# JACARANDA MATHS QUEST 7

AUSTRALIAN CURRICULUM | FIFTH EDITION

# JACARANDA MATHS QUEST 7

## AUSTRALIAN CURRICULUM | FIFTH EDITION


CATHERINE SMITH

JAMES SMART

GEETHA JAMES

CAITLIN MAHONY

BEVERLY LANGSFORD WILLING

CONTRIBUTING AUTHORS

Michael Sheedy | Kahni Burrows | Paul Menta

Fifth edition published 2023 by
John Wiley & Sons Australia, Ltd
Level 4, 600 Bourke Street, Melbourne, Vic 3000

First edition published 2011
Second edition published 2014
Third edition published 2018
Fourth edition published 2021

Typeset in 10.5/13 pt TimesLT Std



ISBN: 978-1-394-19426-1



Front cover images: © Marish/Shutterstock; © Irina Strelnikova/Shutterstock

Illustrated by various artists, diacriTech and Wiley Composition Services

Typeset in India by diacriTech

A catalogue record for this book is available from the National Library of Australia

Printed in Singapore
M WEP290196 130624

The Publishers of this series acknowledge and pay their respects to Aboriginal Peoples and Torres Strait Islander Peoples as the traditional custodians of the land on which this resource was produced.

This suite of resources may include references to (including names, images, footage or voices of) people of Aboriginal and/or Torres Strait Islander heritage who are deceased. These images and references have been included to help Australian students from all cultural backgrounds develop a better understanding of Aboriginal and Torres Strait Islander Peoples' history, culture and lived experience.

It is strongly recommended that teachers examine resources on topics related to Aboriginal and/or Torres Strait Islander Cultures and Peoples to assess their suitability for their own specific class and school context. It is also recommended that teachers know and follow the guidelines laid down by the relevant educational authorities and local Elders or community advisors regarding content about all First Nations Peoples.

All activities in this resource have been written with the safety of both teacher and student in mind. Some, however, involve physical activity or the use of equipment or tools. **All due care should be taken when performing such activities**. To the maximum extent permitted by law, the author and publisher disclaim all responsibility and liability for any injury or loss that may be sustained when completing activities described in this resource.

The Publisher acknowledges ongoing discussions related to gender-based population data. At the time of publishing, there was insufficient data available to allow for the meaningful analysis of trends and patterns to broaden our discussion of demographics beyond male and female gender identification.

# Contents

About this resource ........ vii
Acknowledgements ........ xiv

**Powering up for Year 7** (online only)

**Week 1** Finding your place
**Week 2** Numbering up
**Week 3** Thinking quickly
**Week 4** Fishing for fractions
**Week 5** Getting to the point
**Week 6** Giving 100%

**NAPLAN practice** (online only)

**Set A** Calculator allowed
**Set B** Non-calculator
**Set C** Calculator allowed
**Set D** Non-calculator
**Set E** Calculator allowed
**Set F** Non-calculator

**1 Positive integers** 1

**1.1** Overview ........ 2
**1.2** Place value ........ 4
**1.3** Strategies for adding and subtracting positive integers ........ 10
**1.4** Algorithms for adding and subtracting positive integers ........ 16
**1.5** Multiplying positive integers ........ 25
**1.6** Dividing positive integers ........ 34
**1.7** Rounding and estimating ........ 41
**1.8** Order of operations ........ 45
**1.9** Review ........ 52
Answers ........ 61

**2 Positive and negative integers** 67

**2.1** Overview ........ 68
**2.2** Integers on the number line ........ 70
**2.3** Integers on the number plane ........ 75
**2.4** Adding integers ........ 80
**2.5** Subtracting integers ........ 85
**2.6** Multiplying and dividing integers ........ 91
**2.7** Order of operations with integers ........ 97
**2.8** Review ........ 102
Answers ........ 109

**3 Factors, multiples, indices and primes** 115

**3.1** Overview ........ 116
**3.2** Factors and multiples ........ 118
**3.3** Lowest common multiple and highest common factor ........ 123
**3.4** Index notation ........ 128
**3.5** Prime and composite numbers ........ 133
**3.6** Squares and square roots ........ 141
**3.7** Divisibility tests ........ 147
**3.8** Cubes and cube roots ........ 150
**3.9** Review ........ 155
Answers ........ 162

**4 Fractions and percentages** 167

**4.1** Overview ........ 168
**4.2** What are fractions? ........ 170
**4.3** Simplifying fractions ........ 182
**4.4** Mixed numbers and improper fractions ........ 186
**4.5** Adding and subtracting fractions ........ 192
**4.6** Multiplying fractions ........ 200
**4.7** Dividing fractions ........ 208
**4.8** Working with mixed numbers ........ 213
**4.9** Percentages as fractions ........ 220
**4.10** Calculating percentages of an amount ........ 225
**4.11** One amount as a percentage of another ........ 230
**4.12** Common percentages and short cuts ........ 236
**4.13** Review ........ 242
Answers ........ 251

**5 Ratios, rates and best buys** 261

**5.1** Overview ........ 262
**5.2** Ratios ........ 265
**5.3** Rates ........ 271
**5.4** The unitary method and best buys ........ 276
**5.5** Review ........ 281
Answers ........ 286

**6 Decimals** 289

**6.1** Overview ........ 290
**6.2** Decimals and place value ........ 292
**6.3** Converting decimals to fractions and fractions to decimals ........ 300
**6.4** Rounding and repeating decimals ........ 305
**6.5** Adding and subtracting decimals ........ 311
**6.6** Multiplying decimals ........ 317
**6.7** Dividing decimals ........ 323
**6.8** Decimals and percentages ........ 330
**6.9** Review ........ 333
Answers ........ 340

7 Algebra 347
7.1 Overview 348
7.2 Introduction to algebra 350
7.3 Substituting and evaluating 358
7.4 Simplifying expressions using like terms 362
7.5 Multiplying and dividing terms 367
7.6 Number laws 371
7.7 Expanding brackets 375
7.8 Review 379
Answers 385

Semester review 1 online only 390

8 Equations 393
8.1 Overview 394
8.2 Introduction to equations 396
8.3 Building up expressions and backtracking 402
8.4 Solving equations using backtracking 408
8.5 Solving equations using inverse operations 415
8.6 Checking solutions 424
8.7 Review 429
Answers 435

9 Geometry 441
9.1 Overview 442
9.2 Measuring angles 445
9.3 Constructing angles with a protractor 453
9.4 Types of angles and naming angles 459
9.5 Triangles 465
9.6 Quadrilaterals and their properties 477
9.7 Parallel and perpendicular lines 487
9.8 Review 498
Answers 507

10 Measurement 517
10.1 Overview 518
10.2 Units of measurement 521
10.3 Reading scales and measuring length 528
10.4 Perimeter 536
10.5 Circles and circumference 544
10.6 Area 553
10.7 Area of composite shapes 566
10.8 Volume of rectangular prisms 571
10.9 Capacity 579
10.10 Drawing solids 583
10.11 Review 594
Answers 605

11 Coordinates and the Cartesian plane 613
11.1 Overview 614
11.2 The Cartesian plane 617
11.3 Linear number patterns 625
11.4 Plotting simple linear relationships 632
11.5 Interpreting graphs 639
11.6 Review 651
Answers 659

12 Transformations 671
12.1 Overview 672
12.2 Line and rotational symmetry 674
12.3 Translations 680
12.4 Reflections 686
12.5 Rotations and combined transformations 693
12.6 Review 705
Answers 711

13 Introduction to probability 723
13.1 Overview 724
13.2 The language of chance 726
13.3 The sample space 731
13.4 Simple probability 736
13.5 Experimental probability 742
13.6 Review 750
Answers 757

14 Representing and interpreting data 763
14.1 Overview 764
14.2 Collecting and classifying data 767
14.3 Displaying data in tables 774
14.4 Measures of centre and spread 780
14.5 Column graphs and dot plots 791
14.6 Stem-and-leaf plots 798
14.7 Pie charts and divided bar graphs 804
14.8 Comparing data 814
14.9 Review 820
Answers 829

Semester review 2 online only 842

15 Algorithmic thinking online only
15.1 Overview
15.2 Variables
15.3 Expressions
15.4 Sequences
15.5 Algorithms
15.6 Review
Answers

Glossary 844
Index 850

# About this resource

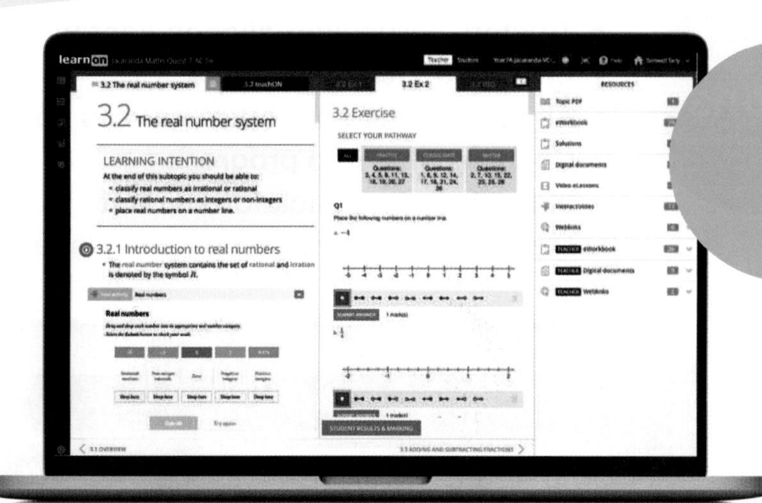

NEW FOR

AUSTRALIAN CURRICULUM V9.0

JACARANDA

MATHS QUEST 7 AUSTRALIAN CURRICULUM

FIFTH EDITION

## Developed by teachers for students

**Tried, tested and trusted. The fifth edition of the *Jacaranda Maths Quest series*, revised fourth edition, continues to focus on helping teachers achieve learning success for every student — ensuring no student is left behind, and no student is held back.**

**Because both** what **and** how **students learn matter**

### Learning is personal

Whether students need a challenge or a helping hand, you'll find what you need to create engaging lessons.

Whether in class or at home, students can get unstuck and progress! Scaffolded lessons, with detailed worked examples, are all supported by teacher-led video eLessons. Automatically marked, differentiated question sets are all supported by detailed worked solutions. And Brand-new Quick Quizzes support in-depth skill acquisition.

### Learning is effortful

Personalise student learning pathways, ensuring that confidence builds and that students push themselves to achieve … all in learnON, Australia's most innovative learning platform.

### Learning is rewarding

Through real-time results data, students can track and monitor their own progress and easily identify areas of strength and weakness.

And for teachers, Learning Analytics provide valuable insights to support student growth and drive informed intervention strategies.

# Learn online with Australia's most

## Everything you need for each of your lessons in one simple view

- Trusted, curriculum-aligned content
- Engaging, rich multimedia
- All the teacher support resources you need
- Deep insights into progress
- Immediate feedback for students
- Create custom assignments in just a few clicks.

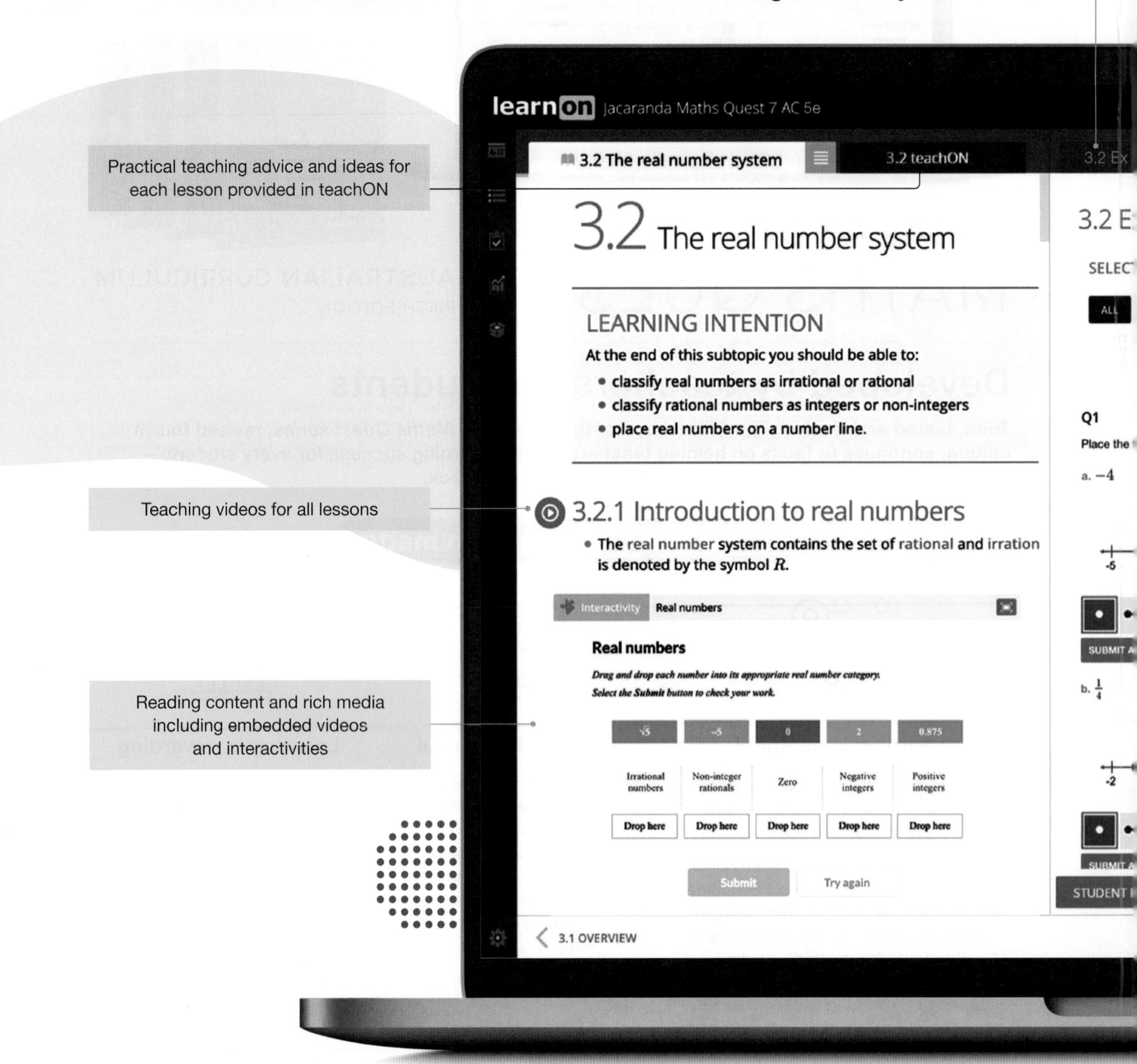

# powerful learning tool, learnON

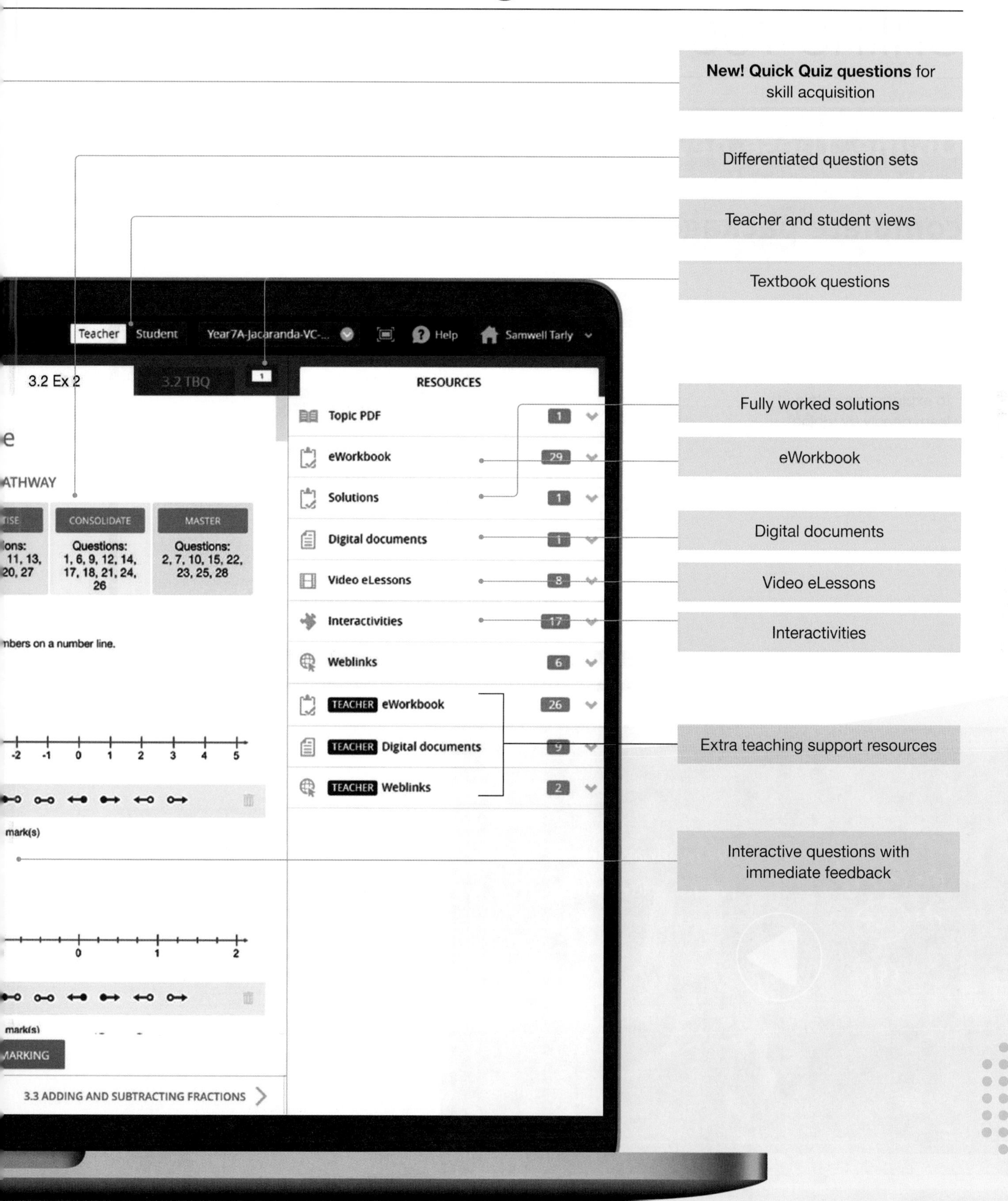

# Get the most from your online resources

## Online, these new editions are the **complete package**

**Trusted Jacaranda theory, plus tools to support teaching and make learning more engaging, personalised and visible.**

Embedded interactivities and videos enable students to explore concepts and learn deeply by 'doing'.

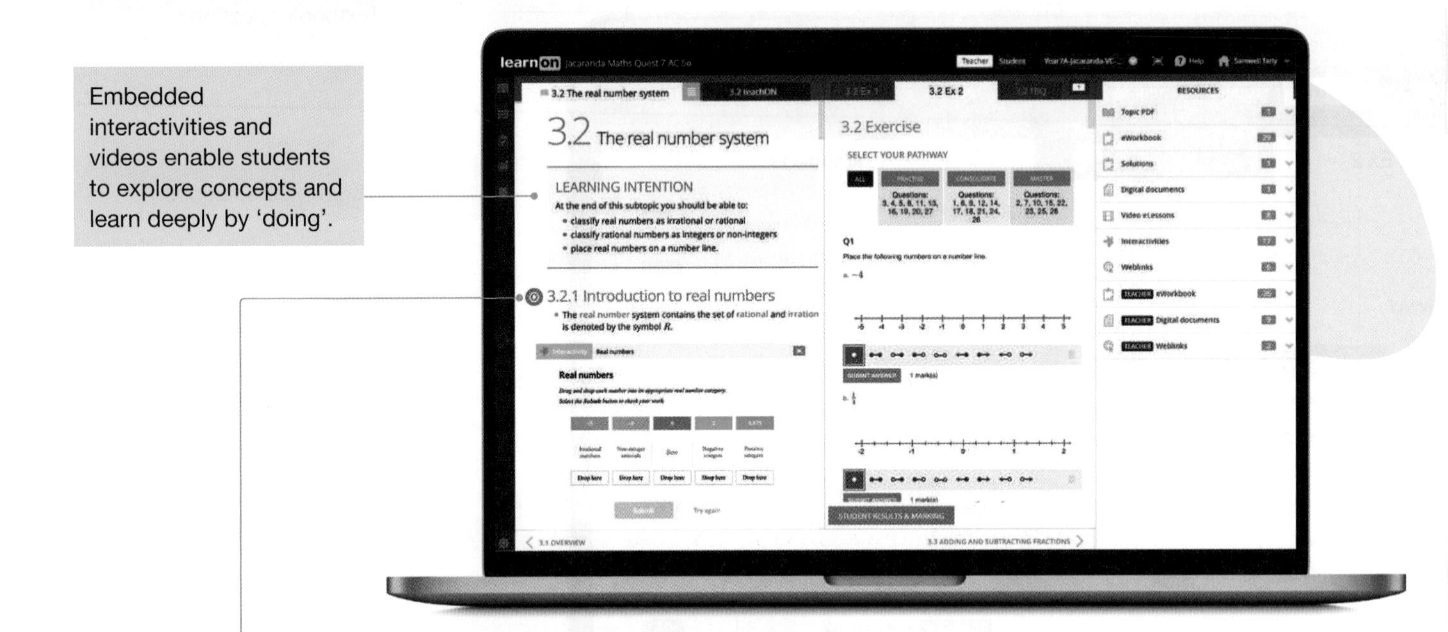

New teaching videos for every lesson are designed to help students learn concepts by having a 'teacher at home', and are flexible enough to be used for pre- and post-learning, flipped classrooms, class discussions, remediation and more.

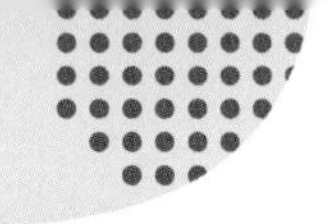

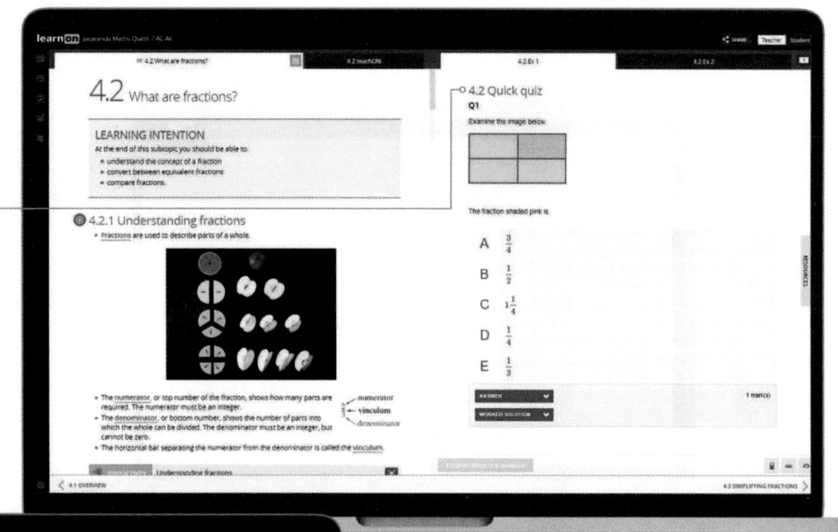

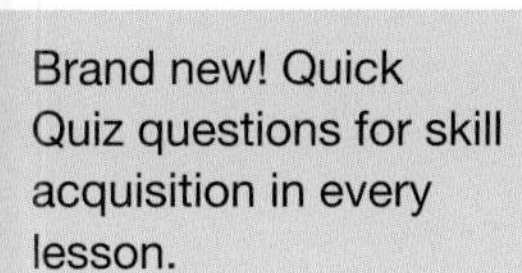

Brand new! Quick Quiz questions for skill acquisition in every lesson.

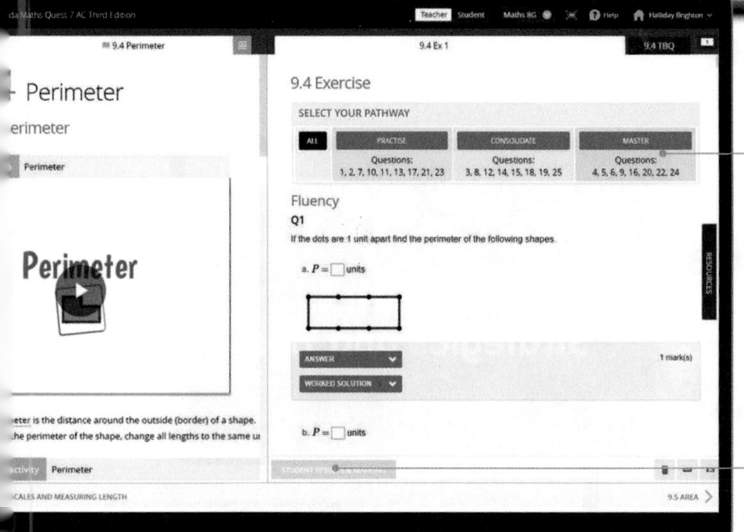

Three differentiated question sets, with immediate feedback in every lesson, enable students to challenge themselves at their own level.

Instant reports give students visibility into progress and performance.

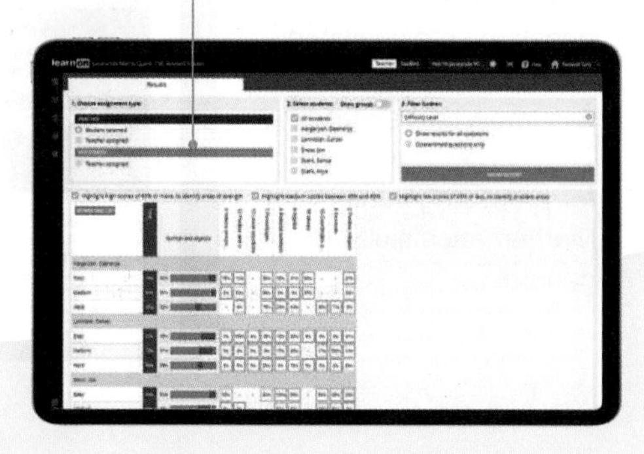

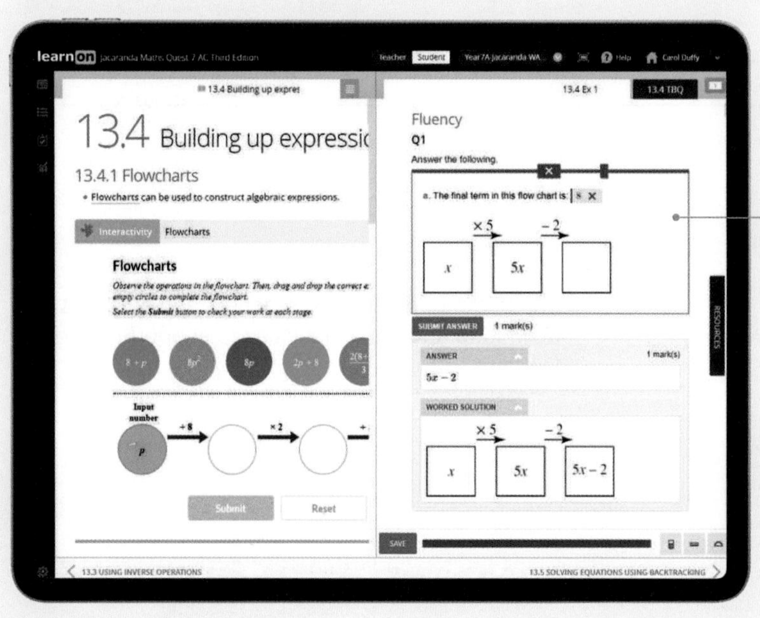

Every question has immediate, corrective feedback to help students overcome misconceptions as they occur and get unstuck as they study independently — in class and at home.

# Powering up for Year 7

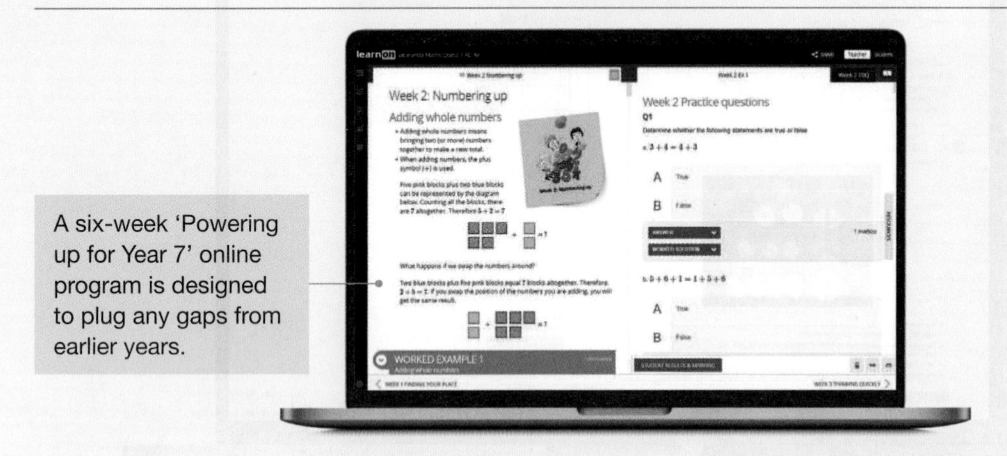


# NAPLAN Online Practice

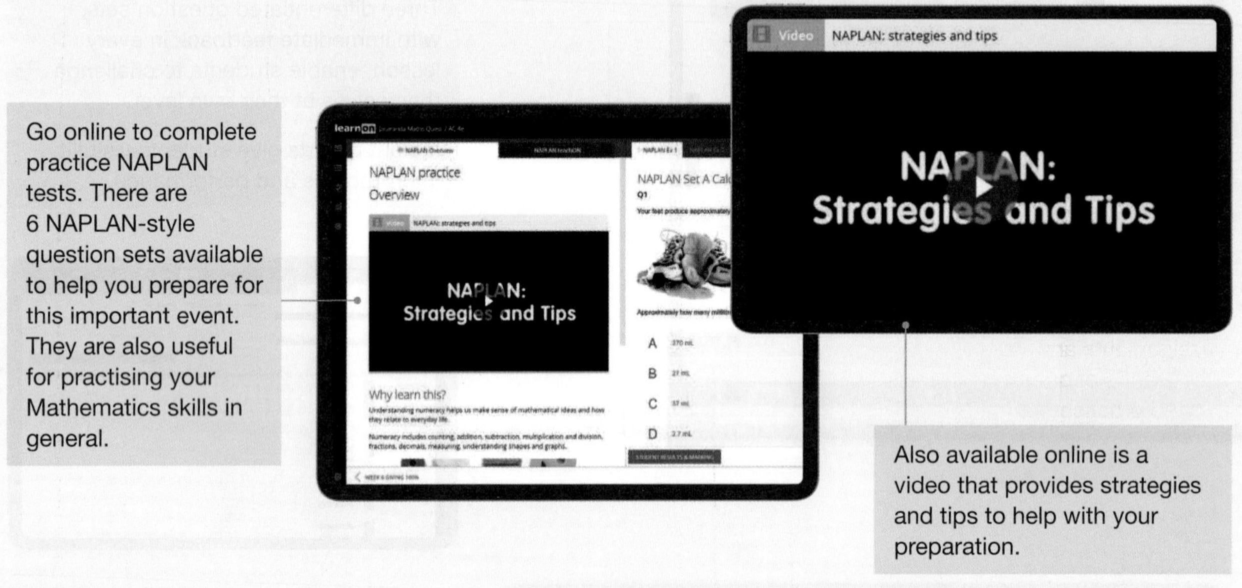


# eWorkbook

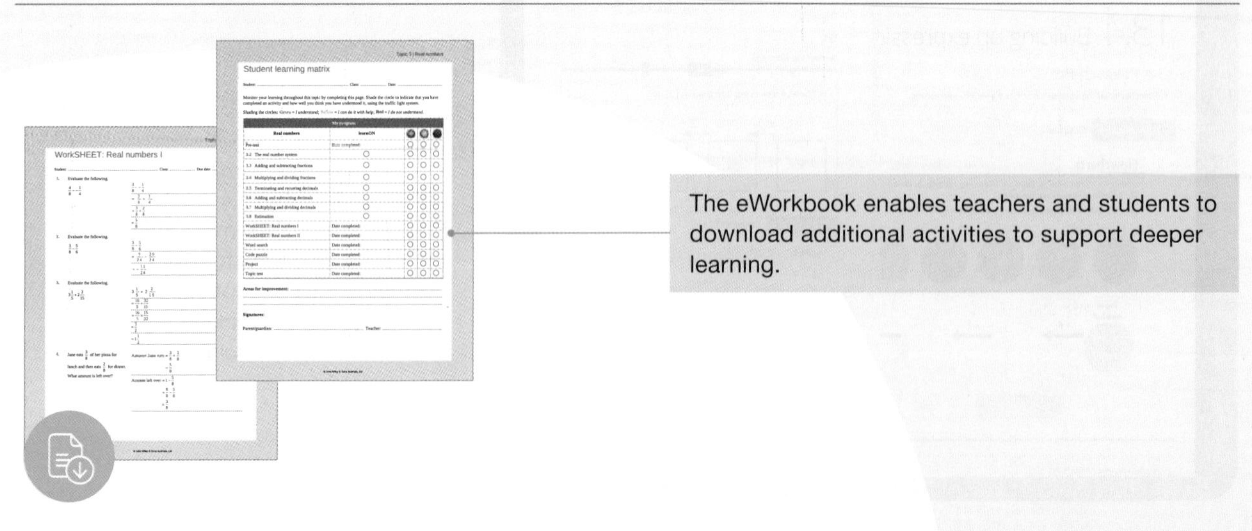

# A wealth of teacher resources

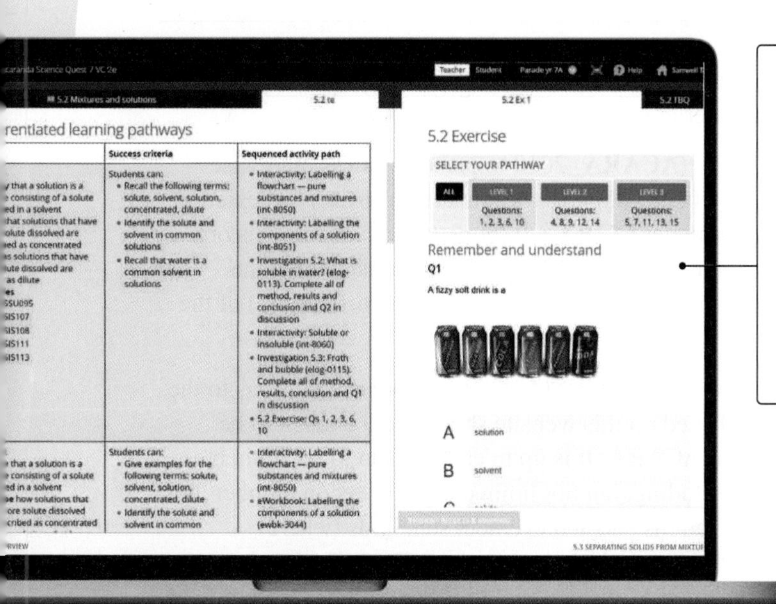


**Enhanced teacher support resources for every lesson, including:**

- work programs and curriculum grids
- practical teaching advice
- three levels of differentiated teaching programs
- quarantined topic tests (with solutions)

# Customise and assign

An inbuilt testmaker enables you to create custom assignments and tests from the complete bank of thousands of questions for immediate, spaced and mixed practice.

# Reports and results

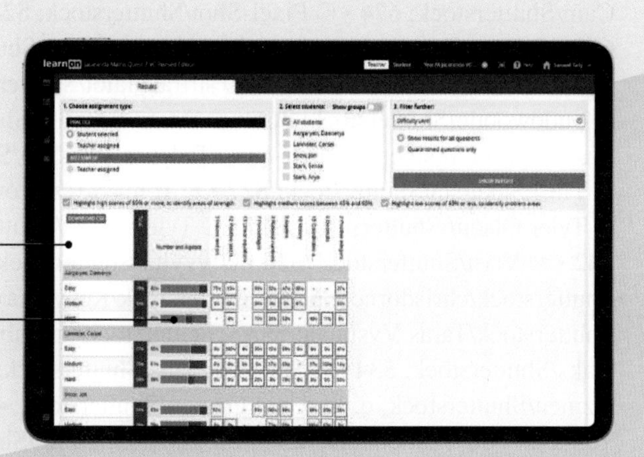

Data analytics and instant reports provide data-driven insights into progress and performance within each lesson and across the entire course.

Show students (and their parents or carers) their own assessment data in fine detail. You can filter their results to identify areas of strength and weakness.

# Acknowledgements

The authors and publisher would like to thank the following copyright holders, organisations and individuals for their assistance and for permission to reproduce copyright material in this book.



## Images

• © Shutterstock: **5** • © Christopher Futcher/Getty Images: **825** • © Getty Images/iStockphoto: **676** • Shutterstock/dikkenss: **596** • Shutterstock/OlekStock: **578** • Shutterstock/pio3: **577** • Shutterstock/Tum3000: **596** • © Africa Studio/Shutterstock: **348, 378** • © Ammit Jack: **649** • © Andrey_Popov/Shutterstock: **768** • © Anton Balazh/Shutterstock: **704** • © Antonio Guillem/Shutterstock: **278** • © Denphumi/Shutterstock: **730** • © docstockmedia/Shutterstock: **672** • © Dragon Images/Shutterstock: **779** • © Dusan Petkovic/Shutterstock: **357** • © Etaphop photo/Shutterstock: **489** • © Fabrik Bilder/Shutterstock: **785** • © Fresnel/Shutterstock: **349** • © Gary Reinwald/Shutterstock: **674** • © George Rudy/Shutterstock: **765** • © Gjermund/Shutterstock: **625** • © goodluz/Shutterstock: **773** • © Hong Vo/Shutterstock: **383** • © Ilja Generalov/Shutterstock: **638** • © Jaroslaw Grudzinski/Shutterstock: **375** • © Jeni Jenny Jeni/Shutterstock: **168** • © Kampol Taepanich/Shutterstock: **356** • © Katilda/Shutterstock: **489** • © Kostyazar/Shutterstock: **384** • © LightField Studios/Shutterstock: **518** • © Luca9257/Shutterstock: **532** • © Mak3t/Shutterstock: **675** • © Minerva Studio/Shutterstock: **310** • © Monkey Business Images/Shutterstock: **780** • © Murni/Shutterstock: **356** • © naito29/Shutterstock: **349** • © Nerthuz/Shutterstock: **284** • © New Africa/Shutterstock: **768** • © Norman Allchin/Shutterstock: **536** • © nortongo/Shutterstock: **378** • © Oez/Shutterstock: **499** • © Olga Kashubin/Shutterstock: **674** • © Orhan Cam/Shutterstock: **674** • © Pixel-Shot/Shutterstock: **828** • © Pixel-ShotShutterstock: **383** • © ppart/Shutterstock: **675** • © Radu Bercan/Shutterstock: **481** • © rfranca/Shutterstock: **352** • © sculpies/Shutterstock: **442** • © Shahjehan/Shutterstock: **790** • © sirtravelalot/Shutterstock: **262** • © Stock-Asso/Shutterstock: **528** • © Sudowoodo/Shutterstock: **366** • © Susan Schmitz/Shutterstock: **828** • © Syda Productions/Shutterstock: **772** • © szefei/Shutterstock: **443, 464** • © Tadeusz Wejkszo/Shutterstock: **357** • © Taras Vyshnya/Shutterstock: **674** • © Thiti Sukapan/Shutterstock: **827** • © Tomasz Czajkowski/Shutterstock: **526** • © tr3gin/Shutterstock: **675** • © Tyler Olson/ Shutterstock: **819** • © Tyler Olson/Shutterstock: **384** • © Viacheslav Nikolaenko/Shutterstock: **542** • © Vixit/Shutterstock: **526** • © vvvita/Shutterstock: **73** • © zhao jiankang/shutterstock: **527** • Shutterstock/chrisdorney: **553** • Shutterstock/Kekyalyaynen: **553** • Shutterstock/nortongo: **571** • Shutterstock/Taras Vyshnya: **553** • Shutterstock/urbanbuzz: **590** • Shutterstock/Zety Akhzar: **553** • © Four Oaks/Shutterstock: **534** • © 300 librarians/Shutterstock: **353** • © a. Chones/Shutterstock; b. M. Unal Ozmen/Shutterstock; c. M. Unal Ozmen/Shutterstock: **402** • © a. Photodisc; b.  Katie Dickinson/Shutterstock; c. Tomasz Pado/Shutterstock: **23** • © a. Elena Shashki /Shutterstock; b. Tulpahn/Shutterstock: **334** • © a. Raphael Christinat/Shutterstock; b. Photodisc: **125** • © Aaron Amat/Shutterstock: **323** • © abriendomundoShutterstock: **9**

• © Adam Middleton/ Shutterstock: **653** • © aekikuis/Shutterstock: **283** • © al7/Shutterstock: **434** • © Alberto Loyo/Shutterstock: **249** • © alessandro0770/Shutterstock: **228** • © Alexandra Lande/Shutterstock: **674** • © Alinute Silzeviciute/ Shutterstock: **818** • © ALPA PROD/Shutterstock: **133** • © alterfalter/Shutterstock: **157** • © alybaba/Shutterstock: **707** • © AmaPhoto/Shutterstock: **250** • © Andrew Burgess/Shutterstock: **6** • © Andrey Armyagov/Shutterstock: **126** • © Andrey Bayda/Shutterstock: **263, 271** • © Andrii_M/Shutterstock: **100** • © Annches/Shutterstock: **398** • © Anthony Hall/Shutterstock: **465** • © Aptyp_koK/Shutterstock: **653** • © ArliftAtoz2205/Shutterstock: **434** • © auddmin/Shutterstock: **749** • © AustralianCamera/ Shutterstock: **235** • © Axel Bueckert/Shutterstock: **292** • © bernashafo/Shutterstock: **336** • © Bilanol/Shutterstock: **40** • © BIRTHPIX/Shutterstock: **768** • © Blend Images/Shutterstock: **33** • © Bloomicon/Shutterstock: **329** • © c John Wiley & Sons Australia/Jennifer Wright: **179** • © CapturePB/Shutterstock: **740** • © Carlos Gutierrez PhotoShutterstock: **74** • © Carsten Reisinger/Shutterstock: **178** • © Celso Pupo/Shutterstock: **431** • © ChameleonsEye/Shutterstock: **24** • © Chekyravaa/Shutterstock: **170** • © Chris Jenner/Shutterstock: **653** • © clearviewstock/Shutterstock: **756** • © Corbis Royalty Free: **535** • © Corepics VOF/Shutterstock: **23** • © corund/Shutterstock: **685** • © cpaulfell/Shutterstock: **434** • © Damian Pankowiec/Shutterstock: **33** • © Daniel Jedzura/Shutterstock: **218** • © debra hughes/Shutterstock: **119** • © Denphumi/Shutterstock: **732** • © Desmos: **697** • © Digital Stock/Corbis Corporation: **90, 462** • © Digital Vision: **643** • © dim.po: **270** • © DisobeyArt/Shutterstock: **100** • © Dmitry Eagle Orlov/Shutterstock: **381** • © Dmytro Zinkevych/Shutterstock: **411** • © dolomite-summits/Shutterstock: **159** • © DOUG RAPHAEL/Shutterstock: **220** • © Dragon ImagesShutterstock: **85** • © eAlisa/Shutterstock: **244** • © Elartico/Shutterstock: **116** • © Elena Veselova/Shutterstock: **264** • © ESB Professional/Shutterstock: **234, 777** • © ET-ARTWORKS/Getty Images: **693** • © Eva Speshneva/Shutterstock: **362** • © f11photo/Shutterstock: **50** • © fckncg/Shutterstock: **178** • © Fedor Selivanov/Shutterstock: **53** • © Fertas/Shutterstock: **433** • © freevideophotoagency/Shutterstock: **160** • © frenchiestravel/Shutterstock: **58** • © fresher/Shutterstock: **34** • © fritz16/Shutterstock: **8** • © garagestock/Shutterstock: **278** • © Gentian Polovina/Shutterstock: **284** • © Gino Santa Maria/Shutterstock: **755** • © givaga/Shutterstock: **279** • © goodluz/Shutterstock: **132** • © GorodenkoffShutterstock: **75** • © Gringoann/Shutterstock: **284** • © grum_l/Shutterstock: **739** • © gualtiero boffi/Shutterstock: **230** • © Halfpoint/Shutterstock: **126** • © hddigital/Shutterstock: **316** • © HG Photography/Shutterstock: **192** • © Hquality/Shutterstock: **305, 754** • © Hugo Felix/Shutterstock: **154** • © Iakov Filimonov: **212** • © Image Addict: **229** • © Image Disk Photography: **7** • © imageBROKER/Alamy Stock Photo: **245** • © Imagenet/Shutterstock: **764** • © ingehogenbijl/Shutterstock: **8** • © Ingrid Balabanova/Shutterstock: **336** • © Irina Strelnikova; © Marish: **1, 67, 115, 167, 261, 289, 347, 393, 441, 517, 613, 671, 723, 763** • © IT Stock: **462** • © Ivan MarjanovicShutterstock: **79** • © Jan Hopgood/Shutterstock: **394** • © Jasni/Shutterstock: **740** • © John Wiley & Sons Australia: **34, 464** • © John Wiley & Sons Australia/ Taken by Kari-Ann Tapp: **196** • © John Wiley & Sons Australia/Carolyn Mews: **450, 486** • © John Wiley & Sons Australia/Jennifer Wright: **179** • © John Wiley & Sons Australia/Jo Patterson: **532** • © John Wiley/Jo Patterson: **532** • © jokerpro/Shutterstock: **44** • © Jorge Casais/Shutterstock: **645** • © Joshua Resnick/Shutterstock: **336** • © JPL Designs/Shutterstock: **706** • © JR-stock/Shutterstock: **361** • © Julinzy: **269** • © Kari-Ann Tapp: **742** • © Krakenimages.com/Shutterstock: **755** • © Kutlayev Dmitry/Shutterstock: **264** • © l i g h t p o e t/Shutterstock: **33** • © Lance Bellers/Shutterstock: **730** • © Lesya Dolyuk/Shutterstock: **191** • © LightField Studios/Shutterstock: **264** • © M. Unal OzmenShutterstock: **58** • © Madeleine R/Shutterstock: **279** • © Maks Narodenko/Shutterstock: **189** • © Maliutina AnnaShutterstock: **96** • © MaraZe/Shutterstock: **45** • © Maridav/Shutterstock: **219, 646** • © mashurov/Shutterstock: **39** • © matimix/Shutterstock: **734** • © Maxisport/Shutterstock: **315, 430** • © MCruzUA/Shutterstock: **35** • © Mega Pixel/Shutterstock: **737** • © Michael Rosskothen/Shutterstock: **653** • © MichaelJayBerlin/Shutterstock: **138** • © michaeljung/Shutterstock: **185** • © Mila Supinskaya Glashchenko/Shutterstock: **39** • © Milagli/Shutterstock: **55** • © Mind Pro Studio/Shutterstock: **232** • © Minerva Studio/ Shutterstock: **789** • © MiniDoodle/Shutterstock: **751** • © mirtmirt/Shutterstock: **735** • © Mitch Hutchinson/Shutterstock: **724** • © Monkey Business Images/Shutterstock: **40, 122, 191, 223, 224, 245, 401, 407, 778** • © monticello/Shutterstock: **240** • © monticello/Shutterstock.com: **741** • © mooinblack/Shutterstock: **240** • © Natali Glado/Shutterstock: **788** • © National Council of Teachers of Mathematics: **745** • © Neale Cousland/Shutterstock: **59, 229, 730** • © Neirfy/Shutterstock: **381** • © Nejron PhotoShutterstock: **85** • © Nerthuz/Shutterstock: **280** • © New Africa/Shutterstock: **125, 246** • © News Ltd/Newspix: **480** • © Nicholas Piccillo/Shutterstock: **414** • © NicoElNino/Shutterstock: **405** • © nobeastsofierce/Shutterstock: **133** •

© oksana2010/Shutterstock: **280** • © Oleksiy Mark/Shutterstock: **250** • © PA Images/Alamy Stock Photo: **2, 290, 316** • © PANAGIOTIS KARAPANAGIOTIS/Alamy Stock Photo: **54** • © PanyaStudio/Shutterstock: **358** • © Pepgooner/Shutterstock: **90** • © Petrenko Andriy/Shutterstock.com: **181** • © Phatthanit/Shutterstock: **230** • © Phils Mommy/Shutterstock: **428** • © Photoagriculture/Shutterstock: **421** • © Photodisc: **198, 207, 463, 479, 486, 529, 772, 774, 801** • © PhotoDisc: Inc, **525** • © Photodisc: Inc, **470** • © photokup/Shutterstock: **427** • © pilipphoto/Shutterstock: **637** • © Ppstock/Shutterstock: **160** • © Pressmaster/Shutterstock: **219, 357** • © ProStockStudio/Shutterstock: **396** • © Qilins prance FilmmakerShutterstock: **101** • © Radu Bercan/Shutterstock: **276** • © ratselmeister/Shutterstock: **263** • © Rawpixel.com/Shutterstock: **249** • © Renovacio/Shutterstock: **741** • © Richard Semik/Shutterstock: **421** • © Rido/Shutterstock: **126** • © Robyn Mackenzie/Shutterstock: **339** • © Roma Koshel/Shutterstock: **748** • © Roman SamborskyiShutterstock: **32** • © ROMSVETNIK/Shutterstock: **726** • © Ronald Sumners/Shutterstock: **50** • © Rostislav Král/Shutterstock: **217** • © Ruth Black/Shutterstock: **50** • © safakcakir/Shutterstock: **217** • © Sandarina/Shutterstock: **40** • © Sean Donohue Photo/Shutterstock: **315** • © Serg64/Shutterstock: **332** • © Sergey Nivens/Shutterstock: **72** • © Sergey Novikov/ Shutterstock: **790** • © Sergey Novikov/Shutterstock: **234** • © Shuang Li/Shutterstock: **276** • © Shutterstock: **304, 315, 339** • © Shvaygert Ekaterina/Shutterstock: **789** • © sirtravelalot/Shutterstock: **401, 465** • © Songsak P/Shutterstock: **486** • © Sonia Alves-Polidor/Shutterstock: **728** • © Sonulkaster/Shutterstock: **104** • © stockcreations/Shutterstock: **90** • © StockPhotoAstur/Shutterstock: **686** • © Sudowoodo: **363** • © Sundraw Photography/Shutterstock: **645** • © Swapan Photography/Shutterstock: **748** • © Syda Productions/Shutterstock: **100, 181** • © technotr/ Getty Images: **22** • © Tetyana Kokhanets/Shutterstock: **727** • © THPStock/Shutterstock: **735** • © TK Kurikawa/Shutterstock: **332** • © TORWAISTUDIO/Shutterstock: **650** • © tr3gin/Shutterstock: **218** • © trigga/Getty Images: **74** • © Tyler Olson/Shutterstock: **282** • © Tyler Olson/Shutterstock.com: **310** • © UnknownLatitude Images/Shutterstock: **7** • © upthebanner/Shutterstock: **217** • © v74/Shutterstock: **728** • © Velychko/Shutterstock: **199** • © Vereshchagin Dmitry/Shutterstock: **772** • © Viewfinder Australia Photo Library: **535** • © Viktor Fedorenko/Shutterstock: **732, 738** • © VisualArtStudioShutterstock: **22** • © vkilikovShutterstock: **17** • © VladisChern/Shutterstock: **70** • © Volodymyr TVERDOKHLIB: **122** • © Volt Collection/Shutterstock: **296** • © VovanIvanovich/Shutterstock: **68** • © VP Photo Studio/Shutterstock: **729** • © wavebreakmedia/Shutterstock: **140, 736** • © wolfman57/Shutterstock: **225** • © Yeti Studio/Shutterstock: **739** • © Yevgenij_D/Shutterstock: **614** • © You Touch Pix of EuToch/Shutterstock: **229** • © Zbigniew Guzowski/Shutterstock: **740** • © zbindere/Getty Images: **7** • © zentilia/Shutterstock: **734** • ©Dragon Images/Shutterstock: **133** • ©Michael D Brown/Shutterstock: **753** • ©Nataliya Peregudova/Shutterstock: **225** • © Andrey Arkusha/Shutterstock: **521** • © Benny Marty/Shutterstock: **521** • © Bruno Herdt/Getty Images: **449** • © Corbis: **524** • © CurrywurstmitPommes/Shutterstock: **521** • © Digital Stock: **449** • © Digital Vision: **484** • © Dimedrol68/Shutterstock: **452** • © Dimedrol68/Shutterstock: **452** • © diskoVisnja/Shutterstock: **497** • © Eugene Buchko/Shutterstock: **483** • © Haali/Shutterstock: **156, 751** • © HaaliShutterstock: **53** • © John Wiley & Sons Australia: **532** • © John Wiley & Sons Australia/Photo by Kari-Ann Tapp: **483** • © John Wiley & Sons Australia/Photo by Theresa Pagon: **484** • © JUGVOLUME1250ML: **532** • © Julenochek/Shutterstock: **521** • © Kitch Bain/Shutterstock.com: **350** • © Kokkai Ng/Getty Images: **524** • © Lilyana Vynogradova/Shutterstock: **450** • © Olesia Bilkei/Shutterstock: **452** • © Photo Melon/Shutterstock: **483** • © PhotoDisc: **450, 475** • © PhotoDisc: Inc, **524** • © PhotoDisc: Inc., **524** • © pornvit_v/Shutterstock: **483** • © Ruud Morijn Photographer/Shutterstock: **497** • © Sheila Fitzgerald/Shutterstock: **353** • © Shutterstock: **103** • © totajla/Shutterstock: **524**

Every effort has been made to trace the ownership of copyright material. Information that will enable the publisher to rectify any error or omission in subsequent reprints will be welcome. In such cases, please contact the Permissions Section of John Wiley & Sons Australia, Ltd.

# Powering up for Year 7

We care deeply about ensuring that all students feel confident in knowing they are equipped with the essential core skills in Mathematics as they transition into Year 7.

*Powering up for Year 7* is a brand-new online resource developed to fill learning gaps in key mathematical areas to ensure that all students have the building blocks needed to move smoothly into Year 7. The course is organised around key skills over a period of six weeks:

Students will learn key skills such as as place value, operations with whole numbers, mental strategies for calculations, fractions, decimals and percentages.

All questions are automatically marked, with instant feedback and worked solutions. This incredible resource is available online only in our digital learnON platform.

# NAPLAN practice

Go online to complete practice NAPLAN tests. There are 6 NAPLAN-style question sets available to help you prepare for this important event. They are also useful for practising your Mathematics skills in general.

Also available online is a video that provides strategies and tips to help with your preparation.

**SET A**
Calculator allowed

**SET B**
Non-calculator

**SET C**
Calculator allowed

**SET D**
Non-calculator

**SET E**
Calculator allowed

**SET F**
Non-calculator

# 1 Positive integers

## LESSON SEQUENCE


1.1 Overview ........ 2
1.2 Place value ........ 4
1.3 Strategies for adding and subtracting positive integers ........ 10
1.4 Algorithms for adding and subtracting positive integers ........ 16
1.5 Multiplying positive integers ........ 25
1.6 Dividing positive integers ........ 34
1.7 Rounding and estimating ........ 41
1.8 Order of operations ........ 45
1.9 Review ........ 52

# LESSON
# 1.1 Overview

## Why learn this?

Positive integers are whole numbers. We use positive integers every day. Addition, subtraction, multiplication and division of positive integers are the building blocks of Mathematics.

When you count the number of runs you make in a game of cricket, you are using integers. When you add your runs to the runs your teammates scored, the total number of runs will be a positive integer. This result is found by the addition of integers.

Many daily activities depend on knowing how to answer simple questions like, 'When you went for that walk, how far did you go?' or 'How many people live in your house?'

Every day we see integers displayed on screens — for example when we look at the weather forecast or check our unread messages. We normally don't even think about integers — they are just the numbers we see all around us.

Understanding integers and their addition, subtraction, multiplication and division is important for everyday life and work. Many jobs, including hospitality, banking, construction, design, engineering, nursing, teaching, finance and medicine all require an understanding of the use of integers.

## Exercise 1.1 Pre-test

learnon

1. **MC** Choose the largest of the following numbers.
   **A.** 5656 **B.** 5665 **C.** 5566 **D.** 5556 **E.** 5666
2. State how many three-digit numbers you can make from the digits 6, 9 and 1 if you are allowed to repeat digits.
3. Calculate the following additions.
   **a.** $1366 + 948 + 97$
   **b.** $654 + 937 + 23 + 68\,941$
   **c.** $3085 + 38 + 20\,389 + 3000 + 235$
4. Calculate the following subtractions.
   **a.** $987 - 365$ **b.** $9432 - 2175$ **c.** $3001 - 1739$
5. Calculate the following using a mental strategy.
   $2 \times 8 \times 6 \times 5$
6. Evaluate the following.
   $3\overline{)1248}$
7. A large chicken farm sorts its eggs into cartons of 12 eggs. If they have 3852 eggs to sort, calculate the number of complete cartons that can be sorted.
8. A landscape gardener charges an initial fee of $165, plus $52 per hour. Determine how much the landscape gardener will charge for a job that takes 35 hours if a customer has taken advantage of a special offer of $120 off the total.
9. If an order was made for 25 boxes of Fanta and each box contains 24 cans, use factors to calculate the total amount of cans.
10. Round 32 895 to the first digit.
11. Calculate the following.
    **a.** $48 \div 4$ **b.** $168 \div 8$ **c.** $625 \div 25$
12. Complete each statement by placing < , > or = in the empty box so that the statement is true.
    **a.** $2 \times 4 \square 8 \times 2$ **b.** $16 \div 2 \square 8$ **c.** $16 \times 2 \square 32 \div 8$
13. Evaluate the following.
    **a.** $7 \times 6 \div 3 \div 7$ **b.** $3 \times 8 \div 6 + 4$ **c.** $7 + [66 - (5 \times 5)] \times 3$
14. Evaluate the following using long division.
    **a.** $2752 \div 16$ **b.** $7548 \div 17$ **c.** $4935 \div 21$
15. Aziz takes 3 minutes to complete one lap of a running course, while Jani takes 4 minutes to complete one lap. Not counting the start of the race, determine when Aziz will next be running beside Jani.

# LESSON
# 1.2 Place value

## LEARNING INTENTION

At the end of this lesson you should be able to:
- understand that the value of each digit in a number depends on its position or place value.

## 1.2.1 Place value

eles-3643

- Numbers are made up of the digits 0, 1, 2, 3, 4, 5, 6, 7, 8 and 9.
- The position of a digit in a number gives a different value to the digit. The table below shows the value of the digit 6 in some different numbers.

| Number | Name of number | Value of the digit 6 |
|---|---|---|
| 1**6** | Sixteen | 6 |
| 35**6**2 | Three thousand, five hundred and sixty-two | 60 |
| 18 **6**34 | Eighteen thousand, six hundred and thirty-four | 600 |

- Each position in a number has its own **place value**. The number 59 376 can be represented in a place-value table as shown.

| Ten thousands | Thousands | Hundreds | Tens | Ones |
|---|---|---|---|---|
| 5 | 9 | 3 | 7 | 6 |

### WORKED EXAMPLE 1 Expanded form

**Write 59 376 in expanded form.**

**THINK**

1. 59 376 is the same as 5 ten thousands, 9 thousands, 3 hundreds, 7 tens and 6 ones.
2. Multiply each of the digits by its place value.
3. Perform each of the multiplications to show how the expanded form is written.

**WRITE**

| Ten thousands | Thousands | Hundreds | Tens | Ones |
|---|---|---|---|---|
| 5 | 9 | 3 | 7 | 6 |

$59\,376 = 5 \times 10\,000 + 9 \times 1000 + 3 \times 100 + 7 \times 10 + 6 \times 1$

$= 50\,000 + 9000 + 300 + 70 + 6$

### Place-holding zeros

- On 6 November 2020, at 10:15 am, the Australian Bureau of Statistics' population clock estimated Australia's population to be 25 718 205. In a place-value table, this number appears as follows.

| Ten millions | Millions | Hundred thousands | Ten thousands | Thousands | Hundreds | Tens | Ones |
|---|---|---|---|---|---|---|---|
| 2 | 5 | 7 | 1 | 8 | 2 | 0 | 5 |

- The zero (0) in the place-value table means that there are no tens. The zero must be written to hold the place value, otherwise, the number would be written as 2 571 825 and would no longer have the same value.

### Reading numbers

- To make numbers easier to read and name, they are written in groups of three digits with a space between each group. The only exception to this rule is that four-digit numbers are usually written as a group of four digits.

| Number | Name of number |
|---|---|
| 4357 | Four thousand, three hundred and fifty-seven |
| 12 345 | Twelve thousand, three hundred and forty-five |
| 102 345 | One hundred and two thousand, three hundred and forty-five |
| 123 456 789 | One hundred and twenty-three million, four hundred and fifty-six thousand, seven hundred and eighty-nine |

## 1.2.2 Ordering and comparing numbers

eles-3644

- Numbers in **ascending order** are placed from smallest to largest, starting on the left. The numbers, 1, 25, 192, 908 and 1115 are in ascending order.

1 25 192 908 1115

- Numbers in **descending order** are placed from largest to smallest, starting on the left. The numbers 8532, 934, 105, 53 and 5 are in descending order.

8532 934 105 53 5

- Numbers are ordered according to their place values. For whole numbers, the number with the most digits is the greatest in value because the first digit will have the highest place value.
- If two numbers have the same number of digits, then the digits with the highest place value are compared. If they are equal, the next highest place values are compared, and so on.

### WORKED EXAMPLE 2 Ordering numbers

**Write the following numbers in descending order.**
**858, 58, 85, 8588, 5888, 855**

| THINK | WRITE |
|---|---|
| 1. Write the numbers with the most digits. | 8588 and 5888 |
| 2. There are two numbers with 4 digits. The number with the higher digit in the thousands column is larger. The other number is placed second. | 8588, 5888 |
| 3. Compare the two numbers with 3 digits. Both have the same hundreds and tens values, so compare the value of the units. | 858, 855 |
| 4. Compare the two 2-digit numbers. | 85, 58 |
| 5. Write the answer. | 8588, 5888, 858, 855, 85, 58 |

7. Using words, write the value of the 3 in the distance to Coober Pedy.

8. Using words, write the value of the 5 in the distance to Alice Springs.

### Understanding

9. **MC** Select the largest of the following numbers.

   **A.** 4884 **B.** 4488 **C.** 4848 **D.** 4844 **E.** 4888

10. **MC** Select the smallest of the following numbers.

   **A.** 4884 **B.** 4488 **C.** 4848 **D.** 4844 **E.** 4888

11. **WE2** Write the following numbers in descending order.
   8569, 742, 48 987, 28, 647

12. Rearrange the following numbers so that they are in descending order.
   47 890, 58 625, 72 167, 12 947, 32 320

13. Organise the following numbers into descending order.
   6477, 7647, 7476, 4776, 6747

14. Change the order of the following numbers to put them in descending order.
   8088, 8800, 8080, 8808, 8008, 8880

15. Write the following numbers in ascending order.
   58, 9, 743, 68 247, 1 258 647

16. Rearrange the following numbers so that they are in ascending order.
   78 645, 58 610, 60 000, 34 108, 84 364

17. Organise the following numbers into ascending order.
   9201, 2910, 1902, 9021, 2019, 1290

18. Change the order of the following numbers so that they are in ascending order.
211, 221, 212, 1112, 222, 111

19. WE3 Copy and complete the following number sentences by placing the < or > symbol in each box.
a. 345 ☐ 567
b. 89 ☐ 98
c. 234 596 ☐ 23 459
d. 7765 ☐ 7756

20. We can use the abbreviation K to represent \$1000. For example, \$50 000 can be written as \$50K. Using this rule, determine what amounts each of the following represent.
a. \$6K
b. \$340K
c. \$58K

21. Write the following using K as an abbreviation, as shown in **Question 20**.
a. \$430 000
b. \$7000
c. \$800 000

## Reasoning

22. Determine the largest five-digit number you can write if each digit must be different and no digit may be prime. Show your working.

23. A new mobile phone company is investigating the habits of phone users. From the following sample of phone calls, determine the most common length of call. Start by rounding each call time to the nearest half-minute (30 seconds).

| | | | | | |
|---|---|---|---|---|---|
| 132 s | 10 s | 43 s | 71 s | 243 s | 52s |
| 142 s | 332 s | 6 s | 38 s | 63 s | 32 s |
| 132 s | 32 s | 43 s | 52 s | 352 s | 101 s |
| 124 s | 28 s | 153 s | 10 s | 243 s | 34 s |

24. Astronomers and other scientists use scientific notation when working with very large or very small numbers. For example, the 'astronomical unit' (AU) is the average distance of the Earth from the Sun; it is equal to150 million km or $1.5 \times 10^{13}$ cm. Explain why scientists use scientific notation.

## Problem solving

25. Determine how many two-digit numbers you can make using the digits 1 and 2 if:
a. you are not allowed to repeat any digits
b. you can repeat the digits.

26. Determine how many three-digit numbers you can make using the digits 1, 2 and 3 if:
a. you are not allowed to repeat any digits
b. you can repeat the digits.

Explain whether there is a relationship between the initial number of digits and the number of arrangements when:
c. repetition is not allowed
d. repetition is allowed.

27. Without actually counting to one million, determine a way of estimating the time it would take to count out loud from 0 to 1 000 000.

# LESSON
# 1.3 Strategies for adding and subtracting positive integers

**LEARNING INTENTION**

At the end of this lesson you should be able to:
- add and subtract positive integers using mental strategies.

## 1.3.1 Mental strategies for addition

eles-3645

- Mental strategies for addition are different ways to add up large numbers or integers in your head without a calculator.
- There are lots of different methods that can help you add 2- and 3-digit numbers in your head.

### Jump strategy for addition

- The jump strategy for addition involves breaking the number being added into smaller parts.

  For example:

$$535 + 87 =$$

Break 87 into 80 and 7.

$$535 + 80 = 615$$

$$615 + 7 = 622$$

+ 80
+ 7
510 520 530 540 550 560 570 580 590 600 610 620 630 640
535 615 622

### Split strategy for addition

- The split strategy for addition involves splitting both numbers being added into their expanded forms and then adding the parts that have the same place value.

  For example:

$$165 + 432 =$$

Split 165 into 1 hundred, 6 tens and 5 ones: $165 = 100 + 60 + 5$
Split 432 into 4 hundreds, 3 tens and 2 ones: $432 = 400 + 30 + 2$

$$\begin{aligned} 165 + 432 &= 100 + 60 + 5 + 400 + 30 + 2 \\ &= 100 + 400 + 60 + 30 + 5 + 2 \\ &= 500 + 90 + 7 \\ &= 597 \end{aligned}$$

**WORKED EXAMPLE 4 Using mental strategies to add numbers**

**a. Calculate $514 + 88$ using the jump strategy.**
**b. Calculate $759 + 412$ using the split strategy.**

| THINK | WRITE |
|---|---|
| a. 1. Write the question. | $514 + 88$ |
| 2. Break 88 into 80 plus 8. | $88 = 80 + 8$ |
| 3. Add 80 to 514 to get 594. | $514 + 80 = 594$ |
| 4. Add 8 to 594 to get 602. | $594 + 8 = 602$ |
| 5. Write the answer. | $514 + 88 = 602$ |
| b. 1. Write the question. | $759 + 412$ |
| 2. Split each number up by writing it in expanded form. | $759 = 700 + 50 + 9$<br>$412 = 400 + 10 + 2$ |
| 3. Add the numbers with the same place value together. | $700 + 400 + 50 + 10 + 9 + 2 = 1100 + 60 + 11$<br>$= 1171$ |
| 4. Write the answer. | $759 + 412 = 1171$ |

## Compensation strategy for addition

- The compensation strategy for addition involves making one of the numbers being added larger by rounding up to the nearest ten or hundred, completing the addition, and then subtracting the added amount.

  For example:

$$243 + 78 =$$

1. Round 78 up to 80.

$$243 + 80 = 323$$

2. Now subtract 2, since 78 was rounded up to 80 by adding 2.

$$323 - 2 = 321$$

## Rearrange strategy for addition

- The rearrange strategy for addition involves rearranging the numbers being added to form numbers that add up to multiples of 10.

  For example:

$$16 + 17 + 14 =$$

Swap the order and add 16 and 14 first, since this will result in a multiple of 10.

$$16 + 14 + 17 =$$

$$30 + 17 = 47$$

**WORKED EXAMPLE 5 Using mental strategies to add numbers**

**a. Calculate $576 + 86$ using the compensation strategy.**
**b. Calculate $38 + 13 + 22$ using the rearrange strategy.**

| THINK | WRITE |
|---|---|
| **a. 1.** Write the question. | $576 + 86$ |
| **2.** Round 86 up to the nearest 10. | 86 becomes 90 when rounded up by adding 4. |
| **3.** Add 576 to the rounded-up number (90). | $576 + 90 = 666$ |
| **4.** Subtract 4, since 86 was rounded up to 90 by adding 4. | $666 - 4 = 662$ |
| **5.** Write the answer. | $576 + 86 = 662$ |
| **b. 1.** Write the question. | $38 + 13 + 22$ |
| **2.** Rearrange the numbers so that a multiple of 10 can be formed. $38 + 22 = 60$, so add these numbers first. | $= 38 + 22 + 13$<br>$= 60 + 13$ |
| **3.** Add 13 to 60. | $= 73$ |
| **4.** Write the answer. | $38 + 13 + 22 = 73$ |

**WORKED EXAMPLE 6 Using mental strategies to add numbers**

**Calculate $713 + 143$ using an appropriate mental strategy.**

| THINK | WRITE |
|---|---|
| **1.** Write the question. | $713 + 143$ |
| **2.** Select a strategy. Since the numbers are both large numbers, it may be easiest to split them up. | Split strategy |
| **3.** Use the strategy to add numbers.<br>• Write each number in expanded form.<br>• Add numbers with the same place values.<br>• Complete the final addition. | $713 = 700 + 10 + 3$<br>$143 = 100 + 40 + 3$<br>$= 700 + 100 + 10 + 40 + 3 + 3$<br>$= 800 + 50 + 6$<br>$= 856$ |
| **4.** Write the answer. | $713 + 143 = 856$ |

## 1.3.2 Mental strategies for subtraction

eles-3646

- Mental strategies for subtraction are different ways to subtract numbers without doing lots of working.
- There are lots of different methods that can help you subtract two- and three-digit numbers in your head.

### Jump strategy for subtraction

- The jump strategy for subtraction is like the jump strategy for addition, but you jump backward instead of forward.
- The number being subtracted is broken into smaller parts and then each part is subtracted one by one.

For example:

$$635 - 127 =$$

Break 127 into 100, 20 and 7.

$$635 - 100 = 535$$

$$535 - 20 = 515$$

$$515 - 7 = 508$$

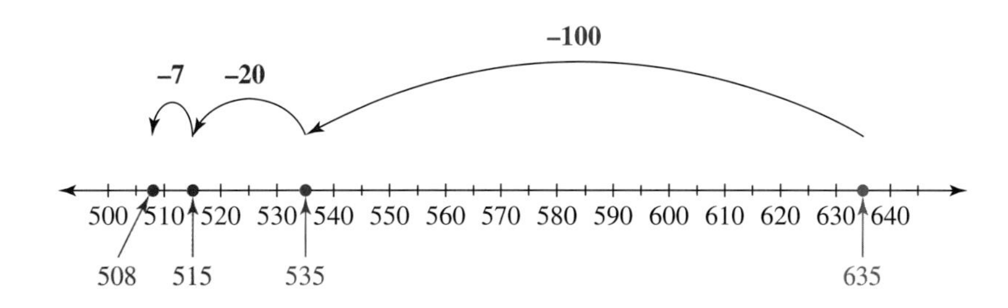


## Adding up to find the difference

- Sometimes, even though you are subtracting, it can be helpful to use addition when trying to subtract numbers mentally.
- Start with the smaller number and add to that number in parts until you get to the larger number.

For example:

$$\begin{aligned} 672 - 613 &= \\ 613 + 7 &= 620 \\ 620 + 50 &= 670 \\ 670 + 2 &= 672 \end{aligned}$$

The numbers added are 7, 50 and 2, which give a total of 59.

Therefore $672 - 613 = 59$.

### WORKED EXAMPLE 7 Using mental strategies to subtract numbers

**Calculate the following.**

**a. 854 – 84 using the jump strategy.**

**b. 436 – 381 using the adding strategy.**

| THINK | WRITE |
|---|---|
| **a.** 1. Write the question. | $854 - 84$ |
| 2. Break 84 into 80 plus 4. | $84 = 80 + 4$ |
| 3. Subtract 80 from 854 to get 774. | $854 - 80 = 774$ |
| 4. Subtract 4 from 774 to get 770. | $774 - 4 = 770$ |
| 5. Write the answer. | $854 - 84 = 770$ |
| **b.** 1. Write the question. | $436 - 381$ |
| 2. Starting at the smaller number, determine what needs to be added to get to the larger number. | $381 + 9 = 390$<br>$390 + 40 = 430$ |
| 3. Add up all the parts. | $430 + 6 = 436$ |
| 4. Write the answer. | $9 + 40 + 6 = 55$<br>$436 - 381 = 55$ |

## Compensation strategy for subtraction

- The compensation strategy for subtraction is like the compensation strategy for addition, except that you will round up the number being subtracted to the nearest ten or hundred and then add the amount used to do the rounding up.

  For example:

$$94 - 49 =$$

Round up 49 to 50 by adding 1.

$$94 - 50 = 44$$

Add 1 to the answer, since 49 was rounded up to 50 by adding 1.

$$44 + 1 = 45$$

$$94 - 49 = 45$$

### WORKED EXAMPLE 8 Using the compensation strategy to subtract numbers

**Calculate 174 – 38 using the compensation strategy.**

| THINK | WRITE |
|---|---|
| 1. Write the question. | $174 - 38$ |
| 2. Round 38 up to the nearest 10 by adding 2. | 38 becomes 40. |
| 3. Subtract 40 from 174. | $174 - 40 = 134$ |
| 4. Add 2 since, 38 was rounded up to 40 by adding 2. | $134 + 2 = 136$ |
| 5. Write the answer. | $174 - 38 = 136$ |

**Checking subtraction by adding**

**Remember, a subtraction can always be checked by adding the two smaller numbers to see if they add up to the larger number.**

**DISCUSSION**

What techniques have you used to mentally add numbers? What shortcuts have you used?

Resources

**eWorkbook** Topic 1 Workbook (worksheets, code puzzle and project) (ewbk-1902)

**Interactivity** Individual pathway interactivity: Strategies for adding and subtracting positive integers (int-8731)

## Exercise 1.3 Strategies for adding and subtracting positive integers

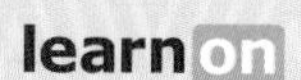

**1.3 Quick quiz** on | **1.3 Exercise**

Individual pathways

| ■ PRACTISE | ■ CONSOLIDATE | ■ MASTER |
|---|---|---|
| 1, 2, 5, 7, 12, 13, 16 | 3, 6, 8, 11, 14, 17 | 4, 9, 10, 15, 18 |

### Fluency

1. **WE4a** Use the jump strategy to calculate the following.
   a. $143 + 85$  b. $1537 + 266$
2. **WE4b** Use the split strategy to calculate the following.
   a. $645 + 261$  b. $370 + 78$
3. **WE5a** Use the compensation strategy to calculate the following.
   a. $471 + 89$  b. $74 + 28$
4. **WE5b** Use the rearrange strategy to calculate the following.
   a. $27 + 36 + 23$  b. $14 + 34 + 16$

### Understanding

5. State whether the following statement is true or false. Show your working.
   $346 + 451$ calculated using the split strategy becomes $300 + 400 + 40 + 50 + 6 + 1$.
6. **MC** $56 + 12 + 14$ is equal to:
   A. 60  B. 82  C. 70  D. 92  E. 72
7. **WE6** Calculate $421 + 54 + 372$ using an appropriate mental strategy.
8. **WE7a** Use the jump strategy to calculate the following.
   a. $88 - 43$  b. $674 - 323$
9. **WE7b** Add up the difference to calculate the following.
   a. $351 - 287$  b. $79 - 46$
10. **WE8** Use the compensation strategy to calculate the following.
    a. $65 - 48$  b. $956 - 729$
11. Calculate $953 - 675$ using an appropriate mental strategy.
12. Fill in the gaps for the following sentences.
    a. All the numbers in the 5-times table end in _________ or _________.
    b. The numbers in the _________-times tables all end in 0.

### Reasoning

13. A student calculates $145 + 671 - 472$. They give their answer as 344. State whether you agree with their answer. Explain your response.

14. A student has shown the following working on a test.

$$641+357=600+10+4+300+70+5=989$$

a. Determine where the student made an error.
b. Determine the correct answer.

15. Twenty children were playing in the playground. Eight of the children left for home, then six more children came to the playground. Determine how many children are in the playground now. Show your working.

### Problem solving

16. There are 763 students at a senior school campus. Of these, 213 students are currently completing exams. Calculate how many students are not completing exams.

17. Harriet has $243 in savings. She adds $93 to her savings account. The next day she spends $175 from her account. Determine how much is left in her savings account.

18. The sum of two numbers is 19 and their difference is 7. Evaluate the two numbers.

# LESSON
# 1.4 Algorithms for adding and subtracting positive integers

**LEARNING INTENTION**

At the end of this lesson you should be able to:
- add and subtract positive integers using algorithms.

## 1.4.1 Addition of positive integers using algorithms

eles-3647

- To add numbers or positive integers, write them in columns according to place value and then add them, starting at the ones column.
- The **sum** is the result obtained when numbers are added together.

**WORKED EXAMPLE 9 Adding integers in columns**

**Arrange these numbers in columns, then add them.**

$$\mathbf{1462+78+316}$$

**THINK**

1. Set out the numbers in columns according to place value.
2. Add the digits in the ones column in your head ($2+8+6=16$). Write the 6 in the ones column of your answer and carry the 1 to the tens column, as shown in the plum colour.
3. Add the digits in the tens column ($1+6+7+1=15$). Write the 5 in the tens column of your answer and carry the 1 to the hundreds column, as shown in pink.

**WRITE**

| | Thousands | Hundreds | Tens | Ones |
|---|---|---|---|---|
| | 1 | $^{1}4$ | $^{1}6$ | 2 |
| | | | 7 | 8 |
| + | | 3 | 1 | 6 |
| | 1 | 8 | 5 | 6 |

4. Add the digits in the hundreds column ($1 + 4 + 3 = 8$). Write 8 in the hundreds column of your answer, as shown in pink. There is nothing to carry.
5. There is only a 1 in the thousands column. Write 1 in the thousands column of your answer.
6. State the answer. — The sum of 1462, 78 and 316 is 1856.

**WORKED EXAMPLE 10 Adding by making a multiple of 10**

**Perform the addition $27 + 19 + 141 + 73$ by finding suitable pairs of numbers to make multiples of 50.**

| THINK | WRITE |
|---|---|
| 1. Write the question. | $27 + 19 + 141 + 73$ |
| 2. Look for pairs of numbers that can be added to make a multiple of 10. Reorder the sum, pairing these numbers. | $= (27 + 73) + (141 + 19)$ |
| 3. Add the number pairs. | $= 100 + 160$ |
| 4. Complete the addition. | $= 260$ |

eles-3648

## 1.4.2 Commutative and associative laws for addition

- The **Commutative Law for addition** means that you can add numbers in any order. For example, $3 + 4$ gives the same result as $4 + 3$.
- The **Associative Law for addition**, simply stated, means that the order in which additions are calculated is not important and does not change the result. For example, $2 + 3 + 7$ could be calculated as $5 + 7 = 12$ or $2 + 10 = 12$
  This is a very useful property when performing mental calculations.

**ACTIVITY: There's something about Gauss**

In the late 1780s, a German teacher gave his class the task of adding the numbers from 1 to 100. One student came up with the correct answer in less than a minute. That student, Johann Carl Friedrich Gauss, used grouping in pairs to work out the sum. He paired the smallest number with the largest number, then the second smallest with the second largest, and so on. Gauss went on to become one of the world's most famous mathematicians.

Try the technique that Gauss used with the numbers from 1 to 10, then 1 to 20, then 1 to 50 and finally 1 to 100. Did you notice any patterns?

## 1.4.3 Subtraction of positive integers using algorithms

eles-3649

- Subtraction is the opposite (or 'inverse') of addition. To work out 800 – 360 you could ask the question, 'What do I need to add to 360 to get 800?' The answer is 440.
- The **difference** is the result when one number is subtracted from another number. For example, the difference between 800 and 360 is 440.
- One of the main distinctions between addition and subtraction is the fact that the order in which you subtract numbers is very important. For example, 6 – 3 does not give the same result as 3 – 6.
- This means that subtraction is neither commutative nor associative.

### Subtraction of positive integers with no borrowing

- To subtract two numbers, write them in columns according to their place values and then subtract each column in order, starting with the ones column.

**WORKED EXAMPLE 11 Subtracting numbers without borrowing**

**Perform the subtraction 395 – 174.**

| THINK | WRITE |
|---|---|
| 1. Set out the difference in columns according to place value. | Hundreds Tens Ones<br>$\begin{array}{r} 3\ 9\ 5 \\ -\ 1\ 7\ 4 \\ \hline \\ \hline \end{array}$ |
| 2. Subtract the digits in the ones column (5 – 4 = 1). Write the 1 in the ones column of your answer. | $\begin{array}{r} 3\ 9\ 5 \\ -\ 1\ 7\ 4 \\ \hline 1 \\ \hline \end{array}$ |
| 3. Subtract the digits in the tens column (9 – 7 = 2). Write the 2 in the tens column of your answer. | $\begin{array}{r} 3\ 9\ 5 \\ -\ 1\ 7\ 4 \\ \hline 2\ 1 \\ \hline \end{array}$ |
| 4. Subtract the digits in the hundreds column (3 – 1 = 2). Write the 2 in the hundreds column of your answer. | $\begin{array}{r} 3\ 9\ 5 \\ -\ 1\ 7\ 4 \\ \hline 2\ 2\ 1 \\ \hline \end{array}$ |
| 5. State the answer. | The difference between 395 and 174 is 221. |

Not all subtraction problems are as simple as the one shown in Worked example 11. For example, working out 32 – 14 requires a different approach because you can't subtract 4 from 2 in the ones column. A subtraction like this requires borrowing.

$$\begin{array}{r} 3\ 2 \\ -\ 1\ 4 \\ \hline ? \\ \hline \end{array}$$

## Subtraction of positive integers with borrowing

- The most common method of subtraction is to use borrowing, which is called the **decomposition method**.
- It's called 'decomposition' because the larger number is decomposed, or taken apart.
- A visual representation of the decomposition method is shown here to help you understand how it works.

## Visual representation of subtracting using decomposition

Evaluate $32 - 14$.

1. Decompose each number by breaking it up into place value parts (tens and ones in this case).

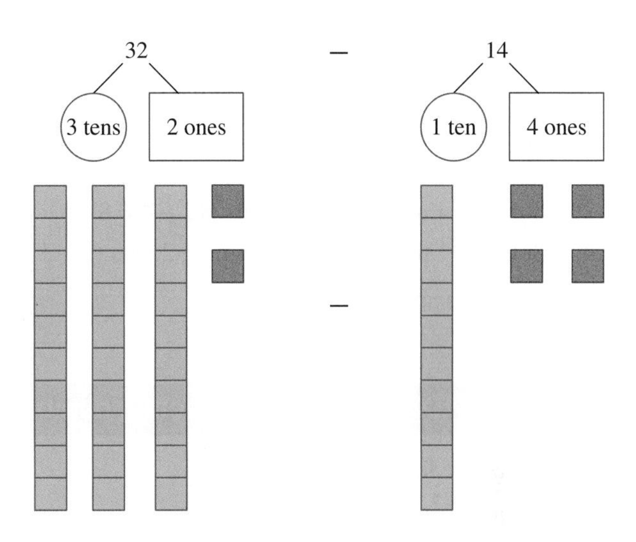


2. Since we can't take 4 ones from 2 ones, we instead borrow a block of 10 and split them into ones.

Borrowing step:

3. Rewrite 32 as 2 tens and 12 ones.
*Note:* Effectively we have borrowed a ten from the tens column and given it to the ones column so that we can work out the subtraction.

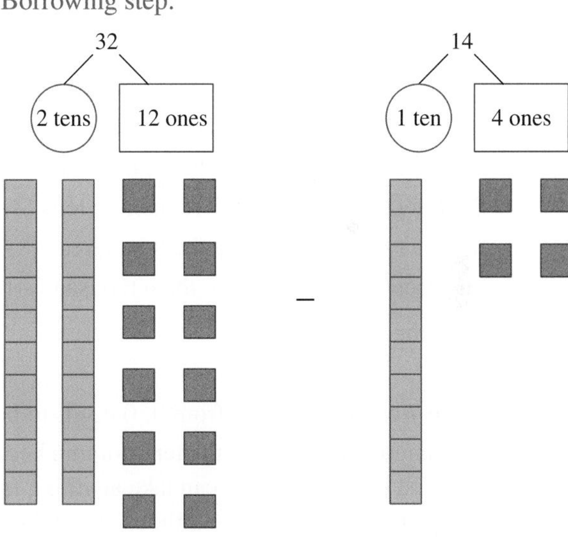


4. We can now complete the subtraction.
   - Subtracting the ones first gives:
     $12 - 4 = 8$ ones
   - Subtracting the tens next gives:
     $2 - 1 = 1$ ten
   - This gives us 1 ten and 8 ones left over after the subtraction.
   - This is the same as 18 individual blocks.

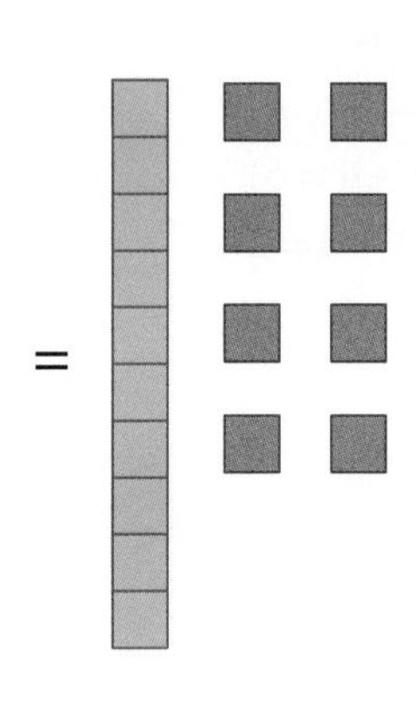


5. State the answer.

$32 - 14 = 18$

## Algorithmic approach to subtraction using decomposition

The subtraction 32 – 14 can be solved using the following algorithm (set of steps).

1. Set out the difference in columns according to place value.

$$\begin{array}{r} 3\ 2 \\ -\ 1\ 4 \\ \hline \\ \hline \end{array}$$

2. Because 4 can't be subtracted from 2, take one 10 from the tens column of the larger number and add it to the ones column of the same number. So 2 becomes 12, and the 3 tens become 2 tens.

$$\begin{array}{r} {}^{2}\not{3}\ \ {}^{1}2 \\ -\ 1\ \ \ 4 \\ \hline 1\ \ \ 8 \\ \hline \end{array}$$

3. Subtract 4 units from 12 units (12 – 4 = 8).
4. Subtract 1 ten from the remaining 2 tens (2 – 1 = 1).
5. Write the answer. 32 – 14 = 18

### WORKED EXAMPLE 12 Subtracting numbers using decomposition

**Subtract the following:**

a. **6892 – 467**

b. **3000 – 467**

**THINK**

a. 1. Since 7 cannot be subtracted from 2, take one ten from the tens column of the larger number and add it to the units column of the same number. The 2 becomes 12 and the 9 tens become 8 tens.
2. Subtract the 7 units from the 12 units (12 – 7 = 5).
3. Subtract 6 tens from the 8 remaining tens (8 – 6 = 2).
4. Subtract 4 hundreds from the 8 hundreds (8 – 4 = 4).
5. Subtract 0 thousands from the 6 thousands (6 – 0 = 6).
6. State the answer.

**WRITE**

a.

$$\begin{array}{r} 6\ 8\ {}^{8}\not{9}\ {}^{1}2 \\ -\ \ \ 4\ 6\ 7 \\ \hline 6\ 4\ 2\ 5 \\ \hline \end{array}$$

6892 – 467 = 6425

**THINK**

b. 1. Since 7 cannot be taken from 0, 0 needs to become 10.
2. We cannot take 10 from the tens column because it is also 0. The first column that we can take anything from is the thousands, so 3000 is decomposed to 2 thousands, 9 hundreds, 9 tens and 10 units.
3. Subtract the units (10 – 7 = 3).
4. Subtract the tens (9 – 6 = 3).
5. Subtract the hundreds (9 – 4 = 5).
6. Subtract the thousands (2 – 0 = 2).
7. State the answer.

**WRITE**

b.

$$\begin{array}{r} {}^{2}\not{3}\ {}^{9}\not{0}\ {}^{9}\not{0}\ {}^{1}0 \\ -\ \ \ 4\ 6\ 7 \\ \hline 2\ 5\ 3\ 3 \\ \hline \end{array}$$

3000 – 467 = 2533

## Digital technology

Scientific calculators can evaluate sums of two or more numbers.

DEG
17245+378+936
18559

Scientific calculators can evaluate differences between two numbers.

DEG
25487-8954
16533

**on** Resources

**Workbook** Topic 1 Workbook (worksheets, code puzzle and project) (ewbk-1902)

**Interactivities** Individual pathway interactivity: Algorithms for adding and subtracting positive integers (int-8732)
Addition of positive integers (int-3922)
Subtraction of positive integers (int-3924)

## Exercise 1.4 Algorithms for adding and subtracting positive integers

**learn on**

**1.4 Quick quiz** 

**1.4 Exercise**

### Individual pathways

| ■ PRACTISE | ■ CONSOLIDATE | ■ MASTER |
|---|---|---|
| 1, 2, 6, 9, 12, 13, 18, 22, 25 | 3, 5, 7, 14, 15, 17, 19, 23, 26 | 4, 8, 10, 11, 16, 20, 21, 24, 27, 28 |

### Fluency

1. Answer these questions by doing the working in your head.
   a. $7+8$
   b. $20+17$
   c. $195+15$
   d. $227+13$
   e. $1000+730$

2. Answer these questions by doing the working in your head.
   a. $17\,000+1220$
   b. $125\,000+50\,000$
   c. $2+8+1+9$
   d. $6+8+9+3+2+4+1+7$
   e. $12+5+3+7+15+8$

3. Add the following numbers, setting them out in columns as shown. Check your answers using a calculator.
   a. $\begin{array}{r} 34 \\ +\ 65 \\ \hline \\ \hline \end{array}$
   b. $\begin{array}{r} 68\,069 \\ 317 \\ 8 \\ +\,4254 \\ \hline \\ \hline \end{array}$
   c. $\begin{array}{r} 399 \\ 1489 \\ 2798 \\ +\,8943 \\ \hline \\ \hline \end{array}$

4. **WE9** Arrange the following numbers in columns, then add them.

   **a.** $137 + 841$
   **b.** $149 + 562 + 55$
   **c.** $376 + 948 + 11$
   **d.** $8 + 12\,972 + 59 + 1423$
   **e.** $1700\,245 + 378 + 930$

   Check your answers with a calculator.

5. **WE10** Mentally perform each of the following additions by pairing suitable numbers together.

   a. $56 + 87 + 24 + 13$
   b. $74 + 189 + 6 + 11$
   c. $98 + 247 + 305 + 3 + 95 + 42$
   d. $180 + 364 + 59 + 141 + 47 + 20 + 16$

6. Answer the following questions without using a calculator.

   a. $11 - 5$
   b. $53 - 30$
   c. $100 - 95$
   d. $150 - 25$
   e. $1100 - 200$

7. Work out the following without using a calculator.

   a. $1700 - 1000$
   b. $100 - 20 - 10$
   c. $1000 - 50 - 300 - 150$
   d. $24 - 3 - 16$
   e. $54 - 28$

8. Calculate the following.

   **a.** $10 + 8 - 5 + 2 - 11$
   **b.** $40 + 15 - 35$
   **c.** $120 - 40 - 25$
   **d.** $15 + 45 + 25 - 85$
   **e.** $100 - 70 + 43$
   **f.** $1000 - 400 + 250 + 150 + 150$

9. **WE11&12** Work out the following subtractions.

   a. $167 - 132$
   b. $47\,836 - 12\,713$
   c. $642\,803 - 58\,204$

## Understanding

10. Solve the following subtractions.

    **a.** $664 - 397$
    **b.** $12\,900 - 8487$
    **c.** $69\,000 - 3561$
    **d.** $2683 - 49$

    Check your answers using a calculator.

11. Work out the following subtractions.

    **a.** $70\,400 - 1003$
    **b.** $27\,321 - 25\,768$
    **c.** $812\,741 - 462\,923$
    **d.** $23\,718\,482 - 4\,629\,738$

    Check your answers using a calculator.

12. Hella was performing in a ballet and needed to buy a tutu, ballet shoes and white tights. Based on the prices shown in the photograph, calculate how much she spent in total on her costume.

$150
$12
$85

13. A print dictionary is split into two volumes. There are 1544 pages in the A–K volume and 1488 pages in the L–Z volume. Calculate how many pages the dictionary has in total.

14. Nathan has taken his sister and her friends to lunch for her birthday. Determine what the total cost of the items shown will be.

15. Of all the world's rivers, the Amazon in South America and the Nile in Africa are the two longest. The Amazon is 6437 kilometres in length and the Nile is 233 kilometres longer than the Amazon. Calculate the length of the Nile.

16. An **arithmagon** is a triangular figure in which the two numbers at the end of each line add to the number along the line, like in the example shown.
Solve each of these arithmagons.

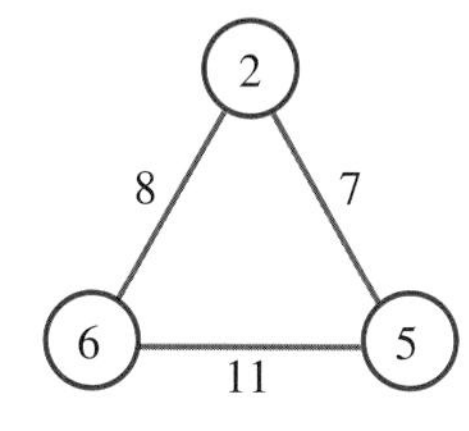


a.
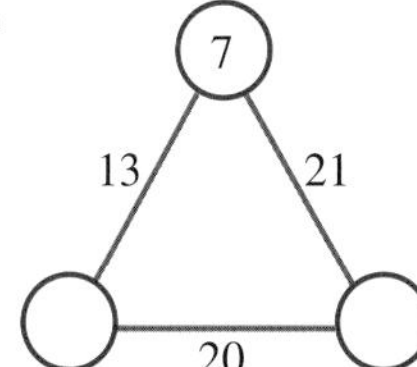


b.
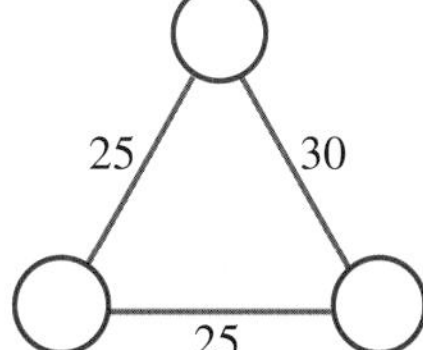


c.
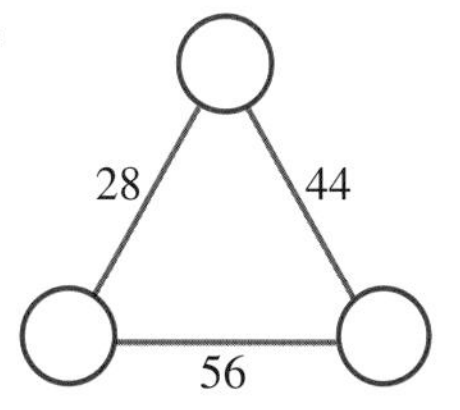


17. The following photographs show three of the highest waterfalls in the world.
Determine how much higher:
    a. Victoria Falls is than Iguazu Falls
    b. Iguazu Falls is than Niagara Falls
    c. Victoria Falls is than Niagara Falls.

18. Prithvi received a box of 36 chocolates. They ate 3 on Monday, 11 on Tuesday and gave 7 away on Wednesday. Calculate how many they had left.

19. A crowd of 24 083 attended an NRL match between Canterbury-Bankstown Bulldogs and St George Illawarra Dragons. If 14 492 people supported the Bulldogs and the rest supported the Dragons, calculate how many supporters the Dragons had.

**20.** A school bus left Laurel High School with 31 students aboard. Thirteen of these passengers disembarked at Hardy Railway Station. The bus collected 24 more students at Hardy High School and a further 11 students disembarked at Laurel Swimming Pool. Calculate how many students were still on the bus.

**21.** Shu-Ling and Ty were driving from Melbourne to Sydney for a holiday. The distance between Melbourne and Sydney via the Hume Highway is 867 kilometres, but they chose the more scenic Princes Highway even though the distance is 1039 kilometres. They drove to Lakes Entrance on the first day (339 kilometres), a further 347 kilometres to Narooma on the second day, and arrived in Sydney on the third day.

**a.** Calculate how much further it is from Melbourne to Sydney via the Princes Highway than via the Hume Highway.

**b.** Determine how far Shu-Ling and Ty travelled on the third day.

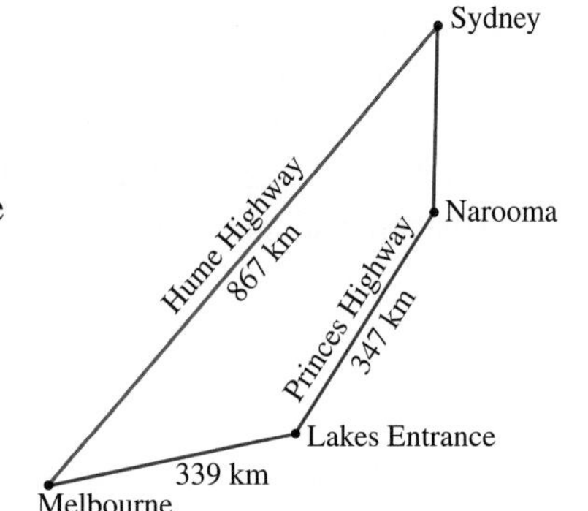


## Reasoning

**22.** Fill in the missing numbers. The * can represent any digit.

**a.**
$$\begin{array}{r} 6*8*2* \\ -488417 \\ \hline *499*4 \\ \hline \end{array}$$

**b.**
$$\begin{array}{r} 3*9* \\ -*6*5 \\ \hline 1*07 \\ \hline \end{array}$$

**23.** In less than 10 seconds, without using a calculator, calculate the answer to 6 849 317 − 999 999. Explain how you reached your answer.

**24.** A beetle has fallen into a hole that is 15 metres deep. It is able to climb a distance of 3 metres during the day, but at night the beetle is tired and must rest. However, during the night it slides back 1 metre. How many days will it take the beetle to reach the top of the hole to freedom? Explain your answer.

## Problem solving

**25.** Five termites start munching through a log.
Pixie is 19 mm ahead of Bitsie.
Trixie is 6 mm ahead of Mixie.
Mixie is twice as far ahead as Itsie.
Itsie is 10 mm behind Bitsie.
Mixie has eaten through 52 mm.

Determine which termite has eaten the most. Show your working.

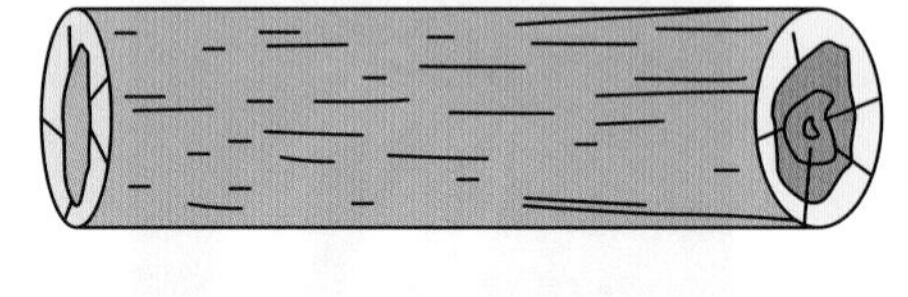

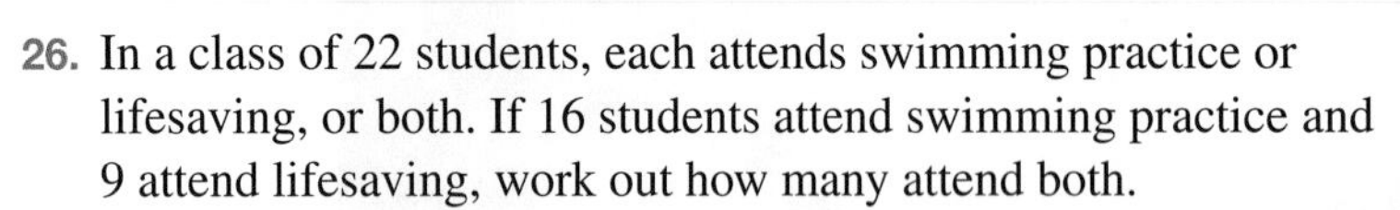

**26.** In a class of 22 students, each attends swimming practice or lifesaving, or both. If 16 students attend swimming practice and 9 attend lifesaving, work out how many attend both.

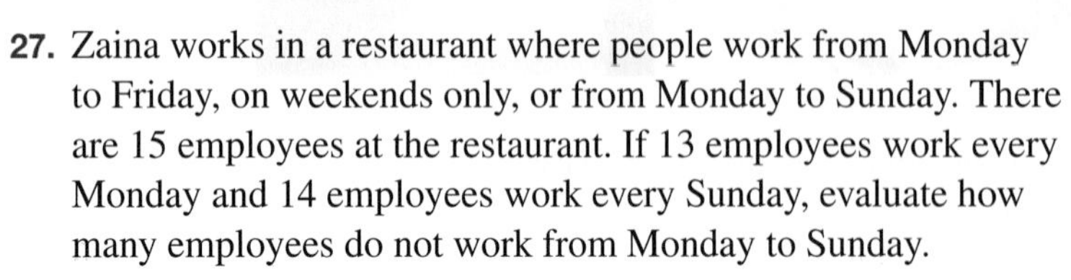

**27.** Zaina works in a restaurant where people work from Monday to Friday, on weekends only, or from Monday to Sunday. There are 15 employees at the restaurant. If 13 employees work every Monday and 14 employees work every Sunday, evaluate how many employees do not work from Monday to Sunday.

**28.** There are 20 chocolate bars on the table all containing only fruit, only nuts, or fruit and nuts together. If 12 chocolate bars contain nuts and 14 chocolate bars contain fruit, evaluate the number of chocolate bars that contain both fruit and nuts.

# LESSON
# 1.5 Multiplying positive integers

**LEARNING INTENTION**

At the end of this lesson you should be able to:
- multiply positive integers using mental strategies and algorithms.

## 1.5.1 Multiplying positive integers

eles-3650

- **Multiplication** is a short way of performing repeated addition of the same number.
  For example, $3+3+3+3+3$ represents 5 groups of 3 and can be written as $5 \times 3$.
- The **product** is the answer you get when numbers are multiplied.
  For example, the product of 5 and 3 is 15 because $5 \times 3 = 15$.
- Multiplication is commutative because the order in which numbers are multiplied is not important.
  For example, $3 \times 8 = 8 \times 3$.

### Multiplying using diagrams

- The following diagram shows 2 groups of balls with 4 balls in each group. There are 8 balls in total.
  This can be written as a multiplication.

$$2 \times 4 = 8$$

Group 1 Group 2

- Instead, what if we looked at the picture as 4 groups of balls with 2 balls in each group? Would we get the same answer? Count the balls and see. There are still 8 balls in total.

$$4 \times 2 = 8$$

Group 1 Group 2 Group 3 Group 4

**Multiplying two numbers**

**When two numbers are multiplied, the order does not matter; the answer will be the same.**

**WORKED EXAMPLE 13 Using a diagram to multiply numbers**

**a. Draw a diagram to show $2 \times 6$.** **b. State what $2 \times 6$ is equal to.**

| THINK | WRITE/DRAW |
|---|---|
| a. Draw a diagram with 2 groups of 6. | Group 1 Group 2 |
| b. Count how many circles there are altogether. | There are 12 circles. |
| Write the answer. | $2 \times 6 = 12$ |

- To multiply small numbers, we learn the times tables. We can also use diagrams.
- To multiply larger numbers, we need to use other methods. These methods are discussed later in this topic.

## Multiplying using expanded form

- In this method of multiplication the larger number being multplied is broken up into expanded form first, then each component is multiplied by the smaller number.
  For example, here is how the expanded form can be used to calculate $215 \times 4$.
  The expanded form of $215 = 200 + 10 + 5$

$$\begin{aligned} 215 \times 4 &= (200 \times 4) + (10 \times 4) + (5 \times 4) \\ &= 800 + 40 + 20 \\ &= 860 \end{aligned}$$

## Multiplying using the area model

- The area model of multiplication involves using the expanded form of the larger number and constructing rectangles for each component, arranged according to place value.
  This diagram shows how to calculate $215 \times 4$ using an area model. (*Note:* This graph is not drawn to scale.)

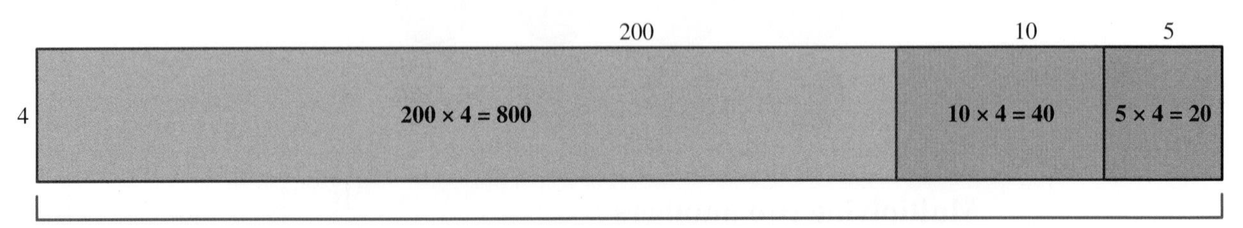

## 1.5.2 The short multiplication algorithm

eles-3651

- The short multiplication algorithm can be used when multiplying a large number by a single digit number.
- Start by lining up the numbers in columns, with one number under the other number. Place the larger number on top.
- Next, multiply each place value in the larger number by the smaller number, starting with the units column.
- Write the answer to each multiplication underneath the smaller number in the appropriate column.
- If an answer has two place values, carry the larger place value to the next column on the left and add it to the result of the next multiplication.

$$\begin{array}{r} {}^{2} \\ 215 \\ \times \quad 4 \\ \hline 860 \end{array}$$

1. $5 \times 4 = 20$
2. Write the 0 in the units column and carry the 2 over to the tens column.
3. Perform the next multiplication and add the carried number to the result.
   $1 \times 4 = 4;\ 4 + 2 = 6$
4. Write the 6 in the tens column.
5. $2 \times 4 = 8$
6. Write the 8 in the hundreds column.

### WORKED EXAMPLE 14 Multiplying using short multiplication

**Calculate $1456 \times 5$.**

| THINK | WRITE |
|---|---|
| 1. Write the numbers in columns according to place value, with the larger number on top. | $\begin{array}{r} 1\,4\,5\,6 \\ \times \quad 5 \\ \hline \\ \hline \end{array}$ |
| 2. Multiply the units by 5 ($5 \times 6 = 30$). Write down the 0 and carry the 3 to the tens column.<br>3. Multiply the tens by 5 and add the carried number ($5 \times 5 = 25; 25 + 3 = 28$). Write the 8 in the tens column and carry the 2 to the hundreds column.<br>4. Multiply the hundreds by 5 and add the carried number ($5 \times 4 = 20; 20 + 2 = 22$). Write the last 2 in the hundreds column and carry the other 2 to the thousands column.<br>5. Multiply the thousands by 5 and add the carried number ($5 \times 1 = 5;\ 5 + 2 = 7$). Write 7 in the thousands column of the answer. | $\begin{array}{r} {}^{2}1\,{}^{2}4\,{}^{3}5\,6 \\ \times \quad 5 \\ \hline 7\,2\,8\,0 \\ \hline \end{array}$ |
| 6. Write the answer. | $1456 \times 5 = 7280$ |

## 1.5.3 The long multiplication algorithm

eles-3652

- The long multiplication algorithm is used to multiply numbers that both have more than one digit. The process is the same as that for short multiplication, but it is repeated for each additional digit.
- When multiplying two numbers that both have more than one digit, a 0 is added in the second row.

**WORKED EXAMPLE 15 Multiplying whole numbers**

**Evaluate the following.**

a. **$547 \times 6$**

b. **$35 \times 62$**

| THINK | WRITE |
|---|---|
| a. 1. Set up the multiplication in columns, with the smaller number on the bottom. | $\begin{array}{r} 547 \\ \times\ 6 \\ \hline \end{array}$ |
| 2. Multiply each digit in 547 by 6.<br>• $7 \times 6 = 42$. Write 2 and carry 4.<br>• $4 \times 6 + 4 = 28$. Write 8 and carry 2.<br>• $5 \times 6 + 2 = 32$. Write 32. | $\begin{array}{r} {}^{2}5\,{}^{4}4\ 7 \\ \times\ 6 \\ \hline 3\ 2\ 8\ 2 \end{array}$ |
| 3. Write the answer. | $547 \times 6 = 3282$ |
| b. 1. Set up the multiplication in columns. | $\begin{array}{r} 35 \\ \times\ 62 \\ \hline \end{array}$ |
| 2. Multiply 35 by 2.<br>• $5 \times 2 = 10$. Write 0 and carry 1.<br>• $3 \times 2 + 1 = 7$. 6+1= 7. Write 7. | $\begin{array}{r} {}^{1}3\ 5 \\ \times\ 6\ 2 \\ \hline 7\ 0 \\ \hline \end{array}$ |
| 3. Add a 0 to the second row because you are multiplying by a number in the tens column. | $\begin{array}{r} {}^{3}3\ 5 \\ \times\ 6\ 2 \\ \hline 7\ 0 \\ 2\ 1\ 0 \\ \hline \end{array}$ |
| 4. Multiply 35 by 6.<br>• $5 \times 6 = 30$. Write 0 and carry 3.<br>• $3 \times 6 + 3 = 21$. Write 21.<br>5. Add 70 and 2100 to get the answer. | $\begin{array}{r} {}^{3}3\ 5 \\ \times\ 6\ 2 \\ \hline 7\ 0 \\ +2\ 1\ 0\ 0 \\ \hline 2\ 1\ 7\ 0 \\ \hline \end{array}$ |
| 6. Write the answer. | $35 \times 62 = 2170$ |

## WORKED EXAMPLE 16 Multiplying using long multiplication

**Calculate $1456 \times 132$ using long multiplication.**

| THINK | WRITE |
|---|---|
| 1. Set the product up in columns according to place value, writing the larger number on top. | $\begin{array}{r} 1456 \\ \times\ 132 \\ \hline \end{array}$ |
| 2. Multiply the larger number by the units digit in the smaller number using short multiplication ($1456 \times 2 = 2912$). Write the answer directly below the problem. | $\begin{array}{r} 1\,{}^{1}4\,{}^{1}5\,6 \\ \times\ 1\,3\,2 \\ \hline 2\,9\,1\,2 \end{array}$ |
| 3. Place a zero in the units column, as shown in pink, when multiplying the larger number by the tens digit of the smaller number ($1456 \times 3 = 4368$) This is because you are really working out $1456 \times 30 = 43\,680$. Write the answer directly below the previous answer. | $\begin{array}{r} {}^{1}1\,{}^{1}4\,{}^{1}5\,6 \\ \times\ 1\,3\,2 \\ \hline 2\,9\,1\,2 \\ 4\,3\,6\,8\,0 \end{array}$ |
| 4. Place zeros in the units and tens columns, as shown in pink, when multiplying the larger number by the hundreds digit of the smaller number ($1456 \times 1 = 1456$). This is because you are really working out $1456 \times 100 = 145\,600$. Write the answer directly below the previous answer. | $\begin{array}{r} 1456 \\ \times\ 132 \\ \hline 2912 \\ 43680 \\ +145600 \\ \hline 192192 \end{array}$ |
| 5. Add the numbers in each column of the three rows to determine the answer. | |
| 6. Write the answer. | $1456 \times 132 = 192\,192$ |

- The **Distributive Law** applies only to multiplication. It states that $a\,(b + c) = ab + ac$. The Distributive Law helps with mental calculation because it means, for example, that $8 \times 13$ can be thought of as:

$$\begin{aligned} 8 \times 13 &= 8 \times (3 + 10) \\ &= 8 \times 3 + 8 \times 10 \\ &= 24 + 80 \\ &= 104 \end{aligned}$$

### Digital technology

Scientific calculators can evaluate the multiplication of large numbers.

DEG
57835*2074
119949790

## 1.5.4 Mental strategies for multiplication

eles-4838

- In many cases it is not practical to use pen and paper (or even a calculator) to perform multiplication.
- There are mental strategies that can make multiplication easier.
- Multiplication is associative — this means smaller numbers can be paired up for multiplication.

  For example, $\begin{aligned} 2 \times 17 \times 5 &= 17 \times (2 \times 5) \\ &= 17 \times 10 \\ &= 170 \end{aligned}$

  This calculation has been simplified by finding the pair of numbers that multiply to equal 10.

### WORKED EXAMPLE 17 Using mental strategies for multiplication

**Use mental strategies to calculate $4 \times 23 \times 25$.**

| THINK | WRITE |
|---|---|
| 1. Write the question. | $4 \times 23 \times 25$ |
| 2. Look for a number pair that makes a simpler multiplication and rearrange. $4 \times 25 = 100$, and multiplying by 100 is simpler than multiplying by either 4 or 25. | $= 23 \times (4 \times 25)$ |
| 3. Mentally calculate $4 \times 25$. | $= 23 \times 100$ |
| 4. Mentally calculate the final answer. | $= 2300$ |

### WORKED EXAMPLE 18 Using mental strategies for multiplication

**Use a mental strategy to calculate $34 \times 200$.**

| THINK | WRITE |
|---|---|
| 1. Write the question. | $34 \times 200$ |
| 2. Write 200 as $2 \times 100$. | $= 34 \times 2 \times 100$ |
| 3. Mentally calculate $34 \times 2$. | $= 68 \times 100$ |
| 4. Mentally calculate $68 \times 100$. | $= 6800$ |

## Multiplying multiples of 10 or 100

**If both numbers are multiples of 10, 100 and so on, ignore the zeros, multiply the remaining numbers, then add the total number of zeros to the answer.**

**For example, $900 \times 6000 = 5\,400\,000$.**

- Consider the multiplication $9 \times 58$. This multiplication can be regarded as $10 \times 58 - 1 \times 58$. Using this way of writing the multiplication, the answer can be mentally calculated by multiplying 58 by 10 and then subtracting 58 from the answer. This can be thought of as the 'multiply by 10' strategy.

### WORKED EXAMPLE 19 Using mental strategies for multiplication

**Use a mental strategy to calculate $77 \times 9$.**

| THINK | WRITE |
|---|---|
| 1. Write the question. | $77 \times 9$ |
| 2. Use the strategy of 'multiply by 10'. | $= 77 \times 10 - 77 \times 1$ |
| 3. Calculate $77 \times 10$ and then subtract 77. | $= 770 - 77$<br>$= 693$ |

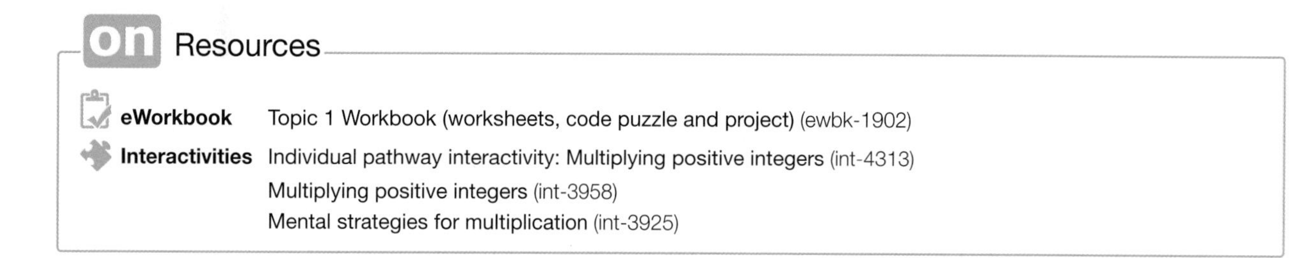


## Exercise 1.5 Multiplying positive integers

learnon

1.5 Quick quiz on | 1.5 Exercise

### Individual pathways

| ■ PRACTISE | ■ CONSOLIDATE | ■ MASTER |
|---|---|---|
| 1, 2, 6, 8, 11, 13, 16, 17, 22, 23, 24, 27, 30, 33 | 3, 4, 9, 12, 14, 18, 19, 25, 26, 31, 34 | 5, 7, 10, 15, 20, 21, 28, 29, 32, 35, 36 |

### Fluency

1. **WE13** Represent each of the following as a diagram and then use the diagram to calculate the product.
   a. $8 \times 3$ b. $6 \times 9$ c. $12 \times 5$ d. $13 \times 2$ e. $15 \times 3$
2. Calculate the product of each of the following without using a calculator.
   a. $25 \times 4$ b. $45 \times 2$ c. $16 \times 2$ d. $21 \times 3$ e. $54 \times 2$
3. Calculate the product of each of the following without using a calculator.
   a. $25 \times 3$ b. $3 \times 4 \times 6$ c. $3 \times 3 \times 3$ d. $5 \times 6 \times 3$ e. $8 \times 5 \times 2$
4. Calculate $267 \times 3$ by rewriting the larger number in expanded form.
5. Use an area model to:
   a. show $523 \times 4$
   b. calculate $523 \times 4$.
6. **WE14** Calculate the following using short multiplication.
   a. $16 \times 8$ b. $137 \times 9$ c. $857 \times 3$ d. $4920 \times 5$
7. Calculate the following using short multiplication.
   a. $7888 \times 8$ b. $2015 \times 8$ c. $10\,597 \times 6$ d. $41\,060 \times 12$
   Check your answers using a calculator.
8. **WE16** Calculate the following using long multiplication.
   a. $52 \times 44$ b. $97 \times 31$ c. $59 \times 28$
9. Calculate the following using long multiplication.
   a. $16 \times 57$ b. $850 \times 76$ c. $407 \times 53$
10. Calculate the following using long multiplication.
    a. $80\,055 \times 27$ b. $57\,835 \times 476$ c. $8027 \times 215$
    Check your answers using a calculator.

11. WE17 Use mental strategies to calculate the following.

a. $2 \times 8 \times 5$ b. $4 \times 19 \times 25$ c. $50 \times 45 \times 2$ d. $4 \times 67 \times 250$

12. WE18 Use mental strategies to calculate the following.

a. $45 \times 20$ b. $62 \times 50$ c. $84 \times 200$ d. $86 \times 2000$

13. Calculate each of the following.

a. $200 \times 40$ b. $600 \times 800$ c. $1100 \times 5000$ d. $900\,000 \times 7000$

14. Calculate each of the following.

a. $90 \times 80$ b. $800 \times 7000$ c. $9000 \times 6000$ d. $12\,000 \times 1100$

15. WE19 Use mental strategies to calculate each of the following.

a. $34 \times 9$ b. $628 \times 9$ c. $75 \times 99$ d. $26 \times 8$

16. a. Calculate $56 \times 100$.
b. Calculate $56 \times 10$.
c. Use your answers to parts a and b to calculate the answer to $56 \times 90$.

17. Use the method demonstrated in question 16 to calculate each of the following.

a. $48 \times 90$ b. $125 \times 90$ c. $32 \times 900$

18. a. Calculate $25 \times 6$.
b. Multiply your answer to part a by 2.
c. Now calculate $25 \times 12$.

19. Use one of the methods demonstrated in this lesson to mentally calculate the value of each of the following.

a. $15 \times 12$ b. $70 \times 12$ c. $40 \times 16$ d. $34 \times 20$

20. Answer the following questions.
a. Calculate the value of $9 \times 10$.
b. Calculate the value of $9 \times 3$.
c. Calculate the value of $9 \times 13$.

21. Use one of the methods demonstrated in this lesson to mentally calculate the value of each of the following.

a. $25 \times 13$ b. $24 \times 13$

## Understanding

22. A school has eight Year 7 classes with 26 students in each class. Calculate how many students there are in Year 7 in total.

23. A shop owner sells 84 bananas each day. Calculate how many bananas are sold in 2 weeks.

24. Ezra wants to make a telephone call to their friend Aasuka, who lives in San Francisco. The call will cost \$3 per minute. If Ezra speaks to Aasuka for 24 minutes:

a. calculate what the call will cost
b. calculate what Ezra would pay if they made this call every month for 2 years.

25. Santilla is buying some generators. The generators cost $12 000 each. She needs 11 of them. Calculate how much they will cost her in total.

26. Julie was saving money to buy a digital camera. She was able to save $75 each month.
   a. Calculate how much she saved over 9 months.
   b. Calculate how much she saved over 16 months.
   c. If Julie continues to save money at the same rate,calculate how much she will save over a period of 3 years.

27. A car can travel 14 kilometres using 1 litre of fuel. Calculate how far it could travel with 35 litres of fuel.

28. In 1995 a team of British soldiers in Hamelin, Germany, constructed a bridge in the fastest time ever. The bridge spanned an 8-metre gap and took the soldiers 8 minutes and 44 seconds to build. Determine how many seconds in total it took to build the bridge.

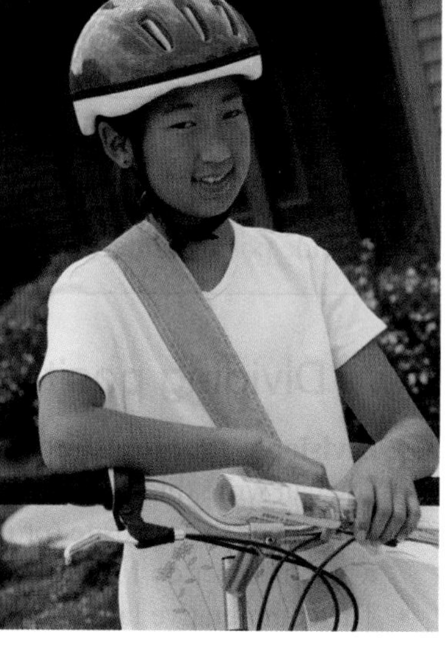

29. Narissa does a paper round each morning before school. She travels 2 kilometres each morning on her bicycle, delivers 80 papers and is paid $35. She does her round each weekday.
   a. Calculate how far she travels in 1 week.
   b. Determine how much she gets paid in 1 week.
   c. Calculate how far she travels in 12 weeks.
   d. Calculate how much she would be paid over 52 weeks.
   e. Calculate how many papers she would deliver in 1 week.
   f. Determine how many papers she would deliver in 52 weeks.

### Reasoning

30. Jake says the product of 42 and 35 is 1428. His friend Ben worked out the answer to be 1470. State who is correct, Jake or Ben. Explain your answer.

31. Explain whether the mathematical statement $23 \times 4 > 20 \times 7$ is true.

32. Imagine a simplified form of a car numberplate that consists of one letter followed by one number, from A0 up to Z9. Determine how many numberplates could be issued under this system.
   Remember that all numberplates issued have to be different.

### Problem solving

33. Evaluate the product of 4, 16 and 9 using expanded form.

34. Consider numbers with two identical digits multiplied by 99. Determine:

$$11 \times 99,\ 22 \times 99,\ 33 \times 99$$

   Can you see a pattern? Explain it.
   Without using a calculator or long multiplication, write down the answers to the following.

| | | |
|---|---|---|
| a. $44 \times 99$ | b. $55 \times 99$ | c. $66 \times 99$ |
| d. $77 \times 99$ | e. $88 \times 99$ | f. $99 \times 99$ |

## WORKED EXAMPLE 20 Dividing using short division

**Evaluate the following.**

**a.** $834 \div 2$ **b.** $466 \div 3$

| THINK | WRITE |
|---|---|
| **a. 1.** Write the equation as a short division. | $2\overline{)834}$ |
| **2.** Divide each digit by 2.<br>• $8 \div 2 = 4$. Write 4 on top.<br>• $3 \div 2 = 1$ r 1. Write 1 on top and carry the remainder of 1 to the next digit. The next digit becomes 14.<br>• $14 \div 2 = 7$. Write 7 on top. | $\begin{array}{r} 417 \\ 2\overline{)83^{1}4} \end{array}$ |
| **3.** Write the answer. | $834 \div 2 = 417$ |
| **b. 1.** Write the equation as a short division. | $3\overline{)466}$ |
| **2.** Divide each digit by 3.<br>• $4 \div 3 = 1$ r 1. Write 1 on top and carry the remainder of 1 to the next digit. The next digit becomes 16.<br>• $16 \div 3 = 5$ r 1. Write 5 on top and carry the remainder of 1 over to the next digit. The next digit becomes 16.<br>• $16 \div 3 = 5$ r 1. Write 5 r 1 on top. | $\begin{array}{r} 1\ 5\ 5 \text{ r } 1 \\ 3\overline{)4^{1}6^{1}6^{1}} \end{array}$ |
| **3.** Write the answer. | $466 \div 3 = 155$ r 1 |

## WORKED EXAMPLE 21 Dividing using short division

**Calculate $89\,656 \div 8$ using short division.**

| THINK | WRITE |
|---|---|
| **1.** Write the equation as a short division. | $8\overline{)89\,656}$ |
| **2.** Divide 8 into the first digit and carry the remainder to the next digit. 8 divides into 8 once. Write 1 above the 8 as shown. There is no remainder. | $\begin{array}{r} 1\ 1\ 2\ 0\ 7 \\ 8\overline{)8\ 9\ ^{1}6\ 5\ ^{5}6} \end{array}$ |
| **3.** Divide 8 into the second digit and carry the remainder to the next digit. 8 divides into 9 once, with 1 left over. Write 1 above the 9 and carry 1 to the hundreds column. | |
| **4.** Divide 8 into the third digit and carry the remainder to the next digit. 8 divides into 16 twice, with no remainder. Write 2 above the 6. | |
| **5.** Divide 8 into the fourth digit and carry the remainder to the next digit. 8 doesn't divide into 5. Write 0 above the 5. Carry 5 to the next digit. | |
| **6.** Divide 8 into 56. 8 divides into 56 seven times. Write 7 above the 6. There is no remainder. | |
| **7.** Write the answer. | $89\,656 \div 8 = 11\,207$ |

## 1.6.3 Dividing numbers that are multiples of 10

eles-3655

- When we are dividing numbers that are multiples of 10, an equivalent number of zeros can be cancelled from both the **dividend (numerator)** and the **divisor (denominator)**.
- The dividend is the quantity to be divided.
- The divisor is the number by which the dividend is to be divided.
- The **quotient** is the answer obtained when one number is divided by another number.

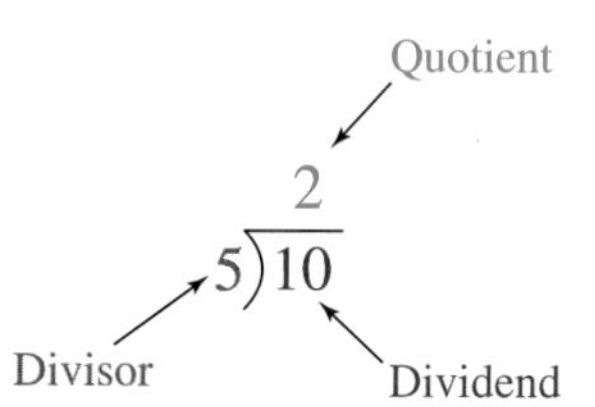


### WORKED EXAMPLE 22 Dividing numbers that are multiples of 10

**Calculate 48 000 ÷ 600. Note that both numbers are multiples of 10.**

| THINK | WRITE |
|---|---|
| 1. Write the question. | $48\,000 \div 600$ |
| 2. Write the question as a fraction. | $= \dfrac{48\,0\not{0}\not{0}}{6\not{0}\not{0}}$ |
| 3. Cancel as many zeros as possible, crossing off the same number of zeros in both numerator and denominator. | $= \dfrac{480}{6}$ |
| 4. Perform the division. | $\begin{array}{r} 0\,8\,0 \\ 6\overline{)4^{6}80} \end{array}$ |
| 5. Write the answer. | $48\,000 \div 600 = 80$ |

*Note:* The principle of associativity is not true for division. This means that, for example, $(80 \div 8) \div 2$ is not equivalent to $80 \div (8 \div 2)$. This is shown by the following.

$$\begin{aligned}(80 \div 8) \div 2 &= 10 \div 2 \\ &= 5\end{aligned} \qquad \begin{aligned}80 \div (8 \div 2) &= 80 \div 4 \\ &= 20\end{aligned}$$

### Digital technology

Scientific calculators can evaluate the division of large numbers.

DEG
27768÷24 1157

### ACTIVITY: Magic division

1. a. Choose a digit from 2 to 9. Write it 6 times to form a 6-digit number. For example, if you choose 4 your 6-digit number will be 444 444.
   b. Divide your 6-digit number by 33.
   c. Divide your result from part **b** by 37.
   d. Divide your result from part **c** by 91.
   e. What do you notice about your result from part **d**?
2. Repeat question **1** with a different 6-digit number and explain how the ‘magic division’ works.

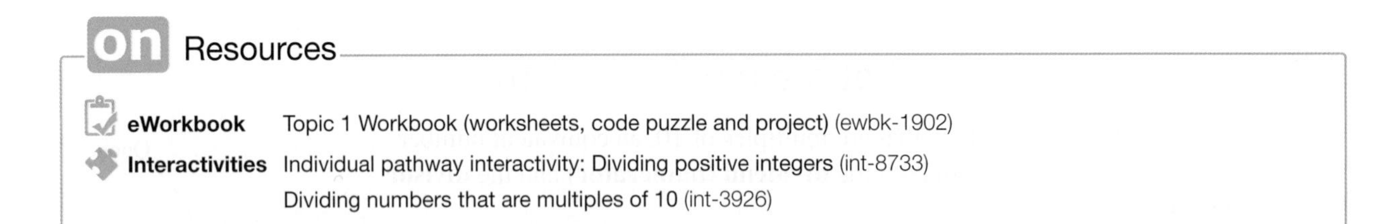


## Exercise 1.6 Dividing positive integers

learnon

1.6 Quick quiz | 1.6 Exercise

### Individual pathways

| ■ PRACTISE | ■ CONSOLIDATE | ■ MASTER |
|---|---|---|
| 1, 6, 8, 12, 15, 18 | 2, 4, 7, 9, 13, 14, 16, 19 | 3, 5, 10, 11, 17, 20 |

### Fluency

1. Calculate these divisions without using a calculator. There should be no remainder.
   a. $24 \div 6$  b. $36 \div 9$  c. $96 \div 12$
   d. $56 \div 7$  e. $26 \div 2$  f. $45 \div 15$
2. Calculate these divisions without using a calculator. There should be no remainder.
   a. $27 \div 3 \div 3$  b. $96 \div 8 \div 6$  c. $48 \div 12 \div 2$
   d. $56 \div 7 \div 4$  e. $100 \div 2 \div 10$  f. $90 \div 3 \div 2$
   Check your answers using multiplication.
3. Perform the following calculations, which involve a combination of multiplication and division. Remember to work from left to right.
   a. $4 \times 5 \div 2$  b. $80 \div 10 \times 7$  c. $144 \div 12 \times 7$
   d. $120 \div 10 \times 5$  e. $121 \div 11 \times 4$
4. **WE20** Calculate the following using short division.
   a. $3\overline{)1455}$  b. $7\overline{)43\,456}$  c. $11\overline{)30\,371}$  d. $8\overline{)640\,360}$
5. **WE21** Calculate the following using short division.
   a. $3\overline{)255\,194}$  b. $6\overline{)516\,285}$  c. $7\overline{)6\,328\,520}$  d. $8\overline{)480\,594}$
   Check your answers using a calculator.
6. Divide the following numbers, which are multiples of 10.
   a. $4200 \div 6$  b. $210 \div 30$  c. $720\,000 \div 800$
7. Divide the following numbers, which are multiples of 10.
   a. $4\,000\,000 \div 8000$  b. $600\,000 \div 120$  c. $480\,000 \div 600$
8. Calculate $144 \div 2$ by breaking 144 up into expanded form.
9. Calculate $642 \div 3$ by breaking 642 up into expanded form.
10. Calculate $4256 \div 8$ by breaking 4256 up into expanded form.

## Understanding

11. Kelly works part time at the local pet shop. Last year she earned $2496.
    a. Calculate how much Kelly earned each month.
    b. Calculate how much Kelly earned each week.
12. At the milk processing plant, the engineer asked Farid how many cows he had to milk each day. Farid said he milked 192 cows, because he had 1674 litres of milk at the end of each day and each cow produced 9 litres. Determine whether Farid really milks 192 cows each day. If not, calculate how many cows he does milk.

13. When Juan caters for a celebration like a wedding, he confirms the arrangements with the client by filling out a form. Juan has been called to answer the phone, so it has been left to you to fill in the missing details. Copy and complete this planning form.

| Celebration type | Wedding |
|---|---|
| Number of guests | 152 |
| Number of people per table | 8 |
| Number of tables required | |
| Number of courses for each guest | 4 |
| Total number of courses to be served | |
| Number of courses each waiter can serve | 80 |
| Number of waiters required | |
| Charge per guest | $55 |
| Total charge for catering | |

14. Janet is a land developer. She has bought 10 450 square metres of land. She intends to subdivide the land into 11 separate blocks.
    a. Calculate how many square metres each block will be.
    b. If she sells each block for $72 250, determine how much she will receive for the subdivided land.

## Reasoning

15. Kayla divides 243 by 9 and works out the answer to be 27. Her friend Tristan calculates the answer to be 26. State whether either Kayla or Tristan calculated the correct answer.
16. Explain whether this statement is true: $1148 \div 4 = 282$. Justify your answer by breaking 1148 up into expanded form.

17. Some people like to keep mice as pets. Female mice give birth to about six litters of babies per year, with about six babies in each litter.

   a. If you own a female mouse, how many pet mice will you have at the end of one year if none of the female's babies have any litters of their own? Explain how you reached your answer.

   b. Explain how many female mice you had at the start of the year if, at the end of the year, you had 111 mice and none of your females' babies had any litters of their own. Show your working.

### Problem solving

18. Danh has booked a beach house for a week over the summer period for a group of 12 friends. The house costs $1344 for the week. If all 12 people stayed for 7 nights, determine how much each person will pay per night.

19. A total of 62 people have been invited to a dinner party. Each table can seat 8 people. Determine how many tables are needed.

20. List the first four whole numbers that will have a remainder of 2 when you divide by 8. Describe the pattern and explain how you found these numbers.

# LESSON
# 1.7 Rounding and estimating

## LEARNING INTENTION

At the end of this lesson you should be able to:
- provide an estimate to a problem by rounding to the first digit.

## 1.7.1 Rounding integers

eles-3656

- Numbers or integers can be rounded to different degrees of accuracy.

### Rounding to the nearest 10

- To round to the nearest 10, think about which multiple of 10 the number is closest to. For example, if 34 is rounded to the nearest 10, the result is 30 because 34 is closer to 30 than it is to 40.

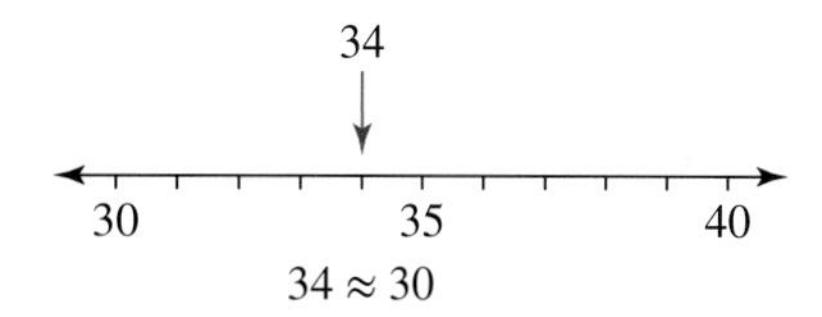


(*Note:* The symbol ≈ means 'is approximately equal to'.)

### Rounding to the nearest 100

- To round to the nearest 100, think about which multiple of 100 the number is closest to. For example, if 177 is rounded to the nearest 100, the result is 200 because 177 is closer to 200 than it is to 100.

### Rounding to the first (or leading) digit

- To round to the first (or leading) digit, use the following guidelines:
  - Consider the digit after the leading one (i.e. the second digit).
  - If the second digit is 0, 1, 2, 3 or 4, the first digit stays the same and all the following digits are replaced with zeros.
  - If the second digit is 5, 6, 7, 8 or 9, the first digit is raised by 1 (rounded up) and all the following digits are replaced with zeros.
- For example, if 2345 is rounded to the first digit, the result is 2000 because 2345 is closer to 2000 than it is to 3000.

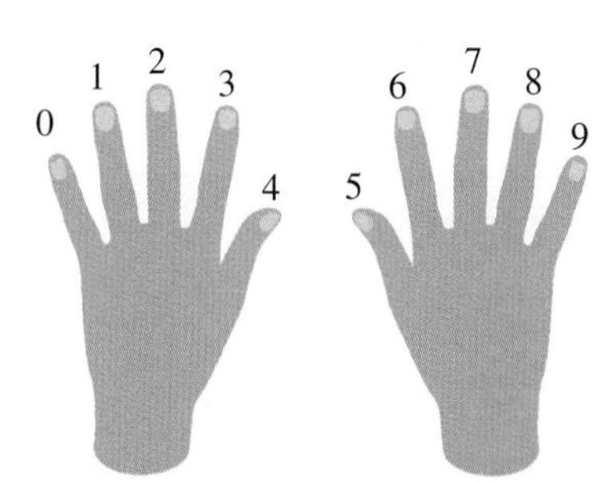


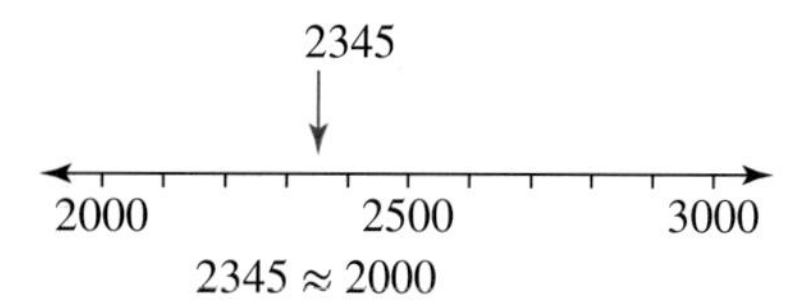

**WORKED EXAMPLE 23 Rounding to the first digit**

**Round the following numbers to the first (or leading) digit.**

a. **2371**

b. **872**

| THINK | WRITE |
|---|---|
| a. Since the second digit (3) is less than 5, leave the leading digit unchanged and replace all other digits with zeros. | a. $2371 \approx 2000$ |
| b. Since the second digit (7) is greater than 5, add 1 to the leading digit and replace all other digits with zeros. | b. $872 \approx 900$ |

## 1.7.2 Estimation

eles-3657

- An estimate is not the same as a guess, because an estimate is based on information.
- When you do not need to know an exact amount, an estimate or approximation is enough.
- To estimate the answer to a mathematical problem, round the numbers to the first digit and find an approximate answer.
- Estimations can be made when multiplying, dividing, adding or subtracting. They can also be used when there is more than one operation in the same question.

**WORKED EXAMPLE 24 Estimating by rounding to the first digit**

**Estimate $48\,921 \times 823$ by rounding to the first digit.**

| THINK | WRITE |
|---|---|
| 1. Write the question. | $48\,921 \times 823$ |
| 2. Round each part of the question to the first digit. | $\approx 50\,000 \times 800$ |
| 3. Multiply | $\approx 40\,000\,000$ |
| 4. Write the answer. | $48\,921 \times 823 \approx 40\,000\,000$ |

**DISCUSSION**

Is the estimated answer to the multiplication in Worked example 24 higher or lower than the actual answer? Justify your response.

**on Resources**

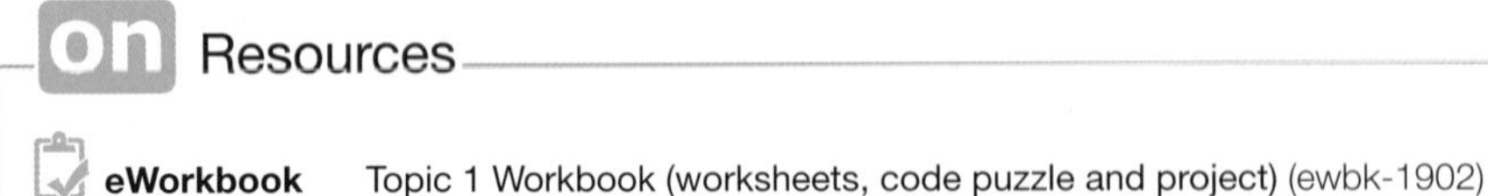

**eWorkbook** Topic 1 Workbook (worksheets, code puzzle and project) (ewbk-1902)

**Interactivities** Individual pathway interactivity: Estimation (int-4318)

Rounding (int-3932)

## Exercise 1.7 Rounding and estimating

learnon

1.7 Quick quiz on | 1.7 Exercise

### Individual pathways

| ■ PRACTISE | ■ CONSOLIDATE | ■ MASTER |
|---|---|---|
| 1, 4, 5, 6, 15, 18 | 2, 7, 8, 13, 14, 16, 19 | 3, 9, 10, 11, 12, 17, 20 |

### Fluency

1. **WE23** Round each of the following to the first (or leading) digit.
   a. 6 b. 45 c. 1368
   d. 12 145 e. 168 879 f. 4 985 452
2. **WE24** Estimate $67\,451 \times 432$ by rounding to the first digit.
3. Copy and complete the following table by rounding the numbers to the first digit. The first row has been completed as an example.
   - In the 'Estimate' column, round each number to the first digit.
   - In the 'Estimated answer' column, calculate the answer.
   - In the 'Prediction' column, guess whether the actual answer will be higher or lower than your estimate.
   - Use a calculator to work out the actual answer and record it in the 'Calculation' column to determine whether it is higher or lower than your estimate.

| | | Estimate | Estimated answer | Prediction | Calculation |
|---|---|---|---|---|---|
| Example | $4129 \div 246$ | $4000 \div 200$ | 20 | Lower | 16.784 553, so lower |
| **a.** | $487 + 962$ | | | | |
| **b.** | $33\,041 + 82\,629$ | | | | |
| **c.** | $184\,029 + 723\,419$ | | | | |
| **d.** | $93\,261 - 37\,381$ | | | | |
| **e.** | $321 - 194$ | | | | |
| **f.** | $468\,011 - 171\,962$ | | | | |
| **g.** | $36 \times 198$ | | | | |
| **h.** | $623 \times 12\,671$ | | | | |
| **i.** | $29\,486 \times 39$ | | | | |
| **j.** | $31\,690 \div 963$ | | | | |
| **k.** | $63\,003 \div 2590$ | | | | |
| **l.** | $69\,241 \div 1297$ | | | | |

4. **MC** Select which of the following is the best estimate of $4372 + 2587$.
   A. 1000 B. 5527 C. 6000 D. 7000 E. 7459
5. **MC** Choose which of the following is the best estimate of $672 \times 54$.
   A. 30 000 B. 35 000 C. 36 000 D. 40 000 E. 42 000
6. **MC** Select which of the following is the best estimate of $67\,843 \div 365$.
   A. 150 B. 175 C. 200 D. 230 E. 250

### Understanding

7. Estimate the answers to each of the following by rounding to the first digit.

   a. $5961 + 1768$
   b. $48\,022 \div 538$
   c. $9701 \times 37$
   d. $98\,631 + 608\,897$
   e. $6501 + 3790$

8. Estimate the answers to each of the following by rounding to the first digit.

   a. $11\,890 - 3642$
   b. $112\,000 \times 83$
   c. $66\,501 \div 738$
   d. $392 \times 113\,486$
   e. $12\,476 \div 24$

**Questions 9 to 12 relate to the following information.**

Su-Lin was using her calculator to answer some mathematical questions, but she got a different answer each time she performed the same calculation. Using your estimation skills, predict which of Su-Lin's answers is most likely to be correct.

9. **MC** $217 \times 489$

   A. 706 B. 106 113 C. 13 203 D. 19 313 E. 105 203

10. **MC** $89\,344 \div 256$

    A. 39 B. 1595 C. 89 088 D. 349 E. 485

11. **MC** $78 \times 6703$

    A. 522 834 B. 52 260 C. 6781 D. 56 732 501 E. 51 624

12. **MC** $53\,669 \div 451$

    A. 10 B. 1076 C. 53 218 D. 119 E. 183

13. Kody is selling tickets for his school's theatre production of *South Pacific*. So far he has sold 439 tickets for Thursday night's performance, 529 for Friday's and 587 for Saturday's. The costs of the tickets are \$9.80 for adults and \$4.90 for students.

    a. Round the figures to the first digit to estimate the number of tickets Kody has sold so far.
    b. If approximately half the tickets sold were adult tickets and the other half were student tickets, estimate how much money has been received so far by rounding the cost of the tickets to the first digit.

14. For the following calculations:

    i. estimate the answer by rounding to the first digit
    ii. calculate the answer using a written method for parts a and b and a calculator for part c.

    a. $6650 - 1310$
    b. $36 \times 223$
    c. $18\,251 \div 391$

### Reasoning

15. Ava wants to buy three items that cost \$5.58, \$15.92 and \$7.22. Estimate the total cost of the three items by rounding each number to the nearest whole number, and then adding.

16. Estimate the number of graduating students shown in the photograph. (Do not count all of the students in the photograph.) If the hall where they are graduating holds 12 times this number, estimate the total capacity of the hall.
    Show your working and write a sentence explaining how you solved this problem.

17. At Kooboora Secondary College there are 127 students in Year 7, 152 students in Year 8 and 319 students in Years 9 and 10. Estimate the total number of students in Years 7–10.

### Problem solving

18. Describe a situation in everyday life where it is *not* appropriate to estimate values.
19. Shari bought a car five years ago for $21 798. The car's value has decreased to one-third of its initial value. Determine the approximate value of the car today.
20. During the intermission of the school production of *South Pacific*, Jia is planning to run a stall selling hamburgers to raise money for the school. She has priced the items she needs and made a list in order to estimate her expenses.
    a. By rounding the item price to the first digit, use the following table to estimate how much each item will cost Jia for the quantity she requires.
    b. Estimate what Jia's total shopping bill will be.
    c. If Jia sells 300 hamburgers over the 3 nights for $2 each, determine how much money she will receive for the hamburgers.
    d. Determine approximately how much money Jia will raise through selling hamburgers over the 3 nights.

| Item | Item price | Quantity required | Estimated cost |
|---|---|---|---|
| Bread rolls | $2.90/dozen | 25 packets of 12 | |
| Hamburgers | $2.40/dozen | 25 packets of 12 | |
| Tomato sauce | $1.80/litre | 2 litres | |
| Margarine | $2.20/tub | 2 tubs | |
| Onions | $1.85/kilogram | 2 kilograms | |
| Tomatoes | $3.50/kilogram | 2 kilograms | |
| Lettuce | $1.10 each | 5 lettuces | |

# LESSON
# 1.8 Order of operations

### LEARNING INTENTION

At the end of this lesson you should be able to:
- understand how the order of operations is to be applied
- apply the order of operations to solve problems.

## 1.8.1 BIDMAS

eles-3658

- There are rules in mathematics that ensure everyone has the same understanding of mathematical operations. These rules are known as the **order of operations**.
- The acronym **BIDMAS** (**B**rackets, **I**ndices or roots, **D**ivision and **M**ultiplication, **A**ddition and **S**ubtraction) can be used to remember the correct order of operations.

*Note*: Division and multiplication have the same order as each other so are to be performed left to right. Addition and subtraction have the same order as each other and also should be performed left to right.

## BIDMAS

**BIDMAS helps us to remember the correct order in which we perform the various operations, working from left to right.**

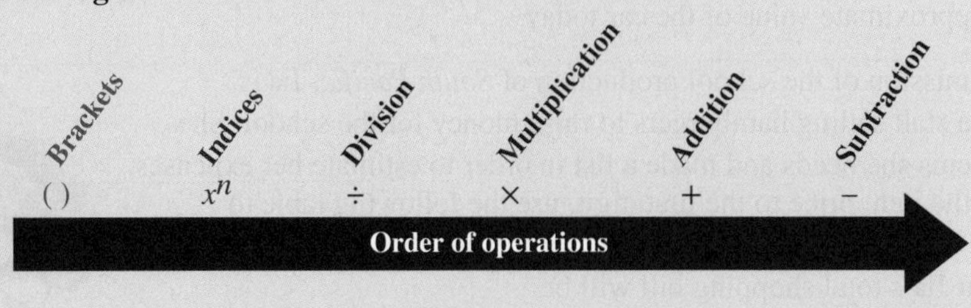


### WORKED EXAMPLE 25 Applying the order of operations

**Calculate $6 + 12 \div 4$.**

| THINK | WRITE |
|---|---|
| 1. Write the question. | $6 + 12 \div 4$ |
| 2. Follow BIDMAS.<br>• Brackets: there are none.<br>• Indices or roots: there are none.<br>• Division and multiplication: there is one division.<br>• Addition and subtraction: there is one addition. | |
| 3. Apply BIDMAS by working out the division before the addition. | $6 + 12 \div 4 = 6 + 3$ |
| 4. Calculate the answer. | $= 9$ |

### WORKED EXAMPLE 26 Applying the order of operations

**Calculate the following.**

**a.** $12 \div 2 + 4 \times (4 + 6)$ **b.** $80 \div [((11 - 2) \times 2) + 2]$

| THINK | WRITE |
|---|---|
| a. 1. Write the question. | a. $12 \div 2 + 4 \times (4 + 6)$ |
| 2. Apply BIDMAS. Remove the brackets by working out the addition inside. | $= 12 \div 2 + 4 \times 10$ |
| 3. Perform the division and multiplication next, doing whichever occurs first as you work from left to right. Work out $12 \div 2$ first, then work out $4 \times 10$. | $= 12 \div 2 + 4 \times 10$<br>$= 6 + 40$ |
| 4. Complete the addition last and calculate the answer. | $= 46$ |
| b. 1. Write the question. | $80 \div [((11 - 2) \times 2) + 2]$ |
| 2. Remove the innermost brackets first by working out the subtraction inside them. | $= 80 \div [(9 \times 2) + 2]$ |
| 3. Remove the next pair of brackets by working out the multiplication inside them. | $= 80 \div [18 + 2]$ |
| 4. Remove the final pair of brackets by working out the addition inside them. | $= 80 \div 20$ |
| 5. Perform the division last and calculate the answer. | $= 4$ |

### ACTIVITY: Order of operations

Using the expressions from Worked example 26, investigate whether the order of operations is applied on different devices such as your mobile phone or a computer. Do you have to take any special steps to ensure that the order of operations is followed?

### WORKED EXAMPLE 27 Inserting brackets to make a statement true

**Insert one set of brackets in the appropriate place to make the following statement true.**
$3 \times 10 - 8 \div 2 + 4 = 7$

| THINK | WRITE |
|---|---|
| 1. Write the left-hand side of the equation. | $3 \times 10 - 8 \div 2 + 4$ |
| 2. Place one set of brackets around the first two values. | $= (3 \times 10) - 8 \div 2 + 4$ |
| 3. Perform the multiplication inside the bracket. | $= 30 - 8 \div 2 + 4$ |
| 4. Perform the division. | $= 30 - 4 + 4$ |
| 5. Perform the subtraction and addition working from left to right. *Note*: Since this is not the answer, the above process must be repeated. | $= 26 + 4$ $= 30$ Since this is not equal to 7 we must place the brackets in a different position. |
| 6. Place one set of brackets around the second and third values. | $3 \times 10 - 8 \div 2 + 4$ $= 3 \times (10 - 8) \div 2 + 4$ |
| 7. Perform the subtraction inside the bracket. | $= 3 \times 2 \div 2 + 4$ |
| 8. Perform the multiplication and division working from left to right. | $= 6 \div 2 + 4$ $= 3 + 4$ |
| 9. Perform the addition last and calculate the answer. | $= 7$ |
| 10. Write the answer showing the bracket placement that makes the statement true. | $3 \times (10 - 8) \div 2 + 4 = 7$ |

### WORKED EXAMPLE 28 Using the order of operations

**Calculate $58 - (2 \times 8 + 3^2)$ using the order of operations.**

| THINK | WRITE |
|---|---|
| 1. Write the question. | $58 - (2 \times 8 + 3^2)$ |
| 2. Apply BIDMAS. | |
| • Remove the brackets first. | |
| • Calculate the indices inside the brackets. | $= 58 - (2 \times 8 + 9)$ |
| • Perform the multiplication inside the brackets. | $= 58 - (16 + 9)$ |
| • Perform the addition inside the brackets. | $= 58 - (25)$ |
| • Perform the subtraction. | |
| 3. Write the answer. | $= 33$ |

### Digital technology

Scientific calculators can evaluate the expression using the order of operations.

```
DEG
64+(4*6-2²)    84
```

### DISCUSSION

Use a calculator to work out the answer to Worked example 28. Do you think it is easier or harder to perform the calculation using a calculator? What do you think the advantages and disadvantages are of using a calculator compared to calculating using mental strategies?

Resources

**eWorkbook** Topic 1 Workbook (worksheets, code puzzle and project) (ewbk-1902)

**Video eLesson** BIDMAS (eles-2425)

**Interactivities** Individual pathway interactivity: Order of operations (int-4315)
Order of operations (int-3707)

## Exercise 1.8 Order of operations

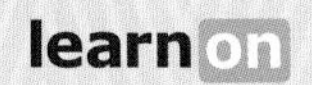

**1.8 Quick quiz**  **1.8 Exercise**

### Individual pathways

| ■ PRACTISE | ■ CONSOLIDATE | ■ MASTER |
|---|---|---|
| 1, 2, 6, 9, 14, 17, 20, 21 | 3, 7, 10, 12, 13, 16, 18, 22, 23 | 4, 5, 8, 11, 15, 19, 24, 25 |

### Fluency

1. **WE25&26** Follow the order of operations rules and calculate each of the following.
   a. $3+4\div 2$
   b. $8+1\times 1$
   c. $24\div(12-4)$
   d. $15\times(17-15)$
   e. $11+6\times 8$
2. Follow the order of operations rules and calculate each of the following.
   a. $30-45\div 9$
   b. $56\div(7+1)$
   c. $12\times(20-12)$
   d. $(7+5)-(10+2)$
3. Follow the order of operations rules and calculate each of the following.
   a. $3\times 4+23-10-5\times 2$
   b. $42\div 7\times 8-8\times 3$
   c. $10+40\div 5+14$
   d. $81\div 9+108\div 12$
   e. $16+12\div 2\times 10$
   f. $(18-15)\div 3\times 27$

4. Follow the order of operations rules and calculate each of the following.

a. $4+(6+3\times 9)-11$
b. $52\div 13+75\div 25$
c. $88\div(24-13)\times 12$
d. $(4+5)\times(20-14)\div 2$
e. $\{[(16+4)\div 4]-2\}\times 6$
f. $60\div\{[(12-3)\times 2]+2\}$

## Understanding

5. **WE27** Insert one set of brackets in the appropriate place to make these statements true.

a. $12-8\div 4=1$
b. $4+8\times 5-4\times 5=40$
c. $3\times 10-2\div 4+4=10$
d. $12\times 4+2-12=60$
e. $10\div 5+5\times 9\times 9=81$
f. $18-3\times 3\div 5=9$

6. Calculate the following, using the order of operations.

a. $6+3\times 4$
b. $18-12\div 3$
c. $2-4-6$
d. $17-3+8$

7. Calculate the following, using the order of operations.

a. $6\times 3\div 9$
b. $72\div 8\times 3$
c. $7-(3+4)$
d. $(6+3)\div 9$

8. Evaluate each of the following.

a. $3\times 2-3\times 1$
b. $6\times 5-2\times 6$
c. $4\times(6+4)$
d. $(8+3)\times 7$

9. **WE28** Evaluate each of the following.

a. $4+7\times 3-2$
b. $6-4+(4)^2$
c. $(2)^3-3\times 2$
d. $3+(8-2)+6$

10. Evaluate each of the following.

a. $8\div 2+(2)^2$
b. $4\times 8-\left[2+(3)^2\right]$
c. $(7+5)+24\div 6$
d. $30\div(5-2)-10$

11. Evaluate each of the following.

a. $54\div 6+8\times 9\div 4$
b. $(9+6)\div 5-8\times 0$
c. $7+7\div 7\times 7+7$
d. $9\times 5-(3+2)-48\div 6$

12. **MC** $20-6\times 3+28\div 7$ is equal to:

A. 46
B. 10
C. 6
D. 4
E. 2

13. **MC** Choose which of the following is the correct order of the two signs that can replace * in the equation $7*2*4-3=12$.

A. $-, +$
B. $+, \times$
C. $=, \div$
D. $\div, \times$
E. $\times, \div$

14. **MC** Select which of the following the expression $6-2\times 5+10\div 2$ is equal to:

A. 25
B. 40
C. 30
D. 1
E. 3

15. Rewrite the left side of each equation, inserting brackets if necessary, to make each of the following statements true.

a. $6+2\times 4-3\times 2=10$
b. $6+2\times 4-3\times 2=26$
c. $6+2\times 4-3\times 2=16$
d. $6+2\times 4-3\times 2=8$

16. Model each situation with integers, then calculate the result.

a. Jemma starts out with \$274 in the bank. She makes 2 withdrawals of \$68 each, then 3 deposits of \$50 each. Calculate how much money Jemma now has in the bank.
b. If 200 boxes of apples were each 3 short of the stated number of 40 apples, calculate the overall shortfall in the number of apples.
c. A person with a mass of 108 kg wants to reduce his mass to 84 kg in 3 months. Determine what average mass reduction is needed per month.

### Reasoning

17. Discuss a real-life situation in which performing calculations in the wrong order could cause problems.

18. Keenan and Amona discovered that they had different answers to the same question, which was to calculate $6 + 6 \div 3$. Keenan thought the answer was 8. Amona thought the answer was 4. State who was correct, Keenan or Amona. Explain your response.

19. I have a pile of pebbles. Arranging them in piles of 7 leaves 1 extra pebble, and arranging them in piles of 5 leaves 3 extra pebbles. If there are more than 10 pebbles in my pile, determine the smallest number of pebbles possible for my pile to have in it. Show your working.

### Problem solving

20. The student lockers at Jacaranda College are to be numbered consecutively from 1 to 800 using plastic digits. Each digit costs 10c. Calculate the total cost of all the digits.

21. Xanthe bakes and sells cupcakes. A customer ordered 3 lemon cupcakes at \$4.25 each, 3 chocolate cupcakes at \$4.75 each and 2 original cupcakes at \$3.75 each.
    a. Write an expression to help Xanthe calculate the amount the customer has to pay.
    b. Use this expression to calculate the amount that the customer has to pay.

22. Local time in Sydney is 3 hours ahead of Singapore time, which is 5 hours behind Auckland (NZ) time. Auckland is 11 hours ahead of Berlin (Germany) time. Determine the time difference between:
    a. Sydney and Berlin
    b. Singapore and Berlin.

23. Minh was asked to buy enough ice creams for everyone in his class. There are 31 people in the class. At the supermarket, ice creams come in packets of 3 or packets of 5. Minh is told that the cost of one ice cream is the same from either packet. Determine how many packets (or what combination of packets) Minh should buy.

**24.** Use the digits 1, 2, 3 and 4 and the operators +, −, × and ÷ to construct equations that result in the numbers 1 to 5 (the numbers 2 and 4 are already done for you). You must use each digit in each expression, and you may not use the digits more than once. You may combine the digits to form larger numbers (such as 21). You may also use brackets to make sure the operations are done in the correct order.

$$\begin{aligned}
1 &= \\
2 &= 4 - 3 + 2 - 1 \\
3 &= \\
4 &= 4 - 2 + 3 - 1 \\
5 &=
\end{aligned}$$

**25.** Use the digits 1, 2, 3 and 4 and the operators +, −, × and ÷ to construct equations that result in the numbers 5 to 10 (the numbers 7 and 10 have already been done for you). You must use each digit in each expression and you may not use the digits more than once. You may combine the digits to form larger numbers (such as 21). You may also use brackets to make sure the operations are done in the correct order.

$$\begin{aligned}
5 &= \\
6 &= \\
7 &= 24 \div 3 - 1 \\
8 &= \\
9 &= \\
10 &= 1 + 2 + 3 + 4
\end{aligned}$$

# LESSON
# 1.9 Review

## 1.9.1 Topic summary

**POSITIVE INTEGERS**

**Place value**

The position of a digit in a number gives a different value to the digit.

For example, the place value of each digit in the number 613 452 is:

| Hundred thousands | Ten thousands | Thousands | Hundreds | Tens | Ones |
|---|---|---|---|---|---|
| 6 | 1 | 3 | 4 | 5 | 2 |

- The number 613 452 is read as '*six hundred and thirteen thousand, four hundred and fifty-two*'.

**Rounding and estimating**

To round to the first digit, use the following guidelines:

- If the second digit is 0, 1, 2, 3 or 4, the first digit stays the same and all the following digits are replaced with zeros.
- If the second digit is 5, 6, 7, 8 or 9, the first digit is increased by 1 (rounded up) and all the following digits are replaced with zeros.

**Positive integers**

Positive integers are whole numbers including 0 without decimals or fractions.

- The integers 1, 2, 3,... are called positive integers.
- 0 is neither positive nor negative....

**Addition of positive integers**

Some mental strategies and algorithms used to add up numbers without a calculator are:

- jump strategy
- split strategy
- compensation strategy
- rearrange strategy
- algorithm to add positive integers.

e.g. 462 + 78 + 316

$$\begin{array}{r} {}^{1}4\ {}^{1}6\ 2 \\ 7\ 8 \\ 3\ 1\ 6 \\ \hline 8\ 5\ 6 \end{array}$$

**Subtraction of positive integers**

Some mental strategies and algorithms used to subtract numbers without a calculator are:

- jump strategy
- adding up to find the difference
- compensation strategy
- algorithm to subtract positive integers.

e.g. 892 − 467

$$\begin{array}{r} 8^{8}\ \not{9}^{1}\ 2 \\ 4\ 6\ 7 \\ \hline 4\ 2\ 5 \end{array}$$

**Multiplication of positive integers**

- A short way of performing repeated addition of the same number
- Expanded form, area model, short and long multiplication algorithms are used to multiply.

e.g.

Short multiplication

$$\begin{array}{r} {}^{2}1{}^{2}456 \\ \times\quad 5 \\ \hline 7280 \end{array}$$

Long multiplication

$$\begin{array}{r} 1456 \\ \times\ 132 \\ \hline 2912 \\ 43680 \\ 145600 \\ \hline 192192 \end{array}$$

**Division of positive integers**

- The process of sharing or dividing a number into equal parts.
- Expanded form and short division algorithm are used to divide.

e.g.

$$\begin{array}{r} 417 \\ 2\overline{)83^{1}4} \end{array}$$

$$\begin{array}{r} 2 \\ 5\overline{)10} \end{array}$$

Quotient (2), Divisor (5), Dividend (10)

**Order of operations**

- The order of operations is a set of rules we must follow so that we all have a common understanding of mathematical operations.
- The set order in which we calculate problems is:
  1. **Brackets** ( ) or [ ]
  2. **Indices or roots** $a^x$ or $\sqrt[n]{x}$
  3. **Division and Multiplication** (working left to right) ÷ or ×
  4. **Addition and Subtraction** (working left to right) + or −
- The acronym **BIDMAS** can be used to remember the correct order of operations.

## 1.9.2 Success criteria

Tick the column to indicate that you have completed the lesson and how well you think you have understood it using the traffic light system.

(**Green:** I understand; **Yellow:** I can do it with help; **Red:** I do not understand)

| Lesson | Success criteria | Green | Yellow | Red |
|---|---|---|---|---|
| 1.2 | I understand that the value of each digit in a number depends on its position or place value. | | | |
| 1.3 | I can add and subtract positive integers using mental strategies. | | | |
| 1.4 | I can add and subtract positive integers using algorithms. | | | |
| 1.5 | I can multiply positive integers using mental strategies and algorithms. | | | |
| 1.6 | I can divide positive integers using mental strategies and the short division algorithm. | | | |
| 1.7 | I can provide an estimate to a problem by rounding to the first digit. | | | |
| 1.8 | I understand how the order of operations is to be applied. | | | |
| | I can apply the order of operations to solve problems. | | | |

## 1.9.3 Project

### Ancient number systems

All ancient civilisations developed methods to count and use numbers. The methods they used to represent numbers did not involve the ten digits 1, 2, 3, 4, 5, 6, 7, 8, 9 and 0 that we use today. Instead they involved different systems like letters, other symbols, pebbles, tying knots in ropes, or cutting notches in sticks. As the need to use and represent larger numbers became more and more important, many old systems fell out of use.

### Egyptian numbers

The ancient Egyptians were one of the oldest known civilisations to have a recorded number system. In about 5000 BCE, Egyptians used a system of symbols called hieroglyphs. An example of hieroglyphs can be seen in the photograph.

The hieroglyphs used for numbers are shown in this table.

| **Number** | 1 | 10 | 100 | 1000 | 10 000 | 100 000 | 1 000 000 | 10 000 000 |
|---|---|---|---|---|---|---|---|---|
| **Symbol** | | | | | | | | |

The following example shows how the symbols are used to represent numbers. As the numbers get larger in value, the symbols are simply repeated. This set of symbols represents 241 513.

1. What number is represented by the following Egyptian hieroglyphs?

2. Show how the ancient Egyptians would have represented the number 2147.

The ancient Egyptian system of numbers was not limited to simply representing numbers. Addition and subtraction could also be performed when it was required. Perform the following additions and subtractions, giving your answers as Egyptian numbers.

3. +

4. +

5. −

6. −

7. Comment on any difficulties you encountered when performing the calculations using Egyptian hieroglyphs.

### Greek numbers

The ancient Greeks used an alphabet that had 27 letters (the modern Greek alphabet has only 24 letters). These letters were also used to represent numbers.

The first nine letters of the alphabet represented the numbers 1 to 9. The middle nine letters of the alphabet represented the tens from 10 to 90. The last nine letters of the alphabet represented the hundreds from 100 to 900. Numbers were written by combining these letters.

8. Search online to find the 27 symbols used in the ancient Greek number system and the number that each symbol represented.
9. How did the ancient Greeks represent numbers greater than or equal to 1000?

10. Search online to find out about number systems used by other ancient civilisations. Try to find out what symbols were used to represent numbers and how those symbols were used in calculations. Discuss any advantages of these systems and suggest why the system may have fallen out of use. Present your findings on a separate sheet of paper.

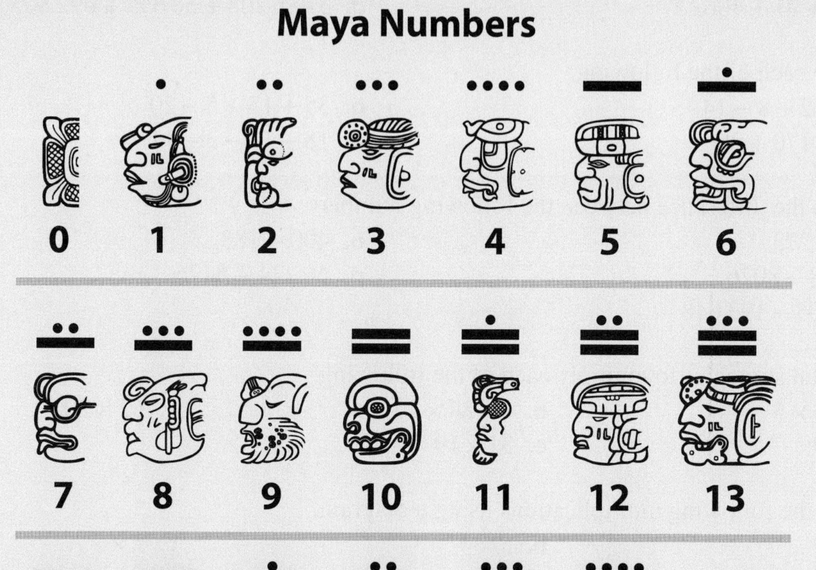


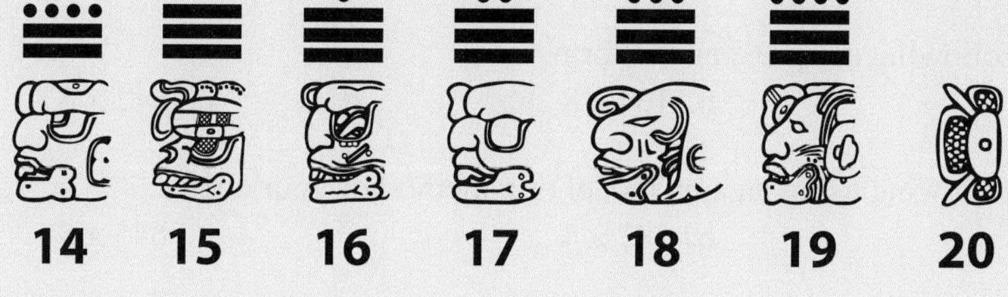

Resources

**eWorkbook** Topic 1 Workbook (worksheets, code puzzle and project) (ewbk-1902)

**Interactivities** Crossword (int-2586)

Sudoku puzzle (int-3163)

## Exercise 1.9 Review questions

**learnon**

### Fluency

1. State the place value of the digit shown in pink in each of the following.

   a. 74 037 b. 541 910 c. 1 904 000 d. 290

2. Write each of the following numbers in expanded form.

   a. 392 b. 4109 c. 42 001 d. 120 000

3. List the numbers 394, 349, 943, 934, 3994, 3499 in ascending order.

4. List the numbers 1011, 101, 110, 1100, 1101 in descending order.

5. Calculate the sum of these numbers.
   a. $43 + 84$
   b. $139 + 3048$
   c. $3488 + 91 + 4062$
   d. $3\,486\,208 + 38\,645 + 692\,803$

6. Calculate each of the following.
   a. $20 - 12 + 8 - 14$
   b. $35 + 15 + 5 - 20$
   c. $300 - 170 + 20$
   d. $18 + 10 - 3 - 11$

7. Calculate the difference between the following numbers.
   a. $688 - 273$
   b. $400 - 183$
   c. $68\,348 - 8026$
   d. $46\,234 - 8476$
   e. $286\,005 - 193\,048$

8. Use mental strategies to multiply each of the following.
   a. $2 \times 15 \times 5$
   b. $4 \times 84 \times 25$
   c. $62 \times 20$
   d. $67 \times 9$
   e. $31 \times 19$

9. Evaluate the following multiplications using a diagram.
   a. $3 \times 6$
   b. $5 \times 4$
   c. $2 \times 7$

10. Calculate the following using expanded form.
    a. $24 \times 4$
    b. $134 \times 8$
    c. $642 \times 5$

11. Calculate the following using an area model or a method of your choice.
    a. $236 \times 6$
    b. $347 \times 4$
    c. $675 \times 5$

12. Calculate the following using the short multiplication algorithm.
    a. $128 \times 9$
    b. $762 \times 3$
    c. $3840 \times 5$

13. Calculate the following using the long multiplication algorithm.
    a. $64 \times 72$
    b. $542 \times 57$
    c. $50\,426 \times 364$

14. Calculate each of the following using expanded form.
    a. $645 \div 5$
    b. $1040 \div 4$
    c. $15\,063 \div 3$

15. Calculate each of the following.
    a. $6 \times 4 \div 3$
    b. $4 \times 9 \div 12$
    c. $49 \div 7 \times 12$
    d. $81 \div 9 \times 5$
    e. $6 \times 3 \div 9 \div 2$

16. Divide the following multiples of 10.
    a. $84\,000 \div 120$
    b. $4900 \div 700$
    c. $12\,300 \div 30$

17. Evaluate the following using short division.
    a. $4704 \div 3$
    b. $6867 \div 7$
    c. $3762 \div 7$
    d. $5702 \div 17$
    e. $11\,852 \div 13$

18. By rounding each number to its first digit, estimate the answer to each of the following calculations.

a. $6802 + 7486$
b. $8914 - 3571$
c. $5304 \div 143$
d. $5706 \times 68$
e. $49\,581 + 73\,258$
f. $289 \times 671$

19. **MC** Choose which of the following is an integer.

A. $2\frac{1}{2}$
B. 0.45
C. 0
D. 201.50
E. $\frac{7}{2}$

20. Complete each statement by inserting =, < or > in the box to make the statement correct.

a. $6 - 4 \square 3 + 2$
b. $1 \square 7 \div 7$
c. $5 - 4 \square 0 \times 5$
d. $100 - 50 \square 9 \times 5$

21. Calculate the following using a mental strategy.

a. $315 + 12$
b. $1010 + 80$
c. $735 + 54$
d. $154 + 62$

22. Calculate the following using a mental strategy.

a. $152 - 25$
b. $135 - 54$
c. $745 - 34$
d. $354 - 76$

23. Evaluate the following.

a. $(6 - 2) \times 7$
b. $4 \times (4 + 2)$
c. $2 \times 5$
d. $(10 - 6) \times 2$
e. $7 + 8 \div 2$
f. $3 \times 5 + (6 \times 2)$

24. Calculate the following.

a. $36 \div 3$
b. $210 \div 7$
c. $45 \div 9$
d. $1800 \div 2$
e. $672 \div 48$
f. $625 \div 25$

25. Complete the following by inserting =, < or > in the box to make the statement correct.

a. $3.5 \square 3.75$
b. $22 \times 2 \square 44$
c. $4 \times 3 \square 8 \times 0.5$
d. $5 \times 2 \square 15$

26. Complete the following by inserting =, < or > in the box to make the statement correct.

a. $12 \div -4 \square 3$
b. $5 \times (2 + 10) \div 4 \square 25$
c. $10 \times 6.5 \square 50$
d. $5 \times 2 - 6 \div 4 \square 2.5$

27. The lowest temperature recorded on a particular day in Melbourne was 4 °C and the highest was 27.5 °C. Calculate the difference between these two temperatures.

28. Follow the rules for the order of operations to calculate each of the following.

a. $35 \div (12 - 5)$
b. $11 \times 3 + 5$
c. $8 \times 3 \div 4$
d. $5 \times 12 - 11 \times 5$
e. $(6 + 4) \times 7$
f. $5 + [21 - (5 \times 3)] \times 4$

## Problem solving

**29.** Uluru is a sacred Aboriginal site. The following map shows some roads between Uluru and Alice Springs. The distances (in kilometres) along particular sections of road are indicated.

**a.** Calculate how far Kings Canyon Resort is from Ayers Rock Resort near Uluru.

**b.** Determine the shortest distance by road from Kings Canyon Resort to Alice Springs.

**c.** If you are travelling in a hire car in this part of Australia, you must travel only on sealed roads. Work out the distance you need to travel if you are driving in a hire car from Kings Canyon Resort to Alice Springs.

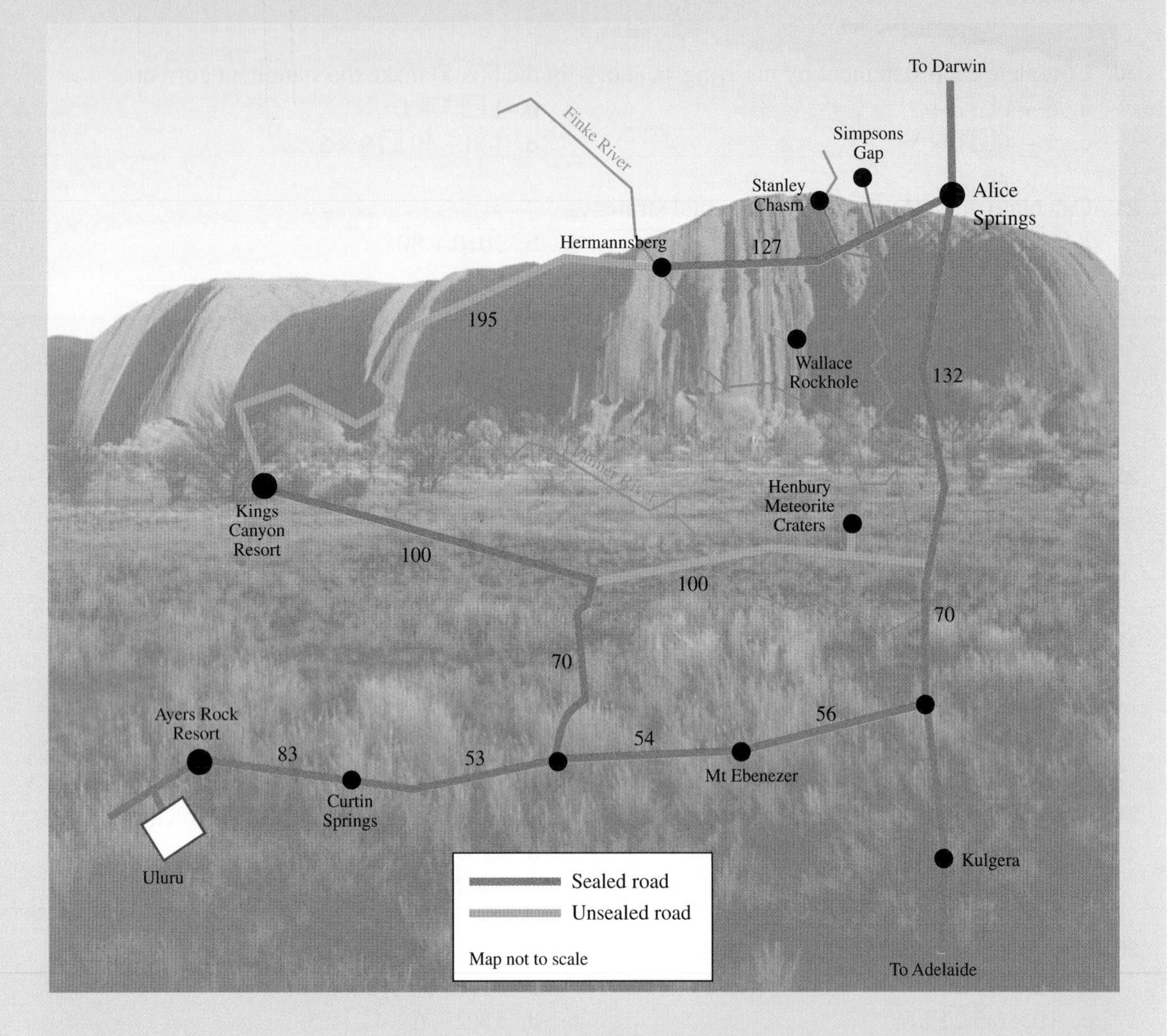


**30.** In summer, an ice cream factory operates 16 hours a day and makes 28 ice creams an hour.

**a.** Calculate how many ice creams are made each day.

**b.** If the factory operates 7 days a week, determine how many ice creams are made in one week.

**c.** If there are 32 staff who run the machines over a week, calculate how many ice creams each person would make.

**31.** When you add two even positive integers, the answer is even. The sum of an even and an odd positive integer is an odd integer. Two odd positive integers add to give an even integer.
Determine what happens if you perform multiplication on these types of integers. Give examples to support your answers. Write a general statement to summarise what would happen if you multiplied more than two positive integers.

**32.** Complete the following subtraction by adding the missing digits to the boxes.

$$\begin{array}{r} 8\ \ 2\ \ \square\ \ 0 \\ -\ 3\ \ \square\ \ 8\ \ \square \\ \hline \square\ \ 5\ \ 5\ \ 9 \\ \hline \end{array}$$

**33.** A lolly factory machine dispenses 760 lollies into 8 jars. Assuming that an equal number of lollies is dispensed into each jar, evaluate how many lollies there are in each jar.

**34.** Julie sells 8 bottles of soft drink for $3 each and 12 bottles of water for $2 each.
**a.** Write a calculation that will find the total value of Julie's sales.
**b.** Calculate the total value of Julie's sales.

**35.** Dara estimates the crowd at a football match to be 20 000 people. If Dara's estimate is correct to the nearest 1000 people, evaluate the greatest number of people that could possibly be at the match.

**36. a.** Complete the addition table below.

| + | 6 | | 3 |
|---|---|---|---|
| | 11 | 15 | |
| 12 | | | |
| | 15 | | |

**b.** Complete the multiplication table below.

| × | 4 | | 6 |
|---|---|---|---|
| | 12 | 24 | 18 |
| 5 | | | |
| | 8 | | |

**37.** Minimum and maximum temperatures are recorded at many locations around the world.

**a.** In Melbourne the temperature often changes rapidly during the day. On one very hot day the temperature reached 42 °C. A cool change arrived during the afternoon and the temperature dropped by 18 °C. Determine the temperature after the cool change.

**b.** In Alaska the weather often changes suddenly as storms sweep across the frozen plains. On one day, the temperature was 3 °C, but a storm caused the temperature to drop by 24 °C. Determine the temperature during the storm.

**38.** You have 30 questions to answer on a multiple choice test. Each correct answer scores one mark (+1). Each incorrect answer loses two marks (−2). Any unmarked questions will be counted as incorrect. Evaluate the lowest number of questions that you can answer correctly and still obtain a score greater than 0. Determine how many marks you would earn. Explain your reasoning.

**39.** Place the digits 1–9 (with no repeats) in the grid shown so that the equations reading across and down are true.

| | | | | | | |
|---|---|---|---|---|---|---|
| | + | | × | | = | 32 |
| + | | × | | + | | |
| | − | | + | | = | 7 |
| × | | ÷ | | ÷ | | |
| | − | | − | | = | 4 |
| = | | = | | = | | |
| 29 | | 10 | | 15 | | |

**40.** The following table lists a business's profit over the past four years. Evaluate which two consecutive years had the greatest difference. Show your working.

| Year | 2016 | 2017 | 2018 | 2019 |
|---|---|---|---|---|
| $ Profit/loss | 1020 | 2620 | 3808 | 8222 |

on To test your understanding and knowledge of this topic, go to your learnON title at www.jacplus.com.au and complete the **post-test**.

# Answers

## Topic 1 Positive integers

### 1.1 Pre-test

1. E
2. 27
3. a. 2411 b. 70 555 c. 26 747
4. a. 622 b. 7257 c. 1262
5. 480
6. 416
7. 321
8. $1865
9. 600
10. 30 000
11. a. 12 b. 21 c. 25
12. a. $<$ b. $=$ c. $>$
13. a. 2 b. 8 c. 130
14. a. 172 b. 444 c. 235
15. 12 minutes after the race has started.

### 1.2 Place value

1. a. $900 + 20 + 5$
   b. $20\,000 + 8000 + 400 + 60 + 9$
   c. $800\,000 + 2$
   d. $1\,000\,000 + 80\,000 + 100$
2. a. Seven hundred and sixty-five
   b. Nine thousand, one hundred and five
   c. Ninety thousand, four hundred and fifty
   d. One hundred thousand, two hundred and thirty-six
3. a. 495 b. 2670 c. 109 605
4. Forty
5. Two hundred
6. Zero
7. Three hundred
8. Fifty
9. E
10. B
11. 48 987, 8569, 742, 647, 28
12. 72 167, 58 625, 47 890, 32 320, 12 947
13. 7647, 7476, 6747, 6477, 4776
14. 8880, 8808, 8800, 8088, 8080, 8008
15. 9, 58, 743, 68 247, 1 258 647
16. 34 108, 58 610, 60 000, 78 645, 84 364
17. 1290, 1902, 2019, 2910, 9021, 9201
18. 111, 211, 212, 221, 222, 1112
19. a. $345 < 567$ b. $89 < 98$ c. $234\,596 > 23\,459$
   d. $7765 > 7756$
20. a. $6000 b. $340 000 c. $58 000
21. a. $430K b. $7K c. $800K
22. 98 641
23. Half a minute (30 s)
24. Astronomers and other scientists work with very small or very large numbers. Scientific notation is a system that makes working with very small or very large numbers easier.
25. a. 2
   b. 4
26. a. 6
   b. 27
   c. In part a, the number of two-digit numbers is the same as the number of digits. In part b the number of three-digit numbers is double the number of digits because the number can begin with any one of the three digits, and then there are two digits to choose from for the second place and one digit for the third place; this makes six possible numbers.
   d. In part a, the number of two-digit numbers is $2^2$. In part b the number of three-digit numbers is $3^3$.
27. About 12 days at a rate of 1 number per second

### 1.3 Strategies for adding and subtracting positive integers

1. a. 228 b. 1803
2. a. 906 b. 448
3. a. 560 b. 102
4. a. 86 b. 64
5. True
6. B
7. 847
8. a. 45 b. 351
9. a. 64 b. 33
10. a. 17 b. 227
11. 278
12. a. 5 or 0 b. 10
13. Yes. Working shown:
    $145 + 671$: Break into 600, 70 and 1
    $145 + 600 = 745$
    $745 + 70 = 815$
    $815 + 1 = 816$
    $816 - 472$: Break into 400, 70 and 2
    $816 - 400 = 416$
    $416 - 70 = 346$
    $346 - 2 = 344$
14. a. The student has not listed the tens and ones correctly.
   b. 998
15. There are 18 children on the playground now.
16. 550 students are not completing exams.
17. Harriet has $161 left in her savings account.
18. The two numbers are 13 and 6.

## 1.4 Algorithms for adding and subtracting positive integers

1. a. 15 b. 37 c. 210
   d. 240 e. 1730
2. a. 18 220 b. 175 000 c. 20
   d. 40 e. 50
3. a. 99 b. 72 648 c. 13 629
4. a. 978 b. 766 c. 1335
   d. 14 462 e. 1 701 553
5. a. 180 b. 280 c. 790
   d. 827
6. a. 6 b. 23 c. 5
   d. 125 e. 900
7. a. 700 b. 70 c. 500
   d. 5 e. 26
8. a. 4 b. 20 c. 55
   d. 0 e. 73 f. 1150
9. a. 35 b. 35 123 c. 584 599
10. a. 267 b. 4413 c. 65 439
    d. 2634
11. a. 69 397 b. 1553 c. 349 818
    d. 19 088 744
12. $247
13. 3032 pages
14. $71
15. 6670 km
16. a. 7, 6, 14 b. 15, 10, 15 c. 8, 20, 36
17. a. 26 m b. 26 m c. 52 m
18. 15
19. 12 149
20. 31
21. a. 172 km b. 353 km
22. a. 638 321 − 488 417 = 149 904
    b. 3492 − 1685 = 1807 (other answers possible)
23. 5 849 318
24. 7 days
25. Trixie
26. Three students attend both swimming practice and lifesaving.
27. Three people do not work Monday to Sunday.
28. 6

## 1.5 Multiplying positive integers

1. a. 24 b. 54 c. 60
   d. 26 e. 45
2. a. 100 b. 90 c. 32
   d. 63 e. 108
3. a. 75 b. 72 c. 27
   d. 90 e. 80
4. 801
5. a.

|   | 500 | 20 | 3 |
|---|---|---|---|
| 4 | 2000 | 80 | 12 |

2000 + 80 + 12 = 2092

   b. 2092
6. a. 128 b. 1233
   c. 2571 d. 24 600
7. a. 63 104 b. 16 120
   c. 63 582 d. 492 720
8. a. 2288 b. 3007 c. 1652
9. a. 912 b. 64 600 c. 21 571
10. a. 2 161 485 b. 27 529 460 c. 1 725 805
11. a. 80 b. 1900
    c. 4500 d. 67 000
12. a. 900 b. 3100
    c. 16 800 d. 172 000
13. a 8000 b 480 000
    c 5 500 000 d 6 300 000 000
14. a. 7200 b. 5 600 000
    c. 54 000 000 d. 13 200 000
15. a. 306 b. 5652
    c. 7425 d. 208
16. a. 5600 b. 560 c. 5040
17. a 4320 b 11 250 c 28 800
18. a. 150 b. 300 c. 300
19. a. 180 b. 840 c. 640 d. 680
20. a. 90 b. 27 c. 117
21. a. 325 b. 312
22. 208
23. 1176
24. a. $72 b. $1728
25. $132 000
26. a. $675 b. $1200 c. $2700
27. 490 km
28. 524 seconds
29. a. 10 km b. $175 c. 120 km
    d. $9100 e. 400 papers f. 20 800 papers
30. The correct answer is 1470 so Ben is correct.
31. 92 > 140 False
32. 260
33. 576
34. 1089; 2178; 3267
    a. 4356 b. 5445 c. 6534
    d. 7623 e. 8712 f. 9801
35. 46
36. The two numbers are 34 and 35. Their sum is 69.

## 1.6 Dividing positive integers

1. a. 4 b. 4 c. 8
   d. 8 e. 13 f. 3
2. a. 3 b. 2 c. 2
   d. 2 e. 5 f. 15
3. a. 10 b. 56 c. 84
   d. 60 e. 44
4. a. 485 b. 6208
   c. 2761 d. 80 045
5. a. 85 064 remainder 2 b. 86 047 remainder 3
   c. 904 075 remainder 5 d. 60 074 remainder 2
6. a. 700 b. 7 c. 900
7. a. 500 b. 5000 c. 800
8. 72
9. 214
10. 532
11. a. \$208 b. \$48
12. 186 cows
13. Number of tables required: 19; total number of courses to be served: 608; number of waiters required: 8; total charge for catering: \$8360
14. a. 950 $m^2$ b. \$794 750
15. Kayla: $243 \div 9 = 27$
    $$9\overline{)2^{2}4^{6}3}\quad 027$$
16. The statement is false.
17. a. 37 b. 3
18. \$16 per night
19. There needs to be 8 tables.
20. 10, 18, 26, 34. Students should mention that the pattern in the solution differs by the divisor and that it is found by adding the remainder to each of the multiples.

## 1.7 Rounding and estimating

1. a. 10 b. 50 c. 1000
   d. 10 000 e. 200 000 f. 5 000 000
2. 28 000 000
3. Estimation table:

| | Estimate | Estimated answer | Actual answer |
|---|---|---|---|
| a. | 500 + 1000 | 1500 | 1449 |
| b. | 30 000 + 80 000 | 110 000 | 115 670 |
| c. | 200 000 + 700 000 | 900 000 | 907 448 |
| d. | 90 000 − 40 000 | 50 000 | 55 880 |
| e. | 300 − 200 | 100 | 127 |
| f. | 500 000 − 200 000 | 300 000 | 296 049 |
| g. | 40 × 200 | 8000 | 7128 |
| h. | 600 × 10 000 | 6 000 000 | 7 894 033 |
| i. | 30 000 × 40 | 1 200 000 | 1 149 954 |
| j. | 30 000 ÷ 1000 | 30 | 32.907 58 |
| k. | 60 000 ÷ 3000 | 20 | 24.325 483 |
| i. | 70 000 ÷ 1000 | 70 | 53.385 505 |

4. D
5. B
6. B
7. a. 8000 b. 100 c. 400 000
   d. 700 000 e. 11 000
8. a. 6000 b. 8 000 000 c. 100
   d. 40 000 000 e. 500
9. E
10. D
11. A
12. D
13. a. 1500 tickets b. \$11 250
14. a. i. 6000 ii. 5340
    b. i. 8000 ii. 8028
    c. i. 50 ii. 46.677 749
15. \$29
16. One way to estimate the number of students is to count (to your best ability) the number of students in 1 row and in 1 column and multiply. It appears to be about $15 \times 20 = 300$ students in the photograph. If the hall holds 12 times this number, the capacity of the hall can be estimated as $30 \times 12 = 3600$.
17. 600 students
18. It is important not to estimate calculations by rounding when a very accurate answer is needed.
19. \$7270
20. a.

| Item | Estimated cost |
|---|---|
| Bread rolls | \$75 |
| Hamburgers | \$50 |
| Tomato sauce | \$4 |
| Margarine | \$4 |
| Onions | \$4 |
| Tomatoes | \$8 |
| Lettuce | \$5 |

   b. \$150 c. \$600 d. \$450

## 1.8 Order of operations

1. a. 5 b. 9 c. 3
   d. 30 e. 59
2. a. 25 b. 7 c. 96
   d. 0

3. a. 15 b. 24 c. 32
 d. 18 e. 76 f. 27

4. a. 26 b. 7 c. 96
 d. 27 e. 18 f. 3

5. a. $(12-8) \div 4 = 1$
 b. $(4+8) \times 5 - 4 \times 5 = 40$
 c. $3 \times (10-2) \div 4 + 4 = 10$
 d. $12 \times (4+2) - 12 = 60$
 e. $10 \div (5+5) \times 9 \times 9 = 81$
 f. $(18-3) \times 3 \div 5 = 9$

6. a. 18 b. 14 c. −8 d. 22

7. a. 2 b. 27 c. 0 d. 1

8. a. 3 b. 18 c. 40 d. 77

9. a. 23 b. 18 c. 2 d. 15

10. a. 8 b. 21 c. 16 d. 0

11. a. 27 b. 3 c. 21 d. 32

12. C

13. B

14. D

15. a. $6 + 2 \times (4-3) \times 2 = 10$
 b. $(6+2) \times 4 - 3 \times 2 = 26$
 c. $6 + (2 \times 4 - 3) \times 2 = 16$
 d. $6 + 2 \times 4 - 3 \times 2 = 8$, no brackets required

16. a. $274 + 2 \times -68 + 3 \times 50 = \$288$
 b. 600 apples
 c. 8 kg

17. When dealing with money.

18. Keenan $(6 + 6 \div 3 = 6 + 2 = 8)$

19. 43 pebbles

20. $229.20

21. a. $3 \times (4.25 + 4.75) + 2 \times 3.75$
 b. $34.50

22. a. Sydney 9h ahead
 b. Singapore 6h ahead

23. Two possible solutions: 2 packets with 3 ice creams and 5 packets with 5 ice creams, or 7 packets with 3 ice creams and 2 packets with 5 ice creams.
 With further investigation of the number pattern obtained by summing 3s and 5s, it appears that there are combinations of 3 packs and 5 packs that will amount to any quantity of ice creams equal to or larger than 8.

24. An example is shown below.
 $1 = (4-3) \div (2-1)$
 $3 = 21 \div (3+4)$
 $5 = 3 + 4 - (2 \times 1)$

25. An example is shown below.
 $5 = 4 + 3 - (2 \times 1)$
 $6 = 3 \times 4 \div (2 \times 1)$
 $8 = 4 \times (3 - 2 + 1)$
 $9 = 4 + 3 + 2 \div 1$

## 1.9 Review

### Project

1. 1 530 430

2. 𓆼𓆼𓍢𓎆𓎆𓎆𓎆𓏺𓏺𓏺𓏺𓏺𓏺𓏺

3. 𓍢𓍢𓎆𓎆𓎆𓎆𓎆𓏺𓏺

4. 𓂭𓆼𓆼𓆼𓍢𓍢𓍢𓍢𓍢𓍢𓎆𓎆𓎆𓎆𓎆𓎆𓎆𓎆𓏺𓏺

5. 𓎆𓎆𓏺𓏺𓏺𓏺𓏺𓏺

6. 𓍢𓍢𓍢𓍢𓎆𓎆𓎆𓎆𓎆𓎆𓎆𓎆𓏺𓏺𓏺𓏺

7. List any difficulties you encountered.

8.

| Symbol | Value | Symbol | Value |
|---|---|---|---|
| $\alpha$ | 1 | $\xi$ | 60 |
| $\beta$ | 2 | $o$ | 70 |
| $\gamma$ | 3 | $\pi$ | 80 |
| $\delta$ | 4 | $\varphi$ or $\zeta$ | 90 |
| $\varepsilon$ | 5 | $\rho$ | 100 |
| $\zeta$ or $F$ | 6 | $\sigma$ | 200 |
| $\xi$ | 7 | $\tau$ | 300 |
| $\eta$ | 8 | $\upsilon$ | 400 |
| $\theta$ | 9 | $\phi$ | 500 |
| $\iota$ | 10 | $\chi$ | 600 |
| $\kappa$ | 20 | $\psi$ | 700 |
| $\lambda$ | 30 | $\omega$ | 800 |
| $\mu$ | 40 | ϶̄ | 900 |
| $\nu$ | 50 | | |

9. Adding a subscript or superscript *iota* (Greek letter for 10) with the symbols used for 1 to 9 represented the numbers 1000 to 9000. *M* was used to represent 10 000. When symbols were placed on top of the *M*, this meant that the value of the symbols was multiplied by 10 000. This enabled the ancient Greeks to represent larger numbers.

10. Sample response can be found in the worked solutions in the online resources.

### Exercise 1.9 Review questions

1. a. 7 b. 40 000
 c. 1 000 000 d. 90

2. a. $300 + 90 + 2$ b. $4000 + 100 + 9$
 c. $40\,000 + 2000 + 1$ d. $100\,000 + 20\,000$

3. 349, 394, 934, 943, 3499, 3994

4. 1101, 1100, 1011, 110, 101

5. a. 127 b. 3187
 c. 7641 d. 4 217 656

6. a. 2 b. 35 c. 150 d. 14

7. a. 415 b. 217 c. 60 322
 d. 37 758 e. 92 957

8. a. 150 b. 8400 c. 1240
 d. 603 e. 589

9. a. $3 \times 6$

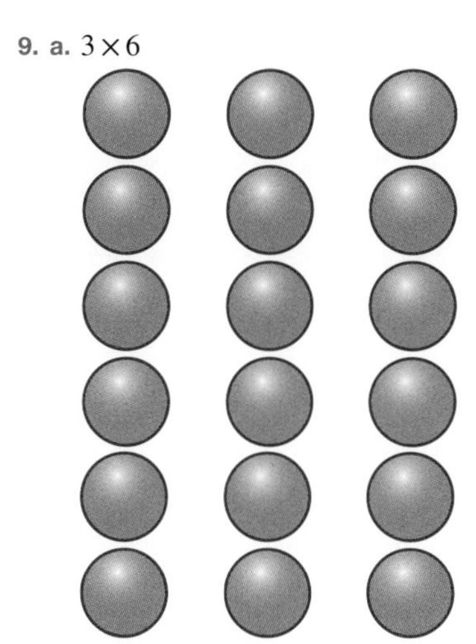

There are 18 balls = 18

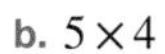

b. $5 \times 4$

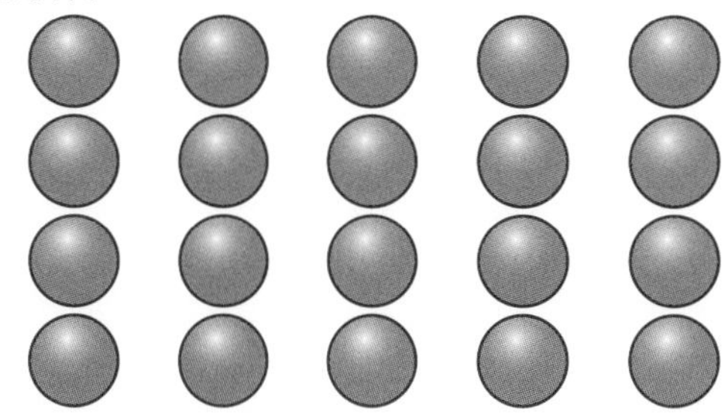

There are 20 balls = 20

c. $2 \times 7$

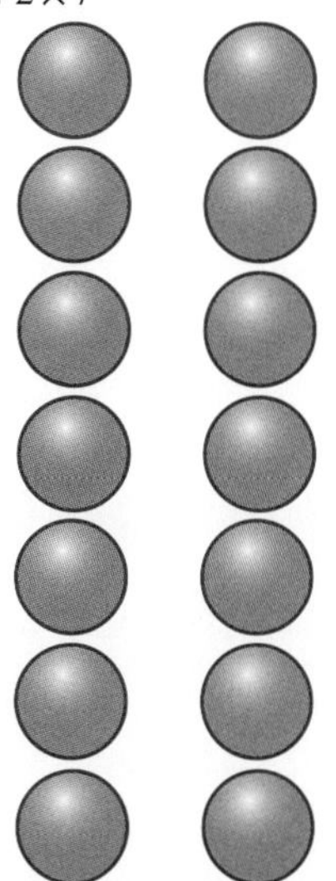

There are 14 balls = 14

10. a. 96 b. 1072 c. 3210
11. a. 1416 b. 1388 c. 3375
12. a. 1152 b. 2286 c. 19 200
13. a. 4608 b. 30 894 c. 18 355 064
14. a. 129 b. 260 c. 5021
15. a. 8 b. 3 c. 84 d. 45 e. 1
16. a. 700 b. 7 c. 410
17. a. 1568 b. 981 c. 537 r 3 d. 335 r 7 e. 911 r 9
18. a. 14 000 b. 5000 c. 50 d. 420 000 e. 120 000 f. 210 000
19. C
20. a. $<$ b. $=$ c. $>$ d. $>$
21. a. 327 b. 1090 c. 789 d. 216
22. a. 127 b. 81 c. 711 d. 278
23. a. 28 b. 24 c. 10 d. 8 e. 11 f. 27
24. a. 12 b. 30 c. 5 d. 900 e. 14 f. 25
25. a. $<$ b. $=$ c. $>$ d. $<$
26. a. $<$ b. $<$ c. $>$ d. $>$
27. 23.5
28. a. 5 b. 38 c. 6 d. 5 e. 70 f. 29
29. a. 306 km b. 322 km c. 482 km
30. a. 448 b. 3136 c. 98
31. As long as one of the factors in a multiplication is even, the result will be even. Otherwise, the answer will be odd.
32.

$$\begin{array}{r} 8240 \\ -3681 \\ \hline 4559 \\ \hline \end{array}$$

33. 95
34. a. $8 \times 3 + 12 \times 2$ b. \$48
35. 20 499
36. a.

| + | 6 | 10 | 3 |
|---|---|---|---|
| 5 | 11 | 15 | 8 |
| 12 | 18 | 22 | 15 |
| 9 | 15 | 19 | 12 |

b.

| × | 4 | 8 | 6 |
|---|---|---|---|
| 3 | 12 | 24 | 18 |
| 5 | 20 | 40 | 30 |
| 2 | 8 | 16 | 12 |

37. a. 24 °C b. −21 °C
38. 21 questions for 3 marks
39.

| 8 | + | 4 | × | 6 | = | 32 |
|---|---|---|---|---|---|---|
| + | | × | | × | | |
| 3 | − | 5 | + | 9 | = | 7 |
| × | | ÷ | | ÷ | | |
| 7 | − | 2 | − | 1 | = | 4 |
| = | | = | | = | | |
| 29 | | 10 | | 15 | | |

40. 2019 and 2018

# 2 Positive and negative integers

## LESSON SEQUENCE

**2.1** Overview ........ 68
**2.2** Integers on the number line ........ 70
**2.3** Integers on the number plane ........ 75
**2.4** Adding integers ........ 80
**2.5** Subtracting integers ........ 85
**2.6** Multiplying and dividing integers (extending) ........ 91
**2.7** Order of operations with integers (extending) ........ 97
**2.8** Review ........ 102

# LESSON
# 2.1 Overview

## Why learn this?

Integers are whole numbers. They can be positive, negative or equal to zero. Positive integers have values above zero and negative integers have values below zero.

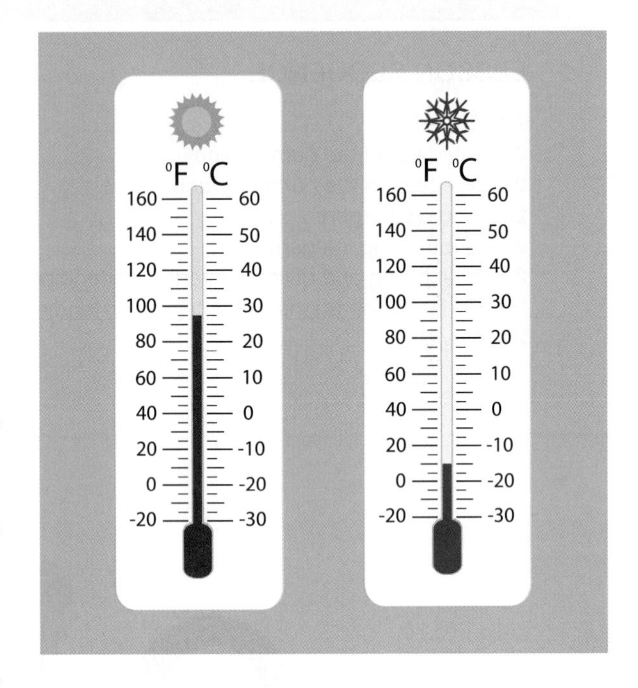


If you live in the southern parts of Australia, overnight temperatures often fall below zero in winter. These very cold temperatures are written as negative integers. In summer the generally very hot temperatures are written as positive integers.

Using your understanding of the basic operations of addition, subtraction, multiplication and division, you can measure the temperature changes throughout a single day. You can also measure the average daily temperature over a month and use those measurements to compare your home town to other parts of Australia, or other parts of the world.

In finance, positive numbers are used to represent the amount of money in someone's bank account, while negative numbers are used to represent how much money someone owes, for example how much they have to pay back after borrowing money or taking out a loan from a bank.

## Exercise 2.1 Pre-test

learnon

1. Write an integer to express the following.
   a. A debt of \$250
   b. 15 metres below sea level
   c. The temperature is 20°C

2. Complete each statement by placing $<$, $>$ or $=$ in the empty box so that the statement is true.
   a. $-3 \square 2$  b. $-5 \square -7$  c. $-21 \square -17$  d. $0 \square -8$

3. MC Select which of the following options has the numbers −5, 8, 2, 0, −1, −8, 3 arranged in ascending order.
   A. 8, −5, −1, 0, 2, 3, −8  B. −8, −5, −1, 0, 2, 3, 8  C. 0, −1, 2, 3, −5, −8, 8
   D. −8, 5, −1, 0, 2, 3, 8  E. −1, 0, 2, 3, −5, −8, 8

4. Determine the coordinates of point A in the graph.

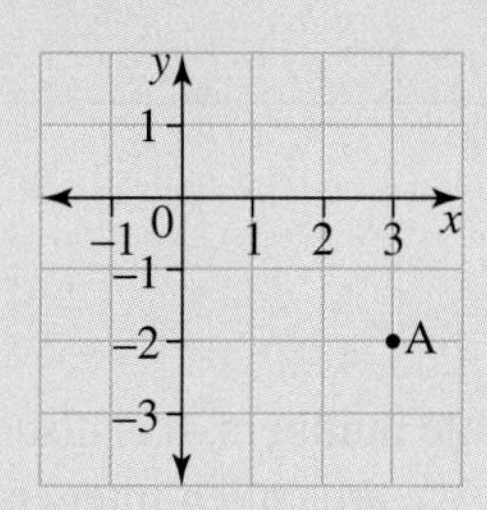


5. MC Select the quadrant in which the point (−7, −1) is located.
   A. First  B. Second  C. Third  D. Fourth  E. On the $x$-axis

6. MC If the points (−2, −2), (−2, 2), (2, 2) and (2, −2) were plotted on a Cartesian plane and joined by straight lines, select the shape this would make.
   A. Rectangle  B. Circle  C. Square  D. Trapezium  E. Kite

7. Calculate the following.
   a. $3 + -5 =$  b. $7 - -6 =$  c. $-7 + -12 =$  d. $-16 + 5 =$

8. Write the number that is 12 more than −5.

9. Narelle went for a hike at Mount Baw Baw. From her starting point, she climbed up 72 m and then descended 15 m, before ascending a further 21m. Determine how far above her starting point she is now.

10. Evaluate $-8 \times -6$.

11. Evaluate the following.
    a. $7 \times 6 \div 3 \div 7$  b. $3 \times 8 \div 6 + 4$  c. $7 + [66 - (5 \times 5)] \times 3$

12. MC Select the simplified form of the algebraic expression $5g \times -3h \times -6k$.
    A. $-15\,ghk$  B. $18\,ghk$  C. $-90\,ghk$  D. $90\,gh$  E. $90\,ghk$

13. If $a = -5$, $b = 4$ and $c = -2$, evaluate $4(a - b) \div 3c$.

14. Hang has \$340 in the bank. She makes a deposit of \$75 each month for the next three months, before withdrawing \$185 to buy a new dress. Determine the amount of money Hang has left in her account.

15. To improve goal kicking a new training drill was developed. In this training run, kicking a goal scores you 6 points, but missing a goal loses you 2 points. If a team had 30 shots at goal and kicked 21 goals, determine how many points they scored.

# LESSON
# 2.2 Integers on the number line

### LEARNING INTENTION

At the end of this lesson you should be able to:
- place integers on a number line
- compare integer values.

## 2.2.1 Integers

eles-3695

- **Whole numbers** are the numbers 0, 1, 2, 3, 4 ...
- **Integers** are positive and negative whole numbers. 0 is also an integer.
  - The numbers ... − 3, −2, −1, 0, 1, 2, 3 ... are called integers.
  - The integers −1, −2, −3 ... are **negative integers**.
  - The integers 1, 2, 3 ... are **positive integers**. They can also be written as +1, +2, +3 ...
  - 0 is an integer. It is neither positive nor negative.
  - The direction of an integer indicates whether it is positive or negative. The + sign can be used to represent either an addition (if it appears between two quantities) or the direction of an integer (if it appears in front of an integer). In the same way, the − sign can be used to represent either a subtraction or the direction of an integer.

### DISCUSSION

Can you think of an example of negative numbers that we encounter in everyday life?

### WORKED EXAMPLE 1 Converting a description into an integer

**Write the integer that is suggested by the following descriptions.**
**a. My flat is on the 5th floor.**
**b. This town is 20 m below sea level.**

| THINK | WRITE |
|---|---|
| a. Numbers above 0 are positive, so place a + sign before the number. | a. +5 (or 5) |
| b. Numbers below 0 are negative, so place a − sign before the number. | b. −20 |

## 2.2.2 The number line

eles-3696

- Integers can be represented on a number line. On a number line, positive integers are written to the right of 0, while negative integers are written to the left of 0.

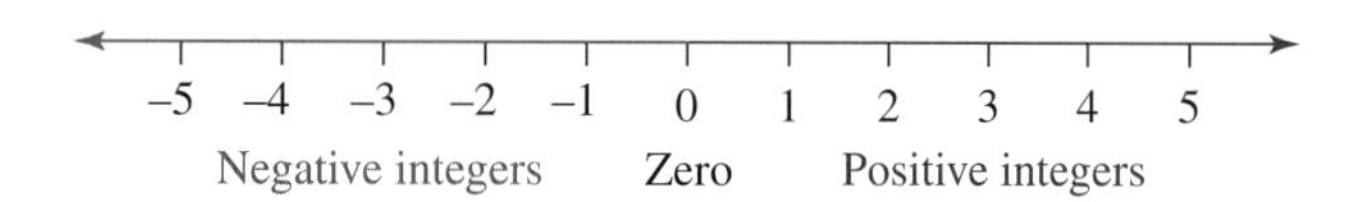


- The **magnitude** of an integer shows the size of the integer in terms of its distance from 0. For example, the number −3 is negative and has a magnitude of 3. This means it is a distance of 3 units from 0.
- The further a number is to the right of any number on a number line, the larger it is. For example, +5 is greater than +2, while + 2 is greater than 0 and 0 is greater than −5.
- **Opposite integers** are the same distance from 0, but on opposite sides of 0. For example, 3 and −3 are opposite integers because they are both 3 units from 0.

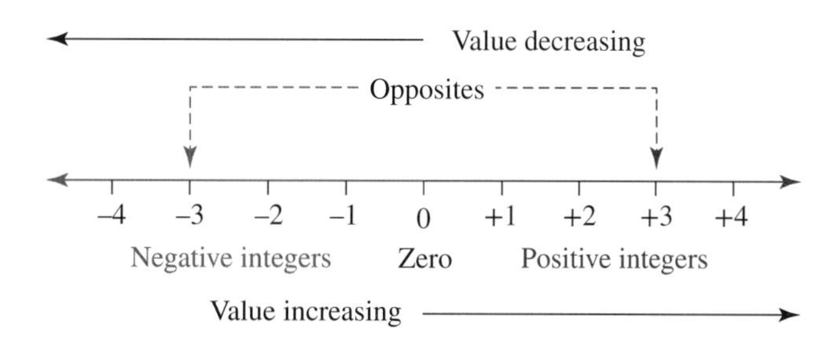


### WORKED EXAMPLE 2 Placing integers on a number line

**Place the integers 9, − 2, − 7, 5 on a number line.**

| THINK | WRITE |
|---|---|
| 1. Draw a number line from −10 to 10 so that you can include all of the numbers. | 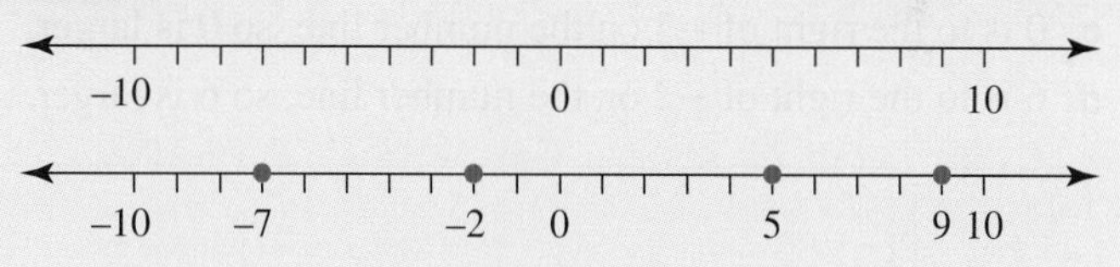 |
| 2. Mark the values 9, −2, −7, 5 on the number line. | |



### WORKED EXAMPLE 3 Writing integers in numerical order

**Use a number line to help list the integers −4, 2, −3, 1 in numerical order.**

| THINK | WRITE |
|---|---|
| 1. Draw a number line from −5 to 5 so that you can include all of the numbers. | 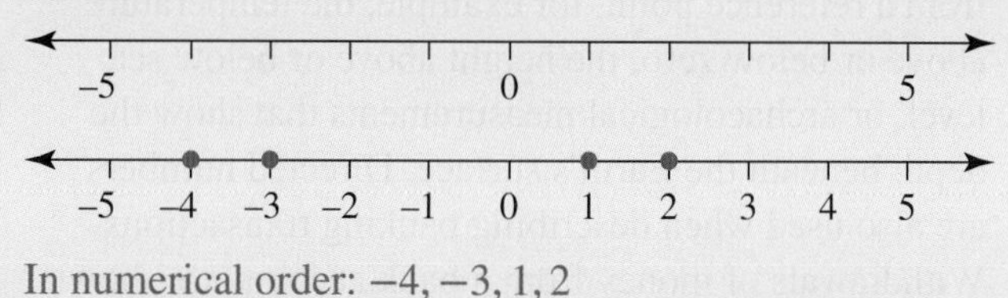 |
| 2. Mark the values. As you move from left to right on the number line, the numbers become larger. | In numerical order: −4, −3, 1, 2 |

**WORKED EXAMPLE 4 Determining the opposite of an integer**

**Write opposites for the following integers.**

a. $-2$ b. 3 c. 4

| THINK | WRITE |
|---|---|
| a. An opposite integer is the same distance from 0, but has the opposite sign. | a. The opposite of $-2$ is 2. |
| b. An opposite integer is the same distance from 0, but has the opposite sign. | b. The opposite of 3 is $-3$. |
| c. An opposite integer is the same distance from 0, but has the opposite sign. | c. The opposite of 4 is $-4$. |

### Comparing integers

- The following inequality symbols can be used when comparing integers.
  $>$ means 'greater than'
  $<$ means 'less than'
  $\geq$ means 'greater than or equal to'
  $\leq$ means 'less than or equal to'
- Remember that inequality signs look a little bit like a crocodile's mouth. When placing an inequality symbol between two numbers, the crocodile's mouth always opens up to eat the larger number.

**WORKED EXAMPLE 5 Comparing integers**

**Complete each statement by inserting the correct symbol: < or >.**

a. $2 \square 5$ b. $-4 \square -1$ c. $0 \square -3$ d. $6 \square -2$

| THINK | WRITE |
|---|---|
| a. 2 sits to the left of 5 on the number line, so 2 is smaller. | a. $2 < 5$ |
| b. $-4$ sits to the left of $-1$ on the number line, so $-4$ is smaller. | b. $-4 < -1$ |
| c. 0 is to the right of $-3$ on the number line, so 0 is larger. | c. $0 > -3$ |
| d. 6 is to the right of $-2$ on the number line, so 6 is larger. | d. $6 > -2$ |

## 2.2.3 Using positive and negative numbers in daily life

eles-3697

- Because positive and negative numbers show both their size and their direction away from zero, they are often referred to as **directed numbers**.
- Directed numbers are often used to measure objects from a reference point, for example, the temperature above or below zero, the height above or below sea level, or archaeological measurements that show the depth beneath the Earth's surface. Directed numbers are also used when describing banking transactions. Withdrawals of money from a bank account are shown using negative signs while deposits into a bank account are shown using positive signs.

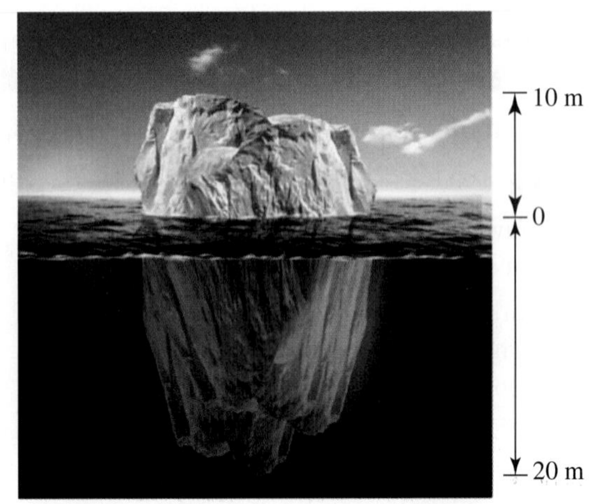

## WORKED EXAMPLE 6 Converting a statement into a directed number

**Write the integer that is suggested by each of the following situations.**

**a. The temperature was 1 degree below zero.**

**b. Romie has $500 in her bank account.**

| THINK | WRITE |
|---|---|
| a. 'Below zero' means negative. | a. $-1$ |
| b. Romie has money in her account, so the integer is positive. | b. $+500$ |

### Resources

**eWorkbook** Topic 2 Workbook (worksheets, code puzzle and project) (ewbk-1903)

**Interactivities** Individual pathway interactivity: Integers on the number line (int-4367)

Number line (int-4033)

## Exercise 2.2 Integers on the number line

learnon

2.2 Quick quiz on | 2.2 Exercise

### Individual pathways

| ■ PRACTISE | ■ CONSOLIDATE | ■ MASTER |
|---|---|---|
| 1, 4, 7, 10 | 2, 5, 8, 11 | 3, 6, 9, 12 |

### Fluency

1. WE1&6 Write the integer suggested by each of the following descriptions.
   a. The temperature today is 7 degrees below zero.
   b. The penthouse is on the 10th floor.
   c. The Dead Sea is 422 m below sea level.
   d. Mount Kosciuszko is 2228 m above sea level.
   e. The Dow Jones dropped 3 points overnight.
   f. Chloe lost $30.

2. WE2 Draw a number line from $-10$ to 10. Mark the following values on your number line.

   a. 2  b. $-4$  c. $-9$  d. 7
   e. 5  f. $-1$  g. 0  h. $-3$

### Understanding

3. WE3 Use a number line to place the following numbers in numerical order.

   a. $5, -2, 3, -6, 7$  b. $0, -1, 3, -4, -6$  c. $-5, 7, -2, -3, 1$  d. $-4, 2, 1, -3, 5$

4. WE4 Write opposite integers for the following.

   a. $-6$  b. 7  c. 1  d. $-8$

5. **WE5** Complete the following by inserting $<$ or $>$ in the box.

a. $2 \square -5$  b. $3 \square 7$  c. $-2 \square -6$
d. $6 \square -2$  e. $-1 \square 3$  f. $-10 \square -6$

6. From the following lists, select:

i. the smallest number  ii. the largest number.

a. $-3, 7, -5$  b. $2, -4, 5, 3, -2$  c. $7, -10, 5, -2, -4$  d. $-4, -1, 3, 0, -2$

### Reasoning

7. The temperature in Oslo dropped to −6 °C overnight. During the day, the temperature rose by 4 °C. Determine the maximum temperature on that day. Show full working.

8. A new apartment building is being built. It will include a ground floor reception area. If there will be 10 floors of apartments and 3 floors of underground parking, how many buttons will be needed in the lift? Explain your answer.

9. Lake Mungo, part of Mungo National Park in New South Wales, is a World Heritage listed area. Archaeologists estimate that the lake dried out 1800 years ago and, before that, Aboriginal people lived on the shores of the lake. Lake Mungo became one of the world's most significant archaeological sites following the discovery there of some of the oldest remains of human civilisation on Earth.
Explain what −60 000 could mean in terms of human occupation in the area around Lake Mungo.

### Problem solving

10. Raj and Vanessa are going home from school in opposite directions. Raj lives 1.2 km south of the school and Vanessa lives 1.6 km north of the school.
    a. Draw a number line to represent the three locations: Raj's house, Vanessa's house and the school. Label the three locations with directed numbers.
    b. Determine how far apart Raj and Vanessa live from each other.

11. An aeroplane flying at 12 km above sea level flies directly over a submarine travelling at 500 m below sea level.
    a. Draw a number line to represent the three locations: the aeroplane, the submarine and the ocean level. Label the three locations with directed numbers.
    b. Evaluate the distance between the submarine and the aeroplane.

12. a. Historians use BCE (Before Common Era) and CE (Common Era) when discussing dates. Explain what is meant by 2000 CE.
    b. If the millennium year 2000 CE (+2000) was considered to be zero, determine the:
        i. integer value that the year 2010 would represent
        ii. integer value that the year 1894 would represent.

c. Human exploration of space started in 1957, which led to new discoveries. The successful orbit of Earth by Sputnik I gave hope to humans that one day they would enter space and even walk on the Moon. If we consider 1957 as being zero, state the directed numbers that could be used to describe the following information.

   i. Cosmonaut Yuri Gagarin was the first man to enter space — 1961
   ii. Neil Armstrong and Edwin 'Buzz' Aldrin of Apollo 11 were the first humans to walk on the Moon — 1969
   iii. Sally Ride was the first American woman in space — 1983
   iv. *Discovery* was the first shuttle to dock with the International Space Station — 1999
   v. *Rosetta* was the first man-made probe to make a soft landing on a comet — 2014

d. Draw a number line from −10 to 60 to show your answers to part **c**.

# LESSON
# 2.3 Integers on the number plane

**LEARNING INTENTION**

At the end of this lesson you should be able to:
- understand the number plane (also called the Cartesian plane), origin and coordinates
- infer the location of a point described by its coordinates
- plot one or more points using their coordinates.

## 2.3.1 Positive integers and zero on the number plane

eles-3698

- The **number plane**, also called the Cartesian plane, has two axes: the horizontal axis, also called the $x$-axis, and the vertical axis, also called the $y$-axis.
- Every point on the number plane can be described by its position relative to the $x$- and $y$-axes.
- A pair of coordinates, also called an ordered pair $(x, y)$, fixes the position of a point. The $x$ value shows the point's distance along the $x$-axis, while the $y$ value shows the point's distance along the $y$-axis.

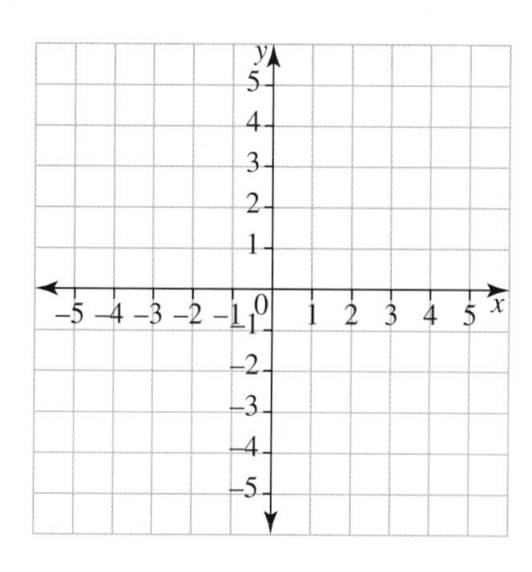

## WORKED EXAMPLE 7 Determining the coordinates

**Determine the coordinates of the following points in the number plane shown.**

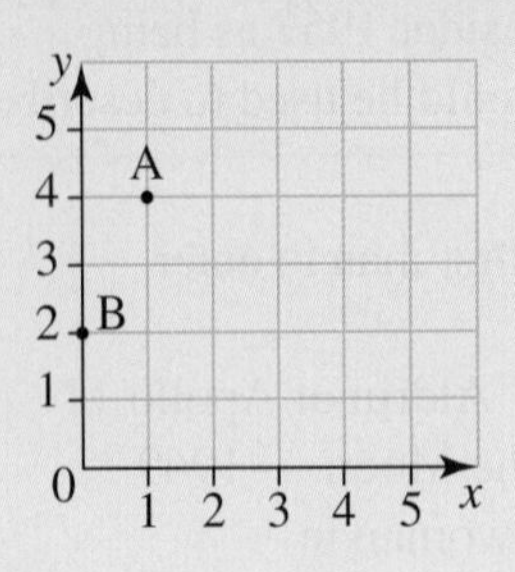


**a.** A **b.** B

| THINK | WRITE |
|---|---|
| **a.** Point A is 1 unit to the right of zero along the $x$-axis and 4 units above zero along the $y$-axis. | **a.** A (1, 4) |
| **b.** Point B lies on the $y$-axis, so it is 0 units to the right of zero along the $x$-axis. It is 2 units above zero along the $y$-axis. | **b.** B (0, 2) |

## 2.3.2 Integers on the number plane

eles-3699

- The origin (0, 0) is the point where the $x$- and $y$-axes intersect to divide the number plane into 4 quadrants, as shown. The origin is labelled as '0'.
- The first quadrant has positive $x$ and positive $y$ values.
- The second quadrant has negative $x$ and positive $y$ values.
- The third quadrant has negative $x$ and negative $y$ values.
- The fourth quadrant has positive $x$ and negative $y$ values.
- The illustration demonstrates the position of points on the number plane.
  - A (3, 2) — 3 right and 2 up
  - B (−3, 2) — 3 left and 2 up
  - C (−3, −2) — 3 left and 2 down
  - D (3, −2) — 3 right and 2 down

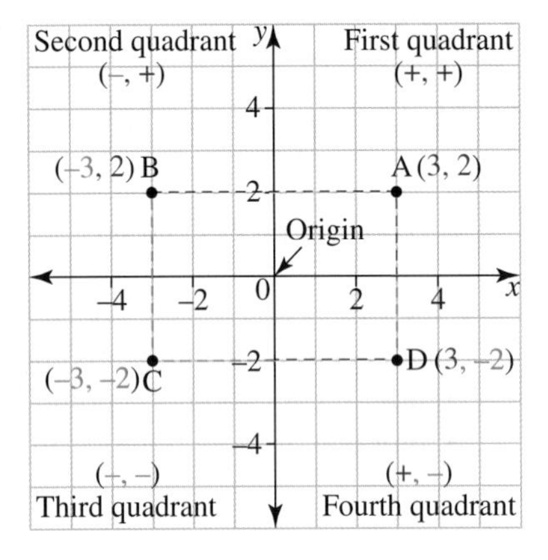


## WORKED EXAMPLE 8 Determining the coordinates, quadrant and axis

**Write the coordinates and state the quadrant or axis of each point on this number plane.**

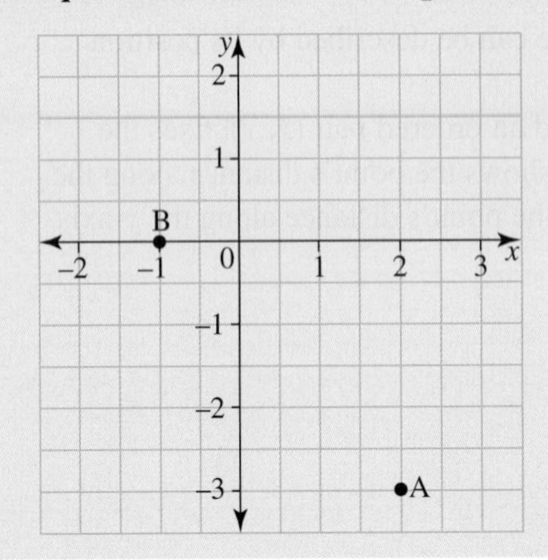

| THINK | WRITE |
|---|---|
| **a. 1.** A is 2 units to the right of the origin along the $x$-axis and 3 units down from the origin along the $y$-axis. | **a.** A (2, −3) |
| **2.** A is in the lower right-hand corner. | A is in the fourth quadrant. |
| **b. 1.** B is 1 unit to the left of the origin along the $x$-axis and 0 units up or down from the origin along the $y$-axis. | **b.** B (−1, 0) |
| **2.** B is on the $x$-axis. | B is on the $x$-axis. |

Resources

**eWorkbook** Topic 2 Workbook (worksheets, code puzzle and project) (ewbk-1903)

**Interactivities** Individual pathway interactivity: Integers on the number plane (int-4368)

Integers on the number plane (int-4035)

Positive integers and zero on the number plane (int-4034)

## Exercise 2.3 Integers on the number plane

**learnon**

**2.3 Quick quiz**  **2.3 Exercise**

### Individual pathways

| ■ PRACTISE | ■ CONSOLIDATE | ■ MASTER |
|---|---|---|
| 1, 2, 4, 7, 10, 13, 16, 19 | 3, 6, 8, 12, 15, 17, 20 | 5, 9, 11, 14, 18, 21 |

### Fluency

Questions 1 and 2 refer to the graph shown.

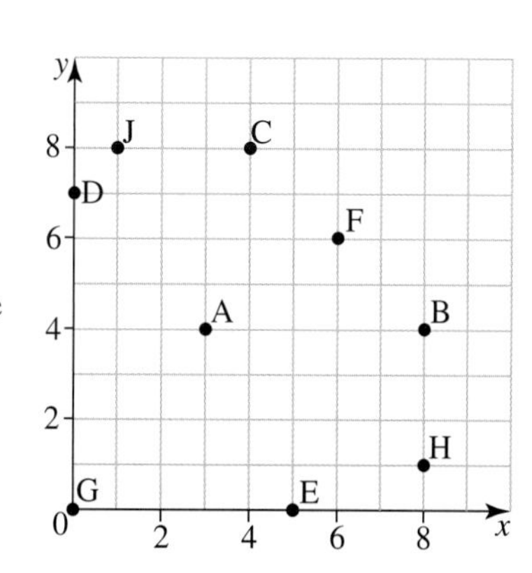


1. **WE7** Determine the coordinates of the following.
   a. A  b. B  c. C  d. D  e. E
2. Match a letter to the points with the following coordinates.
   a. (0, 0)  b. (6, 6)  c. (8, 1)  d. (1, 8)
3. **MC** Select the quadrant in which all the points have negative $x$ and negative $y$ coordinates.
   **A.** Quadrant 1  **B.** Quadrant 2  **C.** Quadrant 3
   **D.** Quadrant 4  **E.** None of the above
4. **MC** Select the quadrant in which all the points have positive $x$ and negative $y$ coordinates.
   **A.** Quadrant 1  **B.** Quadrant 2  **C.** Quadrant 3
   **D.** Quadrant 4  **E.** None of the above

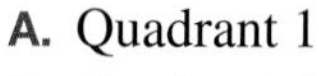

5. If the $x$-axis represents the direction from west to east, and the $y$-axis represents the direction from south to north, calculate the coordinates if a person starts at the origin and walks 5 m west and then 2 m north.

## Understanding

Use the graph shown for questions **6** through **10**.

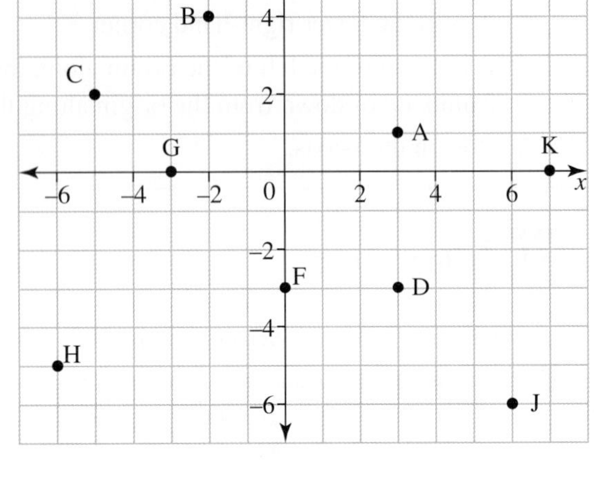


6.  **WE8** Write the coordinates and state the quadrant or axis of each of the following points.

   a. A b. B c. H
   d. F e. J

7. Match the following points to a letter and state their quadrant or axis.

   a. $(-5, 2)$ b. $(0, 5)$ c. $(3, -3)$
   d. $(2, 5)$ e. $(-3, 0)$

8. Determine the $x$-coordinate of the following points.

   a. A b. D c. K d. L

9. Determine the $y$-coordinate of the following points.

   a. C b. J c. G d. F

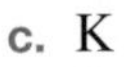

10. State all the points lying in the third quadrant in the number plane.

11. Determine whether the following statements are true or false.

    a. The origin has coordinates $(0, 0)$.
    b. The point at $(3, 5)$ is the same point as at $(5, 3)$.
    c. The point at $(-5, 4)$ is in the third quadrant.
    d. The point at $(0, 2)$ must lie on the $y$-axis.

12. **MC** The point $(5, -2)$ lies:

    A. in the first quadrant. B. in the second quadrant.
    C. in the third quadrant. D. in the fourth quadrant.
    E. on the$x$-axis.

13. **MC** The point $(0, 4)$ lies:

    A. in the first quadrant. B. on the $y$-axis.
    C. in the third quadrant. D. in the fourth quadrant.
    E. on the $x$-axis

14. **MC** The point $(-4, 5)$ lies:

    A. in the first quadrant. B. in the second quadrant.
    C. in the third quadrant. D. in the fourth quadrant.
    E. on the $x$-axis.

15. Draw a number plane with both axes scaled from $-6$ to 6. Plot the points listed and join them with straight lines in the order given. State the name for each completed shape.

    a. $(5, 5), (3, 2), (-2, 2), (0, 5), (5, 5)$ b. $(4, -1), (4, -5), (-1, -3), (4, -1)$
    c. $(-4, 4), (-2, 1), (-4, -5), (-6, 1), (-4, 4)$ d. $(-2, 1), (1, 1), (1, -2), (-2, -2), (-2, 1)$

## Reasoning

16. Consider the graph relating to question **6** and explain whether the points F and D have the same $y$-coordinate, and whether points A and D have the same $x$-coordinate.

17. Consider the graph shown.

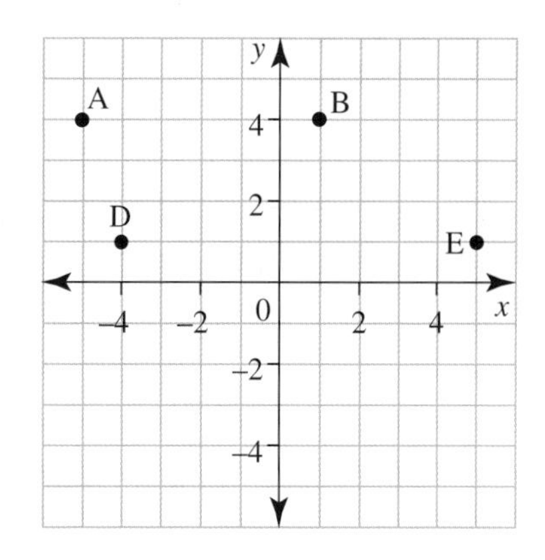


a. Determine the coordinates of a point, C, so that ABCD is a parallelogram.
b. Determine the coordinates of a point, F, so that DBEF is a kite shape.
c. Show that the point (4, −1) lies on the line that passes through D and the origin.
d. List 2 points on the line joining D to E.
e. Give the coordinates of a point, T, in the third quadrant that would complete the isosceles triangle ADT.

18. a. If you were moving east and then you did a U-turn, determine the direction you would then be moving in. Assume that a U-turn is a 180 degree turn, or going back the way you came and is the same as putting a negative sign in front of you.
b. Determine the direction you would be facing if you had two negative signs in front of you.
c. Determine the direction you would be facing if you had three negative signs in front of you.
d. Determine the direction you would be facing if you had ten negative signs in front of you.
e. Determine the direction you would be facing if you had 101 negative signs in front of you. Explain your answer.
f. Explain what would happen if you were moving north and then you made a U-turn (which would mean putting a negative sign in front of you). Determine the direction you would be moving in after you did the U-turn.
g. If you were moving north, explain which direction you would be facing after an odd number of negative signs. Explain which direction you would be facing after an even number of negative signs.

## Problem solving

19. Eugenia uses a remote control to move a robot car on a grid map. The robot car moves from point (7, 5) to point (7, −4). Determine the distance travelled by the robot car.

20. The tennis court shown is enclosed in an area that has a post in each corner.
a. Draw a number plane with the origin in the centre of the court.
b. Determine the coordinates of the four posts on the number plane drawn in part a.

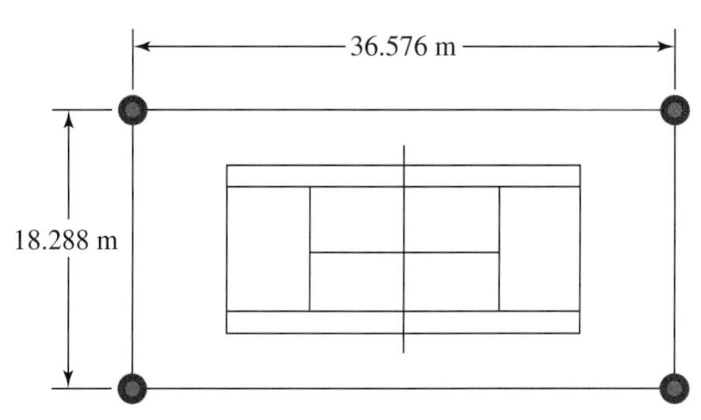


21. If you swap the $x$- and $y$-coordinates of a point, the position of the point may change to another quadrant. Determine what happens to points from each quadrant if you swap their coordinates.

# LESSON
# 2.4 Adding integers

## LEARNING INTENTION

At the end of this lesson you should be able to:
- understand how a number line can be used to add integers
- use a number line to add integers
- add integers without the aid of a number line.

## 2.4.1 Addition of integers

eles-3700

- A number line can be used to add integers.
- When adding a positive integer, move right along the number line.
- When adding a negative integer, move left along the number line.

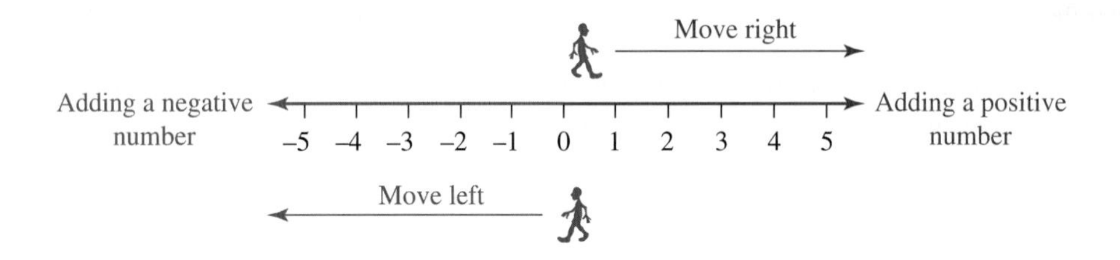


- Consider the addition of $-1+(+4)$.

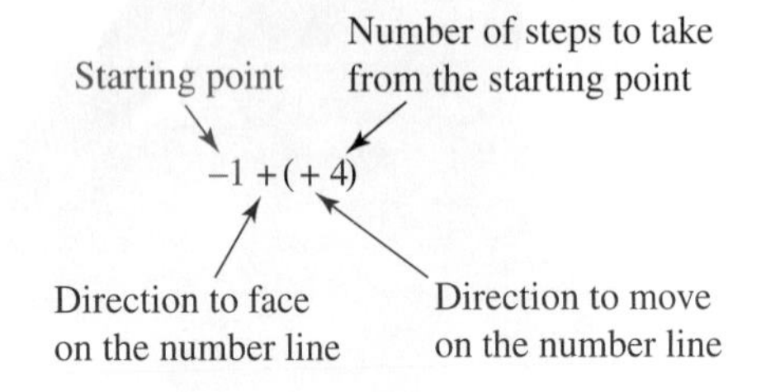


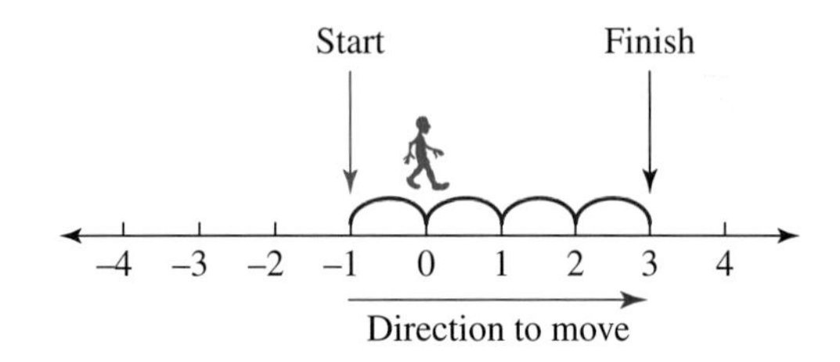


| Symbol | Meaning | Action to take |
|---|---|---|
| $-1$ | Starting point | Start at $-1$. |
| $+$ | Direction to face on the number line | Face right for addition $+$. |
| $(+)$ | Direction to move on the number line | Move forward for positive $(+)$. |
| $(4)$ | Number of steps to take from the starting point | Move 4 places from $-1$. |
| $=$ | Finishing point | Stop at $+3$. |

Therefore, $-1+(+4)=+3$

## WORKED EXAMPLE 9 Adding integers using a number line

**Calculate the value of each of the following, using a number line.**

a. $-5 + (+3)$

b. $-5 + (-3)$

**THINK**

a. 1. Start at −5 and face right (because this is an addition). Move 3 steps forward (because this number is positive). The finishing point is −2.

2. Write the answer.

b. 1. Start at −5 and face right (because this is an addition). Take 3 steps backwards (because this number is negative). The finishing point is −8.

2. Write the answer.

**WRITE**

a.

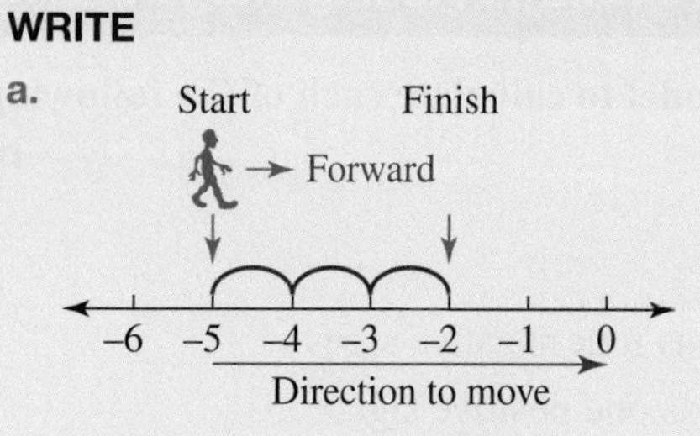


$-5 + (+3) = -2$

b.

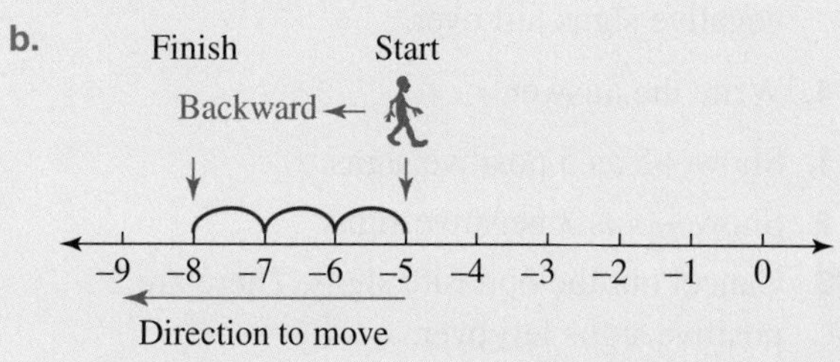


$-5 + (-3) = -8$

## WORKED EXAMPLE 10 Writing a number sentence from a number line

**Write number sentences to show the addition problems in the following diagrams.**

a.

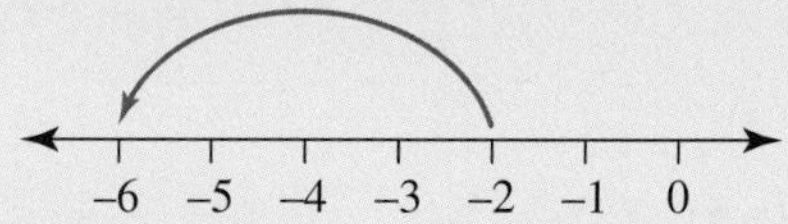


b.

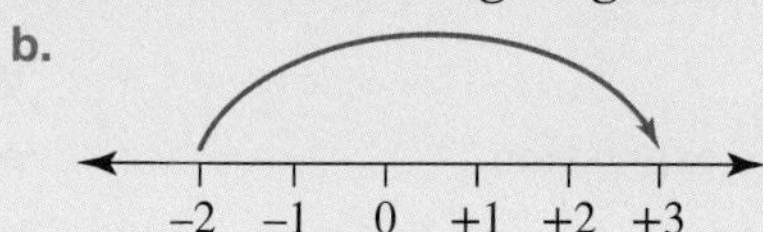


**THINK**

a. 1. The first integer is 2 units to the left of zero, so start at −2. The second integer is 4 units to the left of −2, finishing at −6.

2. Write the number sentence.

b. 1. The first integer is 2 units to the left of zero, so start at −2. The second integer is 5 units to the right of −2, finishing at +3.

2. Write the number sentence.

**WRITE**

a.

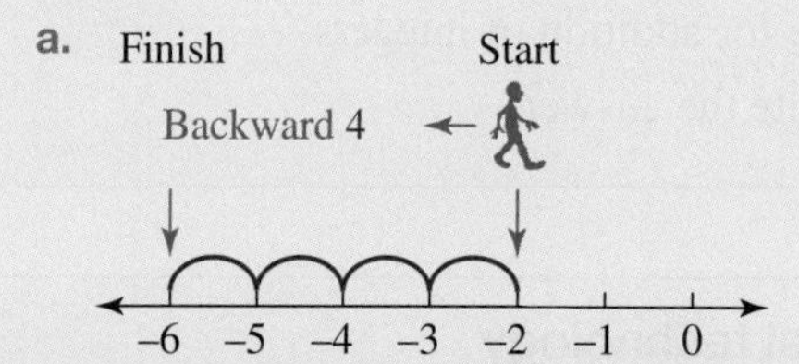


$-2 + (-4) = -6$

b. Start

→ Forward 5

Finish

−2 −1 0 +1 +2 +3

$-2 + (+5) = +3$

## Addition of integers using the sign model

- In the sign model, a − sign and a + sign cancel each other out. This model can be used to add integers, as shown in Worked example 11.

### WORKED EXAMPLE 11 Adding integers using the sign model

**Use the sign model to calculate each of the following.**

**a.** $-4+(+1)$ **b.** $+5+(-3)$

| THINK | WRITE |
|---|---|
| **a. 1.** Show −4 as four negative signs. | **a.** − − − − |
| **2.** Show +1 as one positive sign. | + |
| **3.** Cancel out the opposite signs. There are 3 negative signs left over. | ~~−~~ − − −<br>~~+~~ |
| **4.** Write the answer. | $-4+(+1)=-3$ |
| **b. 1.** Show +5 as 5 positive signs. | **b.** + + + + + |
| **2.** Show −3 as 3 negative signs. | − − − |
| **3.** Cancel out the opposite signs. There are 2 positive signs left over. | ~~+~~ ~~+~~ ~~+~~ + +<br>~~−~~ ~~−~~ ~~−~~ |
| **4.** Write the answer. | $+5+(-3)=+2$ |

### WORKED EXAMPLE 12 Evaluating an algebraic expression

**Evaluate the algebraic expression $a+2b+c$, if $a=-2$, $b=1$ and $c=-5$.**

| THINK | WRITE |
|---|---|
| **1.** Write the algebraic expression. | $a+2b+c$ |
| **2.** Substitute each pronumeral with the appropriate integer. | $=-2+2\times1+(-5)$ |
| **3.** Evaluate the expression by using the addition rule for addition of integers. | $=-2+2-5$ |
| **4.** Write the answer. | $=-5$ |

### Digital technology

Scientific calculators can evaluate the addition of positive and negative numbers.

```
DEG
-350+82+(-156)
              -424
```

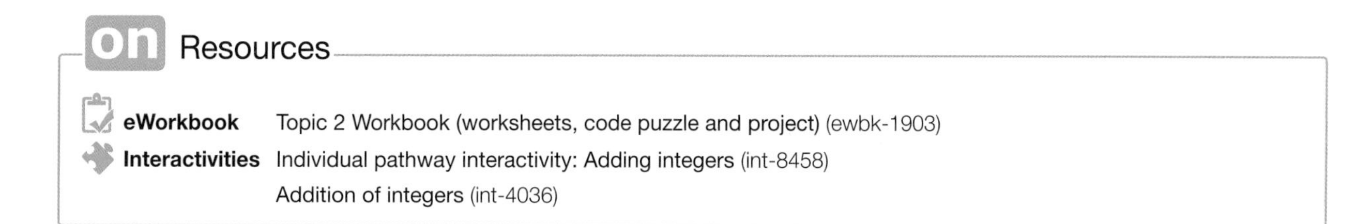


## Exercise 2.4 Adding integers

learnon

2.4 Quick quiz 

2.4 Exercise

### Individual pathways

| ■ PRACTISE | ■ CONSOLIDATE | ■ MASTER |
|---|---|---|
| 1, 3, 6, 10, 13, 16 | 2, 4, 7, 11, 14, 17 | 5, 8, 9, 12, 15, 18, 19 |

### Fluency

1. **WE9** Calculate the value of each of the following.
   a. $-25+(+10)$ b. $-7+(-3)$ c. $6+(-7)$
   d. $-8+(-5)$ e. $13+(+6)$ f. $16+(-16)$

2. **WE10** Write number sentences to show the addition problems in the following diagrams.

a. 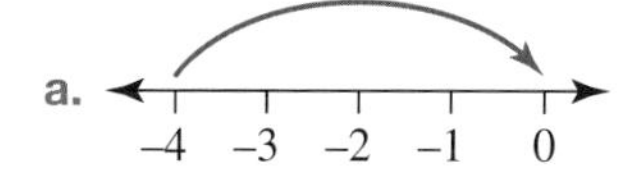


b. 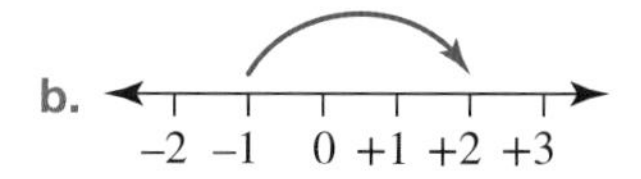


c. –6 –5 –4 –3 –2 –1 0

d. 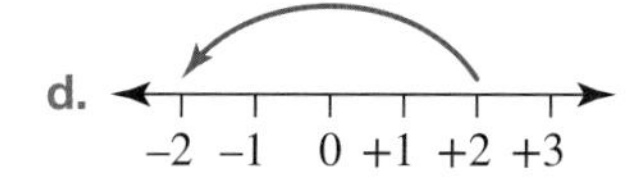


e. 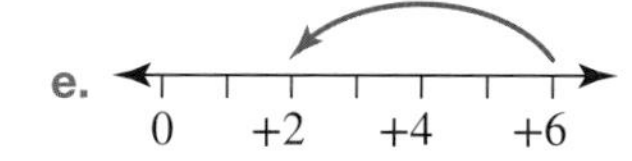


f. 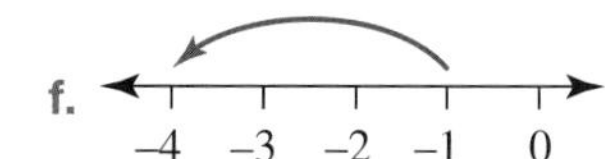


3. **WE11** Use the sign model to calculate each of the following.
   a. $+2+(-3)$ b. $+3+(-4)$ c. $+4+(-4)$
   d. $+3+(-2)$ e. $-4+(+2)$

4. Calculate each of the following using number sentences. (Draw a number line or other model if you wish.)
   a. $5+(-2)$ b. $-3+(-4)$ c. $-2+2$
   d. $6+(-5)$ e. $-5+5$

5. Calculate each of the following using number sentences. (Draw a number line or other model if you wish.)
   **a.** $4+(-6)$ **b.** $-5+7$ **c.** $6+(-9)$
   **d.** $3+(-3)$ **e.** $0+(-6)$

6. Write the answer for each of the following.
   a. $-5+(-2)$ b. $-6+4$ c. $-8+8$
   d. $3+(-7)$ e. $-3+7$ f. $-3+(-7)$

7. Write the answer for each of the following.

a. $-8+12$
b. $19+(-22)$
c. $-64+(-36)$
d. $-80+90$
e. $-2+4$
f. $-15+(-7)$

8. Copy and complete this addition table.

| + | −13 | 5 |
|---|---|---|
| 21 | | |
| −18 | | |

9. **WE12** Evaluate each algebraic expression, if $a=3, b=2, c=-4$.

a. $a+b$
b. $b+c$
c. $a+b+c$
d. $c+2b+a$

## Understanding

10. Determine the missing number in each of these incomplete number sentences.

a. $8+$ ____ $=0$
b. $-2+$ ____ $=-8$
c. ______ $+(-6)=4$
d. ______ $+5=-2$
e. ______ $+(-5)=-2$

11. a. Copy and complete the addition table shown.

b. i. Determine the pattern shown along the leading (dotted) diagonal.

ii. Determine the pattern shown along the other (unmarked) diagonal.

iii. State whether the chart is symmetrical about the leading diagonal.

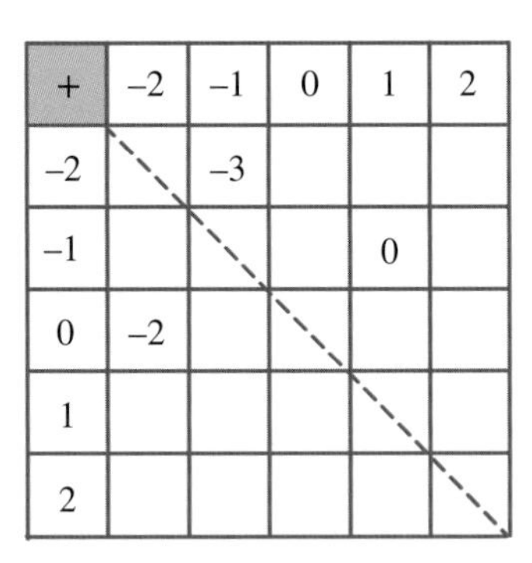

| + | −2 | −1 | 0 | 1 | 2 |
|---|---|---|---|---|---|
| −2 | | −3 | | | |
| −1 | | | | 0 | |
| 0 | −2 | | | | |
| 1 | | | | | |
| 2 | | | | | |

12. Model each situation with an integer number sentence that shows the result of the following.

a. A lift went down 2 floors, then up another 3 floors.
b. A lift went down 3 floors, then up 5 floors.
c. Australia was 50 runs behind, then made another 63 runs.
d. A submarine at sea level, dived 50 metres then rose 26 m.
e. An account with a balance of \$200 had \$350 withdrawn from it.

## Reasoning

13. Describe a situation to fit each of the following number sentences.

a. $-3+(-2)=-5$
b. $-10+(-40)=-50$
c. $2+(-6)=-4$
d. $-20+20=0$
e. $-8+10=2$

14. a. Write a rule starting with 'When adding two negative numbers…'

b. Write a rule starting with 'When adding a negative number and a positive number…'

15. Arrange the numbers in this magic square so that all rows, columns and diagonals add up to the same value. No numbers can be used twice.

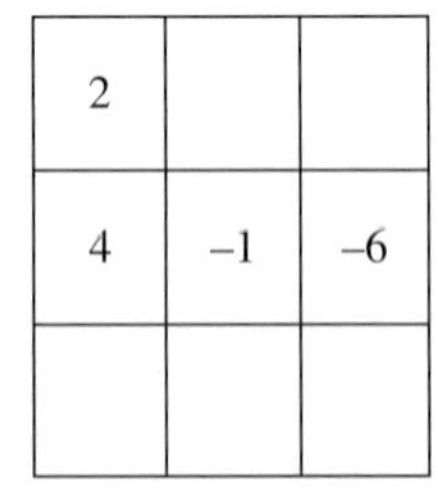

| | | |
|---|---|---|
| 2 | | |
| 4 | −1 | −6 |
| | | |

### Problem solving

**16.** Kuni has a sporting equipment business. Due to a financial crisis, she lost \$2500 after six months and then lost \$3075 by the end of the year. Calculate the total amount that Kuni lost by the end of the year.

**17.** In an AFL match, Dustin Martin scored two goals in the first quarter and was awarded six points for each. He scored three behinds in the fourth quarter and was awarded 1 point for each. Calculate the total points gained by Dustin for his team.

**18.** Avery parks her car in a car park two floors below ground level. She takes a lift to the shopping centre and goes up seven floors. Then she takes a lift down three floors to catch up with her friend Lucy. Determine the floor at which Avery gets off to meet Lucy.

**19.** Determine the effect of adding a number to its opposite integer.

# LESSON
# 2.5 Subtracting integers

**LEARNING INTENTION**

At the end of this lesson you should be able to:
- understand how a number line can be used to subtract integers
- use a number line to subtract integers
- subtract integers without using a number line.

## 2.5.1 Subtraction of integers

eles-3701

- A number line can be used to subtract integers.
- Subtracting a number gives the same result as adding its opposite integer.
- When subtracting a positive integer, move left along the number line.
- When subtracting a negative integer, move right along the number line.

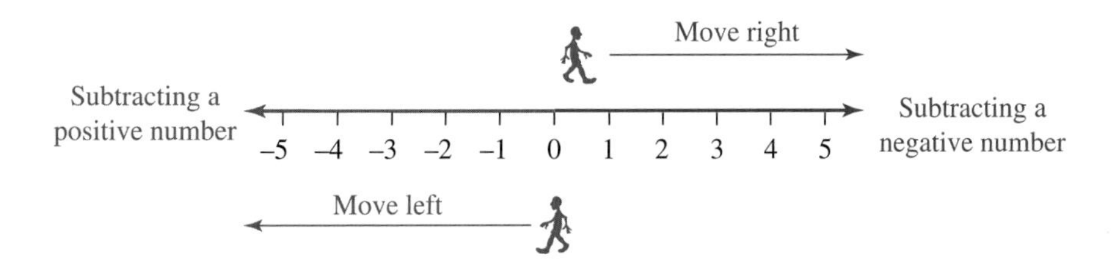


- Consider the subtraction of $12 - (+4)$.

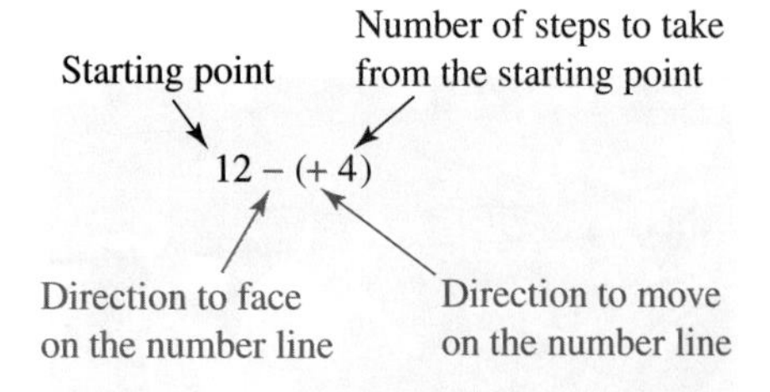


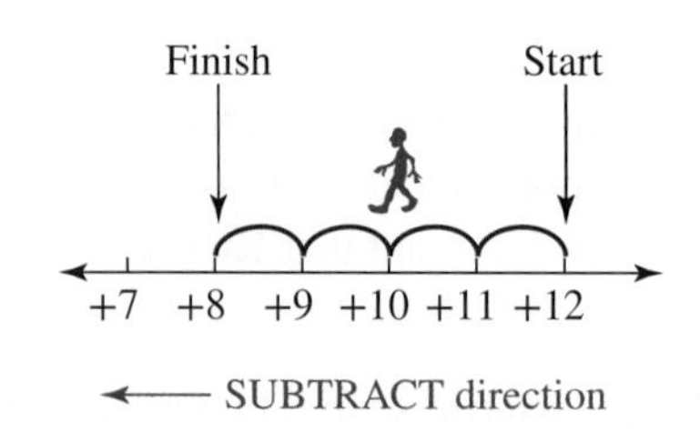


| Symbol | Meaning | Action to take |
|---|---|---|
| 12 | Starting point | Start at 12. |
| − | Direction to face on the number line | Face left for subtraction −. |
| (+) | Direction to move on the number line | Move forward for positive (+). |
| (4) | Number of steps to take from the starting point | Move 4 places from 12. |
| = | Finishing point | Stop at +8. |

Therefore, $12-(+4)=+8$

## WORKED EXAMPLE 13 Subtracting integers using a number line

**Calculate the value of each of the following, using a number line.**

**a.** $\mathbf{-2-(+3)}$ **b.** $\mathbf{-2-(-3)}$

**THINK**

**a. 1.** Start at −2 and face left (because this is subtraction). Take 3 steps forwards (because the number being subtracted is positive). The finishing point is −5.

**2.** Write the answer.

**b. 1.** Start at −2 and face left (because this is subtraction). Take 3 steps backwards (because the number being subtracted is negative). The finishing point is +1.

**2.** Write the answer.

**WRITE**

a.

SUBTRACT

Finish Start

forward 3

−5 −4 −3 −2 −1 0 +1 +2 +3

$-2-(+3)=-5$

b.

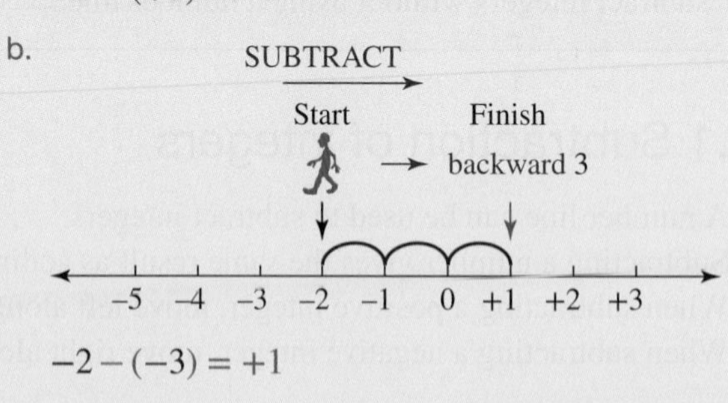


$-2-(-3)=+1$

## 2.5.2 Subtracting integers by adding opposites

eles-4091

- By developing and extending a pattern, we can show that subtracting negatives has the same effect as adding a positive. This means that subtracting a negative integer is the same as adding the opposite integer. For example, as demonstrated in part **b** of Worked example 13:

$$\begin{aligned}-2-(-3) &= -2+(+3)\\ &= -2+3\\ &= +1\end{aligned}$$

**WORKED EXAMPLE 14 Subtracting integers**

**Calculate the following.**

**a.** $2-5$ **b.** $-3-6$ **c.** $5-(-3)$ **d.** $-5-(-4)$

| THINK | WRITE |
|---|---|
| **a. 1.** Write the question. | **a.** $2-5=2-(+5)$ |
| **2.** Rewrite the equation, changing subtraction to addition of the opposite integer. | $=2+(-5)$ |
| **3.** Add, using the addition rule for addition of integers. <br>• Start at +2. <br>• Face right for addition. <br>• Move backwards 5 units. | $=-3$ |
| **b. 1.** Write the question. | **b.** $-3-6=-3-(+6)$ |
| **2.** Rewrite the equation, changing subtraction to addition of the opposite integer. | $=-3+(-6)$ |
| **3.** Add, using the addition rule for addition of integers. <br>• Start at −3. <br>• Face right for addition. <br>• Move backwards 6 units. | $=-9$ |
| **c. 1.** Write the question. | **c.** $5-(-3)=5+(+3)$ |
| **2.** Rewrite the equation, changing subtraction to addition of the opposite integer. | $=5+3$ |
| **3.** Add, using the addition rule for addition of integers. <br>• Start at +5. <br>• Face right for addition. <br>• Move forward 3 units. | $=8$ |
| **d. 1.** Write the question. | **d.** $-5-(-4)=-5+(+4)$ |
| **2.** Rewrite the equation, changing subtraction to addition of the opposite integer. | $=-5+4$ |
| **3.** Add, using the addition rule for addition of integers. <br>• Start at −5. <br>• Face right for addition. <br>• Move forward 4 units. | $=-1$ |

## WORKED EXAMPLE 15 Application of adding or subtracting integers

**A news flash in Freezonia announced that there had been a record drop in temperature overnight. At 6 pm the temperature was 10 °C and by 4 am it had fallen by 25 °C. What was the temperature at 4 am?**

| THINK | WRITE |
|---|---|
| 1. The original temperature is 10 °C, so write the number 10. The temperature fell by 25 °C, so write this as 10 – 25. Write the number sentence. Rewrite as the addition of the opposite. | $10 - 25$<br>$= 10 + (-25)$ |
| 2. Complete the addition by starting at +10, facing right, then moving backwards by 25 units. | $= -15$ |
| 3. Write the answer in a sentence. | The temperature in Freezonia at 4 am was −15°C. |

### Digital technology

Scientific calculators can evaluate subtraction of positive and negative numbers.

DEG
247-57-(-68)
258

### ACTIVITY: Radial diagrams

Insert the integers from −6 to +2 into the circles on the diagram so that each straight line of three circles has each of the following totals.

a. −6
b. −3
c. −9

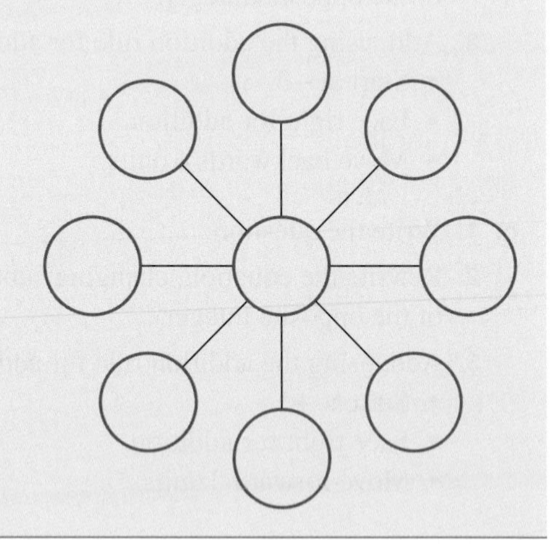

## Resources

 **eWorkbook** Topic 2 Workbook (worksheets, code puzzle and project) (ewbk-1903)

 **Video eLesson** Addition and subtraction of positive and negative numbers (eles-1869)

 **Interactivities** Individual pathway interactivity: Subtracting integers (int-8341)
Subtraction of integers (int-4037)

# Exercise 2.5 Subtracting integers

2.5 Quick quiz on | 2.5 Exercise

## Individual pathways

| PRACTISE | CONSOLIDATE | MASTER |
| --- | --- | --- |
| 1, 5, 9, 12, 14, 17 | 2, 4, 6, 8, 11, 15, 18 | 3, 7, 10, 13, 16, 19, 20 |

### Fluency

1. **WE13** Calculate the value of each of the following.

   a. $7-(+2)$  b. $-18-(+6)$  c. $3-(+8)$
   d. $17-(-9)$  e. $-28-(-12)$  f. $-17-(-28)$

2. **WE14** Calculate the value of each of the following.

   a. $7-5$  b. $8-(-2)$  c. $-4-6$
   d. $-6-(-8)$  e. $1-10$

3. Calculate the value of each of the following.

   a. $-5-5$  b. $-8-(-8)$  c. $0-4$
   d. $0-(-3)$  e. $-10-(-20)$  f. $-5-(-5)$

4. Write number sentences to show the subtraction problems in the following diagrams.

   a.
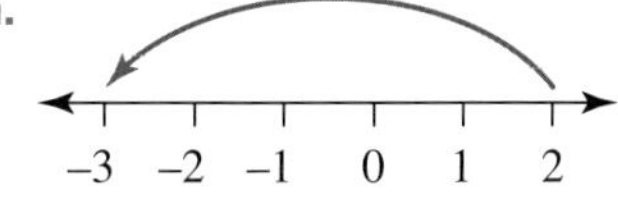


   b.
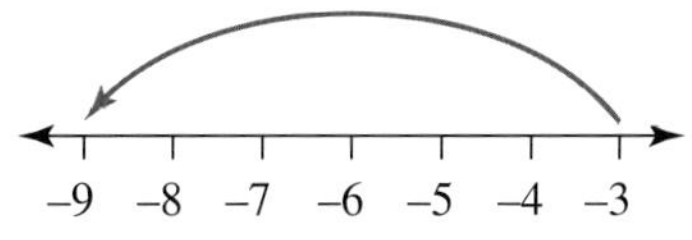


   c.
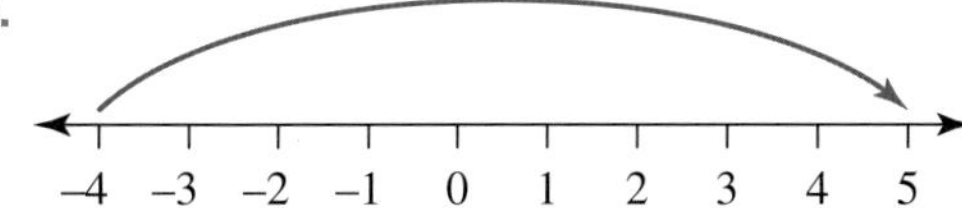


   d.
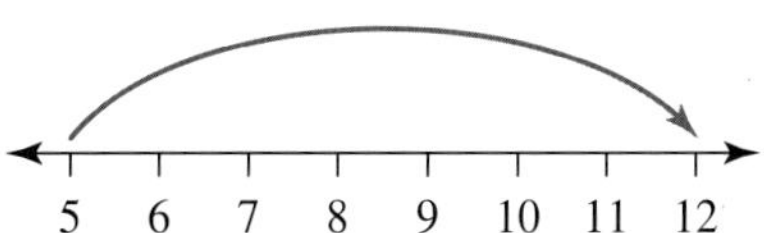


### Understanding

5. Mentally calculate the following and write the answer.

   a. $-7-3$  b. $8-(-5)$  c. $-6-(-9)$  d. $-0-(-12)$

6. Mentally calculate the following and write the answer.

   a. $-8-8$  b. $3-20$  c. $20-(-3)$  d. $-4-8$

7. Determine the missing number in these incomplete number sentences.

   a. ______ $-7=-6$  b. $-8-$ ______ $=-17$
   c. $-8-$ ______ $=17$  d. ______ $-(-2)=7$

8. Write each of the numbers that can be described by the following.

   a. 6 less than $-2$  b. 5 less than $-8$
   c. 8 °C below $-1$ °C  d. 3 °C below 2 °C
   e. 3 to the left of $+4$  f. 4 to the left of $-3$

9. **MC** Select which of the following is equal to $7+(-4)-(-2)$.

   A. 9  B. 1  C. 13
   D. 5  E. $-5$

10. Model each situation with an integer number sentence that shows the result of the following.
 a. From ground level, a lift went down 2 floors, then down another 3 floors.
 b. An Olympian dived down 5 metres from a board 3 metres above water level.
 c. At 5:00 pm in Falls Creek the temperature was 1 °C. It then fell 6 °C by 11:00 pm.
 d. A bank account with a balance of $500 had $150 withdrawn from it.

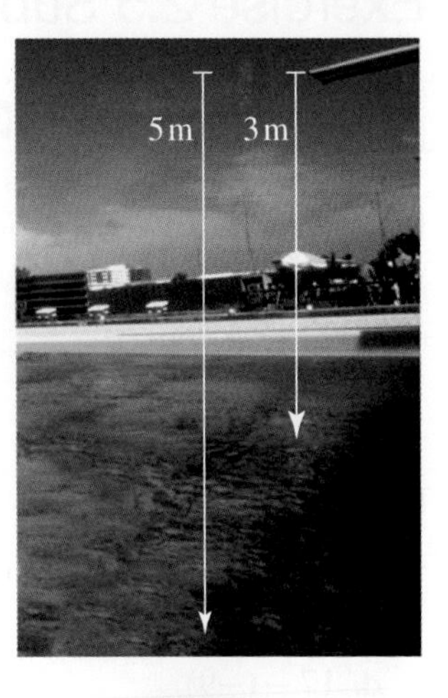


11. **MC** From ground level, a lift went down 2 floors, then down another 3 floors to a level 5 floors below the ground. Select which of the following number sentences describes this situation.
 A. $2+3=5$ B. $-2+(-3)=-5$ C. $-2+3=1$
 D. $2+(-3)=-1$ E. $-3+(-2)=5$

12. State whether the following number sentences are true or false. Show your full working.
 a. $7-9=7+-9$ b. $0-8=-8-0$
 c. $8-12=-12+8$ d. $0-p=-p$

13. **WE15** The temperature in the freezer was −20 °C. Just before he went to bed, Chen had a spoonful of ice cream and left the freezer door open all night. The temperature in the freezer rose by 18 °C and the ice cream melted. Determine the temperature in the freezer when Chen's mother found the ice cream in the morning.

### Reasoning

14. Nihal has to hand in an essay for English class with a maximum of 400 words. Currently his essay has 520 words. Calculate the number of words Nihal has to cut out of his essay.

15. Jill is climbing up a steep and slippery path to fetch a bucket of water. When she is 6 m above her starting point, she slips back 1 m, grabs onto some bushes by the side of the path and climbs 7 m more to a flat section of the path. Determine how far above her starting point Jill is when she reaches the flat section. Show your working.

16. Sharyn opened a bank account when she started working at her local fish and chip shop. Below is a week-by-week breakdown of her transactions.
 Week 1: deposited $56.00, then withdrew $45.00
 Week 2: deposited $44.00, then withdrew $75.00
 Week 3: deposited $52.80, then withdrew $22.00
 Week 4: deposited $39.20, then withdrew $50.00
 Sharyn's bank account was opened with a $10 deposit.
 a. Determine how much money was in Sharyn's account at the end of the month.
 b. Show full working to determine in which week the balance of Sharyn's account was negative.

### Problem solving

17. Stuart is diving in the ocean. He starts at a depth of 5 m below the water's surface, then rises 2 m. He then dives a further 12 m down. Calculate the final depth that he reached.

18. A hospital building has 6 floors above ground level and 4 floors below ground level. Over a 10-minute period one busy morning, the lift started from ground level (reception) and made the following moves: up 2 floors, up 1 floor, down 4 floors, down 2 floors, up 3 floors, down 4 floors and up 6 floors.
   a. Determine the floor on which the lift finished.
   b. Determine the number of floors the lift travelled in the whole 10 minutes.
19. Rendah has a maths trick. For her trick she asks people to do the following mental maths steps:
   1. Pick a number between 1 and 10.
   2. Multiply the number by 9.
   3. Add the digits of the number. For example, if you chose the number 25 you would add 2 and 5 to get 7.
   4. ________________________________
   5. Match your number to a letter of the alphabet; for example, 1 is A, 2 is B, 3 is C and so on.
   6. Think of an animal that starts with your letter.

   When the trick is done, Rendah asks people if they are thinking of a dog, and they almost always say yes. If they say no, it's usually because they picked a donkey, a dolphin or a dugong. Determine the missing step in Rendah's trick.
20. Determine the effect of subtracting a number from its opposite integer.

# LESSON
# 2.6 Multiplying and dividing integers

### LEARNING INTENTION

At the end of this lesson you should be able to:
- understand the patterns for multiplying and dividing integers
- multiply positive and negative integers
- divide positive and negative integers.

## 2.6.1 Multiplication of integers

eles-3702

- Patterns from multiplication tables can be used to determine the product when two directed numbers are multiplied.
- Consider the following multiplication table – we already know how to complete part of the table.
  - Notice that multiplying by 0 always results in 0.
  - Notice that multiplying by 1 does not change the value of the number.

| × | −3 | −2 | −1 | 0 | 1 | 2 | 3 |
|---|---|---|---|---|---|---|---|
| 3 | | | | 0 | 3 | 6 | 9 |
| 2 | | | | 0 | 2 | 4 | 6 |
| 1 | | | | 0 | 1 | 2 | 3 |
| 0 | | | | 0 | 0 | 0 | 0 |
| −1 | | | | | | | |
| −2 | | | | | | | |
| −3 | | | | | | | |

- Using a calculator to help with the multiplication of negative numbers, we can complete the table, as shown here.

| × | −3 | −2 | −1 | 0 | 1 | 2 | 3 |
|---|---|---|---|---|---|---|---|
| 3 | −9 | −6 | −3 | 0 | 3 | 6 | 9 |
| 2 | −6 | −4 | −2 | 0 | 2 | 4 | 6 |
| 1 | −3 | −2 | −1 | 0 | 1 | 2 | 3 |
| 0 | 0 | 0 | 0 | 0 | 0 | 0 | 0 |
| −1 | 3 | 2 | 1 | 0 | −1 | −2 | −3 |
| −2 | 6 | 4 | 2 | 0 | −2 | −4 | −6 |
| −3 | 9 | 6 | 3 | 0 | −3 | −6 | −9 |

- From this table, we can develop the following rules for multiplying integers.

### Rules for multiplying integers

- **When multiplying two integers with the *same* sign, the answer is always *positive*.**

  $+ \times + = +$ **or** $- \times - = +$

- **When multiplying two integers with *different* signs, the answer is always *negative*.**

  $+ \times - = -$ **or** $- \times + = -$

### WORKED EXAMPLE 16 Multiplying integers

**Evaluate:**

a. $-5 \times +2$

b. $-4 \times -6$

| THINK | WRITE |
|---|---|
| a. 1. Write the question. | a. $-5 \times +2$ |
| 2. Negative × positive = negative. | $= -10$ |
| b. 1. Write the question. | b. $-4 \times -6$ |
| 2. Negative × negative = positive. | $= 24$ |

- Brackets can also be used to represent multiplication. For example, the expression $(-3)(5)$ is the same as $-3 \times 5$. Likewise, the expression $(2-4)(11-3)$ is the same as $(2-4) \times (11-3)$.

## 2.6.2 Division of integers

eles-3703

- Division is the inverse, or opposite, of multiplication.
- The multiplication rules for directed numbers can be used to work out the division rules for directed numbers.

  Since $3 \times 2 = 6$ then $6 \div 3 = 2$ and $6 \div 2 = 3$.

  Since $3 \times -2 = -6$ then $-6 \div 3 = -2$ and $-6 \div -2 = 3$

  Since $-2 \times -3 = 6$ then $6 \div -3 = -2$ and $6 \div -2 = -3$.

From this, we can develop the following rules for dividing integers.

### Rules for dividing integers

- **When dividing two integers with the *same* sign, the answer is always *positive*.**

$$+ \div + = + \text{ or } - \div - = +$$

- **When dividing two integers with *different* signs, the answer is *negative*.**

$$+ \div - = - \text{ or } - \div + = -$$

**WORKED EXAMPLE 17 Dividing integers**

**Evaluate the following.**

a. $10 \div -2$ b. $-12 \div 4$ c. $-20 \div -5$

| THINK | WRITE |
|---|---|
| a. 1. Write the question. | a. $10 \div -2$ |
| 2. The two integers have different signs, so the answer will be negative. | $= -5$ |
| b. 1. Write the question. | b. $-12 \div 4$ |
| 2. The two integers have different signs, so the answer will be negative. | $= -3$ |
| c. 1. Write the question. | c. $-20 \div -5$ |
| 2. Both numbers have the same sign, so the answer will be positive. | $= 4$ |

**WORKED EXAMPLE 18 Multiplying and dividing integers**

**Evaluate:**

a. $\dfrac{-16}{+2}$ b. $-5 \times \dfrac{4}{20}$

| THINK | WRITE |
|---|---|
| a. 1. Write the question. *Note:* $\dfrac{-16}{+2}$ is the same as $-16 \div +2$. | a. $\dfrac{-16}{+2}$ |
| 2. Evaluate the expression. *Note:* negative ÷ positive = negative | $= -8$ |
| b. 1. Write the question. | b. $-5 \times \dfrac{4}{20}$ |
| 2. Write the integer as a fraction with a denominator of 1 and simplify by cancelling. | $= \dfrac{-5}{1} \times \dfrac{\cancel{4}^{1}}{\cancel{20}^{5}}$ |

3. Multiply the numerators then multiply the denominators and simplify.
*Note:* negative ÷ positive = negative

$= \frac{-5}{5}$

4. Write the answer.

$= -1$

### Digital technology

Scientific calculators can evaluate multiplication of positive and negative numbers.

```
DEG
(2-4)*(11-3)
                -16
```

Scientific calculators can evaluate division of positive and negative numbers.

```
DEG
-5*4/20         -1
```

**on** Resources

**eWorkbook** Topic 2 Workbook (worksheets, code puzzle and project) (ewbk-1903)

**Interactivities** Individual pathway interactivity: Multiplication and division of integers (int-4370)
Multiplication of integers (int-4038)
Division of integers (int-4039)

## Exercise 2.6 Multiplying and dividing integers

learn**on**

| 2.6 Quick quiz **on** | 2.6 Exercise |
|---|---|

### Individual pathways

| ■ PRACTISE | ■ CONSOLIDATE | ■ MASTER |
|---|---|---|
| 1, 4, 10, 15, 16, 19 | 2, 5, 7, 9, 11, 12, 17, 20 | 3, 6, 8, 13, 14, 18, 21 |

### Fluency

1. **WE16** Calculate the following.

a. $2 \times -5$
b. $-6 \times 3$
c. $-7 \times 9$
d. $6 \times -5$
e. $-2 \times -3$

2. Evaluate the following.

a. $-6 \times -3$ b. $-5 \times -5$ c. $0 \times -7$ d. $-8 \times -1$

3. Calculate the following.

a. $10 \times -1$ b. $-15 \times 2$ c. $-20 \times -10$ d. $-6 \times -6$

4. **WE17** Evaluate the following.

a. $-8 \div -2$ b. $-8 \div 2$ c. $12 \div -3$
d. $-16 \div -8$ e. $-90 \div -10$

5. Calculate the following.

a. $88 \div -11$ b. $-6 \div 1$ c. $0 \div -4$
d. $-84 \div 4$ e. $-184 \div 2$

6. Evaluate the following.

a. $-125 \div -5$ b. $-67 \div -1$ c. $129 \div -3$
d. $-284 \div 4$ e. $336 \div -6$

## Understanding

7. **WE18a** Calculate the following.

a. $\frac{-6}{2}$ b. $\frac{-24}{-8}$ c. $\frac{-8}{8}$

8. **WE18b** Evaluate the following.

a. $3 \times \frac{-2}{-6}$ b. $4 \times \frac{-5}{10}$ c. $-9 \times \frac{-3}{18}$

9. Calculate the value of each of the following.

a. $2 \times -3 \times 4$ b. $-4 \times -3 \times 3$ c. $-8 \times 9 \times -2$

10. Fill in the missing numbers in these number sentences.

a. $6 \times$ _______ $= -18$ b. _______ $\times 3 = -18$ c. $-8 \times$ _______ $= -8$
d. $-8 \times$ _______ $= 8$ e. $-8 \times$ _______ $= 0$

11. Fill in the missing numbers in these number sentences.

a. $-21 \div$ _______ $= -7$ b. _______ $\div -8 = -4$ c. _______ $\div -9 = 8$
d. $-11 \div$ _______ $= 1$ e. _______ $\div -7 = 0$ f. $-150 \div -25 =$ _______

12. Evaluate the following expressions.

a. $-4 + (-4)$
b. $2 \times -4$
c. $-2 + (-2) + (-2)$
d. $3 \times -2$
e. $-5 + (-5) + (-5) + (-5)$
f. $4 \times -5$

13. **MC** Select which of the following could be the missing numbers in the number sentence $16 \div$ ____ $=$ ____.

A. $2, -8$ B. $-2, -8$ C. $-4, 4$ D. $-2, 8$ E. $1, -16$

14. **MC** Select which of the following could be the missing numbers in the number sentence ____ $\div$ ____ $= -5$.

A. $-15, 3$ B. $15, 5$ C. $25, 5$ D. $-30, -6$ E. $-25, -5$

15. **MC** Six people each owe the bank \$50. Calculate the combined total of their six bank accounts.

A. \$300 B. −\$50 C. \$50 D. $-\$\frac{6}{5}$ E. −\$300

### Reasoning

16. Evaluate the following multiplications.
    a. i. $(-1)(-2)$ ii. $(-3)(-2)(-1)$
    iii. $(-2)(-3)(-1)(-4)$ iv. $(-4)(-1)(-5)(-2)(-1)$
    b. Write a rule for multiplying more than two negative numbers.
    c. Does this rule also apply to the division of negative numbers? Explain your answer.

17. When you multiply two integers, is the result an integer? What about when you divide two integers? Explain your answer.

18. A spider is running down the stairs from the first floor of an old lady's house to the basement below. It stops every 5 steps to catch a fly. If there are 26 steps above the ground floor and 14 steps below the ground floor, determine how many flies the spider catches.

### Problem solving

19. Blake is jumping down the stairs, stepping on every second step only.
    He jumps a total of ten times.

    a. Determine the number of steps he has skipped.
    b. Determine the number of steps there are altogether.
    c. If each step is 25 cm high, determine how far he has travelled in total.
    d. If going up the stairs is considered going in a positive direction and going down the stairs is considered going in a negative direction, rewrite your answer to part c.

20. Manu is working on the expression $8 \div 2 \times (2 + 2)$ and says the answer is 1. Her friend Kareen says the answer should be 16. Show, with full working, who is correct.

21. On a test, each correct answer scores 5 points, each incorrect answer scores −2 points, and each question left unanswered scores 0 points. A student answers 16 questions on the test correctly, 3 incorrectly, and does not answer 1 question. Write an expression for the student's score and calculate the score.

# LESSON
# 2.7 Order of operations with integers

## LEARNING INTENTION

At the end of this lesson you should be able to:
- understand how to apply the order of operations with integers when more than one operation is involved
- apply the order of operations with positive and negative integers to solve expressions that involve more than one operation.

## 2.7.1 Applying BIDMAS to positive and negative integers

eles-3704

- Lesson 1.8 in topic 1 explored how the order of operations is applied with positive integers when more than one mathematical operation is involved.
- The same order of operations is also applied for negative integers.
- The order of operations involves solving the brackets first, then solving powers or indices, then multiplication and division (working from left to right) and finally addition and subtraction (working from left to right).
- The order of operations can be written as BIDMAS (**B**rackets, **I**ndices and roots, **D**ivision and **M**ultiplication, **A**ddition and **S**ubtraction).

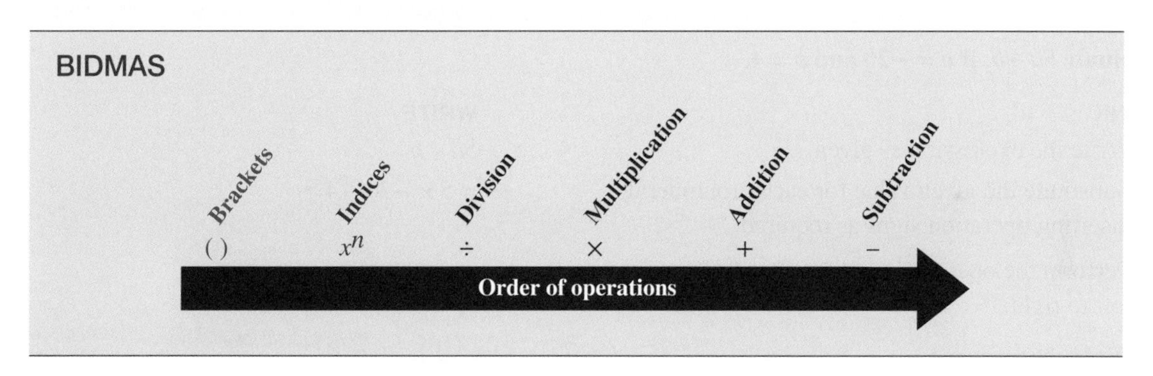


- Examples of these operations are shown in this table.

| B | Brackets | $8 \times (3+5) = 8 \times (8) = 64$ |
|---|---|---|
| I | **Indices** | $5 + 3^2 = 5 + 9 = 14$ |
| D | **Division** | $17 - 8 \div 4 = 17 - 2 = 15$ |
| M | **Multiplication** | $7 + 6 \times 3 = 7 + 18 = 25$ |
| A | **Addition** | $6 \times 2 + 5 = 12 + 5 = 17$ |
| S | **Subtraction** | $21 \div 3 - 4 = 7 - 4 = 3$ |

### WORKED EXAMPLE 19 Calculating an expression using order of operations

**Calculate $58 - (2 \times -8 + 3^2)$ using the correct order of operations.**

| THINK | WRITE |
|---|---|
| 1. Write the question. | $58 - \left(2 \times -8 + 3^2\right)$ |
| 2. Using BIDMAS, start inside the brackets and simplify the squared term (indices). | $= 58 - \left(2 \times -8 + 3^2\right)$<br>$= 58 - (2 \times -8 + 9)$ |
| 3. Perform the multiplication within the brackets. | $= 58 - (2 \times -8 + 9)$<br>$= 58 - (-16 + 9)$ |
| 4. Perform the addition within the brackets. | $= 58 - (-16 + 9)$<br>$= 58 - (-7)$ |
| 5. When the brackets have been removed, perform the subtraction that had been outside the brackets. | $= 58 - -7$<br>$= 58 + 7$ |
| 6. Write the answer. | $= 65$ |

### WORKED EXAMPLE 20 Evaluating an expression using order of operations

**Evaluate $5a \div b$, if $a = -20$ and $b = 4$.**

| THINK | WRITE |
|---|---|
| 1. Write the expression as given. | $5a \div b$ |
| 2. Substitute the given value for each pronumeral, inserting operation signs as required. | $= 5 \times -20 \div 4$ |
| 3. Perform the operations as they appear, from left to right. | $= -100 \div 4$ |
| 4. Write the answer. | $= -25$ |

### Digital technology

Scientific calculators can calculate expressions using order of operations.

```
DEG
(-4)²-5*-3          31
```

 Resources

 **eWorkbook** Topic 2 Workbook (worksheets, code puzzle and project) (ewbk-1903)

**Interactivities** Individual pathway interactivity: Combined operations (int-4371)

Combined operations (int-4040)

## Exercise 2.7 Order of operations with integers

learnon

| 2.7 Quick quiz  | 2.7 Exercise |
|---|---|

### Individual pathways

| ■ PRACTISE | ■ CONSOLIDATE | ■ MASTER |
|---|---|---|
| 1, 3, 7, 9, 11, 14, 17 | 2, 4, 8, 10, 12, 15, 18 | 5, 6, 13, 16, 19, 20 |

### Fluency

1. **WE19** Calculate the following using the correct order of operations.
   a. $6 + 3 \times -4$ b. $18 - 12 \div -3$ c. $8 + -4 - 10$ d. $17 - 3 + -8$
2. Evaluate each of the following expressions using the correct order of operations.
   a. $6 \times -3 \div 9$ b. $72 \div 8 \times -3$ c. $7 + (-3 - 4)$ d. $(6 + 3) \div -9$
3. Calculate each of the following expressions using the correct order of operations.
   a. $-3 \times -2 + 3 \times -1$ b. $-6 \times 5 - 2 \times -6$
   c. $-4 \times (6 - -4)$ d. $(-8 + 3) \times -7$
4. Evaluate each of the expressions using the correct order of operations.
   a. $4 + 7 \times -3 - 2$ b. $-6 - 4 + (-3)^2$
   c. $(-2)^3 - 3 \times -2$ d. $3 + (2 - 8) + -6$
5. Calculate each of the expressions using the correct order of operations.
   a. $-8 \div 2 + (-2)^2$ b. $-4 \times -8 - [2 + (-3)^2]$
   c. $(-7 + 5) - -24 \div 6$ d. $-15 \div (2 - 5) - 10$
6. Evaluate each of the expressions using the correct order of operations.
   a. $54 \div -6 + 8 \times -9 \div -4$ b. $(9 - -6) \div -5 + -8 \times 1$
   c. $-7 + -7 \div -7 \times -7 - -7$ d. $-9 \times -5 - (3 - -2) + -48 \div 6$
7. **WE20** Evaluate each of the following expressions.
   a. $2x + 3x$, if $x = -4$ b. $5 + 3d$, if $d = -2$ c. $-5b - 3$, if $b = -7$
8. Evaluate each of the following expressions.
   a. $a(b + c)$, if $a = 6, b = -2, c = -4$
   b. $x^3 - y$, if $x = -4, y = 4$
   c. $2a(3b - 24)$, if $a = -2,\ b = 6$

9. **MC** The expression $6 + 2 \times -5 - -10 \div 2$ is equal to:

A. $-15$ B. $-35$ C. $-60$ D. 1 E. 3

10. **MC** The expression $(9 - -6) \div -5 + -8 \times 0$ is equal to:

A. $-15$ B. $-3$ C. $-60$ D. 0 E. 30

### Understanding

11. A submarine dives 100 m below sea level, rises 60 m, then dives 25 m. Calculate its final position.

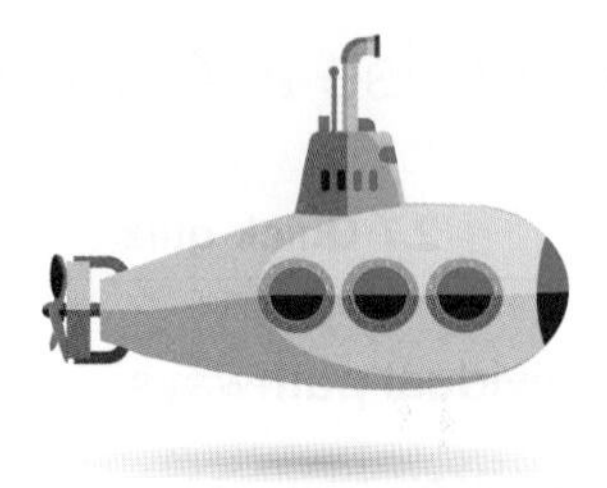

12. Jini has \$2075 in the bank. She makes 2 withdrawals of \$150 each, then 5 deposits of \$60 each. Calculate the amount of money Jini now has in the bank.

13. Sami weighs 156 kg and wants to reduce his weight to 64 kg in 4 months. Calculate the average weight loss per month Sami will need to achieve in order to meet his weight loss goal.

### Reasoning

14. Darcy says there is no difference between the expressions $8 - 2 \times (4 + 2)$ and $8 - 2 \times 4 + 2$. Do you agree with Darcy? Justify your answer with correct working.

15. Are the following expressions true or false? Justify your answer with correct working.

a. $5 \times 6 - 3 + (5 - 2) < 5 \times (6 - 3) + 5 - 2$
b. $2 + (15 \div 3) \times 1 = 2 + 15 \div (3 \times 1)$

16. Merlin is riding his bike east at a steady pace of 10 km/h, while Morgan is riding her bike west at a steady 8 km/h. They pass each other on Backpedal Bridge at 12 noon. Assume that east is the positive direction and west is negative, and that time before noon is negative and after noon is positive.

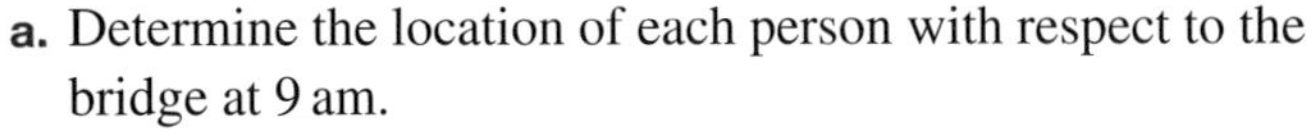

a. Determine the location of each person with respect to the bridge at 9 am.
b. Determine their locations with respect to the bridge at 2 pm.
c. Determine how far apart they were at 10 am.
d. Determine how far apart they will be at 4 pm.

### Problem solving

17. Ryan buys 5 vanilla cupcakes at \$5.75 each, 2 cinnamon donuts at \$3.25 each, and 3 chocolate chip cookies at \$2.50 each. Determine the total amount Ryan has to pay.

18. A snail begins to climb up the side of a bucket. It climbs 3 cm and slips back 2 cm, then climbs a further 4 cm and slips back 1 cm. Write a number sentence to help you determine how far the snail is from the bottom of the bucket.

19. At many locations around the world, minimum and maximum temperatures are recorded. On one particular day, the minimum temperature in the Arctic circle was recorded as −24 °C and the maximum temperature was recorded as −7 °C. In London the maximum temperature was 10 °C and the minimum temperature was −2 °C. In Mexico City the maximum temperature was 38 °C and the minimum temperature was 16 °C. Calculate the difference in maximum and minimum temperatures at each of these three places.

20. In a Maths competition marks are awarded as follows: 3 marks for a hard question (H), 2 marks for a medium question (M) and 1 mark for an easy question (E).
If an answer is incorrect (I), 1 mark is deducted.
The top three students had the following scores:
Student 1: 10H, 3M, 5E, 2I
Student 2: 7H, 7M, 3E, 3I
Student 3: 9H, 6M, 5E, 0I

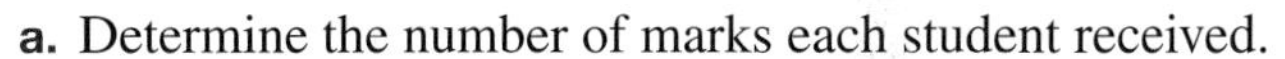

a. Determine the number of marks each student received.
b. Determine who won the competition.

# LESSON
# 2.8 Review

## 2.8.1 Topic summary

### Number line

Integers can be represented on a number line.

- Positive integers are written to the right of 0.
- Negative integers are written to the left of 0.

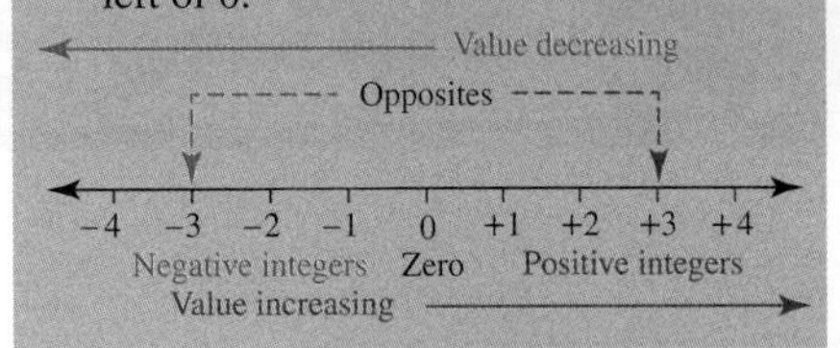


### Number plane

- The number plane, or Cartesian plane, has two axes:
  - the horizontal axis ($x$-axis)
  - the vertical axis ($y$-axis).
- The origin (0, 0) is where the two axes intersect.

e.g. A (3, 2) move 3 right and 2 up
B (−3, 2) move 3 left and 2 up
C (−3, −2) move 3 left and 2 down
D (3, −2) move 3 right and 2 down

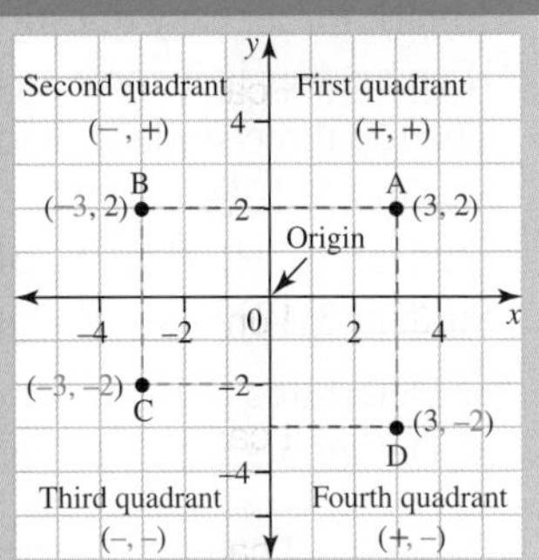


## POSITIVE AND NEGATIVE INTEGERS

### Integers

Integers are positive and negative whole numbers including 0.

- The integers −1, −2, −3.. are called negative integers.
- The integers 1, 2, 3... are called positive integers.
- 0 is neither positive nor negative.

### Addition of integers

A number line can be used to add integers.

- To add a positive integer, move to the right.
  e.g. $-4 + 1 = -3$
- To add a negative integer, move to the left.
  e.g. $5 + (-3) = 2$

Move right when adding a positive integer

Move left when adding a negative integer

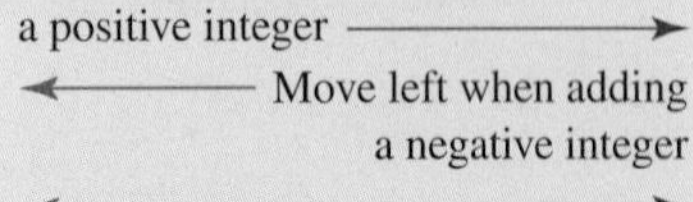

### Subtraction of integers

- Subtracting a number gives the same result as adding its opposite.
  e.g. $2 - (5) = 2 + (-5) = -3$ or $-6 - (-11) = -6 + (+11) = 5$

### Division of integers

- When dividing integers with the same sign, the answer is positive.
  e.g. $-20 \div -4 = 5$ or $24 \div 6 = 4$
- When dividing integers with different signs, the answer is negative.
  e.g. $10 \div -5 = -2$ or $\frac{-35}{7} = -5$

### Multiplication of integers

- When multiplying integers with the same sign, the answer is positive.
  e.g. $-4 \times -6 = 24$ or $7 \times 8 = 56$
- When multiplying integers with different signs, the answer is negative.
  e.g. $-5 \times 2 = -10$ or $9 \times -3 = -27$

### Order of operations

- The order of operations is a set of rules that is followed so there is a common understanding of the order in which to solve mathematical operations.
- The set order in which we calculate problems is:
  1. **Brackets** ( ) or [ ]
  2. **Indices or roots** $a^x$ or $\sqrt[n]{x}$
  3. **Division and Multiplication** (working left to right) ÷ or ×
  4. **Addition and Subtraction** (working left to right) + or −
- The acronym **BIDMAS** can be used to remember the correct order of operations.

## 2.8.2 Success criteria

Tick the column to indicate that you have completed the lesson and how well you think you have understood it using the traffic light system.

(**Green:** I understand; **Yellow:** I can do it with help; **Red:** I do not understand)

| Lesson | Success criteria | Green | Yellow | Red |
|---|---|---|---|---|
| 2.2 | I can place integers on a number line. | | | |
| | I can compare integer values. | | | |
| 2.3 | I understand the number plane (Cartesian plane), origin and coordinates. | | | |
| | I can work out the location of a point described by its coordinates. | | | |
| | I can plot one or more points given their coordinates. | | | |
| 2.4 | I understand how a number line can be used to add integers. | | | |
| | I can use a number line to add integers. | | | |
| | I can add integers without using a number line. | | | |
| 2.5 | I understand how a number line can be used to subtract integers. | | | |
| | I can use a number line to subtract integers. | | | |
| | I can subtract integers without using a number line. | | | |
| 2.6 | I understand the patterns for multiplying and dividing integers. | | | |
| | I can multiply positive and negative integers. | | | |
| | I can divide positive and negative integers. | | | |
| 2.7 | I understand how to apply the order of operations with integers when more than one operation is involved. | | | |
| | I can apply the order of operations with positive and negative integers to solve expressions with more than one operation. | | | |

## 2.8.3 Project

### Directed numbers dice game

In this game, the winner is the first to complete a row of three numbers, horizontally, vertically or diagonally.

**Equipment:** 2 players, 2 standard dice, Connect-3 board (shown), dice placement sheet (shown), 9 blue counters, 9 red counters

**Instructions:**

- Roll the two dice and decide whether to add or subtract the numbers on the dice.
- Put the dice on one of the squares on the dice placement sheet to produce a total shown on the Connect-3 board shown.
- Once you have worked out the answer, place your counter on the appropriate number on the Connect-3 board. You should check each other's answers.
- You cannot cover a number that has already been covered. If you are unable to find a total that has not been covered, you must pass. The winner is the first to complete a row of three numbers either horizontally, vertically or diagonally.

Make your own board game using dice with negative numbers on each face and put different numbers on the board. The dice placement sheet could also be changed to include × or ÷ symbols.

**Connect-3 board**

| | –5 | –4 | –3 | –2 |
|---|---|---|---|---|
| –1 | 0 | 1 | 2 | 3 |
| 4 | 5 | 6 | 7 | 8 |
| 9 | 10 | 11 | 12 | |

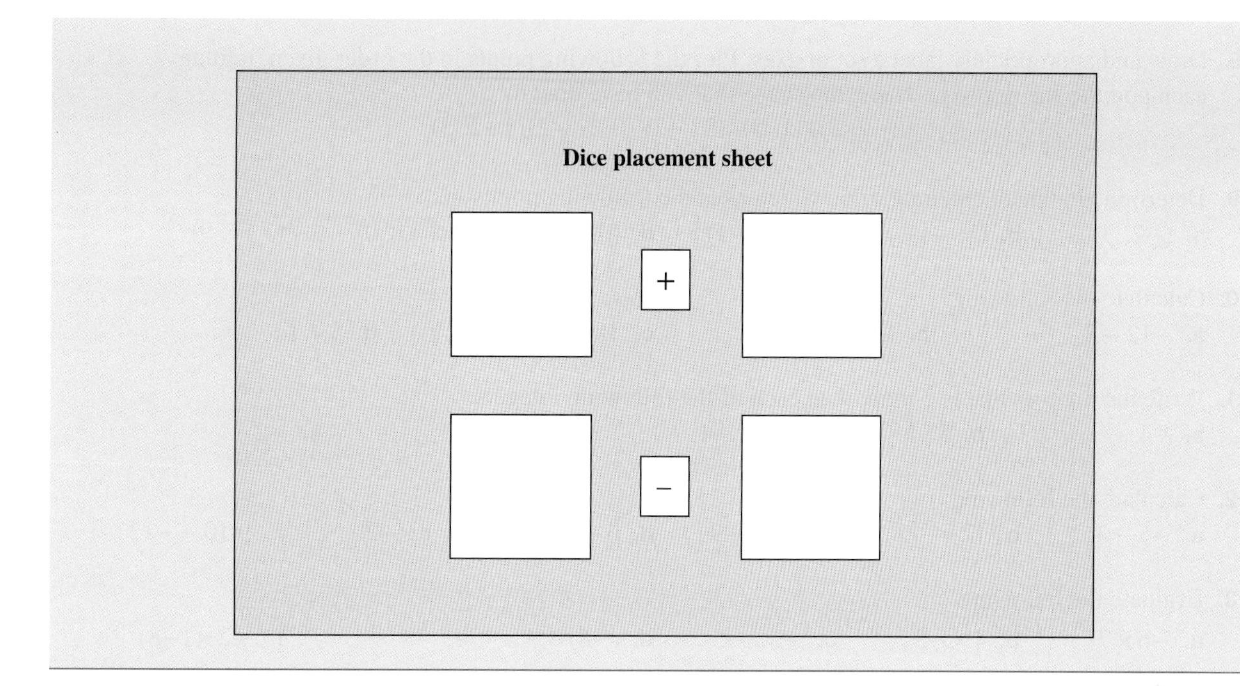

 Resources

eWorkbook Topic 2 Workbook (worksheets, code puzzle and project) (ewbk-1903)

Interactivities Crossword (int-2604)
Sudoku puzzle (int-3172)

## Exercise 2.8 Review questions

learnon

### Fluency

1. Select which of the following are integers.
   a. $-2\frac{1}{2}$ b. 0.45 c. 0 d. −201

2. Complete each statement by inserting the correct symbol:>, < or =.
   a. $-6\square-2$ b. $-7\square7$ c. $0\square-5$ d. $-100\square9$

3. List the integers between −21 and −15.

4. Arrange the numbers −3, 2, 0, −15 in descending order.

5. Describe the integers shown on each number line.
   a. 

   b. 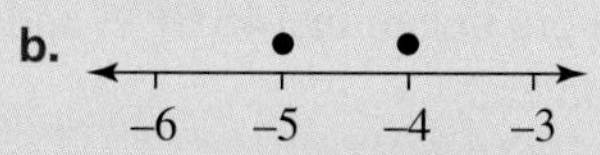
   
   c. 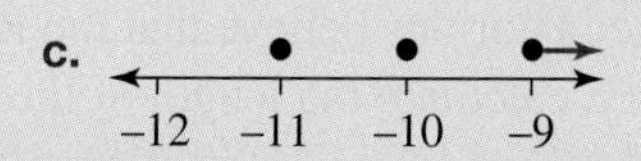
   

6. Graph each of the following sets of integers on a number line.
   a. integers between −7 and −2 b. integers $>-3$ c. integers $\leq -4$

7. State whether the following points are on the $x$-axis, the $y$-axis, both axes or in the first quadrant.
   a. (0, 0) b. (0, 5) c. (3, 0) d. (3, 2)

**8.** Draw and appropriately label a set of axes. Plot the following points in the order given, joining each point to the next one. Name the shape that you have drawn.

$$(-2, 3), (1, 3), (2, -2), (-1, -2), (-2, 3)$$

**9.** Determine in which quadrant or on which axes the following points lie.

**a.** $(-2, 3)$ **b.** $(3, -1)$ **c.** $(-4, -1)$ **d.** $(0, 2)$ **e.** $(-1, 0)$ **f.** $(7, 9)$

**10.** Calculate the following.

**a.** $-12+7$ **b.** $-9+-8$ **c.** $18+-10$ **d.** $5+1$

**11.** Write the number that is 2 more than each of the following integers.

**a.** $-4$ **b.** $5$ **c.** $-1$ **d.** $0$ **e.** $-2$

**12.** Calculate the following.

**a.** $-5-3$ **b.** $17--9$ **c.** $-6--9$ **d.** $6-8$ **e.** $12-20$ **f.** $-10--12$

**13.** Evaluate the following.

**a.** $-6\times7$ **b.** $4\times-8$ **c.** $-2\times-5$ **d.** $(-8)^2$ **e.** $-8^2$ **f.** $-2\times(-8)^2$

**14.** Calculate the following.

**a.** $-36\div3$ **b.** $-21\div-7$ **c.** $45\div-9$ **d.** $\dfrac{-18}{-2}$ **e.** $64\div-4$ **f.** $-100\div25$

**15.** Evaluate the following expressions.

**a.** $10-6\times2$ **b.** $-7--8\div2$ **c.** $-3\times-5--6\times2$
**d.** $(-2)^3$ **e.** $(-3-12)\div(-10+7)$

**16.** Calculate the following expressions.

**a.** $2c+3c$, if $c=-4$ **b.** $-2x(x+5)$, if $x=-2$
**c.** $2a^2+a$, if $a=-3$ **d.** $b^3$, if $b=-5$
**e.** $2a-5b+c$, if $a=-2$, $b=2$, $c=5$

**17.** Replace the box with =, < or > to make each of the following statements correct.

**a.** $-5\square-3$ **b.** $-22\times-2\square44$ **c.** $4\square2$
**d.** $-5\times-3\square-15$ **e.** $0\square-7$ **f.** $-2(-4+7)\square6$

**18.** Replace the box with =, < or > to make each of the following statements correct.

**a.** $-2\times5\square-9$ **b.** $5\times(-2-18)\div-4\square25$ **c.** $12\div-4\square3$
**d.** $5\times(-2-18)\div4\square-25$ **e.** $-10\times-5\square50$ **f.** $5\times-2-18\div4\square2.5$

**19.** Model the following situation with integers, then find the result. A scuba diver at 52 metres below sea level made her ascent in 3 stages of 15 metres each. Determine how far below sea level she was then.

**20.** Some historians believe that the Roman era began in 146 BCE and ended in 455 CE. According to this theory, calculate the length of the Roman era.

**21.** The lowest temperature recorded on Earth was −90 °C in Antarctica and the highest was 58 °C in Africa. Determine the difference between these two temperatures.

**22.** Nazeem's bank account has a balance of −\$43. He pays a bill of \$51 using direct debit. Calculate his bank balance now.

## Problem solving

**23.** For the following expressions, determine if the answer will be negative or positive. Explain your answers using mathematical reasoning.

**a.** $-17\,489 - 25\,636$ **b.** $-65\,234 + 123\,468$ **c.** $-44\,322 + 31\,212$

**24.** Examine the graph shown.

**a.** Determine the coordinates of a point, D, that can be added to the graph so that ABCD is a rectangle.

**b.** Name the combination of points that make:

**i.** a right-angled triangle

**ii.** an isosceles triangle that doesn't include a right angle. (There may be more than one solution.)

**c.** Determine the coordinates of a point, F, that can be added to the graph so that ACFE is a parallelogram. Mark point F on the grid and draw the parallelogram.

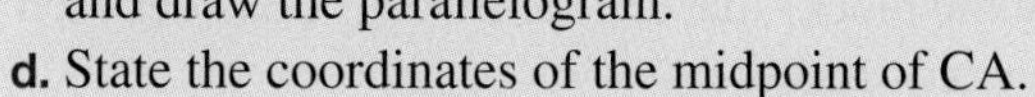

**d.** State the coordinates of the midpoint of CA.

**e.** State the coordinates of the midpoint of AE.

**f.** Estimate the coordinates of G.

**25.** Answer the following questions.

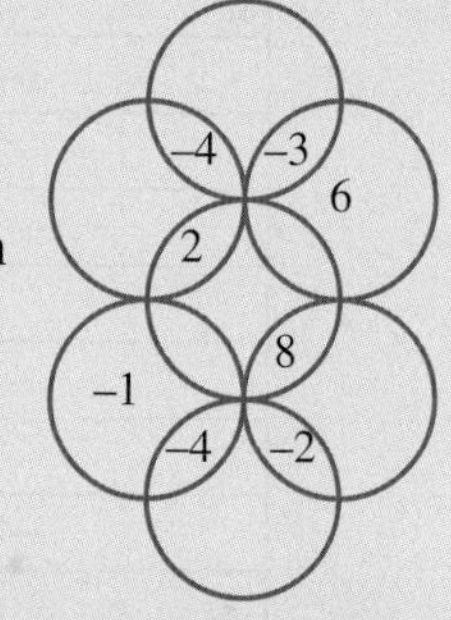


**a.** Add numbers to the empty sections in this diagram so that the numbers in each circle add up to zero.

**b.** Add numbers to the empty sections in a different way so that the numbers in each circle add up to −2.

**26.** If $(X - Y > X)$, $(Y - X < Y)$ and $X$ and $Y$ are integers, determine whether $X$ and $Y$ are positive or negative values.

**27.** The negative integers −2 to −1000 are to be arranged in 5 columns as shown.

| Column 1 | Column 2 | Column 3 | Column 4 | Column 5 |
|---|---|---|---|---|
| | −2 | −3 | −4 | −5 |
| −9 | −8 | −7 | −6 | |
| | −10 | −11 | −12 | −13 |
| −17 | −16 | −15 | −14 | |

Identify the column in which −1000 should be placed.

**28. a.** Copy this table and use your calculator to complete it.

| Integer ($x$) | $x^2$ | $x^3$ | $x^4$ | $x^5$ |
|---|---|---|---|---|
| 2 | | | | |
| −2 | | | | |
| 3 | | | | |
| −3 | | | | |
| 4 | | | | |
| −4 | | | | |

**b.** Look at your results in the $x^2$ column. State the sign of all the numbers.

**c.** Look at the sign of the numbers in the $x^3$ column. What do you notice?
**d.** Describe the resulting sign of the numbers in the $x^4$ and $x^5$ columns.
**e.** Is the resulting sign in the $x^2$ column the same as the sign in the $x^4$ column? Comment about the signs of the numbers in the $x^3$ and $x^5$ columns.
**f.** Now consider the inverse of raising a number to a power: taking the root of a number. You will notice that $2^2 = 4$ and $(-2)^2 = 4$. It follows that, if we take the square root of 4, we can get $+2$ or $-2$. (Your calculator will only give you the positive answer.) The shorthand way of writing this is as $\sqrt{4} = \pm 2$. Similarly, you will notice that $\sqrt[4]{16} = \pm 2$. This pattern only applies to even roots. Now that you know this, write a statement showing the square root of 100.
**g.** It is not possible to take the even root of a negative number, because no number raised to an even power can produce a negative number. Explain what happens when you try to evaluate $\sqrt{-144}$ on the calculator.
**h.** Notice that with odd-numbered roots the sign of the answer is the same as the sign of the original number. For example, $\sqrt[3]{8} = 2$, but $\sqrt[3]{-8} = -2$. Now that you know this, calculate $\sqrt[3]{-125}$.
**i.** Copy and complete the following table by filling in the blocks that are not shaded. Consider each answer carefully — some of these calculations are not possible.

| Integer ($x$) | $\sqrt{x}$ | $\sqrt[3]{x}$ | $\sqrt[4]{x}$ | $\sqrt[5]{x}$ |
|---|---|---|---|---|
| 16 | ±4 | | | |
| −16 | | | | |
| 27 | | | | |
| −27 | | | | |
| 32 | | | | |
| −32 | | | | |
| 81 | | | | |
| −81 | | | | |
| 64 | | | | |
| −64 | | | | |

**j.** In your own words, describe the sign that results from taking odd and even roots of positive and negative numbers.

on To test your understanding and knowledge of this topic, go to your learnON title at www.jacplus.com.au and complete the **post-test**.

# Answers

## Topic 2 Positive and negative integers

### 2.1 Pre-text

1. a. $-250$ b. $-15$ c. $+20\,°C$
2. a. $-3 < 2$ b. $-5 > -7$ c. $-21 < -17$ d. $0 > -8$
3. B
4. A $(3, -2)$
5. C
6. C
7. a. $-2$ b. 13 c. $-19$ d. $-11$
8. 7
9. 78 m
10. 48
11. a. 2 b. 8 c. 130
12. E
13. 6
14. $380
15. 108

### 2.2 Integers on the number line

1. a. $-7$ b. 10 or $+10$ c. $-422$ d. 2228 or $+2228$ e. $-3$ f. $-30$
2. 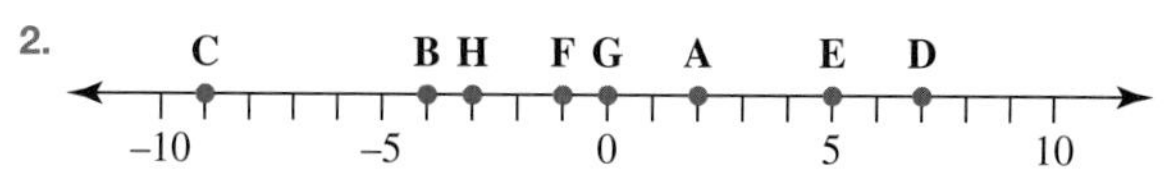

3. a. $-6, -2, 3, 5, 7$ b. $-6, -4, -1, 0, 3$ c. $-5, -3, -2, 1, 7$ d. $-4, -3, 1, 2, 5$
4. a. 6 b. $-7$ c. $-1$ d. 8
5. a. $>$ b. $<$ c. $>$ d. $>$ e. $<$ f. $<$
6. a. i. $-5$ ii. 7
   b. i. $-4$ ii. 5
   c. i. $-10$ ii. 7
   d. i. $-4$ ii. 3
7. $-2\,°C$
8. 14
9. Sample response: There were human settlements here 60 thousand years ago.
10. a. 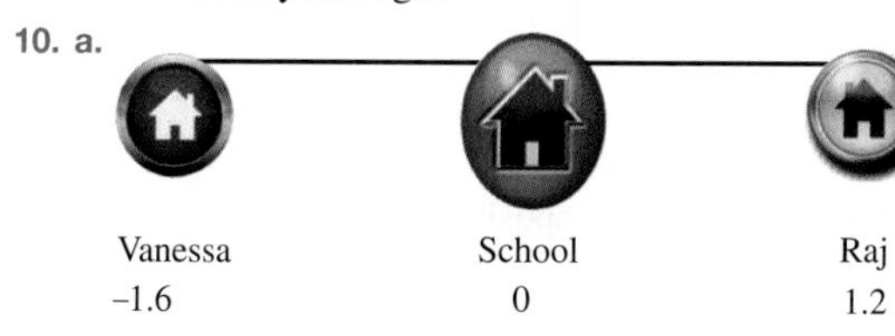

   b. 2.8 km
11. a. 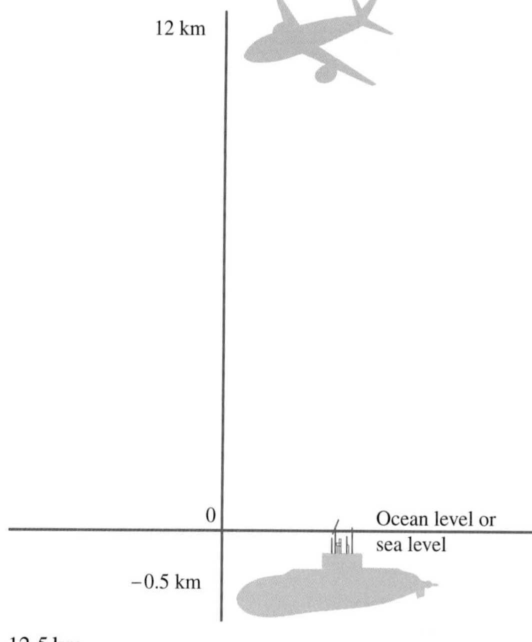

   b. 12.5 km
12. a. 2000 years in Common Era.
   b. i. 10 ii. $-106$
   c. i. 4 ii. 12 iii. 26 iv. 42 v. 57
   d. 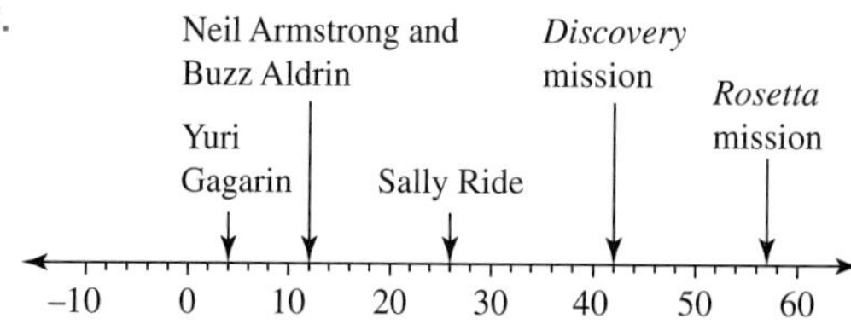


### 2.3 Integers on the number plane

1. a. $(3, 4)$ b. $(8, 4)$ c. $(4, 8)$ d. $(0, 7)$ e. $(5, 0)$
2. a. G b. F c. H d. J
3. C
4. D
5. $(-5, 2)$
6. a. $(3, 1)$; 1st quadrant
   b. $(-2, 4)$; 2nd quadrant
   c. $(-6, -5)$; 3rd quadrant
   d. $(0, -3)$; $y$-axis
   e. $(6, -6)$; 4th quadrant
7. a. C; 2nd quadrant b. L; $y$-axis c. D; 4th quadrant d. E; 1st quadrant e. G; $x$-axis
8. a. 3 b. 3 c. 7 d. 0
9. a. 2 b. $-6$ c. 0 d. $-3$
10. H
11. a. True b. False c. False d. True
12. D
13. B
14. B

**15.** 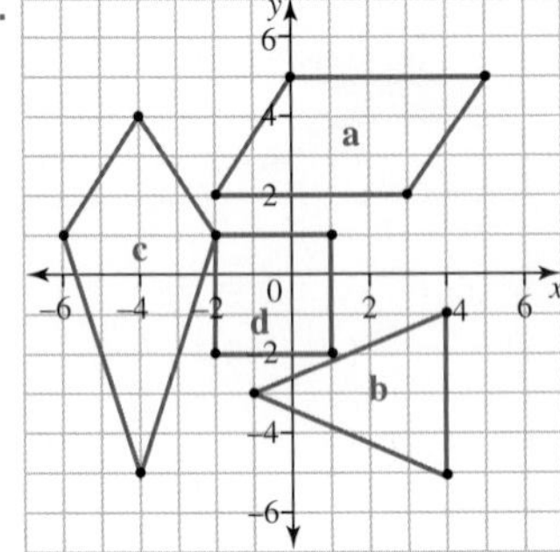


a. Parallelogram b. Isosceles triangle

c. Kite d. Square

**16.** According to the graph, both points F and D are 3 units down from the origin along the $y$-axis. Therefore they have the same $y$-coordinate. Similarly, both A and D are 3 units to the right along the $x$-axis. Therefore, they have the same $x$-coordinate.

**17.** a. (2, 1)
b. (1, −2)
c. Draw line DO
d. (−3, 1), (−2, 1), (−1, 1) etc. Any two points on the line DE are correct.
e. (−5, −2)

**18.** a. West b. East
c. West d. East
e. An odd number of negative signs turns you west, and an even number of negative signs turns you east.
f. South
g. An odd number of negative signs turns you south, and an even number of negative signs turns you north.

**19.** 9 units

**20.** a.

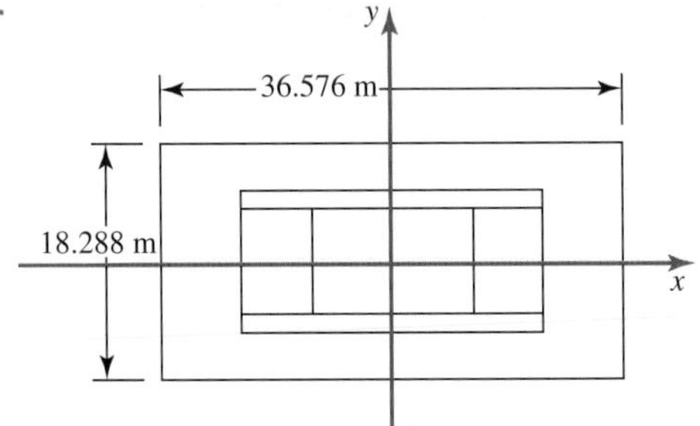


b.

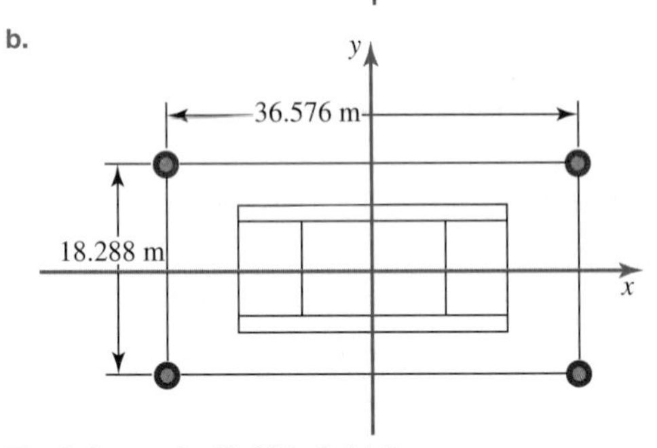


Top left post (−18.288, 9.144)
Top right post (18.288, 9.144)
Bottom right post (18.288, −9.144)
Bottom left post (−18.288, −9.144)

**21.** Points in the first quadrant are in the form (+, +). Swapping the $x$- and $y$-coordinates creates a point in the form (+, +), so the point will remain in the first quadrant.
Points in the second quadrant are in the form (−, +). Swapping the $x$- and $y$-coordinates creates a point in the form (+, −), so the point will change to the fourth quadrant.
Points in the third quadrant are in the form (−, −). Swapping the $x$- and $y$-coordinates creates a point in the form (−, −), so the point will remain in the third quadrant.
Points in the fourth quadrant are in the form (+, −). Swapping the $x$- and $y$-coordinates creates a point in the form (−, +), so the point will change to the second quadrant.

## 2.4 Adding integers

**1.** a. −15 b. −10 c. −1
d. −13 e. 19 f. 0

**2.** a. $-4+(+4)=0$ b. $-1+(+3)=+2$
c. $-2+(-4)=-6$ d. $+2+(-4)=-2$
e. $+6+(-4)=+2$ f. $-1+(-3)=-4$

**3.** a. −1 b. −1 c. 0
d. +1 e. −2

**4.** a. $5+(-2)=3$ b. $-3+(-4)=-7$
c. $-2(+2)=0$ d. $6+(-5)=1$
e. $-5(+5)=0$

**5.** a. $4+(-6)=-2$ b. $-5+7=2$
c. $6+(-9)=-3$ d. $3+(-3)=0$
e. $0+(-6)=-6$

**6.** a. −7 b. −2 c. 0
d. −4 e. 4 f. −10

**7.** a. 4 b. −3 c. −100
d. 10 e. 2 f. −22

**8.**

| + | −13 | 5 |
|---|---|---|
| 21 | 8 | 26 |
| −18 | −31 | −13 |

**9.** a. 5 b. −2 c. 1 d. 3

**10.** a. −8 b. −6 c. 10
d. −7 e. 3

**11.** a.

| + | −2 | −1 | 0 | 1 | 2 |
|---|---|---|---|---|---|
| −2 | −4 | −3 | −2 | −1 | 0 |
| −1 | −3 | −2 | −1 | 0 | 1 |
| 0 | −2 | −1 | 0 | 1 | 2 |
| 1 | −1 | 0 | 1 | 2 | 3 |
| 2 | 0 | 1 | 2 | 3 | 4 |

b. i. Even numbers, increase by increments of 2
ii. Zeros (addition of opposites)
iii. Yes

**12.** a. $-2+3=1$ b. $-3+5=2$
c. $-50+63=13$ d. $-50+26=-24$
e. $200+(-350)=-150$

13. a. Sample response: A diver is 3 m deep and dives a further 2 m deeper until she is 5 m deep.
b. Sample response: A water pipe was drilled to 10 m deep, but a further 40 m was required to reach the water, which was 50 m deep.
c. Sample response: The lift was on the second floor and then travelled six floors down, stopping at the fourth-floor basement car park.
d. Sample response: Billy borrowed $20 from his mum. One week later he paid the loan back in full.
e. Sample response: Shalini gave eight lollies to her brother and later she got ten more lollies from her dad, leaving her with two extra lollies.

14. a. 'When adding two negative numbers, add the two numbers as if they were positive numbers and then write a negative sign in front of the answer.'
b. 'When adding a negative numbers and a positive number, ignore the signs, subtract the smaller number and then write the sign the large number has.'

15.

| | | |
|---|---|---|
| 2 | –12 | 7 |
| 4 | –1 | –6 |
| –9 | 10 | –4 |

16. $5575
17. 15 points
18. 2nd floor
19. If you add a number to its opposite, the answer will always be zero. For example: $-2+2=0$.

## 2.5 Subtracting integers

1. a. 5 b. −24 c. −5 d. 26 e. −16 f. 11
2. a. 2 b. 10 c. −10 d. 2 e. −9
3. a. −10 b. 0 c. −4 d. 3 e. 10 f. 0
4. a. $2-(+5)=-3$ b. $-3-(+6)=-9$ c. $-4-(-9)=5$ d. $5-(-7)=12$
5. a. −10 b. 13 c. 3 d. 12
6. a. −16 b. −17 c. 23 d. −12
7. a. 1 b. 9 c. −25 d. 5
8. a. $-2-6=-8$ b. $-8-5=-13$ c. $-1-8=-9$ d. $2-3=-1$ e. $4-3=1$ f. $-3-4=-7$
9. D
10. a. $-2-3=-5$ b. $3-5=-2$ c. $1-6=-5$ d. $500 − $150 = $350
11. B
12. a. True b. True c. True d. True
13. −2 °C
14. $400-520=-120$. Nihal has to cut 120 words.
15. 12 m
16. a. Balance at end of month: $10
b. During the second week her balance was negative.
17. Depth of 5 m below the surface = −5
An ascent of 2 m = +2
A descent of 12 m = −12
Total depth = $-5+2-12=-15$
−15 = 15 m below the surface
18. a. 2nd floor
b. It travelled 22 floors in 10 minutes.
19. Subtract 5 from the number: $9-5=4$
20. If you subtract a number from its opposite, the answer will be double the original number value, but it can be positive or negative. For example: $2-(-2)=4$, or $-2-2=-4$.

## 2.6 Multiplying and dividing integers

1. a. −10 b. −18 c. −63 d. −30 e. 6
2. a. 18 b. 25 c. 0 d. 8
3. a. −10 b. −30 c. 200 d. 36
4. a. 4 b. −4 c. −4 d. 2 e. 9
5. a. −8 b. −6 c. 0 d. −21 e. −92
6. a. 25 b. 67 c. −43 d. −71 e. −56
7. a. −3 b. 3 c. −1
8. a. 1 b. −2 c. $1\frac{1}{2}$
9. a. −24 b. 36 c. 144
10. a. −3 b. −6 c. 1 d. −1 e. 0
11. a. 3 b. 32 c. −72 d. −11 e. 0 f. 6
12. a. −8 b. −8 c. −6 d. −6 e. −20 f. −20
13. B
14. A
15. E
16. a. i. 2 ii. −6 iii. 24 iv. −40
b. When multiplying an even number of negative numbers, the answer is always positive. When multiplying an odd number of negative numbers, the answer is always negative.
c. Yes, because multiplication and division have the same rules when working with negative numbers.
17. Agree. When you multiply two integers, the result is always an integer. When you divide two integers, the answer is not always an integer. If the integers do not divide equally, the answer will be a fraction.
18. 8 flies
19. a. 10 b. 20 c. 5 m d. −5 m

20. According to Kareen: $8 \div 2 \times (2 + 2)$ applying BIDMAS to this expression

$$= 8 \div 2 \times (4)$$
$$= 4 \times 4$$
$$= 16$$

Kareen is correct. We do need to apply BIDMAS when more than one operation is used in an expression.

21. $16 \times 5 + 3 \times -2 + 0 = 80 - 6 = 74$ points

## 2.7 Order of operations with integers

1. a. $-6$ b. 22 c. $-6$ d. 6
2. a. $-2$ b. $-27$ c. 0 d. $-1$
3. a. 3 b. $-18$ c. $-40$ d. 35
4. a. $-19$ b. $-1$ c. $-2$ d. $-9$
5. a. 0 b. 21 c. 2 d. $-5$
6. a. 9 b. $-11$ c. $-7$ d. 32
7. a. $-20$ b. $-1$ c. 32
8. a. $-36$ b. $-68$ c. 24
9. D
10. B
11. $-100 + 60 + -25 = -65$ m
12. \$2075
13. 23 kg per month
14. The two expressions are different due to the placement of the bracket in the first expression. See the working below.

First expression: $8 - 2 \times (4 + 2)$
$$= 8 - 2 \times (6)$$
$$= 8 - 12 = -4$$
Second expression: $8 - 2 \times 4 + 2 = 8 - 8 + 2$
$$= 0 + 2$$
$$= +2$$

15. a. False. See the working below.
$5 \times 6 - 3 + (5-2) < 5 \times (6 - 3) + 5 - 2$
Apply BIDMAS to both of the expressions.
$5 \times 6 - 3 + (5 - 2) = 5 \times 6 - 3 + (3)$
$= 30 - 3 + 3 = 33 - 3 = 30$ (larger)
$5 \times (6 - 3) + 5 - 2 = 5 \times (3) + 5 - 2$
$= 15 + 5 - 2 = 20 - 2 = 18$ (smaller)
$5 \times 6 - 3 + (5 - 2) >$
$5 \times (6 - 3) + 5 - 2.$
Therefore, the statement given is not true.

b. True. See the working below.
$2 + (15 \div 3) \times 1 = 2 + 15 \div (3 \times 1)$
Apply BIDMAS to both of the expressions.
$2 + (15 \div 3) \times 1 = 2 + (5) \times 1 = 2 + 5 = 7$
$2 + 15 \div (3 \times 1) = 2 + 15 \div (3) = 2 + 5 = 7$
Because both the expressions are equal to 7, the statement is true.

16. a. Merlin: $-3 \times 10 = -30$, 30 km west of bridge
Morgan: $-3 \times -8 = 24$, 24 km east of bridge

b. Merlin: $2 \times 10 = 20$, 20 km east of bridge
Morgan: $2 \times -8 = -16$, 16 km west of bridge

c. 36 km

d. 72 km

17. The total amount Ryan paid to the bakery is \$42.75.
18. $3 + -2 + 4 + -1 = 4$
The snail is 4 cm from the bottom of the bucket.
19. a. Arctic Circle 17 °C,
London 12 °C,
Mexico 22 °C
20. a. Student 1: 39 marks; Student 2: 35 marks; Student 3, 44 marks

b. Student 3 won the competition with 44 marks.

## Project

Students should play the game to understand how it works. Then students should present a game they have created, which should include positive and negative numbers and any operation.

## 2.8 Review questions

1. a. No b. No c. Yes d. Yes
2. a. $<$ b. $<$ c. $>$ d. $<$
3. $-20, -19, -18 - 17, -16$
4. $2, 0, -3, -15$
5. a. $< 1$ or $\leq 0$ b. Between $-6$ and $-3$
c. $> -12$ or $\geq -11$
6. a. $-7\ -6\ -5\ -4\ -3\ -2$

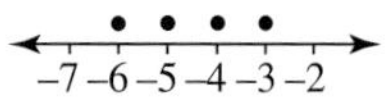

b. $-3\ -2\ -1\ 0$

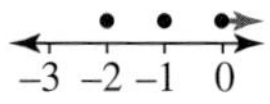

c. $-6\ -5\ -4\ -3$

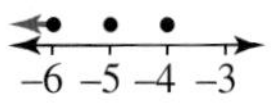

7. a. Both axes b. $y$-axis
c. $x$-axis d. First quadrant
8. (−2, 3) (1, 3) (−1, −2) (2, −2)

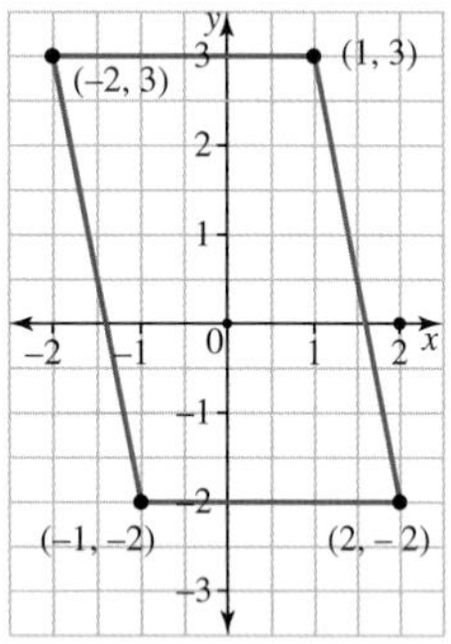

Parallelogram

9. a. Second quadrant b. Fourth quadrant
c. Third quadrant d. $y$-axis
e. $x$-axis f. First quadrant
10. a. $-5$ b. $-17$ c. 8 d. 6
11. a. $-2$ b. 7 c. 1 d. 2
e. 0
12. a. $-8$ b. 26 c. 3 d. $-2$
e. $-8$ f. 2
13. a. $-42$ b. $-32$ c. 10 d. 64
e. $-64$ f. $-128$

14. a. −12 b. 3 c. −5 d. 9
e. −16 f. −4

15. a. −2 b. −3 c. 27 d. −8
e. 5

16. a. −20 b. 12 c. 15
d. −125 e. −9

17. a. $-5 < -3$ b. $-22 \times -2 = 44$
c. $4 > 2$ d. $-5 \times 3 > -15$
e. $0 > -7$ f. $-2(-4 + 7) < 6$

18. a. $-2 \times 5 < -9$ b. $5 \times (-2 - 18) \div -4 = 25$
c. $12 \div -4 < 3$ d. $5 \times (-2 - 18) \div 4 = -25$
e. $-10 \times -5 = 50$ f. $5 \times -2 - 18 \div 4 < 2.5$

19. $-52 + 3 \times 15 = -7$

20. 601 years

21. 148 °C

22. −$94

23. a. Negative
b. Positive
c. Negative
Two negative numbers added together will always be negative. When a positive and a negative number are added together, if the larger number is positive the result will be positive. If the larger number is negative the result will be negative.

24. a. D(−1, 0)
b. i. CBA, BAD (Others are possible).
ii. ADE or BCE
c. F( − 3, −1)

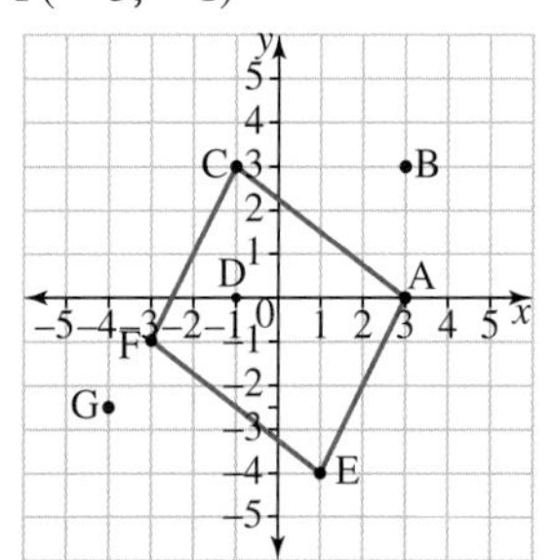


d. (1, 1.5)
e. (2, −2)
f. G( − 4, −2.5)

25. a.

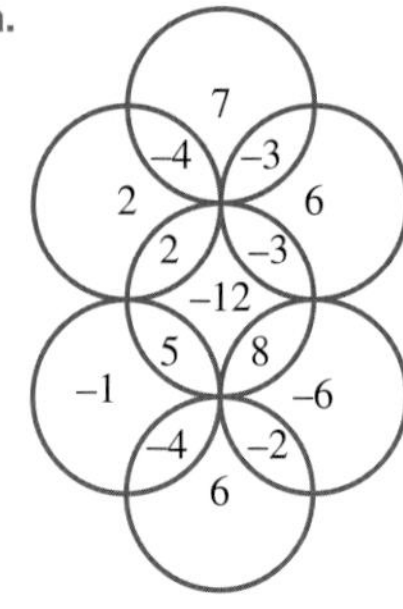


b.

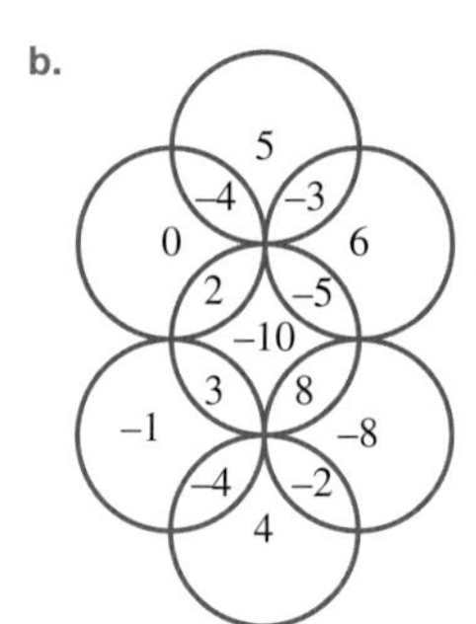


26. *X* is positive and *Y* is negative.

27. Column 2

28. a.

| Integer ($x$) | $x^2$ | $x^3$ | $x^4$ | $x^5$ |
|---|---|---|---|---|
| 2 | 4 | 8 | 16 | 32 |
| −2 | 4 | − 8 | 16 | − 32 |
| 3 | 9 | 27 | 81 | 243 |
| −3 | 9 | −27 | 81 | − 243 |
| 4 | 16 | 64 | 256 | 1024 |
| −4 | 16 | −64 | 256 | −1024 |

b. Positive

c. If $x$ was negative, then $x^3$ is negative.

d. For $x^4$, all the results are positive; for $x^5$, if $x$ is negative then $x^5$ is negative.

e. The sign in the $x^2$ column is the same as the sign in the $x^4$. The sign in the $x^3$ is the same as the matching term in the $x^5$ column.

f. The square root of 100 is 10 or −10.

g. It will give an error message.

h. −5

i.

| Integer ($x$) | $\sqrt{x}$ | $\sqrt[3]{x}$ | $\sqrt[4]{x}$ | $\sqrt[5]{x}$ |
|---|---|---|---|---|
| 16 | 4 | | 2 | |
| −16 | Not possible | | Not possible | |
| 27 | | 3 | | |
| −27 | | −3 | | |
| 32 | | | | 2 |
| −32 | | | | −2 |
| 81 | 9 | | 3 | |
| −81 | Not possible | | Not possible | |
| 64 | 8 | 4 | | |
| −64 | Not possible | −4 | | |

j. When taking an even root of a positive number, the answer can be positive or negative.
It is not possible to take the even root of a negative number.
When taking the odd root of a positive number, the answer is always positive.
When taking the odd root of a negative number, the answer is always negative.

# 3 Factors, multiples, indices and primes

## LESSON SEQUENCE


**3.1** Overview ..... 116
**3.2** Factors and multiples ..... 118
**3.3** Lowest common multiple and highest common factor ..... 123
**3.4** Index notation ..... 128
**3.5** Prime and composite numbers ..... 133
**3.6** Squares and square roots ..... 141
**3.7** Divisibility tests ..... 147
**3.8** Cubes and cube roots (extending) ..... 150
**3.9** Review ..... 155

# LESSON
# 3.1 Overview

## 3.1.1 Why learn this?

Whole numbers can be written in different ways that can make calculations easier if you do not have your calculator handy. You have been using multiplication tables already without necessarily realising that the they give you numbers that are multiples of each other. Knowing how to write numbers in different ways helps in understanding how numbers are connected.

Indices are short ways of writing a repeated multiplication and are very useful in everyday life, because they allow us to write very large and very small numbers more easily. Indices can simplify calculations involving these very large or very small numbers. Astronomy is a branch of science in which very large distances are involved. The stars we see in the Southern Hemisphere night sky are different to the stars people see in the Northern Hemisphere. Alpha Crucis, the brightest and closest star to us in the Southern Cross constellation, is approximately 3000 million billion kilometres from Earth. That is a huge number — 3 followed by 15 zeros! Other scientists and engineers also need to be able to communicate and work with numbers of all sizes.

## Exercise 3.1 Pre-test

**learnon**

1. **MC** The first 3 multiples of 8 are:

   **A.** 2, 4, 8 **B.** 8, 16, 24 **C.** 4, 8, 16 **D.** 1, 8, 16 **E.** 1, 2, 4

2. Kate earned $7 pocket money each week for 5 weeks for emptying the dishwasher. Given she saves all her money, calculate how much she would have saved at the end of the 5 weeks.

3. Determine the highest common factor of 42 and 66.

4. Fred takes 3 minutes to complete one lap of a running course, while Sam takes 4 minutes to complete one lap. After the start, determine when Fred will next be running beside Sam.

5. Write $5 \times 3 \times 5 \times 3 \times 3 \times 3$ in simplified index notation.

6. Select all the prime numbers from the following.

   **A.** 21 **B.** 109 **C.** 33 **D.** 81 **E.** 17 **F.** 31 **G.** 91

7. Select two pairs of prime numbers whose sum is 20.

   **A.** 1, 19 **B.** 2, 18 **C.** 3, 17 **D.** 5, 15 **E.** 7, 13 **F.** 9, 11

8. Select all the prime factors of 40.

   **A.** 1 **B.** 2 **C.** 3 **D.** 4 **E.** 5
   **F.** 8 **G.** 10 **H.** 20 **I.** 40

9. Determine the lowest common multiple (LCM) and highest common factor (HCF) of 54 and 96 by first expressing each number as a product of its prime factors.

10. A chessboard is a large square with sides that are 8 small squares in length. Calculate the total number of small squares on a chessboard.

11. Evaluate the following.

    $3^2 + \sqrt{16} \times 5^2$

12. Determine between which two integers $\sqrt{70}$ lies.

13. Evaluate the following.

    $\sqrt[3]{27}$

14. Sarah wants to organise a party with her friends at short notice. She decides to send a text to three of her friends, who contact another three friends each, and then they all contact a further three friends each.

    Determine the number of friends that are invited to the party.

15. Evaluate the following.

    $$\frac{11^2 - \sqrt[3]{216} \times \sqrt{144}}{\sqrt{49}}$$

# LESSON
# 3.2 Factors and multiples

## LEARNING INTENTION

At the end of this lesson you should be able to:
- list multiples and factors of a whole number
- determine multiples and factor pairs.

## 3.2.1 Multiples

eles-3725

- A **multiple** of a number is the result of **multiplying** that number by another whole number. For example, all numbers in the 5 times table are multiples of 5: so 5, 10, 15, 20, 25, ... are all multiples of 5.

| × | 1 | 2 | 3 | 4 | 5 |
|---|---|---|---|---|---|
| 1 | 1 | 2 | 3 | 4 | 5 |
| 2 | 2 | 4 | 6 | 8 | 10 |
| 3 | 3 | 6 | 9 | 12 | 15 |
| 4 | 4 | 8 | 12 | 16 | 20 |
| 5 | 5 | 10 | 15 | 20 | 25 |

### WORKED EXAMPLE 1 Listing multiples of a whole number

**List the first five multiples of 7.**

| THINK | WRITE |
|---|---|
| 1. The first multiple is the number × 1, that is, $7 \times 1$. | 7 |
| 2. The second multiple is the number × 2, that is, $7 \times 2$. | 14 |
| 3. The third multiple is the number × 3, that is, $7 \times 3$. | 21 |
| 4. The fourth multiple is the number × 4, that is, $7 \times 4$. | 28 |
| 5. The fifth multiple is the number × 5, that is, $7 \times 5$. | 35 |
| 6. Write the answer as a sentence. | The first five multiples of 7 are 7, 14, 21, 28 and 35. |

### WORKED EXAMPLE 2 Determining multiples

**State which numbers in the following list are multiples of 8.**

**18, 8, 80, 100, 24, 60, 9, 40**

| THINK | WRITE |
|---|---|
| 1. The biggest number in the list is 100. List multiples of 8 using the 8 times table just past 100; that is, $8 \times 1 = 8$, $8 \times 2 = 16$, $8 \times 3 = 24$, $8 \times 4 = 32$, $8 \times 5 = 40$, and so on. | 8, 16, 24, 32, 40, 48, 56, 64, 72, 80, 88, 96, 104 |
| 2. Write any multiples that appear in the given list. | Numbers in the given list that are multiples of 8 are 8, 24, 40 and 80. |

### 3.2.2 Factors

eles-3726

- A **factor** of a whole number **divides** into that whole number exactly. For example, the number 4 is a factor of 8 because 4 divides into 8 twice or $8 \div 4 = 2$.
- Factors of a number can be written as factor pairs. For example, since $4 \times 2 = 8$, it follows that 2 and 4 are a factor pair of 8.

**Unique factors**

**Except for the number 1, every whole number has at least two unique factors, 1 and itself.**

**For example, the factors of 5 are 1 and 5.**

**WORKED EXAMPLE 3 Listing factors of a whole number**

**List all the factors of 14.**

| THINK | WRITE |
| --- | --- |
| 1. 1 is a factor of every number, and the number itself is a factor of itself, that is, $1 \times 14 = 14$. | 1, 14 |
| 2. 14 is an even number, so 14 is divisible by 2; therefore, 2 is a factor. Divide the number by 2 to find the other factor $(14 \div 2 = 7)$. There are no other whole numbers that divide evenly into 14. | 2, 7 |
| 3. Write the answer as a sentence, placing the factors in order from smallest to largest. | The factors of 14 are 1, 2, 7 and 14. |

**WORKED EXAMPLE 4 Determining factor pairs**

**List the factor pairs of 30.**

| THINK | WRITE |
| --- | --- |
| 1. 1 and the number itself are factors, that is, $1 \times 30 = 30$. | 1, 30 |
| 2. 30 is an even number, so 2 and 15 are factors, that is, $2 \times 15 = 30$. | 2, 15 |
| 3. Divide the next smallest number into 30. Therefore, 3 and 10 are factors, that is, $3 \times 10 = 30$. | 3, 10 |
| 4. 30 ends in 0 so 5 divides evenly into 30, that is, $5 \times 6 = 30$. | 5, 6 |
| 5. Write the answer as a sentence by listing the factor pairs. | The factor pairs of 30 are: 1, 30; 2, 15; 3, 10 and 5, 6. |

## COLLABORATIVE TASK: Sorting multiples and factors

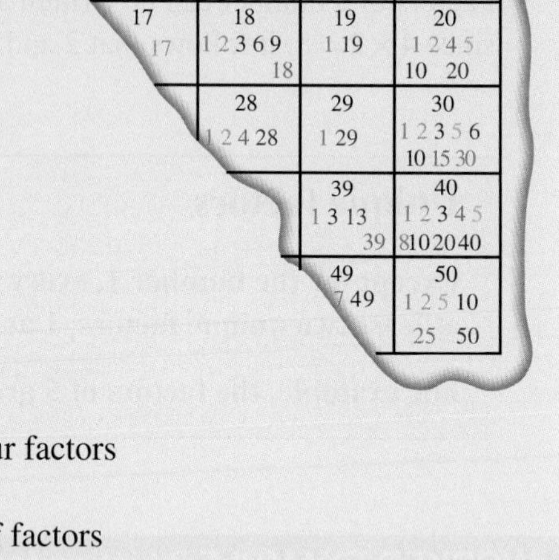


1. As a class, draw a large $10 \times 5$ number chart on the board and number the boxes from 1 to 50.
2. Your teacher will share the numbers 1 to 50 among the class. For each of your numbers, list the multiples up to and including 50.
3. Your teacher will call out the numbers between 1 and 50 one at a time. When one of your numbers is called, write it in each box on the chart that is a multiple of your number. For example, if your number is 10, write 10 in each box that corresponds to a multiple of 10.
4. At the end of the activity, the numbers written in each of the boxes are the factors of the box numbers. You can see that different numbers have different numbers of factors.
5. As a class, investigate:
   a. which numbers have one factor, two factors, three factors or four factors
   b. which numbers have more than four factors
   c. whether there are more numbers with an odd or even number of factors
   d. how factors and multiples are related.

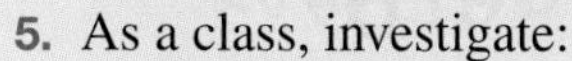

### on Resources

**eWorkbook** Topic 3 Workbook (worksheets, code puzzle and project) (ewbk-1904)

**Video eLesson** Factors and multiples (eles-2310)

**Interactivities** Individual pathway interactivity: Factors and multiples (int-4316)
Multiples (int-3929)
Factors (int-3930)

## Exercise 3.2 Factors and multiples

learnon

3.2 Quick quiz  | 3.2 Exercise

### Individual pathways

| ■ PRACTISE | ■ CONSOLIDATE | ■ MASTER |
| --- | --- | --- |
| 1, 4, 6, 8, 10, 12, 15, 18, 22, 23, 26 | 2, 5, 9, 13, 16, 19, 24, 27 | 3, 7, 11, 14, 17, 20, 21, 25, 28 |

### Fluency

1. **WE1** List the first five multiples of the following numbers.
   a. 3 b. 6 c. 100 d. 11
2. List the first five multiples of the following numbers.
   a. 120 b. 45 c. 72 d. 33

3. **WE2** Select the numbers in the following list that are multiples of 10.
10, 15, 20, 100, 38, 62, 70

4. Select the numbers in the following list that are multiples of 7.
17, 21, 7, 70, 47, 27, 35

5. Write the numbers in the following list that are multiples of 16.
16, 8, 24, 64, 160, 42, 4, 32, 1, 2, 80

6. List the multiples of 9 that are less than 100.

7. List the multiples of 6 between 100 and 160.

8. **MC** The first three multiples of 9 are:
**A.** 1, 3, 9 **B.** 3, 6, 9 **C.** 9, 18, 27
**D.** 9, 18, 81 **E.** 18, 27, 36

9. **MC** The first three multiples of 15 are:
**A.** 15, 30, 45 **B.** 30, 45, 60 **C.** 1, 15, 30
**D.** 45 **E.** 3, 5, 15

10. **WE3** List all the factors of the following numbers.
a. 12 b. 40

11. A list of numbers is: 2, 7, 5, 20, 25, 15, 10, 3, 1.
From this list, select all the numbers that are factors of each of the following values.
**a.** 28 **b.** 60 **c.** 100

### Understanding

12. List all factors of the following numbers in ascending order.
a. 72 b. 250

13. A list of numbers is: 21, 5, 11, 9, 1, 33, 3, 17, 7.
From this list, select all the numbers that are factors of each of the following values.
a. 85 b. 99 c. 51

14. List the factor pairs of 20.

15. List the factor pairs of 132.

16. **MC** Select all the factor pairs of 18.
**A.** 1, 9 **B.** 3, 6 **C.** 2, 9
**D.** 6, 12 **E.** 1, 18

17. Three is a factor of 12. State the smallest number greater than 12 that has 3 as one of its factors.

18. **MC** A factor pair of 24 is:
**A.** 2, 4 **B.** 4, 6 **C.** 6, 2
**D.** 2, 8 **E.** 3, 9

19. **MC** A factor pair of 42 is:
**A.** 6, 7 **B.** 20, 2 **C.** 21, 1
**D.** 16, 2 **E.** 6, 8

20. Determine which of the numbers 3, 4, 5 and 11 are factors of 2004.

**21.** Alex and Nadia were racing down a flight of stairs. Nadia took the stairs two at a time while Alex took the stairs three at a time. In each case, they reached the bottom with no steps left over.

**a.** Determine how many steps there are in the flight of stairs. List three possible answers.

**b.** Determine the smallest number of steps there could be.

**c.** If Alex can also take the stairs five at a time with no steps left over, determine the smallest number of steps in the flight of stairs.

## Reasoning

**22.** Place each of the first six multiples of 3 into the separate circles around the triangle so the numbers along each side of the triangle add up to 27. Use each number once only.

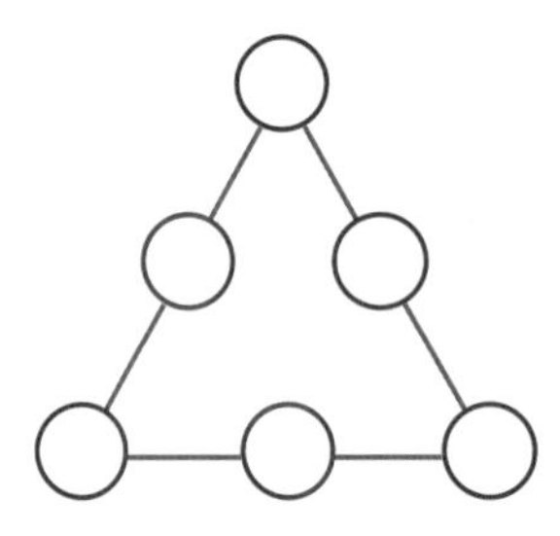

**23.** Ms Pythagoras is trying to organise their Year 4 class into rows for their class photograph.
If Ms Pythagoras wishes to organise the 20 students into rows containing equal numbers of students, determine what possible arrangements they can have.
*Note:* Ms Pythagoras will not be in the photograph.

**24.** My age is a multiple of 3 and a factor of 60. The sum of the digits of my age is 3. Determine my age. (There are three possible answers.)

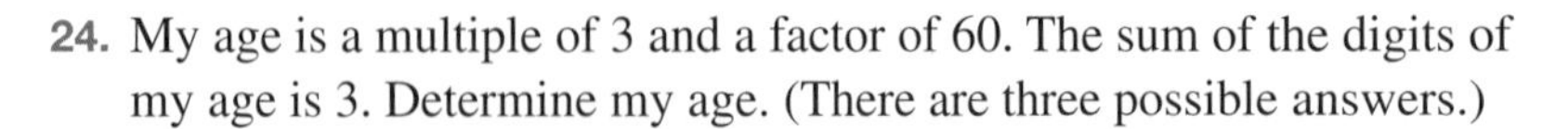

**25.** A man and his grandson share the same birthday. For exactly six consecutive years the grandson's age is a factor of his grandfather's age. Determine how old each of them is at the sixth of these birthdays, given that the grandson is no more than 10.

## Problem solving

**26.** I am a two-digit number that can be divided by 3 with no remainder. The sum of my digits is a multiple of 4 and 6. My first digit is double my second digit. Evaluate the number. Show your working.

**27.** Determine the following number. I am a multiple of 5 with factors of 6, 4 and 3. The sum of my digits is 6. Show your working.

**28.** Identify a two-digit number such that if you subtract 3 from it, the result is a multiple of 3; if you subtract 4 from it, the result is a multiple of 4; and if you subtract 5 from it, the result is a multiple of 5. Explain how you reached the answer.

# LESSON
# 3.3 Lowest common multiple and highest common factor

**LEARNING INTENTION**

At the end of this lesson you should be able to:

- determine the lowest common multiple and the highest common factor of two or more whole numbers.

## 3.3.1 The LCM and HCF

eles-3727

- **Common multiples** are numbers that are multiples of more than one number.
  Multiples of 3: 3, 6, 9, **12**, 15, 18, 21, **24**, …
  Multiples of 4: 4, 8, **12**, 16, 20, **24**, 28, …
  The first two common multiples of 3 and 4 are 12 and 24.
- The **lowest common multiple (LCM)** is the smallest multiple that is common to two or more numbers. The LCM of 3 and 4 is 12.
- **Common factors** are numbers that are factors of more than one number.
  Factors of 16: **1**, **2**, **4**, **8** and 16
  Factors of 24: **1**, **2**, 3, **4**, 6, **8**, 12 and 24
  The common factors of 16 and 24 are 1, 2, 4 and 8.
- The common factors of 16 and 24 can be seen in the overlapping region of a Venn diagram, as shown.
- The **highest common factor (HCF)** is the largest of the common factors. The HCF of 16 and 24 is 8.

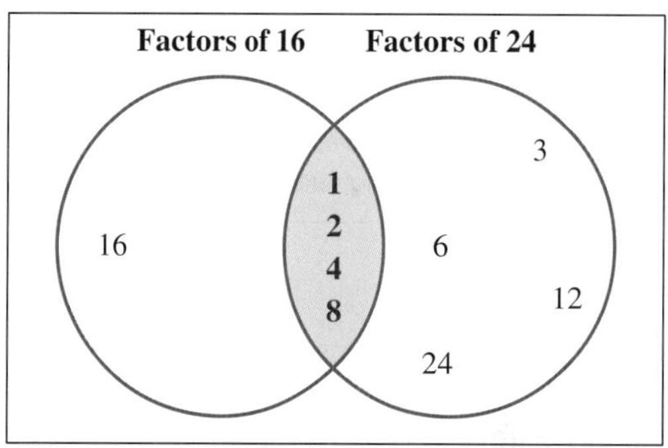


**WORKED EXAMPLE 5 Determining the lowest common multiple of two numbers**

a. **List the first nine multiples of 8.**
b. **List the first nine multiples of 12.**
c. **List the multiples that 8 and 12 have in common.**
d. **State the lowest common multiple of 8 and 12.**

| THINK | WRITE |
|---|---|
| a. Write out the first nine numbers of the 8 times table. | 8, 16, 24, 32, 40, 48, 56, 64, 72 |
| b. Write out the first nine numbers of the 12 times table. | 12, 24, 36, 48, 60, 72, 84, 96, 108 |
| c. Choose the numbers that appear in both lists. | 24, 48, 72 |
| d. State the lowest of these numbers to give the lowest common multiple (LCM). | LCM is 24. |

**WORKED EXAMPLE 6 Determining the highest common factor**

**a. Determine the common factors of 8 and 24 by:**
  **i. listing the factors of 8**
  **ii. listing the factors of 24**
  **iii. listing the factors common to both 8 and 24.**
**b. State the highest common factor of 8 and 24.**

| THINK | WRITE |
|---|---|
| a. i. 1. List the pairs of factors of 8. | a. i. 1, 8, 2, 4 |
| 2. Write them in order. | Factors of 8 are 1, 2, 4, 8. |
| ii. 1. List the pairs of factors of 24. | ii. 1, 24; 2, 12; 3, 8; 4, 6 |
| 2. Write them in order. | Factors of 24 are 1, 2, 3, 4, 6, 8, 12, 24. |
| iii. Write the common factors. | iii. Common factors are 1, 2, 4, 8. |
| b. State the highest common factor, that is, the largest of the common factors. | b. HCF is 8. |

**on** Resources

 **eWorkbook** Topic 3 Workbook (worksheets, code puzzle and project) (ewbk-1904)

 **Interactivities** Individual pathway interactivity: Lowest common multiple and highest common factor (int-4317)
Lowest common multiple (int-3931)
Highest common factor (int-4545)

## Exercise 3.3 Lowest common multiple and highest common factor

**learnon**

**3.3 Quick quiz**  | **3.3 Exercise**

### Individual pathways

| ■ PRACTISE | ■ CONSOLIDATE | ■ MASTER |
|---|---|---|
| 1, 2, 7, 10, 11, 12, 16, 19 | 3, 4, 8, 13, 14, 17, 20 | 5, 6, 9, 15, 18, 21 |

### Fluency

1. a. **WE5** List the first ten multiples of 4.
   b. List the first ten multiples of 6.
   c. In your lists, circle the multiples that 4 and 6 have in common (that is, circle the numbers that appear in both lists).
   d. State the lowest multiple that 4 and 6 have in common. This is the lowest common multiple (LCM) of 4 and 6.

2. State the LCM for each of the following pairs of numbers.
   a. 3 and 9   b. 6 and 15   c. 7 and 10   d. 12 and 16   e. 4 and 15

3. State whether the following statement is true or false.
   20 is a multiple of 10 and 2 only.

4. State whether the following statement is true or false.
   15 and 36 are both multiples of 3.

5. State whether the following statement is true or false.
   60 is a multiple of 2, 3, 6, 10 and 12.

6. State whether the following statement is true or false.
   100 is a multiple of 2, 4, 5, 10, 12 and 25.

7. WE6 By listing the factors of each number, determine the highest common factor (HCF) for each of the following pairs of numbers.
   a. 21 and 56 b. 7 and 28 c. 48 and 30

8. Determine the HCF for each of the following pairs of numbers.
   a. 9 and 36 b. 42 and 77

9. Determine the HCF of the following numbers.
   **a.** 36 and 64 **b.** 45, 72 and 108

### Understanding

10. Kate goes to the gym every second evening, while Ian goes every third evening.
    a. On Monday evening, Kate and Ian are at the gym together. Determine how many days it will be before both attend the gym again on the same evening.
    b. Explain how this answer relates to the multiples of 2 and 3.

11. Vinod and Elena are riding around a mountain bike trail. Vinod completes one lap in the time shown on the stopwatch on the left, and Elena completes one lap in the time shown on the stopwatch on the right.
    a. If they both begin cycling from the starting point at the same time, determine how long it will be before they pass the starting point again at exactly the same time.
    b. Relate your answer to the multiples of 5 and 7.

12. A warehouse owner employs Bob and Charlotte as security guards. Each security guard checks the building at midnight. Bob then checks the building every 4 hours, and Charlotte checks every 6 hours.
    a. Determine how long it will be until both Bob and Charlotte are next at the warehouse at the same time.
    b. Relate your answer to the multiples of 4 and 6.

13. Two smugglers, Bill Bogus and Sally Seadog, have set up signal lights that flash continuously across the ocean. Bill's light flashes every 5 seconds and Sally's light flashes every 4 seconds. If they start together, determine how long it will take for both lights to flash again at the same time.

14. Twenty students in Year 7 were each given a different number from 1 to 20 and then asked to sit in numerical order in a circle. Three older girls — Milly, Molly and Mandy — came to distribute jelly beans to the class. Milly gave red jelly beans to every second student, Molly gave green jelly beans to every third student and Mandy gave yellow jelly beans to every fourth student.
    a. Determine which student had jelly beans of all 3 colours.
    b. Calculate how many students received exactly 2 jelly beans.
    c. Calculate how many students did not receive any jelly beans.

15. Wah needs to cut tubing into the largest pieces of equal length that he can, without having any offcuts left over. He has three sections of tubing: one 6 metres long, another 9 metres long and the third 15 metres long.
    a. Calculate how long each piece of tubing should be.
    b. Determine how many pieces of tubing Wah will end up with.

### Reasoning

16. Two candles of equal length are lit at the same time. The first candle takes 6 hours to burn out and the second takes 3 hours to burn out. Determine how many hours it will be until the slower-burning candle is twice as long as the faster-burning candle.

17. Mario, Luigi, Zoe and Daniella are playing a video game. Mario takes 2 minutes to play one level of the game, Luigi takes 3 minutes, Zoe takes 4 minutes and Daniella takes 5 minutes. They have 12 minutes to play.
    a. If they play continuously, determine which player would be in the middle of a game as time ran out.
    b. Determine after how many minutes this player began the last game.

18. Six church bells ring repeatedly at intervals of 2, 4, 6, 8, 10 and 12 minutes respectively. Determine how many times three or more bells will ring together in 30 minutes.

### Problem solving

19. Carmen types 30 words per minute on her laptop while Evan types 40 words per minute on his laptop. They have to type the same number of words and they both end up typing for a whole number of minutes.
    a. Determine how long Carmen has been typing for.
    b. Determine how long Evan has been typing for.

20. Alex enjoys running around the park after school. It takes him 10 minutes to complete one lap. His mother Claire walks the family dog at the same time. It takes her 15 minutes to complete one lap. They decide to go home when they meet next. Determine how long it will take them to meet again.

**21.** The runners in a 100 m race are lined up next to each other while those in a 400 m race are staggered around the track. Look at the following diagram of a standard 400 m athletic track and take note of both the starting position and finish line for a 100 m race.

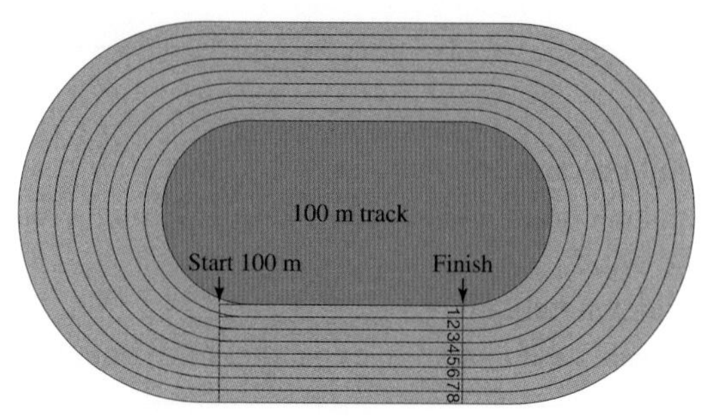


**a.** State the shapes of the two curved ends of the track.
**b.** Explain why the starting line for a 100 m race has been positioned in the straight section.
**c.** Explain why the finish line for a 100 m race has been positioned where it is.

Track events such as the 200 m, 400 m and 800 m races have staggered starting positions. The following diagram illustrates the position of the starting blocks $S_1$–$S_8$ for each lane in the 400 m race.

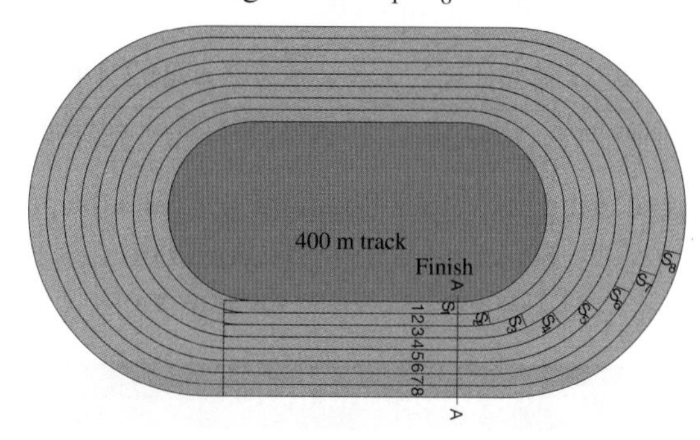


**d.** Explain why there is a need to stagger the starting blocks in the 200 m, 400 m and 800 m track events.
**e.** A runner completing one lap of the inside lane (lane 1) runs a distance of 400 m, while if there is no staggered start a runner completing one lap in the next lane (lane 2) runs a distance of 408 m. If this pattern continues, determine how far runners in lanes 3–8 run. Enter these results in the appropriate column in the table.

| Lane number | Distance travelled (m) | Difference |
|---|---|---|
| 1 | 400 m | |
| 2 | 408 m | |
| 3 | | |
| 4 | | |
| 5 | | |
| 6 | | |
| 7 | | |
| 8 | 456 m | |

**f.** Calculate the difference between the distances travelled by the runners in each of the lanes compared to the distance travelled by the runner in lane 1. Enter these results in the appropriate column in the table.
**g.** Comment on what you notice about the values obtained in part **f**.

# LESSON
# 3.4 Index notation

## LEARNING INTENTION

At the end of this lesson you should be able to:
- understand that index or exponent notation is a short way of writing a repeated multiplication
- write expressions in index notation
- simplify expressions using index notation
- use place value to write and evaluate numbers in expanded form with index notation.

## 3.4.1 Introduction to indices

eles-3728

- An **index** (or **exponent** or **power**) is a short way of writing a repeated multiplication.
- The **base** is the number that is being repeatedly multiplied and the index (plural *indices*) is the number of times it is multiplied.

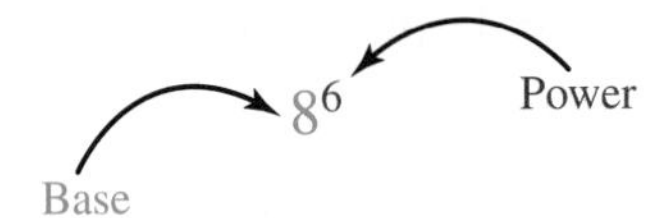


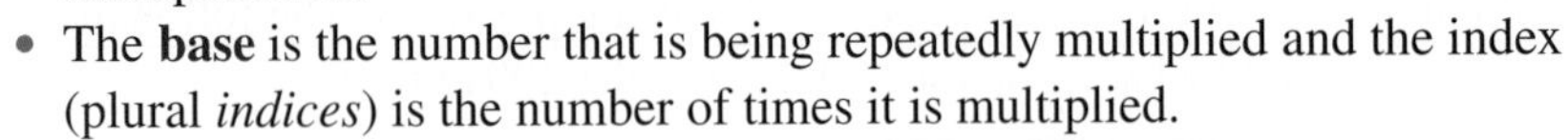


- Numbers in index form are read using the value of both the base and the power.

$8^6$ is read as '8 to the power of 6'.
$12^4$ is read as '12 to the power of 4'.

### WORKED EXAMPLE 7 Identifying the base and index

**For the following expressions, state:**
**i. the number or pronumeral that is the base**
**ii. the number or pronumeral that is the index.**

**a.** $6^7$ **b.** $x^9$ **c.** $3^a$ **d.** $y^b$

| THINK | WRITE |
|---|---|
| The base is the number or pronumeral that is repeatedly multiplied, and the index is the number of times that it is multiplied. | **a.** $6^7$<br>**i.** Base is 6<br>**ii.** Index is 7 |
| | **b.** $x^9$<br>**i.** Base is $x$<br>**ii.** Index is 9 |
| | **c.** $3^a$<br>**i.** Base is 3<br>**ii.** Index is $a$ |
| | **d.** $y^b$<br>**i.** Base is $y$<br>**ii.** Index is $b$ |

## WORKED EXAMPLE 8 Writing an expression using index notation

**Write the following expressions using index notation.**

**a.** $5 \times 5 \times 5 \times 5 \times 5 \times 5 \times 5$ **b.** $3 \times 3 \times 3 \times 3 \times 7 \times 7$

| THINK | WRITE |
|---|---|
| a. 1. Write the multiplication. | a. $5 \times 5 \times 5 \times 5 \times 5 \times 5 \times 5$ |
| 2. Write the number being repeatedly multiplied as the base, and the number of times it is written as the index (the number 5 is written 7 times). | $= 5^7$ |
| b. 1. Write the multiplication. | b. $3 \times 3 \times 3 \times 3 \times 7 \times 7$ |
| 2. Write the number being multiplied as the base, and the number of times it is written as the index (the number 3 is written 4 times, and 7 is written 2 times). | $= 3^4 \times 7^2$ |

## WORKED EXAMPLE 9 Simplifying expressions using index notation

**Simplify each of the following expressions by first writing each expression as a repeated multiplication and then in index notation.**

**a.** $3^4 \times 3^6$ **b.** $(4^3)^3$ **c.** $(3 \times 5)^2$

| THINK | WRITE |
|---|---|
| a. 1. Write the expression. | a. $3^4 \times 3^6$ |
| 2. Write the expression using repeated multiplication (that is, in expanded form). | $= (3 \times 3 \times 3 \times 3) \times (3 \times 3 \times 3 \times 3 \times 3 \times 3)$<br>$= 3 \times 3 \times 3 \times 3 \times 3 \times 3 \times 3 \times 3 \times 3 \times 3$ |
| 3. Write the repeated multiplication using index notation. The number being repeatedly multiplied is 3 (base), and the number of times it is written is 10 (index). | $= 3^{10}$ |
| b. 1. Write the expression. | b. $(4^3)^3$ |
| 2. Write the expression using repeated multiplication (that is, in expanded form). | $= 4^3 \times 4^3 \times 4^3$<br>$= (4 \times 4 \times 4) \times (4 \times 4 \times 4) \times (4 \times 4 \times 4)$ |
| 3. Write the repeated multiplication using index notation. The number being repeatedly multiplied is 4 (base) and the number of times it is written is 9 (index). | $= 4 \times 4 \times 4 \times 4 \times 4 \times 4 \times 4 \times 4 \times 4$<br>$= 4^9$ |
| c. 1. Write the expression. | c. $(3 \times 5)^2$ |
| 2. Write the expression using repeated multiplication (that is, in expanded form). | $= (3 \times 5) \times (3 \times 5)$<br>$= 3 \times 5 \times 3 \times 5$ |
| 3. Write the repeated multiplication using index notation. The numbers being repeatedly multiplied are 3 and 5 (base) and the number of times they are written is 2 (index) in each case. | $= 3 \times 3 \times 5 \times 5$<br>$= 3^2 \times 5^2$ |

## 3.4.2 Indices and place values

eles-3729

- By using place value, you can write numbers in expanded form with index notation.
  For example:

$$\begin{aligned}2700 &= 2000 + 700\\ &= 2\times10\times10\times10+7\times10\times10\\ &= 2\times10^3+7\times10^2\end{aligned}$$

**WORKED EXAMPLE 10 Expressing numbers in expanded form using index notation**

**Write the following numbers in expanded form using index notation.**

**a. 59 176** **b. 108 009**

| THINK | WRITE |
|---|---|
| a. 1. Write the number as the sum of each place value. | a. $59\,176 = 50\,000 + 9000 + 100 + 70 + 6$ |
| 2. Write each place value in multiples of 10. | $59\,176 = 5\times10\times10\times10\times10 + 9\times10\times10\times10 + 1\times10\times10 + 7\times10 + 6$ |
| 3. Write each place value in index notation. | $59\,176 = 5\times10^4 + 9\times10^3 + 1\times10^2 + 7\times10^1 + 6$ |
| b. 1. Write the number as the sum of each place value. | b. $108\,009 = 100\,000 + 8000 + 9$ |
| 2. Write each place value in multiples of 10. | $108\,009 = 1\times10\times10\times10\times10\times10 + 8\times10\times10\times10 + 9$ |
| 3. Write each place value in index notation. | $108\,009 = 1\times10^5 + 8\times10^3 + 9$ |

**WORKED EXAMPLE 11 Evaluating indices**

**Evaluate each of the following.**

**a.** $2^5$ **b.** $3^2\times2^3$ **c.** $3^2+5^2-2^4$

| THINK | WRITE |
|---|---|
| a. 1. Write in expanded form. | $2^5 = 2\times2\times2\times2\times2$ |
| 2. Multiply the terms. | $= 32$ |
| 3. Write the answer. | $2^5 = 32$ |
| b. 1. Write both terms in expanded form. | $3^2\times2^3 = (3\times3)\times(2\times2\times2)$ |
| 2. Calculate the product of the numbers in the brackets. | $= 9\times8$ |
| 3. Multiply the terms. | $= 72$ |
| 4. Write the answer. | $3^2\times2^3 = 72$ |
| c. 1. Write all terms in expanded form. | $3^2+5^2-2^4 = (3\times3)+(5\times5)-(2\times2\times2\times2)$ |
| 2. Calculate the product of the numbers in the brackets. | $= 9+25-16$ |

3. Remember the order of operations. Since the operations are addition and subtraction, work left to right. $= 34 - 16$
$= 18$

4. Write the answer. $3^2 + 5^2 - 2^4 = 18$

### Digital technology

Scientific calculators can evaluate expressions in index form where the base is a number.

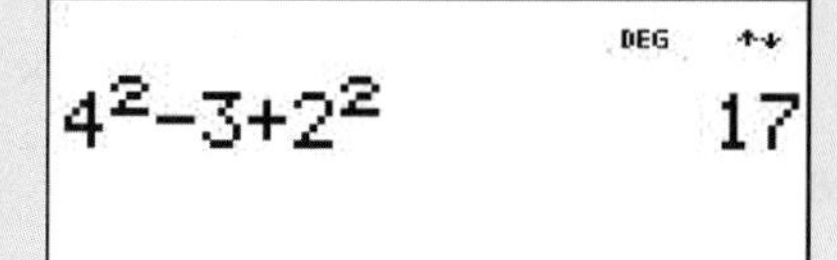

 Resources

**eWorkbook** Topic 3 Workbook (worksheets, code puzzle and project) (ewbk-1904)

**Interactivities** Individual pathway interactivity: Index notation (int-4319)
Index notation (int-3957)

## Exercise 3.4 Index notation

**learnon**

| 3.4 Quick quiz on | 3.4 Exercise |
|---|---|

### Individual pathways

| ■ PRACTISE | ■ CONSOLIDATE | ■ MASTER |
|---|---|---|
| 1, 2, 5, 7, 10, 12, 13, 16 | 3, 6, 8, 14, 17, 18 | 4, 9, 11, 15, 19, 20, 21 |

### Fluency

1. **WE7** In each of the following expressions, enter the missing number into the box and state what the base and index are.
   a. $7^{\square} = 49$  b. $4^{\square} = 4096$  c. $\square^4 = 81$  d. $\square^3 = 1000$
2. **WE8a** Write each of the following expressions in index notation.
   a. $7 \times 7 \times 7 \times 7$  b. $8 \times 8 \times 8 \times 8 \times 8 \times 8$
   c. $3 \times 3 \times 3 \times 3 \times 3 \times 3 \times 3 \times 3 \times 3$  d. $13 \times 13 \times 13$
3. **WE8b** Write the following expressions in index notation.
   a. $4 \times 4 \times 4 \times 4 \times 4 \times 6 \times 6 \times 6$  b. $2 \times 2 \times 3$
   c. $5 \times 5 \times 2 \times 2 \times 2 \times 2$  d. $3 \times 3 \times 2 \times 2 \times 5 \times 5 \times 5$
4. Write the following using repeated multiplication (that is, in expanded form).
   a. $6^5$  b. $11^3$

5. **WE9** Simplify each of the following expressions by first writing each expression as a repeated multiplication and then using index notation.

a. $5^8 \times 5^3$ b. $4^8 \times 4^5$ c. $(2^7)^3$

6. **WE10** Write the following numbers in expanded form using index notation.

a. 300 b. 4500 c. 6705 d. 10 000

## Understanding

7. **WE9** Simplify each of the following expressions by first writing each expression as a repeated multiplication and then using index notation.

a. $(9^5)^3$ b. $(3 \times 13)^6$

8. **MC** The value of $4^4$ is:

A. 8 B. 16 C. 64 D. 256 E. 484

9. **WE11** Evaluate each of the following.

a. $7^3$ b. $6^2 \times 2^3$ c. $5^3 - 4^3 + 2^4$

10. Evaluate each of the following.

a. $2^3 \times 3^2$ b. $3^4 \times 4^3$ c. $3^5 \times 9^3$

11. Evaluate each of following.

a. $6^4 - 9^3$ b. $5^3 + 2^5 \times 9^2$ c. $2^7 - 4^5 \div 2^6$

## Reasoning

12. a. **MC** Which of the following expressions has the greatest value?

A. $2^8$ B. $8^2$ C. $3^4$ D. $4^3$ E. $9^2$

b. Justify your answer using mathematical reasoning.

13. We know that $12^2 = 144$ and $21^2 = 441$. It is also true that $13^2 = 169$ and $31^2 = 961$. If $14^2 = 196$, is $41^2 = 691$? Try to justify your answer without calculating $41^2$.

14. Write < (less than) or > (greater than) in the boxes below to make each statement true.

a. $1^3 \square 3^1$ b. $2^3 \square 3^2$ c. $5^3 \square 3^5$ d. $4^4 \square 3^5$

15. You have a choice of how your weekly allowance is increased for 15 weeks.
Option 1: Start at 1 cent and double your allowance every week.
Option 2: Start at $1 and increase your allowance by $1 every week.

a. Explain which option you would choose and why.
b. Explain whether powers can be used to help with these calculations.

## Problem solving

16. Jessica has a 'clean-up' button on her mobile phone that will clear half of the read messages in her inbox. She pressed the button six times, causing just one message to be left in her inbox. Calculate how many messages were in her inbox before the clean-up.

17. A student sent an email to eight of his cousins, who then each forwarded the email to eight different friends, who in turn forwarded the email to another eight friends each.

a. Calculate how many emails in total were sent during this process.
b. Assuming that everyone received the email only once, determine how many people in total received the email by the end of this process.

18. You received a text message from your friend. After 5 minutes, you forward the text message to two of your other friends. After 5 more minutes, those two friends forward it to two more friends. If the text message is sent every 5 minutes in this way, calculate how many people have received it in 30 minutes.

19. Bacteria such as those shown in the image grow at an alarming rate. Each bacterium can split itself in half, forming two new cells each hour.
   a. If you start with one bacterium cell, calculate how many you will have after 12 hours. Show your working to include the number of bacteria after each hour.
   b. If a bacterium could successfully divide into three new cells each hour, calculate how many bacteria you would have after 12 hours.
   c. Show how powers can be applied to the calculations in parts **a** and **b**.

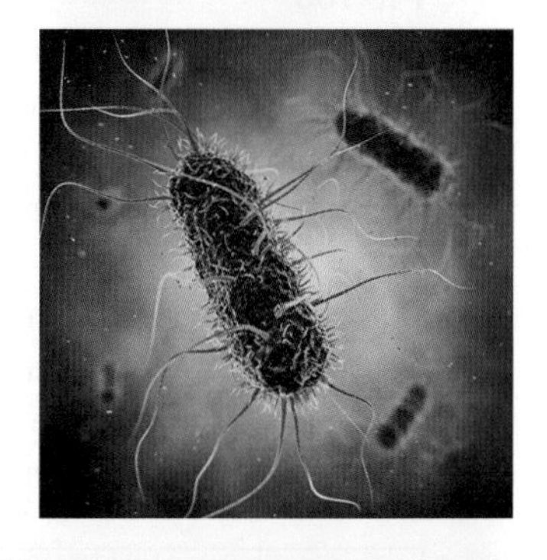

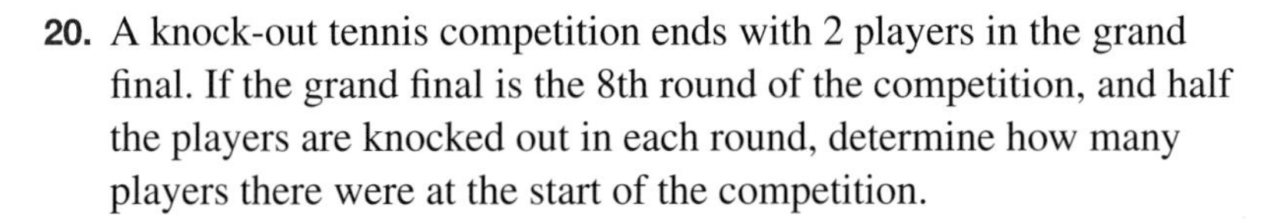

20. A knock-out tennis competition ends with 2 players in the grand final. If the grand final is the 8th round of the competition, and half the players are knocked out in each round, determine how many players there were at the start of the competition.

21. A particular rule applies to each of these three sets of numbers.

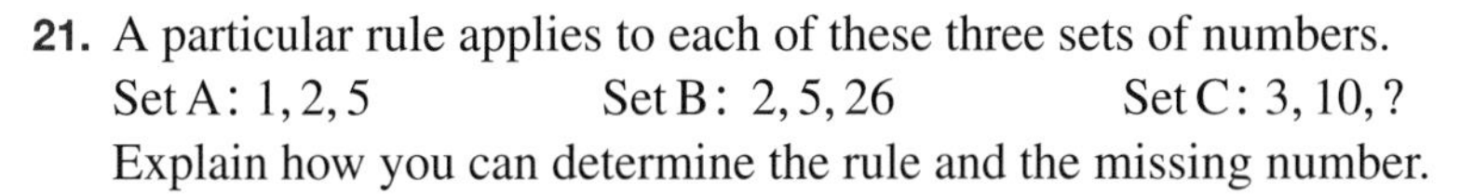

Set A: 1, 2, 5    Set B: 2, 5, 26    Set C: 3, 10, ?

Explain how you can determine the rule and the missing number.

# LESSON
# 3.5 Prime and composite numbers

**LEARNING INTENTION**

At the end of this lesson you should be able to:
- understand the difference between prime and composite numbers
- determine prime factors by drawing a factor tree
- write a composite number in index form
- determine the highest common factor (HCF) and lowest common multiple (LCM) of large numbers as products of their prime factors.

## 3.5.1 Prime numbers

eles-3730

- A **prime number** is a counting number that has exactly 2 unique factors: itself and 1.
  (Counting numbers are 1, 2, 3, 4, …)
  The number 3 is a prime number; its only factors are 3 and 1.
  The number 2 is the only even prime number; its only factors are 2 and 1.
- A **composite number** is one that has more than two factors.
  The number 4 is a composite number; its factors are 1, 2 and 4.
- The number 1 is a special number. It is neither a prime number nor a composite number because it has only one factor, 1.

### COLLABORATIVE TASK: The sieve of Eratosthenes

Eratosthenes (pronounced 'e-rah-toss-thee-knees') was a Greek mathematician who lived around 200 BC. He worked out a way to sort prime numbers from composite numbers.

1. As a class, draw a $10 \times 10$ grid on the board and number the boxes from 1 to 100.
2. Shade the number 1. It is not a prime or composite number.
3. Circle the number 2. Cross out all other multiples of 2.
4. Circle the next smallest uncrossed number, 3. Cross out all other multiples of 3.
5. Continue this process until all the numbers in the grid are either circled or crossed out.
6. Describe the circled numbers.
7. Describe the crossed out numbers.
8. a. Are there more prime numbers or composite numbers?
   b. What connections can you make between composite numbers and multiples?

16 4 48 24 49 8 2 11 5 17 31

| 1 | 2 | 3 | 4 | 5 | 6 | 7 | 8 | 9 | 10 |
|---|---|---|---|---|---|---|---|---|---|
| 11 | 12 | 13 | 14 | 15 | 16 | 17 | 18 | 19 | 20 |
| 21 | 22 | 23 | 24 | 25 | 26 | 27 | 28 | 29 | 30 |
| 31 | 32 | 33 | 34 | 35 | 36 | 37 | 38 | 39 | 40 |
| 41 | 42 | 43 | 44 | 45 | 46 | 47 | 48 | 49 | 50 |
| 51 | 52 | 53 | 54 | 55 | 56 | 57 | 58 | 59 | 60 |
| 61 | 62 | 63 | 64 | 65 | 66 | 67 | 68 | 69 | 70 |
| 71 | 72 | 73 | 74 | 75 | 76 | 77 | 78 | 79 | 80 |
| 81 | 82 | 83 | 84 | 85 | 86 | 87 | 88 | 89 | 90 |
| 91 | 92 | 93 | 94 | 95 | 96 | 97 | 98 | 99 | 100 |

### WORKED EXAMPLE 12 Listing prime numbers

**List the prime numbers between 50 and 70.**

**THINK**

1. The only even prime number is 2. The prime numbers between 50 and 70 will be odd. Numbers ending in 5 are divisible by 5 so 55 and 65 are not primes.
2. Check the remaining odd numbers between 50 and 70:
   $51 = 3 \times 17$ and $1 \times 51$. It has four factors so it is composite.
   $53 = 1 \times 53$. It has two unique factors only so it is prime.
   $57 = 3 \times 19$ and $1 \times 57$. It has four factors so it is composite.
   $59 = 1 \times 59$. It has two unique factors only so it is prime.
   $61 = 1 \times 61$. It has two unique factors only so it is prime.
   $63 = 7 \times 9$, $3 \times 21$ and $1 \times 63$. It has six factors so it is composite.
   $67 = 1 \times 67$. It has two unique factors only so it is prime.
   $69 = 3 \times 23$ and $1 \times 69$. It has four factors so it is composite.

**WRITE**

The prime numbers are: 53, 59, 61, 67.

### WORKED EXAMPLE 13 Checking for composite and prime numbers

**State whether the following numbers are prime or composite.**

**a. 45** **b. 37** **c. 86**

| THINK | WRITE |
| --- | --- |
| a. Factors of 45 are 1, 3, 5, 9, 15 and 45. | a. 45 is composite. |
| b. The only factors of 37 are 1 and 37. | b. 37 is prime. |
| c. Factors of 86 are 1, 2, 43 and 86. All even numbers except 2 are composite. | c. 86 is composite. |

### DISCUSSION

Explain your method of working out whether a number is a prime number.

## 3.5.2 Composite numbers and factor trees

eles-3731

- Every composite number can be written as the product of its prime factors.
- Writing any number as a product of its prime factors simply means to rewrite a number using **multiplication** of **prime numbers** only. For example:

$$12 = 2 \times 2 \times 3 = 2^2 \times 3$$

Prime numbers that multiply to give 12

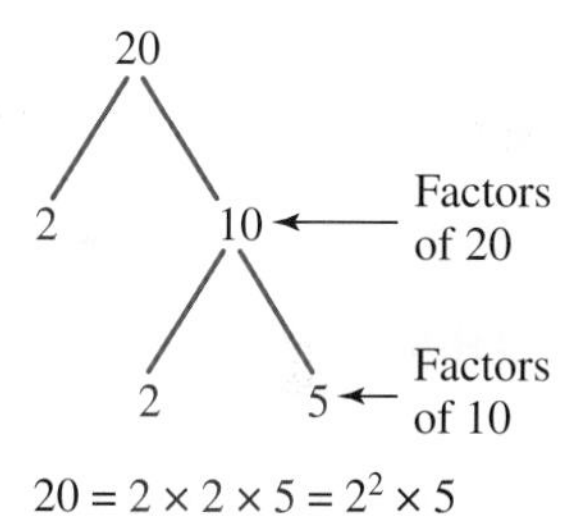


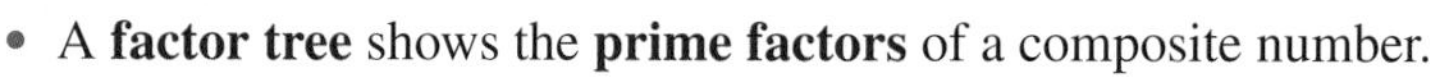

- A **factor tree** shows the **prime factors** of a composite number.
- Each branch shows a factor of the number above it.
- Branches stop at factors that are prime numbers, as shown at right in pink.
- In the factor tree shown, 20 can be written as $2 \times 2 \times 5$; this is known as writing a number as a product of its prime factors.

### WORKED EXAMPLE 14 Determining the prime factors by drawing a factor tree

**a. Determine the prime factors of 50 by drawing a factor tree.**
**b. Write 50 as a product of its prime factors in index form.**

**THINK**

a. 1. Write a factor pair of the given number and draw the factor tree ($50 = 5 \times 10$).

2. If a branch is prime, no other factors can be found (5 is prime). If a branch is composite, write factors of that number: 10 is composite, so $10 = 5 \times 2$.

**WRITE**

a. 50
5 10

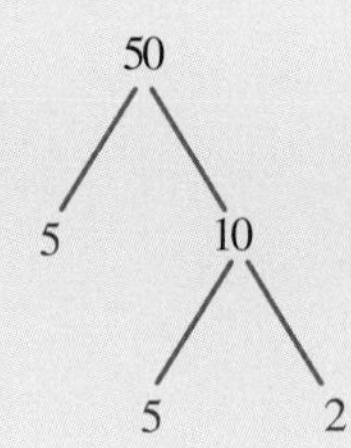

3. Continue until all branches end in a prime number, then stop.

4. Write the prime factors. The prime factors of 50 are 2 and 5.

b. Write 50 as a product of prime factors in index form found in part a.

b. $50 = 5 \times 5 \times 2$

$= 5^2 \times 2$

## WORKED EXAMPLE 15 Writing composite numbers in index form

**Write 72 as a product of its prime factors in index form.**

**THINK**

1. Draw a factor tree. When all factors are prime numbers you have found the prime factors.

**WRITE**

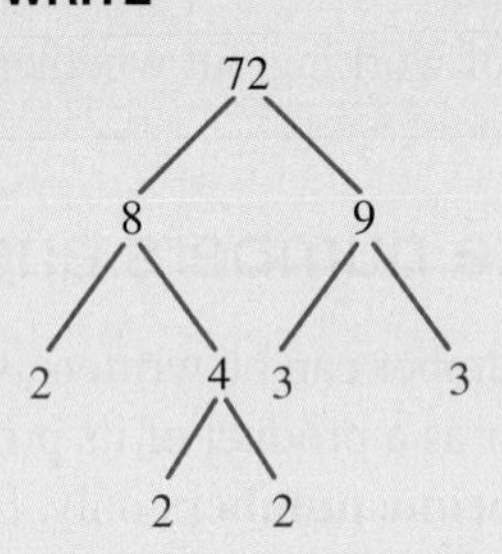


2. Write 72 as a product of its prime factors in index form.

$72 = 2 \times 2 \times 2 \times 3 \times 3$

$= 2^3 \times 3^2$

## WORKED EXAMPLE 16 Writing composite numbers in index form

**Write 360 as a product of prime factors using index notation.**

**THINK**

1. Draw a factor tree. If the number on the branch is a prime number, stop. If not, continue until a prime number is reached.

**WRITE**

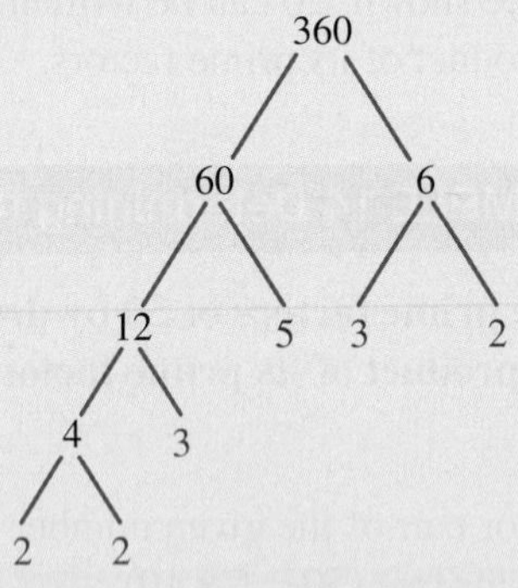


2. Write the number as a product of prime factors.

$360 = 2 \times 2 \times 2 \times 3 \times 3 \times 5$

3. Write your answer using index notation.

$360 = 2^3 \times 3^2 \times 5$

There can often be more than one way of drawing a factor tree for a number, as shown below.

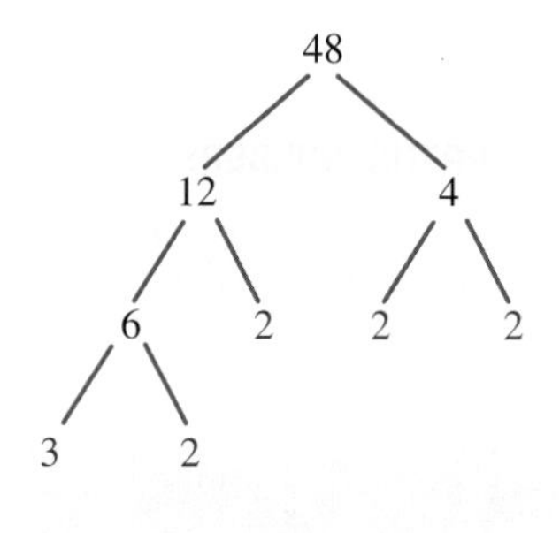

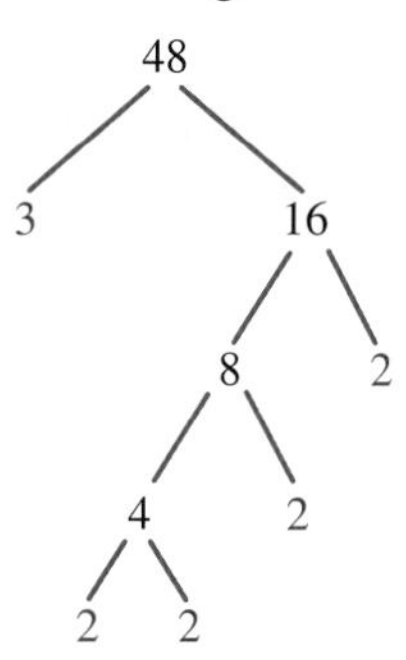

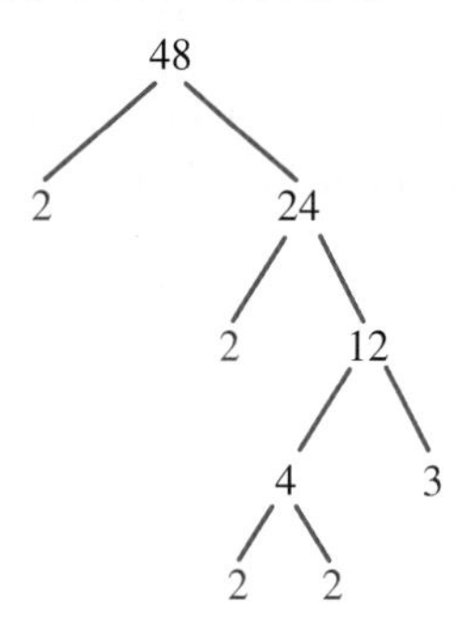

$48 = 12 \times 4$
$= 2 \times 3 \times 2 \times 2 \times 2$
$= 2^4 \times 3$

$48 = 3 \times 16$
$= 3 \times 2 \times 2 \times 2 \times 2$
$= 2^4 \times 3$

$48 = 2 \times 24$
$= 2 \times 3 \times 2 \times 2 \times 2$
$= 2^4 \times 3$

## 3.5.3 Lowest common multiple and highest common factor using prime factors

eles-3732

- Expressing large numbers as products of their prime factors makes it easier to determine their lowest common multiple (LCM) or highest common factor (HCF).

### Lowest common multiple (LCM)

**The LCM of two numbers expressed as the products of their prime factors is the product of the prime factors that are factors of either number.**

### Highest common factor (HCF)

**The HCF of two numbers expressed as the products of their prime factors is the product of the prime factors common to both numbers.**

### WORKED EXAMPLE 17 Determining HCF and LCM using prime factors

**Determine the highest common factor and lowest common multiple of 270 and 900.**

| THINK | WRITE |
|---|---|
| 1. Write 270 and 900 as products of their prime factors. Alternatively, a factor tree may be used to determine the prime factors. | $270 = 2 \times 3 \times 3 \times 3 \times 5$<br>$900 = 2 \times 2 \times 3 \times 3 \times 5 \times 5$ |
| 2. Circle the prime factors common to both numbers. | $270 = \quad (2) \times (3) \times (3) \times 3 \times (5)$<br>$900 = 2 \times (2) \times (3) \times (3) \times 5 \times (5)$ |
| 3. For the HCF, multiply the prime factors common to both numbers. | $2 \times 3 \times 3 \times 5 = 90$.<br>The HCF of 270 and 900 is 90. |
| 4. For the LCM, multiply the prime factors that are factors of either number.<br>*Hint:* Only include one circled pair. | $2 \times 2 \times 3 \times 3 \times 3 \times 5 \times 5 = 2700$<br>The LCM of 270 and 900 is 2700. |

### COLLABORATIVE TASK: Goldbach's conjecture

In 1742, mathematician Christian Goldbach suggested that every even number greater than 2 could be written as the sum of two prime numbers.

In pairs, test this for the even numbers greater than 2 and less than 40.

In pairs, choose 5 even numbers between 100 and 300. Test these numbers to see if Goldbach's conjecture is true.

Do you agree with Christian Goldbach?

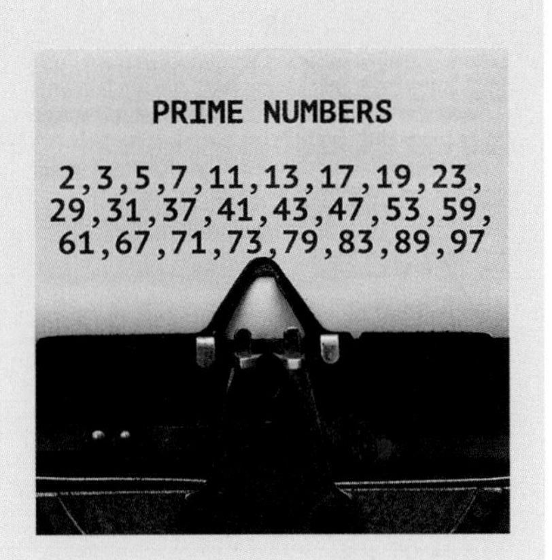


### Resources

**eWorkbook** Topic 3 Workbook (worksheets, code puzzle and project) (ewbk-1904)

**Interactivities** Individual pathway interactivity: Prime numbers and composite numbers (int-4320)
Prime numbers (int-3933)
Composite numbers and factor trees (int-3934)

## Exercise 3.5 Prime and composite numbers

learnon

**3.5 Quick quiz**  **3.5 Exercise**

### Individual pathways

| ■ PRACTISE | ■ CONSOLIDATE | ■ MASTER |
|---|---|---|
| 1, 2, 4, 8, 10, 13, 15, 16, 21, 24, 25, 28 | 5, 7, 9, 12, 17, 19, 20, 22, 26, 29 | 3, 6, 11, 14, 18, 23, 27, 30 |

### Fluency

1. **WE12** List four prime numbers that are between 20 and 40.
2. **WE13** State whether each of the following numbers is prime or composite.
   a. 9 b. 13 c. 55 d. 41
3. State whether each of the following numbers is prime or composite.
   a. 64 b. 79 c. 98 d. 101
4. i. **WE14** Determine the prime factors of each of the following numbers by drawing a factor tree.
   ii. Write each one as a product of its prime factors in index form.
   a. 15 b. 30 c. 100 d. 49
5. i. Determine the prime factors of each of the following numbers by drawing a factor tree.
   ii. Write each one as a product of its prime factors in index form.
   a. 72 b. 56 c. 45

6. i. Determine the prime factors of each of the following numbers by drawing a factor tree.
   ii. Write each one as a product of its prime factors in index form.
   a. 84 b. 112 c. 40

7. i. WE15 Determine the prime factors of the following numbers by drawing a factor tree.
   ii. Express the number as a product of its prime factors in index form.
   a. 96 b. 32 c. 3000

8. Determine the prime factors of each of the following numbers.
   a. 48 b. 200 c. 81 d. 18

9. Determine the prime factors of each of the following numbers.
   a. 300 b. 60 c. 120 d. 80

### Understanding

10. WE16 Write the following as a product of prime factors using index notation.
    a. 60 b. 75 c. 220

11. Write the following as a product of prime factors using index notation.
    a. 192 b. 124

12. WE17 By expressing the following pairs of numbers as products of their prime factors, determine their lowest common multiple and their highest common factor.
    a. 36 and 84 b. 48 and 60 c. 120 and 400 d. 220 and 800

13. State whether each of the following is true or false.
    a. All odd numbers are prime numbers.
    b. No even numbers are prime numbers.
    c. 1, 2, 3 and 5 are the first four prime numbers.
    d. A prime number has two factors only.
    e. 2 is the only even prime number.

14. State whether each of the following is true or false.
    a. The sum of two prime numbers is always even.
    b. The product of two prime numbers is always odd.
    c. There are no consecutive prime numbers.

15. MC The number of primes less than 10 is:
    A. 4 B. 3 C. 5 D. 2 E. 1

16. MC The first three prime numbers are:
    A. 1, 3, 5 B. 2, 3, 4 C. 2, 3, 5 D. 3, 5, 7 E. 2, 5, 7

17. MC The number 15 can be written as the sum of two prime numbers. These are:
    A. $3 + 12$ B. $1 + 14$ C. $13 + 2$ D. $7 + 8$ E. $9 + 6$

18. MC Factors of 12 that are prime numbers are:
    A. 1, 2, 3, 4 B. 2, 3, 6 C. 2, 3 D. 2, 4, 6, 12 E. 1, 2, 3, 4, 6, 12

19. The following numbers are not primes. Each of them is the product of two primes. Determine the two primes in each case.
    a. 365 b. 187

20. **MC** A factor tree for 21 is:

**A.** 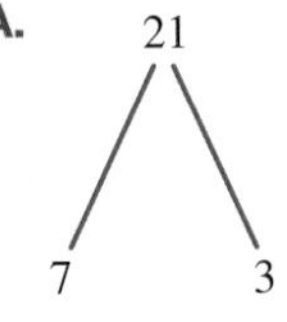


**B.** 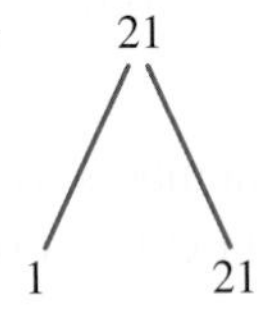


**C.** 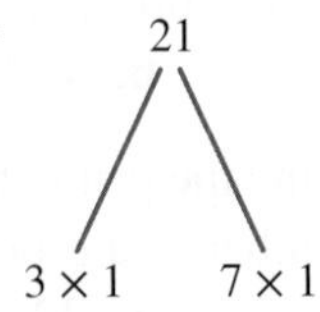


**D.** 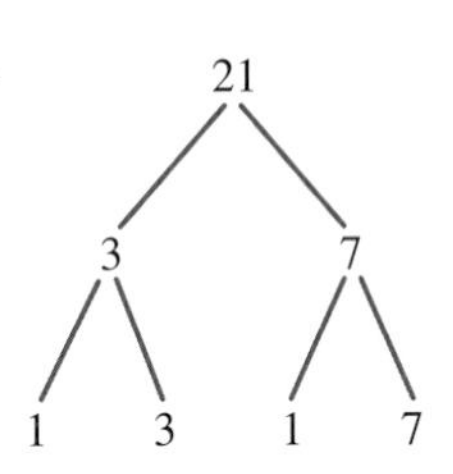


**E.** 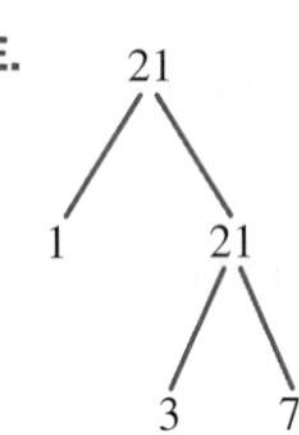


21. **MC** A factor tree for 36 is:

**A.** 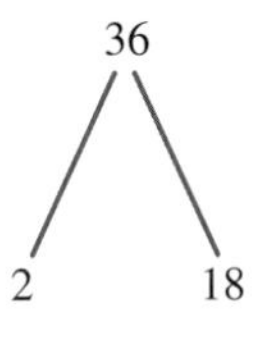


**B.** 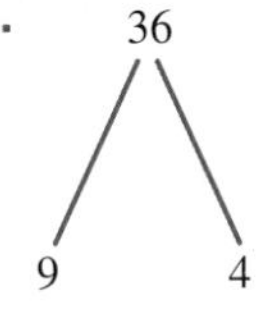


**C.** 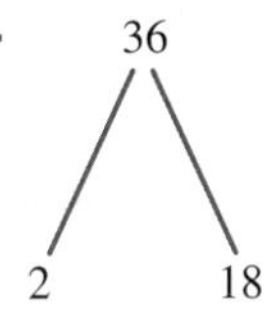


**D.** 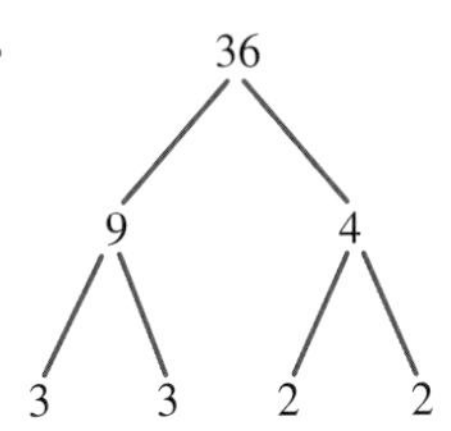


**E.** 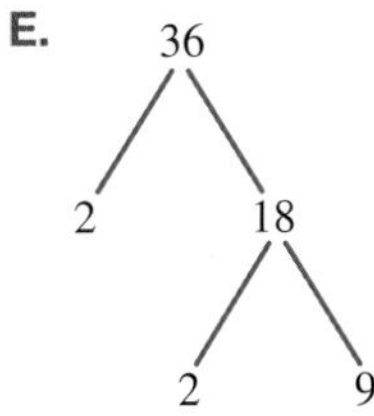


22. **MC** The prime factors of 16 are:

**A.** 1, 2 **B.** 1, 2, 4 **C.** 2 **D.** 1, 2, 4, 8, 16 **E.** 2, 4, 8

23. **MC** The prime factors of 28 are:

**A.** 1, 28 **B.** 2, 7 **C.** 1, 2, 14 **D.** 1, 2, 7 **E.** 2, 7, 14

## Reasoning

24. Determine the largest three-digit prime number in which each digit is a prime number. Prove that this number is a prime number.

25. Is the sum of two prime numbers always a prime number? Explain your answer.

26. Determine a prime number greater than 10 with a sum of digits that equals 11. Show your working.

27. My age is a prime number. I am older than 50. The sum of the digits in my age is also a prime number. If you add a multiple of 13 to my age the result is 100. Determine my age.

28. Determine two prime numbers with a product of:

a. 21 b. 26 c. 323.

29. Determine how many even integers between 2 and 20 can be the sum of two different prime numbers.

30. Twin primes are pairs of prime numbers that differ by 2. Except for the pair of primes 2 and 3, this is the smallest difference between two prime numbers. The first twin primes are 3 and 5, followed by 5 and 7, then 11 and 13. Identify other twin primes below 100.

# LESSON
# 3.6 Squares and square roots

## LEARNING INTENTION

At the end of this lesson you should be able to:
- determine values of squares and square roots
- estimate the value of the square root of other numbers by using the perfect squares that lie on either side of the number
- evaluate squares or square roots using BIDMAS.

## 3.6.1 Square numbers

eles-3733

- The process of multiplying a number by itself is known as **squaring a number**.
- **Square numbers** or **perfect squares** are numbers that can be arranged in a square, as shown.

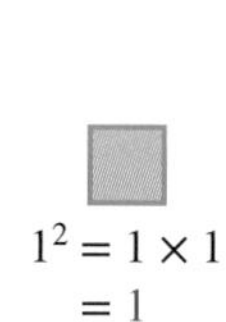

$1^2 = 1 \times 1$
$= 1$

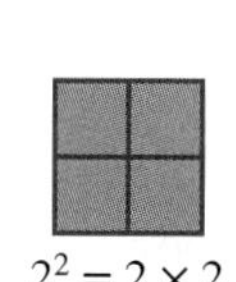

$2^2 = 2 \times 2$
$= 4$

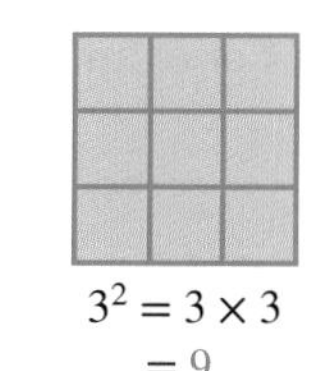

$3^2 = 3 \times 3$
$= 9$

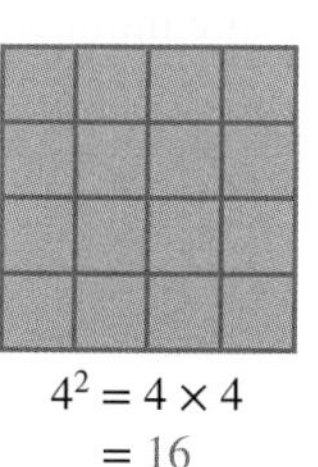

$4^2 = 4 \times 4$
$= 16$

- A square number is the number we get after multiplying a whole number by itself.
- The first four square numbers are: 1, 4, 9 and 16.
- All square numbers can be written in index notation, for example, $4^2 = 16$.

### WORKED EXAMPLE 18 Determining square numbers

**Determine the square numbers between 90 and 150.**

| THINK | WRITE |
|---|---|
| 1. Use your knowledge of multiplication tables to determine the first square number after 90. | $10^2 = 10 \times 10 = 100$ |
| 2. Determine the square numbers that come after that one, but before 150. | $11^2 = 11 \times 11 = 121$<br>$12^2 = 12 \times 12 = 144$<br>$13^2 = 13 \times 13 = 169$ (too big) |
| 3. Write the answer in a sentence. | The square numbers between 90 and 150 are 100, 121 and 144. |

**WORKED EXAMPLE 19 Determining approximate values of square numbers**

**Write the two whole square numbers between which $5.7^2$ will lie.**

| THINK | WRITE |
|---|---|
| 1. Write the whole numbers either side of 5.7. | 5.7 is between 5 and 6. |
| 2. Consider the square of each whole number. | $5.7^2$ is between $5^2$ and $6^2$. |
| 3. Simplify $5^2$ and $6^2$, then write the answer in a sentence. | So $5.7^2$ is between 25 and 36. |
| 4. Verify your answer with a calculator. | $5.7^2 = 32.49$, which lies between 25 and 36. |
| 5. To determine the value of $5.7^2$ press (5) (.) (7) [$x^2$] (enter). | 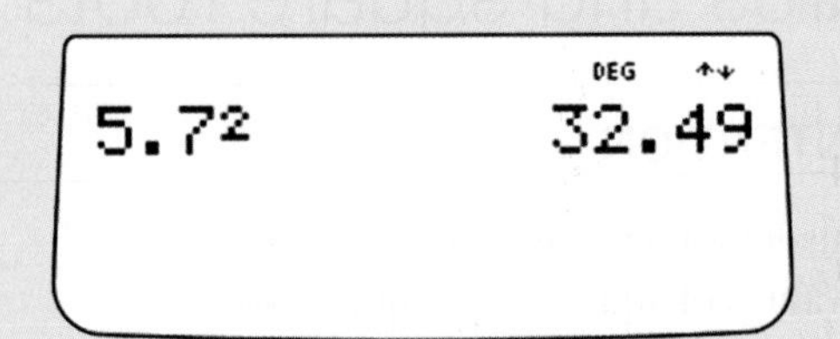  |
| | We have confirmed that $5.7^2$ lies between 25 and 36. |

**DISCUSSION**

Can you predict whether the perfect square of a number will be odd or even? Give some examples.

- When a composite number is squared, the result is equal to the product of the squares of its factors.

$$\text{For example: } 10^2 = (2 \times 5)^2 = 2^2 \times 5^2 = 4 \times 25 = 100$$

## 3.6.2 Square roots

eles-3734

- Evaluating the **square root** of a number is the opposite of squaring the number; for example, since $4^2 = 16$, then $\sqrt{16} = 4$.
- The symbol for square root is called the radical symbol. It is written as $\sqrt{\ }$.
- Using the illustrations of the squares shown earlier, a square of area 16 square units must have a side length of 4 units. This may help understand why $\sqrt{16} = 4$.

**Square roots**

**When determining the square root of a number, you are determining the number that, when multiplied by itself, equals the number underneath the radical symbol.**

**For example, to determine $\sqrt{9}$, identify the number that when multiplied by itself gives 9.**

**Since $3 \times 3 = 9$, we can conclude that $\sqrt{9} = 3$.**

- To determine the square roots of larger numbers, it helps to break the number up as a product of two smaller square roots with which we are more familiar. For example:

$$\begin{aligned}\sqrt{900} &= \sqrt{9}\times\sqrt{100}\\ &= 3\times 10\\ &= 30\end{aligned}$$

### WORKED EXAMPLE 20 Determining square roots

**Evaluate the following square roots.**

**a.** $\sqrt{49}$ **b.** $\sqrt{3600}$

| THINK | WRITE |
|---|---|
| **a.** Determine a number that when multiplied by itself gives 49. | **a.** $\sqrt{49} = 7$ $(7\times 7 = 49)$ |
| **b. 1.** Write 3600 as the product of two smaller numbers for which we can calculate the square root. | **b.** $\sqrt{3600} = \sqrt{36}\times\sqrt{100}$ |
| **2.** Take the square root of each of these numbers. | $= 6\times 10$ |
| **3.** Determine the product and write the answer. | $\sqrt{3600} = 60$ |

- Only perfect squares have square roots that are whole numbers.
- The value of the square root of other numbers can be estimated by using the perfect squares that lie on either side of the number.

### WORKED EXAMPLE 21 Estimating square roots

**Between which two numbers do the following numbers lie?**

**a.** $\sqrt{74}$ **b.** $\sqrt{342}$

| THINK | WRITE |
|---|---|
| **a. 1.** Write the square numbers on either side of 74. | **a.** 74 is between 64 and 81. |
| **2.** Consider the square root of each number. | $\sqrt{74}$ is between $\sqrt{64}$ and $\sqrt{81}$. |
| **3.** Simplify $\sqrt{64}$ and $\sqrt{81}$. | So $\sqrt{74}$ is between 8 and 9. |
| **4.** Verify your answer with a calculator. To determine the value of $\sqrt{74}$ press [2nd] [$x^2$] (7) (4) [ ) ] (enter). | $\sqrt{74}\approx 8.6023$<br>√(74) 8.602325267<br>We have confirmed that $\sqrt{74}$ lies between 8 and 9. |
| **b. 1.** Write the square numbers on either side of 342. | **b.** 342 is between 324 $(18^2)$ and 361 $(19^2)$. |
| **2.** Consider the square root of each number. | $\sqrt{342}$ is between $\sqrt{324}$ and $\sqrt{361}$. |

3. Simplify $\sqrt{324}$ and $\sqrt{361}$. So $\sqrt{342}$ is between 18 and 19.

4. Verify your answer with a calculator. $\sqrt{342} \approx 18.4932$

To determine the value of $\sqrt{342}$ press [2nd] [$x^2$] (3) (4) (2) [)] (enter).

√(342)
18.49324201

We have confirmed that $\sqrt{342}$ lies between 18 and 19.

## Order of operations

- The square or square root is an index represented by the letter I in BIDMAS, which determines the order in which operations are evaluated in a mathematical expression (see topic 2).

### WORKED EXAMPLE 22 Evaluating square roots using the order of operations

**Evaluate:**

a. $\sqrt{16+9}$

b. $\sqrt{16}+9$

| THINK | WRITE |
| --- | --- |
| a. 1. Simplify the expression inside the square root and write the answer. | a. $\sqrt{16+9} = \sqrt{25}$ |
| 2. Complete the calculation by evaluating the square root and write the answer. | $= 5$ |
| b. 1. Evaluate the square root. *Note:* There is no expression inside the square root to evaluate first. | b. $\sqrt{16}+9 = 4+9$ |
| 2. Perform the addition to complete the calculation and write the answer. | $= 13$ |

### COLLABORATIVE TASK: How many squares?

Work in pairs to answer the following question.

How many squares of any size can be found in the following diagram?

*Hint:* Count the number of single squares, count the number of $2 \times 2$ squares, then count the number of $3 \times 3$ squares. Look for a pattern in these numbers, and use the pattern to help you.

### Digital technology

Scientific calculators can evaluate squares of decimal numbers.

```
54.25²
        2943.0625
```

Scientific calculators can evaluate square roots of decimal numbers.

```
√248.64
      15.76832268
```

**on Resources**

**eWorkbook** Topic 3 Workbook (worksheets, code puzzle and project) (ewbk-1904)

**Interactivities** Individual pathway interactivity: Squares and square roots (int-4321)
Square numbers (int-3936)
Square roots (int-3937)

## Exercise 3.6 Squares and square roots

**learnon**

**3.6 Quick quiz** on | **3.6 Exercise**

Individual pathways

| ■ PRACTISE | ■ CONSOLIDATE | ■ MASTER |
|---|---|---|
| 1, 4, 6, 11, 13, 16, 19, 22 | 2, 5, 7, 9, 12, 17, 20, 23 | 3, 8, 10, 14, 15, 18, 21, 24, 25 |

### Fluency

1. Evaluate the following and verify your answers with a calculator.
   a. $8^2$  b. $11^2$  c. $15^2$  d. $25^2$
2. **WE18** State the square numbers between 50 and 100.
3. State the square numbers between 160 and 200.
4. **WE19** Write two whole square numbers between which each of the following will lie.
   a. $6.4^2$  b. $7.8^2$  c. $9.2^2$  d. $12.5^2$
5. **WE20** Evaluate:
   a. $\sqrt{25}$  b. $\sqrt{81}$  c. $\sqrt{144}$  d. $\sqrt{400}$
6. Evaluate the following and verify your answers with a calculator.
   a. $\sqrt{4900}$  b. $\sqrt{14\,400}$  c. $\sqrt{360\,000}$  d. $\sqrt{160\,000}$
7. **WE21** Write the two whole numbers between which each of the following will lie.
   a. $\sqrt{60}$  b. $\sqrt{14}$  c. $\sqrt{200}$  d. $\sqrt{2}$

8. **WE22** Evaluate:

a. $\sqrt{144+25}$ b. $\sqrt{144}+25$

9. State the even square numbers between 10 and 70.

10. State the odd square numbers between 50 and 120.

## Understanding

11. **MC** For which of the following square roots can we calculate an exact answer?

A. $\sqrt{10}$ B. $\sqrt{25}$ C. $\sqrt{50}$ D. $\sqrt{75}$ E. $\sqrt{82}$

12. **MC** For which of the following square roots can we not calculate the exact value?

A. $\sqrt{160}$ B. $\sqrt{400}$ C. $\sqrt{900}$ D. $\sqrt{2500}$ E. $\sqrt{3600}$

13. Evaluate the following. Verify your answers with a calculator.

a. $2^2+\sqrt{25}$ b. $9^2-\sqrt{36}$ c. $5^2\times2^2\times\sqrt{49}$ d. $3^2+2^2\times\sqrt{16}$

14. Evaluate the following. Verify your answers with a calculator.

a. $3^2-2^2\div\sqrt{4}+\sqrt{49}$ b. $\sqrt{9}\times4^2-\sqrt{144}\div2^2$

15. a. Determine between which two whole numbers $\sqrt{150}$ will lie.
b. Use your answer to part **a** to estimate the value of $\sqrt{150}$ and check your answer with a calculator.

## Reasoning

16. a. Evaluate $15^2$. b. Evaluate $3^2\times5^2$.
c. Are your answers to parts a and b equal? Explain why or why not.

17. a. Evaluate $\sqrt{225}$. b. Evaluate $\sqrt{25}\times\sqrt{9}$.
c. Are your answers to parts a and b equal? Explain why or why not.

18. a. Express 196 as a product of its prime factors.
b. Express 200 as a product of its prime factors.
c. Use the answers to parts **a** and **b** to explain how you can use prime factors to determine whether the square root of a number is an integer.

19. a. Evaluate $\sqrt{36}+\sqrt{64}$. b. Evaluate $\sqrt{36+64}$.
c. Are your answers to parts a and b equal? Explain why or why not.

20. a. Evaluate $\sqrt{25}+\sqrt{144}$. b. Evaluate $\sqrt{169}$.
c. $169=25+144$. Does $\sqrt{169}=\sqrt{25}+\sqrt{144}$? Explain your answer.

21. Megan has 3 different game scores that happen to be square numbers. The first 2 scores have the same three digits. The total of the 3 scores is 590.
Determine the 3 scores. Explain how you solved this question.

## Problem solving

22. a. Evaluate each of the following using a calculator.
i. $25^2-24^2$ ii. $24^2-23^2$ iii. $23^2-22^2$ iv. $22^2-21^2$ v. $21^2-20^2$
b. Comment on any pattern in your answers.
c. Discuss whether this pattern occurs with other numbers that are squared. Try some examples.

23. Guess the number that matches the following:
It is odd and it has an odd number of factors.
The sum of the digits is a two-digit prime.
The number is less than $\sqrt{10\,000}$ but greater than $\sqrt{100}$.

24. Rewrite the following in ascending order.
a. $8^2$, $\sqrt{56}$, 62, $\sqrt{72}$, 8, 7, $3^2$, $\sqrt{100}$
b. 100, $\sqrt{121}$, $\sqrt{10}$, $3\times 4$, $9^2$, $\sqrt{169}$, $2\times 7$

25. The sum of two numbers is 20 and the difference between their squares is 40. Determine the two numbers.

# LESSON
# 3.7 Divisibility tests

## LEARNING INTENTION

At the end of this lesson you should be able to:
- test for factors of a given number using divisibility tests.

## 3.7.1 Tests of divisibility

eles-3736

- Divisible means that one number divided by another gives a result that is a whole number.
- **Divisibility tests** are a quick way of testing for factors of a given number.
- These tests save time compared with dividing by each possible factor to determine whether the number is divisible by the chosen divisor.
- All positive integers are divisible by 1.

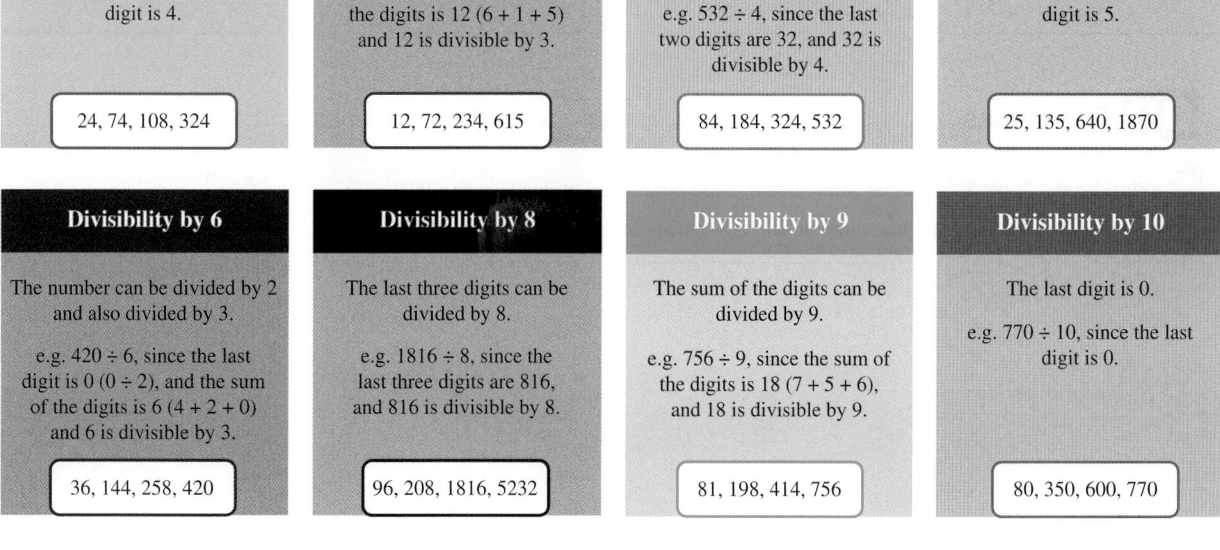

### WORKED EXAMPLE 23 Applying divisibility tests

**Determine whether the results of the following divisions are whole numbers.**

a. **$3454 \div 2$** b. **$6381 \div 9$**

| THINK | WRITE |
|---|---|
| a. 1. To be divisible by 2, the last digit must be 0, 2, 4, 6 or 8. | a. In the number 3454, the last digit is 4. |
| 2. Write the answer in a sentence. | Therefore, 3454 is divisible by 2. |
| b. 1. To be divisible by 9, the sum of all the digits must by divisible by 9. | b. The sum of all the digits: $6+3+8+1=18$ 18 is divisible by 9. |
| 2. Write the answer in a sentence. | Therefore, 6381 is divisible by 9. |

### WORKED EXAMPLE 24 Checking divisibility

**Check the divisibility of the following numbers.**

a. **$816 \div 4$** b. **$7634 \div 6$**

| THINK | WRITE |
|---|---|
| a. 1. To be divisible by 4, the last two digits together must be divisible by 4. | a. In the number 816, the last two digits are 16. 16 is divisible by 4. |
| 2. Write the answer in a sentence. | Therefore, 816 is divisible by 4. |
| b. 1. To be divisible by 6, the number must be divisible by 2 and also by 3. | |
| 2. To be divisible by 2, the last digit must be 0, 2, 4, 6 or 8. | b. In the number 7634, the last digit is 4. Therefore, 7634 is divisible by 2. |
| 3. To be divisible by 3, the sum of all the digits must be divisible by 3. | In the number 7634, the sum of all the digits: $7+6+3+4=20$ 20 is **not** divisible by 3. |
| 4. Write the answer in a sentence. | The number 7634 is divisible by 2 but not divisible by 3. Therefore, 7634 is not divisible by 6. |

**on** Resources

**eWorkbook** Topic 3 Workbook (worksheets, code puzzle and project) (ewbk-1904)

**Interactivity** Individual pathway interactivity: Divisibility tests (int-8342)

# Exercise 3.7 Divisibility tests

learnon

3.7 Quick quiz on

3.7 Exercise

## Individual pathways

| ■ PRACTISE | ■ CONSOLIDATE | ■ MASTER |
|---|---|---|
| 1, 4, 6, 8, 11, 14 | 2, 5, 9, 12, 15 | 3, 7, 10, 13, 16 |

### Fluency

1. State the divisibility tests for 2, 4 and 6 with an example for each.
2. State the divisibility tests for 3, 5 and 9 with an example for each.
3. State how the divisibility tests for 3 and 9 are similar.
4. **WE23** Determine whether or not the results of the following divisions are whole numbers.
   a. $54 \div 2$  b. $150 \div 9$  c. $295 \div 5$  d. $1293 \div 3$
5. **WE24** Check the divisibility of the following.
   a. $1816 \div 4$  b. $2638 \div 6$  c. $62\,873 \div 9$  d. $876 \div 8$

### Understanding

6. State the factors (from 1 to 10) of the following numbers.
   a. 144  b. 264  c. 1456  d. 567 844
7. Use the divisibility rules to determine the factors (from 1 to 10) of the following numbers.
   a. 760  b. 535  c. 3457  d. 234 568
8. Apply the divisibility rules to determine the factors (from 1 to 10) of the following numbers.
   a. 76  b. 30  c. 458  d. 227
9. Without using a calculator, use divisibility tests to determine whether the following are True or False.
   a. 2 is a factor of 65  b. 3 is a factor of 96
   c. 4 is a factor of 22  d. 5 is a factor of 44
   e. 6 is a factor of 72  f. 10 is a factor of 85
10. a. Use the tests of divisibility to determine whether the following numbers are divisible by 2, 3, 4, 5, 6 or 10.
    i. 415  ii. 600  iii. 935
    iv. 1010  v. 4321  vi. 55 112
    
    b. Use a calculator to check your answers to part **a**.

### Reasoning

11. **WE24b** Jack says that 34 281 is divisible by 6, but Julie says that Jack is incorrect. Explain whether you agree with Jack or Julie, and why.
12. If a number is divisible by 9, determine another factor (2, 3, 4, 5, 6, 7, 8, or 10) for that number.
13. 6 is a factor of 4572 and 2868. Is 6 a factor of $4572 + 2868$ and $4572 - 2868$? Show full working to justify your answer.

### Problem solving

14. Determine the missing digit to create the number.
    a. Divisible by 3 : ______316
    b. Divisible by 6 : 9______32
    c. Divisible by 8 : 59______16
    d. Divisible by 9 : 1543______
15. Determine 2 two-digit numbers that are divisible by 4, 5 and 8.
16. A number between 550 and 560 is divisible by 6 and 4. Determine the number.

# LESSON
# 3.8 Cubes and cube roots

**LEARNING INTENTION**

At the end of this lesson you should be able to:
- determine values of cubes and cube roots
- evaluate cubes or cube roots using BIDMAS.

## 3.8.1 Cube numbers

eles-3737

- **Cube numbers** are numbers that can be arranged in a cube, as shown.

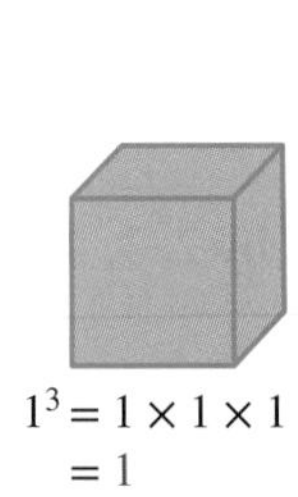

$1^3 = 1 \times 1 \times 1$
$= 1$

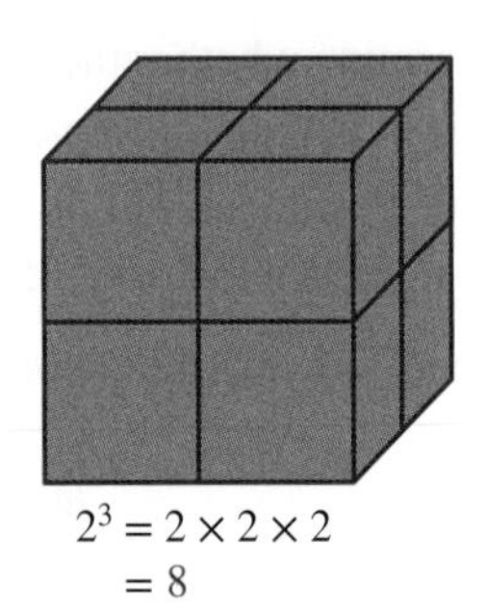

$2^3 = 2 \times 2 \times 2$
$= 8$

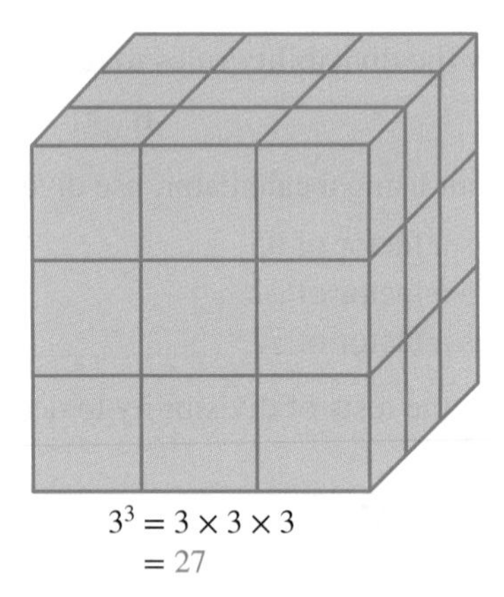

$3^3 = 3 \times 3 \times 3$
$= 27$

- The first cube number, 1, equals $1 \times 1 \times 1$.
- The second cube number, 8, equals $2 \times 2 \times 2$.
- The third cube number, 27, equals $3 \times 3 \times 3$.
- Each new perfect cube is found by multiplying a number by itself three times.
  This is known as *cubing a number* and is written using an index (or power) of 3. For example, $4^3 = 4 \times 4 \times 4 = 64$.

## WORKED EXAMPLE 25 Calculating the value of a cube

**Calculate the value of $5^3$.**

| THINK | WRITE |
|---|---|
| 1. Write $5^3$ as the product of three lots of 5. | $5^3 = 5 \times 5 \times 5$ |
| 2. Evaluate and write the answer. | $= 125$ |

### Order of operations

- When working with squares and cubes, remember to evaluate the index as part of the I in BIDMAS.

## WORKED EXAMPLE 26 Calculating cubes using the order of operations

**Calculate the value of $3^3 + 4 \times 5^2$.**

| THINK | WRITE |
|---|---|
| 1. According to the order of operations (BIDMAS), first simplify both the cubed and the square terms. | $3^3 + 4 \times 5^2 = (3 \times 3 \times 3) + 4 \times (5 \times 5)$<br>$= 27 + 4 \times 25$ |
| 2. Then complete the multiplication. | $= 27 + 100$ |
| 3. Write the answer. | $= 127$ |

## 3.8.2 Cube roots

eles-3738

- Evaluating the **cube root** of a number is the opposite of cubing a number.
- The cube root symbol is similar to the square root symbol but with a small 3 written in front, and is written as $\sqrt[3]{\ }$.

  For example, since $2^3 = 8$, then $\sqrt[3]{8} = 2$.
- Using the illustrations of the cubes in section 3.8.1, a cube of 8 cubic units has a side length of 2 units.

## WORKED EXAMPLE 27 Calculating cube roots

**Calculate $\sqrt[3]{27}$.**

| THINK | WRITE |
|---|---|
| Look for a number that when multiplied by itself three times gives 27, and write the answer. | $27 = 3 \times 3 \times 3$<br>$\sqrt[3]{27} = 3$ |

## WORKED EXAMPLE 28 Calculating cube roots using the order of operations

**Calculate:**

a. $\sqrt[3]{64+61}$ b. $\sqrt[3]{64}+61.$

| THINK | WRITE |
|---|---|
| a. 1. First simplify the expression inside the cube root. | a. $\sqrt[3]{64+61}=\sqrt[3]{125}$ |
| 2. To complete the calculation, evaluate the cube root by identifying the number that when multiplied by itself three times gives 125, and write the answer. | $=5$ |
| b. 1. Evaluate the cube root first. (Identify the number that when multiplied by itself three times gives 64.) | b. $\sqrt[3]{64}+61=4+61$ |
| 2. Perform the addition to complete the calculation, and write the answer. | $=65$ |

### Digital technology

Scientific calculators can evaluate cubes of decimal numbers.

DEG
$34.84^3$
42289.6839

Scientific calculators can evaluate cube roots of decimal numbers.

DEG
$\sqrt[3]{548.69}$
8.18670262

 Resources

**eWorkbook** Topic 3 Workbook (worksheets, code puzzle and project) (ewbk-1904)

**Interactivities** Individual pathway interactivity: Cubes and cube roots (int-4322)
Cube numbers (int-3959)
Cube roots (int-3960)

# Exercise 3.8 Cubes and cube roots

learn on

3.8 Quick quiz on

3.8 Exercise

## Individual pathways

| ■ PRACTISE | ■ CONSOLIDATE | ■ MASTER |
|---|---|---|
| 1, 4, 7, 10, 12, 15 | 2, 5, 13, 16 | 3, 6, 8, 9, 11, 14, 17 |

## Fluency

1. **WE25** Evaluate:
   a. $4^3$
   b. $2^3$
   c. $3^3$
2. Evaluate:
   a. $6^3$
   b. $10^3$
3. Write the first eight cube numbers.
4. **WE26** Evaluate $3^2 + 4^3 \times 2$.
5. Evaluate the following.
   a. $10^3 - 5^2 \times 2^3$
   b. $4^3 - 3^3 - 2^3$
6. **WE27** Evaluate:
   a. $\sqrt[3]{8}$
   b. $\sqrt[3]{64}$

## Understanding

7. Evaluate each of the following. Verify your answers with a calculator.
   a. $\sqrt[3]{216}$
   b. $\sqrt[3]{343}$
   (*Hint:* Use your answer to question **3**.)
8. **WE28** Determine the value of:
   a. $\sqrt[3]{125+91}$
   b. $\sqrt[3]{125}+91$
9. Evaluate each of the following:
   a. $\sqrt[3]{343+169}$
   b. $\sqrt[3]{343}+\sqrt{169}$
10. a. Determine between which two whole numbers $\sqrt[3]{50}$ will lie.
    b. Use your answer to part **a** to estimate the value of $\sqrt[3]{50}$ and check your answer with a calculator.
11. a. Determine between which two whole numbers $\sqrt[3]{90}$ will lie.
    b. Use your answer to part **a** to estimate the value of $\sqrt[3]{90}$ and check your answer with a calculator.

## Reasoning

12. Determine the first four numbers that could be arranged as a triangle-based pyramid (where all triangles are equilateral).
13. a. Express 216 as a product of its prime factors.
    b. Express 135 as a product of its prime factors.
    c. Use the answers to parts **a** and **b** to explain how you can use prime factors to determine whether the cube root of a number is an integer.

**14.** The first five square numbers are $1, 4, 9, 16, 25$. If we find the difference between these numbers, we get $4 - 1 = 3$, $9 - 4 = 5$, $16 - 9 = 7$ and $25 - 16 = 9$.
These numbers all differ by 2. Representing this in a table, we get:

| **Square numbers** | **1** | | **4** | | **9** | | **16** | | **25** |
|---|---|---|---|---|---|---|---|---|---|
| First difference | | 3 | | 5 | | 7 | | 9 | |
| Second difference | | | 2 | | 2 | | 2 | | |

Repeat this process for the first six cube numbers. Explain how many times you needed to determine the difference until they were equal.
If you look at $1^4, 2^4, 3^4, 4^4, \ldots$, explain how many differences you would need to find until they are equal.

## Problem solving

**15.** Dave said that $a^3 = \sqrt[a]{a}$ is a true statement for all numbers. Show your working to prove or disprove this statement.

**16.** Jack was organising a get-together of all his football mates from the last three years. He decided to phone 4 footballer mates and ask them to phone 4 footballers each, who in turn would phone 4 more footballers. Each team member was given a list of 4 names so that no-one received more than one call.

**a.** Calculate how many footballers would receive a phone call about the party.
**b.** Jack's friend Karin liked the idea, so she decided to contact 258 people about a 10-year reunion using Jack's method.
Assuming there were three rounds of calls, calculate how many people each person should call.

**17.** It's an interesting fact that the sum of the digits of a cube of an integer, when reduced to a single digit, can only be 1, 8 or 9.
For example:

$$12^3 = 1728$$
$$1 + 7 + 2 + 8 = 18$$
$$1 + 8 = 9$$

There is a pattern in the sequence of these numbers.
**a.** Use this rule with the cubes of the first 10 integers.
**b.** Determine the single-digit sum of $35^3$.
**c.** Explain your answer.

# LESSON
# 3.9 Review

## 3.9.1 Topic summary

### Multiples

- A multiple of a number is the result of multiplying that number by another whole number.
- The times table gives lists of multiples.

e.g.
Multiples of 4 are: 4, 8, 12, 16, 20, …
Multiples of 6 are: 6, 12, 18, 24, 30, …

### LCM and HCF

- The lowest common multiple (LCM) is the smallest multiple that is common to two or more numbers.
  e.g.
  Consider the numbers 8 and 12.
  Multiples of 8: 8, 16, **24**, 32, …
  Multiples of 12: 12, **24**, 36, 48, …
  LCM(8,12) = 24
- The highest common factor (HCF) is the largest number that is a factor of both numbers.
  e.g.
  Factors of 8: 1, 2, **4**, 8
  Factors of 12: 1, 2, 3, **4**, 6, 12
  HCF(8,12) = 4

### Factors

- A factor of a whole number divides that number exactly.
- Factors are also numbers that multiply together to get another number.

e.g.
Since $20 = 4 \times 5$, or $2 \times 10$, or $1 \times 20$, then the factors of 20 are 1, 2, 4, 5, 10 and 20.

## FACTORS, MULTIPLES, INDICES AND PRIMES

### Index notation

- Index notation is a short way of writing a repeated multiplication.
  e.g. $2 \times 2 \times 2 \times 2 \times 2 \times 2$ can be written as $2^6$, which is read as '2 to the power of 6'.
- The base is the number that is being repeatedly multiplied, and the index is the number of times it is multiplied.
  $2^6 = 2 \times 2 \times 2 \times 2 \times 2 \times 2 = 64$

### Divisibility tests

Divisibility tests determine whether or not the number is divisible by the chosen divisor.
2: last digit is 0, 2, 4, 6 or 8
3: sum of the digits is divisible by 3
4: last 2 digits together are divisible by 4
5: last digit is either 0 or 5
6: divisible by 2 and divisible by 3
8: last 3 digits together are divisible by 8
9: sum of the digits is divisible by 9
10: last digit is 0

### Squares

- The process of multiplying a number by itself is known as squaring a number.
  e.g.
  $1 \times 1 = 1^2 = 1$
  $2 \times 2 = 2^2 = 4$
- A square number is the number we get after multiplying a whole number by itself.
- The first four square numbers are 1, 4, 9 and 16.

### Cubes

- A cube number is found by multiplying a number by itself three times.
  e.g.
  $1 \times 1 \times 1 = 1^3 = 1$
  $2 \times 2 \times 2 = 2^3 = 8$
  $3 \times 3 \times 3 = 3^3 = 27$
- The first four cube numbers are 1, 8, 27 and 64.

### Prime and composite numbers

- A *prime* number has exactly two unique factors: 1 and itself.
- The first six prime numbers are 2, 3, 5, 7, 11 and 13.
- A *composite* number has more than two unique factors.
- Some examples of composite numbers are 4, 6, 8, 9 and 10.

### Square roots

- The radical or square root symbol is $\sqrt{\ }$.
- The square root of a number is the opposite of squaring the number.
  e.g. If $4^2 = 16$ then $\sqrt{16} = 4$.

### Cube roots

- The cube root symbol is $\sqrt[3]{\ }$.
- The cube root of a number is the opposite of cubing the number.
  e.g. If $5^3 = 125$ then $\sqrt[3]{125} = 5$.

### One — a special number

The number 1 has only one unique factor: itself. It is neither prime nor composite but a special number.

## 3.9.2 Success criteria

Tick the column to indicate that you have completed the lesson and how well you think you have understood it using the traffic light system.

(**Green:** I understand; **Yellow:** I can do it with help; **Red:** I do not understand)

| Lesson | Success criteria | Green | Yellow | Red |
|---|---|---|---|---|
| 3.2 | I can list multiples and factors of a whole number. | | | |
| | I can determine multiples and factor pairs. | | | |
| 3.3 | I can determine the lowest common multiple and the highest common factor of two or more whole numbers. | | | |
| 3.4 | I understand that index or exponent notation is a short way of writing a repeated multiplication. | | | |
| | I can write expressions in index notation. | | | |
| | I can simplify expressions using index notation. | | | |
| | I can use place value to write and evaluate numbers in expanded form with index notation. | | | |
| 3.5 | I understand the difference between prime and composite numbers. | | | |
| | I can determine prime factors by drawing a factor tree. | | | |
| | I can write a composite number in index form. | | | |
| | I can determine the highest common factor (HCF) and lowest common multiple (LCM) of large numbers as products of their prime factors. | | | |
| 3.6 | I can determine values of squares and square roots. | | | |
| | I can estimate the value of the square root of other numbers by using the perfect squares that lie on either side of the number. | | | |
| | I can evaluate squares or square roots using BIDMAS. | | | |
| 3.7 | I can test for factors of a given number using divisibility tests. | | | |
| 3.8 | I can determine values of cubes and cube roots. | | | |
| | I can evaluate cubes or cube roots using BIDMAS. | | | |

## 3.9.3 Project

### Alphabet sizes

Bob's business, Reading Resources, produces individual letters of the alphabet printed on cardboard. He caters for the market of visually impaired students who are learning to read.

Teachers use the cards to form words for students to read. Because the students have varying difficulties with their sight, the letters are printed in heights of 1 cm, 2 cm, 3 cm, …, 10 cm. Bob has a set of all the letters at 1 cm and uses these as the basis for forming the larger letters.

Bob's factory has 10 machines. Each machine performs only one specific job, as indicated:

- Machine 1 makes letters 1 cm high.
- Machine 2 enlarges the letters so they are twice as high.
- Machine 3 enlarges the letters so they are 3 times as high.
- Machine 4 enlarges the letters so they are 4 times as high.
- Machine 10 enlarges the letters so they are 10 times as high.

Bob's business is thriving and he relies heavily on all his machines working. This morning he arrived to process a large order of letters that need to be 6 cm high, only to find that machine 6 was not working. He gathered his staff together to discuss the problem.

'No problem!' said Ken. 'As long as the other machines are working we can still get this order done.'

1. Suggest a solution that Ken might have proposed.
2. If Bob can do without machine 6, are there others he can also do without?
3. What is the minimum number of machines that Bob needs to make letters up to 10 cm high?

As Bob's business has become more successful, there has been a demand for letters of a greater height.

4. For Bob to make letters of every whole number up to a height of 20 cm, would he need twice as many machines as he currently has? Explain.
5. Having become very excited about needing fewer machines than he first thought, Bob considered expanding his business to make advertising signs using letters of all whole number heights up to 100 cm. To succeed in this new market, what is the minimum number of machines he would need?
6. Explain why it is necessary to have some specific machine numbers, while some others are not necessary.

**on** Resources

eWorkbook Topic 3 Workbook (worksheets, code puzzle and project) (ewbk-1904)

Interactivities Crossword (int-2588)

Sudoku puzzle (int-3164)

## Exercise 3.9 Review questions

learn**on**

### Fluency

1. List the first 5 multiples of each number.

   a. 11 b. 100 c. 5 d. 20 e. 35

2. Calculate the lowest common multiple (LCM) of the following pairs of numbers.

   a. 3 and 12 b. 6 and 15 c. 4 and 7 d. 5 and 8

3. List all the factors of each of the following numbers.

   a. 27 b. 50

4. List all the factors of each of the following numbers.

   a. 16 b. 42 c. 72

5. List the factor pairs of the following numbers.

   a. 24 b. 40 c. 48 d. 21 e. 99 f. 100

6. Evaluate:

   a. $6^2$ b. $14^2$ c. $19^2$ d. $80^2$

7. Write the two whole square numbers between which each of the following will lie.

   a. $3.8^2$ b. $5.1^2$ c. $10.6^2$ d. $15.2^2$

8. Evaluate:

   a. $\sqrt{49}$ b. $\sqrt{256}$ c. $\sqrt{900}$ d. $\sqrt{1369}$

9. Calculate:

   a. $4^3$ b. $7^3$ c. $15^3$ d. $30^3$

10. Evaluate each of the following.

    a. $\sqrt[3]{27}$ b. $\sqrt[3]{125}$ c. $\sqrt[3]{1000}$ d. $\sqrt[3]{8000}$

11. Write the following using index notation.

    a. $2 \times 3 \times 3 \times 3$

    b. $5 \times 5 \times 6 \times 6 \times 6 \times 6$

    c. $2 \times 5 \times 5 \times 5 \times 9 \times 9 \times 9$

12. Use your calculator to evaluate:
    a. $3^5$ b. $7^3$ c. $8^4$ d. $11^5$

13. List all of the prime numbers less than 30.

14. List all of the single-digit prime numbers.

15. State the prime number that comes next after 50.

16. State the prime factors of:
    a. 99 b. 63 c. 125 d. 124.

17. Express the following numbers as products of their prime factors, in index form.
    a. 280 b. 144

18. Write the following numbers in expanded form using index notation.
    a. 1344 b. 30 601

19. Determine the highest common factor and lowest common multiple of 120 and 384 by first expressing each number as a product of its prime numbers.

20. Determine the highest common factor and lowest common multiple of 72 and 224 by first expressing each number as a product of its prime numbers.

## Problem solving

21. Hung and Frank are cyclists who are riding around a track. They ride past the finish line together, and from that point Hung takes 25 seconds to complete a lap and Frank takes 40 seconds.
    a. Calculate how long it will be until they next pass the finish line together.
    b. Determine how many laps each will have ridden when this occurs.

22. Joe, Claire and Daniela were racing up and down a flight of stairs. Joe took the stairs three at a time, Daniela took the stairs two at a time, while Claire took the stairs four at a time. In each case, they reached the bottom with no steps left over.
    a. Calculate how many steps are in the flight of stairs. List three possible answers.
    b. Determine the smallest number of steps there could be.
    c. If Joe can also take the stairs five at a time with no steps left over, determine the smallest number of steps in the flight of stairs.

23. A perfect number is one whose factors (all except the number itself) add up to the number. For example, 6 is a perfect number because $1 + 2 + 3$ (the sum of its factors, excluding 6) is equal to 6. Show why 496 is a perfect number.

24. In a race, one dirt bike rider completes each lap in 40 seconds while another completes a lap in 60 seconds. Determine how long after the start of the race the two bikes will pass the starting point together.

25. A three-digit number is divisible by 6. The middle digit is a prime number. The sum of the digits is 9. The number is between 400 and 500. The digits are in descending order. Determine the number.

26. Explain why we cannot determine the exact value of $\sqrt{20}$.

27. Write the following numbers in the order of size, starting with the smallest.

$$5, \sqrt{64}, 2, \sqrt[3]{27}$$

28. Calculate the smallest possible number which, when multiplied by 120, will give a result that is a square number. Now calculate the smallest possible number that will give a result that is a cube number.

29. List the prime numbers up to 100. Some prime numbers are separated by a difference of 2 or 6. (For example, $5 - 3 = 2$ and $19 - 17 = 2$.)
    **a.** Calculate the prime number pairs up to 100 that have a difference of 2.
    **b.** Calculate the prime number pairs up to 100 that have a difference of 6.
    **c.** Explain why there are no prime number pairs with a difference of 7.

30. Determine the smallest number that is both a square number and a cube number. Identify the next smallest number.

31. Complete the following sequence: $2, 3, 5, 7, 11, 13,$ ___, ___, ___.

32. Write each of the numbers 250, 375 and 625 as a product of its prime factors and:
    **a.** determine the highest common *prime* factor
    **b.** determine the HCF.

33. A motor boat requires an engine service every 5000 nautical miles, refuelling every 300 nautical miles and an oil change after 2250 nautical miles. Determine how many nautical miles the motor boat will have travelled before all three services are required simultaneously.

**34.** By folding a sheet of paper in half, you divide it into 2 regions.

**a.** If you fold it in half again, determine how many regions there are.

**b.** Set up a table similar to the one shown and fill it in.

**c.** Explain how the number of regions increases with each new fold.

**d.** Write a formula for the number of regions ($R$) for $n$ folds.

| $n$ (number of folds) | $R$ (number of regions made) |
|---|---|
| 0 | 1 |
| 1 | 2 |
| 2 | 4 |
| 3 | |
| 4 | |
| 5 | |
| 6 | |

**35.** Look at the following table. As you read down each column, you'll notice that the equations follow a pattern. Continue the patterns to complete the table.

| | |
|---|---|
| $2^5 = 32$ | $3^5 = 243$ |
| $2^4 = 16$ | $3^4 = 81$ |
| $2^3 = 8$ | $3^? = 27$ |
| $2^2 =$ | $3^2 =$ |
| $2^1 =$ | $3^1 =$ |
| $2^0 =$ | $3^0 =$ |
| $2^{-1} =$ | $3^{-1} =$ |

**a.** State what $5^0$ equals.

**b.** State what $2^{-2}$ is equal to.

**on** To test your understanding and knowledge of this topic, go to your learnON title at www.jacplus.com.au and complete the **post-test**.

# Answers

## Topic 3 Factors, multiples, indices and primes

### 3.1 Pre-test

1. B
2. $35
3. 6
4. 12 minutes
5. $5^2 \times 3^4$
6. 17, 31, 109
7. C, E
8. B, E
9. LCM = 864; HCF = 6
10. 64
11. 109
12. 8 and 9
13. 3
14. 39
15. 7

### 3.2 Factors and multiples

1. a. 3, 6, 9, 12, 15 b. 6, 12, 18, 24, 30
   c. 100, 200, 300, 400, 500 d. 11, 22, 33, 44, 55
2. a. 120, 240, 360, 480, 600 b. 45, 90, 135, 180, 225
   c. 72, 144, 216, 288, 360 d. 33, 66, 99, 132, 165
3. 10, 20, 100, 70
4. 21, 7, 70, 35
5. 16, 64, 160, 32, 80
6. 9, 18, 27, 36, 45, 54, 63, 72, 81, 90, 99
7. 102, 108, 114, 120, 126, 132, 138, 144, 150, 156
8. C
9. A
10. a. 1, 2, 3, 4, 6, 12 b. 1, 2, 4, 5, 8, 10, 20, 40
11. a. 1, 2, 7
    b. 1, 2, 3, 5, 10, 15, 20
    c. 1, 2, 5, 10, 20, 25
12. a. 1, 2, 3, 4, 6, 8, 9, 12, 18, 24, 36, 72
    b. 1, 2, 5, 10, 25, 50, 125, 250
13. a. 1, 5, 17 b. 1, 3, 9, 11, 33 c. 1, 3, 17
14. 1, 20; 2, 10; 4, 5
15. 1, 132; 2, 66; 3, 44; 4, 33; 6, 22; 11, 12
16. B, C, E
17. 15
18. B
19. A
20. 3, 4
21. a. 6, 12, 18 (or any other multiple of 6)
    b. 6
    c. 30
22. 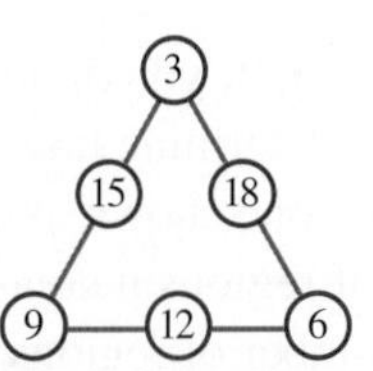

23. 4 rows of 5 students or 5 rows of 4 students; 2 rows of 10 students or 10 rows of 2 students; 1 row of 20 students or 20 rows of 1 student
24. 3, 12 or 30
25. Boy 6, grandfather 66
26. 84
27. 60
28. 60

### 3.3 Lowest common multiple and highest common factor

1. a. 4, 8, 12, 16, 20, 24, 28, 32, 36, 40
   b. 6, 12, 18, 24, 30, 36, 42, 48, 54, 60
   c. 12, 24, 36
   d. 12
2. a. 9 b. 30 c. 70 d. 48 e. 60
3. False
4. True
5. True
6. False
7. a. HCF = 7 b. HCF = 7 c. HCF = 6
8. a. 9 b. 7
9. a. 4 b. 9
10. a. 6 days
    b. 6 is the lowest common multiple of 2 and 3.
11. a. 35 minutes
    b. 35 is the lowest common multiple of 5 and 7.
12. a. 12 hours
    b. 12 is the lowest common multiple of 4 and 6.
13. 20 seconds
14. a. 12 b. 6 c. 7
15. a. 3 metres b. 10
16. 2 hours
17. a. Daniella b. 10 minutes
18. 6 times
19. a. 4 minutes and any multiple of 4 thereafter
    b. 3 minutes and any multiple of 3 thereafter
20. 30 minutes
21. a. Two semicircles
    b. The starting line for a 100 m race has been positioned in the extended section so the athletes can all start from the same starting line.

c. The finish line for a 100 m race has been positioned so that all athletes can run in a direct line without having to have staggered starting positions.

d. There is a need to stagger the starting blocks in the 200 m, 400 m and 800 m track events so that the outside runners will not be running further than the inside runners.

e. and f.

| Lane number | Distance travelled | Difference |
|---|---|---|
| 1 | 400 m | 0 |
| 2 | 408 m | 8 |
| 3 | 416 m | 16 |
| 4 | 424 m | 24 |
| 5 | 432 m | 32 |
| 6 | 440 m | 40 |
| 7 | 448 m | 48 |
| 8 | 456 m | 56 |

g. The runner in lane 8 is running 56 metres further than the runner in lane 1.

## 3.4 Index notation

1. a. $7^2 = 49$ Base is 7 and index is 2.
   b. $4^5 = 4096$ Base is 4 and index is 5.
   c. $3^4 = 81$ Base is 3 and index is 4.
   d. $10^3 = 1000$ Base is 10 and index is 3.
2. a. $7^4$ b. $8^6$ c. $3^9$ d. $13^3$
3. a. $4^5 \times 6^3$ b. $2^2 \times 3$
   c. $2^4 \times 5^2$ d. $2^2 \times 3^2 \times 5^3$
4. a. $6 \times 6 \times 6 \times 6 \times 6$ b. $11 \times 11 \times 11$
5. a. $5^{11}$ b. $4^{13}$ c. $2^{21}$
6. a. $300 = 3 \times 10^2$
   b. $4500 = 4 \times 10^3 + 5 \times 10^2$
   c. $6705 = 6 \times 10^3 + 7 \times 10^2 + 0 \times 10^1 + 5$
   d. $10\,000 = 1 \times 10^4$
7. a. $9^{15}$ b. $3^6 \times 13^6$
8. D
9. a. 343 b. 288 c. 77
10. a. 72 b. 5184 c. 177 147
11. a. 567 b. 2717 c. 112
12. a. A
    b. At first the base number 2 would appear to be the smallest, but it has the highest power. Although the other numbers have a higher base, they have relatively small powers — therefore $2^8$ will be the largest number. This can easily be verified by evaluating each number.
13. $41^2 \neq 691$. Notice that $31^2 = 961$, therefore $41^2$ will need to be larger than 961.
14. a. < b. < c. < d. >
15. a. Option 1: \$327.68; Option 2 : \$120 after the 15 weeks
    b. Option 1: $2^{15}$ cents after week 15
16. 64
17. a. $8^3 = 584$ b. 584
18. 127
19. a. $2^{12} = 4096$
    b. $3^{12} = 531\,441$
    c. Using powers is a quick way of getting answers that are really large.
20. 256
21. Square the first number and add one to calculate the second number. Square the second number and add one to calculate the third number. The missing number is 101.

## 3.5 Prime and composite numbers

1. 23, 29, 31, 37
2. a. Composite b. Prime
   c. Composite d. Prime
3. a. Composite b. Prime
   c. Composite d. Prime
4. a. i.

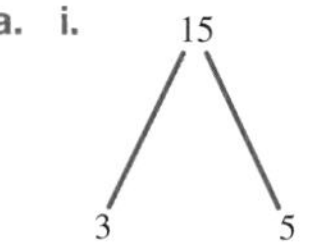


3, 5

ii. $15 = 3 \times 5$

b. i.

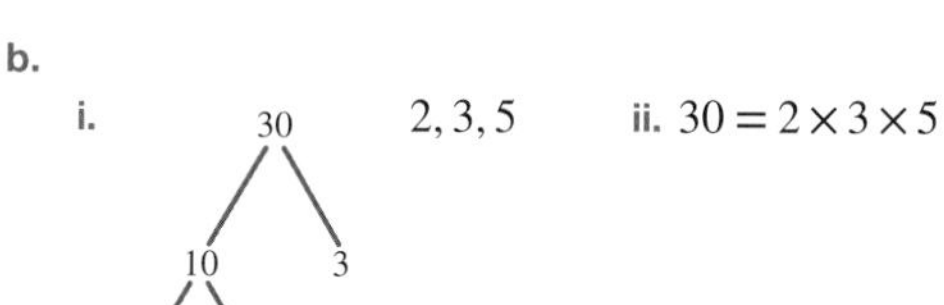


2, 3, 5

ii. $30 = 2 \times 3 \times 5$

c. i. 100 → 10, 10; 10 → 2, 5; 10 → 2, 5

2, 5

ii. $100 = 2^2 \times 5^2$

d. i. 49 → 7, 7

7

ii. $49 = 7^2$

5. a. i. 72 → 9, 8; 9 → 3, 3; 8 → 4, 2; 4 → 2, 2

2, 3

ii. $72 = 2^3 \times 3^2$

b. i. 56 → 8, 7; 8 → 2, 4; 4 → 2, 2

2, 7

ii. $56 = 2^3 \times 7$

c. i. 3, 5

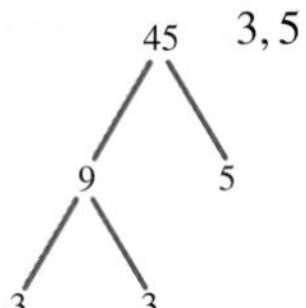


ii. $45 = 3^2 \times 5$

6. a. i. 2, 3, 7

84
7 12
3 4
2 2

ii. $84 = 2^2 \times 3 \times 7$

b. i. 2, 7

112
4 28
2 2 4 7
2 2

ii. $112 = 2^4 \times 7$

c. i. 2, 5

40
4 10
2 2 2 5

ii. $40 = 2^3 \times 5$

7. a. i. 2, 3

96
8 12
2 4 3 4
2 2 2 2

ii. $96 = 2^5 \times 3$

b. i. 2

32
4 8
2 2 2 4
2 2

ii. $32 = 2^5$

c. i. 2, 3, 5

3000
3 1000
10 100
2 5 10 10
2 5 2 5

ii. $3000 = 2^3 \times 3 \times 5^3$

8. a. 2, 3 b. 2, 5 c. 3 d. 2, 3

9. a. 2, 3, 5 b. 2, 3, 5 c. 2, 3, 5 d. 2, 5

10. a. $2^2 \times 3 \times 5$ b. $3 \times 5^2$ c. $2^2 \times 5 \times 11$

11. a. $2^6 \times 3$ b. $2^2 \times 31$

12. a. LCM is 252; HCF is 12.
b. LCM is 240; HCF is 12.
c. LCM is 1200; HCF is 40.
d. LCM is 8800; HCF is 20.

13. a. False b. False c. False
d. True e. True

14. a. False b. False c. False

15. A

16. C

17. C

18. C

19. a. $365 = 5 \times 73$ b. $187 = 11 \times 17$

20. A

21. D

22. C

23. B

24. 773

25. No. $3 + 7 = 10$, 3 and 7 are prime numbers, but their sum 10 is not a prime number.

26. 29 or 47 or 83 or 137. Others may be possible.

27. 61

28. a. 3, 7 b. 2, 13 c. 17, 19

29. 6 integers

30. 17 and 19, 29 and 31, 41 and 43, 59 and 61, 71 and 73

## 3.6 Squares and square roots

1. a. 64 b. 121 c. 225 d. 625

2. 64, 81

3. 169, 196

4. a. 36 and 49 b. 49 and 64
c. 81 and 100 d. 144 and 169

5. a. 5 b. 9 c. 12 d. 20

6. a. 70 b. 120 c. 600 d. 400

7. a. 7 and 8 b. 3 and 4
c. 14 and 15 d. 1 and 2

8. a. 13 b. 37

9. 16, 36, 64

10. 81

11. B

12. A

13. a. 9 b. 75 c. 700 d. 25

14. a. 14 b. 45

15. a. 12 and 13 b. 12.2

16. a. 225
b. 225
c. Yes, if two squares are multiplied together, then the result is equal to the square of the base numbers multiplied together.

17. a. 15
b. 15
c. Yes, if two square roots are multiplied together, then the result is equal to the square root of the base numbers multiplied together.

18. a. $2^2 \times 7^2$
b. $2^3 \times 5^2$
c. If all of the prime factors of a number are squares, then the square root of that number will be an integer.

19. a. 14
b. 10

c. No, the square root of two numbers that have been added together is not equal to adding the square roots of these two numbers together.

20. a. 17

b. 13

c. No. You have to evaluate the individual square roots first before adding.

21. 169, 196, 225

22. a. i. 49 ii. 47 iii. 45 iv. 43 v. 41

b. Answer is the sum of the two numbers.

c. No

23. 49

24. a. $7, \sqrt{56}, 8, \sqrt{74}, 3^2, \sqrt{100}, 62, 8^2$

b. $\sqrt{10}, \sqrt{121}, 3 \times 4, \sqrt{169}, 2 \times 7, 9^2, 100$

25. 9 and 11

## 3.7 Divisibility tests

1. A number is divisible by 2 if the last digit is 0, 2, 4, 6 or 8. For example, 56 is divisible by 2 because the last digit is 6.
A number is divisible by 4 if the last two digits together (as a two-digit number) are divisible by 4. For example, 3416 is divisible by 4 because the last two digits are 16, which is divisible by 4.
A number is divisible by 6 if it is divisible by 2 and divisible by 3. For example, 354 is divisible by 2 because the last digit is 6; and the sum of the digits $(3 + 5 + 4)$ is 12, which is divisible by 3. Therefore, 354 is divisible by 6.

2. A number is divisible by 3 if the sum of the digits is divisible by 3. For example, 156 is divisible by 3 because the sum of the digits $(1 + 5 + 6)$ is 12, which is divisible by 3.
A number is divisible by 5 if the last digit is either 0 or 5. For example, 845 is divisible by 5 because the last digit is 5.
A number is divisible by 9 if the sum of the digits is divisible by 9. For example, 54 is divisible by 9 because the sum of the digits $(5 + 4)$ is 9, which is divisible by 9.

3. The sum of the digits is calculated to test for divisibility by 3 and by 9.

4. a. 54 is divisible by 2 as the last digit is 4.

b. 150 is not divisible by 9 as the sum of the digits is $6 (1 + 5 + 0)$, which is not divisible by 9.

c. 295 is divisible by 5 as the last digit is 5.

d. 1293 is divisible by 3 as the sum of the digits is $15 (1 + 2 + 9 + 3)$, which is divisible by 3.

5. a. Divisible by 4 as the last two digits are divisible by 4

b. Not divisible by 6 as the number is divisible by 2 but not by 3

c. Not divisible by 9 as the sum of digits is not divisible by 9

d. Not divisible by 8; incorrect rule

6. a. 1, 2, 3, 4, 6, 8, 9 b. 1, 2, 3, 4, 6, 8
c. 1, 2, 4, 7, 8 d. 1, 2, 4

7. a. 1, 2, 4, 5, 8, 10 b. 1, 5
c. 1 d. 1, 2, 4, 8

8. a. 1, 2, 4 b. 1, 2, 3, 5, 6
c. 1, 2 d. 1

9. a. False b. True
c. False d. False
e. True f. False

10. a. i. 5 only ii. 2, 3, 4, 5, 6 and 10
iii. 5 only iv. 2, 5 and 10
v. None vi. 2 and 4

11. Julie is correct. 34 281 is not divisible by 6.

12. 3

13. Yes, $4572 + 2868$ and $4572 - 2868$ are divisible by 6.

14. a. 2, 5 or 8 b. 1, 4 or 7
c. 0, 2, 4, 6 or 8 d. 5

15. The two 2-digit numbers that are divisible by 4, 5 and 8 are 40 and 80.

16. 552

## 3.8 Cubes and cube roots

1. a. 64 b. 8 c. 27

2. a. 216 b. 1000

3. 1, 8, 27, 64, 125, 216, 343, 512

4. 137

5. a. 800 b. 29

6. a. 2 b. 4

7. a. 6 b. 7

8. a. 6 b. 96

9. a. 8 b. 20

10. a. 3 and 4 b. 3.7

11. a. 4 and 5 b. 4.5

12. 3, 6, 10, 15 (not including 1, which would not really be a triangle at all).

13. a. $2^3 \times 3^3$

b. $3^3 \times 5$

c. If all of the prime factors of a number are cubes, then the cube root of that number will be an integer.

14. For cube numbers, it was necessary to find the third difference. If we look at the power 4, it will be necessary to find the fourth difference.

15. Dave is incorrect.

16. a. 84 footballers would receive a call.
b. Each person would call 6 people.

17. a. 1, 8, 9, 1, 8, 9, 1, 8, 9, 1

b. 8

c. The single-digit sums of cube numbers repeat in the order 1, 8, 9.

## Project

1. Enlarge the letters using machine 2, then machine 3.

2. 4, 6, 8, 9 and 10

3. 5

4. No. Numbers not required can be made using their prime factors.

5. Machine 1 and the prime numbers less than 100. Total of 26 machines.
6. Machine 1 is needed to make letters one centimetre high. Machines whose numbers are prime numbers can make all the sizes up to 100 centimetres.

## 3.9 Review questions

1. a. 11, 22, 33, 44, 55
   b. 100, 200, 300, 400, 500
   c. 5, 10, 15, 20, 25
   d. 20, 40, 60, 80, 100
   e. 35, 70, 105, 140, 175
2. a. 12 b. 30 c. 28 d. 40
3. a. 1, 3, 9, 27 b. 1, 2, 5, 10, 25, 50
4. a. 1, 2, 4, 8, 16
   b. 1, 2, 3, 6, 7, 14, 21, 42
   c. 1, 2, 3, 4, 6, 8, 9, 12, 18, 24, 36, 72
5. a. 1, 24; 2, 12; 3, 8; 4, 6
   b. 1, 40; 2, 20; 4, 10; 5, 8
   c. 1, 48; 2, 24; 3, 16; 4, 12; 6, 8
   d. 1, 21; 3, 7
   e. 1, 99; 3, 33; 9, 11
   f. 1, 100; 2, 50; 4, 25; 5, 20; 10, 10
6. a. 36 b. 196 c. 361 d. 6400
7. a. 9 and 16 b. 25 and 36
   c. 100 and 121 d. 225 and 256
8. a. 7 b. 16 c. 30 d. 37
9. a. 64 b. 343 c. 3375 d. 27 000
10. a. 3 b. 5 c. 10 d. 20
11. a. $2\times3^3$ b. $5^2\times6^4$ c. $2\times5^3\times9^3$
12. a. 243 b. 343 c. 4096 d. 161 051
13. 2, 3, 5, 7, 11, 13, 17, 19, 23, 29
14. 4
15. 53
16. a. 3, 11 b. 3, 7 c. 5 d. 2, 31
17. a. $280 = 2\times2\times2\times5\times7 = 2^3\times5\times7$
    b. $144 = 2^4\times3^2$
18. a. $1344 = 1\times10^3 + 3\times10^2 + 4\times10^1 + 4$
    b. $30\,601 = 3\times10^4 + 0\times10^3 + 6\times10^2 + 0\times10^1 + 1$
19. HCF is 24; LCM is 1920.
20. HCF is 8; LCM is 2016.
21. a. 200s
    b. Hung 8 laps
    Frank 5 laps
22. a. 12, 24, 36 b. 12 c. 60
23. The sum of its factors (except for 496) add up to give 496.
24. 120 seconds or 2 minutes
25. 432
26. 20 is not a perfect square, so we can only find an approximate answer. It will be between 4 and 5.
27. 2, $\sqrt[3]{27}$, 5, $\sqrt{64}$
28. 30, 225
29. a. 3, 5; 5, 7; 11, 13; 17, 19; 29, 31; 41, 43; 59, 61; 71, 73
    b. 5, 11; 7, 13; 11, 17; 13, 19; 17, 23; 23, 29; 31, 37; 37, 43; 41, 47; 47, 53; 53, 59; 61, 67; 67, 73; 73, 79; 83, 89
    c. For 2, the number that's 7 higher is 9, which is not prime. For odd primes, the number that's 7 higher is even and therefore not prime.
30. 1, 64
31. 2, 3, 5, 7, 11, 13, $\underline{17}$, $\underline{19}$, $\underline{23}$
32. a. 5 b. 125
33. 45 000
34. a. 4 regions
    b.

| *n* (number of folds) | 0 | 1 | 2 | 3 | 4 | 5 | 6 |
|---|---|---|---|---|---|---|---|
| *R* (number of regions made) | 1 | 2 | 4 | 8 | 16 | 32 | 64 |

    c. Each time the page is folded the number of regions is multiplied by 2 (doubled).
    d. $R = 2^n$
35.

| $2^5 = 32$ | $3^5 = 243$ |
|---|---|
| $2^4 = 16$ | $3^4 = 81$ |
| $2^3 = 8$ | $3^3 = 27$ |
| $2^2 = 4$ | $3^2 = 9$ |
| $2^1 = 2$ | $3^1 = 3$ |
| $2^0 = 1$ | $3^0 = 1$ |
| $2^{-1} = \frac{1}{2}$ | $3^{-1} = \frac{1}{3}$ |

    a. $5^0 = 1$ b. $2^{-2} = \frac{1}{4}$

# 4 Fractions and percentages

## LESSON SEQUENCE


4.1 Overview ... 168
4.2 What are fractions? ... 170
4.3 Simplifying fractions ... 182
4.4 Mixed numbers and improper fractions ... 186
4.5 Adding and subtracting fractions ... 192
4.6 Multiplying fractions ... 200
4.7 Dividing fractions ... 208
4.8 Working with mixed numbers ... 213
4.9 Percentages as fractions ... 220
4.10 Calculating percentages of an amount ... 225
4.11 One amount as a percentage of another ... 230
4.12 Common percentages and short cuts ... 236
4.13 Review ... 242

# LESSON
# 4.1 Overview

## Why learn this?

There are many everyday situations in which we do not work with whole numbers. You may share a pizza with a friend, so you would eat part of the pizza. Fractions are a way of expressing parts of whole things, amounts or quantities. They are important in many situations, such as cooking, shopping and telling the time. Did you know that approximately one quarter $\left(\frac{1}{4}\right)$ of the population of Australia lives in Victoria?

Percentages, like fractions, are a way of expressing parts of a whole. They are commonly used in advertising, statistics and shopping. Did you know that the area of New South Wales is approximately 10%, or one tenth $\left(\frac{1}{10}\right)$, of the size of Australia? Fractions and percentages can be used interchangeably to represent the same part of a whole. Many different professions use fractions and percentages extensively in their work, including hospitality, finance, statistics and journalism. When you are next in a shopping centre, check out how many percentage (%) signs you see. You might be surprised.

## Exercise 4.1 Pre-test

learnon

1. **MC** Select the equivalent fraction to $\frac{2}{3}$:

   **A.** $\frac{3}{2}$  **B.** $\frac{30}{45}$  **C.** $\frac{30}{40}$  **D.** $\frac{12}{15}$  **E.** $\frac{21}{30}$

2. Luke's football team scored 16 goals. Luke scored 6 of the team's goals. Answer the following questions, giving your answers in simplified form.
   **a.** Calculate the fraction of the team's goals that Luke scored.
   **b.** Calculate the fraction of the team's goals that the rest of the team scored.

3. A bowl of apples was handed out to kindergarten students. If 23 students ate $\frac{1}{4}$ of an apple each, calculate how many apples the students ate in total. Give your answer as a mixed number in simplified form.

4. Evaluate the following, giving your answer as a simplified fraction.

$$\frac{2}{3} - \frac{1}{6} + \frac{3}{12}$$

5. An AFL match went for 120 minutes. If Patrick Cripps played $\frac{5}{6}$ of the game time, calculate how many minutes he played.

6. **MC** Select the simplified answer to $\frac{18}{56} \div \frac{27}{64}$.

   **A.** $\frac{16}{21}$  **B.** $\frac{18}{21}$  **C.** $\frac{2}{3}$  **D.** $\frac{15}{24}$  **E.** $\frac{243}{1792}$

7. Yuto gets $16 a week for cleaning his room, but if he also empties the dishwasher and feeds the dogs he can get $2\frac{1}{4}$ times his pay. If Yuto completes all his chores in the week, calculate how much he earned.

8. When Thomas started school he was the shortest in his class, at 85 cm tall. By the time he finished grade 12 he was the tallest in his class. His height had increased by $2\frac{1}{5}$ times. Calculate how tall he was at the end of grade 12.

9. Newcastle East Public School is having a cupcake sale to raise money for a local charity. They have 250 cupcakes to sell. Of these, $\frac{1}{5}$ are chocolate and the remainder are vanilla. At the end of the day, $\frac{24}{25}$ of the chocolate cupcakes were sold, and $\frac{7}{10}$ of all the cupcakes were sold. Calculate what fraction of the vanilla cupcakes were sold.

10. **MC** Max spends 78% of his weekly pocket money on his mobile phone plan. Calculate what fraction of his pocket money is spent on the phone plan.

    **A.** $\frac{38}{50}$  **B.** $\frac{37}{50}$  **C.** $\frac{39}{50}$  **D.** $\frac{18}{25}$  **E.** $\frac{19}{25}$

11. The cost of the new model PlayStation has increased by 12% compared to the previous model. Write the percentage increase as a:

a. simplified fraction  b. decimal.

12. At a particular soccer game, 62% of the crowd supported the Matildas. If 17 800 people attended the game, determine how many supported the Matildas.

13. Alexander got 24 marks out of a possible 32. Calculate his percentage score to the nearest whole number.

14. MC A survey of 20 students found that 9 like Coke, 6 like Solo, 3 like Fanta and 2 like lemonade. Select the percentage of students who like Fanta.

A. 3%  B. 6%  C. 10%  D. 15%  E. 20%

15. Sarah purchased a dress that was on sale for a discount of 10%.
If the dress originally cost $155.75, calculate how much she saved, to the nearest cent.

# LESSON
# 4.2 What are fractions?

## LEARNING INTENTION

At the end of this lesson you should be able to:

- understand the concept of a fraction
- convert between equivalent fractions
- compare fractions.

## 4.2.1 Understanding fractions

eles-3798

- **Fractions** are used to describe parts of a whole.

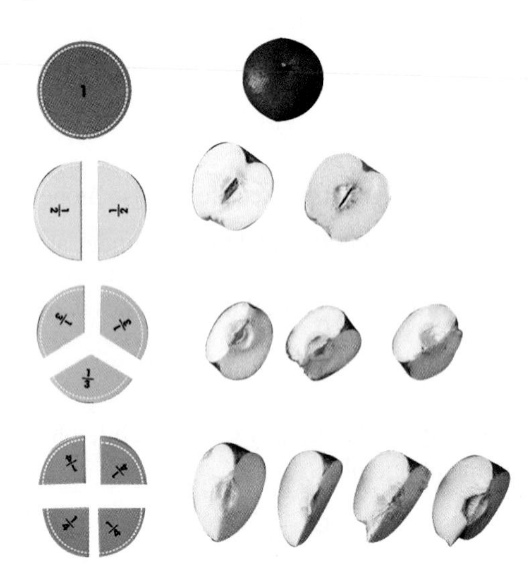

- The **numerator**, or top number of the fraction, shows how many parts are required. The numerator must be an integer.
- The **denominator**, or bottom number, shows the number of parts into which the whole can be divided. The denominator must be an integer, but cannot be zero.
- The horizontal bar separating the numerator from the denominator is called the **vinculum**.

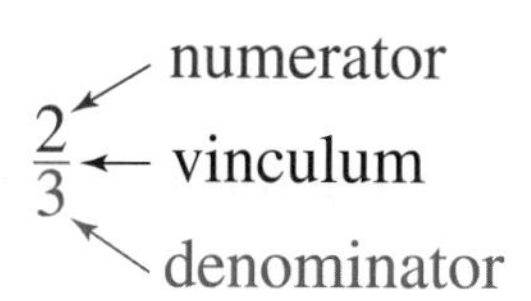


| Proper fractions | Improper fractions | Mixed numbers |
|---|---|---|
| **Proper fractions** have a numerator *less than* the denominator. | **Improper fractions** have a numerator *greater than* the denominator. | **Mixed numbers** consist of a whole number and a proper fraction. |
| $\frac{1}{4}, \frac{2}{3}, \frac{27}{68}$ | $\frac{5}{3}, \frac{3}{2}, \frac{123}{17}$ | $1\frac{1}{2}, 5\frac{7}{8}, 14\frac{152}{986}$ |

## Understanding fractions

**A fraction can be understood as numerator *out of* denominator.**

**For example: the fraction $\frac{6}{10}$ means 6 *out of* 10.**

**When the numerator is equal to the denominator, we have a 'whole', which is equivalent to 1.**

**For example: $\frac{10}{10}$ means 10 *out of* 10 and is one 'whole'.**

- To express one quantity as a fraction of another, write the first quantity as a fraction of the second quantity.

### WORKED EXAMPLE 1 Identifying fractions of a quantity

**a. Identify the fraction of the rectangle that has been shaded.**
**b. Express the number of unshaded squares as a fraction of the number of squares in total.**

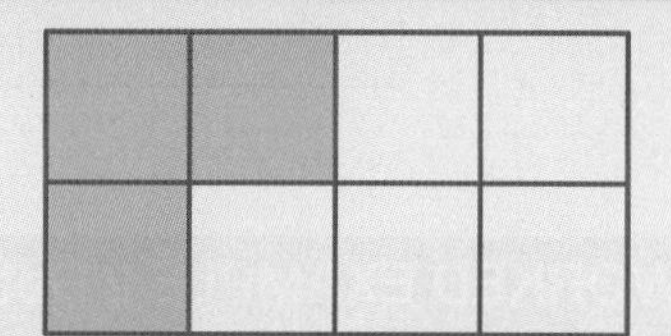

| THINK | WRITE |
|---|---|
| a. 1. Count how many equal parts the rectangle has been divided into. | a. Total number of parts = 8 |
| 2. State how many parts are shaded. | 3 parts are shaded. |
| 3. Write the number of shaded parts as a fraction of the total number of parts. | $\frac{3}{8}$ of the rectangle has been shaded. |
| b. 1. Count the number of unshaded squares. | b. There are 5 unshaded squares. |
| 2. State the number of squares in total. | There are 8 squares in total. |
| 3. Write the number of unshaded squares as a fraction of the total number of squares. | $\frac{5}{8}$ of the squares are unshaded. |

## 4.2.2 Equivalent fractions

eles-3799

- **Equivalent fractions** are equal fractions. For example, the fractions $\frac{1}{2}, \frac{2}{4}, \frac{5}{10}, \frac{8}{16}$ are all equivalent because they are all equal to a half.
- Equivalent fractions can be shown using diagrams. Equivalent fractions for $\frac{2}{3}$ are shown in the following diagrams.

Note that the same portion, $\frac{2}{3}$ of the rectangle, has been shaded in each case.

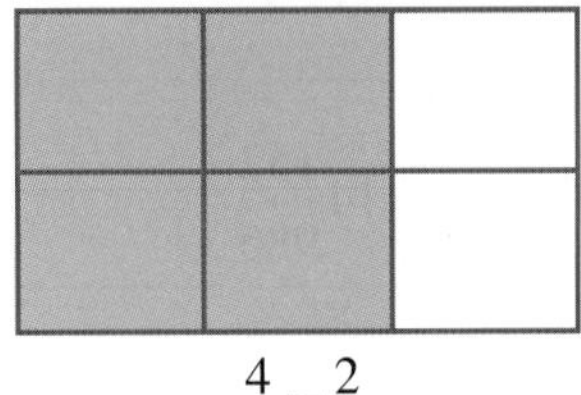

$\frac{4}{6} = \frac{2}{3}$

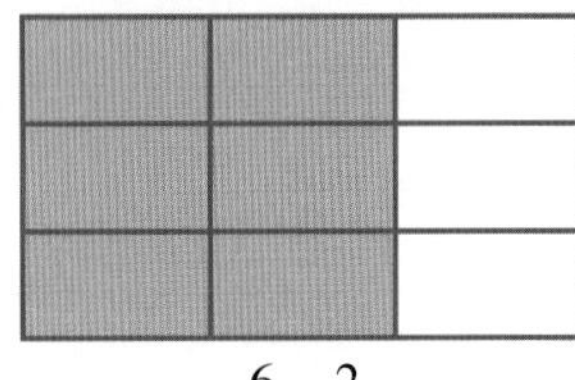

$\frac{6}{9} = \frac{2}{3}$

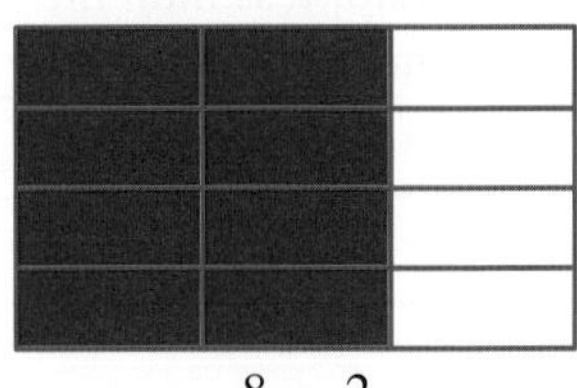

$\frac{8}{12} = \frac{2}{3}$

### Equivalent fractions

**Equivalent fractions are equal fractions. Equivalent fractions can be found by multiplying or dividing the numerator and denominator by the same number.**

**For example:**

$$\frac{2}{3} \overset{\times 2}{=} \frac{4}{6} \qquad \frac{10}{20} \overset{\div 10}{=} \frac{1}{2}$$

### WORKED EXAMPLE 2 Determining the multiplier

**For the equivalent fractions $\frac{4}{5} = \frac{8}{10}$, determine the number (multiplier) that has been used to multiply both numerator and denominator of the first fraction in order to get the second.**

| THINK | WRITE |
| --- | --- |
| 1. What number is 4 multiplied by to equal 8? ($4 \times \_ = 8$) What number is 5 multiplied by to equal 10? ($5 \times \_ = 10$) | $\frac{4}{5} = \frac{4 \times 2}{5 \times 2} = \frac{8}{10}$ |
| 2. Write the answer. | The number used is 2. |

## WORKED EXAMPLE 3 Determining equivalent fractions

**From the list, determine those fractions that are equivalent to $\frac{1}{2}$.**

$$\frac{3}{9}, \frac{3}{6}, \frac{9}{18}, \frac{10}{15}, \frac{7}{14}, \frac{17}{34}$$

**THINK**

1. Multiply the numerator and denominator of $\frac{1}{2}$ by the numerator of the first fraction in the list (3) to check whether the new fraction is in the list.
2. Multiply the numerator and denominator of $\frac{1}{2}$ by the next different numerator in the list (9) and check whether the new fraction is in the list.
3. Continue until all fractions have been considered.
4. Write the equivalent fractions.

**WRITE**

1. $\frac{1 \times 3}{2 \times 3} = \frac{3}{6}$
2. $\frac{1 \times 9}{2 \times 9} = \frac{9}{18}$
3. $\frac{1 \times 10}{2 \times 10} = \frac{10}{20}$
   $\frac{1 \times 7}{2 \times 7} = \frac{7}{14}$
   $\frac{1 \times 17}{2 \times 17} = \frac{17}{34}$
4. From the given list, the equivalent fractions of $\frac{1}{2}$ are $\frac{3}{6}, \frac{9}{18}, \frac{7}{14}, \frac{17}{34}$.

## WORKED EXAMPLE 4 Writing a sequence of equivalent fractions

**Write the sequence of the first three equivalent fractions for $\frac{2}{3}$.**

**THINK**

1. Write the first three equivalent fractions in the sequence by multiplying both the numerator and denominator by 2, 3 and 4.
2. Write the equivalent fractions.
3. Write your answer.

**WRITE**

1. $\frac{2}{3} = \frac{2 \times 2}{3 \times 2} = \frac{4}{6}$
   $= \frac{2 \times 3}{3 \times 3} = \frac{6}{9}$
   $= \frac{2 \times 4}{3 \times 4} = \frac{8}{12}$
2. $\frac{2}{3} = \frac{4}{6} = \frac{6}{9} = \frac{8}{12}$
3. The first three equivalent fractions for $\frac{2}{3}$ are $\frac{4}{6}, \frac{6}{9}$ and $\frac{8}{12}$.

## 4.2.3 Fractions on a number line

eles-3800

- The space between each whole number on a number line can be divided into equal parts. The number of equal parts is determined by the denominator of the fraction.

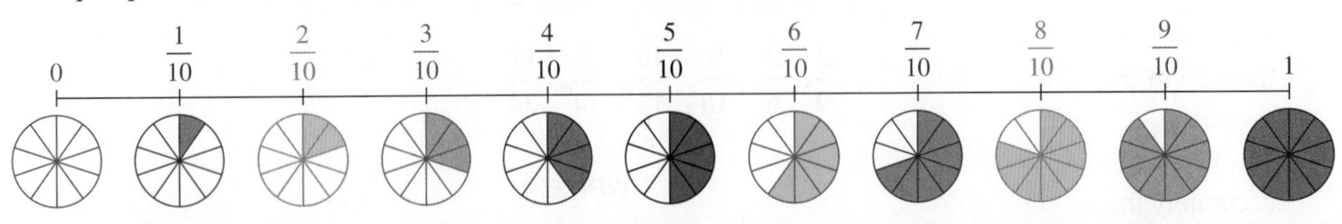


### WORKED EXAMPLE 5 Placing fractions on a number line

**Show the positions of $\frac{5}{6}$ and $1\frac{1}{6}$ on a number line.**

| THINK | WRITE |
|---|---|
| 1. Draw a number line from 0 to 2. | 0 1 2 |
| 2. The denominator of both fractions is 6. Divide the sections between 0 and 1 and between 1 and 2 into 6 equal parts each. | |
| 3. To place $\frac{5}{6}$, count 5 marks from 0. To place $1\frac{1}{6}$ count 1 mark from the whole number 1. | 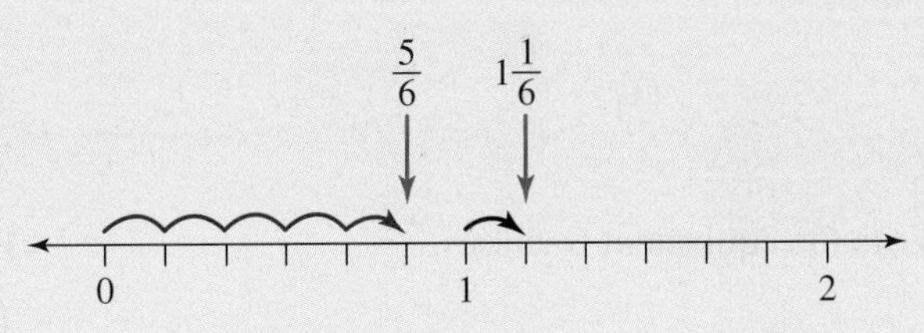 |



- When choosing a scale for your number lines, take into account the number of equal parts each whole number has been divided into. For example, when plotting thirds it is easier to use a scale of 3 cm for each whole number (for each part) as opposed to a scale of 1cm for each whole number.

### COLLABORATIVE TASK: Peg the fraction

**Equipment:** one long piece of string, one smaller piece of string for each group, laundry pegs, sticky notes

Your task is to place a peg on the string to show a fraction, following the steps below.

1. Two students hold the ends of a long piece of string across the front of the classroom.
2. The teacher calls out a fraction and, with the help of the class, another student places a peg on the string to show the position of that fraction.
3. The two students holding the ends of the string then fold the string into equal lengths to see how accurately the peg was placed.
4. Repeat steps **1** to **3** in small groups using the smaller lengths of string.
5. As a class, discuss the following.
   a. How do you visually estimate fractions when the numerator is 1?
   b. How can this strategy be used to estimate the position of a fraction in which the numerator is more than 1?

6. Two students divide the long piece of string in half and place a peg at the halfway mark. They hold the ends of the string so that it stretches across the front of the classroom. Another student writes the number 1 on a sticky note and attaches it to the peg.
7. The teacher calls out a fraction, which will be a mixed or an improper fraction, such as $1\frac{1}{2}$ or $\frac{7}{4}$. With the help of the class, another student places a peg on the string to show the position of that fraction on the string.
8. The two students holding the ends of the string then fold the string into equal lengths to see how accurately the peg was placed.
9. Repeat steps 6 to 8 in small groups using the smaller lengths of string.
10. As a class, discuss the effect the denominator has on determining the position of the peg.

## 4.2.4 Comparing fractions

eles-3801

- If we cut a block into 4 equal pieces, then 1 piece is less than 2 pieces. This means that $\frac{1}{4}$ of the block is smaller than $\frac{2}{4}$.
- Fraction walls, or number lines, are useful tools when comparing fractions. The following figure shows an example of a fraction wall.

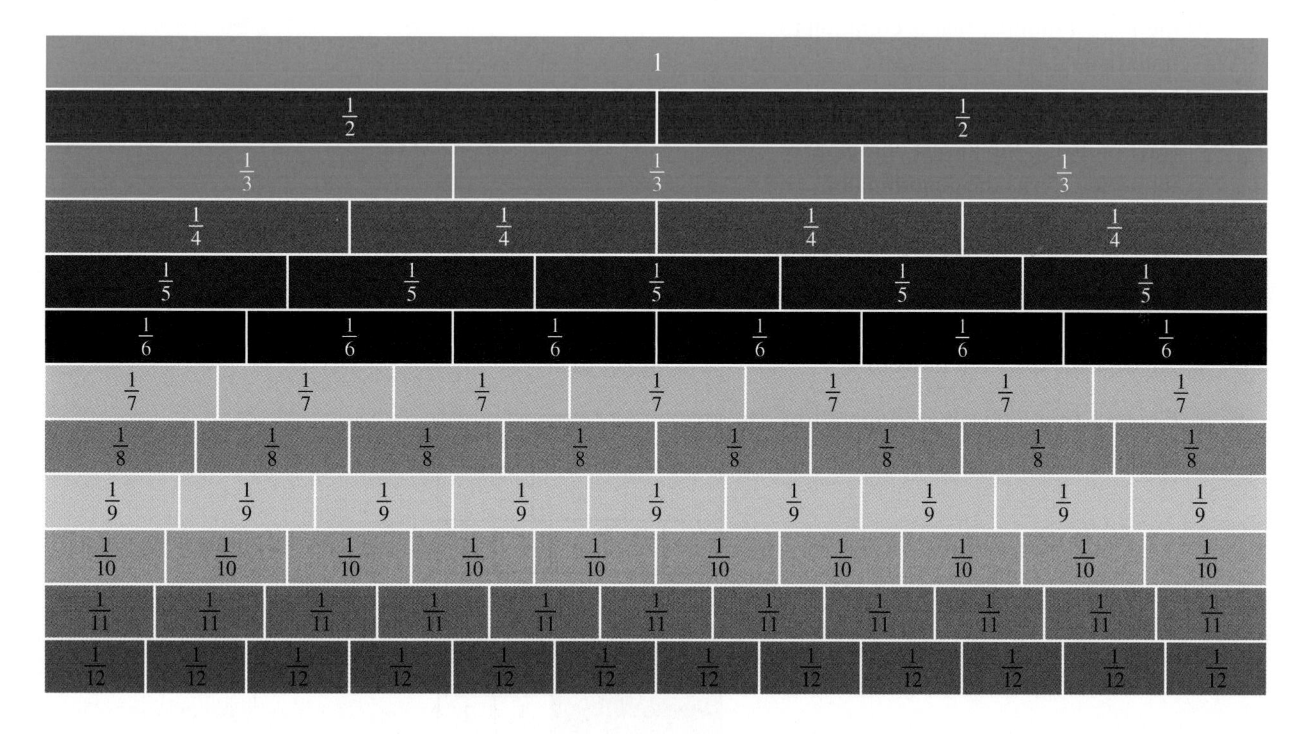


- With a diagram such as a fraction wall or a number line, we can see that $\frac{1}{4}$ is less than $\frac{1}{3}$. That is, if we divided a block into 4 equal parts, each part would be smaller than if we divided it into 3 equal parts.
- If the numerators are the same, the smaller fraction is the one with the larger denominator. For example, $\frac{1}{7}$ is less than $\frac{1}{6}$, and $\frac{3}{10}$ is less than $\frac{3}{7}$.

## Comparing fractions

**If the fractions you wish to compare do not have the same denominator:**

- **use equivalent fractions to convert both fractions so that they have the same denominator**
- **use the lowest common multiple of the two denominators (which is the best choice for the denominator).**

**Once the fractions have the same denominator, compare the numerators. The fraction with the largest numerator is bigger.**

### WORKED EXAMPLE 6 Comparing fractions

a. **Identify which is the bigger fraction, $\frac{2}{3}$ or $\frac{3}{5}$.**

b. **Justify your answer using a fraction wall and a number line.**

| THINK | WRITE |
| --- | --- |
| a. 1. Determine the lowest common multiple of the denominators. First, list the multiples of 3 and 5. Identify the lowest number that is common to both lists. | a. Multiples of 3 are 3, 6, 9, 12, (15), 18, ...<br>Multiples of 5 are 5, 10, (15), 20, ...<br>The lowest common multiple is 15. |
| 2. Write each fraction as an equivalent fraction using the lowest common multiple (15) as the denominator. | $\frac{2}{3}=\frac{2\times 5}{3\times 5}=\frac{10}{15}$ and $\frac{3}{5}=\frac{3\times 3}{5\times 3}=\frac{9}{15}$ |
| 3. Decide which is bigger by comparing the numerators of the equivalent fractions. | $\frac{10}{15}=\frac{2}{3}$ is bigger than $\frac{9}{15}=\frac{3}{5}$. |
| 4. Answer the question in words. | $\frac{2}{3}$ is bigger than $\frac{3}{5}$. |
| b. 1. Create a fraction wall showing thirds and fifths. | b. 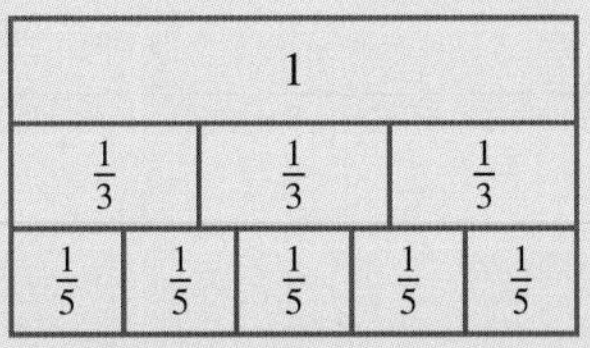 Fraction wall: 1; $\frac{1}{3}$ $\frac{1}{3}$ $\frac{1}{3}$; $\frac{1}{5}$ $\frac{1}{5}$ $\frac{1}{5}$ $\frac{1}{5}$ $\frac{1}{5}$ |
| 2. Shade $\frac{2}{3}$ and $\frac{3}{5}$. | 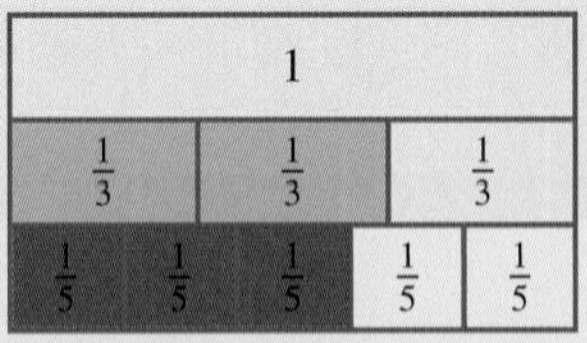 Fraction wall: 1; $\frac{1}{3}$ $\frac{1}{3}$ (shaded) $\frac{1}{3}$; $\frac{1}{5}$ $\frac{1}{5}$ $\frac{1}{5}$ (shaded) $\frac{1}{5}$ $\frac{1}{5}$ |
| 3. Compare the lengths of the shaded areas to compare the fractions. Answer the question in words. | $\frac{2}{3}$ is bigger than $\frac{3}{5}$. |

4. Draw a number line showing 0 to 1 in intervals of $\frac{1}{15}$ (found by using the lowest common multiple of the denominators).

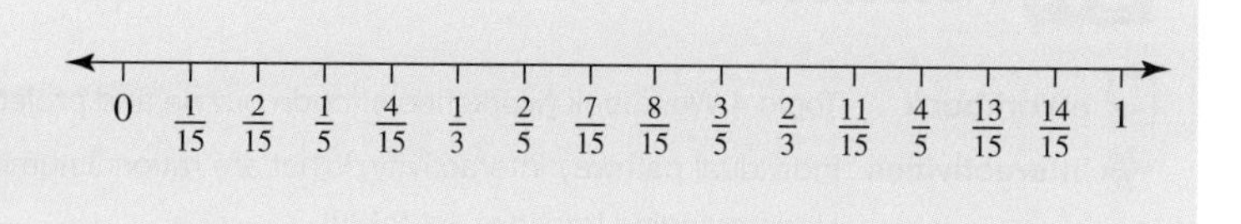


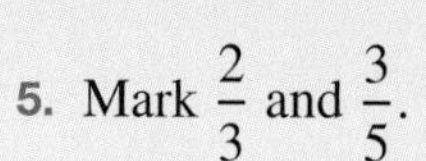

5. Mark $\frac{2}{3}$ and $\frac{3}{5}$.

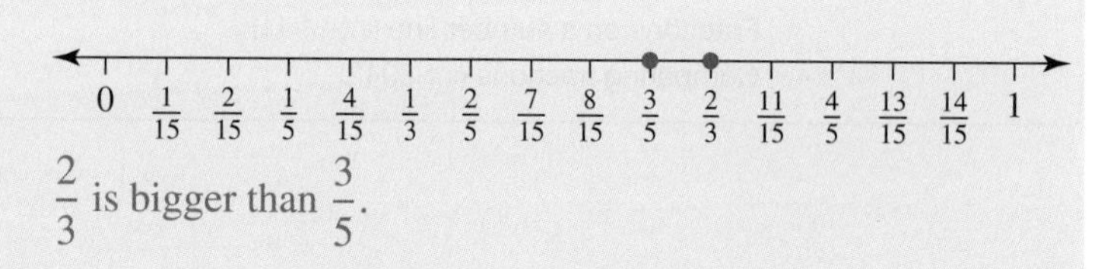


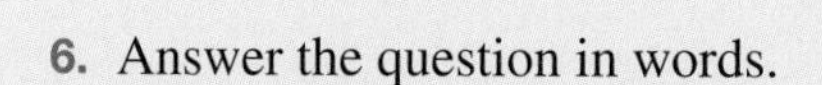

6. Answer the question in words.

$\frac{2}{3}$ is bigger than $\frac{3}{5}$.

## Comparison symbols

- The symbol '>' means *is bigger than* or *is greater than.*
- The symbol '<' means *is smaller than* or *is less than.*
- When we write $\frac{1}{2} < \frac{3}{4}$, it means that $\frac{1}{2}$ is less than $\frac{3}{4}$.

### WORKED EXAMPLE 7 Comparing fractions

**Insert the appropriate symbol, < or >, between each pair of fractions to make a true statement.**

a. $\frac{6}{7}$ $\frac{7}{8}$

b. $\frac{7}{12}$ $\frac{5}{9}$

| THINK | WRITE |
|---|---|
| a. 1. Determine the lowest common multiple of the denominators. | a. Multiples of 7 are: 7, 14, 21, 28, 35, 42, 49, (56), 63, ...<br>Multiples of 8 are: 8, 16, 24, 32, 40, 48, (56), 64, ...<br>The lowest common multiple is 56. |
| 2. Write each fraction as an equivalent fraction using the lowest common multiple (56) as the denominator. | $\frac{6}{7} = \frac{6 \times 8}{7 \times 8} = \frac{48}{56}$ and $\frac{7}{8} = \frac{7 \times 7}{8 \times 7} = \frac{49}{56}$ |
| 3. Decide which fraction is bigger by comparing the numerators of the equivalent fractions. | $\frac{48}{56} = \frac{6}{7}$ is less than $\frac{49}{56} = \frac{7}{8}$. |
| 4. Answer the question. | $\frac{6}{7} < \frac{7}{8}$ |
| b. 1. Determine the lowest common multiple of the denominators. | b. Multiples of 12 are: 12, 24, (36), 48, …<br>Multiples of 9 are: 9, 18, 27, (36), 45, ...<br>The lowest common multiple is 36. |
| 2. Write each fraction as an equivalent fraction using the lowest common multiple (36) as the denominator. | $\frac{7}{12} = \frac{7 \times 3}{12 \times 3} = \frac{21}{36}$ and $\frac{5}{9} = \frac{5 \times 4}{9 \times 4} = \frac{20}{36}$ |
| 3. Decide which fraction is bigger by comparing the numerators of the equivalent fractions. | $\frac{21}{36} = \frac{7}{12}$ is greater than $\frac{20}{36} = \frac{5}{9}$. |
| 4. Answer the question. | $\frac{7}{12} > \frac{5}{9}$ |

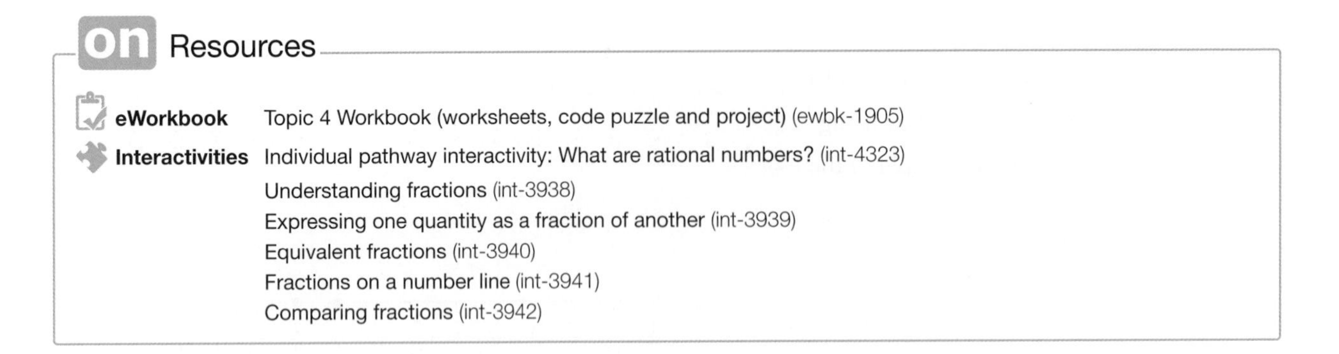
**on** Resources

**eWorkbook** Topic 4 Workbook (worksheets, code puzzle and project) (ewbk-1905)

**Interactivities** Individual pathway interactivity: What are rational numbers? (int-4323)
Understanding fractions (int-3938)
Expressing one quantity as a fraction of another (int-3939)
Equivalent fractions (int-3940)
Fractions on a number line (int-3941)
Comparing fractions (int-3942)

## Exercise 4.2 What are fractions?

learn**on**

4.2 Quick quiz  | **4.2 Exercise**

### Individual pathways

| ■ PRACTISE | ■ CONSOLIDATE | ■ MASTER |
|---|---|---|
| 1, 2, 4, 7, 9, 12, 14, 16, 19, 25, 28 | 3, 5, 8, 10, 13, 17, 20, 21, 23, 26, 29 | 6, 11, 15, 18, 22, 24, 27, 30 |

### Fluency

1. **WE1** Identify the fraction of each of the following rectangles that has been shaded.

a.

b. 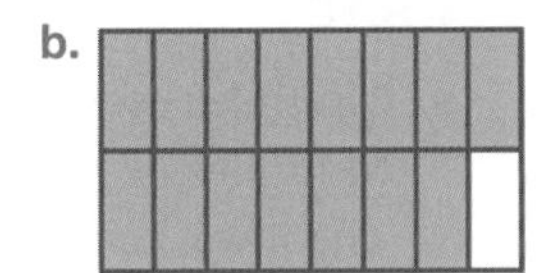

c. 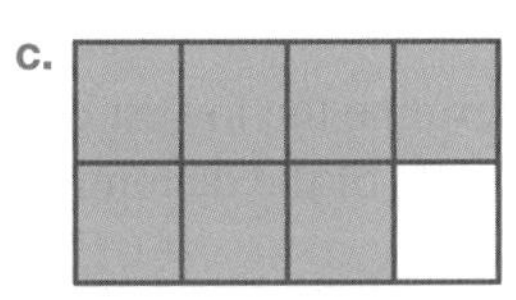

2. Identify the fraction of this flag that is coloured red.

3. Identify the fraction of this flag that is coloured red.

4. Calculate the fraction of the total number of pieces of fruit in the photograph that is made up of:
   a. bananas  b. oranges.

5. Calculate the fraction of chocolate biscuits in this packet that has been eaten.

6. Express 17 minutes as a fraction of an hour.
7. WE2 For the following equivalent fractions, determine the number that has been used to multiply both numerator and denominator.
   a. $\frac{3}{8} = \frac{9}{24}$  b. $\frac{1}{4} = \frac{3}{12}$  c. $\frac{3}{5} = \frac{12}{20}$  d. $\frac{2}{3} = \frac{10}{15}$
8. For the following equivalent fractions, determine the number that has been used to multiply both numerator and denominator.
   a. $\frac{5}{6} = \frac{30}{36}$  b. $\frac{9}{10} = \frac{81}{90}$  c. $\frac{7}{8} = \frac{77}{88}$  d. $\frac{7}{8} = \frac{84}{96}$
9. WE3 From the list, identify the fractions that are equivalent to the fraction marked in pink.
   $\frac{2}{3}$  $\frac{20}{30}$  $\frac{5}{8}$  $\frac{12}{16}$  $\frac{14}{21}$  $\frac{40}{60}$
10. From the list, identify the fractions that are equivalent to the fraction marked in pink.
   $\frac{4}{5}$  $\frac{12}{15}$  $\frac{15}{20}$  $\frac{36}{45}$  $\frac{16}{20}$  $\frac{28}{35}$  $\frac{80}{100}$
11. From the list, identify the fractions that are equivalent to the fraction marked in pink.
   $\frac{7}{10}$  $\frac{18}{25}$  $\frac{35}{50}$  $\frac{14}{21}$  $\frac{21}{30}$  $\frac{14}{20}$  $\frac{140}{200}$
12. WE4 Write a sequence of the first three equivalent fractions for $\frac{5}{6}$.
13. Write a sequence of the first three equivalent fractions for $\frac{7}{8}$.
14. WE5 Draw a number line from 0 to 2. Show the position of each of the following fractions on the number line.
   a. $\frac{1}{5}$  b. $\frac{4}{5}$  c. $\frac{7}{5}$  d. $1\frac{3}{5}$
15. Draw a number line from 0 to 3. Show the position of each of the following fractions on the number line.
   a. $\frac{1}{3}$  b. $\frac{4}{3}$  c. $\frac{8}{3}$  d. $\frac{3}{3}$

## Understanding

16. **WE6** Identify which is the bigger fraction. Justify your answer using a fraction wall or number line.

a. $\frac{2}{5}$ or $\frac{3}{5}$ b. $\frac{5}{8}$ or $\frac{7}{8}$ c. $\frac{1}{5}$ or $\frac{1}{6}$ d. $\frac{1}{8}$ or $\frac{1}{10}$

17. Identify which is the bigger fraction. Justify your answer using a fraction wall or number line.

a. $\frac{3}{4}$ or $\frac{3}{5}$ b. $\frac{2}{5}$ or $\frac{5}{8}$ c. $\frac{3}{5}$ or $\frac{5}{8}$ d. $\frac{4}{5}$ or $\frac{5}{8}$

18. **WE7** Insert the appropriate symbol, < or >, between each pair of fractions to make a true statement.

a. $\frac{5}{8} \quad \frac{3}{8}$ b. $\frac{3}{8} \quad \frac{1}{4}$ c. $\frac{3}{4} \quad \frac{3}{5}$ d. $\frac{3}{10} \quad \frac{2}{5}$

19. **MC** Select which fraction is smaller than $\frac{5}{8}$.

A. $\frac{7}{8}$ B. $\frac{17}{24}$ C. $\frac{11}{12}$ D. $\frac{3}{5}$ E. $\frac{13}{16}$

20. **MC** Select which fraction is equivalent to $\frac{3}{4}$.

A. $\frac{3}{8}$ B. $\frac{6}{8}$ C. $\frac{9}{16}$ D. $\frac{12}{24}$ E. $\frac{15}{24}$

21. Write the following fractions with the same denominator and then write them in ascending order (smallest to largest).

$$\frac{3}{8}, \frac{1}{2}, \frac{1}{3}$$

22. Write the following fractions with the same denominator and then write them in descending order (largest to smallest).

$$\frac{3}{4}, \frac{2}{3}, \frac{7}{15}$$

23. Fill in the gaps.

$$\frac{3}{4} = \frac{\phantom{00}}{40} = \frac{18}{\phantom{00}} = \frac{\phantom{00}}{36}$$

24. Fill in the gaps.

$$\frac{5}{6} = \frac{15}{\phantom{00}} = \frac{\phantom{00}}{42} = \frac{40}{\phantom{00}}$$

## Reasoning

25. Explain why a scale of 2 cm per whole number would not be the most appropriate scale if you are displaying sevenths on a number line.

26. On the number line shown, the fractions have the same numerator but the denominator is always double the previous one. Explain, with reasons, the relationship between these denominators and the distance between the numbers.

27. Four hungry teenagers ordered individual pizzas, but none could eat the whole pizza. Sally ate $\frac{6}{8}$ of her seafood pizza. Hanna ate $\frac{1}{2}$ of her Hawaiian pizza. Sonya ate $\frac{3}{4}$ of her special pizza. Vanessa ate $\frac{5}{6}$ of her vegetarian pizza.

a. Draw the four pizzas and shade the amount that each girl ate.
b. Determine who ate the most pizza.
c. State who ate more pizza, Sally or Sonya.

## Problem solving

28. Five students came in at lunchtime to get help with mathematics. Macy used $\frac{1}{4}$ of the time asking questions, Lee Yin used $\frac{1}{3}$ of the time and Lachlan used $\frac{1}{6}$ of the time. Abbey also asked questions for $\frac{2}{12}$ of the time and Tao $\frac{1}{12}$ of the time. Determine which of the students used most of the lunchtime asking questions.

29. Meilin puts four equally spaced points on a number line. The first point is at the $\frac{3}{10}$ mark and the last point is at the $\frac{7}{20}$ mark. Determine where she placed the two points in between.

30. Shirly is sitting $2\frac{5}{7}$ m from the classroom door and Brent is $2\frac{2}{11}$ m from the classroom door. Place Shirly's and Brent's seats on the number line given.

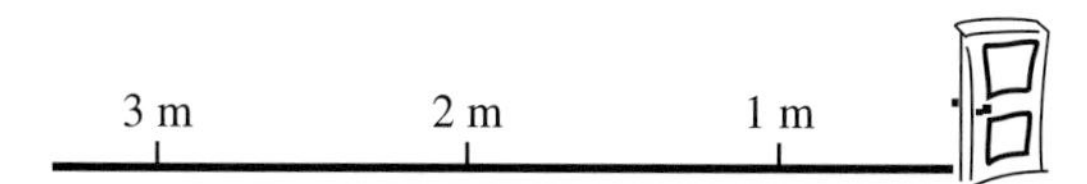

# LESSON
# 4.3 Simplifying fractions

## LEARNING INTENTION

At the end of this lesson you should be able to:
- simplify fractions
- simplify mixed numbers.

## 4.3.1 Simplifying fractions

eles-3802

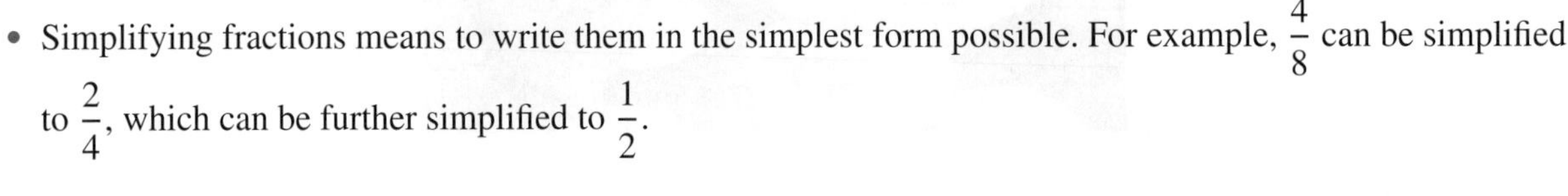

- Simplifying fractions means to write them in the simplest form possible. For example, $\frac{4}{8}$ can be simplified to $\frac{2}{4}$, which can be further simplified to $\frac{1}{2}$.

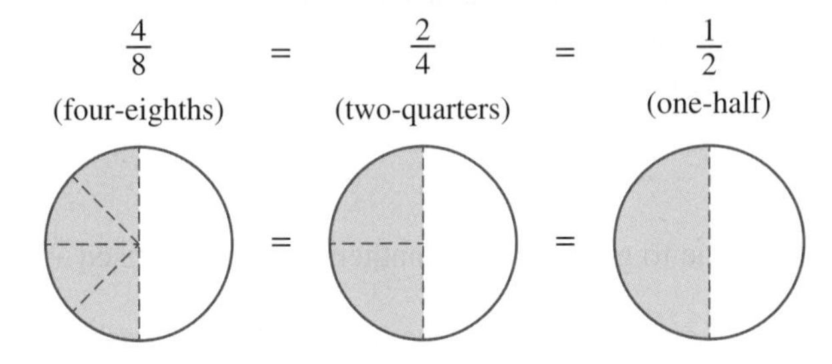


- Generally, fractions are written in their simplest form. In other words, we reduce the fraction to its lowest equivalent form.
- To simplify a fraction, divide the numerator and the denominator by the same number.

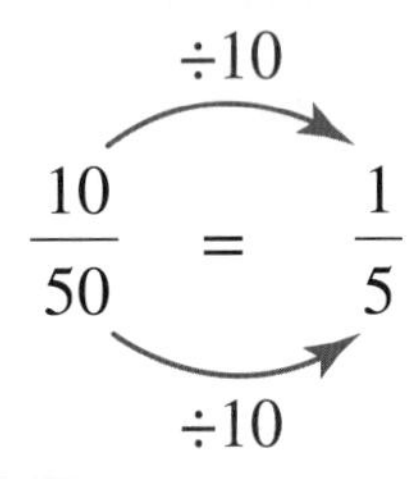


- The highest common factor (HCF) of two numbers is the largest factor of both numbers.
- A fraction is in simplest form when it cannot be simplified further. This occurs when the numerator and denominator share no common factor.

### Simplifying fractions

**There are two ways to convert a fraction into its simplest form:**
1. **Divide the numerator and denominator by the highest common factor (HCF).**
2. **Simplify repeatedly, by whichever common factor you like, until there are no common factors between the numerator and denominator.**

## WORKED EXAMPLE 8 Simplifying fractions

**Write $\frac{3}{6}$ in simplest form.**

| THINK | WRITE |
| --- | --- |
| 1. Write the fraction and determine the highest common factor (HCF) or the largest number that is a factor of both the numerator and the denominator. | $\frac{3}{6}$, HCF = 3 |
| 2. Divide the numerator and denominator by this factor ($3 \div 3 = 1, 6 \div 3 = 2$). | $\frac{3}{6} \overset{\div 3}{=} \frac{1}{2}$ |
| 3. Write the answer in simplest form. | $= \frac{1}{2}$ |

### Simplifying mixed numbers

- To simplify a mixed number, leave the whole number part and simplify the fraction part.

## WORKED EXAMPLE 9 Simplifying mixed numbers

**Simplify $7\frac{40}{64}$.**

| THINK | WRITE |
| --- | --- |
| 1. Write the mixed number and determine the HCF of the numerator and denominator of the fraction part. | $7\frac{40}{64}$, HCF = 8 |
| 2. Divide both numerator and denominator by this factor ($40 \div 8 = 5, 64 \div 8 = 8$). | $\frac{40}{64} \overset{\div 8}{=} \frac{5}{8}$ |
| 3. Write your answer as a mixed number in simplest form. | $= 7\frac{5}{8}$ |

### DISCUSSION

When simplifying a fraction, how do you know when it has been fully simplified? What method do you use to simplify fractions?

Resources

eWorkbook Topic 4 Workbook (worksheets, code puzzle and project) (ewbk-1905)

Interactivities Individual pathway interactivity: Simplifying rational numbers (int-4324)
Simplifying fractions (int-3943)

## Exercise 4.3 Simplifying fractions

learnon

4.3 Quick quiz on | 4.3 Exercise

Individual pathways

| ■ PRACTISE | ■ CONSOLIDATE | ■ MASTER |
|---|---|---|
| 1, 4, 6, 7, 11, 14 | 2, 5, 8, 10, 12, 15 | 3, 9, 13, 16 |

### Fluency

1. **WE8** Write the following fractions in simplest form.
   a. $\frac{5}{10}$  b. $\frac{8}{12}$  c. $\frac{21}{24}$  d. $\frac{48}{60}$
2. Write the following fractions in simplest form.
   a. $\frac{100}{120}$  b. $\frac{48}{50}$  c. $\frac{63}{72}$  d. $\frac{49}{70}$
3. Write the following fractions in simplest form.
   a. $\frac{33}{36}$  b. $\frac{22}{50}$  c. $\frac{21}{56}$  d. $\frac{49}{56}$
4. **WE9** Simplify.
   a. $4\frac{21}{35}$  b. $7\frac{45}{54}$  c. $10\frac{10}{20}$  d. $1\frac{44}{48}$
5. Simplify.
   a. $6\frac{21}{28}$  b. $5\frac{16}{48}$  c. $3\frac{11}{55}$  d. $2\frac{16}{64}$
6. **MC** Select the equivalent fraction to $\frac{26}{28}$.
   A. $\frac{2}{3}$  B. $\frac{2}{6}$  C. $\frac{3}{8}$  D. $\frac{9}{12}$  E. $\frac{13}{14}$

### Understanding

7. Kylie's netball team scored 28 goals. Kylie scored 21 of her team's goals. Calculate the fraction of the team's goals that Kylie scored. Simplify the answer.
8. Angelo's basketball team scored 36 points. Angelo scored 20 of the team's points.
   a. Calculate the fraction of the team's points that Angelo scored. Simplify the answer.
   b. Calculate the fraction of the team's points that the rest of the team scored. Simplify the answer.

9. Year 7 students at Springfield High School ran a car wash to raise money for the local hospital. They raised a total of \$1000 and drew up a table showing how much money each class raised.

| 7A | 7B | 7C | 7D | 7E |
|---|---|---|---|---|
| \$200 | \$150 | \$320 | \$80 | \$250 |

Express as a simple fraction how much of the total each class raised.

10. Below are the results of a book-reading competition.

Mark 5 books
Ken 10 books
Samantha 8 books
Jules 6 books
Ahmed 4 books
Darren 7 books

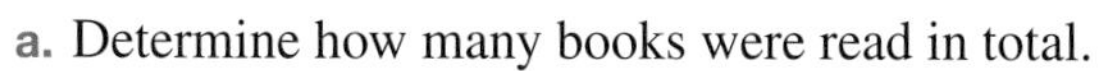

a. Determine how many books were read in total.
b. For each contestant, record the number of books read as a fraction of the total number of books read. Where possible, reduce the fractions to their simplest forms.

### Reasoning

11. The Newcastle to Sydney train runs on time 3 out of every 5 trains. The Sydney to Newcastle train runs on time 4 out of every 7 trains. Determine which train is more reliable.

12. a. Draw three lines to divide the triangle shown into 4 equal parts.
b. State what fraction of the original triangle is one of the new triangles.
c. Halve each smaller triangle.
d. State what fraction of the smaller triangle is one of these even smaller triangles.
e. Using all the triangles available, explain why $\frac{2}{8} = \frac{1}{4}$.

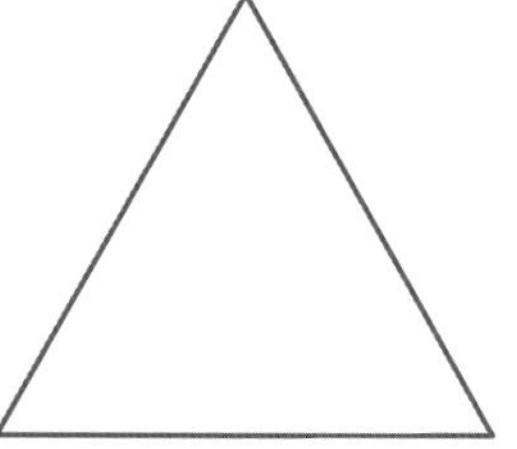

13. The fraction $\frac{13}{n}$ is not written in its simplest form. Explain what number the denominator could be.

### Problem solving

14. Sarina orders pizza for 22 friends. Each friend must get $\frac{1}{4}$ of a pizza.
a. Determine the number of pizzas Sarina needs to order.
b. Calculate the amount of leftover pizza.

15. a. Arrange the digits 1, 2, 3, 4, 5, 6, 7, 8 and 9 to form a fraction that is equivalent to $\frac{1}{2}$.
For example, one arrangement is $\frac{7329}{14\,658}$. Can you find two others?
*Hint:* Try starting with a numerator beginning with 72 for one of them and 67 for the other.
b. Arrange the nine digits so that they form a fraction equivalent to $\frac{1}{3}$.

16. One of the designs for a patchwork motif is made up of squares and right-angled isosceles triangles as shown. Determine what fraction of the whole has been shaded.

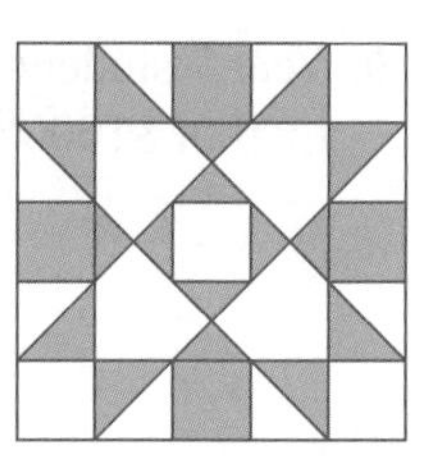

# LESSON
# 4.4 Mixed numbers and improper fractions

## LEARNING INTENTION

At the end of this lesson you should be able to:

- convert between mixed numbers and improper fractions.

## 4.4.1 Converting improper fractions to mixed numbers

eles-3803

- An improper fraction has a numerator larger than the denominator: for example, $\frac{3}{2}$.
- An improper fraction can be changed to a mixed number by dividing the denominator into the numerator and writing the remainder over the same denominator.

**Mixed numbers and improper fractions**

**Mixed numbers and improper fractions are different ways of writing the same quantity.**

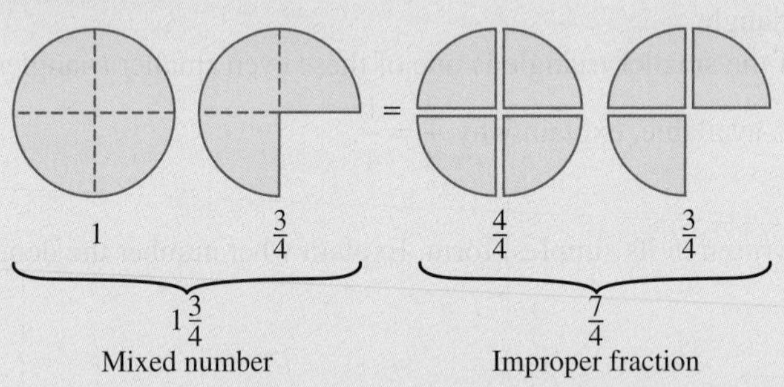


**WORKED EXAMPLE 10 Converting improper fractions to mixed numbers**

**Draw a diagram to show $\frac{5}{4}$ as parts of a circle, then write the improper fraction as a mixed number.**

| THINK | WRITE |
|---|---|
| 1. Draw a whole circle and divide it into the number of parts shown by the denominator. | 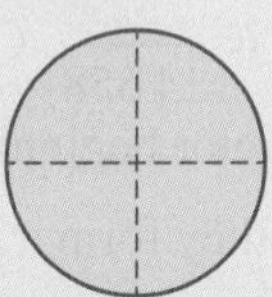 This is 4 quarters or $\frac{4}{4}$. |

2. Determine the number of parts left over and draw them.

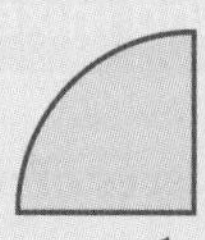

Extra $\frac{1}{4}$ needed.

3. Write the improper fraction as a mixed number. $\frac{5}{4} = 1\frac{1}{4}$

**WORKED EXAMPLE 11 Converting improper fractions to mixed numbers**

**Express $\frac{11}{5}$ as a mixed number.**

| THINK | WRITE |
|---|---|
| 1. Write the improper fraction. | $\frac{11}{5}$ |
| 2. Determine how many times the denominator can be divided into the numerator and what the remainder is. The whole number is part of the answer and the remainder becomes the numerator of the fractional part with the same denominator as the original improper fraction. | $= 11 \div 5$<br>$= 2$ remainder 1<br>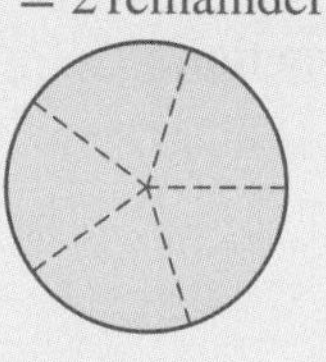 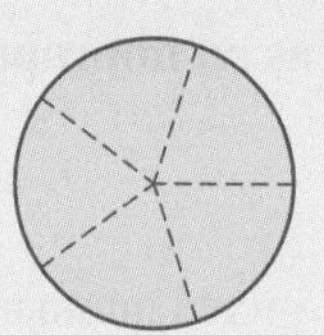 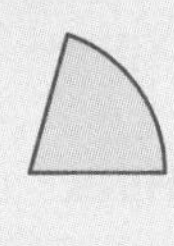 |
| 3. Write the answer. | $= 2\frac{1}{5}$ |

## 4.4.2 Converting mixed numbers to improper fractions

eles-3804

- A mixed number can be changed to an improper fraction by first multiplying the whole number by the denominator and then adding the numerator. The denominator stays the same.
- You may wish to draw a diagram to help visualise the conversion.

**Mixed numbers to improper fractions**

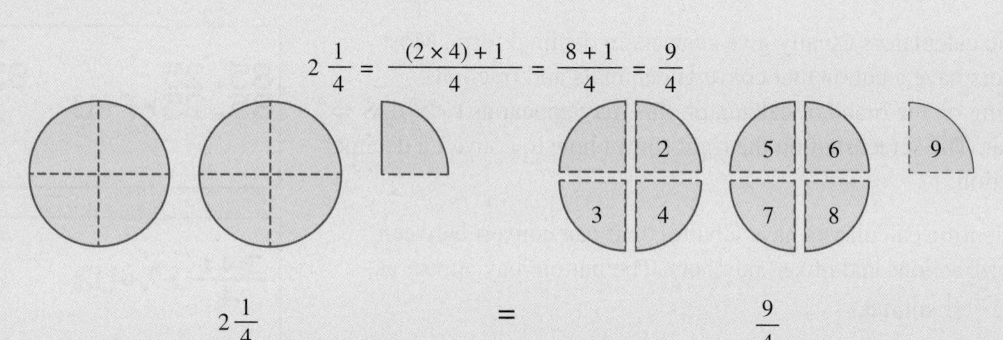

## WORKED EXAMPLE 12 Converting mixed numbers to improper fractions

**Draw a diagram to illustrate $2\frac{1}{3}$ as pieces of a circle, then write $2\frac{1}{3}$ as an improper fraction.**

| THINK | WRITE |
|---|---|
| 1. Draw two whole circles and $\frac{1}{3}$ of a circle. <br> 2. Divide the whole circles into thirds. | 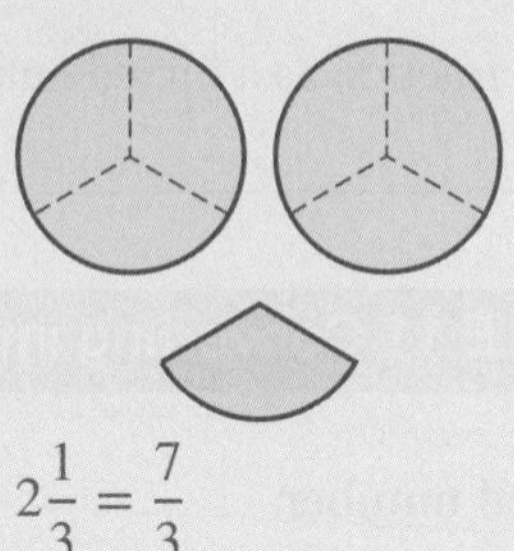 |
| 3. Count the number of thirds and write the mixed number as an improper fraction. | $2\frac{1}{3}=\frac{7}{3}$ |

## WORKED EXAMPLE 13 Converting mixed numbers to improper fractions

**Express $2\frac{3}{4}$ as an improper fraction.**

| THINK | WRITE |
|---|---|
| 1. Write the mixed number. | $2\frac{3}{4}$ |
| 2. Multiply the whole number by the denominator, then add the numerator. The result becomes the numerator, and the denominator stays the same. | $=\frac{2\times4+3}{4}$ |
| 3. Evaluate the top line of the fraction. | $=\frac{8+3}{4}$ |
| 4. Write the answer. | $=\frac{11}{4}$ |

### Digital technology

Scientific calculators usually give answers in decimal form. Most calculators have a button that converts decimals and fractions. Depending on the brand of calculator, this may appear as f ◁▷ d, S ⇔ or similar. The screenshot on the right shows how to convert a decimal to a fraction.

DEG
85.25 85.25
85.25▸f◂▸d $\frac{341}{4}$

Most scientific calculators have a button that can convert between improper fractions and mixed numbers. The button may appear as $\frac{n}{d}$ ◁▷ $U\frac{n}{d}$ or similar.

DEG
$\frac{341}{4}$▸%◂▸U% $85\frac{1}{4}$

Resources

eWorkbook Topic 4 Workbook (worksheets, code puzzle and project) (ewbk-1905)

Interactivities Individual pathway interactivity: Mixed numbers and improper fractions (int-4325)
Mixed numbers to improper fractions (int-3944)

## Exercise 4.4 Mixed numbers and improper fractions

learnon

4.4 Quick quiz 

4.4 Exercise

### Individual pathways

| ■ PRACTISE | ■ CONSOLIDATE | ■ MASTER |
|---|---|---|
| 1, 2, 4, 7, 10, 12, 14, 17, 20, 22, 25 | 5, 9, 11, 15, 18, 21, 23, 26 | 3, 6, 8, 13, 16, 19, 24, 27, 28 |

### Fluency

1. WE10 Draw a diagram to show the following improper fractions as pieces of a circle, then write each improper fraction as a mixed number.

   a. $\frac{5}{2}$ b. $\frac{4}{3}$ c. $\frac{13}{8}$

2. Aamira ate $\frac{3}{2}$ or 3 halves of an apple. Draw the amount of apple that Aamira ate and express $\frac{3}{2}$ as a mixed number.

3. Dean ate $\frac{9}{4}$ or 9 quarters of pizza. Draw the amount of pizza that Dean ate and express $\frac{9}{4}$ as a mixed number.

4. WE11 Express these improper fractions as mixed numbers.

   a. $\frac{7}{5}$ b. $\frac{11}{4}$ c. $\frac{21}{2}$ d. $\frac{51}{12}$ e. $\frac{92}{11}$

5. Express these improper fractions as mixed numbers.

a. $\frac{29}{13}$ b. $\frac{100}{3}$ c. $\frac{25}{2}$ d. $\frac{20}{3}$ e. $\frac{8}{5}$

6. Change these improper fractions to mixed numbers.

a. $\frac{27}{8}$ b. $\frac{11}{2}$ c. $\frac{58}{7}$ d. $\frac{117}{10}$ e. $\frac{67}{8}$

7. **MC** Select which of the following is the same as $\frac{61}{8}$.

A. $5\frac{5}{8}$ B. $6\frac{5}{8}$ C. $7\frac{5}{8}$ D. $8\frac{5}{8}$ E. $9\frac{5}{8}$

8. **MC** Select which of the following is the same as $\frac{74}{10}$.

A. $3\frac{7}{10}$ B. $7\frac{2}{5}$ C. $7\frac{7}{10}$ D. $7\frac{4}{5}$ E. $7\frac{4}{10}$

9. **WE12** Draw a diagram to show the following mixed numbers as pieces of a circle and then write each one as an improper fraction.

a. $1\frac{2}{3}$ b. $3\frac{1}{4}$ c. $2\frac{3}{5}$

10. **WE13** Express the following mixed numbers as improper fractions.

a. $1\frac{1}{8}$ b. $3\frac{3}{4}$ c. $2\frac{1}{6}$ d. $1\frac{1}{7}$ e. $1\frac{4}{5}$

11. Express the following mixed numbers as improper fractions.

a. $3\frac{1}{3}$ b. $6\frac{1}{4}$ c. $4\frac{1}{9}$ d. $11\frac{1}{2}$ e. $7\frac{4}{5}$

12. Express the following mixed numbers as improper fractions.

a. $6\frac{3}{5}$ b. $9\frac{2}{5}$ c. $3\frac{1}{2}$ d. $4\frac{1}{5}$ e. $6\frac{5}{7}$

13. Express the following mixed numbers as improper fractions.

a. $2\frac{9}{10}$ b. $3\frac{7}{12}$ c. $9\frac{4}{7}$ d. $1\frac{10}{11}$ e. $8\frac{5}{6}$

14. **MC** Select which of the following is the same as $10\frac{2}{7}$.

A. $\frac{27}{10}$ B. $\frac{72}{10}$ C. $\frac{10}{27}$ D. $\frac{72}{7}$ E. $\frac{72}{2}$

15. **MC** Select which of the following is the same as $8\frac{5}{6}$.

A. $\frac{85}{6}$ B. $\frac{58}{6}$ C. $\frac{56}{8}$ D. $\frac{53}{6}$ E. $\frac{68}{5}$

16. **MC** Select which of the following is the same as $7\frac{9}{11}$.

A. $\frac{86}{11}$ B. $\frac{99}{7}$ C. $\frac{79}{11}$ D. $\frac{97}{11}$ E. $\frac{77}{9}$

### Understanding

17. Kim and Carly arrived home from school quite hungry and cut up some fruit. Kim ate 9 quarters of apple and Carly ate 11 quarters. Calculate how many apples they each ate.

18. Thanh was selling pieces of quiche at the school fete. Each quiche was divided into 8 pieces. If Thanh sold 52 pieces, calculate how many quiches Thanh sold.

19. It was Cecilia's responsibility to supply oranges for her soccer team at half-time. If 19 players ate $\frac{1}{4}$ of an orange each, calculate how many oranges were eaten.

20. Insert the appropriate $<$ or $>$ sign between each pair of fractions to make a true statement.
    a. $\frac{8}{5} \quad 1\frac{2}{5}$
    b. $\frac{19}{2} \quad 8\frac{1}{2}$
    c. $\frac{33}{32} \quad 1\frac{3}{32}$
    d. $\frac{15}{2} \quad 5\frac{1}{2}$

21. Insert the appropriate $<$ or $>$ sign between each pair of fractions to make a true statement.
    a. $7\frac{1}{7} \quad \frac{65}{7}$
    b. $2\frac{1}{3} \quad \frac{8}{3}$
    c. $1\frac{16}{17} \quad \frac{35}{17}$
    d. $4\frac{3}{8} \quad \frac{36}{8}$

### Reasoning

22. A catering company provided 12 quiches that were each cut into 8 pieces for a luncheon. At the end of the luncheon, 12 pieces of quiche were left over. Calculate how many whole quiches were eaten. Show your full working.

23. a. Convert the following numbers to improper fractions.
    i. $3\frac{2}{7}$
    ii. $5\frac{3}{4}$

    b. Explain with the use of diagrams why $3\frac{2}{7} = \frac{23}{7}$.

24. A bakery makes birthday cakes to serve 24 people. Arabella takes one to school to celebrate her birthday, but there are 30 people in her class. Determine what fraction of the original slices each student will receive.

### Problem solving

25. a. Nasira made apple pies for her Grandma's 80th birthday party. She divided each pie into 6 pieces, and after the party she noted that $4\frac{1}{6}$ pies had been eaten. Calculate how many pieces were eaten.
    b. Nasira's cousin Ria provided cordial for the same party and calculated that she could make 20 drinks from each bottle. At the end of the party, $3\frac{17}{20}$ bottles had been used. Calculate how many drinks were consumed.

26. Baljinder walks to school four days a week and her father drives her to school on Fridays. If she lives $2\frac{1}{3}$ km from school, calculate:
    a. the distance Baljinder walks to school and back home each day
    b. the total distance walked by Baljinder over the four days
    c. the total distance she travels to school either walking or by car over the whole week.
    State your answers in improper fractions and mixed numbers if possible.

27. The following mixed numbers are missing their denominators. The missing denominators are 3, 4, 6, 8 and 14. The mixed numbers are placed in order from smallest to largest. Determine the missing numbers.

$$1\frac{1}{a},\quad 2\frac{1}{b},\quad 2\frac{1}{c},\quad 3\frac{1}{d},\quad 3\frac{1}{e}$$

28. Explain the strategies you use to write an improper fraction as a mixed number.

# LESSON
# 4.5 Adding and subtracting fractions

**LEARNING INTENTION**

At the end of this lesson you should be able to:
- add and subtract fractions.

## 4.5.1 Adding and subtracting proper fractions

eles-3805

- Fractions can be added and subtracted if they have the same denominator.

## Adding and subtracting fractions with equal denominators

**To add or subtract fractions of the same denominator, simply add or subtract the numerators. Leave the denominator unchanged.**

**For example:**

$$\frac{1}{5}+\frac{2}{5}=\frac{1+2}{5}=\frac{3}{5} \qquad \frac{5}{8}-\frac{2}{8}=\frac{5-2}{8}=\frac{3}{8}$$

- Addition and subtraction of fractions can be visualised using areas of shapes.

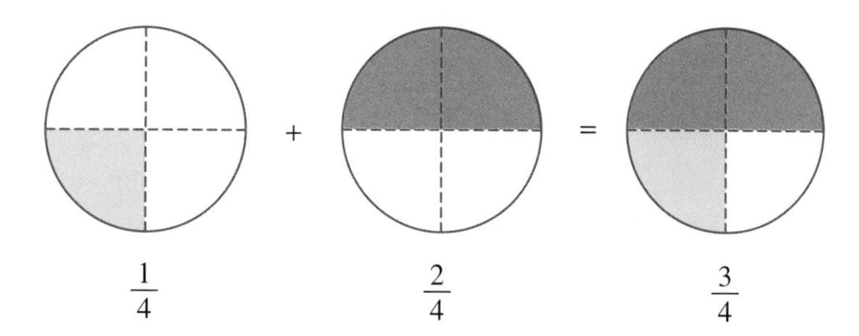


$\frac{1}{4}$ $\qquad \frac{2}{4}$ $\qquad \frac{3}{4}$

### WORKED EXAMPLE 14 Adding proper fractions

Evaluate $\frac{5}{9}+\frac{1}{9}$.

| THINK | WRITE |
|---|---|
| 1. Write the question. | $\frac{5}{9}+\frac{1}{9}$ |
| 2. Since the denominators are the same, add the numerators. | $=\frac{6}{9}$ |
| 3. Simplify the answer if possible by cancelling. Divide by the highest common factor $(6 \div 3 = 2, 9 \div 3 = 3)$. | $=\frac{\cancel{6}^2}{\cancel{9}^3}$ <br> $=\frac{2}{3}$ |

## Lowest common denominator

- The lowest common denominator (LCD) is the lowest number that is a multiple of the denominators. This is also known as the lowest common multiple (LCM) of the numbers in the denominators.

## Adding and subtracting fractions with different denominators

**To add or subtract fractions with different denominators, convert to equivalent fractions with the lowest common denominator (LCD), then add the numerators.**

**For example:** $\frac{1}{2}+\frac{1}{3}$

**The LCD of 2 and 3 is 6.**

**Add the intermediate step to the equation and calculate the answer:**

$$\frac{1}{2}+\frac{1}{3}=\frac{3}{6}+\frac{2}{6}=\frac{3+2}{6}=\frac{5}{6}$$

- Let us visualise the previous example using areas of circles. We can divide both circles into the same number of parts and then add the parts.

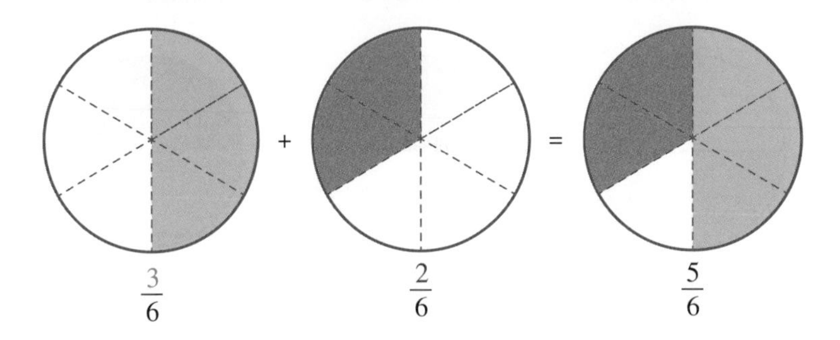


$\frac{3}{6}$ $\frac{2}{6}$ $\frac{5}{6}$

## WORKED EXAMPLE 15 Subtracting proper fractions

**Evaluate $\frac{5}{6}-\frac{1}{12}$, expressing the answer in simplest form.**

| THINK | WRITE |
|---|---|
| 1. List multiples of each denominator. | Multiples of 6 are 6, 12, ...<br>Multiples of 12 are 12, ... |
| 2. Identify the lowest common denominator. | LCD is 12. |
| 3. Rewrite the fractions as equivalent fractions with 12 as the LCD. | $\frac{5}{6}-\frac{1}{12}=\frac{5\times 2}{6\times 2}-\frac{1}{12}=\frac{10}{12}-\frac{1}{12}$ |
| | 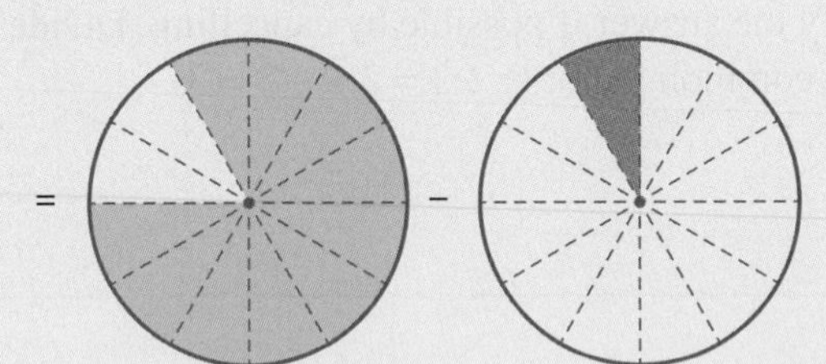 |
| 4. Subtract the numerators. | $=\frac{9}{12}$ |
| | 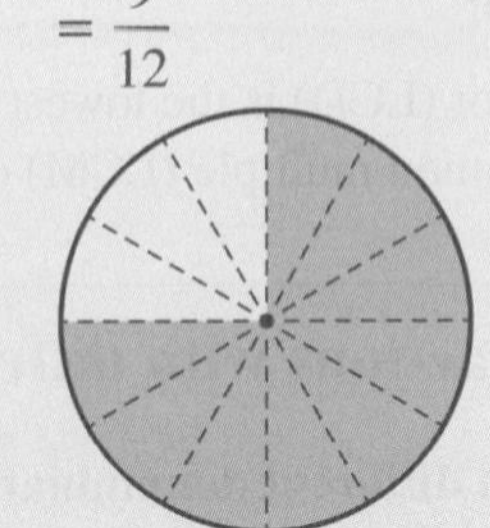 |

5. Simplify the answer if possible by cancelling. Divide by the highest common factor ($9 \div 3 = 3, 12 \div 3 = 4$).

$$= \frac{\cancel{9}^{3}}{\cancel{12}^{4}}$$

$$= \frac{3}{4}$$

## WORKED EXAMPLE 16 Subtracting a fraction from a whole number

**Evaluate $4 - \frac{1}{3}$, expressing the answer as a mixed number.**

**THINK** | **WRITE**

1. Express the whole number as a fraction over 1.

$$4 - \frac{1}{3} = \frac{4}{1} - \frac{1}{3}$$

2. Equate the denominators.

$$= \frac{12}{3} - \frac{1}{3}$$

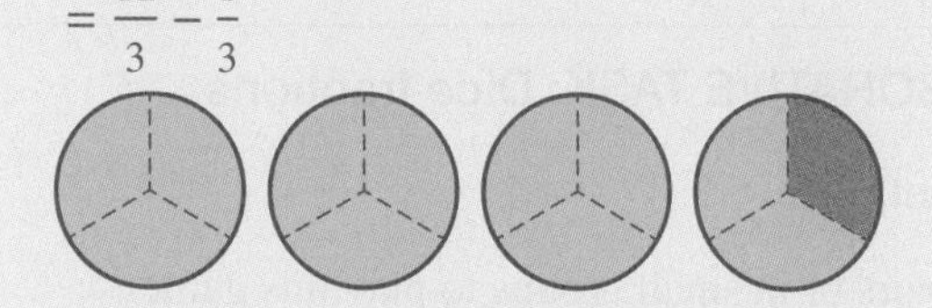

3. Perform the subtraction.

$$= \frac{11}{3}$$

1 2 3 | 4 5 6 | 7 8 9 | 10 11

4. Convert to a mixed number.

$$= 3\frac{2}{3}$$

## WORKED EXAMPLE 17 Application of adding and subtracting fractions

**A farmer picks the fruit from $\frac{2}{3}$ of her orange trees and her partner collects fruit from another $\frac{1}{6}$ of the trees. Determine the fraction of the total number of trees that still need to have their fruit picked.**

**THINK** | **WRITE**

1. Calculate the total fraction of trees that have had fruit picked, by adding the fractions.

$$\frac{2}{3} + \frac{1}{6} = \frac{4}{6} + \frac{1}{6}$$

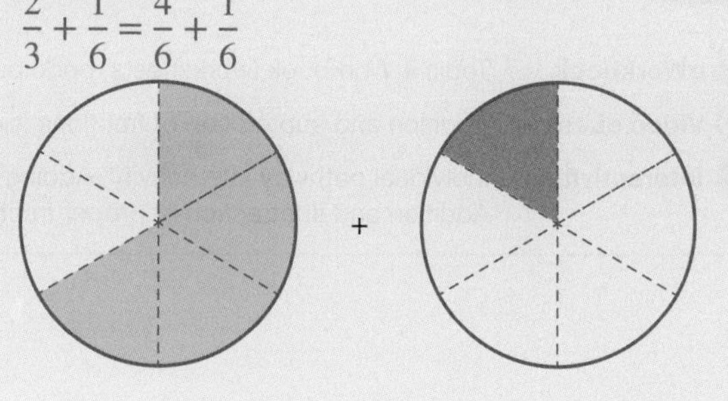

$$= \frac{5}{6}$$

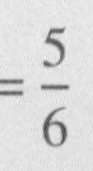

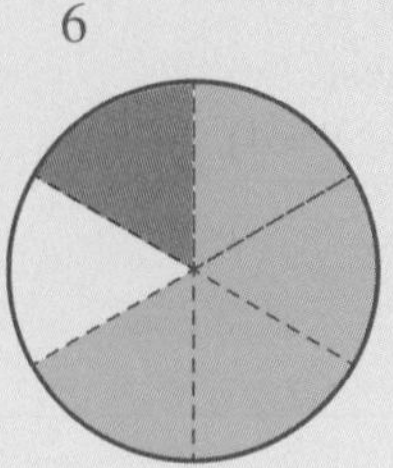

2. Using the diagram, determine the remaining fraction of trees that have not yet had the fruit picked.

$\frac{1}{6}$

Or, alternatively

$$1 - \frac{5}{6} = \frac{6}{6} - \frac{5}{6}$$
$$= \frac{1}{6}$$

3. Write the answer.

$\frac{1}{6}$ of the orange trees still have fruit that needs to be picked.

---

### COLLABORATIVE TASK: Dice fractions

**Equipment:** dice

Work in pairs or in small groups to play this game.

1. The first player rolls both dice. One number rolled is the numerator of the fraction and the other number rolled is the denominator. The player makes the smallest possible fraction with the numbers rolled.
2. The first player rolls the dice again to make another fraction, which should be as close as possible to the first fraction.
3. The first player works out the difference between the two fractions rolled by subtracting one fraction from the other. This amount is recorded as their result for this round.
4. Repeat steps 1 to 3 until every player has a result for the first round.
5. Compare the results for the round. The player with the smallest difference between their two fractions receives 1 point. If two players have the smallest difference, they both receive 1 point. The game is over when a player has collected 5 points.
6. What strategy did you use to make the smallest difference between your fractions?

---

**on** Resources

**eWorkbook** Topic 4 Workbook (worksheets, code puzzle and project) (ewbk-1905)

**Video eLesson** Addition and subtraction of fractions (eles-1866)

**Interactivities** Individual pathway interactivity: Adding and subtracting rational numbers (int-4326)
Addition and subtraction of proper fractions (int-3945)

# Exercise 4.5 Adding and subtracting fractions

learnon

4.5 Quick quiz 

4.5 Exercise

Individual pathways

| ■ PRACTISE | ■ CONSOLIDATE | ■ MASTER |
| --- | --- | --- |
| 1, 3, 5, 7, 11, 14, 16, 20, 23 | 2, 6, 8, 9, 12, 15, 17, 21, 24 | 4, 10, 13, 18, 19, 22, 25, 26 |

## Fluency

1. **WE14** Evaluate:
   a. $\frac{1}{5}+\frac{3}{5}$ b. $\frac{9}{11}-\frac{3}{11}$ c. $\frac{3}{6}+\frac{2}{6}$ d. $\frac{20}{50}+\frac{11}{50}$
2. Evaluate the following, expressing the answers in simplest form.
   a. $\frac{5}{12}+\frac{1}{12}$ b. $\frac{7}{16}-\frac{3}{16}$ c. $\frac{9}{15}-\frac{4}{15}$ d. $\frac{37}{100}+\frac{13}{100}$
3. Evaluate:
   a. $\frac{15}{25}-\frac{6}{25}$ b. $\frac{12}{12}-\frac{1}{12}$ c. $\frac{5}{8}-\frac{1}{8}+\frac{3}{8}$ d. $\frac{4}{7}+\frac{5}{7}-\frac{3}{7}$
4. Evaluate the following, expressing the answers in simplest form.
   a. $\frac{4}{7}+\frac{5}{7}-\frac{2}{7}$ b. $\frac{13}{16}+\frac{9}{16}-\frac{10}{16}$ c. $\frac{11}{28}+\frac{10}{28}$ d. $\frac{21}{81}+\frac{21}{81}+\frac{30}{81}$
5. **WE15** Evaluate the following, expressing the answers in simplest form.
   a. $\frac{1}{2}+\frac{1}{4}$ b. $\frac{1}{2}-\frac{1}{4}$ c. $\frac{1}{3}+\frac{2}{21}$ d. $\frac{1}{8}+\frac{20}{32}$
6. Evaluate the following, expressing the answers in simplest form.
   a. $\frac{9}{18}+\frac{1}{6}$ b. $\frac{7}{9}-\frac{8}{27}$ c. $\frac{31}{35}-\frac{3}{7}$ d. $\frac{8}{10}-\frac{28}{70}$
7. a. Determine the lowest common multiple of each of the following pairs of numbers.
      i. 2 and 3 ii. 3 and 5 iii. 5 and 10 iv. 4 and 6
   b. Use the answers that were found in part a as the lowest common denominators to add or subtract these fractions.
      i. $\frac{1}{2}-\frac{1}{3}$ ii. $\frac{2}{3}-\frac{2}{5}$ iii. $\frac{3}{5}+\frac{3}{10}$ iv. $\frac{3}{4}-\frac{1}{6}$
8. a. Determine the lowest common multiple of each of the following pairs of numbers.
      i. 9 and 6 ii. 4 and 8 iii. 6 and 8 iv. 7 and 5
   b. Use the answers from part a as the lowest common denominators to add or subtract these fractions.
      i. $\frac{1}{9}+\frac{5}{6}$ ii. $\frac{1}{4}+\frac{3}{8}$ iii. $\frac{5}{6}-\frac{3}{8}$ iv. $\frac{5}{7}-\frac{2}{5}$
9. Answer the following by first identifying the lowest common denominator. Simplify your answer if necessary.
   a. $\frac{7}{10}+\frac{2}{15}$ b. $\frac{5}{12}+\frac{2}{9}$ c. $\frac{7}{11}+\frac{1}{4}$ d. $\frac{3}{10}-\frac{1}{20}$

10. Answer the following by first identifying the lowest common denominator. Simplify your answer if necessary.

a. $\frac{5}{9}-\frac{2}{7}$  b. $\frac{2}{3}-\frac{1}{13}$  c. $\frac{4}{5}-\frac{2}{3}$  d. $\frac{6}{11}-\frac{1}{3}$

11. **MC** The lowest common denominator of $\frac{1}{3}$, $\frac{1}{4}$ and $\frac{1}{6}$ is:

**A.** 16  **B.** 24  **C.** 18  **D.** 12  **E.** 72

### Understanding

12. Answer the following:

a. $\frac{1}{2}+\frac{1}{3}-\frac{1}{4}$  b. $\frac{3}{10}+\frac{4}{15}-\frac{1}{6}$  c. $\frac{7}{8}-\frac{3}{4}+\frac{4}{6}$  d. $\frac{11}{12}-\frac{8}{15}-\frac{3}{20}$

13. Answer the following:

a. $\frac{1}{16}+\frac{1}{8}+\frac{1}{4}$  b. $\frac{5}{6}-\frac{7}{18}-\frac{1}{9}$  c. $\frac{1}{2}+\frac{2}{3}+\frac{3}{4}$  d. $\frac{3}{4}+\frac{1}{3}+\frac{5}{6}$

14. **WE16** Evaluate the following, expressing the answers as proper fractions or mixed numbers.

a. $1-\frac{1}{4}$  b. $1-\frac{3}{7}$  c. $1-\frac{11}{71}$  d. $2-\frac{2}{5}$

15. Evaluate the following, expressing the answers as proper fractions or mixed numbers.

a. $3-\frac{7}{15}$  b. $5-\frac{5}{6}$  c. $10-\frac{1}{3}$  d. $23-\frac{9}{10}$

16. Melissa and Antonio went berry picking. Melissa picked $\frac{3}{10}$ of a bucket of berries and Antonio picked $\frac{1}{6}$ of a bucket.

a. Calculate how much they picked altogether.
b. If they needed 1 full bucket of berries, calculate what fraction more they still need to pick.

17. A loaf of bread needed to be cooked at 200 °C for $\frac{3}{4}$ hour then a further $\frac{1}{2}$ hour at 150 °C. Calculate the total cooking time in hours.

18. Robert, Jason and Luke play football in the same team. Last Saturday, Robert kicked $\frac{1}{4}$ of the team's goals, Jason kicked $\frac{1}{3}$ and Luke kicked $\frac{5}{12}$. Calculate what fraction of the team's goals were kicked by:

a. Jason and Luke
b. Robert and Jason
c. the three together.

**19.** Suraya and her friend decided to climb the tree in her front yard. It took Suraya 10 minutes to climb halfway up, 10 more minutes to climb another sixth of the way and yet another 10 minutes to reach a further tenth of the way. Here she sat admiring the view for 5 minutes before beginning her descent.

**a.** Calculate what fraction of the height of the tree Suraya had climbed after 20 minutes.

**b.** Calculate what fraction of the tree Suraya climbed in total.

## Reasoning

**20.** Yuna sold $\frac{2}{3}$ of the cupcakes she made. She had 45 cupcakes left. Calculate how many cupcakes she made.

**21.** Explain why $\frac{1}{3}+\frac{2}{5}\neq\frac{3}{8}$.

**22.** Chris lives on a farm and has a horse he rides every day. He fills up the horse's trough with $\frac{2}{3}$ hay, $\frac{1}{6}$ carrots, $\frac{1}{12}$ oats and the rest maize. Calculate the fraction of the trough that is filled with maize.

## Problem solving

**23. a.** Without using a calculator, perform the following additions:

i. $\frac{1}{2}+\frac{1}{3}$ ii. $\frac{1}{3}+\frac{1}{4}$ iii. $\frac{1}{4}+\frac{1}{5}$

**b.** Explain any pattern you noticed.

**24. a.** Without using a calculator, perform the following subtractions:

i. $\frac{1}{2}-\frac{1}{3}$ ii. $\frac{1}{3}-\frac{1}{4}$ iii. $\frac{1}{4}-\frac{1}{5}$

**b.** Explain any pattern you noticed.

**c.** Using your answer to part **b**, state the answer to the subtraction $\frac{1}{n}-\frac{1}{n+1}$.

**25.** Explain the difficulties you might experience when adding and subtracting fractions with different denominators and how you could overcome them.

**26.** Arrange the following fractions in the circles in the diagram so that each row adds up to the same value.

$$\frac{1}{12}, \frac{1}{4}, \frac{1}{2}, \frac{1}{6}, \frac{3}{4}, \frac{1}{3}, \frac{2}{3}$$

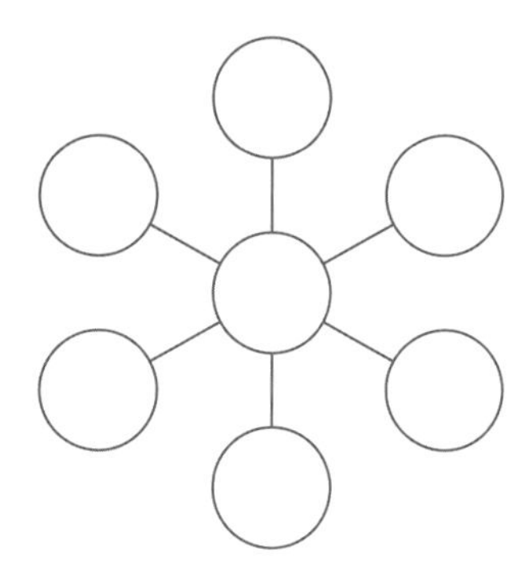

# LESSON
# 4.6 Multiplying fractions

## LEARNING INTENTION

At the end of this lesson you should be able to:
- multiply fractions by a whole number
- multiply fractions by another fraction
- calculate fractions of a fraction
- calculate fractions of an amount.

## 4.6.1 Multiplying fractions

eles-3806

- Multiplication by a whole number can be thought of as repeated addition.
  For example, $3 \times 4 = 4 + 4 + 4 = 12$.
  The same principle holds when multiplying a fraction by a whole number.
  For example, $3 \times \frac{1}{4} = \frac{1}{4} + \frac{1}{4} + \frac{1}{4} = \frac{3}{4}$

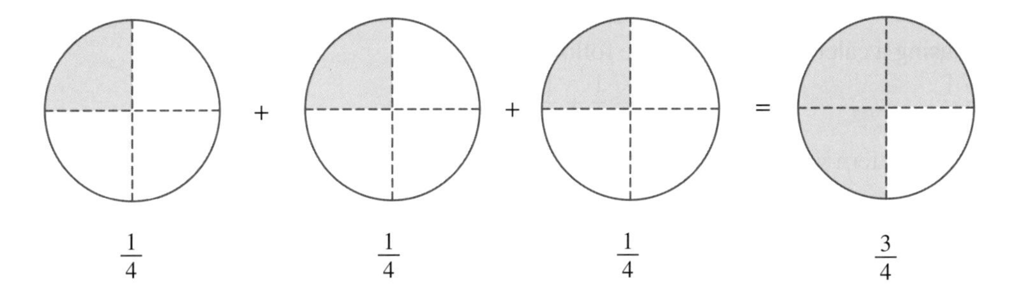


$\frac{1}{4}$ $\frac{1}{4}$ $\frac{1}{4}$ $\frac{3}{4}$

- A quicker way of multiplying a fraction by a whole number is to simply multiply the numerator by the whole number.

### Multiplying a fraction by a whole number

**To multiply a fraction by a whole number, simply multiply the numerator by the whole number and keep the denominator the same.**

**For example:**

$$\frac{1}{5} \times 6 = \frac{1 \times 6}{5} = \frac{6}{5}$$

### WORKED EXAMPLE 18 Multiplying a fraction by a whole number

**Calculate $6 \times \frac{1}{3}$.**

| THINK | WRITE |
|---|---|
| 1. Write the question. | $6 \times \frac{1}{3}$ |

2. Multiply by a whole number using repeated addition. Add $\frac{1}{3}$ six times.

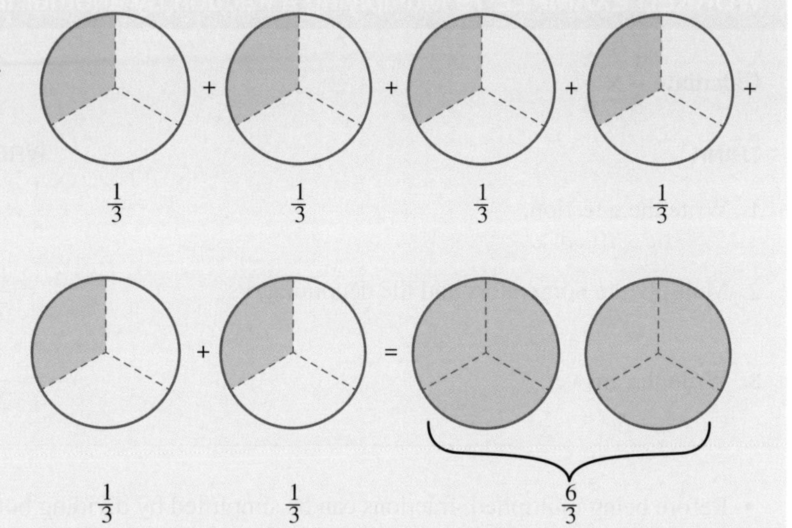


3. Alternatively, multiply the numerator by the whole number.

$$= \frac{6 \times 1}{3}$$

$$= \frac{6}{3}$$

4. Simplify and write the answer. $= 2$

- The multiplication $\frac{1}{3} \times \frac{1}{2}$ is the same as one-third of a half; this is the same as one-sixth of a whole unit, as shown here.

| $\frac{1}{2}$ | | | | | |
|---|---|---|---|---|---|
| $\frac{1}{3}$ of $\frac{1}{2}$ | | | | | |

$$\frac{1}{3} \times \frac{1}{2} = \frac{1}{6} \text{ (of a whole unit)}$$

## Multiplying a fraction by another fraction

**To multiply a fraction by another fraction, simply multiply the numerators and then multiply the denominators.**

**For example:**

$$\frac{1}{2} \times \frac{3}{4} = \frac{1 \times 3}{2 \times 4} = \frac{3}{8}$$

**Fractions don't need to have the same denominator to be multiplied together.**

### WORKED EXAMPLE 19 Multiplying a fraction by another fraction

**Calculate $\frac{1}{5} \times \frac{3}{4}$.**

| THINK | WRITE |
|---|---|
| 1. Write the question. | $\frac{1}{5} \times \frac{3}{4}$ |
| 2. Multiply the numerators and the denominators. | $= \frac{1 \times 3}{5 \times 4}$ |
| 3. Write the answer. | $= \frac{3}{20}$ |

- Before being multiplied, fractions can be simplified by dividing both the numerator and denominator by their highest common factor. This is often called *cancelling*.
- When multiplying fractions, cancelling can only occur:
  - vertically, for example, $\frac{\cancel{3}^1}{\cancel{24}^8} \times \frac{5}{7} = \frac{1 \times 5}{8 \times 7} = \frac{5}{56}$
  - diagonally, for example, $\frac{\cancel{2}^1}{3} \times \frac{5}{\cancel{6}^3} = \frac{1 \times 5}{3 \times 3} = \frac{5}{9}$

### DISCUSSION

Why is cancelling pairs of numerators or denominators not allowed?

### WORKED EXAMPLE 20 Multiplying fractions

**Calculate $\frac{2}{3} \times \frac{9}{10}$.**

| THINK | WRITE |
|---|---|
| 1. Write the question. | $\frac{2}{3} \times \frac{9}{10}$ |
| 2. Simplify by dividing the numbers in the numerator and denominator by their highest common factor. The HCF of 2 and 10 is 2, and the HCF of 3 and 9 is 3.<br>*Note*: This may be done vertically or diagonally. | $= \frac{\cancel{2}^1 \times \cancel{9}^3}{\cancel{3}^1 \times \cancel{10}^5}$ |
| 3. Multiply the numerators, then the denominators and simplify the answer if appropriate. | $\frac{1 \times 3}{1 \times 5} = \frac{3}{5}$ |

COLLABORATIVE TASK: Fractions of fractions

**Equipment:** string, pegs, counters

1. Form small groups. One student from each group should collect a piece of string and two pegs.
2. Your teacher will give each group a multiplication of two fractions.
3. As a group, place a peg on the length of string to show the position of one of your fractions.
4. Using this peg as a new marker for the end of the string, find the second fraction of this new string length and place a second peg in the correct position on the string.
5. Check the position of the second peg by calculating the product of the fractions and then folding the complete length of string into the appropriate number of parts to find the position of the product.
6. Repeat steps 2 to 5 with other fraction multiplications.
7. One student from each group should collect a handful of counters.
8. Your teacher will give each group a fraction of an amount.
9. As a group, divide up the counters to show your allocated fraction.
10. Write a description of the process that your group used to find your fraction of an amount.

## 4.6.2 Using the word *of*

eles-3807

- Sometimes the word *of* means 'multiply'. For example, $\frac{1}{2}$ of $\frac{1}{3}$ can be visualised as

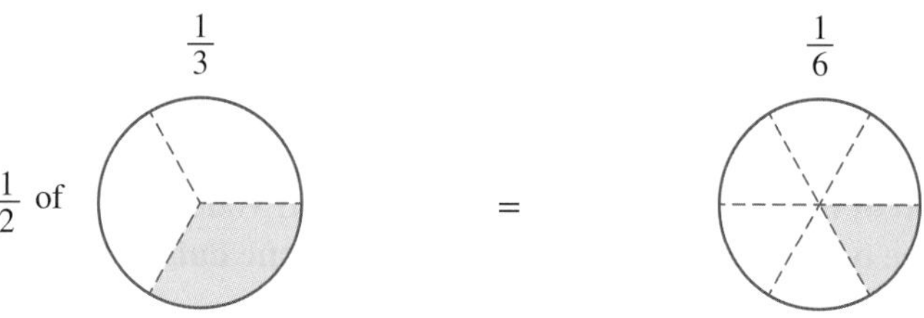


Half of $\frac{1}{3} = \frac{1}{2} \times \frac{1}{3} = \frac{1}{6}$

- A third of an hour, or $\frac{1}{3}$ of 60 minutes, is the same as $\frac{1}{3} \times \frac{60}{1} = 20$ minutes.
- If a customer wants to buy $\frac{1}{4}$ of the remaining 8 pizzas, it is the same as $\frac{1}{4} \times \frac{8}{1} = 2$ pizzas. The customer buys 2 pizzas.
- A diagram can also be used to display multiplication of a fraction by another fraction. The following model shows how to calculate $\frac{2}{3}$ of $\frac{4}{5}$.

The rectangle is divided vertically into fifths; $\frac{4}{5}$ is shaded blue.

The rectangle is divided horizontally into thirds; $\frac{2}{3}$ of the blue rectangles are now shaded purple.

The purple shading represents $\frac{2}{3}$ of $\frac{4}{5}$.

$$\frac{2}{3} \times \frac{4}{5} = \frac{8}{15}$$

## WORKED EXAMPLE 21 Calculating a fraction of a fraction

**Calculate $\frac{2}{3}$ of $\frac{1}{4}$.**

| THINK | WRITE |
|---|---|
| 1. Write the question. | $\frac{2}{3}$ of $\frac{1}{4}$ |
| | $\frac{1}{4}$ <br> $\frac{2}{3}$ of $\frac{1}{4}$ |
| 2. Change 'of' to × and cancel if appropriate. | $= \frac{\cancel{2}^{1}}{3} \times \frac{1}{\cancel{4}^{2}}$ |
| 3. Perform the multiplication and write the answer. | $= \frac{1}{6}$ |

## WORKED EXAMPLE 22 Calculating fractions of an amount

**Mum put a 2-litre carton of flavoured milk in the refrigerator. When the children came home from school, Joanna drank one quarter of the milk, Kieren drank one half of it and Daisy drank one sixth of it. Calculate how much milk, in litres, each person drank.**

| THINK | WRITE |
|---|---|
| 1. Write the fraction of the milk that Joanna drank and simplify the answer. | Joanna: $\frac{1}{4}$ of $2 = \frac{1}{4} \times 2$ |
| 2. Perform the multiplication. | $= \frac{1}{{}_{2}\cancel{4}} \times \frac{\cancel{2}^{1}}{1}$ <br> $= \frac{1}{2}$ |
| 3. Write the fraction of the milk that Kieren drank and simplify the answer. | Kieren: $\frac{1}{2}$ of $2 = \frac{1}{2} \times 2$ |
| 4. Perform the multiplication. | $= \frac{1}{{}_{1}\cancel{2}} \times \frac{\cancel{2}^{1}}{1}$ <br> $= 1$ |

5. Write the fraction of the milk that Daisy drank and simplify the answer.

Daisy: $\frac{1}{6}$ of $2 = \frac{1}{6} \times 2$

6. Perform the multiplication.

$= \frac{1}{{}^{3}\cancel{6}} \times \frac{\cancel{2}^{1}}{1}$

$= \frac{1}{3}$

7. Write a sentence giving the answer in litres.

Joanna drank half a litre, Kieren drank 1 litre and Daisy drank a third of a litre of milk.

## WORKED EXAMPLE 23 Calculating a fraction of a fraction of an amount

**Before a road trip John completely fills up his car's fuel tank with 72 litres of petrol. On the first day of the trip, the car uses $\frac{3}{8}$ of the full tank. On the second day, the car uses $\frac{2}{5}$ of the remaining amount. Determine how much petrol was used on the second day.**

| THINK | WRITE |
| --- | --- |
| a. 1. Determine the fraction left after $\frac{3}{8}$ of the petrol is used on the first day. | a. $1 - \frac{3}{8} = \frac{5}{8}$<br>$\frac{5}{8}$ remains after the first day. |
| 2. Calculate the total fraction of the tank of petrol that has been used on day 2. | $\frac{2}{5}$ of $\frac{5}{8}$ of a tank $= \frac{2 \times 5}{5 \times 8}$<br>$= \frac{1}{4}$ of the full tank is used on day 2. |
| 3. Calculate the amount of petrol used on day 2 as the answer. Remember to state the units. | $\frac{1}{4} \times 72 = \frac{72}{4} = 18$<br>18 litres of petrol were used on the second day. |

## DISCUSSION

Is repeated addition an efficient method of multiplying fractions?

## Digital technology

Scientific calculators can evaluate the multiplication of a fraction by another fraction or a whole number.

DEG

$\frac{12}{23} * \frac{45}{84}$ $\frac{45}{161}$

 Resources

eWorkbook Topic 4 Workbook (worksheets, code puzzle and project) (ewbk-1905)

Video eLesson Multiplication and division of fractions (eles-1867)

Interactivities Individual pathway interactivity: Multiplying rational numbers (int-4327)
Repeated addition (int-2738)
Multiplication of rational numbers (int-3946)
Using the word *of* (int-3947)

## Exercise 4.6 Multiplying fractions

learnon

4.6 Quick quiz 

4.6 Exercise

### Individual pathways

| ■ PRACTISE | ■ CONSOLIDATE | ■ MASTER |
|---|---|---|
| 1, 2, 5, 8, 10, 13, 15, 18 | 3, 6, 9, 11, 16, 19 | 4, 7, 12, 14, 17, 20 |

### Fluency

1. **WE18** Calculate the following.
   a. $\frac{2}{3} \times 3$ b. $\frac{5}{14} \times 7$ c. $\frac{15}{22} \times 11$ d. $\frac{5}{16} \times 4$

2. **WE19** Calculate the following using written methods, then check your answers with a calculator.
   a. $\frac{1}{2} \times \frac{1}{4}$ b. $\frac{1}{3} \times \frac{2}{3}$ c. $\frac{5}{9} \times \frac{4}{3}$ d. $\frac{11}{13} \times \frac{1}{2}$

3. Calculate the following using written methods, then check your answers with a calculator.
   a. $\frac{6}{7} \times \frac{9}{11}$ b. $\frac{11}{12} \times \frac{11}{12}$ c. $\frac{8}{9} \times \frac{5}{9}$ d. $\frac{5}{8} \times \frac{1}{3}$

4. Calculate the following using written methods, then check your answers with a calculator.
   a. $\frac{5}{6} \times \frac{11}{12}$ b. $\frac{5}{11} \times \frac{5}{12}$ c. $\frac{2}{3} \times \frac{4}{13}$ d. $\frac{4}{7} \times \frac{2}{3}$

5. **WE20** Calculate the following using written methods, then check your answers with a calculator.
   a. $\frac{4}{8} \times \frac{3}{9}$ b. $\frac{2}{5} \times \frac{1}{2}$ c. $\frac{3}{7} \times \frac{5}{6}$ d. $\frac{3}{8} \times \frac{4}{15}$

6. Calculate the following using written methods, then check your answers with a calculator.
   a. $\frac{10}{11} \times \frac{22}{25}$ b. $\frac{24}{27} \times \frac{9}{8}$ c. $\frac{18}{32} \times \frac{64}{72}$ d. $\frac{48}{56} \times \frac{24}{60}$

7. Calculate the following using written methods, then check your answers with a calculator.
   a. $\frac{15}{27} \times \frac{36}{45}$ b. $\frac{8}{49} \times \frac{14}{16}$ c. $\frac{4}{9} \times \frac{6}{8}$ d. $\frac{5}{25} \times \frac{12}{48}$

8. **WE21** Calculate:
   a. $\frac{1}{2}$ of $\frac{1}{4}$ b. $\frac{3}{4}$ of $\frac{2}{3}$ c. $\frac{5}{6}$ of $\frac{5}{6}$ d. $\frac{8}{9}$ of $\frac{1}{4}$

9. Calculate:
   a. $\frac{7}{8}$ of 32 b. $\frac{9}{10}$ of 50 c. $\frac{3}{5}$ of 25 d. $\frac{4}{5}$ of 20

## Understanding

10. WE22 Joshua's dad made 6 litres of cordial for his birthday party. Emily drank $\frac{1}{3}$ of the cordial, Tracy drank $\frac{1}{12}$ and Jonathan drank $\frac{1}{18}$. Calculate how much cordial, in litres, each person drank.

11. Zoe and Jimin play basketball with the Sharp Shooters. Each game lasts 40 minutes. Zoe played $\frac{4}{5}$ of last week's game and Jimin played $\frac{7}{8}$. Calculate how many minutes of the game:

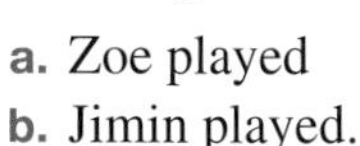

a. Zoe played
b. Jimin played.

12. Mark's uncle gave him a family-size block of chocolate. He divided it into thirds to share with his two sisters. His friends Tao and Nick then arrived to visit, and Mark shared his portion of the chocolate with them, so that they each had the same amount.
   a. Draw a block of chocolate with $9 \times 6$ pieces, and shade Mark's third of the block. Then draw lines to represent Tao's share.
   b. Identify the fraction of the block of chocolate that Nick received.

13. WE23 A car's petrol tank holds 48 litres of fuel. The tank was full at the start of a trip.
   a. By examining the gauge shown, determine what fraction of the tank of petrol has been used.
   b. Determine how many litres of petrol have been used.

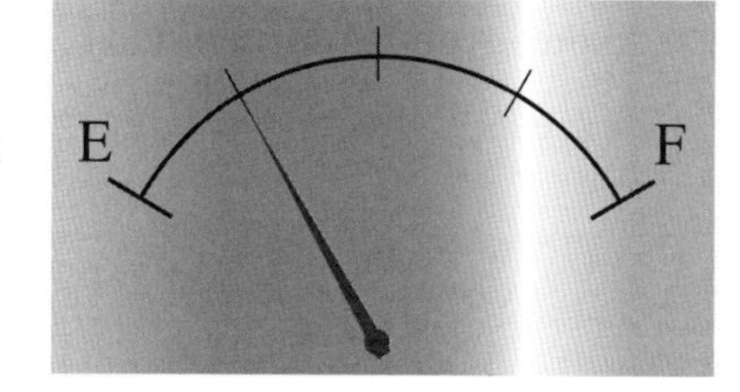


14. Answer the following questions.
   a. State how many half pizzas are in 1 pizza.
   b. State how many half pizzas are in 2 whole pizzas.
   c. State how many half pizzas are in 5 whole pizzas.

## Reasoning

15. a. Draw a quarter of a pie.
   b. Shade half of this piece of pie.
   c. Determine what fraction of the whole pie is shaded.
   d. Complete the mathematical sentence $\frac{1}{4} \times \frac{1}{2} =$ __ to show how the answer was found.

16. To make a loaf of fruit bread in a breadmaker, the fruit needs to be added after the second rise stage and the top of the bread needs to be glazed two-thirds of the way through the baking stage. The recipe shows how long each stage of the total cycle takes.

| Stage | Length of time |
|---|---|
| First knead | 5 minutes |
| Second knead | 20 minutes |
| First rise then punch down | 40 minutes |
| Second rise then punch down | 25 minutes |
| Third rise | 50 minutes |
| Bake | 40 minutes |

a. State when the fruit should be added.
b. Express this length of time as a fraction of the total time in the breadmaker.
c. Calculate the length of time during which the bread should be glazed.

**17. a.** Without using a calculator, evaluate the following:

i. $\frac{1}{2} \times \frac{2}{3}$ ii. $\frac{1}{2} \times \frac{2}{3} \times \frac{3}{4}$ iii. $\frac{1}{2} \times \frac{2}{3} \times \frac{3}{4} \times \frac{4}{5}$

**b.** Describe any pattern you see in part **a**.

**c.** Hence, evaluate the solution to:

$$\frac{1}{2} \times \frac{2}{3} \times \frac{3}{4} \ldots \times \frac{98}{99} \times \frac{99}{100}$$

### Problem solving

**18.** Gustave's monthly take-home pay is \$2400. From this he spends a quarter on his home loan payments, one half on food and drink and one sixth on clothing. One half of the remainder goes into his savings account. Determine how much money Gustave deposits into his savings account each month.

**19.** On a given map, one kilometre is represented by $\frac{2}{9}$ cm. If a distance on the map is 27 cm, determine how many kilometres this represents.

**20.** Andrea requires $5\frac{2}{7}$ m of fabric to make a pattern on a quilt. One third of this fabric has to be red and $\frac{1}{4}$ must be green. Determine how many metres of red and green fabric she needs.

# LESSON
# 4.7 Dividing fractions

**LEARNING INTENTION**

At the end of this lesson you should be able to:
- divide fractions.

## 4.7.1 Dividing fractions

eles-3808

- To divide fractions you need to know what their reciprocal is.
- The **reciprocal** of a fraction is found by turning the fraction upside down. So $\frac{3}{5}$ is the reciprocal of $\frac{5}{3}$.
- A whole number can be written as a fraction by putting it over 1. Therefore, the reciprocal of a whole number is 1 over the number. For example, since $4 = \frac{4}{1}$, the reciprocal of 4 is $\frac{1}{4}$.

## WORKED EXAMPLE 24 Reciprocal of a proper fraction

**Identify the reciprocal of $\frac{2}{3}$.**

| THINK | WRITE |
|---|---|
| Turn the fraction upside down and write the answer in a sentence. | The reciprocal of $\frac{2}{3}$ is $\frac{3}{2}$. |

- To determine the reciprocal of a mixed number, first express it as an improper fraction.

## WORKED EXAMPLE 25 Reciprocal of a mixed number

**Determine the reciprocal of $1\frac{2}{3}$.**

| THINK | WRITE |
|---|---|
| 1. Write $1\frac{2}{3}$ as an improper fraction. | $1\frac{2}{3} = \frac{5}{3}$ |
| 2. Turn the improper fraction upside down and write the answer in a sentence. | The reciprocal of $1\frac{2}{3}$ is $\frac{3}{5}$. |

- Try to think of division as a question. For example, $10 \div 4$ can be interpreted as 'How many groups of 4 are there in 10?'

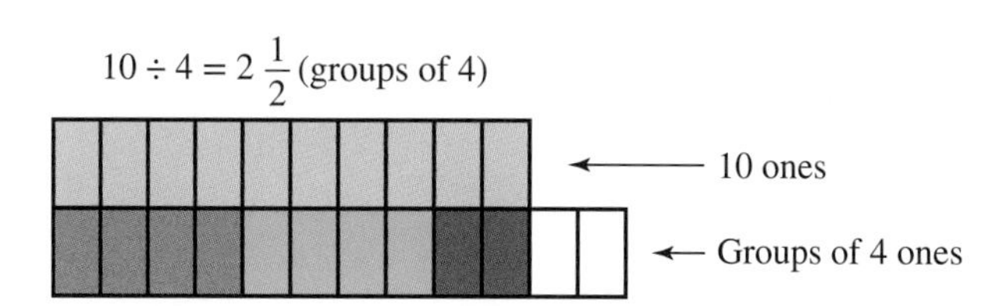


- Division of fractions follows the same logic. For example, $3 \div \frac{1}{2}$ can be interpreted as 'How many halves are there in 3?'

$3 \div \frac{1}{2} = 6\left(\text{groups of } \frac{1}{2}\right)$

| 1 | | 1 | | 1 | | ← 3 ones |
|---|---|---|---|---|---|---|
| $\frac{1}{2}$ | $\frac{1}{2}$ | $\frac{1}{2}$ | $\frac{1}{2}$ | $\frac{1}{2}$ | $\frac{1}{2}$ | ← Groups of 1 half |

- This method is great for understanding the concept of division of fractions, but is not so good for evaluating many divisions involving fractions.

## Dividing fractions

**To divide fractions, multiply the first fraction by the reciprocal of the second fraction.**

**For example:** $\frac{1}{3} \div \frac{2}{5} = \frac{1}{3} \times \frac{5}{2} = \frac{5}{6}$

**This process can be remembered easily by the saying KEEP, CHANGE, FLIP.**

$$\frac{3}{4} \quad \div \quad \frac{2}{7}$$

**KEEP** — **CHANGE** — **FLIP**

$$\frac{3}{4} \quad \times \quad \frac{7}{2}$$

### WORKED EXAMPLE 26 Dividing fractions

**Evaluate** $\frac{2}{3} \div \frac{4}{9}$.

| THINK | WRITE |
|---|---|
| 1. Write the question. | $\frac{2}{3} \div \frac{4}{9}$ |
| 2. Keep the first fraction the same. Change the division sign to a multiplication sign and flip the second fraction. | $= \frac{2}{3} \times \frac{9}{4}$ |
| 3. Cancel diagonally. | $= \frac{\cancel{2}^{1}}{\cancel{3}^{1}} \times \frac{\cancel{9}^{3}}{\cancel{4}^{2}}$ |
| 4. Perform the multiplication. | $= \frac{3}{2}$ |

### DISCUSSION

Explore what happens when you multiply and divide positive numbers by a fraction between 0 and 1. Discuss the results.

### Digital technology

Scientific calculators can evaluate division of fractions.

DEG

$\frac{1}{3} \div \frac{5}{6}$ $\quad \frac{2}{5}$

Resources

**eWorkbook** Topic 4 Workbook (worksheets, code puzzle and project) (ewbk-1905)

**Interactivities** Individual pathway interactivity: Dividing rational numbers (int-4328)
Division of rational numbers (int-3948)

## Exercise 4.7 Dividing fractions

learnon

**4.7 Quick quiz** 

**4.7 Exercise**

### Individual pathways

| ■ PRACTISE | ■ CONSOLIDATE | ■ MASTER |
|---|---|---|
| 1, 4, 7, 10, 13 | 2, 6, 8, 11, 14 | 3, 5, 9, 12, 15 |

### Fluency

1. **WE24** Identify the reciprocals of each of the following.
   a. $\frac{3}{4}$ b. $\frac{2}{7}$ c. $\frac{6}{5}$ d. $\frac{5}{3}$ e. 5
2. Identify the reciprocals of each of the following.
   a. $\frac{5}{12}$ b. $\frac{9}{2}$ c. $\frac{5}{3}$ d. $\frac{10}{3}$ e. 20
3. Identify the reciprocals of each of the following.
   a. $\frac{1}{12}$ b. $\frac{3}{15}$ c. $\frac{2}{22}$ d. $\frac{1}{6}$ e. 1
4. **WE25** Determine the reciprocals of these mixed numbers.
   a. $1\frac{2}{3}$ b. $4\frac{3}{7}$ c. $9\frac{2}{7}$ d. $3\frac{3}{8}$ e. $6\frac{9}{10}$
5. Determine the reciprocals of these mixed numbers.
   a. $3\frac{6}{7}$ b. $5\frac{1}{3}$ c. $7\frac{1}{2}$ d. $10\frac{2}{9}$ e. $2\frac{5}{9}$
6. Multiply each of these numbers by its reciprocal.
   a. $\frac{5}{7}$ b. $\frac{3}{5}$ c. $2\frac{1}{8}$

### Understanding

7. **WE26** Evaluate the following using written methods, then check your answers with a calculator.
   a. $\frac{1}{4} \div \frac{2}{3}$ b. $\frac{8}{9} \div \frac{7}{6}$ c. $\frac{1}{12} \div \frac{2}{3}$ d. $\frac{4}{11} \div \frac{5}{6}$ e. $\frac{10}{3} \div \frac{15}{21}$

8. Evaluate the following using written methods, then check your answers with a calculator.

a. $\frac{2}{7} \div \frac{20}{21}$ b. $\frac{4}{9} \div \frac{4}{9}$ c. $\frac{4}{7} \div 10$ d. $\frac{4}{3} \div 12$ e. $\frac{15}{11} \div 5$

9. Evaluate the following using written methods, then check your answers with a calculator.

a. $\frac{5}{6} \div \frac{3}{2}$ b. $\frac{3}{11} \div 6$ c. $\frac{9}{10} \div \frac{5}{7}$ d. $\frac{20}{17} \div \frac{5}{3}$ e. $\frac{50}{7} \div 5$

### Reasoning

10. Lucy says that the following statement is correct.

$$3\frac{1}{2} \div \frac{4}{6} > \frac{4}{3} \div \frac{2}{15}$$

However, Lucy's friend Lily does not agree. Explain who is correct.

11. Mai is making sushi. The recipe requires $\frac{1}{4}$ cup of rice. She has a 2 kg bag of rice. If 1 cup of rice weighs 250 g, calculate the number of batches of the recipe she can make.

12. Maytreyi bought $1\frac{1}{2}$ kg of chocolates for her birthday party. She is going to give each friend attending her birthday $\frac{1}{10}$ kg. Calculate the number of friends attending her birthday party.

### Problem solving

13. A plumber has a piece of pipe that is $\frac{26}{4}$ cm long. He divides the pipe into pieces that are $\frac{3}{2}$ cm long. Determine the number of pieces he has now.

14. Calculate the average of $\frac{1}{4}, \frac{1}{5}, \frac{1}{10}$ and $\frac{1}{20}$.

15. Determine the value of the following expression.

$$\frac{1}{2+\frac{1}{2\frac{1}{2}}}$$

# LESSON
# 4.8 Working with mixed numbers

**LEARNING INTENTION**

At the end of this lesson you should be able to:
- add, subtract, multiply and divide mixed numbers.

## 4.8.1 Adding and subtracting mixed numbers

eles-3809

- To add or subtract mixed numbers, first convert to improper fractions, and then add and subtract as usual.
- To perform operations on mixed numbers, first convert them to improper fractions, and then use the rules for operations on fractions to evaluate.

**Working with mixed numbers**

**To add, subtract, multiply or divide mixed numbers, convert them to improper fractions and then perform the operation as usual.**

**WORKED EXAMPLE 27 Adding mixed numbers**

**Evaluate $2\frac{3}{4}+1\frac{1}{2}$.**

| THINK | WRITE |
|---|---|
| 1. Write the question. | $2\frac{3}{4}+1\frac{1}{2}$ |
| 2. Convert each mixed number into an improper fraction. | $=\frac{11}{4}+\frac{3}{2}$ |
| 3. Write both fractions with the lowest common denominator (LCD). | $=\frac{11}{4}+\frac{3\times2}{2\times2}$ <br> $=\frac{11}{4}+\frac{6}{4}$ |
| 4. Add the numerators. | $=\frac{17}{4}$ |
| 5. Convert to a mixed number if appropriate and write the answer. | $=4\frac{1}{4}$ |

**WORKED EXAMPLE 28 Subtracting mixed numbers**

**Evaluate $3\frac{1}{5}-1\frac{3}{4}$.**

**THINK** | **WRITE**

1. Write the question. $3\frac{1}{5}-1\frac{3}{4}$
2. Convert each mixed number into an improper fraction. $=\frac{16}{5}-\frac{7}{4}$
3. Write both fractions with the lowest common denominator (LCD). $=\frac{16\times4}{5\times4}-\frac{7\times5}{4\times5}$

   $=\frac{64}{20}-\frac{35}{20}$
4. Subtract the numerators. $=\frac{29}{20}$
5. Convert to a mixed number if appropriate and write the answer. $=1\frac{9}{20}$

## Estimation

- Estimation can be used to determine approximate answers to addition and subtraction of mixed numbers.
- This allows you to check easily whether your answer to a calculation is reasonable.
- To estimate, add or subtract the whole number parts. The answer should be close to the result obtained by this method.

**WORKED EXAMPLE 29 Checking answers by estimation**

**Evaluate $3\frac{1}{3}+2\frac{1}{2}$ and check the answer by estimation.**

**THINK** | **WRITE**

1. Write the question. $3\frac{1}{3}+2\frac{1}{2}$
2. Convert each mixed number into an improper fraction. $=\frac{10}{3}+\frac{5}{2}$
3. Write both fractions with the lowest common denominator (LCD). $=\frac{10\times2}{3\times2}+\frac{5\times3}{2\times3}$

   $=\frac{20}{6}+\frac{15}{6}$
4. Add the numerators. $=\frac{35}{6}$

5. Convert to a mixed number if appropriate. $= 5\frac{5}{6}$

6. Check the answer by adding the whole numbers to determine an approximation.

$3\frac{1}{3} + 2\frac{1}{2} \simeq 3 + 2$

$3\frac{1}{3} + 2\frac{1}{2} \simeq 5$

The approximation shows that the answer we obtain should be close to 5.

## 4.8.2 Multiplying and dividing mixed numbers

eles-3810

- To multiply mixed numbers, first convert to improper fractions, and then multiply as usual.
- To divide mixed numbers, first convert them into improper fractions, and then divide as usual.

### WORKED EXAMPLE 30 Multiplying mixed numbers

**Evaluate $1\frac{1}{2} \times 3\frac{3}{5}$.**

| THINK | WRITE |
|---|---|
| 1. Write the question. | $1\frac{1}{2} \times 3\frac{3}{5}$ |
| 2. Change mixed numbers to improper fractions and cancel if possible. | $= \frac{3}{{}_{1}\cancel{2}} \times \frac{\cancel{18}^{9}}{5}$ |
| 3. Multiply the numerators and then the denominators. | $= \frac{27}{5}$ |
| 4. Convert to a mixed number if appropriate and write the answer. | $= 5\frac{2}{5}$ |

### WORKED EXAMPLE 31 Dividing mixed numbers

**Evaluate $1\frac{1}{2} \div 3\frac{2}{5}$.**

| THINK | WRITE |
|---|---|
| 1. Write the question. | $1\frac{1}{2} \div 3\frac{2}{5}$ |
| 2. Convert the mixed numbers to improper fractions. | $= \frac{3}{2} \div \frac{17}{5}$ |

3. Change the ÷ to × and flip the second fraction. $= \frac{3}{2} \times \frac{5}{17}$

4. Multiply the numerators, then the denominators and simplify the answer if necessary. $= \frac{15}{34}$

 Resources

 **eWorkbook** Topic 4 Workbook (worksheets, code puzzle and project) (ewbk-1905)

 **Interactivities** Individual pathway interactivity: Working with mixed numbers (int-4330)

Addition and subtraction of mixed numbers (int-3951)

## Exercise 4.8 Working with mixed numbers

learnon

**4.8 Quick quiz**  **4.8 Exercise**

### Individual pathways

| ■ PRACTISE | ■ CONSOLIDATE | ■ MASTER |
|---|---|---|
| 1, 3, 5, 6, 8, 10, 14, 17, 20, 23 | 2, 7, 11, 12, 16, 18, 21, 24, 25 | 4, 9, 13, 15, 19, 22, 26 |

### Fluency

1. **WE27&28** Evaluate the following, giving answers as mixed numbers in simplest form.
   a. $1\frac{1}{5} + 3\frac{1}{2}$ b. $3\frac{2}{3} - 2\frac{1}{2}$ c. $5\frac{2}{3} - 2\frac{1}{4}$ d. $9\frac{3}{8} + 4$
2. Evaluate the following, giving answers as mixed numbers in simplest form.
   a. $2\frac{3}{5} - 1\frac{1}{2}$ b. $3\frac{1}{2} + 2\frac{3}{4}$ c. $7 - 5\frac{2}{3}$ d. $2\frac{1}{3} - \frac{5}{6}$
3. Evaluate the following, giving answers as mixed numbers in simplest form.
   a. $3\frac{1}{2} - 1\frac{7}{8}$ b. $4\frac{3}{5} - 2\frac{9}{10}$ c. $5\frac{1}{4} - 1\frac{7}{12}$ d. $5\frac{3}{4} - 3\frac{5}{6}$
4. Evaluate the following, giving answers as mixed numbers in simplest form.
   a. $\frac{1}{5} + 2\frac{2}{7}$ b. $\frac{5}{8} + 3\frac{3}{4} - \frac{1}{2}$ c. $2\frac{2}{9} + 1\frac{1}{3} + \frac{1}{6}$ d. $4\frac{2}{3} + 1\frac{3}{8} - 3$
5. **WE29** Evaluate the following and check the answers by estimation.
   a. $5\frac{1}{3} + 2\frac{1}{4}$ b. $1\frac{1}{6} + 2\frac{1}{2}$ c. $3\frac{1}{3} + 1\frac{1}{6}$ d. $6\frac{1}{2} - 2\frac{1}{4}$
6. **WE30** Evaluate:
   a. $1\frac{1}{4} \times \frac{2}{3}$ b. $1\frac{1}{2} \times 1\frac{1}{2}$ c. $2\frac{5}{8} \times 1\frac{3}{4}$ d. $2\frac{1}{5} \times 3\frac{1}{2}$
7. Evaluate:
   a. $3\frac{5}{6} \times 9\frac{5}{7}$ b. $5\frac{1}{5} \times 6\frac{2}{3}$ c. $3\frac{3}{4} \times 2\frac{2}{3}$ d. $\frac{7}{8} \times 3\frac{1}{3}$

8. WE31 Evaluate:

a. $1\frac{1}{4} \div \frac{3}{4}$
b. $\frac{9}{4} \div 3\frac{1}{2}$
c. $\frac{10}{3} \div 3\frac{1}{3}$
d. $7\frac{4}{9} \div \frac{1}{3}$

9. Evaluate:

a. $1\frac{1}{2} \div \frac{1}{2}$
b. $10\frac{9}{10} \div 4$
c. $3\frac{2}{7} \div 2\frac{1}{7}$
d. $3\frac{5}{8} \div 1\frac{3}{4}$

### Understanding

10. One-third of a litre of cordial is mixed with $1\frac{1}{2}$ litres of water. Calculate how many litres of drink have been made.

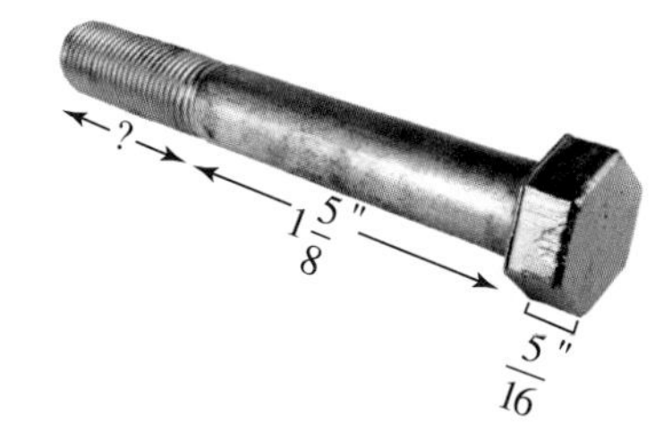


11. The lengths of bolts and nails are often measured in inches. For the bolt shown, calculate the length of the threaded section if the total length of the bolt is $3\frac{1}{4}$ inches.

12. Julia is planning a holiday to the US. She has 5 months and has worked out the following itinerary. She will be in California for $1\frac{1}{2}$ months, in Colorado for $1\frac{2}{3}$ months and in Florida for $\frac{3}{4}$ of a month. The other state she will be visiting is New York. Determine how long she will spend in New York.

13. Theo is building a house. Before the frame of the house can be assembled, the footings need to be dug. The digging will take $3\frac{1}{2}$ days. The concrete slab that is then needed will take $2\frac{3}{4}$ days to pour and set. Determine how long it will be before Theo can start assembling the frame.

14. Felicity is dividing $1\frac{1}{2}$ kilograms of minced steak into 8 hamburger portions. Calculate how many kilograms of mince are in each portion.

15. Ned spends his school holidays helping his father with shearing sheep.
    a. It takes Ned $\frac{1}{4}$ hour to shear one sheep. Calculate how many sheep Ned will shear in $5\frac{1}{2}$ hours.
    b. Ned's father, Wesley, shears a sheep every $\frac{1}{12}$ of an hour. If Wesley worked continuously for 8 hours, calculate how many sheep he would shear.

16. Lachlan was looking for something to do one Saturday afternoon and his dad suggested he cook something for afternoon tea. Lachlan found a recipe for peanut butter muffins.

**Ingredients:**

$\frac{1}{4}$ cup sugar | $\frac{1}{4}$ cup margarine
$\frac{1}{2}$ cup peanut butter | 2 eggs
$1\frac{1}{2}$ cups milk | $2\frac{1}{2}$ cups self-raising flour
$\frac{1}{4}$ teaspoon baking soda

**Method:**

Blend the sugar, margarine and peanut butter. Beat in the eggs and milk. Add the self-raising flour and baking soda. Place the mixture in greased muffin pans and sprinkle with cinnamon sugar. Bake for 15–20 minutes in a 200 °C oven.

This recipe makes 10 muffins, but Lachlan wants enough mixture to make 15. He needs to multiply the quantities by $1\frac{1}{2}$. Write the quantities of each ingredient Lachlan needs to make 15 muffins.

17. Anh bought $7\frac{3}{4}$ kg of prawns for a party and $\frac{3}{5}$ of them were eaten. Calculate how many kilograms of prawns were left over.

18. Teagan buys $8\frac{3}{4}$ kg of sugar. She wants to store it in containers that can hold $1\frac{1}{4}$ kg. Calculate how many containers she needs.

19. Max works as a train driver and is paid $21 per hour. If Max works on a public holiday, he is paid double time and a half, which means he earns $2\frac{1}{2}$ times his normal hourly rate. Calculate what Max earns per hour working on a public holiday.

## Reasoning

20. To make the casing for a simple collapsible telescope, you need two cylinders, one of which fits inside the other.

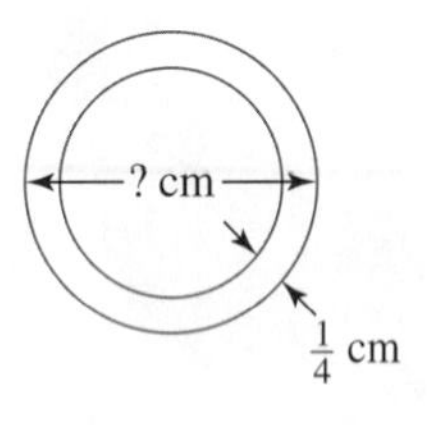


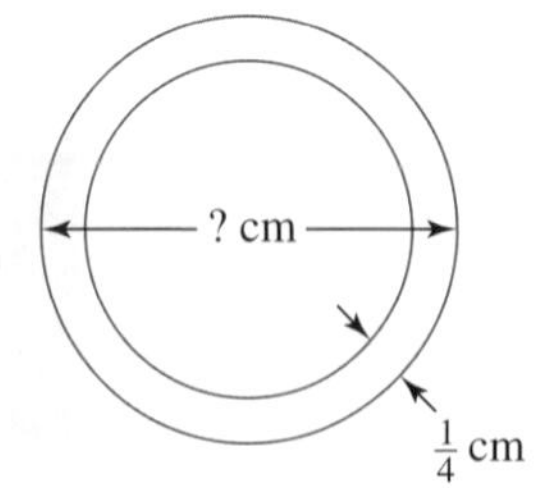


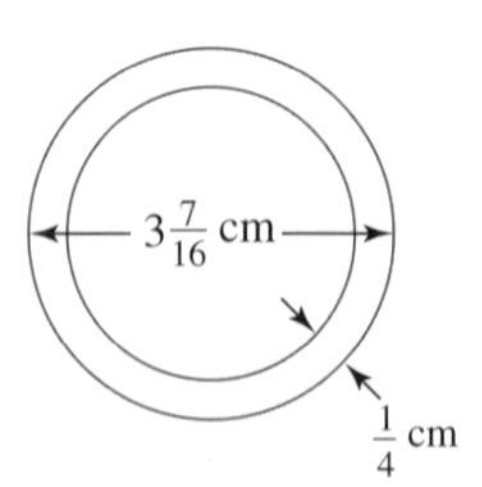


a. Calculate the width of the cylinder in the middle diagram that would slide over the one shown at right.
b. Calculate the width of the cylinder in the left diagram that would slide into the one shown at right.

21. Worldwide figures indicate that $\frac{1}{10}$ of men are left-handed, while $\frac{2}{25}$ of women are left-handed. If there are 400 men in a room, calculate how many women would need to be there so that there are the same number of left-handed men as left-handed women.

22. Explain how you could you convince someone that $\frac{1}{2}$ of $3\frac{1}{3}$ is $1\frac{2}{3}$.

### Problem solving

23. Five-eighths of a class were girls and $\frac{2}{3}$ of the girls were brunette.

    a. Calculate what fraction of the class were brunette girls.
    b. If there were 24 people in the class, calculate how many were brunette girls.

24. Michelle can run one lap of a cross-country course in $\frac{1}{5}$ hour.

    a. Calculate how many laps she completed in $1\frac{1}{3}$ hours. (Assume she can keep up the same pace.)
    b. Calculate the number of minutes it takes for Michelle to run one lap.

25. A skydiver opens his parachute at 846 m above ground level. He has already fallen five-sevenths of the distance to the ground. Calculate the height of the plane when he jumped.

26. A new vending machine stocking water and sports drinks was installed at the local gymnasium. The number of bottles sold each day over the first 5 weeks after its installation is shown.

    4, 39, 31, 31, 50, 43, 60, 45, 57, 61, 18, 26, 3, 52, 51, 59, 33,
    51, 27, 62, 30, 40, 3, 30, 37, 9, 33, 44, 53, 16, 22, 6, 42, 33, 19

    The gymnasium realises that the number of daily sales varies, depending particularly on the weather. The machine is stocked with 60 bottles at the start of each day. The supplier advises that, because of electricity costs and other running expenses, the machine is not worthwhile having unless at least three-quarters of the bottles are sold on average 2 days per week.
    Round these sales to the nearest 5 and analyse the figures to determine whether the gymnasium should keep the vending machine.

# LESSON
# 4.9 Percentages as fractions

## LEARNING INTENTION

At the end of this lesson you should be able to:

- convert percentages to fractions.

## 4.9.1 Definition of a percentage

eles-3812

- The term **per cent** means 'per hundred' or 'out of a hundred'.
- The symbol for per cent is %. For example, 7% means 7 out of 100.

### WORKED EXAMPLE 32 Writing a percentage as a fraction

**Write 47% as a fraction.**

| THINK | WRITE |
|---|---|
| Recall the definition of a percentage. 47% means 47 out of 100. | $47\% = \frac{47}{100}$ |

### Percentages as fractions

- All percentages can be expressed as fractions.
  - To convert a whole number percentage into a fraction, write it over 100 and simplify.
    For example, $10\% = \frac{10}{100} = \frac{1}{10}$

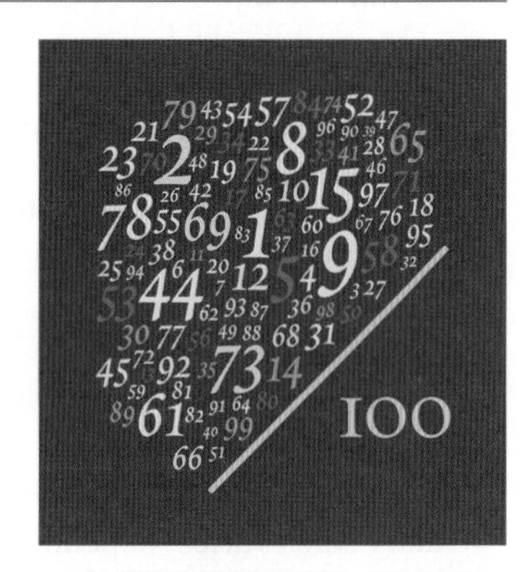


### WORKED EXAMPLE 33 Converting a percentage into a fraction

**Write 20% as a fraction in simplest form.**

| THINK | WRITE |
|---|---|
| 1. Write the percentage as a fraction with a denominator of 100. | $20\% = \frac{20}{100}$ |
| 2. Simplify the fraction. | $= \frac{1}{5}$ |

- To convert a percentage containing a decimal number into a fraction:
  - write it as a fraction over 100
  - change the numerator to a whole number by multiplying both the numerator and denominator by the appropriate power of 10
  - simplify if possible.

  For example: $28.2\% = \frac{28.2}{100} = \frac{282}{1000} = \frac{141}{500}$

### WORKED EXAMPLE 34 Converting percentages to fractions

**Write 36.4% as a fraction in simplest form.**

**THINK** / **WRITE**

1. Write the percentage as a fraction out of 100.

$$36.4\% = \frac{36.4}{100}$$

2. Change the numerator to a whole number by multiplying it by an appropriate power of 10, and multiply the denominator by the same power of 10.

$$= \frac{36.4 \times 10}{100 \times 10}$$

$$= \frac{364}{1000}$$

3. Simplify the fraction.

$$= \frac{91}{250}$$

- To convert a percentage that is a proper fraction into a fraction, multiply the denominator by 100 and simplify if possible.
- For example: $\frac{5}{8}\% = \frac{5}{800} = \frac{1}{160}$
- To convert a mixed number percentage into a fraction:
  - change the mixed number into an improper fraction
  - multiply the denominator by 100
  - simplify if possible.

  For example: $12\frac{1}{2}\% = \frac{25}{2}\% = \frac{25}{200} = \frac{1}{8}$

## Converting percentages into fractions

**In simple terms, to convert a percentage into a fraction, place the percentage as a fraction over 100, then manipulate so that both the numerator and denominator are whole numbers with no common factors.**

$$x\,\% = \frac{x}{100}$$

### WORKED EXAMPLE 35 Converting percentages to fractions

**Write the following percentages as fractions in simplest form.**

**a.** $\frac{1}{4}\%$ **b.** $15\frac{1}{3}\%$

**THINK** / **WRITE**

a. 1. Write the percentage and then multiply the denominator by 100.

$$\text{a. } \frac{1}{4}\% = \frac{1}{4 \times 100}$$

2. Simplify.

$$= \frac{1}{400}$$

b. 1. Convert the mixed number to an improper fraction.

2. Multiply the denominator by 100.

3. Simplify the fraction.

b. $15\frac{1}{3}\% = \frac{46}{3}\%$

$= \frac{46}{3 \times 100}$

$= \frac{46}{300}$

$= \frac{23}{150}$

## Resources

**eWorkbook** Topic 4 Workbook (worksheets, code puzzle and project) (ewbk-1905)

**Interactivities** Individual pathway interactivity: Percentages as fractions (int-4343)

Percentages as fractions (int-3992)

# Exercise 4.9 Percentages as fractions

learnon

**4.9 Quick quiz** 

**4.9 Exercise**

### Individual pathways

| ■ PRACTISE | ■ CONSOLIDATE | ■ MASTER |
|---|---|---|
| 1, 3, 6, 9, 12, 15, 16, 17, 19, 22 | 2, 4, 7, 10, 13, 20, 23, 24 | 5, 8, 11, 14, 18, 21, 25 |

### Fluency

1. **WE32** Write the following percentages as fractions.
   a. 17 %  b. 29 %  c. 79 %  d. 99 %  e. 3 %
2. Write the following percentages as fractions.
   a. 19%  b. 33%  c. 9%  d. 243%  e. 127%
3. **WE33** Write the following percentages as fractions in simplest form.
   a. 50%  b. 80%  c. 35%  d. 60%  e. 10%
4. Write the following percentages as fractions in simplest form.
   a. 45%  b. 12%  c. 5%  d. 74%  e. 2%
5. Write the following percentages as fractions in simplest form. Change the answer to a mixed number where appropriate.
   a. 150%  b. 180%  c. 200%  d. 500%  e. 112%
6. **WE34** Write the following percentages as fractions in simplest form.
   a. 3.5%  b. 7.2%  c. 19.7%  d. 32.4%

7. Write the following percentages as fractions in simplest form.

a. 15.5% b. 16.2% c. 28.3% d. 41.38%

8. Write the following percentages as fractions in simplest form.

a. 18.14% b. 57.99% c. 12.15% d. 0.05%

9. WE35 Write the following percentages as fractions in simplest form.

a. $\frac{1}{2}\%$ b. $\frac{1}{5}\%$ c. $\frac{2}{3}\%$ d. $\frac{1}{10}\%$ e. $\frac{6}{11}\%$

10. Write the following percentages as fractions in simplest form.

a. $8\frac{1}{4}\%$ b. $20\frac{2}{3}\%$ c. $9\frac{2}{3}\%$ d. $15\frac{1}{2}\%$ e. $11\frac{1}{5}\%$

11. Write the following percentages as fractions in simplest form. Leave the answer as a mixed number where appropriate.

a. $10\frac{3}{8}\%$ b. $11\frac{2}{3}\%$ c. $150\frac{1}{2}\%$ d. $120\frac{1}{2}\%$ e. $33\frac{1}{3}\%$

## Understanding

12. MC 40% as a fraction in simplest form is:

A. $\frac{1}{4}$ B. $\frac{2}{5}$ C. $\frac{4}{100}$ D. $\frac{4000}{1}$ E. $\frac{0.4}{1000}$

13. MC $10\frac{1}{2}\%$ as a fraction is:

A. $\frac{21}{2}$ B. $\frac{10.5}{100}$ C. $\frac{2100}{2}$ D. $\frac{42}{100}$ E. $\frac{21}{200}$

14. MC 1.5% as a fraction in simplest form is:

A. $\frac{1.5}{100}$ B. $\frac{3}{100}$ C. $\frac{3}{200}$ D. $\frac{15}{100}$ E. $\frac{15}{1000}$

15. Imran saves 20% of his pocket money each week. Calculate the fraction that he saves.

16. In one game, 35% of a football team is injured. Calculate what fraction of the team is injured.

17. Each week Jodie spends 45% of her wages at the supermarket. Calculate what fraction of Jodie's wages is spent at the supermarket.

18. Calculate what fraction of a class were born overseas if 68% were born in Australia.

## Reasoning

19. Of the students in a Year 7 class, 80% have pets and 20% have blue eyes.
    a. Calculate what fraction of students have pets.
    b. Calculate what fraction of students have blue eyes.
    c. Explain whether this means that none of the students with blue eyes have pets.

20. a. In a molecule of methane (natural gas), 80% of the atoms are hydrogen. Calculate the fraction of the methane molecule that is composed of hydrogen atoms.

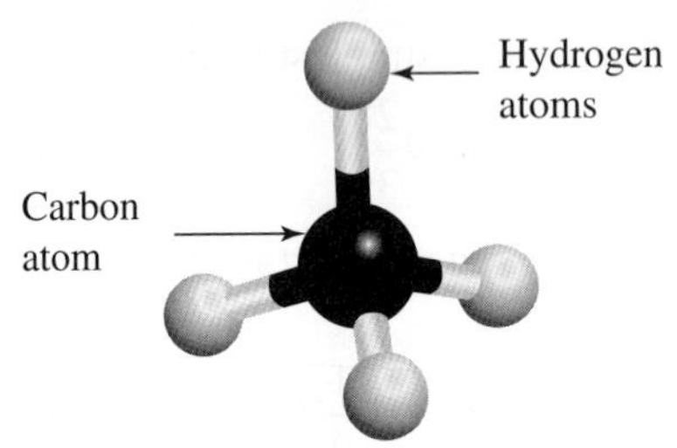

Methane molecule

b. The diagram represents a molecule of glycolaldehyde, a simple form of sugar found in the gas surrounding a young star called IRAS 16293-2422, located around 400 light-years away from Earth. Hydrogen atoms make up 50% of the atoms of this molecule. Calculate the fraction of the glycolaldehyde molecule that is composed of hydrogen atoms.

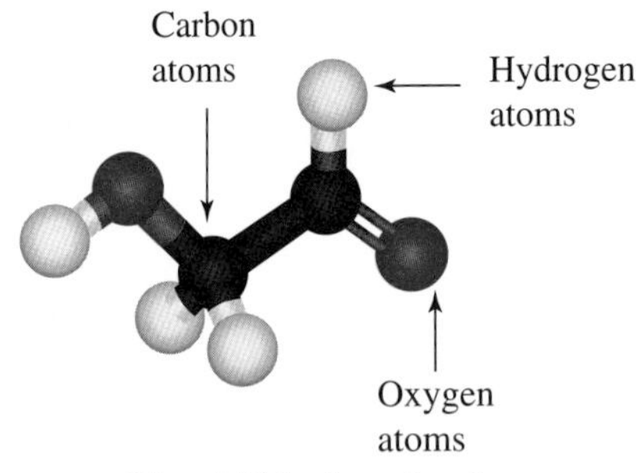

Glycolaldehyde molecule

c. Although both molecules have the same number of hydrogen atoms, explain why they represent different fractions.

21. Shirley plays basketball in her school's team. In the last game she scored 50 baskets from 75 shots but in the previous game she scored 45 baskets from 60 shots.
    a. Compare the percentage of baskets scored relative to the number of shots for each game to determine which game she performed best in.
    b. Explain how the percentage is affected by:
        i. increasing the number of baskets when the number of shots stays the same
        ii. increasing the number of shots when the number of baskets stays the same
        iii. increasing both the number of baskets and the number of shots at the same percentage rate.

### Problem solving

22. Less than 50% of the estimated 200 000 Australian invertebrate species have been described. Determine what fraction this is.
23. If the unemployment rate in Australia is 8%, calculate:
    a. what fraction of the population is unemployed
    b. what fraction of the population is employed
    c. out of 100 people, the number you would expect to be unemployed.
24. Determine what fraction remains if:
    a. 65% of winnings have been spent
    b. 19% of an audience hated a movie
    c. all stock was discounted by 15%
    d. 93.5% of the school population supports the uniform policy.
25. Seventeen per cent of visitors to Australia are from New Zealand.
    a. State the fraction of visitors to Australia from New Zealand.
    b. Calculate the fraction of visitors to Australia not from New Zealand.
    c. Out of 100 visitors to Australia, state how many you would expect to be from New Zealand.

# LESSON
# 4.10 Calculating percentages of an amount

### LEARNING INTENTION

At the end of this lesson you should be able to:
- calculate percentages of amounts using fractions.

## 4.10.1 Calculating percentages of an amount using fractions

eles-3814

- To calculate a percentage of an amount using fractions, follow these steps.
  Step 1: Convert the percentage into a fraction.
  Step 2: Multiply by the amount.
- When calculating a percentage of an amount, the unit of the answer will always be the unit of the initial quantity. For example, the percentage of an amount of kilograms will be a certain number of kilograms.

### COLLABORATIVE TASK: Elastic percentages

**Equipment:** scissors, wide rubber bands, rulers, A4 paper

1. Working in pairs, cut a rubber band in half to make one long rubber strip.
2. Use your ruler to draw an even scale on the rubber strip from 0 to 100, using intervals of 10. Make sure the rubber strip is not stretched in any way during this process.

3. Select five different objects around the classroom for which you will identify a percentage of their length, width or height; an object may be a book, a pencil or even someone's arm.
4. On an A4 sheet of paper, write the percentage that you will measure for each object.
   Your teacher may allocate your pair a percentage, or you may like to choose your own. Measure the same percentage for each of your objects; for example, you may choose to calculate 60% of the heights of a chair, a table, a computer screen, a poster and a bookshelf.
5. For each of your five objects, work with your partner to record the measurement of your given percentage using the following steps.
   - Stretch the rubber strip along the edge of the object that you wish to measure. The 0 mark on the strip must be at one end of the object you are measuring, and the 100 mark must be at the other end of the object. Hold onto the rubber strip tightly.
   - Place a pencil point on the object next to the position of your percentage on the rubber strip.
   - Use a ruler to measure the distance from the end of the object to the pencil point.
   - Draw a line of this length on your sheet of paper, and then label the line with a description of the object that you measured.
6. Each line on your paper represents the same percentage of different lengths. With your partner, discuss why the lines are different lengths.
7. As a class, discuss the relationship between the size of a percentage and the size of the whole object being measured. For example, is 60% of the length of a book more or less than 60% of the height of a door? Why?

### WORKED EXAMPLE 36 Calculating a percentage of an amount using fractions

**Calculate 40% of $135.**

| THINK | WRITE |
|---|---|
| 1. Write the question. | 40% of $135 |
| 2. Write the percentage as a fraction with a denominator of 100, change *of* to × and write the amount as a fraction over 1. | $= \frac{40}{100} \times \frac{135}{1}$ |
| 3. Simplify as appropriate. | $= \frac{40}{20} \times \frac{27}{1}$ <br> $= \frac{2}{1} \times \frac{27}{1}$ |
| 4. Perform the multiplication. | $= \frac{54}{1}$ |
| 5. State the answer. | = $54 |

### WORKED EXAMPLE 37 Calculating a percentage of an amount using fractions

**Of the 250 students selected at random to complete a survey, 18% were in Year 11. Calculate how many students were in Year 11.**

| THINK | WRITE |
|---|---|
| 1. Decide what percentage of the total is required and write an expression to determine the percentage of the total. | 18% of 250 |
| 2. Write the percentage as a fraction, change *of* to × and write the total as a fraction. | $= \frac{18}{100} \times \frac{250}{1}$ |

3. Simplify as appropriate.

$$= \frac{18}{2} \times \frac{5}{1}$$

$$= \frac{9}{1} \times \frac{5}{1}$$

4. Perform the multiplication.

$$= \frac{45}{1}$$

$$= 45$$

5. Answer the question by writing a sentence.

45 of the 250 students were in Year 11.

**DISCUSSION**

What method would you use to calculate 60% of an amount? Are there other ways? Which is the most efficient?

Resources

**eWorkbook** Topic 4 Workbook (worksheets, code puzzle and project) (ewbk-1905)

**Video eLesson** Percentages of an amount (eles-1882)

**Interactivities** Individual pathway interactivity: Calculating percentages of an amount (int-4346)

Percentages of an amount using fractions (int-3996)

## Exercise 4.10 Calculating percentages of an amount

learnon

**4.10 Quick quiz** 

**4.10 Exercise**

### Individual pathways

| ■ PRACTISE | ■ CONSOLIDATE | ■ MASTER |
|---|---|---|
| 1, 3, 6, 9, 12, 16, 18, 23, 25, 28, 31 | 2, 4, 7, 10, 13, 15, 19, 21, 24, 27, 29, 32 | 5, 8, 11, 14, 17, 20, 22, 26, 30, 33 |

### Fluency

1. Copy each of the following problems and then calculate the answers by completing the working.

   a. $90\% \text{ of } 200 = \frac{90}{100} \times \frac{200}{1}$
   b. $8\% \text{ of } 50 = \frac{8}{100} \times \frac{50}{1}$
   c. $50\% \text{ of } 120 = \frac{50}{100} \times \frac{120}{1}$

2. Copy each of the following problems and then calculate the answers by completing the working.

   a. $5\% \text{ of } 30 = \frac{5}{100} \times \frac{30}{1}$
   b. $15\% \text{ of } 70 = \frac{15}{100} \times \frac{70}{1}$
   c. $65\% \text{ of } 120 = \frac{65}{100} \times \frac{120}{1}$

3. WE36 Calculate the following by converting the percentage to a fraction.

a. 50% of 20 b. 20% of 80 c. 10% of 30 d. 9% of 200 e. 40% of 15

4. Calculate the following by converting the percentage to a fraction.

a. 12% of 50 b. 70% of 110 c. 80% of 5000 d. 44% of 150 e. 68% of 25

5. Calculate the following by converting the percentage to a fraction.

a. 52% of 75 b. 38% of 250 c. 110% of 50 d. 150% of 8 e. 125% of 20

6. Calculate the following by converting the percentage to a fraction. Write the answer as a mixed number.

a. 18% of 20 b. 16% of 30 c. 11% of 70 d. 74% of 25 e. 66% of 20

7. Calculate the following by converting the percentage to a fraction. Write the answer as a mixed number.

a. 55% of 45 b. 15% of 74 c. 95% of 62 d. 32% of 65 e. 82% of 120

8. MC 45% written as a fraction is:

A. $\frac{45}{100}$ B. $\frac{45}{1}$ C. $\frac{450}{1}$ D. $\frac{1}{45}$ E. $\frac{90}{45}$

9. MC To evaluate 17% of 22, *of* should be changed to:

A. $\div$ B. of C. $+$ D. $\times$ E. $-$

10. MC Select which of the following would determine 15% of 33.

A. 15 of 33 B. $\frac{15}{1} \times 33$ C. $\frac{15}{100} \times \frac{33}{100}$ D. $\frac{15}{1} \times \frac{33}{100}$ E. $\frac{15}{100} \times \frac{33}{1}$

11. MC 60% of 30 is:

A. $19\frac{4}{5}$ B. $\frac{31}{5}$ C. 186 D. 19 E. 18

12. MC 8% of 50 equals:

A. 40 B. 4 C. 140 D. 14 E. 16

13. MC 40% of 160 equals:

A. 400 B. 6400 C. 640 D. 0.64 E. 64

14. MC 60% of 30 equals:

A. 180 B. 1.8 C. 400 D. 18 E. 300

## Understanding

15. WE37 Of the 300 Year 7 students selected to complete a questionnaire on ice cream, 70% said that their favourite flavour was chocolate chip. Calculate how many students favoured chocolate chip ice cream.

16. A thunder day at a given location is a calendar day on which thunder is heard at least once. About 20% of days near Darwin are thunder days.
Calculate the number of thunder days you would expect in Darwin in one year.

17. Two per cent of Australians play lawn bowls. In a group of 50 people, state how many you would expect to play lawn bowls.

18. Two thousand people entered a marathon. Some walked and the rest jogged. If 20% walked, calculate:

a. what percentage jogged
b. how many people jogged.

19. William earns $550 per week. He has just received a pay rise of 4%.

a. Calculate how much more William will earn per week.
b. Calculate William's total pay after the pay rise.

20. The lotto jackpot for Saturday night is $20 million. If you win 8% of the jackpot, calculate how much money you would win.

21. The Australian cricket team was fined 15% of their match payments for a slow over rate. If the team was paid $80 000 for the match, calculate how much the team was fined.

22. Australia has 315 species of mammals. Of these, 15% are threatened. Calculate how many threatened species of mammals Australia has. Round your answer to the nearest whole number.

23. Two million people attended at least one AFL game last year. Of these people, 30% attended 10 games or more. Calculate how many people attended 10 or more AFL games.

24. The water content of a particular brand of shampoo is 35%. After performing a calculation, Peter claims that an 800 mL bottle of this particular shampoo contains 28 mL of water.

a. Explain why Peter's calculation is incorrect.
b. Calculate how much of the shampoo in the 800 mL bottle is actually water.

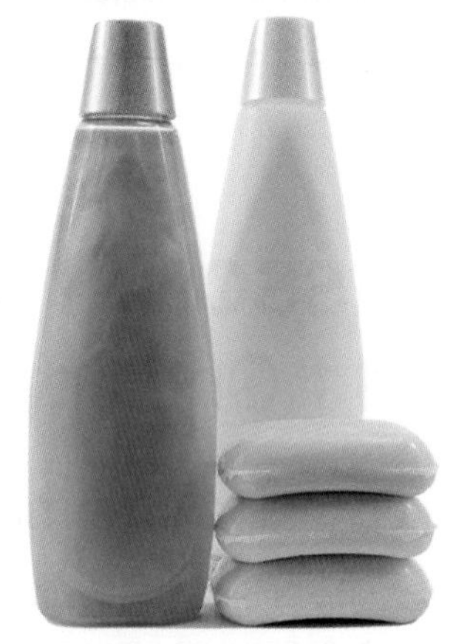

25. In Sydney, 14% of adults use a telephone to pay bills and 38% use the internet. If 50 adults need to pay bills, calculate how many will use the:

a. telephone
b. internet.

26. In Weburbia, 85% of all households have access to the internet. If there were 34 houses in Website Street, calculate how many you would expect to have access to the internet. Write the answer to the nearest whole number.

27. Terry runs a factory that makes parts for trucks. He has increased his staff by 12%. If Terry originally had 34 workers, calculate how many workers he has after the increase. Write the answer to the nearest whole number.

## Reasoning

28. In the year 2020, 60% of households in Geelong had a smoke alarm. Of these smoke alarms, 94% worked. A street in Geelong has 20 houses.

a. Calculate how many houses you would expect to have a smoke alarm.

b. Calculate how many would have a smoke alarm that works. Write the answer to the nearest whole number.

29. In some parts of Africa, 6% of women die during pregnancy or childbirth. Of 50 women living in these parts who are pregnant, calculate how many are likely to die during pregnancy or childbirth.

30. For a radioactive substance, a half-life is the time it takes for half of the radioactive atoms in the substance to decay (lose their radioactivity). For 400 g of a given radioactive substance, the decay rate is 50% every 20 days.
   a. Calculate the quantity of radioactive substance left after 20 days, 40 days, 60 days, 80 days and 100 days.
   b. Calculate the percentage of the initial quantity remaining after 20 days, 40 days, 60 days, 80 days and 100 days.
   c. The quantity of the radioactive substance halves every 20 days. Explain why the percentage decrease is smaller after every 20 days.

### Problem solving

31. Squeezy fruit drink is 36% pure orange juice and 64% water. If 1 litre of Squeezy fruit drink is mixed with 600 mL of water, determine the concentration of juice in the new mixture.

32. The students in Year 7A were voting for class captain. There were only two candidates. Rachel received 84% of the votes and Susi received 28% of the votes. It was then discovered that some students had voted twice. If the class consisted of 25 students, determine how many of them voted twice.

33. A department store stocked a selection of denim jeans. At the start of December, the store had 320 of these jeans. During December, it sold 50% of the jeans. In January, it sold 25% of the remaining denim jeans. Determine how many were left at the end of January.

# LESSON
# 4.11 One amount as a percentage of another

**LEARNING INTENTION**

At the end of this lesson you should be able to:
- express an amount as a percentage of another amount.

## 4.11.1 One amount as a percentage of another

eles-3815

- To express one amount as a percentage of another, write the amounts as a fraction $\frac{\text{amount}}{\text{total amount}}$, then convert the fraction to a percentage.
  (*Remember:* To change a fraction into a percentage, multiply by 100 and add the % sign.)
- When expressing one amount as a percentage of another, make sure that both amounts are in the same unit.

## WORKED EXAMPLE 38 Expressing one amount as a percentage of another

**Express:**

a. **15 as a percentage of 20**

b. **9 as a percentage of 33. Write the answer as a mixed number.**

| THINK | WRITE |
|---|---|
| a. 1. Write the amount as a fraction of the total: $\frac{\text{amount}}{\text{total amount}}$. | a. $\frac{15}{20}$ |
| 2. Multiply by $\frac{100}{1}$ and include the % sign. | $= \frac{15}{20} \times \frac{100}{1}\%$ |
| 3. Simplify and then multiply. | $= \frac{15}{1} \times \frac{5}{1}\%$ <br> $= \frac{75}{1}\%$ |
| 4. State the answer. | $= 75\%$ |
| b. 1. Write the amount as a fraction of the total: $\frac{\text{amount}}{\text{total amount}}$. | b. $\frac{9}{33}$ |
| 2. Multiply by $\frac{100}{1}$ and include the % sign. | $= \frac{9}{33} \times \frac{100}{1}\%$ |
| 3. Simplify and then multiply. | $= \frac{3}{11} \times \frac{100}{1}\%$ <br> $= \frac{300}{11}\%$ |
| 4. Write as a mixed number by dividing the denominator into the numerator. | $= 27\frac{3}{11}\%$ |

## WORKED EXAMPLE 39 Calculating percentages with different units

**Express 45c as a percentage of $2.**

| THINK | WRITE |
|---|---|
| 1. Write the larger amount using the smaller unit. | $2 = 200 cents |
| 2. Write the first amount as a fraction of the second amount: $\frac{\text{amount}}{\text{total amount}}$. | $\frac{45}{200}$ |
| 3. Multiply by $\frac{100}{1}$ and include the % sign. | $= \frac{45}{200} \times \frac{100}{1}\%$ |
| 4. Simplify and then multiply. | $= \frac{45}{2} \times \frac{1}{1}\%$ <br> $= \frac{45}{2}\%$ |

5. Convert to a mixed number. $= 22\frac{1}{2}\%$

6. Write the answer. 45c is $22\frac{1}{2}\%$ of $2.

## WORKED EXAMPLE 40 Calculating percentages of test marks

**Kye obtained 17 out of 30 on their Science test. Calculate their percentage score. Round your answer to the nearest whole number.**

| THINK | WRITE |
|---|---|
| 1. Write the amount as a fraction of the total: $\frac{\text{amount}}{\text{total amount}}$. | $\frac{17}{30}$ |
| 2. Multiply by $\frac{100}{1}$ and include the % sign. | $= \frac{17}{30} \times \frac{100}{1}\%$ |
| 3. Simplify and then multiply. | $= \frac{17}{3} \times \frac{10}{1}\%$ <br> $= \frac{170}{3}\%$ |
| 4. Convert to a mixed number. | $= 56\frac{2}{3}\%$ |
| 5. Round to the nearest whole number. (If the fraction is a half or more, add one to the number.) | $\approx 57\%$ |
| 6. Write the answer as a sentence. | Kye obtained 57% for their Science test. |

### Digital technology

Scientific calculators can calculate one amount as a percentage of another. The answer could be either in fraction or mixed number format.

DEG

$\frac{17}{30} * 100$ $\frac{170}{3}$

$\frac{170}{3}$ ▸%↔U% $56\frac{2}{3}$

Resources

**eWorkbook** Topic 4 Workbook (worksheets, code puzzle and project) (ewbk-1905)

**Interactivities** Individual pathway interactivity: One amount as a percentage of another (int-4347)
One amount as a percentage of another (int-3998)

# Exercise 4.11 One amount as a percentage of another

4.11 Quick quiz 

4.11 Exercise

Individual pathways

| ■ PRACTISE | ■ CONSOLIDATE | ■ MASTER |
|---|---|---|
| 1, 4, 7, 10, 12, 14, 21, 23, 26 | 2, 5, 8, 11, 15, 17, 19, 22, 24, 27, 28 | 3, 6, 9, 13, 16, 18, 20, 25, 29, 30 |

## Fluency

1. **WE38a** Express:
   a. 2 as a percentage of 10
   b. 13 as a percentage of 52
   c. 8 as a percentage of 80
   d. 8 as a percentage of 20
   e. 15 as a percentage of 60.
2. Express:
   a. 120 as a percentage of 150
   b. 8 as a percentage of 25
   c. 12 as a percentage of 30
   d. 40 as a percentage of 200
   e. 10 as a percentage of 200.
3. **WE38b** Express (writing the answers as mixed numbers):
   a. 180 as a percentage of 720
   b. 63 as a percentage of 200
   c. 25 as a percentage of 400
   d. 32 as a percentage of 500
   e. 620 as a percentage of 3000.
4. Express (writing the answers as mixed numbers):
   a. 70 as a percentage of 80
   b. 12 as a percentage of 42
   c. 52 as a percentage of 78
   d. 11 as a percentage of 30
   e. 4 as a percentage of 15.
5. Express (writing the answers as mixed numbers):
   a. 5 as a percentage of 70
   b. 4 as a percentage of 18
   c. 3 as a percentage of 7
   d. 64 as a percentage of 72
   e. 15 as a percentage of 21.
6. Express (rounding all answers to the nearest whole number):
   a. 7 as a percentage of 42
   b. 18 as a percentage of 54
   c. 95 as a percentage of 150
   d. 7 as a percentage of 65.
7. Express (rounding all answers to the nearest whole number):
   a. 30 as a percentage of 36
   b. 14 as a percentage of 18
   c. 80 as a percentage of 450
   d. 8 as a percentage of 68.
8. **WE39** Express the first amount as a percentage of the second amount.
   a. 30c and $3
   b. 200 m and 6 km
   c. 5 mm and 4 cm
9. Express the first amount as a percentage of the second amount.
   a. 6 days and 3 weeks
   b. 15 minutes and 1 hour
   c. 25 mL and 2 L
10. **MC** 4 as a percentage of 50 is:
    A. 2%
    B. 12.5%
    C. 46%
    D. 8%
    E. 4%
11. **MC** 9 out of 36 as a percentage is:
    A. 25%
    B. 9%
    C. 400%
    D. 27%
    E. 36%

12. **MC** 14 as a percentage of 35 is:

**A.** 14% **B.** 40% **C.** 20% **D.** 35% **E.** $2\frac{1}{2}\%$

13. **MC** Alice got 42 out of 60 for her Mathematics test. Her mark as a percentage is:

**A.** 58% **B.** 60% **C.** 30% **D.** 42% **E.** 70%

### Understanding

14. **WE40** Ash Barty has won 7 out of her last 10 singles matches. Calculate the number of wins in her last 10 matches as a percentage.

15. Three out of five people prefer chocolate ice cream to vanilla. Calculate what percentage of people prefer chocolate ice cream to vanilla.

16. Caillan's pocket money increased from $15 per week to $20 per week.
   **a.** Calculate how much Caillan's pocket money has increased by.
   **b.** Calculate this as a percentage increase.

17. Daniel earns $500 per week. He spends $50 on petrol, $70 on rent and $60 on food.
   **a.** Calculate the percentage Daniel spends on:
      **i.** petrol **ii.** rent **iii.** food.
   **b.** Calculate what percentage of his total wage Daniel spends on petrol, rent and food combined.
   **c.** Calculate the percentage Daniel has left.

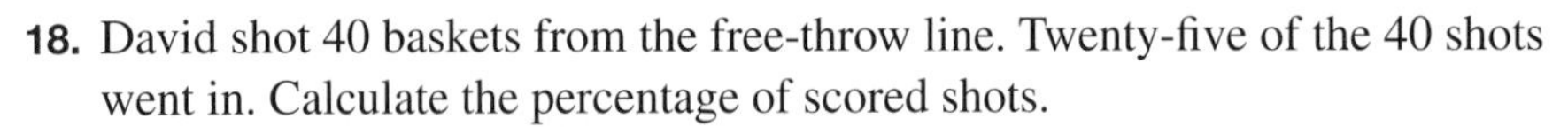

18. David shot 40 baskets from the free-throw line. Twenty-five of the 40 shots went in. Calculate the percentage of scored shots.

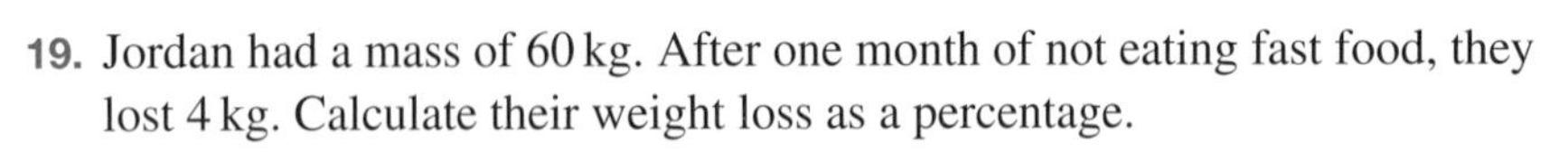

19. Jordan had a mass of 60 kg. After one month of not eating fast food, they lost 4 kg. Calculate their weight loss as a percentage.

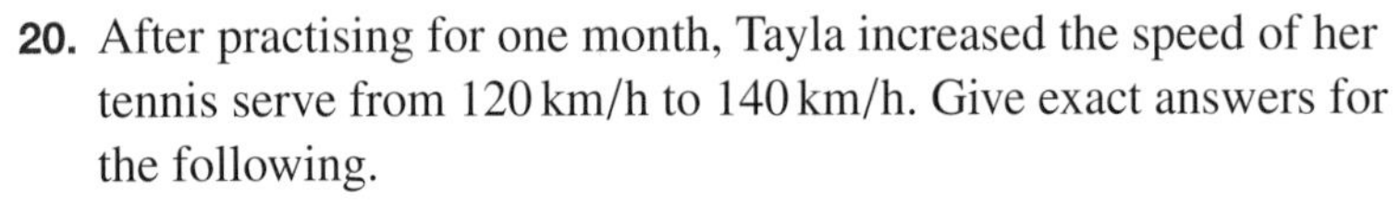

20. After practising for one month, Tayla increased the speed of her tennis serve from 120 km/h to 140 km/h. Give exact answers for the following.
   **a.** Calculate how much her serve speed increased by.
   **b.** Calculate this as a percentage increase.

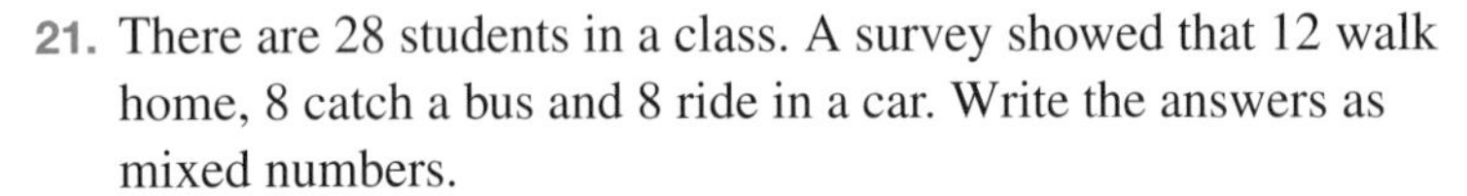

21. There are 28 students in a class. A survey showed that 12 walk home, 8 catch a bus and 8 ride in a car. Write the answers as mixed numbers.
   **a.** Calculate the percentage of students who walk.
   **b.** Calculate the percentage of students who catch a bus.
   **c.** Calculate the total percentage who walk or catch a bus.
   **d.** Calculate the percentage of students who do not walk.

22. During one AFL season, the GWS Giants won 16 out of 22 games. Calculate the percentage of games the Giants lost.

## Reasoning

23. A baker uses 200 g of flour for 500 g of dough in one recipe, and 250 g of flour for 500 g of dough in another recipe.
    a. Identify which recipe contains the higher percentage of flour. Explain your choice.
    b. Explain what happens when the quantity of flour is increased but the quantity of dough remains constant.

24. a. If you increase $100 by 25% and then decrease the new amount by 25%, will you end up with more than $100, less than $100 or exactly $100? Give reasons for your answer.
    b. Choose another amount and another percentage to test your reasoning.

25. My mother is four times older than I am. My sister is 75% of my age. She is also 10% of my grandfather's age. My father is 50, which is 2 years older than my mother. Calculate how old my sister and my grandfather are.

## Problem solving

26. The teacher's mark book shows Karina has achieved the following test results in Maths.

| Topic | Probability | Measurement | Algebra | Equations | Geometry |
|---|---|---|---|---|---|
| Score | $\frac{15}{20}$ | $\frac{13}{15}$ | $\frac{27}{30}$ | $\frac{70}{80}$ | $\frac{95}{100}$ |
| Percentage | | | | | |

    a. Copy and complete the table to show Karina's percentage for each test. Round answers to the nearest whole number.
    b. Compare the percentage results to determine which test Karina performed best in.
    c. Calculate Karina's average percentage mark by adding up the percentages and dividing by 5, the number of tests.

27. Determine which of the following quantities represents:
    a. the highest percentage
    b. the lowest percentage
    c. the same percentage.

    200 g out of 500 g
    300 g out of 450 g
    420 g out of 840 g
    350 g out of 700 g

28. Suggest why we need to be able to express one quantity as a percentage of another quantity.

29. A type of concrete uses 10 kg of cement for a quantity of 65 kg of concrete. Another type of concrete uses 10 kg of cement for 85 kg of concrete.
    a. Explain which type of concrete contains the higher percentage of cement.
    b. Explain what happens when the quantity of cement is constant but the quantity of concrete increases.

30. Roseanne is planning a tour of Australia. She calculates that 70% of the distance will be on sealed roads and the remainder on dirt roads. If there are 1356 km of dirt roads, determine how long Roseanne's journey will be.

# LESSON
# 4.12 Common percentages and short cuts

## LEARNING INTENTION

At the end of this lesson you should be able to:
- recognise commonly used percentages
- use common percentages as short cuts to calculate related percentages.

## 4.12.1 Common percentages and short cuts

eles-3816

- Common percentages that are often used in everyday life include 10%, 20%, 25%, 50%, 75%.
- It is useful to remember the fraction equivalents of the commonly used percentages to make calculations involving them quicker and easier.
- The table below shows the fractional equivalents of some commonly used fractions and how to use these to speed up calculations.

| Percentage | Fraction | How to calculate this percentage |
|---|---|---|
| 1% | $\frac{1}{100}$ | Divide by 100 |
| 5% | $\frac{1}{20}$ | Divide by 20 |
| 10% | $\frac{1}{10}$ | Divide by 10 |
| $12\frac{1}{2}\%$ | $\frac{1}{8}$ | Divide by 8 |
| 20% | $\frac{1}{5}$ | Divide by 5 |

| Percentage | Fraction | How to calculate this percentage |
|---|---|---|
| 25% | $\frac{1}{4}$ | Divide by 4 |
| $33\frac{1}{3}\%$ | $\frac{1}{3}$ | Divide by 3 |
| 50% | $\frac{1}{2}$ | Divide by 2 |
| $66\frac{2}{3}\%$ | $\frac{2}{3}$ | Divide by 3, then multiply by 2 |
| 75% | $\frac{3}{4}$ | Divide by 4, then multiply by 3 |

- Calculating 10% without a calculator is easy. As it is equivalent to dividing by 10, the decimal place just needs to be moved 1 space to the left. Once you have found 10% of the amount, it is easy to calculate some other percentages. For example:
  - to calculate 5%, halve 10%
  - to calculate 20%, double 10%.

**Short cut for calculating whole number percentages**

**Calculate 1% by dividing the amount by 100 (move the decimal point 2 spaces to the left), then multiply 1% by the whole number that you are looking for.**

**For example, to calculate 12%, calculate 1%, then multiply by 12.**

### WORKED EXAMPLE 41 Calculating 10% of an amount

**Calculate 10% of each of the following, rounding the answer to the nearest 5 cents.**

a. **$37** b. **$120**

| THINK | WRITE |
|---|---|
| a. Write the question and move the position of the decimal point 1 place to the left for the answer. Remember that if there is no decimal point, put it at the end of the number. ($37 = $37.00) | a. 10% of $37 = $3.70 |
| b. Write the question and move the position of the decimal point 1 place to the left for the answer. | b. 10% of $120 = $12 |

### WORKED EXAMPLE 42 Calculating common percentages of an amount

**Calculate:**

a. **5% of $180** b. **20% of $7** c. **25% of $46**

| THINK | WRITE |
|---|---|
| a. 1. Calculate 10% of the amount. | a. 10% of $180 = $18 |
| 2. Calculate 5% by halving the amount obtained in step 1. Write the answer. | 5% of $180 = $18 ÷ 2<br>= $9 |
| b. 1. Calculate 10% of the amount. | b. 10% of $7 = $0.70 |
| 2. Calculate 20% by doubling the amount obtained in step 1. Write the answer. | 20% of $7 = 2 × $0.70<br>= $1.40 |
| c. 1. To calculate 25% of an amount divide it by 4. | 25% of $46 = $46 ÷ 4 |
| 2. Perform the division. | = $11.50 |
| 3. Write the answer. | 25% of $46 = $11.50 |

### WORKED EXAMPLE 43 Calculating percentages using short cuts

**Calculate:**

a. **12% of $53** b. **43% of $120**

**Round your answers to the nearest 5 cents.**

| THINK | WRITE |
|---|---|
| a. 1. Break up the percentage into lots of 10% and 1%. | a. 12% = 10% + 2 × 1% |
| 2. Calculate 10% of the amount. | 10% of $53 = $5.30 |
| 3. Calculate 1% of the amount and then double. | 1% of $53 = $0.53<br>2% of $53 = 2 × $0.53<br>= $1.06 |
| 4. Determine 12% of the amount by adding 10% and 2 × 1%. | 12% of 53% = $5.30 + $1.06<br>= $6.36 |
| 5. Round the answer to the nearest 5 cents. | = $6.35 |

Alternatively, you could use the short-cut method shown below.

1. First calculate 1% of \$53 by moving the decimal point 2 places to the left. — $1\% \text{ of } \$53 = \$0.53$
2. Multiply by 12. — $12\% \text{ of } \$53 = 12 \times 1\% \text{ of } \$53 = 12 \times \$0.53 = \$6.36$
3. Round the answer to the nearest 5 cents. — $= \$6.35$

**b.** 1. Break up the percentage into lots of 10% and 1%. — **b.** $43\% = 4 \times 10\% + 3 \times 1\%$

2. Calculate 10% and 1% of the amount. — $10\% \text{ of } \$120 = \$12$, $1\% \text{ of } \$120 = \$1.20$

3. Use the values of 10% and 1% to calculate 43% of \$120. — $43\% \text{ of } \$120 = 4 \times \$12 + 3 \times \$1.20 = \$48 + \$3.60 = \$51.60$

Alternatively, you could use the short-cut method shown below.

1. First calculate 1% of \$120 by moving the decimal point 2 places to the left. — $1\% \text{ of } \$120 = \$1.20$
2. Multiply by 43 and write the answer. — $43\% \text{ of } \$120 = 43 \times 1\% \text{ of } \$120 = 43 \times \$1.20 = \$51.60$

Resources

eWorkbook Topic 4 Workbook (worksheets, code puzzle and project) (ewbk-1905)

Interactivities Individual pathway interactivity: Common percentages and short cuts (int-4348)

Common percentages and short cuts (int-3999)

## Exercise 4.12 Common percentages and short cuts

learnon

4.12 Quick quiz  | 4.12 Exercise

### Individual pathways

| ■ PRACTISE | ■ CONSOLIDATE | ■ MASTER |
|---|---|---|
| 1, 4, 7, 10, 13, 16, 19, 22, 24, 27, 30, 33 | 2, 5, 8, 11, 14, 17, 20, 23, 25, 28, 31, 34 | 3, 6, 9, 12, 15, 18, 21, 26, 29, 32, 35 |

### Fluency

1. **WE41** Calculate 10% of each of the following, rounding the answer to the nearest 5 cents.
   a. \$10.00 b. \$18.00 c. \$81.00 d. \$150.00 e. \$93.00

2. Calculate 10% of each of the following, rounding the answer to the nearest 5 cents.

a. $22.00 b. $17.20 c. $12.60 d. $1.50 e. $47.80

3. Calculate 10% of each of the following, rounding the answer to the nearest 5 cents.

a. $192.40 b. $507.00 c. $1926.00 d. $3041.50 e. $1999.90

4. Calculate 10% of the following. Round your answers to the nearest 5 cents.

a. $15 b. $51 c. $9 d. $137 e. $4.29

5. Calculate 10% of the following. Round your answers to the nearest 5 cents.

a. $6.37 b. $39.17 c. $74.81 d. $102.75 e. $67.87

6. Calculate 10% of the following. Round your answers to the nearest 5 cents.

a. $517.83 b. $304.77 c. $628.41 d. $100.37 e. $207.08

7. **WE42** Calculate 5% of each of the following. Round the answers to the nearest 5 cents.

a. $8.20 b. $6.40 c. $2.20 d. $140.20 e. $42.40

8. Calculate 5% of each of the following. Round the answers to the nearest 5 cents.

a. $10.60 b. $304.80 c. $1000 d. $103.27 e. $31.70

9. Calculate 20% of the following. Round the answers to the nearest 5 cents.

a. $21.50 b. $42.30 c. $3.30 d. $74.10 e. $0.79

10. Calculate 20% of the following. Round the answers to the nearest 5 cents.

a. $16.40 b. $135.80 c. $261.70 d. $1237 e. $5069

11. Calculate the following. Round the answers to the nearest 5 cents.

a. 15% of $12 b. 15% of $8.00 c. 15% of $60.00

12. Calculate the following. Round the answers to the nearest 5 cents.

a. 25% of $30.00 b. 25% of $90.00 c. 25% of $220.00

13. Calculate the following. Round the answers to the nearest 5 cents.

a. 30% of $15.00 b. 30% of $25.00 c. 30% of $47.50

14. Calculate 1% of the following. Round the answers to the nearest 5 cents.

a. $268 b. $713 c. $604 d. $5.60 e. $13

15. Calculate 1% of the following. Round the answers to the nearest 5 cents.

a. $14.80 b. $81.75 c. $19.89 d. $4.25 e. $6.49

16. **WE43** Calculate the following. Round the answers to the nearest 5 cents.

a. 12% of $11 b. 21% of $50 c. 3% of $22 d. 6% of $40 e. 13% of $14

17. Calculate the following. Round the answers to the nearest 5 cents.

a. 35% of $210 b. 9% of $17 c. 2% of $53 d. 7% of $29 e. 45% of $71.50

18. Calculate the following. Round the answers to the nearest 5 cents.

a. 33% of $14.50 b. 42% of $3.80 c. 31% of $1.45 d. 64% of $22.50 e. 41% of $1200

### Understanding

19. **MC** 10% of $7.25 equals:

A. $725 B. $7.30 C. $72.50 D. $0.73 E. $0.72

20. **MC** 1% of $31.48 equals:

A. $3.14  B. $0.31  C. $0.32  D. $31.50  E. $3.15

21. **MC** 15% of $124 equals:

A. $12.40  B. $1.24  C. $6.20  D. $13.64  E. $18.60

22. **MC** 22% of $5050 equals:

A. $60.60  B. $50.50  C. $1111  D. $43.56  E. $55.55

23. Maria is buying a new set of golf clubs. The clubs are marked at $950, but if Maria pays cash, the shop will take 10% off the marked price. Determine how much the clubs will cost if Maria pays cash.

24. Thirty per cent of residents in a district are over the age of 65. If there are 180 000 residents, determine how many are over the age of 65.

25. Jay is buying a new lounge suite worth $2150. Jay has to leave a 15% deposit and then pay the balance in monthly instalments.
    a. Calculate the deposit Jay needs to pay.
    b. Determine how much Jay will have to pay each month if he plans to pay the balance off in one year.

26. Ninety per cent of students at a school were present for school photographs. If the school has 1100 students, calculate how many were absent on the day that the photographs were taken.

27. Jim can swim 50 m in 31 seconds. If he improves his time by 10%, calculate Jim's new 50 m time.

28. In a survey, 40 people were asked if they liked or disliked Vegemite. Of the people surveyed, 5% said they disliked Vegemite. Calculate how many people:
    a. disliked Vegemite  b. liked Vegemite.

29. 35 800 went to the SCG to watch a Swans versus Giants football match. Of the crowd, 30% went to the game by car and 55% caught public transport. Calculate how many people:
    a. arrived by car

    b. caught public transport
    c. did not arrive at the game by car or public transport.

### Reasoning

30. When I am 5% older than I am now, I will be 21 years old. How old am I now?

31. The price of bread is now 250% of what it cost 20 years ago, which is an increase of 150%. If a loaf of bread costs $2.00 now, determine how much it cost 20 years ago.

32. I am six months old. If I gain 10% of my current mass I will be three times my birth mass. If my birth mass was 3 kg, determine my current mass. Round your answer to 1 decimal place.

### Problem solving

**33.** I am 33 years old. I have lived in England for 8 years. If I stay in England, determine how old I will be when the number of years I have lived here is 50% of my age. Show how you reached your answer.

**34.** Kate is a carpenter and has made a cutting board as shown.

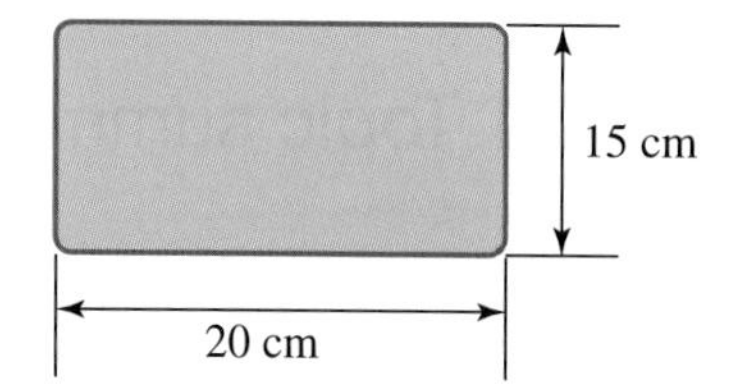


a. One of her customers asked Kate to make her a similar cutting board but to double its width. Calculate the percentage increase of the width of the board.

b. A second customer asked Kate to make her a cutting board with dimensions 30 cm by 30 cm. Calculate:

i. the percentage increase of the width of the board
ii. the percentage increase of the length of the board
iii. the percentage increase of the perimeter of the board.

c. Explain whether there is a connection between the percentage increase of a side and the percentage increase of the perimeter.

**35.** Ten per cent of the messages received by a business were from customers dissatisfied with the product bought. Of these customers, 32% were male.

a. Calculate the overall percentage of dissatisfied male customers.
b. Calculate the overall percentage of dissatisfied female customers.
c. Explain what happens when you apply a percentage to another percentage.

# LESSON
# 4.13 Review

## 4.13.1 Topic summary

### FRACTIONS AND PERCENTAGES

#### Working with fractions

- When working with fractions, mixed numbers provide the user with a good understanding of the value.
- However, when computing operations with fractions (addition, subtraction, multiplication and division), it is best to have all fractions expressed as proper fractions.

#### Types of fractions

- Fractions are used to represent parts of a whole.
- **Proper fractions** have a numerator that is less than the denominator, e.g. $\frac{2}{7}$.
- **Improper fractions** have a numerator that is greater than the denominator, e.g. $\frac{10}{3}$.
- A **mixed number** consists of an integer and a proper fraction, e.g. $2\frac{1}{4}$.

#### Finding a percentage of a quantity

- Finding a percentage of a number or a quantity is the same as finding an equivalent fraction of that number or quantity.
- Remember that in mathematics the word 'of' means multiply.

e.g. Determine 15% of $80.

$$15\% \text{ of } \$80 = \frac{15}{100} \times 80 = \frac{15}{10\not{0}} \times \frac{8\not{0}}{1} = \frac{120}{10} = 12$$

#### Simplifying fractions

- Fractions can be simplified by dividing the numerator and denominator by the same number.
- A fraction is in its simplest form when there are no factors (other than 1) that are common to both the numerator and denominator.

e.g. Simplify $\frac{80}{200}$.

Divide the numerator and denominator by 10.

$$\frac{80^{\div 10}}{200^{\div 10}} = \frac{8}{20}$$

There are still numbers that divide into both 8 and 20.

Divide the numerator and denominator by 4.

$$\frac{80^{\div 10}}{200^{\div 10}} = \frac{8^{\div 4}}{20^{\div 4}} = \frac{2}{5}$$

#### Multiplying fractions

To multiply fractions together, follow the three steps below:

1. Multiply the numerators.
2. Multiply the denominators.
3. Simplify the fraction if possible.

e.g. Evaluate $\frac{1}{2} \times \frac{4}{9}$.

$$\frac{1}{2} \times \frac{4}{9} = \frac{1 \times 4}{2 \times 9} = \frac{4}{18} = \frac{2}{9}$$

#### Dividing fractions

To divide two fractions, follow the KEEP, CHANGE, FLIP method outlined below and then multiply the fractions together.

1. **KEEP** the fraction to the left of the ÷ symbol the same.
2. **CHANGE** the division sign to a multiplication sign.
3. **FLIP** (invert) the fraction to the right of the symbol.

e.g. Evaluate $\frac{3}{4} \div \frac{2}{7}$.

| KEEP | CHANGE | FLIP |
|---|---|---|
| $\frac{3}{4}$ | $\times$ | $\frac{7}{2}$ |

$$\frac{3}{4} \div \frac{2}{7} = \frac{3}{4} \times \frac{7}{2} = \frac{3 \times 7}{4 \times 2} = \frac{21}{8}$$

#### Addition and subtraction of fractions

To add or subtract fractions, follow the steps below:

1. Examine the denominators of all fractions and determine the lowest common multiple (LCM).
2. Use the method of equivalent fractions to change each fraction so that the denominators are all the same.
3. Depending on the symbol in the question, add or subtract the numerators leaving the denominator unchanged.

e.g. Evaluate $\frac{2}{3} + \frac{7}{4}$.

LCM = 12

$$\frac{2}{3} + \frac{7}{4} = \frac{2}{3} \times \frac{4}{4} + \frac{7}{4} \times \frac{3}{3} = \frac{8}{12} + \frac{21}{12}$$

$$= \frac{8 + 21}{12} = \frac{29}{12}$$

#### Expressing a percentage as a fraction

- To convert a percentage to a fraction, simply divide by 100 and remove the percentage symbol.

e.g. Express 80% as a common fraction.

$$80\% = 80 \div 100 = \frac{80^{\div 20}}{100^{\div 20}} = \frac{4}{5} = 0.54$$

#### Equivalent fractions

- Equivalent fractions are fractions that have the same value but look different.

e.g. $\frac{1}{2}, \frac{2}{4}$ and $\frac{50}{100}$ are equivalent fractions.
The value of each fraction is 0.5.

- Equivalent fractions can be created by multiplying both the numerator and denominator by the same number.

e.g.

$$\frac{1}{2} \times \frac{2}{2} = \frac{2}{4}$$

$$\frac{1}{2} \times \frac{50}{50} = \frac{50}{100}$$

#### Converting fractions to percentages

- To express a fraction as a percentage, simply multiply by 100 and include the percentage symbol.

e.g. Express $\frac{9}{20}$ as a percentage.

$$\frac{9}{20} \times 100\% = \frac{9}{2\not{0}} \times \frac{10\not{0}}{1}\% = \frac{90}{2}\% = 45\%$$

## 4.13.2 Success criteria

Tick the column to indicate that you have completed the lesson and how well you think you have understood it using the traffic light system.

(**Green:** I understand; **Yellow:** I can do it with help; **Red:** I do not understand)

| Lesson | Success criteria | Green | Yellow | Red |
|---|---|---|---|---|
| 4.2 | I understand the concept of a fraction. | | | |
| | I can convert between equivalent fractions. | | | |
| | I can compare fractions. | | | |
| 4.3 | I can simplify fractions. | | | |
| | I can simplify mixed numbers. | | | |
| 4.4 | I can convert between mixed numbers and improper fractions. | | | |
| 4.5 | I can add and subtract fractions. | | | |
| 4.6 | I can multiply fractions by a whole number. | | | |
| | I can multiply fractions by another fraction. | | | |
| | I can calculate fractions of a fraction. | | | |
| | I can calculate fractions of an amount. | | | |
| 4.7 | I can divide fractions. | | | |
| 4.8 | I can add, subtract, multiply and divide mixed numbers. | | | |
| 4.9 | I can convert percentages to fractions. | | | |
| 4.10 | I can calculate percentages of amounts using fractions. | | | |
| 4.11 | I can express an amount as a percentage of another amount. | | | |
| 4.12 | I can recognise commonly used percentages. | | | |
| | I can use common percentages as short cuts to calculate related percentages. | | | |

## 4.13.3 Project

### The Memory game

The game Memory is played with a standard deck of playing cards. All cards are placed face-side down on a table and each player takes turns to reveal 2 cards. If the 2 cards contain the same number or suit, they are a pair and the player gets another turn. If they are not a pair, they are returned to their original spot and turned face-down. The object of the game is to remember where the same cards are and collect as many pairs as you can. The winner is the person with the most pairs.

Your task is to develop a version of Memory with a fractions flavour. Each of your cards will have a fraction written on it and its pair will be an equivalent fraction.

1. As an introduction to the game, match the pairs of equivalent fractions on the cards below.

| $\frac{6}{8}$ | $\frac{3}{9}$ | $\frac{1}{3}$ | $\frac{3}{4}$ | $\frac{4}{8}$ | $\frac{1}{2}$ |
|---|---|---|---|---|---|
| A | B | C | D | E | F |

The requirements for your game are as follows.

- There are 15 pairs of equivalent fractions in this game — that makes 30 cards.
- One of the fractions of each pair must be in simplest form.
- For example, $\frac{1}{3}$ and $\frac{4}{12}$ are an acceptable pair, but $\frac{10}{30}$ and $\frac{4}{12}$ are not.
- Fractions don't have to be proper fractions. Equivalent mixed numbers and improper fractions are allowed.

2. Using the information in step 1, write down the 15 pairs of fractions that will be a part of your memory game.

When the game is packaged for sale, additional information accompanies the cards. This information includes a list of the fractions enclosed and instructions for the game. The list of fractions is given in order from smallest to largest.

3. List your fraction pairs in order from smallest to largest.

When items are placed for sale, presentation of the packaging is important. Eye-catching designs will attract attention, and mistakes must be avoided at all costs. You are now the designer for a company that is going to market the new game.

4. In your workbook or on a blank piece of paper, draw two rectangles the size of playing cards. In the first rectangle, design the back of your cards, and in the second rectangle, design how the fractions will appear on the face of the cards.

The playing instructions are very important pieces of information in games. They explain the number of players allowed, how the game is to be played and who wins.

5. Taking into consideration the information that has been provided for you throughout this task, write a set of instructions that show how your Memory game is played.
6. Now construct your own game cards and present them in a suitable package. Remember, presentation of the game is important. When you have finished, try the game out with your friends.

 Resources

 **eWorkbook** Topic 4 Workbook (worksheets, code puzzle and project) (ewbk-1905)

 **Interactivities** Crossword (int-2590)

Sudoku puzzle (int-3165)

## Exercise 4.13 Review questions

learnon

### Fluency

**1.** State the fraction of the eggs in this carton that are white.

**2.** Write the sequence of the first three equivalent fractions for the following.

**a.** $\frac{3}{8}$ **b.** $\frac{4}{9}$ **c.** $\frac{5}{11}$ **d.** $\frac{6}{7}$

**3.** Fill the gaps to make equivalent fractions.

**a.** $\frac{1}{2} = \frac{}{4} = \frac{3}{}$ **b.** $\frac{1}{3} = \frac{3}{} = \frac{}{15}$ **c.** $\frac{3}{4} = \frac{9}{} = \frac{}{28}$ **d.** $\frac{2}{5} = \frac{}{10} = \frac{16}{}$

**4.** Ricky, Leo and Gabriel were the highest goal scorers for their local football team throughout the season. Ricky scored $\frac{2}{7}$ of the team's goals, Leo scored $\frac{1}{3}$ of the goals and Gabriel scored $\frac{1}{4}$ of the goals.

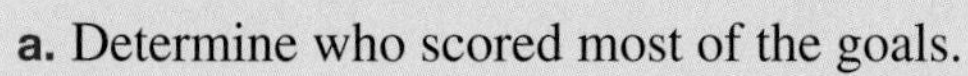

**a.** Determine who scored most of the goals.
**b.** State who scored more goals: Ricky or Gabriel.

**5.** Simplify these fractions.

**a.** $\frac{42}{49}$ **b.** $\frac{81}{108}$ **c.** $\frac{18}{20}$ **d.** $\frac{21}{28}$

**6.** Simplify these fractions.

**a.** $\frac{50}{150}$ **b.** $\frac{75}{125}$ **c.** $\frac{2}{80}$ **d.** $\frac{36}{72}$

7. Simplify the following mixed numbers.

a. $6\frac{8}{10}$ b. $3\frac{6}{9}$ c. $4\frac{15}{20}$ d. $1\frac{20}{100}$

8. Determine what fraction of months of the year begin with the letter J. Express your answer in simplest form.

9. Convert these improper fractions to mixed numbers.

a. $\frac{16}{3}$ b. $\frac{21}{5}$ c. $\frac{12}{7}$ d. $\frac{11}{2}$

10. Convert these improper fractions to mixed numbers.

a. $\frac{5}{4}$ b. $\frac{80}{7}$ c. $\frac{55}{9}$ d. $\frac{72}{10}$

11. Express these mixed numbers as improper fractions.

a. $2\frac{3}{4}$ b. $9\frac{7}{8}$ c. $3\frac{5}{7}$ d. $5\frac{5}{6}$

12. Express these mixed numbers as improper fractions.

a. $3\frac{2}{3}$ b. $9\frac{9}{10}$ c. $3\frac{11}{12}$ d. $8\frac{1}{2}$

13. Marcella was in charge of baking and selling lemon slice at the school fete. Each tray she baked contained 15 pieces of lemon slice. If Marcella sold 68 slices, determine how many empty trays there were.

14. Evaluate:

a. $\frac{5}{6}+\frac{1}{8}$ b. $\frac{7}{9}-\frac{2}{5}$ c. $\frac{12}{5}-\frac{4}{3}$

15. Evaluate:

a. $\frac{4}{7}\times\frac{5}{9}$ b. $\frac{2}{5}\times\frac{3}{5}$ c. $\frac{25}{36}\times\frac{18}{20}$ d. $\frac{56}{81}\times\frac{18}{64}$

16. Evaluate:

a. $\frac{4}{5}$ of 55 b. $\frac{2}{11}$ of 121 c. $\frac{7}{9}$ of $\frac{18}{84}$ d. $\frac{1}{12}$ of 200

17. Write the reciprocals of:

a. $\frac{5}{2}$ b. $\frac{1}{9}$ c. $\frac{1}{4}$ d. $5\frac{3}{4}$

18. Evaluate:

a. $\frac{4}{5}\div\frac{8}{15}$ b. $\frac{9}{10}\div\frac{27}{40}$ c. $\frac{7}{9}\div 4$ d. $\frac{3}{7}\div 10$

**19.** Evaluate:

**a.** $1\frac{7}{8}+2\frac{5}{6}$ **b.** $10\frac{3}{7}-8\frac{1}{2}$ **c.** $6\frac{1}{3}-2\frac{5}{9}$ **d.** $5\frac{1}{2}-2\frac{5}{6}$

**20.** Evaluate:

**a.** $8\frac{2}{9}-4\frac{5}{8}$ **b.** $12\frac{6}{11}-11\frac{7}{9}$ **c.** $1\frac{1}{5}+3\frac{1}{3}-\frac{13}{15}$ **d.** $4\frac{7}{9}-1\frac{5}{6}+3\frac{1}{3}$

**21.** Evaluate:

**a.** $10\frac{2}{7}\times 6\frac{2}{9}$ **b.** $\frac{4}{9}\times 2\frac{1}{4}\times\frac{3}{5}$ **c.** $3\frac{5}{12}\times 1\frac{3}{5}\times 2\frac{1}{2}$ **d.** $\frac{6}{17}\times 1\frac{1}{4}\times 3\frac{2}{5}$

**22.** Evaluate:

**a.** $1\frac{1}{2}\div 4\frac{4}{5}$ **b.** $6\frac{2}{3}\div 4\frac{1}{6}$ **c.** $4\frac{1}{5}\div 2\frac{7}{8}$ **d.** $10\frac{1}{3}\div 12\frac{5}{7}$

**23.** Write the following percentages as fractions in simplest form.

**a.** 13% **b.** 70% **c.** 26% **d.** 132%

**24.** Change the following fractions to percentages.

**a.** $\frac{15}{100}$ **b.** $\frac{11}{20}$ **c.** $\frac{3}{5}$ **d.** $\frac{18}{36}$

**25.** Change the following fractions to percentages.

**a.** $\frac{30}{80}$ **b.** $\frac{8}{15}$ **c.** $\frac{5}{6}$ **d.** $\frac{4}{11}$

**26.** Calculate the following percentages using fractions.

**a.** 5% of 120 **b.** 60% of 20 **c.** 75% of 52 **d.** 130% of 30

**27.** Calculate the following percentages using fractions.

**a.** 14% of 45 **b.** 7% of 530 **c.** 32% of 15 **d.** 17% of 80

**28.** Express:

**a.** 30 as a percentage of 50 **b.** 18 as a percentage of 60
**c.** 32 as a percentage of 80 **d.** 28 as a percentage of 70.

**29.** Express:

**a.** 4 days as a percentage of 8 weeks
**b.** 20 minutes as a percentage of 3 hours.

**30.** Calculate 5% of the following by finding 10% and halving your answer. Round your answer to the nearest 5 cents.

**a.** \$8.00 **b.** \$21.00 **c.** \$64.00 **d.** \$104.00

**31.** Calculate the following using short cuts. Round your answer to the nearest 5 cents.

**a.** 1% of \$16.00 **b.** 1% of \$28.00 **c.** 12% of \$42.00 **d.** 30% of \$90.00

## Problem solving

**32.** Year 7 students at Camberwell Secondary College raised a total of \$1600 for the Royal Children's Hospital Good Friday Appeal. The table below shows how much money each class raised.

| Class | 7.1 | 7.2 | 7.3 | 7.4 | 7.5 | 7.6 |
|---|---|---|---|---|---|---|
| Amount raised (\$) | \$200 | \$400 | \$250 | \$250 | \$280 | \$220 |

Express as a simple fraction how much of the total each class raised.

**33.** Anthony's monthly take-home pay is \$2800. He spends a quarter on his home loan payments, one-half on food and drink, and one-seventh on clothing. One-half of the remainder goes into his savings account. Calculate how much money Anthony saves each month.

**34.** On a balance scale, a brick exactly balances $\frac{2}{3}$ of a brick and a $\frac{2}{3}$ kg weight. How heavy is the brick?

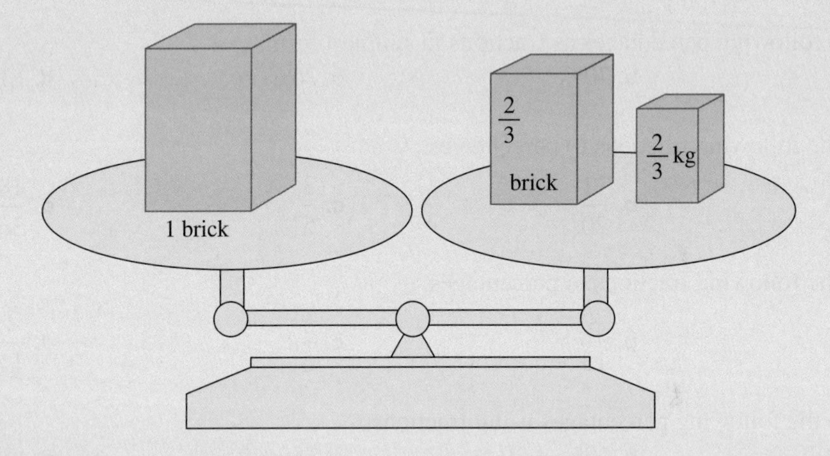


**35.** Complete the following:

**a.** $\frac{5}{12} = \frac{1}{4} + ?$ **b.** $\frac{5}{12} = \frac{2}{3} - ?$ **c.** $\frac{5}{12} = \frac{3}{4} \times ?$ **d.** $\frac{5}{12} = \frac{4}{5} \div ?$

**36.** Explain how you can determine a fraction whose numerator and denominator have a product of 252, and that is equivalent to $\frac{4}{7}$.

**37.** Tomas buys a 24-piece pizza and eats 8 pieces. Drew buys a 16-piece pizza of the same size, and eats 5 pieces. Identify who ate the most pizza and explain your choice.

**38.** An Egyptian fraction is one that can be written as the sum of a set of different unit fractions: for example, $\frac{3}{4} = \frac{1}{2} + \frac{1}{4}$. (*Note:* The numerator of each unit fraction is 1.) Write $\frac{4}{5}$ as the sum of a set of different unit fractions.

**39.** Mr Thompson earned \$2400 per month. Recently, he had a pay increase. He now earns $1\frac{1}{4}$ times his old salary. Calculate his new salary.

Mr Thompson's colleague, Mr Goody, earned \$3000 per month and had a pay cut. He now gets $\frac{1}{6}$ less of his old salary. Calculate Mr Goody's new salary.

**40.** A shopping centre has 250 parking spaces. Of these, $\frac{3}{10}$ are suitable for small cars only. The remaining spaces are for standard cars. Last Friday small cars took up $\frac{14}{15}$ of their allocated spots, but overall the car park was $\frac{7}{10}$ full.
Determine the fraction of standard car spaces that were filled.

**41.** Byron needs $\frac{2}{5}$ kg of catalyst mixed with $4\frac{1}{2}$ kg of resin to create one surfboard.
**a.** If he has $2\frac{1}{4}$ kg of catalyst and $20\frac{1}{4}$ kg of resin, how many whole surfboards can Byron build?
**b.** Calculate how much more catalyst and resin Byron needs to build one more surfboard. Explain your reasoning.

**42.** Calculate the percentage that remains if:
**a.** 38% of the winnings have been spent
**b.** all stock was discounted 25%.

**43.** Eighty per cent of a Year 7 Maths class got question 1 correct on their percentages test. If there are 30 students in the class, calculate how many got the question correct.

**44.** Forty per cent of primary-school children can sing the national anthem. In a group of 675 primary-school children, calculate how many you would expect to be able to sing the national anthem.

**45.** Lake Eyre in South Australia has an annual rainfall of around 100 mm. If 8 mm fell in one week, give this as a percentage of the annual rainfall.

**46.** A tennis player agreed to pay her coach 8% of her winnings from tournaments. If she wins $145 000 in a tournament, calculate how much her coach receives.

**47.** At Queens High School, there are 900 students.
45% walk to school.
55% don't walk to school.
12% of students wear glasses.

Identify which of the following statements is definitely false. (There may be more than one answer.)
**a.** There are 100 students who don't walk to school and who wear glasses.
**b.** 10% of the students who don't walk to school and 2% of the students who do walk to school wear glasses.
**c.** 20% of the students who walk to school wear glasses.
**d.** More than 22% of the students who don't walk to school wear glasses.

**48.** The label on a 500 g pack of butter states that it contains 81% fat. A milk bottle label says that the milk contains 3 grams of fat per 100 mL. Determine the number of litres of this milk that contains the same amount of fat as there is in the butter.

**49.** In a recent survey of house occupancy in a town, it was discovered that 40% of the houses contained 2 or more people. Of those houses that contain only 1 person, 25% of them contained a male. Of all the houses in a town, calculate the percentage that contains only 1 female and no males.

**50.** A television set has a ticketed price of $2500. A 10% discount is allowed on the set.

**a.** Calculate the negotiated price of the television set.
**b.** If a further 10% is negotiated off the price of the set, calculate the price paid.
**c.** Calculate the total discount as a percentage of the original ticketed price.
**d.** Explain why two successive 10% discounts do not equal a 20% discount.

**51.** Joseph wants to spend $2500 on a new barbecue. Since he works at a BBQ store he can get a staff discount of 10%. He can also get a discount of 30% in the stocktake sale.
Joseph wants to know whether he should ask the salesman to apply his staff discount or the sale discount first to achieve the cheapest price.

**a.** Calculate how much the barbecue will cost if the staff discount is taken off first and then the sale discount is taken off.
**b.** Calculate how much it will cost if done the opposite way.
**c.** Decide which method Joseph should select to get the best price.

**52.** Two triangles have the same area. Show that if one triangle has a 25% wider base than the other, then it must be 20% shorter (that is, its height must be 20% less).

**on** To test your understanding and knowledge of this topic, go to your learnON title at www.jacplus.com.au and complete the **post-test**.

# Answers

## Topic 4 Fractions and percentages

### 4.1 Pre-test

1. B
2. a $\frac{3}{8}$ b $\frac{5}{8}$
3. $5\frac{3}{4}$
4. $\frac{3}{4}$
5. 100 minutes
6. A
7. 36
8. 187 cm
9. $\frac{127}{200}$
10. C
11. a. $\frac{3}{25}$ b. 0.12
12. 11 036
13. 75%
14. D
15. $15.58

### 4.2 What are fractions?

1. a. $\frac{3}{4}$ b. $\frac{15}{16}$ c. $\frac{7}{8}$
2. $\frac{1}{3}$
3. $\frac{3}{4}$
4. a. $\frac{3}{11}$ b. $\frac{2}{11}$
5. $\frac{3}{11}$
6. $\frac{17}{60}$ of an hour
7. a. 3 b. 3 c. 4 d. 5
8. a. 6 b. 9 c. 11 d. 12
9. $\frac{20}{30}, \frac{14}{21}, \frac{40}{60}$
10. $\frac{12}{15}, \frac{36}{45}, \frac{16}{20}, \frac{28}{35}, \frac{80}{100}$
11. $\frac{35}{50}, \frac{21}{30}, \frac{14}{20}, \frac{140}{200}$
12. $\frac{5}{6} = \frac{10}{12} = \frac{15}{18} = \frac{20}{24}$
13. $\frac{7}{8} = \frac{14}{16} = \frac{21}{24} = \frac{28}{32}$
14. Number line: $0$, $\frac{1}{5}$, $\frac{4}{5}$, $1$, $\frac{7}{5}$, $1\frac{3}{5}$, $2$
15. Number line: $0$, $\frac{1}{3}$, $\frac{3}{3}$, $\frac{4}{3}$, $2$, $\frac{8}{3}$, $3$
16. a. $\frac{3}{5}$ b. $\frac{7}{8}$ c. $\frac{1}{5}$ d. $\frac{1}{8}$
17. a. $\frac{3}{4}$ b. $\frac{5}{8}$ c. $\frac{5}{8}$ d. $\frac{4}{5}$
18. a. $>$ b. $>$ c. $>$ d. $<$
19. D
20. B
21. $\frac{9}{24}, \frac{12}{24}, \frac{8}{24}; \frac{1}{3}, \frac{3}{8}, \frac{1}{2}$
22. $\frac{45}{60}, \frac{40}{60}, \frac{28}{60}; \frac{7}{15}, \frac{2}{3}, \frac{3}{4}$
23. $\frac{30}{40} = \frac{18}{24} = \frac{27}{36}$
24. $\frac{15}{18} = \frac{35}{42} = \frac{40}{48}$
25. 7 does not divide into 2 evenly, so it would be difficult to mark the equal parts. It would be more appropriate to use a scale of 3.5 cm or 7 cm per whole number.
26. As the denominator doubles, the distance between fractions halves, because doubling the denominator is the same as dividing the fraction by 2.
27. a.

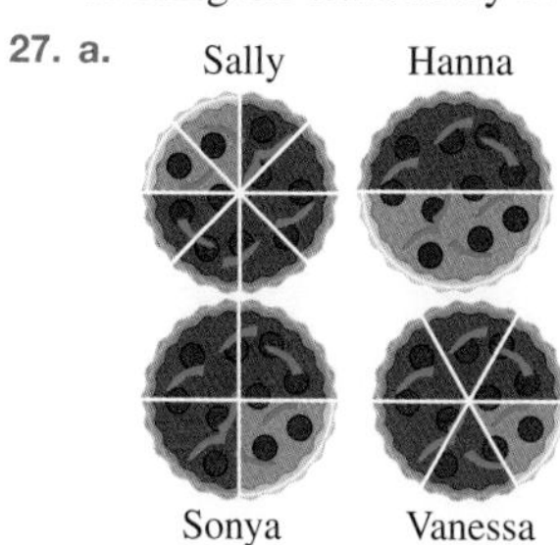


b. Vanessa
c. They ate the same amount.

28. Lee Yin
29. $\frac{19}{60}, \frac{1}{3}$
30.

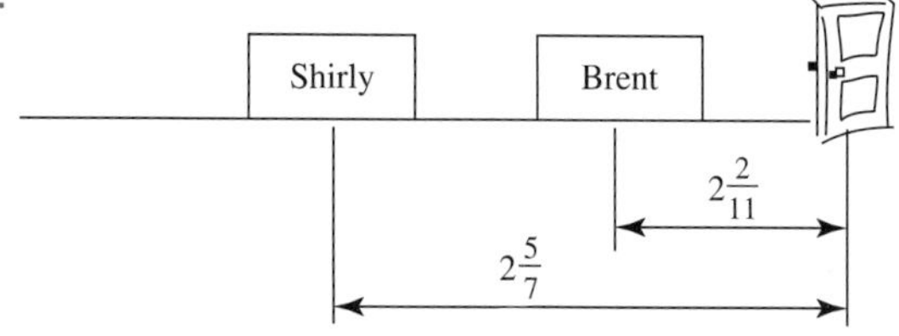


### 4.3 Simplifying fractions

1. a. $\frac{1}{2}$ b. $\frac{2}{3}$ c. $\frac{7}{8}$ d. $\frac{4}{5}$

2. a. $\frac{5}{6}$  b. $\frac{24}{25}$  c. $\frac{7}{8}$  d. $\frac{7}{10}$

3. a. $\frac{11}{12}$  b. $\frac{11}{25}$  c. $\frac{3}{8}$  d. $\frac{7}{8}$

4. a. $4\frac{3}{5}$  b. $7\frac{5}{6}$  c. $10\frac{1}{2}$  d. $1\frac{11}{12}$

5. a. $6\frac{3}{4}$  b. $5\frac{1}{3}$  c. $3\frac{1}{5}$  d. $2\frac{1}{4}$

6. E

7. $\frac{21}{28} = \frac{3}{4}$

8. a. $\frac{20}{36} = \frac{5}{9}$  b. $\frac{16}{36} = \frac{4}{9}$

9. 7A: $\frac{1}{5}$, 7B: $\frac{3}{20}$, 7C: $\frac{8}{25}$, 7D: $\frac{2}{25}$, 7E: $\frac{1}{4}$

10. a. 40

b. Mark: $\frac{5}{40} = \frac{1}{8}$, Ken: $\frac{10}{40} = \frac{1}{4}$, Samantha: $\frac{8}{40} = \frac{1}{5}$, Jules: $\frac{6}{40} = \frac{3}{20}$, Ahmed: $\frac{4}{40} = \frac{1}{10}$, Darren: $\frac{7}{40}$

11. Newcastle to Sydney is more reliable.

12. a.

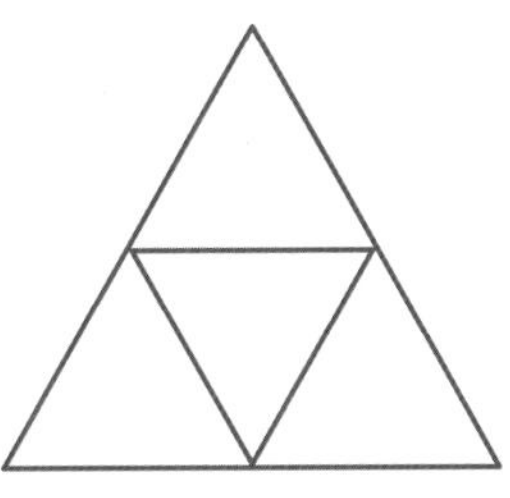

b. $\frac{1}{4}$

c.

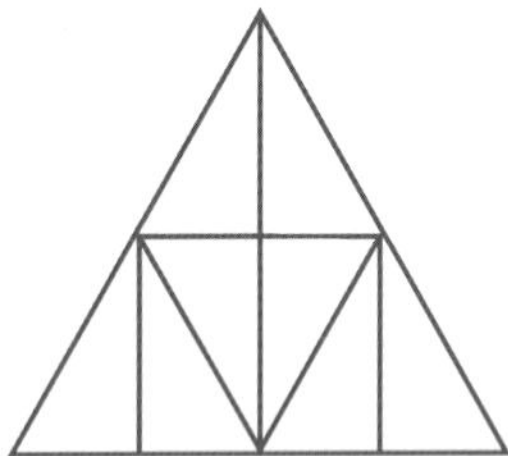

d. $\frac{1}{8}$

e. Because two smallest triangles form one middle size triangle

13. The denominator could be any multiple of 13 because the fraction is not written in its simplest form and 13 is a prime number.

14. a. Six pizzas

b. Two quarters (one half) of a pizza are left over.

15. a. $\frac{7293}{14\,586}, \frac{6792}{13\,584}$  b. $\frac{5823}{17\,469}$

16. $\frac{2}{5}$

## 4.4 Mixed numbers and improper fractions

1. a.

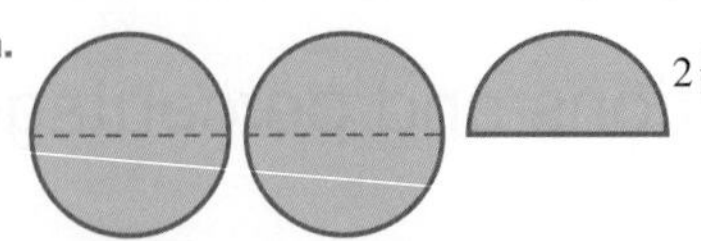


b.

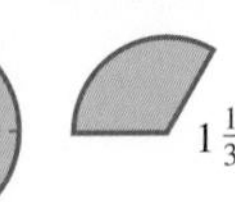


c. $1\frac{5}{8}$

2. $\frac{3}{2} = 1\frac{1}{2}$

3.

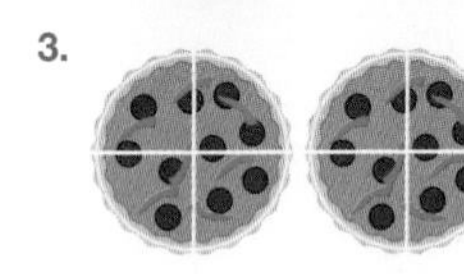


4. a. $1\frac{2}{5}$  b. $2\frac{3}{4}$  c. $10\frac{1}{2}$  d. $4\frac{1}{4}$  e. $8\frac{4}{11}$

5. a. $2\frac{3}{13}$  b. $33\frac{1}{3}$  c. $12\frac{1}{2}$  d. $6\frac{2}{3}$  e. $1\frac{3}{5}$

6. a. $3\frac{3}{8}$  b. $5\frac{1}{2}$  c. $8\frac{2}{7}$  d. $11\frac{7}{10}$  e. $8\frac{3}{8}$

7. C

8. B

9. a.

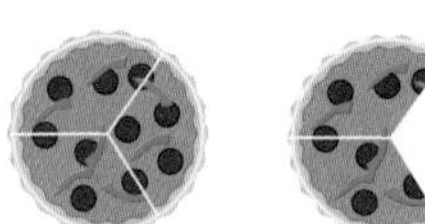


b.

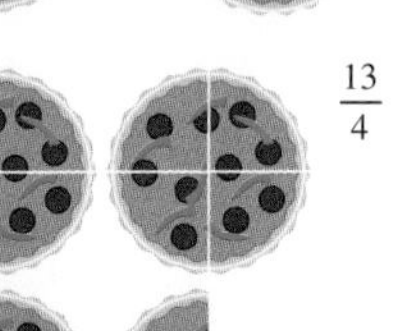


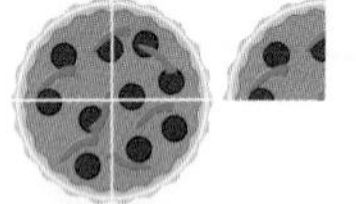

c.

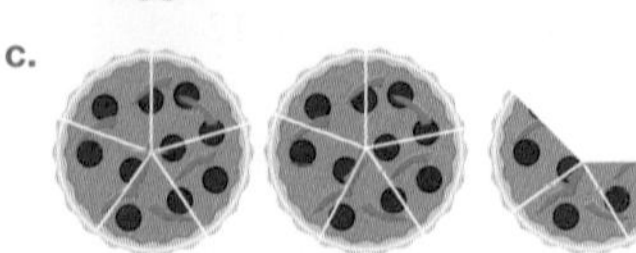


10. a. $\frac{9}{8}$  b. $\frac{15}{4}$  c. $\frac{13}{6}$  d. $\frac{8}{7}$  e. $\frac{9}{5}$

11. a. $\frac{10}{3}$  b. $\frac{25}{4}$  c. $\frac{37}{9}$  d. $\frac{23}{2}$  e. $\frac{39}{5}$

12. a. $\frac{33}{5}$  b. $\frac{47}{5}$  c. $\frac{7}{2}$  d. $\frac{21}{5}$  e. $\frac{47}{7}$

13. a. $\frac{29}{10}$ b. $\frac{43}{12}$ c. $\frac{67}{7}$ d. $\frac{21}{11}$ e. $\frac{53}{6}$

14. D

15. D

16. A

17. Kim: $2\frac{1}{4}$, Carly: $2\frac{3}{4}$

18. $6\frac{1}{2}$

19. $4\frac{3}{4}$

20. a. > b. > c. < d. >

21. a. < b. < c. < d. <

22. $10\frac{1}{2}$

23. a. i. $\frac{23}{7}$ ii. $\frac{23}{4}$

b.

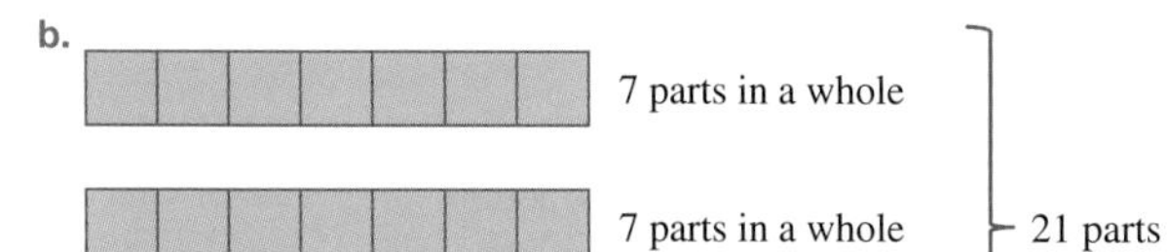
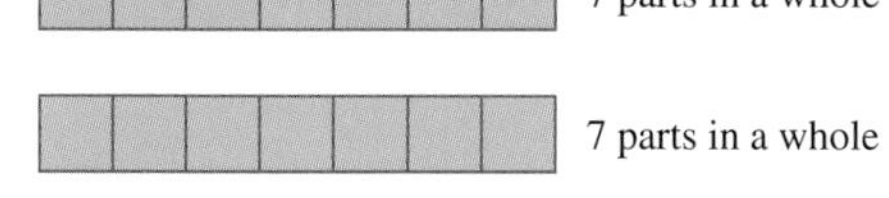
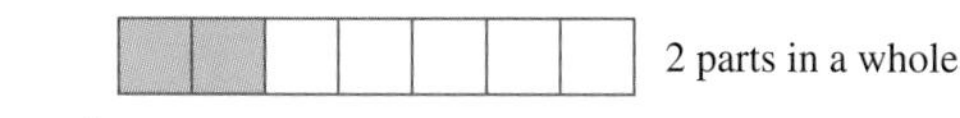


24. $\frac{4}{5}$

25. a. 25 b. 77

26. a. $\frac{14}{3}$ or $4\frac{2}{3}$ km

b. $\frac{56}{3}$ or $18\frac{2}{3}$ km

c. $\frac{70}{3}$ or $23\frac{1}{3}$ km

27. $1\frac{1}{4}, 2\frac{1}{12}, 2\frac{1}{6}, 3\frac{1}{8}, 3\frac{1}{3}$

28. To write an improper fraction as a mixed number we divide the denominator into the numerator and write the remainder over the same denominator. Alternatively, we can draw circles to determine how many whole numbers we have, and the parts left over will be the fraction part of the mixed number.

## 4.5 Adding and subtracting fractions

1. a. $\frac{4}{5}$ b. $\frac{6}{11}$ c. $\frac{5}{6}$ d. $\frac{31}{50}$

2. a. $\frac{1}{2}$ b. $\frac{1}{4}$ c. $\frac{1}{3}$ d. $\frac{1}{2}$

3. a. $\frac{9}{25}$ b. $\frac{11}{12}$ c. $\frac{7}{8}$ d. $\frac{6}{7}$

4. a. 1 b. $\frac{3}{4}$ c. $\frac{3}{4}$ d. $\frac{8}{9}$

5. a. $\frac{3}{4}$ b. $\frac{1}{4}$ c. $\frac{3}{7}$ d. $\frac{3}{4}$

6. a. $\frac{2}{3}$ b. $\frac{13}{27}$ c. $\frac{16}{35}$ d. $\frac{2}{5}$

7. a. i. 6 ii. 15 iii. 10 iv. 12

b. i. $\frac{1}{6}$ ii. $\frac{4}{15}$ iii. $\frac{9}{10}$ iv. $\frac{7}{12}$

8. a. i. 18 ii. 8 iii. 24 iv. 35

b. i. $\frac{17}{18}$ ii. $\frac{5}{8}$ iii. $\frac{11}{24}$ iv. $\frac{11}{35}$

9. a. $\frac{5}{6}$ b. $\frac{23}{36}$ c. $\frac{39}{44}$ d. $\frac{1}{4}$

10. a. $\frac{17}{63}$ b. $\frac{23}{39}$ c. $\frac{2}{15}$ d. $\frac{7}{33}$

11. D

12. a. $\frac{7}{12}$ b. $\frac{2}{5}$ c. $\frac{19}{24}$ d. $\frac{7}{30}$

13. a. $\frac{7}{16}$ b. $\frac{1}{3}$ c. $1\frac{11}{12}$ d. $1\frac{11}{12}$

14. a. $\frac{3}{4}$ b. $\frac{4}{7}$ c. $\frac{60}{71}$ d. $1\frac{3}{5}$

15. a. $2\frac{8}{15}$ b. $4\frac{1}{6}$ c. $9\frac{2}{3}$ d. $22\frac{1}{10}$

16. a. $\frac{7}{15}$ b. $\frac{8}{15}$

17. $1\frac{1}{4}$ hours

18. a. $\frac{3}{4}$ b. $\frac{7}{12}$ c. 1

19. a. $\frac{2}{3}$ b. $\frac{23}{30}$

20. 135

21. To add fractions, you must convert to equivalent fractions with the same denominators. You cannot simply add the numerators and add the denominators.

22. $\frac{1}{12}$

23. a. i. $\frac{5}{6}$ ii. $\frac{7}{12}$ iii. $\frac{9}{20}$

b. When adding two fractions, both with the numerator of 1 and denominators that are consecutive numbers, the answer is $\frac{\text{sum of denominators}}{\text{product of denominators}}$.

24. a. i. $\frac{1}{6}$ ii. $\frac{1}{12}$ iii. $\frac{1}{20}$

b. When subtracting two fractions that both have a numerator of 1 and denominators that are consecutive numbers, the answer is $\dfrac{1}{\text{product of denominators}}$.

c. $\dfrac{1}{n(n+1)}$

25. Fractions cannot be added or subtracted unless they have the same denominator. This is because we need to be adding or subtracting like fractions. If the denominators are different, we can make the two fractions into equivalent fractions by finding the lowest common denominator (LCD). The LCD is the smallest multiple of all the denominators in a set of fractions.

26.

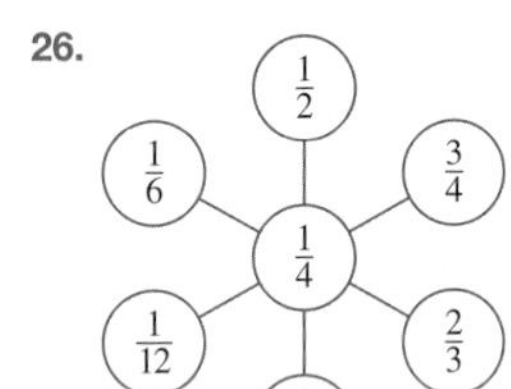


## 4.6 Multiplying fractions

1. a. 2 b. $2\frac{1}{2}$ c. $7\frac{1}{2}$ d. $1\frac{1}{4}$

2. a. $\frac{1}{8}$ b. $\frac{2}{9}$ c. $\frac{20}{27}$ d. $\frac{11}{26}$

3. a. $\frac{54}{77}$ b. $\frac{121}{144}$ c. $\frac{40}{81}$ d. $\frac{5}{24}$

4. a. $\frac{55}{72}$ b. $\frac{25}{132}$ c. $\frac{8}{39}$ d. $\frac{8}{21}$

5. a. $\frac{1}{6}$ b. $\frac{1}{5}$ c. $\frac{5}{14}$ d. $\frac{1}{10}$

6. a. $\frac{4}{5}$ b. 1 c. $\frac{1}{2}$ d. $\frac{12}{35}$

7. a. $\frac{4}{9}$ b. $\frac{1}{7}$ c. $\frac{1}{3}$ d. $\frac{1}{20}$

8. a. $\frac{1}{8}$ b. $\frac{1}{2}$ c. $\frac{25}{36}$ d. $\frac{2}{9}$

9. a. 28 b. 45 c. 15 d. 16

10. Emily drank 2 L. Tracy drank $\frac{1}{2}$ L. Jonathan drank $\frac{1}{3}$ L.

11. a. 32 minutes b. 35 minutes

12. a.

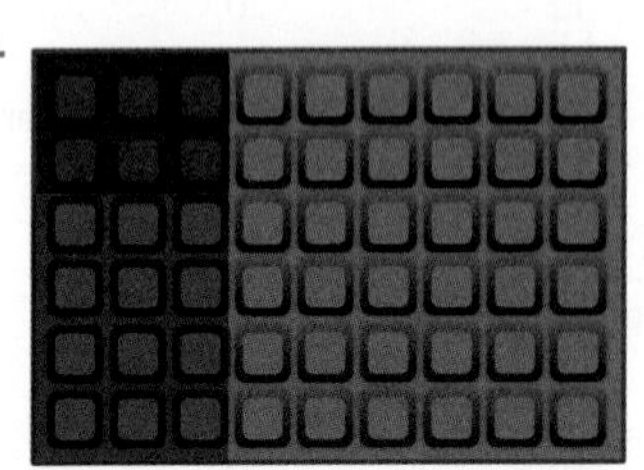

Here is one example answer.

b. $\frac{1}{9}$

13. a $\frac{3}{4}$ b 36 litres

14. a. 2 b. 4 c. 10

15. a.

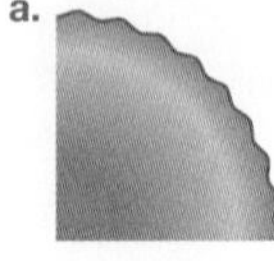

b.

c. $\frac{1}{8}$ d. $\frac{1}{2} \times \frac{1}{4} = \frac{1}{8}$

16. a. 90 minutes

b. $\frac{90}{180} = \frac{1}{2}$

c. $140 + 26\frac{2}{3} = 166\frac{2}{3}$ minutes

17. a. i. $\frac{1}{3}$ ii. $\frac{1}{4}$ iii. $\frac{1}{5}$

b. The result from the numerator of the first fraction and the denominator of the last fraction

c. $\frac{1}{100}$

18. $\frac{1}{24} \times 2400 = \$100$

19. $121\frac{1}{2}$ km

20. Red $= 1\frac{16}{21}$ m, green $= 1\frac{9}{28}$ m

## 4.7 Dividing fractions

1. a. $\frac{4}{3}$ b. $\frac{7}{2}$ c. $\frac{5}{6}$ d. $\frac{3}{5}$ e. $\frac{1}{5}$

2. a $\frac{12}{5}$ b $\frac{2}{9}$ c $\frac{3}{5}$ d $\frac{3}{10}$ e $\frac{1}{20}$

3. a. 12 b. 5 c. 11 d. 6 e. 1

4. a. $\frac{3}{5}$ b. $\frac{7}{31}$ c. $\frac{7}{65}$ d. $\frac{8}{27}$ e. $\frac{10}{69}$

5. a. $\frac{7}{27}$ b. $\frac{3}{16}$ c. $\frac{2}{15}$ d. $\frac{9}{92}$ e. $\frac{9}{23}$

6. a. 1 b. 1 c. 1

7. a. $\frac{3}{8}$ b. $\frac{16}{21}$ c. $\frac{1}{8}$ d. $\frac{24}{55}$ e. $4\frac{2}{3}$

8. a. $\frac{3}{10}$ b. 1 c. $\frac{2}{35}$ d. $\frac{1}{9}$ e. $\frac{3}{11}$

9. a. $\frac{5}{9}$ b. $\frac{1}{22}$ c. $\frac{63}{50}$ d. $\frac{12}{17}$ e. $\frac{10}{7}$

10. Lily is correct.

11. 32

12. 15 people

13. $\frac{13}{3}$

14. $\frac{3}{20}$

15. $\frac{5}{12}$

## 4.8 Working with mixed numbers

1. a. $4\frac{7}{10}$ b. $1\frac{1}{6}$ c. $3\frac{5}{12}$ d. $13\frac{3}{8}$

2. a. $1\frac{1}{10}$ b. $6\frac{1}{4}$ c. $1\frac{1}{3}$ d. $1\frac{1}{2}$

3. a. $1\frac{5}{8}$ b. $1\frac{7}{10}$ c. $3\frac{2}{3}$ d. $1\frac{11}{12}$

4. a. $2\frac{17}{35}$ b. $3\frac{7}{8}$ c. $3\frac{13}{18}$ d. $3\frac{1}{24}$

5. a. $7\frac{7}{12}$; 7 b. $3\frac{2}{3}$; 3 c. $4\frac{1}{2}$; 4 d. $4\frac{1}{4}$; 4

6. a. $\frac{5}{6}$ b. $2\frac{1}{4}$ c. $4\frac{19}{32}$ d. $7\frac{7}{10}$

7. a. $37\frac{5}{21}$ b. $34\frac{2}{3}$ c. 10 d. $2\frac{11}{12}$

8. a. $1\frac{2}{3}$ b. $\frac{9}{14}$ c. 1 d. $22\frac{1}{3}$

9. a. 3 b. $2\frac{29}{40}$ c. $1\frac{8}{15}$ d. $2\frac{1}{14}$

10. $1\frac{5}{6}$ L

11. $1\frac{5}{16}$ inches

12. $1\frac{1}{12}$ months

13. $6\frac{1}{4}$ days

14. $\frac{3}{16}$ kg

15. a. 22 b. 96

16. $\frac{3}{8}$ cup sugar
$\frac{3}{8}$ cup margarine
$\frac{3}{4}$ cup peanut butter
3 eggs
$2\frac{1}{4}$ cups milk
$3\frac{3}{4}$ cups self-raising flour
$\frac{3}{8}$ teaspoon baking soda

17. $3\frac{1}{10}$ kg

18. 7 containers

19. $52.50

20. a. $3\frac{15}{16}$ cm b. $2\frac{15}{16}$ cm

21. 500 women

22. We could convert $3\frac{1}{3}$ to an improper fraction, $\frac{10}{3}$. We would then multiply it by $\frac{1}{2}$.

$\frac{10}{3} \times \frac{1}{2} = \frac{10}{6} = \frac{5}{3} = 1\frac{2}{3}$

Alternatively, we could draw a picture of $3\frac{1}{3}$ circles and divide each by $\frac{1}{2}$, and then add these fractions together, giving us a result of $1\frac{2}{3}$.

23. $\frac{5}{12}$, 10

24. a. $6\frac{2}{3}$ laps b. 12 minutes

25. 2961 m

26. The gymnasium should keep the drink machine.

## 4.9 Percentages as fractions

1. a. $\frac{17}{100}$ b. $\frac{29}{100}$ c. $\frac{79}{100}$ d. $\frac{99}{100}$ e. $\frac{3}{100}$

2. a. $\frac{19}{100}$ b. $\frac{33}{100}$ c. $\frac{9}{100}$ d. $\frac{243}{100}$ e. $\frac{127}{100}$

3. a. $\frac{1}{2}$ b. $\frac{4}{5}$ c. $\frac{7}{20}$ d. $\frac{3}{5}$ e. $\frac{1}{10}$

4. a. $\frac{9}{20}$ b. $\frac{3}{25}$ c. $\frac{1}{20}$ d. $\frac{37}{50}$ e. $\frac{1}{50}$

5. a. $1\frac{1}{2}$ b. $1\frac{4}{5}$ c. $\frac{2}{1} = 2$ d. $\frac{5}{1} = 5$ e. $1\frac{3}{25}$

6. a. $\frac{7}{200}$ b. $\frac{9}{125}$ c. $\frac{197}{1000}$ d. $\frac{81}{250}$

7. a. $\frac{31}{200}$ b. $\frac{81}{500}$ c. $\frac{283}{1000}$ d. $\frac{2069}{5000}$

8. a. $\frac{907}{5000}$ b. $\frac{5799}{10\,000}$ c. $\frac{243}{2000}$ d. $\frac{1}{2000}$

9. a. $\frac{1}{200}$ b. $\frac{1}{500}$ c. $\frac{1}{150}$ d. $\frac{1}{1000}$ e. $\frac{3}{550}$

10. a. $\frac{33}{400}$ b. $\frac{31}{150}$ c. $\frac{29}{300}$ d. $\frac{31}{200}$ e. $\frac{14}{125}$

11. a. $\frac{83}{800}$ b. $\frac{7}{60}$ c. $1\frac{101}{200}$ d. $1\frac{41}{200}$ e. $\frac{1}{3}$

12. B

13. E

14. C

15. $\frac{20}{100} = \frac{1}{5}$

16. $\frac{35}{100} = \frac{7}{20}$

17. $\frac{45}{100} = \frac{9}{20}$

18. 32% were born overseas $\frac{8}{25}$

19. a. $\frac{4}{5}$

b. $\frac{1}{5}$

c. Not necessarily. There could be an overlap of students with blue eyes who have pets.

20. a. $\frac{4}{5}$

b. $\frac{1}{2}$

c. Although the two molecules contain the same number of hydrogen atoms, the total number of atoms differs between the two molecules.

21. a. She scored a higher percentage in the second game (75%).

b. i. Increased percentage
ii. Unchanged
iii. Decreased percentage

22. Less than $\frac{1}{2}$

23. a. $\frac{2}{25}$ b. $\frac{23}{25}$ c. 8 people

24. a. $\frac{7}{20}$ b. $\frac{81}{100}$ c. $\frac{17}{20}$ d. $\frac{13}{200}$

25. a. $\frac{17}{100}$ b. $\frac{83}{100}$ c. 17

## 4.10 Calculating percentages of an amount

1. a. 180 b. 4 c. 60

2. a. $1\frac{1}{2}$ b. $10\frac{1}{2}$ c. 78

3. a. 10 b. 16 c. 3 d. 18 e. 6

4. a. 6 b. 77 c. 4000 d. 66 e. 17

5. a. 39 b. 95 c. 55 d. 12 e. 25

6. a. $3\frac{3}{5}$ b. $4\frac{4}{5}$ c. $7\frac{7}{10}$ d. $18\frac{1}{2}$ e. $13\frac{1}{5}$

7. a. $24\frac{3}{4}$ b. $11\frac{1}{10}$ c. $58\frac{9}{10}$ d. $20\frac{4}{5}$ e. $98\frac{2}{5}$

8. A

9. D

10. E

11. E

12. B

13. E

14. D

15. 210 students

16. 73 days

17. 1 person

18. a. 80% b. 1600 people

19. a. $22 b. $572

20. $1 600 000

21. $12 000

22. 47 species

23. 600 000 people

24. a. Peter calculated 35% of 80 mL rather than 800 mL.
b. 280 mL

25. a. 7 adults b. 19 adults

26. 29 houses

27. 38 workers

28. a. 12 houses b. 11 houses

29. 3

30. a. 200 g, 100 g, 50 g, 25 g and 12.5g
b. 50%, 25%, 12.5%, 6.25% and 3.125%
c. Every 20 days the quantity of the radioactive substance is smaller so there is a 50% reduction of an already reduced quantity.

31. 22.5%

32. 3 students voted twice.

33. 120

## 4.11 One amount as a percentage of another

1. a. 20% b. 25% c. 10% d. 40% e. 25%

2. a. 80% b. 32% c. 40% d. 20% e. 5%

3. a. 25% b. $31\frac{1}{2}$% c. $6\frac{1}{4}$% d. $6\frac{2}{5}$% e. $20\frac{2}{3}$%

4. a. $87\frac{1}{2}$% b. $28\frac{4}{7}$% c. $66\frac{2}{3}$% d. $36\frac{2}{3}$% e. $26\frac{2}{3}$%

5. a. $7\frac{1}{7}$% b. $22\frac{2}{9}$% c. $42\frac{6}{7}$% d. $88\frac{8}{9}$% e. $71\frac{3}{7}$%

6. a. 17% b. 33% c. 63% d. 11%

7. a. 83% b. 78% c. 18% d. 12%

8. a. 10% b. $3\frac{1}{3}$% c. $12\frac{1}{2}$%

9. a. $28\frac{4}{7}$% b. 25% c. $1\frac{1}{4}$%

10. D

11. A

12. B

13. E

14. 70%

15. 60%

16. a. $5 b. $33\frac{1}{3}\%$

17. a. i. 10% ii. 14% iii. 12%
b. 36%
c. 64%

18. $62\frac{1}{2}\%$

19. $6\frac{2}{3}\%$

20. a. 20 km/h b. $16\frac{2}{3}\%$

21. a. $42\frac{6}{7}\%$ b. $28\frac{4}{7}\%$ c. $71\frac{3}{7}\%$ d. $57\frac{1}{7}\%$

22. $27\frac{3}{11}\%$

23. a. Second recipe. The second recipe contains 50% flour compared to 40% in the first recipe.
b. The percentage of flour increases

24. a. Less than $100 because 25% of the new amount is larger than the amount added. So, a higher amount is subtracted.
b. Sample responses can be found in the worked solutions in the online resources.

25. Sister is 9; grandfather is 90.

26. a.

| Topic | Prob. | M'ment | Algebra | Eqns | Geo. |
|---|---|---|---|---|---|
| Score | $\frac{15}{20}$ | $\frac{13}{15}$ | $\frac{27}{30}$ | $\frac{70}{80}$ | $\frac{95}{100}$ |
| Percentage | 75 | 87 | 90 | 88 | 95 |

b. Geometry
c. 87%

27. a. 300 g out of 450 g (300 g out of 450 g represents 67%.)
b. 200 g out of 500 g (200 g out of 500 g represents 40%.)
c. 420 g out of 840 g and 350 g out of 700 g (420 g out of 840 g and 350 g out of 700 g both represent 50%.)

28. Expressing one quantity as a percentage of another quantity allows us to compare the two quantities.

29. a. First recipe. The first type of concrete contains 15.4% cement, and the second type of concrete contains 11.8% cement.
b. The percentage of cement decreases.

30. 4520 km

## 4.12 Common percentages and short cuts

1. a. $1.00 b. $1.80 c. $8.10 d. $15.00 e. $9.30
2. a. $2.20 b. $1.70 c. $1.25 d. $0.15 e. $4.80
3. a. $19.25 b. $50.70 c. $192.60
d. $304.15 e. $200.00
4. a. $1.50 b. $5.10 c. $0.90 d. $13.70 e. $0.45
5. a. $0.65 b. $3.90 c. $7.50 d. $10.30 e. $6.80
6. a. $51.80 b. $30.50 c. $62.85 d. $10.05 e. $20.70
7. a. $0.40 b. $0.30 c. $0.10 d. $7.00 e. $2.10
8. a. $0.55 b. $15.25 c. $50 d. $5.15 e. $1.60
9. a. $4.30 b. $8.45 c. $0.65 d. $14.80 e. $0.15
10. a. $3.30 b. $27.15 c. $52.35
d. $247.40 e. $1013.80
11. a. $1.80 b. $1.20 c. $9.00
12. a. $7.50 b. $22.50 c. $55.00
13. a. $4.50 b. $7.50 c. $14.25
14. a. $2.70 b. $7.15 c. $6.05 d. $0.05 e. $0.15
15. a. $0.15 b. $0.80 c. $0.20 d. $0.05 e. $0.05
16. a. $1.30 b. $10.50 c. $0.65 d. $2.40 e. $1.80
17. a. $73.50 b. $1.55 c. $1.05 d. $2.05 e. $32.20
18. a. $4.80 b. $1.60 c. $0.45 d. $14.40 e. $492
19. D
20. B
21. E
22. C
23. $855
24. 54 000 residents
25. a. $322.50 b. $152.30
26. 110 students
27. 27.9 seconds
28. a. 2 people b. 38 people
29. a. 10 740 people
b. 19 690 people
c. 5370 people
30. 20 years old
31. $0.80
32. 8.2kg
33. 50 years old
34. a. 100%
b. i. 100% ii. 50% iii. 71.4%
c. No, because the two lengths have changed at different percentages
35. a. 3.2%
b. 6.8%
c. The overall percentage is a lot smaller.

## Project

1. A and D, B and C, E and F
2–6. An individual response is required by the student.

## 4.13 Review questions

1. $\frac{3}{20}$

2. a. $\frac{6}{16}, \frac{9}{24}, \frac{12}{32}$ b. $\frac{8}{18}, \frac{12}{27}, \frac{16}{36}$
c. $\frac{10}{22}, \frac{15}{33}, \frac{20}{44}$ d. $\frac{12}{14}, \frac{18}{21}, \frac{24}{28}$

3. a. $\frac{2}{4}=\frac{3}{6}$ b. $\frac{3}{9}=\frac{5}{15}$

c. $\frac{9}{12}=\frac{21}{28}$ d. $\frac{4}{10}=\frac{16}{40}$

4. a. Leo b. Ricky

5. a. $\frac{6}{7}$ b. $\frac{3}{4}$ c. $\frac{9}{10}$ d. $\frac{3}{4}$

6. a. $\frac{1}{3}$ b. $\frac{3}{5}$ c. $\frac{1}{40}$ d. $\frac{1}{2}$

7. a. $6\frac{4}{5}$ b. $3\frac{2}{3}$ c. $4\frac{3}{4}$ d. $1\frac{1}{5}$

8. $\frac{3}{12}=\frac{1}{4}$

9. a. $5\frac{1}{3}$ b. $4\frac{1}{5}$ c. $1\frac{5}{7}$ d. $5\frac{1}{2}$

10. a. $1\frac{1}{4}$ b. $11\frac{3}{7}$ c. $6\frac{1}{9}$ d. $7\frac{1}{5}$

11. a. $\frac{11}{4}$ b. $\frac{79}{8}$ c. $\frac{26}{7}$ d. $\frac{35}{6}$

12. a. $\frac{11}{3}$ b. $\frac{99}{10}$ c. $\frac{47}{12}$ d. $\frac{17}{2}$

13. 4 trays

14. a. $\frac{23}{24}$ b. $\frac{17}{45}$ c. $1\frac{1}{15}$

15. a. $\frac{20}{63}$ b. $\frac{6}{25}$ c. $\frac{5}{8}$ d. $\frac{7}{36}$

16. a. 44 b. 22

c. $\frac{1}{6}$ d. $\frac{50}{3}$ or $16\frac{2}{3}$

17. a. $\frac{2}{5}$ b. 9 c. 4 d. $\frac{4}{23}$

18. a. $\frac{3}{2}$ or $1\frac{1}{2}$ b. $\frac{4}{3}$ or $1\frac{1}{3}$

c. $\frac{7}{36}$ d. $\frac{3}{70}$

19. a. $4\frac{17}{24}$ b. $1\frac{13}{14}$ c. $3\frac{7}{9}$ d. $2\frac{2}{3}$

20. a. $3\frac{43}{72}$ b. $\frac{76}{99}$ c. $3\frac{2}{3}$ d. $6\frac{5}{18}$

21. a. 64 b. $\frac{3}{5}$ c. $13\frac{2}{3}$ d. $1\frac{1}{2}$

22. a. $\frac{5}{16}$ b. $1\frac{3}{5}$ c. $1\frac{53}{115}$ d. $\frac{217}{267}$

23. a. $\frac{13}{100}$ b. $\frac{7}{10}$

c. $\frac{13}{50}$ d. $\frac{33}{25}=1\frac{8}{25}$

24. a. 15% b. 55% c. 60% d. 50%

25. a. $37\frac{1}{2}\%$ b. $53\frac{1}{3}\%$ c. $83\frac{1}{3}\%$ d. $36\frac{4}{11}\%$

26. a. 6 b. 12 c. 39 d. 39

27. a. $6\frac{3}{10}$ b. $37\frac{1}{10}$ c. $4\frac{4}{5}$ d. $13\frac{3}{5}$

28. a. 60% b. 30% c. 40% d. 40%

29. a. $7\frac{1}{7}\%$ b. $11\frac{1}{9}\%$

30. a. \$0.40 b. \$1.05 c. \$3.20 d. \$5.20

31. a. \$0.15 b. \$0.30 c. \$5.05 d. \$27.00

32. $7.1=\frac{1}{8}, 7.2=\frac{1}{4}, 7.3=\frac{5}{32}, 7.4=\frac{5}{32}, 7.5=\frac{7}{40}, 7.6=\frac{11}{80}$

33. \$150

34. 2 kg

35. a. $\frac{1}{6}$ b. $\frac{1}{4}$

c. $\frac{5}{9}$ d. $\frac{48}{25}=1\frac{23}{25}$

36. All fractions that are equivalent to $\frac{4}{7}$ will be of the form $\frac{4n}{7n}$ where $n$ is some number. The product of the numerator and denominator is therefore $4nx7n=28n^2$. Equating this product to 252:

$252=28n^2$

$n^2=9$

$n=\pm 3$

Substituting $n=+3$ gives a fraction of $\frac{12}{21}$, which is equivalent to $\frac{4}{7}$ and for which the product of the numerator and denominator is 252.

$\frac{12}{21}$

37. Tomas

38. $\frac{1}{2}+\frac{1}{4}+\frac{1}{20}$ or $\frac{1}{2}+\frac{1}{5}+\frac{1}{10}$

39. Mr Thompson \$3000, Mr Goody \$2500

40. $\frac{3}{5}$ of the standard spots were filled.

41. a. 4 complete surfboards

b. $2\frac{1}{4}$ kg of resin, 0 catalyst

42. a. 62% b. 75%

43. 24 students

44. 270 children

45. 8%

46. \$11 600

47. b, d

48. 13.5 litres of milk would be needed to make the fat of 500 g of butter.

49. 45%

50. a. $2250

 b. $2025

 c. 19%

 d. The second 10% discount is not 10% of the original price but 10% of an amount, which is only 90% of the original.

51. a. $1575

 b. $1575

 c. It doesn't matter which discount is applied first; the final price will be the same.

52. Sample responses can be found in the worked solutions in the online resources.

# 5 Ratios, rates and best buys

## LESSON SEQUENCE

**5.1** Overview ........ 262
**5.2** Ratios ........ 265
**5.3** Rates ........ 271
**5.4** The unitary method and best buys ........ 276
**5.5** Review ........ 281

# LESSON
# 5.1 Overview

## 5.1.1 Why learn this?

Ratios and rates are used to compare different quantities. Ratios compare quantities of the same kind. For example, in a class of Year 7 students, if 20% walk to school and 80% don't, this can be expressed as a ratio of 1 to 4. This means that for every 1 student who walks to school, 4 students don't. Ratios are used in design or architectural drawings, with a drawing being an exact 'miniature' or scale diagram of the item to be made. Plans for a new home would be drawn at a ratio of 1 : 1000, which means that every 1 mm on the plan represents 1000 mm in the new home. In comparison, rates enable you to compare quantities expressed in different units. Speed is an example of a rate; it compares distance travelled with the time taken, and is measured in km/h or m/s. You will see speed signs whenever you go for a drive. Rates and ratios help you make choices, too, such as when you decide which box of chocolates to purchase. Civil engineers examine rates of water flow, which helps them to design pipes that ensure the water gets to your home. Electrical engineers examine rates for stepping up and stepping down voltage, which ensures electricity is able to get to your home. Business managers explore various types of costed rates to ensure they always get the 'best deal'. Carpenters evaluate different rates to ensure they order enough timber to complete a job, and doctors and pharmacists calculate different rates to ensure you take the right amount of medicine based upon your height and weight. By comparing price and quantity, you can choose the 'best buy', which is the item with the lowest price per unit. Architects, chefs, engineers, fashion designers and builders all use rates and ratios in their field of work.

## Exercise 5.1 Pre-test

learnon

1. A cake mixture contains 1 cup of sugar and 2 cups of flour. Express this statement as a ratio.

2. For the diagram shown, write the following ratios in simplest terms.
   a. shaded parts: unshaded parts
   b. shaded parts: total number of parts

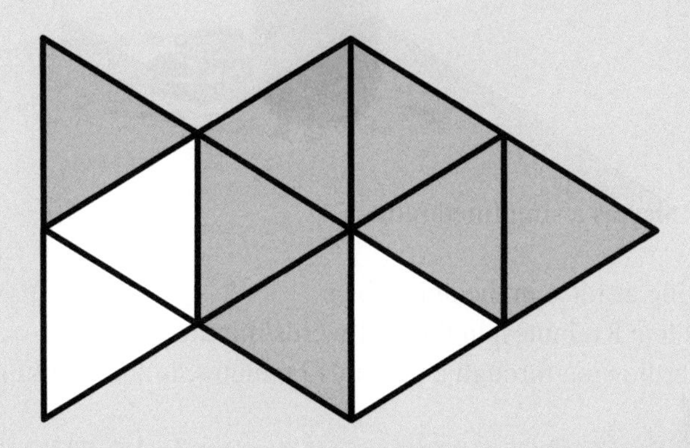

3. Express each of the following as a fraction.
   a. 5 pears to 7 oranges
   b. 500 m to 2 km
   c. 7 basketballs to 35 tennis balls
   d. 8 hours to 120 minutes

4. **MC** Select which of the following options is the simplest form of the ratio 8 : 32.
   A. 4 : 1  B. 1 : 4  C. 4 : 2  D. 2 : 8  E. 4 : 16

5. A school has 2 students who wear glasses to every 5 students who don't wear glasses. A particular class where this ratio applies has 8 students who wear glasses. Calculate how many students don't wear glasses in this class.

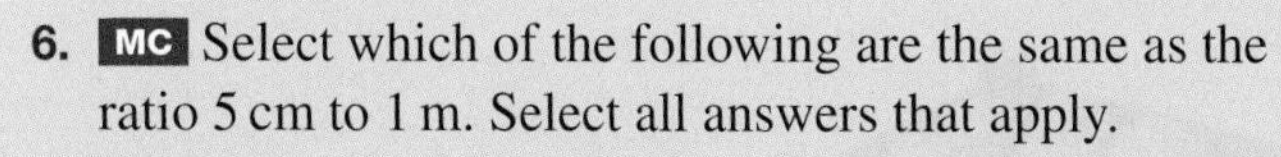

6. **MC** Select which of the following are the same as the ratio 5 cm to 1 m. Select all answers that apply.
   A. 1 : 20  B. 5 : 100  C. 2 : 50
   D. 2 : 25  E. 25 : 500

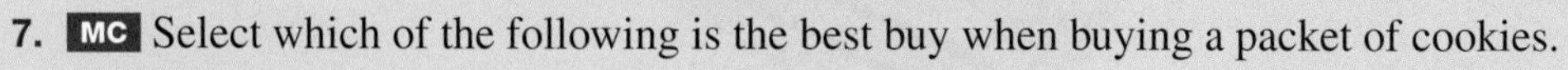

7. **MC** Select which of the following is the best buy when buying a packet of cookies.
   A. 150 g for $1.75  B. 200 g for $2.30  C. 250 g for $2.75
   D. 350 g for $3.55  E. 500 g for $5.45

8. A bus is travelling at a speed of 60 km per hour. Calculate how long it takes to travel 145 km.

9. Hot chocolate can be bought in 250 g jars for $9.50 or in 100 g jars for $4.10. Calculate the cheaper way to buy the hot chocolate, and how much cheaper it is per 100 g.

10. Write 16 hours to 1 day as a simplified ratio.

11. Express the following as rates in the units given.
   a. 464 words written in 8 minutes, in the unit words/minute
   b. 550 litres of water flowing through a pipe in 12 minutes, in the unit kL/h

12. The price of an item is reduced by 10%, then increased by 11.1%. This takes the item back to its original price (to the nearest dollar). True or false?

13. Out of 100 customers selected for a mobile internet usage survey, 63 customers were aged less than 25 years, 28 customers were aged 25–65 years and the rest were aged over 65 years.
   a. Write the ratio of customers aged 25–65 years to those aged less than 25 years.
   b. Write the ratio of customers aged over 65 years to those aged 25–65 years.
   c. Write the ratio of customers aged less than 25 years to the total number of customers surveyed.
   d. Write the number of customers aged less than 25 years as a fraction of the total number of customers surveyed.

14. If 12 oranges costs $6.00, explain how much it will cost to buy 5 oranges.

15. If Eva takes 3 hours to clean a room and Will takes 5 hours to clean a room, determine how long it will take to clean a room if they work together.

# LESSON
# 5.2 Ratios

## LEARNING INTENTION

At the end of this lesson you should be able to:
- compare quantities using ratios
- simplify ratios.

## 5.2.1 Introduction to ratios

eles-3886

- **Ratios** are used to compare quantities of the same kind in the same unit.
  That is, a ratio says how much of one value there is compared to another value.
  For example, in the diagram below there is one blue square to 4 pink squares.

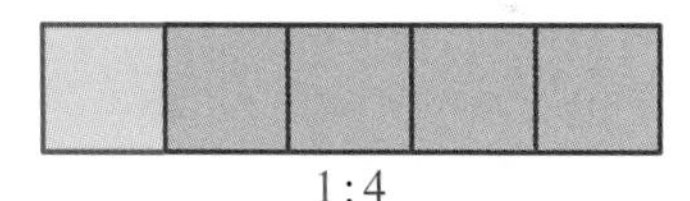

1 : 4

### Expressing ratios

- **A ratio of 1 to 4 is written as 1 : 4 (the quantities are separated by a colon) and is read as 'one to four'.**
- **Ratios can also be written in fraction form:**

$$\mathbf{1:4 \longleftrightarrow \frac{1}{4}}$$

- **Ratios do not have names or units of measurement.**
- **Ratios contain only whole numbers.**

### WORKED EXAMPLE 1 Expressing quantities as ratios

**Express each of the following statements as a ratio.**
**a. The fertiliser contains 3 parts of phosphorus to 2 parts of potassium.**
**b. Mix 1 teaspoon of salt with 3 teaspoons of flour.**

| THINK | WRITE |
|---|---|
| a. Write the 3 parts of phosphorus to 2 parts of potassium as a ratio (the number of parts of phosphorus should be written first). | a. 3 : 2 |
| b. Write the 1 teaspoon of salt with 3 teaspoons of flour as a ratio (the number of teaspoons of salt should be written first). | b. 1 : 3 |

WORKED EXAMPLE 2 Expressing quantities as fractions

**Express each of the following as a fraction.**
a. **5 cm to 7 cm**
b. **3000 m to 2 km**

| THINK | WRITE |
|---|---|
| a. 1. Write the 5 cm to 7 cm as a ratio by replacing 'to' with ':' (5 cm should be written first). As the two numbers have the same unit, units are not included. | a. 5 : 7 |
| 2. Write the ratio as a fraction by placing the first number as a numerator and the second number as a denominator (no unit required). | $\frac{5}{7}$ |
| b. 1. Express both quantities in the same units. Convert 3000 m to km or 2 km to m. | b. 3000 m to 2 km<br>3000 m to 2000 m |
| 2. Write the 3000 m to 2000 m as a ratio by replacing 'to' with ':' (3000 m should be written first). Omit the units and simplify. | 3000 : 2000<br>3 : 2 |
| 3. Write the ratio as a fraction by placing the first number as a numerator and the second number as a denominator. | $\frac{3}{2}$ |

## 5.2.2 Simplifying ratios

eles-3887

- Equivalent ratios are ratios that are the same when we compare them.
  For instance, the ratios 2 : 3 and 4 : 6 are equivalent, as the second ratio can be obtained by multiplying both parts of the first ratio by 2 and therefore they are the same ratios when compared.
  The ratios 10 : 5 and 2 : 1 are also equivalent, as the second ratio is obtained by dividing both parts of the first ratio by 5.
- Ratios are usually written in simplest form — that is, reduced to lowest terms. This is achieved by dividing each number in the ratio by the highest common factor (HCF).

**Simplification of ratios**

**Equivalent ratios are always multiplied or divided by the same number. They work in the same way as equivalent fractions.**

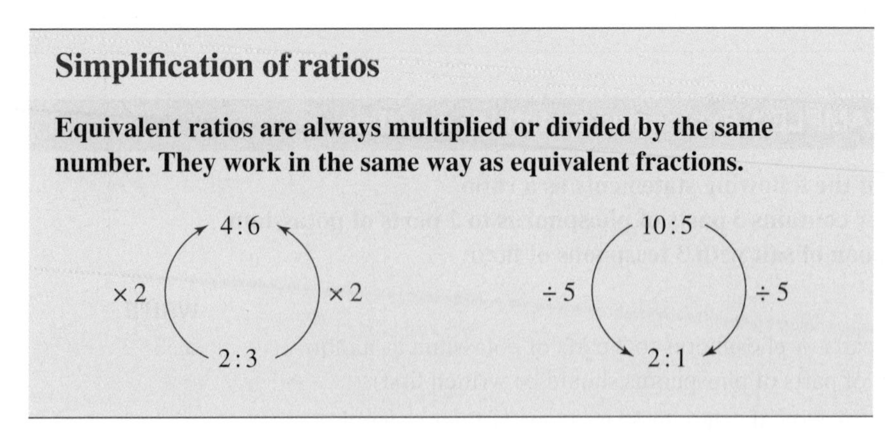

## WORKED EXAMPLE 3 Simplifying ratios

**Express the ratio 9 : 15 in simplest form.**

| THINK | WRITE |
|---|---|
| 1. Determine the largest number by which both 9 and 15 can be divided (that is, the highest common factor, HCF). It is 3. | The HCF is 3. |
| 2. To go from the ratio with the larger values to the ratio with the smaller values, divide both 9 and 15 by 3 to obtain an equivalent ratio in simplest form. | 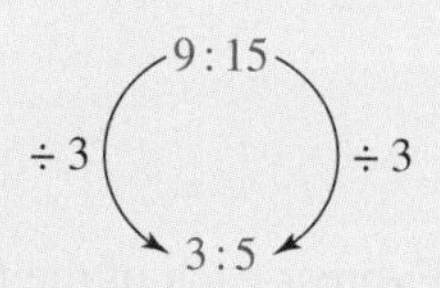 |
| 3. Write the answer. | 3 : 5 |

## Digital technology

Since simplifying ratios follows a similar process to simplifying fractions, a calculator can be used to help check your answer. For example, in Worked example 3 the ratio 9 : 15 was simplified to 3 : 5.

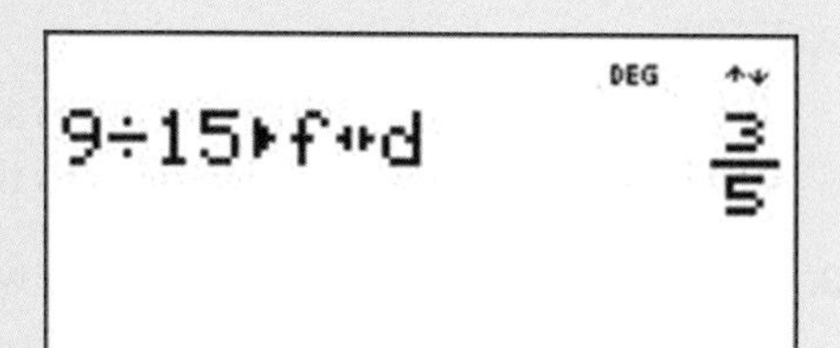


Using a calculator, input 9 ÷ 15, then use the decimal to fraction button to display the answer as a fraction in simplest form.

## WORKED EXAMPLE 4 Simplifying ratios of quantities expressed in different units

**Write the ratio of 10 cm to 30 mm in simplest form.**

| THINK | WRITE |
|---|---|
| 1. Write the question. | 10 cm to 30 mm |
| 2. Express both quantities in the same units by changing 10 cm into mm (1 cm = 10 mm). | 100 mm to 30 mm |
| 3. Omit the units and write the two numbers as a ratio. | 100 : 30 |
| 4. Determine the HCF by which both 100 and 30 can be divided. It is 10. | The HCF is 10. |
| 5. Simplify the ratio by dividing both sides by the HCF. | 100 : 30<br>÷ 10 ÷ 10<br>10 : 3 |
| 6. Write the answer. | 10 : 3 |

**on** Resources

**eWorkbook** Topic 5 Workbook (worksheets, code puzzle and project) (ewbk-1906)

**Interactivities** Individual pathway interactivity: Introduction to ratios (int-4412)

Introduction to ratios (int-3956)

Simplification of ratios (int-3949)

## Exercise 5.2 Ratios

learnon

5.2 Quick quiz 

5.2 Exercise

### Individual pathways

| ■ PRACTISE | ■ CONSOLIDATE | ■ MASTER |
|---|---|---|
| 1, 4, 5, 10, 12, 15, 18 | 2, 6, 8, 11, 13, 16, 19 | 3, 7, 9, 14, 17, 20 |

### Fluency

1. **WE1** Express each of the following statements as a ratio.
   a. The cake mixture contains 2 cups of flour and 5 cups of water.
   b. Mix 1 cup of sugar with 3 cups of milk.
   c. Sebastian Vettel's F1 car is twice as fast as Mick's delivery van.
   d. The Tigers team finished the season with a win to loss ratio of 5 to 2.
2. For the rectangle shown:
   a. write the ratio of unshaded parts to shaded parts
   b. write the ratio of unshaded parts to the total number of parts.
3. Explain why order is important when writing ratios.
4. **WE2** Express each of the following as a fraction.
   a. 5 wins to 7 losses
   b. 3500 m to 25 km
   c. 3 apples to 4 oranges
   d. 10 basketballs to 3 tennis balls
5. **WE3** Express the following ratios in simplest form.
   a. 75 : 60 b. 154 : 24 c. 36 : 6 d. 48 : 42
6. **WE4** Express the following ratios in simplest form.
   a. 7 cm : 9 cm
   b. 120 minutes : 2 hours
   c. 240 g : 2 kg
   d. 5 weeks to 2 months
7. Express the following ratios in simplest form.
   a. 5 cm : 10 mm
   b. 600 g : 8 kg
   c. 24 hours : 16 minutes
   d. 21 boys : 14 girls
8. Draw rectangles to represent the following ratios.
   a. Shaded area to unshaded area is 1 : 1
   b. Shaded area to unshaded area is 2 : 7
   c. Unshaded area to shaded area is 2 : 3
   d. Unshaded area to shaded area is 3 : 1
9. Complete the patterns of equivalent ratios.

| a. | b. | c. | d. |
|---|---|---|---|
| 1 : 4 | 2 : 3 | 64 : 32 | 48 : 64 |
| 2 : 8 | 4 : 6 | ______ : 16 | ______ : 32 |
| ______ : 12 | 8 : ______ | ______ : 8 | ______ : 16 |
| ______ : 16 | ______ : 15 | 8 : ______ | 6 : ______ |
| 5 : ______ | ______ : 21 | ______ : 1 | ______ : 4 |

## Understanding

10. The length to width ratio of the Australian flag is 2 : 3. If the length of the flag is 60 cm, determine the width of the flag.

11. Compare the following, using a mathematical ratio (in simplest form).
   a. The Hawks beat the Cats by 81 points to 60 points.

   b. The cost of a bag of Maltesers chocolate is \$3, but Darrell Lea chocolate costs \$5.
   c. In the movie audience, there were 160 children and 24 adults.
   d. During the race, Rebecca's average speed was 300 km/h while Donna's average speed was 120 km/h.

12. Copy and complete each of the following equivalent ratios.
   a. 1 : 3 = ___ : 12   b. 2 : 3 = 4 : ___   c. 66 : 48 = : ___ : 8   d. 36 : 72 = 12 : ___

13. Copy and complete each of the following equivalent ratios.
   a. 20 : 12 = : __ : 3   b. __ : 6 = 8 : 24   c. 8 : 6 = __ : 72   d. 40 : __ = 5 : 6

**14.** The table shows the nutritional information from a box of cereal.

| Nutritional information | Average quantity per serving |
|---|---|
| Energy | 660 kJ |
| Protein | 4.5 g |
| Fat | 0.1 g |
| Carbohydrates | 22 g |
| Sugars | 2 g |
| Fibre | 3 g |
| Sodium | 55 mg |

Write the following ratios in simplest form.

**a.** Sugar to total carbohydrate
**b.** Fat to protein
**c.** Protein to fibre
**d.** Sodium to protein

### Reasoning

**15.** The cost of tickets to two different concerts is in the ratio 4 : 7. If the more expensive ticket costs \$560, then determine the cost of the cheaper ticket.

**16.** A student received a score of $\frac{77}{100}$ for a Maths test.

**a.** Write the ratio of the total marks received compared with the total marks lost.
**b.** Write the ratio of the total marks lost compared with the total possible marks.

**17.** The total attendance at a concert was 62 400 people. Of this total, 24 800 people were male.

**a.** Calculate how many females attended the concert.
**b.** Compare the number of females to males by writing the ratio in simplest form.
**c.** Write the number of females as a fraction of the number of males.

### Problem solving

**18.** The ratio of gold coins to silver coins in a purse is 2 : 5. If there are 10 gold coins in the purse, determine the smallest number of coins that needs to be added so that the ratio of gold coins to silver coins changes to 3 : 4.

**19.** Out of 200 people selected for a school survey, 168 were students, 20 were teachers and the rest were support staff.

**a.** Write the ratio of teachers to students.
**b.** Write the ratio of support staff to teachers.
**c.** Write the ratio of students to people surveyed.
**d.** Write the number of teachers as a fraction of the number of people surveyed.

**20.** You drop a rubber ball from a height of 10 m. The ball rebounds to a height of 8 m and then to a height of 6.4 m. The ball continues to bounce in the same ratio.

**a.** Calculate the height of the ball when the top of its bounce is less than 3 m.
**b.** Determine on which number bounce this occurs.

# LESSON
# 5.3 Rates

## LEARNING INTENTION

At the end of this lesson you should be able to:
- identify and express a rate from a written context
- simplify a rate and express the rate in its simplest form
- apply a rate to calculate the total value of an unknown quantity.

## 5.3.1 Identifying rates

eles-3888

- A **rate** is used to compare two measurements of different kinds that are in different units.
- Rates have units. An example of a rate is speed (measured in km/h or m/s). A speed limit sign tells the driver that the maximum speed they are allowed to drive is 25 kilometres per hour.
- A slash (/), which is the mathematical symbol for 'per', is used to separate two different units.
- A rate is in its simplest form if it is per one unit.
- Rates are often written as one quantity **per** another quantity.

**Expressing rates**

**First quantity per second quantity**

**or**

$$\frac{\textbf{first quantity}}{\textbf{second quantity}}$$

**WORKED EXAMPLE 5 Expressing rate from a worded problem**

**Jennifer goes for a run every Monday morning and knows that she can run 3 kilometres in 15 minutes. Express, as a rate, the speed at which she can run.**

| THINK | WRITE |
|---|---|
| 1. Speed is a rate the measures distance over time. From the question, determine the values for distance and time. | Distance = 3 km<br>Time = 15 minutes |
| 2. Express the rate as first quantity (distance) per second quantity (time) and write the answer. | 3 km per 15 min |

**WORKED EXAMPLE 6 Expressing rate from a worded problem**

**Max notices on his food packaging there is a warning that tells him the sugar content in one serving is equal to 25 g. Rewrite this information in the form of a rate.**

| THINK | WRITE |
| --- | --- |
| 1. Read the question carefully and determine two quantities. In this question, quantities are sugar and serving size. | Sugar = 25 g<br>Serving size = 1 serve |
| 2. Express rate as first quantity (sugar) per second quantity (serve) and write the answer. | 25 g per 1 serve |
| 3. As there is only 1 serving size, this can be written more simply as 25 g per serve. | 25 g per serve |

## 5.3.2 Simplifying rates

eles-3889

- It is often most useful to express rates in their simplest form.
  For instance, in Worked example 5, Jennifer knows she can run 3 kilometres in 15 minutes — speeds are usually given in kilometres per hour.
  To **convert** this rate of 3 km per 15 minutes into a rate of kilometres per hour (km/h), we multiply minutes by 4. That is because there are four lots of 15 minutes in every hour.

3 km : 15 minutes
× 4 — There are 60 minutes in an hour (15 × 4 = 60).
3 km : 60 minutes

To keep the rate the same, what we do to one side of the rate is what we also have to do to the other. This ensures that we do not change the rate's **value**.
This only works for multiplying and dividing, and not for adding or subtracting.

We also multiply the left side by 4 (3 × 4 = 12).

3 km : 15 minutes
× 4    × 4
12 km : 60 minutes

Thus, the rate in kilometres per hour (km/h) is written as 12 km/h.

**Converting units**

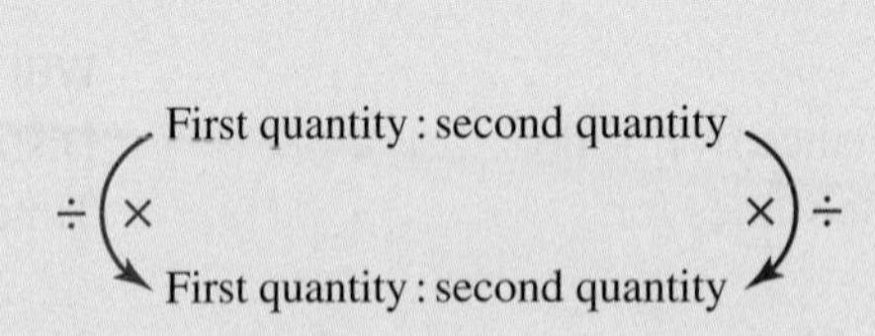

## WORKED EXAMPLE 7 Simplifying rates

**Madison works for 4 hours at her local bakery. At the end of the week, she has earned $100 for her work. Calculate how much Madison earns in dollars per hour.**

| THINK | WRITE |
|---|---|
| 1. Read the question carefully and determine two quantities. In this question, the quantities are dollars and hours. | Dollars = 100<br>Hours = 4 |
| 2. Express as a ratio of the first quantity (dollars) to the second quantity (hours). Remember to write out the units. | 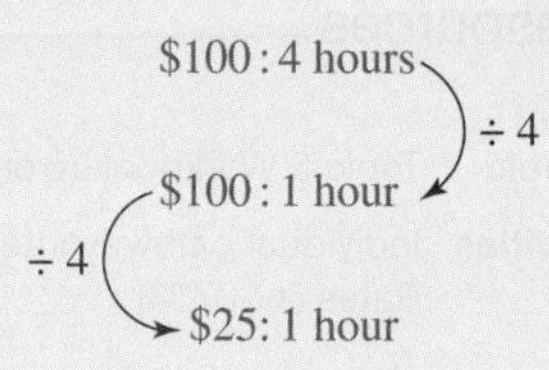 |
| 3. To convert the rate unit to dollars per hour we need to divide the hours by 4, to reduce 4 hours to 1 hour. Remember that what we do to one side of the rate, we must also do to the other side. Therefore, we also divide dollars by 4. | |
| 4. Write the answer as a rate in the correct units. | Madison earns $25 per hour. |

## 5.3.3 Applications of rates

eles-3890

- Rates can be applied to find the total quantity of a measure being compared.
  For example, rate (speed) and time (second quantity) can be applied in order to calculate the total distance (first quantity) covered.

### Arranging rate formula

$$\textbf{Rate} = \frac{\textbf{first quantity}}{\textbf{second quantity}}$$

**First quantity = rate × second quantity**

**Second quantity = first quantity ÷ rate**

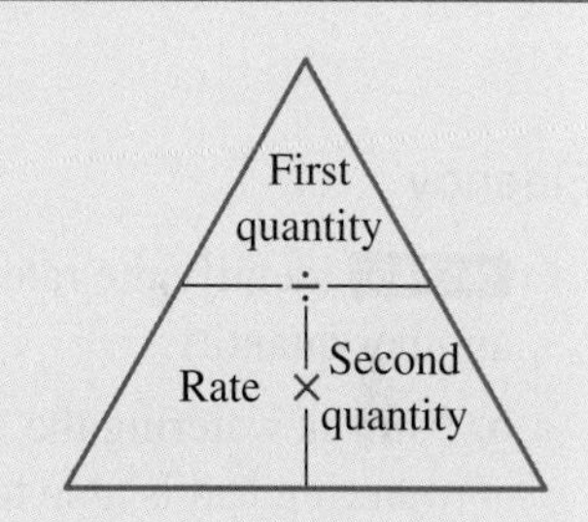


## WORKED EXAMPLE 8 Applying rate to calculate an unknown quantity

**Rick has a job at the local cinema, where he is able to earn $22 per hour. If Rick works for 12 hours this week, calculate how much money he will earn in total.**

| THINK | WRITE |
|---|---|
| 1. Express the quantities given in the question as rate. The first quantity is dollars and the second quantity is hours. | Rate = dollars per hour<br>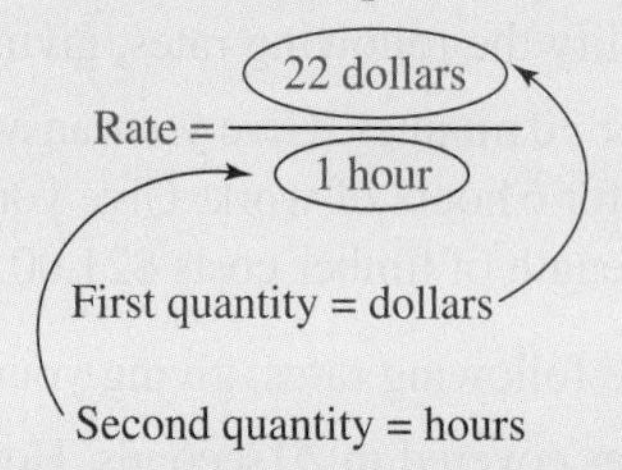 |
| 2. To calculate the first quantity (money earned), rearrange the rate formula. | First quantity = rate × second quantity |
| 3. State the number of hours worked. | Hours worked = 12 |

4. To calculate the total payment, multiply the hourly payment by the total number of hours worked. — Total payment = $22 \times 12$ = $264

5. Write the answer with correct units (in this case the unit is dollars). — Rick will earn $264 in 12 hours.

**on** Resources

**eWorkbook** Topic 5 Workbook (worksheets, code puzzle and project) (ewbk-1906)

**Interactivities** Individual pathway interactivity: Rates (int-4417)
Rates (int-3738)

## Exercise 5.3 Rates

learn**on**

| 5.3 Quick quiz on | 5.3 Exercise |
|---|---|

### Individual pathways

| ■ PRACTISE | ■ CONSOLIDATE | ■ MASTER |
|---|---|---|
| 1, 4, 7, 10, 13 | 2, 5, 8, 11, 14 | 3, 6, 9, 12, 15 |

### Fluency

1. **WE5&6** Identify the rates in the following circumstances and write them out in the form of one quantity per another quantity.
   a. Amir is watering the garden. He notices that he can fill his watering can up with water in one minute. The watering can is able to hold 7 litres of water. Express this as the rate at which the water is filling the can.
   b. James is training for his upcoming cricket game. If his current rate of runs per ball continues, he will make 33 runs from 7 balls. Write this as the rate of runs per ball.
   c. Tilly pays $6.00 for a bunch of bananas. State the cost of bananas per bunch.
   d. Cherese works at the local supermarket where, in a four-hour shift, she earns $80.00. Calculate the rate at which she is being paid.
   e. Jessica swims every second morning. She knows that, in a 25-metre pool, she is able to swim 20 lengths in 15 minutes. Write her rate of swimming. Give your answer in metres per minute. You do not need to simplify your answer.

2. **WE7** Simplify the following rates, giving your answer as the unit rate where possible.
   a. 15 km per 30 minutes. Give your answer in the units km per hour.
   b. $150.00 for 6 hours of work. Give your answer in the units dollars per hour.
   c. A 14 m length of timber costs $21.00. Give your answer in the units dollars per metre.

3. Simplify the following rates, giving your answer as the unit rate where possible.
   a. 100 m was covered in 20 seconds. How many metres were covered per second?
   b. It cost $180.00 to wallpaper a room with a wall space of 50 $m^2$. How much did the wallpaper cost per square metre?
   c. If a 5.5 kg bag of almonds cost $19.25, how much do almonds cost per kilogram?
   d. It costs $300.00 to carpet a room that is 8 $m^2$. What is the cost of carpet per square metre?

4. **WE8** Patrick cycles for 2 hours, at a speed of 18 km/h. Calculate how far he has cycled.

5. Jackson works a 5-hour shift at the local bakery. If he is paid $85.00 for this shift, calculate his hourly pay rate.
6. A train is travelling at a speed of 50 km per hour. Calculate how long it takes to travel 150 km.

### Understanding

7. Larah is working in the garden and buys 50 kg of premium potting mix for $75.00. Calculate how much the potting mix costs per kilogram.
8. Jacob was driving at a constant speed for 30 minutes. If he travelled for 35 km, calculate his speed.
9. Anthony was riding his bike at an average speed of 25 km/h. If he managed to cover a distance of 40 km, calculate how long his bike ride was (to the nearest minute).

### Reasoning

10. If 10 pears costs $15.00, explain how much it will cost to buy 5 pears.
11. If Michael can run 30 km in 3 hours, explain how long it will take him to run 20 km at a constant speed.
12. If Jaye can swim 400 metres in a time of 6 minutes and 45 seconds, calculate how long it will take her to swim 800 metres if she is swimming at a constant speed.

### Problem solving

13. Water is poured into a canister at a rate of 6 L/ min. If the canister can hold 24 litres of water, calculate:
    a. how long will it take to fill the canister
    b. how long will it take to fill two canisters of the same size
    c. how long will it take until the canister is half full
    d. how long will it take to fill the canister with 18 litres of water.
14. Kelly is paid $14.25 per hour for her job working in retail. She is looking to save $240 to buy a new monitor for her computer and she knows she can only be rostered on for 4-hour shifts at work.
    a. Calculate how many hours she will need to work in order to buy the monitor. Round your answer to the nearest hour.
    b. Calculate how many 4-hour shifts she will need to complete in order to achieve this.
    c. After working enough 4-hour shifts, determine how much money is left over if she were to buy the monitor.
15. Charlee works two jobs, one at a supermarket and another as a basketball coach at weekends. She earns an hourly wage of $13.75 for her job at the supermarket and is paid a flat rate of $25.00 per week for coaching basketball. At the supermarket, the work roster is such that each shift goes for three hours and she cannot do more than one shift per day.
    a. If she is rostered on to work for 4 days this week, calculate how much money she will earn for the entire week, including her work as a basketball coach.
    b. At this rate, calculate how much money Charlee will earn for 3 weeks of work.
    c. Charlee worked for 24 hours at the supermarket over a 3-week period. Calculate how much money she earned in this time.

# LESSON
# 5.4 The unitary method and best buys

## LEARNING INTENTION

At the end of this lesson you should be able to:
- use the unitary method to determine the best value option
- apply the unitary method to find amounts from a partial amount.

## 5.4.1 Unit price

eles-3891

- The **unit price** is the price per unit. Examples include the price per 100 mL, the price per kg or the price of a single item.
- The **unitary method** can be used to work out the best buy for an item by comparing the unit price.
- Supermarkets in Australia are now obliged to show the unit price on most products.
- To determine the 'best buy' when comparing similar items:
  - decide what unit you will use in your comparison
  - divide the price of each item by its number of units
  - identify the item with the smallest unit price. This is the best value item.

**Price per unit**

$$\textbf{Price per unit} = \frac{\textbf{price}}{\textbf{number of units}}$$

**Remember to represent the same units when you are making comparisons.**

### WORKED EXAMPLE 9 Determining the best buy using the unitary method

**Three shampoos are sold in the following quantities.**

**Brand A: 200 mL for \$5.38**
**Brand B: 300 mL for \$5.98**
**Brand C: 400 mL for \$8.04**

**Determine which shampoo is the best buy.**

**THINK** | **WRITE**

1. Determine the number of 100-mL units for each shampoo.

$$\text{Brand A} = \frac{200}{100} = 2 \text{ units}$$
$$\text{Brand B} = \frac{300}{100} = 3 \text{ units}$$
$$\text{Brand C} = \frac{400}{100} = 4 \text{ units}$$

2. Determine the price per unit for each shampoo.

$$\text{Price per unit} = \frac{\text{price}}{100\text{-mL units}}$$

$$\text{Brand A} = \frac{5.38}{2} = \$2.69 \text{ per } 100 \text{ mL}$$
$$\text{Brand B} = \frac{5.98}{3} = \$1.99 \text{ per } 100 \text{ mL}$$
$$\text{Brand C} = \frac{8.04}{4} = \$2.01 \text{ per } 100 \text{ mL}$$

3. Write the answer.

Brand B: 300 mL of shampoo for \$5.98 is the best buy.

## 5.4.2 Unitary method in percentages

eles-3892

- The unitary method can be used to determine the whole amount when given a partial amount and the percentage of the total that it corresponds to.

**Determining the whole amount from a partial amount**

- **Step 1: Find 1% of the total amount.**
- **Step 2: Multiply the value of the 1% by 100 to get the whole amount (100).**

### WORKED EXAMPLE 10 Determining total price from a partial amount

**Calculate the whole amount if 5% is represented by \$25.**

**THINK** | **WRITE**

1. Five per cent of a number equals 25. Calculate 1% of the \$25 by dividing the number by 5.

$5\% = \$25$
$1\% = \$25 \div 5$
$= \$5$

2. Calculate 100% by multiplying the amount by 100.

$100\% = \$5 \times 100$
$= \$500$

3. Write the answer.

The whole amount is \$ 500.

- The same method can also be applied when given a percentage that is greater than 100%.

## WORKED EXAMPLE 11 Determining values from an increased amount

**Marcie receives a pay rise of 12%, taking her salary to \$72 000 a year. Calculate her annual salary before the pay rise, to the nearest dollar.**

| THINK | WRITE |
|---|---|
| 1. A pay rise of 12% means that Marcie's new salary is 112% of her old salary. | 112% of previous salary = \$72 000 |
| 2. Calculate what 1% of her previous salary was by dividing by 112. | $1\%$ of previous salary $= \dfrac{\$72\,000}{112}$ <br> $= \$642.8571$ |
| 3. Calculate 100% of her previous salary by multiplying by 100. | $100\%$ of previous salary $= \$642.8571 \times 100$ <br> $= \$64\,285.71$ |
| 4. Write the answer, rounding to the nearest dollar. | Marcie's previous salary was \$64 286. |

### ACTIVITY: Price comparison websites

Use the internet to research different price comparison websites. Investigate what goods and services price comparison websites display and present your research to your class.

 Resources

 **eWorkbook** Topic 5 Workbook (worksheets, code puzzle and project) (ewbk-1906)

 **Interactivities** Individual pathway interactivity: The unitary method and best buys (int-4396)
Unitary method (int-4154)
Unitary method in percentages (int-4155)

# Exercise 5.4 The unitary method and best buys

learnon

5.4 Quick quiz on

5.4 Exercise

### Individual pathways

| ■ PRACTISE | ■ CONSOLIDATE | ■ MASTER |
|---|---|---|
| 1, 4, 8, 9, 11, 17, 21 | 2, 5, 6, 12, 13, 15, 18, 22 | 3, 7, 10, 14, 16, 19, 20, 23 |

### Fluency

1. WE9 Determine which of the following is the best buy.

| Chocolate weight | Cost |
|---|---|
| 150 g | $3.25 |
| 250 g | $4.75 |
| 325 g | $5.50 |

2. Determine which of the following is the best buy.

| No. of pages | Cost |
|---|---|
| 80 | $1.98 |
| 160 | $3.38 |
| 200 | $3.98 |

3. Compare the cost of 400 g of biscuits for $2.48 and 500 g for $3.10 to identify which is the better buy.
4. WE10 Calculate the original amount if:
   a. 10% is $18 b. 20% is $6 c. 25% is $60 d. 40% is $900
5. Calculate the original amount if:
   a. 90% is $380 b. 200% is $800 c. 120% is $420 d. 9% is $54.
6. Calculate:
   a. 60%, if 40% is $120 b. 38%, if 20% is $6 c. 50%, if 25% is $60
7. Calculate:
   a. 150%, if 50% is $900 b. 12%, if 75% is $250 c. 86%, if 14% is $4200.
8. Calculate how much you would pay for 4.5 kg of apples, given that 2 kg cost $3.80.

### Understanding

9. You spend 40% of this month's allowance on a pair of used inline skates. Determine your monthly allowance if you spend $20 on the skates.
10. Hannah saves 32% of her wages each week. If she saves $220, calculate her weekly wage.
11. WE11 Alex receives a 15% pay rise. He now receives $97 290 a year. Calculate his annual income before the pay rise.

12. Compare the following to identify the best buy:
500 g of cheese for \$12.80, 200 g of cheese for \$5.62 or 80 g of cheese for \$2.10.

13. After a discount of 15%, a pair of wireless headphones is worth \$183. Calculate its value before the discount.

14. During a sale, a retailer allows a discount of 15% off the marked price. The sale price of \$60 still gives the retailer a profit of 10%.
    a. Calculate the cost of the article.
    b. Calculate the marked price.

15. A discount of 15% reduced the price of a pen by \$3.20.
    a. Calculate the original price of the pen.
    b. Calculate its reduced selling price.

16. Supermarkets are obliged by law to display the unit price when selling the same items in different quantities. Calculate the price per 100 g for the following amounts of chocolate. Give your answers correct to the nearest cent.
    a. 250 g for \$5.95 b. 400 g for \$8.50 c. 60 g for \$2.00

### Reasoning

17. In some situations it is not always practical to go for the best buy. Suggest two reasons this might occur.

18. Explain some of the different situations in which you would use the unitary method.

19. While shopping in a sale, Carly spots a sign that says 'Buy 2 items, get the third half price (the least expensive item will be counted as the third item)'. If she finds a shirt for \$25.99, a skirt for \$87.99 and a belt for \$15.97 and pays cash for the items, determine how much Carly will pay and justify your answer.

20. Penny loves to make bracelets. To make them, she needs to buy some beads. Brand A costs \$7.50 for a box that contains 1000 beads; Brand B costs \$17.50 for a box that contains 2500 beads. Determine which brand is the better buy.

### Problem solving

21. Ice-cream sticks are sold in the following multi-packs. Compare the following to identify the best buy.

| Cost (\$) | Number of ice-cream sticks |
|---|---|
| \$3.25 | 6 |
| \$4.99 | 10 |
| \$7.50 | 15 |

22. Some players from a soccer club went to a café to celebrate their win. Each had a burger and a drink. The bill came to \$80. When it came time to pay the bill, 3 people had left without paying, so the remaining members had to pay an extra \$2 each to cover the bill. If there were originally $n$ people in the group, and the burger and drink deal cost $\$b$, write an equation using these pronumerals to show how the bill was settled. Do not attempt to solve your equation.

23. The price of entry into a theme park has increased by 10% every year since the theme park opened. If the latest price rise increased the tickets to \$8.80, determine the price of a ticket 2 years ago.

# LESSON
# 5.5 Review

## 5.5.1 Topic summary

**RATIOS, RATES AND BEST BUYS**

**Ratios**

- Ratios compare quantities in the same unit.
- 1 : 3 is read as 'the ratio of 1 to 3'.
- Ratios contain only integer values without units.
- They can be converted into fractions.

  e.g. $2:5 \Rightarrow \frac{2}{7}$ and $\frac{5}{7}$

**Equivalent ratios**

- Ratios are equivalent if one can be converted into the other by multiplying or dividing by a factor.

  e.g.

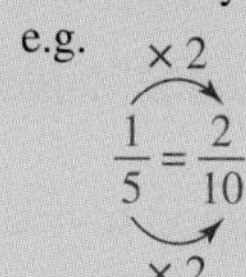


$$\frac{1}{5} = \frac{2}{10}$$

- Ratios are generally written in simplest form.

**Rates**

- A rate compares two different types of quantities in different units.
  e.g. 12 km/h or 12 cm/day
- A rate is in simplified form if it is per one unit.

  $100 : 4 hours (÷ 4) → $100 : 1 hour (÷ 4) → $25 : 1 hour

- A rate can be applied to find the total quantity of a measure being compared.

$$\text{Rate} = \frac{\text{first quantity}}{\text{second quantity}}$$

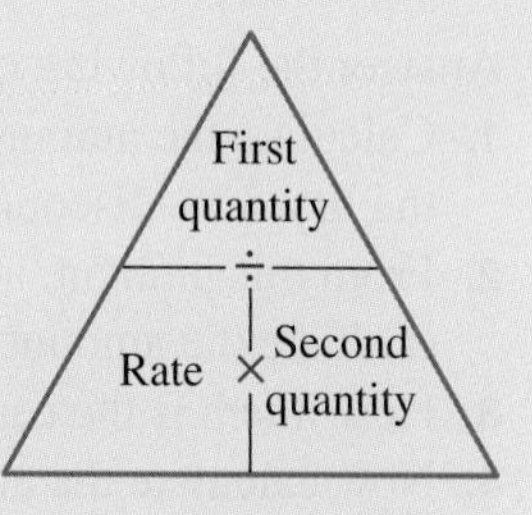


**Unitary method**

- The unitary method is used to calculate the whole amount when a partial percentage amount is known.
  - First find 1% of the total amount.
  - Multiply this value of 1% by 100 to get the whole amount.

**Discount**

- A discount is a reduction in price.

$$\text{Percentage discount} = \frac{\text{discount}}{\text{original selling price}} \times \frac{100\%}{1}$$

- The new sale price can be obtained either by subtracting the discount amount from the original price, or by calculating the remaining percentage.
  e.g. A 10% discount means there is 90% remaining of the original selling price.

**Unit price**

- The unit price is the price per single unit, such as kg, m or item.

$$\text{Price per unit} = \frac{\text{price}}{\text{number of units}}$$

- The 'best buy' is the option with the lowest cost per unit.

## 5.5.2 Success criteria

Tick the column to indicate that you have completed the lesson and how well you have understood it using the traffic light system.

(**Green:** I understand; **Yellow:** I can do it with help; **Red:** I do not understand.)

| Lesson | Success criteria | Green | Yellow | Red |
|---|---|---|---|---|
| 5.2 | I can compare quantities using ratios. | | | |
| | I can simplify ratios. | | | |
| 5.3 | I can identify and express a rate from a written context. | | | |
| | I can simplify a rate and express the rate in its simplest form. | | | |
| | I can apply a rate to calculate the total value of an unknown quantity. | | | |
| 5.4 | I can use the unitary method to determine the best value option. | | | |
| | I can apply the unitary method to find amounts from a partial amount. | | | |

## 5.5.3 Project

### Successive discounts

Tony is a mechanic who has bought $250 worth of equipment at a tool store. Tony received 15% off the marked price of all items and received a further 5% trade discount. When two discounts are given on the same item or group of items, they are called successive discounts. In this case, Tony received successive discounts of 15% and 5%. Is this equal to a single discount of 20%?

Answer the following questions.

1. Calculate the amount that was due after Tony was given the first 15% discount.
2. From this amount, apply the trade discount of 5% to evaluate the amount that was due.
3. How much is the cash discount that Tony received?
4. Now calculate the amount that would have been due had Tony received a single discount of 20%. Is this the same answer?
5. Calculate the amount of cash discount that Tony received as a percentage of the original bill.
6. Explain whether the discount is the same if the 5% discount is applied before the 15% discount.

**on** Resources

**eWorkbook** Topic 5 Workbook (worksheets, code puzzle and project) (ewbk-1906)

**Interactivities** Crossword (int-8415)

Sudoku puzzle (int-3168)

## Exercise 5.5 Review questions

learnon

### Fluency

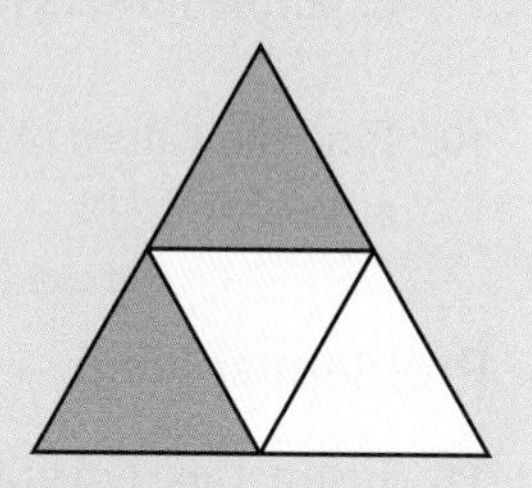

1. For the diagram shown, write the following ratios in simplest terms.
   a. shaded parts: unshaded parts
   b. shaded parts: total number of parts

2. Express in simplest form.
   a. 8 : 10 b. 14 : 21 c. 16 : 64 d. 24 : 18

3. Express the following in simplest form.
   a. $\frac{1}{2} : 4$ b. $\frac{3}{5} : \frac{1}{5}$ c. $1\frac{1}{2} : \frac{3}{4}$
   d. 0.8 : 10 e. $10^2 : 10^3$ f. $\frac{3}{5} : \frac{3}{10}$

4. Rewrite each of the following ratios in simplest terms.
   a. 4 m to 10 cm
   b. 3 oranges to 1 dozen (12) oranges
   c. 8 months to 1 year
   d. 3 cm to 12 mm
   e. 70 cents to $2.00

5. Rewrite each of the following ratios in simplest terms.
   a. 3.5 kilograms to 500 grams
   b. 1.2 cm to 10 mm
   c. $\frac{1}{2}$ of a melon to 3 melons
   d. 2 hours and 40 minutes to 30 minutes
   e. $3\frac{1}{5}$ cm to $\frac{2}{5}$ m

6. Calculate the unit rate for each of the following.
   a. If 35 copies can be made in 7 minutes, how many copies can be made in one minute?
   b. If 9 crates can hold 216 bottles, how many bottles can one crate hold?
   c. If James can drive 180 km in 3 hours and 20 minutes, how many kilometres can he drive in one hour?
   d. If 6 movie tickets cost $50.40, how much does each movie ticket cost?

7. Calculate the given rates for each of the following.
   a. The total cost of 1.5 kg of potatoes at $2.99/ kg
   b. The distance travelled at 65 km/h for 3.75 hours
   c. The time taken to travel 110 km at 80 km/h

8. When the AFL Grand Final was held in Queensland in 2020, 30 000 people attended. If there were 9872 women, 11 300 Richmond supporters, 8400 Geelong supporters and 2100 children in attendance, choose the best approximation of the ratio for each of the following.
   a. **MC** The number of Geelong supporters to the number of Richmond supporters
      **A.** 8400 : 1130 **B.** 113 : 84 **C.** 11 300 : 8400 **D.** 84 : 113
   b. **MC** The number of children to the number of adults
      **A.** 2100 : 30 000 **B.** 2100 : 9872 **C.** 7 : 93 **D.** 2100 : 19 700
   c. **MC** The number of women to the total number of people in attendance
      **A.** 1234 : 3750 **B.** 9872 : 3000 **C.** 9872 : 39 872 **D.** 617 : 1870

9. Look at the bunch of flowers and write the ratios of:
   a. red flowers to yellow flowers
   b. red flowers to non-red flowers
   c. purple flowers to white flowers.

10. Find the values of the pronumeral in the following ratios.
   a. $a : 7 = 3 : 14$ b. $12 : 4 = 9 : b$

## Problem solving

11. Justin uses 375 grams of cream cheese to make a cheesecake that serves 12 people, using the recipe shown.

   **Ingredients**
   250 g sweet plain biscuits
   125 g unsalted butter
   375 g cream cheese, softened
   1 lemon, zested
   2 tsp vanilla essence

   a. Write the ratio of cream cheese to butter.
   b. Calculate how many grams of cream cheese Justin needs to make a cheesecake that serves 8 people.

12. If it takes 15 people 12 days to complete a job, explain how long it would take 20 people to complete the same job.

13. After a 10% discount, a PlayStation costs \$320. Calculate the original cost of the PlayStation.

14. Identify which is the better value — 400 g of biscuits costing \$2.98, or a pack of biscuits with 400 g + 25% extra, costing \$3.28.

15. The rate of ascent for the Discovery space shuttle was 114 000 m in 8.0 minutes.
   a. Write this speed in km/min.
   b. Write this speed in km/h.

16. Mitch is a very fast runner who can run 3 km in 9 minutes. If Mitch is able to maintain this rate, calculate how long it will take him to run 9 km.

17. The dimensions (length and width) of rectangle B are double those of rectangle A. Explain whether the area of rectangle B is double of the area of rectangle A.

A

B

**18.** A rectangle has dimensions (length and width) that are in the ratio 3 : 5. Complete the following table to find the dimensions of the rectangle if it has an area of 135 $cm^2$. The first row has been done for you.

| Length (cm) | Width (cm) | Area ($cm^2$) |
|---|---|---|
| 3 | 5 | 15 |
| | | |
| | | |
| | | |

**19.** Tammy walks to school at an average rate (speed) of 6 km/h.

**a.** It takes her 30 minutes to walk to school. Calculate the distance from her house to the school.

**b.** If the total time she spends walking to school and back home is 70 minutes, calculate the rate at which she is walking on her way home.

**20.** At a Year 7 social, the ratio of females to males was 3 : 2. Four additional females arrived and 4 males left, and the ratio was then 2 : 1. Determine the number of people at the social.

**on** To test your understanding and knowledge of this topic, go to your learnON title at www.jacplus.com.au and complete the **post-test**.

# Answers

## Topic 5 Ratios, rates and best buys

### 5.1 Pre-test

1. 1 : 2
2. a. 7 : 3 b. 7 : 10
3. a. $\frac{5}{7}$ b. $\frac{1}{4}$ c. $\frac{1}{5}$ d. $\frac{4}{1}$
4. B
5. 20 students
6. A, B, E
7. D
8. 2 hours 25 minutes
9. The 250 g jar is cheaper by 30 cents per 100 g.
10. 2 : 3
11. a. 58 words/minute, b. 2.75 kL/h
12. True
13. a. 28 : 63 b. 9 : 28
c. 63 : 100 d. $\frac{63}{100}$
14. \$2.50
15. $1\frac{7}{8}$ hours *or* 1 hour 52 minutes 30 seconds

### 5.2 Ratios

1. a. 2 : 5 b. 1 : 3 c. 2 : 1 d. 5 : 2
2. a. 7 : 5 b. 7 : 12
3. The order is very important in ratios. By changing the order of a ratio, the meaning changes. For example, if the ratio of losses to wins is 2 : 3, the 2 corresponds to the losses and the 3 corresponds to the wins. If we change the order of the ratio to 3 : 2 then the 3 corresponds to the losses and the 2 corresponds to the wins, which is not correct.
4. a. $\frac{5}{7}$ b. $\frac{7}{50}$ c. $\frac{3}{4}$ d. $\frac{10}{3}$
5. a. 5 : 4 b. 77 : 12 c. 6 : 1 d. 8 : 7
6. a. 7 : 9 b. 1 : 1 c. 3 : 25 d. 5 : 8
7. a. 5 : 1 b. 3 : 40 c. 90 : 1 d. 3 : 2
8. 

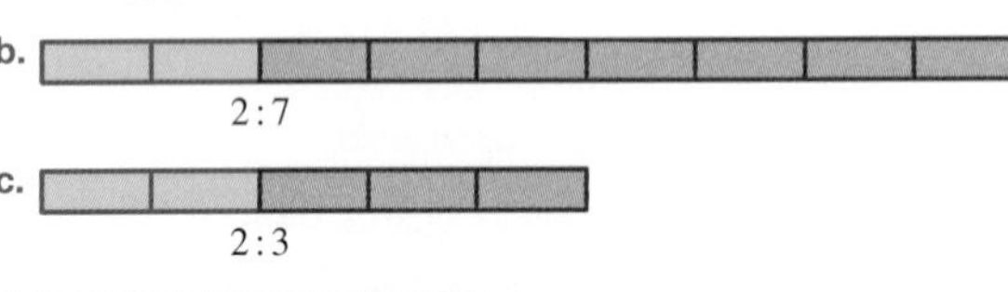


9. a. 1 : 4
2 : 8
3 : 12
4 : 16
5 : 20
b. 2 : 3
4 : 6
8 : 12
10 : 15
14 : 21
c. 64 : 32
32 : 16
16 : 8
8 : 4
2 : 1
d. 48 : 64
24 : 32
12 : 16
6 : 8
3 : 4
10. The width of the flag is 90 cm.
11. a. 27 : 20 b. 3 : 5 c. 20 : 3 d. 5 : 2
12. a. 1 : 3 = 4 : 12 b. 2 : 3 = 4 : 6
c. 66 : 48 = 11 : 8 d. 36 : 72 = 12 : 24
13. a. 20 : 12 = 5 : 3 b. 2 : 6 = 8 : 24
c. 8 : 6 = 96 : 72 d. 40 : 48 = 5 : 6
14. a. 1 : 11 b. 1 : 45 c. 3 : 2 d. 11 : 900
15. The cost of the cheaper ticket is \$320.
16. a. 77 : 23 b. 23 : 100
17. a. 37 600 female b. 47 : 31 c. $\frac{47}{31}$
18. Fourteen coins need to be added to the purse. Three silver coins and 11 gold coins will change the ratio of gold to silver.
19. a. 5 : 42 b. 3 : 5 c. 21 : 25 d. $\frac{1}{10}$
20. a. 2.6 m b. Bounce 6

### 5.3 Rates

1. a. 7 litres per minute
b. 4.71 runs per ball
c. 6 dollars per bunch
d. 20 dollars per hour
e. $\frac{100}{3}$ metres per minute
2. a. 30 km/h
b. \$25.00 per hour
c. \$1.50 per metre
3. a. 5 m/s b. \$3.60 per $m^2$
c. \$3.50 per kg d. \$37.50 per $m^2$
4. 36 km
5. \$17.00 per hour
6. 3 hours
7. \$1.50 per kg
8. 70 km/h

9. 1 hour and 36 minutes
10. \$7.50
11. 2 hours
12. 13.5 minutes or 13 minutes and 30 seconds
13. a. 4 minutes b. 8 minutes
c. 2 minutes d. 3 minutes
14. a. 17 hours b. 5 shifts c. \$45.00
15. a. \$190.00 b. \$570.00 c. \$405.00

## 5.4 The unitary method and best buys

1. The 325-g option is the best buy.
2. The 200-page option is the best buy.
3. Neither is better (they are both 62 cents per 100 g).
4. a. \$180 b. \$30 c. \$240 d. \$2250
5. a. \$422.22 b. \$400 c. \$350 d. \$600
6. a. \$180 b. \$11.40 c. \$120
7. a. \$2700 b. \$40 c. \$25 800
8. \$8.55
9. \$50
10. \$687.50
11. \$84 600
12. 500 g of cheese is the best buy.
13. \$215.29
14. a. \$54.55 b. \$70.59
15. a. \$21.33 b. \$18.13
16. a. \$2.38 b. \$2.13 c. \$3.33
17. The amount required should always be considered, as the best buy will often be for the largest quantity. Waste due to spoilage and the best before date of the item should also be taken into account. The size of the item and storage space may also need to be considered. The total cost should also be considered, particularly if shopping on a budget.
18. There are various situations in which it would be used. Generally they involve comparing different types of a certain item.
19. Brand B. The cost for each bead for Brand B is cheaper (7c compared to 7.5c).
a. \$21.33 b. \$18.13
20. \$121.97
21. Options are 50 c per ice-cream stick, 49.9 c per ice-cream stick and 54.17 c per ice-cream stick. Therefore the \$4.99 pack is the best buy.
22. $(n - 3)(b + 2) = 80$
23. \$7.27

## Project

1. \$212.50
2. \$201.88
3. \$48.13
4. No
5. 19.25%
6. Yes

## 5.5 Review questions

1. a. $2:2 = 1:1$ b. $2:4 = 1:2$
2. a. $4:5$ b. $2:3$ c. $1:4$ d. $4:3$
3. a. $1:8$ b. $3:1$ c. $2:1$
d. $2:25$ e. $1:10$ f. $2:1$
4. a. $40:1$ b. $1:4$ c. $2:3$
d. $5:2$ e. $7:20$
5. a. $7:1$ b. $6:5$ c. $1:6$
d. $16:3$ e. $2:25$
6. a. 5 copies b. 24 bottles
c. 54 km d. \$8.40
7. a. \$4.485 ≈ \$4.49 b. 243.75 km
c. 1.375 hours
8. a. D b. C c. A
9. a. $5:3$ b. $5:7$ c. $3:1$
10. a. $\frac{3}{2}$ b. 3
11. a. $3:1$ b. 250 g
12. 9 days
13. The original cost of the PlayStation was \$355.55.
14. 400 g + 25% extra
15. a. 14.25 km/minutes b. 855 km/h
16. It will take Mitch 27 minutes to run 9 km.
17. No, the area of rectangle *B* will not be double. It will be four times greater.
18.

| Length (cm) | Width (cm) | Area (cm$^2$) |
|---|---|---|
| **3** | **5** | **15** |
| 6 | 10 | 60 |
| 9 | 15 | 135 |

19. a. The school is 3 km away from her house.
b. The rate is 4.5 km/h.
20. The total number of people at the social is 60.

# 6 Decimals

LESSON SEQUENCE


6.1 Overview ..... 290
6.2 Decimals and place value ..... 292
6.3 Converting decimals to fractions and fractions to decimals ..... 300
6.4 Rounding and repeating decimals (extending) ..... 305
6.5 Adding and subtracting decimals ..... 311
6.6 Multiplying decimals ..... 317
6.7 Dividing decimals ..... 323
6.8 Decimals and percentages ..... 330
6.9 Review ..... 333

# LESSON
# 6.1 Overview

## Why learn this?

You have already seen that whole numbers cannot be used to measure everything. How much did you pay for that burger and chips? $12.60? Our money system is in dollars and cents and uses decimals to represent parts of the dollar. Decimals play an important part in our daily lives. You could have been using decimals without realising how important they are. Since our lives are so dependent on money, knowing how to work with decimals is very important. We all need to be skilled in the basic operations of addition, subtraction, multiplication and division with decimals.

Decimals are used not only with money but also with time and measurement. They are also used as another way of expressing a fraction or percentage of something. Did you know that Cathy Freeman won the gold medal in the 400-metre event at the 2000 Sydney Olympics in a time of 49.11 seconds, a winning margin of just 0.02 seconds? Usain Bolt set the men's 100-metre world record in 2009 with a time of just 9.58 seconds. Every person will use decimals in some aspect of their lives, including when checking their pay or their bank balance. Understanding and using decimals is vital in retail, hospitality, design, commerce, construction, the health industry and all areas of business.

## Exercise 6.1 Pre-test

learnon

1. **MC** Select which of the following is the largest number.
   **A.** 3.098 **B.** 3.089 **C.** 3.100 **D.** 2.999 **E.** 3.079

2. The cost of the new PlayStation has increased by 12% compared to the previous model. Write the percentage increase as a decimal.

3. **MC** The decimal 0.56 in simplest fractional form is:
   **A.** $\frac{56}{100}$ **B.** $\frac{12}{25}$ **C.** $\frac{28}{50}$ **D.** $\frac{15}{24}$ **E.** $\frac{14}{25}$

4. Josh swam his race in 22.370 seconds and Ren swam his in $22\frac{3}{8}$ seconds. State who was the fastest.

5. Write the number 23.8951 rounded to 2 decimal places.

6. Walter wants to withdraw all the money from his bank account. He has a total of $378.93, but can only withdraw an amount to the nearest 5 cents. State how much he can withdraw.

7. **MC** The third largest number from the following choices is:
   **A.** $0.276$ **B.** $0.27\dot{6}$ **C.** $0.2\dot{6}\dot{9}$ **D.** $0.2\dot{6}7\dot{9}$ **E.** $0.27\dot{5}$

8. Calculate:

$$\begin{array}{r} 2\,.\,3\;7\;5 \\ +\;3\,.\,4\;4\;2 \\ \hline \end{array}$$

9. **MC** Michael works at McDonald's. If a customer had an order of $23.75 and he handed over a $50 note, calculate the change Michael gave to the customer.
   **A.** $25.25 **B.** $26.75 **C.** $26.25 **D.** $23.75 **E.** $36.25

10. Calculate $8.2 \times 400$.

11. **MC** Callum coaches tennis on Sundays and he gets 1.5 times his usual hourly pay of $18.70. His Sunday hourly rate is:
    **A.** $37.40 **B.** $28.05 **C.** $25.70 **D.** $27.70 **E.** $27.05

12. Kat bought 2.6 kg of sausages for her BBQ at a cost of $5.25 per kg. Calculate how much Kat paid for her sausages.

13. Calculate $34.792 \div 0.4$.

14. If Mei filled her car with petrol and it cost $53.68 at $1.22 per litre, calculate how many litres of petrol she put in her car.

15. Leah pours drinks for a group of people at a volume of 0.32 litres each. If she has 40 litres of the beverage to use, calculate how many drinks she can pour.

# LESSON
# 6.2 Decimals and place value

### LEARNING INTENTION

At the end of this lesson you should be able to:
- understand the place values used in decimal numbers
- determine the value of digits in decimal numbers
- compare the sizes of decimal numbers.

## 6.2.1 Whole numbers and decimal parts

eles-3940

- Each position in a number has its own place value.

| Hundred thousands | Ten thousands | Thousands | Hundreds | Tens | Units |
|---|---|---|---|---|---|
| 100 000 | 10 000 | 1000 | 100 | 10 | 1 |

- Each place to the left of another position has a value that is 10 times larger.
- Each place to the right of another position has a value that is $\frac{1}{10}$ of the previous position.
- The position of a digit in a number gives the value of the digit.
- The table shows the value of the digit 3 in some numbers.

| Number | Value of 3 in number |
|---|---|
| 132 | 3 tens or 30 |
| 3217 | 3 thousands or 3000 |
| 4103 | 3 units (ones) or 3 |

- In a decimal number, the whole number part and the fractional part are separated by a **decimal point.**

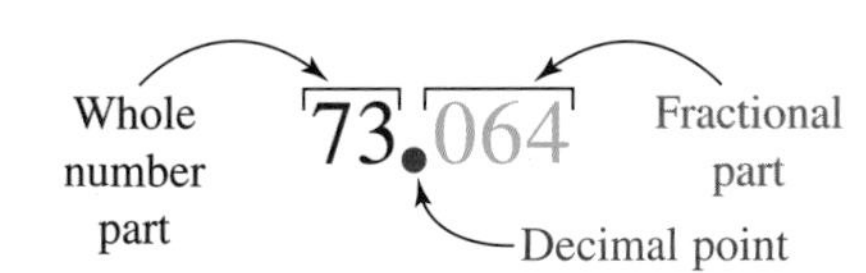


- A place value table can be extended to include decimal place values. It can be continued to an infinite number of **decimal places**.

| Thousands | Hundreds | Tens | Units | . | Tenths | Hundredths | Thousandths | Ten thousandths |
|---|---|---|---|---|---|---|---|---|
| 1000 | 100 | 10 | 1 | . | $\frac{1}{10}$ | $\frac{1}{100}$ | $\frac{1}{1000}$ | $\frac{1}{10\,000}$ |

- The table shows the value of the digit 3 in some decimal numbers.

| Number | Value of 3 in the number |
|---|---|
| 14.32 | The 3 is 3 tenths or $\frac{3}{10}$. |
| 106.013 | The 3 is 3 thousandths or $\frac{3}{1000}$. |
| 0.000 03 | The 3 is 3 hundred thousandths or $\frac{3}{100\,000}$. |

- The number of decimal places in a decimal is the number of digits after the decimal point. The number 73.064 has 3 decimal places.
- The zero (0) in 73.064 means that there are no tenths. The zero must be written to hold the place value; otherwise the number would be written as 73.64, which does not have the same value.

### DISCUSSION

How do you know how many zeros the denominator should have when calculating the value of a digit to the right of the decimal point?

### WORKED EXAMPLE 1 Stating the place value

**State the value of the 7 in each of the following.**

a. **10.74** b. **0.173** c. **321.037**

| THINK | WRITE |
|---|---|
| a. The value of the first place to the right of the decimal point is tenths, so the value of the 7 is seven tenths. | a. $\frac{7}{10}$ |
| b. The second place after the decimal point is hundredths, so the value of the 7 is seven hundredths. | b. $\frac{7}{100}$ |
| c. The third place after the decimal point is thousandths, so the value of the 7 is seven thousandths. | c. $\frac{7}{1000}$ |

### WORKED EXAMPLE 2 Expressing place values in words

**For the number 76.204, write the value of each digit in words and numbers.**

| THINK | WRITE |
|---|---|
| 1. 7 is in the tens position. | Seventy, 70 |
| 2. 6 is in the units position. | Six, 6 |
| 3. 2 is in the first position after the decimal point, so it is tenths. | Two tenths, $\frac{2}{10}$ |
| 4. 0 is in the hundredths position. | Zero hundredths, $\frac{0}{100}$ |
| 5. 4 is in the thousandths position. | Four thousandths, $\frac{4}{1000}$ |

### Expanded notation

**A number can be written in expanded notation by adding the values of each digit.**

**The number 76.204 can be written in expanded notation as:**

$$(7 \times 10) + (6 \times 1) + \left(2 \times \frac{1}{10}\right) + \left(4 \times \frac{1}{1000}\right)$$

### WORKED EXAMPLE 3 Writing in expanded notation

**Write 3.4501 in expanded notation.**

| THINK | WRITE |
|---|---|
| 1. Write the decimal. | 3.4501 |
| 2. Determine the place value of each digit.<br>3 : 3 units<br>4 : 4 tenths<br>5 : 5 hundredths<br>0 : 0 thousandths<br>1 : 1 ten thousandth | = 3 units + 4 tenths + 5 hundredths<br>+0 thousandths + 1 ten thousandth |
| 3. Write the number in expanded notation. | $= (3 \times 1) + \left(4 \times \frac{1}{10}\right) + \left(5 \times \frac{1}{100}\right) + \left(1 \times \frac{1}{10\,000}\right)$ |

## 6.2.2 Comparing decimals

eles-3941

- Decimals are compared using digits with the same place value.

### Comparing decimals

- **The decimal with the largest number in the highest place-value column is the largest decimal, *regardless of the number of decimal places.***
  **For example, 15.71 is larger than 15.702 because the first place value with different digits (moving from left to right) is hundredths, and 1 is greater than 0; that is, 15.71 > 15.702.**

| Tens | Units (ones) | . | Tenths | Hundredths | Thousandths |
|---|---|---|---|---|---|
| 1 | 5 | . | 7 | 1 | |
| 1 | 5 | . | 7 | 0 | 2 |

## WORKED EXAMPLE 4 Determining the largest number with a decimal

**State the largest number in each of the following.**

a. **0.126, 0.216, 0.122** b. **2.384, 2.388, 2.138** c. **0.506, 0.605, 0.612**

| THINK | WRITE |
|---|---|
| a. 1. As the units digit is 0 in each number, compare the tenths. The number 0.216 has 2 tenths, the others have 1 tenth so 0.216 is the largest number. | a. 0.216 is larger than 0.126 and 0.122. |
| 2. Answer the question. | The largest number is 0.216. |
| b. 1. As the units digits are the same, compare the tenths. 2.138 has 1 tenth, the others have 3 tenths, so 2.138 is the smallest. | b. 2.384 and 2.388 are both larger than 2.138. |
| 2. The hundredths are the same, so compare the thousandths. 2.388 has 8 thousandths, 2.384 has 4 thousandths, so 2.388 is the largest. | 2.388 is larger than 2.384 and 2.138. |
| 3. Answer the question. | The largest number is 2.388. |
| c. 1. As the units digits are the same, compare the tenths. 0.506 has 5 tenths, the others have 6 tenths, so 0.506 is the smallest. | c. 0.605 and 0.612 are larger than 0.506. |
| 2. Compare the hundredths. 0.612 has 1 hundredth, 0.605 has 0 hundredths, so 0.612 is the largest. | 0.612 is larger than 0.605 and 0.506. |
| 3. Answer the question. | The largest number is 0.612. |

- Decimals can also be compared using the *area model*. For example, we could use the area model to determine whether 0.67 or 0.7 is larger. From the following area model, it can be seen that 0.7 covers a greater area than 0.67 and is therefore larger.

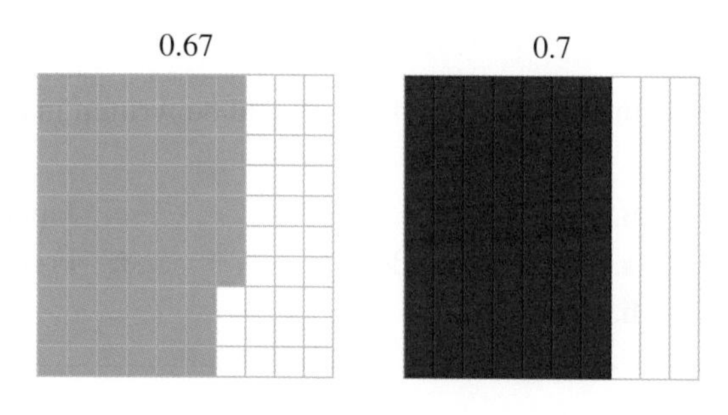


- When comparing two numbers, it is easier to use symbols instead of words, as shown in topic 4 on fractions.
- In Worked example 4a, the result '0.216 is larger than 0.126' could be written as $0.216 > 0.126$. We could also say that 0.126 is less than 0.216, or write it as $0.126 < 0.216$.

## WORKED EXAMPLE 5 Comparing numbers with decimals

**Insert the appropriate < or > sign between the following pairs of numbers to make true statements.**

a. **0.312 0.318**  b. **0.0246 0.0168**

| THINK | WRITE |
|---|---|
| a. Compare the numbers. Both numbers have the same number of tenths and the same number of hundredths, so compare thousandths and insert the correct sign. | a. $0.312 < 0.318$ |
| b. Compare the numbers. Both numbers have no tenths, so compare hundredths and insert the correct sign. | b. $0.0246 > 0.0168$ |

## DISCUSSION

You may read statements in the media quoting decimals that are not achievable every day. Some examples are 'The average number of tries scored by the Roosters is 3.6' or 'The number of children in the average family is 2.3'. Do you think these statements are reasonable?

## COLLABORATIVE TASK: Some points about decimals

In English-speaking countries, the decimal point is a full stop and spaces are left between groups of three digits: for example, 12 345.67. In many other countries, a comma is used as a decimal point and full stops are used to separate groups of three digits; for example, 12.345, 67.

In groups discuss the following:

- What are some of the problems associated with these different methods of writing decimal numbers?
- What could be done to fix these problems?
- Which method of displaying decimals do you think we should use and why?

Use the internet to investigate how the number 12 345.67 would be written in each of the following countries.

a. China  b. India  c. Japan
d. United States of America  e. United Kingdom  f. France
g. Turkey  h. Saudi Arabia

Resources

 **eWorkbook** Topic 6 Workbook (worksheets, code puzzle and project) (ewbk-1907)

 **Interactivities** Individual pathway interactivity: Place value and comparing decimals (int-4337)
Whole numbers (int-3874)
Decimal parts (int-3975)
Comparing decimals (int-3976)
Comparing decimals using the area model (int-3977)

# Exercise 6.2 Decimals and place value

learnon

6.2 Quick quiz on | 6.2 Exercise

Individual pathways

| ■ PRACTISE | ■ CONSOLIDATE | ■ MASTER |
|---|---|---|
| 1, 4, 7, 10, 13, 16, 19, 21, 24, 27 | 2, 5, 8, 11, 14, 17, 20, 23, 25, 28 | 3, 6, 9, 12, 15, 18, 22, 26, 29, 30 |

## Fluency

1. **WE1** State the value of the 2 in each of the following.
   a. 5.2  b. 19.12  c. 100.29  d. 0.982 047
2. State the value of the 2 in each of the following.
   a. 6.1342  b. 90.0002  c. 12.14  d. 1.8902
3. State the value of the 9 in each of the following.
   a. 0.9  b. 14.98  c. 18.89  d. 12.090
4. State the value of the 9 in each of the following.
   a. 3.4629  b. 1.276 89  c. 900.76  d. 9.612
5. For each of the following numbers write the value of each digit in *numbers*.
   a. 0.4  b. 2.7  c. 5.23  d. 0.763
6. For each of the following numbers write the value of each digit in *numbers*.
   a. 0.1101  b. 7.2964  c. 0.330 24  d. 300.03
7. **WE2** For the following numbers write the value of each digit in *words* and *numbers*.
   a. 1.85  b. 0.271  c. 16.001  d. 3.402 07
8. **WE3** Write the following numbers in expanded notation.
   a. 1.25  b. 56.01  c. 39.01  d. 5.987
9. Write the following numbers in expanded notation.
   a. 13.482  b. 0.3062  c. 0.5002  d. 2.47
10. **MC** Five hundredths, 2 thousandths and 7 ten thousandths equals:
    A. 527  B. 52.7  C. 5.27  D. 0.0527  E. 0.527
11. Copy and complete the table by putting only one digit in each box.

| | | Tens | Units | | Tenths | Hundredths | Thousandths |
|---|---|---|---|---|---|---|---|
| Example | 37.684 | 3 | 7 | . | 6 | 8 | 4 |
| a | 0.205 | | | . | | | |
| b | 1.06 | | | . | | | |
| c | 74.108 | | | . | | | |
| d | 0.108 | | | . | | | |
| e | 50.080 | | | . | | | |

12. **WE4** State the largest number in each of the following.
    a. 0.24, 0.32, 0.12  b. 0.57, 0.51, 0.59  c. 0.192, 0.191, 0.901

13. State the largest number in each of the following.

a. 0.0392, 0.039 90, 0.0039 b. 2.506, 2.305, 2.559 c. 0.110 43, 0.110 49, 0.110 40

14. **WE5** Insert the appropriate $<$ or $>$ sign between each of the following pairs of numbers to make true statements.

a. 3.2  2.9 b. 8.6  8.9 c. 0.64  0.67
d. 0.29  0.39 e. 13.103  13.112

15. Insert the appropriate $<$ or $>$ sign between each of the following pairs of numbers to make true statements.

a. 0.427  0.424 b. 0.580  0.508 c. 0.0101  0.0120
d. 0.048 01  0.4801 e. 1.3830  1.3824

### Understanding

16. a. State how many decimal places we use in our currency.
b. Write the amount of money shown in the photograph:
i. in words ii. in numbers.

17. Write the following in order from smallest to largest (ascending order).

a. 0.21, 0.39, 0.17, 0.45, 0.33
b. 0.314, 0.413, 0.420, 0.391, 0.502
c. 0.821, 0.803, 0.811, 0.807, 0.902

18. Write the following in order from smallest to largest (ascending order).

a. 0.9864, 0.9812, 0.9943, 0.9087, 0.9189
b. 4.6249, 4.5097, 4.802, 4.6031, 4.0292
c. 0.004 65, 0.005 02, 0.003, 0.0056, 0.009

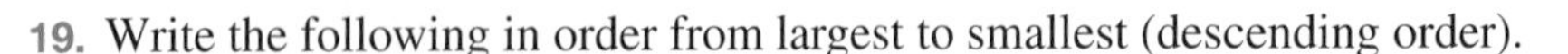

19. Write the following in order from largest to smallest (descending order).

a. 0.36, 0.31, 0.39, 0.48, 0.19
b. 0.91, 0.97, 0.90, 0.95, 0.99
c. 1.264, 1.279, 1.273, 1.291, 1.288

20. Write the following in order from largest to smallest (descending order).

a. 0.372, 0.318, 0.390, 0.309, 0.317
b. 0.8207, 0.8889, 0.8823, 0.8217, 0.8448
c. 1.349 54, 1.486 59, 1.702 96, 1.843 21, 1.486 13

21. **MC** The largest number in the list 0.4261, 0.4265, 0.4273, 0.4199, 0.3999 is:

A. 0.4261 B. 0.4199 C. 0.4265 D. 0.3999 E. 0.4273

22. **MC** The smallest number in the list 0.4261, 0.4265, 0.4273, 0.4199, 0.3999 is:

A. 0.4261 B. 0.4199 C. 0.4265 D. 0.3999 E. 0.4273

23. **MC** The list 0.4261, 0.4265, 0.4273, 0.4199, 0.3999 when arranged from smallest to largest is:

A. 0.4273, 0.4265, 0.4261, 0.4199, 0.3999
B. 0.4273, 0.4261, 0.4265, 0.4199, 0.3999
C. 0.3999, 0.4199, 0.4265, 0.4261, 0.4273
D. 0.3999, 0.4199, 0.4261, 0.4273, 0.4265
E. 0.3999, 0.4199, 0.4261, 0.4265, 0.4273

## Reasoning

24. Write True (T) or False (F) for each of the following and justify your response.
   a. 76.34 has 4 decimal places.
   b. $\frac{6}{10} + \frac{3}{100} + \frac{4}{10\,000}$ is the same as 0.6304.
   c. 4.03 has the same value as 4.3.

25. Write True or False for each of the following and justify your response.
   a. 29.60 has the same value as 29.6.
   b. 1.2804 could be written as $1 + \frac{2}{10} + \frac{8}{100} + \frac{4}{1000}$.
   c. 1090.264 51 has 5 decimal places.

26. a. Calculate the decimal values of the following fractions.

$$\frac{1}{7}, \frac{2}{7}, \frac{3}{7}, \frac{4}{7}, \frac{5}{7}, \frac{6}{7}$$

   Write as many decimals as possible.
   b. Is there a pattern? Explain.

## Problem solving

27. For each of the numbers listed from a to e:
   i. state the place value of the zero
   ii. state whether the value of the number would change if the zero wasn't there. (Write yes or no.)
   a. 6.02   b. 7.360   c. 0.65   d. 20   e. 108.62

28. Year 7 students competing in their school swimming sports recorded the following times in the 50-metre freestyle, backstroke and breaststroke events.

| | **Time (seconds) recorded by contestants** | | | | | | |
|---|---|---|---|---|---|---|---|
| **Event** | **Carolyn** | **Jessica** | **Mara** | **Jenika** | **Robyn** | **Shelley** | **Kyah** |
| Freestyle | 37.23 | 39.04 | 40.90 | 38.91 | 37.45 | 37.02 | 37.89 |
| Backstroke | 40.23 | 43.87 | 44.08 | 42.65 | 41.98 | 40.29 | 41.05 |
| Breaststroke | 41.63 | 42.70 | 41.10 | 41.21 | 42.66 | 41.33 | 41.49 |

   a. State who won the freestyle event and calculate how much they won by.
   b. State who won the backstroke event and calculate how much they won by.
   c. State who won the breaststroke event and calculate how much they won by.
   d. List the first 3 placings of the freestyle event.
   e. List the first 3 placings of the backstroke event.
   f. List the first 3 placings of the breaststroke event.
   g. Determine whether any students obtained a placing in all three events.

29. Determine my number.
My number contains four different even digits and has 3 decimal places. The digit in the thousandths position is half the value of the digit in the units position. The sum of the digits in the units and tenths positions is the same as the sum of the digits in the hundredths and thousandths positions.
The digit in the thousandths position is bigger than the digit in the tenths position.

30. Your friend said, 'You don't have to worry about the zeros in decimals; you can just leave them out.' Explain why your friend is correct or incorrect.

# LESSON
# 6.3 Converting decimals to fractions and fractions to decimals

## LEARNING INTENTION

At the end of this lesson you should be able to:
- convert decimals to fractions and fractions to decimals.

## 6.3.1 Converting decimals to fractions

eles-3943

- Decimal numbers can be written as fractions by applying knowledge of place values.

| Number | Ones | . | Tenths | Hundredths |
|---|---|---|---|---|
| 0.3 | 0 | . | 3 | 0 |
| 1.25 | 1 | . | 2 | 5 |

- The decimal number 0.3 can be written as $\frac{3}{10}$.
- The decimal number 1.25 can be thought of as $1 + \frac{2}{10} + \frac{5}{100} = 1 + \frac{20}{100} + \frac{5}{100} = 1\frac{25}{100}$.

### Converting decimals to fractions

- **The digits after the decimal point become the numerator, and the place value of the last digit gives the denominator.**
- **Simplify fractions by dividing the denominator and numerator by a common factor, e.g. $1\frac{25}{100} = 1\frac{1}{4}$.**

### WORKED EXAMPLE 6 Converting decimals to fractions

**Write the following decimals as fractions, then simplify where appropriate.**

a. **0.2** b. **0.86** c. **0.6021**

| THINK | WRITE |
|---|---|
| a. 1. Write the decimal. | a. 0.2 |
| 2. The numerator is 2 and the last decimal place is tenths so the denominator is 10. | $= \frac{2}{10}$ |
| 3. Simplify the fraction and write the answer. | $= \frac{1}{5}$ |
| b. 1. Write the decimal. | b. 0.86 |
| 2. The numerator is 86. The last decimal place is hundredths so the denominator is 100. | $= \frac{86}{100}$ |
| 3. Simplify the fraction and write the answer. | $= \frac{43}{50}$ |
| c. 1. Write the decimal. | c. 0.6021 |
| 2. The numerator is 6021. The last place is tens of thousandths so the denominator is 10 000. | $= \frac{6021}{10\,000}$ |

### WORKED EXAMPLE 7 Converting mixed numbers to fractions

**Write each of the following as a mixed number in its simplest form.**

a. **3.041** b. **7.264**

| THINK | WRITE |
|---|---|
| a. 1. Write the decimal. | a. 3.041 |
| 2. Write the whole number part and change the decimal part to a fraction. The numerator is 41. The last decimal place is thousandths so the denominator is 1000. | $= 3\frac{41}{1000}$ |
| b. 1. Write the decimal. | b. 7.264 |
| 2. Write the whole number part and change the decimal part to a fraction. The numerator is 264 and the denominator is 1000. | $= 7\frac{264}{1000}$ |
| 3. Simplify the fraction. | $= 7\frac{264 \div 8}{1000 \div 8}$ |
| 4. Write the answer in simplest form. | $= 7\frac{33}{125}$ |

## 6.3.2 Converting fractions to decimals

eles-3944

- Fractions can be written as decimal numbers by dividing the numerator by the denominator.
  For example, to convert $\frac{1}{4}$ into a decimal, divide 1 by 4 : $4\overline{)1.00}$ with quotient $0.25$
- Add trailing zeros to the numerator where required.

**WORKED EXAMPLE 8 Converting fractions to decimals**

**Change the following fractions into decimals.**

a. $\frac{2}{5}$ b. $\frac{1}{8}$

| THINK | WRITE |
| --- | --- |
| a. 1. Set out the question as for division of whole numbers, adding a decimal point and the required number of zeros. | a. $5\overline{)2.0}$ |
| 2. Divide, writing the answer with the decimal points aligned. | $5\overline{)2.0}$ with quotient $0.4$ |
| 3. Write the answer. | $\frac{2}{5} = 0.4$ |
| b. 1. Set out the question as for division of whole numbers, adding a decimal point and the required number of zeros. *Note:* $\frac{1}{8} = 1 \div 8$. | b. $8\overline{)1.000}$ with quotient $0.125$ |
| 2. Divide, writing the answer with the decimal point exactly in line with the decimal point in the question, and write the answer. | $\frac{1}{8} = 0.125$ |

- By knowing the decimal equivalent of any fraction, it is possible to determine the equivalent of any multiple of that fraction. The following worked example illustrates this.

**WORKED EXAMPLE 9 Converting multiples of a known fraction to decimals**

**Use the results of Worked example 8 to determine decimal equivalents for:**

a. $\frac{3}{8}$ b. $4\frac{5}{8}$

| THINK | WRITE |
| --- | --- |
| a. 1. Write the decimal equivalent for the fraction with 1 as the numerator. | a. $\frac{1}{8} = 0.125$ |
| 2. Multiply both sides of this equation by the appropriate multiple (3 in this case). | $\frac{1}{8} \times 3 = 0.125 \times 3$ |
| 3. Write the answer. | $\frac{3}{8} = 0.375$ |
| b. 1. Consider only the fraction part of the mixed number. Write the decimal equivalent of this fraction with 1 as the numerator. | b. $\frac{1}{8} = 0.125$ |
| 2. Multiply both sides of this equation by the appropriate multiple (5 in this case). | $\frac{1}{8} \times 5 = 0.125 \times 5$ |

3. Simplify both sides. $\frac{5}{8} = 0.625$

4. Combine with the whole number and write the answer. $4\frac{5}{8} = 4.625$

Resources

**eWorkbook** Topic 6 Workbook (worksheets, code puzzle and project) (ewbk-1907)

**Interactivities** Individual pathway interactivity: Converting decimals to fractions and fractions to decimals (int-4338)
Conversion of decimals to fractions (int-3978)
Conversion of fractions to decimals (int-3979)

## Exercise 6.3 Converting decimals to fractions and fractions to decimals

learnon

**6.3 Quick quiz** 

**6.3 Exercise**

### Individual pathways

| ■ PRACTISE | ■ CONSOLIDATE | ■ MASTER |
|---|---|---|
| 1, 4, 7, 10, 12, 15, 18 | 2, 5, 8, 11, 13, 16, 19 | 3, 6, 9, 14, 17, 20 |

### Fluency

1. **WE6** Write the following decimals as fractions, then simplify where appropriate.
   a. 0.3 b. 0.5 c. 0.21 d. 0.4 e. 0.44
2. Write the following decimals as fractions, then simplify where appropriate.
   a. 0.49 b. 0.502 c. 0.617 d. 0.30 e. 0.64
3. Write the following decimals as fractions, then simplify where appropriate.
   a. 0.28 b. 0.9456 c. 0.9209 d. 0.4621 e. 0.120
4. **WE7** Write the following decimals as mixed numbers in simplest form.
   a. 1.3 b. 2.7 c. 9.4 d. 1.2 e. 4.2
5. Write the following decimals as mixed numbers in simplest form.
   a. 8.5 b. 5.27 c. 19.182 d. 3.15 e. 6.25
6. Write the following decimals as mixed numbers in simplest form.
   a. 9.140 b. 16.682 c. 2.4917 d. 4.3386 e. 100.0048

### Understanding

7. **MC** 0.13 as a fraction is:
   A. $\frac{13}{10}$ B. $\frac{13}{100}$ C. $\frac{13}{1000}$ D. $\frac{1.3}{100}$ E. $1\frac{3}{10}$

8. **MC** 0.207 as a fraction is:

A. $\frac{207}{1000}$ B. $\frac{207}{100}$ C. $2\frac{7}{10}$ D. $20\frac{7}{10}$ E. 207

9. **MC** 0.52 as a fraction in simplest form is:

A. $\frac{52}{100}$ B. $\frac{26}{50}$ C. $\frac{13}{25}$ D. $\frac{26}{100}$ E. $\frac{13}{50}$

10. **MC** 0.716 as a fraction in simplest form is:

A. $\frac{716}{10\,000}$ B. $\frac{368}{560}$ C. $\frac{716}{1000}$ D. $\frac{179}{250}$ E. $\frac{358}{1000}$

11. **MC** 5.325 as a fraction in simplest form is:

A. $\frac{5325}{1000}$ B. $\frac{325}{1000}$ C. $5\frac{325}{1000}$ D. $5\frac{65}{200}$ E. $5\frac{13}{40}$

12. **WE8** Change the following fractions to decimals.

a. $\frac{3}{4}$ b. $\frac{1}{2}$ c. $\frac{4}{5}$ d. $\frac{1}{20}$

13. Change the following fractions to decimals.

a. $\frac{3}{12}$ b. $\frac{1}{50}$ c. $\frac{8}{25}$ d. $\frac{3}{15}$

14. Write $\frac{1}{8}$ as a decimal. Using this value, write:

a. $\frac{3}{8}$ as a decimal b. $\frac{7}{8}$ as a decimal c. $\frac{1}{16}$ as a decimal.

### Reasoning

15. The twin sisters in the photograph are keen tennis players. One of the twins has won 37 of her last 51 games and claims to be a much better player than her sister, who has won 24 out of her last 31 games.
    a. Compare the twins' results to determine who is the better player.
    b. Explain your answer to the twins.

16. In a survey, 24 people were asked: 'What is the maximum distance you would walk to a train station?' A person who says '800 m' will walk up to 800 m, but no more. The survey responses are:

| | | | | | |
|---|---|---|---|---|---|
| 100 m | 200 m | 250 m | 1 km | $\frac{1}{2}$ km | $\frac{3}{4}$ km |
| 600 m | 450 m | 500 m | 100 m | 1.2 km | 800 m |
| 1.5 km | 1.4 km | 200 m | 300 m | 1.2 km | 350 m |
| 900 m | 750 m | 300 m | 650 m | 320 m | 100 m |

Determine the longest distance (as a multiple of 100 m) that at least $\frac{3}{4}$ of the people would walk to the train station. Justify your response.

17. You are competing in a long jump contest. Your first jump was 3.78 m, your second jump was $3\frac{8}{9}$ m and your third jump is $3\frac{4}{5}$ m. You are congratulated on your third jump, as being your best. Explain whether you agree.

**Problem solving**

18. A student received the following scores in his last three tests: History $\frac{12}{17}$, English $\frac{7}{10}$ and Science $\frac{71}{101}$. Place his test scores in order from highest to lowest score (by first converting each score to a decimal).

19. Place the following fractions in order from smallest to largest by first converting the fractions to decimals:

$$\frac{3}{7}, \frac{4}{15}, \frac{8}{25}, \frac{3}{11}, \frac{30}{100}, \frac{4}{12}, \frac{325}{1000}$$

20. The decimal equivalents of $\frac{1}{7}, \frac{2}{7}, \frac{3}{7} \dots \frac{6}{7}$ can be found in the sequence 142 857 142 857 142 857 …
If $\frac{1}{7} = 0.\dot{1}42\,85\dot{7}$ and $\frac{2}{7} = 0.\dot{2}85\,71\dot{4}$, without using a calculator, write $\frac{3}{7}, \frac{4}{7}, \frac{5}{7}$ and $\frac{6}{7}$ as decimals.

# LESSON
# 6.4 Rounding and repeating decimals

**LEARNING INTENTION**

At the end of this lesson you should be able to:
- round decimal numbers to the required number of decimal places.

## 6.4.1 Rounding

eles-3945

- When rounding decimals, look at the first digit after the number of decimal places required.

**Rules for rounding**

- **If the first digit after the number of decimal places required is 0, 1, 2, 3 or 4, write the number without any change.**
- **If the first digit after the number of decimal places required is 5, 6, 7, 8 or 9, add one to the digit in the last required decimal place.**

- We can use the ≈ symbol to represent approximation when rounding has occurred, as shown in Worked example 10.

### WORKED EXAMPLE 10 Rounding decimal numbers

**Round the following to 2 decimal places.**

**a. 3.641 883** **b. 18.965 402 0**

| THINK | WRITE |
|---|---|
| **a. 1.** Write the number and underline the required decimal place. | **a.** 3.64̲1 883 |
| **2.** Circle the next digit and round according to the rule. *Note:* Since the circled digit is less than 5, we leave the number as it is. | = 3.64(1)883 <br> ≈ 3.64 |
| **b. 1.** Write the number and underline the required decimal place. | **b.** 18.96̲5 402 0 |
| **2.** Circle the next digit and round according to the rule. *Note:* Since the circled digit is greater than or equal to 5, add 1 to the last decimal place that is being kept. | = 18.96(5)4020 <br> ≈ 18.97 |

- If you need to add 1 to the last decimal place and the digit in this position is a 9, the result is 10. The 0 is put in the last required place and the 1 is added to the digit in the next place to the left. For example, 0.298 rounded to 2 decimal places is 0.30.

### WORKED EXAMPLE 11 Rounding decimals to a given number of decimal places

**Round the following to the number of decimal places shown in the brackets.**

**a. 27.462 973 (4)** **b. 0.009 94 (3)**

| THINK | WRITE |
|---|---|
| **a. 1.** Write the number and underline the required decimal place. | **a.** 27.462 9̲73 |
| **2.** Circle the next digit and round according to the rule. *Note:* Since the circled digit is greater than 5, add 1 to the last decimal place that is being kept. As 1 is being added to 9, write 0 in the last place and add 1 to the previous digit. | = 27.462 9(7)3 <br> ≈ 27.4630 |
| **b. 1.** Write the number and underline the required decimal place. | **b.** 0.009̲ 94 |
| **2.** Circle the next digit and round according to the rule. *Note:* Since the circled digit is greater than 5, add 1 to the last decimal place that is being kept. As 1 is being added to 9, write 0 in the last place and add 1 to the previous digit. | = 0.009(9)4 <br> ≈ 0.010 |

### WORKED EXAMPLE 12 Rounding decimals to the nearest unit

**Round 8.672 to the nearest unit.**

| THINK | WRITE |
|---|---|
| **1.** Write the decimal. | 8.672 |
| **2.** Look at the first digit after the decimal point. Since it is greater than 5, add 1 to the whole number. | ≈ 9 |

- When trying to answer Worked example 12, you can think of the question this way: 'Is 8.672 closer to 8 or 9?'

## 6.4.2 Rounding to the nearest 5 cents

eles-3946

- As the smallest denomination of physical money in Australia is the 5-cent coin, prices must be rounded to the nearest 5 cents upon payment.
- Rounding to the nearest 5 cents works in the same way as rounding to the nearest 10: determine whether the number is closest to 0, 5 or 10, then round accordingly.

| **Rounding guidelines for cash transactions** | |
|---|---|
| **Final cash amount** | **Round to:** |
| 1 and 2 cents | nearest 10 |
| 3 and 4 cents | nearest 5 |
| 6 and 7 cents | nearest 5 |
| 8 and 9 cents | nearest 10 |

*Source:* www.accc.gov.au

**WORKED EXAMPLE 13 Rounding to the nearest 5 cents**

**Melinda had $51.67 in her bank account. She wanted to withdraw all her money, so the bank rounded the amount to the nearest 5 cents. State how much money the teller gave Melinda.**

| THINK | WRITE |
|---|---|
| 1. Write the actual amount she had in her account. | $51.67 |
| 2. Determine whether the last digit is closer to 5 or closer to 10, then rewrite the approximate value. *Note:* Alternatively it can be seen that 67 cents is closer to 65 cents than 70 cents. | $\approx$ $51.65 |
| 3. Write the answer in a sentence. | Melinda will receive $51.65 from the bank. |

## 6.4.3 Terminating and recurring (repeating) decimals

eles-3947

- A **terminating decimal** is a decimal number that ends (terminates) after any number of decimal places; for example, $\frac{9}{20} = 0.45$ is a terminating decimal.
- A **recurring decimal**, or a repeating decimal, occurs when we divide the denominator into the numerator and the answer keeps repeating.

  For example, $\frac{1}{3} = 0.333\cdots$

  $\frac{1}{6} = 0.1666\cdots$

**WORKED EXAMPLE 14 Repeating decimals**

**Convert $\frac{1}{11}$ to a decimal. Continue dividing until a pattern emerges, then round the answer to 2 decimal places.**

| THINK | WRITE |
|---|---|
| 1. Set out the question as for division of whole numbers, adding a decimal point and enough zeros to see a pattern emerging. | $11\overline{)1.0000}$ |

2. Divide, writing the answer with the decimal point exactly in line with the decimal point in the question. (The amount left over each time is 10 then 1, then 10 then 1 again. The decimal answer is also repeating.)

$$\begin{array}{r} 0.0\ 9\ 0\ 9\ 0\ 9\ 0\ 9\ldots \\ 11\overline{\big)1.{}^{1}0{}^{10}0{}^{1}0{}^{10}0{}^{1}0{}^{10}0{}^{1}0{}^{10}0} \end{array}$$

3. Write the approximate answer rounded to 2 decimal places. $\frac{1}{11} \approx 0.09$

- Recurring decimals can be written in one of the following shorter ways for an exact answer.
  - 4.6666 ... could be written as $4.\dot{6}$ (with a dot above the repeating part of the decimal).
  - 3.512 512 ... could be written as $3.\dot{5}1\dot{2}$ (with a dot above the first and last digits of the repeating part) or as $3.\overline{512}$ (with a line above the repeating part of the decimal).
  - Like terminating decimals, the decimal equivalent of a fraction can be used to determine the decimal equivalent of any multiple of that fraction.

## WORKED EXAMPLE 15 Equivalent repeating fractions

**Use the result from Worked example 14 to calculate the decimal equivalent for $\frac{6}{11}$.**

**THINK** / **WRITE**

1. Write the decimal equivalent for the fraction with 1 as the numerator. In this case, it is an infinite recurring decimal. Therefore 0.090 909 ... can be written as $0.\overline{09}$.

$$\frac{1}{11} = 0.090\,909\ldots$$

2. Multiply both sides of this equation by the appropriate multiple (6 in this case).

$$\frac{1}{11} \times 6 = 0.090\,909\ldots \times 6$$
$$= 0.545\,454\ldots$$

3. Simplify and write the answer.

$$\frac{6}{11} = 0.\overline{54}$$

## DISCUSSION

When using a calculator to perform a calculation, for example $2 \div 3$, the answer may show an approximation to a recurring decimal, for example 0.666 666 667. Why do you think this may be?

DEG

2÷3 0.666666667

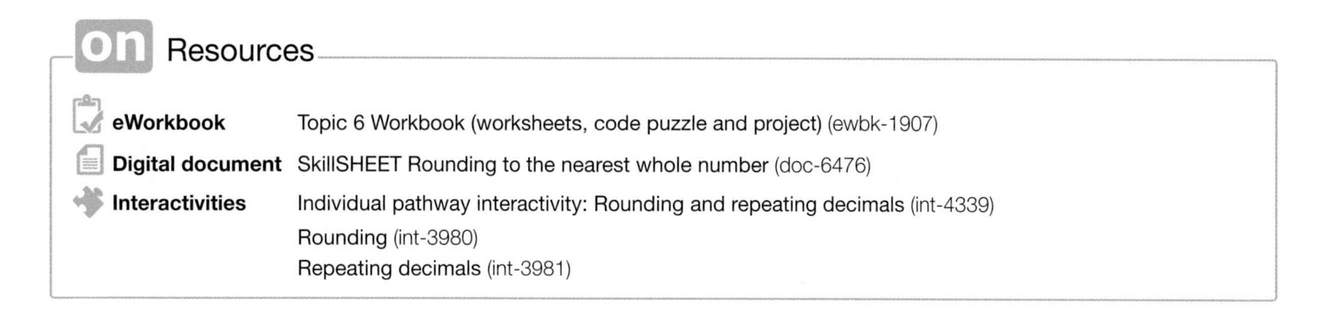


## Exercise 6.4 Rounding and repeating decimals

learnon

6.4 Quick quiz | 6.4 Exercise

### Individual pathways

| PRACTISE | CONSOLIDATE | MASTER |
|---|---|---|
| 1, 4, 7, 9, 13, 15, 18, 21, 22, 26 | 2, 5, 8, 10, 11, 16, 19, 25, 27 | 3, 6, 12, 14, 17, 20, 23, 24, 28 |

### Fluency

1. **WE10** Round the following to 2 decimal places.
   a. 0.3241 b. 0.863 c. 1.246 10 d. 13.049 92 e. 3.0409
2. Round the following to 2 decimal places.
   a. 7.128 63 b. 100.813 82 c. 71.260 39 d. 0.0092 e. 2.386 214
3. Round the following to 2 decimal places.
   a. 0.185 00 b. 19.6979 c. 0.3957 d. 0.999 e. 8.902
4. Round the following to 1 decimal place.
   a. 0.410 b. 0.87 c. 9.27 d. 25.25 e. 3.92
5. Round the following to 1 decimal place.
   a. 300.06 b. 12.82 c. 99.91 d. 8.88 e. 14.034 59
6. Round the following to 1 decimal place.
   a. 17.610 27 b. 0.8989 c. 93.994 d. 0.959 027 e. 96.280 49
7. **MC** 13.179 rounded to 2 decimal places is:
   A. 13.17 B. 13.20 C. 13.18 D. 13.27 E. 13.19
8. **MC** 2.998 rounded to 1 decimal place is:
   A. 3.0 B. 2.9 C. 2.8 D. 3.1 E. 3.9
9. **WE12** Round the following to the nearest unit.
   a. 10.7 b. 8.2 c. 3.6 d. 92.7 e. 112.1
10. Round the following to the nearest unit.
    a. 21.76 b. 42.0379 c. 2137.50 d. 0.12 e. 0.513

11. Write the following infinite recurring decimals using one of the short forms.

a. 2.555 ... b. 0.666 ⋯ c. 49.111 11 ... d. 0.262 626 ...

12. Write the following infinite recurring decimals using one of the short forms.

a. 0.913 913 ... b. 8.641 864 18 ... c. 133.946 246 2 ... d. 0.127 777 7 ...

## Understanding

13. **WE11** Convert each of the following to a decimal. Continue dividing until a pattern emerges, then round the answer to the number of decimal places indicated in the brackets.

a. $\frac{1}{3}$ (1) b. $\frac{1}{9}$ (1) c. $\frac{5}{12}$ (3)

14. Convert each of the following to a decimal. Continue dividing until a pattern emerges, then round the answer to the number of decimal places indicated in the brackets.

a. $\frac{1}{6}$ (2) b. $\frac{2}{11}$ (3) c. $\frac{7}{15}$ (2)

15. **MC** 1.8888 ... written as an exact answer is:

A. $1.\dot{8}$ B. 1.888 C. 1.88 D. 1.9 E. 1.889

16. **MC** $\frac{3}{7}$ as a decimal in exact form is:

A. $0.\dot{4}2857\dot{1}$ B. $0.\overline{428}$ C. 0.4 D. 0.428 517 4 E. $2.\dot{3}$

17. Determine decimal equivalents for the following fractions.

a. $\frac{7}{2}$ b. $4\frac{2}{3}$ c. $\frac{3}{4}$

18. Round the following to the nearest ten.

a. 13 b. 76 c. 138 d. 262 e. 175

19. Round the following to the nearest hundred.

a. 320 b. 190 c. 138 d. 6751 e. 9750.051

20. Round the following to the nearest thousand.

a. 3426 b. 12 300 c. 18 640 d. 9462 e. 1098

## Reasoning

21. **WE13** In the supermarket, Christine's shopping bill came to $27.68. As there are no 1- or 2-cent pieces, this amount must be rounded to the nearest 5 cents. State how much Christine will pay in cash for her shopping.

22. Rank the following decimals from smallest to largest:
$0.295, 0.29\dot{5}, 0.2\dot{9}\dot{5}, 0.\dot{2}9\dot{5}$

23. Write a decimal that is greater than $0.\dot{5}432\dot{1}$ and less than 0.543 22.

24. Using a calculator, Greg worked out that the piece of timber required to finish making a support for a gate should be 3.567 82 metres. Realistically, the timber can be measured only to the nearest millimetre (nearest thousandth of a metre). What measurement should be used for the length of the timber? Explain why 3.567 82 m is unreasonable as a measurement for timber.

25. Barney rounds a decimal to 0.6 correct to 1 decimal place. Fred then says that the number before being rounded off could not have been bigger than 0.64. Explain whether Fred is correct.

**Problem solving**

26. The maximum temperature was recorded as 24.7°C. In the news broadcast, the presenter quoted this to the nearest degree. Write the temperature quoted.

27. a. Round the decimal number 0.8375 to the nearest:
   i. tenth   ii. hundredth   iii. thousandth.

   b. Explain whether the rules for rounding a decimal up or down depend on the number of decimal places.

28. Divide 100 by the first 10 prime numbers. Write the answers correct to 4 decimal places.

# LESSON
# 6.5 Adding and subtracting decimals

**LEARNING INTENTION**

At the end of this lesson you should be able to:
- add and subtract decimal numbers.

## 6.5.1 Adding decimals

eles-3948

- Decimals can be added using a method similar to that for whole numbers.
  Step 1: Set out the addition in vertical columns and line up the decimal points so that the digits with the same place value are underneath each other.
  Step 2: If the question is not written in columns, it is necessary to rewrite it with the decimal points lined up.
  Step 3: Fill the empty places with zeros.

**WORKED EXAMPLE 16 Adding numbers with decimals**

**Calculate:**

a. $\begin{array}{r} 1.3 \\ +0.5 \\ \hline \end{array}$   b. $\begin{array}{r} 12.84 \\ +2.33 \\ \hline \end{array}$   c. $\begin{array}{r} 1.25 \\ 3.146 \\ +7.0 \\ \hline \end{array}$

| THINK | WRITE |
| --- | --- |
| a. Copy the question exactly and add the digits as for whole numbers, working from right to left. Write the decimal point directly below the decimal points in the question. | a. $\begin{array}{r} 1.3 \\ +0.5 \\ \hline 1.8 \\ \hline \end{array}$ |
| b. Copy the question exactly and add the digits as for whole numbers, working from right to left. Write the decimal point directly below the decimal points in the question. | b. $\begin{array}{r} 12.48 \\ +2.33 \\ \hline 14.81 \\ \hline \end{array}$ |

c. Write the question, replacing the spaces with zeros.
Add the digits as for whole numbers, working from right to left.
Write the decimal point directly below the decimal points in the question.

c.
$$\begin{array}{r} 1.250 \\ 3.146 \\ +7.000 \\ \hline 11.396 \\ \hline \end{array}$$

### WORKED EXAMPLE 17 Adding numbers with decimals

**Rewrite in columns, then add 0.26 + 1.8 + 12.214.**

**THINK**

1. Write the question in columns with the decimal points directly beneath each other with the zeros included.
2. Add the digits as for whole numbers. Write the decimal point directly below the decimal points in the question.

**WRITE**

$$\begin{array}{r} 0.260 \\ 1.800 \\ +12.214 \\ \hline 14.274 \\ \hline \end{array}$$

## 6.5.2 Subtracting decimals

eles-3949

- Decimals can be subtracted using a method similar to that for whole numbers.
- Set out the subtraction in vertical columns and line up the decimal points so that the digits with the same place value are underneath each other.
- If the question is not written in columns, it is necessary to rewrite it with the decimal points lined up.
- Fill the empty places with zeros.

### WORKED EXAMPLE 18 Subtracting numbers with decimals

**Calculate:** $\begin{array}{r} \mathbf{0.56} \\ \mathbf{-0.14} \\ \hline \end{array}$

**THINK**

Copy the question exactly and subtract the digits as for whole numbers, working from right to left. Write the decimal point directly below the decimal points in the question.

**WRITE**

$$\begin{array}{r} 0.56 \\ -0.14 \\ \hline 0.42 \\ \hline \end{array}$$

### WORKED EXAMPLE 19 Subtracting numbers with decimals

**Rewrite the following in columns, then subtract.**

a. **1.82 – 0.57**

b. **2.641 – 0.85**

**THINK**

a. Write in columns with the decimal points directly under each other. Subtract, and insert the decimal point directly below the other decimal points in the question.

**WRITE**

a.
$$\begin{array}{r} 1.\overset{7}{\cancel{8}}{}^{1}2 \\ -0.57 \\ \hline 1.25 \\ \hline \end{array}$$

**b.** Write in columns with the decimal points directly under each other, adding zeros as appropriate. Subtract as for whole numbers and insert the decimal point directly below the other decimal points.

**b.**
$$\begin{array}{r} \overset{5}{\cancel{6}} \\ 2.\cancel{6}{}^{1}41 \\ -0.850 \\ \hline 1.791 \\ \hline \end{array}$$

## Checking by estimating

- Answers to decimal addition and subtraction can be checked by estimating. Round each decimal to the nearest whole number or to a similar number of decimal places and then add or subtract them.
  For example, in Worked example 17, you could check the answer by rounding to get an estimate by adding $0.3 + 2 + 12 = 14.3$, which is close to 14.274.
  In Worked example 18, you could check the answer by rounding to get an estimate $0.6 - 0.1 = 0.5$, which is close to 0.42.

**DISCUSSION**

What is your method for estimating the answer to a decimal addition or subtraction?

 Resources

 **eWorkbook** Topic 6 Workbook (worksheets, code puzzle and project) (ewbk-1907)

 **Interactivities** Individual pathway interactivity: Adding and subtracting decimals (int-4340)
Addition of decimals (int-3982)
Checking by estimating (int-3983)
Subtraction of decimals (int-3984)

# Exercise 6.5 Adding and subtracting decimals

learnon

**6.5 Quick quiz**   **6.5 Exercise**

### Individual pathways

| ■ PRACTISE | ■ CONSOLIDATE | ■ MASTER |
| --- | --- | --- |
| 1, 3, 6, 8, 11, 15, 18, 21, 24, 25, 28, 29 | 2, 4, 9, 12, 16, 19, 22, 26, 30, 31, 32 | 5, 7, 10, 13, 14, 17, 20, 23, 27, 33, 34, 35 |

### Fluency

1. **WE16a,b** Calculate the following.

   a. $\begin{array}{r} 1.2 \\ +2.3 \\ \hline \\ \hline \end{array}$ b. $\begin{array}{r} 1.67 \\ +1.02 \\ \hline \\ \hline \end{array}$ c. $\begin{array}{r} 8.062 \\ +5.177 \\ \hline \\ \hline \end{array}$ d. $\begin{array}{r} 10.0364 \\ +92.1494 \\ \hline \\ \hline \end{array}$

2. **WE16c** Calculate the following, after filling the blank spaces.

   a. $\begin{array}{l} 6.27 \\ +0.5 \\ \hline \\ \hline \end{array}$ b. $\begin{array}{l} 3.26 \\ +18.6460 \\ \hline \\ \hline \end{array}$ c. $\begin{array}{l} \quad\;\; 4.2 \\ \;\; 62.013 \\ +1946.12 \\ \hline \\ \hline \end{array}$ d. $\begin{array}{l} \;\; 48.129\,06 \\ \quad\;\; 9 \\ +204.32 \\ \hline \\ \hline \end{array}$

3. **WE17** Rewrite the following in columns, then add. Check your answer by rounding to get an estimate.

a. $1.4+3.2$ b. $6.5+0.4$ c. $3.261+0.21$ d. $15.987+1.293$

4. Rewrite the following sums, then add. Check your answer by rounding to get an estimate.

a. $10.8271+6.5$ b. $1.8+18.6329$
c. $0.24+3.16+8.29$ d. $14.23+1.06+86.29+3.64$

5. Rewrite the following sums, then add. Check your answer by rounding to get an estimate.

a. $5.27+1.381+12.3$ b. $100+4.3+0.298+1.36$
c. $82.3+100.6+0.9949+9$ d. $126+372.8+100.0264+2020.13$

6. **WE18** Calculate the following, filling the spaces with zeros as required. Check your answer by rounding to get an estimate.

a. $\begin{array}{r} 6.87 \\ -6.27 \\ \hline \end{array}$ b. $\begin{array}{r} 12.231 \\ -8.026 \\ \hline \end{array}$ c. $\begin{array}{r} 0.6301 \\ -0.5495 \\ \hline \end{array}$ d. $\begin{array}{r} 3.0091 \\ -1.6723 \\ \hline \end{array}$

7. Calculate the following, filling the spaces with zeros as required. Check your answer by rounding to get an estimate.

a. $\begin{array}{r} 31.02 \\ -26 \\ \hline \end{array}$ b. $\begin{array}{r} 98.26 \\ 9.07 \\ \hline \end{array}$ c. $\begin{array}{r} 146 \\ -58.91 \\ \hline \end{array}$ d. $\begin{array}{r} 3.2 \\ -0.467 \\ \hline \end{array}$

8. **WE19** Rewrite the following in columns, then subtract. Check your answer by rounding to get an estimate.

a. $5.64-2.3$ b. $12.07-6.14$ c. $0.687-0.36$ d. $15.226-11.08$

9. Rewrite the following in columns, then subtract. Check your answer by rounding to get an estimate.

a. $6.734-4.8$ b. $12.2-8.911$ c. $100.562-86.0294$ d. $38-21.234$

10. Rewrite the following in columns, then subtract. Check your answer by rounding to get an estimate.

a. $47-8.762$ b. $5-0.8864$ c. $0.2-0.0049$ d. $3.279-2.506\,84$

## Understanding

11. **MC** $1.6+4.8$ equals:

A. 5.4 B. 6.4 C. 0.54 D. 0.64 E. 64

12. **MC** $3.26+0.458$ equals:

A. 3.718 B. 0.784 C. 0.037 18 D. 3.484 E. 7.84

13. **MC** $1.84+0.61+4.07$ equals:

A. 6.52 B. 6.42 C. 5.42 D. 5.52 E. 0.652

14. **MC** $216+1.38+0.002\,64$ equals:

A. 217.4064 B. 0.618 C. 217.644 D. 217.382 64 E. 21.7644

15. **MC** $0.39-0.15$ equals:

A. 0.0024 B. 0.024 C. 0.24 D. 2.4 E. 24

16. **MC** $1.4-0.147$ would be rewritten as:

A. $\begin{array}{r} 1.4 \\ -0.417 \\ \hline \end{array}$ B. $\begin{array}{r} 1.400 \\ -0.417 \\ \hline \end{array}$ C. $\begin{array}{r} 1.40 \\ -1.47 \\ \hline \end{array}$ D. $\begin{array}{r} 1.004 \\ -0.147 \\ \hline \end{array}$ E. $\begin{array}{r} 1.040 \\ -0.147 \\ \hline \end{array}$

17. **MC** 0.3 – 0.024 equals:

A. 0.06 B. 0.276 C. 0.7 D. 0.76 E. 76

18. **MC** 150.278 – 0.99 equals:

A. 150.728 B. 149.288 C. 1.49288 D. 159.388 E. 1.593 88

19. Josh deposited $27.60 into his bank account. If his balance before the deposit was $139.40, determine Josh's new bank balance.

20. Jessica bought the following items at the school canteen: 1 can of Coke for $1.60, 1 sausage roll for $1.20, 1 packet of chips for $1.50 and 2 red frogs for $0.40 (red frogs cost 20 cents each). Calculate how much Jessica spent.

21. A triathlon consists of a 0.5-kilometre swim, a 15.35-kilometre bike ride and a 4.2-kilometre run. Calculate how far the competitors travel altogether.

22. Amy walked 3.6 kilometres to school, 0.8 kilometres from school to the shops, 1.2 kilometres from the shops to a friend's house and finally 2.5 kilometres from her friend's house to her their home. Calculate how far Amy walked.

23. For lunch Paula ordered 1 potato cake, 1 dim sim, minimum chips and a milkshake from the menu below. Calculate how much Paula spent on her lunch.

| MENU | | | |
|---|---|---|---|
| Flake | $3.50 | Coffee | $2.20 |
| Whiting | $3.50 | Tea | $2.20 |
| Dim sims | $0.60 | Soft drinks | $1.80 |
| Potato cakes | $0.50 | Milkshakes | $3.00 |
| Minimum chips | $2.50 | Water | $1.80 |

24. Ryan works in a newsagency. A customer buys $9.65 worth of goods and gives Ryan a $20 note. Calculate the change Ryan gives the customer.

### Reasoning

25. A jockey has a mass of 52.3 kilograms. After exercising and training for 2 days and spending time in a sauna, the jockey has lost 1.82 kilograms. Determine the jockey's mass now.

26. If 1.27 metres is cut from a piece of material that is 13 metres long, calculate how much material is left.

27. Gary and Liz are replacing the skirting boards in their lounge room. They know the perimeter of the room is 34.28 metres. If there is a door 0.82 metres wide and a fireplace 2.18 metres wide that do not require skirting boards, explain (with working) how much wood they will need to buy for their lounge room.

## Problem solving

28. Cathy Freeman won a particular 400 metre race in 51.35 seconds. In her next race, her time was 2.97 seconds faster than this. Determine Cathy's time for the later race.

29. Without using a calculator, determine the average of the numbers 0.1, 0.11 and 0.111.

30. Explain how rounding is used in estimation.

31. Jo and Anton made a certain number of telephone calls over a 2-month period. There is a charge of 25 cents for each call, and the monthly rental of the phone line is $13.60. The bill for the 2 months comes to $39.45. Determine how many calls they made in the 2-month period.

32. The following table shows the times recorded for each swimmer in the under-13, 50-metre freestyle relay for 6 teams.

| | Times for each swimmer (seconds) | | | |
|---|---|---|---|---|
| **Team** | **Swimmer 1** | **Swimmer 2** | **Swimmer 3** | **Swimmer 4** |
| 1 | 36.7 | 41.3 | 39.2 | 35.8 |
| 2 | 38.1 | 46.5 | 38.8 | 35.9 |
| 3 | 34.6 | 39.2 | 39.9 | 35.2 |
| 4 | 41.6 | 40.8 | 43.7 | 40.5 |
| 5 | 37.9 | 40.2 | 38.6 | 39.2 |
| 6 | 38.3 | 39.1 | 40.8 | 37.6 |

a. Determine the total time for each team. Put your results in a table.
b. Identify which team won the relay.
c. Determine the difference in time between the first and second placed teams.

33. Australian coins, which are minted by the Royal Australian Mint, have the following masses:

| | | | | | |
|---|---|---|---|---|---|
| 5c | 2.83 g | 10c | 5.65 g | 20c | 11.3 g |
| 50c | 15.55 g | $1 | 9.0 g | $2 | 6.6 g |

I have 6 coins in my pocket, with a total value of $2.45. What could the total mass of these coins be?

34. You purchased some shares over a 4-month period. The price fell by $4.23 in the first month, rose by $6.67 in the second month, rose by $1.35 in the third month and fell by $3.28 in the fourth month. Explain whether the shares increased or decreased in value over the four months, and by how much.

35. Arrange the digits 1 to 9 into the boxes to make a true decimal addition.

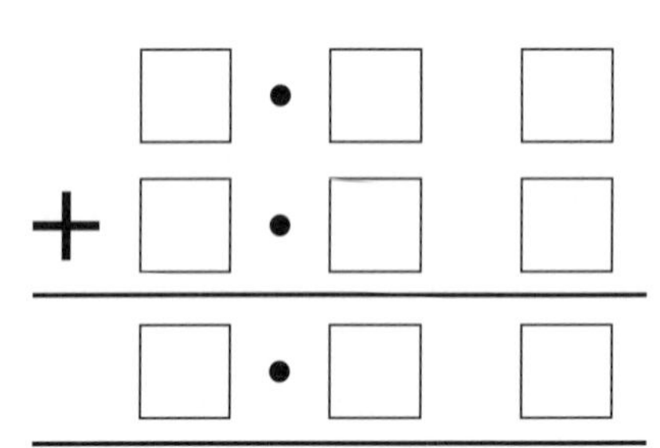

# LESSON
# 6.6 Multiplying decimals

## LEARNING INTENTION

At the end of this lesson you should be able to:
- multiply decimals, including by multiples of 10
- square decimal numbers.

## 6.6.1 Multiplying decimals

eles-3950

- The following calculation shows the multiplication $1.7 \times 2.3$. The diagram is a visual representation of each step in the calculation. There are 1.7 rows of 2.3, or 1.7 groups of 2.3.

$$\begin{array}{r l} 1.7 & \\ \times 2.3 & \\ \hline 0.51 & \leftarrow 0.3 \times 1.7 = 0.51 \\ 3.40 & \leftarrow 2.0 \times 1.7 = 3.40 \\ \hline 3.91 & \leftarrow \textbf{Total} \\ \hline \end{array}$$

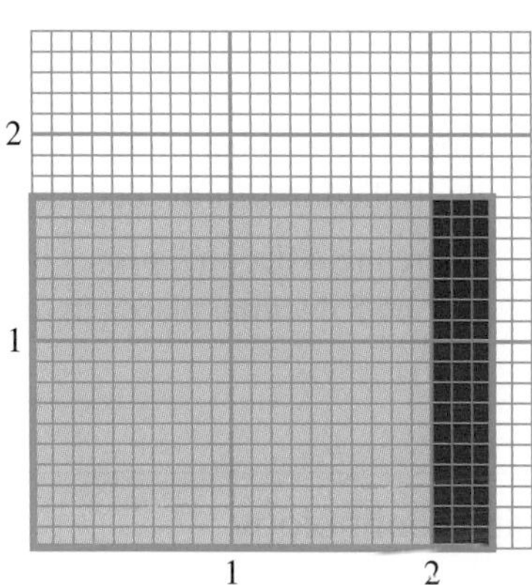


- The smallest place value in the answer is determined by multiplying the smallest place values from each of the decimal numbers. The first multiplication is $\frac{3}{10} \times \frac{7}{10} = \frac{21}{100}$, so the smallest place value in the answer will be hundredths.

### Multiplying decimals

**The number of decimal places in the answer is equal to the total number of decimal places in the numbers being multiplied.**

### WORKED EXAMPLE 20 Multiplying numbers with decimals

**Calculate the following.**

a. **$12.6 \times 7$**

b. $\begin{array}{r} \mathbf{3.26} \\ \mathbf{\times 0.4} \\ \hline \end{array}$

c. $\begin{array}{r} \mathbf{0.4629} \\ \mathbf{\times 2.6} \\ \hline \end{array}$

| THINK | WRITE |
|---|---|
| a. 1. Rewrite and multiply the digits, ignoring the decimal places. | a. $\begin{array}{r} 126 \\ \times 7 \\ \hline 882 \\ \hline \end{array}$ |
| 2. Count the number of decimal places altogether (1) and put in the decimal point. | $12.6 \times 7$<br>$= 88.2$ |

3. Check the answer by rounding: $10 \times 7 = 70$, which is close to 88.2.

**b.** 1. Multiply, ignoring the decimal places.

$$\begin{array}{r} 326 \\ \times 4 \\ \hline 1304 \\ \hline \end{array}$$

2. Count the number of digits after the point in both the decimals being multiplied and insert the decimal point in the answer. There are 2 decimal places in 3.26 and 1 in 0.4 so there will be 3 decimal places in the answer.

$3.26 \times 0.4$
$= 1.304$

3. Check the answer by rounding: $3 \times 0.4 = 1.2$, which is close to 1.304.

**c.** 1. Multiply, ignoring the decimal places.

$$\begin{array}{r} 4629 \\ \times 26 \\ \hline 27\,774 \\ 92\,580 \\ \hline 120\,354 \\ \hline \end{array}$$

2. Count the number of digits after the point in both the decimals being multiplied. Insert the decimal point in that position in the answer.

There are 4 decimal places in 0.4629 and 1 decimal place in 2.6, so there will be 5 decimal places in the answer.

$0.4629 \times 2.6$
$= 1.203\,54$

### Digital technology

Scientific calculators can evaluate multiplication of numbers with decimals.

```
DEG
0.5364*3.7
             1.98468
```

### DISCUSSION

Does multiplication always result in a larger number?

## 6.6.2 Squaring decimals

eles-3951

- To square a decimal, multiply the number by itself.

### Squaring decimals

- **The number of decimal places in the square is twice the number of decimal places in the original number.**

- The following diagrams show how squaring decimal numbers can be represented visually.

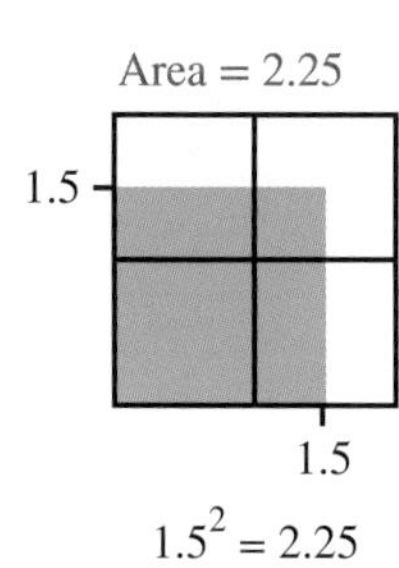


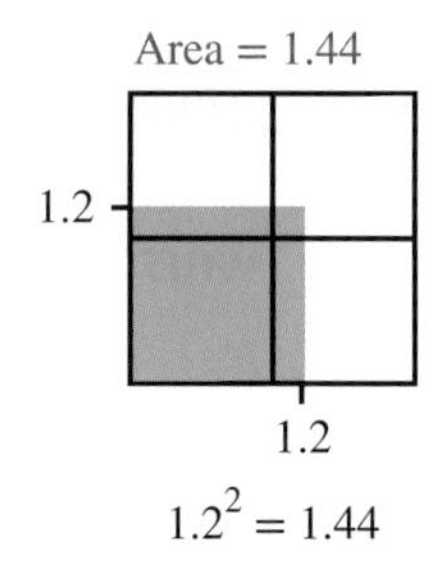


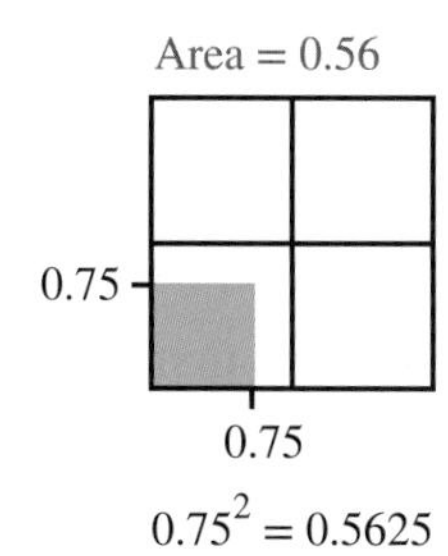


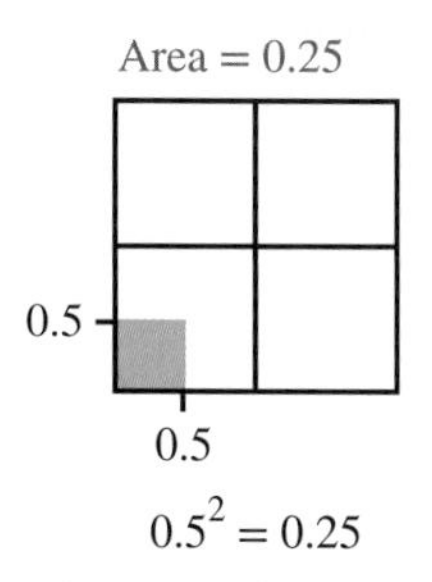


**WORKED EXAMPLE 21 Squaring decimal numbers**

**Calculate the following.**

a. $0.5^2$ b. $1.2^2$

| THINK | WRITE |
|---|---|
| a. 1. Multiply the number by itself, ignoring the decimal places. | a. $5 \times 5 = 25$ |
| 2. Count the number of digits after the point in both the decimals being multiplied and insert the decimal point in the answer. There will be 2 decimal places in the answer. | $0.5 \times 0.5 = 0.25$ |
| 3. Write the answer. | $0.5^2 = 0.25$ |
| b. 1. Multiply the number by itself, ignoring the decimal places. | b. $12 \times 12 = 144$ |
| 2. Count the number of digits after the point in both the decimals being multiplied and insert the decimal point in the answer. There will be 2 decimal places in the answer. | $1.2 \times 1.2 = 1.44$ |
| 3. Write the answer. | $1.2^2 = 1.44$ |

## 6.6.3 Multiplying by multiples of 10

eles-3952

- The multiples of 10 are 10, 20, 30, 40, ... 120, ... 1000, ... 1500, ... 20 000 and so on.
- Multiplying by a multiple of 10 can be carried out by writing the multiple of 10 as the product of two factors.
- The factors used are the non-zero digits from the multiple of 10 and the corresponding power of 10.
  For example, $8700 = 87 \times 100$, so multiplying by 8700 is the same as multiplying by 87 and then multiplying by 100.
- To multiply a number with a decimal by a power of 10, move the decimal point to the right by the number of zeros.
  For example:
  - to multiply by 10, move the decimal point one place to the right
  - to multiply by 100, move the decimal point two places to the right
  - to multiply by 1000, move the decimal point three places to the right.

### WORKED EXAMPLE 22 Multiplying by multiples of 10

**Calculate:**

a. $\mathbf{5.1 \times 600}$

b. $\mathbf{0.0364 \times 24\,000}$

| THINK | WRITE |
|---|---|
| a. 1. Multiplying by 600 is the same as first multiplying by 6 then multiplying by 100. Calculate $5.1 \times 6$. | a. $\begin{array}{r} 5.1 \\ \times\ 6 \\ \hline 30.6 \\ \hline \end{array}$ |
| 2. Multiply the result by 100. Move the position of the decimal point 2 places to the right. | $30.6 \times 100 = 3060$ |
| 3. Write the answer. | $5.1 \times 600 = 3060$ |
| b. 1. Multiplying by 24 000 is the same as first multiplying by 24 then multiplying by 1000. Calculate $0.0364 \times 24$. | b. $\begin{array}{r} 0.0364 \\ \times\ 24 \\ \hline 1456 \\ 7280 \\ \hline 0.8736 \\ \hline \end{array}$ |
| 2. Multiply the result by 1000. Move the position of the decimal point 3 places to the right. | $0.8736 \times 1000 = 873.6$ |
| 3. Write the answer. | $0.0364 \times 24\,000 = 873.6$ |

## Recognising short cuts

- When multiplying by some common decimals, such as 0.1 or 0.25, you can use short cuts to help you complete the calculation more efficiently.

  For example, the fractional equivalent of 0.1 is $\frac{1}{10}$, so instead of multiplying by 0.1 you can multiply by $\frac{1}{10}$, which is the same as dividing by 10. Similarly, the fractional equivalent of 0.25 is $\frac{1}{4}$, so 0.25 of \$60 is equivalent to $\frac{1}{4}$ of \$60, which is equivalent to $\$60 \div 4$.

**eWorkbook** Topic 6 Workbook (worksheets, code puzzle and project) (ewbk-1907)

**Video eLesson** Multiplication of decimals (eles-2311)

**Interactivities** Individual pathway interactivity: Multiplying decimals (including by multiples of 10) (int-4341)
Multiplication of decimals (int-3985)

## Exercise 6.6 Multiplying decimals

learnon

6.6 Quick quiz on | 6.6 Exercise

Individual pathways

| ■ PRACTISE | ■ CONSOLIDATE | ■ MASTER |
|---|---|---|
| 1, 4, 7, 10, 13, 15, 20, 22, 24, 27 | 2, 5, 8, 11, 14, 16, 19, 25, 28 | 3, 6, 9, 12, 17, 18, 21, 23, 26, 29 |

### Fluency

1. Calculate the following.
   a. $3.5 \times 4$ b. $15.7 \times 8$ c. $10.2 \times 6$ d. $22.34 \times 5$
2. Calculate the following.
   a. $27.18 \times 7$ b. $64.87 \times 8$ c. $1.064 \times 6$ d. $0.264\,81 \times 3$
3. **WE20a** Calculate the following.
   a. $1.4 \times 0.6$ b. $4.2 \times 0.7$ c. $0.8 \times 0.4$ d. $9.7 \times 0.8$
4. Calculate the following.
   a. $0.35 \times 0.4$ b. $0.64 \times 0.3$ c. $0.77 \times 0.5$ d. $0.49 \times 0.9$
5. Calculate the following.
   a. $5.38 \times 0.8$ b. $0.347 \times 0.6$ c. $4.832 \times 0.6$ d. $12.2641 \times 0.4$
6. Calculate the following.
   a. $0.002 \times 0.05$ b. $0.003 \times 0.004$ c. $0.7 \times 0.09$ d. $0.037 \times 0.006$
7. Calculate the following.
   a. $0.000\,061 \times 0.04$ b. $0.004 \times 0.09$ c. $0.56 \times 0.7$ d. $0.030\,31 \times 0.02$
8. **WE20b** Calculate the following.
   a. $0.25 \times 1.2$ b. $0.37 \times 2.3$ c. $0.47 \times 5.4$ d. $0.05 \times 3.5$
9. Calculate the following.
   a. $0.79 \times 8.3$ b. $4.68 \times 3.6$ c. $8.04 \times 7.5$ d. $5.06 \times 6.2$
10. **MC** When calculating $8.32 \times 0.64$, the number of decimal places in the answer is:
    A. 0 B. 1 C. 2 D. 3 E. 4
11. **MC** $0.2 \times 0.2$ equals:
    A. 0.004 B. 0.04 C. 0.4 D. 4 E. 40
12. **MC** $1.4 \times 0.8$ equals:
    A. 1.12 B. 8.2 C. 82 D. 11.2 E. 112
13. **MC** $0.0312 \times 0.51$ equals:
    A. 0.001 591 2 B. 0.015 912 C. 0.156 312 D. 0.159 12 E. 1.5912

### Understanding

14. **WE21** Calculate the following.
    a. $0.02^2$ b. $1.3^2$ c. $2.05^2$

6. Divide 4 into the fourth digit ($9 \div 4 = 2$ remainder 1). Write the 2 above the 9, and write the remainder beside the next digit, as shown in purple.

$$\begin{array}{r} 0.622 \\ 4\overline{)2.^{2}489^{1}6} \end{array}$$

7. Divide 4 into the fifth digit, which includes the carried 1 ($16 \div 4 = 4$). Write the 4 above the 6, as shown in orange.

$$\begin{array}{r} 0.6224 \\ 4\overline{)2.^{2}489^{1}6} \end{array}$$

8. Write the answer.

$2.4896 \div 4 = 0.6224$

---

- Sometimes, when you are dividing numbers, there will be a remainder. For example $15.3 \div 4$:

$$\begin{array}{r} 3.8 \\ 4\overline{)15.^{3}3} \end{array} \quad \text{remainder } 1$$

- Instead of leaving a remainder, add zeros to the end of the decimal and keep dividing until there is no remainder.

$$\begin{array}{r} 3.\ 8\ 2\ 5 \\ 4\overline{)15.^{3}3^{1}0^{2}0} \end{array}$$

**WORKED EXAMPLE 24 Dividing a decimal by a whole number**

**Calculate $21.76 \div 5$. Add zeros and keep dividing until there is no remainder.**

**THINK**

1. Set up the division. Write the decimal point in the answer directly above the decimal point in the question and divide as for short division, adding zeros as required.
2. Check the answer by rounding: $20 \div 5 = 4$, which is close to 4.352.

**WRITE**

$$\begin{array}{r} 4\ .\ 3\ 5\ 2 \\ 5\overline{)21.\ ^{1}7^{2}6^{1}0} \end{array}$$

## 6.7.2 Dividing a decimal number by a multiple of 10

eles-3955

- When dividing by a multiple of 10, factorise the multiple to give a power of 10 and its other factor. Divide by the other factor first, and then by the power of 10.
  For example, to divide by 500, divide first by 5 and then by 100.
- To divide by a power of 10, move the decimal point to the left by the number of zeros.
  For example, to divide by 10, move the decimal point one place to the left;
  to divide by 100, move the decimal point two places to the left;
  to divide by 1000, move the decimal point three places to the left, and so on.

**WORKED EXAMPLE 25 Dividing a decimal by a multiple of 10**

**Calculate:**

**a. $4.8 \div 40$** **b. $19.2 \div 6000$**

**THINK**

a. 1. Dividing by 40 is the same as first dividing by 4 then dividing by 10. First, divide by 4.

2. To divide by 10, move the position of the decimal point place to the left.

3. Write the answer.

b. 1. Dividing by 6000 is the same as dividing by 6 then dividing by 1000. First, divide by 6.

**WRITE**

a.
$$\begin{array}{r} 1.2 \\ 4\overline{)4.8} \end{array}$$

$1.2 \div 10 = 0.12$

$4.8 \div 40 = 0.12$

b.
$$\begin{array}{r} 3\ .2 \\ 6\ \overline{)19.2} \end{array}$$

2. To divide by 1000, move the position of the decimal point 3 places to the left. $3.2 \div 1000 = 0.0032$
3. Write the answer. $19.2 \div 6000 = 0.0032$

**COLLABORATIVE TASK: Decimal division patterns**

1. As a class, work out the following divisions using a calculator and write each part with its answer on the board.
   a. $10 \div 0.1$ b. $10 \div 0.2$ c. $10 \div 0.3$ d. $10 \div 0.4$ e. $10 \div 0.5$
   f. $10 \div 0.6$ g. $10 \div 0.7$ h. $10 \div 0.8$ i. $10 \div 0.9$ j. $10 \div 1.0$
2. Have you noticed any pattern in your answers? Discuss.
3. Now repeat the calculations, this time using $20 \div 0.1, 20 \div 0.2$ and so on.
4. Discuss with your class what is happening when you divide by a decimal.
5. Make up some divisions of your own and check if your assumptions about dividing by a decimal hold true.

## 6.7.3 Dividing a decimal number by another decimal number

eles-3956

- A visual representation of $2.724 \div 0.4$ is shown below.

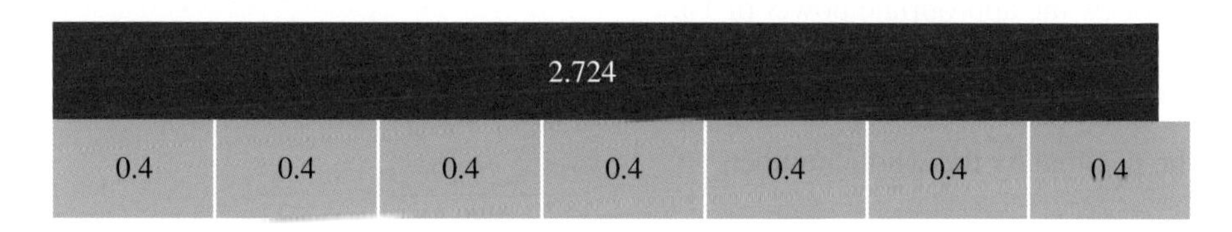


$2.724 \div 0.4$ can be interpreted as 'how many times does 0.4 divide into 2.724?'

$$2.724 \div 0.4 = 6.81$$

### Dividing one decimal number by another

- **When dividing one decimal by another, multiply the decimal you are dividing by (divisor) by a power of 10 to make it a whole number.**
  **Multiply the other decimal *by the same power of* 10, as shown in the following example.**

First multiply this number by 10 to make a whole number.

$$\mathbf{2.724 \div 0.4}$$

Then multiply this number by 10 also.

- **This is the same as writing an equivalent fraction with a whole number as the denominator:**

$$\frac{\mathbf{2.724}}{\mathbf{0.4}} \times \frac{\mathbf{10}}{\mathbf{10}} = \frac{\mathbf{27.24}}{\mathbf{4}}$$

## WORKED EXAMPLE 26 Dividing a decimal by a decimal

**Calculate:**

**a. 26.724 ÷ 0.4** **b. 3.0276 ÷ 0.12**

| THINK | WRITE |
|---|---|
| **a. 1.** Rewrite the question as a fraction. | **a.** $26.724 \div 0.4 = \dfrac{26.724}{0.4}$ |
| **2.** Multiply both the numerator and the denominator by the appropriate power of 10. | $= \dfrac{26.724}{0.4} \times \dfrac{10}{10}$ <br> $= \dfrac{267.24}{4}$ |
| **3.** Divide the decimal by the whole number. | $\begin{array}{r} 6\ 6.\ 81 \\ 4\overline{)26^{2}7.^{3}24} \end{array}$ |
| **4.** Write the answer. | $26.724 \div 0.4 = 66.81$ |
| **b. 1.** Rewrite the question as a fraction. | **b.** $3.0276 \div 0.12 = \dfrac{3.0276}{0.12}$ |
| **2.** Multiply both the numerator and the denominator by the appropriate power of 10. | $= \dfrac{3.0276}{0.12} \times \dfrac{100}{100}$ <br> $= \dfrac{302.76}{12}$ |
| **3.** Divide the decimal by the whole number. | $\begin{array}{r} 2\ 5.\ 2\ 3 \\ 12\overline{)30^{6}2.^{2}7^{3}6} \end{array}$ |
| **4.** Write the answer. | $3.0276 \div 0.12 = 25.23$ |

### Digital technology

Scientific calculators can evaluate division of numbers with decimals.

DEG

34.854÷2.4

14.5225

## WORKED EXAMPLE 27 Dividing a decimal by a decimal

**Calculate how many litres of petrol could be purchased for $50.88 if 1 litre costs $1.06.**

| THINK | WRITE |
|---|---|
| **1.** Write the question. | $50.88 \div 1.06$ |
| **2.** Rewrite the question as a fraction. | $= \dfrac{50.88}{1.06}$ |

| | |
|---|---|
| 3. Multiply the numerator and denominator by the appropriate multiple of 10, in this case 100. Alternatively, the decimal point could be moved twice to the right in both numbers so that the divisor is a whole number ($50.88 \div 1.06 = 5088 \div 106$). | $= \frac{50.88}{1.06} \times \frac{100}{100}$<br>$= \frac{5088}{106}$ |
| 4. Divide the decimal by the whole number. Alternatively, use a calculator. | $= 106\overline{)508^{84}8}$ with quotient $48$<br>$50.88 \div 1.06 = 48$ |
| 5. Write the answer in a sentence. | Forty-eight litres of petrol could be purchased for \$50.88. |

Resources

**eWorkbook** Topic 6 Workbook (worksheets, code puzzle and project) (ewbk-1907)

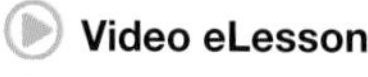

**Video eLesson** Division of decimals (eles-1877)

**Interactivities** Individual pathway interactivity: Dividing decimals (including by multiples of 10) (int-4342)
Division of decimals by a multiple of 10 (int-3990)
Division of a decimal by another decimal (int-3991)

## Exercise 6.7 Dividing decimals

learnon

**6.7 Quick quiz** 

**6.7 Exercise**

### Individual pathways

| ■ PRACTISE | ■ CONSOLIDATE | ■ MASTER |
|---|---|---|
| 1, 4, 10, 13, 15, 17, 20, 23, 24, 25 | 2, 5, 6, 8, 11, 14, 18, 21, 26, 27 | 3, 7, 9, 12, 16, 19, 22, 28, 29, 30 |

### Fluency

1. **WE23** Calculate:
   a. $3.6 \div 6$  b. $21.7 \div 7$  c. $4.86 \div 9$  d. $8.05 \div 5$  e. $9.68 \div 4$
2. Calculate:
   a. $1.576 \div 2$  b. $8.029 \div 7$  c. $32.5608 \div 8$  d. $20.5782 \div 3$  e. $126.4704 \div 4$
3. Calculate:
   a. $46.80 \div 15$  b. $24.541 \div 11$  c. $17.108 \div 14$  d. $77.052 \div 12$  e. $121.3421 \div 11$
4. **WE24** Calculate the following. In each case, add zeros and keep dividing until there is no remainder.
   a. $3.7 \div 2$  b. $9.5 \div 2$  c. $7.3 \div 5$
5. Calculate the following. In each case, add zeros and keep dividing until there is no remainder.
   a. $9.8 \div 4$  b. $7.5 \div 6$  c. $55.6 \div 8$

6. Calculate the following by changing the position of the decimal point.

a. $14.07 \div 10$
b. $968.13 \div 100$
c. $620.8 \div 1000$
d. $3592.87 \div 1000$
e. $2349.78 \div 100\,000$

7. Calculate the following by changing the position of the decimal point.

a. $9.0769 \div 100\,000$
b. $802\,405.6 \div 1\,000\,000$
c. $152.70 \div 1\,000\,000$
d. $0.7205 \div 10\,000$
e. $0.0032 \div 1\,000\,000$

8. **WE25** Calculate the following.

a. $15.9 \div 60$
b. $23.7 \div 30$
c. $238 \div 400$
d. $8.79 \div 6000$
e. $5.22 \div 3000$

9. **WE26a** Calculate each of the following.

a. $2.5 \div 0.5$
b. $4.2 \div 0.6$
c. $8.1 \div 0.9$
d. $2.8 \div 0.7$
e. $0.248 \div 0.8$

10. Calculate each of the following.

a. $1.32 \div 0.6$
b. $39.6 \div 0.6$
c. $57.68 \div 0.8$
d. $0.2556 \div 0.3$
e. $0.5468 \div 0.4$

11. **WE26b** Calculate:

a. $172.0488 \div 0.11$
b. $0.510\,48 \div 0.12$
c. $21.470\,10 \div 0.15$

12. Calculate:

a. $142.888 \div 0.08$
b. $0.028\,692 \div 0.06$
c. $32.619 \div 0.02$

## Understanding

13. **MC** To calculate $9.84 \div 0.8$, it can be rewritten as:

A. $9.84 \div 8$
B. $0.984 \div 0.8$
C. $98.4 \div 0.8$
D. $98.4 \div 8$
E. $984 \div 8$

14. **MC** To calculate $151.368 \div 1.32$, it can be rewritten as:

A. $151.368 \div 132$
B. $151.368 \div 13.2$
C. $1513.68 \div 132$
D. $15\,136.8 \div 132$
E. $151\,368 \div 132$

15. **MC** $0.294 \div 0.7$ equals:

A. 0.042
B. 0.42
C. 4.2
D. 42
E. 420

16. **MC** $21.195 \div 0.15$ equals:

A. 0.1413
B. 1.413
C. 14.13
D. 141.3
E. 1413

17. For the following calculations:

i. Estimate the answer by rounding each number to the first digit.
ii. Calculate the answers by using written methods.
iii. Check your answer by using a calculator.

a. $6600 \div 24$
b. $1044 \div 4.5$
c. $384 \div 0.16$

18. Change the following to dollars ($) by dividing by 100.

a. 365 cents
b. 170 cents
c. 5685 cents
d. 75 cents
e. 90 cents
f. 6350 cents

19. Calculate $100\,987.5412 \div 10^4$.
Round the answer to the nearest hundredth.

## Reasoning

20. The carat is a unit of measure used to weigh precious stones. A 2.9-carat diamond weighs 8.9494 grains. Calculate how many grains one carat is equal to.

21. You are helping to beautify the gardens at your school and have been given the responsibility for paving the path. The path is 6.715 m (or 671.5 cm) long and the width of only two pavers. The pavers you will use are shown.

a. Calculate how many pavers you should buy to minimise waste.
b. If each paver costs $2.32, calculate how much the pavers cost in total.

22. A jug is $\frac{1}{5}$ full of water. Jack added more water to the jug until it was $\frac{3}{4}$ full. At that stage there was 1.5 L of water in the jug. Show with working how much water was originally in the jug.

## Problem solving

23. Stephanie spent $6.95 on these chocolates. Determine the cost of each chocolate. Give your answer to the nearest 5 cents.

24. WE27 Calculate how many 1.25-litre bottles of water could be poured into a 25-litre drink dispenser.
25. Calculate how many burgers could be bought for $562.80 if each burger costs $2.80.
26. If you have $22.50 for bus fares to school for the week, calculate how much you would spend on each of the 5 days.
27. The area of an office floor is 85.8 square metres. Calculate how many people could fit in the office if each person takes up 1.2 square metres.
28. Emily wants to make 10 cushions from 6.75 metres of material that she found on a table of remnants at a fabric shop. Determine how much material she would have for each cushion.
29. Calculate how many books can be stacked on a shelf that is 28.6 centimetres high if each book is 1.1 centimetres high.

30. Anna has a wooden rod that is 2.4 metres long. She needs to cut 3 lengths, each of 0.4 metres, and 2 lengths, each of 0.2 metres. Determine the fraction of the original rod left over after she has cut the required lengths.

# LESSON
# 6.8 Decimals and percentages

## LEARNING INTENTION

At the end of this lesson you should be able to:
- convert percentages to decimals.

## 6.8.1 Percentages as decimals

eles-3957

- Percentages can be expressed as decimals.
- To convert a percentage to a decimal, divide the percentage by 100; that is, move the decimal point 2 spaces to the left.

### WORKED EXAMPLE 28 Converting percentages to decimals

**Write the following percentages as decimals.**

**a. 81%** **b. 16.8%**

| THINK | WRITE |
|---|---|
| **a. 1.** To convert to a decimal, divide by 100. | **a.** $81\% = 81 \div 100$ |
| **2.** Move the position of the decimal point 2 places to the left and place a zero in the units column. | $= \frac{81}{100}$<br>$= 0.81$ |
| **b. 1.** To convert to a decimal divide by 100. | **b.** $16.8\% = 16.8 \div 100$ |
| **2.** Move the position of the decimal point 2 places to the left and place a zero in the units column. | $= \frac{16.8}{100}$<br>$= 0.168$ |

**on** Resources

**eWorkbook** Topic 6 Workbook (worksheets, code puzzle and project) (ewbk-1907)

**Interactivities** Individual pathway interactivity: Percentages as decimals (int-4344)

Percentages as decimals (int-3993)

# Exercise 6.8 Decimals and percentages

6.8 Quick quiz on | 6.8 Exercise

Individual pathways

| PRACTISE | CONSOLIDATE | MASTER |
|---|---|---|
| 1, 4, 8, 10, 12, 15, 18, 21 | 2, 5, 7, 13, 16, 19, 22 | 3, 6, 9, 11, 14, 17, 20, 23, 24 |

## Fluency

1. **WE28a** Write the following percentages as decimals.
   a. 36% b. 14% c. 28% d. 73% e. 66%
2. Write the following percentages as decimals.
   a. 59% b. 99% c. 9% d. 4% e. 1%
3. Write the following percentages as decimals.
   a. 25% b. 200% c. 150% d. 360% e. 100%
4. **WE28b** Write the following percentages as decimals.
   a. 12.3% b. 31.6% c. 59.2% d. 84.9% e. 42.1%
5. Write the following percentages as decimals.
   a. 37.6% b. 21.9% c. 16.9% d. 11.1% e. 3.1%
6. Write the following percentages as decimals.
   a. 9.2% b. 5.9% c. 8.8% d. 14.25% e. 31.75%
7. Write the following percentages as decimals.
   a. 23.55% b. 45.75% c. 0.05% d. 4.01% e. 0.02%

## Understanding

8. **MC** 41% as a decimal is:
   A. 41 B. 0.41 C. 4.100 D. 4.1 E. 0.041
9. **MC** 8% as a decimal is:
   A. 8 B. 0.008 C. 0.08 D. 0.8 E. 800
10. **MC** 43.64% as a decimal is:
    A. 0.4364 B. 4.364 C. 43.6400 D. 436.4 E. 4364.0
11. **MC** 110% as a decimal is:
    A. $\frac{110}{100}$ B. 1100 C. 0.11 D. 1.10 E. 11.0
12. Car prices dropped by 17% during a recent 8-year period. Write this percentage as a:
    a. fraction b. decimal.
13. The maximum legal blood alcohol concentration (BAC) for drivers in New South Wales is 0.05%. Write the BAC as a:
    a. fraction b. decimal.
14. In 2020, the number of visitors to Australia was 80.7% lower than the previous year. Write this percentage as a decimal.

15. Over the past decade, insurance premiums have risen by 218%. Write this percentage as a decimal.

16. In the three months following a drought, the price of vegetables fell by 13.8%. Write this percentage as a decimal.

17. Government funding to state schools has risen by 8.35%. Write this percentage as a decimal.

### Reasoning

18. a. Round the following percentages to the nearest whole number: 23.2%, 23.5%, 23.8%, 23.9%
    b. Round the following decimal numbers to 2 decimal places: 0.232, 0.235, 0.238, 0.239
    c. The percentages in part a are the same as the decimal numbers in part b. Explain whether it matters if you round the percentages or the decimal values up or down.

19. Consider the following percentages: 23.6%, $23.\dot{6}7\%$, 23.66%
    a. Which one of the three percentages is the largest? Justify your answer.
    b. Convert these percentages to decimals. State your answer to 4 decimal places.

20. Consider the following percentages: 37.8%, 39.6%, 30.9%, 34.5%, 32.8%
    a. Convert these numbers to decimals.
    b. Arrange the percentages in order from highest to lowest.
    c. Arrange the decimal numbers in order from highest to lowest.
    d. Explain whether it was easier to arrange the numbers in the order required using decimals or percentages.

### Problem solving

21. Brent buys a used car for $7500. The dealer requires a 15% deposit. Calculate the value of the deposit they must pay the dealer.

22. The price of entry into a theme park has increased by 10% every year since the theme park opened.
    If the latest price rise increased the ticket price to $8.80, determine the cost of a ticket 2 years ago.

23. Consider the following percentages:
    7.291%, 72.91%, 729.1%.
    a. Convert the following percentages to decimal numbers.
       i. 7.291% ii. 72.91% iii. 729.1%
    b. Explain the effect of the conversion on the numbers.

24. Suggest when it is more convenient to express a percentage as a decimal rather than as a fraction.

# LESSON
# 6.9 Review

## 6.9.1 Topic summary

### Decimal numbers and place value

- Decimal numbers can be identified easily because the whole number part and the fractional part are separated by a decimal point (.).
  e.g. 11.234 is a decimal number.
- Decimal numbers can also be written in expanded form.
  e.g. $11.234 = 11 + \frac{2}{10} + \frac{3}{100} + \frac{4}{1000}$

### Rounding decimals

When rounding decimal numbers, follow the steps below.

1. Decide which digit is the last digit to keep (underline it).
2. Leave this underlined digit the same if the **next digit** is 0, 1, 2, 3 or 4 (this is called rounding down).

- Increase the underlined digit by 1 if the **next digit** is 5, 6, 7, 8 or 9 (this is called rounding up).

e.g. $2.\underline{3}456 = 2.3$ correct to 1 decimal place.
$2.3\underline{4}56 = 2.35$ correct to 2 decimal places.

### Types of decimals

- **Terminating (finite) decimals** are decimal numbers that finish (terminate).
  e.g. $\frac{3}{8} = 0.375$
- **Recurring (repeating) decimals** are decimal numbers that keep on going forever and never finish.
  e.g. $\frac{1}{3} = 0.3333333\ldots = 0.\dot{3}$
  (The 3 never stops.)
  $\frac{5}{11} = 0.45454545\ldots = 0.\overline{45}$
  (The 45 pattern continues forever.)

## DECIMALS

### Addition and subtraction

- To add or subtract decimal numbers, set the problem out in columns according to place value and make sure the decimal points are aligned.
- Add and subtract using the same method as for whole numbers.

e.g. Calculate 1.23 + 7.894.

$$\begin{array}{r} 1.230 \\ +\,7.894 \\ \hline 9.124 \\ \hline \end{array}$$

### Multiplying decimal numbers

- To multiply decimal numbers, use the method for multiplying whole numbers.
- To know where to place the decimal point in the answer, *count the total number of decimal places in the entire question.*
- The answer must have the same number of decimal places as your count.

### Multiplying by powers of 10

- Count the number of zeros in the power of 10 being multiplied.
  e.g. 1000 has three zeros.
- Move the decimal point to the *right* by the same number.

e.g. $23.4567 \times 1000 = 23456.7$
(1000 has three zeros, so move the decimal point three places to the right.)

### Dividing by powers of 10

- Count the number of zeros in the power of 10 being divided.
  e.g. 100 has two zeros.
- Move the decimal point to the *left* by the same number.

e.g. $23.4567 \div 100 = 0.234567$
(100 has two zeros, so move the decimal point two places to the left.)

### Converting percentages to decimal

- To convert a percentage to a fraction or a decimal, simply divide by 100 and remove the percentage symbol.

e.g. Express 54% as a decimal. number

$54\% = 54 \div 100$
$= 0.54$

(Follow the rules for dividing by powers of 10; that is, move the decimal point two places to the left.)

### Division by a whole number

- When dividing a decimal number by a whole number, use the same method as for division with whole numbers.

e.g. $5\overline{)10.375}$ = 2.075

### Dividing decimal numbers

- When dividing by a decimal number, convert the denominator to a whole number by multiplying by a power of 10.
  (Put simply, just move the decimal point to the right as many places as needed to make the denominator a whole number.)
- Multiply the numerator by the same power of 10 as the denominator to keep it balanced.
  (Move the decimal point in the numerator exactly the same number of places to the right as you did in the denominator.)
- Divide as normal.

e.g. Calculate $\frac{0.135}{0.5}$.
To make the denominator a whole number, multiply it by 10 (or move the decimal point one place to the right). Do the same to the numerator.

$\frac{0.135}{0.5} = \frac{1.35}{5} = 0.27$

## 6.9.2 Success criteria

Tick the column to indicate that you have completed the lesson and how well you think you have understood it using the traffic light system.

(**Green:** I understand; **Yellow:** I can do it with help; **Red:** I do not understand)

| Lesson | Success criteria | Green | Yellow | Red |
|---|---|---|---|---|
| 6.2 | I can understand the place value of digits in decimal numbers. | | | |
| | I can the value of digits in decimal numbers. | | | |
| | I can compare the sizes of decimal numbers. | | | |
| 6.3 | I can convert decimals to fractions and fractions to decimals. | | | |
| 6.4 | I can round decimal numbers to the required number of decimal places. | | | |
| 6.5 | I can add and subtract decimal numbers. | | | |
| 6.6 | I can multiply decimals, including by multiples of 10. | | | |
| | I can square decimal numbers. | | | |
| 6.7 | I can divide a decimal number by a whole number or a decimal number. | | | |
| | I can divide a decimal number by a multiple of 10. | | | |
| 6.8 | I can convert percentages to fractions. | | | |

## 6.9.3 Project

### Eating out

When we eat out or buy takeaway food, a number of factors are taken into consideration before we make our purchases. What types of foods are available? What is the size of each serving? What is the cost of the meal? These are examples of things we consider when we try to decide what and where we eat.

In a group of four, you and your friends have decided to eat at a local restaurant. Your parents have given each of you $30.00 to spend on your meal.

# EAT EASY RESTAURANT MENU

## SALADS

| | |
|---|---|
| Tossed Green Salad | $5.50 |
| Caesar Salad | $7.50 |
| Chicken Caesar Salad | $8.95 |
| Greens with Tomato | $6.50 |

## STARTERS

| | |
|---|---|
| Garlic or Herb Bread | $3.50 |
| Bruschetta | $4.50 |
| Dips & Bread | $3.50 |
| Potato Skins – served with sour cream | $6.50 |
| Basket of fries | $4.25 |
| Nachos with cheese, salsa and sour cream | $6.50 |
| Italian style baby potatoes | $6.50 |

## SANDWICHES

All served with salad and fries on your choice of bread

| | |
|---|---|
| Bacon, Lettuce & Tomato | $5.50 |
| Ham & Cheese | $5.00 |
| Ham, Cheese & Tomato | $5.50 |
| Grilled Chicken & Cheese | $6.00 |
| Egg & Lettuce | $5.50 |
| Top Quality Steak & Onions | $9.95 |

## BURGERS AND HOT DOGS

All served with salad and fries

| | |
|---|---|
| Jumbo Hot Dog | $5.95 |
| Gourmet Dog – includes smoked sausage | $6.95 |
| Double Beef Burger | $5.95 |
| Grilled Chicken Burger | $6.95 |
| Fresh Fish Burger | $7.55 |
| Vegetable Burger | $4.95 |

Additional items

| | |
|---|---|
| Bacon | $0.50 |
| Grilled onion | $0.50 |
| Olives | $0.75 |
| Cheddar | $0.75 |

## HOUSE SPECIAL

Today's Special is 250 gram Rib Eye Steak served with fries, vegetables with your choice of sauce $18.95

## PASTA

Your choice of Spaghetti, Fettuccine, Gnocchi, Ravioli or Tagliatelle with the sauces below

| | |
|---|---|
| Mama's Meaty Bolognaise | $9.95 |
| Creamy Chicken | $9.95 |
| Fresh Seafood Marinara | $10.95 |
| Pesto | $9.95 |
| Tomato and Hot Chilli | $8.95 |
| Vegetable Lasagne | $9.95 |
| Rich and Meaty Lasagne | $9.95 |

## DESSERTS

| | |
|---|---|
| Homemade Fruit Pie | $3.75 |
| Brownie | $3.50 |
| Cheesecake | $3.95 |
| Frozen Yoghurt – Chocolate or Vanilla | $2.50 |
| Jumbo Banana Split | $4.50 |

1. Using the menu provided, work out how much each of the following meals would cost. On a separate piece of paper, set out each order as it would appear on the final bill. An example is shown.
   a. Tossed green salad and Homemade fruit pies.
   b. Chicken Caesar salad, Jumbo hot dog and Vanilla-flavoured yoghurt.
   c. Garlic bread, Potato skins, House special and Jumbo banana split.
   d. Vegetable burger with cheddar, Basket of fries and Brownie.

**EAT EASY RESTAURANT**

| Item | Cost |
|---|---|
| Tossed green salad | $5.50 |
| Homemade fruit pies | $3.75 |
| | |
| | |
| | |
| Total | $9.25 |

Your group has decided to order a three-course meal. The first course consists of items from the Salads and Starters sections and the third course is from the Desserts section. The second course is ordered from the remaining sections.

The group also decides to buy a large jug of orange juice to drink, and plans to share its cost of $4.60 evenly.

2. Calculate your share of the cost of the drinks.
3. How much does this leave you to spend on your food order?
4. Select your three-course meal, keeping in mind the amount of money you have to spend. Write down your order, and set it out as you did for question 1.

Obtain the orders from three of your classmates. This will represent your group of four.

5. Write down each group member's order on a separate sheet of paper. Present the information as you would like it to appear on your final bill. Include the total cost of each person's order.
6. Calculate how much change each group member received from their $30.
7. Imagine that the restaurant's final bill for your group showed only the total amount owing; that is, there was no information on the breakdown of charges. Comment on how you would feel about this and how you think bills should be presented for group bookings.

**Resources**

**eWorkbook** Topic 6 Workbook (worksheets, code puzzle and project) (ewbk-1907)

**Interactivities** Crossword (int-2594)
Sudoku puzzle (int-3167)

## Exercise 6.9 Review questions

learnon

### Fluency

1. Give the value of the 7 in each of the following.
   **a.** 1.719 **b.** 3.0726 **c.** 0.2078
2. Give the vaue of the 7 in each of the following.
   **a.** 23.1487 **b.** 0.002 57 **c.** 17.592
3. Write the following numbers in expanded notation.
   **a.** 2.64 **b.** 18.406 **c.** 96.3428
4. Add 2 tenths to the following.
   **a.** 6.2 **b.** 0.743 **c.** 3.91
5. Add 3 thousandths to the following.
   **a.** 0.456 **b.** 1.6 **c.** 2.79

**6.** Put $<$ or $>$ between the following.

**a.** 8.72 8.27 **b.** 0.35 0.37 **c.** 10.214 10.219

**7.** Put $<$ or $>$ between the following.

**a.** 13.0496 13.149 **b.** 0.804 06 0.804 17 **c.** 0.000 879 0.000 876

**8.** Write the following decimals in order from smallest to largest.

**a.** 0.13, 0.86, 0.34, 0.71, 0.22 **b.** 0.247, 0.274, 0.124, 0.258, 0.285

**9.** Write the following decimals in order from smallest to largest.

**a.** 0.834, 0.826, 0.859, 0.888, 0.891 **b.** 0.356, 0.358, 0.365, 0.385, 0.217

**10.** Write the following decimals as fractions in simplest form.

**a.** 0.8 **b.** 0.17 **c.** 0.36 **d.** 0.187

**11.** Write the following decimals as fractions in simplest form.

**a.** 0.125 **b.** 0.568 **c.** 0.205 **d.** 0.950

**12.** Write the following decimals as mixed numerals in simplest form.

**a.** 1.5 **b.** 4.60 **c.** 3.48 **d.** 5.25

**13.** Write the following decimals as mixed numerals in simplest form.

**a.** 2.75 **b.** 2.625 **c.** 1.56 **d.** 8.32

**14.** Round the following to 2 decimal places.

**a.** 2.047 **b.** 13.8649 **c.** 17.898 193

**15.** Round the following to the number of decimal places indicated in the brackets.

**a.** 1.29 (1) **b.** 0.0482 (3) **c.** 1.925 96 (4)

**16.** Round the following to the nearest whole number.

**a.** 13.6 **b.** 29.02 **c.** 86.99 **d.** 100.09

**17.** Calculate the following.

**a.** $1.8 + 7.3$ **b.** $4.21 + 5.88$ **c.** $6.75 + 0.243$

**18.** Calculate the following.

**a.** $12.047 + 3.6$ **b.** $194 + 18.62 + 3.1$ **c.** $34.1 + 7.629 + 0.008\,45$

**19.** Calculate the following.

**a.** $9.6 - 4.3$ **b.** $18.25 - 9.18$ **c.** $3.92 - 1.88$

**20.** Calculate the following.

**a.** $100 - 9.341$ **b.** $4.876 - 3.927$ **c.** $1.6 - 0.025$

**21.** Calculate the following.

**a.** $6.2 \times 3$ **b.** $4.67 \times 9$ **c.** $13.2036 \times 5$ **d.** $0.7642 \times 7$

**22.** Calculate the following.

**a.** $0.23 \times 11$ **b.** $16.28 \times 41$ **c.** $182.94 \times 28$ **d.** $0.028\,94 \times 32$

**23.** Calculate the following.

**a.** $0.26 \times 10$ **b.** $1.345 \times 10$ **c.** $0.0645 \times 100$ **d.** $1.8294 \times 100$

**24.** Calculate the following.

**a.** $146.6281 \times 100$ **b.** $0.048\,064\,3 \times 1000$ **c.** $0.839\,204 \times 1000$ **d.** $0.368 \times 1000$

**25.** Calculate the following.

**a.** $3.2 \times 0.41$ **b.** $1.72 \times 0.3$ **c.** $0.87 \times 0.9$ **d.** $0.03 \times 0.006$

**26.** Calculate the following.

**a.** $0.58 \times 1.5$ **b.** $2.83 \times 0.96$ **c.** $11.468 \times 1.3$ **d.** $1.248 \times 0.82$

**27.** Calculate each of the following.

**a.** $2.5^2$ **b.** $0.03^2$ **c.** $0.64^2$ **d.** $0.0025^2$

**28.** Calculate each of the following.

**a.** $1.64 \div 4$ **b.** $12.48 \div 6$ **c.** $147.24 \div 2$ **d.** $1.76 \div 11$

**29.** Calculate each of the following.

**a.** $14.623 \div 10$ **b.** $102.36 \div 10$ **c.** $9612.347 \div 1000$

**30.** Calculate each of the following.

**a.** $20.032 \div 100$ **b.** $264\,983.0026 \div 1000$ **c.** $3462.94 \div 100$

**31.** Write each of these fractions as a terminating decimal or as an infinite recurring decimal.

**a.** $\frac{1}{20}$ **b.** $\frac{5}{8}$ **c.** $\frac{5}{16}$ **d.** $\frac{2}{16}$

**32.** Write these infinite recurring decimals using a short form.

**a.** 4.555 ... **b.** 0.8282 ... **c.** 19.278 127 81 ... **d.** 83.016 262 62 ...

**33.** Calculate the following.

**a.** $4.8 \div 0.6$ **b.** $35.7 \div 0.7$ **c.** $12.1 \div 1.1$ **d.** $13.72 \div 0.4$

**34.** Calculate the following.

**a.** $17.8946 \div 0.02$ **b.** $372.045\,72 \div 0.06$ **c.** $0.289\,56 \div 0.12$ **d.** $3214.0170 \div 0.15$

**35.** Write the following percentages as decimals.

**a.** 42% **b.** 5% **c.** 94% **d.** 139%

36. Write the following percentages as decimals.

a. 6.7% b. 19.7% c. 58.03% d. 0.8%

37. Calculate the following percentages by converting the percentage to a decimal.

a. 20% of 25 b. 12% of 31 c. 4.5% of 50 d. 9.2% of 75

38. Calculate the following percentages by converting the percentage to a decimal.

a. 21.4% of 90 b. 32.3% of 120 c. 76.5% of 8 d. 42.3% of 96.2

## Problem solving

39. Jim saved the following amounts of pocket money to take away on holiday: $12.50, $15.00, $9.30, $5.70 and $10.80.
Calculate how much money Jim had to spend while on holiday.

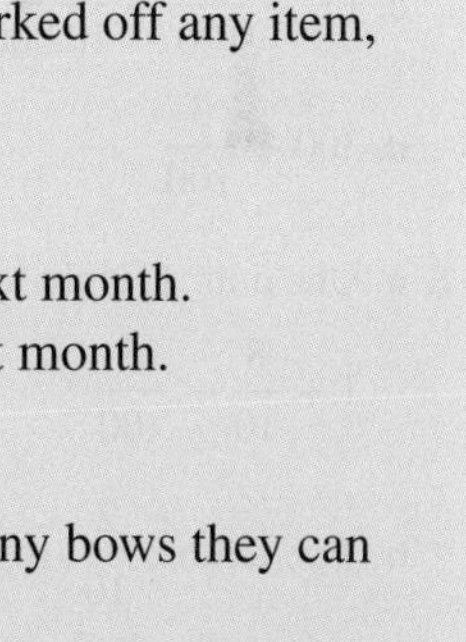

40. Mandie poured 0.375 litres from a 1.5-litre bottle of juice. Calculate how much juice was left in the bottle.

41. Tara bought 0.350 kilograms of shaved ham at $10.50 per kilogram. Calculate how much Tara paid for the ham.

42. The decimal equivalent of $\frac{3}{13}$ is $0.\overline{230\,769}$.

a. Determine the decimal equivalent of $\frac{1}{13}$.

b. Determine the decimal equivalent of $\frac{5}{13}$.

43. Lena saved $32 when she purchased a pair of jeans at a sale. If the sale had 40% marked off any item, calculate the original price for the pair of jeans.

44. You are given two options at work:
*Option 1:* Your salary can be raised by 10% this month and then reduced by 10% next month.
*Option 2:* Your salary will be reduced by 10% this month and increased by 10% next month.
Compare the two options to determine which is the better option.

45. Leesa has 3 m of ribbon. It takes 0.15 m of ribbon to make a bow. Calculate how many bows they can make with the ribbon.

46. Steven lives $\frac{3}{4}$ km from school. He can walk at 6.25 km/h.
Determine how long it takes him to walk to school.

**on** To test your understanding and knowledge of this topic, go to your learnON title at www.jacplus.com.au and complete the **post-test**.

# Answers

## Topic 6 Decimals

### 6.1 Pre-test

1. C
2. 0.12
3. E
4. Josh
5. 23.90
6. \$378.95
7. E
8. 5.817
9. C
10. 3280
11. \$28.05
12. \$13.65
13. 86.98
14. 44 litres
15. 125

### 6.2 Decimals and place value

1. a. $\frac{2}{10}$ b. $\frac{2}{100}$ c. $\frac{2}{10}$ d. $\frac{2}{1000}$
2. a. $\frac{2}{10\,000}$ b. $\frac{2}{10\,000}$ c. 2 d. $\frac{2}{10\,000}$
3. a. $\frac{9}{10}$ b. $\frac{9}{10}$ c. $\frac{9}{100}$ d. $\frac{9}{100}$
4. a. $\frac{9}{10\,000}$ b. $\frac{9}{100\,000}$ c. 900 d. 9
5. a. $\frac{4}{10}$ b. $2+\frac{7}{10}$
   c. $5+\frac{2}{10}+\frac{3}{100}$ d. $\frac{7}{10}+\frac{6}{100}+\frac{3}{1000}$
6. a. $\frac{1}{10}+\frac{1}{100}+\frac{1}{10\,000}$
   b. $7+\frac{2}{10}+\frac{9}{100}+\frac{6}{1000}+\frac{4}{10\,000}$
   c. $\frac{3}{10}+\frac{3}{100}+\frac{2}{10\,000}+\frac{4}{100\,000}$
   d. $300+\frac{3}{100}$
7. a. One unit, 1; eight tenths, $\frac{8}{10}$; five hundredths, $\frac{5}{100}$; $1+\frac{8}{10}+\frac{5}{100}$
   b. Two tenths, $\frac{2}{10}$; seven hundredths, $\frac{7}{100}$; one thousandth, $\frac{1}{1000}$; $\frac{2}{10}+\frac{7}{100}+\frac{1}{1000}$
   c. One ten, 10; six units, 6; one thousandth, $\frac{1}{1000}$; $16+\frac{1}{1000}$
   d. Three units, 3; four tenths, $\frac{4}{10}$; two thousandths, $\frac{2}{1000}$; seven hundred thousandths, $\frac{7}{100\,000}$; $3+\frac{4}{10}+\frac{2}{1000}+\frac{7}{100\,000}$
8. a. $(1\times 1)+\left(2\times\frac{1}{10}\right)+\left(5\times\frac{1}{100}\right)$
   b. $(5\times 10)+(6\times 1)+\left(1\times\frac{1}{100}\right)$
   c. $(3\times 10)+(9\times 1)+\left(1\times\frac{1}{100}\right)$
   d. $(5\times 1)+\left(9\times\frac{1}{10}\right)+\left(8\times\frac{1}{100}\right)+\left(7\times\frac{1}{1000}\right)$
9. a. $(1\times 10)+(3\times 1)+\left(4\times\frac{1}{10}\right)+\left(8\times\frac{1}{100}\right)+\left(2\times\frac{1}{1000}\right)$
   b. $\left(3\times\frac{1}{10}\right)+\left(6\times\frac{1}{1000}\right)+\left(2\times\frac{1}{10\,000}\right)$
   c. $\left(5\times\frac{1}{10}\right)+\left(2\times\frac{1}{10\,000}\right)$
   d. $(2\times 1)+\left(4\times\frac{1}{10}\right)+\left(7\times\frac{1}{100}\right)$
10. D
11. See the table at the foot of the page.*
12. a. 0.32 b. 0.59 c. 0.901
13. a. 0.039 90 b. 2.559 c. 0.110 49
14. a. > b. < c. < d. < e. <
15. a. > b. > c. < d. < e. >
16. a. 2
    b. i. Seventy-eight dollars and fifty-five cents
       ii. \$78.55
17. a. 0.17, 0.21, 0.33, 0.39, 0.45
    b. 0.314, 0.391, 0.413, 0.420, 0.502
    c. 0.803, 0.807, 0.811, 0.821, 0.902
18. a. 0.9087, 0.9189, 0.9812, 0.9864, 0.9943
    b. 4.0292, 4.5097, 4.6031, 4.6249, 4.802
    c. 0.003, 0.004 65, 0.005 02, 0.0056, 0.009

*11.

| | | Tens | Units | | Tenths | Hundredths | Thousandths |
|---|---|---|---|---|---|---|---|
| **a.** | 0.205 | 0 | 0 | · | 2 | 0 | 5 |
| **b.** | 1.06 | 0 | 1 | · | 0 | 6 | 0 |
| **c.** | 74.108 | 7 | 4 | · | 1 | 0 | 8 |
| **d.** | 0.108 | 0 | 0 | · | 1 | 0 | 8 |
| **e.** | 50.080 | 5 | 0 | · | 0 | 8 | 0 |

19. a. $0.48, 0.39, 0.36, 0.31, 0.19$
 b. $0.99, 0.97, 0.95, 0.91, 0.90$
 c. $1.291, 1.288, 1.279, 1.273, 1.264$

20. a. $0.390, 0.372, 0.318, 0.317, 0.309$
 b. $0.8889, 0.8823, 0.8448, 0.8217, 0.8207$
 c. $1.843\,21, 1.702\,96, 1.486\,59, 1.486\,13, 1.349\,54$

21. E

22. D

23. E

24. a. False b. True c. False

25. a. True b. False c. True

26. a. 0.142 857 142 9
 i. 0.285 714 285 7
 ii. 0.428 571 428 57
 iii. 0.571 428 571 4
 iv. 0.714 285 714 3
 v. 0.857 142 857 14
 b. Each number has a repeated group of decimals in the same sequence: 142 857.

27. a. i. Tenths ii. Yes
 b. i. Thousandths ii. No
 c. i. Units ii. No
 d. i. Units ii. Yes
 e. i. Tens ii. Yes

28. a. Shelley, 0.21 seconds
 b. Carolyn, 0.06 seconds
 c. Mara, 0.11 seconds
 d. Shelley, Carolyn, Robyn
 e. Carolyn, Shelley, Kyah
 f. Mara, Jenika, Shelley
 g. Yes, Shelley

29. 8.264

30. They are correct for trailing zeros (to the right) of a decimal. They are incorrect for zeros that are enclosed by non-zero digits, as they would indicate a place value of zero.

## 6.3 Converting decimals to fractions and fractions to decimals

1. a. $\frac{3}{10}$ b. $\frac{1}{2}$ c. $\frac{21}{100}$ d. $\frac{2}{5}$ e. $\frac{11}{25}$

2. a. $\frac{49}{100}$ b. $\frac{251}{500}$ c. $\frac{617}{1000}$ d. $\frac{3}{10}$ e. $\frac{16}{25}$

3. a. $\frac{7}{25}$ b. $\frac{591}{625}$ c. $\frac{9209}{10\,000}$ d. $\frac{4621}{10\,000}$ e. $\frac{3}{25}$

4. a. $1\frac{3}{10}$ b. $2\frac{7}{10}$ c. $9\frac{2}{5}$ d. $1\frac{1}{5}$ e. $4\frac{1}{5}$

5. a. $8\frac{1}{2}$ b. $5\frac{27}{100}$ c. $19\frac{91}{500}$ d. $3\frac{3}{20}$ e. $6\frac{1}{4}$

6. a. $9\frac{7}{50}$ b. $16\frac{341}{500}$ c. $2\frac{4917}{10\,000}$
 d. $4\frac{1693}{5000}$ e. $100\frac{3}{625}$

7. B

8. A

9. C

10. D

11. E

12. a. 0.75 b. 0.5 c. 0.8 d. 0.05

13. a. 0.25 b. 0.02 c. 0.32 d. 0.2

14. a. 0.375 b. 0.875 c. 0.0625

15. a. Darya
 b. Compare the fraction of wins as decimals. Darya has a larger fraction of wins.

16. 300 m

17. No, the second jump was best.

18. History, Science, English

19. $\frac{4}{15}, \frac{3}{11}, \frac{30}{100}, \frac{8}{25}, \frac{325}{1000}, \frac{4}{12}, \frac{3}{7}$

20. $\frac{3}{7} = 0.\dot{4}28\,57\dot{1}$, $\frac{4}{7} = 0.\dot{5}71\,42\dot{8}$, $\frac{5}{7} = 0.\dot{7}14\,28\dot{5}$,
 $\frac{6}{7} = 0.\dot{8}57\,14\dot{2}$

## 6.4 Rounding and repeating decimals

1. a. 0.32 b. 0.86 c. 1.25 d. 13.05 e. 3.04

2. a. 7.13 b. 100.81 c. 71.26 d. 0.01 e. 2.39

3. a. 0.19 b. 19.70 c. 0.40 d. 1.00 e. 8.90

4. a. 0.4 b. 0.9 c. 9.3 d. 25.3 e. 3.9

5. a. 300.1 b. 12.8 c. 99.9 d. 8.9 e. 14.0

6. a. 17.6 b. 0.9 c. 94.0 d. 1.0 e. 96.3

7. C

8. A

9. a. 11 b. 8 c. 4 d. 93 e. 112

10. a. 22 b. 42 c. 2138 d. 0 e. 1

11. a. $2.\dot{5}$ b. $0.\dot{6}$ c. $49.\dot{1}$ d. $0.\dot{2}\dot{6}$

12. a. $0.\dot{9}1\dot{3}$ b. $8.\dot{6}41\dot{8}$ c. $133.9\overline{462}$ d. $0.12\dot{7}$

13. a. 0.3 b. 0.1 c. 0.417

14. a. 0.17 b. 0.182 c. 0.47

15. A

16. A

17. a. 3.5 b. $4.\dot{6}$ c. 0.75

18. a. 10 b. 80 c. 140 d. 260 e. 180

19. a. 300 b. 200 c. 100 d. 6800 e. 9800

20. a. 3000 b. 12 000 c. 19 000 d. 9000 e. 1000

21. $27.70

22. $0.295, 0.\dot{2}9\dot{5}, 0.29\dot{5}, 0.2\dot{9}\dot{5}$

23. An example answer is 0.543216.

24. A more reasonable measurement would be 3.568 m. With the instruments we have available, it would not be possible to cut the timber more accurately.

25. Fred is incorrect; any number between 0.5 and 0.65 will round to 0.6.
26. 25°C
27. a. i. 0.8 ii. 0.84 iii. 0.838
b. The rules are the same regardless of the number of decimals.
28. 50.0000, 33.3333, 20.0000, 14.2857, 9.0909, 7.6923, 5.8824, 5.2632, 4.3478, 3.4483

## 6.5 Adding and subtracting decimals

1. a. 3.5 b. 2.69 c. 13.239 d. 102.1858
2. a. 6.77 b. 21.906 c. 2012.333 d. 261.449 06
3. a. 4.6 b. 6.9 c. 3.471 d. 17.28
4. a. 17.3271 b. 20.4329 c. 11.69 d. 105.22
5. a. 18.951 b. 105.958 c. 192.8949 d. 2618.9564
6. a. 0.60 b. 4.205 c. 0.0806 d. 1.3368
7. a. 5.02 b. 89.19 c. 87.09 d. 2.733
8. a. 3.34 b. 5.93 c. 0.327 d. 4.146
9. a. 1.934 b. 3.289 c. 14.5326 d. 16.766
10. a. 38.238 b. 4.1136 c. 0.1951 d. 0.772 16
11. B
12. A
13. A
14. D
15. C
16. B
17. B
18. B
19. $167.00
20. $4.70
21. 20.05 kilometres
22. 8.1 kilometres
23. $6.60
24. $10.35
25. 50.48 kilograms
26. 11.73 metres
27. 31.28 metres
28. 48.38 seconds
29. 0.107
30. When we estimate an answer we use rounding to make the numbers simple to calculate; for example, 11 may be rounded to 10.
31. 49 calls
32. a. See the table at the foot of the page.*
b. Team 3
c. 4.1 seconds
33. 65.53 g, 43.43 g, 32.03 g or 32.04 g
34. Increase 51c
35. One possible solution is $5.83 + 1.46 = 7.29$.

## 6.6 Multiplying decimals

1. a. 14.0 b. 125.6 c. 61.2 d. 111.70
2. a. 190.26 b. 518.96 c. 6.384 d. 0.794 43
3. a. 0.84 b. 2.94 c. 0.32 d. 7.76
4. a. 0.140 b. 0.192 c. 0.385 d. 0.441
5. a. 4.304 b. 0.2082 c. 2.8992 d. 4.905 64
6. a. 0.0001 b. 0.000 012 c. 0.063 d. 0.000 222
7. a. 0.000 002 44 b. 0.000 36
c. 0.392 d. 0.000 606 2
8. a. 0.3 b. 0.851 c. 2.538 d. 0.175
9. a. 6.557 b. 16.848 c. 60.300 d. 31.372
10. E
11. B
12. A
13. B
14. a. 0.0004 b. 1.69 c. 4.2025
15. a. 1016 b. 5973 c. 437 400
16. a. 1092 b. 4548 c. 38 340
17. a. 64.8 b. 1389.6 c. 270.8
d. 217 148.96 e. 82 049.6783
18. a. 3 268 904.3267 b. 984 326.641
c. 278.498 32 d. 460
19. a. i. 24 000 ii. 24 640
b. i. 300 000 ii. 331 602
c. i. 60 000 ii. 50 449.5
20. a. 3500 b. 12 700 c. 1100 d. 2535 e. 5820
21. $750
22. 108 L
23. 260.7569 m$^2$
24. a. $32.00 b. $32.05
25. The 50 extra sheets are 6 mm thick. You might notice the difference.
26. 13
27. $5.55
28. 0.375 L
29. Together they pay $6.30: Sophie pays $3.60 and Hamish pays $2.70.

*32.

| Team | 1 | 2 | 3 | 4 | 5 | 6 |
|---|---|---|---|---|---|---|
| Time (seconds) | 153 | 159.3 | 148.9 | 166.6 | 155.9 | 155.8 |

## 6.7 Dividing decimals

1. a. 0.6 b. 3.1 c. 0.54 d. 1.61 e. 2.42
2. a. 0.788 b. 1.147 c. 4.0701 d. 6.8594 e. 31.6176
3. a. 3.12 b. 2.231 c. 1.222 d. 6.421 e. 11.0311
4. a. 1.85 b. 4.75 c. 1.46
5. a. 2.45 b. 1.25 c. 6.95
6. a. 1.407 b. 9.6813 c. 0.6208
   d. 3.592 87 e. 0.023 497 8
7. a. 0.000 090 769 b. 0.802 405 6 c. 0.000 152 7
   d. 0.000 072 05 e. 0.000 000 0032
8. a. 0.265 b. 0.79 c. 0.595
   d. 0.001 465 e. 0.001 74
9. a. 5 b. 7 c. 9 d. 4 e. 0.31
10. a. 2.2 b. 66 c. 72.1 d. 0.852 e. 1.367
11. a. 1564.08 b. 4.254 c. 143.134
12. a. 1786.1 b. 0.4782 c. 1630.95
13. D
14. D
15. B
16. D
17. a. i. 350 ii. 275
    b. i. 200 ii. 232
    c. i. 2000 ii. 2400
18. a. $3.65 b. $1.70 c. $56.85
    d. $0.75 e. $0.90 f. $63.50
19. 10.10
20. 3.086 grains
21. a. 32 b. $74.24
22. 0.4 L
23. $0.60
24. 20
25. 201
26. $4.50
27. 71
28. 0.675 m
29. 26
30. $\frac{1}{3}$

## 6.8 Decimals and percentages

1. a. 0.36 b. 0.14 c. 0.28 d. 0.73 e. 0.66
2. a. 0.59 b. 0.99 c. 0.09 d. 0.04 e. 0.01
3. a. 0.25 b. 2.00 c. 1.5 d. 3.6 e. 1
4. a. 0.123 b. 0.316 c. 0.592 d. 0.849 e. 0.421
5. a. 0.376 b. 0.219 c. 0.169 d. 0.111 e. 0.031
6. a. 0.092 b. 0.059 c. 0.088 d. 0.1425 e. 0.3175
7. a. 0.2355 b. 0.4575 c. 0.0005 d. 0.0401 e. 0.0002
8. B
9. C
10. A
11. D
12. a. $\frac{17}{100}$ b. 0.17
13. a. $\frac{1}{2000}$ b. 0.0005
14. 0.807
15. 2.18
16. 0.138
17. 0.807
18. a. 23, 24, 24, 24
    b. 0.23, 0.24, 0.24, 0.24
    c. It does not matter as long as both values are rounded up or down at the same digit.
19. a. 23.67%
    b. 0.2360, 0.2367, 0.2366
20. a. 0.378, 0.396, 0.309, 0.345, 0.328
    b. 39.6%, 37.8%, 34.5%, 32.8%, 30.9%
    c. 0.396, 0.378, 0.345, 0.328, 0.309
    d. Sample response: It is easier to arrange the numbers in the order using percentages as you can actually see the numbers and order them without any calculation. Decimals are more difficult to determine order, because there are more place values to look at.
21. $1125
22. $7.27
23. a. i. 0.072 91 ii. 0.7291 iii. 7.291
    b. When converting a percentage to a decimal number, the number becomes 100 times smaller.
24. When dealing with measurements it is easier to use decimals rather than percentages or fractions.

## Project

1. a. $9.25 b. $17.40 c. $33.45 d. $13.45
2. $1.15
3. $28.85
4. Sample response:

| Item | Cost |
|---|---|
| Nachos with cheese, salsa and sour cream | $6.50 |
| Vegetable burger | $4.95 |
| Cheesecake | $3.95 |
| **Total** | **$15.40** |

5. Sample response:

| Your order | |
|---|---|
| **Item** | **Cost** |
| Nachos with cheese, salsa and sour cream | $6.50 |
| Vegetable burger | $4.95 |
| Cheesecake | $3.95 |
| **Total** | **$15.40** |

| Classmate 1 | |
|---|---|
| **Item** | **Cost** |
| Chicken caesar salad | $8.95 |
| Tomato and hot chilli pasta | $8.95 |
| Brownie | $3.50 |
| **Total** | **$21.40** |

| Classmate 2 | |
|---|---|
| **Item** | **Cost** |
| Bruschetta | $4.50 |
| Double beef burger with bacon | $6.45 |
| Jumbo banana split | $4.50 |
| **Total** | **$15.45** |

| Classmate 3 | |
|---|---|
| **Item** | **Cost** |
| Garlic bread | $3.50 |
| Rib eye steak | $18.95 |
| Chocolate frozen yoghurt | $2.50 |
| **Total** | **$24.95** |

| Item | Cost |
|---|---|
| Nachos with cheese, salsa and sour cream | $6.50 |
| Chicken caesar salad | $8.95 |
| Bruschetta | $4.50 |
| Garlic bread | $3.50 |
| Vegetable burger | $4.95 |
| Tomato and hot chilli pasta | $8.95 |
| Double beef burger with bacon | $6.45 |
| Rib eye steak | $18.95 |
| Cheesecake | $3.95 |
| Brownie | $3.50 |
| Jumbo banana split | $4.50 |
| Chocolate frozen yoghurt | $2.50 |
| Jug of orange juice | $4.60 |
| **Total** | **$81.80** |

6. Sample response:
You: $14.60
Classmate 1: $8.60
Classmate 2: $14.55
Classmate 3: $5.05
7. If the bill only showed the total amount and not the breakdown then it would be very difficult to work out how much each person owes individually. To counter this, the restaurant could use a bill similar to that shown in the answer to question **5** above.

## 6.9 Review questions

1. a. 7 tenths b. 7 hundredths c. 7 thousandths
2. a. 7 ten thousandths
b. 7 hundred thousandths
c. 7 units
3. a. $(2 \times 1) + \left(6 \times \frac{1}{10}\right) + \left(4 \times \frac{1}{100}\right)$
b. $(1 \times 10) + (8 \times 1) + \left(4 \times \frac{1}{10}\right) + \left(6 \times \frac{1}{1000}\right)$
c. $(9 \times 10) + (6 \times 1) + \left(3 \times \frac{1}{10}\right) + \left(4 \times \frac{1}{100}\right) + \left(2 \times \frac{1}{1000}\right) + \left(8 \times \frac{1}{10\,000}\right)$
4. a. 6.4 b. 0.943 c. 4.11
5. a. 0.459 b. 1.603 c. 2.793
6. a. > b. < c. <
7. a. < b. < c. >
8. a. 0.13, 0.22, 0.34, 0.71, 0.86
b. 0.124, 0.247, 0.258, 0.274, 0.285
9. a. 0.826, 0.834, 0.859, 0.888, 0.891
b. 0.217, 0.356, 0.358, 0.365, 0.385
10. a. $\frac{4}{5}$ b. $\frac{17}{100}$ c. $\frac{9}{25}$ d. $\frac{187}{1000}$
11. a. $\frac{1}{8}$ b. $\frac{71}{125}$ c. $\frac{41}{200}$ d. $\frac{19}{20}$
12. a. $1\frac{1}{2}$ b. $4\frac{3}{5}$ c. $3\frac{12}{25}$ d. $5\frac{1}{4}$
13. a. $2\frac{3}{4}$ b. $2\frac{5}{8}$ c. $1\frac{14}{25}$ d. $8\frac{8}{25}$
14. a. 2.05 b. 13.86 c. 17.90
15. a. 1.3 b. 0.048 c. 1.9260
16. a. 14 b. 29 c. 87 d. 100
17. a. 9.1 b. 10.09 c. 6.993
18. a. 15.647 b. 215.72 c. 41.737 45
19. a. 5.3 b. 9.07 c. 2.04
20. a. 90.659 b. 0.949 c. 1.575
21. a. 18.6 b. 42.03 c. 66.0180 d. 5.3494
22. a. 2.53 b. 667.48 c. 5122.32 d. 0.926 08
23. a. 2.6 b. 13.45 c. 6.45 d. 182.94
24. a. 14 662.81 b. 48.0643 c. 839.204 d. 368
25. a. 1.312 b. 0.516 c. 0.783 d. 0.000 18
26. a. 0.87 b. 2.7168 c. 14.9084 d. 1.023 36
27. a. 6.25 b. 0.0009 c. 0.4096 d. 0.000 006 25
28. a. 0.41 b. 2.08 c. 73.62 d. 0.16

29. a. 1.4623 b. 10.236 c. 9.612 347

30. a. 0.200 32 b. 264.983 002 6 c. 34.6294

31. a. 0.05 b. 0.625 c. 0.3125 d. 0.125

32. a. $4.\dot{5}$ b. $0.\overline{82}$ c. $19.\overline{2781}$ d. $83.01\dot{6}\dot{2}$

33. a. 8 b. 51 c. 11 d. 34.3

34. a. 894.73 b. 6200.762 c. 2.413 d. 21 426.78

35. a. 0.42 b. 0.05 c. 0.94 d. 1.39

36. a. 0.067 b. 0.197 c. 0.5803 d. 0.008

37. a. 5 b. 3.72 c. 2.25 d. 6.9

38. a. 19.26 b. 38.76 c. 6.12 d. 40.6926

39. \$53.30

40. 1.125 L

41. \$3.70 (rounded to nearest 5c) or \$3.68

42. a. $0.\overline{076\,923}$ b. $0.\overline{384\,615}$

43. \$80

44. No. Overall, after two months your salary will decrease by 1% using either option. However, option 1 gives an extra income (10% of the original salary) in the first month.

45. 20

46. 7 min 12s

# 7 Algebra

## LESSON SEQUENCE


**7.1** Overview ..... 348
**7.2** Introduction to algebra ..... 350
**7.3** Substituting and evaluating ..... 358
**7.4** Simplifying expressions using like terms ..... 362
**7.5** Multiplying and dividing terms ..... 367
**7.6** Number laws ..... 371
**7.7** Expanding brackets (extending) ..... 375
**7.8** Review ..... 379

# LESSON
# 7.1 Overview

## Why learn this?

Algebra is a fundamental building block of mathematics. Algebra is, in fact, a mathematical language in which you are able to use the basic operations of addition, subtraction, multiplication and division. Using algebra, you are able to break down a problem or a puzzle into smaller parts and then find the solution or answer. Sometimes it is hard to imagine how we would apply algebra in our everyday lives, but we use algebra frequently without even realising it. The methods used to solve problems and the reasoning skills learned will be useful no matter what you do in life.

Every five years, on one particular night, every person in Australia completes a census form. The census collects information such as the number of people living at an address, their ages and their incomes. Statisticians use algebraic techniques to interpret the information, which can then be used in policymaking and to make predictions about future requirements for such things as schools, hospitals and aged care services. Scientists, architects, accountants and engineers all use algebra frequently in their work. Without algebra there would be no space travel, no electrical appliances, no television, no smartphones, no iPads, no computers and no computer games.

1. **MC** Given the pattern 8, 13, 18 …, the next three numbers are:
   **A.** 23, 28, 35 **B.** 23, 28, 33 **C.** 22, 26, 30 **D.** 22, 27, 32 **E.** 23, 29, 34

2. Calculate the value of $q$ when the given substitution is made to the following formula: $q = 3p$ when $p = 8$.

3. In a tennis match, Charlie scored 12 more points than twice Lachlan's score. Lachlan scored 36 points. Calculate the number of points Charlie scored.

4. **MC** An expression that shows the difference between $K$ and $L$, where $K$ is bigger than $L$, is:
   **A.** $L - K$ **B.** $K \div L$ **C.** $2K$ **D.** $K + L$ **E.** $K - L$

5. Simplify the expression $3x + 2y - x + 4y$ by collecting like terms.

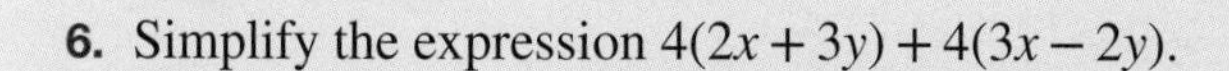

6. Simplify the expression $4(2x + 3y) + 4(3x - 2y)$.

7. Identify two consecutive numbers with a product of 420.

8. Explain what happens when the result of $(5k + 3)$ is doubled.

9. Complete the following: $3x + 7 - \square = 2 + \square - x$.

10. Simplify the following expressions.
    **a.** $3a \times 5b \times 2c$ **b.** $4ab \times 13b^2$ **c.** $\dfrac{15pq}{3p}$ **d.** $\dfrac{36mn}{6n^2}$

11. If $a = 2$, $b = 5$ and $c = 3$, determine the value of the following expression: $c(4a - b) - 3a$.

12. Expand and simplify the following expression: $3(7p - 4q) - 2(2p - 5q)$.

13. A rectangle has a width of $3x$ and a length of $5y$. Determine the simplest expression for the perimeter of the rectangle.

14. Is the expression $a \times a = 2a$ true or false? Justify your answer.

15. Ashton has 2 pairs of basketball shoes and 5 pairs of runners, and Alyssa has 1 pair of basketball shoes and 3 pairs of runners. Given $b$ is the number of basketball shoes and $r$ is the number of runners, determine the simplified expression that shows the total number of basketball shoes and runners they have together.

# LESSON
# 7.2 Introduction to algebra

## LEARNING INTENTION

At the end of this lesson you should be able to:
- understand the use of pronumerals in algebra
- identify algebraic terminology
- construct expressions from given information.

## 7.2.1 Pronumerals

eles-3986

- A **pronumeral** is a letter or symbol that is used to represent a number. For example, $p$ could represent the number of pies in a shop.
- Any letter can be used as a pronumeral.
- A pronumeral can also be called a **variable**.

**Using pronumerals**

- **Pronumerals are used to represent unknown quantities.**
- **For example, without knowing the number of fish in a pond, this number could be represented by the pronumeral $f$.**

## WORKED EXAMPLE 1 Use of pronumerals and constants

**We don't know exactly how many lollies there are in a jar, but we know each jar has the same number. We will refer to this unknown number as $l$, where $l$ is the pronumeral.**

a. **Draw a jar and label the number of lollies in the jar as $l$.**

b. **There is a jar of lollies plus one lolly. Represent this information using pictures and using symbols.**

| THINK | WRITE |
|---|---|
| a. The number of lollies in the jar is unknown. | Let l be the number of lollies in the jar.<br><br>$l$ |
| b. Draw a jar of lollies and one lolly. Use '$l$' in an algebraic expression to represent this information. | <br>$l+1$ |

## 7.2.2 Terms and expressions

eles-3987

- A **term** is a single mathematical expression.
  For example, $2, p, 3x$ and $6xy$ are all terms.
- In terms that contain numbers and pronumerals multiplied together, the number is placed at the front and the multiplication signs are removed.
  For example, $2 \times a \times c$ would be simplified to $2ac$.
- Pronumerals in a term are written in alphabetical order.
  For example, $5 \times y \times x = 5xy$.
- The number in front of the pronumeral(s) is known as the **coefficient**. For example, in the term $3t$, the coefficient of $t$ is 3.
- Terms such as $m \times m$ are expressed as $m^2$. Similarly, $m \times m \times m$ is expressed as $m^3$.
- Numbers without pronumerals are called **constant** terms.
- **Expressions** are mathematical sentences made up of terms separated by + or − signs.
  For example, $3m + 5$, $9 - 2t$ and $a^2 + 2ab + b^2$ are all expressions.
- The definitions are summarised in the following diagram.

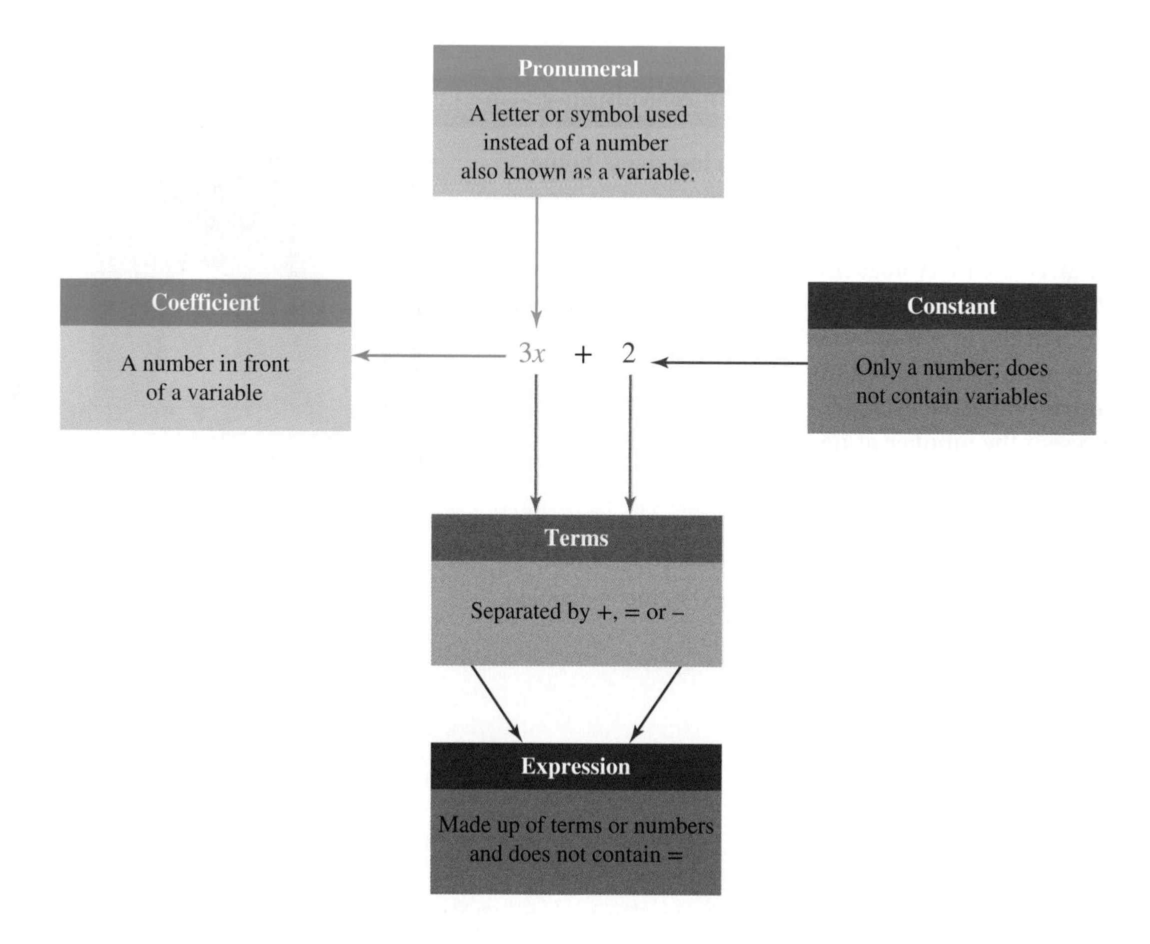

## WORKED EXAMPLE 2 Writing terms and expressions

**If $Y$ represents any number, write terms or expressions for:**

a. **5 times that number**

b. **2 less than that number**

c. **8 more than that number**

d. **the number divided by 4 (or the quotient of $Y$ and 4)**

e. **the next consecutive number (that is, the counting number after $Y$).**

| THINK | WRITE |
|---|---|
| In each case think about which operations are being used and the order in which they occur. | |
| a. An expression with '5 times' means 'multiply by 5'. When multiplying, the multiplication sign is not shown. Remember to put the number first. | a. $5 \times Y = 5Y$ |
| b. An expression with '2 less' means 'subtract 2'. | b. $Y - 2$ |
| c. An expression with '8 more' means 'add 8'. | c. $Y + 8$ |
| d. An expression with 'divided by 4' means 'write as a quotient (fraction)'. | d. $\frac{Y}{4}$ |
| e. An expression with 'the next consecutive number' means 'add 1 to the number'. | e. $Y + 1$ |

## WORKED EXAMPLE 3 Writing simple algebraic expressions

**Suppose there are *d* dogs currently at a lost dogs' home.**

a. **If three dogs find homes, write an expression to represent the total number of dogs at the lost dogs' home.**

b. **Just after the three dogs find homes, the number of dogs at the lost dogs' home is doubled. Write an expression to represent the new number of dogs at the lost dogs' home.**

c. **The dogs are then sorted into 10 pens with the same number of dogs in each pen. Write another expression to represent the number of dogs in each pen.**

| THINK | WRITE |
|---|---|
| a. The original number of dogs ($d$) is decreased by three. | a. $d - 3$ |
| b. The number of dogs left at the lost dogs' home is doubled. | b. $2 \times (d - 3)$ |
| c. The dogs are divided into 10 pens. | c. $2 \times (d - 3) \div 10$ or $\frac{2d - 6}{10}$ |

## WORKED EXAMPLE 4 Simplifying when terms or pronumerals are multiplied

**If one jar of lollies has $l$ lollies, then two jars of lollies will have double the number of lollies, that is $2 \times l = 2l$. In algebra, $2l$ means two lots of $l$, $3l$ means three lots of $l$, and so on.**

**a. There are two jars of lollies and three loose lollies. Each jar has the same number of lollies. Write an expression using symbols.**

**b. There are three jars of lollies, four packets of lollipops and four loose lollies. Each jar has the same number of lollies and each packet of lollipops has the same number of lollipops. Write an expression using symbols.**

| THINK | WRITE |
|---|---|
| a. There are two jars of lollies ($l$) plus (+) the three that are loose. 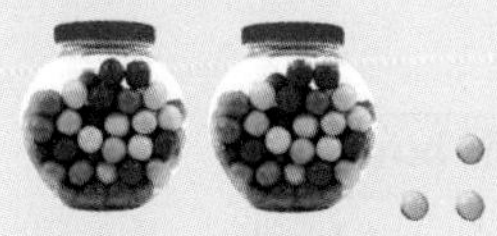 | a. Let $l$ be the number of lollies in the jar.<br>$2 \times l + 3 = 2l + 3$ |
| b. There are three jars of lollies ($l$) and four packets of lollipops ($p$) plus (+) the four loose lollies.   | b. Let $l$ be the number of lollies in each jar and $p$ be the number of lollipops in each packet.<br>$3 \times l + 4 \times p + 4 = 3l + 4p + 4$ |

## COLLABORATIVE TASK: How many shoes?

You own a certain number of pairs of shoes. Your friend owns five pairs more than you own, and your sister owns three pairs fewer than you own. Your brother owns four pairs fewer than your friend owns.

1. Write an expression to individually represent each of the following.
   a. The number of pairs of shoes you own
   b. The number of pairs of shoes your friend owns
   c. The number of pairs of shoes your sister owns
   d. The number of pairs of shoes your brother owns
2. Share your answers with a partner. Discuss any differences between your answers.
3. As a class, discuss any differences that you found.
4. Does it matter if each student in the class uses a different symbol? Why? Why not?

 Resources

 **eWorkbook** Topic 7 Workbook (worksheets, code puzzle and project) (ewbk-1908)

 **Interactivities** Individual pathway interactivity: Using variables (int-4429)

Terms and expressions (int-4004)

# Exercise 7.2 Introduction to algebra

learnon

7.2 Quick quiz on | 7.2 Exercise

## Individual pathways

| ■ PRACTISE | ■ CONSOLIDATE | ■ MASTER |
|---|---|---|
| 1, 3, 4, 7, 11, 12, 13, 18, 19, 24, 25, 26, 28, 31 | 2, 5, 6, 8, 14, 15, 20, 21, 27, 29, 32 | 9, 10, 16, 17, 22, 23, 30, 33, 34 |

### Fluency

1. State what pronumerals are used for in algebra.

2. **WE1** If $b$ represents the total number of bananas in a fruit bowl, write a short paragraph that could explain the following table.

| Day | Number of bananas in the fruit bowl |
|---|---|
| Monday | $b$ |
| Tuesday | $b-4$ |
| Wednesday | $b-6$ |
| Thursday | $b+8$ |
| Friday | $b+5$ |

3. **WE2** Write an expression for the total number of coins in each of the following. The symbol represents a full piggy bank and represents one coin. Use $c$ to represent the number of coins in a full piggy bank.

a. 

b. 

c. 

d. 

4. Write an expression for the total number of coins in each of the following, using $c$ to represent the number of coins in a piggy bank.

a. 

b. 

c. 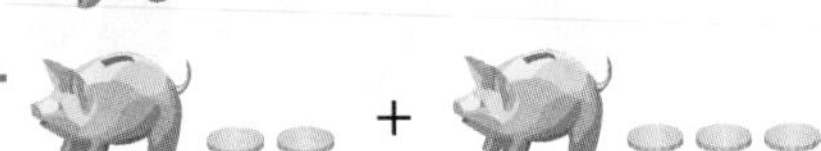

d. 

5. Christie and Jane each had two full piggy banks and seventeen coins. They combined their money and spent every cent on a day out in town.

$2\times($ + $)$

Write an expression to represent the number of coins they had in total. Use $c$ to represent the number of coins in a piggy bank.

6. Luke has three macadamia nut trees in his backyard. He saves takeaway containers to store the nuts in. He has two types of containers, rectangular and round.

Using $m$ to represent the number of nuts in a rectangular container and $n$ to represent the number of nuts in a round container, write expressions for the following.

a.

b.

c.

d. where ● represents one nut.

e. 

7. Using $a$ and $b$ to represent numbers, write an expression for:

a. the sum of $a$ and $b$

b. the difference between $a$ and $b$

c. the product of $a$ and $b$

d. $a$ multiplied by 3 and the result divided by $b$.

8. **WE3** Using $a$ and $b$ to represent numbers, write an expression for:

a. 15 divided by negative $a$

b. five times $b$ subtracted from three times $a$

c. $b$ divided by negative 5

d. seven less than $a$.

9. Write the following expressions in ascending order.

a. $m+3,\ m-2,\ m+1,\ m-8$

b. $n-15,\ n+25,\ n-10.3$

10. Write your own descriptions for what the following expressions could represent.

a. $d+2,\ d,\ d-1$

b. $t,\ 2t,\ t+1$

c. $y,\ 3y,\ 10y$

11. **WE4** Write an expression for each of the following.

a. The sum of $B$ and 2

b. 3 less than $T$

c. 6 added to $D$

d. 5 taken away from $K$

12. Write an expression for each of the following.

a. The sum of $G$, $N$ and $W$

b. $D$ increased by $H$

c. $N$ increased by $N$

d. $H$ added to $C$

13. Write an expression for each of the following.

a. $G$ subtracted from 12

b. The product of $D$ and 4

c. $B$ multiplied by $F$

d. $Y$ added to the product of 3 and $M$

14. If $A$, $B$ and $C$ represent any 3 numbers, write an expression for each of the following.

a. The sum of all 3 numbers

b. The difference between $A$ and $C$

c. The product of $A$ and $C$

d. The product of all 3 numbers

15. If $A$, $B$ and $C$ represent any 3 numbers, write an expression for each of the following.

a. The quotient of $B$ and $C$ (that is, $B$ divided by $C$)

b. The sum of $A$ and $C$, divided by $B$

c. 3 more than $A$

d. The difference between $C$ and the product of $A$ and $B$

16. Write expressions for the following rules.

a. The number of students left in the class if $X$ students leave for the canteen out of a total group of $T$ students

b. The amount of money earned by selling $B$ cakes, where each cake is sold for $4.00

c. The total number of sweets if there are $G$ bags of sweets with 45 sweets in each bag

d. The cost of one concert ticket if 5 tickets cost $$T$

## Understanding

**17.** There are $n$ mice in a cage. If the number of mice doubles each month, write an expression for the number of mice:

**a.** 1 month from now
**b.** 2 months from now
**c.** 3 months from now.

**18.** **MC** $M$ is used to represent an unknown whole number. Choose the answer that matches the description 'six more than the number'.

**A.** $6+M$ **B.** $6M$
**C.** $M-6$ **D.** $M \div 6$
**E.** $6-M$

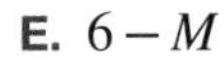

**19.** **MC** $M$ is used to represent an unknown whole number. Choose the answer that matches the description 'a fifth of the number'.

**A.** $5M$ **B.** $M-5$ **C.** $\frac{M}{5}$ **D.** $M+\frac{1}{5}$ **E.** $\frac{1}{5}-M$

**20.** **MC** $M$ is used to represent an unknown whole number. Choose the answer that matches the description 'the number just before that number'.

**A.** $M+1$ **B.** $M-1$ **C.** $M$ **D.** $M+2$ **E.** $M-2$

**21.** **MC** $M$ is used to represent an unknown whole number. Choose the answer that matches the description 'eight more than the product of that number and 10'.

**A.** $10(M+8)$ **B.** $10M+8$ **C.** $8M+10$ **D.** $8(M+10)$ **E.** $88M$

**22.** **MC** $M$ is used to represent an unknown whole number. Choose the answer that matches the description 'five more than three times the number'.

**A.** $5M+3$ **B.** $3M-5$ **C.** $3M+5$ **D.** $5M-3$ **E.** $3(M-5)$

**23.** Answer True or False for each of the statements below.

**a.** $3x$ is a term. **b.** $3\,mn$ is a term.
**c.** $g=23-t$ is an expression. **d.** $g=5t-6$ is an equation.

**24.** Answer True or False for each of the statements below.

**a.** $rt=r\times t$ **b.** $5+A=A+5$
**c.** $3d+5$ is an expression. **d.** $7-B=B-7$

**25.** Answer True or False for each of the statements below.

**a.** $3x=9$ is an expression. **b.** The expression $g+2t$ has two terms.
**c.** $2f+4=4+2f$ **d.** $5a-7t+h$ is an equation.

**26.** There are $a$ apples at your house. At your friend's house there are six more apples than there are at your house. Your friend uses eight apples to make an apple pie. Use this information to write an expression describing the number of apples at your friend's house:

**a.** before making the apple pie
**b.** after making the apple pie.

27. Your friend spends $\$d$ on a bus fare each day.
  a. If she uses the bus three days a week, state how much she spends on bus fares for the week.
  b. There are forty school weeks in a year and she travels three times a week on the bus. Write an expression for how much it will cost her for the bus fare each year.
  c. It costs an adult an extra \$1.50 for the same bus trip. Write an expression for the cost of an adult catching the bus three days per week.

### Reasoning

28. Compare the expressions $\frac{x+1}{3}$ and $\frac{x}{3}+1$. Give possible examples for each expression.

29. Explain the difference between $y+y+y$ and $y \times y \times y$

30. $p$ represents a number between 1 and 2 as shown:

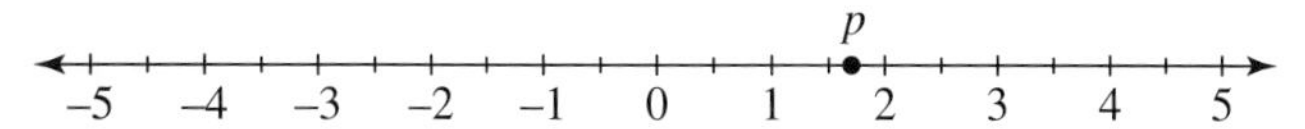


Draw a number line for each of the following:

a. $p+1$ b. $p-2$ c. $3-p$

### Problem solving

31. a. Ali earns \$7 more per week than Halit. If Halit earns $\$x$, write an expression for their total earnings per week.
  b. Sasha is twice as old as Kapila. If Kapila is $x$ years old, write an expression for their total age.
  c. Frank has had 3 more birthdays than James. If James is $2x$ years old, write an expression for their total age.

32. a. Write a rule to describe how you would calculate the total cost of two hamburgers and one drink.
  b. Let $h$ represent the cost of a hamburger. If the drink costs \$3.75 and the total costs \$7.25, write an equation that links all this information.

33. Two friends each had five \$1 coins and a piggy bank full of 20 c coins. They combined their money and spent it all at a games arcade. Assume that both friends had the same amount of money in their piggy banks.
  a. Write an expression for the total amount of money that the two friends had.
  b. If there were 85 coins in each piggy bank, calculate the total amount of money spent.

34. Ten couples go out for dinner. If everyone shakes hands once with everyone else except their partner, determine how many handshakes there are.

# LESSON
# 7.3 Substituting and evaluating

**LEARNING INTENTION**

At the end of this lesson you should be able to:
- substitute a number for a pronumeral to evaluate an algebraic expression.

## 7.3.1 Substituting

eles-3988

- The value of an expression can be determined by substituting in values that are represented by the pronumeral.
- **Substitution** means replacing the pronumeral with the number it represents.

**WORKED EXAMPLE 5 Substituting into an expression**

**If $n = 3$, evaluate the following expressions.**

**a.** $n + 4$ **b.** $2n$ **c.** $4n - 2$

| THINK | WRITE |
|---|---|
| **a. 1.** Substitute $n = 3$ into the expression. | **a.** $n + 4 = 3 + 4$ |
| **2.** Perform the addition and write the answer. | $= 7$ |
| **b. 1.** Substitute $n = 3$ into the expression. | **b.** $2n = 2 \times 3$ |
| **2.** Perform the multiplication and write the answer. | $= 6$ |
| **c. 1.** Substitute $n = 3$ into the expression. | **c.** $4n - 2 = 4 \times 3 - 2$ |
| **2.** Perform the multiplication first. | $= 12 - 2$ |
| **3.** Perform the subtraction and write the answer. | $= 10$ |

**WORKED EXAMPLE 6 Substituting to complete a table of values**

**If $x = 3t - 6$, determine the value of $x$ by substituting the value of $t$ given.**

| $t$ | 1 | 2 | 5 | 12 |
|---|---|---|---|---|
| $x$ | | | | |

| THINK | WRITE |
|---|---|
| **1.** Write the formula. | $x = 3t - 6$ |
| **2.** Substitute $t = 1$ into the formula. | $t = 1,\ x = 3 \times 1 - 6$ |

3. Evaluate the right-hand side. $x = 3 - 6 = -3$

4. Substitute $t = 2$ into the formula. $t = 2, x = 3 \times 2 - 6$

5. Evaluate the right-hand side. $x = 12 - 6 = 6$

6. Substitute $t = 5$ into the formula. $t = 5, x = 3 \times 5 - 6$

7. Evaluate the right-hand side. $x = 15 - 6 = 9$

8. Substitute $t = 12$ into the formula. $t = 12, x = 3 \times 12 - 6$

9. Evaluate the right-hand side. $x = 36 - 6 = 30$

10. Complete the table.

| $t$ | 1 | 2 | 5 | 12 |
|---|---|---|---|---|
| $x$ | −3 | 6 | 9 | 31 |

## WORKED EXAMPLE 7 Substituting into a formula

**When researching holiday destinations, Jack saw that the temperature in Hawaii was 88 °F. They use the formula $C = \frac{5}{9}(F - 32)$ to convert this temperature to Celsius. Calculate the temperature in Hawaii in degrees Celsius.**

**THINK**

1. Write the formula.

2. Substitute $F = 88$ into the equation.

3. Evaluate the right-hand side and write the answer.

**WRITE**

$C = \frac{5}{9}(F - 32)$

$C = \frac{5}{9}(88 - 32)$

$= 31.1$ °C

### Digital technology

Scientific calculators have a fraction button that can be used to compute calculations involving fractions.

DEG

$\frac{5}{9}(88-32)$ ▸f◂▸d

31.11111111

## WORKED EXAMPLE 8 Substituting into a formula

**Calculate the value of $m$ by substituting the given value of $x$ into the following formula.**

$$m = 3(2x + 3)$$
$$x = 4$$

| THINK | WRITE |
|---|---|
| 1. Write the formula. | $m = 3(2x + 3)$ |
| 2. Substitute $x = 4$ into the equation. | $m = 3(2 \times 4 + 3)$ |
| 3. Evaluate the right-hand side of the equation. Use BIDMAS to perform the operations in the correct order. | $m = 3(8 + 3)$<br>$m = 3(11)$<br>$m = 3 \times 11$<br>$m = 33$ |

### on Resources

**eWorkbook** Topic 7 Workbook (worksheets, code puzzle and project) (ewbk-1908)

**Video eLesson** Substitution (eles-2426)

**Interactivities** Individual pathway interactivity: Substitution (int-4351)
Substitution (int-4003)

## Exercise 7.3 Substituting and evaluating

learnon

7.3 Quick quiz on | 7.3 Exercise

### Individual pathways

| ■ PRACTISE | ■ CONSOLIDATE | ■ MASTER |
|---|---|---|
| 1, 4, 7, 8, 12, 17, 20 | 2, 5, 9, 11, 13, 14, 18, 21 | 3, 6, 10, 15, 16, 19, 22 |

### Fluency

1. **WE5** If $a = 4$, evaluate the following expressions.
   a. $a + 7$
   b. $5a$
   c. $2a - 10$
   d. $3 + 3a$

2. If $b = 6$, evaluate the following expressions.
   a. $b - 7$
   b. $11b$
   c. $3b + 1$
   d. $-8b + 7$

3. If $c = 3$, evaluate the following expressions.
   a. $c - 16$
   b. $-4c$
   c. $12 + 7c$
   d. $11 - c$

4. **WE6** Substitute the given values into each formula to determine the value of $d$ if $d = g - 2$.
   a. $g = 4$
   b. $g = 5$
   c. $g = 2$
   d. $g = 102$

5. Substitute the given values into each formula to determine the value of $e$ if $e = 2t - 3$.
   a. $t = 7$
   b. $t = 2$
   c. $t = 100$
   d. $t = 8$

6. Substitute the given values into each formula to determine the value of $f$ if $f = 12h + 7$.

a. $h = 1$ b. $h = 0$ c. $h = 5$ d. $h = 20$

7. Substitute the given values into each formula to determine the value of $g$ if $g = 25 - 4w$.

a. $w = 1$ b. $w = 3$ c. $w = 6$ d. $w = 0$

8. **WE8** Calculate the value of $h$ by substituting the given value of the pronumeral into the formula $h = 2(g + 1)$.

a. $g = 1$ b. $g = 0$ c. $g = 12$ d. $g = 75$

9. Calculate the value of $i$ by substituting the given value of the pronumeral into the formula $i = 5(x - 2)$.

a. $x = 6$ b. $x = 10$ c. $x = 11$ d. $x = 2$

10. Calculate the value of $j$ by substituting the given value of the pronumeral into the formula $j = 4(12 - p)$.

a. $p = 2$ b. $p = 3$ c. $p = 12$ d. $p = 11$

11. Calculate the value of $k$ by substituting the given value of the pronumeral into the formula $k = 5(2g - 3)$.

a. $g = 2$ b. $g = 14$ c. $g = 5$ d. $g = 9$

## Understanding

12. Calculate the value of $l$ by substituting the given value of the pronumeral into the formula $l = 2(d + 2) - 3$.

a. $d = 3$ b. $d = 0$ c. $d = 7$ d. $d = 31$

13. Calculate the value of $m$ by substituting the given value of the pronumeral into the formula $m = 3(f - 1) + 17$.

a. $f = 1$ b. $f = 3$ c. $f = 6$ d. $f = 21$

14. Calculate the value of $n$ by substituting the given value of the pronumeral into the formula $n = 4s - s$.

a. $s = 3$ b. $s = 1$ c. $s = 101$ d. $s = 72$

15. Calculate the value of $o$ by substituting the given value of the pronumeral into the formula $o = 3(y + 5) - 2$.

a. $y = 1$ b. $y = 5$ c. $y = 0$ d. $y = 12$

16. Calculate the value of $p$ by substituting the given value of the pronumeral into the formula $p = 50 - 6v$.

a. $v = 4$ b. $v = 7$ c. $v = 1$ d. $v = 8$

17. The area of a triangle is given by $A = \frac{1}{2} \times w \times h$. Find the height $h$ and width $w$ that gives an area of 12 square units.

## Reasoning

18. **WE7** The formula $M = 0.62K$ can be used to convert distances in kilometres ($K$) into miles ($M$). Use the formula to convert the following distances into miles:

a. 100 kilometres b. 250 kilometres c. 15 kilometres

19. The formula for finding the perimeter ($P$) of a rectangle of length $l$ and width $w$ is $P = 2l + 2w$. Use this formula to calculate the perimeter of the rectangular swimming pool shown.

### Problem solving

20. Determine the values of $x$ when the result of $5x-2$ is a two-digit number.

21. In Australian rules football, a goal is worth 6 points and a behind is worth 1 point. The equation $S=6g+b$ can be used to determine a team's total score ($S$) from the number of goals ($g$) and behinds ($b$) that it scores.
   a. The Giants kicked 16 goals and 13 behinds. Calculate their total score.
   b. The Swans kicked 12 goals and 21 behinds. Calculate their total score.
   c. The Swans had a total of 33 scoring shots. If they'd been more accurate and kicked 15 goals instead of 12, would they have scored more than the Giants?

22. When he was a young man, the mathematician Carl Friedrich Gauss discovered that the sum ($S$) of the first $n$ natural numbers is given by the formula $S=\frac{n(n+1)}{2}$.
   Using this formula, find the answer to the sum of the following series of numbers:
   a. $1+2+3+4+\ldots+46+47+48+49$
   b. $54+55+56+57+\ldots+85+86+87+88$

# LESSON
# 7.4 Simplifying expressions using like terms

**LEARNING INTENTION**

At the end of this lesson you should be able to:
- simplify expressions by adding or subtracting like terms.

## 7.4.1 Simplifying expressions containing like terms

eles-3989

- Terms that contain the exact same pronumerals are called **like terms**.
- Terms that do not contain the exact same pronumerals are called **unlike terms**.
  Examples of like and unlike terms are:

| Like terms | Unlike terms |
|---|---|
| $2a$ and $5a$ | $2a$ and $5b$ |
| $x$ and $13x$ | $3x$ and $5xy$ |
| $2bc$ and $4cb$ | $4abc$ and $3cbd$ |
| $3g^2$ and $45g^2$ | $16g$ and $45g^2$ |

- Note that the order of pronumerals in a term does not matter. The terms $ab$ and $ba$ are like terms because they both contain the exact same pronumerals.
- Like terms can be combined by adding or subtracting coefficients. For example, $3a+2a=5a$.

- Unlike terms can't be combined or simplified in any way. For example $3a + 2b$ cannot be written in a simpler way.

### WORKED EXAMPLE 9 Simplifying expressions

**Where possible, simplify the following expressions by adding or subtracting like terms.**

**a.** $4g + 6g$ **b.** $11ab - ab$ **c.** $6ad + 5da$ **d.** $4t + 7t - 5$ **e.** $8x + 3y$

| THINK | WRITE |
|---|---|
| **a.** The pronumerals are the same. The terms can be added. | **a.** $4g + 6g = 10g$ |
| **b.** The pronumerals are the same. The terms can be subtracted. Note: *ab* is the same as 1*ab* or 1*ba* or *ba*. | **b.** $11ab - ab = 10ab$ |
| **c.** Although the order of the pronumerals is different, the terms are like terms and can be added. | **c.** $6ad + 5da = 6ad + 5ad$ <br> $= 11ad$ |
| **d.** The pronumerals are the same and the first two terms can be added. Do not subtract the constant term. | **d.** $4t + 7t - 5 = 11t - 5$ |
| **e.** The pronumerals are **not** the same. These are not like terms and therefore cannot be simplified. | **e.** $8x + 3y$ cannot be simplified. |

### WORKED EXAMPLE 10 Evaluating expressions containing like terms

**First simplify the following expressions, then determine the value of the expression by substituting $a = 4$.**

**a.** $5a + 2a$ **b.** $7a - a + 5$

| THINK | WRITE |
|---|---|
| **a. 1.** Add the like terms. | **a.** $5a + 2a = 7a$ |
| **2.** Substitute $a = 4$ into the simplified expression. | $= 7 \times 4$ <br> $= 28$ |
| **b. 1.** Subtract the like terms. | **b.** $7a - a + 5 = 6a + 5$ |
| **2.** Substitute $a = 4$ into the simplified expression. | $= 6 \times 4 + 5$ <br> $= 24 + 5$ <br> $= 29$ |

### DISCUSSION

$3x^2 + 4x^2 = 7x^2$

The terms $3x^2$ and $4x^2$ are called *like terms*. Explain why.

$3x^2$ and $4x$ are called *unlike terms*. Explain why.

## Equivalent algebraic expressions

- Terms that are identical but are displayed in different ways are called **equivalent expressions**.
  For example, you can simplify the expression $x+x+y+y-2$ to $2x+2y-2$. This means that $x+x+y+y-2$ and $2x+2y-2$ are equivalent expressions.
- The following are all pairs of equivalent expressions:

$$a+a+a+a=4a \qquad b\times b=b^2 \qquad c\times d=cd \qquad e\div f=\frac{e}{f}$$

### COLLABORATIVE TASK: Silently collecting like terms

**Equipment:** 25–30 cards, each containing an algebraic term. Across the set of cards there should be 5–6 different pronumerals, and each card should have different coefficients.

1. Each student chooses a card with a term written on it. They must not look at their cards or talk.
2. Each student displays their card facing outward so that the rest of the class can see their term, but the student holding the card cannot.
3. Without talking or looking at their own card, students then help each other to form groups around the classroom with other like terms.
4. Once the class is happy that they are all in like-term groups, students can look at their own card and talk.
5. Each group then simplifies by adding and subtracting like terms.
6. Each group then writes their part of the expression on the board to provide the fully simplified class expression.
7. Discuss any difficulties experienced when gathering and simplifying like terms.

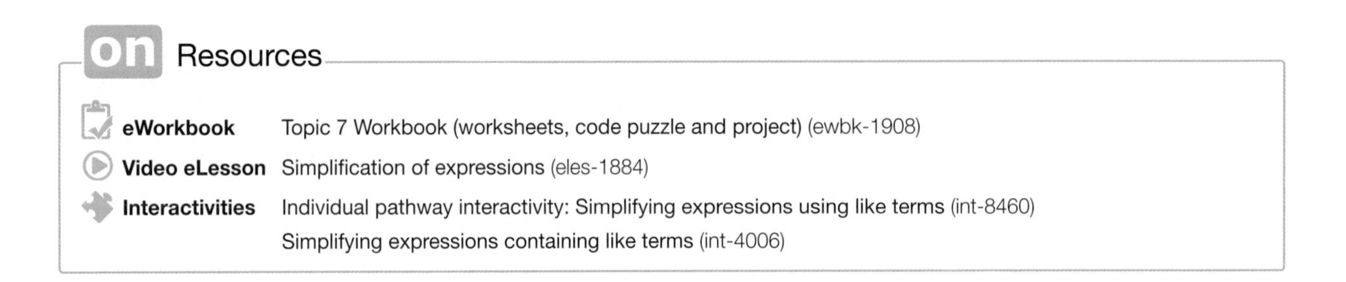

**on** Resources

**eWorkbook** Topic 7 Workbook (worksheets, code puzzle and project) (ewbk-1908)

**Video eLesson** Simplification of expressions (eles-1884)

**Interactivities** Individual pathway interactivity: Simplifying expressions using like terms (int-8460)

Simplifying expressions containing like terms (int-4006)

## Exercise 7.4 Simplifying expressions using like terms

learn**on**

| 7.4 Quick quiz **on** | 7.4 Exercise |
|---|---|

### Individual pathways

| ■ PRACTISE | ■ CONSOLIDATE | ■ MASTER |
|---|---|---|
| 1, 4, 7, 9, 12, 15, 18, 21, 22, 25 | 2, 5, 8, 10, 13, 16, 20, 23, 26 | 3, 6, 11, 14, 17, 19, 24, 27, 28 |

### Fluency

1. **WE9** Where possible, simplify the following expressions by adding or subtracting like terms.
   a. $3a+2a$ b. $9y+5y$ c. $3c+12c$ d. $4u-2u$ e. $7e+13e$
2. Where possible, simplify the following expressions by adding or subtracting like terms.
   a. $7t-2t$ b. $12ab+2ab$ c. $8fg-2fg$ d. $4e-e$ e. $6t+t$

3. Where possible, simplify the following expressions by adding or subtracting like terms.

a. $f+4f$ b. $6y-6y$ c. $4x+14x$ d. $3m+16m-7m$ e. $6a+4a-2a$

4. Simplify the following expressions (if possible).

a. $24ab+ab-7$ b. $5y+y-3y$ c. $5t+5s$ d. $4t+8t-3+2t$ e. $7r+2r+5r-r$

5. Simplify the following expressions (if possible).

a. $7y+6$ b. $4+3g-g$ c. $6t-5t$ d. $18bg-18bg$ e. $11pq+3qp$

6. Simplify the following expressions (if possible).

a. $7t+4t-5$ b. $32t-31t$ c. $7xy-7x$ d. $5t+6t-8$ e. $5t-t+3$

7. Simplify each of the following.

a. $3t-3t$ b. $18r-18r$ c. $12ab-12ab$ d. $5x-5x+8$

8. Simplify each of the following.

a. $6t+7-6t$ b. $13xyz-13xyz$ c. $5x+7-5x$ d. $5y+2y-y$

9. **WE10** Simplify the following expressions, then determine the value of the expression by substituting $a=7$.

a. $3a+2a$ b. $7a+2a$ c. $6a-2a$ d. $9a+a$

10. First simplify the following expressions, then determine the value of the expression by substituting $a=7$.

a. $13a+2a-5a$ b. $3a+7a$ c. $17+5a+3a$ d. $6a-a+2$

11. First simplify the following expressions, then determine the value of the expression by substituting $a=7$.

a. $a+a$ b. $7a-6a$ c. $7a-7a$ d. $12a+5a-16$

## Understanding

12. Answer True or False to each of the following statements.

a. $4t$ and $6t$ are like terms. b. $3x$ and $x$ are like terms. c. $5g$ and $5t$ are like terms.

d. $5a$ and $6a$ are like terms. e. $4a$ and $4ab$ are like terms.

13. Answer True or False to each of the following statements.

a. $6gh$ and $7gk$ are like terms. b. $yz$ and $45zy$ are like terms. c. $3acd$ and $6cda$ are like terms.

d. $5g$ and $5fg$ are like terms. e. $8gefh$ and $3efgh$ are like terms.

14. Answer True or False to each of the following statements.

a. $6ab$ and $3ba$ are like terms. b. $8xy$ and $5xy$ are like terms. c. $7eg$ and $7g$ are like terms.

d. $7y$ and $18yz$ are like terms. e. $12ep$ and $4pe$ are like terms.

15. **MC** Which one of the following is a like term for $7t$?

A. $7g$ B. $5tu$ C. $2$ D. $6f$ E. $5t$

16. **MC** Which one of the following is a like term for $3a$?

A. $5ab$ B. $6a$ C. $3w$ D. $a+2$ E. $3+a$

17. **MC** Which one of the following is a like term for $5ab$?

A. $6abe$ B. $4b$ C. $4ba$ D. $5$ E. $5abe$

18. **MC** Which one of the following is a like term for $7bgt$?

A. $7t$ B. $5gt$ C. $2gb$ D. $btg$ E. $a+b+e$

19. **MC** Which one of the following is a like term for $2fgk$?

A. $fg+k$ B. $5fg$ C. $2fgk$ D. $2f$ E. $12p$

20. **MC** Which one of the following is a like term for $20m$?

A. $11m$ B. $20mn$ C. $20f$ D. $m+2$ E. $20$

21. **MC** Which one of the following is a like term for $9xyz$?

A. $9x$ B. $2yz$ C. $8xz$ D. $xzy$ E. $9z$

## Reasoning

22. Answer True or False to each of the following statements and justify your answer.
    a. The equation $y=4x+3x$ can be simplified to $y=7x$.
    b. The equation $k=8y+4y$ can be simplified to $k=10y$.
    c. The equation $y=4x+3$ can be simplified to $y=7x$.

23. Answer True or False to each of the following statements and justify your answer.
    a. The equation $b=3a-a$ can be simplified to $b=2a$.
    b. The equation $k=7a+4d$ can be simplified to $k=11ad$.
    c. The equation $y=5x-3x$ can be simplified to $y=2x$.

24. Answer True or False to each of the following statements and justify your answer.
    a. The equation $m=7+2x$ can be simplified to $m=9x$.
    b. The equation $t=3h+12h+7$ can be simplified to $t=15h+7$.
    c. The equation $y=16g+6g-7g$ can be simplified to $y=15g$.

## Problem solving

25. There are $x$ chocolates in each box. Ahmed buys 7 boxes while Kevin buys 3 boxes. Write an expression for the total number of chocolates.

26. There are $d$ mangoes in a kilogram. Nora buys 4 kg but they are so good she buys 2 kg more. Write an expression for the total number of mangoes Nora bought.

27. Find the simplest expression for the perimeter of the following shapes:

a.

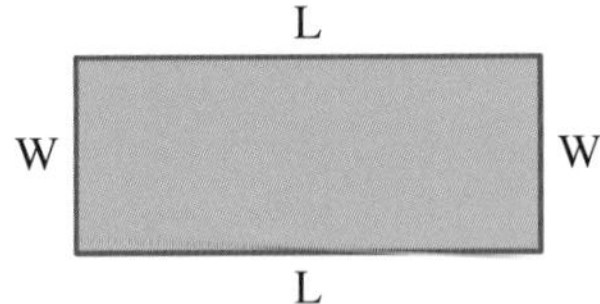


b.

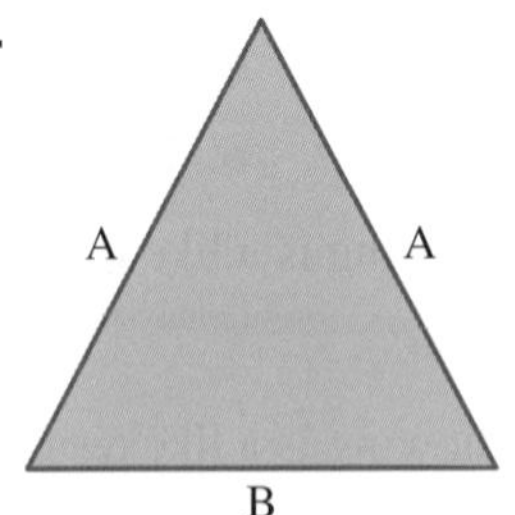


28. Given that OD bisects COE, that $\angle COD=(4x+16)^\circ$ and that $\angle DOE=(6x+4)^\circ$, explain if there is enough information for you to determine the angles for COD and DOE. If so, what are the measures for each?

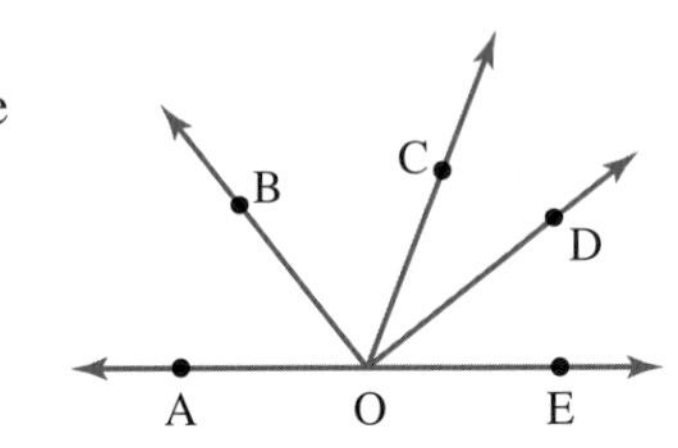

# LESSON
# 7.5 Multiplying and dividing terms

## LEARNING INTENTION

At the end of this lesson you should be able to:
- simplify expressions when multiplying, using indices where appropriate
- simplify expressions when diving by identifying common factors.

## 7.5.1 Multiplying terms

eles-3990

- We multiply algebraic terms using this process:
  - Split each term into its coefficient and its pronumerals.
  - Determine the product of the coefficient of the terms.
  - Combine like factors into a power. For example, $x \times x = x^2$.
  - Combine the coefficient and pronumerals into one term.

Unlike adding and subtracting, when we multiply algebraic terms, we can collect them into one term.

### WORKED EXAMPLE 11 Simplify by using index notation where appropriate

**Simplify:**

**a.** $2 \times k$ **b.** $3 \times 2k$ **c.** $k \times 2kx$ **d.** $3k \times 2k$ **e.** $3k \times 2k^2x$

| THINK | WRITE |
|---|---|
| a. 1. Remove the multiplication sign. | a. $2 \times k$ <br> $= 2k$ |
| b. 1. Rewrite in expanded form. | b. $3 \times 2 \times k$ |
| 2. Multiply the numbers. (the coefficients). | $= 6 \times k$ |
| 3. Remove the multiplication sign. | $= 6k$ |
| c. 1. Rewrite in expanded form. | c. $k \times 2 \times k \times x$ |
| 2. Multiply the like pronumerals. | $= 2 \times k^2 \times x$ |
| 3. Remove the multiplication sign. | $= 2k^2x$ |
| d. 1. Rewrite in expanded form. | d. $3 \times k \times 2 \times k$ |
| 2. Multiply the numbers and write pronumerals with index. | $= 6 \times k^2$ |
| 3. Remove the multiplication sign. | $= 6k^2$ |
| e. 1. Rewrite in expanded form. | e. $3 \times k \times 2 \times k \times k \times x$ |
| 2. Multiply the numbers and write pronumerals with index. | $= 6 \times k^3 \times x$ |
| 3. Remove the multiplication sign. | $= 6k^3x$ |

**Multiplying pronumerals**

**When we multiply pronumerals, order is not important. For example:**

$$3 \times 6 = 6 \times 3 \qquad 6 \times w = w \times 6 \qquad a \times b = b \times a$$

- **The multiplication sign (×) is usually omitted for reasons of convention.**

$$3 \times g \times h = 3gh$$

- **Although order is not important, conventionally the pronumerals in each term are written in alphabetical order. For example:**

$$2 \times b \times a \times c = 2abc$$

## 7.5.2 Dividing terms

eles-3991

- We divide algebraic terms using this process:
  - Split each term into its coefficient and its pronumerals.
  - Determine the quotient of the coefficient of the terms.
  - Cancel any common factors. For example, $\frac{x}{x} = \frac{1x}{x} = 1$.
  - Combine the coefficient and pronumerals into one term.

Unlike adding and subtracting, when we divide algebraic terms, we can collect them into one term.

**WORKED EXAMPLE 12 Simplify by identifying and simplifying common factors**

**Simplify:**

a. $\frac{16f}{4}$ b. $16f \div 2f$ c. $\frac{4f}{16f^2}$ d. $\frac{36fg}{40f^2g}$

| THINK | WRITE |
|---|---|
| a. 1. Write in expanded form. | a. $\frac{16 \times f}{4}$ |
| 2. Divide numerator and denominator by the highest common factor (HCF), 4. | $= \frac{{}^{4}\cancel{16} \times f}{{}_{1}\cancel{4}}$ |
| 3. Write in simplest terms. | $= 4f$ |
| b. 1. Write as a fraction in expanded form. | b. $\frac{16 \times f}{2 \times f}$ |
| 2. Divide numerator and denominator by HCF 2 and $f$. | $= \frac{{}^{8}\cancel{16} \times \cancel{f}}{{}_{1}\cancel{2} \times \cancel{f}}$ |
| 3. Write in simplest terms. | $= 8$ |
| c. 1. Write in expanded form. | c. $\frac{4 \times f}{16 \times f \times f}$ |

2. Divide numerator and denominator by HCF, 4 and $f$. $= \dfrac{{}^{1}\cancel{4} \times \cancel{f}}{{}_{4}\cancel{16} \times \cancel{f} \times f}$

3. Write in simplest terms. Omit the multiplication sign. $= \dfrac{1}{4f}$

d. 1. Write in expanded form. d. $\dfrac{36 \times f \times g}{40 \times f \times f \times g}$

2. Divide numerator and denominator by HCF 4 and $f$. $= \dfrac{{}^{9}\cancel{36} \times \cancel{f} \times \cancel{g}}{{}_{10}\cancel{40} \times \cancel{f} \times f \times \cancel{g}}$

3. Write in simplest terms. Omit the multiplication sign. $= \dfrac{9}{10f}$

**Resources**

eWorkbook Topic 7 Workbook (worksheets, code puzzle and project) (ewbk-1908)

**Interactivity** Individual pathway interactivity: Multiplying and dividing terms (int-8343)

## Exercise 7.5 Multiplying and dividing terms

learnon

**7.5 Quick quiz**  **7.5 Exercise**

### Individual pathways

| ■ PRACTISE | ■ CONSOLIDATE | ■ MASTER |
|---|---|---|
| 1, 2, 6, 9, 11, 14 | 3, 5, 7, 12, 15 | 4, 8, 10, 13, 16 |

### Fluency

1. WE11 Simplify the following.
   a. $2a \times a$ b. $ab \times 7a$ c. $7pq \times 3p$ d. $3b^2 \times 2cd$ e. $20m^{12} \times 2m^3$

2. Simplify the following.
   a. $4 \times 3g$ b. $7 \times 3h$ c. $4d \times 6$ d. $3z \times 5$ e. $6 \times 5r$

3. Simplify the following.
   a. $5t \times 7$ b. $4 \times 3u$ c. $7 \times 6p$ d. $7gy \times 3$ e. $2 \times 11ht$

4. Simplify the following.
   a. $4x \times 6g$ b. $10a \times 7h$ c. $9m \times 4d$ d. $3c \times 5h$ e. $9g \times 2x$

5. WE12 Simplify the following.
   a. $\dfrac{8f}{2}$ b. $\dfrac{6h}{3}$ c. $\dfrac{15x}{3}$ d. $9g \div 3$ e. $10r \div 5$

6. Simplify the following.

a. $4x \div 2x$ b. $8r \div 4r$ c. $\frac{16m}{8m}$ d. $14q \div 21q$ e. $\frac{3x}{6x}$

7. Simplify the following.

a. $\frac{12h}{14h}$ b. $50g \div 75g$ c. $\frac{8f}{24f}$ d. $35x \div 70x$ e. $24m \div 36m$

8. Simplify the following:

a. $\frac{3 \times a}{a^2}$ b. $\frac{25p^{12} \times 4q^7}{15p^2}$ c. $\frac{8x^3 \times 7y^2}{6x \times 14y^3}$ d. $w \times \frac{5}{w^2}$ e. $\frac{12x^3y^2}{6x^4 \times 14y^3}$

## Understanding

9. Simplify the following.

a. $3a \times a$ b. $-2p \times 2p$ c. $5 \times 2x \times x$ d. $ab \times 5a$

10. Simplify the following.

a. $-3xy \times 2 \times 6x$ b. $4pq \times 3p \times 2q$ c. $5m \times n \times 2nt \times -t$ d. $-7 \times xyz \times -3z \times -2y$

## Reasoning

11. A student shows the following working:

$$\frac{\cancel{2+3x}^1}{\cancel{5x}^1}$$

Explain whether you agree with their working.

12. Consider the expression $\frac{24xy^2}{7ab} \times \frac{14a^2}{27y}$.

a. Multiply the numerators and multiply the denominators, then cancel any common factors.
b. Cancel the common factors in the numerators and denominators first, then multiply the remaining factors accordingly.
c. Compare your answer to part a with your answer to part b. Which process do you feel is better to use?

13. Explain why $x \div (yz)$ is equivalent to $\frac{x}{yz}$ and $x \div (y \times z)$, but is not equivalent to $x \div y \times z$ or $\frac{x}{y} \times z$.

## Problem solving

14. Calculate the areas of the following shapes.

a.
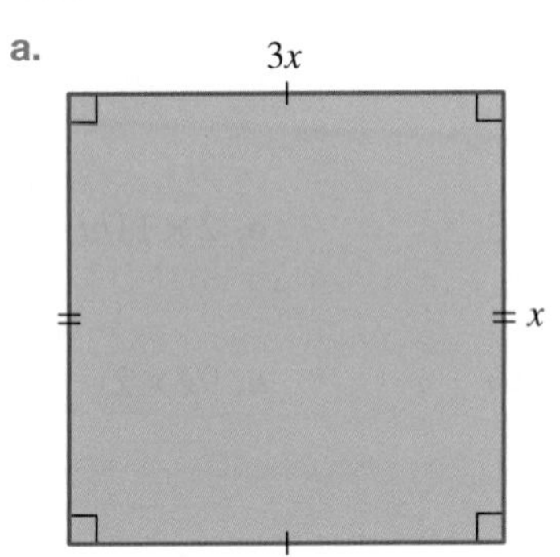


b.
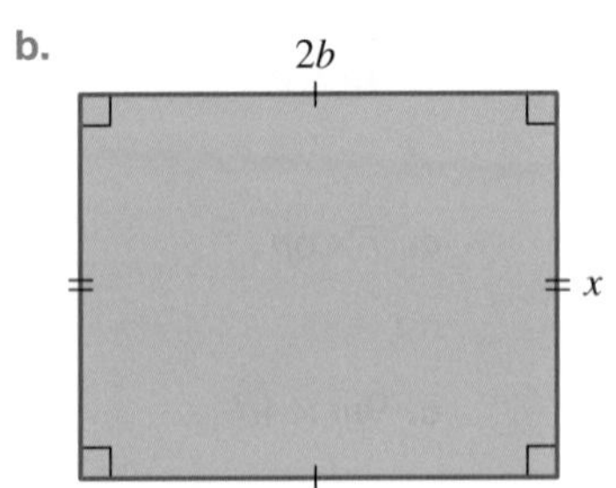


c.
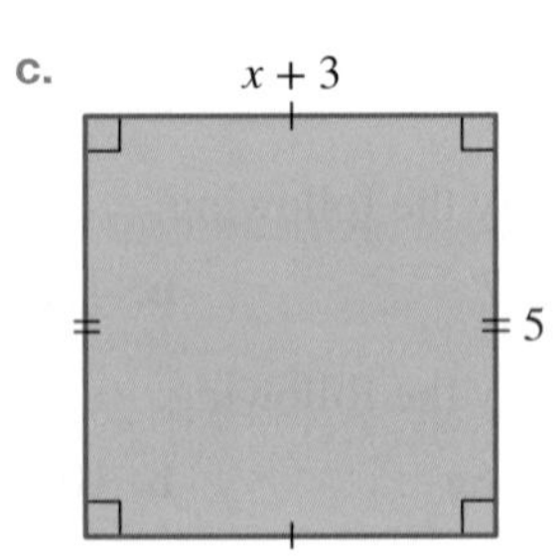

15. Evaluate the expression $\frac{4ht}{3dk} \times \frac{-12hk}{9dt}$, by substituting the values of $h=1$, $d=2$, $k=3$ and $t=4$.

16. A rectangle has been cut out of a corner of a square.

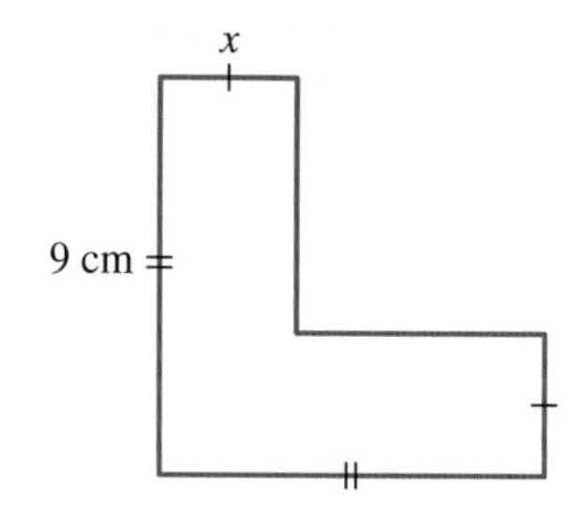


a. Find the perimeter of the shape.
b. Find the area of the shape if:

i. $x = 2$ cm  ii. $x = 3$ cm.

c. Find the area of the shape in terms of $x$.

# LESSON
# 7.6 Number laws

### LEARNING INTENTION

At the end of this lesson you should be able to:
- understand and apply the Commutative Law for addition and multiplication
- understand and apply the Associative Law for addition and multiplication
- understand that the commutative and associative laws do not apply to subtraction or division.

## 7.6.1 The Commutative Law

eles-3992

- The **Commutative Law** says we can swap numbers over and still get the same answer when adding and mutiplying.

**The Commutative Law for addition**

- $a + b = b + a$
  **For example, $5 + 3 = 8$ and $3 + 5 = 8$.**
- **This law *does not hold for subtraction*.**
  **For example, $5 - 3 \neq 3 - 5$.**

**The Commutative Law for multiplication**

- $a \times b = b \times a$
  **For example, $2 \times 3 = 6$ and $3 \times 2 = 6$.**
- **This law *does not hold for division*.**
  **For example, $6 \div 2 \neq 2 \div 6$.**

### WORKED EXAMPLE 13 Testing the Commutative Law

**Evaluate the following expressions if $a = 2$ and $b = 3$. Comment on the results obtained.**

a. i. $a + b$ ii. $b + a$
b. i. $5a - 2b$ ii. $2b - 5a$
c. i. $a \times b$ ii. $b \times a$
d. i. $a \div b$ ii. $b \div a$

| THINK | WRITE |
|---|---|
| a. i. 1. Substitute the correct value for each pronumeral. | a. i. $a + b = 2 + 3$ |
| 2. Evaluate and write the answer. | $= 5$ |
| ii. 1. Substitute the correct value for each pronumeral. | ii. $b + a = 3 + 2$ |
| 2. Evaluate and write the answer. | $= 5$ |
| 3. Compare the result with the answer obtained in part a i. | The same result is obtained; therefore, order is not important when adding two terms. |
| b. i. 1. Substitute the correct value for each pronumeral. | b. i. $5a - 2b = 5 \times 2 - 2 \times 3$ |
| 2. Evaluate and write the answer. | $= 4$ |
| ii. 1. Substitute the correct value for each pronumeral. | ii. $2b - 5a = 2 \times 3 - 5 \times 2$ |
| 2. Evaluate and write the answer. | $= -4$ |
| 3. Compare the result with the answer obtained in part b i. | Two different results are obtained; therefore, order is important when subtracting two terms. |
| c. i. 1. Substitute the correct value for each pronumeral. | c. i. $a \times b = 2 \times 3$ |
| 2. Evaluate and write the answer. | $= 6$ |
| ii. 1. Substitute the correct value for each pronumeral. | ii. $b \times a = 3 \times 2$ |
| 2. Evaluate and write the answer. | $= 6$ |
| 3. Compare the result with the answer obtained in part c i. | The same result is obtained; therefore, order is not important when multiplying two terms. |
| d. i. 1. Substitute the correct value for each pronumeral. | d. i. $a \div b = 2 \div 3$ |
| 2. Evaluate and write the answer. | $= \frac{2}{3} (\approx 0.67)$ |
| ii. 1. Substitute the correct value for each pronumeral. | ii. $b \div a = 3 \div 2$ |
| 2. Evaluate and write the answer. | $= \frac{3}{2} (1.50)$ |
| 3. Compare the result with the answer obtained in part d i. | Two different results are obtained; therefore, order is important when dividing two terms. |

## 7.6.2 The Associative Law

eles-3993

- The **Associative Law** says that it doesn't matter how we group the numbers (i.e. which we calculate first) when adding or multiplying.

## The Associative Law for addition

- $(a+b)+c=a+(b+c)=a+b+c$
  For example, $(9+3)+7=12+7=19$ and $9+(3+7)=9+10=19$.
- This law *does not hold for subtraction.*
  For example, $(8-3)-2=5-2=3$, but $8-(3-2)=8-1=7$.

## The Associative Law for multiplication

- $(a\times b)\times c=a\times(b\times c)=a\times b\times c$
  For example, $(2\times3)\times5=6\times5=30$ and $2\times(3\times5)=2\times15=30$.
- This law *does not hold for division.*
  For example, $(24\div6)\div2=4\div2=2$, but $24\div(6\div2)=24\div3=8$.

### WORKED EXAMPLE 14 Applying the Associative Law

**Use the Associative Law to complete the following.**

a. $2w+(8d+\_\_)=(\_\_+8d)+5h$

b. $(x\times\_\_)\times4g=\_\_(d\times4g)$

| THINK | WRITE |
|---|---|
| a. The Associative Law states that the numbers can be regrouped. $2w$, $8d$ and $5h$ are numbers. $5h$ is missing from the left-hand side and $2w$ is missing from the right-hand side. | a. $2w+(8d+5h)=(2w+8d)+5h$ |
| b. The Associative Law states that the numbers can be regrouped. $d$ is missing from the left-hand side and $x$ is missing from the right-hand side. | b. $(x\times d)\times4g=x\times(d\times4g)$ |

### WORKED EXAMPLE 15 Applying the associative and commutative laws

**Use the associative and commutative laws to complete the following.**

a. $4r+(6t+\_\_)=5g+(\_\_+\_\_)$

b. $a\times(3t\times2w)=(2w\times\_\_)\_\_a$

| THINK | WRITE |
|---|---|
| a. The associative and commutative laws state that the numbers can be regrouped and rearranged. $5g$ is missing from the left-hand side. Both $4r$ and $6t$ are missing from the right-hand side. | a. $4r+(6t+5g)=5g+(4r+6t)$ |
| b. The associative and commutative laws state that the numbers can be regrouped and rearranged. $3t$ and a multiplication sign are missing from the right-hand side. | b. $a\times(3t\times2w)=(2w\times3t)\times a$ |

Resources

**eWorkbook** Topic 7 Workbook (worksheets, code puzzle and project) (ewbk-1908)

**Interactivities** Individual pathway interactivity: Number laws (int-8461)
The associative law for multiplication (int-4008)
The commutative law for multiplication (int-4009)

# Exercise 7.6 Number laws

learnon

7.6 Quick quiz 

7.6 Exercise

## Individual pathways

| ■ PRACTISE | ■ CONSOLIDATE | ■ MASTER |
|---|---|---|
| 1, 2, 5, 9, 12 | 3, 6, 7, 10, 13 | 4, 8, 11, 14 |

### Fluency

1. **WE13** State whether the following are True or False. *Hint:* If you're not sure, try substituting values for the pronumerals to help you decide.
   a. $2s + (3w + 5z) = (2s + 3w) + 5z$
   b. $x \times (d + y) = (x \times d) + y$
   c. $g(jk) = (gj)k$
   d. $4 \div (a \div c) = (4 \div a) \div c$
2. **WE14** Use the Associative Law to complete the following.
   a. $w + (r\_\_6y) = (\_\_ + \_\_) + 6y$
   b. $6t \times (4r \times \_\_) = (6\_\_ \times 4r)\_\_3s$
   c. $(9y + \_\_) + 3w = \_\_ + (2r + 3w)$
   d. $(z + 2p) + \_\_ = z\_\_(\_\_ + 6t)$
3. **WE15** Use the associative and commutative laws to complete the following.
   a. $6t + (3w + \_\_) = \_\_ + (6t + 7v)$
   b. $s \times (9r \times \_\_) = 2c \times (\_\_ \times \_\_)$
   c. $(3c + w) + \_\_ = d + (\_\_ + \_\_)$
   d. $(g \times \_\_) \times 2y = 2y \times (3b \times \_\_)$
4. State whether the following are True or False. *Hint:* If you're not sure, try substituting values for the pronumerals to help you decide.
   a. $3g + (k \div m) = (3g + k) \div m$
   b. $4t - (p + 2b) = (4t - p) + 2b$
   c. $3r + (a + 4c) = a + (3r + 4c)$
   d. $a + (b \times c) = (a \times c) + b$

### Understanding

5. Calculate the values of the following expressions and comment on the results if $a = 3,\ b = 8$ and $c = 2$.
   a. $3a + (2b + 4c)$
   b. $(3a + 2b) + 4c$
6. Calculate the values of the following expressions and comment on the results if $a = 3,\ b = 8$ and $c = 2$.
   a. $9a - (2b - 5c)$
   b. $(9a - 2b) - 5c$
7. Determine the values of the following expressions and comment on the results if $a = 3,\ b = 8$ and $c = 2$.
   a. $a \times (b \times c)$
   b. $(a \times b) \times c$
8. Calculate the values of the following expressions and comment on the results if $a = 3,\ b = 8$ and $c = 2$.
   a. $4a \div (3b \div 2c)$
   b. $(4a \div 3b) \div 2c$

### Reasoning

9. The Commutative Law does not hold for subtraction. What can you say about the results of $x - a$ and $a - x$?
10. Explain whether the Associative Law holds when you are adding fractions.

### Problem solving

11. Use the numbers $1, 5, 6$ and $7$ with any arithmetic operations to give $21$ as the result.

12. a. If $x = 1$, $y = 2$ and $z = 3$, find the value of:
    i. $(x - y) - z$  ii. $x - (y - z)$.
    b. Comment on the answers with special reference to the Associative Law.
13. Janet and Judy want to buy a set of Blu-rays together. Janet has \$47 less than the cost of the Blu-rays and Judy has \$2 less than the cost. Both girls have at least \$1 each. If they pool their money, they still do not have enough to buy the Blu-rays. If the cost of the set is a whole number, calculate how much the Blu-rays cost.

14. Andrew thinks of a number, adds 9 and multiplies the result by 3. Andrea thinks of a number, multiplies it by 3 and adds 9 to the result. Both Andrew and Andrea get an answer of 60. Explain whether Andrew and Andrea began by selecting the same number. Show your working to support your answer.

# LESSON
# 7.7 Expanding brackets

**LEARNING INTENTION**

At the end of this lesson you should be able to:
- write expressions in an expanded form by using the Distributive Law.

## 7.7.1 The Distributive Law

eles-3994

- Expanding brackets means writing an equivalent expression without brackets.
- The Distributive Law can be used to expand or remove brackets.
- In the diagram, the blue rectangles represent $a$, the pink squares represent $b$ and the green rectangles represent $a + b$.
- As can be seen in the diagram, $3(a + b) = 3a + 3b$.
- This can also be determined by collecting like terms:

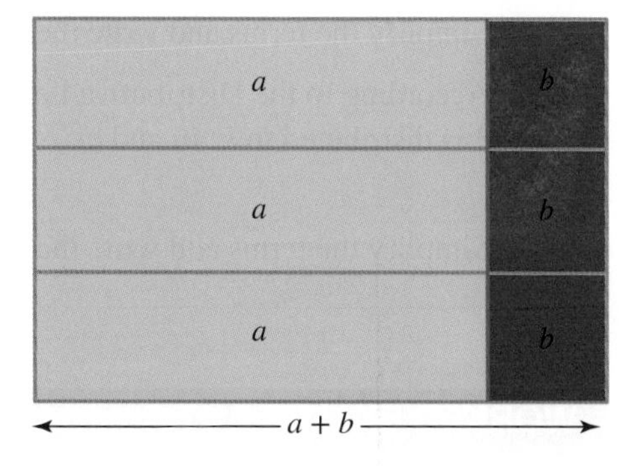


$$\begin{aligned}3(a+b) &= (a+b)+(a+b)+(a+b)\\ &= a+a+a+b+b+b\\ &= 3a+3b\end{aligned}$$

- Expressions such as this may be written in factorised form or expanded form.

$$\overbrace{3(a+b)}^{\text{factorised form}} = \overbrace{3a+3b}^{\text{expanded form}}$$

### The Distributive Law

**The Distributive Law provides a quick method of expanding expressions that are in factorised form.**

$$a(b + c) = ab + ac$$

**The term that is out at the front of the brackets gets distributed to each of the terms inside the brackets.**

**The Distributive Law also applies when the terms inside the brackets are subtracted.**

$$a(b - c) = ab - ac$$

- The Distributive Law holds true no matter how many terms there are inside the brackets. For example, $a(b + c - d) = ab + ac - ad$.
- The Distributive Law is not used when the terms inside the brackets are multiplied or divided. For example, $2(4 \times 5) = 2 \times 4 \times 5$ and not $(2 \times 4) \times (2 \times 5)$.

#### WORKED EXAMPLE 16 Applying the Distributive Law

**Write the following expressions in expanded form, then simplify them.**

a. $4(2a + d)$ b. $a(3x + 2y)$ c. $3(k + 4a - m)$

| THINK | WRITE |
|---|---|
| a. 1. According to the Distributive Law, the number 4 is distributed (or multiplied) to both $2a$ and $d$. | a. $4(2a + d) = 4 \times 2a + 4 \times d$ |
| 2. Simplify the terms and write the answer. | $= 8a + 4d$ |
| b. 1. According to the Distributive Law, the number represented by the pronumeral $a$ is distributed to both $3x$ and $2y$. | b. $a(3x + 2y) = a \times 3x + a \times 2y$ |
| 2. Simplify the terms and write the answer. | $= 3ax + 2ay$ |
| c. 1. According to the Distributive Law, the number 3 is distributed to $k$, $4a$ and $m$. | c. $3(k + 4a - m) = 3 \times k + 3 \times 4a - 3 \times m$ |
| 2. Simplify the terms and write the answer. | $= 3k + 12a - 3m$ |

#### WORKED EXAMPLE 17 Applying the Distributive Law

**Use the Distributive Law to expand and simplify the following.**

a. $4(2s + 3k) - 2k$ b. $a(3 - c) + a$

| THINK | WRITE |
|---|---|
| a. 1. Use the Distributive Law to expand $4(2s + 3k)$. | a. $4(2s + 3k) - 2k = 4 \times 2s + 4 \times 3k - 2k$ |

2. Simplify the terms. $= 8s + 12k - 2k$

3. Collect like terms and write the answer. $= 8s + 10k$

**b.** 1. Use the Distributive Law to expand $a(3-c)$. **b.** $a(3-c) + a = a \times 3 - a \times c + a$

2. Simplify the terms. $= 3a - ac + a$

3. Collect like terms and write the answer. $= 4a - ac$

Resources

**eWorkbook** Topic 7 Workbook (worksheets, code puzzle and project) (ewbk-1908)

**Interactivities** Individual pathway interactivity: Expanding brackets (int-8344)

The distributive law (int-4007)

## Exercise 7.7 Expanding brackets

learnon

**7.7 Quick quiz** on | **7.7 Exercise**

### Individual pathways

| ■ PRACTISE | ■ CONSOLIDATE | ■ MASTER |
|---|---|---|
| 1, 3, 6, 8, 11, 14 | 4, 7, 9, 12, 15 | 2, 5, 10, 13, 16 |

### Fluency

1. **WE16** Write the following expressions in expanded form, then simplify (if possible).
   a. $3(a + 2b)$ b. $5(x - 4z)$ c. $10(2g + 3h)$ d. $b(2c - 3a)$

2. Write the following expressions in expanded form, then simplify (if possible).
   a. $2c(8d - q)$ b. $3b(3b - 2g)$ c. $x(9a + 7b)$ d. $7k(t + 3m)$

3. **WE17** Use the Distributive Law to expand the following expressions.
   a. $5(a + 2b) - 3a$ b. $4(n - 2c) + n$ c. $9(b + 2c) - 9b$ d. $x(a - b) + x$

4. Use the Distributive Law to expand each of the following.
   a. $3(2d + 7e)$ b. $3h(6j - 5) + 4h$ c. $4(k + 5a - 3)$ d. $4x(x + 1) - 2(2x - 3)$

5. Use the Distributive Law to expand the following expressions.
   a. $y(3 + 2z) + 2y$ b. $3x(d - 3) + 4x$ c. $6u(2f - 1) + 20u$ d. $7n(10 + 3h) - 50n$

6. Expand each of the following expressions and then simplify by collecting like terms.
   a. $8(5r + 2) + 14r$ b. $3(t - 6) - 12$ c. $3y + 3(12 - 2y)$ d. $2x - (16 - 3x)$

7. Expand each of the following expressions and then simplify by collecting like terms.

a. $5 - 3t + 6(2t - 8)$
b. $14 + 3ry + 8(r + 2ry)$
c. $12ty + 2t(7y - 8h) + 6h$
d. $16r^2 + 12r(2 - r) - 6r + 3$

8. Expand each of the following expressions and then simplify by collecting like terms.

a. $3(x + 2) + 2(3 - x)$
b. $6(4h - 3) + 7h(4x - 8)$
c. $5(4x + y) - 2(3y + 3)$
d. $6(3 - t^2) - 8(t^2 + 3)$

### Understanding

9. Expand each of the following and then simplify.

a. $3(x - y) + 5(y + 2x)$
b. $5(a - 2y) + a(7 + 6y)$
c. $6c(g + 3d) + g(3c + d)$
d. $y(6d - r) + 2y(3d + 2r)$

10. In expressions such as $a(b + c) = ab + ac$, the left-hand side is in *factor form* and the right-hand side is in *expanded form*. Write each of the following in factor form.

a. $2x + 4$
b. $6x - 6y$
c. $4x + 12$
d. $12fg - 16gh$

### Reasoning

11. The formula for the perimeter ($P$) of a rectangle of length $l$ and width $w$ is $P = 2(l + w)$. Use the rule to calculate the perimeter of a magazine cover with $l = 28.6$ cm and $w = 21$ cm.

12. For each of the following expressions, show three different ways to make the statements true by placing appropriate terms in the boxes.

a. $\square(\square + 2) = 12ab + \square$
b. $\square(\square + 2) = 36kn + \square$

13. Consider the expression $2(1 - 5x)$. If $x$ is a negative integer, explain why the expression will be a positive integer.

### Problem solving

14. Expanding and simplifying the expression $2(8c - 16a)$ gives $16c - 32a$. Determine three expressions that give the same result.

15. a. Expand and simplify $2(x + y) - 3x$.
b. Verify that your simplified expression is correct by substituting $x = 2$ and $y = 3$ into both the original and simplified expressions.

16. The price of a pair of shoes is \$50. During a sale, the price of the shoes is discounted by \$$d$.

a. Write an expression to represent the sale price of the shoes.
b. If you buy three pairs of shoes during the sale, write an expression to represent the total purchase price:
   i. containing brackets
   ii. in expanded form (without brackets).
c. Write an expression to represent the total change you would receive from \$200 for the three pairs of shoes purchased during the sale.

# LESSON
# 7.8 Review

## 7.8.1 Topic summary

### The language of algebra

- **Pronumerals** are letters used to represent numbers.
  e.g. $t$: the number of tickets sold
- A pronumeral is also called a **variable**.
- The number in front of a pronumeral is called the **coefficient**.
  e.g. The coefficient of $7x$ is 7.
  The coefficient of $p$ is 1.
- **Terms** can contain a combination of numbers and one or more pronumerals, or may consist of a number only.
  e.g. $3t$, $2y$, $5gh$, 7 and $m$ are examples of terms.
- A **constant term** is a term containing only a number.
  e.g. 7, 10 and +4 are examples of constant terms.
- **Expressions** are made up of terms separated by + and − signs.
  e.g. $5x + 4 - 2y$ is an example of an expression.
  - This expression has 3 terms: $5x$, +4 and $-2y$ (+4 is called a constant term).
  - The coefficient of $x$ is 5 and the coefficient of $y$ is −2.

# ALGEBRA

### Substitution

- Substitution is the process of replacing a pronumeral with its known value.
- By using substitution, the value of an expression can be determined.

e.g. $2x + 5$ when $x = 6$ becomes:
$2(6) + 5 = 2 \times 6 + 5 = 12 + 5 = 17$

### Like terms

- **Like terms** have identical pronumeral parts.
  e.g. $3x$ and $x$ are like terms.
  $5ab$ and $-6ba$ are like terms.
  $-5xy$ and $11y$ are **not** like terms.
- Like terms can be added together or subtracted.

### Multiplying and dividing expressions

- When multiplying expressions, multiply the numbers and write the pronumerals in alphabetical order. Remove the multiplication sign.
  e.g. $5b \times 2a = 10ab$
- When dividing, write the division as a fraction and simplify by cancelling.
  e.g. $\frac{16f}{4} = 4f$

### Algebraic laws

- Only addition and multiplication are commutative. This means that the order in which we add or multiply does not matter. The **Commutative Law** for addition and multiplication states that:
  $$a + b = b + a \text{ and } a \times b = b \times a$$
- Only addition and multiplication are associative. This means that it does not matter how we group the numbers together when adding or multiplying. The **Associative Law** for addition and multiplication states that:
  $$(a + b) + c = a + (b + c) = a + b + c$$
  $$(a \times b) \times c = a \times (b \times c) = a \times b \times c$$
- The **Distributive Law** states that multiplying a number by a group of numbers added together is the same as doing each multiplication separately. The Distributive Law is useful for expanding brackets.
  The Distributive Law states that:
  $$\begin{aligned} a \times (b + c) &= a \times b + a \times c \\ &= ab + ac \end{aligned}$$

### Equivalent expressions

- Equivalent expressions are always equal when pronumerals are substituted.

e.g. $4 + 6x$ and $11 + 6x - 7$ are equivalent.

## 7.8.2 Success criteria

Tick the column to indicate that you have completed the lesson and how well you think you have understood it using the traffic light system.

(**Green:** I understand; **Yellow:** I can do it with help; **Red:** I do not understand)

| Lesson | Success criteria | Green | Yellow | Red |
|---|---|---|---|---|
| 7.2 | I understand the use of pronumerals in algebra. | | | |
| | I can identify algebraic terminology. | | | |
| | I can construct expressions from given information. | | | |
| 7.3 | I can substitute a number for a pronumeral to evaluate an algebraic expression. | | | |
| 7.4 | I can simplify expressions by adding or subtracting like terms. | | | |
| 7.5 | I can simplify expressions when multiplying, using indices where appropriate. | | | |
| | I can simplify expressions when dividing by identifying common factors. | | | |
| 7.6 | I understand and can apply the Commutative Law for addition and multiplication. | | | |
| | I understand and can apply the Associative Law for addition and multiplication. | | | |
| | I understand that the associative and commutative laws do not apply to subtraction or division. | | | |
| 7.7 | I can write expressions in expanded form by using the Distributive Law. | | | |

## 7.8.3 Project

### Landscaping

A town's local council wants to create more garden beds in the parklands around the town as part of its beautification program. All garden beds are to be square, but can vary in size. They will be surrounded by 1 $m^2$ pavers, as shown in the diagram.

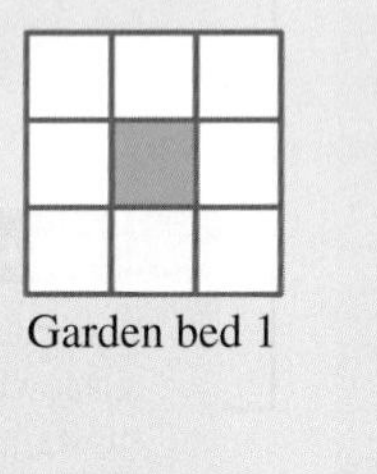

Garden bed 1

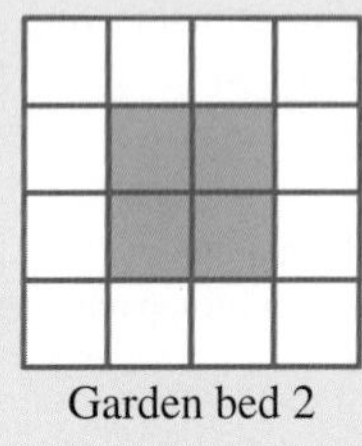

Garden bed 2

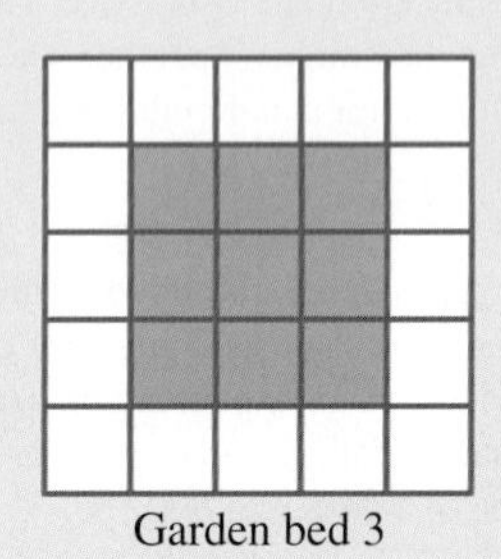

Garden bed 3

- The blue areas represent the square garden beds.
- The white squares represent the 1 $m^2$ pavers.

1. Use these three diagrams of the garden beds to complete the following table.

| Measurements | Garden bed 1 | Garden bed 2 | Garden bed 3 |
|---|---|---|---|
| Side length of the garden bed (m) | | | |
| Area of the garden bed $(m^2)$ | | | |
| Number of pavers used | | | |
| Total length around the outside of pavers (m) | | | |

Let $s$ represent the side length of the garden bed.
Let $a$ represent the area of the garden bed.
Let $p$ represent the number of pavers used.
Let $l$ represent the total length around the outside of the pavers.

2. Examine the patterns formed in your table and write formulas linking:
   - $a$ and $s$
   - $l$ and $s$
   - $p$ and $s$
   - $p$ and $l$.
3. You just got a job as the council's new landscape gardener. Your first task is to design a 10 m × 10 m square garden bed that is to be constructed in the same way as demonstrated in the previous diagram. How many pavers will be needed for this design?
4. The council has decided to put a large, square garden bed on the grounds near the entrance to its offices. The area of ground measures 18 m × 20 m. The council wants you to place the largest possible square garden bed here. You also have to put 1 $m^2$ pavers around it. Draw a design of the council's plan for this area of ground on a sheet of paper, with clear labels showing the measurements involved. Indicate the number of pavers that will be required for the project.

**on** Resources

**eWorkbook** Topic 7 Workbook (worksheets, code puzzle and project) (ewbk-1908)

**Interactivities** Crossword (int-2598)

Sudoku puzzle (int-3169)

## Exercise 7.8 Review questions

### Fluency

1. Simplify the following expressions by adding or subtracting like terms.
   a. $3g+4g$ b. $8y-2y$ c. $4h+5h$ d. $7ag-2ag$ e. $6gy-3yg$
2. Simplify the following expressions by adding or subtracting like terms.
   a. $8r-8r$ b. $6y-2y+y$ c. $4t+6+3t$ d. $12gh+6hg$ e. $8t-2m+3t$
3. Simplify the following expressions by adding or subtracting like terms.
   a. $3m+m$ b. $7g+8g+8+4$ c. $7h+4t-3h$
   d. $2b+7c+8b$ e. $11axy-3axy$
4. First simplify the following expressions, then find the value of the expression if $x=5$.
   a. $7x+3x$ b. $2x+3x-4$ c. $11x+12x$ d. $x+2x$
5. First simplify the following expressions, then find the value of the expression if $x=5$.
   a. $3x-2x+16$ b. $21x-13x+7$ c. $11+2x+5x$ d. $7x-4x+3x$
6. Use the Distributive Law to expand the following expressions.
   a. $7(m-3k)$ b. $w(g+9a)$ c. $2y(8h-7)$
7. Use the Distributive Law to expand the following expressions.
   a. $10m(7r-2p)$ b. $3g(2a+2c)$ c. $4j(6x-3y)$
8. Expand the following and simplify (if possible).
   a. $5(t-2s)-3t$ b. $4(a+3g)-2g$ c. $d(3f+9)-2d$
9. Expand the following and simplify (if possible).
   a. $4z(2-a)+7z$ b. $3p(f-2q)+7p$ c. $10b(a+6)+5b$
10. Use the Associative Law to complete the following equations.
    a. $a+(3b+\_\_)=(\_\_+\_\_)+5c$ b. $\frac{1}{a}+(2b+\_\_)=(\_\_+2b)\_\_7g$
11. Use the Associative Law to complete the following equations.
    a. $(d\times 3y)\times z=\_\_(\_\_\times\_\_)$ b. $\left(\frac{a}{b}\times c\right)\times\_\_=\frac{a}{\_\_}(\_\_\times f)$
12. Use the associative and commutative laws to complete the following equations.
    a. $g+(3m+\_\_)=(2k+\_\_)+3m$ b. $(w+\_\_)+4r=\_\_+(5g+w)$
13. Use the associative and commutative laws to complete the following equations.
    a. $\frac{2}{3}\times(2d\times\_\_)=(r\times 2d)\times\_\_$ b. $(3z\times 5b)\times\_\_=\_\_\times(6m\times 5b)$

14. Write an expression for:
 a. the difference between $M$ and $C$
 b. money earned by selling $B$ cakes for \$3 each
 c. the product of $X$ and $Y$
 d. 15 more than $G$
 e. 1 more than $D$
 f. the cost of 12 bananas at $H$ cents each
 g. $T$ multiplied by 5.

15. Otto works in a warehouse, packing boxes of chocolates for distribution across the country. An order from a store in Wollongong fills 35 boxes, with 18 chocolates left over.
 a. Write an expression to represent the number of chocolates ordered, using the pronumeral $x$ to represent a full box of chocolates.
 b. If each full box contains 30 chocolates, use your expression to calculate how many chocolates the store ordered.

16. Are the following equations True or False for all values of the pronumerals?
 a. $7x - 10y = 10y - 7x$
 b. $9x \times -y = -y \times 9x$
 c. $16 \times 2x \times x = 32x^2$

17. Simplify the following.
 a. $3g \times 7g$
 b. $6xy \times 3y$
 c. $7ad \times 6a^2$
 d. $-3z^2 \times 8zy$

18. Simplify each of the following.
 a. $\frac{16m}{8}$
 b. $\frac{13x}{13}$
 c. $\frac{20d}{60d}$
 d. $\frac{6th}{12h}$
 e. $\frac{63r}{27rt}$

19. Simplify each of the following.
 a. $\frac{15kt^3}{r^2}$
 b. $\frac{8m^2n}{18mn^2}$
 c. $\frac{72def^2}{-18d^2e}$
 d. $\frac{9a^2b}{12ab^2}$
 e. $\frac{33k^2m^2}{11m^2k}$

20. Place the appropriate terms in the boxes below to make the following statements true. Remember to simplify the equation by multiplying by the reciprocal.
 a. $\square \div 3pq = 4p^2$
 b. $21x^2 \div \square = \frac{3x}{y}$
 c. $\square \div \frac{2a}{5b} = \frac{35b^2}{2a^2}$
 d. $\frac{7x}{2y} \div \square = \frac{7y^2}{24}$

## Problem solving

21. Mary has $y$ books. Tom has 4 more books than Mary. Cindy has 5 times as many books as Tom. Write down an expression, in terms of $y$, that can be used to work out the number of books that Cindy has. (Don't actually calculate the number of books.)

22. It's Bianca's birthday today. She is $m$ years old. She asks the ages of people at her party. Write expressions to describe the ages of the following people.
 a. Bianca's cousin Paul is 8 years older than Bianca.
 b. Bianca's sister Kate is 5 years younger than Bianca.
 c. Bianca's aunt Delia is three times Bianca's age.
 d. Delia's friend Nic is 3 years older than Delia.

**23.** Use the pronumeral $x$ to make a general statement in the form ___ × ___ = ___ from the following clues.

$4 \times \frac{1}{4} = 1; \qquad 30 \times \frac{1}{30} = 1; \qquad 1.2 \times \frac{1}{1.2} = 1$

**24.** A couple spends $\$x$ on developing 80 small wedding photos and $\$y$ for 60 large wedding photos. Determine how much more they need to spend if they want an extra dozen photos of each size.

**25.** It costs $\$x$ for four adults' tickets at the cinema. A child's ticket costs \$3 less than an adult's ticket. On family night, all the tickets go for half price. Develop a formula in terms of $x$ to calculate how much it will cost for a family of 2 adults and 3 children on family night.

**on** To test your understanding and knowledge of this topic, go to your learnON title at www.jacplus.com.au and complete the **post-test**.

# Answers

## Topic 7 Algebra

### 7.1 Pre-test

1. B
2. 24
3. 84
4. E
5. $2x + 6y$
6. $20x + 4y$
7. 20, 21
8. $10k + 6$
9. 5, $4x$
10. a. $30abc$ b. $52ab^3$ c. $5q$ d. $\frac{13m}{3n}$
11. 3
12. $17p - 2q$
13. $6x + 10y$
14. False, $a \times a = a^2$
15. $3b + 8r$

### 7.2 Introduction to algebra

1. Pronumerals are used in algebra to represent unknown numbers.
2. As an example, on Monday there were $b$ bananas in the bowl; on Tuesday 4 bananas were taken out; on Wednesday 2 bananas were taken out; on Thursday 14 bananas were added; on Friday 3 bananas were taken out.
3. a. $2 \times c + 4$ b. $3 \times c + 4$ c. $2 \times c + 2$ d. $2 \times c + 3$
4. a. $c + 3$ b. $2c$ c. $2c + 5$ d. $4c + 3$
5. $4c + 34$
6. a. $2m$ b. $2m + n$ c. $4m + 3n$ d. $m + 2n + 5$ e. $3m + 2n + 8$
7. a. $a + b$ b. $a - b$ c. $ab$ d. $\frac{3a}{b}$
8. a. $\frac{-15}{a}$ b. $3a - 5b$ c. $\frac{-b}{5}$ d. $a - 7$
9. a. $m - 8, m - 2, m + 1, m + 3$ b. $n - 15, n - 10.3, n + 25$
10. Examples are given here.
    a. Let $d$ be the number of dogs at my house. My friend has 2 more dogs at her house than I do $(d + 2)$, while another friend has 1 dog less at his house than I do $(d - 1)$.
    b. Let $t$ be the number of textbooks in my schoolbag. My friend has double that number in his schoolbag $(2t)$ while another friend has 1 more textbook than I do, in his schoolbag $(t + 1)$.
    c. Let $y$ be the number of dollars in my wallet. My friend has 3 times the number of dollars as I have, in her wallet $(3y)$, while another friend has 10 times the number of dollars as I do, in her wallet $(10y)$.
11. a. $B + 2$ b. $T - 3$ c. $D + 6$ d. $K - 5$
12. a. $G + N + W$ b. $D + H$ c. $2N$ d. $C + H$
13. a. $12 - G$ b. $4D$ c. $BF$ d. $3M + Y$
14. a. $A + B + C$ b. $A - C$ c. AC d. ABC
15. a. $\frac{B}{C}$ b. $\frac{A + C}{B}$ c. $A + 3$ d. $C - AB$
16. a. $T - X$ b. $\$4B$ c. $45G$ d. $\frac{\$T}{5}$
17. a. $2 \times n$ b. $4 \times n$ c. $8 \times n$
18. A
19. C
20. B
21. B
22. C
23. a. True b. True c. False d. True
24. a. True b. True c. True d. False
25. a. False b. True c. True d. False
26. a. $a + 6$ b. $a - 2$
27. a. $\$3d$ b. $\$120d$ c. $\$3(d + 1.5)$
28. In $\frac{(x + 1)}{3}$, 1 is added to $x$ and then the result is divided by 3. In $\frac{x}{3} + 1$, $x$ is divided by 3 first and then 1 is added.
29. $y + y + y = 3y$ and $y \times y \times y = y^3$
30. 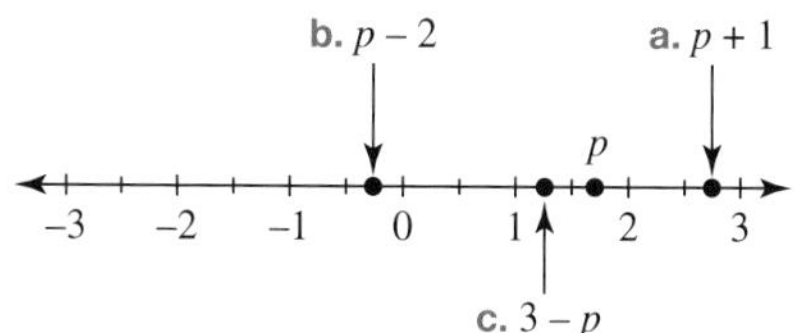

31. a. $\$(2x + 7)$ b. $3x$ c. $4x + 3$
32. a. Cost of a hamburger + cost of a hamburger + cost of a drink = total cost (or cost of the drink + twice the cost of a hamburger = total cost)
    b. $\$3.75 + 2h = \$7.25$
33. a. $10 + \frac{2n}{5}$, where $n$ is the number of coins in each piggy bank.
    b. $44
34. 180

### 7.3 Substituting and evaluating

1. a. 11 b. 20 c. $-2$ d. 15
2. a. $-1$ b. 66 c. 19 d. $-41$
3. a. $-13$ b. $-12$ c. 33 d. 8
4. a. $d = 2$ b. $d = 3$ c. $d = 0$ d. $d = 100$
5. a. $e = 11$ b. $e = 1$ c. $e = 197$ d. $e = 13$
6. a. $f = 19$ b. $f = 7$ c. $f = 67$ d. $f = 247$

6. a. $7m - 21k$ b. $gw + 9aw$ c. $16hy - 14y$

7. a. $70mr - 20mp$ b. $6ag + 6cg$ c. $24jx - 12jy$

8. a. $2t - 10s$ b. $4a + 10g$ c. $3df + 7d$

9. a. $15z - 4az$ b. $3fp - 6pq + 7p$
 c. $10ab + 65b$

10. In this question, the order of the terms is important.
 a. $a + (3b + 5c) = (a + 3b) + 5c$
 b. $t$

11. In this question, the order of the terms is important.
 a. $(d \times 3y) \times z = d \times (3y \times z)$
 b. $\left(\frac{a}{b} \times c\right) \times f = \frac{a}{b}(c \times f)$

12. In this question, the order of the terms is important.
 a. $g + (3m + 2k) = (2k + g) + 3m$
 b. $(w + 5g) + 4r = 4r + (5g + w)$

13. In this question, the order of the terms is important.
 a. $\frac{2}{3} \times (2d + r) = (r \times 2d) \times \frac{2}{3}$
 b. $(3z + 5b) \times 6m = 3z \times (6m \times 5b)$

14. a. $M - C$ b. $\$3B$
 c. $XY$ d. $G + 15$
 e. $D + 1$ f. $12H$ cents
 g. $5T$

15. a. $35x + 18$ b. 1068

16. a. False b. True c. True

17. a. $21g^2$ b. $18xy^2$ c. $42a^3d$ d. $-24yz^3$

18. a. $2m$ b. $x$ c. $\frac{1}{3}$ d. $\frac{t}{2}$ e. $\frac{7}{3t}$

19. a. $15kt$ b. $\frac{4m}{9n}$ c. $-\frac{4f^2}{d}$ d. $\frac{3a}{4b}$ e. $3k$

20. a. $12p^3q$ b. $7xy$ c. $\frac{7b}{a}$ d. $\frac{12x}{y^3}$

21. $5y + 20$

22. a. $m + 8$ b. $m - 5$ c. $3m$ d. $3m + 3$

23. $x \times \frac{1}{x} = 1$

24. $\$\frac{3x + 4y}{20} = \frac{3x}{20} + \frac{y}{5}$

25. $\$\frac{5x - 36}{8}$

# Semester review 1

The learnON platform is a powerful tool that enables students to complete revision independently and allows teachers to set mixed and spaced practice with ease.

## Student self-study

Review the **Course Content** to determine which topics and lessons you studied throughout the year. Notice the green bubbles showing which elements were covered.

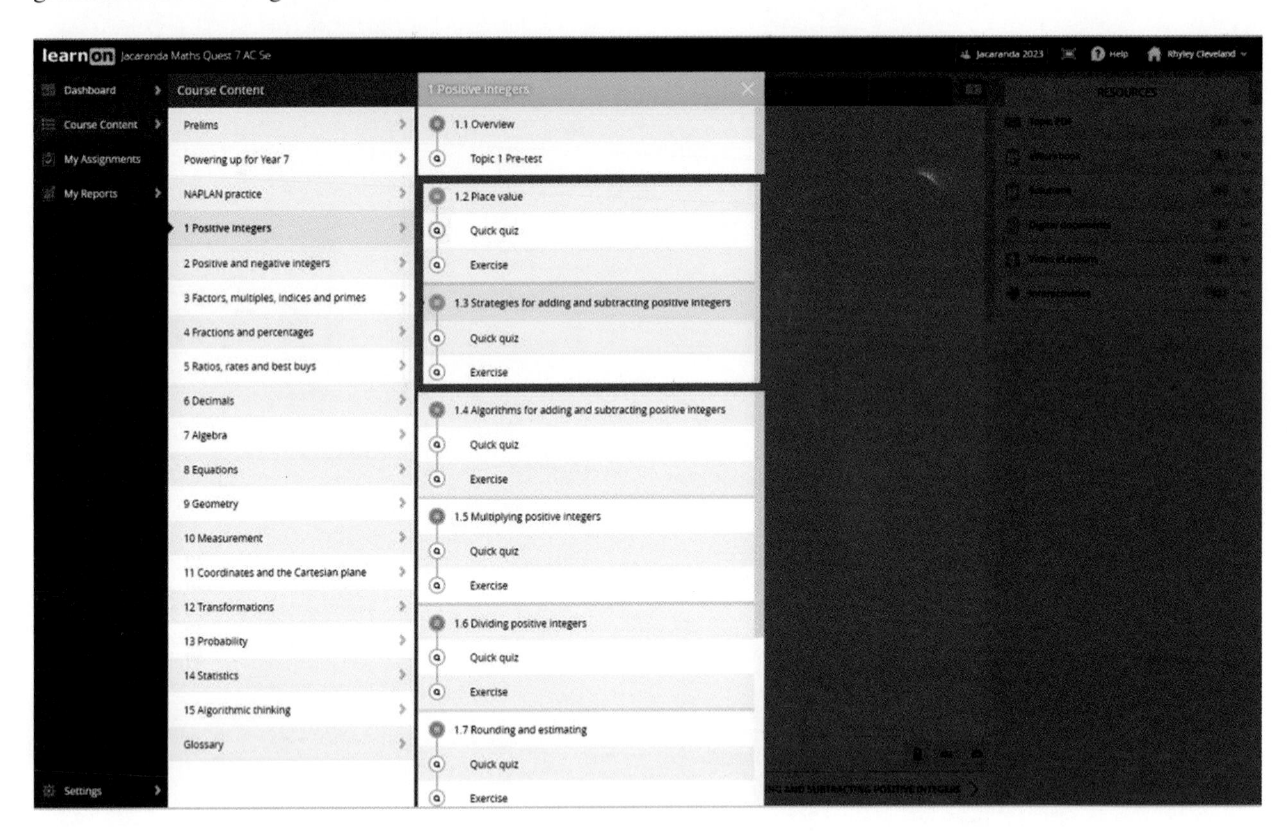


Review your results in **My Reports** and highlight the areas where you may need additional practice.

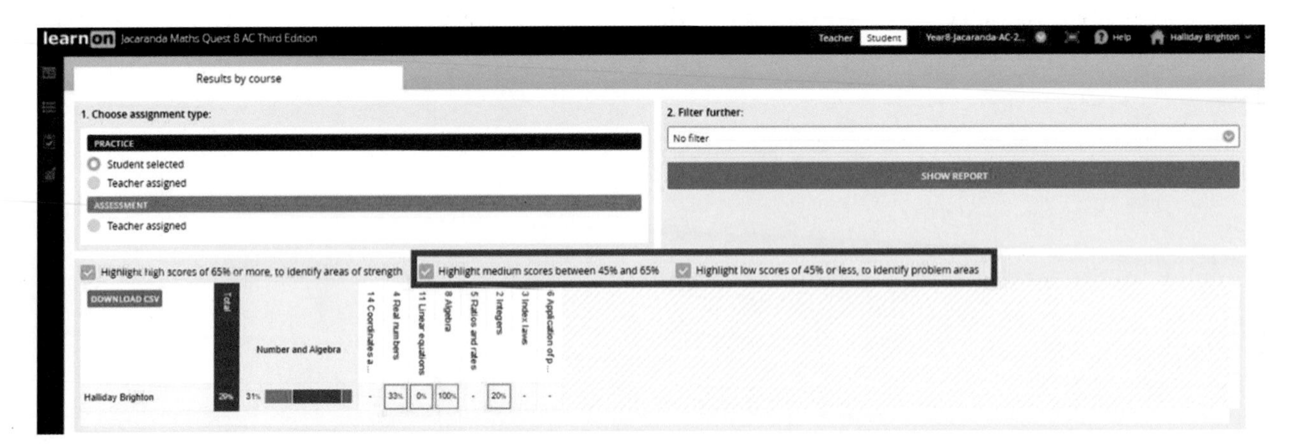


Use these and other tools to help identify areas of strengths and weakness and target those areas for improvement.

## Teachers

It is possible to set questions that span multiple topics. These assignments can be given to individual students, to groups or to the whole class in a few easy steps.

Go to **Menu** and select **Assignments** and then **Create Assignment**. You can select questions from one or many topics simply by ticking the boxes as shown below.

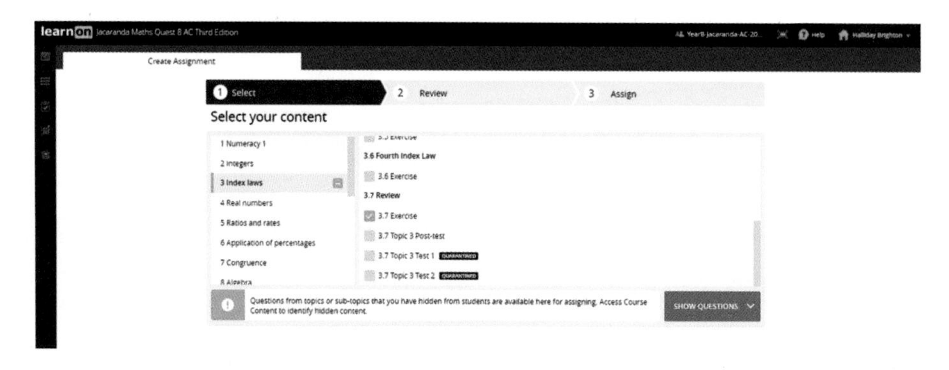


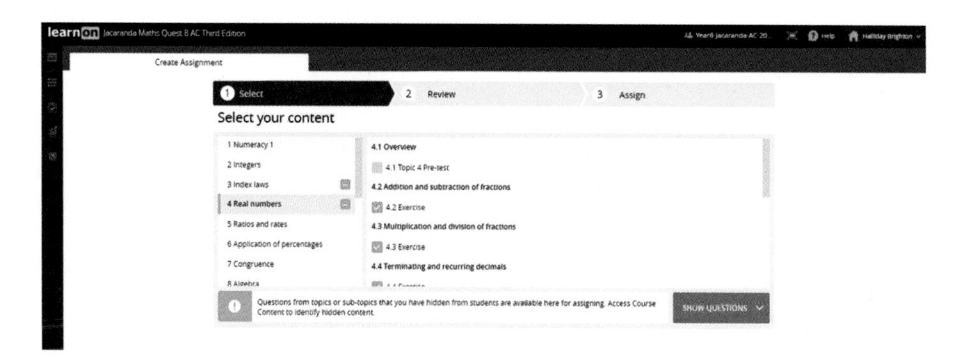


Once your selections are made, you can assign to your whole class or subsets of your class, with individualised start and finish times. You can also share with other teachers.

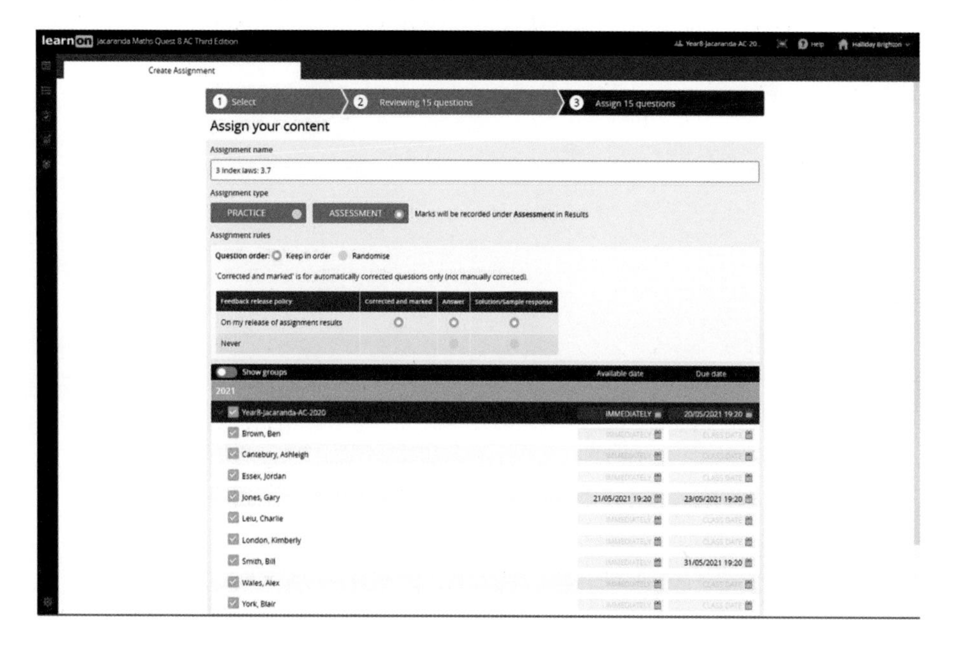


More instructions and helpful hints are available at www.jacplus.com.au.

# 8 Equations

LESSON SEQUENCE


8.1 Overview ........ 394
8.2 Introduction to equations ........ 396
8.3 Building up expressions and backtracking ........ 402
8.4 Solving equations using backtracking ........ 408
8.5 Solving equations using inverse operations ........ 415
8.6 Checking solutions ........ 424
8.7 Review ........ 429

# LESSON
# 8.2 Introduction to equations

**LEARNING INTENTION**

At the end of this lesson you should be able to:
- write an equation to represent a worded problem
- solve equations by inspection
- solve equations using trial and error.

## 8.2.1 Equations

eles-4012

- **Equations** are mathematical statements where two expressions are equal.
- For a mathematical statement to be classified as an equation it must contain an equals sign (=).

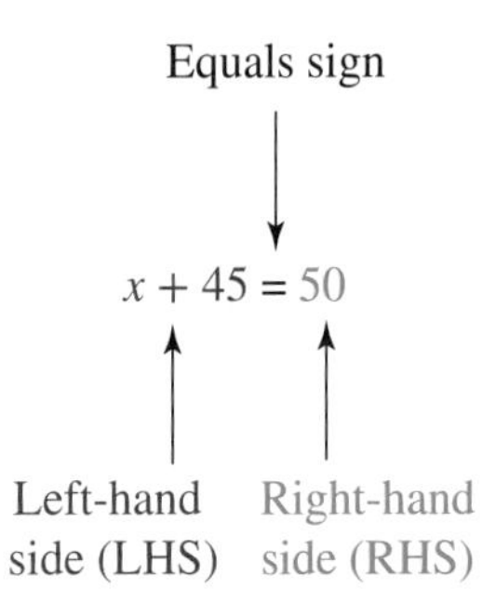


This logic puzzle uses 4 different equations. Can you solve the final equation?

= 28

= 17

= 9

= ?

**Equations**

- **An equation is a mathematical statement containing an equals sign (=).**
- **The LHS of an equation is always equal to (has the same value as) the RHS of the equation.**

### Writing an equation from words

- Mathematical equations can be created based on words or sentences.
- Some English words can be changed into mathematical operations.
  - Words such as *sum, more than, increased, add* or *added* refer to addition and can be replaced with a plus sign (+).
  - Words such as *difference, less than, decreased* or *minus* refer to subtraction and can be replaced with a minus sign (−).
  - Words such as *product, multiply* or *times* refer to multiplication and can be replaced with a multiplication sign (×).
  - Words such as *quotient* or *divide* refer to division and can be replaced with a division sign (÷).

## WORKED EXAMPLE 1 Writing an equation for a worded problem

**Write an equation to represent each of these worded problems.**

a. **When I multiply a number by 8, the answer is 24.**

b. **When I divide a number by 5, the answer is 7.**

c. **When I divide 60 by a number, the answer is 10.**

d. **When I subtract 7 from a number, the answer is 25.**

e. **When a number is increased by 9, the answer is 35.**

f. **When I square a number, the result is 36.**

| THINK | WRITE |
|---|---|
| a. 1. Use a pronumeral to describe the number. | a. Let $m$ be the number. |
| 2. Multiply the number by 8 to form the LHS of the expression. | LHS: $m \times 8 = 8m$ |
| 3. Write the RHS of the expression. | RHS: 24 |
| 4. Write the equation by writing that the LHS is equal to the RHS. | $8m = 24$ |
| b. 1. Use a pronumeral to describe the number. | b. Let $t$ be the number. |
| 2. Divide the number by 5 to form the LHS of the expression. | LHS: $t \div 5 = \frac{t}{5}$ |
| 3. Write the RHS of the expression. | RHS: 7 |
| 4. Write the equation by writing that the LHS is equal to the RHS. | $\frac{t}{5} = 7$ |
| c. 1. Use a pronumeral to describe the number. | c. Let $s$ be the number. |
| 2. Divide 60 by the number to form the LHS of the expression. | LHS: $60 \div s = \frac{60}{s}$ |
| 3. Write the RHS of the expression. | RHS: 10 |
| 4. Write the equation by writing that the LHS is equal to the RHS. | $\frac{60}{s} = 10$ |
| d. 1. Use a pronumeral to describe the number. | d. Let $l$ be the number. |
| 2. Subtract 7 from the number (that is, take 7 away from the number) to form the LHS of the expression. | LHS: $l - 7$ |
| 3. Write the RHS of the expression. | RHS: 25 |
| 4. Write the equation by writing that the LHS is equal to the RHS. | $l - 7 = 25$ |
| e. 1. Use a pronumeral to describe the number. | e. Let $a$ be the number. |
| 2. The word *increased* refers to addition. To increase the number a by 9, add 9 to it. | LHS: $a + 9$ |
| 3. Write the RHS of the expression. | RHS: 35 |
| 4. Write the equation by writing that the LHS is equal to the RHS. | $a + 9 = 35$ |
| f. 1. Use a pronumeral to describe the number. | f. Let $z$ be the number. |
| 2. Square the number (that is, multiply the number by itself) to form the LHS of the expression. | LHS: $z \times z = z^2$ |
| 3. Write the RHS of the expression. | RHS: 36 |
| 4. Write the equation by writing that the LHS is equal to the RHS. | $z^2 = 36$ |

## 8.2.2 Solving an equation by inspection

eles-4013

- To solve an equation you must determine the value of the pronumeral that makes the LHS of the equation equal to the RHS.
- Some basic equations can be solved by **inspection**. This process involves determining the answer by simply inspecting (or looking at) the equation.

### WORKED EXAMPLE 2 Solving an equation by inspection

**Solve the following equations by inspection.**

**a.** $\frac{w}{3} = 4$ **b.** $h - 9 = 10$

| THINK | WRITE |
|---|---|
| **a. 1.** Write the equation. | **a.** $\frac{w}{3} = 4$ |
| **2.** Think of a number that, when divided by 3, gives 4. Try 12. | $12 \div 3 = 4$ |
| **3.** Based on this, $w$ must be 12. | $w = 12$ |
| **b. 1.** Write the equation. | **b.** $h - 9 = 10$ |
| **2.** Think of a number that equals 10 when 9 is subtracted from it. Try 19. | $19 - 9 = 10$ |
| **3.** Based on this, $h$ must be 19. | $h = 19$ |

## 8.2.3 Guess, check and improve

eles-4014

- Sometimes equations cannot be solved easily by inspection.
  In these cases, one possible method that could be used is called 'guess, check and improve'.
  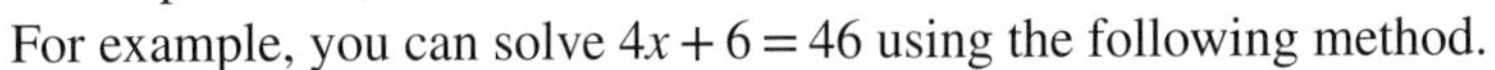
  For example, you can solve $4x + 6 = 46$ using the following method.

| **Solve $4x + 6 = 46$** | | | |
|---|---|---|---|
| **Guess** | **Check** | | **Improve** |
| $x = 3$ | $\text{LHS} = 4 \times 3 + 6$ $= 18$ | $\text{RHS} = 46$ | $x = 3$ makes the LHS too small. Since $x$ appears on the LHS only, choose a larger value for $x$ to make the LHS larger. |
| $x = 11$ | $\text{LHS} = 4 \times 11 + 6$ $= 50$ | $\text{RHS} = 46$ | $x = 11$ makes the LHS too large, but closer than before. Choose a value for $x$ that is less than 11 but greater than 3. |
| $x = 10$ | $\text{LHS} = 4 \times 10 + 6$ $= 46$ | $\text{RHS} = 46$ | LHS = RHS, so $x = 10$ is the solution. |

## WORKED EXAMPLE 3 Solving an equation by guess, check and improve

**Use guess, check and improve to solve the equation $2x + 21 = 4x - 1$.**

**THINK**

1. Set up a table with 4 columns displaying the value of $x$, the value of the LHS and RHS equations (after substitution) and a comment on how these two values compare to the equation.
2. Substitute the first guess, say $x = 1$, into the LHS and RHS equations and comment on the results.
3. Repeat step 2 for the other guesses until the correct answer is found.
4. Write the answer.

**WRITE**

| | Check | | |
|---|---|---|---|
| Guess $x$ | $2x + 21$ | $4x - 1$ | Comment |
| 1 | 23 | 3 | $4x - 1$ is too small. |
| 6 | 33 | 23 | This is closer. |
| 10 | 41 | 39 | Very close. |
| 11 | 43 | 43 | That's it! |

The answer is $x = 11$.

## WORKED EXAMPLE 4 Using guess and check to determine unknown numbers

**Use guess, check and improve to determine two numbers whose sum is 31 and whose product is 150.**

**THINK**

1. The numbers add up to 31, so guess 2 numbers that do this. Then check by finding their product.
2. Guess 1 and 30, then calculate the product.
3. Guess 10 and 21, then calculate the product.
4. Try other numbers between 1 and 10 for the first number. Determine the product of the two numbers.
   Stop when the product of the two numbers is 150.
5. Write the answer.

**WRITE**

| Guess the sum (small number first) | Check the product (P) | Comment |
|---|---|---|
| 1 and 30 | $1 \times 30 = 30$ | P is too low. |
| 10 and 21 | $10 \times 21 = 210$ | P is too high. |
| 5 and 26 | $5 \times 26 = 130$ | P is too low. |
| 8 and 23 | $8 \times 23 = 184$ | P is too high. |
| 6 and 25 | $6 \times 25 = 150$ | That's it! |

The two numbers that add up to 31 and have a product of 150 are 6 and 25.

**on** Resources

**eWorkbook** Topic 8 Workbook (worksheets, code puzzle and project) (ewbk-1909)

**Interactivities** Individual pathway interactivity: Solving equations using trial and error (int-4372)
Writing equations (int-4041)
Guess, check and improve (int-4042)

## Exercise 8.2 Introduction to equations

learnon

8.2 Quick quiz on | 8.2 Exercise

### Individual pathways

| ■ PRACTISE | ■ CONSOLIDATE | ■ MASTER |
|---|---|---|
| 1, 4, 6, 8, 11, 14 | 2, 5, 9, 12, 15 | 3, 7, 10, 13, 16, 17 |

### Fluency

1. **WE1** Write an equation to represent each of these worded problems.
   a. When I add 7 to a number, the answer is 11.
   b. When I add 12 to a number, the answer is 12.
   c. When I subtract 7 from a number, the answer is 1.
   d. When I subtract 4 from a number, the answer is 7.

2. Write an equation to represent each of these worded problems.
   a. When I multiply a number by 2, the answer is 12.
   b. When I multiply a number by 5, the answer is 30.
   c. When I divide a number by 7, the answer is 1.
   d. When I divide a number by 5, the answer is 2.

3. Write an equation to represent each of these worded problems.
   a. When I subtract a number from 15, the answer is 2.
   b. When I subtract a number from 52, the answer is 8.
   c. When I divide 21 by a number, the answer is 7.
   d. When I square a number, the answer is 100.

4. **WE2** Solve the following equations by inspection.
   a. $x+7=18$ b. $y-8=1$ c. $3m=15$ d. $\frac{m}{10}=3$

5. Solve the following equations by inspection.
   a. $\frac{k}{5}=0$ b. $b+15=22$ c. $b-2.1=6.7$ d. $\frac{c}{3}=1.4$ e. $5x=14$

### Understanding

6. **WE3** Use guess, check and improve to solve the following equations.
   a. $3x+11=5x-1$ b. $5x+15=x+27$ c. $x+20=3x$

7. Use guess, check and improve to solve the following equations.
   a. $12x-18=10x$ b. $6(x-2)=4x$ c. $3(x+4)=5x+4$

8. **WE4** Use guess, check and improve to find two numbers whose sum and product are given:
   a. sum $=21$, product $=98$
   b. sum $=26$, product $=165$
   c. sum $=54$, product $=329$
   d. sum $=178$, product $=5712$

9. Use guess, check and improve to find two numbers whose sum and product are given:
   a. sum $=153$, product $=4662$
   b. sum $=242$, product $=14\,065$
   c. sum $=6.1$, product $=8.58$
   d. sum $=978$, product $=218\,957$

**10.** Copy and complete this table by substituting each $x$-value into $x^2 + 4$ and $4x + 1$. The first row has been completed for you. Use the table to determine a solution to the equation $x^2 + 4 = 4x + 1$. (Remember, $x^2$ means $x \times x$.)

| $x$ | $x^2 + 4$ | $4x + 1$ |
|---|---|---|
| 0 | 4 | 1 |
| 1 | | |
| 2 | | 9 |
| 3 | 13 | |
| 4 | | |

## Reasoning

**11.** Explain the thought process you would use to determine the solution to $8x = 48$ by inspection.

**12.** Explain why the inspection method is not suitable to solve the equation $5x = 34$.

**13.** Annisa is trying to find the solution to the equation $3x - 7 = 17$ using the guess and check method. Her first guess is $x = 5$, which is not the solution. Should her next guess be greater than or less than 5?

## Problem solving

**14.** A football team won four more games than it lost. The team played 16 games. Determine how many games the team won.

**15.** A plumber cut a 20-metre pipe into two pieces. One of the pieces is three times as long as the other. Determine the lengths of the two pieces of pipe.

**16.** Lily is half the age of Pedro. Ross is 6 years older than Lily and 6 years younger than Pedro. Determine Pedro's age.

**17.** Angus and his grandfather share a birthday. Both their ages are prime numbers. Angus's age has the same two digits as Grandpa's but in reverse order. In 10 years' time, Grandpa will be three times as old as Angus. Determine how old both Grandpa and Angus will be in 10 years.

# LESSON
# 8.3 Building up expressions and backtracking

**LEARNING INTENTION**

At the end of this lesson you should be able to:
- complete a flowchart to determine the output
- construct a flowchart from a given expression.

## 8.3.1 Flowcharts

eles-4016

- A flowchart is useful for keeping track of the steps in a sequence.

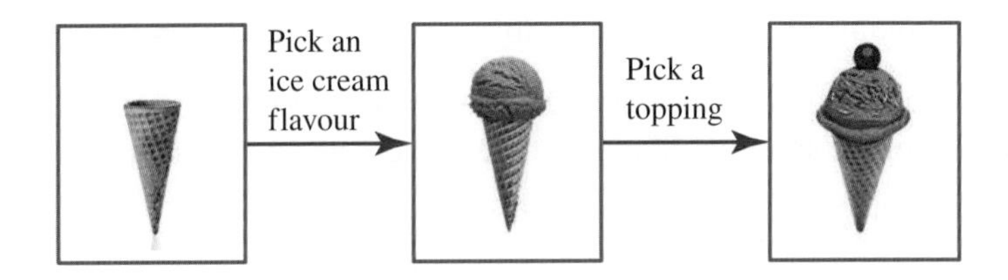


- In the preceding diagram, the instructions for each step are written above the arrows that join the boxes.
  - The first box shows what you start with.
  - The box in the middle shows what happens along the way as you complete the steps.
  - The last box shows what you finish with.
- Flowcharts can be used to keep track of expressions as operations are performed on them.
- In the following flowchart, the starting number is 8. This is called the **input number**. Performing the operations in the order displayed results in 1, the number in the last box. This is called the **output number**.

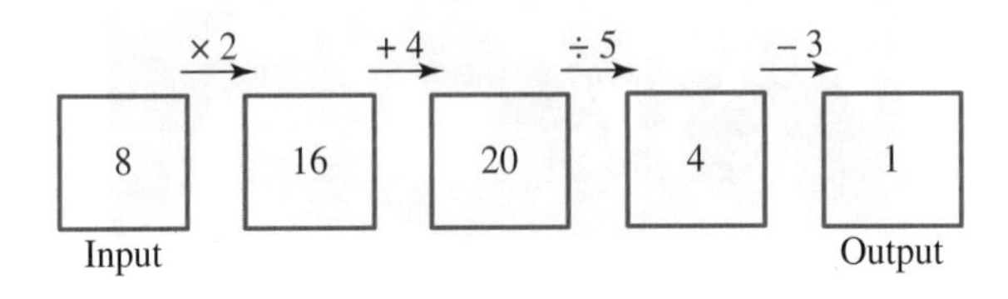


**WORKED EXAMPLE 5 Completing a flowchart**

**Build up an expression by following the instructions on these flowcharts.**

**a.** $m \xrightarrow{\times 3} \square \xrightarrow{+5} \square$

**b.** $m \xrightarrow{+5} \square \xrightarrow{\times 3} \square$

**THINK**

**a. 1.** The instruction above the first arrow says $\times 3$. The result is $m \times 3 = 3m$.

**2.** The instruction above the second arrow says $+5$. Adding 5 to $3m$ gives the result $3m + 5$.

**3.** State the expression (output).

**WRITE**

**a.**

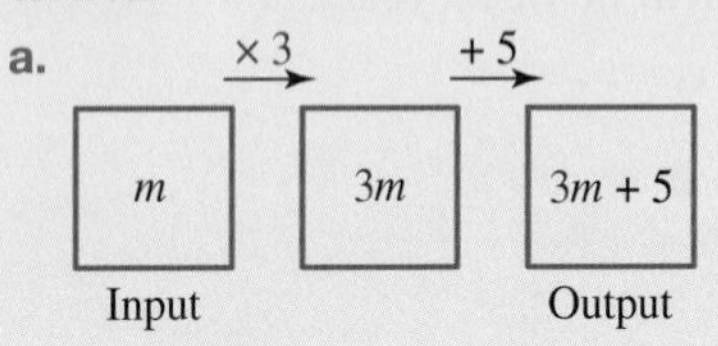


The expression is $3m + 5$.

**b. 1.** The instruction above the first arrow says $+5$. Adding 5 to $m$ gives the result $m+5$.

**2.** The instruction above the second arrow says $\times 3$. That means multiply all of $m+5$ by 3. The result is $3(m+5)$.

**b.** $m \xrightarrow{+5} m+5 \xrightarrow{\times 3} 3(m+5)$

Input Output

**3.** State the expression (output). The expression is $3(m+5)$.

## WORKED EXAMPLE 6 Drawing a flowchart

**Draw a flowchart whose input number is $m$ and whose output number is given by the following expressions.**

**a.** $2m - 11$ **b.** $\dfrac{m+9}{5}$ **c.** $4\left(\dfrac{m}{3}+2\right)$

**THINK**

**WRITE**

**a. 1.** The first step is to obtain $2m$. Do this by multiplying m by 2.

**a.**

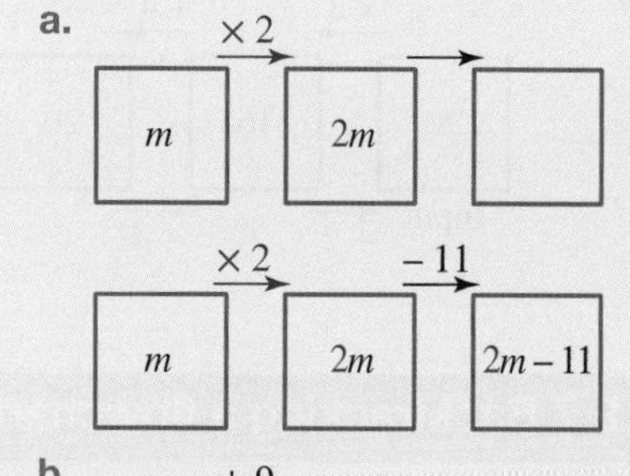


**2.** Next, subtract 11.

**b. 1.** The expression $m+9$ is grouped as though it is in a pair of brackets, so we must work out this part first. This means we add 9 to $m$ first.

**b.**

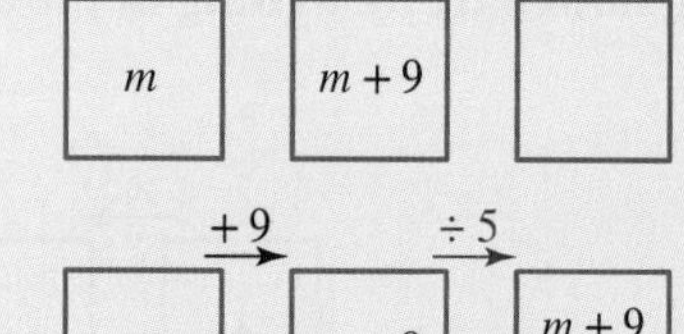


**2.** Next, the whole expression is divided by 5.

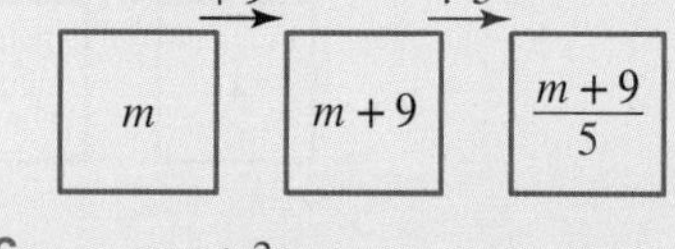


**c. 1.** The brackets indicate we must first work from within the brackets, so:
- divide $m$ by 3
- then add 2 to the result.

**c.** $m \xrightarrow{\div 3} \dfrac{m}{3} \rightarrow \square \rightarrow \square$

$m \xrightarrow{\div 3} \dfrac{m}{3} \xrightarrow{+2} \dfrac{m}{3}+2 \rightarrow \square$

**2.** Multiply the result obtained in step 1 by 4.

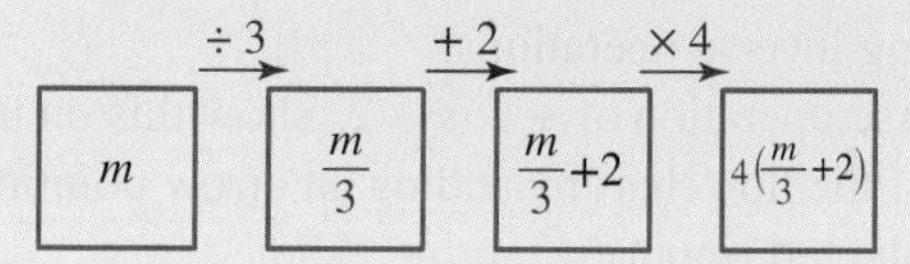


## 8.3.2 Backtracking

eles-4017

- **Backtracking** is a method used to work backwards through a flowchart. It involves moving from the output towards the input.
- When working backwards through a flowchart, use **inverse** (or **opposite**) **operations**.

- A list of operations and their inverses (opposites) is highlighted in the following box.

**Inverse operations**

**$-$ is the inverse operation of $+$**

**$+$ is the inverse operation of $-$**

**$\div$ is the inverse operation of $\times$**

**$\times$ is the inverse operation of $\div$**

- Examine the following flowchart and the operations that were applied to the input to produce the output (working left to right). These are shown in purple.
  Using backtracking we can identify the opposite operations that need to be applied when working backwards through the flowchart. These are shown in pink.

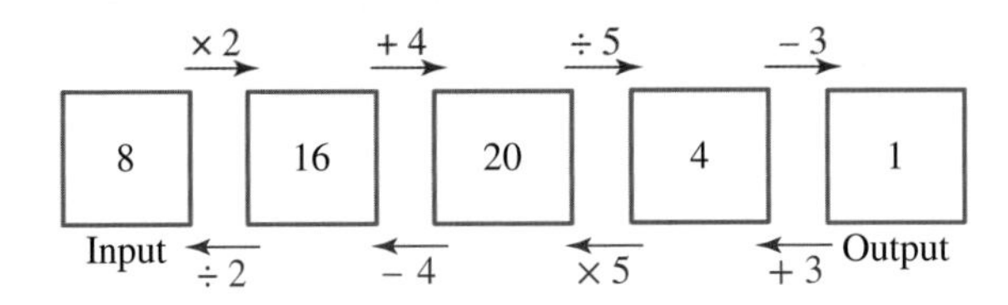


## WORKED EXAMPLE 7 Backtracking a flowchart

**Complete the following flowchart by writing in the operations that need to be carried out in order to backtrack to $x$.**

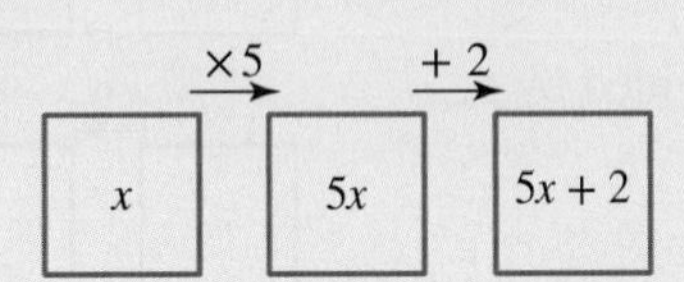


**THINK**

1. Copy the flowchart and look at the operations that have been performed.
   Starting from the input, two operations have been performed: a multiplication by 5 and then the number 2 has been added.
2. Starting with the output, work backwards towards the input using inverse operations.
   The inverse operation of $+2$ is $-2$. Show this on the bottom of the flowchart by adding an arrow pointing towards the left (input).
3. The inverse operation of $\times 5$ is $\div 5$. Show this on the flowchart at the bottom, with an arrow pointing towards the left (input).

**WRITE**

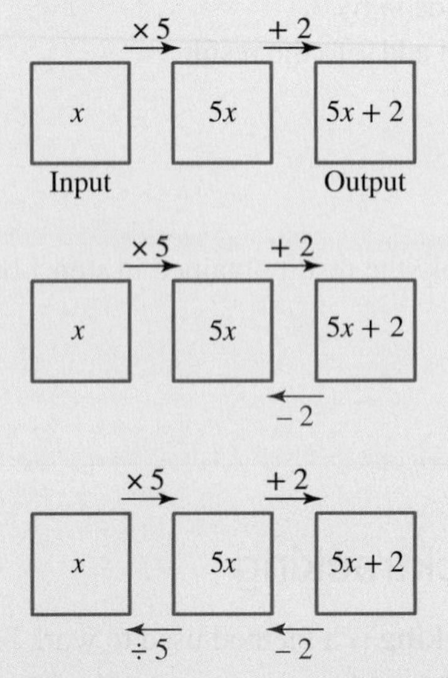

## COLLABORATIVE TASK: Secret numbers

1. Think of a number by yourself and write it in your workbook. Don't tell anyone! Multiply your number by 4 and write down the answer. Next, add 20 and write down the answer. Then divide by 2 and write down the answer. Finally, subtract 5 and write down the answer.
2. As a class, write the flowchart for these instructions on the board.
3. Show how the flowchart works for different starting numbers.
4. Work out some of your classmates' secret numbers by working backwards through the flowchart.

 Resources

 **eWorkbook** Topic 8 Workbook (worksheets, code puzzle and project) (ewbk-1909)

 **Interactivities** Individual pathway interactivity: Building up expressions (int-4374)
Flowcharts (int-4044)

## Exercise 8.3 Building up expressions and backtracking

learnon

8.3 Quick quiz 

8.3 Exercise

### Individual pathways

| ■ PRACTISE | ■ CONSOLIDATE | ■ MASTER |
|---|---|---|
| 1, 3, 7, 10, 13, 16 | 2, 5, 8, 11, 14, 17 | 4, 6, 9, 12, 15, 18, 19 |

### Fluency

1. **WE5** Build up an expression by following the instructions on these flowcharts.

a. 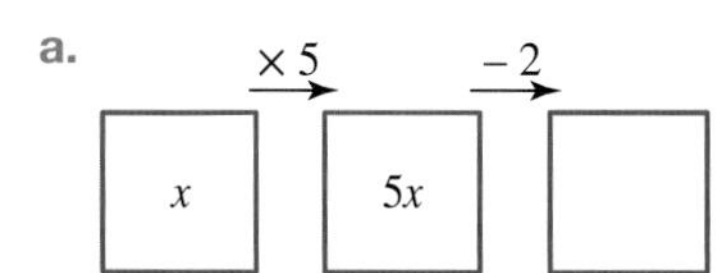


b. 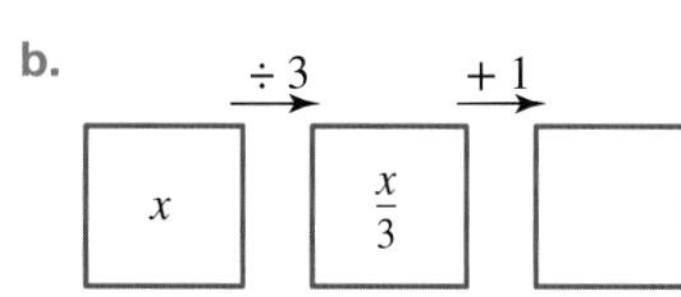


c. 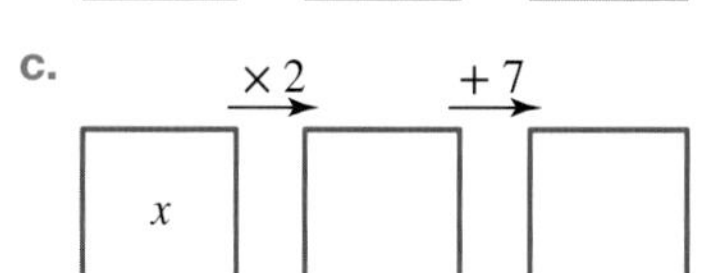


d. 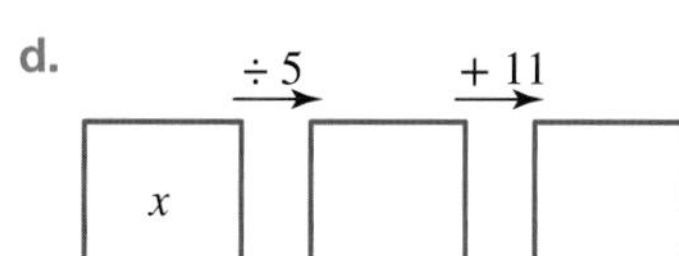

2. Build up an expression by following the instructions on these flowcharts.

a. 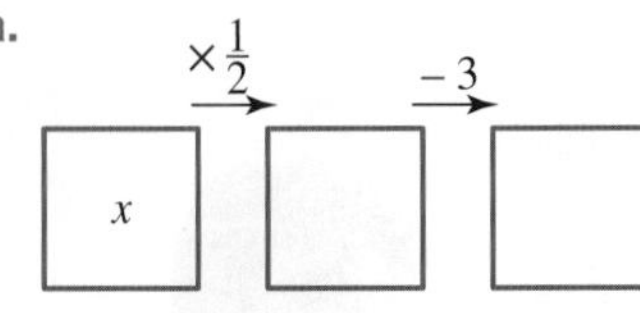


b. 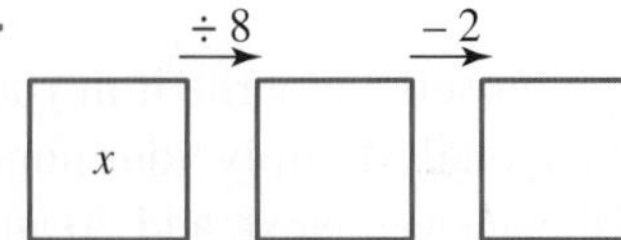


c. 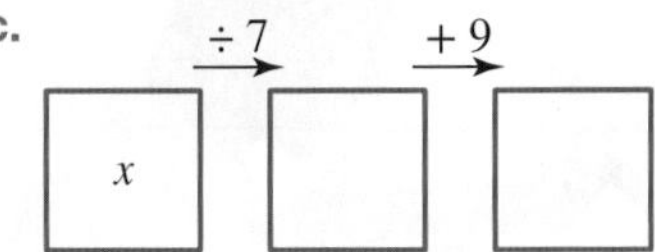


d. 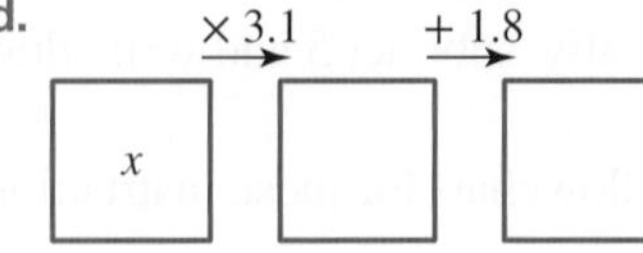


3. Build up an expression by following the instructions on the flowcharts. Use brackets or fractions as a grouping device, for example, $2\,(x+3)$ or $\dfrac{x-5}{4}$.

a. 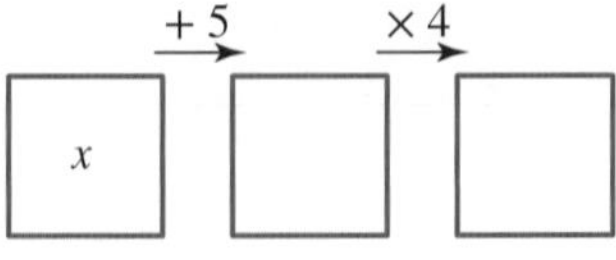


b. 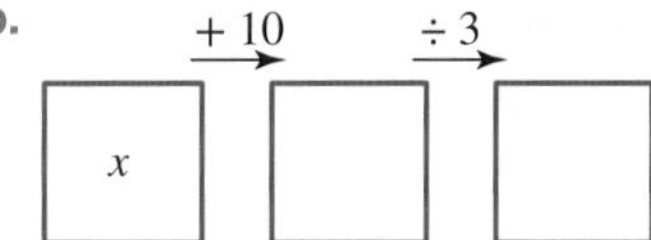


c. 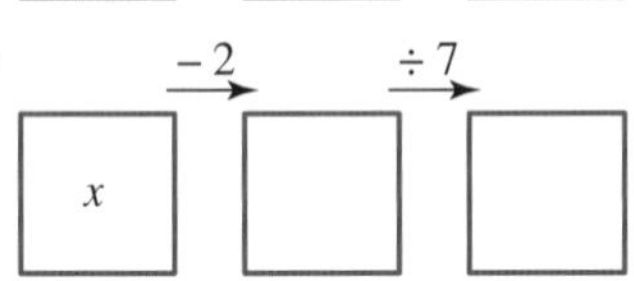


d. 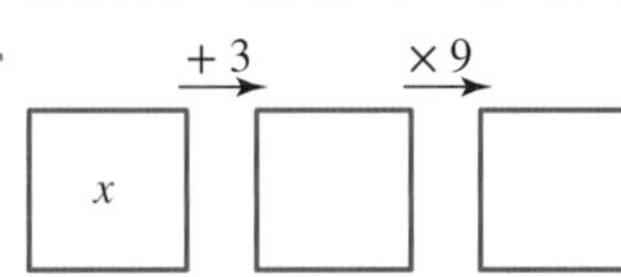


4. Build up an expression by following the instructions on the flowcharts. Use brackets or fractions as a grouping device, for example, $2\,(x+3)$ or $\dfrac{x-5}{4}$.

a. 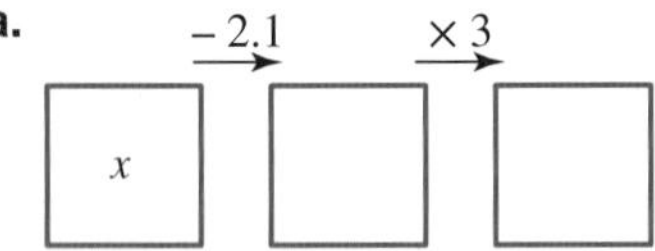


b. 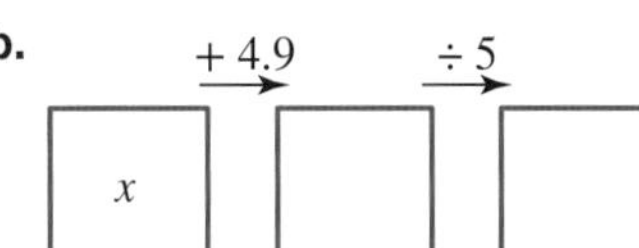


c. 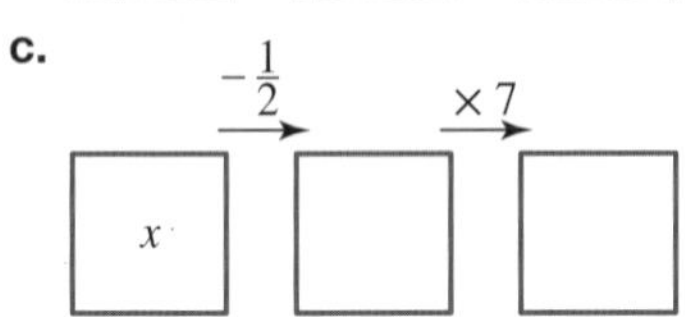


d. 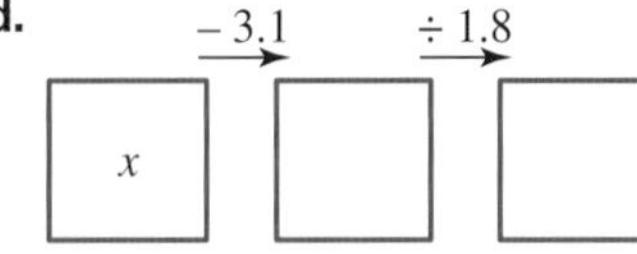


5. Copy and complete the following flowcharts by filling in the missing expressions.

a. 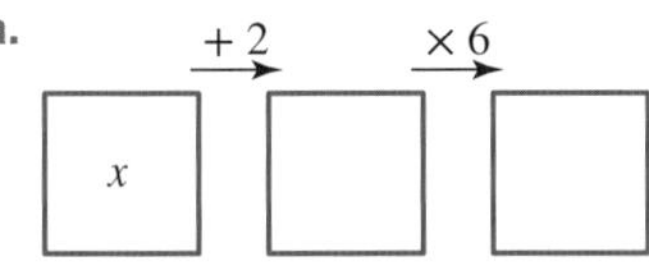


b. 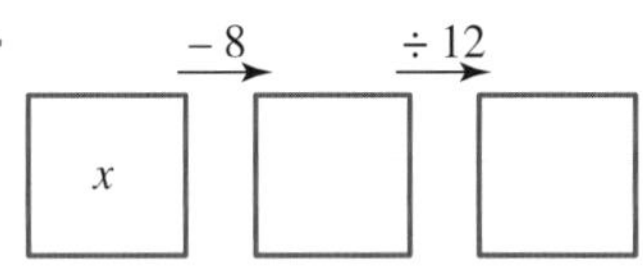


c. 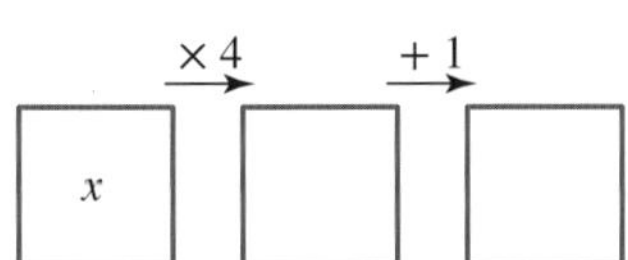


d. 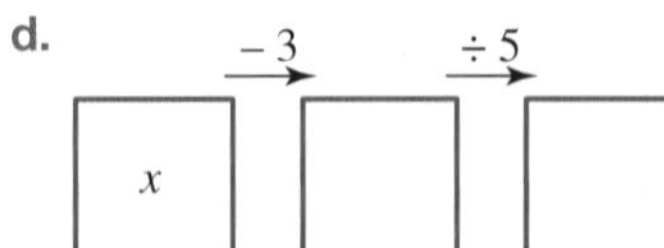


6. Copy and complete the following flowcharts by filling in the missing expressions.

a. 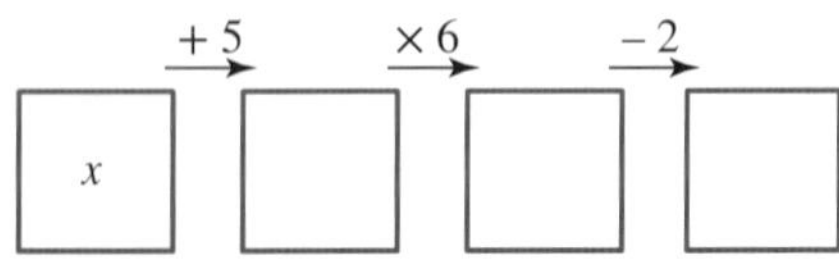


b. 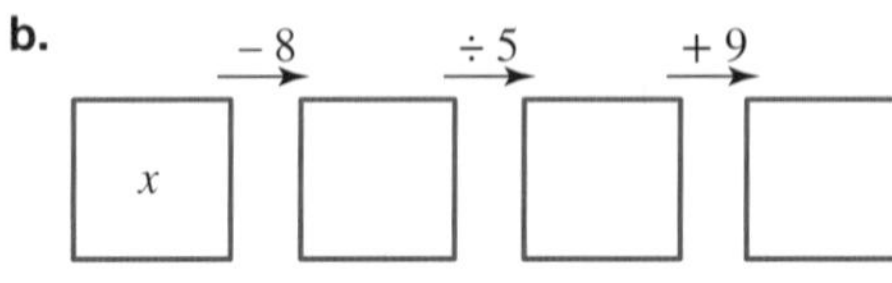


c. 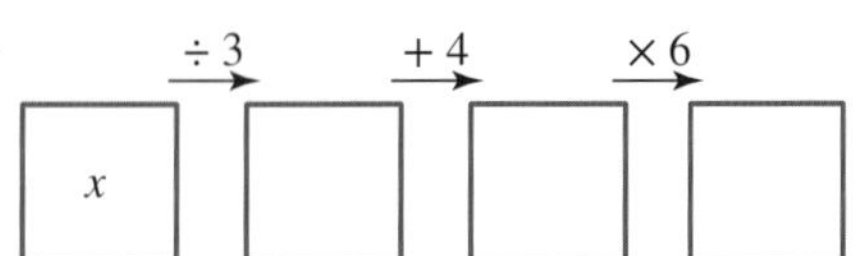


d. 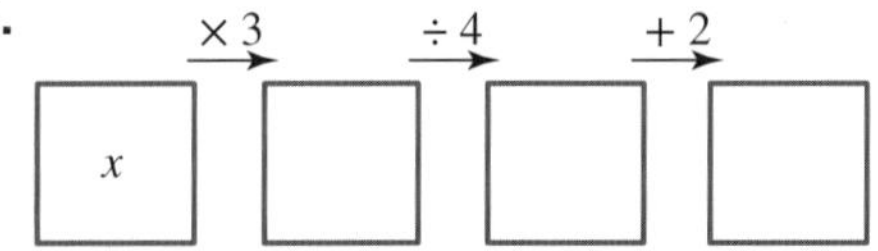

## Understanding

7. **WE6** Draw a flowchart whose input number is $x$ and whose output is given by the following expressions.

   a. $5x+9$ b. $2(x+1)$ c. $\dfrac{x-8}{7}$

8. Draw a flowchart whose input number is $x$ and whose output is given by the following expressions.

   a. $12(x-7)$ b. $\dfrac{x}{5}-2$ c. $\dfrac{x+6}{3}$

9. Draw a flowchart whose input number is $x$ and whose output is given by the following expressions.

   **a.** $3(x+7)-5$ **b.** $\dfrac{3x+7}{2}$ **c.** $4(3x+1)$

10. **WE7** Complete the following flowcharts by writing in the operations that must be carried out in order to backtrack to $x$.

a.

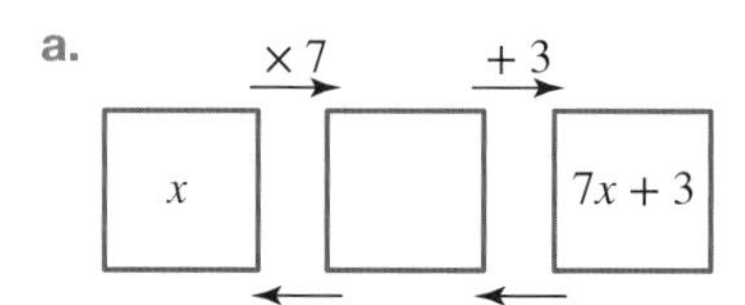


b.

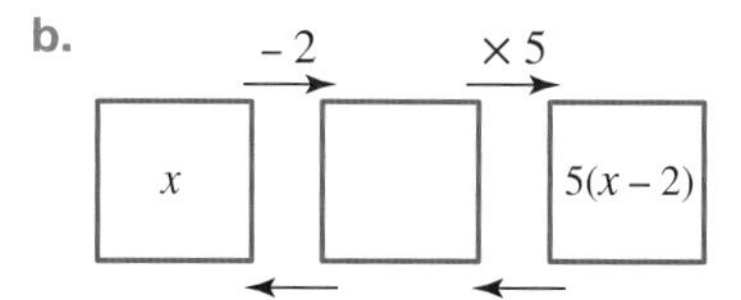


c.

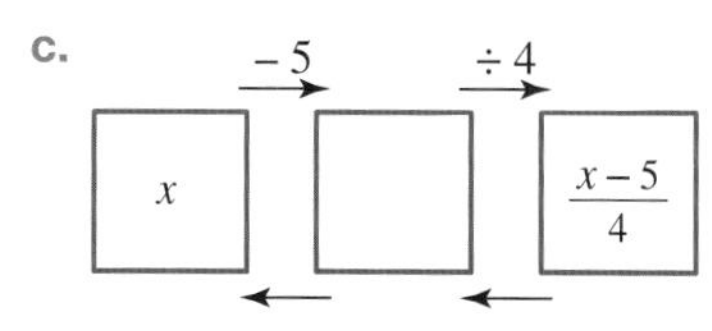


11. Add the operations to complete these flowcharts.

a.

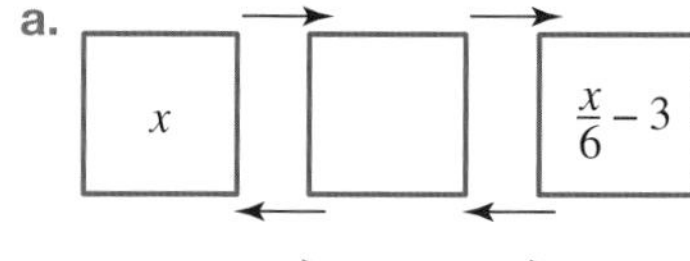


b.

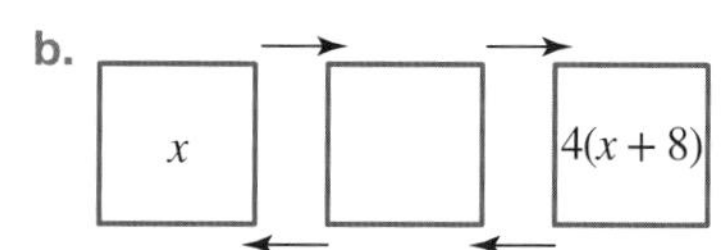


c.

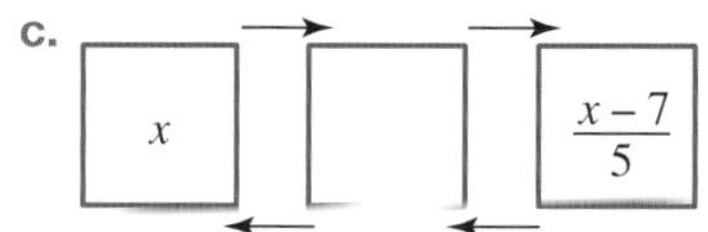


d.

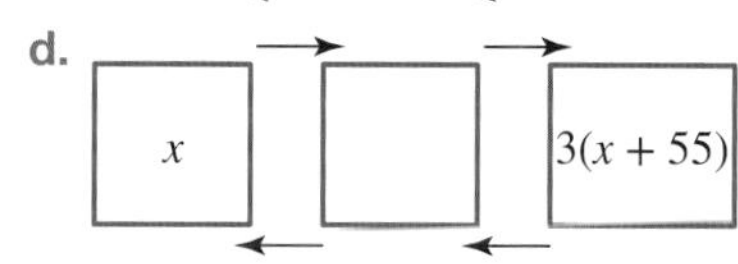


12. Add the operations to complete these flowcharts.

a.

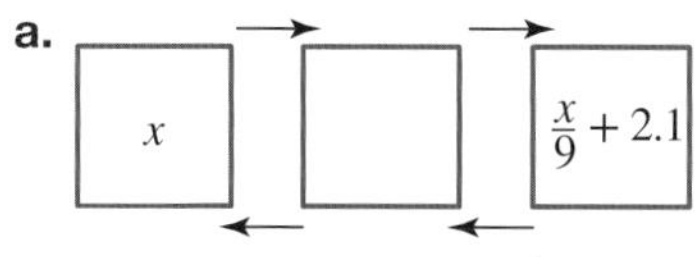


b.

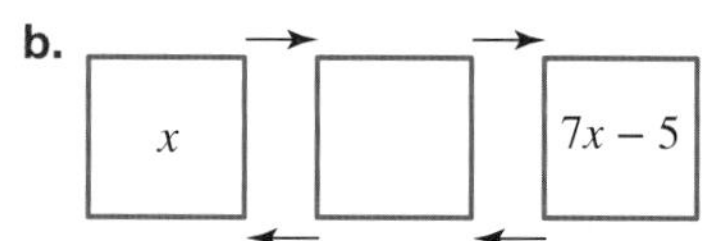


c.

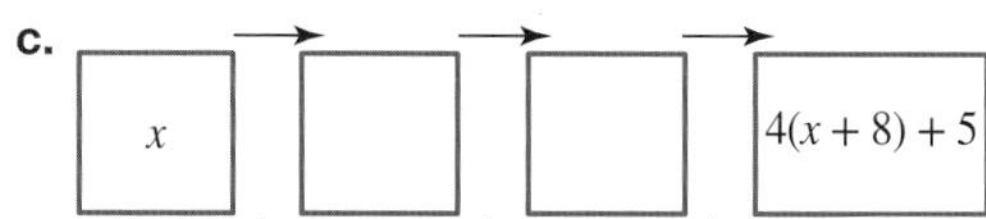


d.

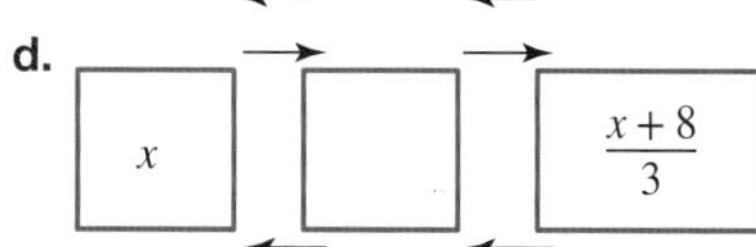


## Reasoning

13. Starting from $p$, the expression $2(p-3)$ is built using the operations $\times 2$ and $-3$. Identify which operation should be performed first.

14. I think of a number and add 8 to it. I multiply the result by 5 and then divide the result by 4. The answer is 30. Use a flowchart to build up an expression and write these steps in the form of an equation.

15. I think of a number, multiply it by 5 and add 15. The result is 3 less than 4 times the original number. If the original number is $n$, write an equation to show the relation. Show all your working.

### Problem solving

16. Rachel and Jackson are 7 years apart in age. Jackson is older than Rachel. The sum of their ages is 51. Determine Rachel's age.

17. The sum of two numbers is 32 and their product is 247. Determine the two numbers.

18. Marcus and Melanie pooled their funds and purchased shares on the stock exchange. Melanie invested \$350 more than Marcus. Together they invested \$2780. Determine how much money Marcus invested.

19. The equation $1+2+3+4+5+6+7+8+9=100$ is clearly incorrect. By removing some + signs, including some − signs and combining some of the digits to make larger numbers, see if you can make the equation correct. Remember, only + and − signs are allowed and the equation must equal 100.
One example is shown below. Determine as many other possible solutions as you can, showing full working.

$$123+45-67+8-9=100$$

# LESSON
# 8.4 Solving equations using backtracking

## LEARNING INTENTION

At the end of this lesson you should be able to:
- solve equations using backtracking.

## 8.4.1 Using backtracking to solve equations

eles-4018

- Backtracking is a process that can be used to solve equations.

### Solving equations using backtracking

**To solve an equation using backtracking, follow these steps.**
1. **Use a flowchart to build up the equation.**
2. **Work backwards through the flowchart using inverse operations to determine the solution.**

For example, if we are asked to solve the equation $2x+9=23$, we must first build up the equation and then use backtracking to determine the solution.

**Build up the equation**

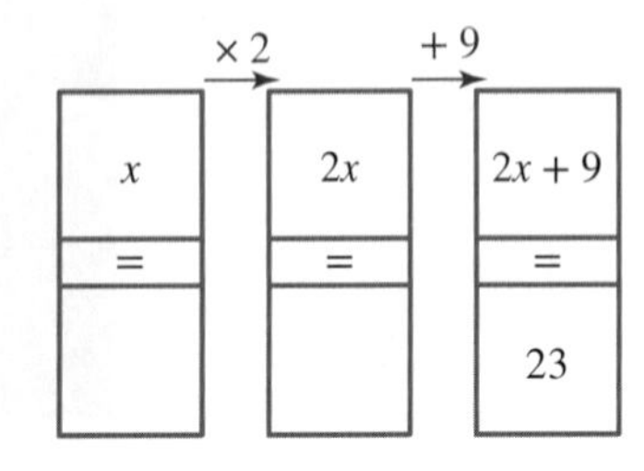


**Backtrack to determine the solution**

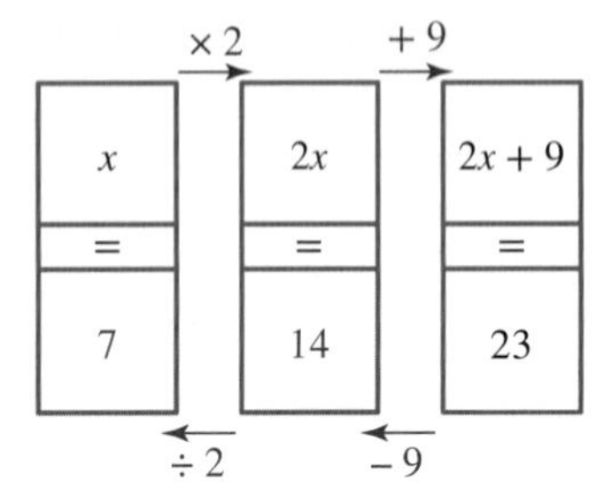


- The solution of the equation $2x+9=23$ is $x=7$.
- We can also check that we have found the correct answer by substitution.
That is, LHS $=2(7)+9=2\times 7+9=14+9=23=$ RHS.

## WORKED EXAMPLE 8 Drawing a flowchart and solving by backtracking

**Draw a flowchart to represent the following worded problem and then solve it by backtracking.**
**I am thinking of a number. When I multiply it by 4 and then add 2, the answer is 14.**

**THINK**

1. Build an expression using $x$ to represent the number. Start with $x$, multiply by 4 and add 2. The output number is 14.

**WRITE**

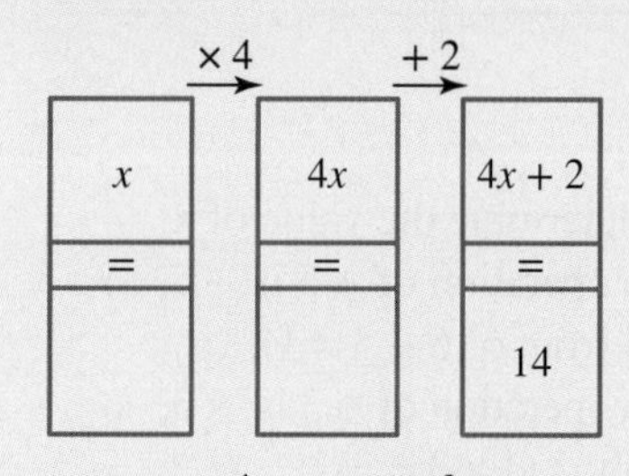


2. Backtrack to determine the value of $x$.
   - The inverse operation of $+2$ is $-2$ $(14-2=12)$.
   - The inverse operation of $\times 4$ is $\div 4$ $(12 \div 4 = 3)$.

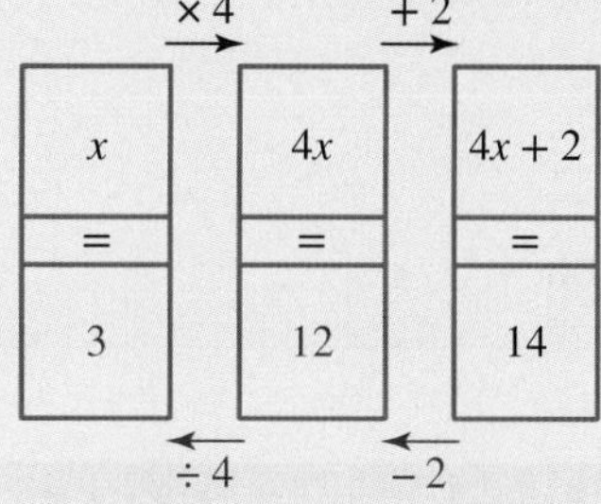


3. Write the solution.

The solution is $x = 3$. Therefore the number is 3.

## WORKED EXAMPLE 9 Solving equations by backtracking

**Solve the following equations by backtracking.**

**a.** $3(x+7) = 24$

**b.** $\dfrac{x}{3} + 5 = 6$

**THINK**

**a.** 1. Build the expression on the left-hand side of the equation. Start with $x$, then add 7 and then multiply by 3. The output number is 24.

**WRITE**

**a.**

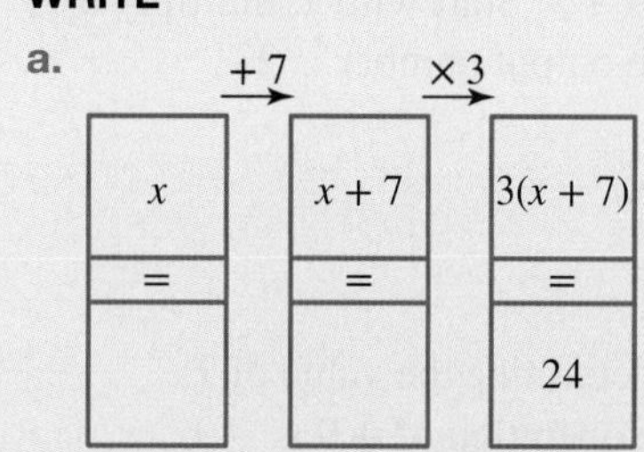


2. Backtrack to determine the value of $x$.
   - The inverse operation of $\times 3$ is $\div 3$ $(24 \div 3 = 8)$.
   - The inverse operation of $+7$ is $-7$ $(8-7=1)$.

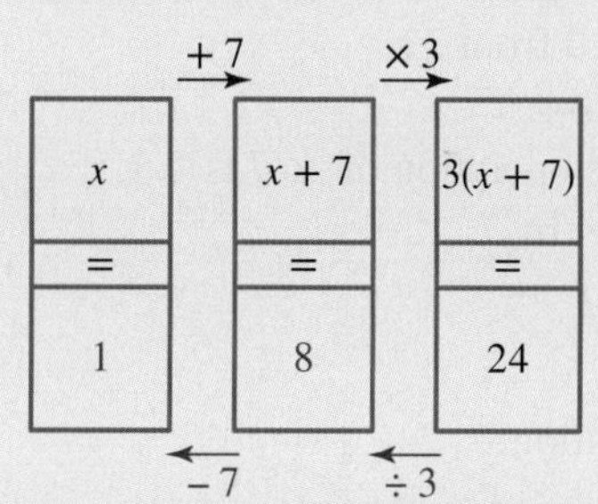


3. Write the solution.

The solution for the equation is $x = 1$.

**b.** 1. Build the expression on the left-hand side of the equation. Start with $x$, then divide by 3 and then add 5. The output number is 6.

**b.**

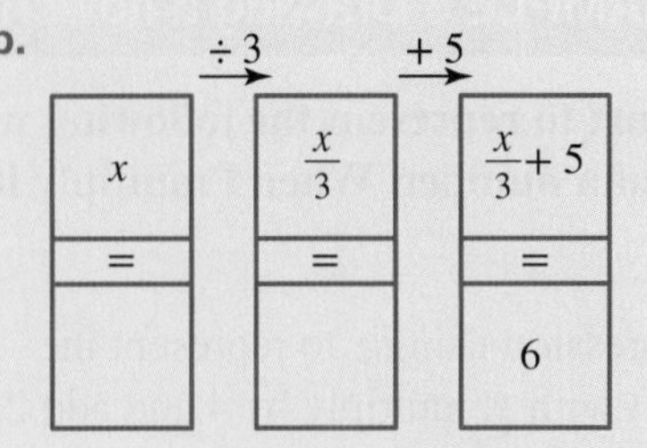


2. Backtrack to determine the value of x.
   - The inverse operation of + 5 is − 5, so subtract −5 from 6 ($6 - 5 = 1$).
   - The inverse operation of ÷ 3 is × 3, so multiply 1 by 3 ($1 \times 3 = 3$).

3. Write the solution.

The solution for the equation is $x = 3$.

## WORKED EXAMPLE 10 Simplifying and solving equations by backtracking

**Simplify and then solve the following equation by backtracking.**

$5x + 13 + 2x - 4 = 23$

**THINK**

1. Simplify by adding the like terms together on the left-hand side of the equation. $5x + 2x = 7x$, $13 - 4 = 9$

2. Draw a flowchart and build the left-hand side of the equation $7x + 9$. Start with $x$, multiply by 7 and add 9. The output number is 23.

3. Backtrack to determine the value of $x$.
   - The inverse operation of +9 is −9, so subtract 9 from 23 ($23 - 9 = 14$).
   - The inverse operation of × 7 is ÷ 7, so divide 14 by 7 ($14 \div 7 = 2$).

4. Write the solution.

**WRITE**

$$5x + 13 + 2x - 4 = 23$$
$$7x + 9 = 23$$

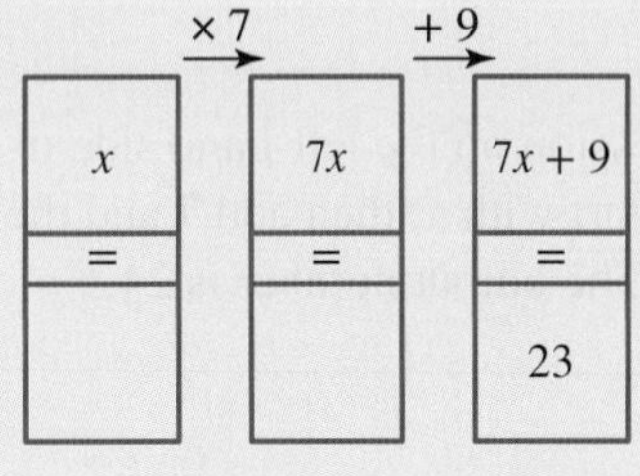


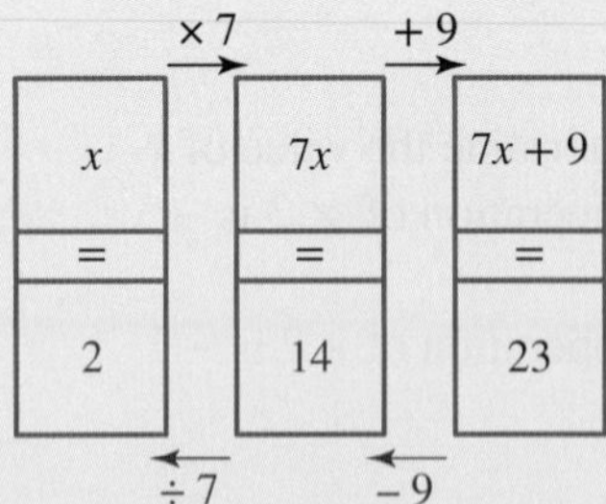


The solution for the equation is $x = 2$.

## COLLABORATIVE TASK: Think of a number

In pairs, discuss and investigate Mel's 'think of a number' puzzle, which is described below.

Mel loves playing 'Think of a number' with friends. Here's an example of one of her puzzles.

- Think of a number.
- Double it.
- Add 10.
- Divide by 2.
- Take away the number you first thought of.
- Your answer is … 5!

Let's investigate to see why the answer is always 5, whatever number you first think of. We can form expressions for each of the steps, using a variable as the starting value.

- Think of a number. $n$
- Double it. $n \times 2 = 2n$
- Add 10. $2n + 10$
- Divide by 2. $(2n + 10) \div 2 = n + 5$
- Take away the number you first thought of. $n + 5 - n = 5$
- Your answer is... $5$

As you can see, the answer will always be 5 for any starting number.

### Now you try

**1.** Write expressions for each step in the following, showing that you can determine the answer in each case.

**a. Puzzle 1: Your birth year**

- Take the year in which you were born.
- Subtract 500.
- Multiply by 2.
- Add 1000.
- Divide by 2.
- Your answer is … your birth year!

**b. Puzzle 2: Your age**

- Take your age (in years).
- Add 4.
- Multiply by 10.
- Subtract 10.
- Divide by 5.
- Subtract your age.
- Take away 6.
- Your answer will be your age!

**c. Puzzle 3: Think of a number…**

- Think of a number.
- Divide it by 2.
- Add 2.
- Multiply by 4.
- Take away your original number.
- Subtract your original number again.
- Your answer should be …

**2.** Write some 'Think of a number' puzzles yourself. Try them out on friends. They will marvel at your mystical powers!

 Resources

eWorkbook Topic 8 Workbook (worksheets, code puzzle and project) (ewbk-1909)

Interactivities Individual pathway interactivity: Solving equations using backtracking (int-4375)

Backtracking (int-4045)

## Exercise 8.4 Solving equations using backtracking

learnon

8.4 Quick quiz 

8.4 Exercise

### Individual pathways

| ■ PRACTISE | ■ CONSOLIDATE | ■ MASTER |
| --- | --- | --- |
| 1, 4, 7, 11, 12, 15, 16 | 2, 5, 8, 10, 13, 17, 18, 19, 20 | 3, 6, 9, 14, 21, 22, 23, 24 |

### Fluency

1. **WE8** Draw a flowchart to represent each of the following worded problems and then solve them by backtracking.
   a. When I multiply a number by 2 and then add 7 the answer is 11.
   b. When I add 3 to a number and then multiply by 5 the answer is 35.
   c. When I divide a number by 4 and then add 12 the answer is 14.
   d. When I add 5 to a number and then divide by 3 the answer is 6.
   e. When I subtract 7 from a number and then multiply by 6 the answer is 18.
2. Draw a flowchart to represent each of the following worded problems and then solve them by backtracking.
   a. When I subtract 4 from a number and then divide by 9 the answer is 7.
   b. When I divide a number by 11 and then subtract 8 the answer is 0.
   c. When I multiply a number by 6 and then add 4 the answer is 34.
   d. When I multiply a number by 5 and then subtract 10 the answer is 30.
   e. When I subtract 3.1 from a number and then multiply by 6 the answer is 13.2.
3. Draw a flowchart to represent each of the following worded problems and then solve them by backtracking.
   a. When I add $\frac{2}{5}$ to a number and then divide by 6 the answer is $\frac{4}{5}$.
   b. When I subtract $\frac{3}{4}$ from a number and then divide by $\frac{2}{3}$ the answer is $\frac{1}{6}$.
4. Draw a flowchart and use backtracking to determine the solution to the following equations.
   a. $5x+7=22$ b. $9y-8=1$ c. $4x+12=32$ d. $8w+2=26$ e. $4w+5.2=28$
5. **WE9a** Solve the following equations by backtracking.
   a. $3(x+7)=24$ b. $2(x-7)=22$ c. $11(x+5)=99$ d. $6(x+9)=72$ e. $4(w+5.2)=26$
6. **WE9b** Solve the following equations by backtracking.
   a. $\frac{x}{3}+5=6$ b. $\frac{x}{9}-2=3$ c. $\frac{x}{2}-11=6$
   d. $\frac{x}{7}-5=6$ e. $\frac{x}{5}+2.3=4.9$

## Understanding

7. Solve the following equations by backtracking.

   a. $\dfrac{x+4}{3}=6$  b. $\dfrac{x-8}{7}=10$  c. $\dfrac{x+11}{2}=6$

   d. $\dfrac{x-5}{7}=0$  e. $\dfrac{x+2.21}{1.4}=4.9$

8. Use backtracking to find the solution to the following equations.

   a. $3x-7=23$

   b. $4(x+7)=40$

   c. $\dfrac{x}{5}-2=8$

   d. $5(x-3)=15$

   e. $6(x-4)=18$

9. Use backtracking to find the solution to the following equations.

   a. $\dfrac{x}{2.1}-1.7=3.6$

   b. $\dfrac{x+5}{3}-3=7$

   c. $4(x-2)+5=21$

   d. $3\left(\dfrac{x}{2}+1\right)=15$

   e. $2(3x+4)-5=15$

10. **WE10** Simplify and then solve the following equations by backtracking.

    a. $2x+7+3x+5=27$

    b. $3x+9+x-4=17$

    c. $3x+5x+2x=40$

    d. $6x+6-x-4=37$

    e. $7x-4x+8-x=10$

11. **MC** The correct flowchart required to solve the equation $6x+5=8$ is:

    **A.**

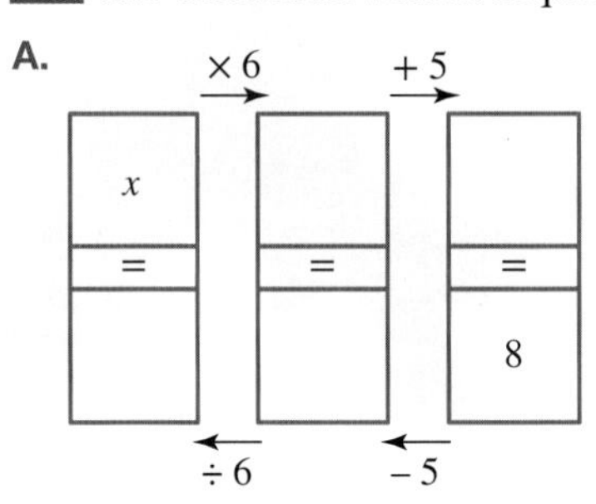


    **B.**

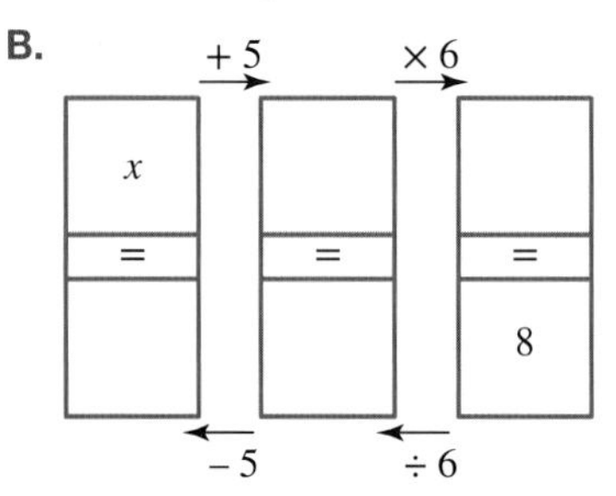


    **C.**

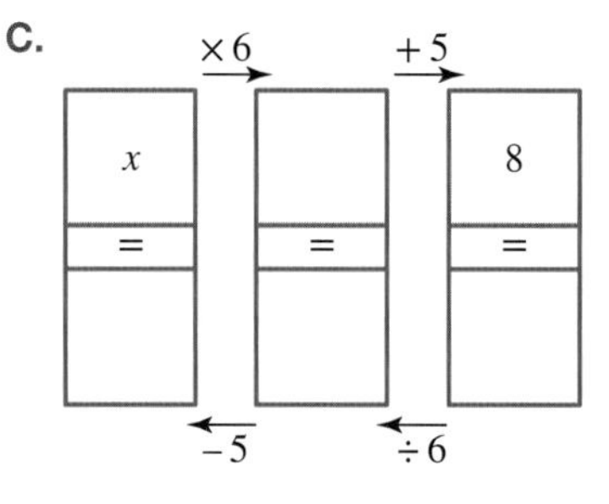


    **D.**

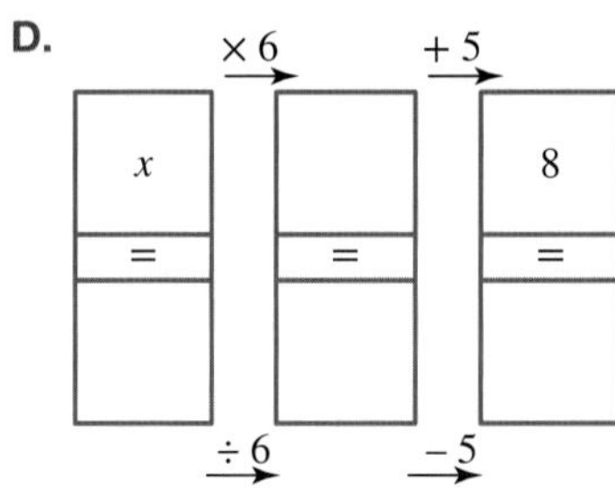


    **E.**

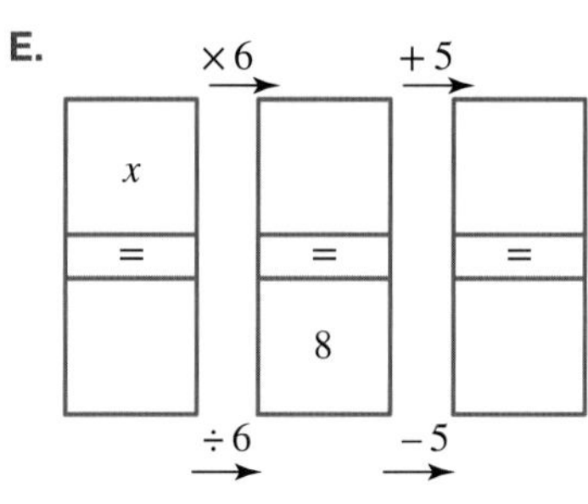

## Reasoning

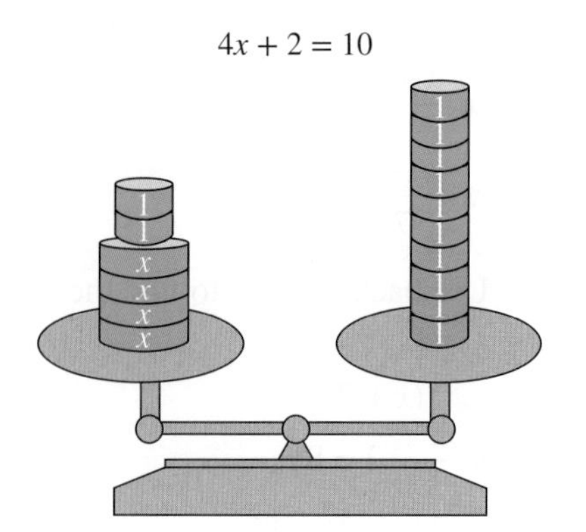


12. Based on the scales in the diagram, identify the operations you need to apply to find the value of $x$. Calculate the value of $x$.

13. Chloe wrote the following explanation to solve the equation$3(x-6)+5=8$ by the backtracking method:
    *To solve this equation, work backwards and do things in the reverse order.*
    *First add 5, then subtract 6, then finally multiply by 3.*
    Explain why her instructions are not correct.

14. The equation $4x-20=8$ can be solved using backtracking by adding 20, then dividing by 4. Explain whether the equation could also be solved by dividing by 4 first.

## Problem solving

15. Kevin is 5 years older than his brother Gareth. The sum of their ages is 31 years. Determine Gareth's age, letting $x$ represent Gareth's age. Show your working.

16. The sum of three consecutive whole numbers is 51. Determine the numbers. (*Hint:* Let the smallest number equal $x$.)

17. Melanie and Callie went tenpin bowling together. Melanie scored 15 more pins than Callie. Their total score was 207. Determine what Callie scored.

18. The sum of three consecutive odd numbers is 27. Determine the 3 numbers.

19. The sum of 3 consecutive odd numbers is 39. Determine the 3 numbers.

20. In the high jump event Chris leapt 12 centimetres higher than Tim. Their two jumps made a total of 3 metres. Determine how high Chris jumped.

21. In three basketball games Karina has averaged 12 points each game. In the first game she scored 11 points. In the second she scored 17 points. In the third game she scored $x$ points.
    **a.** From the given information, calculate the average of 11, 17 and $x$.
    **b.** Write an equation using the answer to part **a**.
    **c.** Solve the equation.
    **d.** State how many points Karina scored in the third game.

22. Three consecutive multiples of 5 add up to 90. Determine the 3 numbers.

23. David is 5 years younger than his twin brothers. If the sum of their ages is 52, then how old is David?

24. Using the six consecutive numbers from 4 to 9, complete the magic square shown so that each row, column and diagonal totals 15.

| | | |
|---|---|---|
| | 1 | |
| 3 | | |
| | | 2 |

# LESSON
# 8.5 Solving equations using inverse operations

## LEARNING INTENTION

At the end of this lesson you should be able to:
- solve equations using the balancing method
- solve equations using inverse operations.

## 8.5.1 Keeping equations balanced

eles-4019

- A balance scale can be used to show whether things have equal mass.

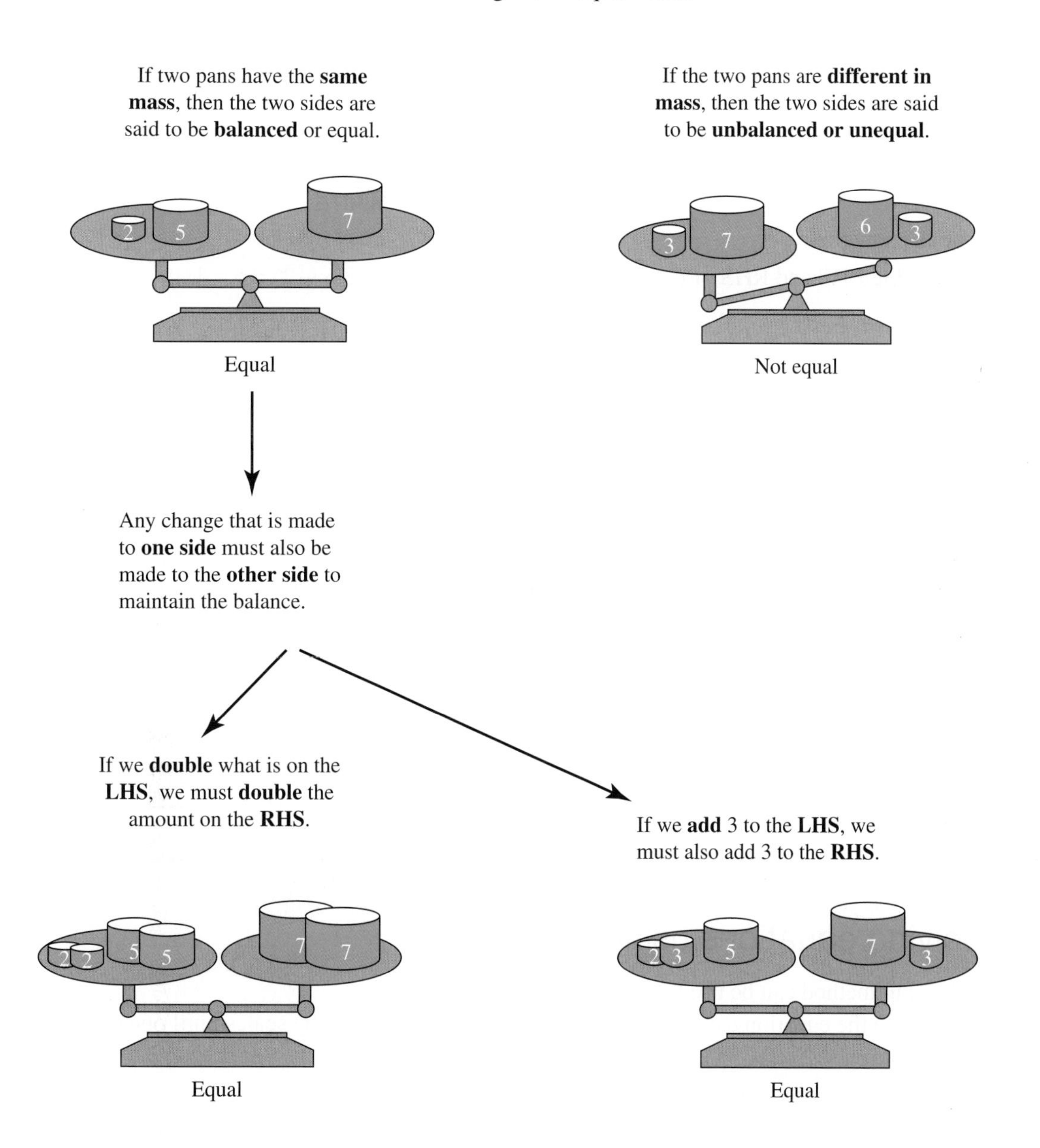

- For example, the scale in this diagram can be described by the equation $x + 3 = 7$.
- Making changes to both sides of the scale lets you work out how many weights are in the bag (the value of the pronumeral).
- If we remove 3 weights from each side of the scale, we can see that the bag with $x$ weights in it weighs the same as 4 weights. That means $x = 4$.

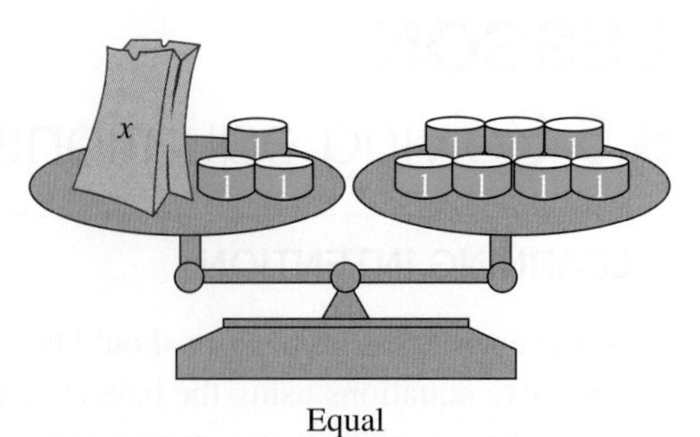


**WORKED EXAMPLE 11 Applying operations to keep an equation balanced**

**Apply the given operation to the LHS and RHS of the following equations to maintain the balance between both sides.**

a. $x + 2 = 5$ **[multiply both sides by 3]**
b. $2x = 8$ **[add 2 to both sides]**
c. $5x = 15$ **[subtract 2 from both sides]**
d. $4x = 16$ **[divide both sides by 4]**

| THINK | WRITE |
|---|---|
| a. 1. Write the equation. | a. $x + 2 = 5$ |
| 2. Multiply the LHS and RHS of the equation by 3. | $3(x + 2) = 5 \times 3$ |
| 3. Write the answer by simplifying both sides. | $3x + 6 = 15$ |
| b. 1. Write the equation. | b. $2x = 8$ |
| 2. Add 2 to the LHS and RHS of the equation. | $2x + 2 = 8 + 2$ |
| 3. Write the answer by simplifying both sides. | $2x + 2 = 10$ |
| c. 1. Write the equation. | c. $5x = 15$ |
| 2. Subtract 2 from the LHS and RHS of the equation. | $5x - 2 = 15 - 2$ |
| 3. Write the answer by simplifying both sides. | $5x - 2 = 13$ |
| d. 1. Write the equation. | d. $4x = 16$ |
| 2. Divide the LHS and RHS of the equation by 4 | $\frac{4x}{4} = \frac{16}{4}$ |
| 3. Write the answer by simplifying both sides. | $x = 4$ |

## 8.5.2 Solving equations using balancing

eles-4020

- The balancing method can be used to solve equations.
- The solution to the equation can be found by removing weights from the scales until one side contains the pronumeral only.

## WORKED EXAMPLE 12 Solving an equation by keeping equations balanced

**For the following balance scale:**

**a. write the equation represented by the scale**

**b. calculate the value of the pronumeral.**

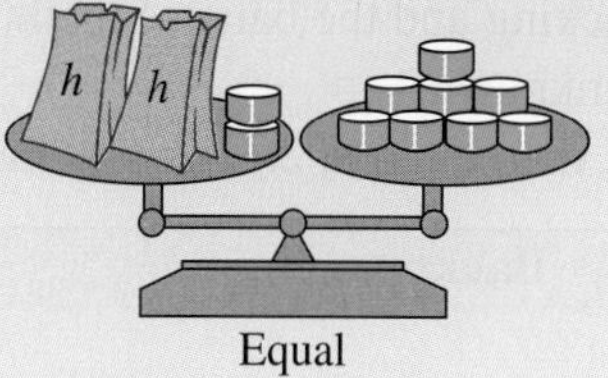


**THINK** / **WRITE**

a. 1. Examine the balance scale. On the LHS there are 2 bags with $h$ weights in each, and 2 weights outside the bags. On the RHS there are 8 weights.

a.

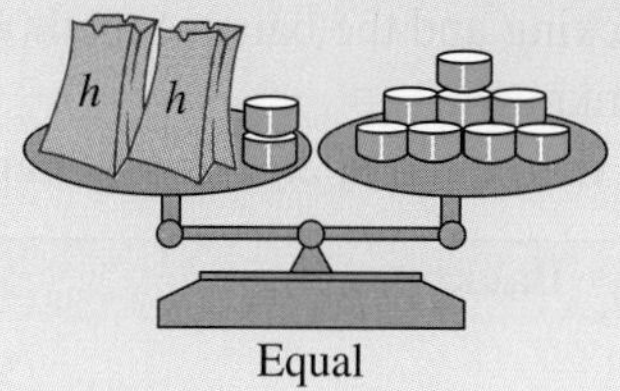


2. Write this as an equation.

$2h + 2 = 8$

b. 1. Remove 2 weights from both sides so that the balance will be maintained (−2). This leaves 2 bags on the LHS and 6 weights on the RHS.

b.

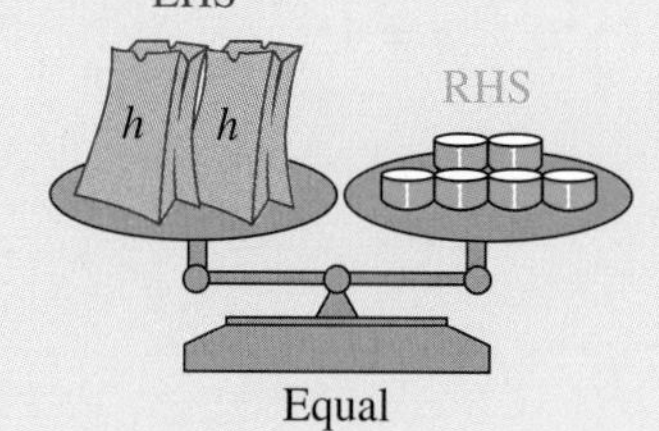


$$2h + 2 - 2 = 8 - 2$$
$$2h = 6$$

2. To work out the value of h (the value of 1 bag), divide both sides by 2 and then simplify.

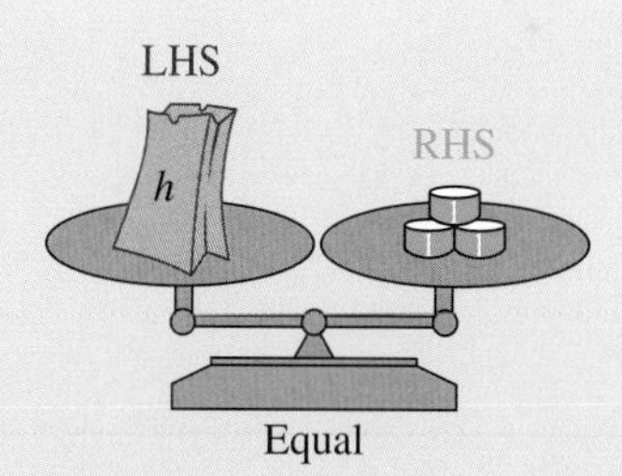


$$\frac{(2h)}{2} = \frac{6}{2}$$
$$h = 3$$

Each bag is equivalent to 3 weights.

That means $h = 3$.

## 8.5.3 Solving equations using inverse operations

eles-4021

- Both backtracking and the balance method of solving equations use inverse operations to calculate the value of an unknown.
For example, the equation $2q + 3 = 11$ can be solved as follows.

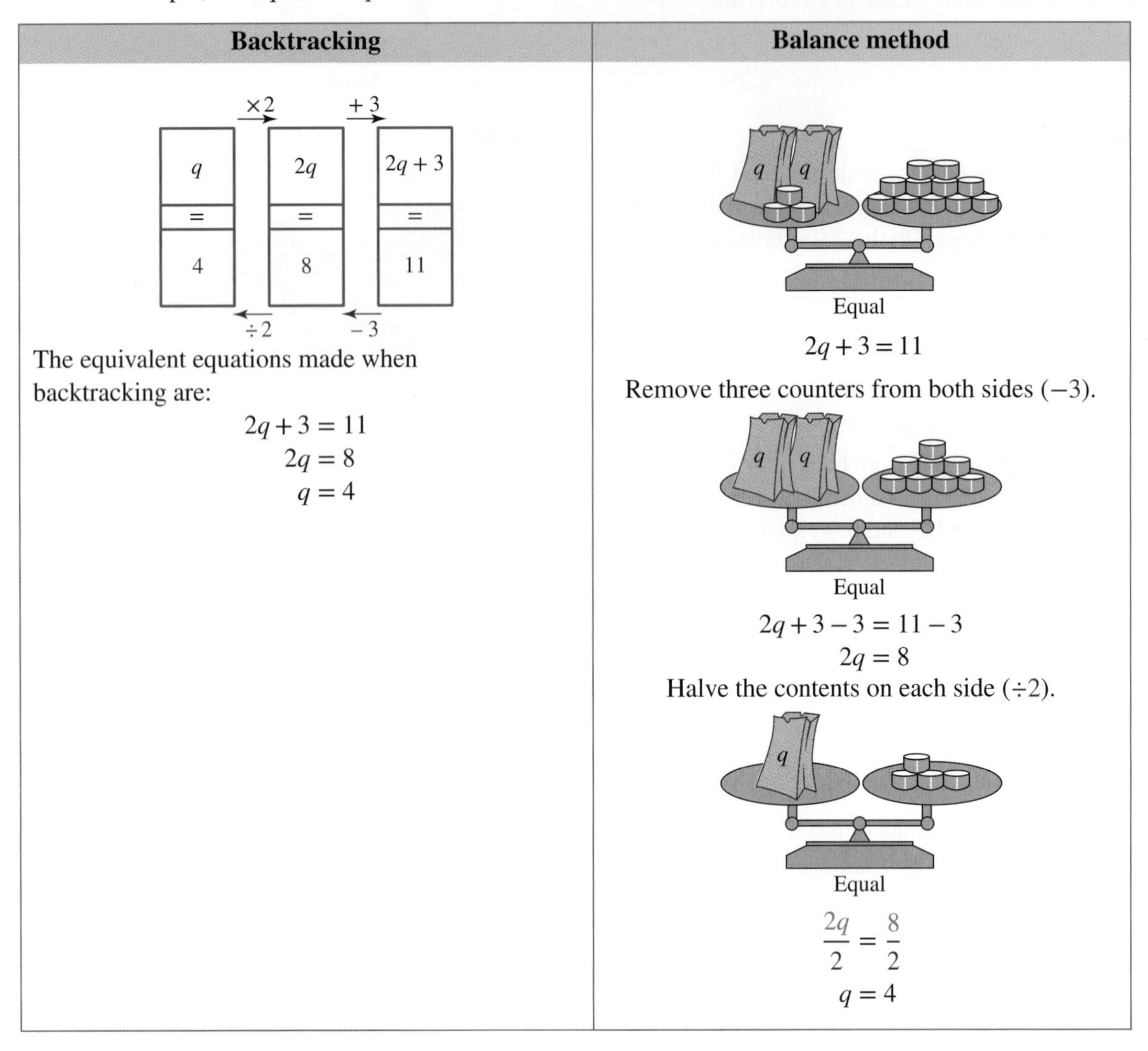

| Backtracking | Balance method |
|---|---|
| $q \xrightarrow{\times 2} 2q \xrightarrow{+3} 2q + 3$; $4 \xleftarrow{\div 2} 8 \xleftarrow{-3} 11$ | Equal<br>$2q + 3 = 11$<br>Remove three counters from both sides ($-3$). |
| The equivalent equations made when backtracking are:<br>$2q + 3 = 11$<br>$2q = 8$<br>$q = 4$ | Equal<br>$2q + 3 - 3 = 11 - 3$<br>$2q = 8$<br>Halve the contents on each side ($\div 2$). |
| | Equal<br>$\frac{2q}{2} = \frac{8}{2}$<br>$q = 4$ |

- Both methods are equivalent, and use inverse operations to simplify the LHS of the equation until only the pronumeral remains.
- The operations that solve the equation and the order in which they are performed are the same.
- The equivalent equations created on the way to the solution are the same.
- The aim is to isolate the unknown value on one side of the equation.
- The last operation that was performed on the unknown value when building the equation is the first operation that is undone.

### How to solve equations using inverse operations

- **To solve an equation using inverse operations:**
  1. **determine the operations used to build the expression containing the pronumeral**
  2. **apply the inverse operations in reverse order to both sides of the equation.**
- **You can use a flowchart or a written list to determine the operations and their inverse.**

*Note:* Building up expressions was covered in lesson 8.3.

## WORKED EXAMPLE 13 Using inverse operations

**Use inverse operations to solve the following equations.**

a. $2y + 3 = 11$

b. $\frac{h}{5} + 1 = 3$

c. $2(k - 4) = 4$

**THINK** / **WRITE**

a. 1. Write the equation.

$2y + 3 = 11$

2. Note that on the LHS, $y$ is multiplied by 2 and 3 is added to it.
Apply inverse operations to:
- remove 3 by subtracting 3 from both sides first
- divide both sides by 2 to remove 2.

$2y + 3 - 3 = 11 - 3$

$2y = 8$

$\frac{2y}{2} = \frac{8}{2}$

3. Simplify.

$y = 4$

4. Write the solution.

The solution for the equation is $y = 4$.

b. 1. Write the equation.

$\frac{h}{5} + 1 = 3$

2. Note that on the LHS, $h$ is divided by 5 and 1 is added to it.
Apply inverse operations to:
- remove 1 by subtracting 1 from both sides first
- multiply both sides by 5 to remove 5.

$\frac{h}{5} + 1 - 1 = 3 - 1$

$\frac{h}{5} = 2$

$\frac{h}{5} \times 5 = 2 \times 5$

3. Simplify.

$h = 10.$

4. Write the solution.

The solution for the equation is $h = 10$.

c. 1. Write the equation.

$2(k - 4) = 4$

2. Note that on the LHS, 2 is multiplied by $k - 4$ and 4 is subtracted from $k$.
Apply inverse operations to:
- remove 2 by dividing both sides by 2 first
- add 4 to both sides to remove −4.

$\frac{2(k-4)}{2} = \frac{4}{2}$

$k - 4 = 2$

$k - 4 + 4 = 2 + 4$

3. Simplify.

$k = 6.$

4. Write the solution.

The solution for the equation is $k = 6$.

### WORKED EXAMPLE 14 Solving equations with non-integer solutions

**Solve the following equations.**

**a.** $5d = 4$ **b.** $3c + 1 = 6$

| THINK | WRITE |
|---|---|
| **a. 1.** Write the equation. | $5d = 4$ |
| **2.** Note that on the LHS, $d$ is multiplied by 5. Apply inverse operation to: • remove 5 by dividing both sides by 5. | $\frac{5d}{5} = \frac{4}{5}$ |
| **3.** Simplify. | $d = \frac{4}{5}$ |
| **4.** State the solution. | The solution for the equation is $d = \frac{4}{5}$. |
| **b. 1.** Write the equation. | $3c + 1 = 6$ |
| **2.** Note that on the LHS, $c$ is multiplied by 3 and 1 is added to it. Apply inverse operations to: • remove 1 by subtracting 1 from both sides first • divide both sides by 3 to remove 3. | $3c + 1 - 1 = 6 - 1$<br>$3c = 5$<br>$\frac{3c}{3} = \frac{5}{3}$ |
| **3.** Simplify. | $c = \frac{5}{3}$ |
| **4.** Write the solution. | The solution for the equation is $c = \frac{5}{3}$ or $c = 1\frac{2}{3}$. |

## 8.5.4 Solving equations by making the pronumeral positive

eles-4022

- When an equation includes a pronumeral that is being subtracted (for example, $28 - 3x = 7$), an effective way to solve the equation is to add the pronumeral part to both sides first, then solve as normal using balancing or backtracking.

### WORKED EXAMPLE 15 Solving equations by making the pronumeral positive

**Solve the following equations.**

**a.** $5 - v = 2$ **b.** $17 - 2a = 11$

| THINK | WRITE |
|---|---|
| **a. 1.** Undo the subtraction of $v$ by adding $v$ to both sides and simplify. | **a.** $5 - v = 2$<br>$5 - v + v = 2 + v$<br>$5 = 2 + v$ |
| **2.** To isolate $v$, subtract 2 from both sides and simplify. | $5 - 2 = 2 + v - 2$<br>$3 = v$ |
| **3.** State the solution. | $v = 3$ |

| | |
|---|---|
| **b. 1.** Undo subtracting $2a$ by adding $2a$ to both sides and simplify. | **b.** $17 - 2a = 11$<br>$17 - 2a + 2a = 11 + 2a$<br>$17 = 11 + 2a$ |
| **2.** The last operation performed on the RHS when building $17 = 11 + 2a$ was $+11$. Subtract 11 from both sides and simplify. | $17 - 11 = 11 + 2a - 11$<br>$6 = 2a$ |
| **3.** To isolate $a$, divide both sides by 2 and simplify. | $\frac{6}{2} = \frac{2a}{2}$ |
| **4.** State the solution. | $3 = a$<br>$a = 3$ |

### COLLABORATIVE TASK: How many was that?

Work in pairs to answer the following questions and then share your ideas with others in your class.

1. A farmer told two friends that they could pick peaches from his tree but that they must not take more than 30 peaches each. They worked for a while, and then the taller one asked her friend, 'Have you picked your limit yet?'
   He replied, 'Not yet, but if I had twice as many as I have now, plus half as many as I have now, I would have reached my limit.'
   How many peaches did he have?

2. Your aunt from overseas sent an enormous box of chocolates for your mother's birthday. You were told not to eat any of the chocolates. Nevertheless, over the next five days you sneaked into the pantry when your mother was out and ate some of the chocolates. Each day you ate six more than the day before. After five days you were caught and your mother found out that you had eaten 100 chocolates. Your mother was extremely cross!
   How many chocolates did you eat on *each* of the five days?

 Resources

 **eWorkbook** Topic 8 Workbook (worksheets, code puzzle and project) (ewbk-1909)

 **Interactivities** Individual pathway interactivity: Keeping equations balanced (int-4377)
Keeping equations balanced (int-4047)
Equations with rational number solutions (int-4048)
Negative integers (int-4049)

## Exercise 8.5 Solving equations using inverse operations

learnon

| 8.5 Quick quiz on | 8.5 Exercise |
|---|---|

### Individual pathways

| ■ PRACTISE | ■ CONSOLIDATE | ■ MASTER |
|---|---|---|
| 1, 3, 6, 9, 14, 17 | 2, 4, 7, 10, 13, 15, 18 | 5, 8, 11, 12, 16, 19, 20, 21 |

### Fluency

1. WE11 Apply the given operations to the LHS and RHS of the following equations to maintain the balance of the two sides.
   a. $x + 3 = 6$ [multiply both sides by 3]
   b. $3x = 15$ [subtract 2 from both sides]
   c. $4x + 12 = 24$ [divide both sides by 4]
2. WE12 For each of the following balance scales:
   i. write the equation represented by the scale
   ii. calculate the value of the pronumeral.

a.

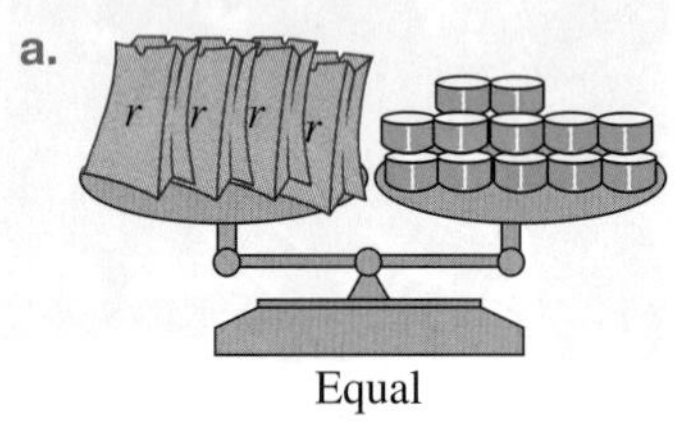


Equal

b.

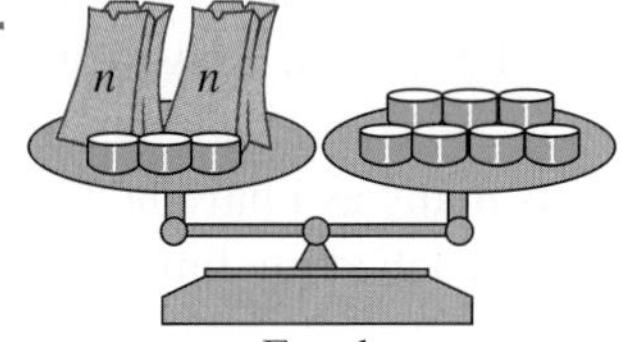


Equal

c.

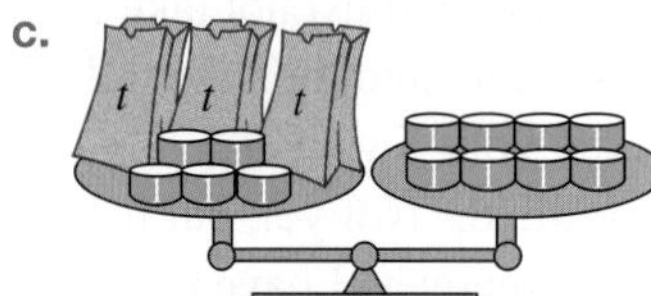


Equal

3. WE13 Use inverse operations to solve the following.
   a. $3g + 7 = 10$ b. $4m - 6 = 6$ c. $\frac{q}{3} + 8 = 11$
4. Use inverse operations to solve the following.
   a. $5b - 2 = 8$ b. $2z + 13 = 37$ c. $\frac{v}{4} + 31 = 59$
5. Use inverse operations to solve the following.
   a. $2(n - 5) = 8$ b. $\frac{g - 2}{4} = 3$ c. $4(y + 1) = 16$
6. WE14 Solve the following equations.
   a. $3h = 7$ b. $2k = 5$ c. $2w + 1 = 8$
7. Solve the following equations.
   a. $3t - 4 = 9$ b. $6h - 3 = 10$ c. $3l + 4 = 8$
8. Solve the following equations.
   a. $5g + 3 = 11$ b. $3h - 10 = 1$ c. $8n - 4 = 0$

### Understanding

9. WE15 Solve the following equations.
   a. $6 - m = 2$ b. $4 - d = 1$ c. $12 - 3v = 6$

10. Solve the following equations.

a. $13 - 2s = 7$

b. $19 - 3g = 4$

c. $30 - 5k = 20$

11. Solve the following equations.

a. $2(x + 3) = 11$

b. $p + 3 = 6\frac{1}{4}$

c. $7 - x = 5$

12. Solve the following equations.

a. $3(5 - y) = 12$

b. $5(8 - 2h) = 15$

c. $2(x + 3) = 7$

13. The formula $C = \frac{5(F - 32)}{9}$ is used to convert degrees Fahrenheit to degrees Celsius. Use the formula to find:

a. 45° Fahrenheit in degrees Celsius

b. 45° Celsius in degrees Fahrenheit.

### Reasoning

14. A taxi company charges a $3.60 flag fall (a fixed fee before the journey starts). An additional $2.19 per km is charged for the journey.

a. Calculate the cost of a 4.6 km journey.

b. If the journey cost $110.91, determine the distance travelled.

15. While shopping for music online, Olivia found an album she liked. She could buy the whole album for $17.95 or buy songs from the album for $1.69 each. Determine the number of individual songs Olivia could buy to make it cheaper to buy the whole album.

16. Explain why the order of inverse operations is important, using the equation $2x + 3 = 15$ as an example. In your explanation, use balanced scales to represent the equation.

### Problem solving

17. A class of 25 students has 7 more boys than girls. Determine how many boys there are.

18. When 12 is subtracted from 7 times a number, the result is 9. Create an equation that represents this statement, then solve your equation to determine the number.

19. Given that the perimeter of the following triangle is 22 cm, calculate the value of $x$.

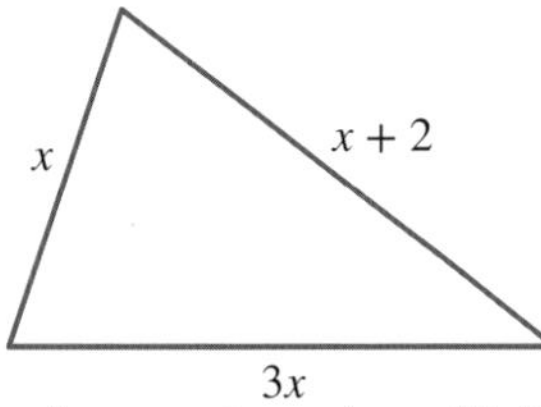


20. A number $x$ becomes the fifth number in the set of numbers $\{3, 7, 9, 13\}$ to make the mean of the five numbers equal to their median. Determine all possible positive values for $x$.

21. These scales show that two bricks are equal to 2 kilograms plus half a brick. Determine the weight of each brick.

# LESSON
# 8.6 Checking solutions

**LEARNING INTENTION**

At the end of this lesson you should be able to:

- check the solution of an equation using substitution.

## 8.6.1 Checking solutions by substituting

eles-4023

- Once the solution to an equation has been determined, the method of **substitution** can be used to check whether that solution is correct.
- Substitution involves:
  - replacing the pronumeral with the numerical value of the solution
  - simplifying the equation to determine the value of the LHS and RHS of the equation.
- If the LHS = RHS after substitution, then the solution is correct.

**WORKED EXAMPLE 16 Checking solutions by substitution**

**For each of the following equations, use substitution to determine whether $x = 7$ is the solution to the equation.**

a. $\dfrac{x+5}{3} = 4$

b. $2x - 8 = 10$

| THINK | WRITE |
|---|---|
| a. 1. Write the equation. | a. $\dfrac{x+5}{3} = 4$ |
| 2. Write the LHS of the equation and substitute $x = 7$. | If $x = 7$, LHS $= \dfrac{x+5}{3}$<br>$= \dfrac{7+5}{3}$ |
| 3. Perform the calculation. | $= \dfrac{12}{3}$<br>$= 4$ |
| 4. Compare the LHS with the RHS of the equation. | LHS = 4, RHS = 4 |
| 5. Comment on the answer. | Since LHS = RHS, $x = 7$ is the solution. |
| b. 1. Write the equation. | b. $2x - 8 = 10$ |
| 2. Write the LHS of the equation and substitute $x = 7$. | If $x = 7$, LHS $= 2x - 8$<br>$= 2(7) - 8$<br>$= 14 - 8$ |
| 3. Perform the calculation. | $= 6$ |
| 4. Compare the LHS with the RHS of the equation. | LHS = 6, RHS = 10 |
| 5. Comment on the answer. | $x = 7$ is not the solution, since LHS $\neq$ RHS. |

**Digital technology**

Scientific calculators can be used to check solutions by substitution. Replace pronumerals with the desired value in brackets.

For example, the solution to the equation $3x - 15 = 12$ is $x = 9$.

| 3(9)−15 | 12 |
|---|---|

**WORKED EXAMPLE 17 Verifying solutions by substitution**

**A solution is given for each equation below. Verify whether the solution is correct.**

a. $\frac{x+2}{3} = 2x - 12, x = 10$

b. $3x - 7 = 2x + 3, x = 10$

| THINK | WRITE |
|---|---|
| a. 1. Write the equation. | a. $\frac{x+2}{3} = 2x - 12$ |
| 2. Write the LHS of the equation and substitute $x = 10$. | If $x = 10$, LHS $= \frac{x+2}{3}$ <br> $= \frac{10+2}{3}$ |
| 3. Perform the calculation. | $= \frac{12}{3} = 4$ |
| 4. Write the RHS of the equation and substitute $x = 10$. | If $x = 10$, RHS $= 2x - 12$ <br> $= 2(10) - 12$ |
| 5. Perform the calculation. | $= 20 - 12 = 8$ |
| 6. Comment on the answer. | $x = 10$ is not the solution, since LHS $\neq$ RHS. |
| b. 1. Write the equation. | b. $3x - 7 = 2x + 3$ |
| 2. Write the LHS of the equation and substitute $x = 10$. | If $x = 10$, LHS $= 3x - 7$ <br> $= 3(10) - 7$ |
| 3. Perform the calculation. | $= 30 - 7 = 23$ |
| 4. Write the RHS of the equation and substitute $x = 10$. | If $x = 10$, RHS $= 2x + 3$ <br> $= 2(10) + 3$ |
| 5. Perform the calculation. | $= 20 + 3 = 23$ |
| 6. Comment on the answer. | Since LHS = RHS, $x = 10$ is the correct solution. |

**on** Resources

**eWorkbook** Topic 8 Workbook (worksheets, code puzzle and project) (ewbk-1909)

**Interactivities** Individual pathway interactivity: Checking solutions (int-4376)
Checking solutions (int-4046)

## Exercise 8.6 Checking solutions

learnon

8.6 Quick quiz on | 8.6 Exercise

### Individual pathways

| ■ PRACTISE | ■ CONSOLIDATE | ■ MASTER |
|---|---|---|
| 1, 4, 6, 10, 13 | 2, 7, 8, 11, 14 | 3, 5, 9, 12, 15 |

### Fluency

1. **WE16** Use substitution to determine whether each of the following is correct.
   a. $x = 3$ is the solution to the equation $x + 2 = 6$
   b. $x = 3$ is the solution to the equation $2x - 1 = 5$
   c. $x = 4$ is the solution to the equation $6x - 6 = 24$

2. Use substitution to determine whether each of the following is correct.
   a. $x = 5$ is the solution to the equation $4(x - 3) = 8$
   b. $x = 7$ is the solution to the equation $3(x - 2) = 25$
   c. $x = 8$ is the solution to the equation $5(x + 1) = 90$

3. Use substitution to determine whether each of the following is correct.
   a. $x = 81$ is the solution to the equation $3x - 53 = 80$
   b. $x = 2.36$ is the solution to the equation $5x - 7 = 4.8$
   c. $x = 4.4$ is the solution to the equation $7x - 2.15 = 18.64$.

4. **WE17** Verify whether the solution given for each of the following equations is correct.
   a. $5x + 1 = 2x - 7, x = 8$
   b. $5x = 2x + 12, x = 4$
   c. $4x = 3x + 8, x = 8$

5. Verify whether the solution given for each of the following equations is correct.
   a. $3x - 1.2 = x + 2.9, x = 1.9$
   b. $6x + 1.5 = 2x + 41.5, x = 10$
   c. $1.2(x + 1.65) = 0.2(x + 9.85), x = 0.45$

### Understanding

6. Complete the following table to determine the value of $2x + 3$ when $x = 0, 1, 2, 3, 4$.

| $x$ | 0 | 1 | 2 | 3 | 4 |
|---|---|---|---|---|---|
| $2x + 3$ | | | | | |

   a. Identify the solution (that is, the value of $x$) for $2x + 3 = 11$.
   b. Identify the solution (that is, the value of $x$) for $2x + 3 = 5$.

7. Complete the following table to determine the value of $5(x - 2)$ when $x = 2, 3, 4, 5, 6$.

| $x$ | 2 | 3 | 4 | 5 | 6 |
|---|---|---|---|---|---|
| $5(x - 2)$ | | | | | |

   a. Identify the solution (that is, the value of $x$) for $5(x - 2) = 10$.
   b. Identify the solution (that is, the value of $x$) for $5(x - 2) = 20$.
   c. Guess the solution (that is, the value of $x$) to $5(x - 2) = 30$. Check your guess.

8. a. Copy and complete the following table.

| $x$ | $2x + 1$ | $3x - 5$ |
|---|---|---|
| 3 | | |
| 4 | | |
| 5 | | 10 |
| 6 | | |
| 7 | | |

b. Identify the solution to $2x + 1 = 3x - 5$.

9. a. Copy and complete the following table.

| $x$ | $\frac{x+3}{2}$ | $2x - 6$ |
|---|---|---|
| 3 | 3 | |
| 5 | | |
| 7 | | 8 |
| 9 | | |
| 11 | | |

b. Identify the solution to $\frac{x+3}{2} = 2x - 6$.

## Reasoning

10. Substitution is a method used to check that a solution to an equation is correct. Explain how you could use substitution to answer a multiple choice question that has potential solutions as the options.

11. Consider the diagram shown.

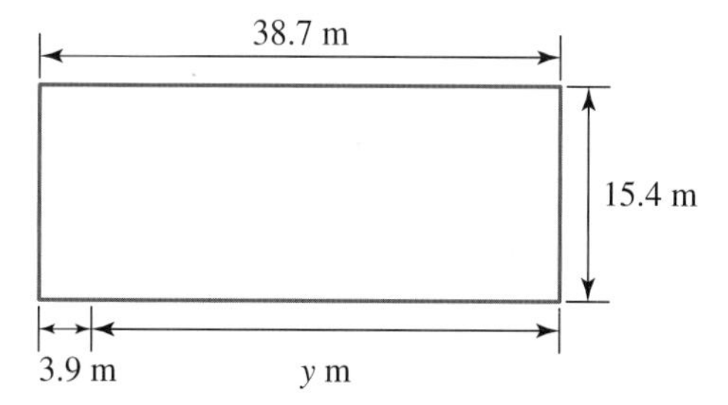


a. Explain whether you need to know both the length and width of this rectangle to be able to calculate the value of $y$.
b. Calculate the value of $y$.
c. Explain how you would check the solution.

12. Carol is making a quilt for her granddaughter. The quilt pattern requires that $\frac{1}{7}$ of the quilt is made of a pink fabric. Carol has $0.5\ \text{m}^2$ of pink fabric. She intends to use all the pink material in the quilt.
Calculate how much more fabric she needs for the entire quilt.

a. Write an equation for this problem.
b. Calculate the value of the unknown variable.
c. Check your solution.

## Problem solving

13. James and Alison share \$125 between them, but Alison gets \$19 more than James. Create an equation to describe this situation. Calculate how much money James and Alison each receive.

14. Aiko and Miyu are sisters who receive pocket money based on their age. Their parents pay them according to the equation $\$P = 3x - 25$, where $\$P$ represents their weekly pocket money and $x$ is their age in years.
    a. Determine at what age Aiko and Miyu start receiving pocket money.
    b. Miyu is 2 years older than Aiko. Determine how much more pocket money Miyu receives each week compared to her sister. (Assume the sisters are both old enough to be receiving pocket money.)
    c. When Aiko is 11 years old, calculate how much pocket money each child receives.

15. Ivan is trying to find the solution to the equation $n + (n + 1) + (n + 2) = 393$. He believes the solution is $n = 133$.
    a. Show that Ivan is incorrect.
    b. Determine the correct solution, and prove that it is the solution using substitution.
    c. Write a worded problem that this equation represents.

# LESSON
# 8.7 Review

## 8.7.1 Topic summary

**EQUATIONS**

### What is an equation?

- An equation is a mathematical statement that says two expressions are equal.

e.g. $7x + 9 = 23$ is an example of an equation.

- The left-hand side (LHS) is $7x + 9$.
- The right-hand side (RHS) is 23.
- This equation states that the LHS and the RHS are equal (have the same value).

### Solving equations

- Solving an equation is the process of finding pronumeral values to make the equation true.
  e.g. The equation $7x + 9 = 23$ is true only when $x = 2$.
  Hence, the solution of the equation is $x = 2$.
- Equations can be solved using many different techniques.

### Solving equations using guess, check and improve

- Solving equations using the guess, check and improve method involves selecting a number for the pronumeral that you think might be a solution to the equation.
- Use substitution to check whether your guess is correct or not.
- After checking your guess, make improvements to your chosen number by either increasing or decreasing it until you find the solution.

e.g. For $2x + 5 = 11$

Guess: $x = 2$

Check: LHS $= 2(2) + 5 = 4 + 5 = 9 \neq$ RHS

Improve: Try $x = 3$

LHS $2(3) + 5 = 6 + 5 = 11 =$ RHS

Solution is $x = 3$.

### Solving equations using balancing

- By performing the same operation on *both* sides of an equation, it remains balanced.
- Equations can be solved using inverse operations with balancing, that is, applying an inverse operation to both sides of the equation.

e.g. $x - 20 = 12$

$x - 20 + 20 = 12 + 20$

$x = 32$

### Developing an equation from words

- When developing an equation from words, some English phrases or words can be turned into mathematical symbols.

e.g.

- Words such as *sum*, *more than*, *increased*, *add* or *added* imply addition and can be replaced with +.
- Words such as *difference*, *less than*, *decreased* or *minus* imply subtraction and can be replaced with −.
- Words such as *product* or *times* imply multiplication and can be replaced with ×.
- Words such as *quotient* or *divide* imply division and can be replaced with ÷.

### Building expressions using flowcharts

- Flowcharts can be used to build an expression from words, steps, instructions or operations.
- In a flowchart, the starting number or position is the *input* and the final number or position is the *output*.

e.g.

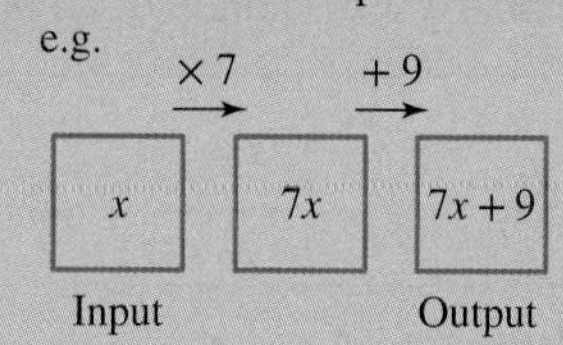


### Inverse operations

- Inverse (or opposite) operations are often used to help solve equations.

+ is the inverse operation of −

− is the inverse operation of +

× is the inverse operation of ÷

÷ is the inverse operation of ×

### Solving equations using backtracking

- Backtracking uses inverse operations to determine the solution of an equation by moving backwards through a flowchart.

e.g. Solving $7x + 9 = 23$ using backtracking gives:

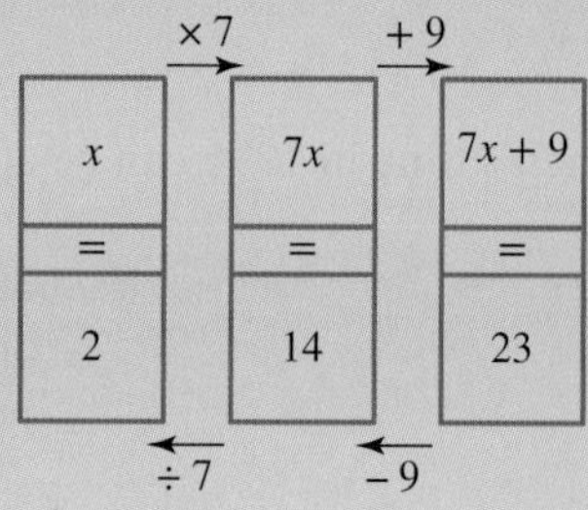


Solution is $x = 2$.

## 8.7.2 Success criteria

Tick the column to indicate that you have completed the lesson and how well you think you have understood it using the traffic light system.

(**Green:** I understand; **Yellow:** I can do it with help; **Red:** I do not understand)

| Lesson | Success criteria | Green | Yellow | Red |
|---|---|---|---|---|
| 8.2 | I can write an equation to represent a worded problem. | | | |
| | I can solve equations by inspection. | | | |
| | I can solve equations using trial and error. | | | |
| 8.3 | I can complete a flowchart to determine the output. | | | |
| | I can construct a flowchart from a given expression. | | | |
| 8.4 | I can solve equations using backtracking. | | | |
| 8.5 | I can solve equations using the balancing method. | | | |
| | I can solve equations using inverse operations. | | | |
| 8.6 | I can check the solution of an equation using substitution. | | | |

## 8.7.3 Project

### Equations at the Olympic Games

The Olympic Games is held every four years between competing nations from all over the world. You might know that Sydney was the host city for the 2000 Olympic Games. A feature of this international sporting event is that records continue to be broken. Every Olympic Games sees competitors run faster, lift heavier weights, and so on. The desire to become better urges competitors to train harder in the hope that they will become record holders.

This project looks at running times for the men's and women's 100-metre running events. The running times for these events can be approximated using the following equations:

- Men's 100-metre event $t = -0.0094y + 28.73$
- Women's 100-metre event $t = -0.0173y + 45.31$

where $t$ represents the running time (in seconds) and $y$ represents the year.

For example, the men's time in 1996 can be approximated by substituting 1996 into the equation as shown.

$$\begin{aligned} t &= -0.0094y + 28.73 \\ &= -0.0094 \times 1996 + 28.73 \\ &= 9.9676 \text{ seconds} \end{aligned}$$

1. Use the equations provided to calculate the men's and women's running times for the 100-metre event in the following years.
   a. 1928  b. 1968  c. 1988  d. 2016
2. The following table shows the winners and their running times for the 100-metre event final at four Olympic Games events. How well do these actual results compare with the times you calculated using the equations? What does this say about the equations?

| Year | Men's winner | Time taken (seconds) | Women's winner | Time taken (seconds) |
|---|---|---|---|---|
| 1928 | Percy Williams | 10.80 | Elizabeth Robinson | 12.20 |
| 1968 | James Hines | 9.95 | Wyomia Tyus | 11.08 |
| 1988 | Carl Lewis | 9.92 | Florence Griffith Joyner | 10.54 |
| 2016 | Usain Bolt | 9.81 | Elaine Thompson | 10.71 |

3. Predict the times for both men and women at the 2024 Olympic Games.

When making predictions about the future, we must remember that these predictions are based on the assumption that the patterns we observe now will continue into the future.

Your answers to question **1** show you that men have historically run faster times than women in the 100-metre event. However, closer inspection of the times shows that women are making greater improvements in their times over the years.

4. If running times continue to follow these patterns in future Olympic Games, decide whether it is possible that women's times will become equal to men's times. Discuss your answer.
5. Suggest reasons that running times are getting shorter. Do you think they will follow this pattern forever?

## on Resources

**eWorkbook** Topic 8 Workbook (worksheets, code puzzle and project) (ewbk-1909)

**Interactivities** Crossword (int-2606)
Sudoku puzzle (int-3173)

## Exercise 8.7 Review questions

### Fluency

1. Solve these equations by inspection.
   a. $m + 7 = 12$  b. $5h = 30$  c. $s - 12 = 7$  d. $\frac{d}{5} = 4$

2. Use guess, check and improve to find two numbers whose sum and product are given.
   a. sum = 83, product = 1632  b. sum = 86, product = 1593

3. Calculate the output number for each of the following flowcharts.

   a.
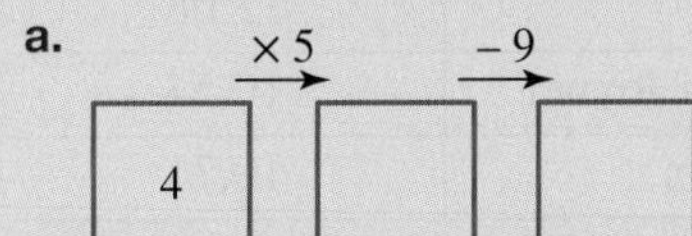


   b.
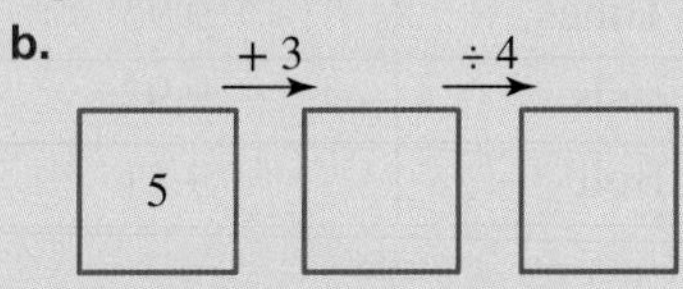


   c. ÷ 2 → + 11

   10

   d.
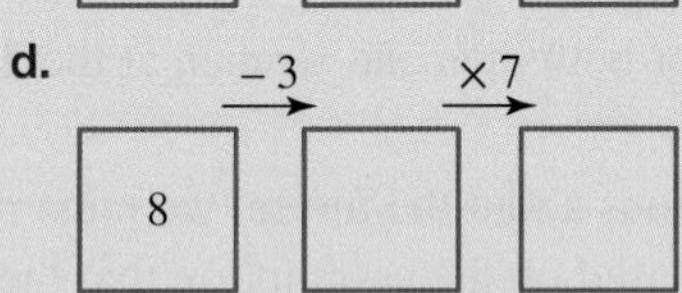


4. Use backtracking and inverse operations to calculate the input number for each of the following flowcharts.

   a.
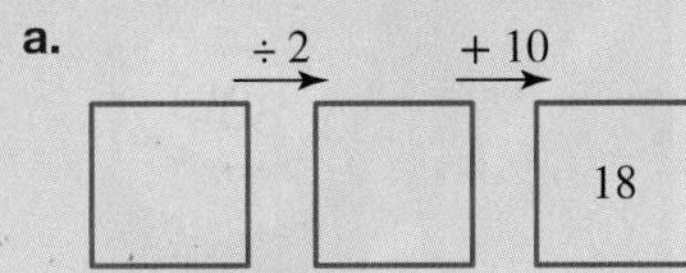


   b.
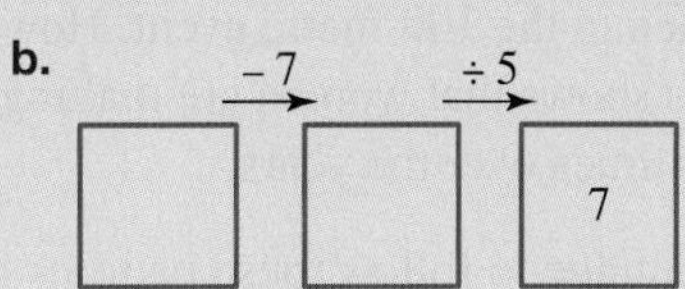


   c. − 6 → × 7

   35

   d.
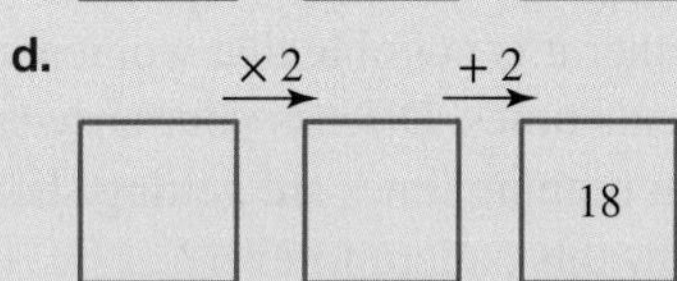


5. Build up an expression by following the instructions on the flowchart.

   a.
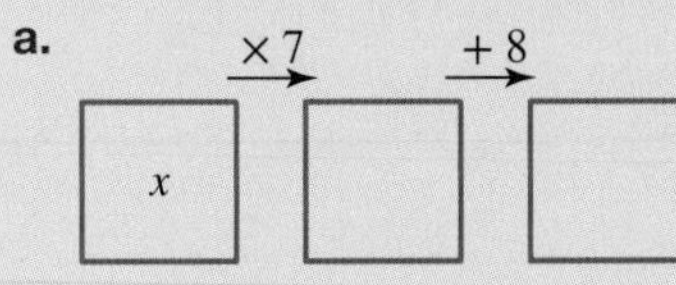


   b.
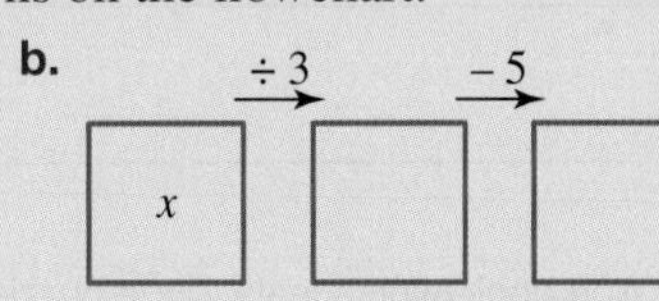


   c. + 2 → × 6

   $x$

   d. − 7 → ÷ 5

   $x$

6. Build up an expression by following the instructions on the flowchart.

   a.
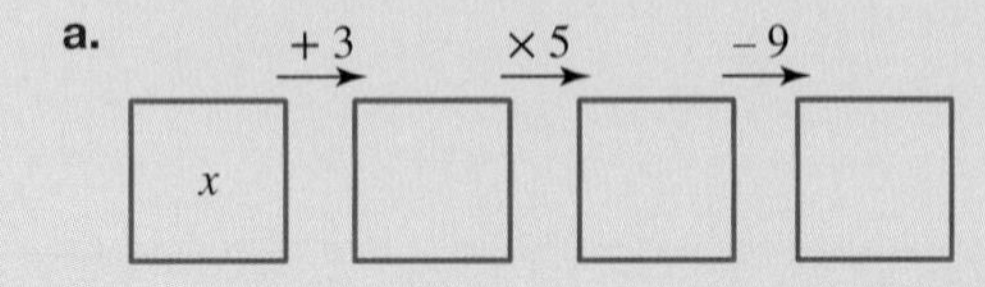


   b.
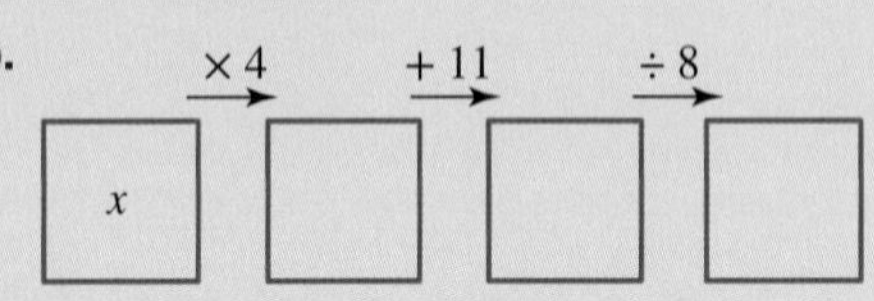


7. Draw a flowchart that has an input of $x$ and an output that is given by the expressions below.
   a. $5(x + 7)$  b. $\frac{x}{4} - 3$  c. $6x - 14$  d. $\frac{x + 2}{5}$

8. Draw a flowchart and use backtracking to calculate the solution to the following equations.
 a. $7x+6=20$  b. $9(y-8)=18$  c. $\frac{m}{5}-3=9$  d. $\frac{s+7}{5}=5$

9. Use backtracking to determine the solution to these equations.
 a. $3(d+1)=15$  b. $\frac{t}{4}-11=14$  c. $6d-3=15$  d. $\frac{a+6}{4}=3$

10. Simplify the expression and then solve the equation for each of the following.
 a. $7v+3+3v+4=37$  b. $6c+15-5c-8=19$

11. For each of the following equations there is a solution given. Is the solution correct?
 a. $5x-7=2x+2, x=3$  b. $\frac{x+9}{2}=2x-7, x=5$

12. For each of the following balance scales:
 i. write the equation represented by the scale
 ii. calculate the value of the pronumeral.

a.

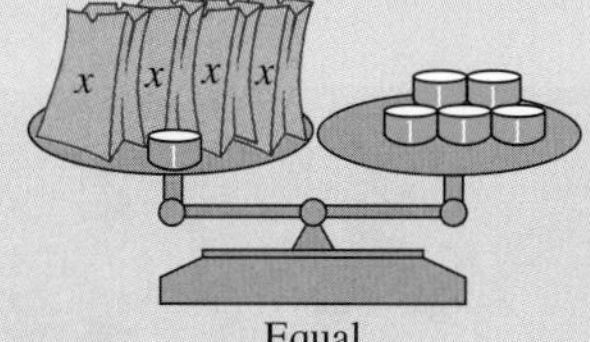


b.

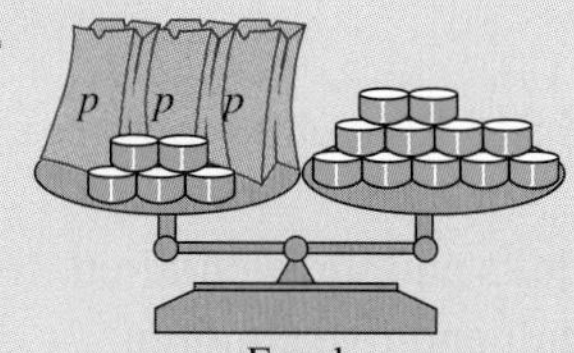


c.

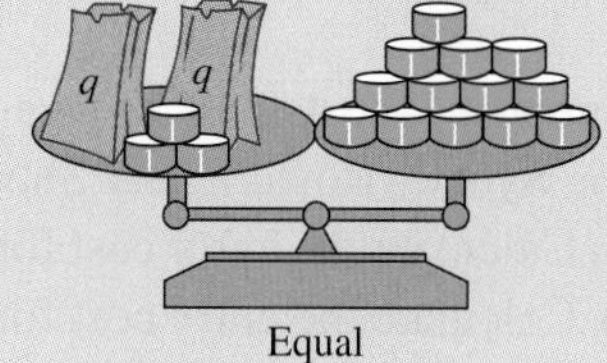


13. Use inverse operations to solve the following equations.
 a. $2v+1=7$  b. $\frac{x}{2}+3=4$  c. $4w-3=9$  d. $2(g+1)=10$

14. Solve the following equations. Make sure you check your solutions.
 a. $4k+3=10$  b. $2p-3=8$  c. $3q+6=7$  d. $5t+1=3$

15. Solve the following equations. Make sure you check your solutions.
 a. $3-x=1$  b. $7-2p=3$  c. $10-4r=2$  d. $13-3z=7$

## Problem solving

16. Sophie and Jackie each have a collection of football cards. Jackie has 5 more cards than Sophie. Together Jackie and Sophie have 67 cards. Create and solve an equation that shows how many cards Sophie owns.

17. Andreas has completed 2 more pieces of homework than Richard, who submitted $x$ pieces of homework for the semester. If the total number of pieces of homework submitted by the two boys is 12, determine how many pieces of homework Andreas submitted.

18. Keith is 6 years younger than his twin brothers. If the sum of the brothers' ages is 48, determine Keith's age.

**19.** Margaret bought six 2-hour tickets and four daily tickets for $28. If the cost of one daily ticket is $4, write and solve an equation to calculate the cost of a 2-hour ticket.

**20.** Jetski hire is $10 per hour, plus a $30 deposit.

**a.** Calculate the cost of hiring a jetski for:

**i.** 1 hour **ii.** 2 hours **iii.** 3 hours.

**b.** Write a rule that could be used to calculate the cost of hiring a jetski for $h$ hours.

**c.** Use your rule to calculate the cost of hiring a jetski for 8 hours.

**d.** You have $85 to spend. Write an equation to help you work out how many hours of jetski hire you can afford. Solve this equation.

**e.** Work out how much money (if any) you would have left over after you paid the hire charge and hire the jetski for 5 hours.

**f.** After spending 5 hours on the water you return the jetski with no damage. Explain whether you would have enough money for a hamburger on the way home.

**21.** Suppose it costs $30 for an adult and $15 for a child to enter the Sydney Royal Easter Show.

**a.** Calculate the entry cost for two adults and three children.

**b.** Calculate the entry cost for one adult and five children.

**c.** Determine the maximum number of children who could get into the Show for $100.

**d.** Identify the possible combinations of adults and children that could get in for exactly $300.

**22.** Judy is watching her daughter play in the park. She notices that some children are riding tricycles and some are riding bicycles. Altogether, 19 children are on cycles. She counts that there are 43 wheels on the cycles. Determine how many children are on tricycles and how many are on bicycles.

**23.** Two angles in a triangle have the same magnitude. The sum of the magnitudes of these angles is the magnitude of the third angle. Determine the magnitudes of the angles and use this information to describe the triangle in this scenario.

**24.** A stage screen is 4 times as long as it is wide. If it were 5 metres wider and 4 metres shorter it would be a square. Determine the dimensions of the stage screen.

**25.** The largest angle in a triangle is 65 degrees more than the smallest angle. The third angle is 10 degrees more than the smallest angle. Calculate the size of the smallest angle.

**on** To test your understanding and knowledge of this topic, go to your learnON title at www.jacplus.com.au and complete the **post-test**.

# Answers

## Topic 8 Equations

### 8.1 Pre-test

1. $6x = 72$
2. 32
3. 11 games
4. 4
5. C
6. 4
7. 32
8. A
9. B
10. 12 years old
11. $x = 3$
12. $x = 10$
13. 26
14. 16 boys
15. 5.5 cm

### 8.2 Introduction to equations

1. a. $x + 7 = 11$ b. $x + 12 = 12$
   c. $x - 7 = 1$ d. $x - 4 = 7$
2. a. $2x = 12$ b. $5x = 30$
   c. $\frac{x}{7} = 1$ d. $\frac{x}{5} = 2$
3. a. $15 - x = 2$ b. $52 - x = 8$
   c. $\frac{21}{x} = 7$ d. $x^2 = 100$
4. a. $x = 11$ b. $y = 9$
   c. $m = 5$ d. $m = 30$
5. a. $k = 0$ b. $b = 7$
   c. $b = 8.8$ d. $c = 4.2$
   e. $x = 2.8$
6. a. $x = 6$ b. $x = 3$ c. $x = 10$
7. a. $x = 9$ b. $x = 6$ c. $x = 4$
8. a. 7 and 14 b. 11 and 15
   c. 7 and 47 d. 42 and 136
9. a. 42 and 111 b. 97 and 145
   c. 2.2 and 3.9 d. 347 and 631
10.

| $x$ | $x^2 + 4$ | $4x + 1$ |
|---|---|---|
| 0 | 4 | 1 |
| 1 | 5 | 5 |
| 2 | 8 | 9 |
| 3 | 13 | 13 |
| 4 | 20 | 17 |

$x = 1$ and $x = 3$ are both solutions.

11. Think of the 8 times table. Determine which multiple of 8 is equal to 48. $6 \times 8 = 48$, so the answer is $x = 6$.
12. 34 does not appear in the 5 times table. Therefore the solution will not be a whole number.
13. Greater than 5
14. 10
15. 5 metres, 15 metres
16. 24 years old
17. Angus will be 27 and his grandfather will be 81.

### 8.3 Building up expressions and backtracking

1. a.

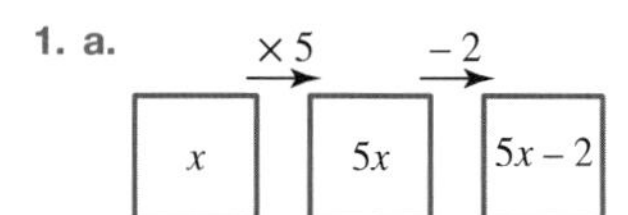


b.

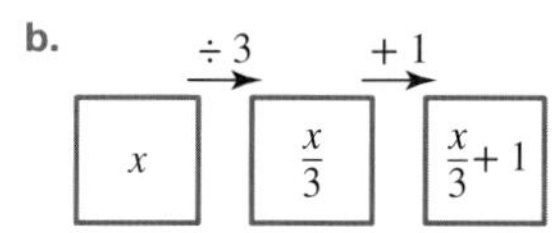


c.

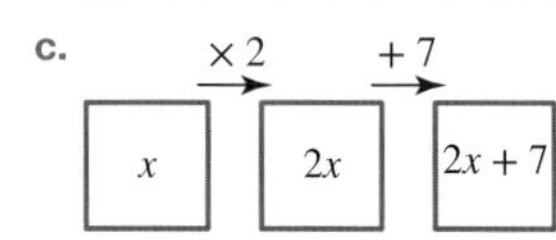


d.

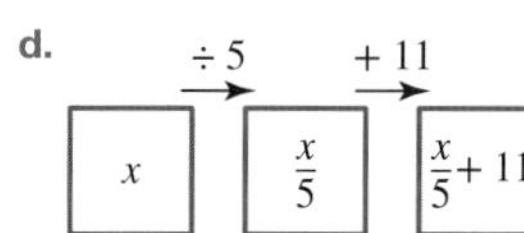


2. a.

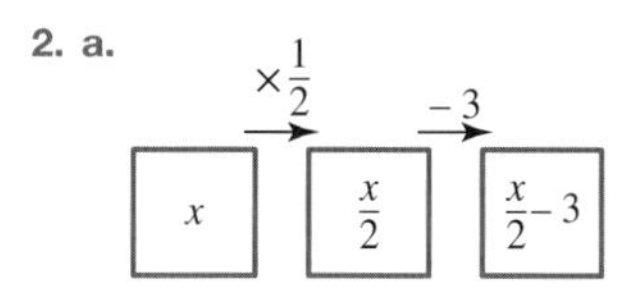


b.

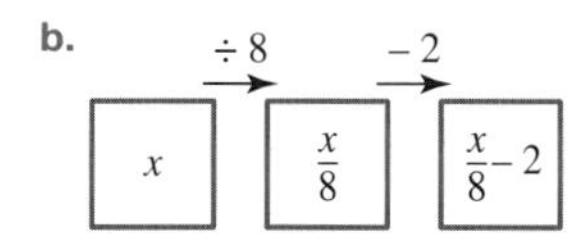


c.

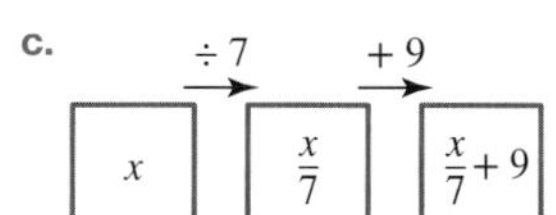


d. $x \xrightarrow{\times 3.1} 3.1x \xrightarrow{+1.8} 3.1x + 1.8$

3. a. $x \xrightarrow{+5} x + 5 \xrightarrow{\times 4} 4(x + 5)$

b. $x \xrightarrow{+10} x + 10 \xrightarrow{\div 3} \frac{x + 10}{3}$

c. $x \xrightarrow{-2} x - 2 \xrightarrow{\div 7} \frac{x - 2}{7}$

d. $x \xrightarrow{+3} x + 3 \xrightarrow{\times 9} 9(x + 3)$

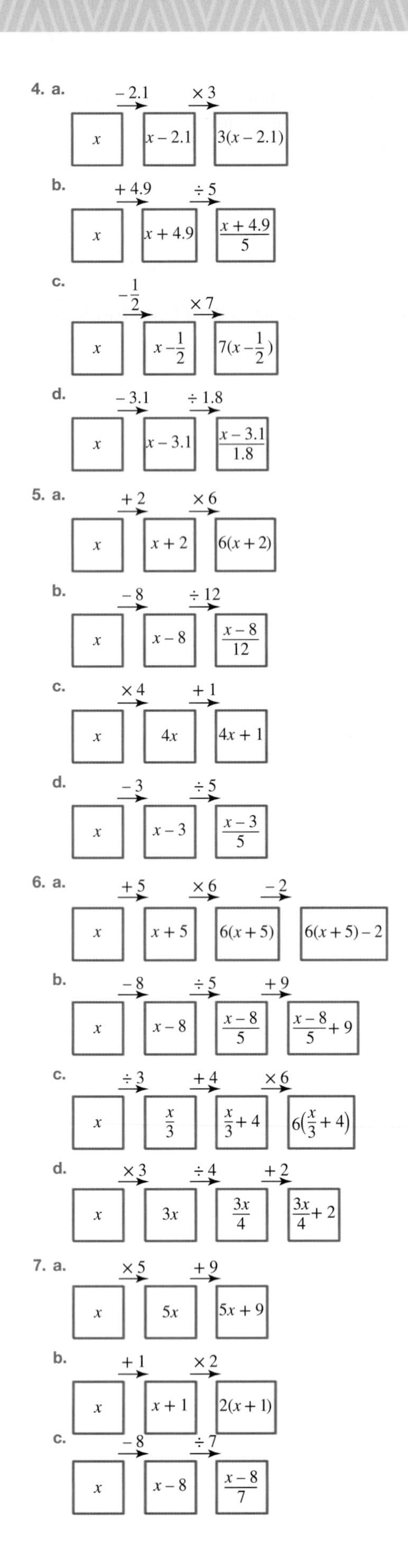
4. a.
− 2.1
× 3
x
x − 2.1
3(x − 2.1)
b.
+ 4.9
÷ 5
x
x + 4.9
(x + 4.9)/5
c.
− 1/2
× 7
x
x − 1/2
7(x − 1/2)
d.
− 3.1
÷ 1.8
x
x − 3.1
(x − 3.1)/1.8
5. a.
+ 2
× 6
x
x + 2
6(x + 2)
b.
− 8
÷ 12
x
x − 8
(x − 8)/12
c.
× 4
+ 1
x
4x
4x + 1
d.
− 3
÷ 5
x
x − 3
(x − 3)/5
6. a.
+ 5
× 6
− 2
x
x + 5
6(x + 5)
6(x + 5) − 2
b.
− 8
÷ 5
+ 9
x
x − 8
(x − 8)/5
(x − 8)/5 + 9
c.
÷ 3
+ 4
× 6
x
x/3
x/3 + 4
6(x/3 + 4)
d.
× 3
÷ 4
+ 2
x
3x
3x/4
3x/4 + 2
7. a.
× 5
+ 9
x
5x
5x + 9
b.
+ 1
× 2
x
x + 1
2(x + 1)
c.
− 8
÷ 7
x
x − 8
(x − 8)/7

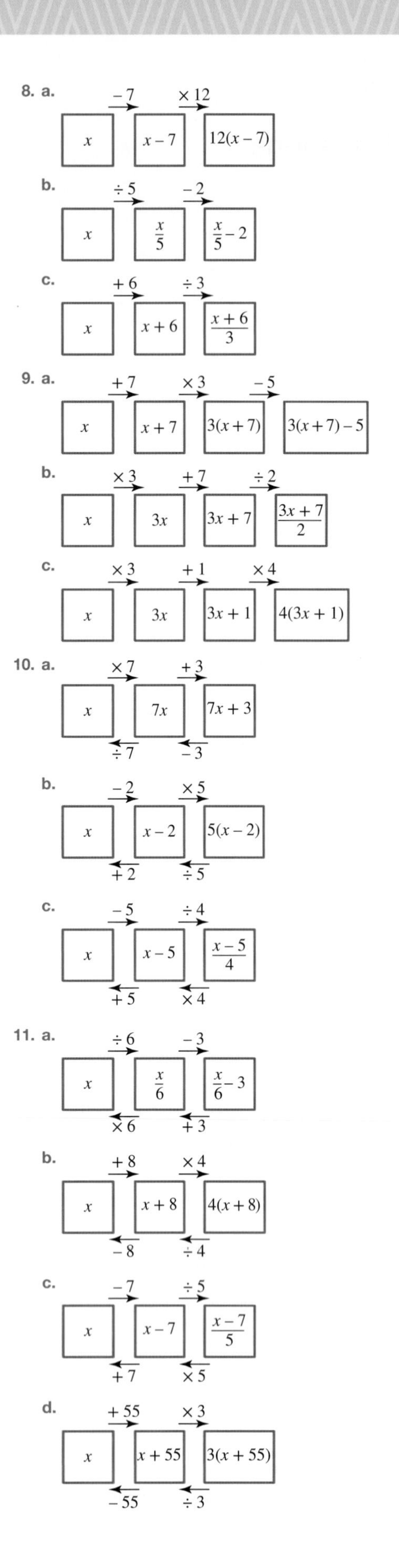
8. a.
− 7
× 12
x
x − 7
12(x − 7)
b.
÷ 5
− 2
x
x/5
x/5 − 2
c.
+ 6
÷ 3
x
x + 6
(x + 6)/3
9. a.
+ 7
× 3
− 5
x
x + 7
3(x + 7)
3(x + 7) − 5
b.
× 3
+ 7
÷ 2
x
3x
3x + 7
(3x + 7)/2
c.
× 3
+ 1
× 4
x
3x
3x + 1
4(3x + 1)
10. a.
× 7
+ 3
x
7x
7x + 3
÷ 7
− 3
b.
− 2
× 5
x
x − 2
5(x − 2)
+ 2
÷ 5
c.
− 5
÷ 4
x
x − 5
(x − 5)/4
+ 5
× 4
11. a.
÷ 6
− 3
x
x/6
x/6 − 3
× 6
+ 3
b.
+ 8
× 4
x
x + 8
4(x + 8)
− 8
÷ 4
c.
− 7
÷ 5
x
x − 7
(x − 7)/5
+ 7
× 5
d.
+ 55
× 3
x
x + 55
3(x + 55)
− 55
÷ 3

12. a.

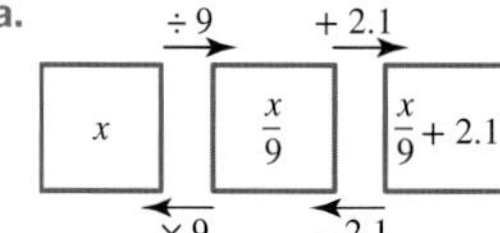


b. × 7, − 5: $x$ → $7x$ → $7x - 5$; ÷ 7, + 5

c.

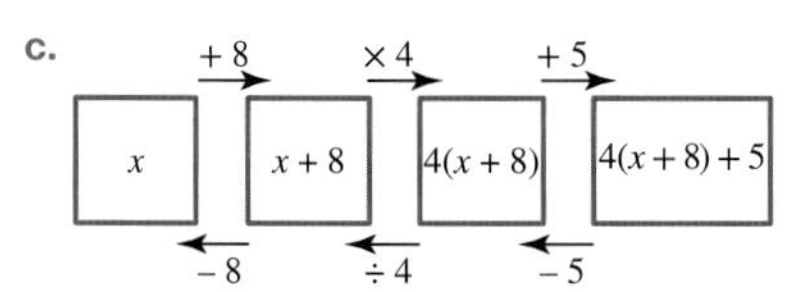


d.

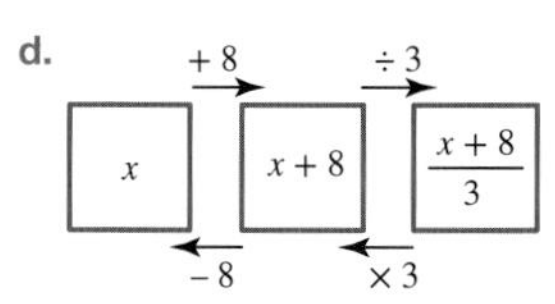


13. $-3$ should be performed first, then $\times 2$.

14. $\frac{5(x+8)}{4} = 30$

15. $5n + 15 = 4n - 3$

16. 22 years old

17. The two numbers are 19 and 13.

18. \$1215

19. Some examples are:
$12 + 3 - 4 + 5 + 67 + 8 + 9 = 100$
$123 + 4 - 5 + 67 - 89 = 100$
$123 - 45 - 67 + 89 = 100$

## 8.4 Solving equations using backtracking

1. a. 2 b. 4 c. 8 d. 13 e. 10

2. a. 67 b. 88 c. 5 d. 8 e. 5.3

3. a. $4\frac{2}{5}$ b. $\frac{31}{36}$

4. a. $x = 3$ b. $y = 1$ c. $x = 5$ d. $w = 3$ e. $w = 5.7$

5. a. $x = 1$ b. $x = 18$ c. $x = 4$ d. $x = 3$ e. $w = 1.3$

6. a. $x = 3$ b. $x = 45$ c. $x = 34$ d. $x = 77$ e. $x = 13$

7. a. $x = 14$ b. $x = 78$ c. $x = 1$ d. $x = 5$ e. $x = 4.65$

8. a. $x = 10$ b. $x = 3$ c. $x = 50$ d. $x = 6$ e. $x = 7$

9. a. $x = 11.13$ b. $x = 25$ c. $x = 6$ d. $x = 8$ e. $x = 2$

10. a. $x = 3$ b. $x = 3$ c. $x = 4$ d. $x = 7$ e. $x = 1$

11. A

12. First subtract 2 from both sides, then divide both sides by 4; $x = 2$.

13. Inverse operations must be performed in the reverse order. The forward order is: First subtract 6, then multiply by 3 and finally add 5. The reverse order is: Subtract 5, divide by 3, then add 6.

14. Yes. Dividing both sides by 4 gives $x - 5 = 2$. By then adding 5 to both sides, the solution $x = 7$ is obtained.

15. Gareth is 13 years old.

16. 16, 17 and 18

17. Callie scored 96 pins.

18. 7, 9 and 11

19. 11, 13 and 15

20. Chris jumped 156 centimetres.

21. a. 12
b. $\frac{x + 28}{3} = 12$
c. $x = 8$
d. Karina scored 8 points in the third game.

22. 25, 30 and 35

23. David is 14 years old.

24.

| 8 | 1 | 6 |
|---|---|---|
| 3 | 5 | 7 |
| 4 | 9 | 2 |

## 8.5 Solving equations using inverse operations

1. a. $3x + 9 = 18$ b. $3x - 2 = 13$ c. $x + 3 = 6$

2. a. i. $4r = 12$ ii. $r = 3$
b. i. $2n + 3 = 7$ ii. $n = 2$
c. i. $3t + 5 = 8$ ii. $t = 1$

3. a. $g = 1$ b. $m = 3$ c. $q = 9$

4. a. $b = 2$ b. $z = 12$ c. $v = 112$

5. a. $n = 9$ b. $g = 14$ c. $y = 3$

6. a. $h = 2\frac{1}{3}$ b. $k = 2\frac{1}{2}$ c. $w = 3\frac{1}{2}$

7. a. $t = 4\frac{1}{3}$ b. $h = 2\frac{1}{6}$ c. $l = 1\frac{1}{3}$

8. a. $g = \frac{8}{5}$ b. $h = \frac{11}{3}$ c. $n = \frac{1}{2}$

9. a. $m = 4$ b. $d = 3$ c. $v = 2$

10. a. $s = 3$ b. $g = 5$ c. $k = 2$

11. a. $x = 2\frac{1}{2}$ b. $p = 3\frac{1}{4}$ c. $x = 2$

12. a. $y = 1$ b. $h = 2\frac{1}{2}$ c. $x = \frac{1}{2}$

13. a. $7\frac{2}{9}$ °C b. 113 °F

14. a. $13.67 b. 49 km

15. 11 songs

16. The removal of the three weights from the right-hand side needs to be done before the weights on each side are halved. If the division is performed first, the odd numbers of weights on each side of the scales would have to be halved, which will not result in whole numbers.

17. 16 boys

18. 3

19. 4 cm

20. 3, 8, 13

21. $1\frac{1}{3}$ kilograms

## 8.6 Checking solutions

1. a. No b. Yes c. No

2. a. Yes b. No c. No

3. a. No b. Yes c. No

4. a. No b. Yes c. Yes

5. a. No b. Yes c. No

6.

| $x$ | 0 | 1 | 2 | 3 | 4 |
|---|---|---|---|---|---|
| $2x + 3$ | 3 | 5 | 7 | 9 | 11 |

a. $x = 4$ b. $x = 1$

7.

| $x$ | 2 | 3 | 4 | 5 | 6 |
|---|---|---|---|---|---|
| $5(x - 2)$ | 0 | 5 | 10 | 15 | 20 |

a. $x = 4$ b. $x = 6$ c. $x = 8$

8. a.

| $x$ | $2x + 1$ | $3x - 5$ |
|---|---|---|
| 3 | 7 | 4 |
| 4 | 9 | 7 |
| 5 | 11 | 10 |
| 6 | 13 | 13 |
| 7 | 15 | 16 |

b. $x = 6$

9. a.

| $x$ | $\frac{x+3}{2}$ | $2x - 6$ |
|---|---|---|
| 3 | 3 | 0 |
| 5 | 4 | 4 |
| 7 | 5 | 8 |
| 9 | 6 | 12 |
| 11 | 7 | 16 |

b. $x = 5$

10. You could substitute all of the options into the equation and see which one makes that equation true.

11. a. No. Length is enough.
b. 34.8 m
c. Adding 3.9 and the value of y. The answer should be 38.7 m.

12. a. $\frac{1}{7}x = 0.5$ b. $x = 3.5$

13. James $53 and Alison $72

14. a. 9
b. $6
c. Aiko $8 , Miyu $14

15. a. $133 + (133 + 1) + (133 + 2) = 402 \neq 393$
b. $n = 130$, $130 + (130 + 1) + (130 + 2) = 393$
c. Sample response: The sum of three consecutive numbers is 393. What is the smallest of these numbers?

## Project

1. Men's: 1928, 10.607 s; 1968, 10.231 s; 1988, 10.043 s; 2016, 9.780 s
Women's: 1928, 11.956 s; 1968, 11.264 s; 1988, 10.918 s; 2016, 10.433 s

2. All calculated times are well within 0.4 seconds of the actual running times.

3. Using the formula: Men's time = 9.7044 s; Women's time = 10.2948 s.

4. Women's times are coming down faster than men's. It is possible only if these patterns continue.

5. Different training programs, fitter athletes, better shoes, etc. There must be some levelling out, as the times cannot keep coming down forever – it will be impossible to run 100 metres in 0 seconds.

## 8.7 Review questions

1. a. $m = 5$
b. $h = 6$
c. $s = 19$
d. $d = 20$

2. a. 32, 51 b. 27, 59

3. a. 11
b. 2
c. 16
d. 35

4. a. 16 b. 42 c. 11 d. 8

5. a. $7x + 8$
b. $\frac{x}{3} - 5$
c. $6(x + 2)$
d. $\frac{x - 7}{5}$

6. a. $5(x + 3) - 9$ b. $\frac{4x + 11}{8}$

7. a.

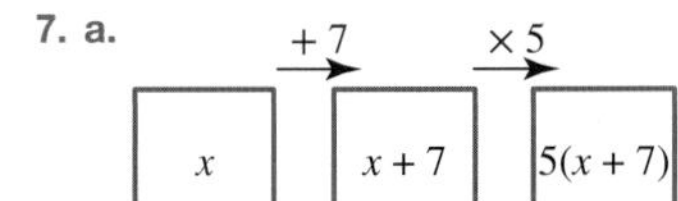


b.

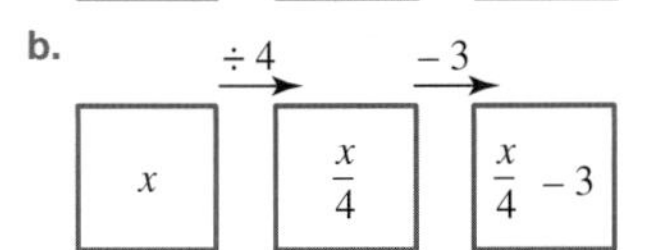


c.

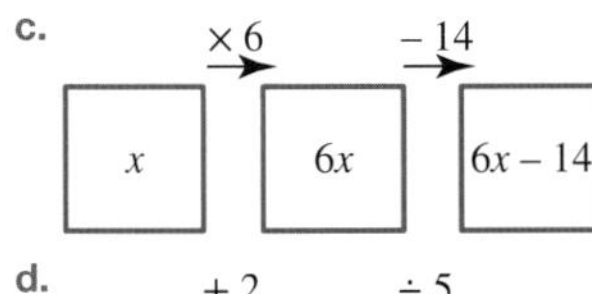


d.

+ 2 → ÷ 5 →

$x$ → $x + 2$ → $\frac{x + 2}{5}$

8. a. $x = 2$ b. $y = 10$ c. $m = 60$ d. $s = 18$

9. a. $d = 4$ b. $t = 100$ c. $d = 3$ d. $a = 6$

10. a. $v = 3$ b. $c = 12$

11. a. Yes b. No

12. a. i. $4x + 1 = 5$
ii. $x = 1$
b. i. $3p + 5 = 11$
ii. $p = 2$
c. i. $2q + 3 = 13$
ii. $q = 5$

13. a. $v = 3$ b. $x = 2$ c. $w = 3$ d. $g = 4$

14. a. $k = 1\frac{3}{4}$ b. $p = 5\frac{1}{2}$ c. $q = \frac{1}{3}$ d. $t = \frac{2}{5}$

15. a. $x = 2$ b. $p = 2$ c. $r = 2$ d. $z = 2$

16. Sophie owns 31 cards.

17. 7 pieces of homework

18. 12 years old

19. $2

20. a. i. $40 ii. $50 iii. $60
b. Cost = $30 + 10h$
c. $110
d. $85 = 30 + 10h$; $h = 5.5$ hours
e. $5
f. Your deposit is returned ($30), which means you could buy a hamburger on the way home.

21. a. $105
b. $105
c. 6 children
d. 10 adults; 9 adults and 2 children; 8 adults and 4 children; 7 adults and 6 children; 6 adults and 8 children; 5 adults and 10 children; 4 adults and 12 children; 3 adults and 14 children; 2 adults and 16 children; 1 adult and 18 children; 20 children

22. 14 bicycles and 5 tricycles

23. 45°, 45°, 90°. This is a right-angled isosceles triangle.

24. Length 12 m, width 3 m

25. 35°

# 9 Geometry

## LESSON SEQUENCE


**9.1** Overview ....442
**9.2** Measuring angles ....445
**9.3** Constructing angles with a protractor ....453
**9.4** Types of angles and naming angles ....459
**9.5** Triangles ....465
**9.6** Quadrilaterals and their properties ....477
**9.7** Parallel and perpendicular lines ....487
**9.8** Review ....498

# LESSON
# 9.1 Overview

## Why learn this?

The word geometry comes from the Greek terms *geo-* (meaning *earth*) and *metron* (meaning *measurement*). Geometry is one of the oldest areas of mathematics, and it allows us to explore our world in a very precise way. Have you ever wondered how early sailors such as Captain Cook navigated around the world? They didn't have computers to guide them. They navigated using the stars and their knowledge of geometry. The famous Greek philosopher Thales of Miletus, who lived around 600 BC, used geometry to calculate the height of the pyramids in Egypt.

The geometry we study today can be traced back to around the year 300 BC and the work of another Greek mathematician, Euclid. Builders, architects, surveyors and engineers all use their knowledge of geometry to ensure buildings are stable and visually pleasing. Navigation, both in the air and at sea, is based on geometric principles. Travel in outer space involves the use of complex geometry. Geometry is also involved in sport. You need to carefully estimate the angle required when you are shooting for a goal in hockey or football. If you have played or watched someone playing pool, you will know that understanding angles is vital to play well. Many professions, including landscape gardening and interior design, use geometry every day.

## Exercise 9.1 Pre-test

learnon

1. State the type of angle that describes an angle of 125°.

2. If an analogue clock is showing 1:30 pm, determine the smaller angle between the big and small hands.

3. State how many degrees are in a right angle.

4. MC A reflex angle is between:

   A. 45° and 180° B. 0° and 90° C. 90° and 120°
   D. 180° and 360° E. 0° and 45°

5. MC Two acute angles that sum to be an acute angle are:

   A. 180° and 360° B. 90° and 180° C. 30° and 70°
   D. 50° and 30° E. 20° and 70°

6. MC An analogue clock is showing 4 am. The obtuse and reflex angles that the big and small hands make respectively are:

   A. 90° and 180° B. 240° and 120° C. 120° and 240°
   D. 20° and 40° E. 100° and 200°

7. Name the acute angle in the following diagram.

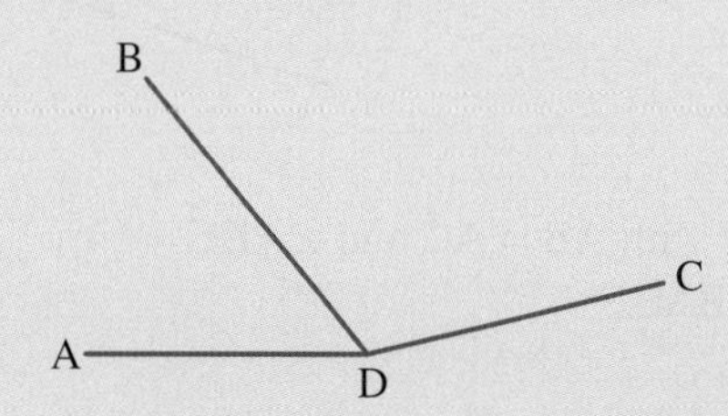


8. MC Identify the values of $p$ and $q$ in the following diagram.

   A. $p = 110°, q = 70°$
   B. $p = 10°, q = 170°$
   C. $p = 100°, q = 80°$
   D. $p = 110°, q = 80°$
   E. $q = 110°, p = 70°$

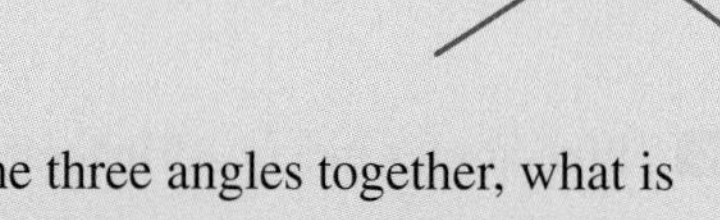


9. MC If you measure the three angles of a triangle and then add the three angles together, what is the result?

   A. 90° B. 120° C. Between 90° and 180°
   D. Between 120° and 240° E. 180°

10. Classify each of the following triangles, based on side length and angle type.

a. 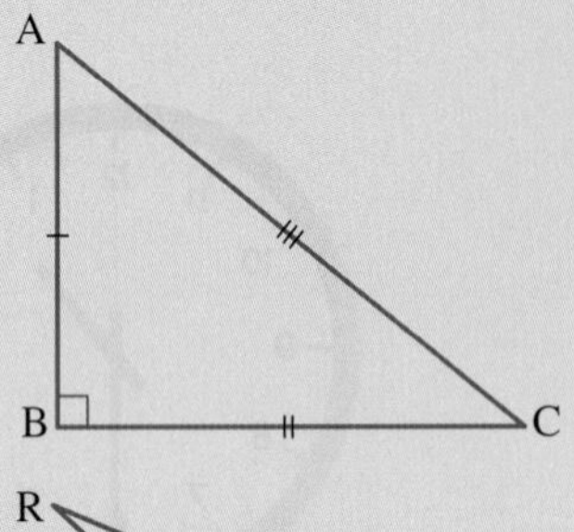


b. 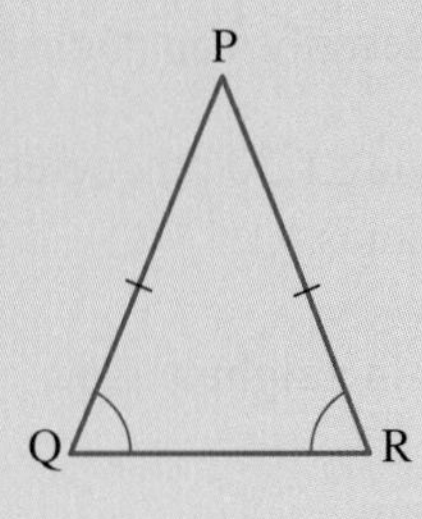


c. 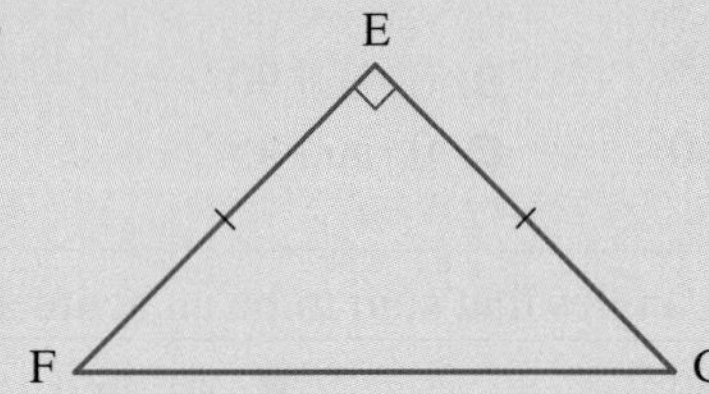


d. 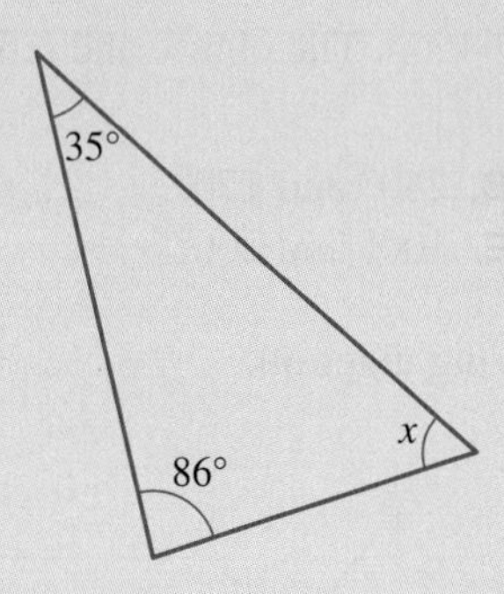


11. Calculate the value of $x$ in the following diagram.

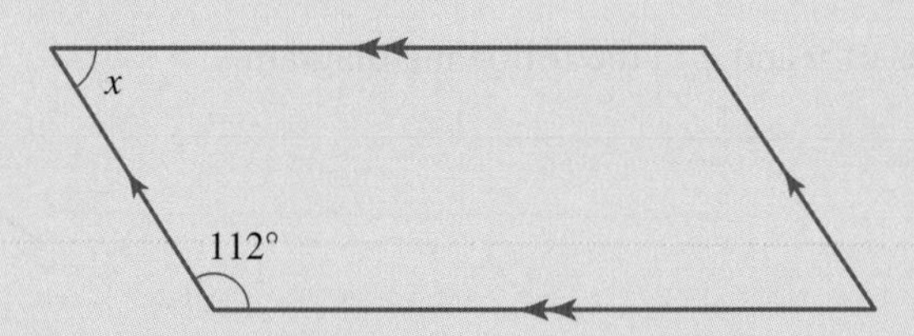


12. $\Delta ABC$ is an isosceles triangle with $AB = AC$ and $\angle ABC = 64°$.
Calculate the magnitude of $\angle BAC$.

13. Calculate the value of $x$ in the following diagram.



14. **MC** Three angles in a quadrilateral are 135°, 72° and 38°. Identify the fourth angle.

**A.** 135° **B.** 38° **C.** 72° **D.** 115° **E.** 105°

15. A driveway inclined at an angle of 18° from the horizontal has vertical posts along the driveway.
Determine the acute angle between the posts and the driveway.

# LESSON
## 9.2 Measuring angles

### LEARNING INTENTION

At the end of this lesson you should be able to:
- understand what an angle is
- estimate angles
- measure angles using a protractor.

### 9.2.1 Understanding angles

eles-4089

- An angle is a measurement of space between two lines that meet at a point.
- The point where the two lines meet is called a **vertex** and the lines are the arms of the angle.
- For *angle*, we use the symbol ∠.
- Angles are measured in degrees. The symbol for degrees is °.

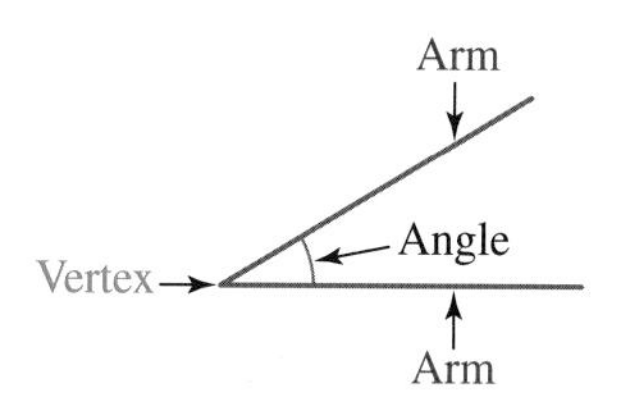


#### Estimating angles

- There are four important angles: 90°, 180°, 270° and 360°.

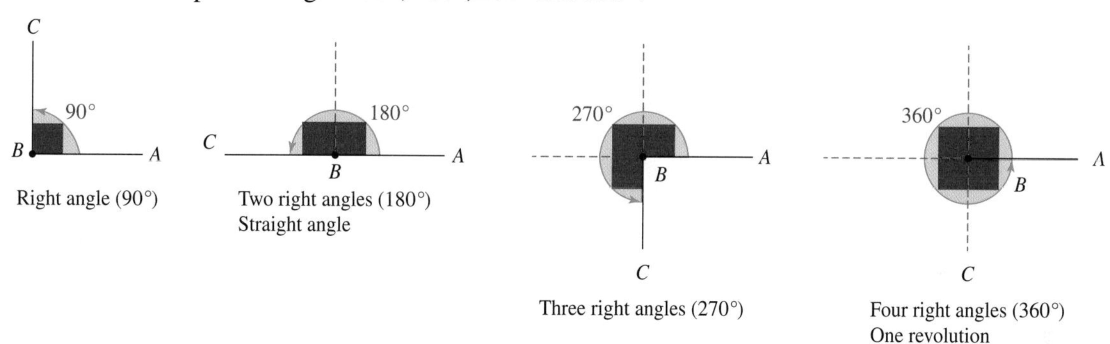


Right angle (90°)

Two right angles (180°)
Straight angle

Three right angles (270°)

Four right angles (360°)
One revolution

**Estimating the size of an angle**

- **First: determine which important angle it would measure closest to**
- **Second: decide whether it is greater or less than that angle**
- **Third: make an estimate.**

### WORKED EXAMPLE 1 Estimating the angle

**Estimate the size of the following angles.**

a.

b.

c.

**THINK**

a. 1. The angle is closest to 90°.
2. The angle is less than 90°.
3. Make an estimate.

**WRITE**

a. The estimate is 60°.

b. 1. The angle is closest to 90°.
2. The angle is more than 90°.
3. Make an estimate.

b. The estimate is 120°.

c. 1. The angle is roughly halfway between 270° and 360°.
2. Halfway between 270° and 360° is $270° + 45° = 315°$
*Note:* You add 45° because 45° is half of 90°.
3. Write your estimate.

c. The estimate is 315°.

## 9.2.2 Measuring angles

eles-4090

- The tool for measuring angles is called a **protractor**.
- There are two types of protractors: circular and semicircular.

**Circular protractor**

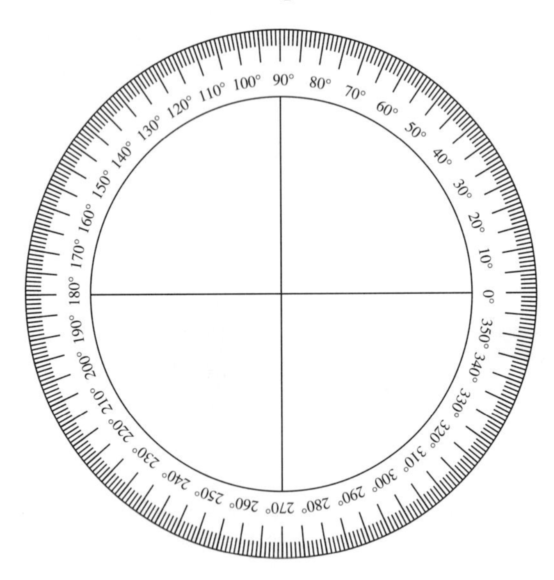

**Semicircular protractor**

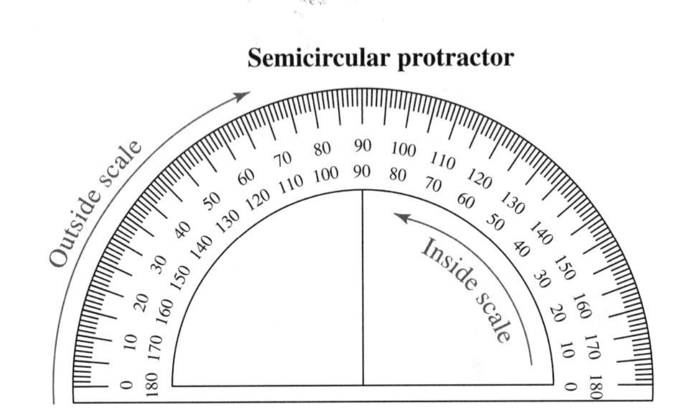


### Using a protractor

**To use a protractor, follow these steps.**

**1. Place the protractor on the angle so that:**
- **the zero line is along one line of the angle**
- **the centre of the protractor lines up with where the two lines meet (the vertex).**

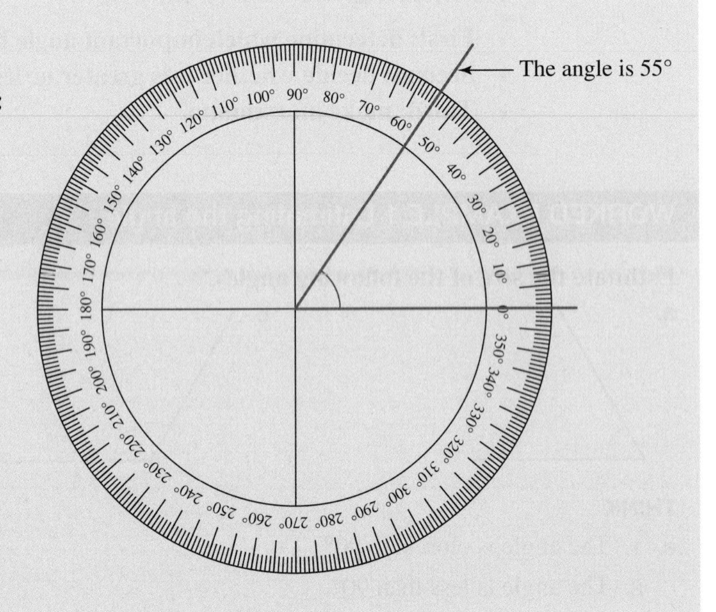

2. **Read the angle by looking at where the other line crosses the protractor numbers.**

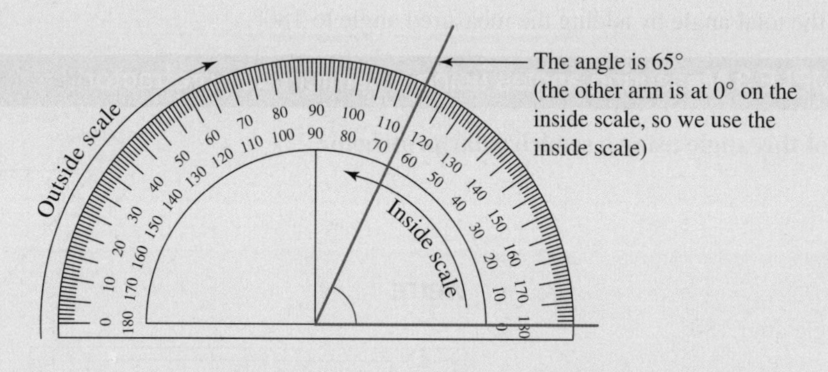


## WORKED EXAMPLE 2 Measuring angles

**Measure the size of each of the following angles.**

a.

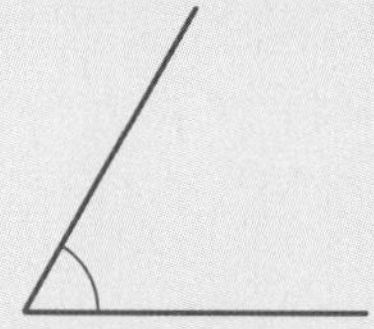

b.

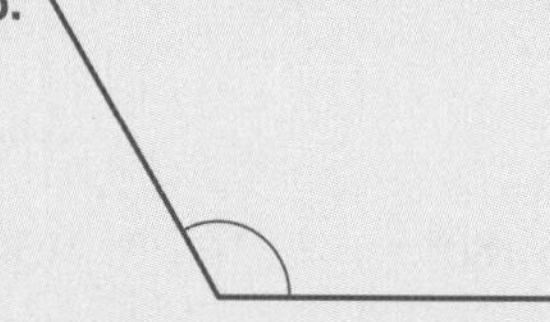

| THINK | WRITE |
|---|---|
| a. 1. Place the centre of the protractor on the vertex of the angle. | a. 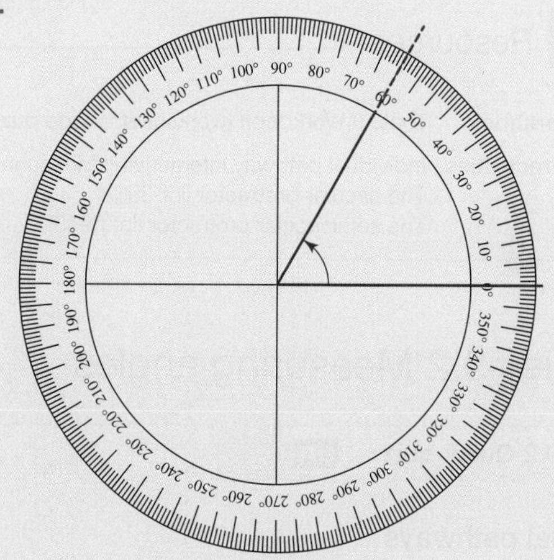 |
| 2. Match the line that passes through the centre of the protractor and points to 0° with one of the arms of the angle. | |
| 3. Read the size of the angle indicated by the other arm. (You may need to extend the line so that you can read it on the protractor.) | The size of this angle is 60°. |
| b. 1. Place the centre of the protractor on the vertex of the angle. | b. 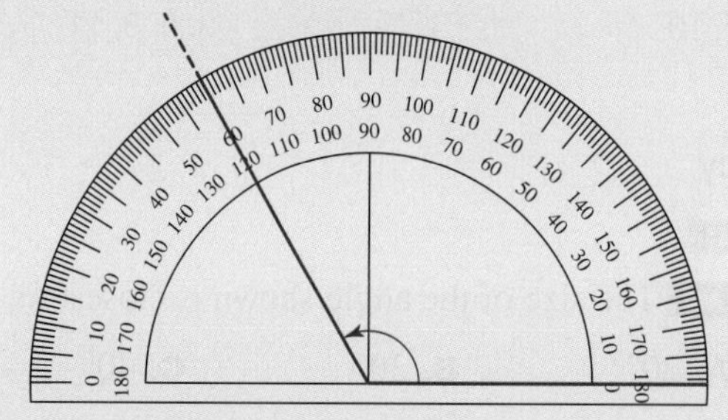 |
| 2. Match the line that passes through the centre of the protractor and points to 0° with one of the arms of the angle. | |
| 3. Read the size of the angle indicated by the other arm. (You may need to extend the line so that you can read it on the protractor.) | The size of this angle is 120°. |

- A semicircular protractor has angles only up to 180°. To measure angles larger than 180°:
  1. measure the part of the angle that is bigger than 180°
  2. determine the total angle by adding the measured angle to 180°.

## WORKED EXAMPLE 3 Measuring angles using a semicircular protractor

**Measure the size of this angle using a semicircular protractor.**

| THINK | WRITE |
|---|---|
| 1. Measure the angle after 180°. | 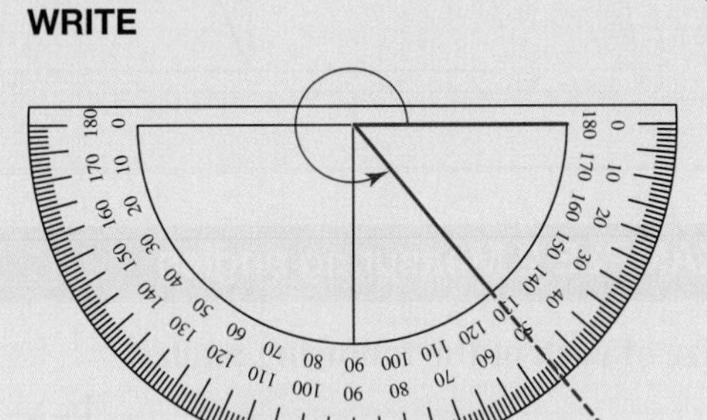 The angle after 180° is 130°. |
| 2. Add the measured angle to 180°. | $130° + 180° = 310°$ |
| 3. Write the answer. | The size of the angle is 310°. |

 Resources

**eWorkbook** Topic 9 Workbook (worksheets, code puzzle and project) (ewbk-1910)

**Interactivities** Individual pathway interactivity: Measuring angles (int-4331)
The circular protractor (int-3952)
The semicircular protractor (int-3953)

# Exercise 9.2 Measuring angles

learnon

**9.2 Quick quiz** 

**9.2 Exercise**

### Individual pathways

| ■ PRACTISE | ■ CONSOLIDATE | ■ MASTER |
|---|---|---|
| 1, 2, 7, 10 | 3, 5, 8, 11 | 4, 6, 9, 12 |

### Fluency

1. **WE1**

   a. **MC** The size of the angle shown is closest to:

   **A.** 20° **B.** 30° **C.** 40° **D.** 150° **E.** 140°

b. The size of the angle shown is closest to:

A. 42° B. 138° C. 142° D. 145° E. 38°

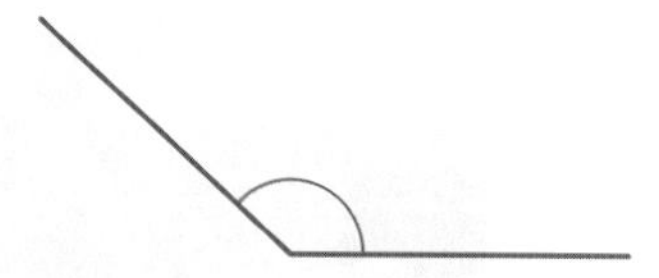

2. WE2,3 Measure the size of each of the following angles.

a. 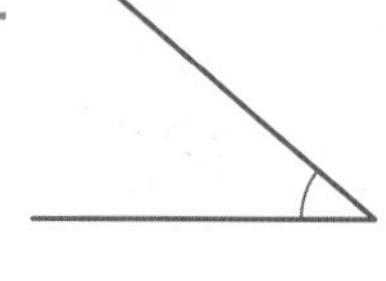

b. 

c. 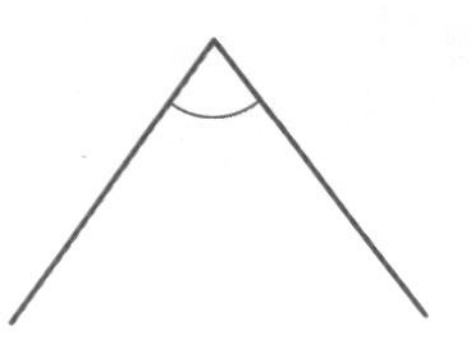

d. 

3. Measure the size of each of the following angles.

a. 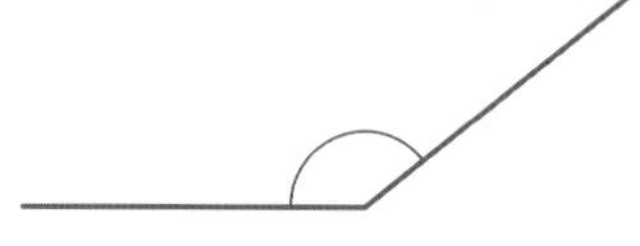

b. 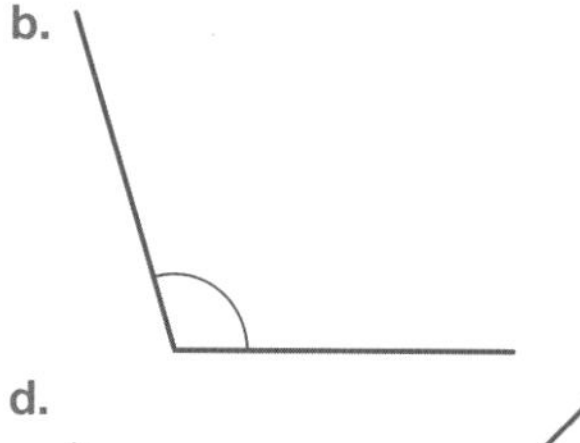

c. 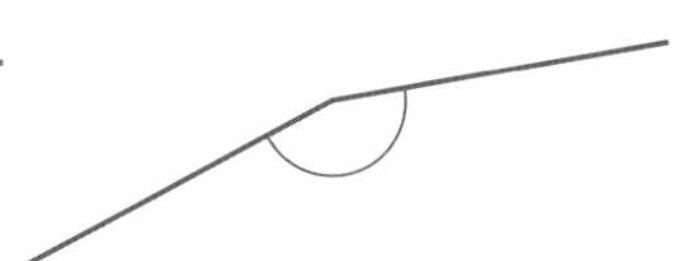

d. 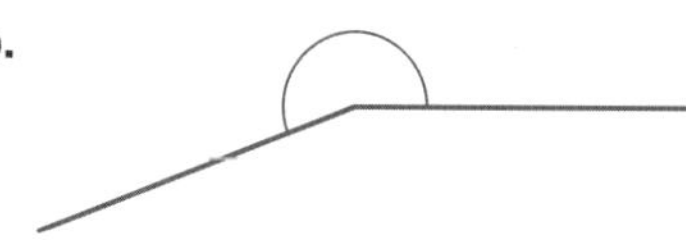

4. Measure the size of each of the following angles.

a. 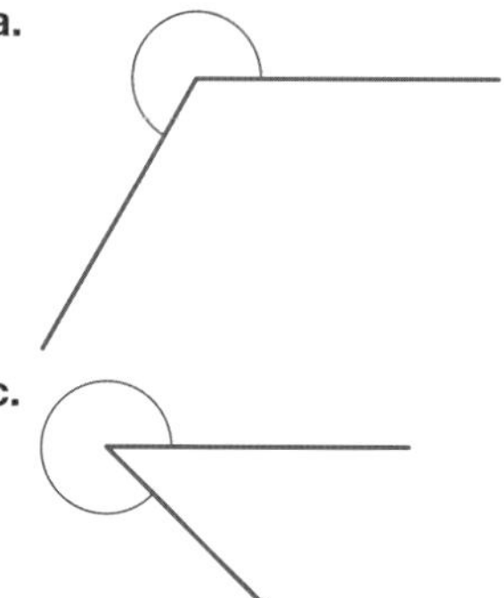

b. 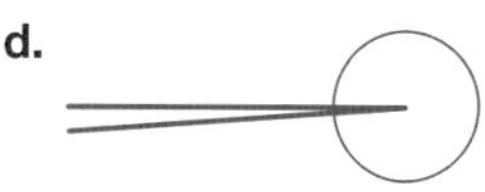

c.

d.

e. Comment on the accuracy of your measurement compared with the answer in the text.

## Understanding

5. In each of the photographs, measure the indicated angle.

a. 

b. 

c. 

d. 

6. The recommended slope for wheelchair access is 4° from the horizontal. An angle greater than this makes it more difficult for the person in the wheelchair to travel up the ramp.
   a. Would you rate the slope of the bottom section of this ramp as 'difficult', 'as recommended' or 'easy'? Justify your rating.
   b. Describe how you would rate the second section of the ramp.

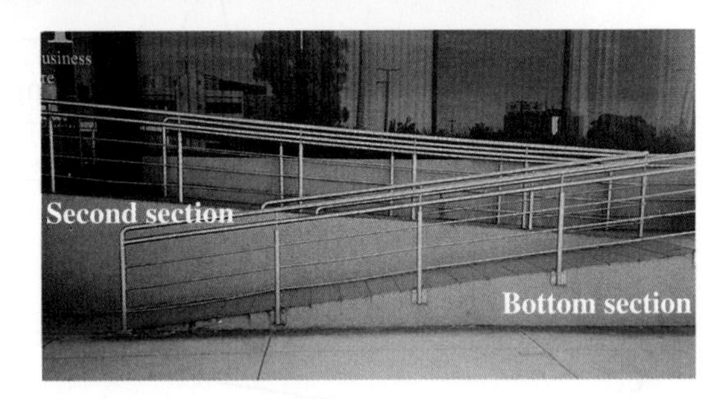


### Reasoning

7. Consider the four different angles shown.
   a. Carefully measure each angle and write the measurements.
   b. Identify which two of these angles combined add to 90°.
   c. Identify which two of these angles combined add to 180°.
   d. Suggest one way of combining these angles to add to 360°. You can repeat angles.

i. 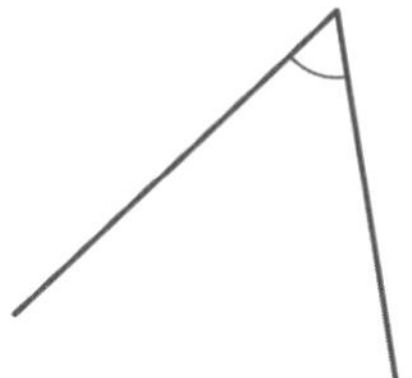

ii.

iii. 

iv. 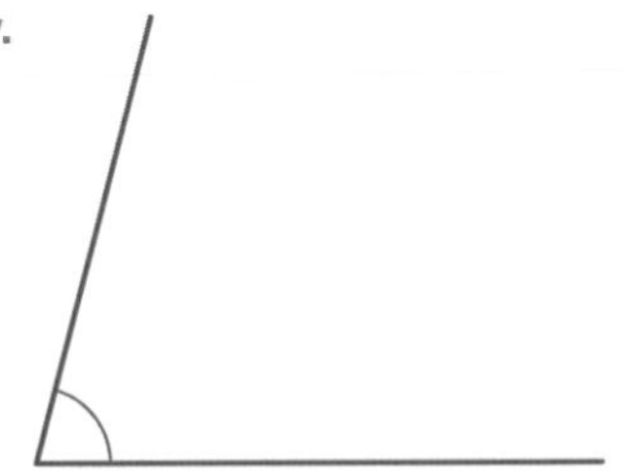

8. Sal is designing a new slide for her water park. She wants the slide to be both fun and safe. Below are side views of four possible designs:

Design 1

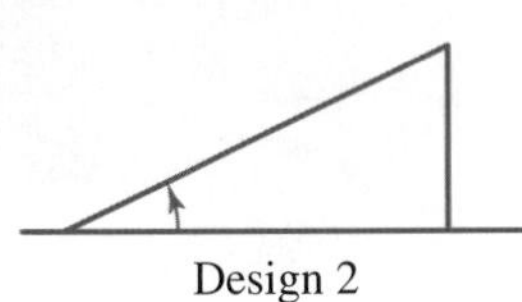
Design 2

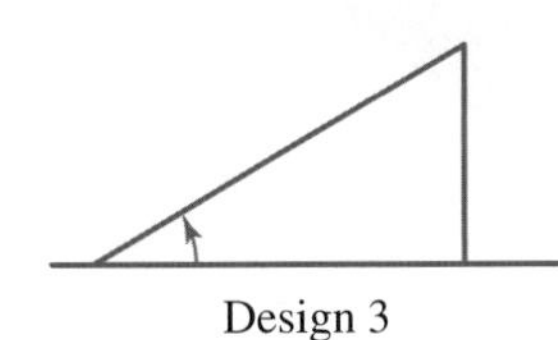
Design 3

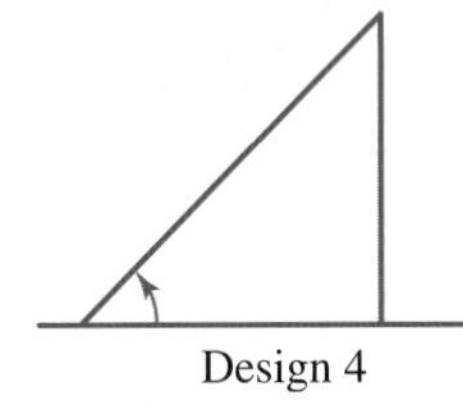
Design 4

a. State which design will give riders the fastest ride. What is the angle of that slide?
b. State which design will give riders the slowest ride. What is the angle of that slide?
c. The safety requirements are that slides for people aged 12 years and older must have an angle of no more than 35°. Identify which slides fit the safety requirements for Sal's park.
d. The safety requirements are that slides for people aged less than 12 years must have an angle of no more than 20°. Identify which slides fit the safety requirements for Sal's park.

Sal needs to choose the designs that give the fastest ride but also fit the safety requirements.

e. Identify which design Sal should use for the slide for people aged 12 and older.
f. Identify which design Sal should use for the slide for people aged less than 12 years.

9. Study the cartoon of the golfer to help you answer the following questions.

Copy the following table to record your answers.

| Diagram | Estimate of angle | Measured angle | Difference |
|---|---|---|---|
| 1 | | | |
| 2 | | | |
| 3 | | | |
| 4 | | | |

a. Without using a protractor, estimate the size of each angle in the first four diagrams of the golf sequence above and complete the second column of the table.
b. Explain how you obtained an estimate. Discuss the steps you followed.
c. Complete the third column by measuring each angle with a protractor.
d. Calculate the difference between your estimate and the actual value of each angle.
e. Describe how you could improve your skills in estimating the size of an angle.
f. Try these improvements for estimating the size of the angles in the remaining four diagrams. Copy and complete the following table.

| Diagram | Estimate of angle | Measured angle | Difference |
|---|---|---|---|
| 5 | | | |
| 6 | | | |
| 7 | | | |
| 8 | | | |

g. Has the difference between the estimated and actual value for each angle become smaller? Comment on whether your estimating skills have improved.

## Problem solving

10. Sean is flying a kite. Depending on the position of the kite, the angle between the string and the horizontal line changes.
Complete the following statements:
    a. As the kite moves from point A to point B, the angle increases from __________ to ____________.
    b. As the kite moves from point B to point C, the angle decreases from __________ to ____________.

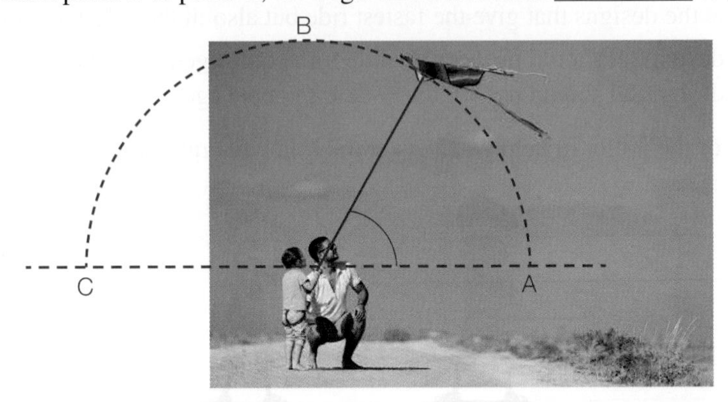


11. a. Using a protractor, measure the angles between the two clock hands shown in the diagrams.

i. 

ii. 

iii. 

iv. 

v. 

vi. 

b. Calculate the angle between the two arms without using a protractor.

12. Abby and Brinda are having an argument. Abby says there isn't a pattern when you measure the angles inside triangles and add them together. Brinda says there is a pattern.
Three triangles are shown.

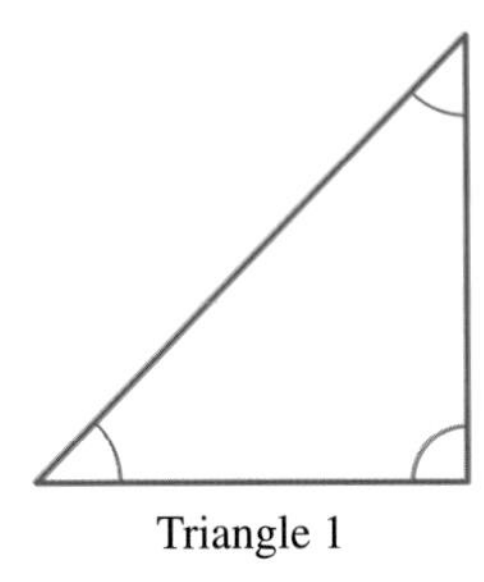

Triangle 1

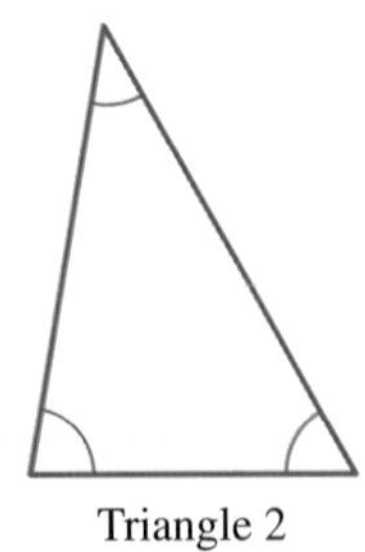

Triangle 2

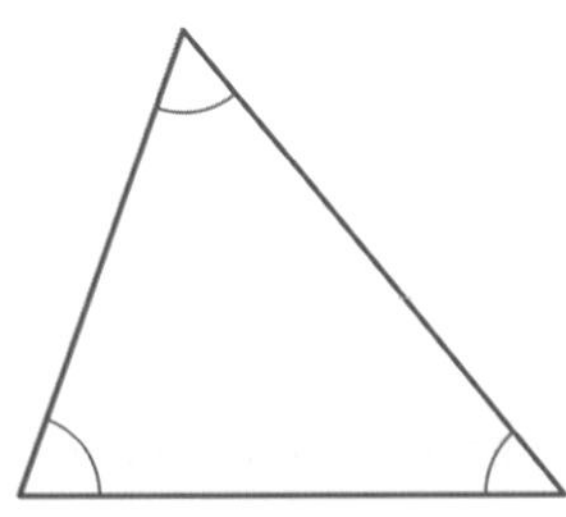

Triangle 3

a. Measure the angles in the corners of each triangle.
b. Add the angles of each triangle.
c. Explain who is right, Abby or Brinda.

# LESSON
# 9.3 Constructing angles with a protractor

## LEARNING INTENTION

At the end of this lesson you should be able to:
- draw an angle using a protractor.

## 9.3.1 Constructing angles with a protractor

eles-4092

- Protractors are used to draw angles.
- To draw an angle, start by drawing a straight line (arm) and putting a dot (vertex) at one end.

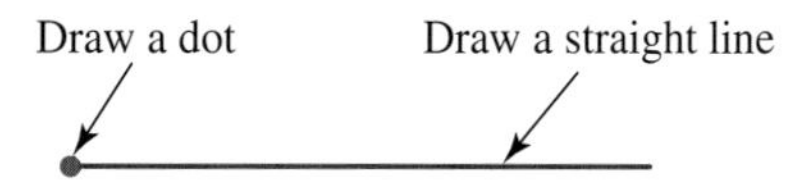


- Next, line up the protractor so that the dot (vertex) is in the centre and the line points to 0°.
- Place a dot for the angle you want to measure (for example 60°).

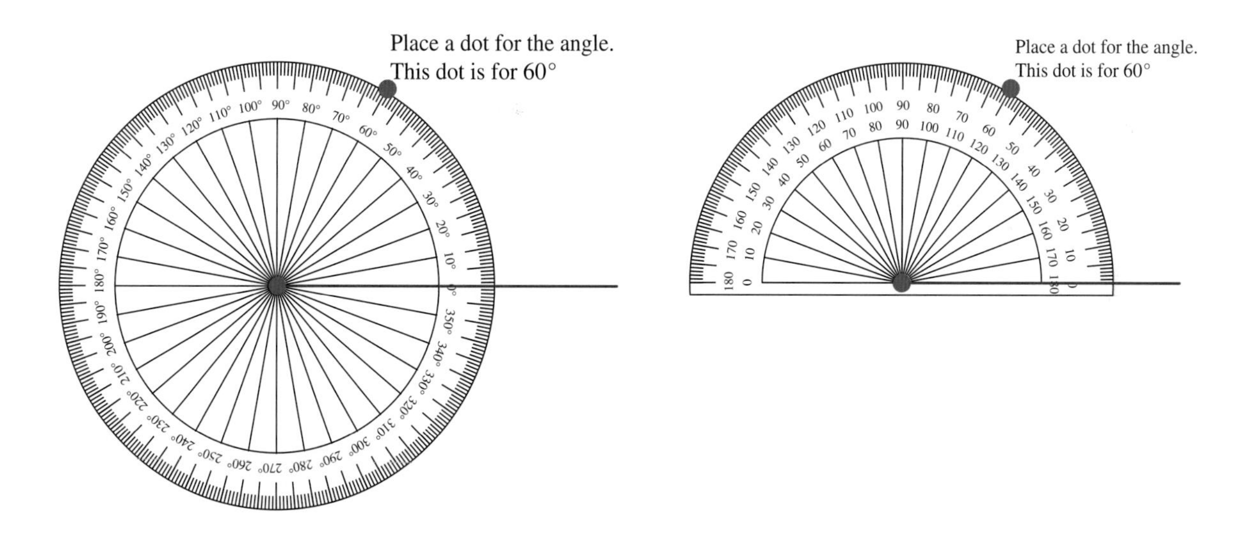


- Lastly, join the dots, and label the angle by drawing an arc between the two lines.

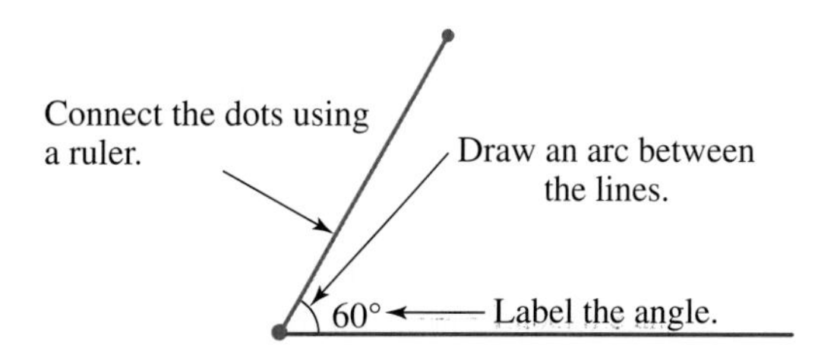

## WORKED EXAMPLE 4 Constructing angles using a protractor

**Construct each of the following angles.**

**a. 50°** **b. 152.5°**

**THINK** **WRITE**

**a. 1.** Draw a horizontal line. Put a dot at one end. This line is the arm and the dot is the vertex.

**a.**

**2.** Place the protractor so that its centre is on the vertex and 0° is on the line.

**3.** Find 50° on the scale and mark a dot at the edge of the protractor.

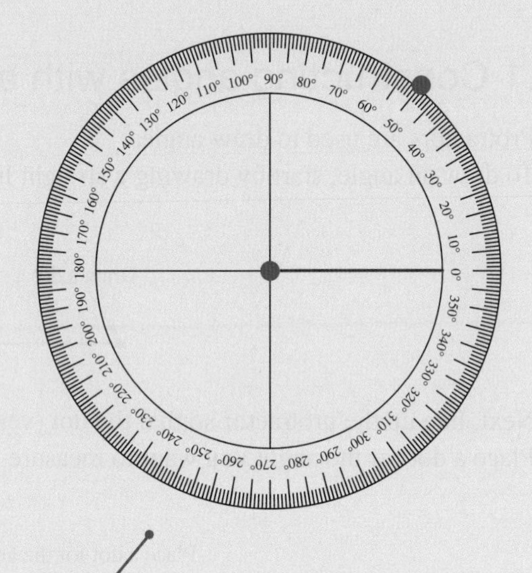


**4.** Join the dot and the vertex with a straight line using a ruler.

**5.** Draw the arc and label the angle.

50°

**b. 1.** Repeat steps **1** and **2** from part **a**.

**2.** Find 152.5° on the scale and mark a dot at the edge of the protractor. Remember to start at 0° and use the inside scale.

**b.**

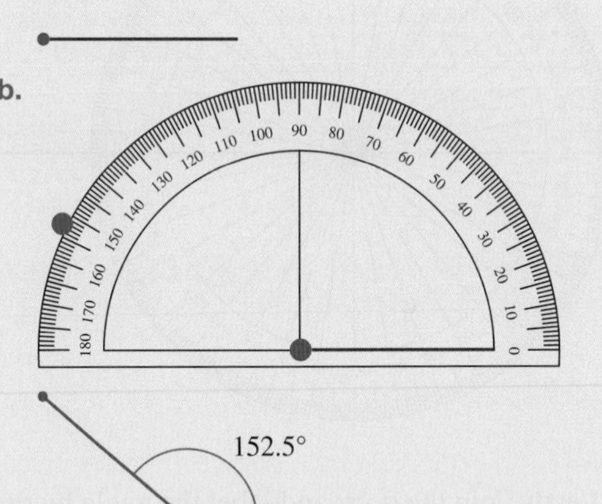


**3.** Join the dots with a straight line using a ruler.

**4.** Draw the arc and label the angle.

152.5°

## 9.3.2 Constructing angles greater than 180° using a semicircular protractor

eles-4093

- To draw an angle greater than 180°, recall that 180° is a straight line.
- Subtract 180° from the given angle to determine how much larger than 180° the angle is.
  For example, for an angle of 220°: 220° − 180° = 40°
  220° is 40° more than 180°.

- To draw the angle, draw the arm (line) and the vertex (dot) as described in section 9.3.1.
- Use the semicircular protractor upside down and start counting at 0° (inner scale) or 180° (outer scale) to measure the given angle (40° in the above example).

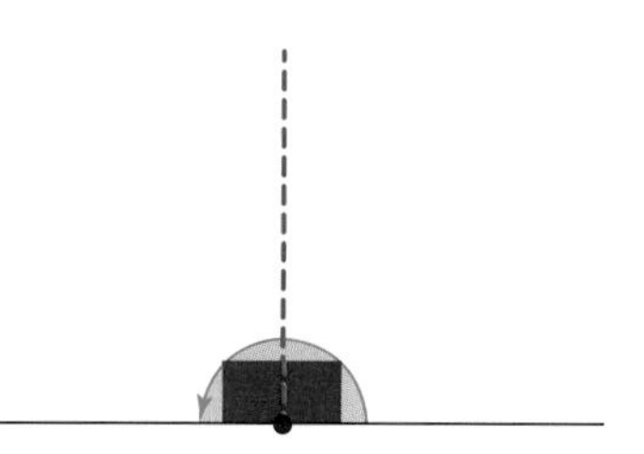

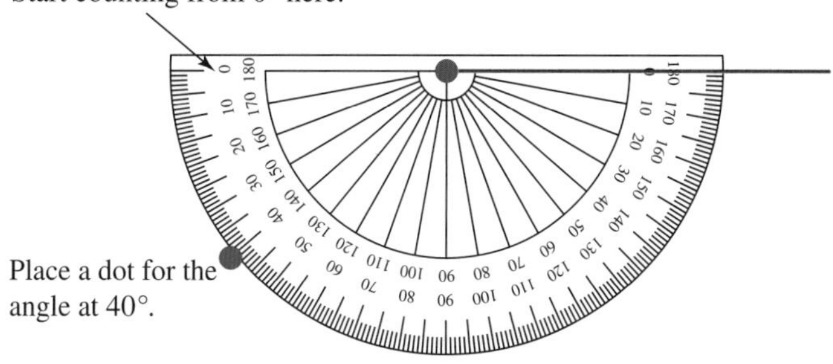


- Lastly, join the dots and label the outer angle by drawing the arc.

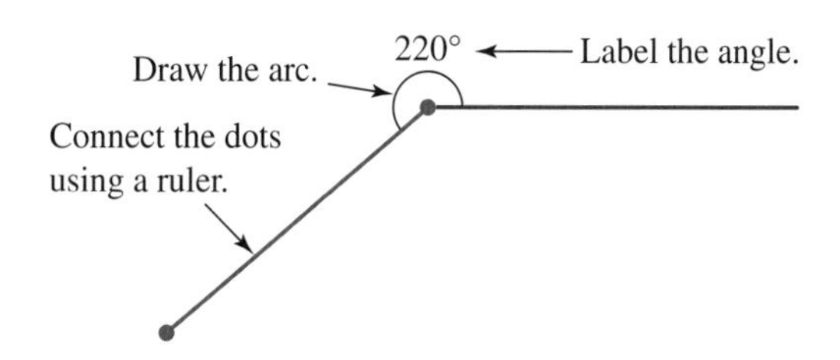


## WORKED EXAMPLE 5 Constructing angles using a semicircular protractor

**Construct an angle of 210° using a semicircular protractor.**

| THINK | WRITE |
|---|---|
| 1. Subtract 180° from 210°. | $210° - 180° = 30°$ |
| 2. Draw a horizontal line. Put a dot at one end. The line is the arm and the dot is the vertex. | |
| 3. Place the protractor upside down on the line so that its centre is on the dot. | |
| 4. Start at 0° on the left-hand side. Mark the position of 30° with a dot. | |
| 5. Remove the protractor and join the dots using a ruler. | |
| 6. Draw the arc and label the outer angle as 210°. | 210° |

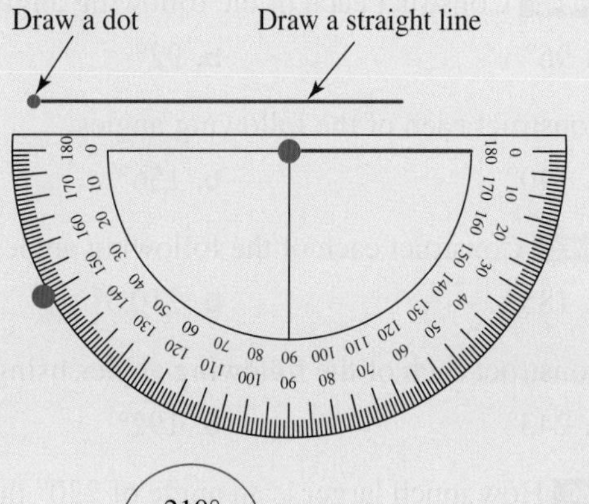

- *Note:* It is possible to draw an angle greater than 180° with the protractor upright. This diagram shows the angle of 210° (the same as in Worked example 5) constructed by placing the protractor in the upright position.

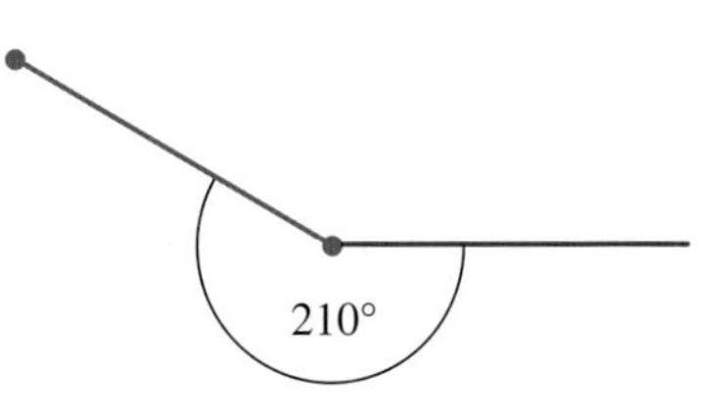

**on Resources**

| | |
|---|---|
| **eWorkbook** | Topic 9 Workbook (worksheets, code puzzle and project) (ewbk-1910) |
| **Interactivities** | Individual pathway interactivity: Constructing angles with a protractor (int-4332)<br>Construction of angles using a circular protractor (int-3954)<br>Construction of angles using a semicircular protractor (int-3955) |

## Exercise 9.3 Constructing angles with a protractor

learnon

**9.3 Quick quiz** on | **9.3 Exercise**

### Individual pathways

| ■ PRACTISE | ■ CONSOLIDATE | ■ MASTER |
|---|---|---|
| 1, 4, 7, 10, 13, 16 | 2, 5, 8, 11, 14, 17 | 3, 6, 9, 12, 15, 18 |

### Fluency

1. **WE4a** Construct each of the following angles.
   a. 15° b. 9° c. 53° d. 75°
2. Construct each of the following angles.
   a. 45° b. 40° c. 88° d. 76°
3. **WE4b** Construct each of the following angles.
   a. 96° b. 92° c. 165° d. 143°
4. Construct each of the following angles.
   a. 140° b. 156° c. 127° d. 149°
5. **WE5** Construct each of the following angles using a semicircular protractor.
   a. 185° b. 210.5° c. 235° d. 260°
6. Construct each of the following angles using a semicircular protractor.
   a. 243° b. 192° c. 249° d. 214°
7. **MC** How much larger is an angle of 220° than an angle of 180°?
   A. 212° B. 58° C. 32° D. 112° E. None of these
8. Construct each of the following angles.
   a. 295° b. 269° c. 307° d. 349°
9. Construct each of the following angles.
   a. 328° b. 300° c. 345° d. 358°

## Understanding

10. a. Draw the following angles.
    i. 45°  ii. 90°  iii. 180°  iv. 270°  v. 360°
  b. Use the angles you have drawn to help fill in the gaps.
    i. The angle _____ is half of the angle 90°.
    ii. The angle _____ is half of the angle 180°.
    iii. The angle _____ is half of the angle 360°.
    iv. The angle _____ is a third of the angle 270°.
    v. The angle _____ is an eighth of the angle 360°.

11. a. Construct a 20° angle.
  b. Draw a 45° angle on one side of this angle.
  c. On the second arm of the angle in part **b** draw a 108° angle.
  d. What is the remaining reflex angle? *Hint:* A reflex angle is more than 180° but less than 360°.
  e. What is the total sum of the angles?

12. Windmills use wind power to produce electricity, pump water and mill grain. The three types of machines shown in the diagram have blades constructed at equal angles. Using a protractor, measure the angles between the blades.

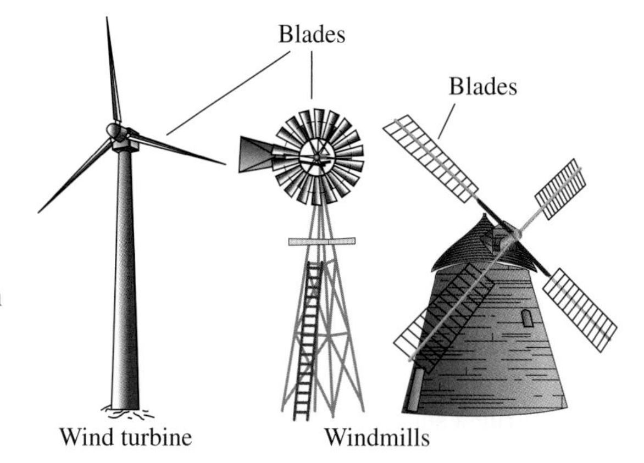


## Reasoning

13. Explain the method you would use to construct a reflex angle if you only had a 180° protractor.

14. It is possible to draw angles larger than 360°.
  a. Draw an angle of 400°.
  b. What smaller (0° to 90°) angle is the same as 400°?

15. We see the objects surrounding us because light reflects off the surface of the objects, as shown in the first diagram. The two angles shown are always equal.
  Imagine you are in a room lit by a light globe as shown in the second diagram.

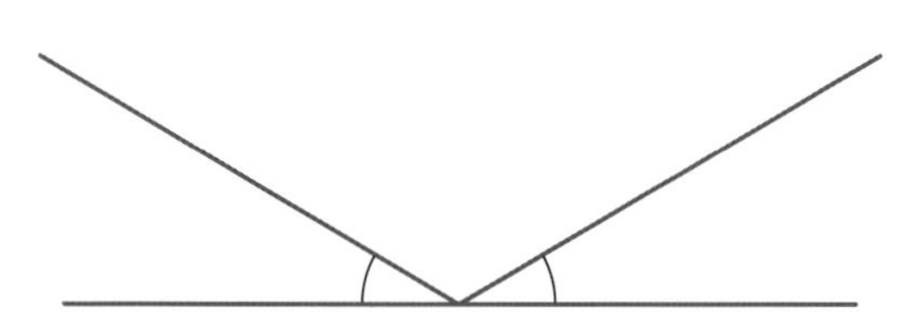

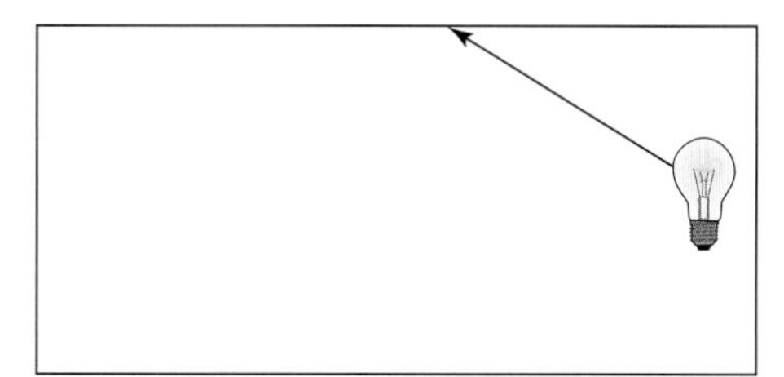

  a. Using a protractor and a ruler, draw the next four paths followed by the ray of light shown in the diagram.
  b. Draw another ray of light in a different direction.
  c. Use the same procedure as in part **a** to draw the first four paths of the line drawn in part **b**.

### Problem solving

16. a. Draw a horizontal line that is 3 cm long. Use that line as the base to draw a triangle with angles of 60° in each corner.
    b. Draw a horizontal line that is 6 cm long. Use that line as the base to draw a triangle with angles of 60° in each corner.
    c. Compare the two triangles and state the similarities and differences.
    d. If you drew a triangle with base length 10 m and two corner angles of 60° each, determine the angle of the third corner.

17. On an A4 sheet of paper, draw a 20 cm long straight horizontal line close to the bottom of the page. Label this line AB.

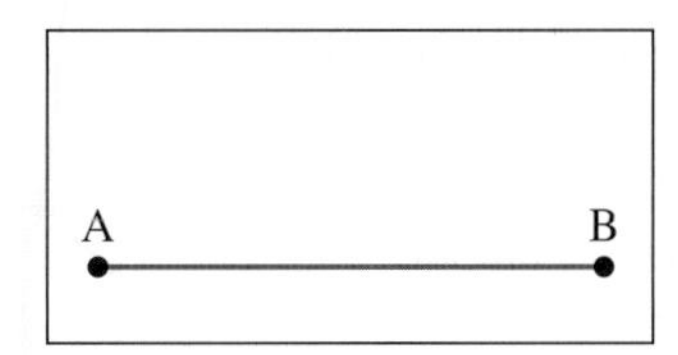


    a. Using a protractor construct an angle of 30° at point A and an angle of 30° at point B. Extend the two lines until they intersect.
    b. Cut out the triangle and compare it with the triangles constructed in this way by other students.
    c. Write a statement about your findings.

18. Draw any triangle on an A4 sheet of paper.
    a. Use a protractor to measure the three angles.
    b. Cut out the triangle and then cut the angles as shown in the diagram.

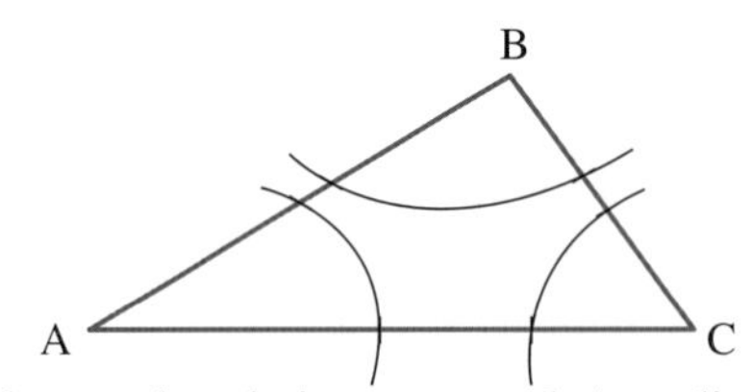


    c. Place the angles next to each other so that their arms touch (see diagram).

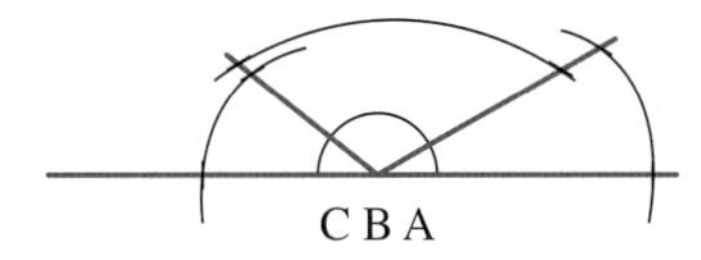


    d. Measure the new angle formed.
    e. Compare your results with those of other students and write a statement to summarise your findings.

# LESSON
# 9.4 Types of angles and naming angles

## LEARNING INTENTION

At the end of this lesson you should be able to:
- classify different types of angles according to size
- name angles using three points.

## 9.4.1 Types of angles

eles-4094

- Angles can be classified according to their size.

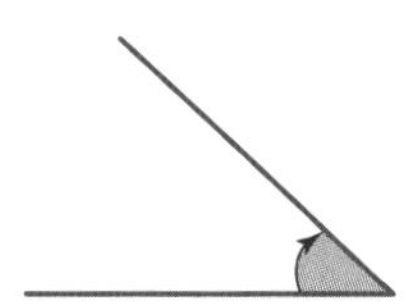

An **acute angle** is greater than 0°, but less than 90°.

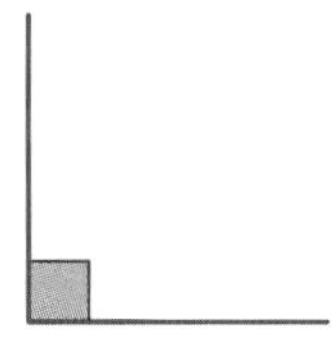

A **right angle** is an angle that equals exactly 90°.

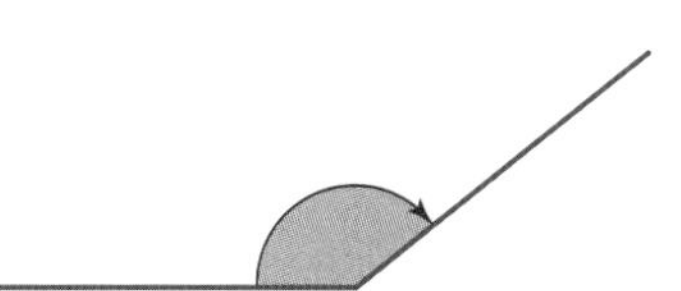

An **obtuse angle** is greater than 90° but less than 180°.

A **straight angle** equals exactly 180°.

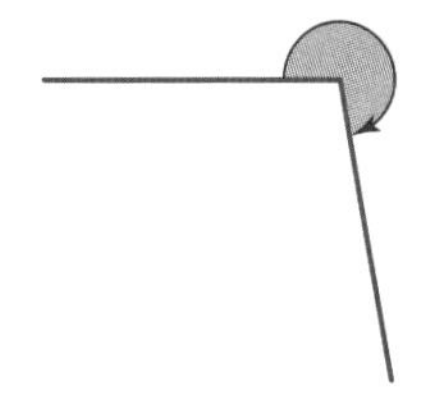

A **reflex angle** is greater than 180° but less than 360°.

An angle of **complete revolution** or a **perigon** is an angle of (a full 360°circle).

### WORKED EXAMPLE 6 Classifying angles

**Classify each of the following angles according to its size.**

**a. 115°** **b. 27°** **c. 300°**

| THINK | WRITE |
|---|---|
| a. The given angle is larger than 90°, but smaller than 180°, so classify it accordingly. | a. 115° is an obtuse angle. |
| b. The given angle is between 0° and 90°, so classify it accordingly. | b. 27° is an acute angle. |
| c. The given angle is larger than 180°, but less 360°, so classify it accordingly. | c. 300° is a reflex angle. |

## 9.4.2 Naming angles

eles-4095

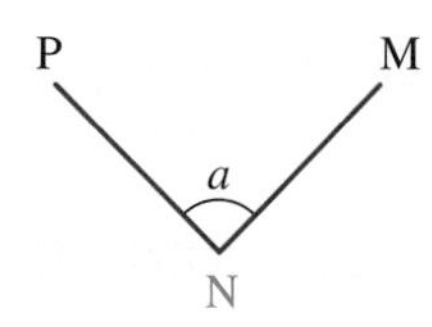


- Angles are named by
  - using three letters: two letters to represent the arms of the angle, and a third letter to represent its vertex. The letter representing the vertex is always placed in the middle. For example, the angle shown can be named either ∠PNM or ∠MNP.
  - writing the letter that represents the vertex only. For example, the angle shown could be named ∠N.
  - writing the pronumeral that represents the angle inside the arms. For example, the angle shown could be named *a*.

### WORKED EXAMPLE 7 Naming angles

**Name each of the following angles.**

**a.**

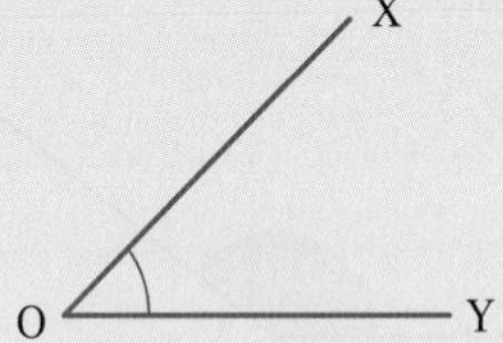


**b.**



**THINK**

**a.** Name the angle by starting either with X or with Y, then placing O (the letter representing the vertex) in the middle. You could also simply use the letter at the vertex. Remember to place the angle symbol (∠) before the letters.

**b.** Name the angle by starting either with A or with C, then placing B (the letter representing the vertex) in the middle. You could also simply use the letter at the vertex. Remember to place the angle symbol (∠) before the letters.

**WRITE**

**a.** ∠XOY or ∠YOX or ∠O

**b.** ∠ABC or ∠CBA or ∠B

### WORKLED EXAMPLE 8 Drawing angles

**Draw an acute angle ∠ADG.**

**THINK**

1. Construct any acute angle (an angle greater than 0°, but less than 90°).
2. Letter D is in the middle, so place it at the vertex.
3. Place letter A at the end of one arm of the angle and letter G at the end of the other arm.
   (*Note:* It does not matter which arm is represented by AD or GD.)

**WRITE**

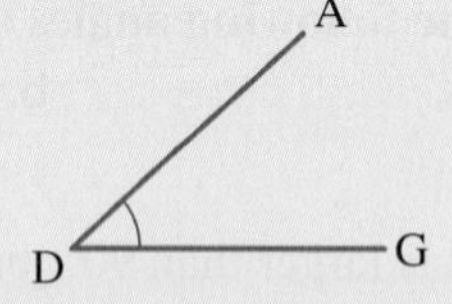

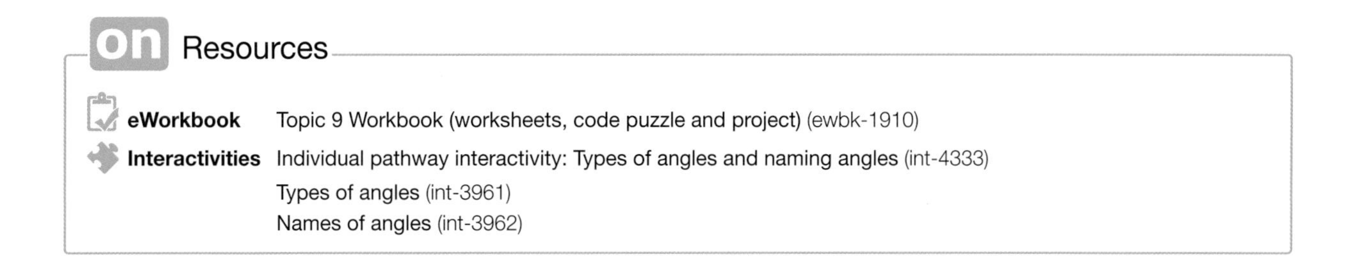


## Exercise 9.4 Types of angles and naming angles

learnon

**9.4 Quick quiz** 

**9.4 Exercise**

### Individual pathways

| ■ PRACTISE | ■ CONSOLIDATE | ■ MASTER |
|---|---|---|
| 1, 4, 7, 10, 12, 15, 20, 23 | 2, 5, 8, 11, 14, 16, 18, 21, 24 | 3, 6, 9, 13, 17, 19, 22, 25 |

### Fluency

1. **WE6** Classify each of the following angles according to its size.
   a. 12°  b. 215.3°  c. 156°  d. 180°  e. 355.2°
2. Classify each of the following angles according to its size.
   a. 90°  b. 4.8°  c. 360°  d. 45°  e. 270°
3. The following list gives values of particular angles.
   $3°, 45°, 65°, 123°, 69°, 234°, 90°, 360°, 300°, 270°, 165°, 210°, 180°$
   **a.** List the acute angles.
   **b.** List the obtuse angles.
   **c.** List the reflex angles.
4. State whether each of the following is true or false. The list of angles in question **3** contains:
   a. at least one right angle
   b. at least one straight angle
   c. at least one full revolution.
5. Consider the following diagrams and write down the type of angle shown in each case.

a. 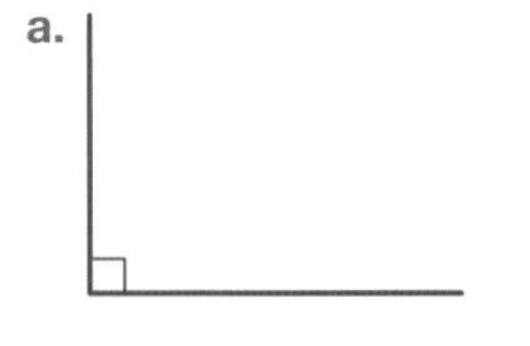

b. 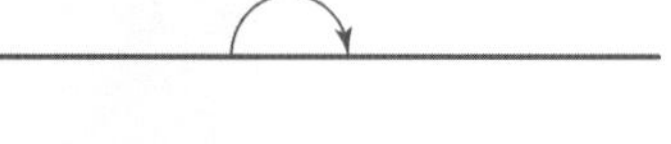

c. 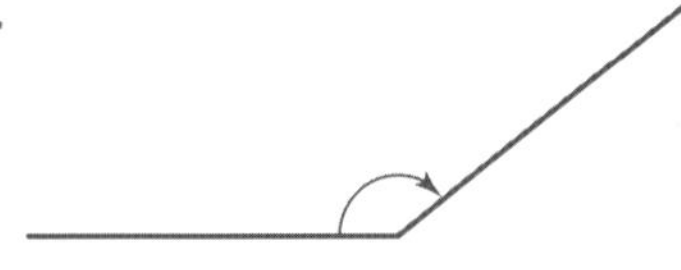

d. 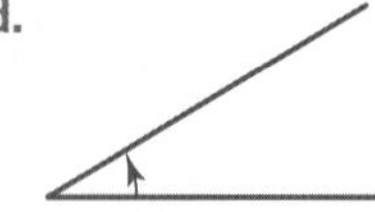

e. 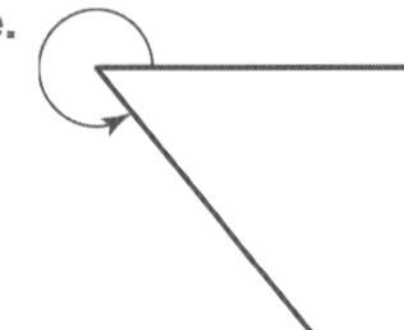

f. 

6. WE7 Name each of the following angles.

a. 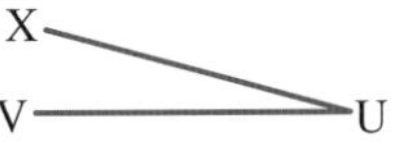


b. 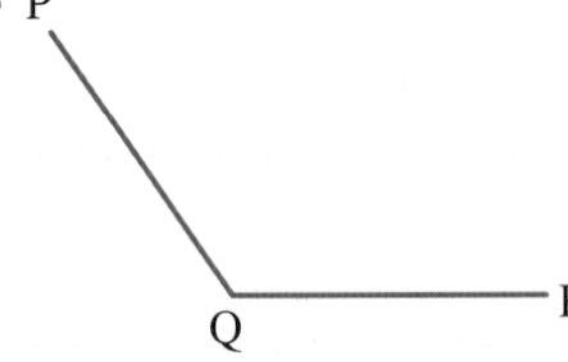


c. 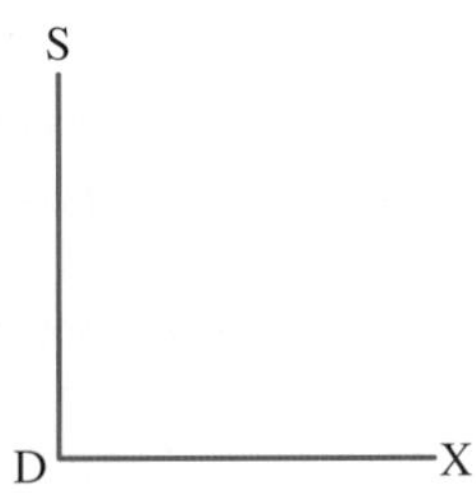


d. 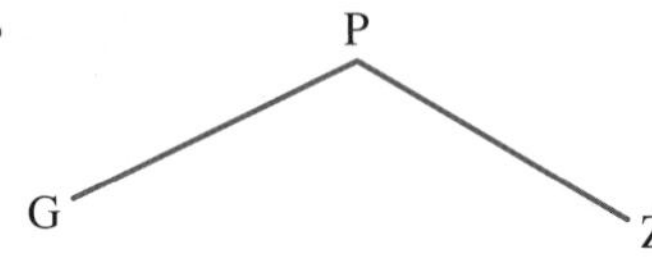


e. P ——•—— H
F

## Understanding

7. WE8 Draw each of the following angles as specified in the brackets.

a. ∠CDE (acute)
b. ∠TRE (obtuse)
c. ∠QAS (straight)

8. Draw each of the following angles as specified in the brackets.

a. ∠FGH (reflex)
b. ∠KJF (right)
c. ∠VBN (acute)

9. Name the type of angle shown in this photograph.

10. Name the type of angle shown in this photograph.

11. Name the type of angle shown in this photograph.

12. Name each acute and obtuse angle in the following diagram.

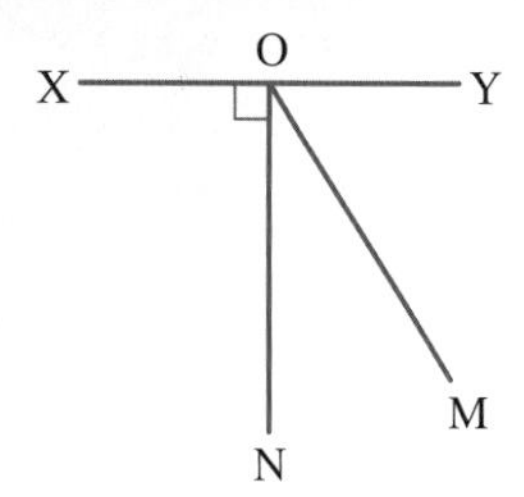


13. Name each acute and obtuse angle in the following diagram.

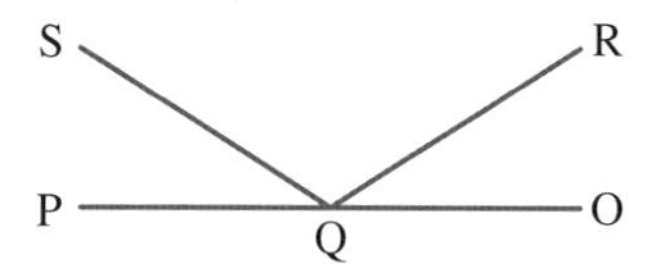


14. Name each acute and obtuse angle in the following diagrams.

a. 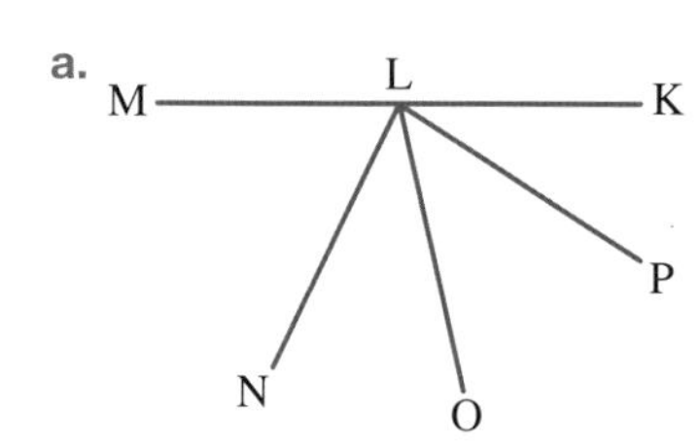


b. 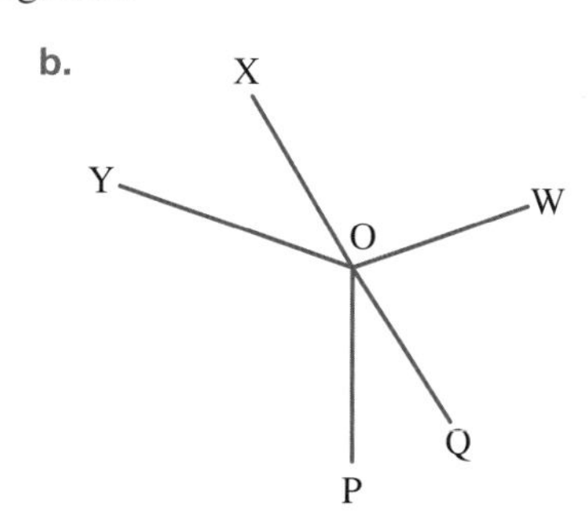


15. State the type of angle that the minute hand of a clock makes when it moves for the following lengths of time.

a. 10 minutes
b. 15 minutes
c. 20 minutes

16. State the type of angle that the minute hand of a clock makes when moving for the following lengths of time.

a. 30 minutes
b. 45 minutes
c. 1 hour

17. At various times of the day the hands of an analogue clock form two angles: the inside angle (the smaller one) and the outside angle (the larger one).
State the type of the inside angle formed by the hands of a clock at the following times.

a. 1 o'clock
b. 20 minutes to 12
c. 6 o'clock
d. 9 o'clock
e. quarter to 4

18. At various times of the day the hands of an analogue clock form two angles: the inside angle (the smaller one) and the outside angle (the larger one).
Write two different times when the inside angle of the clock is:

a. acute
b. obtuse
c. right.

19. In this photograph, acute, right-angled and obtuse angles have been used.

a. Name an acute angle.
b. Name two right angles.
c. Name an obtuse angle.
d. State the name of a reflex angle.

## Reasoning

20. Give reasons for the following statements.
    a. The sum of two acute angles can be an acute angle.
    b. The sum of two acute angles can be a right angle.
    c. The sum of two acute angles can be an obtuse angle.
    d. The sum of two acute angles can be a reflex angle.

21. a. State whether it is possible to construct a complete revolution using one 90° angle, one straight angle and two acute angles.
    b. Construct a diagram to represent the complete revolution in part a.

22. Watchmakers take particular care that the angles between the hands of an analogue clock represent the time accurately. Determine the size of the smaller angle between the hour and minute hands when an analogue clock displays the time as 10 past 9.

### Problem solving

23. Explain why it is important to always put the letter representing the vertex in the middle of an angle's name.

24. Label each angle marked in this photograph with the word describing the type of angle it is.

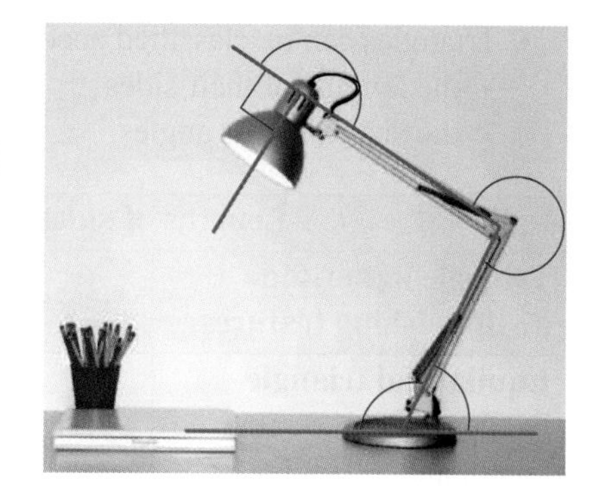

25. a. Consider the figure and identify how many angles are shown.
    b. At midday, the hour hand, the minute hand and the second hand on an analogue clock are all pointing to the 12.
    Determine the size of the smaller angle between the hour hand and the minute hand after the second hand has completed 30 revolutions.
    *Hint:* It is not 180°— draw a separate diagram to help.

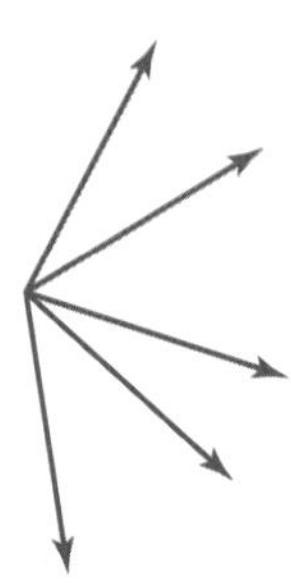

# LESSON
# 9.5 Triangles

**LEARNING INTENTION**

At the end of this lesson you should be able to:
- classify triangles according to the lengths of their sides or the sizes of their angles
- calculate both interior and exterior angles of a triangle.

## 9.5.1 Naming triangles

eles-4096

- **Triangles** are closed shapes that have three edges (or sides) and three vertices (or angles).
- Capital letters are used at each vertex to name a triangle.
  For example, the triangle in the following bicycle frame can be named Δ PQR. (It can also be named Δ RPQ or Δ QRP, and so on.)

## Classifying triangles

- Triangles can be classified according to:
  - the lengths of their sides
  - the sizes of their angles.

| Lengths of sides | | Sizes of angles | |
|---|---|---|---|
| **Triangle name and distinguishing features** | **Shape** | **Triangle name and distinguishing features** | **Shape** |
| **Equilateral triangle** All edges are the same length and all angles are the same size. This is shown by the identical marks on each edge and each angle. | | **Acute-angled triangle** All interior angles are less than 90°. | |
| **Isosceles triangle** Two edges are the same length. The third edge is a different length. It is often called the base of the triangle. The two base angles are the same size. | | **Right-angled triangle** One angle is exactly 90°. This angle is marked by a small square. The edge opposite the right angle is called the hypotenuse. | Hypotenuse |
| **Scalene triangle** All edges are different lengths and all angles are different sizes. This is shown by the unique markings on each edge and each angle. | | **Obtuse-angled triangle** One angle is greater than 90°. | |

### DISCUSSION

1. In the triangle shown, count the number of:
   a. equilateral triangles
   b. right-angled triangles
   c. isosceles triangles.

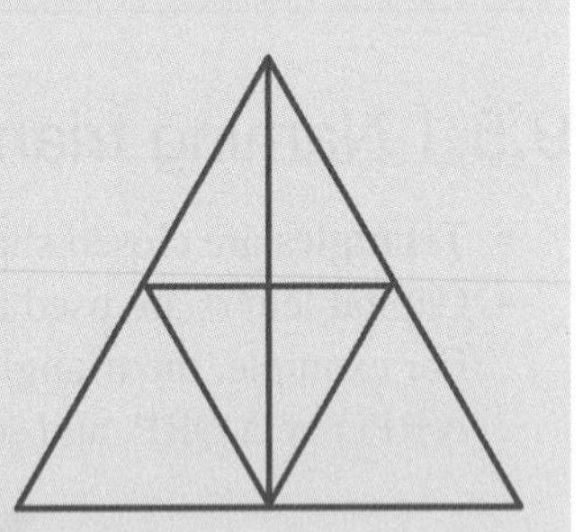

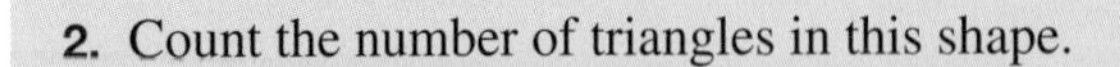

2. Count the number of triangles in this shape.

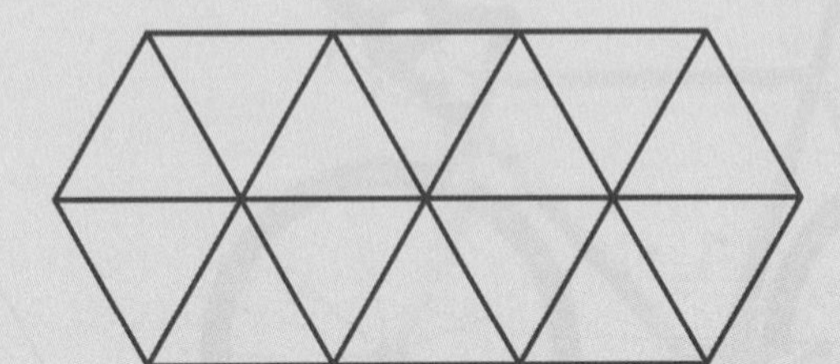

## WORKED EXAMPLE 9 Classifying triangles by sides

**Classify each of these triangles according to the lengths of its sides.**

a. 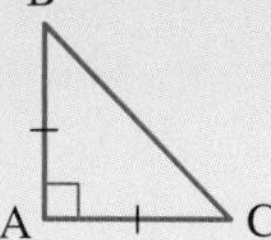


b. 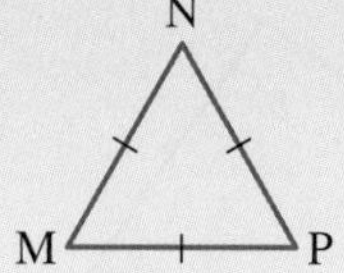


c. 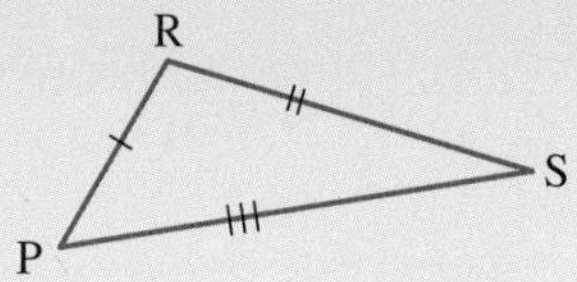


**THINK**

a. Sides AB and AC have identical markings on them, which means that they are equal in length. So ΔABC has 2 equal sides. Classify this triangle based on that infomation.

b. The 3 sides of ΔMNP have identical markings on them, which means that all 3 sides are equal in length. Classify this triangle based on that information.

c. All 3 sides of ΔPRS are marked differently, which means no sides in this triangle are equal in length. Classify this triangle based on that information.

**WRITE**

a. ΔABC is an isosceles triangle.

b. ΔMNP is an equilateral triangle.

c. ΔPRS is a scalene triangle.

## WORKED EXAMPLE 10 Classifying triangles by angles

**Classify each of the triangles in Worked example 9 according to the size of its angles.**

**THINK**

a. In ΔABC, ∠CAB is marked as the right angle. Classify this triangle based on that information.

b. In ΔMNP all angles are less than 90°. Classify this triangle based on that information.

c. In ΔPRS, ∠PRS is greater than 90°; that is, it is obtuse. Classify this triangle based on that information.

**WRITE**

a. ΔABC is a right-angled triangle.

b. ΔMNP is an acute-angled triangle.

c. ΔPRS is an obtuse-angled triangle.

## COLLABORATIVE TASKS

### Sketching triangles from a verbal description

Pair up with a classmate and each of you draw a triangle without the other person seeing. Take turns describing your triangles using only verbal descriptions and asking your partner to draw the triangle you have described. How accurate is their drawing compared to the triangle you described?

### Investigating the angles

Pair up with a classmate and cut out a triangle. Tear off the three angles (the points of your triangle) and pass them to your partner. Can you make a straight line with the three angles your partner has given you?

## 9.5.2 Angles in a triangle

eles-4097

- The sum of the three angles in any triangle is always 180°.

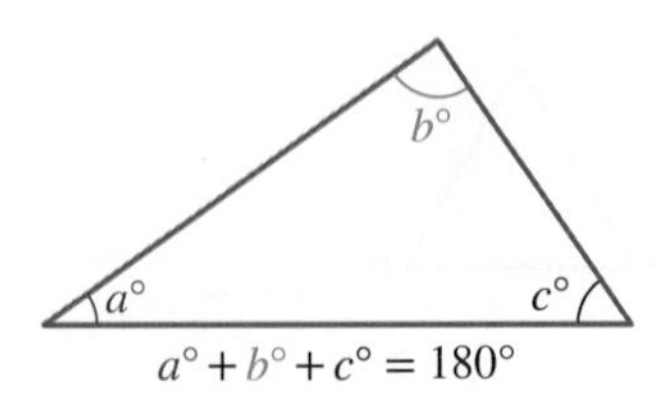


$a° + b° + c° = 180°$

### Sum of angles in a triangle

The three angles of a triangle add up to 180°. That is, $a° + b° + c° = 180°$.

### WORKED EXAMPLE 11 Calculating angles in triangles

**Calculate the values of the pronumeral in each of the following triangles.**

a.

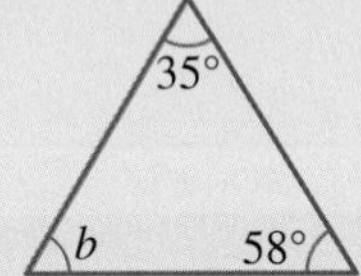


b.

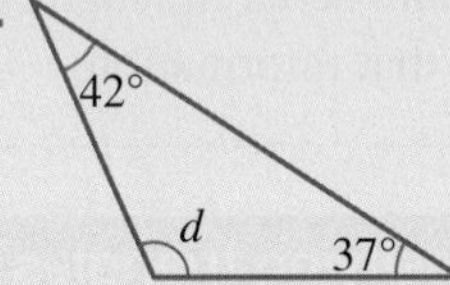


| THINK | WRITE |
|---|---|
| a. 1. The sum of the 3 angles ($b$, 35° and 58°) must be 180°. Write this as an equation. | a. $b + 35° + 58° = 180°$ |
| 2. Simplify by adding 35° and 58° together. | $b + 93° = 180°$ |
| 3. Use inspection or backtracking to solve for $b$. | $b = 180° - 93°$<br>$b = 87°$ |
| b. 1. The sum of the three angles ($d$, 37° and 42°) must be 180°. Write this as an equation. | b. $d + 37° + 42° = 180°$ |
| 2. Simplify by adding 37° and 42° together. | $d + 79° = 180°$ |
| 3. Use inspection or backtracking to calculate the value of $d$. | $d = 180° - 79°$<br>$d = 101°$ |

+ 93°

| $b$ | $b + 93°$ |
|---|---|
| 87° | 180° |

− 93°

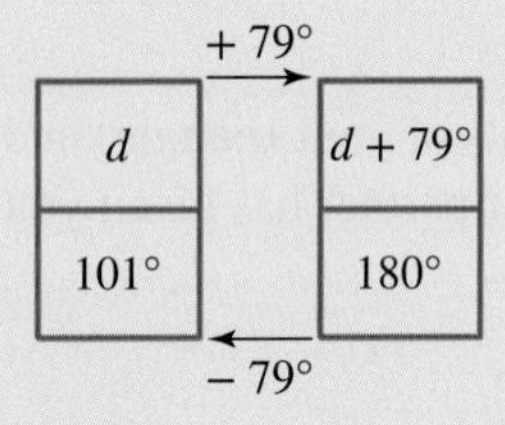

## WORKED EXAMPLE 12 Calculating angles in triangles

**Calculate the value of the pronumeral in the following triangle.**

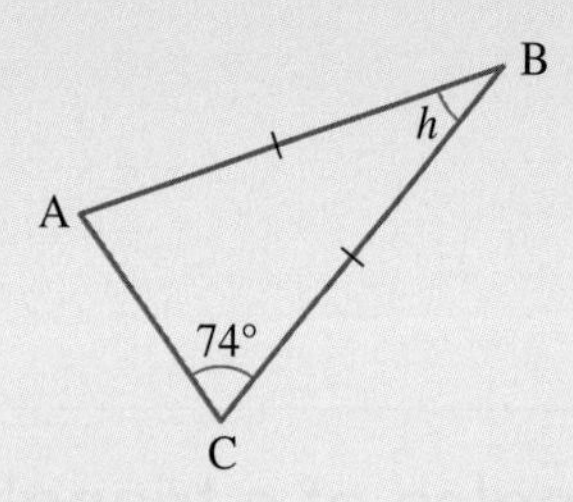


| THINK | WRITE |
|---|---|
| 1. The markings on the diagram indicate that $\Delta ABC$ is isosceles because AB = BC. Therefore, the angles at the base are equal in size; that is, $\angle BCA = \angle BAC = 74°$. Write this as an equation. | $\angle BCA = \angle BAC = 74°$ |
| 2. All 3 angles in a triangle must add up to 180°. Write this as an equation. | $\angle ABC + \angle BAC + \angle BCA = 180°$<br>$h + 74° + 74° = 180°$ |
| 3. Simplify by adding 74° and 74° together. | $h + 148° = 180°$ |
| 4. Subtract 148° from 180° to calculate the value of $h$. | $h = 180° - 148°$<br>$h = 32°$ |

## WORKED EXAMPLE 13 Calculating angles in triangles

**Calculate the value of the pronumeral in the following triangle.**

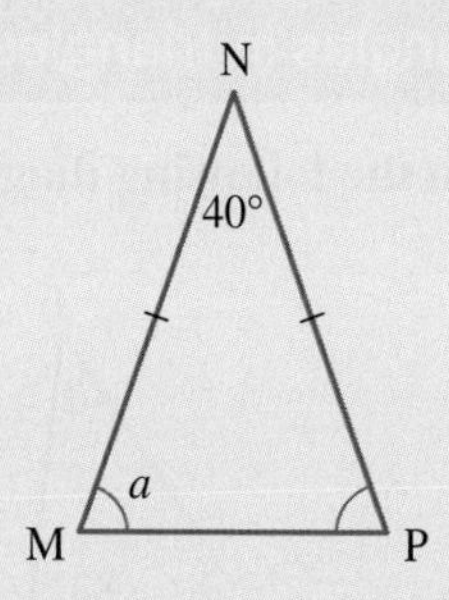


| THINK | WRITE |
|---|---|
| 1. The markings on the diagram indicate that $\Delta MNP$ is isosceles because MN = NP. Therefore the angles at the base are equal in size. Write this as an equation. | $\angle NPM = \angle NMP = a$ |
| 2. All 3 angles in a triangle must add up to 180°. Write this as an equation. | $\angle NMP + \angle NPM + \angle MNP = 180°$<br>$a + a + 40° = 180°$ |
| 3. Simplify by collecting like terms. In this example $a + a = 2a$. | $2a + 40° = 180°$ |

4. Use inspection or backtracking to calculate the value of $a$.

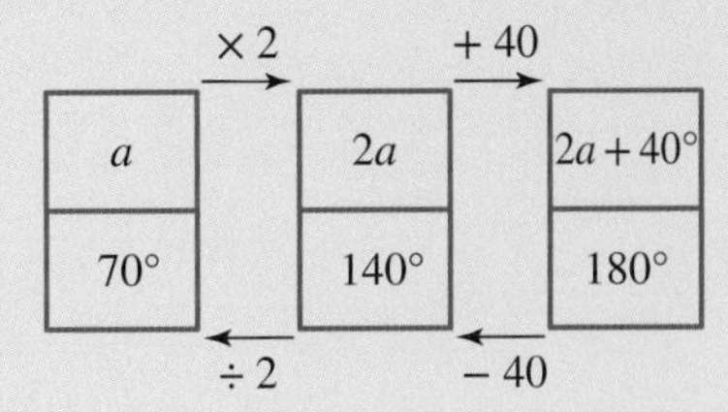


$2a = 180° - 40°$

$2a = 140°$

$a = \frac{140°}{2}$

$a = 70°$

## 9.5.3 Interior and exterior angles of a triangle

eles-4098

- The angles inside a triangle are called **interior angles**.
- When a side of a triangle is extended outwards, the angle formed is called an **exterior angle**.
- The exterior angle and the interior angle adjacent (next) to it add up to 180°.
- The sum of the two interior opposite angles is equal to the exterior angle.

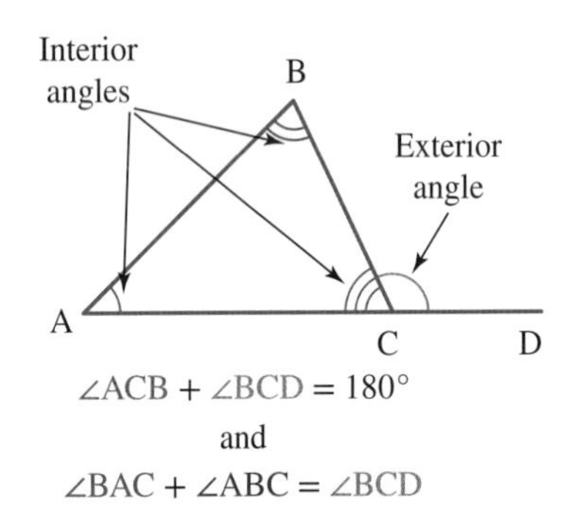

### WORKED EXAMPLE 14 Calculating angles in triangles

**Determine the value of the pronumerals in the following diagram.**

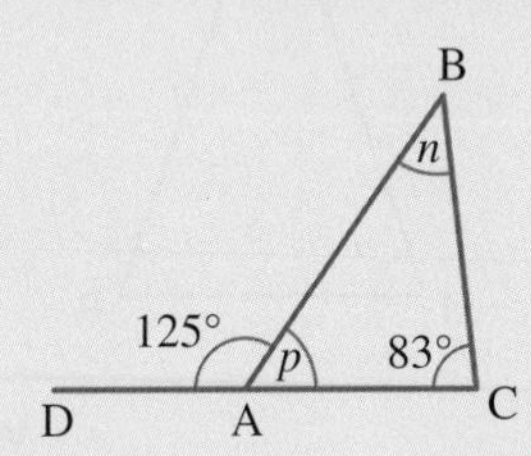


| THINK | WRITE |
|---|---|
| 1. ∠DAB = 125°. ∠BAC (angle $p$) and its adjacent exterior angle (∠DAB) add up to 180°. Write this as an equation. | ∠BAC = $p$; ∠DAB = 125°<br>∠BAC + ∠DAB = 180°<br>$p + 125° = 180°$ |
| 2. Subtract 125° from 180° to determine the value of $p$. | $p = 180° - 125°$<br>$p = 55°$ |
| 3. The interior angles of ΔABC add up to 180°. Identify the values of the angles and write this as an equation. | ∠BCA + ∠BAC + ∠ABC = 180°<br>∠BCA = 83°; ∠BAC = $p$ = 55°; ∠ABC = $n$<br>$83° + 55° + n = 180°$ |

4. Simplify by adding 83° and 55°. $n + 138° = 180°$

5. Subtract 138° from 180° to determine the value of *n*. $n = 180° - 138°$
$n = 42°$

## Resources

**eWorkbook** Topic 9 Workbook (worksheets, code puzzle and project) (ewbk-1910)

**Interactivities** Individual pathway interactivity: Triangles (int-4334)
Classification of triangles — sides (int-3963)
Classification of triangles — angles (int-3964)
Angles in a triangle (int-3965)
Interior and exterior angles of a triangle (int-3966)

# Exercise 9.5 Triangles

learnon

9.5 Quick quiz on

9.5 Exercise

### Individual pathways

| ■ PRACTISE | ■ CONSOLIDATE | ■ MASTER |
| --- | --- | --- |
| 1, 2, 5, 10, 13, 15, 17, 20 | 3, 6, 7, 9, 11, 14, 18, 21 | 4, 8, 12, 16, 19, 22 |

### Fluency

1. **WE9** Name each of the following triangles using the capital letters, then classify each triangle according to the lengths of its sides.

a.
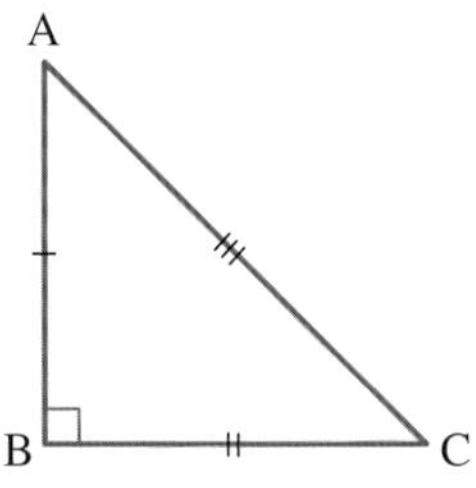


b.
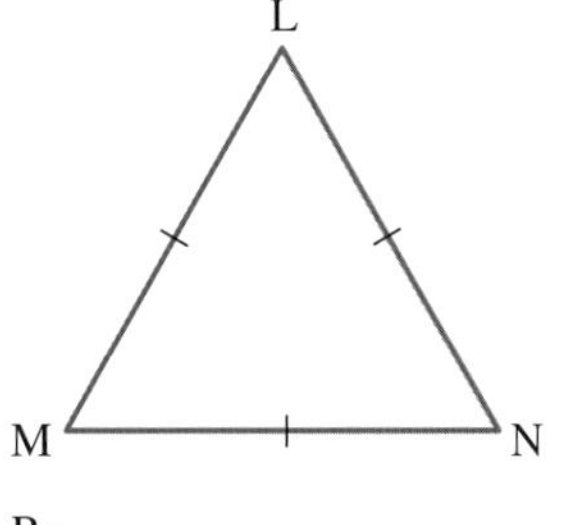


c.
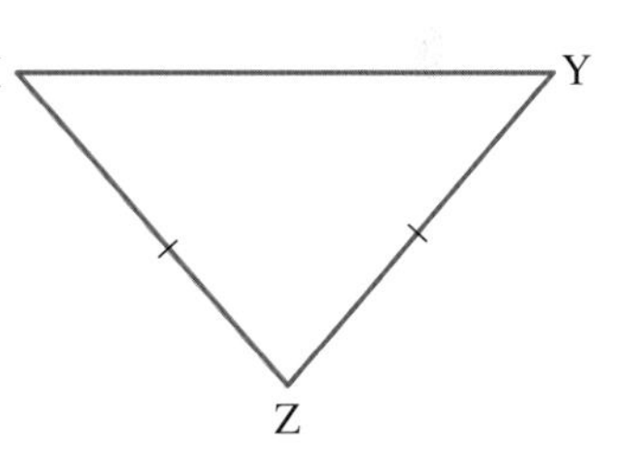


d.
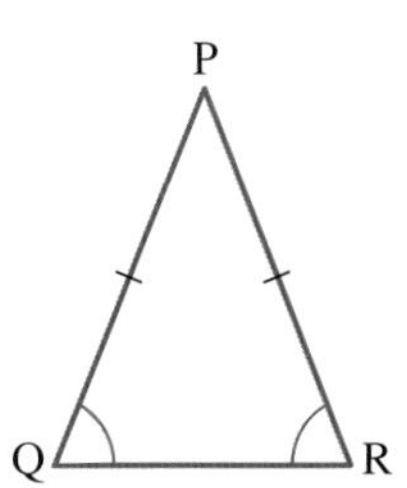


e.
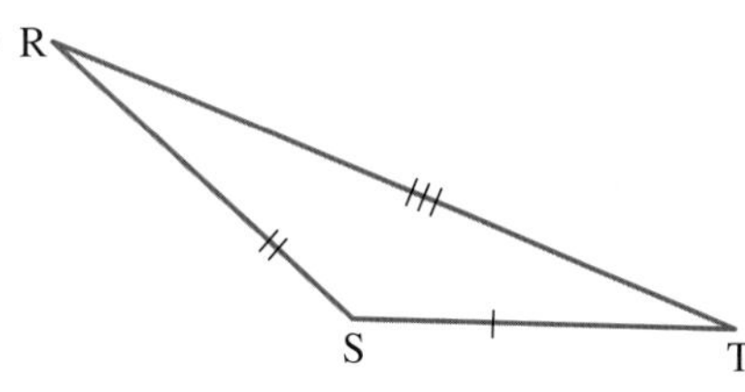


2. **WE10** Classify each of the triangles in question 1 according to the size of its angles.

3. **MC** Select which of these triangles is an equilateral triangle.

A. 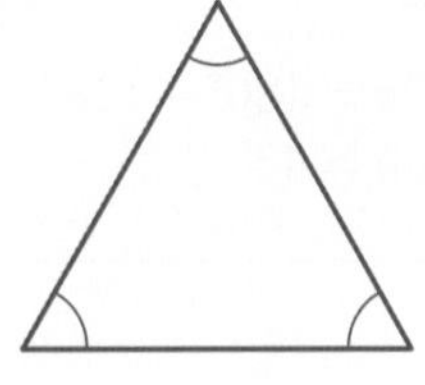

B. 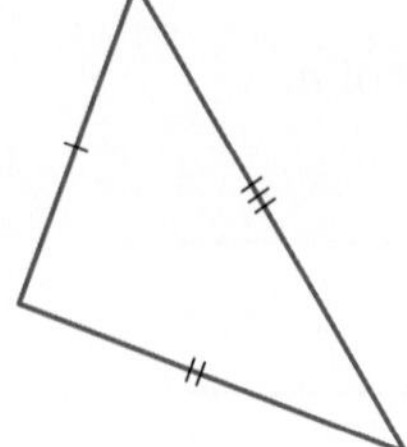

C. 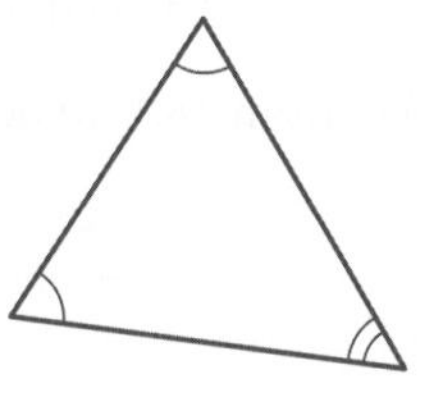

D. 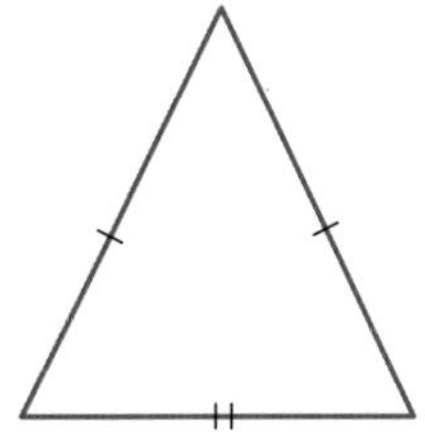

E. 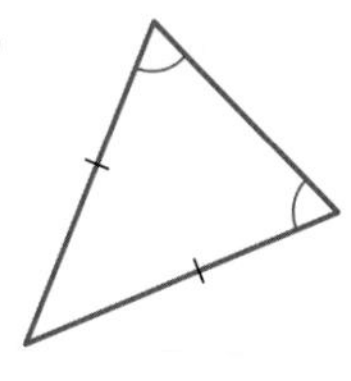

4. **MC** Select which of these triangles is not a scalene triangle.

A. 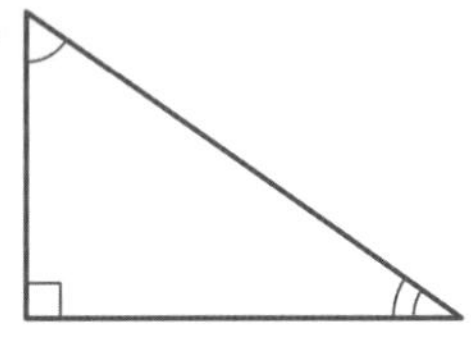

B. 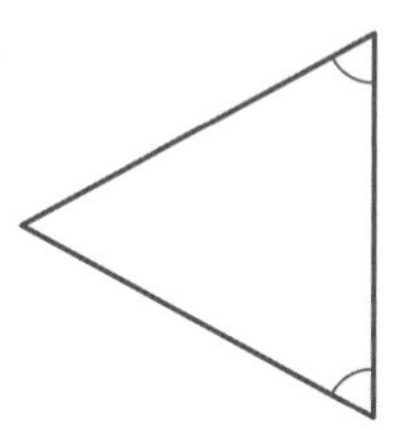

C. 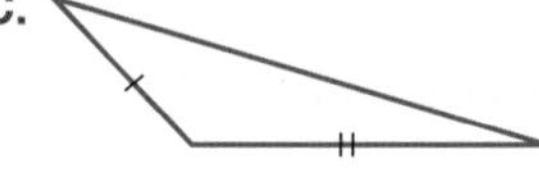

D. 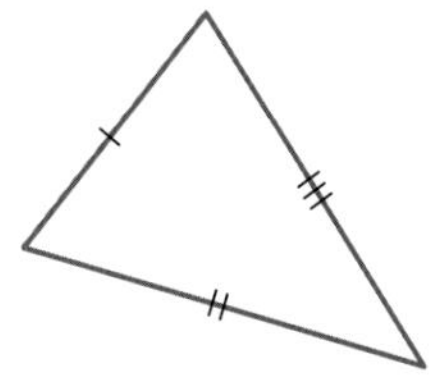

E. 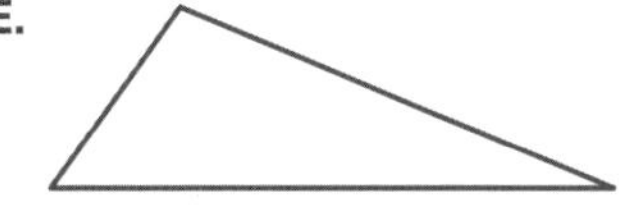

5. **MC** Select the triangle that is both right-angled and scalene.

A. 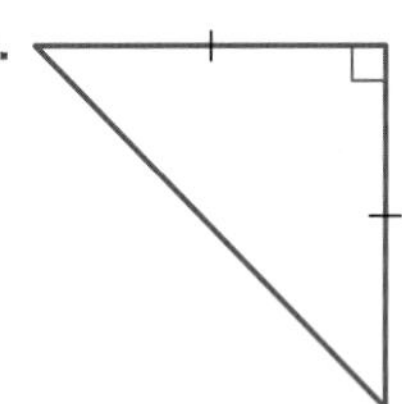

B. 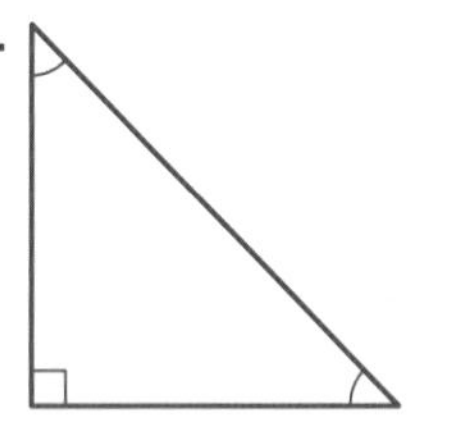

C. 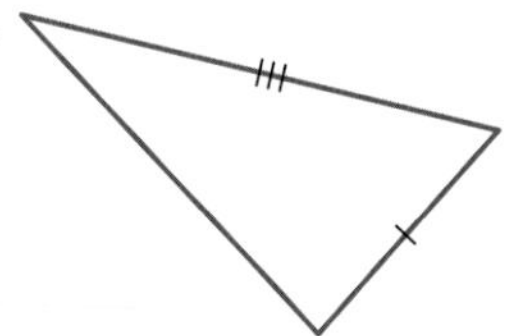

D. 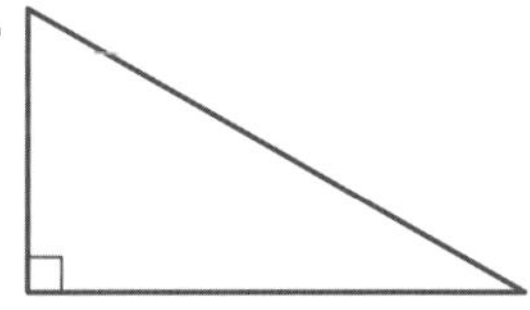

E. 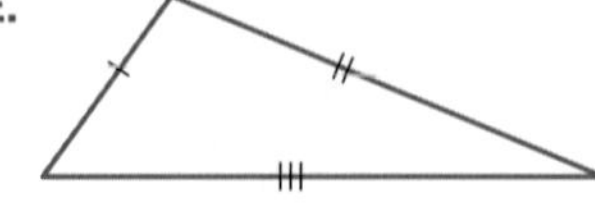

6. **MC** Select which of these triangles is both acute-angled and isosceles.

A. 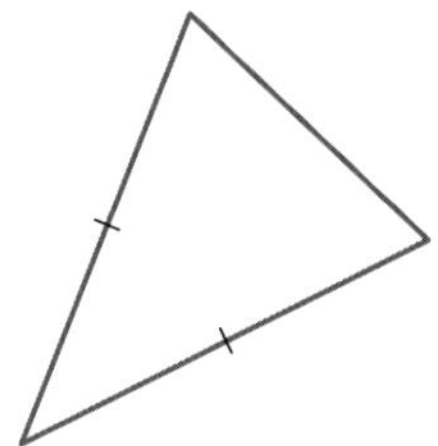

B. 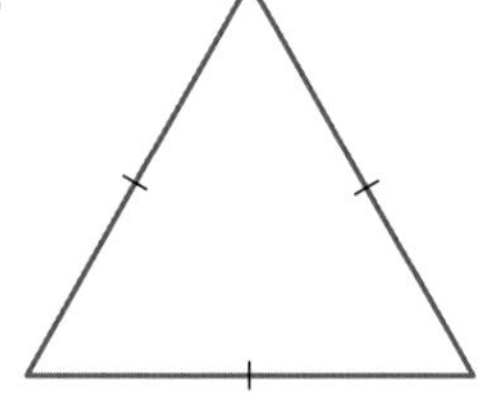

C. 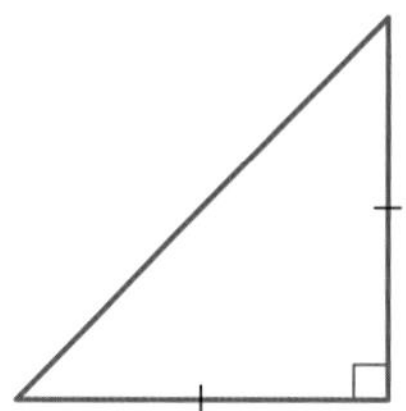

D. 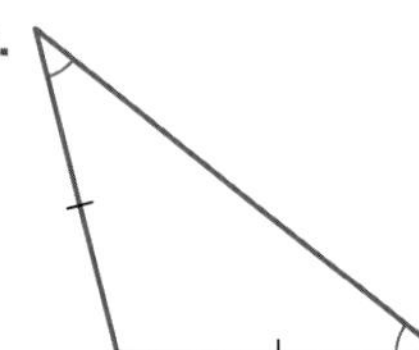

E. 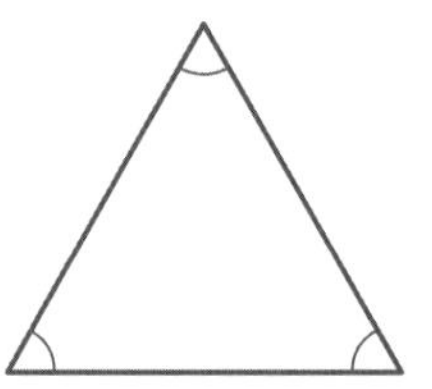

7. **WE11** Calculate the value of the pronumeral in each of the following triangles.

a. 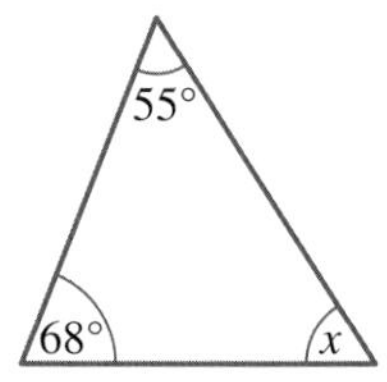


b. 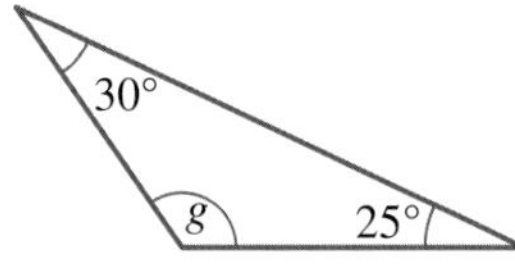


c. 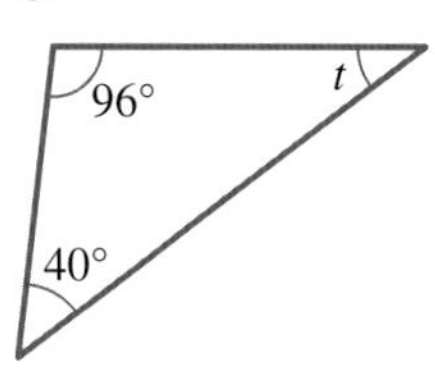


d. 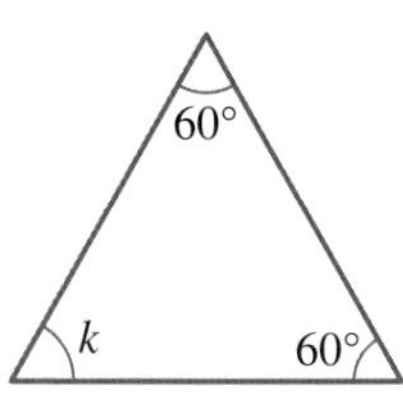


e. 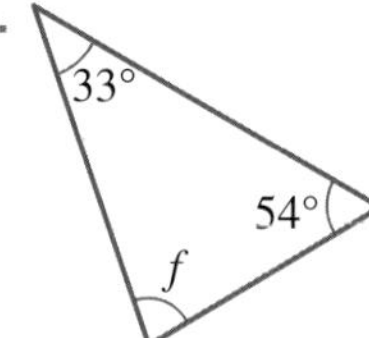


f. 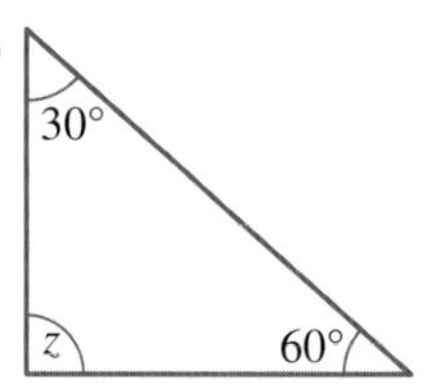


8. Calculate the value of the pronumeral in each of the following right-angled triangles.

a. 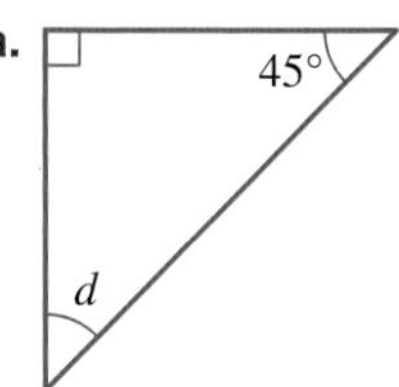


b. 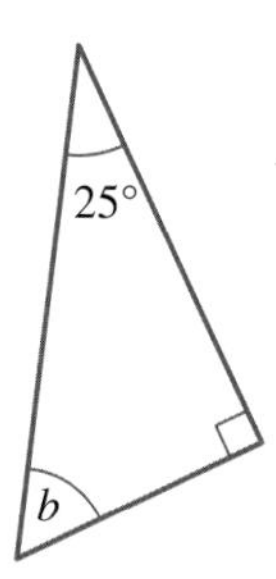


c. 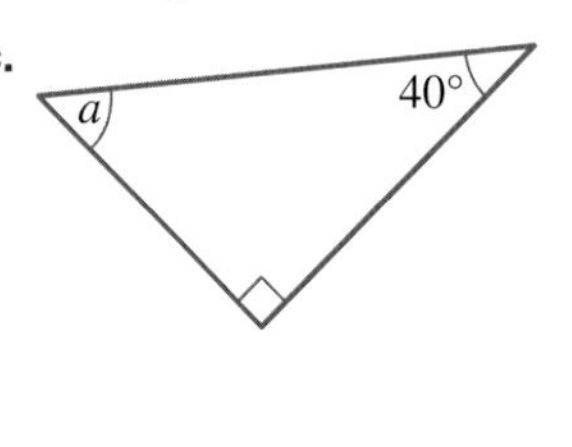


9. **WE12** Calculate the value of the pronumeral in each of the following triangles, giving reasons for your answer.

a. 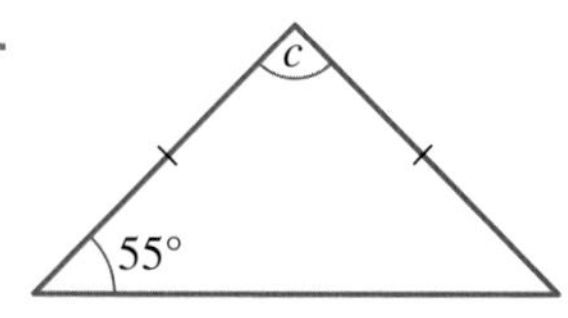


b. 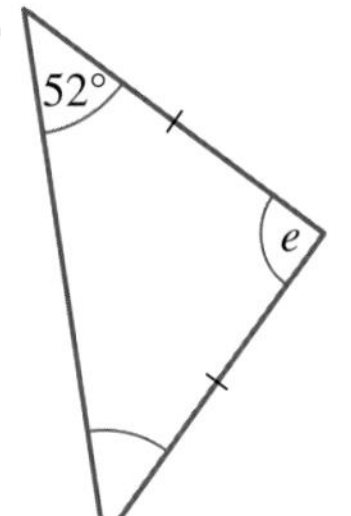


c. 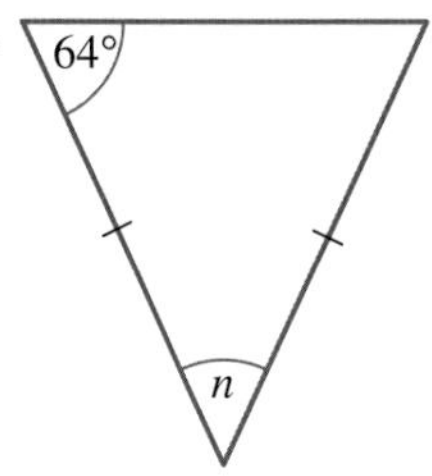

10. **WE13** Calculate the value of the pronumeral in each of the following triangles, giving reasons for your answer.

a.

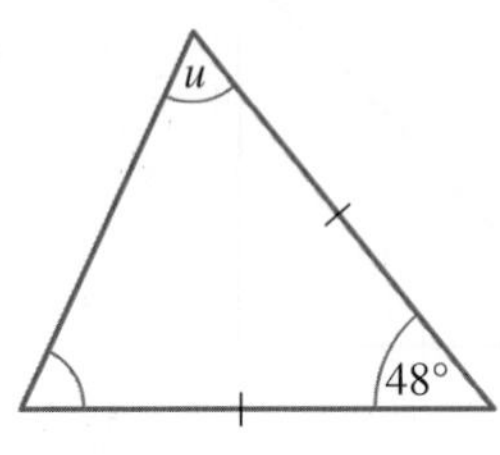


b.

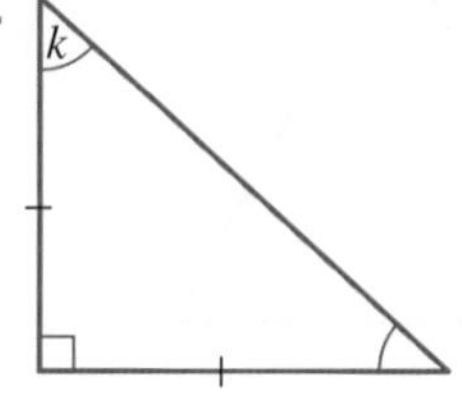


c.

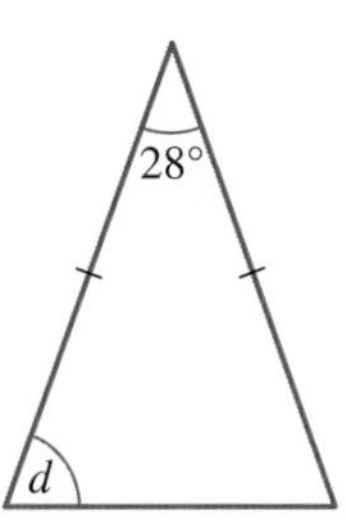


d.

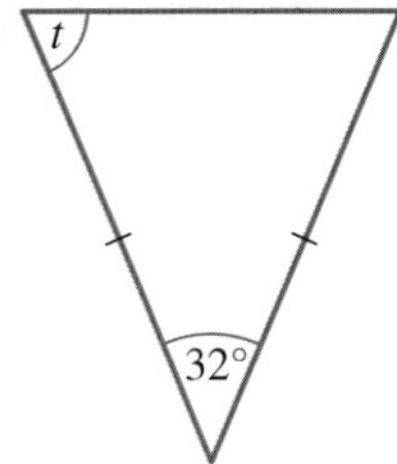


e.

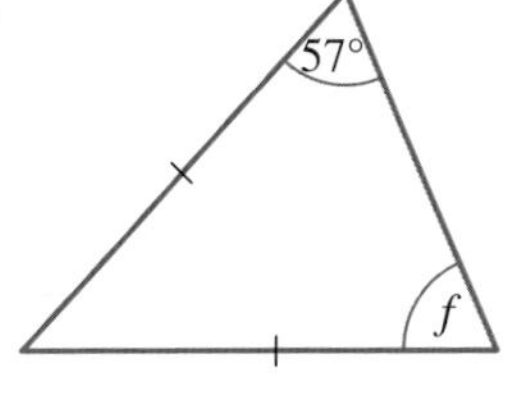


f.

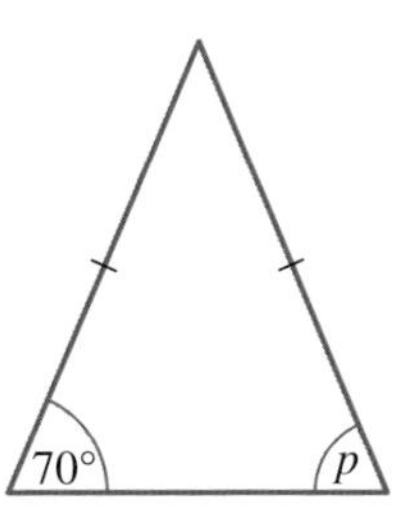


11. **WE14** Calculate the value of the pronumerals in each of the following diagrams.

a.

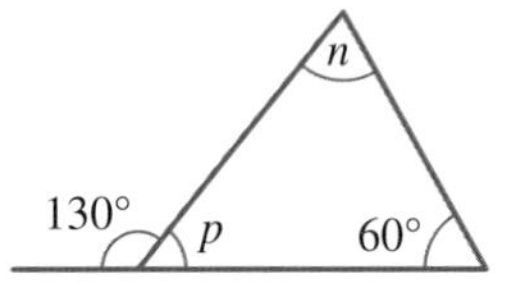


b.

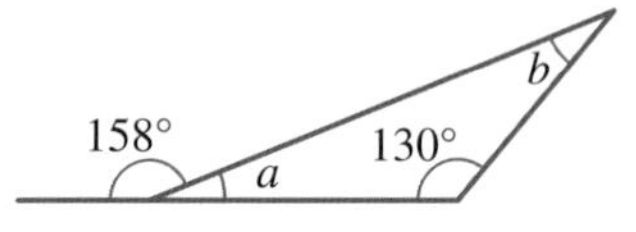


c.

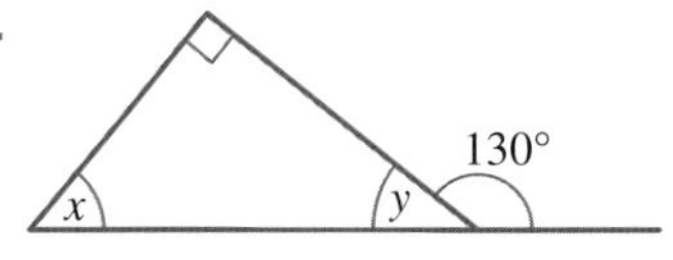


d.

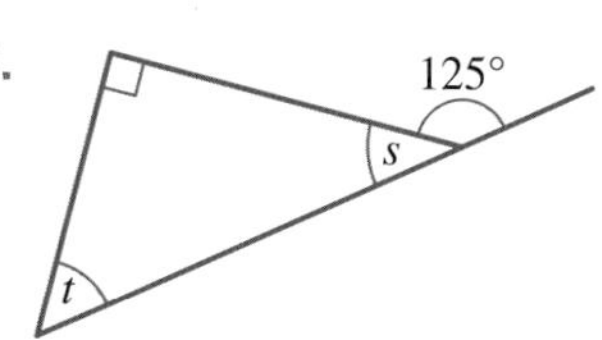


12. Calculate the value of the pronumerals in each of the following diagrams.

a.

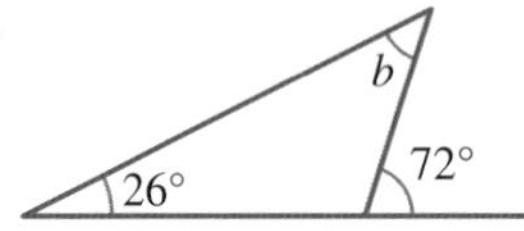


b.

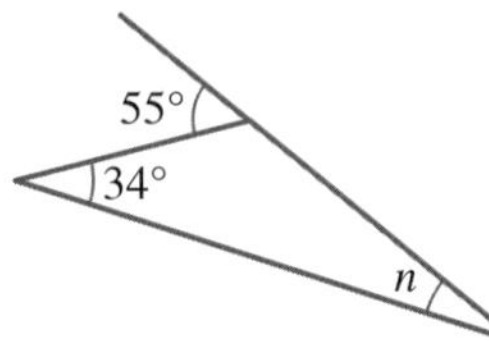


c.

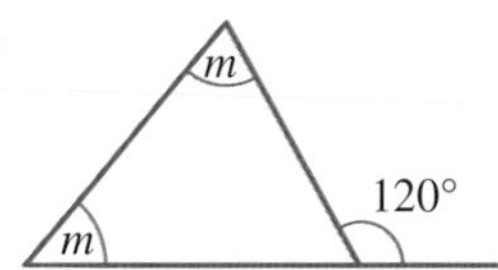


d.

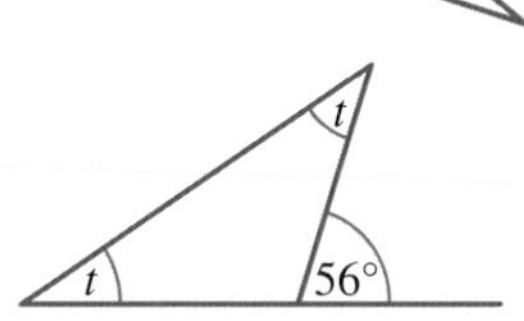


## Understanding

13. Add side and angle markings to these diagrams to show that:

a. ΔRST is an equilateral triangle b. ΔUVW is an isosceles triangle c. ΔPQR is a scalene triangle.

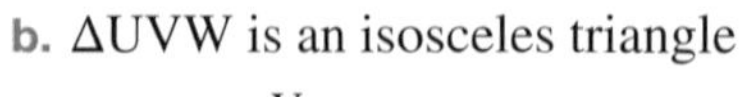

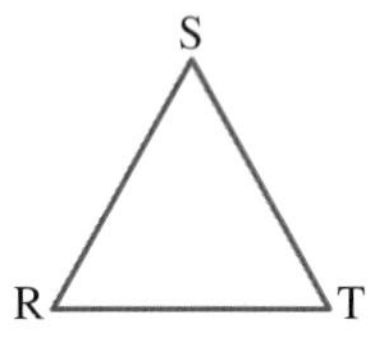


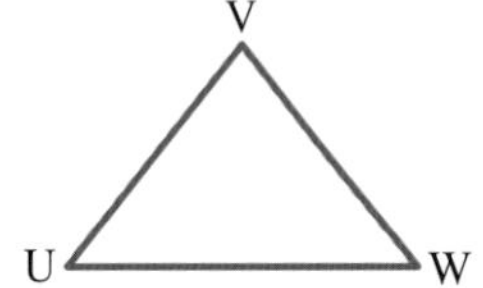


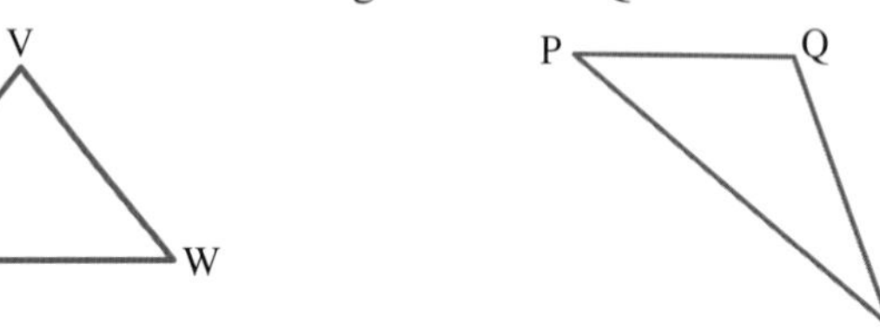

14. Add side and angle markings to these diagrams to show that:

a. $\Delta$MNP is a right-angled triangle

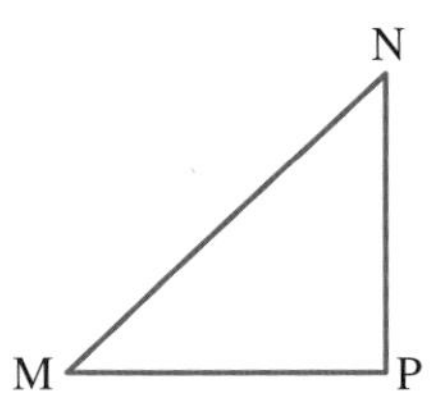


b. $\Delta$ABC is a right-angled and isosceles triangle

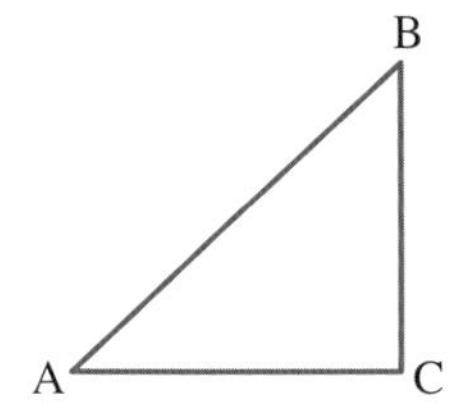


c. $\Delta$MNO is a right-angled and scalene triangle.

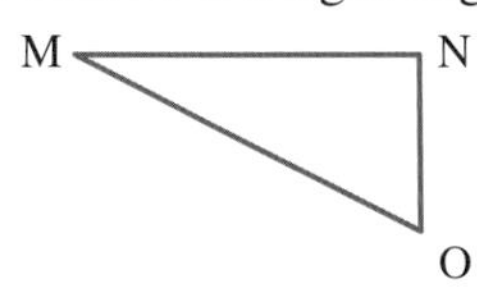


15. Use your ruler, pencil and protractor to accurately draw:

a. an equilateral triangle with side lengths 6 cm and all angles 60°

b. an isosceles triangle with two sides that are 6 cm each with a 40° angle between them

c. a right-angled triangle whose two short sides are 6 cm and 8 cm (How long is the longest side?)

d. a scalene triangle with two of the sides measuring 4 cm and 5 cm and an angle of 70° between the two sides.

16. Calculate the missing angle in each of the triangles marked on the following photographs.

a.

b.

## Reasoning

17. Calculate the magnitudes of the unknown angles in the following triangles.
    a. $\Delta$MNP is an isosceles triangle with MN = MP and $\angle$MNP = 72°. Calculate the sizes of $\angle$MPN and $\angle$NMP.
    b. $\Delta$MNP is an isosceles triangle with MN = MP and $\angle$NMP = 72°. Calculate the sizes of $\angle$MPN and $\angle$MNP

18. a. Explain in your own words why it is impossible to construct a triangle with two obtuse angles.
    b. Explain why the longest side of a triangle is always opposite the largest angle of the triangle.
    c. Explain why the sum of the lengths of two sides of a triangle must be greater than the length of the third side.

19. a. Construct the isosceles triangle ABC with $\angle$B = $\angle$C = 80°.
    b. Extend the line BC to the right and label the other end D where BD = AB and form the $\Delta$ABD.
    c. Calculate the size of $\angle$D.
    d. Extend the line CD to the right and label the other end E where DE = AD and form the $\Delta$ADE.
    e. Calculate the size of $\angle$E.
    f. Repeat this process again and discuss any pattern you observe.

## Problem solving

20. a. An isosceles triangle has two angles of 55° each. Calculate the third angle.
    b. An isosceles triangle has two angles of 12° each. Calculate the third angle.
    c. Two angles of a triangle are 55° and 75°. Calculate the third angle.
    d. Two angles of a triangle are 48° and 68°. Calculate the third angle.

21. For each of the following sets of three angles, state whether it is possible to construct a triangle with these angles. Give a reason for your answer.
    a. 40°, 40°, 100°
    b. 45°, 60°, 70°
    c. 45°, 55°, 85°
    d. 111°, 34.5°, 34.5°

22. Determine the value of the pronumerals in the following shapes.

a.

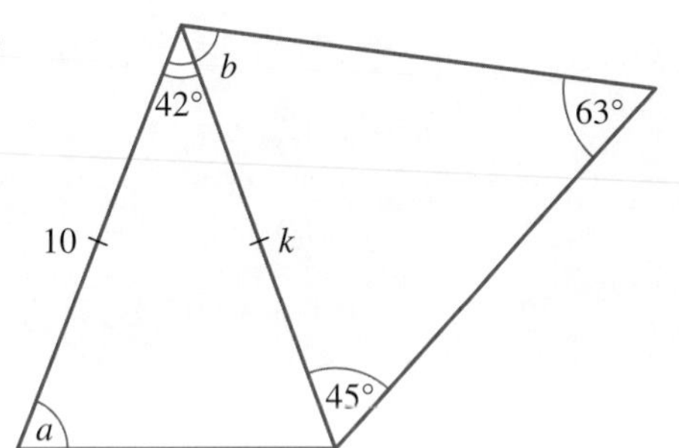


b.

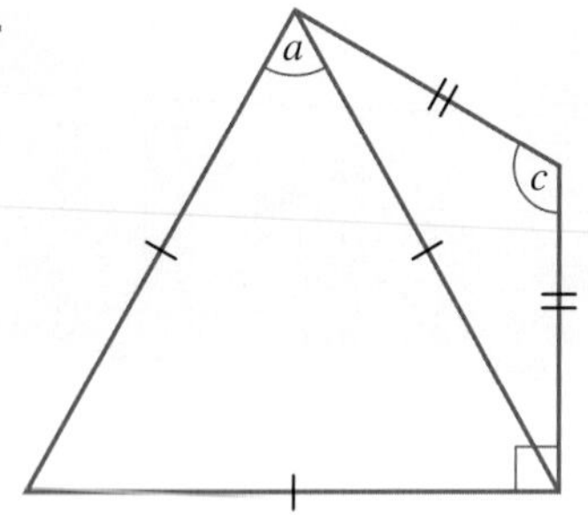

# LESSON
# 9.6 Quadrilaterals and their properties

## LEARNING INTENTION

At the end of this lesson you should be able to:
- classify quadrilaterals from a given diagram
- calculate the angles in a quadrilateral.

## 9.6.1 Types of quadrilaterals

eles-4100

- A **quadrilateral** is a 2-dimensional closed shape with four straight sides.
- Quadrilaterals can be either **convex** or non-convex.
- In a convex quadrilateral, each of the interior angles is less than 180° and the two diagonals lie inside the quadrilateral.
- In a non-convex quadrilateral, one interior angle is more than 180° and one of the two diagonals lies outside of the quadrilateral.
- Quadrilaterals can be divided into two major groups: **parallelograms** and other quadrilaterals.
- Parallelograms are quadrilaterals with:
  1. both pairs of opposite sides equal in length and parallel
  2. opposite angles equal.

  A *rectangle*, a *square* and a *rhombus* (diamond) are all examples of parallelograms.

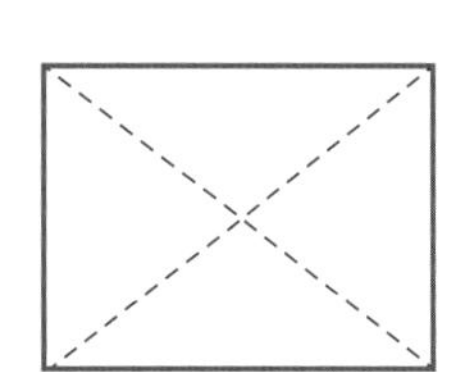
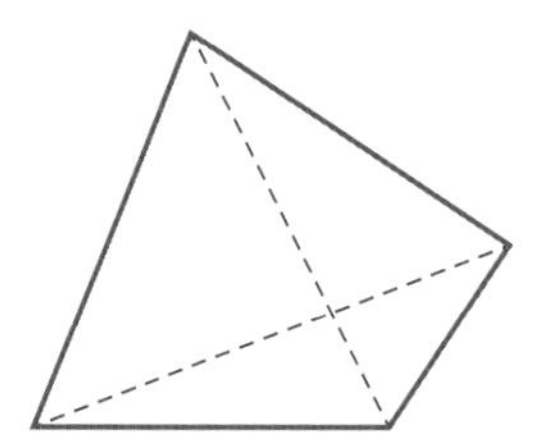

**Convex quadrilaterals**

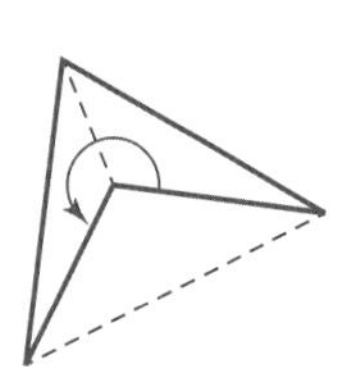
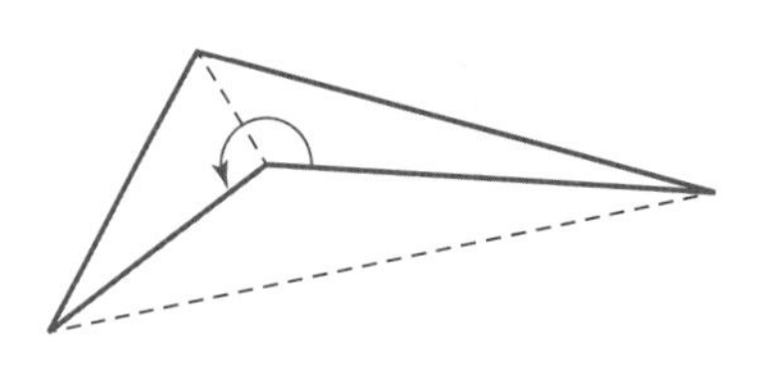

**Non-convex quadrilaterals**

| Parallelograms | Shape | Properties |
|---|---|---|
| Parallelogram | | Opposite sides are equal in length.<br>Opposite angles are equal in size. |
| Rectangle | | Opposite sides are equal in length.<br>Adjacent sides are perpendicular.<br>All angles are the same and equal 90°. |
| Rhombus | | All sides are equal in length.<br>Opposite angles are equal in size. |
| Square | | All sides are equal in length.<br>Adjacent sides are perpendicular.<br>All angles are the same and equal 90°. |

- Examples of other quadrilaterals (non-parallelograms) include *trapeziums*, *kites* and *irregular quadrilaterals*.

| Other quadrilaterals | Shape | Properties |
|---|---|---|
| Trapezium | | One pair of opposite sides is parallel. |
| Kite | | Two pairs of adjacent (next to each other) sides are equal in length.<br>One pair of opposite angles (the ones that are between the sides of unequal length) are equal. |
| Irregular quadrilateral | | This shape does not have any special properties. |

## WORKED EXAMPLE 15 Naming quadrilaterals

**Name the following quadrilaterals, giving reasons for your answers.**

a.

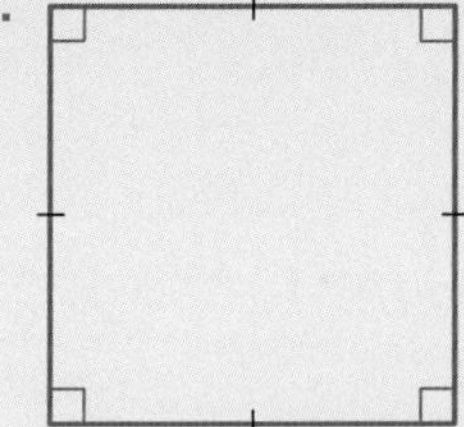

b.

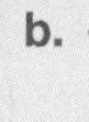

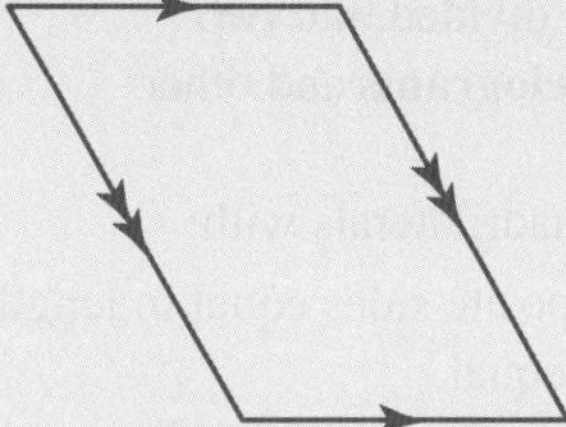

**THINK**

a. The markings on this quadrilateral indicate that all sides are equal in length and all angles equal 90°. Name the quadrilateral by finding the matching description in the table.

b. The arrows on the sides of this quadrilateral indicate that there are two pairs of parallel sides. Therefore it is a parallelogram. Check the descriptions in the table to see if it is a particular type of parallelogram.

**WRITE**

a. This quadrilateral is a square, since all sides are equal and all angles are 90°.

b. This quadrilateral is a parallelogram, since it has two pairs of parallel sides.

## 9.6.2 Angles in a quadrilateral

eles-4101

To investigate the sum of the angles in a quadrilateral, consider the following:

- All quadrilaterals have four internal angles.

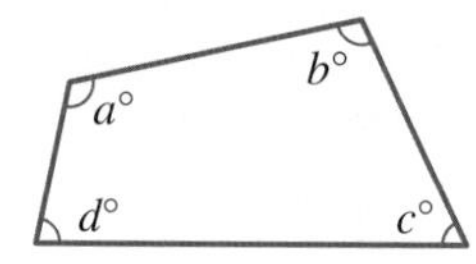


- Quadrilaterals are made up of two triangles.

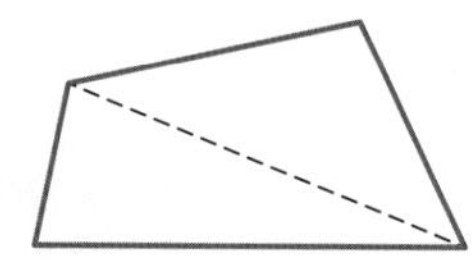

From earlier work, we know that the sum of angles in a triangle = 180°.

Therefore:

$$\begin{aligned} 2\text{ triangles} &= 2 \times 180° \\ &= 360° \end{aligned}$$

### Sum of angles in a quadrilateral

The four angles of a quadrilateral add up to 360°. That is, $a° + b° + c° + d° = 360°$

### WORKED EXAMPLE 16 Determining angles in a quadrilateral

**Calculate the value of the pronumeral in the diagram.**

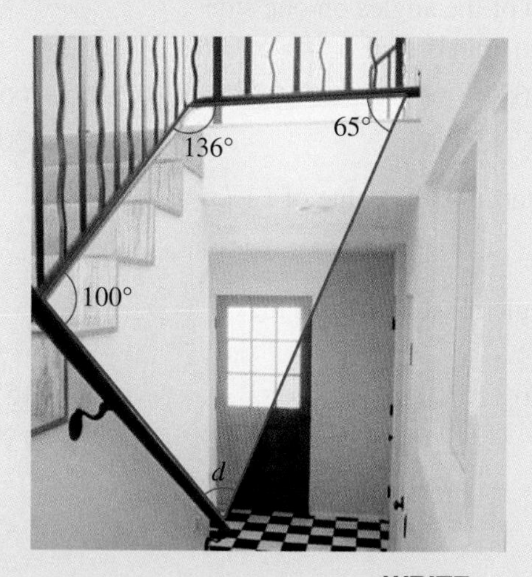


| THINK | WRITE |
|---|---|
| 1. The sum of the angles in a quadrilateral is 360°. Write this as an equation. | $d + 100° + 136° + 65° = 360°$ |
| 2. Simplify by adding 100°, 136° and 65°. | $d + 301° = 360°$ |
| 3. Subtract 301° from 360° to calculate the value of $d$. | $d = 360° - 301°$<br>$d = 59°$ |

WORKED EXAMPLE 17 Evaluating angles in a parallelogram

**Determine the value of the pronomerals in the photograph, giving a reason for your answer.**

| THINK | WRITE |
|---|---|
| 1. According to the markings, the opposite sides of the quadrilateral in the photograph are parallel and equal in length. Therefore, this quadrilateral is a parallelogram. In a parallelogram, opposite angles are equal. Write down the value of the pronumerals. | Opposite angles in a parallelogram are equal in size. Therefore, $j = 65°$ and $i = k$. |
| 2. The interior angles of a quadrilateral add up to 360°. Form an equation by writing the sum of the angles on one side and 360° on the other side of an equals sign. | $65 + j + i + k = 360°$ |
| 3. Replace $j$ in the equation with 65°. | $65 + 65 + i + k = 360°$ |
| 4. Simplify, by adding 65° and 65° together. | $130 + i + k = 360°$ |
| 5. Subtract 130° from 360° to determine the value of $i + k$. | $i + k = 360 - 130°$<br>$i + k = 230°$ |
| 6. We know that $i = k$, so each angle must be half of the remaining 230°. Divide 230° by 2. | $\frac{230°}{2} = 115°$ |
| 7. Angles $i$ and $k$ are both equal to 115°. | $i = 115°$ and $k = 115°$ |
| 8. Write the answer. | $i = 115°, j = 65°, k = 115°$ |

## WORKED EXAMPLE 18 Evaluating angles in a kite

**Calculate the value of the pronumerals in the diagram.**

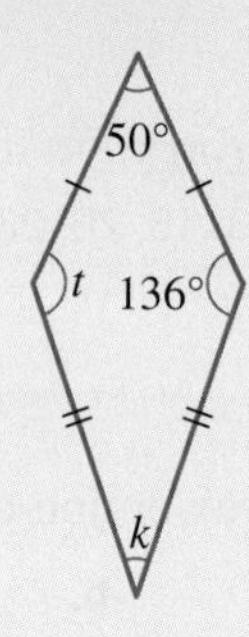


| THINK | WRITE |
|---|---|
| 1. Form an equation by writing the sum of the angles on one side, and 360° on the other side of an equals sign. | $k + t + 50° + 136° = 360°$ |
| 2. The quadrilateral shown in the diagram is a kite. Referring to the description of a kite in the table, angle $t$ and angle 136° are the angles between unequal sides and therefore must be equal in size. | The figure is a kite. $t = 136°$ |
| 3. Replace $t$ in the equation with 136°. | $k + 136° + 50° + 136° = 360°$ |
| 4. Simplify by adding 136°, 50° and 136° together. | $k + 322° = 360°$ |
| 5. Subtract 322° from 360° to calculate the value of $k$. | $k = 360° - 322°$ <br> $k = 38°$ |

## COLLABORATIVE TASK: Sketching quadrilaterals from a verbal description

Pair up with a classmate and each of you draw a quadrilateral without the other person seeing. Take turns describing your quadrilaterals using only verbal descriptions. Your partner must try to draw the quadrilateral you describe. How accurate is the drawing compared to the quadrilateral you described?

Resources

**eWorkbook** Topic 9 Workbook (worksheets, code puzzle and project) (ewbk-1910)

**Interactivities** Individual pathway interactivity: Quadrilaterals and their properties (int-4335)
Types of quadrilaterals (int-4025)
Angles in a quadrilateral (int-3967)

## Exercise 9.6 Quadrilaterals and their properties

learnon

9.6 Quick quiz on | 9.6 Exercise

### Individual pathways

| ■ PRACTISE | ■ CONSOLIDATE | ■ MASTER |
|---|---|---|
| 1, 3, 6, 11, 12, 15, 18, 19, 22 | 2, 4, 7, 10, 13, 16, 20, 23 | 5, 8, 9, 14, 17, 21, 24 |

### Fluency

1. State whether the quadrilaterals shown are convex or non-convex.

a. 

b. 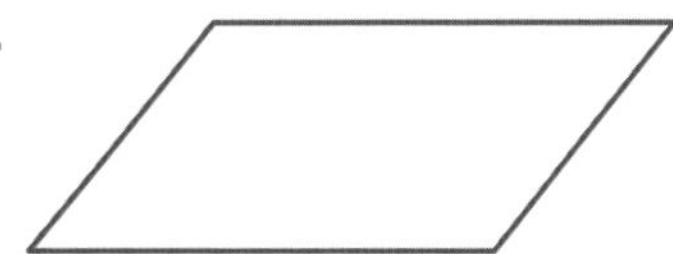

c. 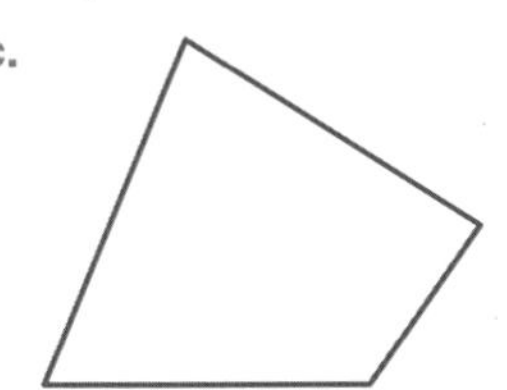

d. 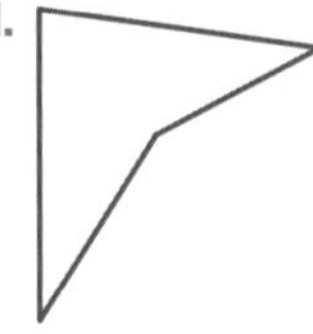

2. **WE15** Name the following quadrilaterals, giving reasons for your answers.

a. i. 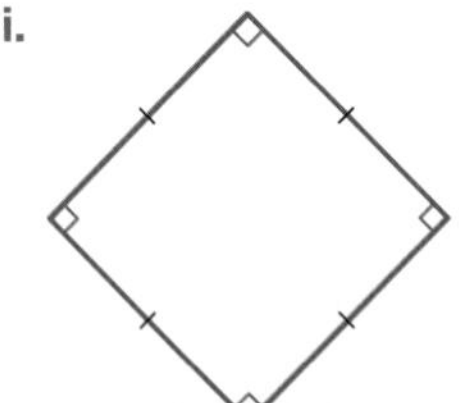

ii. 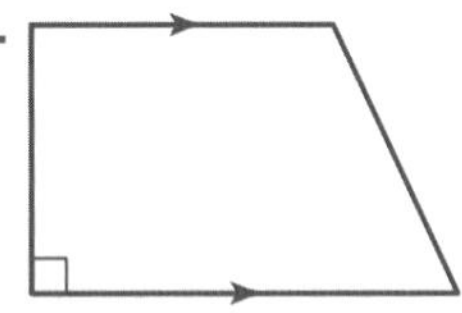

iii. 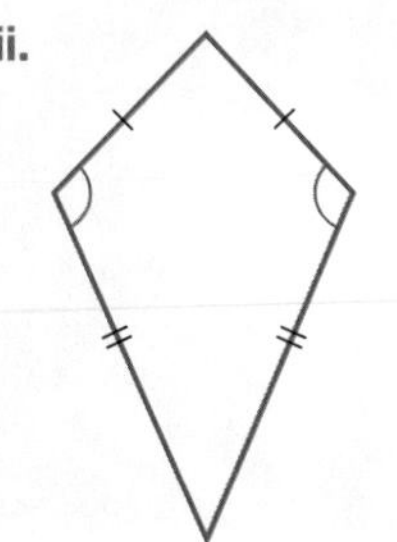

iv. 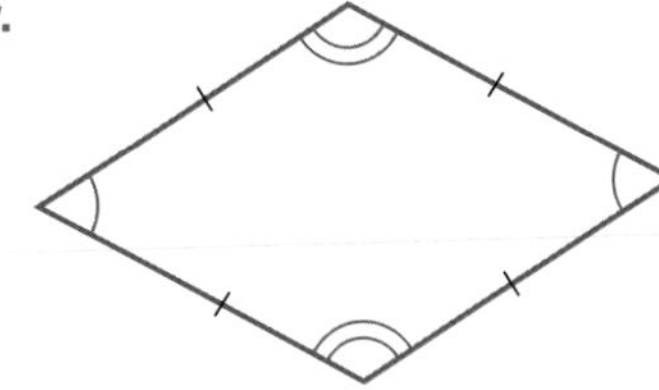

v. 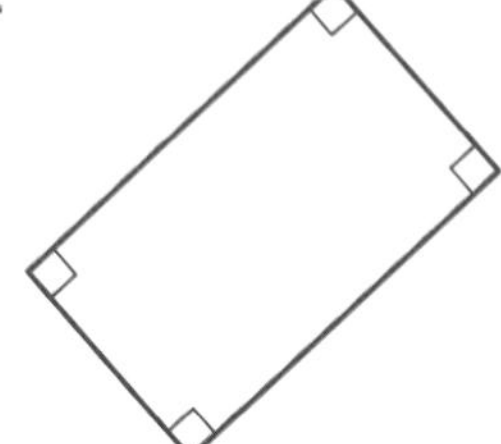

vi. 

**b.** Name each of the quadrilaterals in the photographs, giving reasons for your answers.

**i.**

**ii.**

**iii.**

**iv.**

**3.** MC Identify the quadrilateral shown.

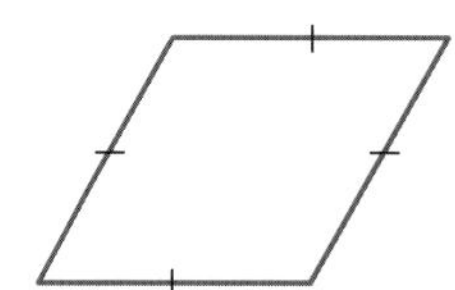

**A.** Square
**B.** Rectangle
**C.** Kite
**D.** Parallelogram
**E.** Rhombus

**4.** MC Identify the quadrilateral shown.

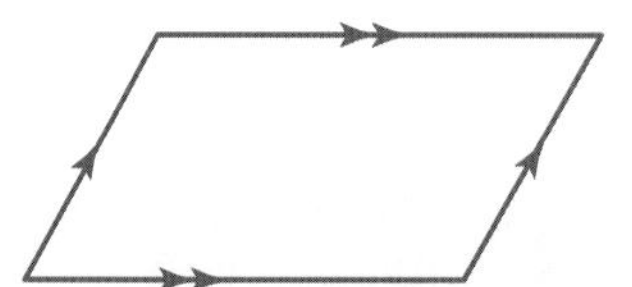

**A.** Trapezium
**B.** Square
**C.** Kite
**D.** Rhombus
**E.** Parallelogram

**5.** MC Identify the quadrilateral shown.

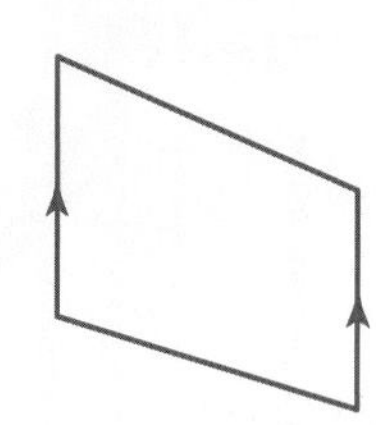

**A.** Trapezium
**B.** Parallelogram
**C.** Rhombus
**D.** Kite
**E.** Square

6. WE16 Calculate the value of the pronumeral in each of the following diagrams.

a.

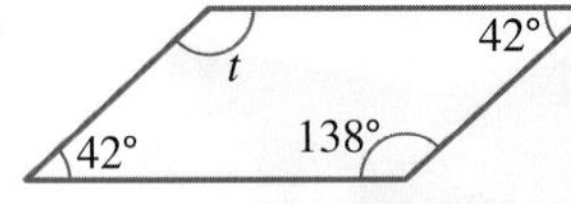


b.

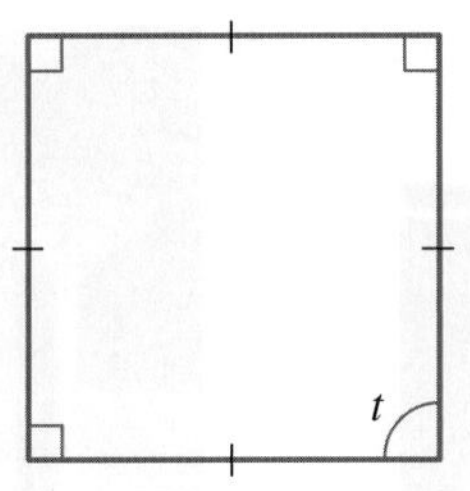


c.

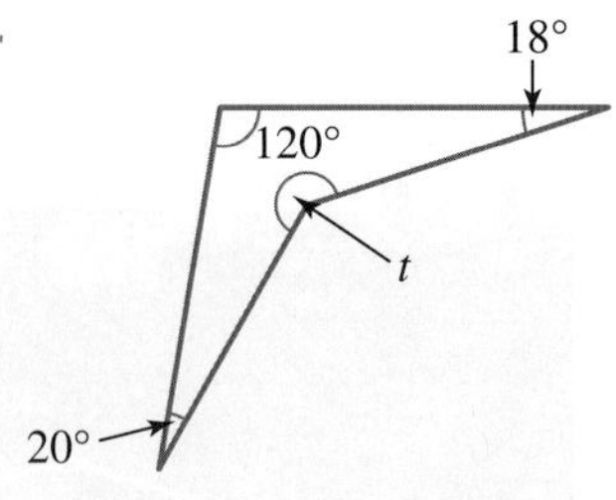


d.

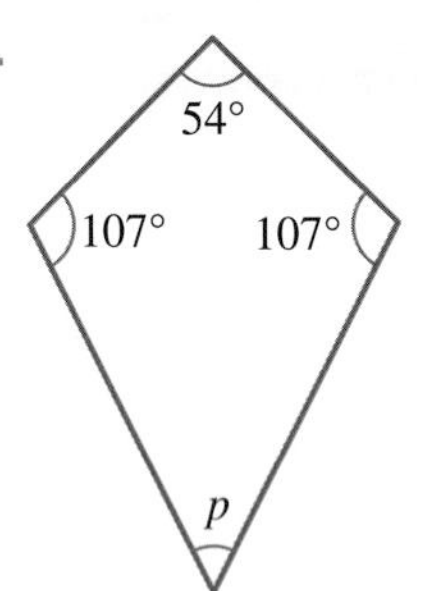


e.

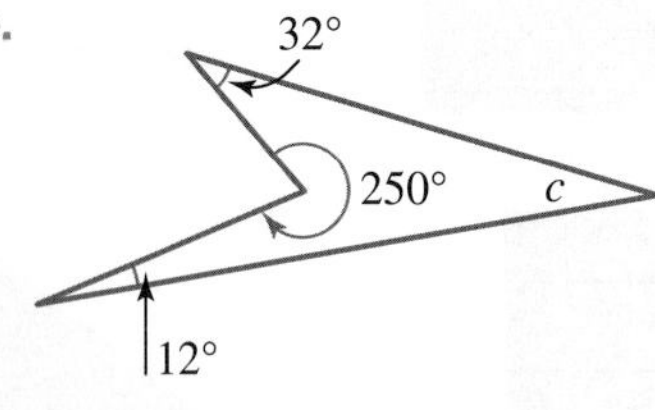


f.

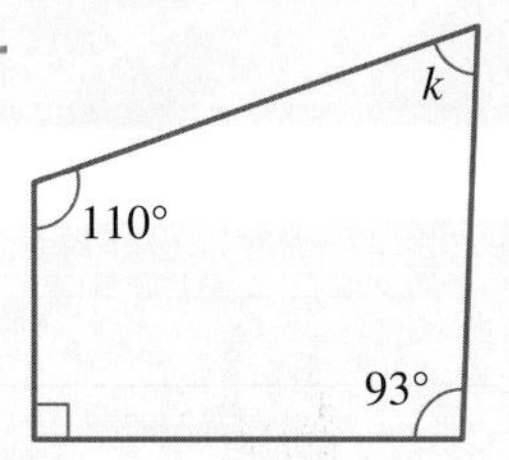


7. WE17 Calculate the value of the pronumeral in each of the following diagrams, giving reasons for your answers.

a.

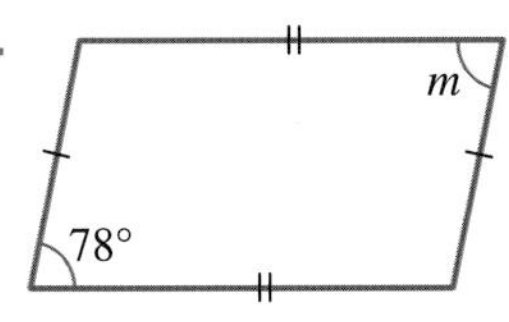


b.

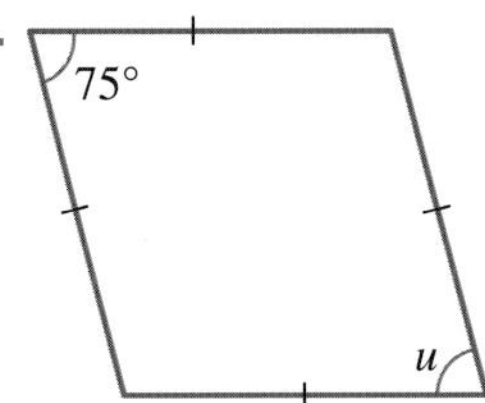


c.

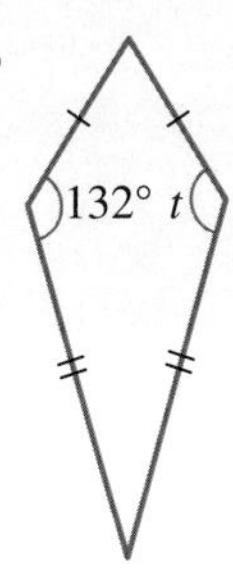


d.

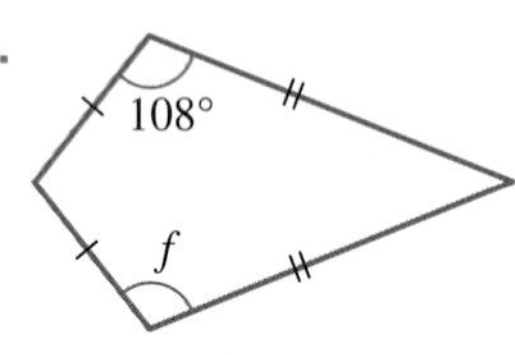


e.

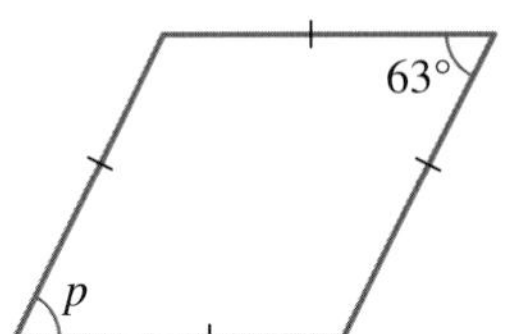


f.

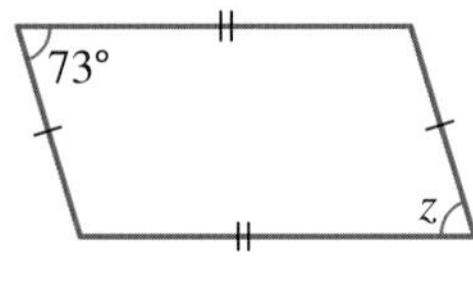


8. Use the following photographs to determine the values of the unknown angles.

a.

b.

9. **WE18** Calculate the value of the pronumerals in each of the following diagrams.

a.

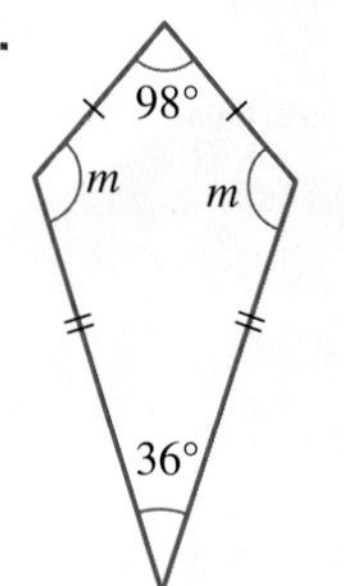


b.

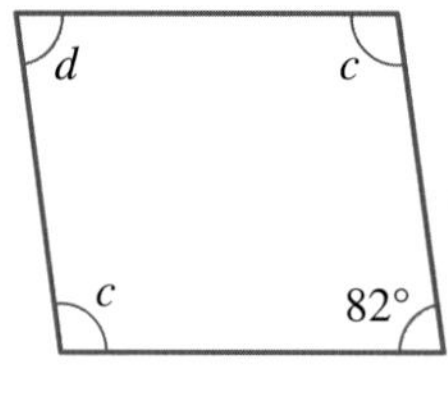


c.

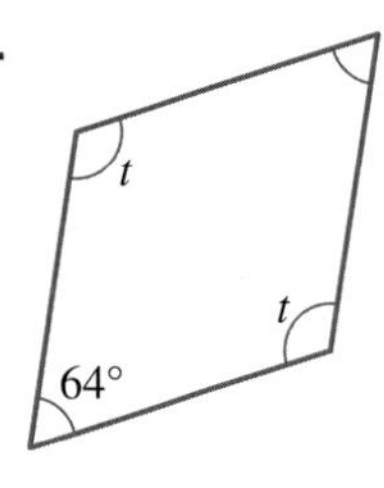


d.

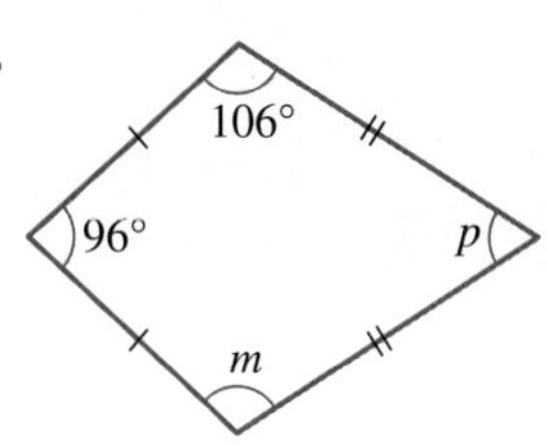


e.

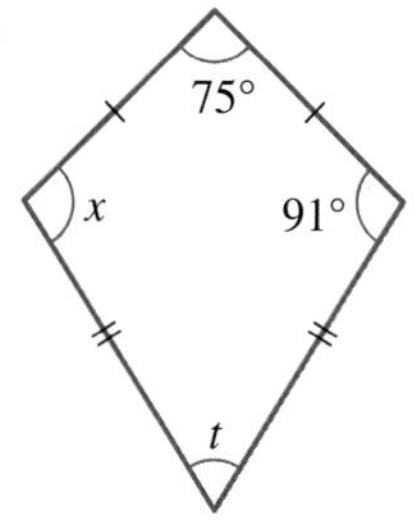


f.

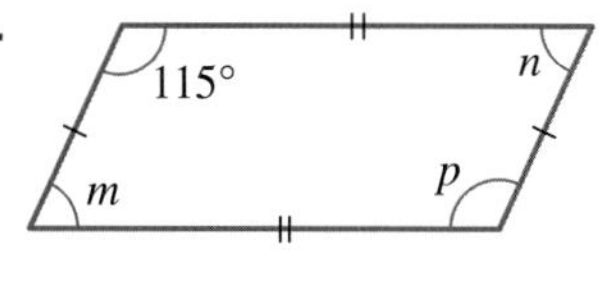


10. **MC** Identify the value of $t$ in the diagram.

A. 360°
B. 112°
C. 222°
D. 138°
E. 180°

11. **MC** Identify the value of $r$ in the diagram.

A. 117°
B. 180°
C. 234°
D. 126°
E. 63°

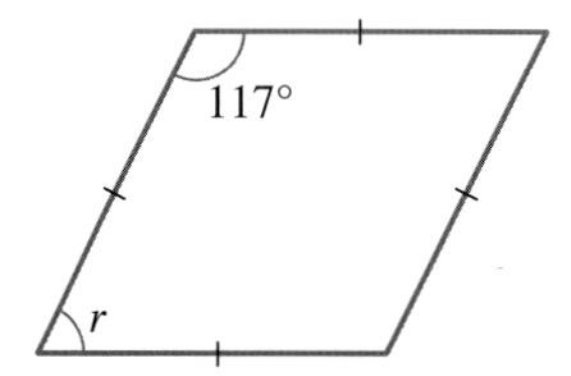


## Understanding

12. State whether each of the following statements is true or false.
    a. All squares are rectangles.
    b. All squares are rhombuses.
    c. All rectangles are squares.
    d. Any rhombus with at least one right angle is a square.
    e. A rectangle is a parallelogram with at least one angle equal to 90°.

13. **MC** A rectangle is a quadrilateral because:

A. it has 4 right angles
B. it has 2 pairs of parallel sides
C. its opposite sides are equal in length
D. it has 4 straight sides
E. it has 2 pairs of parallel sides and 4 right angles.

14. The photograph shows the roof of a gazebo. Calculate the value of $p$.

15. Determine the size of the obtuse angle in the following photograph of a kite.

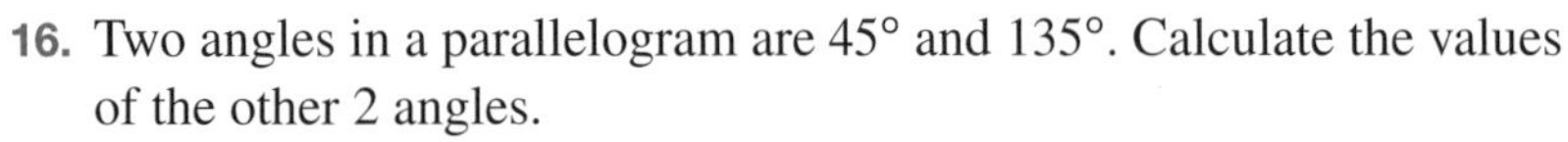

16. Two angles in a parallelogram are 45° and 135°. Calculate the values of the other 2 angles.

17. Tom measures 2 angles of a kite at 60° and 110°, but forgets which angle is which. Draw three different kites that Tom may have measured, showing the size of all angles in each diagram.

18. For each of the following sets of four angles, decide whether it is possible to construct a quadrilateral. Explain your answer.

a. $25°, 95°, 140°, 100°$ b. $40°, 80°, 99°, 51°$

## Reasoning

19. Three angles of a quadrilateral are 60°, 70° and 100°.
   a. Calculate the size of the fourth angle of this quadrilateral.
   b. Determine how many quadrilaterals with this set of angles are possible.
   c. Construct a quadrilateral with the given angle sizes. The choice of the length of the sides is up to you.

20. Val and Peter want to replace their front gate with another of the same design. To have this gate made, they need to supply a diagram of it with all measurements and angles shown. Study the photograph of Val and Peter's gate and use it to help you answer the following questions.

   a. There are 4 different shapes formed by the metal bars of the gate. How many different types of triangles are there? Identify them.
   b. State how many types of quadrilaterals there are. Identify them.
   c. Draw a diagram of the gate showing the length measurements and the one angle that is given.
   d. Use this angle to calculate all the remaining angles in the diagram.

21. a. Consider the diagram. Prove that the sum of all angles in a quadrilateral is 360°.

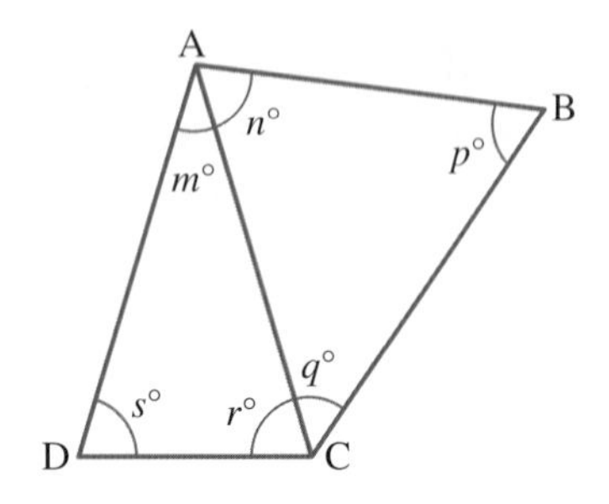


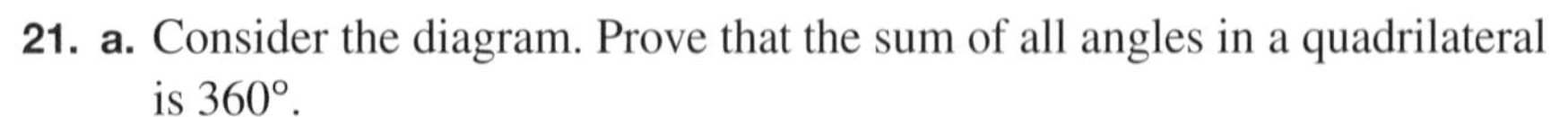

   b. Using a similar method, determine the sum of all angles in the following shapes. *Hint*: Draw the shape first.
      i. Pentagon
      ii. Hexagon
      iii. Octagon
      iv. Decagon

### Problem solving

22. Determine the values of the pronumerals in the following shapes.

a. 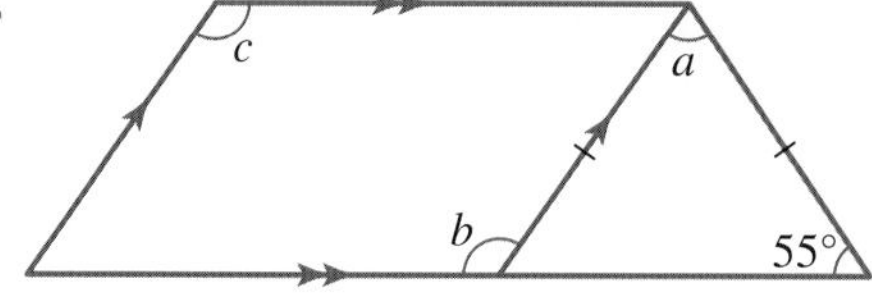


b. 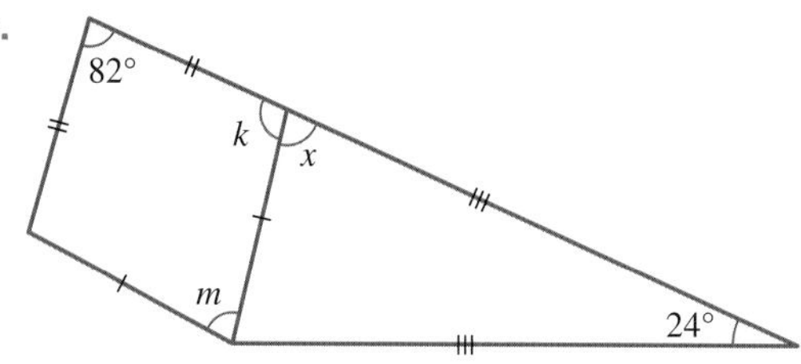


23. Calculate the values of the pronumerals in each of the following quadrilaterals.

a. 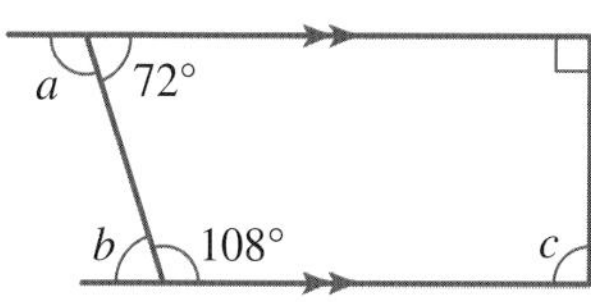


b. 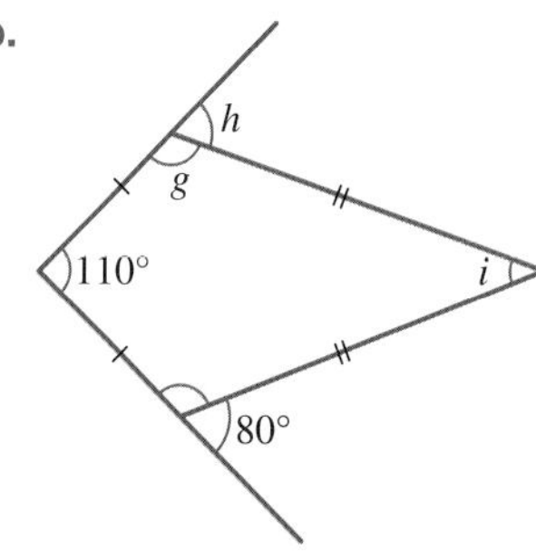


c. 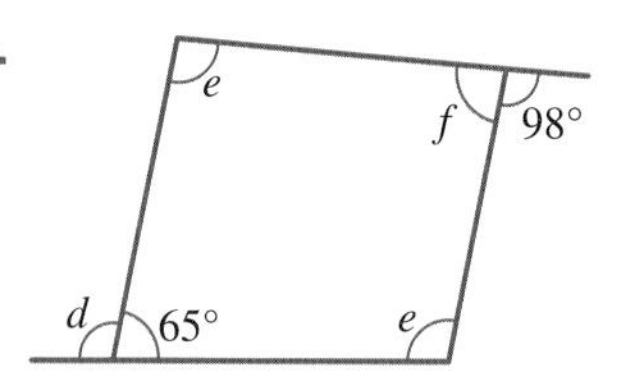


d. 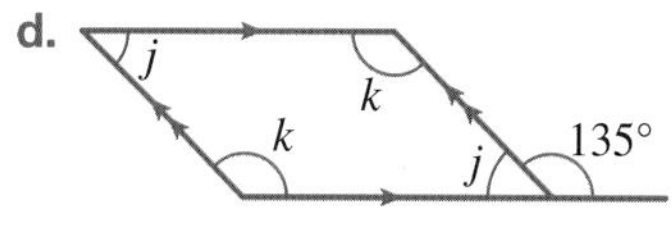


24. Draw a quadrilateral PQRS. Label angle $P$ as 100° and angle $R$ as a right angle. Angle $Q$ is two-thirds of angle $S$. Determine the measurement of angle $S$.

# LESSON
# 9.7 Parallel and perpendicular lines

### LEARNING INTENTION

At the end of this lesson you should be able to:

- identify and determine the size of vertically opposite angles
- identify and determine the size of angles formed when a transversal intersects parallel lines.

## 9.7.1 Vertically opposite and adjacent angles

eles-4102

- When two straight lines intersect they form four angles with a common vertex.
- The angles that are opposite each other are called **vertically opposite angles**.
- Vertically opposite angles are equal in size.
- In this diagram there are two pairs of vertically opposite angles: $\angle AOB = \angle COD$ and $\angle BOC = \angle AOD$.
- **Adjacent angles** share a common arm and a common vertex. In this diagram $\angle AOB$ is adjacent to $\angle BOC$.

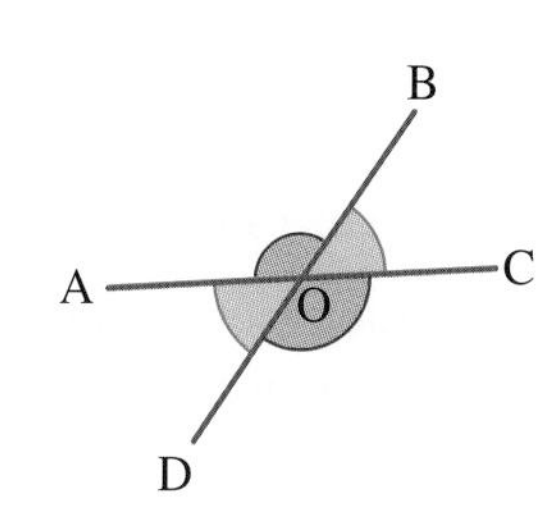

- **Complementary angles** are adjacent angles that add up to 90°.
  In this diagram, $x + y = 90°$.

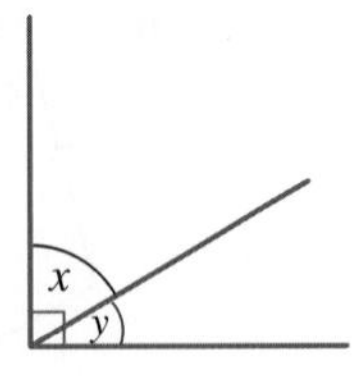


- **Supplementary angles** are adjacent angles that add up to 180°.
  In this diagram, $x + y = 180°$.

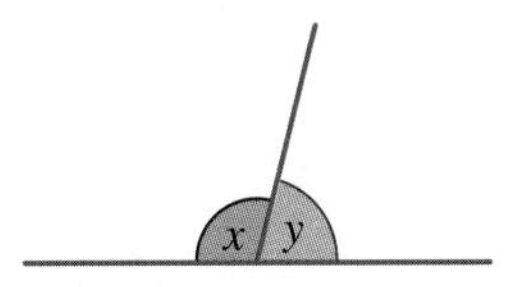


### WORKED EXAMPLE 19 Calculating a complementary angle

**Calculate the value of *x* in the diagram.**

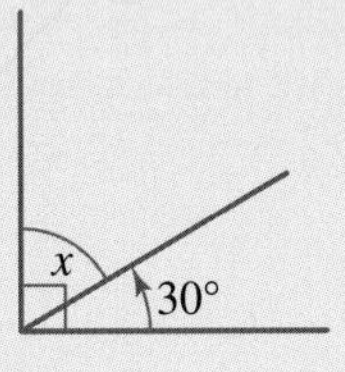


| THINK | WRITE |
|---|---|
| 1. These angles are supplementary because they add up to 90°. | $x + 30° = 90°$ |
| 2. Subtract 30° from 90° to calculate the value of *x*. | $x = 90° - 30°$<br>$x = 60°$ |
| 3. Write the answer. | The value of $x$ is 60°. |

### WORKED EXAMPLE 20 Calculating a supplementary angle

**Calculate the value of *x* in the diagram.**

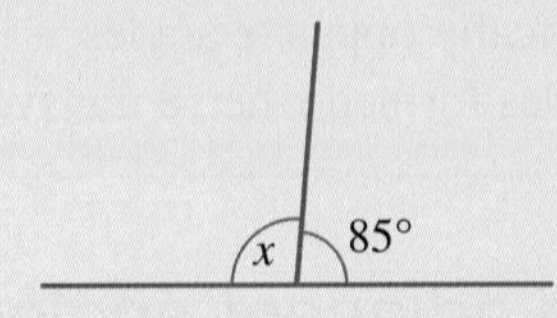


| THINK | WRITE |
|---|---|
| 1. These angles are supplementary because they add up to 180°. | $x + 85° = 180°$ |
| 2. Subtract the known angle from 180° to calculate the value of *x*. | $x = 180° - 85°$<br>$x = 95°$ |
| 3. Write the answer. | The value of $x$ is 95°. |

## WORKED EXAMPLE 21 Determining vertically opposite angles

**Determine the value of the pronumerals in the diagram.**

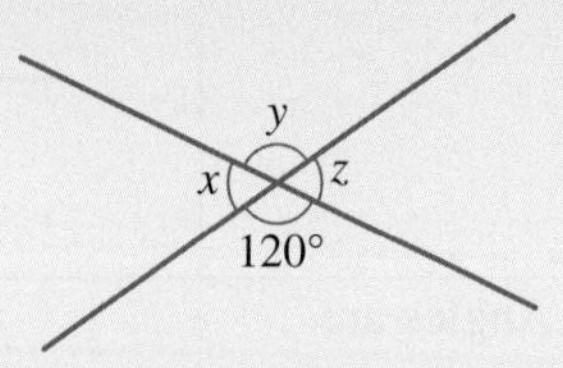


| THINK | WRITE |
|---|---|
| 1. The angle $y$ and 120° are vertically opposite, so they are equal. | $y = 120°$ |
| 2. The angle $z$ and the angle marked 120° form a straight line, so they are supplementary angles (that is, they add to 180°). Subtract 120° from 180° to find the value of $z$. | $z = 180° - 120°$<br>$= 60°$ |
| 3. The angles $x$ and $z$ are vertically opposite, so they are equal. | $x = 60°$ |

## 9.7.2 Parallel and perpendicular lines

eles-4103

- **Parallel lines** are two or more lines that are simple translations of each other.
- The distance between parallel lines is the same across their entire lengths. They never intersect each other. The two rails of the train tracks in this picture are parallel to each other.
- To write 'the line segments AB and CD are parallel' using mathematical symbols, simply write AB || CD.
- The symbol || means 'is parallel to'.
- The term 'transverse' means 'crossways'. A line that intersects with a pair of parallel lines is called a **transversal**.
- When a transversal cuts a set of parallel lines, a number of angles are created. The following table shows how pairs of angles are related.

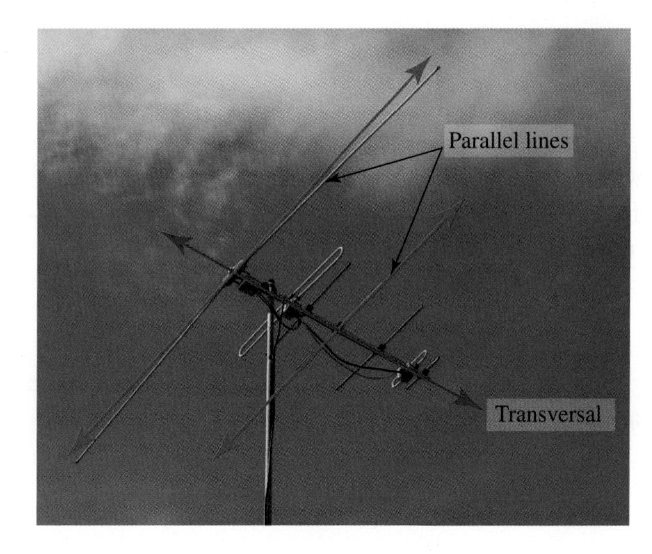

| Angle identification | Relationship | Example |
|---|---|---|
| **Corresponding angles** are positioned on the same side of the transversal and are either both above or both below the parallel lines. Think of them as F-shaped. | $\angle a = \angle b$ | |
| **Co-interior angles** are positioned 'inside' the parallel lines, on the same side of the transversal. Think of them as C-shaped. Co-interior angles are also called **allied angles**. | Angles are supplementary: $\angle a + \angle b = 180°$ | |
| **Alternate angles** are positioned 'inside' the parallel lines on alternate sides of the transversal. think of them as Z-shaped. | $\angle a = \angle b$ | |
| **Vertically opposite angles** are created when two lines intersect. The angles opposite each other are equal in size. Think of them as X-shaped. | $\angle a = \angle a$<br>$\angle b = \angle b$ | |

## 9.7.3 Calculating angles associated with parallel lines

eles-4104

- Angle relationships can be used to calculate the size of missing angles, as shown in the following worked examples.
- Remember that the Z, F or C shapes may be upside-down or backwards in some diagrams.

### WORKED EXAMPLE 22 Calculating angles

**For the diagram shown:**
**a. state the type of angle relationship**
**b. calculate the value of the pronumeral.**

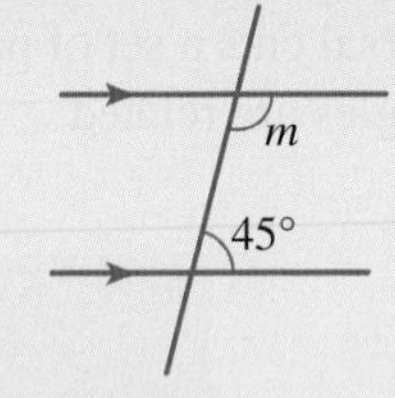


**THINK**

a. 1. Study the diagram: which shape, Z, F or C, would include both angles that are shown? Copy the diagram into your workbook and highlight the appropriate shape.

2. Write the answer by stating the name of the angles suggested by a C shape.

**WRITE**

a. $m$, 45°

The highlighted angles are co-interior.

**b. 1.** Co-interior angles add up to 180°. Write this as an equation.

**b.** $m + 45° = 180°$

**2.** Solve for $m$ by subtracting 45° from 180°.

$m = 180° - 45°$
$m = 135°$

**3.** Write the answer.

The value of $m$ is 135°

## WORKED EXAMPLE 23 Calculating angles

**Calculate the value of the pronumeral in the diagram shown, giving reasons.**

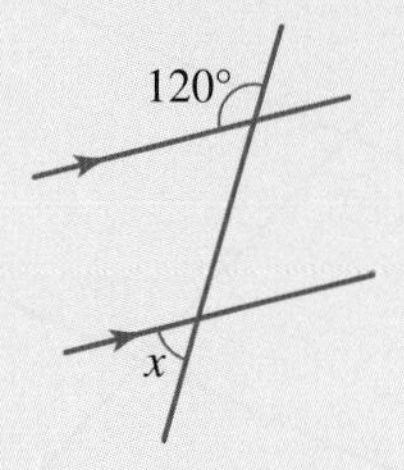


**THINK**

**WRITE**

**1.** The two angles shown are neither C, Z nor F angles. So we must find some other angle first that will enable us to determine the size of angle $x$.

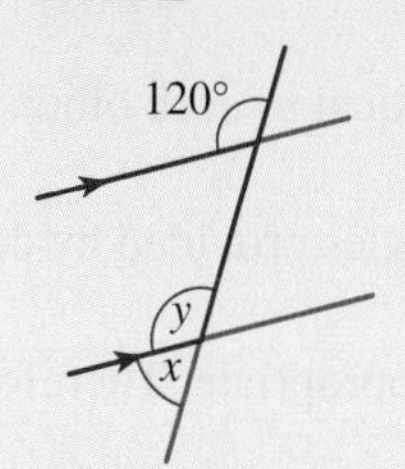


**2.** This other angle must be in a certain relationship with *both* given angles. Draw an F shape that includes a 120° angle.

**3.** The other angle in this F shape is related to both given angles: it corresponds to the 120° angle and is supplementary to angle $x$. Call this angle $y$.

**4.** State the size of angle $y$, specifying the reason (the angle relation).

$y = 120°$ (because corresponding angles are equal)

**5.** The angles $x$ and $y$ are supplementary (add to 180°) as they lie on a straight line. State this as an equation.

$x + y = 180°$ (because the angle sum of a straight line is 180°)

**6.** Substitute the value of $y$ into the equation.

$x + 120° = 180°$

**7.** Solve for $x$ by subtracting 120° from 180°

$x = 180° - 120°$
$x = 60°$

**8.** Write the answer.

The value of $x$ is 60°.

## 9.7.4 Testing for parallel lines

eles-4105

- Two lines are parallel if:
  - corresponding angles are equal
  - alternate angles are equal
  - co-interior angles are supplementary.

**WORKED EXAMPLE 24 Determining parallel lines**

**Are the following lines A and B parallel? Explain your answer.**

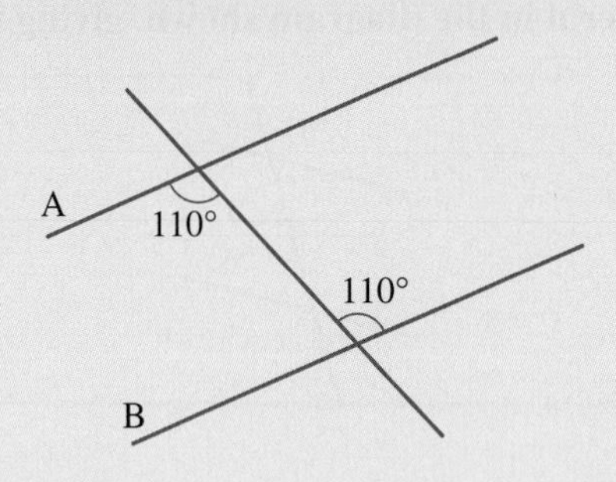


**THINK**

1. The given angles are equal to each other. They are both 110°.
2. Identify the type of angles provided by considering the Z, F or C shapes.
   The Z shape is most appropriate. Therefore the angles are alternate.
3. Since the two angles are alternate and equal, then the lines A and B must be parallel.

**WRITE**

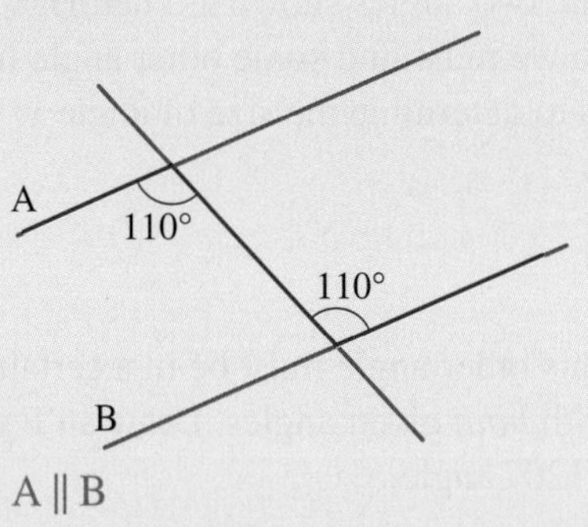


A || B

### Perpendicular lines

eles-4106

- **Perpendicular** means 'at right angles'.
- In this diagram, the angles ∠AOX and ∠XOB are both right angles, which means the lines AB and OX are perpendicular.
- To write 'the line segments AB and OX are perpendicular' using mathematical symbols, simply write AB⊥ OX.
- The symbol ⊥ means 'is perpendicular to'.

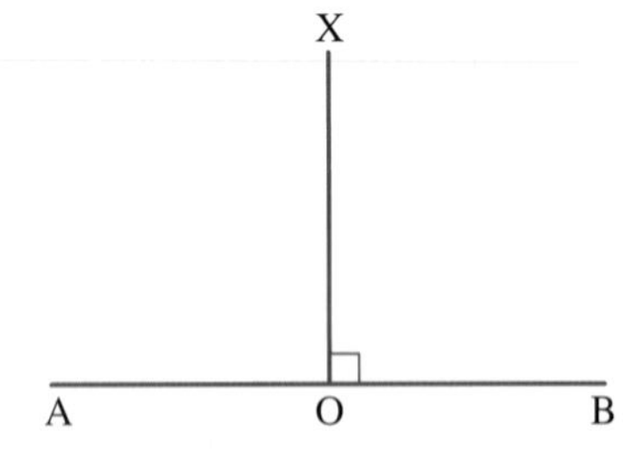

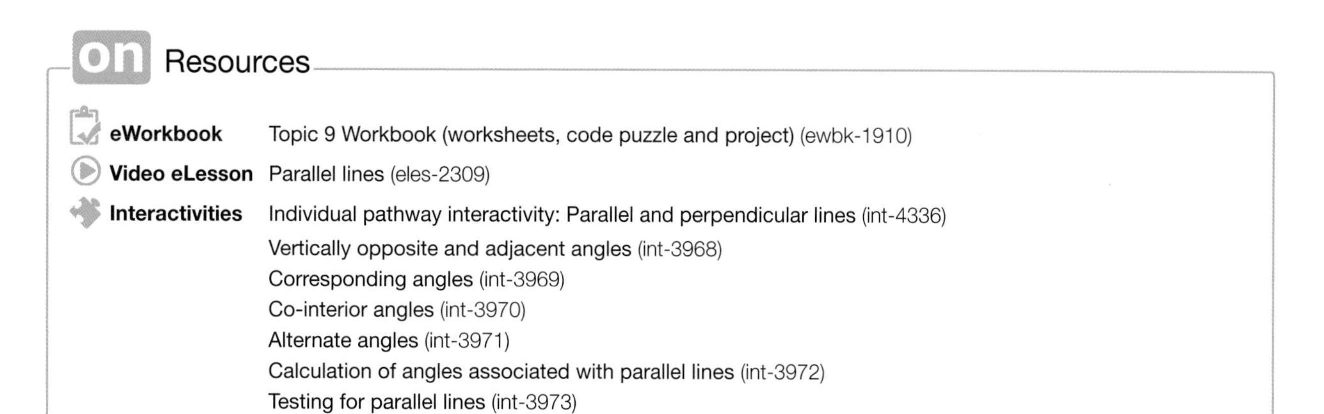

**on** Resources

**eWorkbook** Topic 9 Workbook (worksheets, code puzzle and project) (ewbk-1910)

**Video eLesson** Parallel lines (eles-2309)

**Interactivities** Individual pathway interactivity: Parallel and perpendicular lines (int-4336)
Vertically opposite and adjacent angles (int-3968)
Corresponding angles (int-3969)
Co-interior angles (int-3970)
Alternate angles (int-3971)
Calculation of angles associated with parallel lines (int-3972)
Testing for parallel lines (int-3973)

## Exercise 9.7 Parallel and perpendicular lines

learn**on**

| 9.7 Quick quiz **on** | 9.7 Exercise |
| --- | --- |

### Individual pathways

| ■ PRACTISE | ■ CONSOLIDATE | ■ MASTER |
| --- | --- | --- |
| 1, 4, 7, 8, 11, 15, 18, 20, 23 | 2, 5, 9, 12, 14, 17, 21, 24 | 3, 6, 10, 13, 16, 19, 22, 25 |

### Fluency

1. **WE19** Calculate the value of the pronumeral in the following diagram.

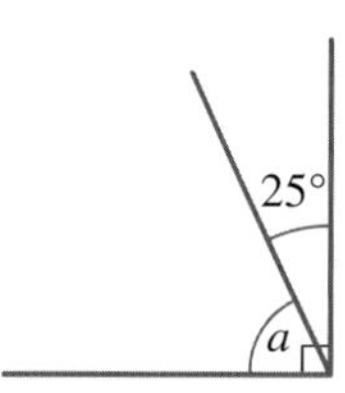


2. Calculate the value of the pronumeral in each of the following diagrams.

a. 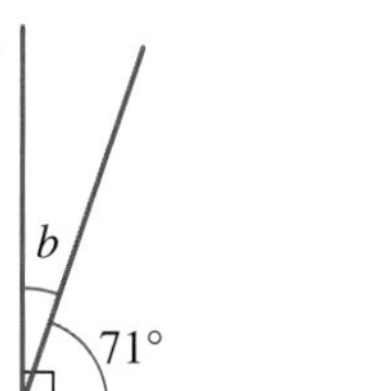


b. 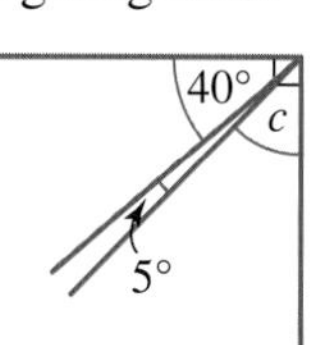


3. **WE20** Calculate the size of $\angle$ BOC.

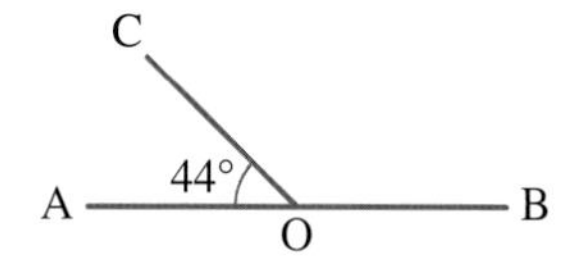

4. Calculate the value of the pronumeral in each of the following diagrams.

a.

b.

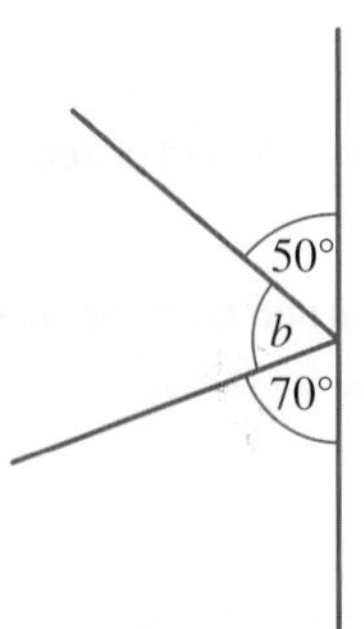


c.

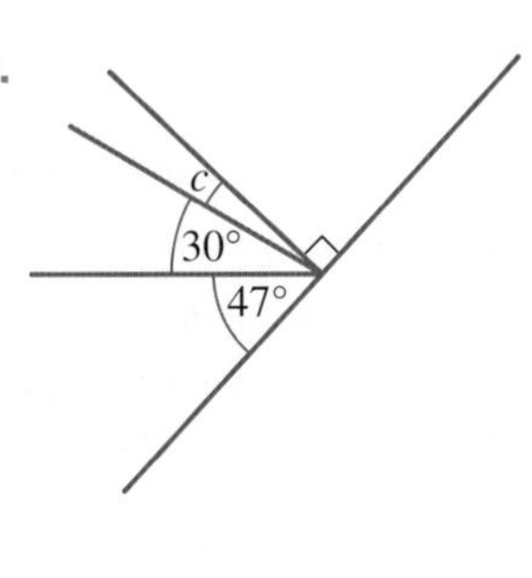


5. **WE21** Calculate the value of the pronumerals in each of the following diagrams.

a.

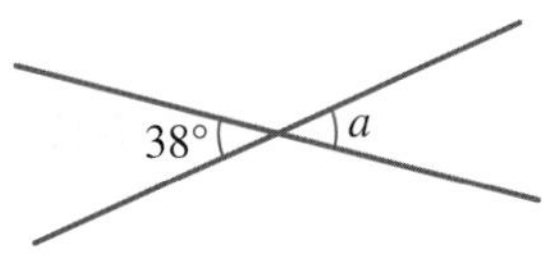


b.

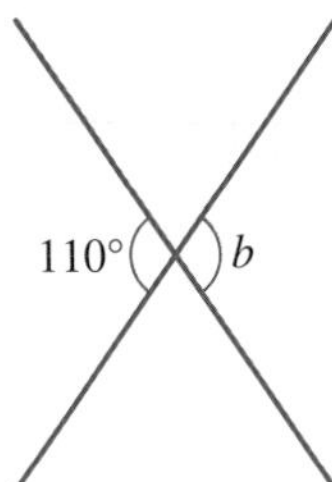


c.

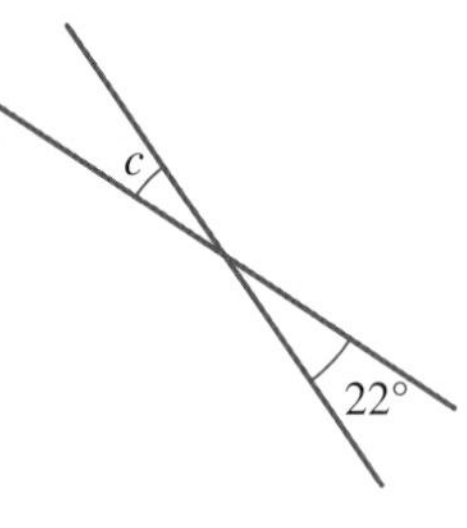


6. Determine the value of the pronumerals in each of the following diagrams.

a.

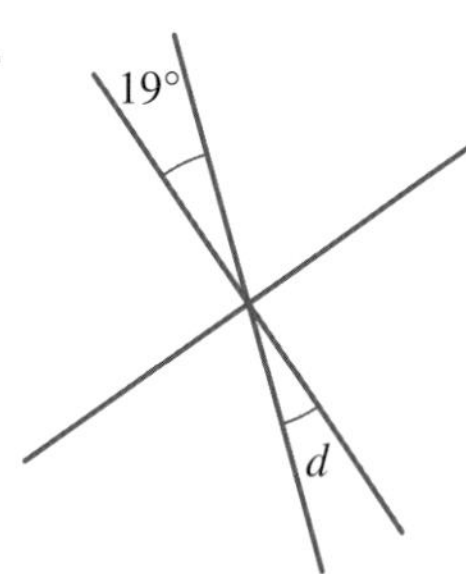


b.

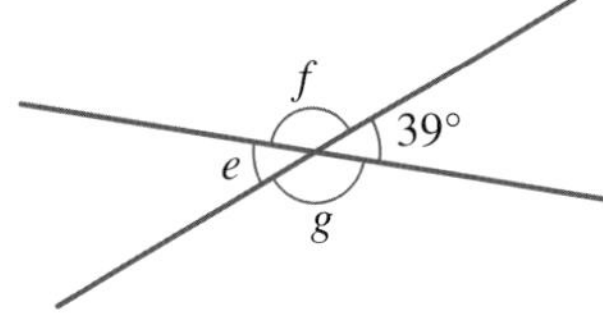


c.

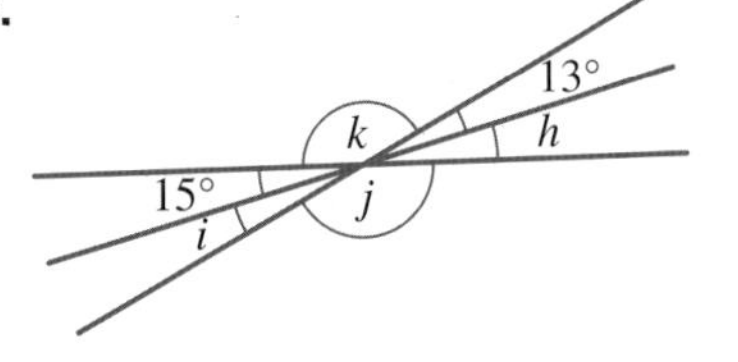


7. **MC** In the diagram:

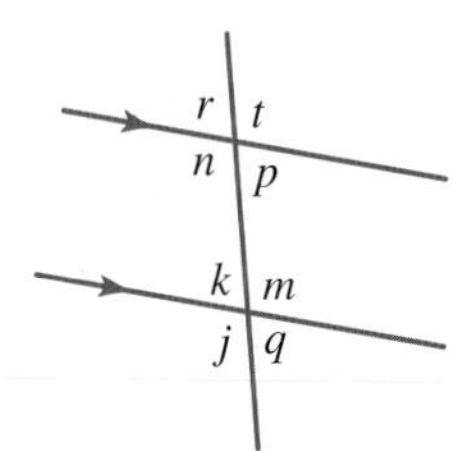


a. which angle is vertically opposite to angle $p$?

A. $k$ B. $m$ C. $r$ D. $q$ E. $t$

b. which angle is corresponding to angle $p$?

A. $k$ B. $m$ C. $r$ D. $q$ E. $t$

c. which angle is co-interior to angle $p$?

A. $k$ B. $m$ C. $r$ D. $q$ E. $n$

d. which angle is alternate to angle $p$?

A. $k$ B. $m$ C. $r$ D. $q$ E. $n$

8. **WE22** For each of the following diagrams:

a. state the type of angle relationship

b. determine the value of the pronumeral.

i.
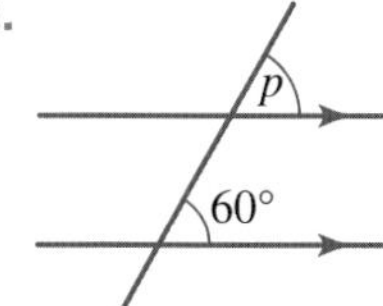


ii.
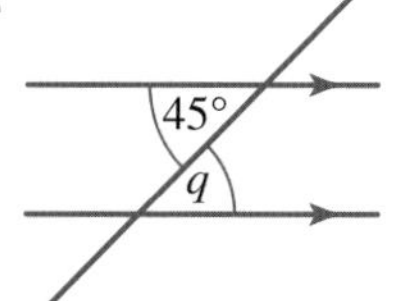


iii.
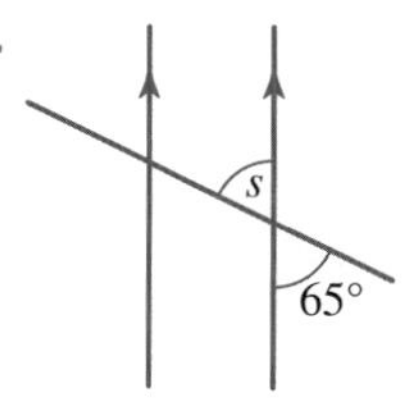


iv.
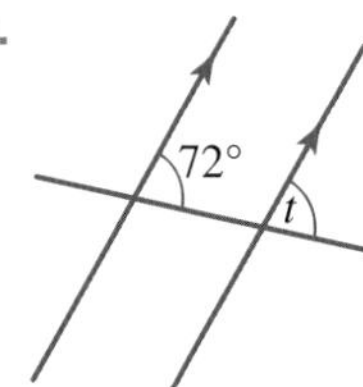


v.
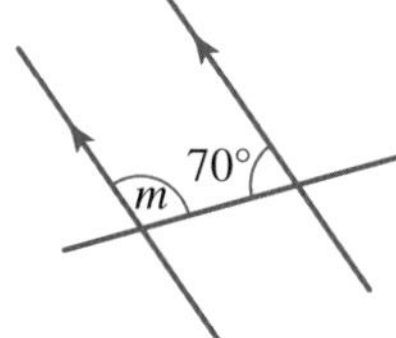


vi.
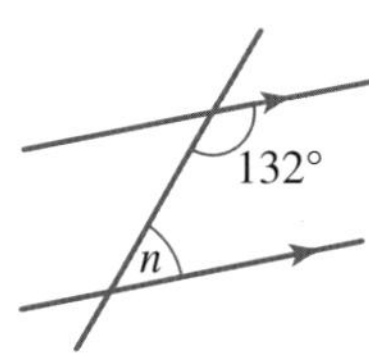


9. **WE23** Calculate the value of the pronumeral in each of the following diagrams. Give reasons for your answers.

a.
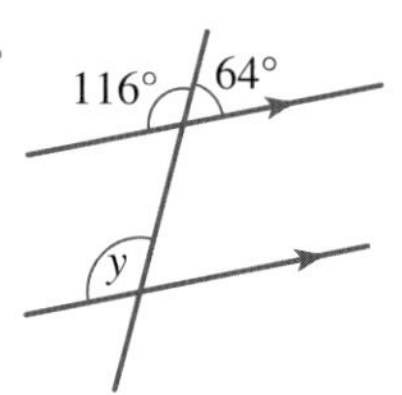


b.
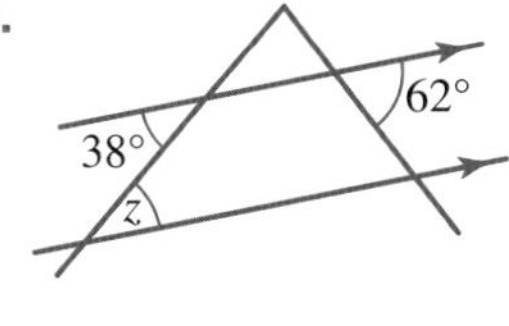


c.
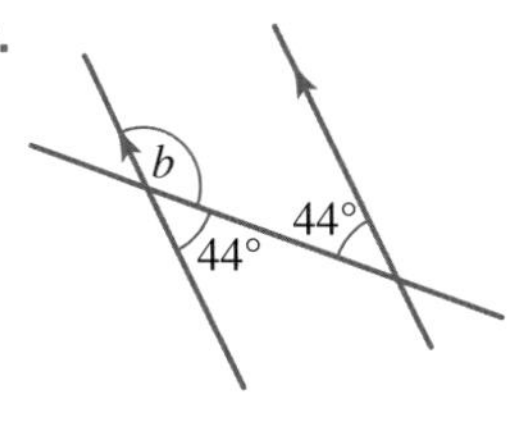


10. Calculate the value of the pronumerals in each of the following diagrams. Give reasons for your answers.

a.
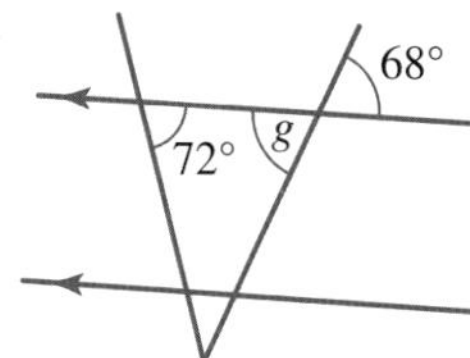


b.
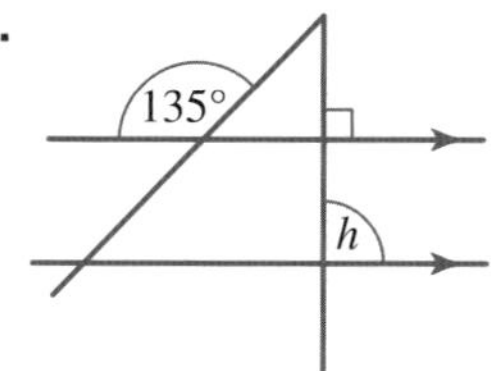


## Understanding

11. Identify which of the following pairs of angles are complementary.
   a. 32° and 60°
   b. 70° and 30°
   c. 45° and 45°
   d. 28° and 72°
   e. 30° and 150°
   f. 42° and 48°

12. Identify which of the following pairs of angles are supplementary.
   a. 30° and 60°
   b. 70° and 110°
   c. 145° and 45°
   d. 118° and 72°
   e. 35° and 145°
   f. 42° and 138°

13. a. Determine the complement of 65°.
   b. Determine the supplement of 123°.

14. a. If the angle allied to $x$ is 135°, calculate the size of angle $x$.
   b. If the angle corresponding to $y$ is 55°, calculate the size of angle $y$.

15. Copy and complete this table.

| Diagram | Type of angle relation | Associated shape | Rule |
|---|---|---|---|
| | Corresponding | | Are equal in size |
| | Alternate | Z | |
| | | | Add up to 180° |

16. Match each diagram with the appropriate angle name from the four options listed.

a. 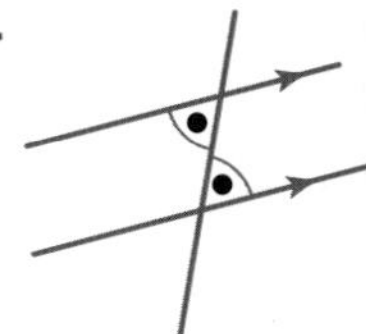

b. 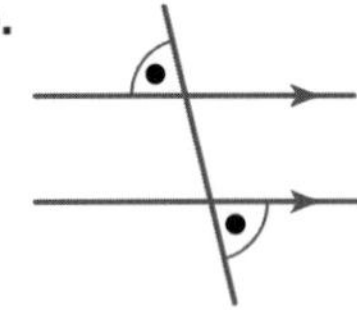

c. 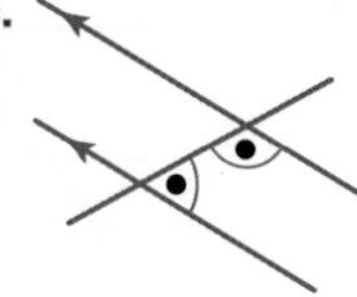

d. 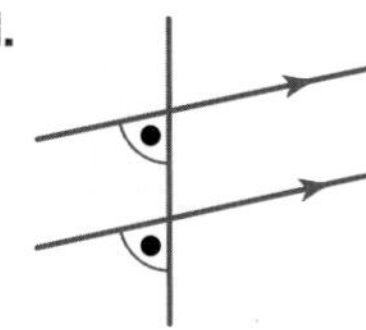

A. Co-interior angles (C)
B. Corresponding angles (F)
C. Alternate angles (Z)
D. None of the above

17. Look at the diagram. List all pairs of:

a. vertically opposite angles
b. corresponding angles
c. co-interior angles
d. alternate angles.

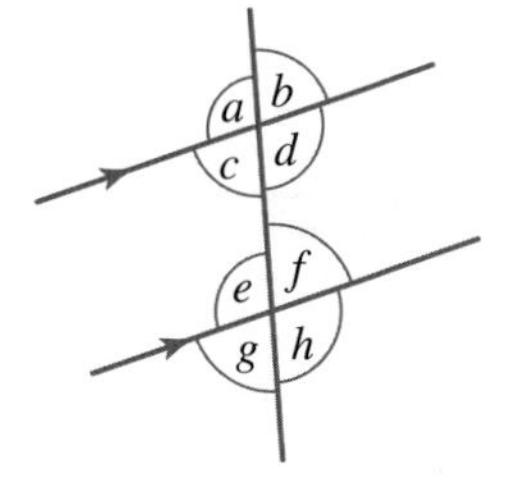


18. Calculate the value of the pronumerals in the following diagrams.

a. 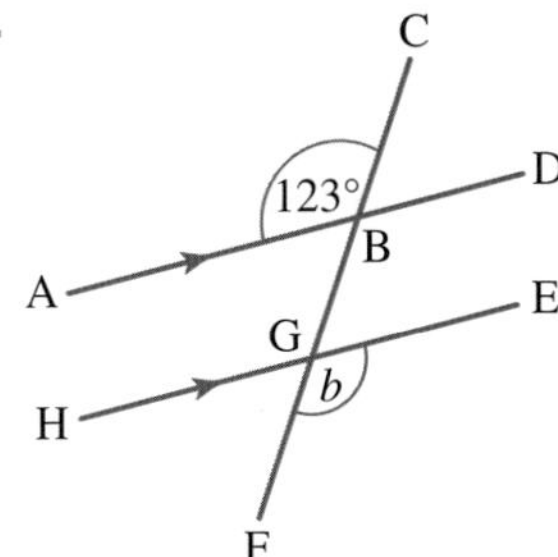


b. 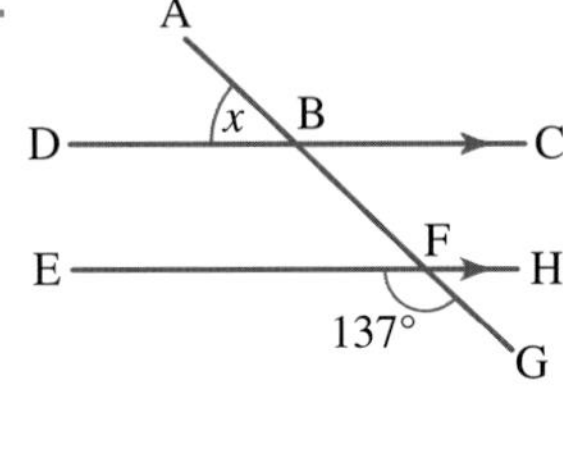


c. 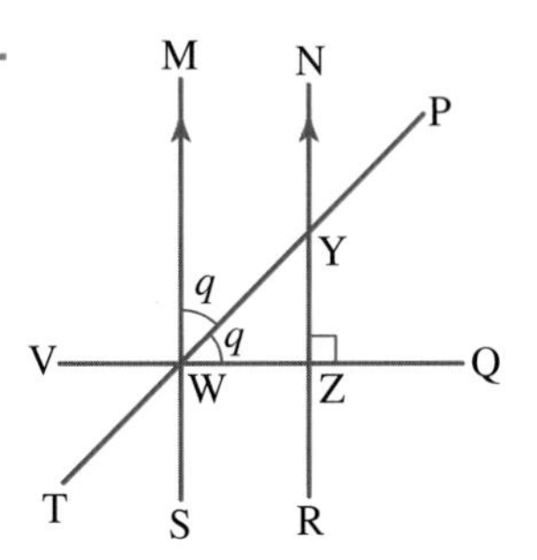


19. Calculate the value of the pronumerals in the following diagrams.

a. 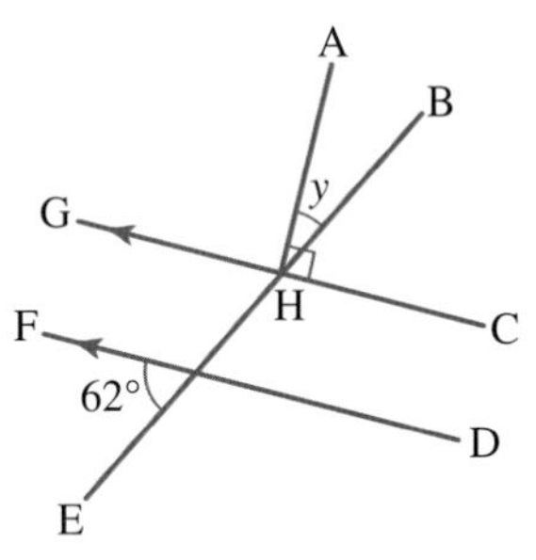


b. 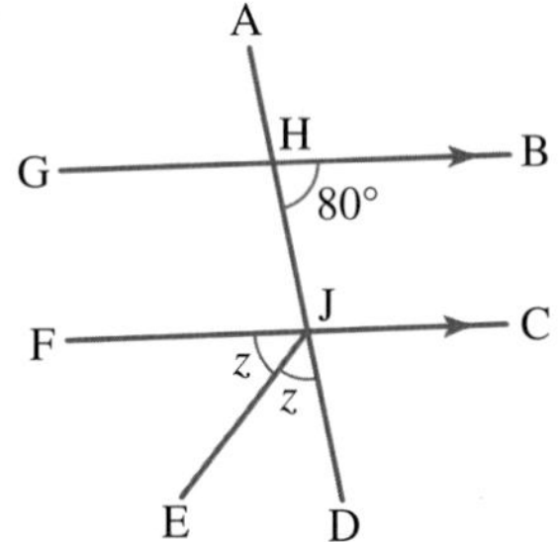


c. 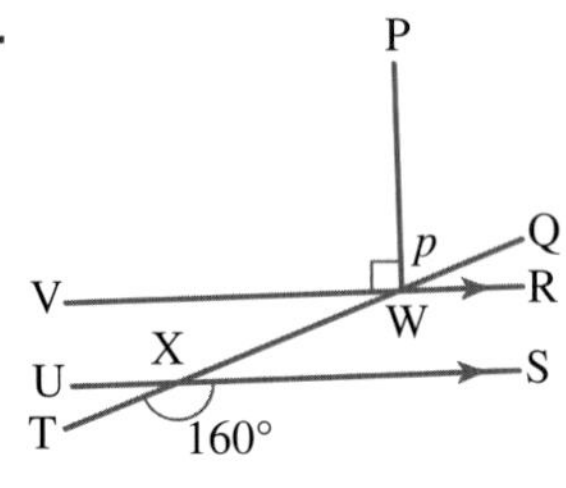


### Reasoning

20. WE24 Is the line AB parallel to the line CD? Explain your answer.

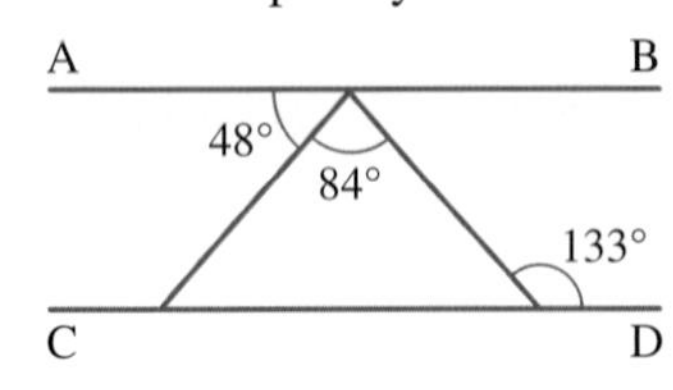

21. a. Calculate the size of angle $a$. Explain how you found it.
b. Is the line KL parallel to MN? Explain your answer.
c. Calculate all of the remaining internal and external angles of the triangle in the centre of the diagram.

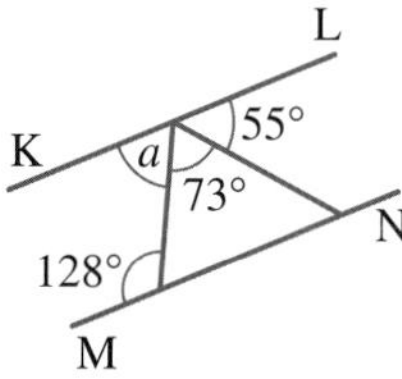


22. Look at the following optical illusions. Are any of the lines parallel? Explain your answer.

a.

b.

## Problem solving

23. A hill is at an angle of 30° to the horizontal. A fence is put in, consisting of a railing parallel to the ground and a series of vertical fence posts. Calculate the angle, $p$, between the top of the fence post and the rail.

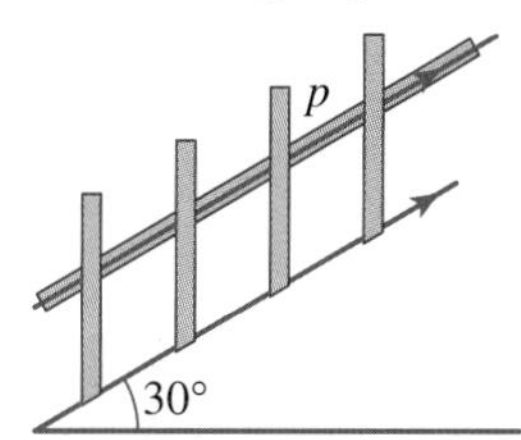


24. Two gates consist of vertical posts, horizontal struts and diagonal beams. Calculate the angle, $a$, as shown in the following gates.

a.
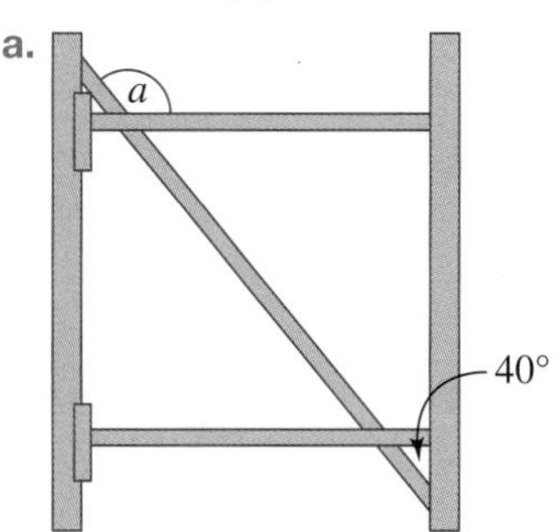


b.

25. In the diagram shown, calculate the magnitude of the acute angle ∠PQR and the reflex angle ∠PQR.

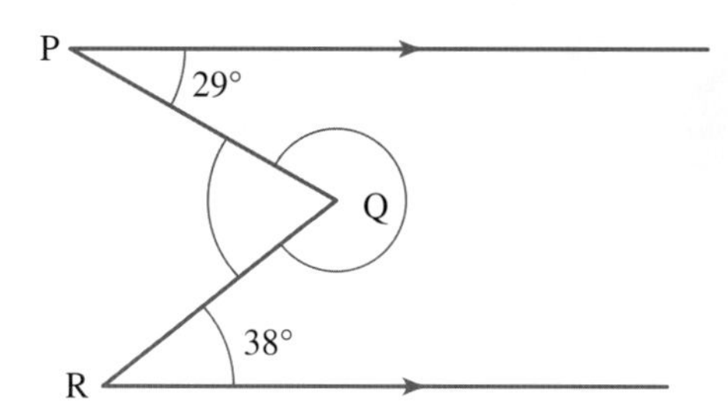

# LESSON
# 9.8 Review

## 9.8.1 Topic summary

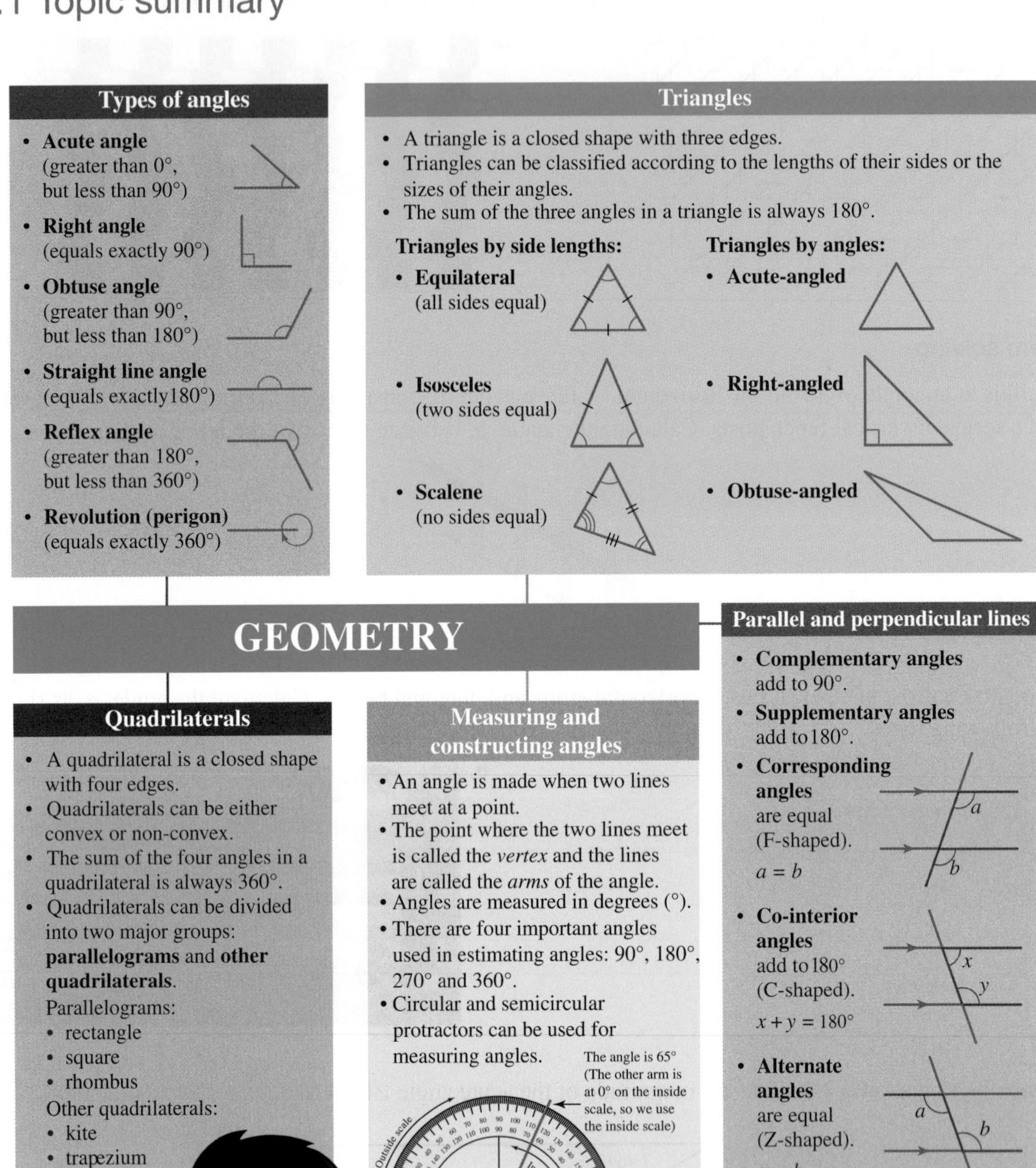

## 9.8.2 Success criteria

Tick the column to indicate that you have completed the lesson and how well you think you have understood it using the traffic light system.

(**Green:** I understand; **Yellow:** I can do it with help; **Red:** I do not understand)

| Lesson | Success criteria | Green | Yellow | Red |
|---|---|---|---|---|
| 9.2 | I understand what an angle is. | | | |
| | I can estimate angles. | | | |
| | I can measure angles using a protractor. | | | |
| 9.3 | I can draw an angle using a protractor. | | | |
| 9.4 | I can classify different types of angles according to size. | | | |
| | I can name angles using three points. | | | |
| 9.5 | I can classify triangles according to the lengths of their sides or the sizes of their angles. | | | |
| | I can calculate both interior and exterior angles of a triangle. | | | |
| 9.6 | I can classify quadrilaterals from a given diagram. | | | |
| | I can calculate the angles in a quadrilateral. | | | |
| 9.7 | I can identify and determine the size of vertically opposite angles. | | | |
| | I can identify and determine the size of angles formed when a transversal intersects parallel lines. | | | |

## 9.8.3 Project

### Tangrams

A tangram is an ancient Chinese puzzle game, also called *ch'i ch'ae pan* or Seven-board of Cunning. The first tangram books were printed in 1813, but the word *ch'i ch'ae* dates back to 740 – 330 BCE.

According to one Chinese legend:

> A servant of a Chinese emperor was carrying a very expensive square ceramic tray and, when he tripped and fell, it was shattered into seven pieces (called *tans*). He was unable to arrange the pieces back into the original shape, but he realised that there were many other shapes that could be built from the pieces.

## Instructions

Trace the following tangram pieces onto cardboard and cut out each piece, then follow the instructions below.

1. Using only the two small triangles, make and draw the following shapes, showing the joins in your book.

a.

Square

b.

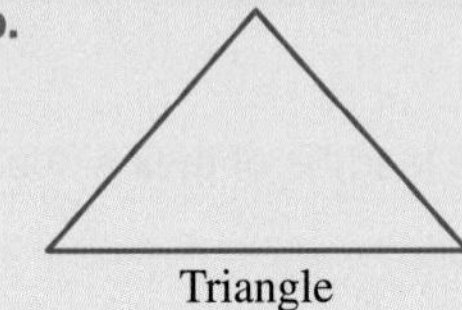

Triangle

c.

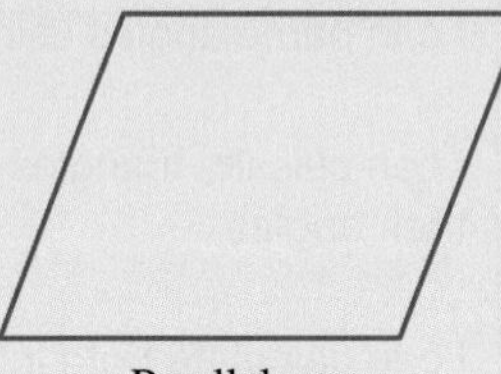

Parallelogram

2. Using the square, the two small triangles, the parallelogram and the medium triangle, make and draw the following shapes.

a.

Square

b.

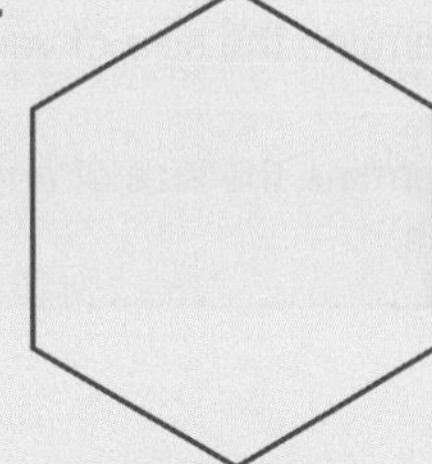

Hexagon

c.

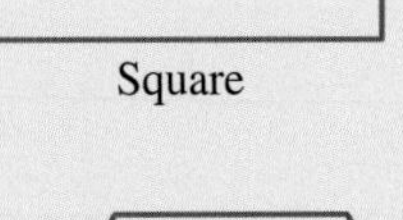

Trapezium

d.

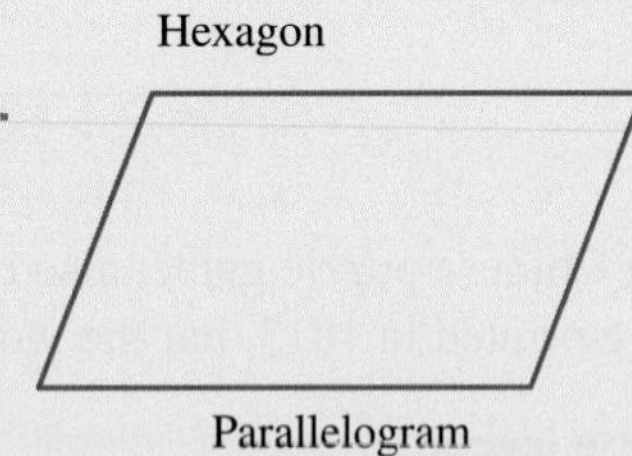

Parallelogram

3. Using all the pieces, make and draw the following shapes.

a. Triangle

b. Rectangle

c. Hexagon

4. Use all of the pieces to make the following shapes and all of the other shapes from the previous questions.

a. A rocket

b. The letter E

**on Resources**

**eWorkbook** Topic 9 Workbook (worksheets, code puzzle and project) (ewbk-1910)

**Interactivities** Crossword (int-2592)

Sudoku puzzle (int-3166)

## Exercise 9.8 Review questions

**learnon**

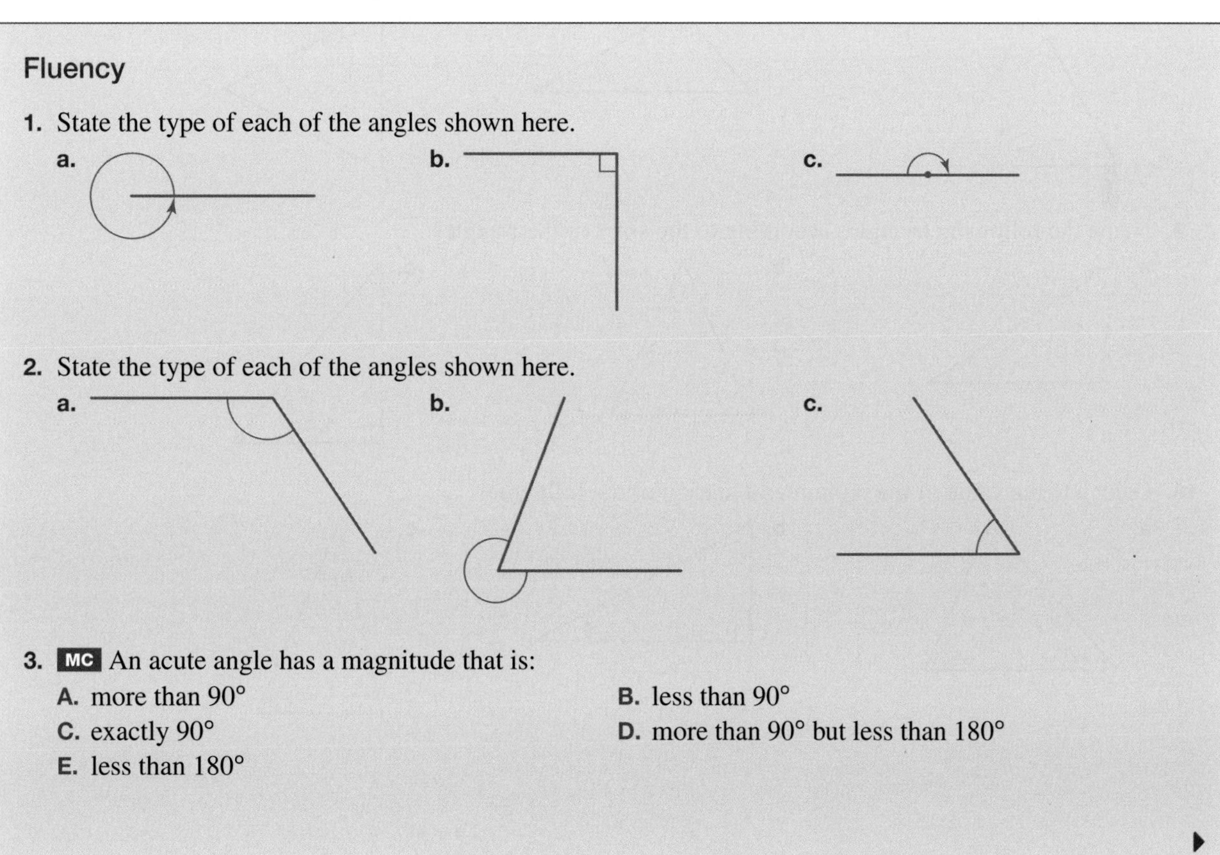

### Fluency

1. State the type of each of the angles shown here.

a. b. c.

2. State the type of each of the angles shown here.

a. b. c.

3. **MC** An acute angle has a magnitude that is:

A. more than 90°
B. less than 90°
C. exactly 90°
D. more than 90° but less than 180°
E. less than 180°

4. **MC** A straight angle has a magnitude of:
**A.** 90° **B.** 270° **C.** 360° **D.** 180° **E.** 60°

**Questions 5 and 6 refer to the following diagram.**

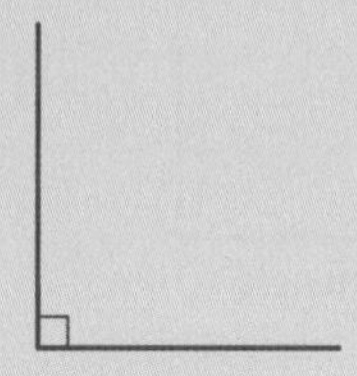

5. **MC** The angle shown in the diagram has a magnitude of:
**A.** 50° **B.** 40° **C.** 90° **D.** 100° **E.** 190°

6. **MC** The angle shown in the diagram is:
**A.** a right angle **B.** a straight angle **C.** an acute angle
**D.** an obtuse angle **E.** a reflex angle

7. Name the angle shown.

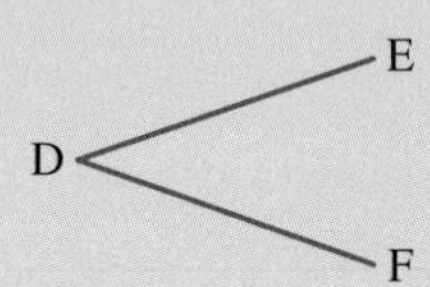


8. Name the following triangles according to the lengths of their sides.

**a.** 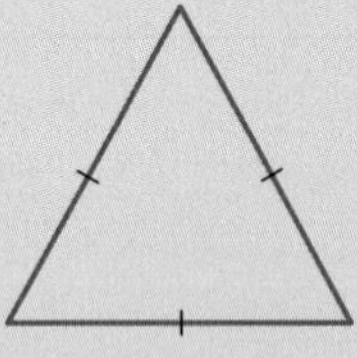

**b.** 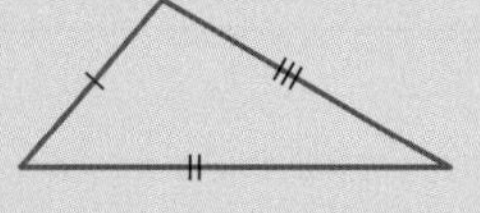

**c.** 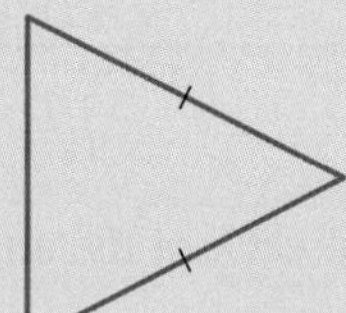

9. Name the following triangles according to the sizes of their angles.

**a.** 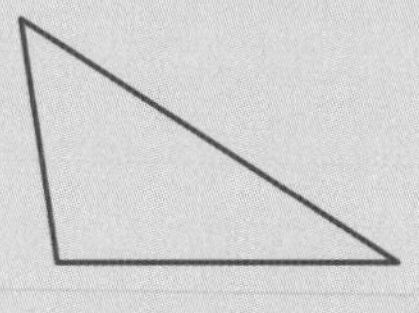

**b.** 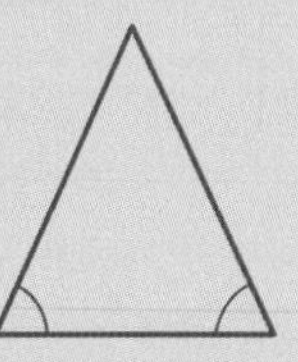

**c.** 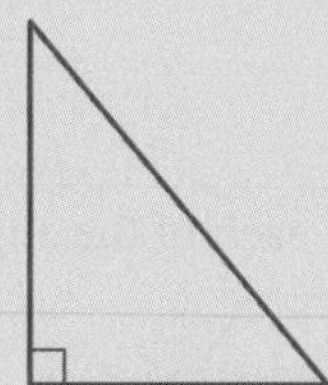

10. Calculate the value of the pronumeral in each of the following.

**a.** 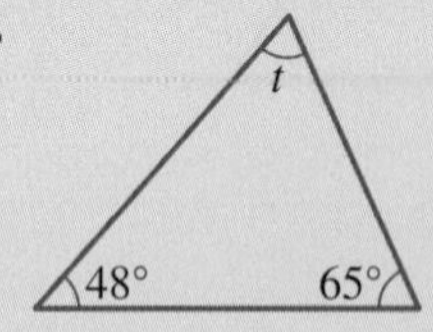


**b.** 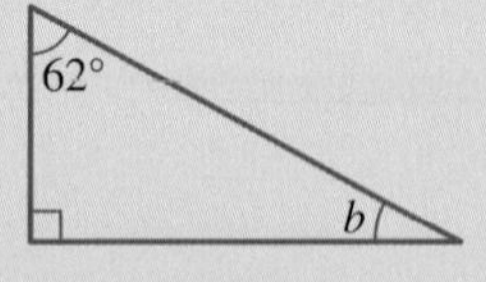


**c.** 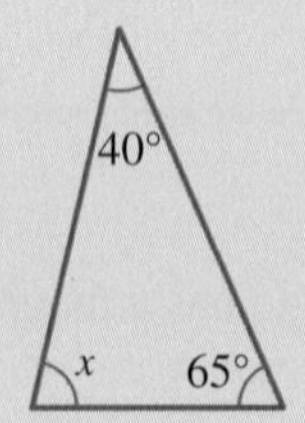

**11.** Calculate the value of the pronumeral in each of the following.

**a.**
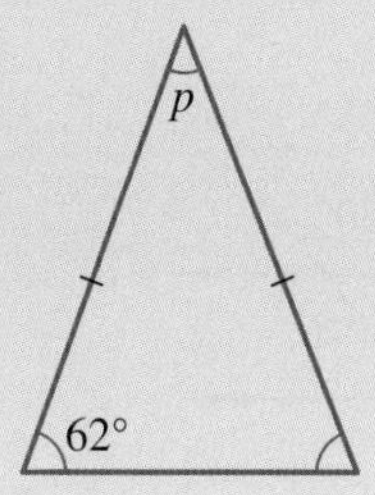


**b.**
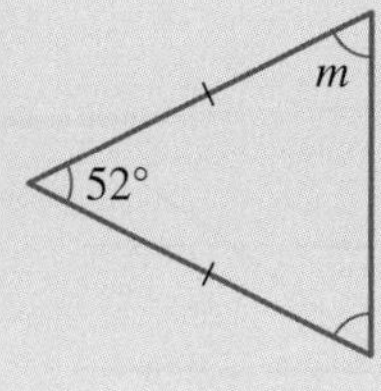


**c.**
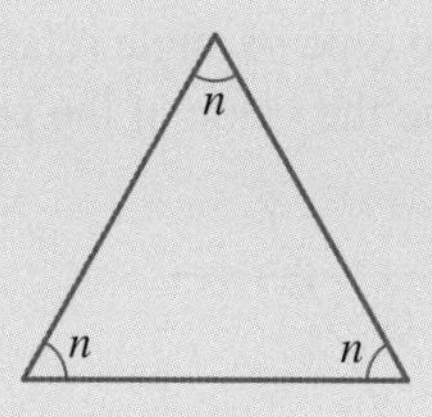


**12.** Name the following quadrilaterals, giving reasons for your answers.

**a.**
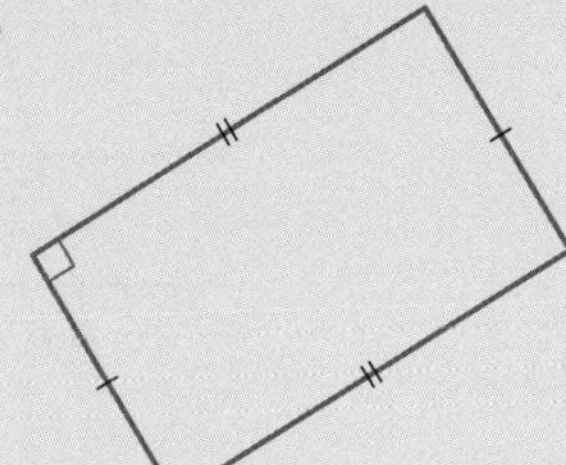

**b.**
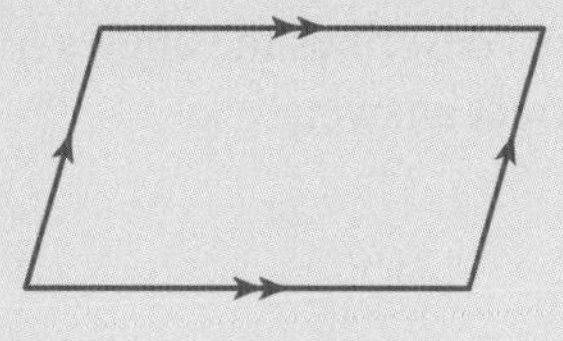

**c.**
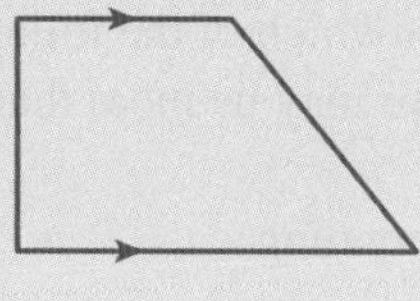

**13.** Name the following quadrilaterals, giving reasons for your answers.

**a.**
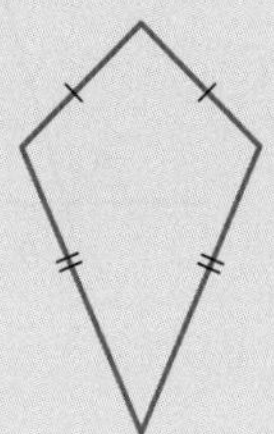

**b.**
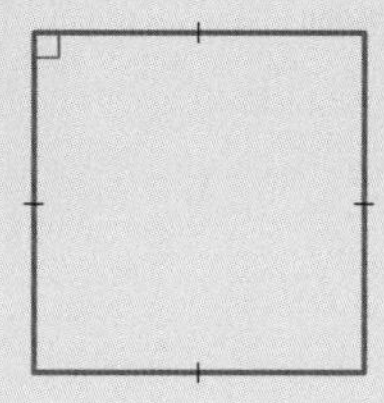

**c.**
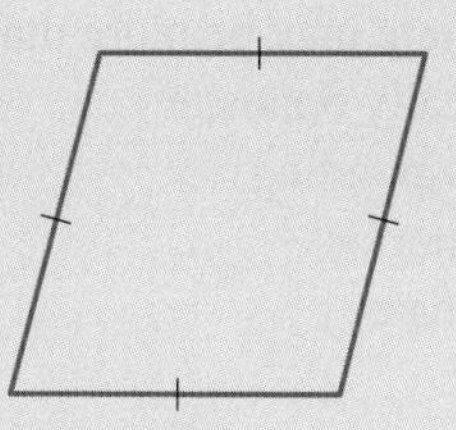

**14.** Calculate the value of the pronumeral in each of the following quadrilaterals.

**a.**
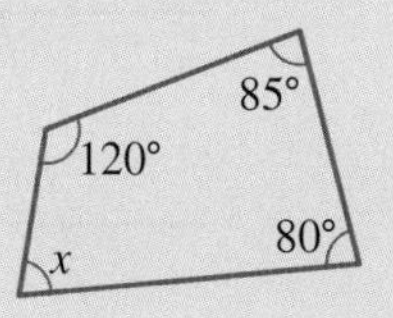


**b.**
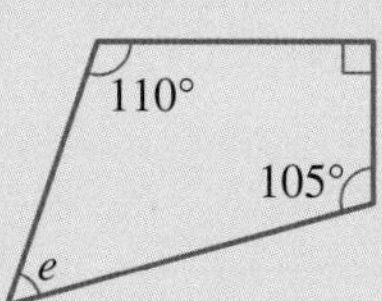


**c.**
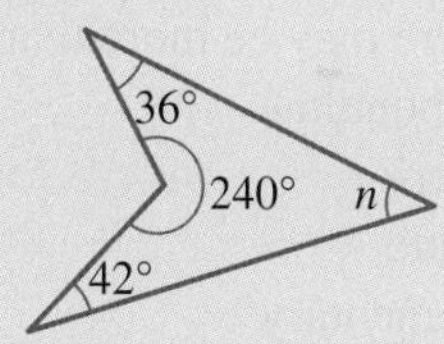


**15.** Calculate the value of the pronumeral in each of the following quadrilaterals.

**a.**
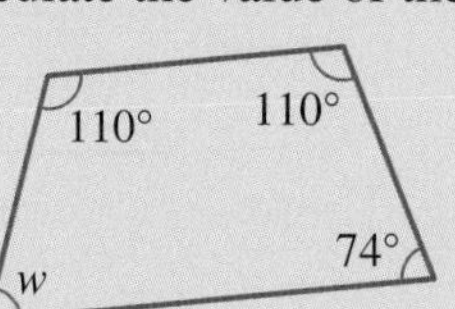


**b.**
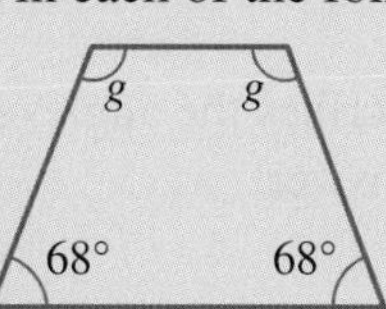


**c.**
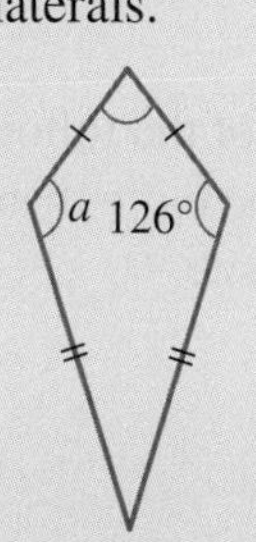


**16.** Determine the value of each of the pronumerals, giving reasons for your answer.

**a.**
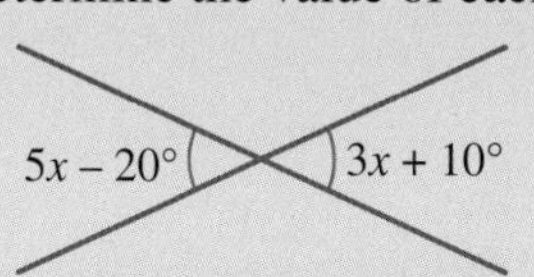


**b.**
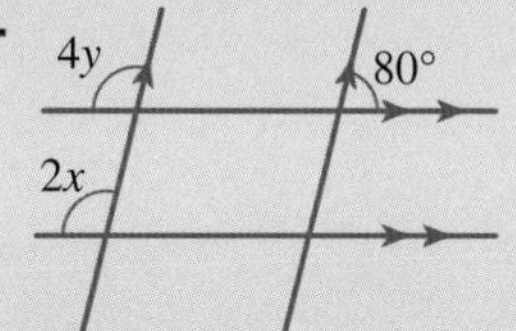


**c.**
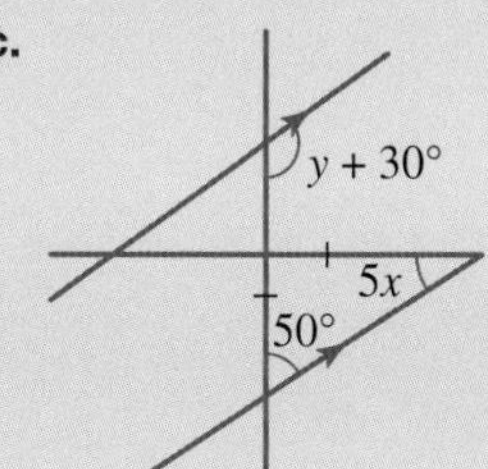

17. For each diagram:
    a. state the type of angle relation that you use
    b. calculate the value of the pronumeral.

i.

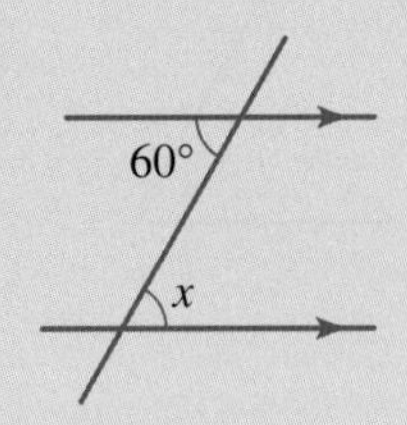


ii.

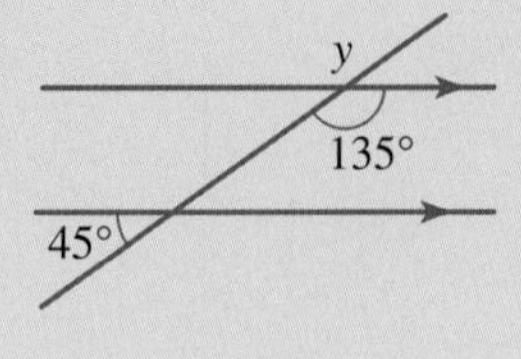


iii.

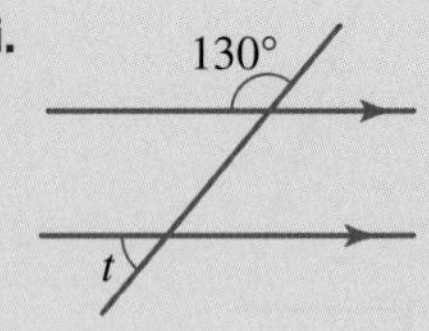


18. **MC** Select which of the following terms describes the angles $a$ and $b$ .
    *Hint:* There may be more than one correct answer.
    A. Equal
    B. Corresponding
    C. Allied
    D. Alternate
    E. Supplementary

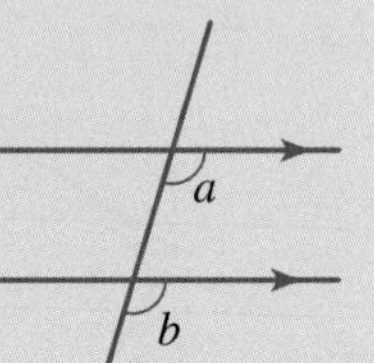


19. **MC** Select which of the following terms describes the angles $a$ and $b$.
    *Hint:* There may be more than one correct answer.
    A. Vertically opposite
    B. Corresponding
    C. Co-interior
    D. Alternate
    E. Equal

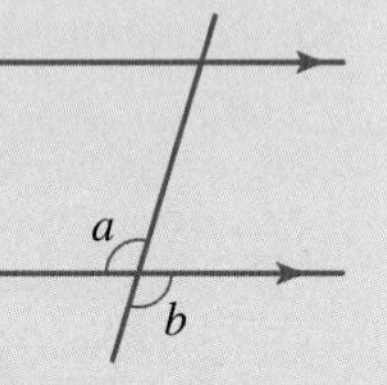


20. **MC** Select which of the following terms describes the angles $a$ and $b$.
    *Hint:* There may be more than one correct answer.
    A. Corresponding
    B. Co-interior
    C. Alternate
    D. Supplementary
    E. Equal

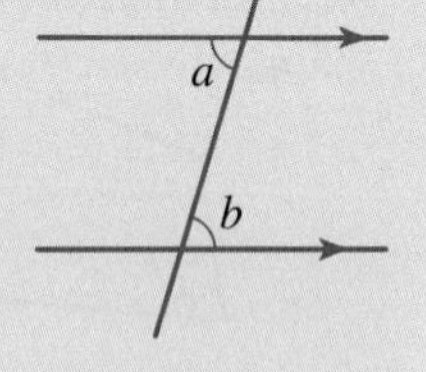


21. **MC** Select which of the following terms describes the angles $a$ and $b$.
    *Hint:* There may be more than one correct answer.
    A. Vertically opposite
    B. Corresponding
    C. Allied
    D. Alternate
    E. Supplementary

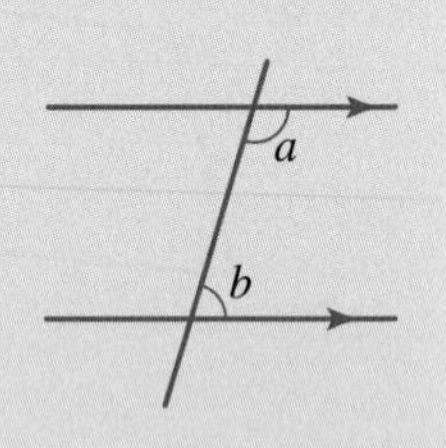

## Problem solving

**22.** The teepee shown has an angle of 46° at its peak. What is the size of the angle, $w$, that the wall makes with the floor?

**23.** A circus trapeze is attached to a rope as shown in the diagram. Calculate the size of angle $t$.

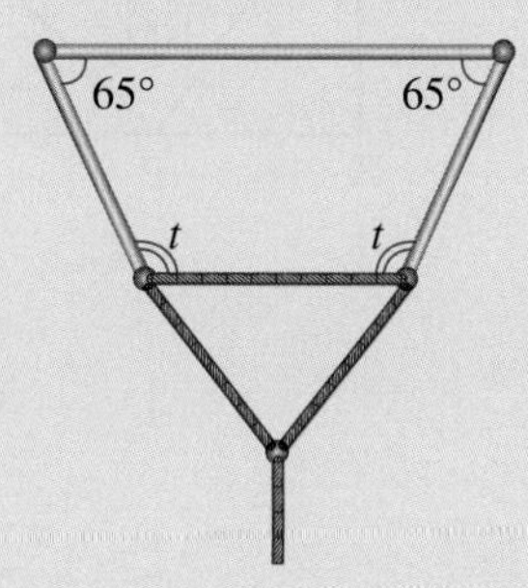


**24.** Triangles can be classified according to the number of sides of equal length (scalene, isosceles or equilateral) and by the type of angles (acute, obtuse or right). Decide which of the following combinations are possible. Draw examples of the combinations that are possible.

- Right–scalene, right–isosceles, right–equilateral
- Acute–scalene, acute–isosceles, acute–equilateral
- Obtuse–scalene, obtuse–isosceles, obtuse–equilateral

**25.** The following triangles are both equilateral triangles, but the side length of triangle ABC is 3 units and the side length of triangle DEF is 6 units. Determine how many of the smaller triangles would you need to fill the larger triangle.

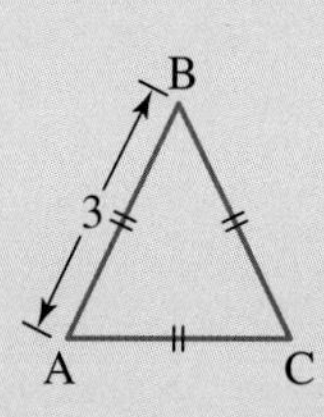


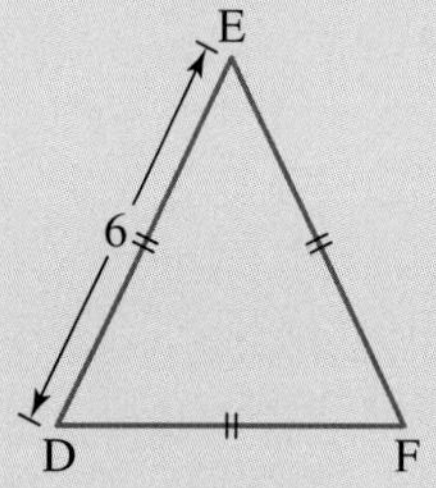


**26.** If angle AOE is obtuse, name all of the acute angles in the diagram below.

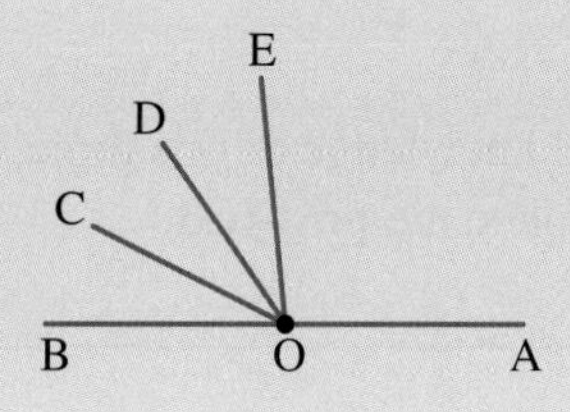

**27.** An equilateral triangle EDC is drawn inside a square ABCD as shown below. Determine the size of $\angle AED$.

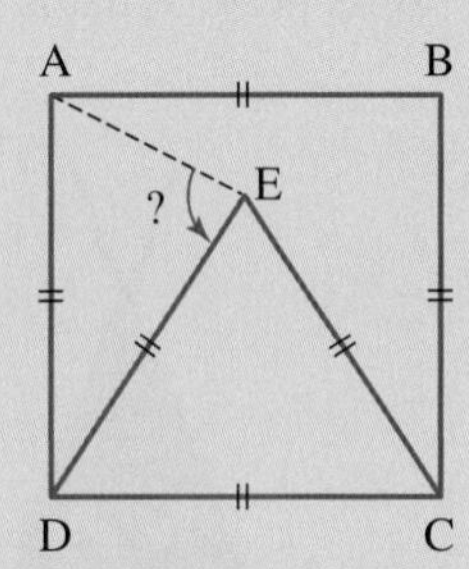


**28.** WXYZ is a rectangle as shown here. Points P and Q are drawn so that $XY = PY$ and $\angle XPQ = 100°$. Determine the size of $\angle PQX$.

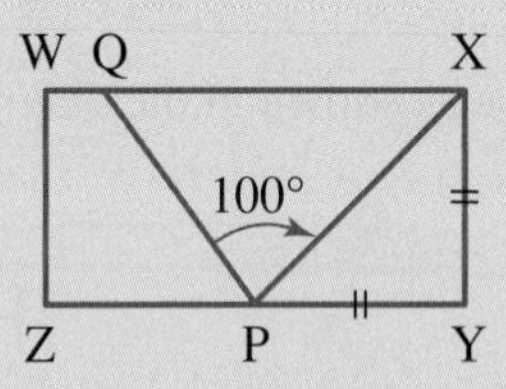


**on** To test your understanding and knowledge of this topic, go to your learnON title at www.jacplus.com.au and complete the **post-test**.

# Answers

## Topic 9 Geometry

### 9.1 Pre-test

1. Obtuse
2. 135°
3. 90°
4. D
5. D
6. C
7. ∠ADB
8. A
9. E
10. a. Scalene, right-angled triangle
    b. Isosceles, acute-angled triangle
    c. Scalene, obtuse-angled triangle
    d. Isosceles, right-angled triangle
11. 59°
12. 52°
13. 68°
14. D
15. 72°

### 9.2 Measuring angles

1. a. B b. B
2. a. 40° b. 81° c. 75° d. 13°
3. a. 142° b. 107° c. 162° d. 103°
4. a. 240° b. 201° c. 316° d. 356°
    e. The accuracy of your measurement will depend on the thickness of your pencil, the position of the protractor and the side from where the angle is read.
5. a. 45° b. 155° c. 180° d. 63°
6. a. Difficult, as angle is approximately 7°
    b. As recommended
7. a. 57°, 105°, 33°, 75°
    b. 57° and 33°
    c. 105°, and 75°
    d. Multiple correct answers: four 57°, and four 33°, two 105° and two 75°, two 57°, two 33°, one 105° and one 75°
8. a. Design 4. 45°
    b. Design 1. 15°
    c. Design 1, Design 2 and Design 3
    d. Design 1
    e. Design 3
    f. Design 1
9. a, b, d, e, f, g. — Responses will vary depending on the estimations made in part a.
    c.

| Diagram | Measured angle |
|---|---|
| 1 | 190° |
| 2 | 130° |
| 3 | 110° |
| 4 | 40° |
| 5 | 300° |
| 6 | 200° |
| 7 | 140° |
| 8 | 290 |

10. a. As the kite moves from point A to point B, the angle increases from 0° to 90°
    b. As the kite moves from point B to point C, the angle decreases from 90° to 0°.
11. a. i. 210° (reflex angle) and 150° (obtuse angle)
    ii. 180° (straight angle) and 180° (straight angle)
    iii. 150° (obtuse angle) and 210° (reflex angle)
    iv. 15° (acute angle) and 345° (reflex angle)
    v. 108° (obtuse angle) and 252° (reflex angle)
    vi. 114° (obtuse angle) and 246°(reflex angle)
    b. 360° ÷ 60 minutes = 6°; 360° ÷ 12 hours = 30°
12. a. Triangle 1: 45°, 45°, 90°, Triangle 2: 80°, 40°, 60°, Triangle 3: 70°, 60°, 50°
    b. 180°, 180°, 180°
    c. Brinda is correct, there is a pattern. The pattern is that the angles within a triangle always add to 180°.

### 9.3 Constructing angles with a protractor

1. a. 15°

b.

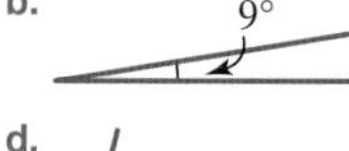


c.

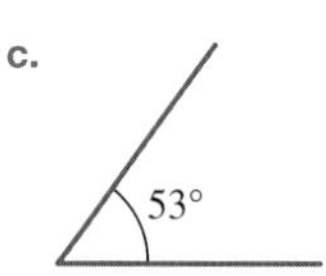


d.

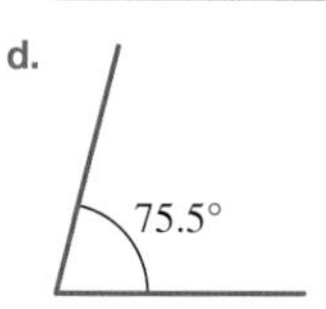


2. a.

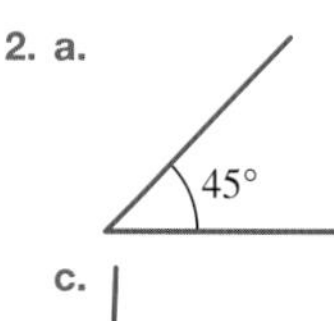


b.

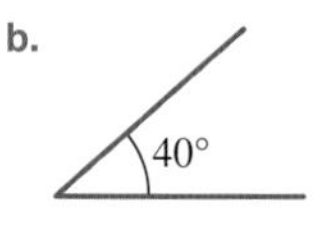


c.

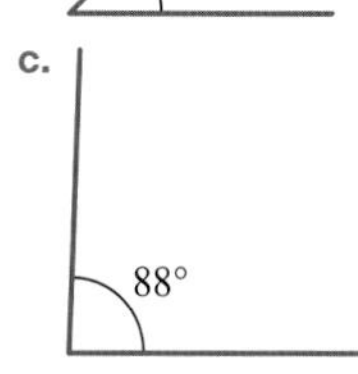


d.

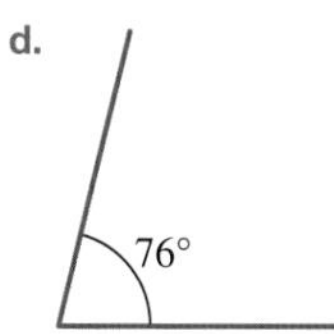

3. a.

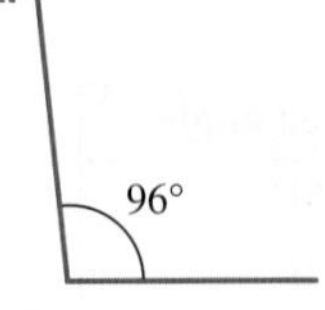


b.

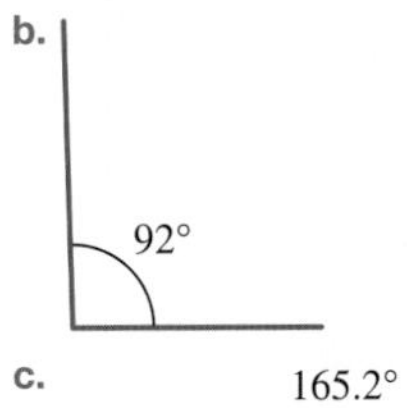


c.

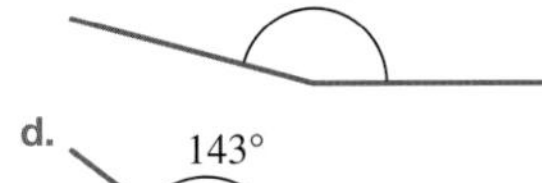


d. 143°

4. a.

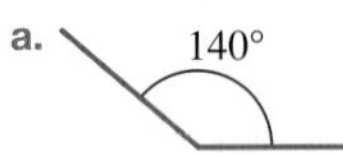


b.

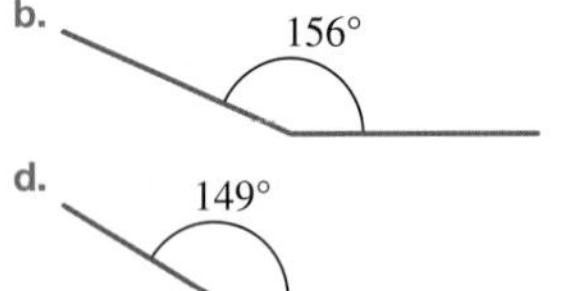


c. 127°

d. 149°

5. a.

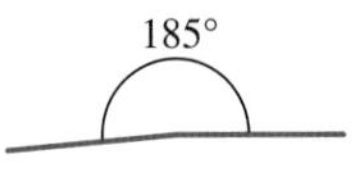


b.

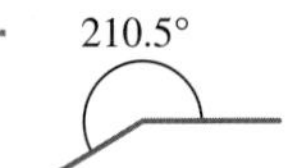


c.

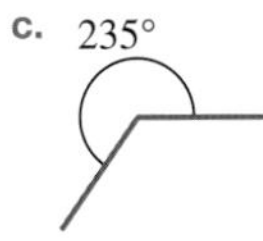


d.

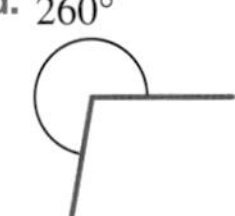


6. a.

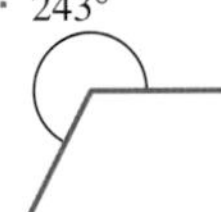


b.

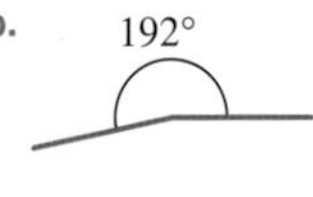


c.

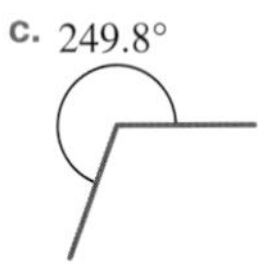


d.

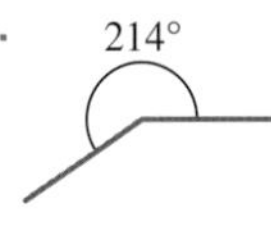


7. E

8. a.

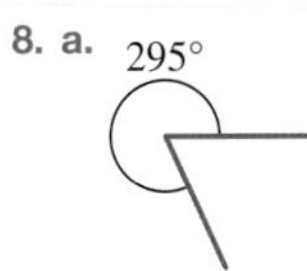


b.

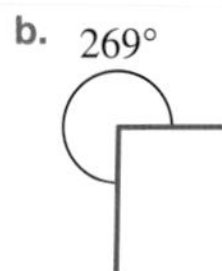


c.

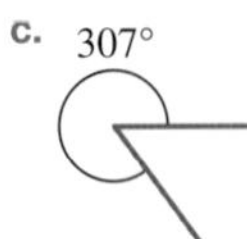


d.

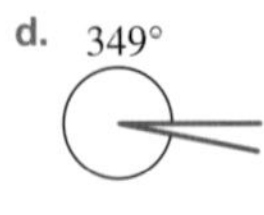


9. a.

b.

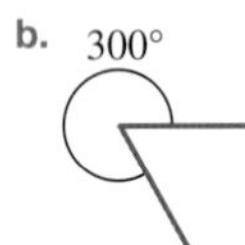


c.

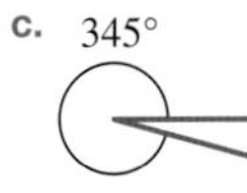


d. 358°

10. a.

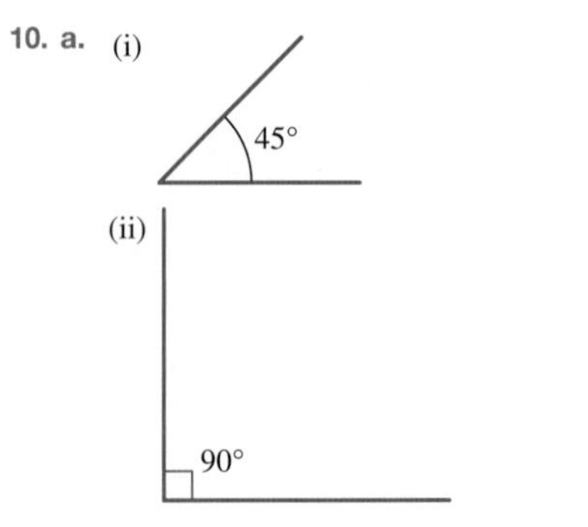


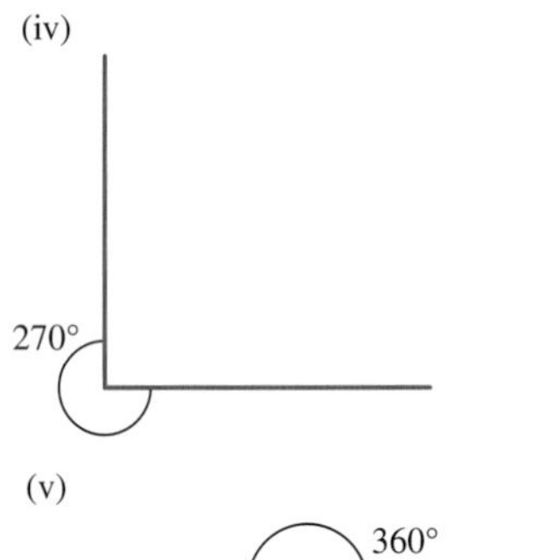


(v) 360°

b. 45°, 90°, 180°, 90°, 45°

11. a.

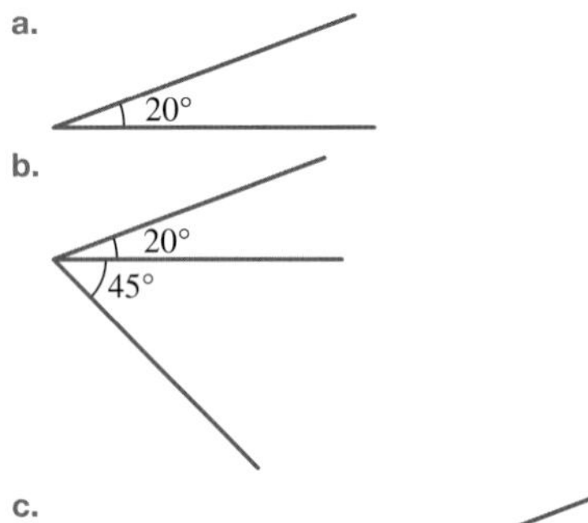


c. 20° 45° 108°

d. 187°

e. 360°

12. 120°, 20° and 90°

13. Construct the corresponding acute angle and then label the reflex angle.

14. a.

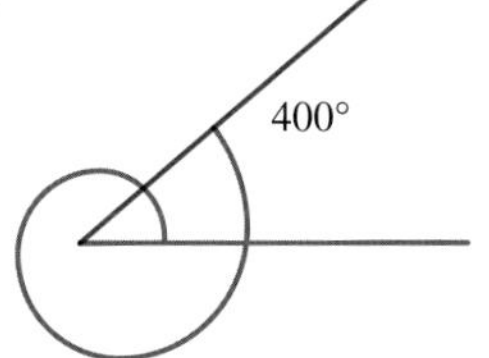


b. 40°

15. a. 

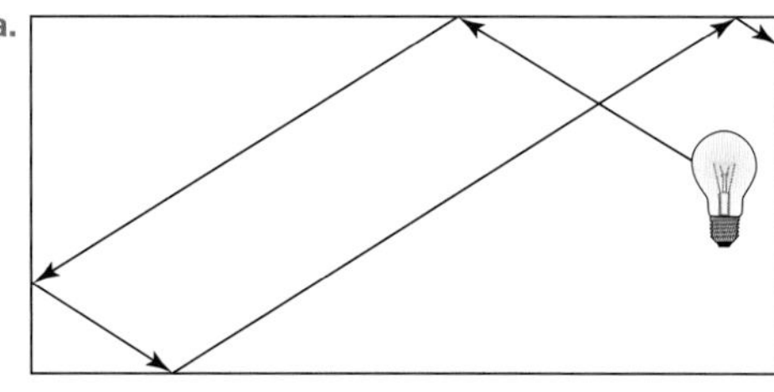

b.

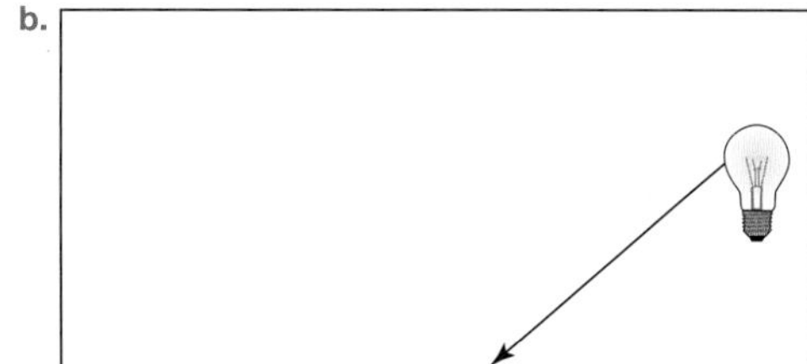

c.

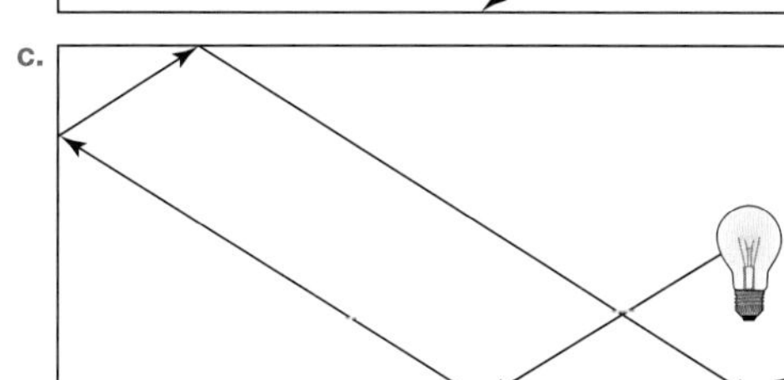

16. a.

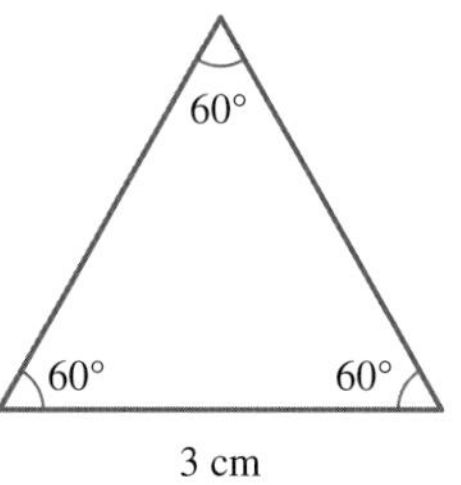


b.

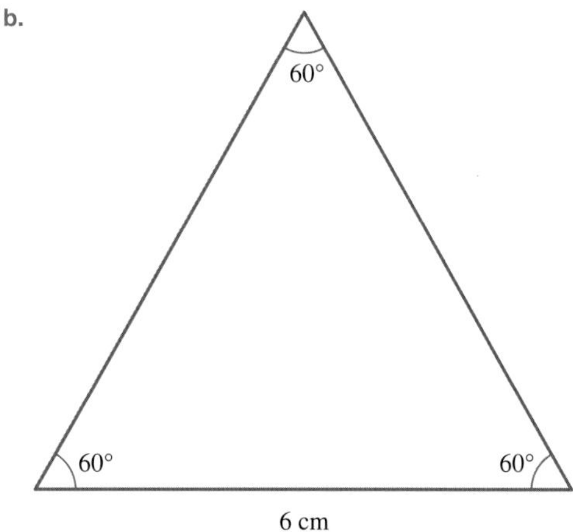


c. Sample response: Similarities include angles, ratios of sides, same shape. Differences include sizes, areas, accuracy.

d. 60°

17. a.

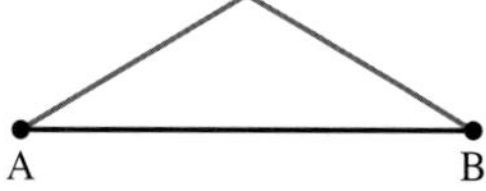


b. All triangles should be identical.

c. If two equal angles are drawn at the ends of a straight line of set length, there is only one size triangle that forms.

18. Draw any triangle on an A4 sheet of paper.

a Student activity; responses will vary.

e. Despite different triangles drawn by students, the sum of the angles of a triangle is always 180°.

## 9.4 Types of angles and naming angles

1. a. Acute b. Reflex c. Obtuse
 d. Straight e. Reflex

2. a. Right b. Acute c. Revolution
 d. Acute e. Reflex

3. a. 4; 3°, 45°, 65°, 69°
 b. 2; 123°, 165°
 c. 4; 234°, 300°, 270°, 210°

4. a. True b. True c. True

5. a. Right b. Straight c. Obtuse
 d. Acute e. Reflex f. Revolution

6. a. ∠XUV or ∠VUX b. ∠PQR or ∠RQP
 c. ∠SDX or ∠XDS d. ∠GPZ or ∠ZPG
 e. ∠PFH or ∠HFP

7. In each case the middle letter should be at the vertex of the angle.

a.

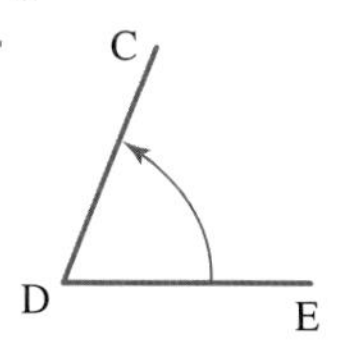


b.

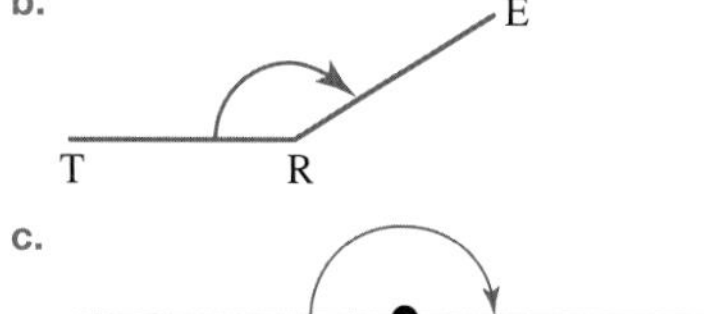


c.



8. a.

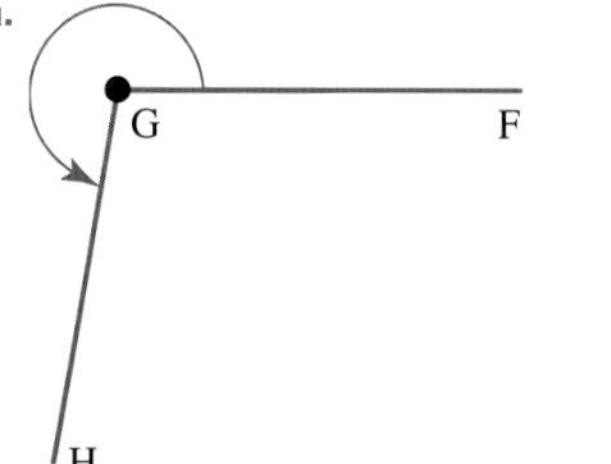


In each case the middle letter should be at the vertex of the angle.

b. 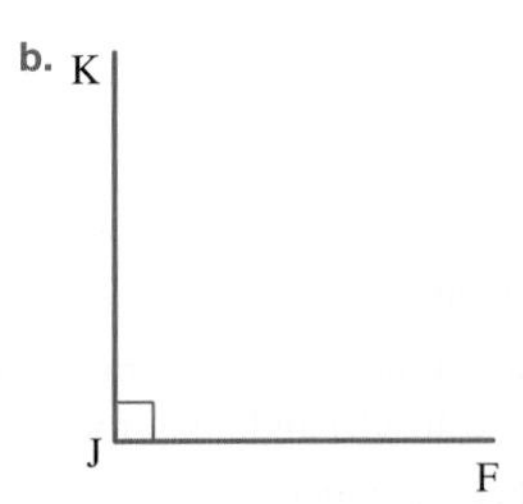


In each case the middle letter should be at the vertex.

c. 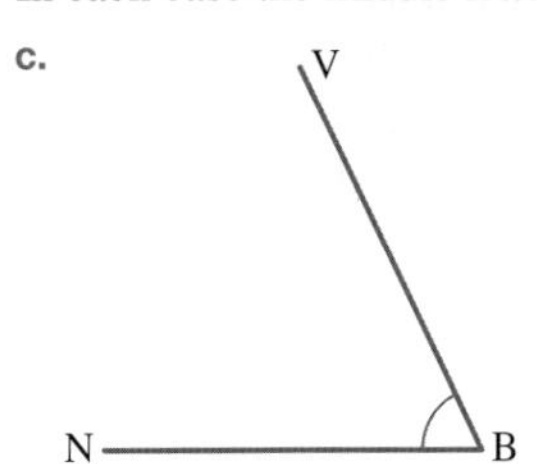


9. Acute
10. Revolution
11. Obtuse
12. Acute: ∠NOM, ∠MOY
    Obtuse: ∠MOX
13. Acute: ∠PQS, ∠RQO
    Obtuse: ∠SQR, ∠SQO, ∠RQP
14. a. Acute: ∠MLN, ∠NLO, ∠OLP, ∠PLK, ∠NLP, ∠OLK
       Obtuse: ∠MLO, ∠MLP, ∠NLK
    b. Acute: ∠YOX, ∠WOQ, ∠POQ
       Obtuse: ∠XOW, ∠YOP, ∠YOW, ∠XOP, ∠YOQ, ∠POW
15. a. Acute b. Right c. Obtuse
16. a. Straight b. Reflex c. Revolution
17. a. Acute b. Obtuse
    c. Straight d. Right
    e. Obtuse
18. a. Sample responses:
       1 o'clock, 10 o'clock
    b. Sample responses:
       4 o'clock, 20 past 11
    c. Sample responses:
       9 o'clock, 3 o'clock
19. In each case the middle letter should be at the vertex of the angle.
    a. ∠DEF
    b. ∠ABC, ∠XWY
    c. ∠PQR
    d. Sample responses:
       outside angle ∠DEF,
       outside angle ∠RQP
20. a. This statement is true if the sum of the two angles is less than 90°.
    b. This statement is true if the sum of the two angles is equal to 90°.
    c. This statement is true if the sum of the two angles is greater than 90°.
    d. This statement is untrue because two acute angles can't add up to more than 180°.
21. a. Yes
    b. 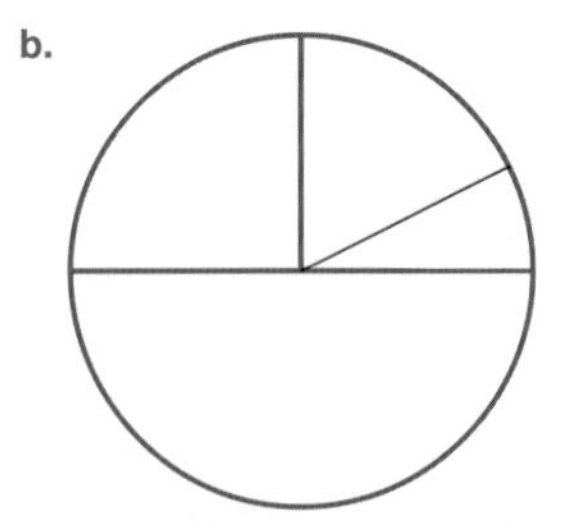
22. 145°
23. It is the naming custom to always put the letter representing the vertex in the middle of the angle's name because this allows anyone to understand the name of the angle.
24. 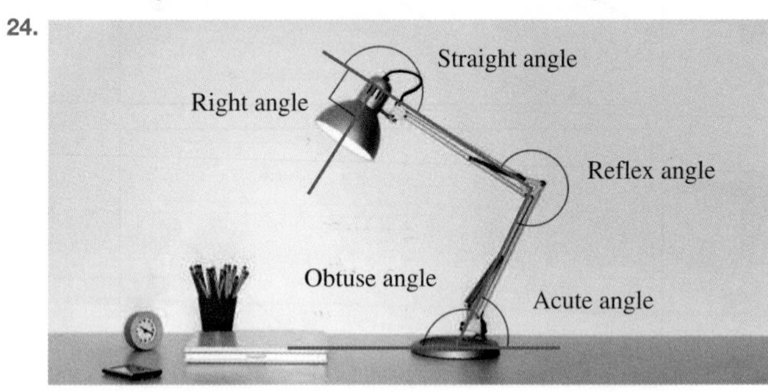

25. a. 25 including revolutions, 20 if revolutions not included
    b. 165°

## 9.5 Triangles

1. a. ΔABC Scalene
   b. ΔLMN Equilateral
   c. ΔXYZ Isosceles
   d. ΔPQR Isosceles
   e. ΔRST Scalene
2. a. Right-angled triangle
   b. Acute-angled triangle
   c. Acute-angled triangle
   d. Acute-angled triangle
   e. Obtuse-angled triangle
3. A
4. B
5. D
6. A
7. a. $x = 57°$ b. $g = 125°$ c. $t = 44°$
   d. $k = 60°$ e. $f = 93°$ f. $z = 90°$
8. a. $d = 45°$ b. $b = 65°$ c. $a = 50°$
9. a. $c = 70°$ b. $e = 76°$ c. $n = 52°$
10. a. $u = 66°$ b. $k = 45°$ c. $d = 76°$
    d. $t = 74°$ e. $f = 57°$ f. $p = 70°$
11. a. $p = 50°; n = 70°$ b. $a = 22°; b = 28°$
    c. $x = 40°; y = 50°$ d. $t = 35°; s = 55°$
12. a. $b = 46°$ b. $n = 21°$
    c. $m = 60°$ d. $t = 28°$

13. a.

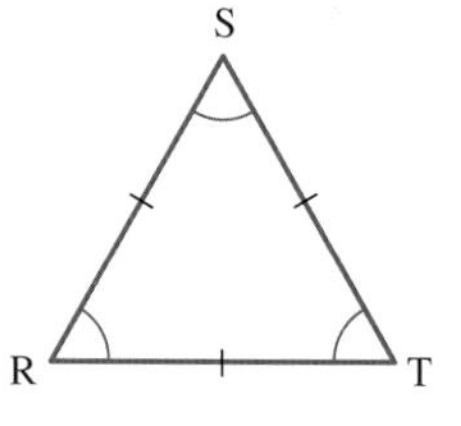


b.

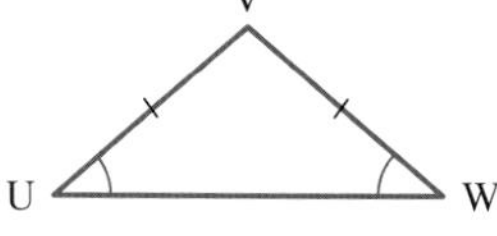


c.

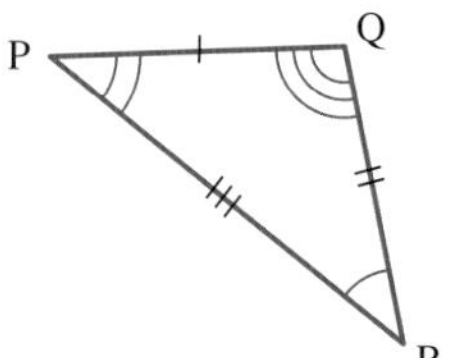


14. a.

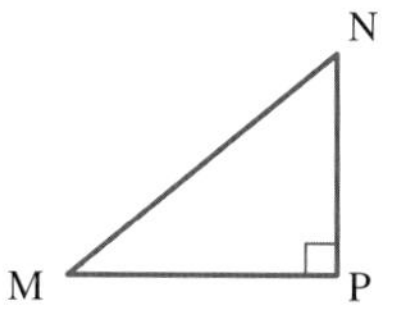


b.

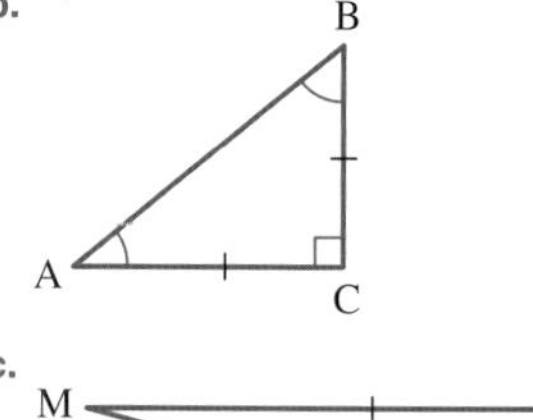


c. M N O

15. c. Please see the worked solutions; the longest side is 10 cm in length. (a, b, d; Please see the worked solutions.)

16. a. $b = 20°$ b. $k = 30°$

17. a. ∠MPN = 72° and ∠NMP = 36°
 b. ∠MPN = 54° and ∠MNP = 54°

18. a. The sum of the angles will be greater than 180°.
 b. Start with an equilateral triangle ABC, equal sides and 3 equal angles. If point A and C are fixed and point B is extended, length AB increases. This in turn means that length BC increases and therefore ∠C increases in size. The longest side (in this case AB) will be opposite the largest angle (in this case ∠C). This is true for all triangles, with the longest side always being opposite the largest angle.
 c. Imagine that the base of a triangle is the longest side. The other two side lengths must be attached to opposite vertices of the base of the triangle. If the sum of these two lengths was shorter than the base length then these two sides could not meet and complete the triangle. Similarly, if the sum of these two lengths was equal to the base length then they could only meet in a straight line along the base, and a triangle would not be created. Therefore the sum of the two lengths must be longer than the base length to create a triangle.

19. a.

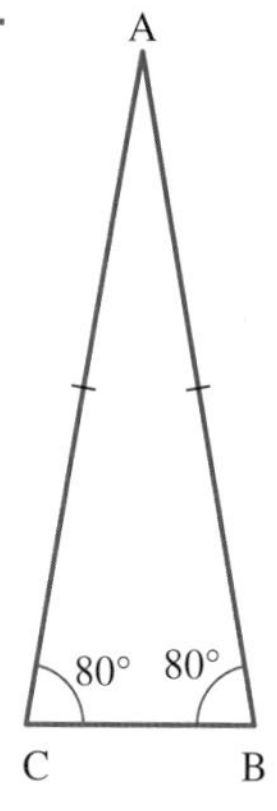


b.

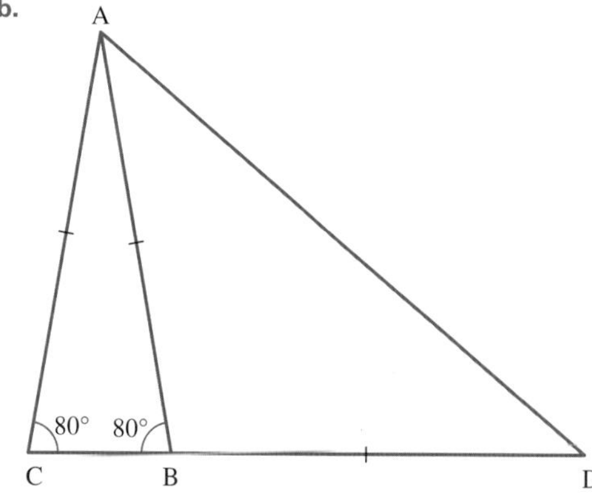


c. 40°

d.

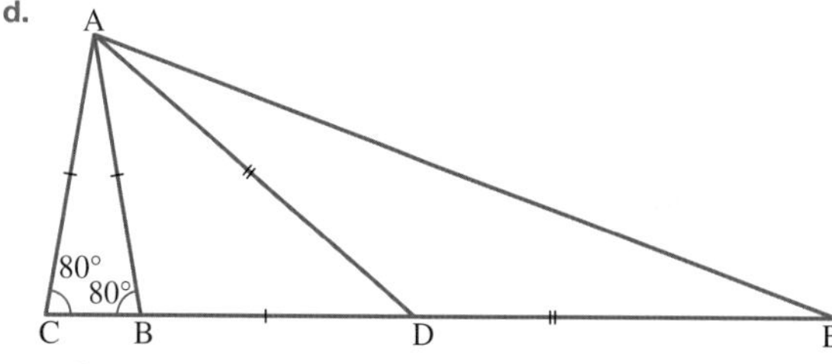


e. 20°

f.

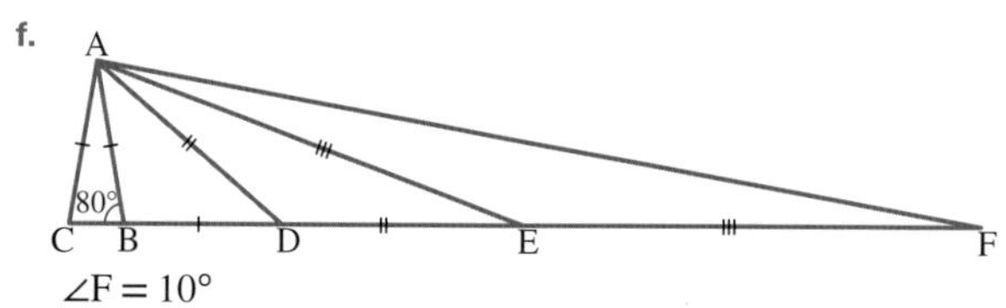


∠F = 10°

The angle is halved every time a new triangle is constructed.

20. a. 70° b. 156° c. 50° d. 64°

21. a. Yes, angles add up to 180°.
 b. No, angles add up to 175°.
 c. No, angles add up to 185°.
 d. Yes, angles add up to 180°.

22. a. $a = 69°, k = 10, b = 114°$
 b. $a = 60°, c = 120°$

## 9.6 Quadrilaterals and their properties

1. a. Non-convex
   b. Convex
   c. Convex
   d. Non-convex
2. a. i. Square
   ii. Trapezium
   iii. Kite
   iv. Rhombus
   v. Rectangle
   vi. Irregular quadrilateral
   b. i. Square: 4 sides equal, 4 angles equal
   ii. Trapezium: one pair of parallel sides
   iii. Square: 4 equal sides and angles
   iv. Kite: adjacent sides equal
3. E
4. E
5. A
6. a. $t = 138°$ b. $t = 90°$ c. $t = 202°$
   d. $p = 92°$ e. $c = 66°$ f. $k = 67°$
7. a. $m = 78°$ (opposite angles of parallelogram equal)
   b. $u = 75°$ (opposite angles of rhombus equal)
   c. $t = 132°$ (opposite angles of a kite equal)
   d. $f = 108°$ (opposite angles of a kite equal)
   e. $p = 63°$ (opposite angles of rhombus equal)
   f. $z = 73°$ (opposite angles of parallelogram equal)
8. a. $a = 129°, b = 51°, c = 129°$
   b. $h = 130°$
9. a. $m = 113°$
   b. $c = 98°, d = 82°$
   c. $t = 116°$
   d. $p = 52°, m = 106°$
   e. $t = 103°$, $x = 91°$
   f. $m = 65°$, $p = 115°, n = 65°$
10. D
11. E
12. a. True b. True
    c. False d. True
    e. True
13. D
14. $p = 61°$
15. 115°
16. 45° and 135°
17.

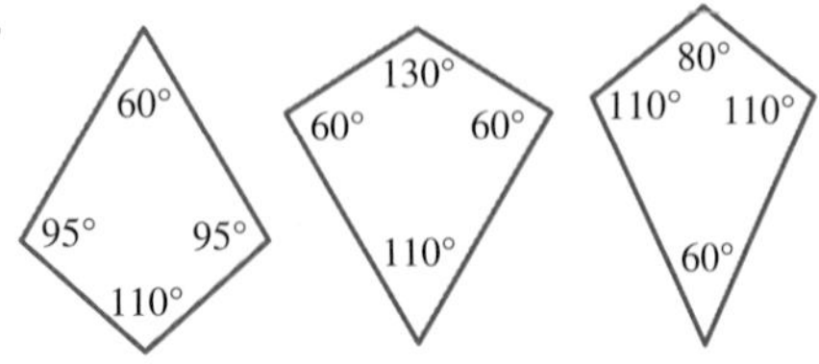


18. a. Yes, angles add up to 360°.
    b. No, angles add up to 270°.
19. a. 130°
    b. Infinite number
    c.

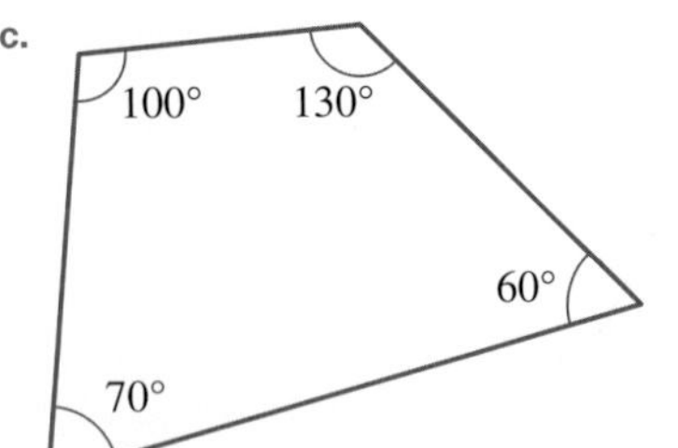


20. a. Two smaller obtuse-angled isosceles triangles, one larger size obtuse-angled isosceles triangle and two right-angled scalene triangles.
    b. Two types: rectangles (of different sizes) and a rhombus.
    c.

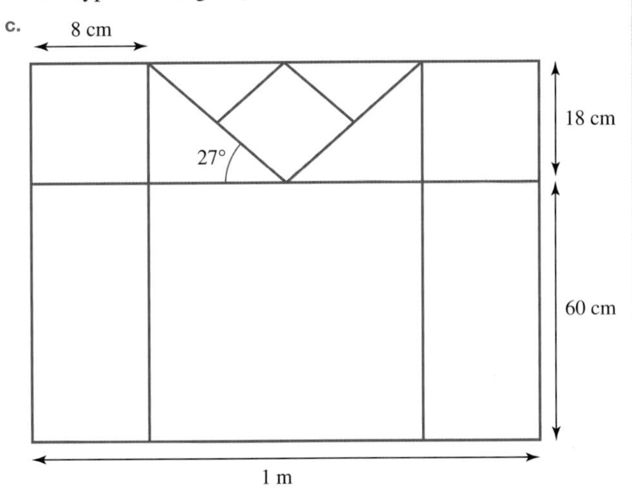


    d. In both right-angled triangles, the angles are: 27°, 63°, 90°; in all isosceles triangles, angles are: 126°, 27°, 27°; in the rhombus, angles are: 126°, 54°, 126°, 54°; in all rectangles, all angles are 90° each.
21. a. Consider the sum of the angles of the two triangles ABC and ACD.
    In $\Delta$ABC : $n + p + q = 180°$
    In $\Delta$ACD : $m + r + s = 180°$
    In quadrilateral ABCD : $n + p + q + m + r + s = 360°$
    b. i. 540°
    ii. 720°
    iii. 1080°
    iv. 1440°
22. a. $a = 70°, b = 125°, c = 125°$
    b. $x = 78°, k = 102°, m = 74°$
23. a. $a = 108°, b = 72°, c = 90°$
    b. $g = 100°, h = 80°, i = 50°$
    c. $d = 115°, e = 106.5°, f = 82°$
    d. $j = 45°, k = 135°$
24. 102°

## 9.7 Parallel and perpendicular lines

1. 65°
2. a. 19° b. 45°
3. 136°
4. a. 10° b. 60° c. 13°
5. a. 38° b. 110° c. 22°

6. a. 19°
   b. $e = 39°, f = 141°, g = 141°$
   c. $i = 13°, h = 15°, k = 152°, j = 152°$

7. a. C   b. D   c. B   d. A

8. a. i. Corresponding angles
      ii. Alternate angles
      iii. Vertically opposite angles
      iv. Corresponding angles
      v. Co-interior angles
      vi. Co-interior angles
   b. i. $p = 60°$
      ii. $q = 45°$
      iii. $s = 65°$
      iv. $t = 72°$
      v. $m = 110°$
      vi. $n = 48°$

9. a. $y = 116°$, corresponding
   b. $z = 38°$, alternate
   c. $b = 136°$, supplementary

10. a. 68°, vertically opposite
    b. 90°, corresponding

11. c, f

12. b, e, f

13. a. 25°   b. 57°

14. a. 45°   b. 55°

15.

| Diagram | Type of angle relation | Associated shape | Rule |
|---|---|---|---|
| | Corresponding | F | Are equal |
| | Alternate | Z | Are equal |
| | Co-interior | C | Add to 180° |

16. a. C   b. D   c. A   d. B

17. a. *a* and *d*, *c* and *b*, *e* and *h*, *f* and *g*
    b. *a* and *e*, *c* and *g*, *b* and *f*, *d* and *h*
    c. *c* and *e*, *d* and *f*
    d. *c* and *f*, *d* and *e*

18. a. 123°   b. 43°   c. 45°

19. a. 28°   b. 50°   c. 70°

20. No, because $48° + 133° \neq 180°$. For parallel lines, co-interior angles need to add to 180° (or alternate angles must be equal).

21. a. 52°. The three angles forming the straight line add to 180°.
    b. Yes. $52° + 128° = 180°$, so lines parallel as co-interior angles add to 180°.
    c. 52°, 55°. Exterior angle: 125°

22. a. The vertical lines are parallel, the horizontal lines are parallel and the oblique lines are parallel because they are equally distant from each other.
    b. The vertical lines are parallel and the horizontal lines are parallel because they are equally distant from each other.

23. 60°

24. a. 130°   b. 50°

25. Acute $\angle PQR = 67°$ and reflex $\angle PQR = 293°$.

## Project

1. a.

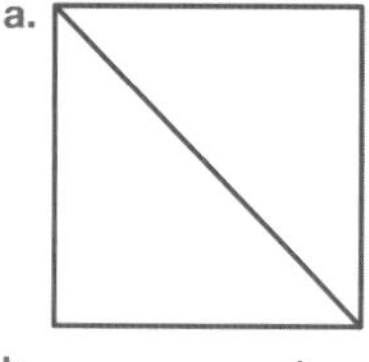

b.

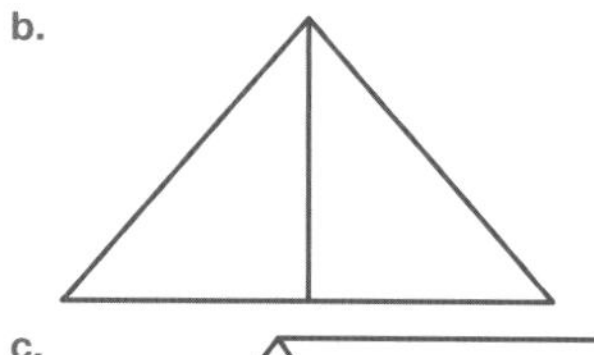

c.

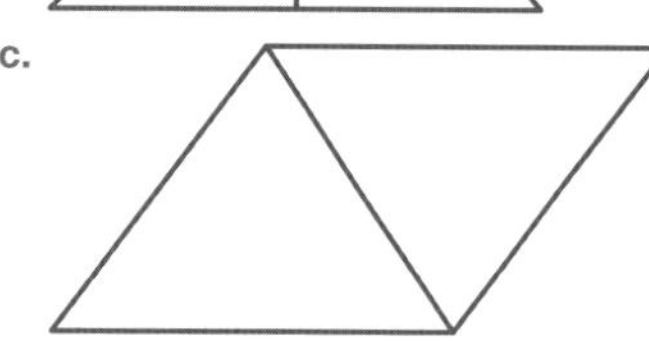

2. a.

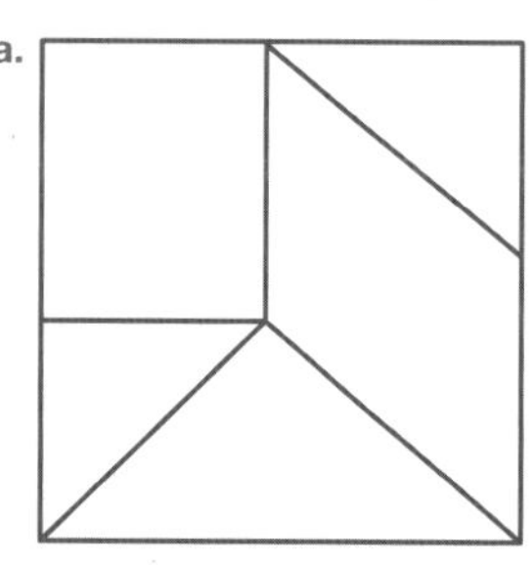

b.

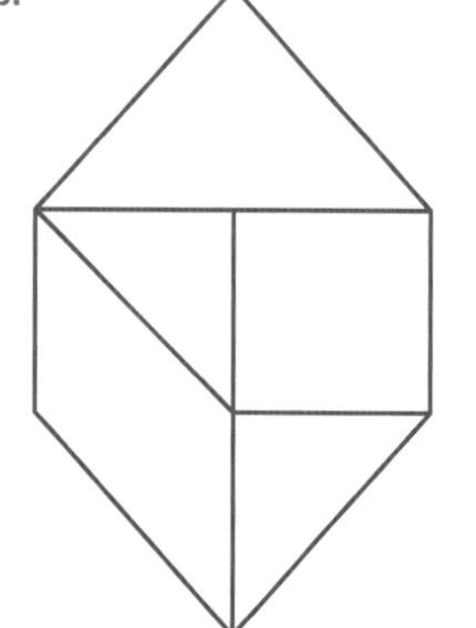

c.

d.

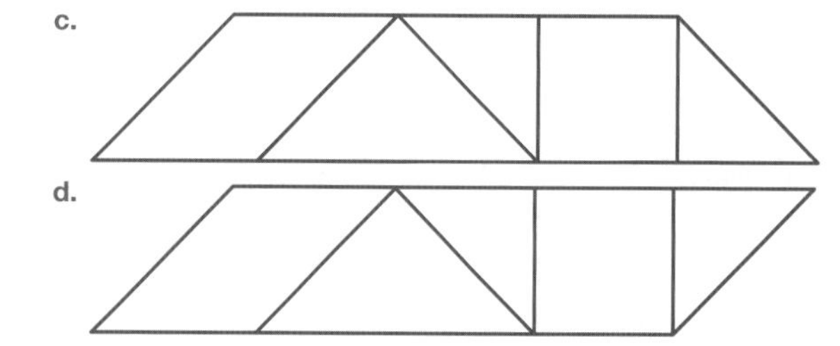

3. a.

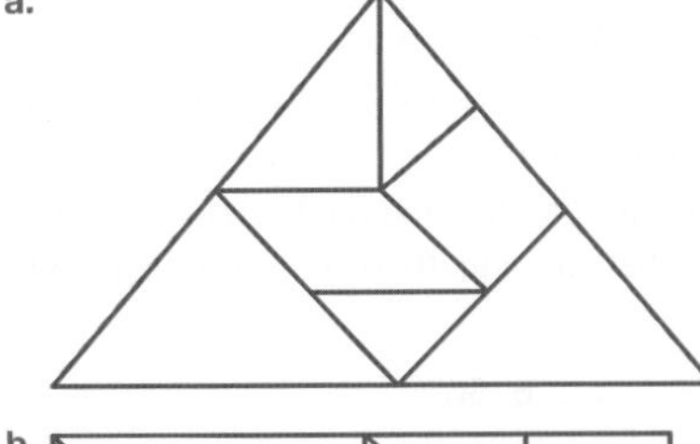

b.

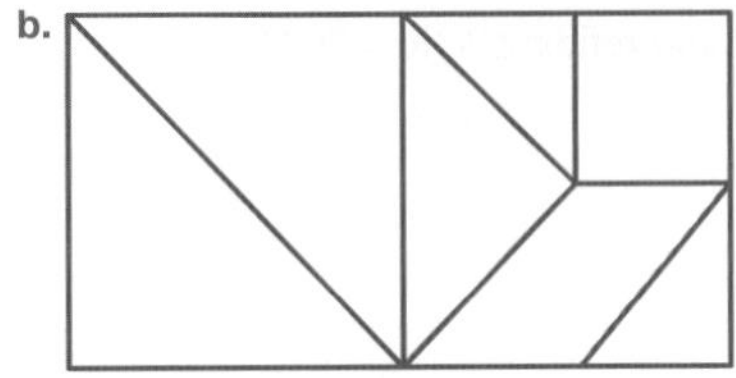

c.

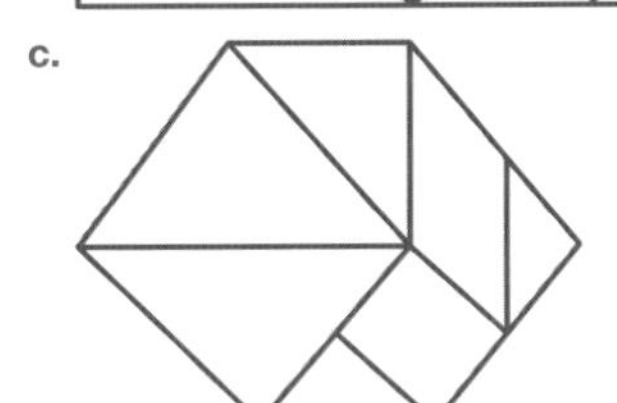

4. a.

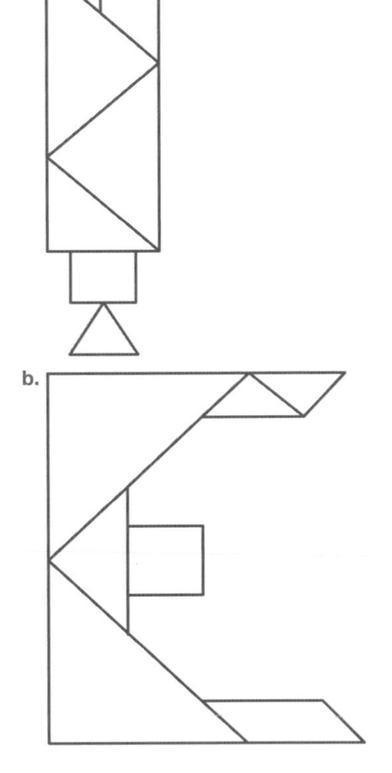

b.

## 9.8 Review questions

1. a. Revolution
   b. Right angle
   c. Straight angle
2. a. Obtuse b. Reflex c. Acute
3. B
4. D
5. C
6. A
7. ∠EDF or ∠FDE
8. a. Equilateral triangle
   b. Scalene triangle
   c. Isosceles triangle
9. a. Obtuse-angled triangle
   b. Acute-angled triangle
   c. Right-angled triangle
10. a. $t = 67°$ b. $b = 28°$ c. $x = 75°$
11. a. $p = 56°$ b. $m = 64°$ c. $n = 60°$
12. a. Rectangle b. Parallelogram c. Trapezium
13. a. Kite b. Square c. Rhombus
14. a. 75° b. 55° c. 42°
15. a. 66° b. 112° c. 126°
16. a. $x = 15°$ because vertically opposite angles are equal.
    b. $x = 50°$ because corresponding angles are equal and the angle sum of a straight line is 180°,
    $y = 25°$ because corresponding angles are equal.
    c. $y = 100°$ because co-interior angles add to 180°, $x = 8°$
    The sum of the interior angles of a triangle equals 180°.
    Since $90 + 50 = 140$ and $180 - 140 = 40$, $5x = 40$.
    Therefore $x = 8$.
17. a. i. Alternate angles
    ii. $x = 60°$
    b. i. Vertically opposite angles
    ii. $y = 135°$
    c. i. Vertically opposite angles, then co-interior angles, then vertically opposite angles. Another solution is supplementary angles, followed by corresponding angles.
    ii. $t = 50°$
18. A, B
19. A, E
20. C, E
21. C
22. 67°
23. 115°
24. Because an equilateral triangle has 3 congruent angles, right-equilateral and obtuse-equilateral are not possible. Examples of the other triangles:

i.

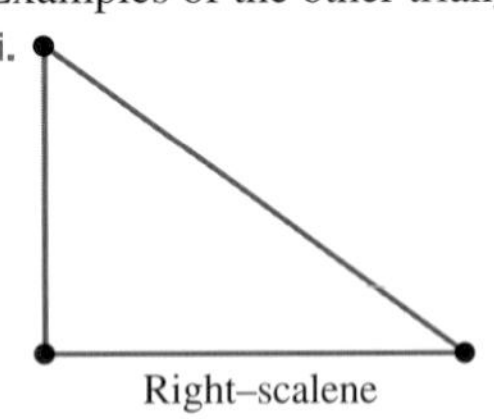

Right–scalene

ii.

Right–isosceles

iii.

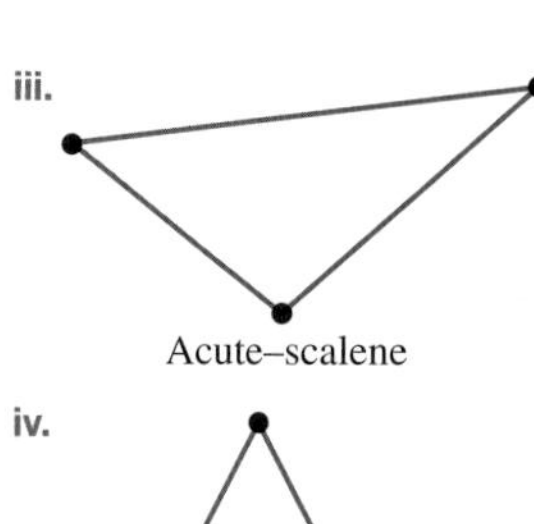

Acute–scalene

iv.

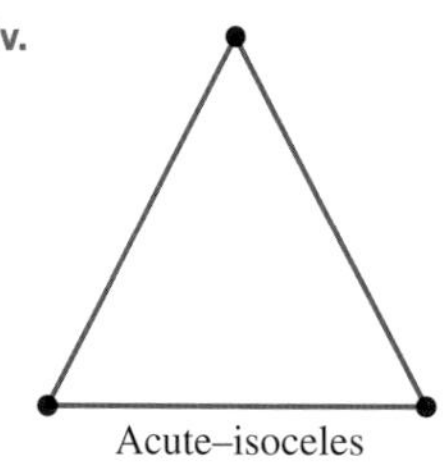

Acute–isoceles

v.

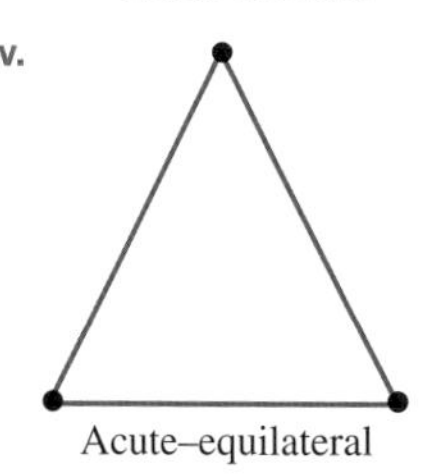

Acute–equilateral

vi.

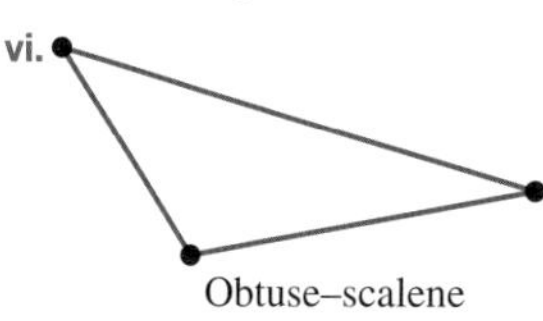

Obtuse–scalene

vii.

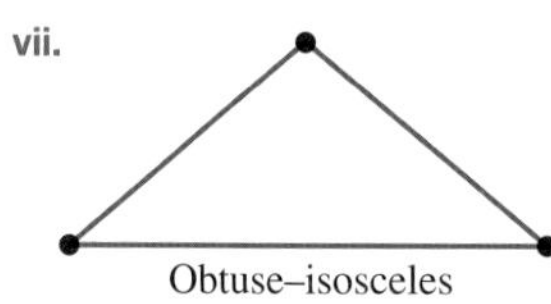

Obtuse–isosceles

25. 4
26. ∠BOE, ∠BOD, ∠BOC, ∠COE, ∠COD, ∠DOE
27. 75°
28. 35°

# 10 Measurement

## LESSON SEQUENCE


**10.1** Overview .....518
**10.2** Units of measurement .....521
**10.3** Reading scales and measuring length .....528
**10.4** Perimeter .....536
**10.5** Circles and circumference .....544
**10.6** Area .....553
**10.7** Area of composite shapes (extending) .....566
**10.8** Volume of right prisms .....571
**10.9** Capacity .....580
**10.10** Drawing solids .....584
**10.11** Review .....595

# LESSON
# 10.1 Overview

## Why learn this?

Measurement, together with geometry, is important in our everyday lives. We are able to describe objects using numbers and units of measurement, such as millimetres, centimetres, grams and kilograms. How far is it from your home to the nearest beach? What is the size of a tennis court or a football ground? How long is a cricket pitch? How much sugar is in your favourite cookie? All of these questions and thousands more are answered using measurement.

When we measure objects we need to understand the units of measurement. Which is longer, a centimetre or a millimetre? Can you change centimetres to millimetres and millimetres to centimetres? From stump to stump, a cricket pitch is 20.12 metres or 2012 centimetres or 20120 millimetres!

Olympic swimming pools need to be 50 metres long and 25 metres wide. Builders must know these dimensions before constructing a pool. Before anything can be made, you need to decide how small or large it will be — that is, its dimensions or measurements. We use measurement to describe length, perimeter, area, volume and capacity every day. Can you imagine trying to build your home without first knowing what its size will be? Many professionals use measurement in their day-to-day work. Among these are property developers, builders, engineers, designers, dressmakers, chefs, architects and construction workers.

### DISCUSSION

Why is it important to have an international standard metric system? Which countries don't follow this system?

## Exercise 10.1 Pre-test

learnon

1. Convert 0.565 m into cm.

2. If the length of a cricket pitch is 20.12 m, calculate its length in km.

3. If the height of a city building is 36 m and it measures 10 cm in a photo, calculate how tall a building that measured 25 cm in a photo would be in real life.

4. Luca went for a training run that was 1200 strides long. If each of his strides was 1.5 m, calculate the total distance he covered.

5. Determine the perimeter of this figure.

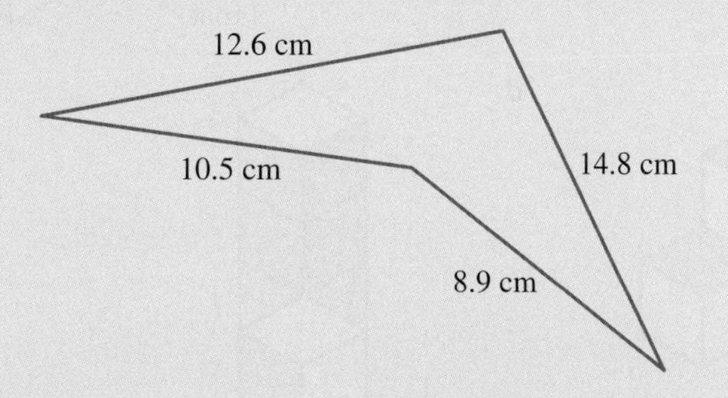


6. A rectangular pool has a perimeter of 34 m. If a concrete border of width 2 m was laid around the pool, calculate the perimeter around the outer section of the concrete border.

7. Determine the area of the triangle at right.

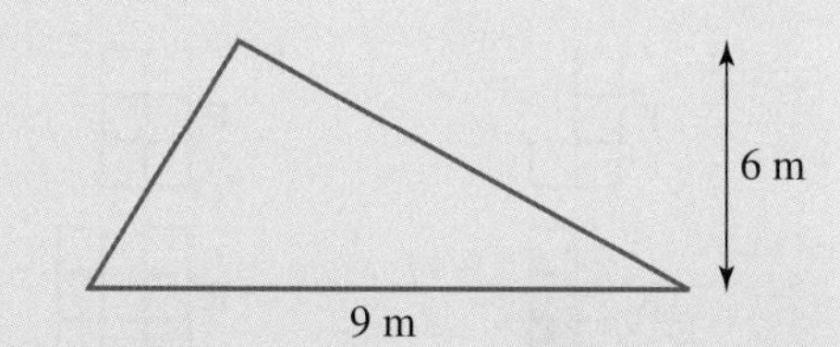


8. A rectangular room with dimensions 3.5 m by 6 m is to be carpeted at a cost of \$22.50 per $m^2$.
   Calculate the total cost to carpet the room.

9. Calculate the area of the figure at right.

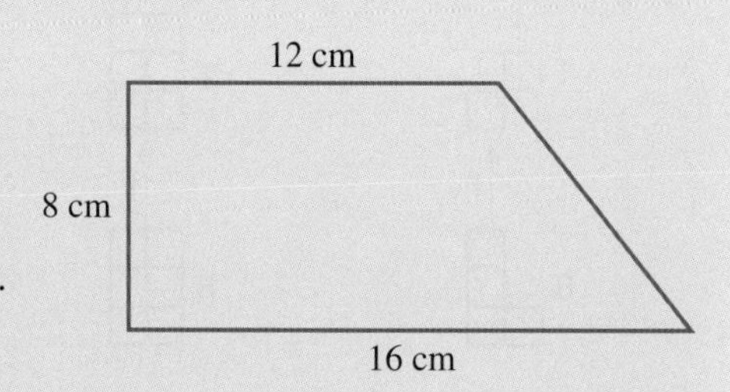


10. A rectangular wall 6.2 m by 2.4 m is to be painted.
    A rectangular cabinet 1.8 m by 1.2 m rests against this wall.
    The area to be painted does not include the area of the cabinet.
    Calculate the area of the wall to be painted.

11. Calculate the volume of a right prism with a length of 8 m, a width of 3 m and a height of 2 m.

12. A right prism has a volume of 112 $cm^3$.
    If it has a height of 2 cm and a width of 8 cm, determine its length.

13. A rectangular sink has a length of 45 cm, width of 40 cm and a depth of 30 cm. Calculate the sink's capacity, in litres.

14. Match the following three-dimensional shapes to their plan views.

a.

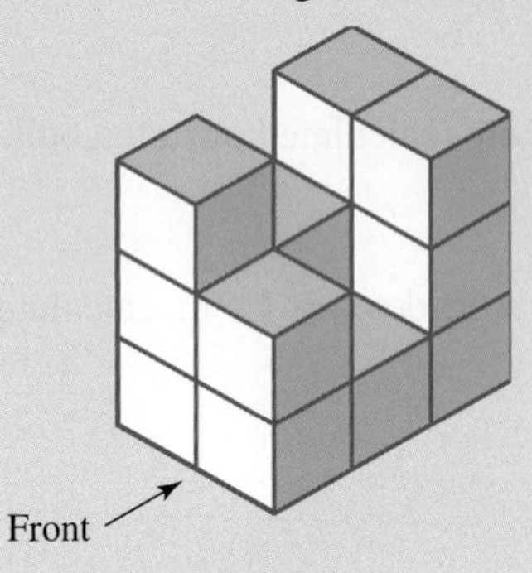


b.

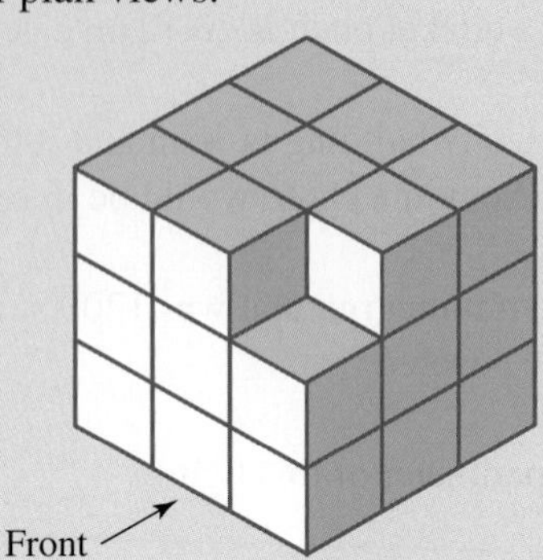


c.

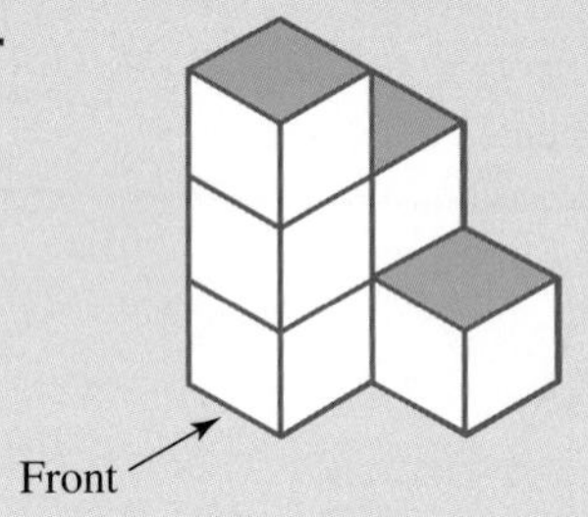


d.

e.

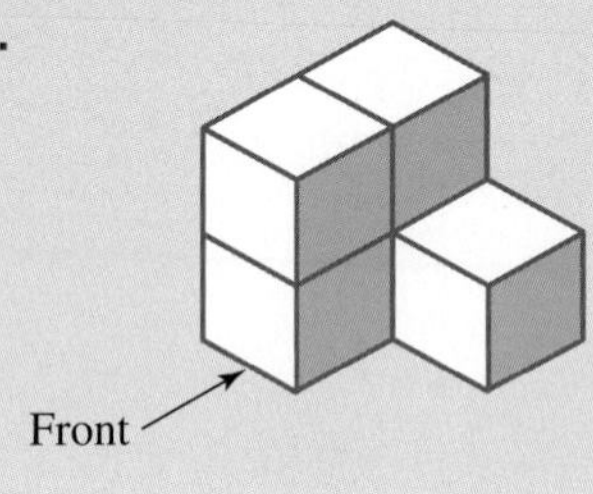


Plan views:

i. F R T F B

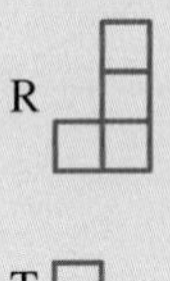

ii. F R T F B

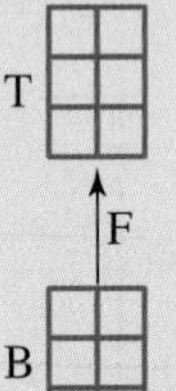

iii. F R T F B

iv. F R T F B

v. F R T F B

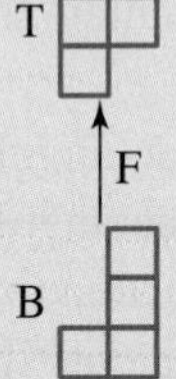

15. Select the correct shape that represents each of the following views.

a. **MC** The top view of a telephone pole
   **A.** Oval **B.** Circle **C.** Trapezium **D.** Rectangle **E.** Triangle

b. **MC** The top view of the Sydney Cricket Ground
   **A.** Oval **B.** Circle **C.** Trapezium **D.** Rectangle **E.** Triangle

c. **MC** The side view of a bucket
   **A.** Oval **B.** Circle **C.** Trapezium **D.** Rectangle **E.** Triangle

d. **MC** The top view of a car
   **A.** Oval **B.** Circle **C.** Trapezium **D.** Rectangle **E.** Triangle

# LESSON
# 10.2 Units of measurement

**LEARNING INTENTION**

At the end of this lesson you should be able to:
- convert between different units of length.

## 10.2.1 Metric units of length

eles-4549

- The metric system is based on the number 10.
- The base unit of length of the metric system is the metre.
- The following figures show the most commonly used units of length along with their abbreviations and photos showing approximate examples.

1. **Kilometre** (**km**)

One kilometre is the distance travelled in one minute by a car travelling at a speed of 60 kilometres per hour.

2. **Metre** (**m**)

The length of an adult's stride

3. **Centimetre** (**cm**)

The width of each of your fingers

4. **Millimetre** (**mm**)

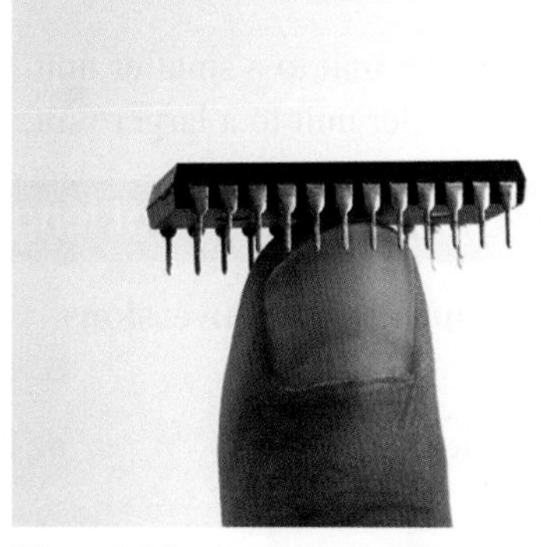

The width of a wire in this computer chip

**DISCUSSION**

Where do you use measurement in your everyday life?

**WORKED EXAMPLE 1 Identifying the most appropriate unit of length**

**You have been given the task of measuring the width of an A4 page. State which metric units of length you would use.**

| THINK | WRITE |
|---|---|
| Estimate the length involved. In this case, it will be less than a metre and more than a centimetre. Consider which unit would be easiest to use. If you use metres, the width would be expressed as a decimal. If you use centimetres, the width would be expressed as a whole number. Therefore, centimetres is the best choice of unit. | Centimetres (or cm) |

## 10.2.2 Converting units of length

eles-4550

- The main metric units of length are related as follows:

$$1\text{ km} = 1000\text{ m} \quad 1\text{ m} = 100\text{ cm} \quad 1\text{ cm} = 10\text{ mm}$$

### Unit conversion

Units of length can be converted as shown in the following diagram. The numbers next to each arrow are called **conversion factors**.

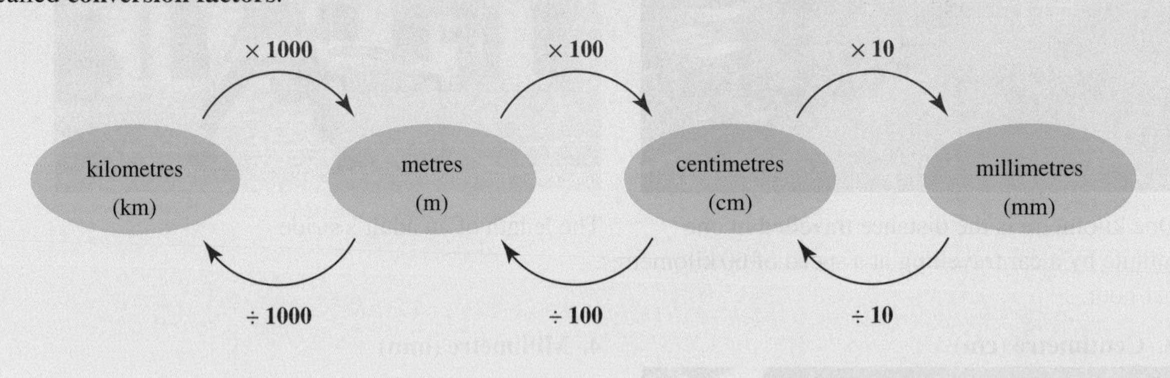


- When converting from a larger unit to a smaller unit, multiply by the conversion factor.
- When converting from a smaller unit to a larger unit, divide by the conversion factor.

**WORKED EXAMPLE 2 Converting units of length**

**Complete each of the following metric conversions.**

**a.** **0.3285 km = ____________ m**
**b.** **560 m = ____________ mm**
**c.** **480 cm = ____________ km**
**d.** $\mathbf{2\frac{3}{5}}$ **m = ________ cm**

**THINK**

**a.** To convert kilometres to metres, multiply by 1000. (Move the decimal point 3 places to the right.)

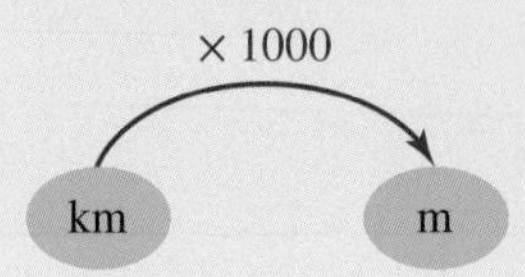


**WRITE**

**a.** $0.3285\text{ km} = 0.3285 \times 1000\text{ m}$
$= 0.3285$
$= 328.5\text{ m}$

**b.** To convert from metres to centimetres, multiply by 100 (since there is no decimal point, place two zeros after the final digit).

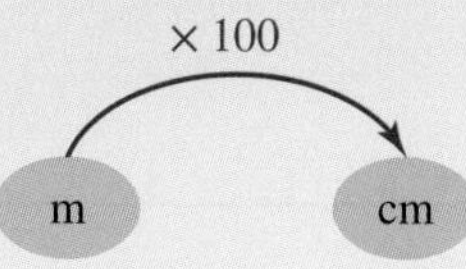


**b.** $560\text{ m} = 560 \times 100\text{ cm}$
$= 56\,000\text{ cm}$

To convert from centimetres to millimetres, multiply by 10 (place one zero after the final digit).

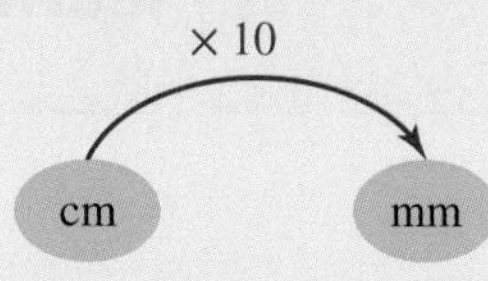


$= 56\,000 \times 10\text{ mm}$
$= 560\,000\text{ mm}$

*Note:* Overall, we need to multiply by $100 \times 10$ or 1000.

**c.** To convert from centimetres to metres, divide by 100 (move the decimal point 2 places to the left).

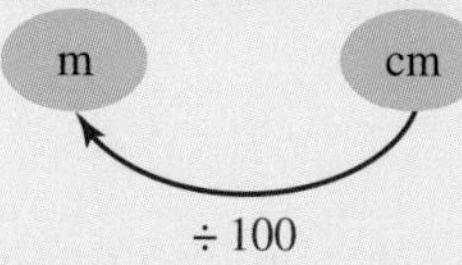


**c.** $480\text{ cm} = 480 \div 100\text{ m}$
$= 4.8\text{ m}$

To convert from metres to kilometres, divide by 1000 (move the decimal point 3 places to the left).

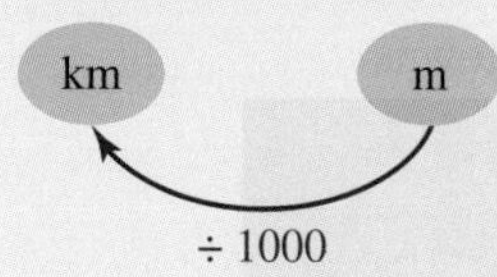


$= 4.8 \div 1000\text{ km}$
$= 0.0048\text{ km}$

**d. 1.** Convert $2\frac{3}{5}$ m to an improper fraction.

**d.** $2\frac{3}{5}\text{ m} = \frac{2 \times 5 + 3}{5}\text{ m}$
$= \frac{13}{5}\text{ m}$

**2.** Convert metres to centimetres by multiplying by 100.

× 100
m → cm

*Note:* Remember when multiplying fractions by a whole number, the number multiplies the numerator.

$= \frac{13}{5} \times 100\text{ cm}$
$= \frac{1300}{5}\text{ cm}$

**3.** Simplify the fraction.

$= 260\text{ cm}$

## Resources

**eWorkbook** Topic 10 Workbook (worksheets, code puzzle and project) (ewbk-1960)

**Interactivities** Individual pathway interactivity: Units of measurement and converting units of measurement (int-4355)
Metric units of length (int-4010)
Converting units of length (int-4011)

# Exercise 10.2 Units of measurement

learnon

**10.2 Quick quiz** on

**10.2 Exercise**

## Individual pathways

| ■ PRACTISE | ■ CONSOLIDATE | ■ MASTER |
| --- | --- | --- |
| 1, 3, 4, 6, 9, 14, 16, 18, 20, 25, 28 | 2, 5, 7, 11, 13, 15, 17, 22, 23, 26, 29, 30 | 8, 10, 12, 19, 21, 24, 27, 31, 32, 33, 34 |

## Fluency

1. WE1 State which metric units (mm, cm, m or km) would be most suitable for measuring the real lengths marked in each photograph.

a. The length of a large kangaroo

b. Large distances between towns

c. The diameter of a DVD

d. The diameter of a planet

e. The height of a tall building

2. State which metric units of length you would use for measuring the following.

a. The length of a netball court
b. The diameter of a netball
c. The distance between Melbourne and Sydney
d. The thickness of a magazine

3. **MC** Select which unit you would use to measure and compare the thicknesses of two different brands of chocolate biscuits.

A. millimetres
B. kilometres
C. metres
D. centimetres
E. kilometres or metres

4. **WE2** Complete each of the following metric conversions.

a. 2.0 km = _______ m
b. 7.0 km = _______ m
c. 5.3 km = _______ m
d. 0.66 km = _______ m

5. Complete each of the following metric conversions.

a. 0.25 m = _______ cm
b. 28.0 cm = _______ mm
c. 200.0 cm = _______ mm
d. 700.0 m = _______ cm

6. Convert to the units indicated.

a. 8000 m = _______ km
b. 6500 m = _______ km
c. 700 m = _______ km
d. 50 m = _______ km

7. Convert to the units indicated.

a. 6000 cm = _______ m
b. 57 cm = _______ m
c. 45 mm = _______ cm
d. 25 600 mm = _______ cm

8. Convert to the units indicated.

a. 8 km = _______ cm
b. 101 m = _______ mm
c. 72.33 m = _______ mm
d. $30\frac{7}{20}$ mm = _______ m

9. **MC** Identify which of these distances is the same as 6.25 km.

A. 625 m
B. 0.006 25 m
C. 625 000 mm
D. 625 000 cm
E. 62.50 cm

10. **MC** Identify which of these distances is the same as 7 860 000 cm.

A. 78.6 km
B. 786 m
C. 786 km
D. 786 000 cm
E. 7.86 km

11. Convert to the units indicated.

a. $45\frac{1}{5}$ m to km
b. 560 mm to m
c. $8\frac{3}{4}$ cm to mm

12. Convert to the units indicated.

a. 0.0006 km to cm
b. $3\frac{9}{100}$ km to cm
c. 48 mm to cm

## Understanding

13. Give an example of a length that each of the following people might measure in their jobs. (For example, a carpet layer would measure the length of a room.)

a. Veterinary surgeon
b. Costume designer
c. Carpenter
d. Landscape gardener
e. Field athlete

14. The longest snake ever held in captivity was a female reticulated python named Colossus. She measured 8.68 m long. The shortest species of snake, the West Indian *Leptotyphlops bilineatus*, only grows to 108 mm. Write, in centimetres:

a. the length of the longest snake
b. the length of the shortest snake.

15. The world's highest mountains are Everest (8848 m) and K2 (8611 m). Convert these heights to kilometres.

16. Arrange the following in order from smallest to largest.

   a. 12.5 m, 150 cm, 0.02 km
   b. 350 cm, 0.445 m, 3000 mm
   c. 50 km, 500 m, 50 000 mm

17. Arrange the following in order from smallest to largest.

   a. 1700 cm, 1.7 m, 0.17 km
   b. 0.052 cm, 0.0052 mm, 0.000 052 m
   c. 990 cm, 0.909 m, 9000 mm

18. Add the following lengths, giving your answer in the specified unit.

   a. 75 cm and 3 m: cm
   b. 2700 m and 7.5 km: m
   c. 1.66 m and 58.2 cm: cm
   d. 0.000 675 km and 87.8 cm: cm

19. Calculate the difference between each of the following lengths, giving your answer in the specified unit.

   **a.** 72 km and 5600 m: m
   **b.** 418 000 mm and 7.6 m: m
   **c.** $34\frac{3}{5}$ cm and $\frac{9}{20}$ m: cm
   **d.** $2\frac{4}{5}$ km and 450 000 cm: km

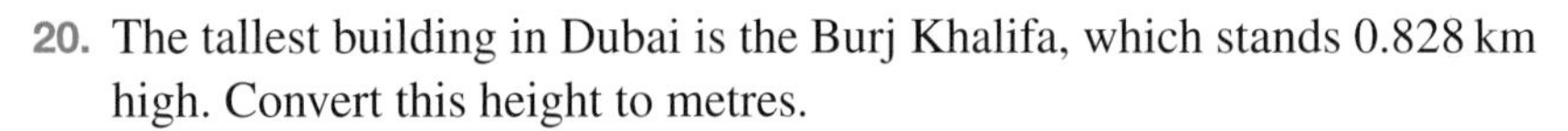

20. The tallest building in Dubai is the Burj Khalifa, which stands 0.828 km high. Convert this height to metres.

21. A builder needs to build a wall 3.5 m high. If each layer of bricks adds 8 cm of height, calculate how many layers of bricks are in the wall.

22. Deanne buys a length of rope and cuts it into three smaller sections, each of length 7200 cm.
Calculate how long the original piece of rope measured.

23. Norbert is 1.53 m tall in bare feet. If he wears shoes with soles that are 6.3 cm thick, calculate how tall he is when wearing these shoes, in metres.

24. Adrian is driving a truck with a rooftop 3.4 m above road level when he approaches an overpass bridge that has a clearance of 376 cm. Determine whether Adrian's truck will fit under the bridge. If so, state the room he has spare, in centimetres.

## Reasoning

25. Zvenglo is stacking identical boxes of height 330 mm. Calculate the height of a stack of six boxes, in centimetres.

26. Finita attaches a trellis that is 0.6 m high to the top of her 180 cm high fence. Calculate the height to the top of the new trellis from ground level. Give your answer in metres.

27. Astronomers use light-years as a measure of distance in the universe. A light-year is the distance travelled by light in one year. If light travels approximately 300 000 km in one second, calculate the distance travelled by light in:
   a. i. one minute
      ii. one hour
      iii. one day
      iv. one year (365 days).
   b. Explain why astronomers use this measurement for distance rather than kilometres.

## Problem solving

28. A pin, 14 mm long, is made of wire. Determine how many whole pins could be made from 1 km of wire.

29. A licorice strap machine takes 3.80 m lengths of licorice and chops them into 10 cm long pieces. Determine how many pieces each 3.80 m length produces.

30. Waldo's noticeboard is 1.5 m long and 1.2 m wide. If he pins a calendar of length 70 cm and width 60 cm exactly in the middle of the board, determine the distance from the top of the calendar to the top of the noticeboard. (*Hint:* Draw a diagram of the situation.)

31. A childcare centre has three large cardboard boxes that the children stack up in various combinations. What stack heights are possible if the boxes' individual heights are 600 mm, 45 cm and 1.1 m? Give your answer in centimetres.

32. A mother and daughter are riding their bikes to the local market. The circumference of the mother's bike wheel is 2 m while the circumference of the daughter's bike wheel is 1 m.
   a. Calculate the number of rotations of the wheel there are in 100 m for the:
      i. mother
      ii. daughter.
   b. Calculate the number of rotations of the wheel there are in 1 km for the:
      i. mother
      ii. daughter.

33. In a 'log cabin' quilt, a pattern is created from a series of squares and rectangles. From a centre square, 4 congruent rectangles are placed to build a larger square. This process is repeated a number of times to build larger and larger squares. Laurel starts with a centre square of side length 2 cm. If each rectangle has a width of 2 cm and a length that depends on its position in the block, give the dimensions of the rectangles that will need to be cut if the final block is to be 18 cm square. (Don't worry about seam allowance.)

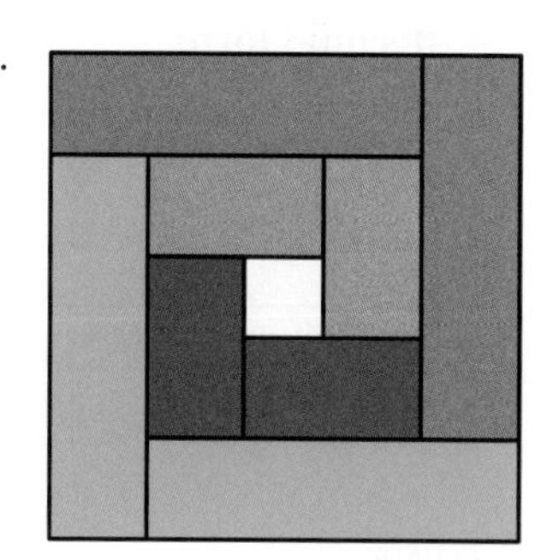

34. It is said that the average person walks the equivalent of four times around the Earth in a lifetime. The circumference of the Earth is about 40 000 km. If you lived to age 80, approximate the number of kilometres per week you would walk to achieve this distance.

# LESSON
# 10.3 Reading scales and measuring length

## LEARNING INTENTION

At the end of this lesson you should be able to:
- read and interpret various scales.

## 10.3.1 Reading scales and measuring length

eles-4551

- A **scale** is a set of levels or numbers used to measure length, mass, temperature or any other quantity.

### Reading scales

When reading scales and measuring lengths:
- check that the scale starts with zero
- check the value of each small division by counting along the scale to the next major mark
- always give units (for example, centimetres or kilograms) with your answer.

### WORKED EXAMPLE 3 Reading a scale

**State the reading indicated by the arrow, giving answers in:**
**i. decimal form** **ii. fractional form where appropriate.**

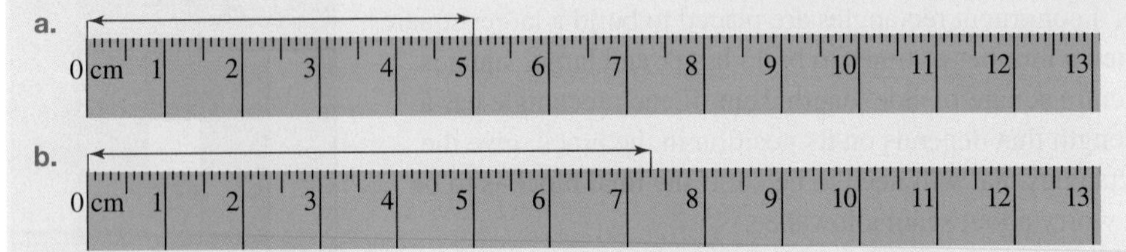


| THINK | WRITE |
|---|---|
| a. i. 1. Check that the line starts at 0. It does. | a. i. 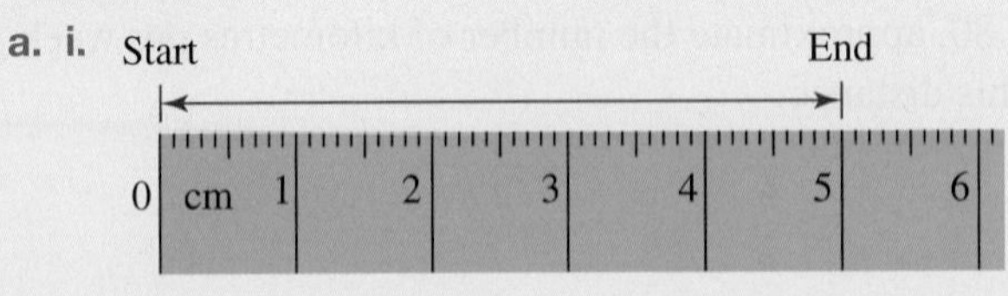 |
| 2. Note the units printed on the ruler (cm). | |
| 3. Read the last centimetre mark (5). | |
| 4. Does the line go past the last centimetre mark? No. | |
| 5. Write the answer with units. | 5 cm |
| ii. There is no need to write in fractional form as the value is a whole number. | |

| | |
|---|---|
| **b. i. 1.** Check that the line starts at 0. It does. | **b. i.** Start — End<br>0 cm 1 2 3 4 5 6 7 8 |
| **2.** Note the units printed on the ruler (cm). | |
| **3.** Read the last centimetre mark (7). | |
| **4.** Consider how many smaller intervals (or divisions) there are between each number. There are 10, so each smaller division represents 0.1 of a unit. | |
| **5.** Count the number of small divisions past the last centimetre mark (3). | |
| **6.** Write the answer with units. | 7.3 cm |
| **ii.** Express the answer obtained in **b i** as a fraction. Recall the conversion of decimals to fractions. | **ii.** $7.3 = \frac{73}{10}$ cm<br>Or, $7\frac{3}{10}$ cm |

## WORKED EXAMPLE 4 Applying a scale

**Use the given length of the climber's lower leg to estimate the labelled length of the rope in the following diagram.**

| THINK | WRITE |
|---|---|
| **1.** Refer to the 0.5 m lengths indicated on the diagram and determine how many of these lengths correspond to the unknown length of the rope above the climber.<br>*Note:* There are approximately $3\frac{1}{2}$ or 3.5 of these lengths in the rope. | From the diagram there are approximately $3\frac{1}{2}$ (or 3.5) of these lengths in the rope. |
| **2.** Multiply the number of lengths by 0.5 m. | $\text{Length} = 3.5 \times 0.5$<br>$= 1.75$ |
| **3.** Write the answer. | The length of the rope is approximately 1.75 m. |

## COLLABORATIVE TASK: Using your body to estimate lengths

To help improve your ability to estimate lengths, you can use different parts of your body for reference.

1. Use a ruler or a tape measure to measure the following distances and complete the following table.

| | Digit width | Handwidth | Fathom | Cubit | Arm length |
|---|---|---|---|---|---|
| Picture of measurement | 1 digit | 1 hand width | 1 fathom | 1 cubit | 1 arm length |
| Your measurement | | | | | |

2. Complete the following sentences.
   a. One metre is ___________.
   b. The bottom of my knee is ___________ off the floor.
3. As a class, select a range of objects around the room and:
   a. estimate the dimensions simply by looking at the object
   b. estimate the dimensions using a body part as a reference
   c. measure the dimensions using a ruler or tape measure.
4. Make a table, like the following one, to record your data for each object. A sample is shown.

| Object and dimension | First estimate | Number and type of body part unit | Estimate using body parts | Length, using a ruler or tape measure |
|---|---|---|---|---|
| Height of door | 190 cm | $23\frac{1}{2}$ handwidths | $23.5 \times 7.5$ cm $= 176.25$ cm | 180 cm |
| Length of pen | | | | |

5. List some other body dimensions that you could use to estimate length: for example, the distance from the tip of your nose to the end of your outstretched arm, or the length of your foot.
6. Use a ruler or tape measure to measure these distances on your own body.
7. As a class, discuss whether the measurements that you made were exact measurements or approximations.

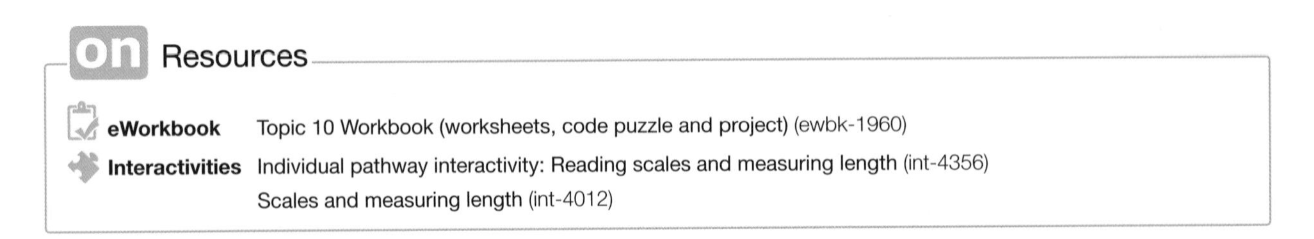

on Resources

eWorkbook Topic 10 Workbook (worksheets, code puzzle and project) (ewbk-1960)

Interactivities Individual pathway interactivity: Reading scales and measuring length (int-4356)

Scales and measuring length (int-4012)

# Exercise 10.3 Reading scales and measuring length

| 10.3 Quick quiz | 10.3 Exercise |
|---|---|

Individual pathways

| ■ PRACTISE | ■ CONSOLIDATE | ■ MASTER |
|---|---|---|
| 1, 4, 8, 9, 12, 15, 18, 21 | 2, 5, 10, 13, 16, 19, 22 | 3, 6, 7, 11, 14, 17, 20 |

## Fluency

1. WE3a State the reading indicated by the arrow.

a.

b.

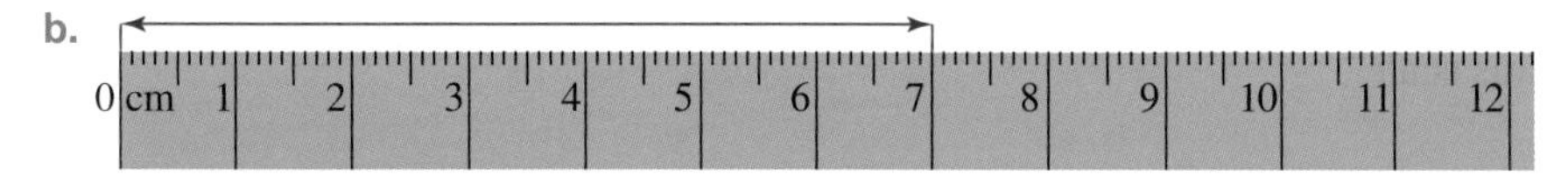


c.

2. WE3b State the reading indicated by the arrow, in decimal form.

a.

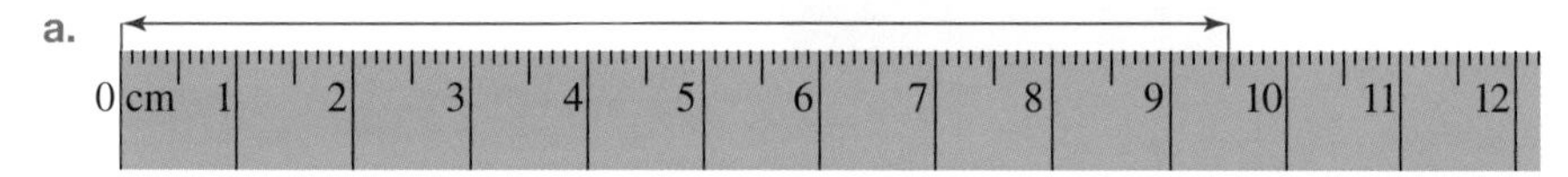


b.

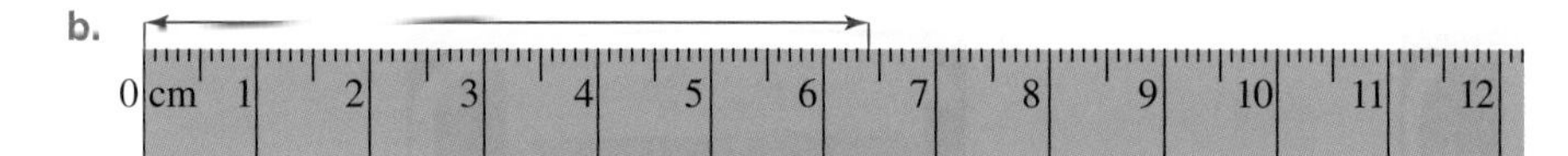


c.

d.

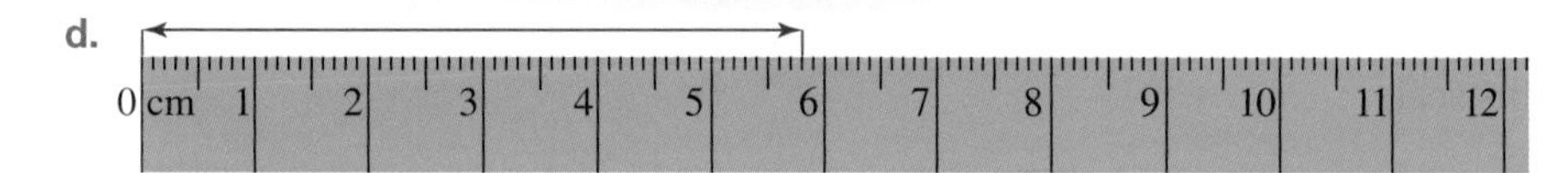


3. State the reading indicated by the arrow, in decimal form.

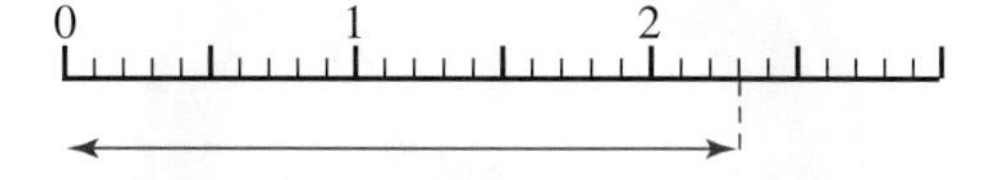

4. State the values shown on this scale.

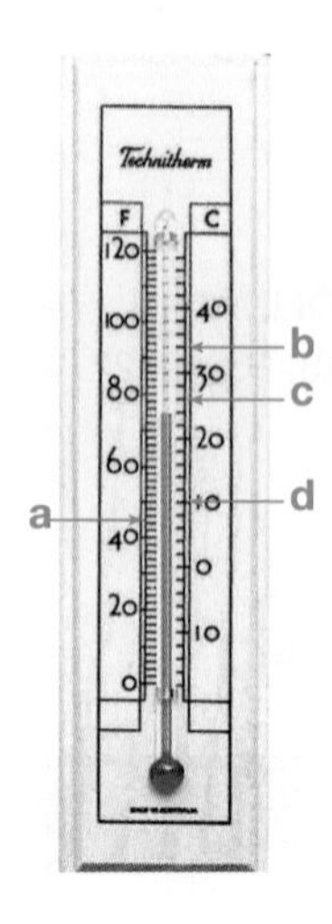


5. State the values shown on this scale.

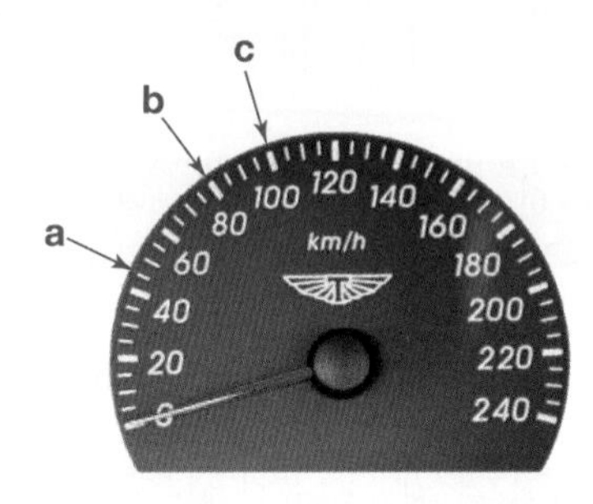


6. State the values shown in the following jugs.

a.

b.

7. State the value shown on this scale.

8. State the values shown on this scale.

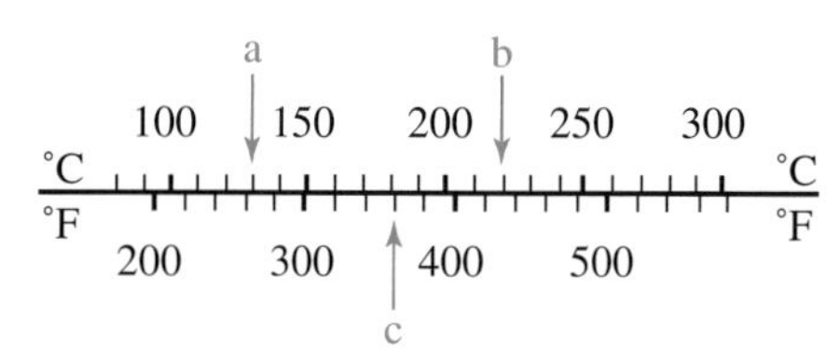


9. **MC** Choose the correct reading for the following scale.

**A.** 2.2 **B.** 2.4 **C.** 2.6 **D.** 2.8 **E.** 3.3

10. **MC** Choose the correct reading for the following scale.

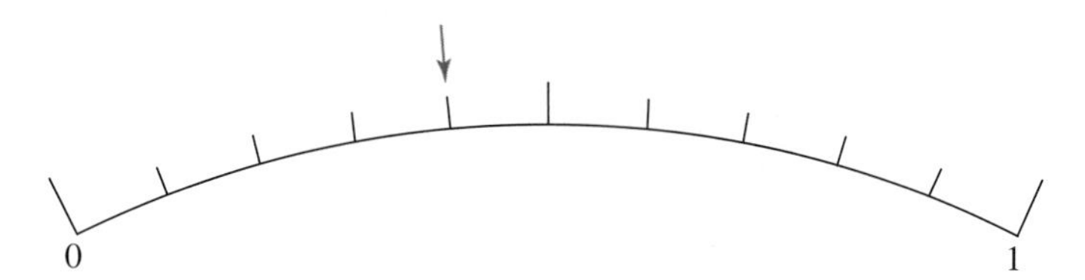


**A.** $\frac{4}{5}$ **B.** 4 **C.** $\frac{2}{5}$ **D.** $\frac{2}{10}$ **E.** $\frac{1}{2}$

## Understanding

For questions **11** to **17**, use the given length to estimate the other length mentioned.

11. **MC** Approximate the distance from the centre of Earth to the centre of the Sun, using the scale bar provided.

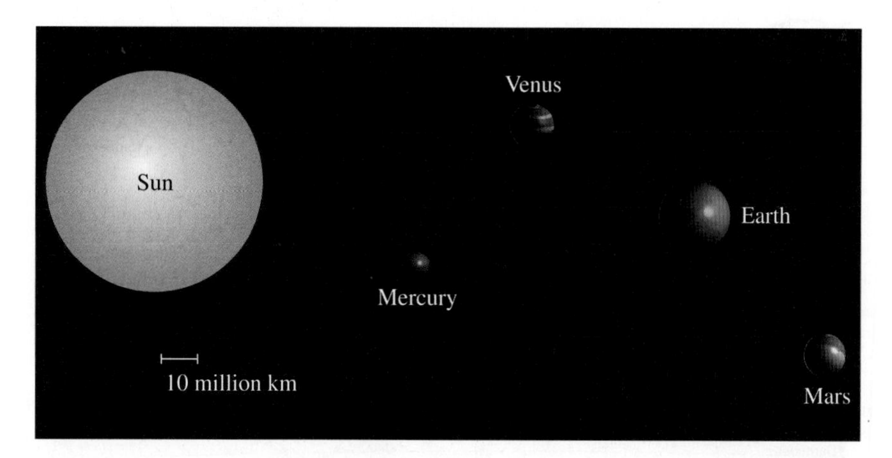


**A.** 10 million km **B.** 50 million km **C.** 100 million km
**D.** 150 million km **E.** 200 million km

12. **MC** If the train pictured has four carriages, identify the length of the train.

**A.** 15 m
**B.** 30 m
**C.** 45 m
**D.** 60 m
**E.** 150 m

13. **MC** If the height of the rhino in the photo is 1.8 m, choose the approximate height of the giraffe.

A. 1.8 m
B. 4.9 m
C. 3.6 m
D. 18 m
E. 7.0 m

14. State the approximate height of the window (to the inside centre of the frame) if the bucket is 30 cm tall.

15. **MC** On the following map, select the approximate straight-line distance from Geelong to Canberra.

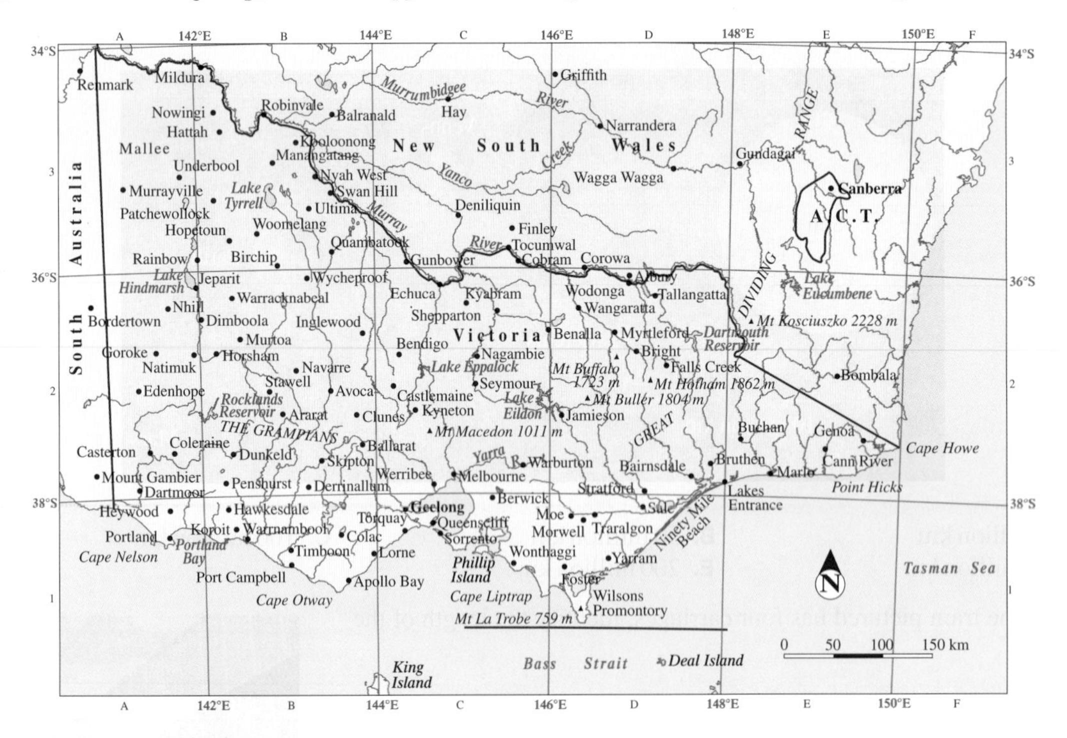


A. 150 km
B. 320 km
C. 540 km
D. 700 km
E. 1000 km

16. **MC** If 2 floors are 10 m tall, estimate the height of the building.

A. 136 m B. 100 m C. 184 m D. 350 m E. 388 m

17. **MC** If the orange ute is 180 cm tall, estimate the height ($x$) of the pole next to the building.

A. 18 m B. 12 m C. 24 m D. 15 m E. 8 m

## Reasoning

18. List three examples of situations in which you would use each of the following measuring tools in everyday life.
    a. A 30 cm ruler that includes millimetre markings
    b. A 100 cm ruler that has only centimetre markings
    c. A 2 m dressmaker's tape that includes centimetre markings
    d. A 100 m measuring tape that has metre and centimetre markings

19. Use the information provided in the following photograph to estimate the height of the lamppost. Explain how you estimated this height.

20. Explain how you could use a normal ruler to determine the thickness of a page in a textbook. Estimate the thickness of a page in your Maths textbook and Science textbook.

### Problem solving

21. Your friend is working on the top floor of a very tall office building. She walks up 726 steps from the ground floor to her office. Each step is 23 cm high. Calculate how high she climbs (in km), to 1 decimal place.

22. A gift box has dimensions 30 cm by 22 cm by 15 cm. If a ribbon is wrapped around this box as shown using 25 cm for a bow, determine the total length of ribbon needed.

# LESSON
# 10.4 Perimeter

**LEARNING INTENTION**

At the end of this lesson you should be able to:
- calculate the perimeter of a given shape
- calculate the perimeter of rectangles and squares.

## 10.4.1 Calculating the perimeter

eles-4552

- A **perimeter** is the distance around the outside (border) of a shape.
- To calculate the perimeter of the shape, convert all lengths to the same unit and then add all the lengths.

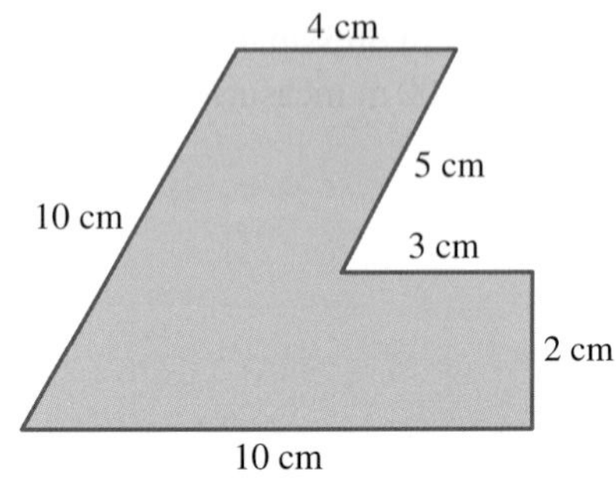


$$\text{Perimeter} = 4 + 5 + 3 + 2 + 10 + 10$$
$$= 34 \text{ cm}$$

## WORKED EXAMPLE 5 Calculating the perimeter of shapes

**Calculate the perimeter of each shape.**

a.

b. 13 cm, 12 cm, 10 cm

c. 30 mm, 6 cm, 45 mm

***Note:* The dots are 1 unit apart.**

| THINK | WRITE |
|---|---|
| a. 1. Count the number of unit intervals around the outside of the shape (16). | a. 3 4 5 6 7, 2 14 13 9 8, 1 15 12 10, 16 11 |
| 2. Write the answer. | The perimeter is 16 units. |
| b. 1. Check that the measurements are in the same unit. All are in centimetres. | b. |
| 2. Add the measurements to determine the perimeter. | $\begin{array}{r} 12 \\ 13 \\ +10 \\ \hline 35 \end{array}$ |
| 3. Write the answer with the correct unit. | The perimeter is 35 cm. |
| c. 1. Notice the measurements are not all the same. Convert to the smaller unit (6 cm = 60 mm). | c. 6 cm = 60 mm |
| 2. Add the measurements that now have the same unit (mm) to determine the perimeter. | $\begin{array}{r} 60 \\ 30 \\ +45 \\ \hline 135 \end{array}$ |
| 3. Write the answer with the correct unit. | The perimeter is 135 mm. |

## WORKED EXAMPLE 6 Calculating perimeters of squares and rectangles

**Calculate the perimeter of:**

**a. a rectangle that is 49.3 cm long and 22.0 cm wide**

**b. a square whose side length is 28 cm.**

| THINK | WRITE |
|---|---|
| a. 1. Draw a diagram and write its measurements. | a. 49.3 cm, 22.0 cm, 22.0 cm, 49.3 cm |

| | |
|---|---|
| **2.** Check that the measurements are in the same unit. All are in centimetres. The perimeter is the distance around the rectangle, so add all the distances together. | $P = 49.3 + 22.0 + 49.3 + 22.0$<br>$= 142.6$ |
| **3.** Write the answer with the correct unit. | The perimeter is 142.6 cm. |
| **b. 1.** Draw a diagram and write its measurements. | **b.** 28 cm<br>28 cm 28 cm<br>28 cm |
| **2.** Check that the measurements are in the same unit. The perimeter is the distance around the square, so add all the distances together. | $P = 28 + 28 + 28 + 28$<br>$= 112$ |
| **3.** Write the answer with the correct unit. | The perimeter is 112 cm. |

**DISCUSSION**

What information does the perimeter give you about a shape? Is it enough to draw the shape?

## 10.4.2 Calculating the perimeter of a rectangle and a square

eles-4553

- The perimeters of rectangles and squares can be found using the following formulas.

**Perimeter of a rectangle**

For a rectangle, the perimeter $P$ is:

$$P = 2(l + w)$$

where $l$ is its length and $w$ is its width.

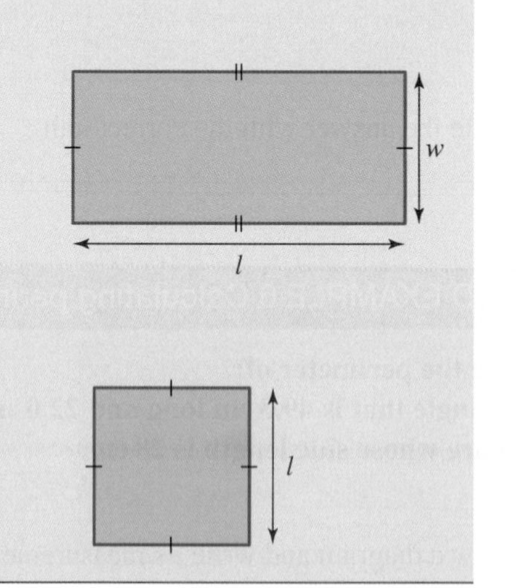


**Perimeter of a square**

For a square, the perimeter P is:

$$P = 4l$$

where $l$ is its side length.

## COLLABORATIVE TASK: Same perimeter, different shape

**Equipment:** a ball of string, A3 paper, marker pens, glue or sticky tape, ruler

1. Cut a piece of string into four equal lengths.
2. Measure and record the lengths of your pieces of string.
3. Make an enclosed shape with one of your pieces of string and stick it onto an A3 sheet of paper.
4. Write a caption about the perimeter of the shape. For example, you might write, 'It is 25 cm around this shape'.
5. Repeat steps **3** and **4** with your other pieces of string. Each time, make a different shape and write a different caption.
6. As a class, discuss the answer to the following question: What information does the perimeter give you about a shape?

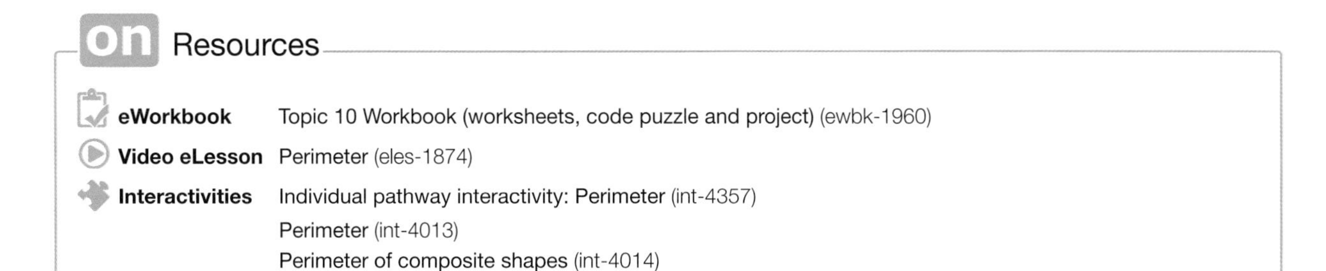

on Resources

| | |
|---|---|
| **eWorkbook** | Topic 10 Workbook (worksheets, code puzzle and project) (ewbk-1960) |
| **Video eLesson** | Perimeter (eles-1874) |
| **Interactivities** | Individual pathway interactivity: Perimeter (int-4357) |
| | Perimeter (int-4013) |
| | Perimeter of composite shapes (int-4014) |

# Exercise 10.4 Perimeter

learnon

**10.4 Quick quiz** on | **10.4 Exercise**

### Individual pathways

| ■ PRACTISE | ■ CONSOLIDATE | ■ MASTER |
|---|---|---|
| 1, 2, 4, 7, 10, 13, 16, 19, 22 | 3, 5, 8, 11, 14, 17, 20, 23 | 6, 9, 12, 15, 18, 21, 24 |

### Fluency

1. **WE5a** Calculate the perimeter of each shape. The dots are 1 unit apart.

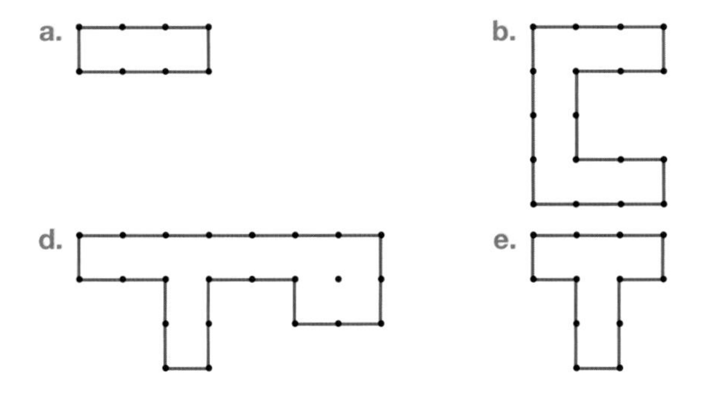


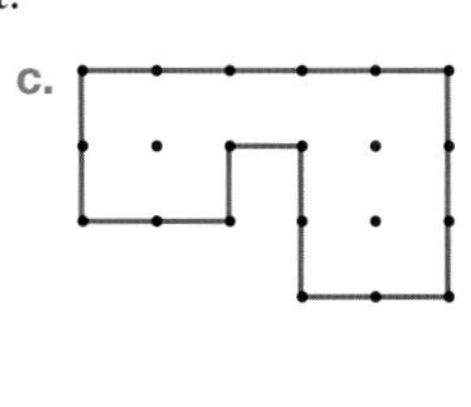

2. WE5b Calculate the perimeter of each of the following.

a.

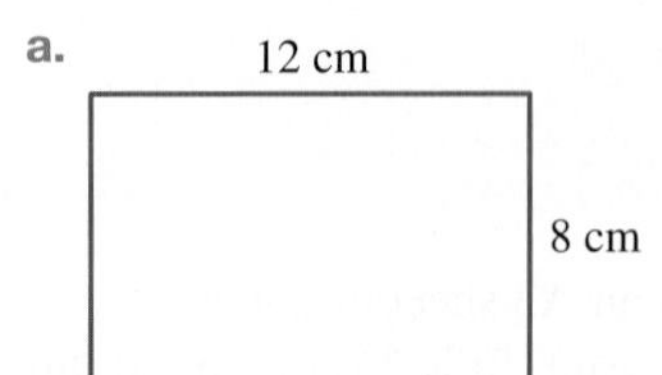


b.

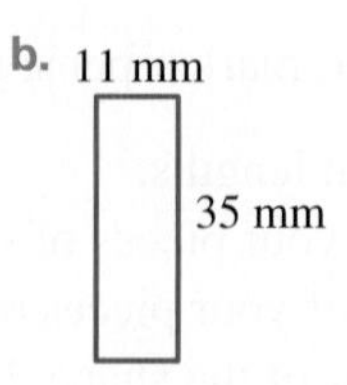


3. Calculate the perimeter of each of the following.

a.

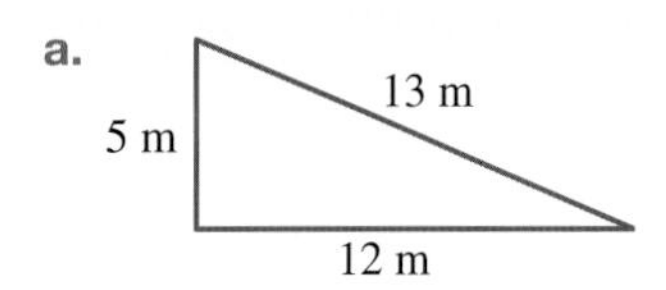


b.

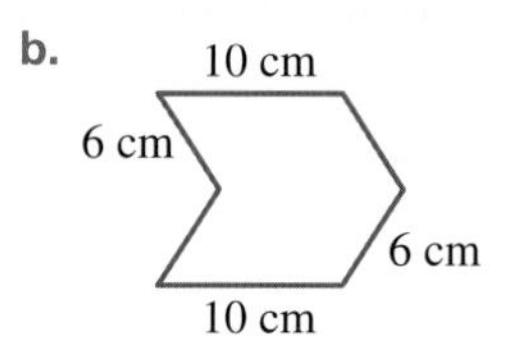


4. Calculate the perimeter of the shape shown.

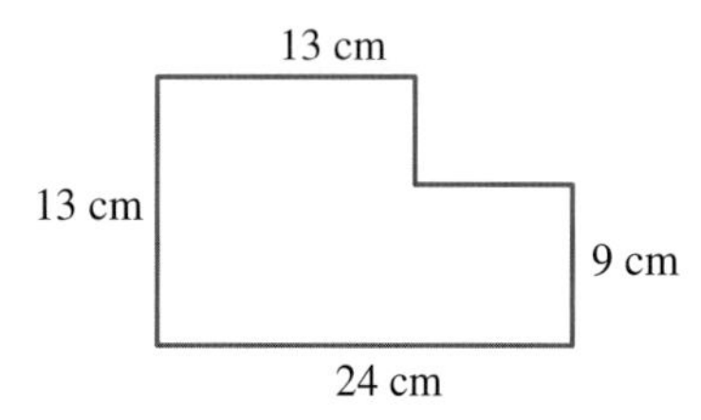


5. Calculate the perimeter of the shape shown.

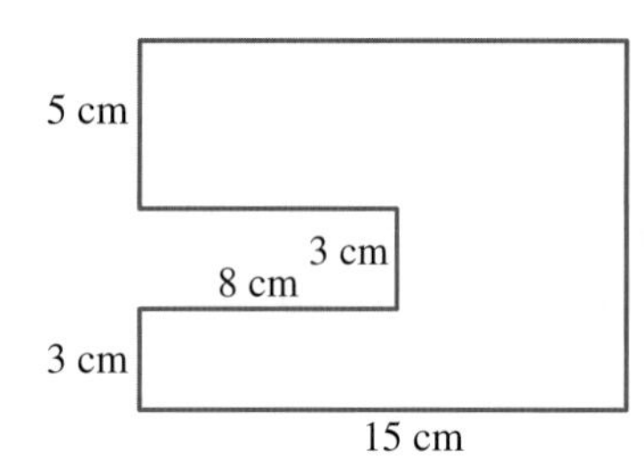


6. Calculate the perimeter of the shape shown.

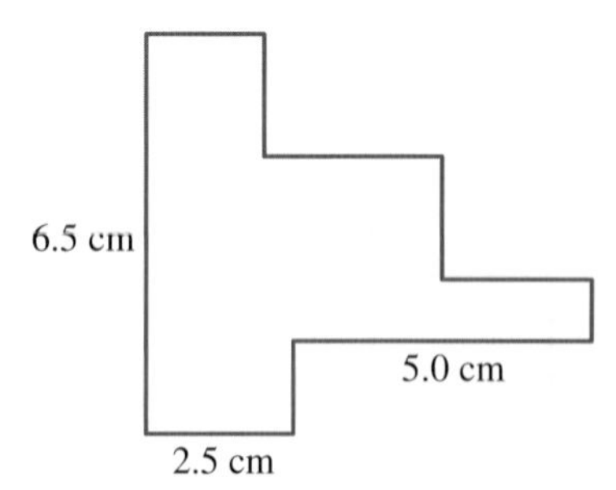

7. **WE5c** Calculate the perimeter of each shape, giving answers in the smaller unit in each case.

a.

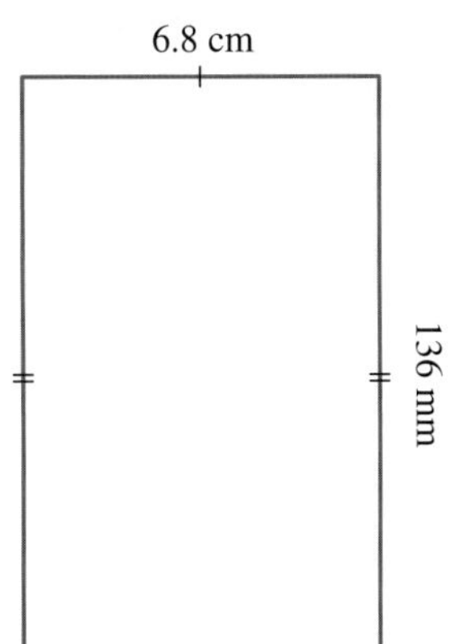


b.

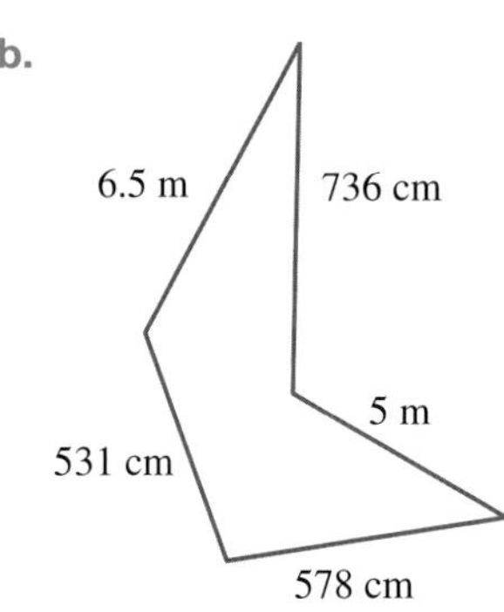


8. Calculate the perimeter of each shape, giving answers in the smaller unit in each case.

a.

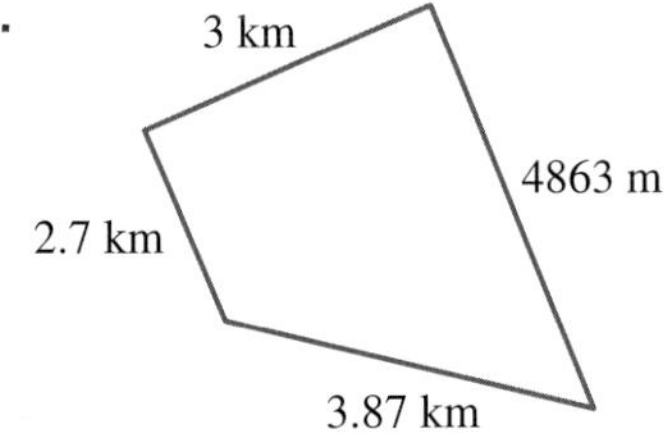


b.

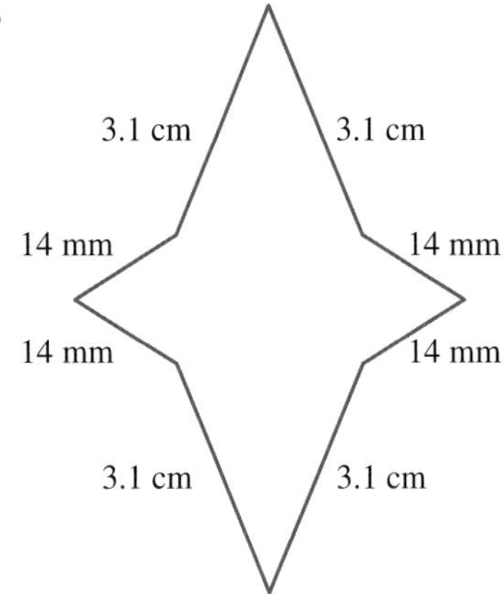


9. Calculate the perimeter of each shape, giving answers in the smaller unit in each case.

a.

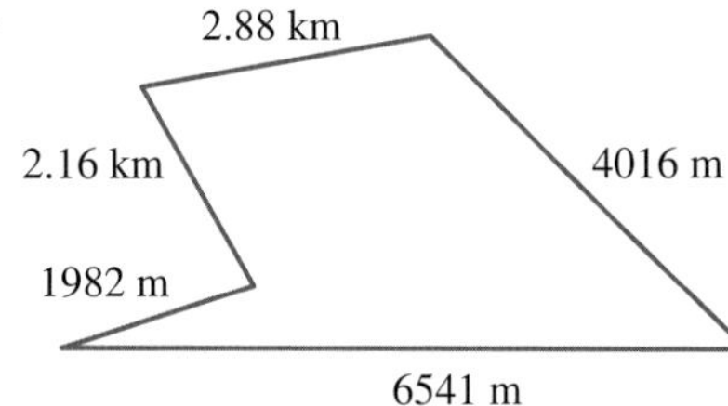


b.



10. **MC** Identify the perimeter of the shape shown.

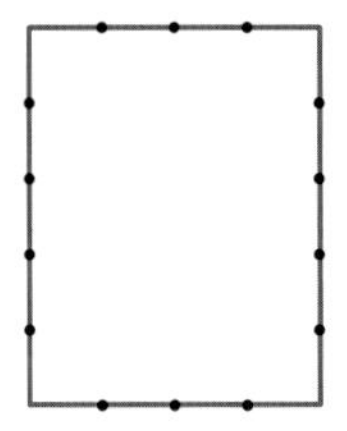

A. 9 units
B. 18 units
C. 16 units
D. 14 units
E. 12 units

*Note:* The dots are 1 unit apart.

11. **WE6**

a. Calculate the perimeter of a rectangle of length 45 cm and width 33 cm.
b. Calculate the perimeter of a rectangle of length 2.8 m and width 52.1 cm.
c. Calculate the perimeter of a square of side length 3.7 cm.

12. a. Calculate the perimeter of a rectangle of length $4\frac{1}{4}$ m and width $2\frac{1}{5}$ m.

b. Calculate the perimeter of a square of side length $8\frac{1}{5}$ mm.

## Understanding

13. Calculate the length of party lights needed to decorate the perimeter of a rectangular tent with dimensions 15.5 m by 8.75 m.

14. Allowing an extra 30 cm for waste, calculate the length of picture frame required to frame the artwork shown in the photo.

15. Zedken wishes to install three strands of barbed wire at the top of the fences around a rectangular work site. The length of the site is 34.5 m, and its width is 19.8 m. Calculate the length of wire needed.

16. A new game, Bop-ball, is played on a triangular field, where each side of the triangle measures 46.6 m. A greenkeeper is marking the field's perimeter using a chalk-dispensing trundle wheel. Determine how far the greenkeeper will walk to mark the field.

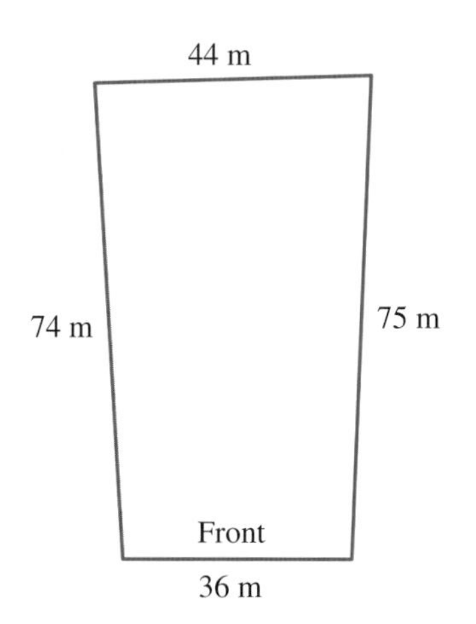


17. Phang's property boundary dimensions are shown in the diagram.
    a. Calculate how many metres of fencing are needed to fence all but the front boundary.
    b. If the fencing costs $19 per metre, calculate the total cost.

18. Lucille has received a quote of $37 per metre for new fencing for a tennis court. The tennis court is 23.77 m long and 10.97 m wide. There should be a 3.6 m space between each side of the court and the fence, and a 6.4 m gap between each end of the court and the fence.
    a. Draw a diagram showing all given measurements.
    b. Determine how many metres of fencing are needed.
    c. Based on the quote, calculate the total cost of the fencing.

## Reasoning

19. Suggest a type of work where people need to calculate perimeters on a daily basis.

20. A piece of paper is 22 cm long, including a 2.2 cm sticky strip at each end. Five strips of paper are stuck together so that the sticky parts overlap exactly to make a loop of paper. Determine the circumference of (distance around) the loop.

21. Marc and Cathy are seeking quotes for the cost of building a fence on three sides of their property. They want to calculate approximate costs for each item to decide whether the quotes supplied sound reasonable.

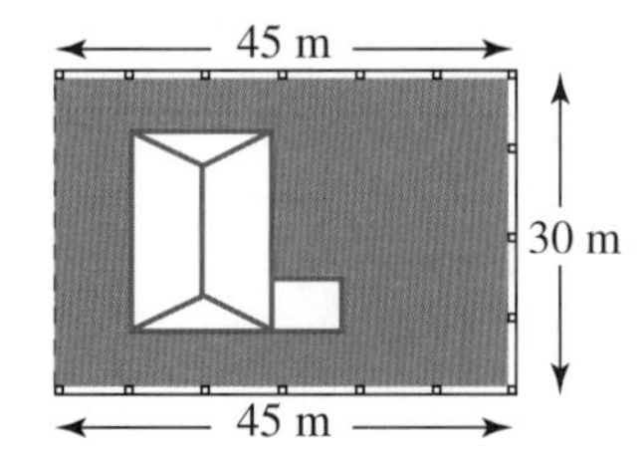


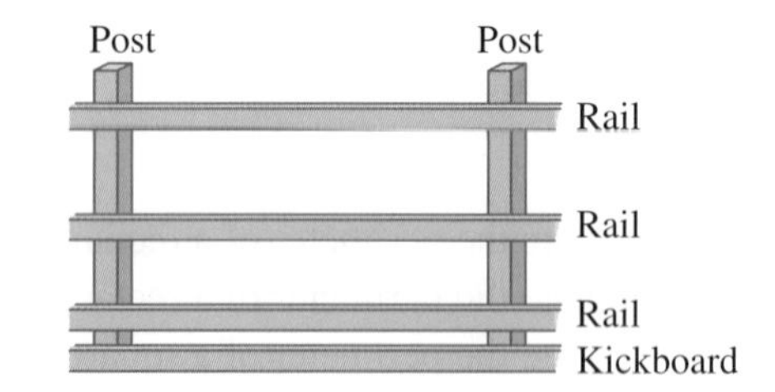


The new fence requires three rails, a kickboard, posts and palings.
    a. Calculate the length of timber needed for the kickboard.
    b. Calculate the cost of the kickboard if the timber required for this costs $1.90 per metre.
    c. If the timber for the rails costs $2.25 per metre, determine the total cost of the timber for the railings.

**d.** Determine how many posts will be needed for the new fence if each post is to be 5 metres apart and there needs to be a post at the end of each straight section of fence.

**e.** Calculate the cost of the posts if the price of each post is $13.65.

Palings are 9 cm wide and are nailed so they overlap each other by 15 mm on each side.

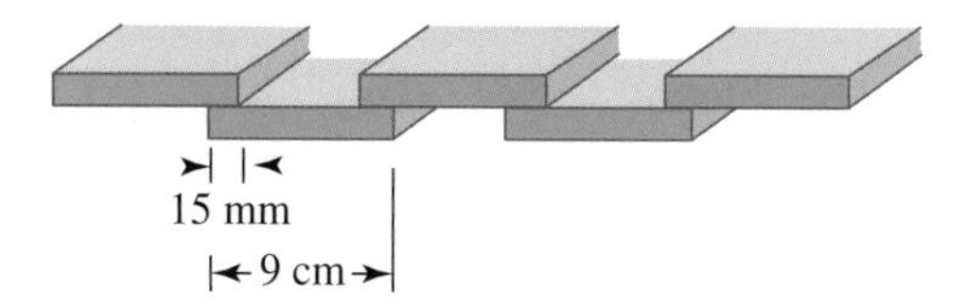


**f. i.** Calculate the approximate number of palings needed for the fence.

**ii.** Palings cost $1.05 each. Determine how much money should be allowed for the total cost of the palings.

**g.** Write an itemised list of all the costs involved. Include an amount to cover labour charges and miscellaneous items like the cost of nails. This amount is around $1000 (two people for two days at approximately $30.00 per hour for an eight-hour day). Estimate the cost of the new fence. This will provide Marc and Cathy with information to use when comparing builders' quotes.

## Problem solving

**22.** Kelly has a piece of string 32 cm long. He uses it to form rectangular shapes that each have a length and a width that are a whole number of centimetres.

a. Identify how many different rectangular shapes he can form.

b. Predict what the sides of the rectangle will be to give the largest possible area. Check by drawing different rectangles to see if your prediction is correct.

**23.** Write an equation for the perimeter of the diagram shown in terms of $a$ and $b$.

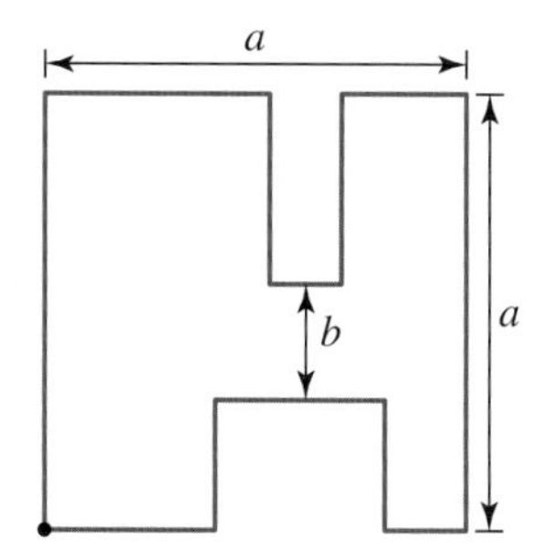


**24.** The rectangle shown is made up of three identical, smaller rectangles that fit inside (as shown). If its length measures 9 cm, work out its width without using a ruler.

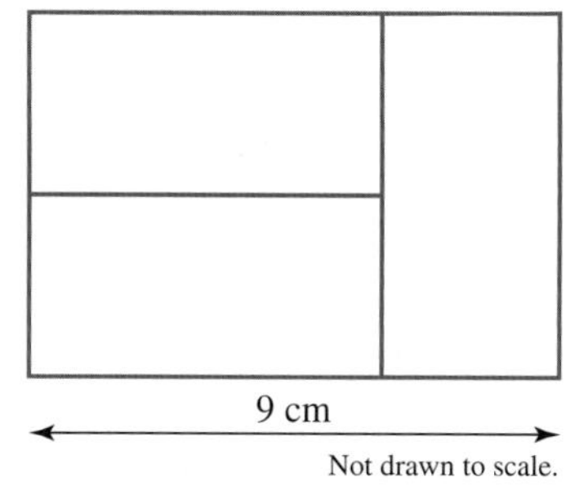

# LESSON
# 10.5 Circles and circumference

## LEARNING INTENTION

At the end of this lesson you should be able to:
- calculate the circumference of a circle in terms of $\pi$, or to an approximate value
- calculate the length of an arc of a circle
- calculate the perimeter of shapes involving circles

## 10.5.1 Circumference

eles-4044

- The distance around a circle is called the **circumference** ($C$).
- The **diameter** ($d$) of a circle is the straight-line distance across a circle though its centre.
- The **radius** ($r$) of a circle is the distance from the centre to the circumference.
- The diameter and radius of a circle are related by the formula $d = 2r$. That is, the diameter is twice as long as the radius.

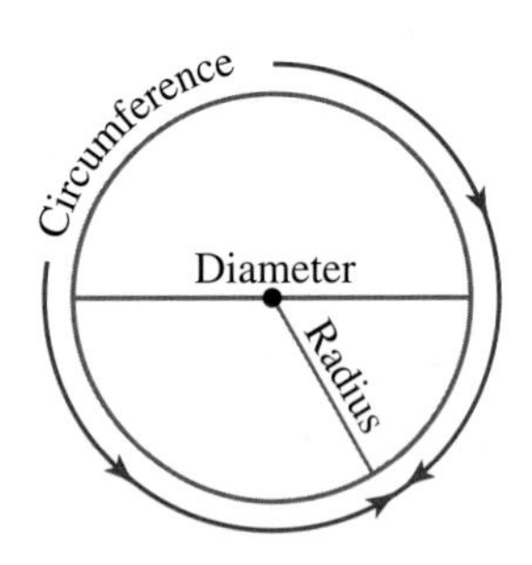


### COLLABORATIVE TASK: The diameter and circumference of a circle

Measure the diameters and circumferences of a variety of different circles and cylinders (for example, coins and drink bottles), and record your results in a table. What do you notice about the value of $\frac{C}{d}$ for each circle you measured?

### Pi ($\pi$)

- The ratio of acircle's circumference to its diameter, $\frac{C}{d}$ gives the same value for any circle no matter how large or small the circle is.
- This special numberis known as pi ($\pi$). That is, $\pi = \frac{C}{d}$.
- Pi ($\pi$) is an irrational number and cannot be written as a fraction.
- When pi is written as a decimal number, the decimal places continue forever with no repeated pattern.
- Pi written to eight decimal places is 3.14159265.
- All scientific calculators have a $\pi$ button and this feature can be used when completing calculations involving pi.

## Digital technology

Scientific calculators can be used to assist with calculations involving pi. Locate the $\pi$ button on your calculator and become familiar with accessing this feature.

DEG

π 3.141592654

### Calculating circumference

**The circumference, $C$ of a circle can be determined using one of the following formulas.**

$$C = 2\pi r, \text{ where } r \text{ is the radius of the circle}$$

**or**

$$C = \pi d, \text{ where } d \text{ is the diameter of the circle.}$$

### WORKED EXAMPLE 7 Calculating the circumference of a circle

**Calculate the circumference of each of the following circles, giving answers:**

**i. in terms of $\pi$**

**ii. correct to 2 decimal places.**

a.

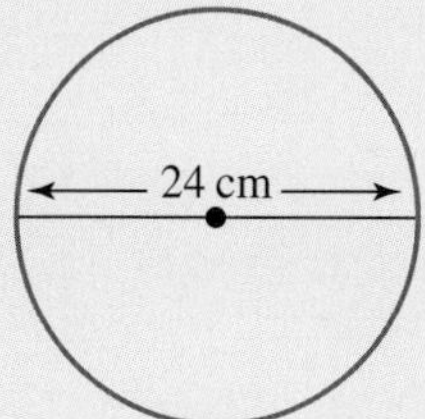


b.

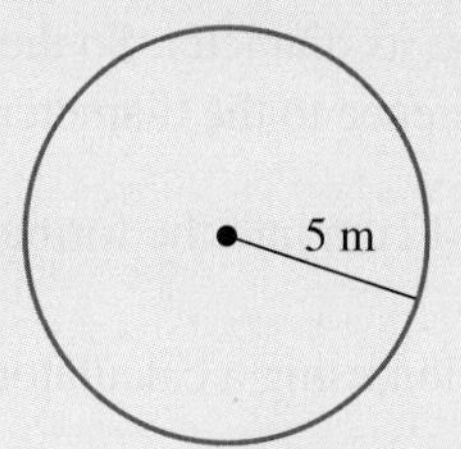


| THINK | WRITE |
|---|---|
| a. i. 1. Write the formula for the circumference of a circle. *Note:* Since the diameter of the circle is given, use the formula that relates the circumference to the diameter. | a. i. $C = \pi d$ |
| 2. Substitute the value $d = 24$ cm into the formula. | $= \pi \times 24$ |
| 3. Write the answer and include the units. | $= 24\pi$ cm |
| ii. 1. Write the formula for the circumference of a circle. | ii. $C = \pi d$ |
| 2. Substitute the value $d = 24$ cm into the formula. | $= \pi \times 24$ |
| 3. Evaluate the multiplication using a calculator and the $\pi$ button. | $= 75.398\ldots$ |
| 4. Round correct to 2 decimal places and include the units. | $= 75.40$ cm |

**b. i.** 1. Write the formula for the circumference of a circle.
*Note:* Since the radius of the circle is given, use the formula that relates the circumference to the radius.

**b. i.** $C = 2\pi r$

2. Substitute the value $r = 5$ m into the formula. $= 2 \times \pi \times 5$

3. Write the answer and include the units. $= 10\pi$ m

**ii.** 1. Write the formula for the circumference of a circle. **ii.** $C = 2\pi r$

2. Substitute the value $r = 5$ m into the formula. $= 2 \times \pi \times 5$

3. Evaluate the multiplication using a calculator and the $\pi$ button. $= 31.415\ldots$

4. Round the answer correct to 2 decimal places and include the units. $= 31.42$ m

## WORKED EXAMPLE 8 Calculating the perimeter of a semicircle

**Calculate the perimeter of the following shape, correct to 2 decimal places.**

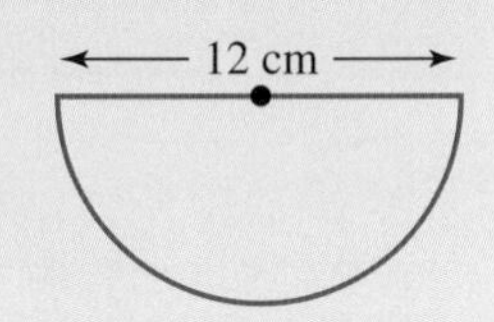


| THINK | WRITE |
| --- | --- |
| 1. Identify the parts that constitute the perimeter of the given shape. | $P = \frac{1}{2}$ circumference + straight-line section |
| 2. Write the formula for the circumference of a circle. *Note:* If the circle were complete, the straight-line segment shown would be its diameter. So the formula that relates the circumference to the diameter is used. | $P = \frac{1}{2}\pi d$ + straight-line section |
| 3. Substitute the value $d = 12$ cm into the formula. | $= \frac{1}{2} \times \pi \times 12 + 12$ |
| 4. Evaluate the multiplication using a calculator and the $\pi$ button. | $= 18.849\ldots + 12$<br>$= 30.849\ldots$ |
| 5. Round the answer correct to 2 decimal places and include the units. | $= 30.85$ cm |

## 10.5.2 Parts of a circle

eles-4045

- A **sector** is the region of a circle between two radii and an arc. It looks like a slice of pizza.
- An **arc** is a section of the circumference of a circle.
- A **chord** is a straight line joining any two points on the circumference of a circle. The diameter is a type of chord.
- A **segment** of a circle is a section bounded by a chord and an arc.

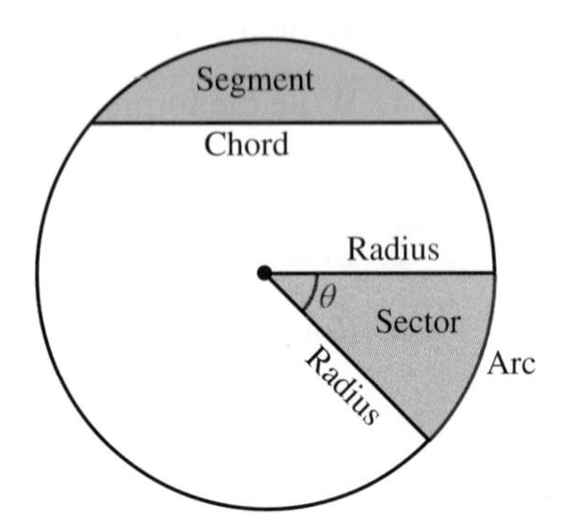

## Calculating arc length

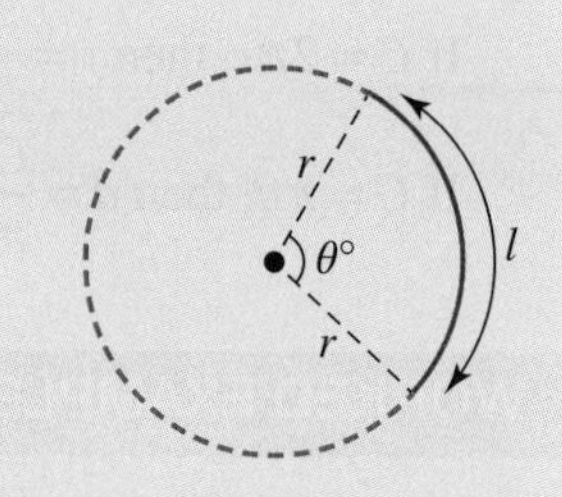


- **An arc is a portion of the circumference of a circle.**
- **The length, $l$, of an arc can be determined using one of the following formulas:**

$$l = \frac{\theta}{360} \times 2\pi r \quad \text{or} \quad l = \frac{\theta}{360} \times \pi d$$

**where: $\theta$ is angle (in degrees) at the centre of the circle**
**$r$ is the radius of the circle**
**$d$ is the diameter of the circle.**

## Calculating the perimeter of a sector

- **A sector consists of an arc and 2 radii.**
- **The perimeter, $P$, of a sector can therefore be calculated using the following formula:**

$$P = l + 2r = \frac{\theta}{360} \times 2\pi r + 2r$$

**where: $\theta$ is angle (in degrees) at the centre of the circle**
**$r$ is the radius of the circle.**

- Recalling that $d = 2r$, the perimeter of a sector could also be determined using the formula:

$$\begin{aligned} P &= l + d \\ &= \frac{\theta}{360} \times \pi d + d \end{aligned}$$

### WORKED EXAMPLE 9 Calculating the perimeter a sector

**Calculate the perimeter of the following sector, correct to 2 decimal places.**

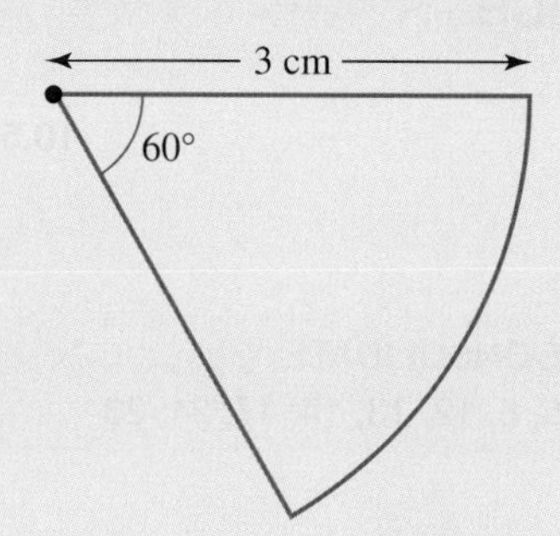


| THINK | WRITE |
|---|---|
| 1. Identify the values of $\theta$ and $r$. | $\theta = 60°, r = 3\text{ cm}$ |
| 2. Write the formula for the perimeter of a sector. | $P = \frac{\theta}{360} \times 2\pi r + 2r$ |
| 3. Substitute these values into the formula for the perimeter of a sector. | $P = \frac{60}{360} \times 2\pi \times 3 + 2 \times 3$ |
| 4. Evaluate the formula. | $= 9.141\ldots$ |
| 5. Round the answer correct to 2 decimal places and include the units. | $= 9.14\text{ cm}$ |

- If the circumference of a circle is known, it is possible to determine its radius or diameter using one of the formulas below.
If $C = 2\pi r$, then $r = \dfrac{C}{2\pi}$.
If $C = \pi d$, then $d = \dfrac{C}{\pi}$.

**WORKED EXAMPLE 10 Determining the radius of a circle from its circumference**

**Calculate the radius of a cylindrical water tank with a circumference of 807 cm, correct to 2 decimal places.**

| THINK | WRITE |
|---|---|
| 1. Write the rule relating circumference and radius. | If $C = 2\pi r$, then $r = \dfrac{C}{2\pi}$. |
| 2. Substitute the value of the circumference into the formula. | $r = \dfrac{807}{2\pi}$ cm |
| 3. Evaluate the expression using a calculator. | $r = 128.438\ldots$ cm |
| 4. Round the answer correct to 2 decimal places. | $r = 128.44$ cm |

eWorkbook Topic 10 Workbook (worksheets, code puzzle and project) (ewbk-1960)

Interactivities Individual pathways interactivity: Circumference (int-4439)
Circumference (int-3792)

## Exercise 10.5 Circumference

learnon

10.5 Quick quiz 

10.5 Exercise

### Individual pathways

| ■ PRACTISE | ■ CONSOLIDATE | ■ MASTER |
|---|---|---|
| 1, 4, 7, 10, 11, 16, 19, 20 | 2, 5, 8, 12, 13, 14, 17, 21, 23 | 3, 6, 9, 15, 18, 22, 24 |

### Fluency

1. WE7a Calculate the circumference of each of these circles, giving answers in terms of $\pi$.

a. 2 cm

b.
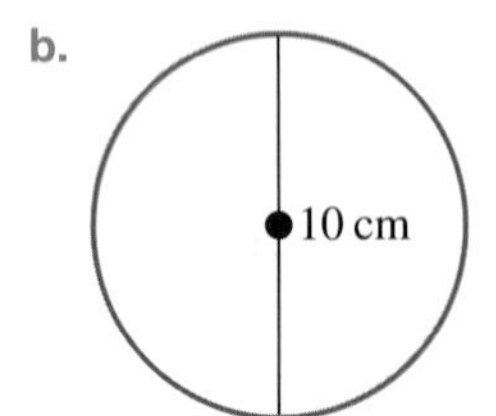


c.
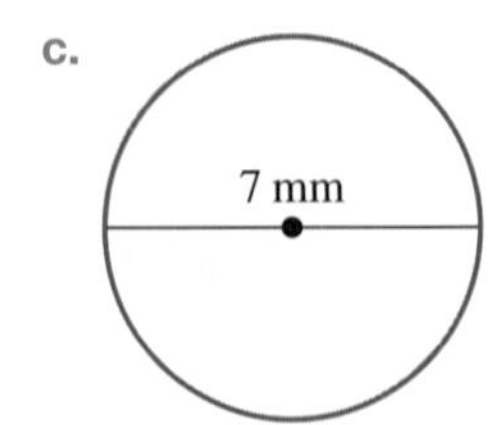

2. Calculate the circumference of each of these circles, giving answers correct to 2 decimal places.

a. 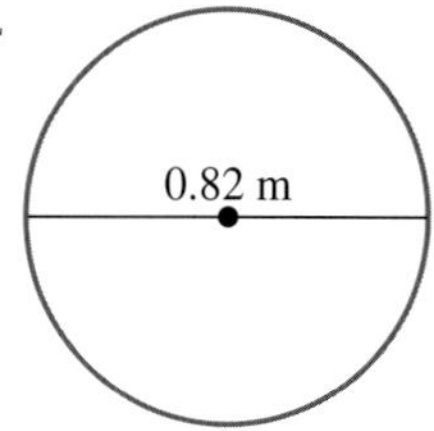


b. 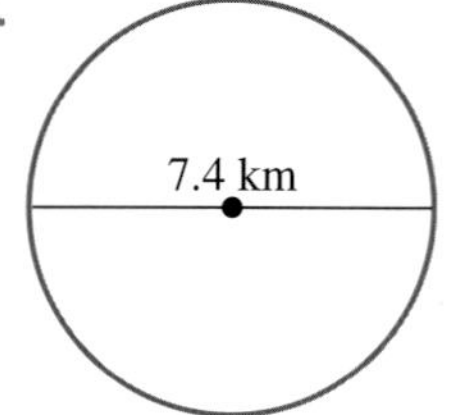


c. 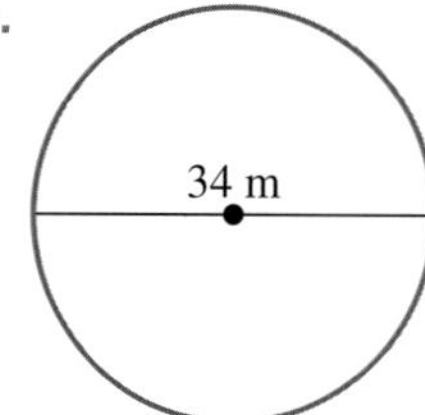


3. **WE7b** Determine the circumference of each of the following circles, giving answers in terms of $\pi$

a. 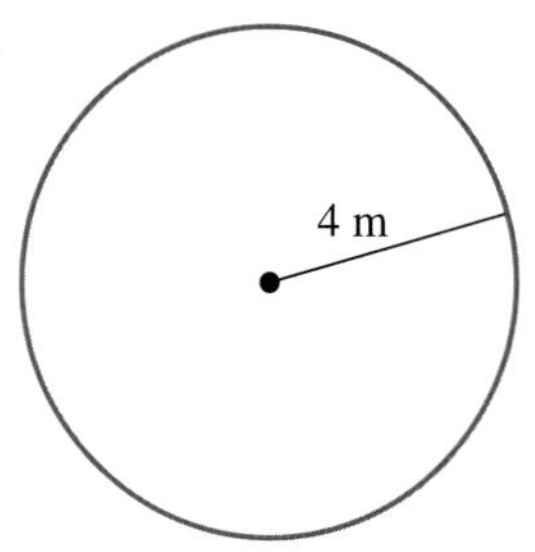


b. 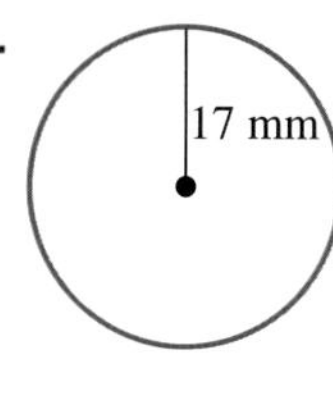


c. 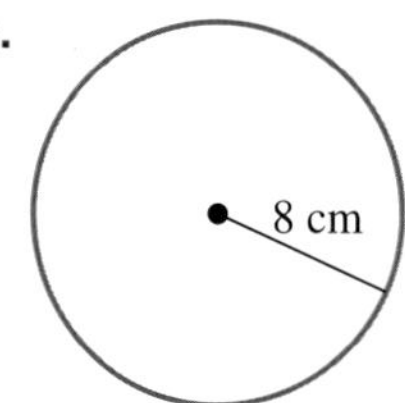


4. Calculate the circumference of each of the following circles, giving answers correct to 2 decimal places.

a. 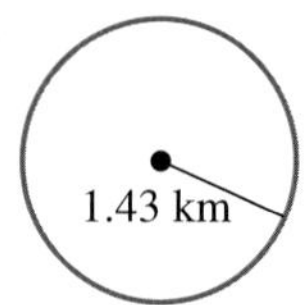


b. 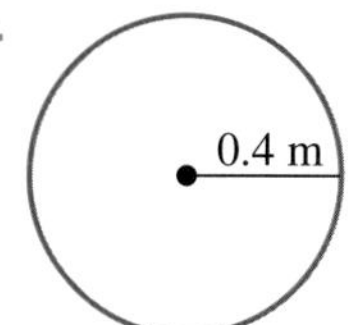


c. 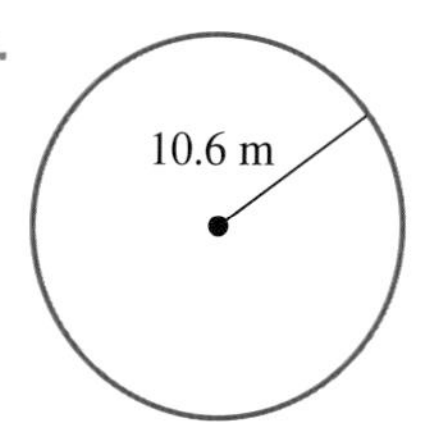


5. **WE8** Determine the perimeter of each of the shapes below. (Remember to add the lengths of the straight sections.) Give your answers to 2 decimal places.

a. 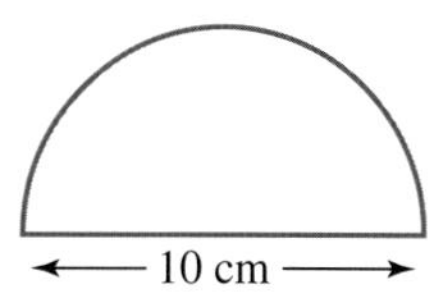


b. 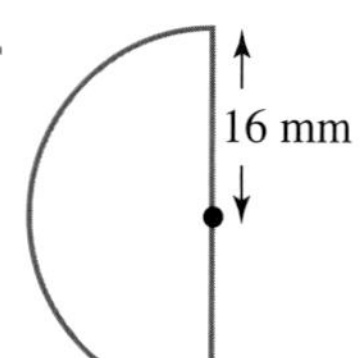


c. 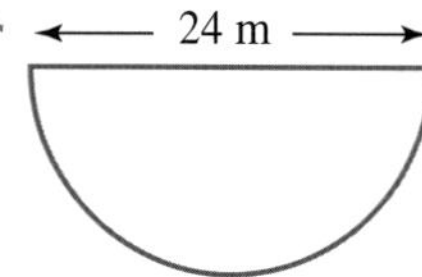


d. 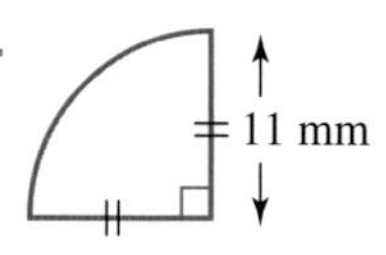


e. 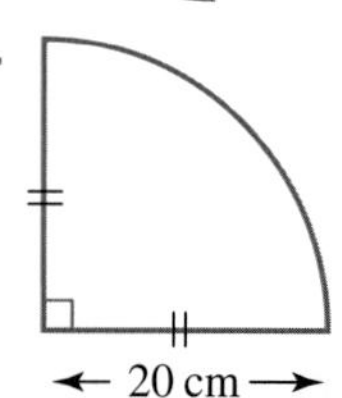


6. Calculate the perimeter of each of the shapes below. (Remember to add the lengths of the straight sections.) Give your answers to 2 decimal places.

a. 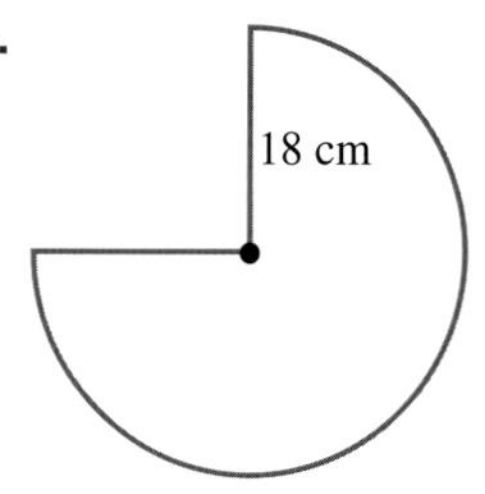


b. 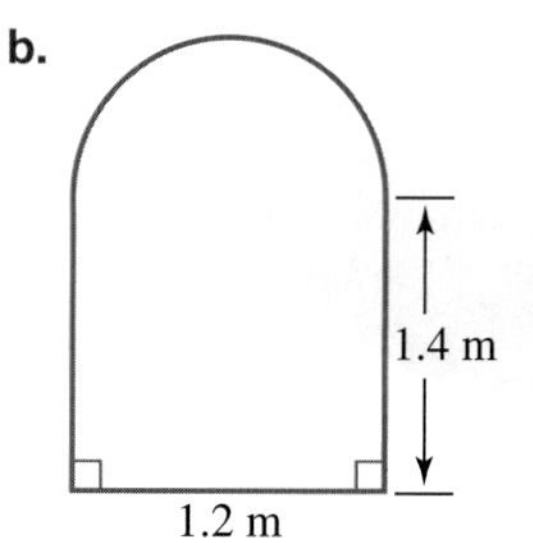


c. 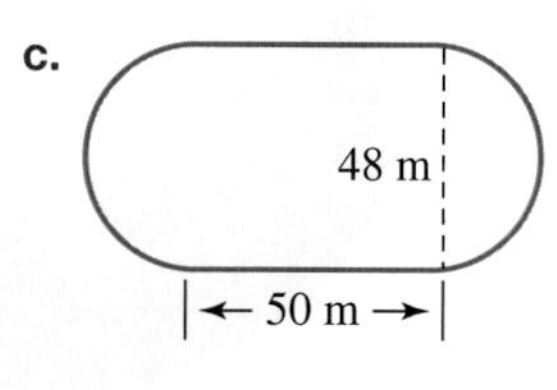


d. 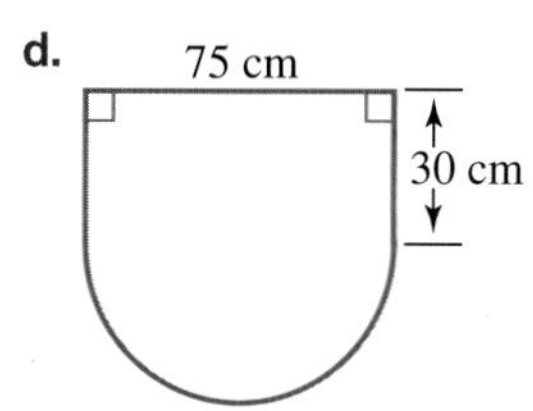

7. **MC** The circumference of a circle with a radius of 12 cm is:

A. $\pi \times 12$ cm
B. $2 \times \pi \times 12$ cm
C. $2 \times \pi \times 24$ cm
D. $\pi \times 6$ cm
E. $\pi \times 18$ cm

8. **MC** The circumference of a circle with a diameter of 55 m is:

A. $2 \times \pi \times 55$ m
B. $\pi \times \frac{55}{2}$ m
C. $\pi \times 55$ m
D. $\pi \times 110 \times 2$ m
E. $2 \times \pi \times 110$ m

9. **WE9** Calculate the perimeter of the following sectors, correct to 2 decimal places.

a. 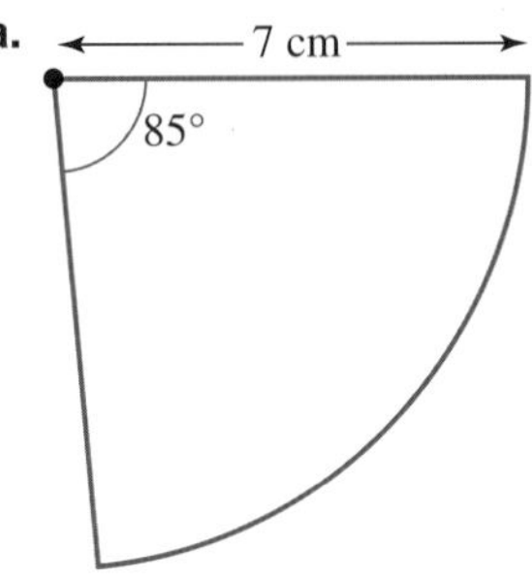


b. 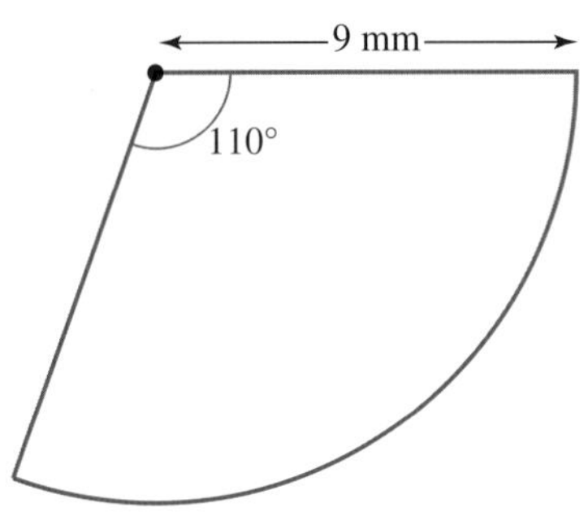


c. 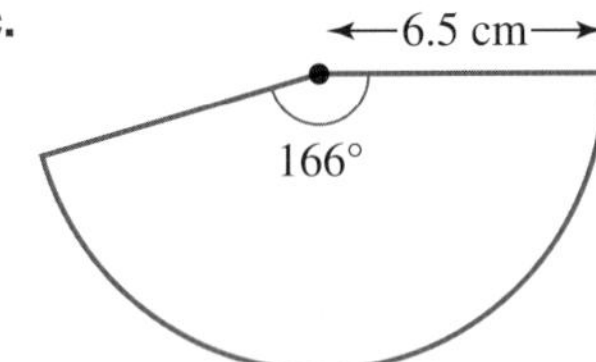


d. 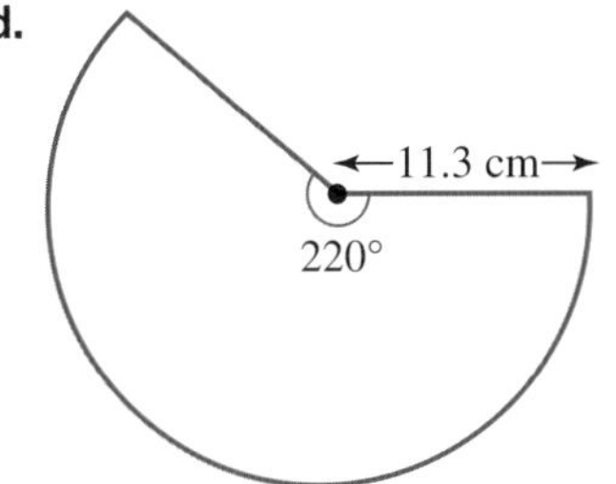


### Understanding

10. Calculate the circumference of the seaweed around the outside of this sushi roll, correct to 2 decimal places.

11. A scooter tyre has a diameter of 32 cm. Determine the circumference of the tyre. Give your answer to 2 decimal places.

12. Calculate the circumference of the Ferris wheel shown at right, correct to 2 decimal places.

13. In a Physics experiment, students spin a metal weight around on the end of a nylon thread. How far does the metal weight travel if it completes 10 revolutions on the end of a 0.88 m thread? Give your answer to 2 decimal places.

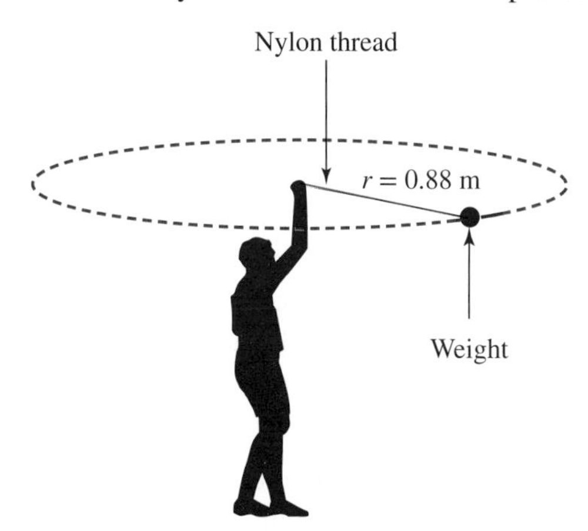


14. **WE10** Calculate the radius of a tyre with a circumference of 135.56 cm. Give your answer to 2 decimal places.

15. Calculate the diameter of a circle (correct to 2 decimal places where appropriate) with a circumference of:

 a. 18.84 m
 b. 64.81 cm
 c. 74.62 mm.

16. Calculate the radius of a circle (correct to 2 decimal places where appropriate) with a circumference of:

 a. 12.62 cm
 b. 47.35 m
 c. 157 mm.

17. Determine the total length of metal pipe needed to assemble the wading pool frame shown at right. Give your answer in metres to 2 decimal places.

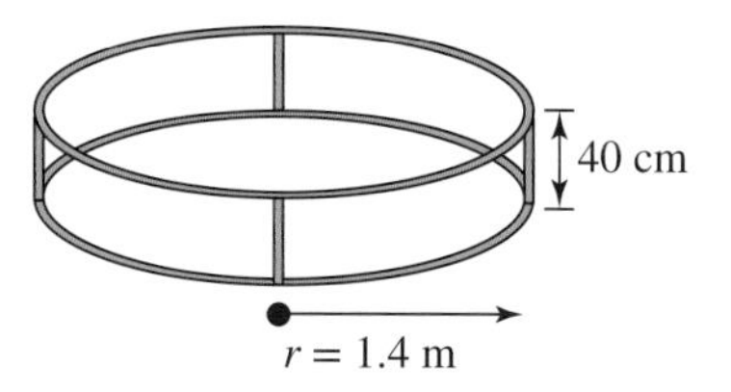


18. Nathan runs around the inside lane of a circular track that has a radius of 29 m. Rachel runs in the outer lane, which is 2.5 m further from the centre of the track. How much longer is the distance Rachel runs each lap. Give your answer to 2 decimal places.

### Reasoning

19. To cover a total distance of 1.5 km, a student needs to run around a circular track three times. Calculate the radius of the track correct to the nearest metre.

20. A shop sells circular trampolines of four different sizes. Safety nets that go around the trampoline are optional and can be purchased separately. The entrance to the trampoline is via a zip in the net. The following table shows the diameters of all available trampolines and their net lengths. Which safety net matches which trampoline?

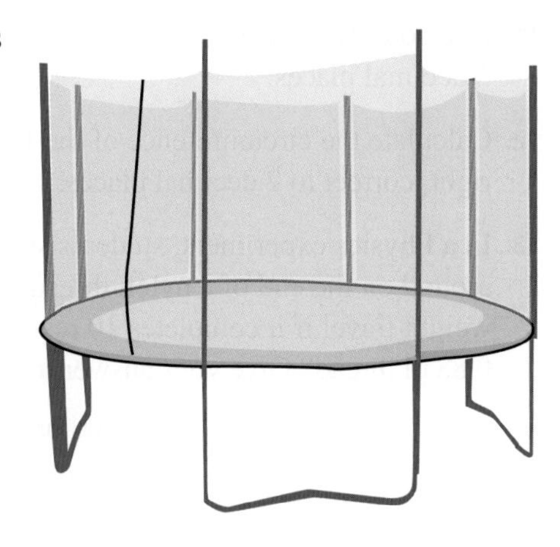

| Diameter of trampoline | | Length of safety net | |
|---|---|---|---|
| a | 1.75 m | i | 6.03 m |
| b | 1.92 m | ii | 9.86 m |
| c | 2.46 m | iii | 5.50 m |
| d | 3.14 m | iv | 7.73 m |

21. In *Around the world in eighty days* by Jules Verne, Phileas Fogg boasts that he can travel around the world in 80 days or fewer. This was in the 1800s, so he couldn't take a plane. What average speed is needed to go around the Earth at the equator in 80 days? Assume you travel for 12 hours each day and that the radius of the Earth is approximately 6390 km. Give your answer in km/h to 2 decimal places.

### Problem solving

22. Liesel's bicycle covers 19 m in 10 revolutions of her bicycle wheel while Jared's bicycle covers 20 m in 8 revolutions of his bicycle wheel. Determine the difference between the radii of the two bicycle wheels. Give your answer in cm to 2 decimal places.

23. Evaluate the perimeter of each of the following shapes. Give your answers correct to 2 decimal places.

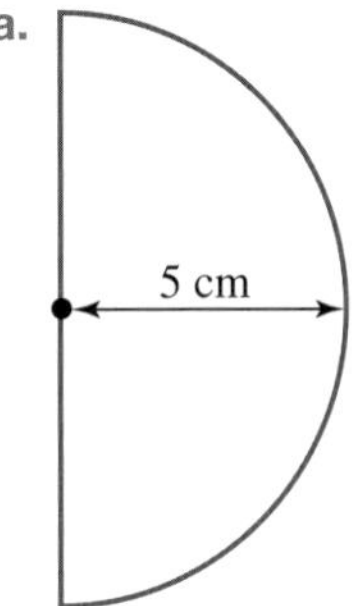


b.

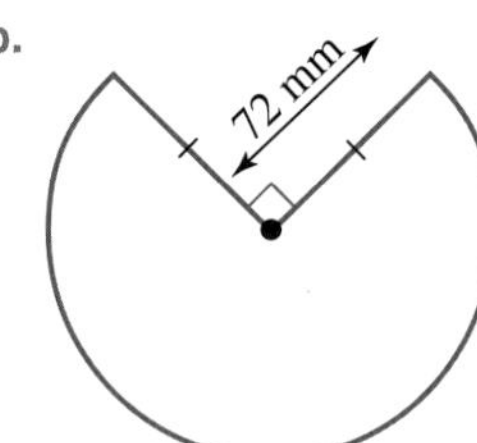

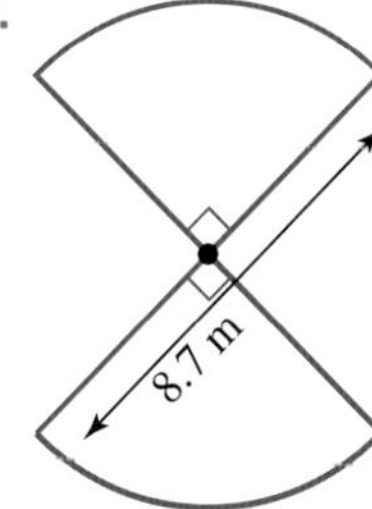


d.

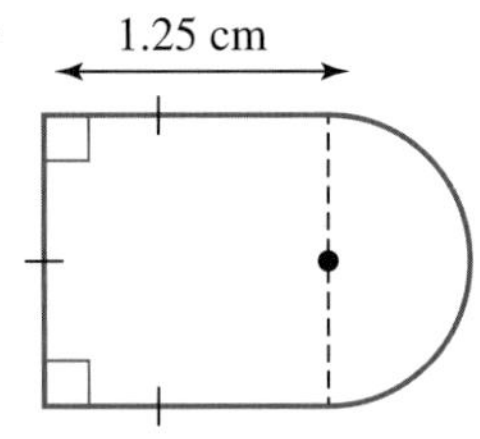


24. Determine the perimeter of the segment shown. Give your answer correct to 2 decimal places.

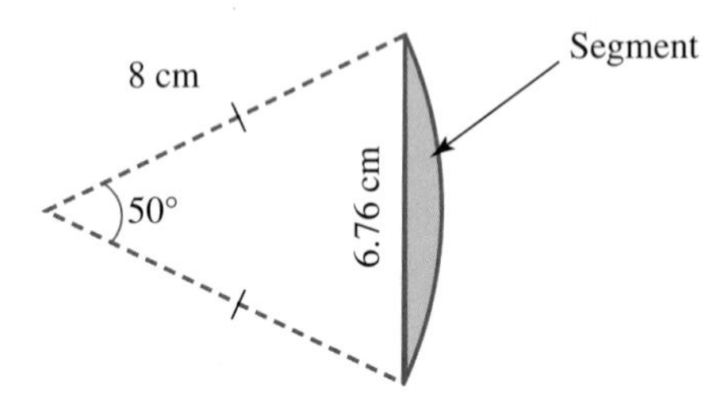

# LESSON
# 10.6 Area

**LEARNING INTENTION**

At the end of this lesson you should be able to:
- calculate, or estimate, the area of a given shape
- calculate the area of rectangles, squares, triangles and parallelograms using formulas.

## 10.6.1 Metric units of area

eles-4554

- The **area** of a shape is the amount of flat surface enclosed by the shape.
- Area is measured in square units such as square centimetres, square metres and square kilometres.
- Commonly used metric units of area, with their abbreviations and examples, are shown in the following figures.

1. **Square kilometres ($km^2$)**

The area of a country or a large city like Sydney is given in square kilometres.

2. **Square metres ($m^2$)**

Square metres are used to measure the area of an object such as a classroom floor, whiteboard or a window.

3. **Square centimetres ($cm^2$)**

Small areas, such as the area of a sheet of A4 paper or a book cover, are measured in square centimetres.

4. **Square millimetres ($mm^2$)**

Very small areas, such as the area of a button or a postage stamp, are measured in square millimetres.

- The area of a farm or a large city park is measured in hectares (ha). $1 \text{ ha} = 10\,000 \text{ m}^2$.
- 1 square metre is the area enclosed by a square with sides that are 1 m long. The same is true for any other unit.
- If a shape is drawn on 1 cm grid paper, its area can be found by counting the number of squares that the shape covers.

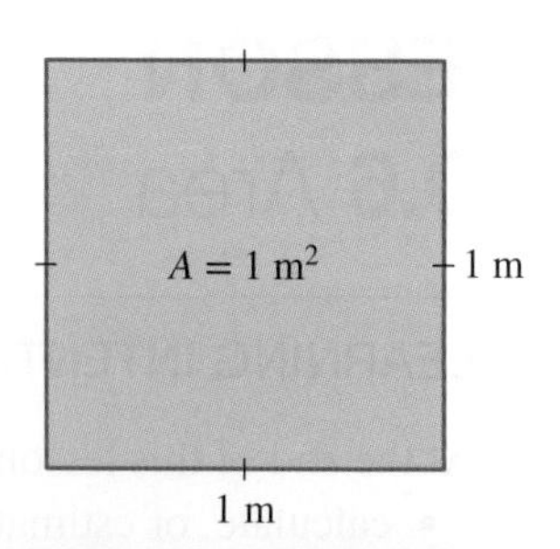


## WORKED EXAMPLE 11 Determining area by counting squares

**The following figures are drawn on centimetre grid paper. Determine the area of each one.**

**a.** 

**b.** 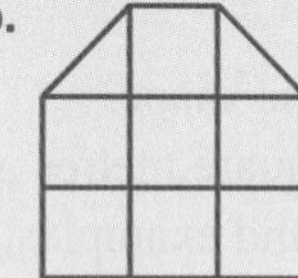

**THINK**

**a.** Count the squares. Remember to include the correct unit $(\text{cm}^2)$ in the answer.

**b.** Some of the squares are cut in half by the diagonal line. It takes two of these to make up one square centimetre. Count the squares. Remember to include the correct unit $(\text{cm}^2)$ in the answer.

**WRITE**

**a.** $8 \text{ cm}^2$

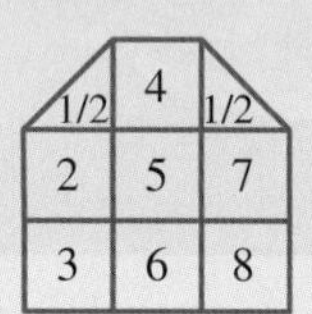

| 1 | 4 | |
|---|---|---|
| 2 | 5 | 7 |
| 3 | 6 | 8 |

**b.** $8 \text{ cm}^2$

| 1/2 | 4 | 1/2 |
|---|---|---|
| 2 | 5 | 7 |
| 3 | 6 | 8 |

### Estimating shaded areas

**If a square of the grid paper is not completely covered by the shape, use the following rule to obtain an estimate of the area.**
- **If more than half the square is covered, count it as a full square.**
- **If less than half the square is covered, do not count it at all.**

## WORKED EXAMPLE 12 Estimating the shaded area

**Estimate the shaded area of the following diagram drawn on centimetre grid paper.**

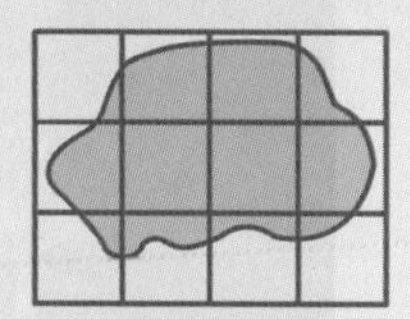

**THINK**

1. Tick the squares that are more than half covered and count them.
2. State the answer, with the correct unit.

**WRITE**

$6 \text{ cm}^2$

## 10.6.2 Calculating the area of a rectangle

eles-4555

- The areas of rectangles and squares can also be calculated by using the following formulas.

### Area of a rectangle

**For a rectangle, the area $A$ is:**

$$A = l \times w$$
$$= lw$$

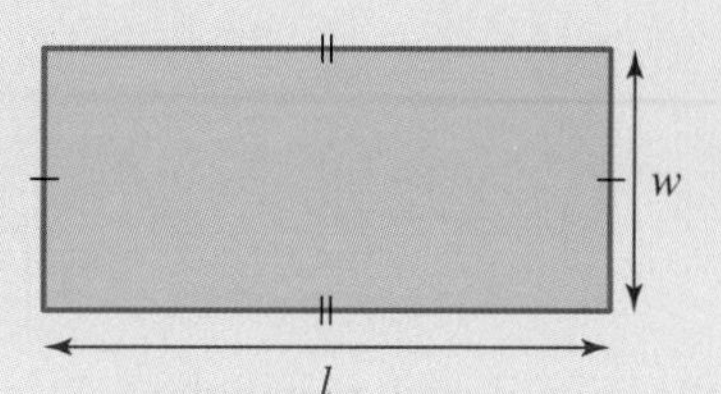


**where $l$ is the length and $w$ is the width of the rectangle.**

### Area of a square

**For a square, the area $A$ is:**

$$A = l \times l$$
$$= l^2$$

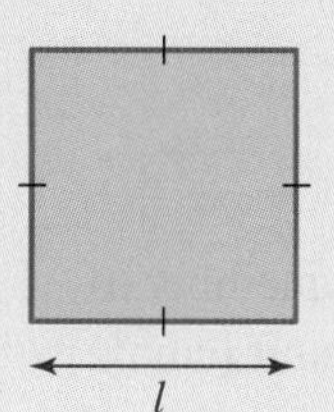


**where $l$ is the length of the square's side.**

### Digital technology

A quick way to calculate the area of a square on a calculator is to use the 'square' button. The button has the symbol $x^2$ and will square the number that you type in.

DEG
11.4²    129.96

### WORKED EXAMPLE 13 Calculating areas of different shapes

**Calculate the areas of the following shapes.**

a.

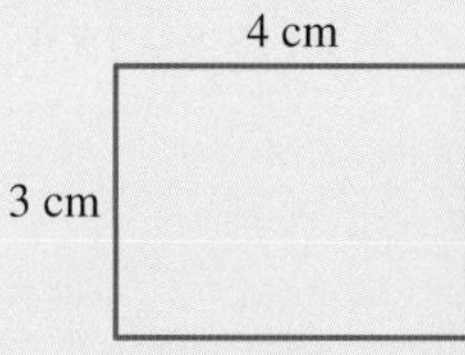


b.

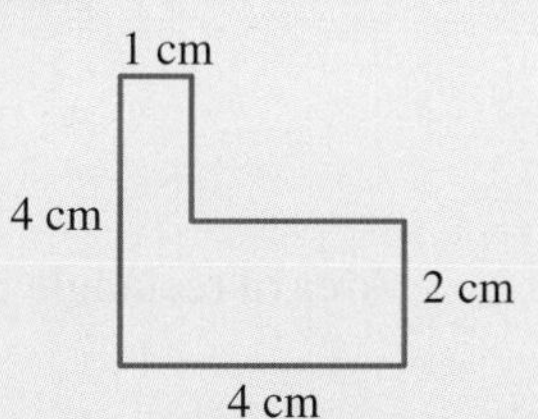


c.

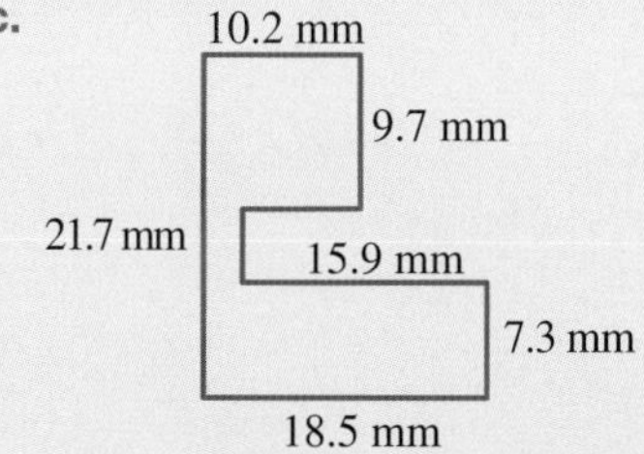


| THINK | WRITE |
|---|---|
| **a. 1.** Write the formula for the area of a rectangle. | **a.** $A = lw$<br>$l = 4$ cm<br>$w = 3$ cm |
| **2.** Substitute the value 4 for $l$ and 3 for $w$ and calculate the area. | $A = 4 \times 3$<br>$= 12\text{ cm}^2$ |

**3.** State the answer in the correct unit ($cm^2$).

The area is $12\,cm^2$.

**b. 1.** Divide the shape into two rectangles.

**b.**

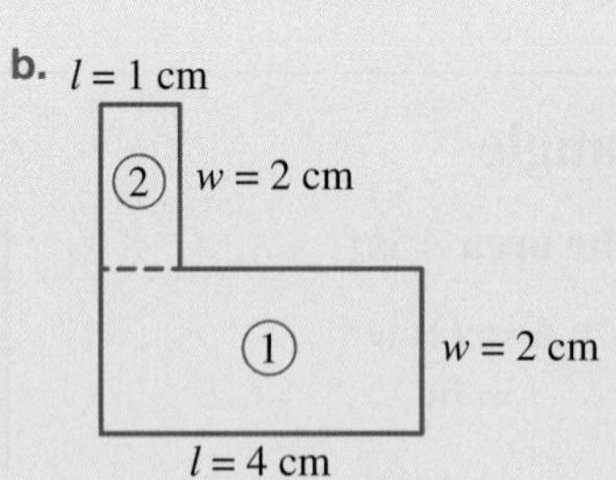


**2.** Calculate the area of each rectangle separately by substituting the correct values of $l$ and $w$ into the formula $A = lw$.

$$\begin{aligned}\text{Area of rectangle 1} &= l \times w \\ &= 4 \times 2 \\ &= 8\,cm^2\end{aligned}$$

$$\begin{aligned}\text{Area of rectangle 2} &= l \times w \\ &= 1 \times 2 \\ &= 2\,cm^2\end{aligned}$$

**3.** Add the two areas. Remember to answer in the correct unit ($cm^2$).

$$\begin{aligned}\text{Area of shape} &= \text{Area of rectangle 1} + \text{Area of rectangle 2} \\ &= 8\,cm^2 + 2\,cm^2 \\ &= 10\,cm^2\end{aligned}$$

**c. 1.** Divide the shape into three rectangles.

**c.**

$l = 10.2$ mm
1
$w = 9.7$ mm
21.7 mm
2
15.9 mm
3
$w = 7.3$ mm
$l = 18.5$ mm

**2.** Calculate the area of each rectangle separately by substituting the correct values of $l$ and $w$ into the formula $A = lw$.

$$\begin{aligned}\text{Area of rectangle 1} &= l \times w \\ &= 10.2 \times 9.7 \\ &= 98.94\,mm^2\end{aligned}$$

$$\begin{aligned}\text{Area of rectangle 2} &= l \times w \\ &= (21.7 - 9.7 - 7.3) \times (18.5 - 15.9) \\ &= 4.7 \times 2.6 \\ &= 12.22\,mm^2\end{aligned}$$

$$\begin{aligned}\text{Area of rectangle 3} &= l \times w \\ &= 18.5 \times 7.3 \\ &= 135.05\,mm^2\end{aligned}$$

**3.** Add the three areas. Remember to answer in the correct unit ($mm^2$).

$$\begin{aligned}\text{Area of shape} &= \text{Area 1} + \text{Area 2} + \text{Area 3} \\ &= 98.94\,mm^2 + 12.22\,mm^2 + 135.05\,mm^2 \\ &= 246.21\,mm^2\end{aligned}$$

**COLLABORATIVE TASK: Rectangles with the same areas and perimeters**

Work in pairs to answer the following questions.

Compare the perimeters of two different rectangles, each with an area of 60 cm$^2$. Which rectangles have the smallest perimeters and which have the largest perimeters?

Compare the areas of different rectangles, each with a perimeter of 24 cm. Which rectangles have the smallest areas and which have the largest areas?

## 10.6.3 Calculating the area of a triangle

eles-4556

- Triangles can be formed by cutting rectangles in half.

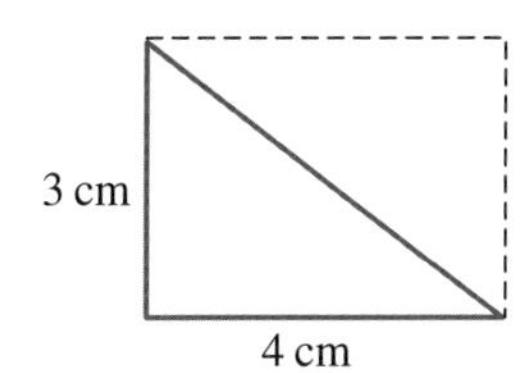


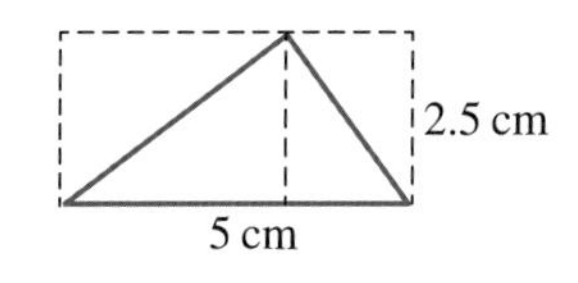


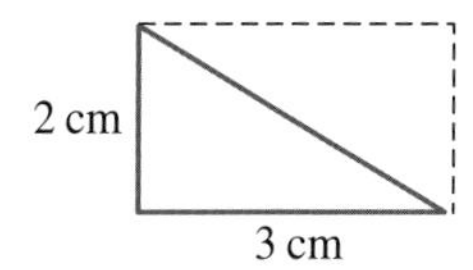


- The area of a triangle is therefore equal to half the area of a rectangle of the same base length and height.

**Area of a triangle**

**To calculate the area A of a triangle:**

$$A = \frac{1}{2}b \times h$$
$$= \frac{1}{2}bh$$

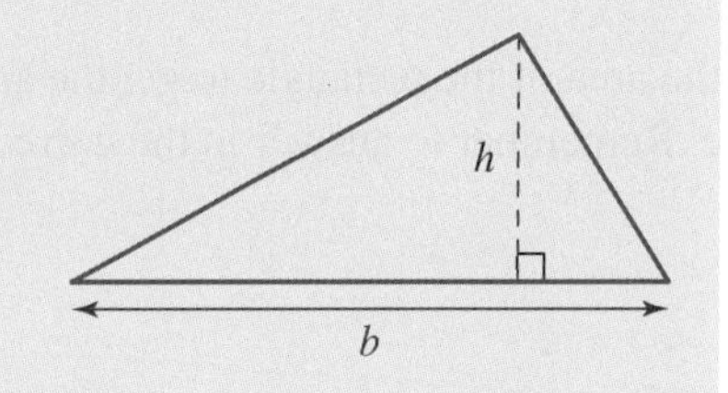


**where *b* is the base and *h* the perpendicular height.**

- To measure the perpendicular height of a triangle, the line must form a right angle with the base of the triangle. Sometimes this line will have to be drawn outside the triangle, as shown in the following diagram.

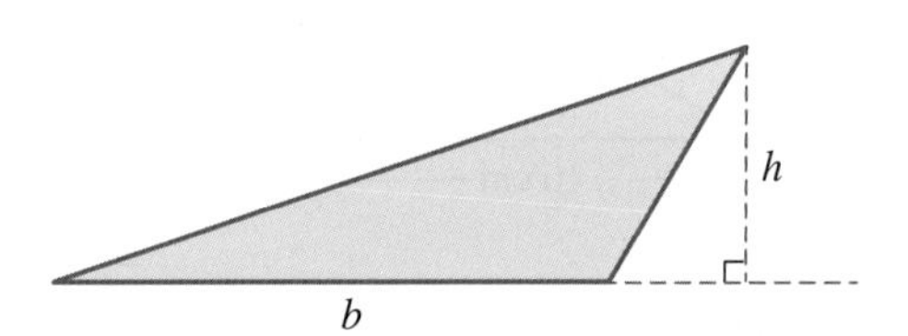


**WORKED EXAMPLE 14 Calculating the area of a right-angled triangle**

**Calculate the area of the following shape.**

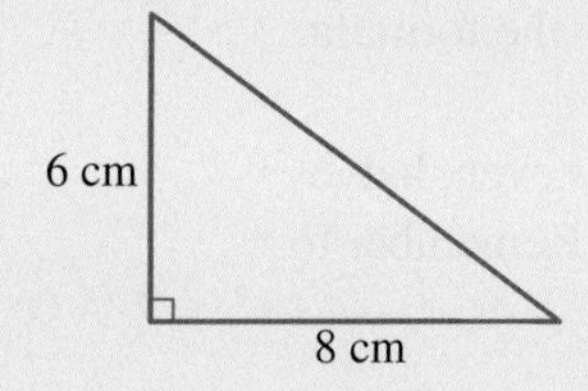

**THINK** | **WRITE**

Method 1: Using the formula for the area of a triangle

1. Write the formula for the area of a triangle. — $A = \frac{1}{2}bh$

2. Identify the values of $b$ and $h$. — $b = 8\text{ cm}, h = 6\text{ cm}$

3. Substitute the values of $b$ and $h$ into the formula. — $A = \frac{1}{2} \times 8 \times 6$

4. Evaluate. (Since one of the values is even, halve it first to make calculations easier.) Remember to include the correct unit (cm$^2$). — $= 4 \times 6$ <br> $= 24\text{ cm}^2$

Method 2: By cutting the rectangle in half

1. Draw an imaginary rectangle that contains the triangle. The rectangle should have the same base length and height as the triangle. Notice that the triangle forms half of the rectangle.

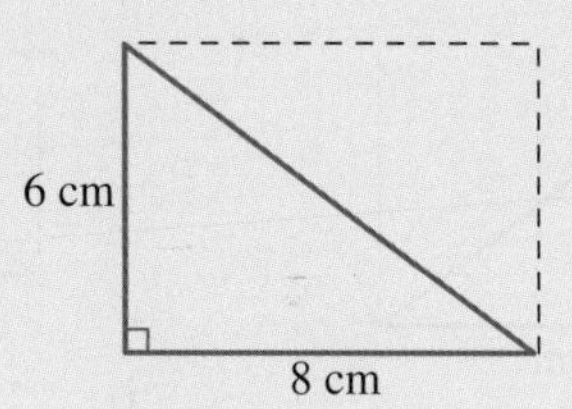


2. Use the formula $A = lw$ to calculate the area of this rectangle. — $A = lw$ <br> $= 8\text{ cm} \times 6\text{ cm}$ <br> $= 48\text{ cm}^2$

3. Halve the area of the rectangle to get the area of the triangle. Remember to answer in the correct unit (cm$^2$). — Area of triangle $= \frac{1}{2}$ of area of rectangle <br> Area of triangle $= \frac{1}{2} \times 48$ <br> $= 24\text{ cm}^2$

## WORKED EXAMPLE 15 Calculating the area of a triangle

**Calculate the area of each of these triangles.**

a.

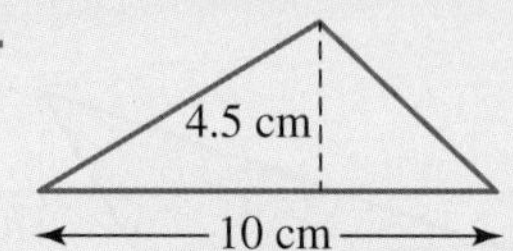


b.

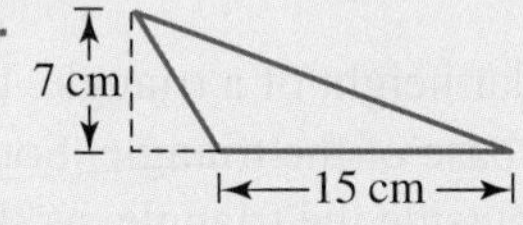


**THINK** | **WRITE**

a. 1. Write the formula for the area of a triangle. — a. $A = \frac{1}{2}bh$

2. Identify the values of $b$ and $h$. — $b = 10, h = 4.5$

3. Substitute the values of $b$ and $h$ into the formula. — $A = \frac{1}{2} \times 10 \times 4.5$

4. Evaluate. (Since one of the values is even, halve it first to make calculations easier.) Remember to include the correct unit (cm$^2$). — $= 5 \times 4.5$ <br> $= 22.5\text{ cm}^2$

**b. 1.** Write the formula for the area of a triangle. | **b.** $A = \frac{1}{2}bh$

**2.** Identify the values of $b$ and $h$. | $b = 15, h = 7$

**3.** Substitute the values of $b$ and $h$ into the formula. | $A = \frac{1}{2} \times 15 \times 7$

**4.** Evaluate. (Since neither value is even, multiply 15 and 7 first, and then divide by 2.) Remember to include the correct unit ($cm^2$). | $= \frac{1}{2} \times 105$
$= 52.5\ cm^2$

## 10.6.4 Area of a parallelogram

eles-4557

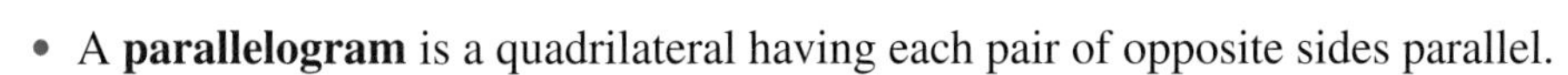

- A **parallelogram** is a quadrilateral having each pair of opposite sides parallel.

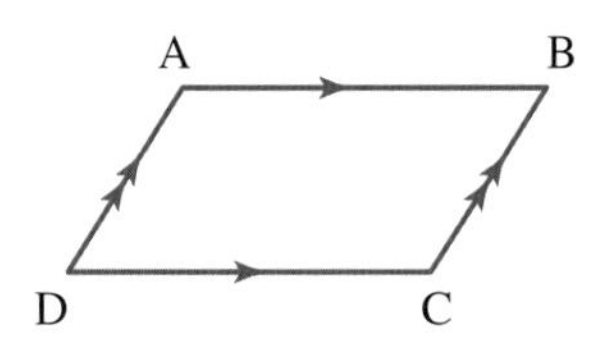


- To calculate the area of the parallelogram, first draw a diagonal from B to D to form two triangles, ABD and BDC. Label the base $b$ and the perpendicular height $h$.

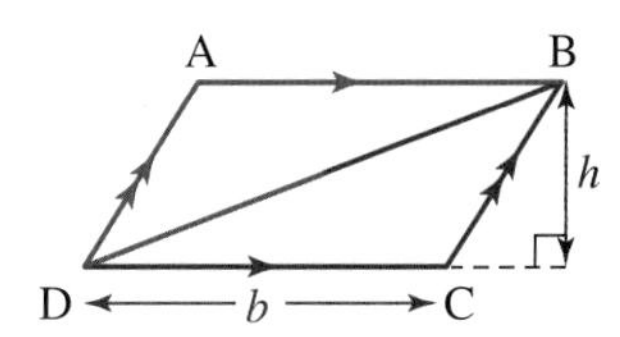


- The area of triangle BCD $= \frac{1}{2}b \times h$ and the area of triangle ABD $= \frac{1}{2}b \times h$.
  So, the area of the parallelogram:

$$\begin{aligned} \text{ABCD} &= \frac{1}{2}b \times h + \frac{1}{2}b \times h \\ &= b \times h \end{aligned}$$

where $b$ is the base and $h$ is the perpendicular height.

### Area of a parallelogram

**The area of a parallelogram is:**

$$A = bh$$

**where $b$ is the base and $h$ is the perpendicular height.**

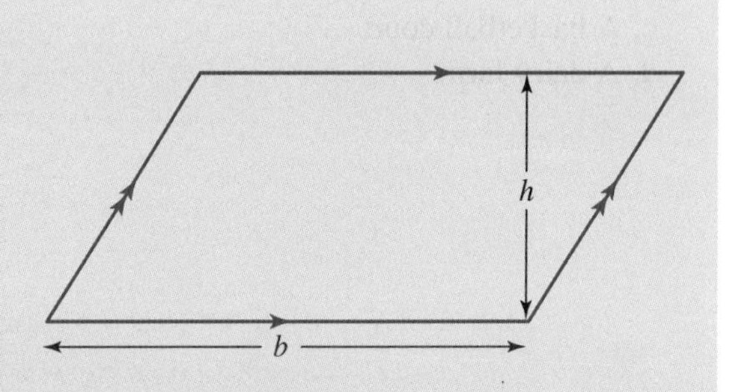

- For example, the area of the parallelogram shown is:

$$\begin{aligned} \text{ABCD} &= b \times h \\ &= 6 \times 4 \\ &= 24\ \text{cm}^2 \end{aligned}$$

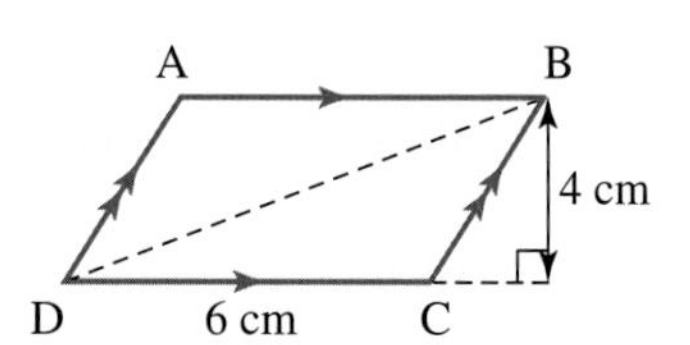

 Resources

 **eWorkbook** Topic 10 Workbook (worksheets, code puzzle and project) (ewbk-1960)

**Interactivities** Individual pathway interactivity: Area (int-4358)
Metric units of area 1 (int-4015)
Metric units of area 2 (int-4016)
Area of a rectangle (int-4017)
Area of a triangle (int-4018)
Area of a parallelogram (int-4019)

## Exercise 10.6 Area

**learn**on

**10.6 Quick quiz**  | **10.6 Exercise**

### Individual pathways

| ■ PRACTISE | ■ CONSOLIDATE | ■ MASTER |
|---|---|---|
| 1, 4, 5, 8, 9, 13, 14, 18, 21, 24, 26, 29, 31 | 2, 6, 10, 11, 15, 17, 19, 22, 25, 27, 30, 32, 34, 35 | 3, 7, 12, 16, 20, 23, 28, 33, 36, 37 |

### Fluency

1. Identify which unit would be most suitable to measure the following areas. Choose from $\text{mm}^2$, $\text{cm}^2$, $\text{m}^2$, ha or $\text{km}^2$.
   a. A computer screen
   b. The Melbourne Cricket Ground
   c. A shirt button
   d. The Brisbane metropolitan area
2. Identify which unit would be most suitable to measure the following areas. Choose from $\text{mm}^2$, $\text{cm}^2$, $\text{m}^2$, ha or $\text{km}^2$.
   a. A house block
   b. The state of Queensland
   c. A basketball court
   d. A dairy farm

3. WE11 The following figures are drawn on centimetre grid paper. Determine the area of each one.

a. 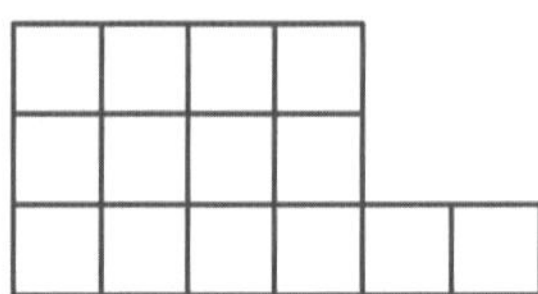

b. 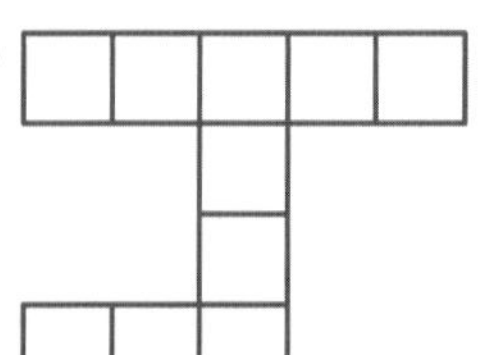

c. 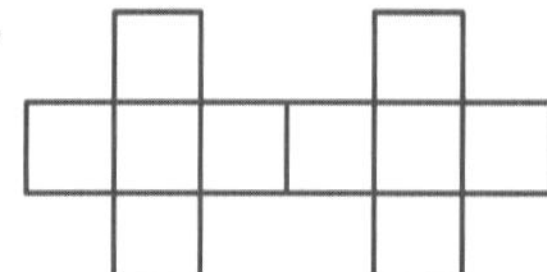

d. 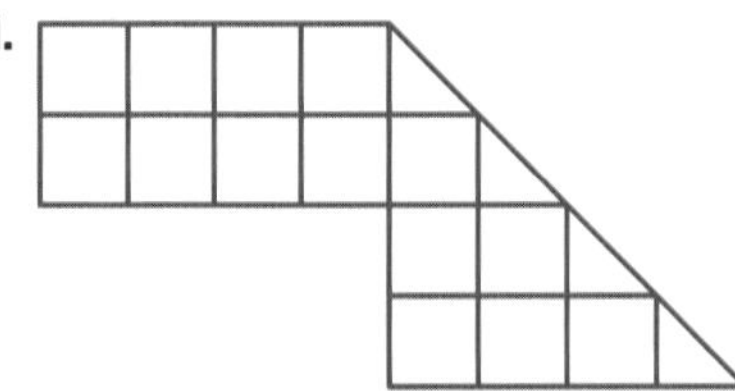

e. 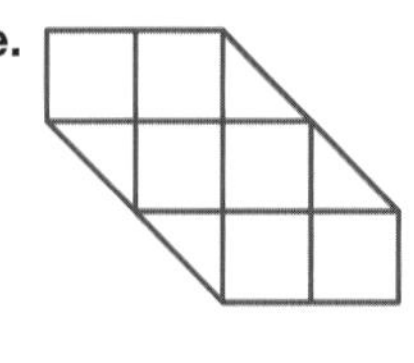

4. Estimate the areas of the following figures, which are drawn on centimetre grid paper.

a. 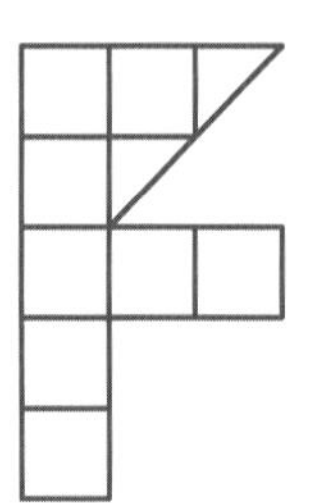

b. 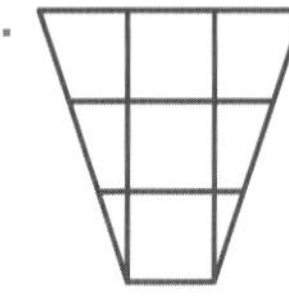

c. 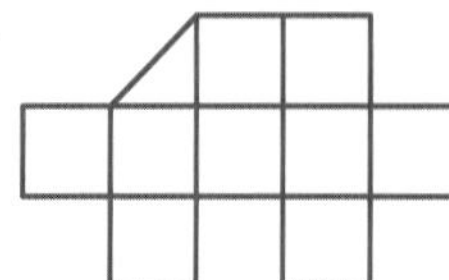

d. 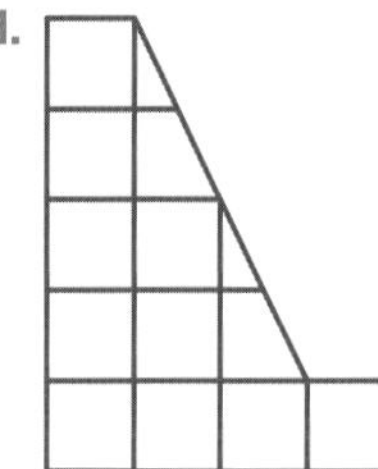

e. 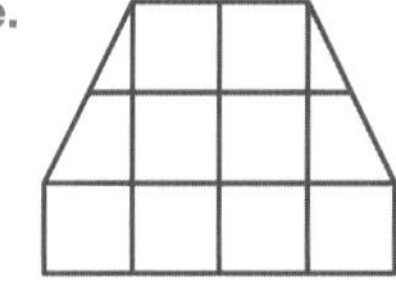

5.  Select from the following an estimate for the area of the shape drawn on centimetre grid paper.

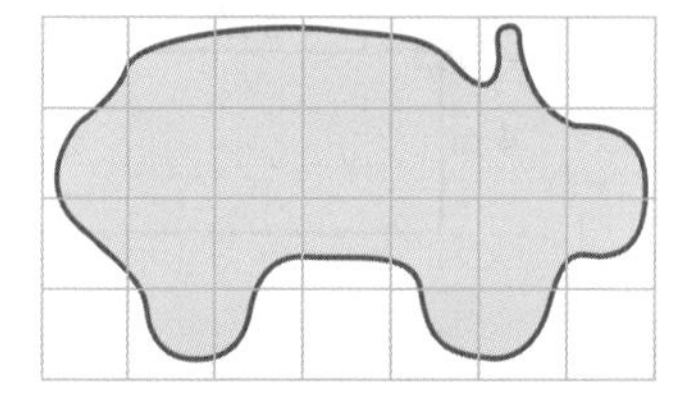

A. $18\,\text{cm}^2$
B. $21\,\text{cm}^2$
C. $16\,\text{cm}^2$
D. $23\,\text{cm}^2$
E. $28\,\text{cm}^2$

6. 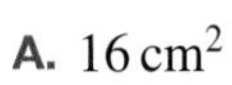 Select from the following an estimate for the area of the shape drawn on centimetre grid paper.

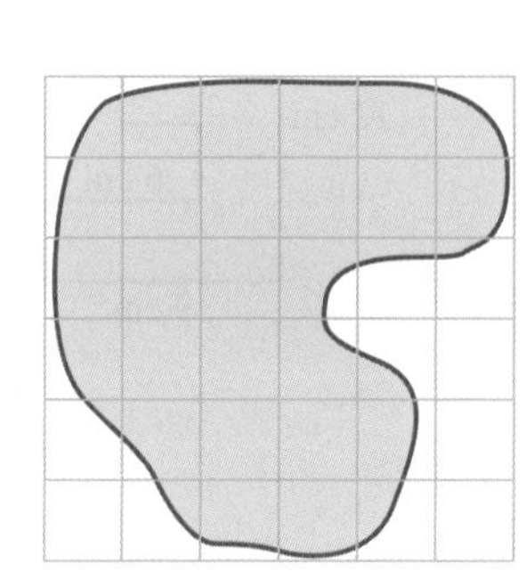

A. $16\,\text{cm}^2$
B. $20\,\text{cm}^2$
C. $25\,\text{cm}^2$
D. $28\,\text{cm}^2$
E. $30\,\text{cm}^2$

7. **MC** Select from the following an estimate for the area of the shape drawn on centimetre grid paper.

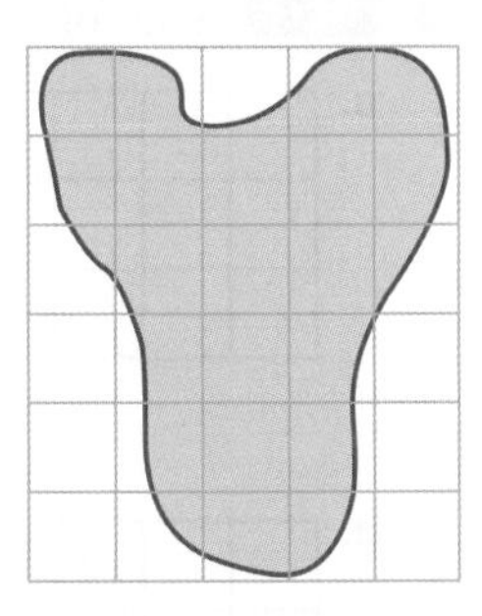

A. $15\text{ cm}^2$
B. $17\text{ cm}^2$
C. $19\text{ cm}^2$
D. $21\text{ cm}^2$
E. $25\text{ cm}^2$

8. **WE13a** Determine the areas of the following rectangles. (*Hint:* Use the formula $A = lw$.)

a. 5 cm, 3 cm

b. 7 km, 2 km

c. 2 m, 5 m

**WE13b,c** For questions **9** to **12**, calculate the areas of the following shapes. (*Hint:* Divide the shapes into rectangles and squares before using the formula $A = lw$.)

9. a.

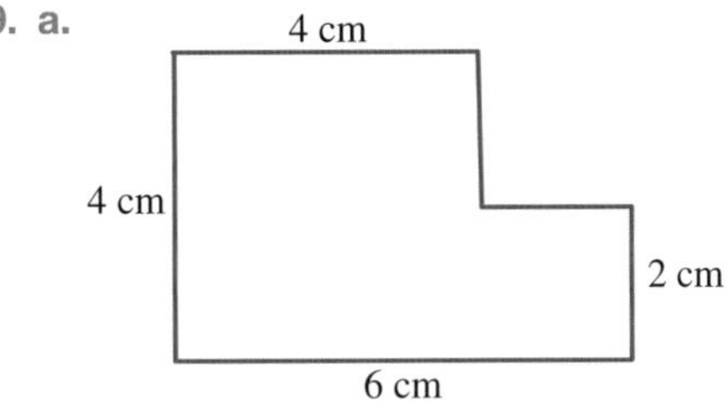


b.

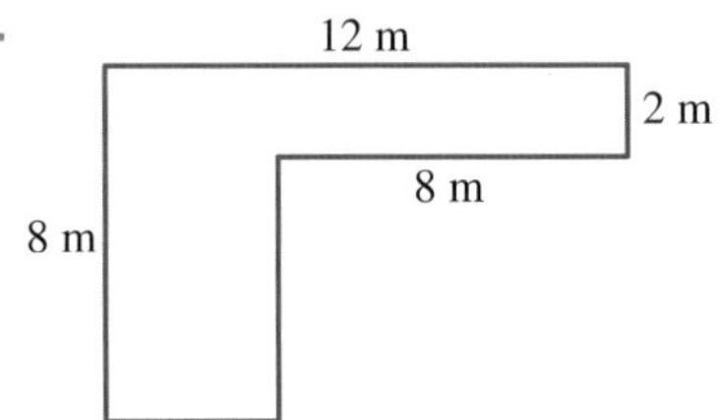


10. a.

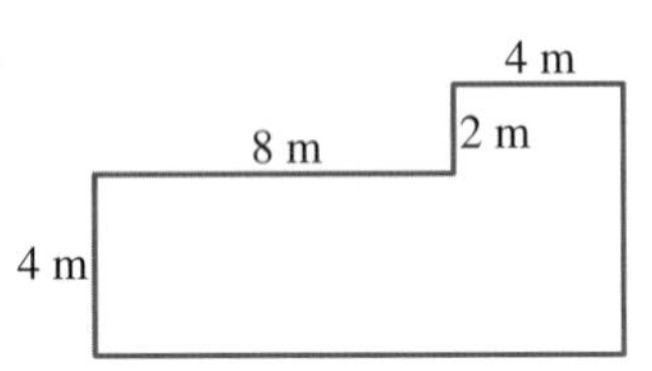


b.

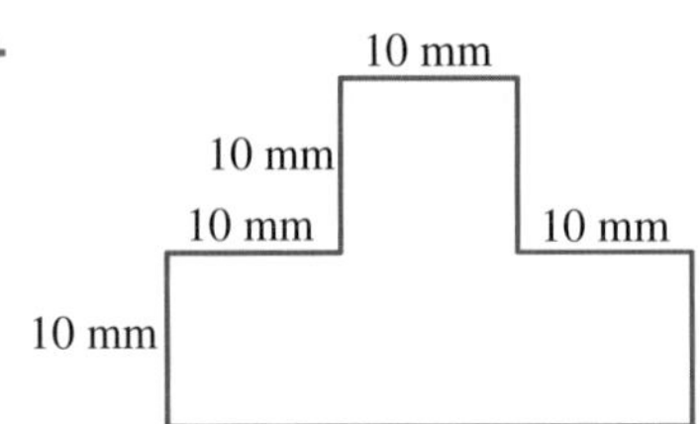


11. a.

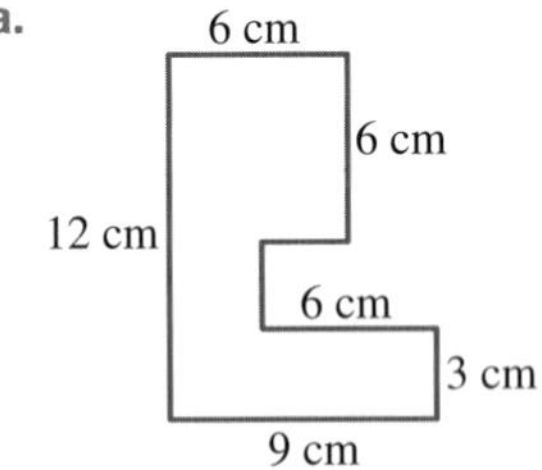


b.

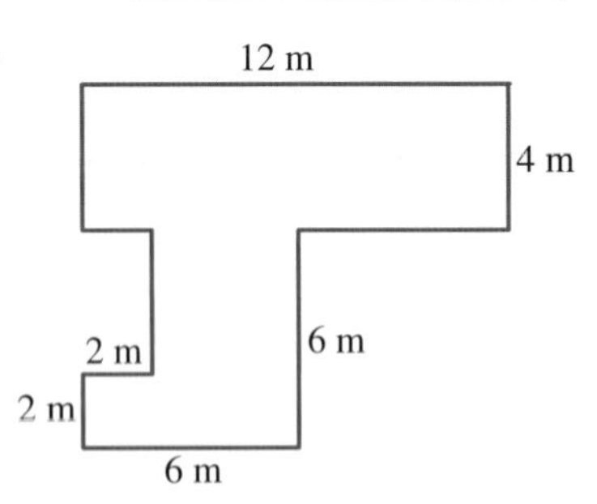

12. a.

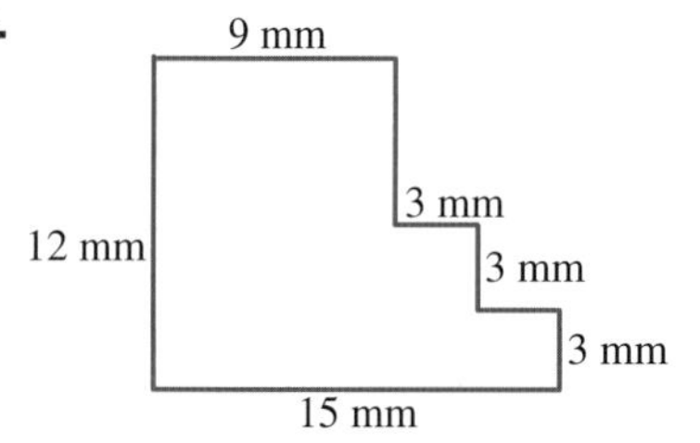


b.

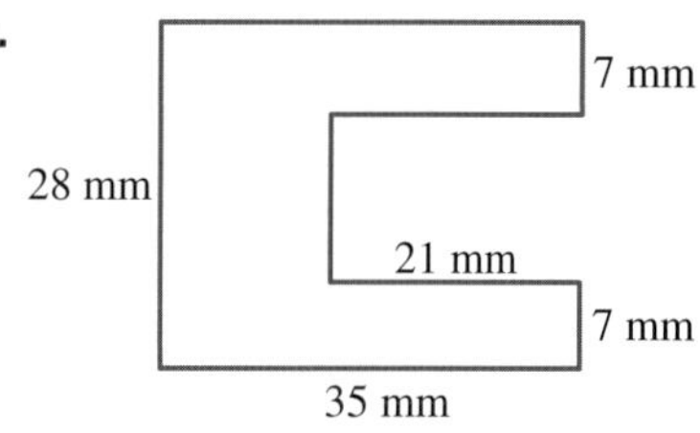


13. **WE14** Calculate the areas of the following triangles.

a.

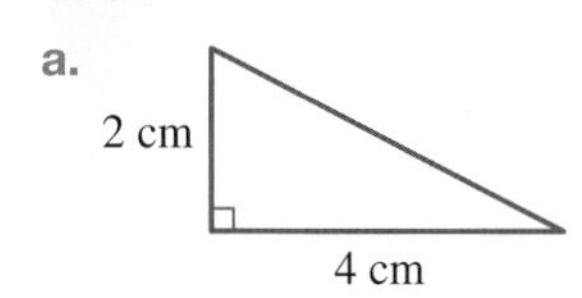


b.

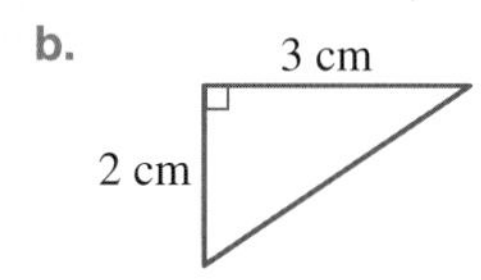


c.

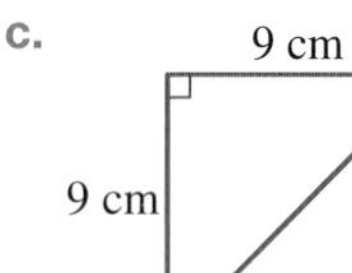


d.

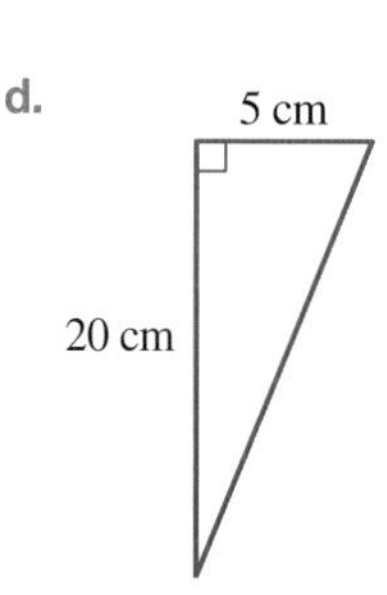


14. **MC** Identify the height of the triangle shown.

A. 60 mm
B. 30 mm
C. 27 mm
D. 53 mm
E. 49 mm

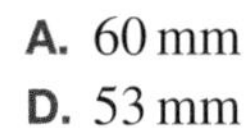

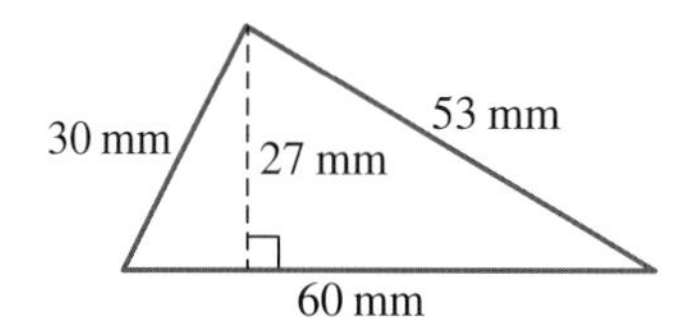


15. **MC** Identify the height of the triangle shown.

A. 2.6 cm
B. 4.0 cm
C. 3.0 cm
D. 4.2 cm
E. 3.8 cm

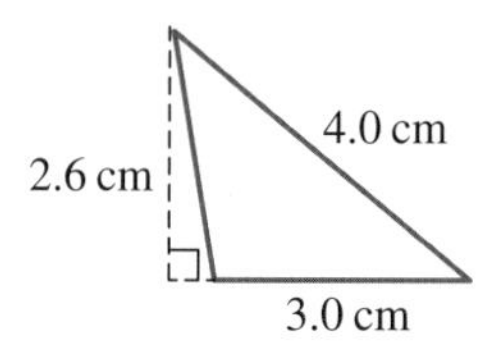


16. **MC** Identify the base length of the triangle shown.

A. 39 cm
B. 58 cm
C. 29 cm
D. 25 cm
E. 43 cm

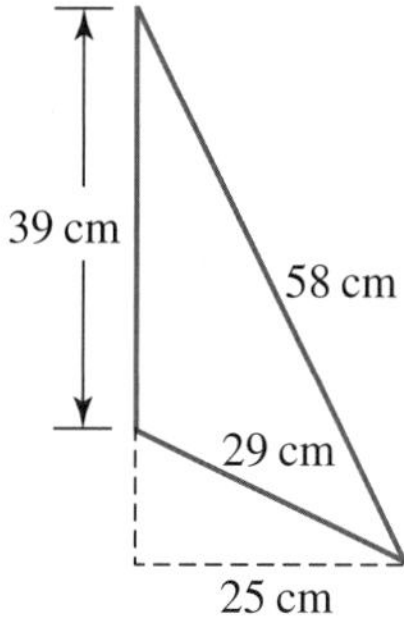


17. **WE15** Calculate the area of each of these triangles.

a.

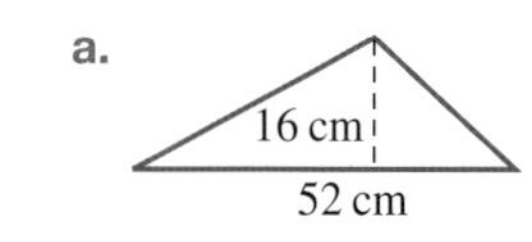


b.

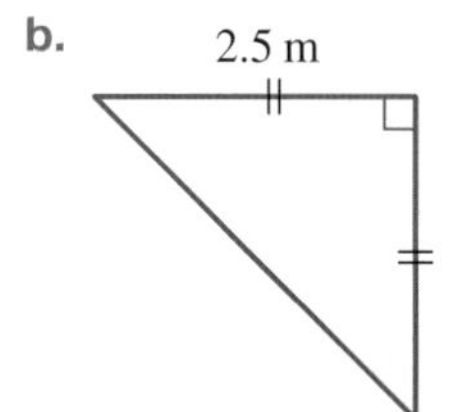


c.

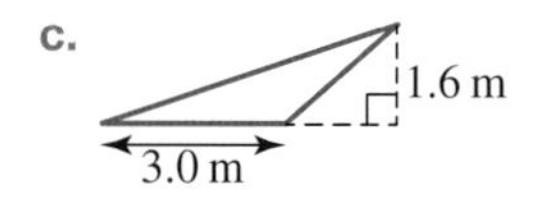


d.

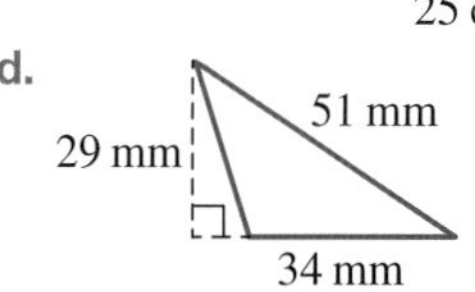


18. Calculate the area of each of these triangles.

a.

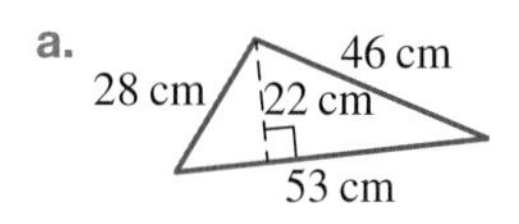


b.

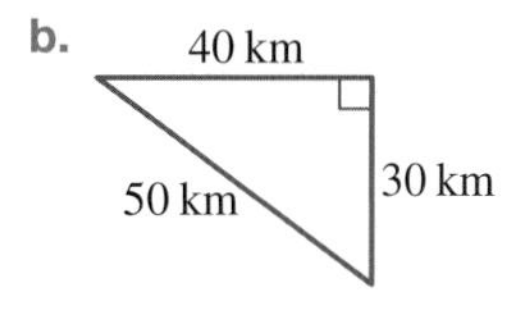


c.

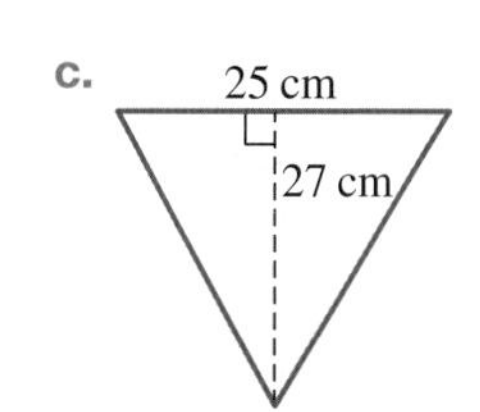


d.

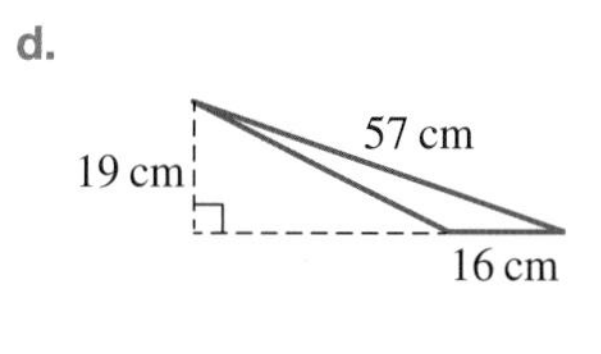

19. Calculate the area of the following parallelogram.

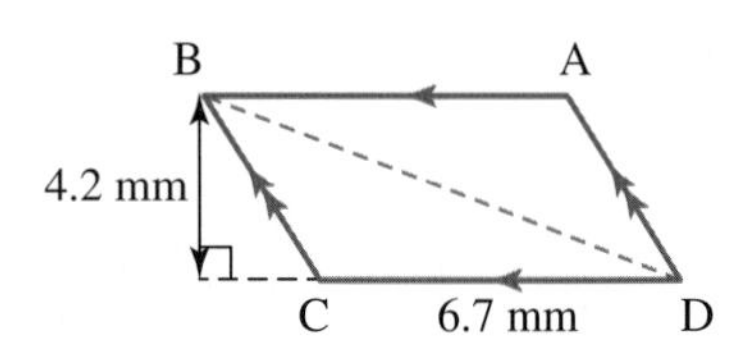


## Understanding

20. **MC** Select the closest estimate of the area of this butterfly if a centimetre grid is placed over it as shown.
    A. $10\,\text{cm}^2$
    B. $18\,\text{cm}^2$
    C. $27\,\text{cm}^2$
    D. $38\,\text{cm}^2$
    E. $51\,\text{cm}^2$

21. One of the largest open-air shopping centres is the Ala Moana centre in Honolulu, Hawaii, USA with more than 200 shops covering an area of 20 hectares. Convert this area to square metres.

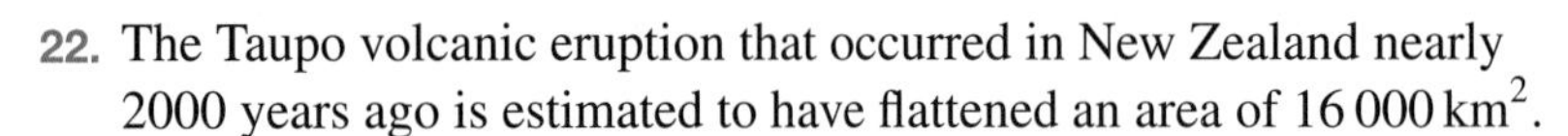

22. The Taupo volcanic eruption that occurred in New Zealand nearly 2000 years ago is estimated to have flattened an area of $16\,000\,\text{km}^2$.
    a. Convert this area to square metres.
    b. Convert this area to hectares.

23. Calculate how many square metres of carpet are needed to cover a rectangular room of length 5 m and width 3.5 m.

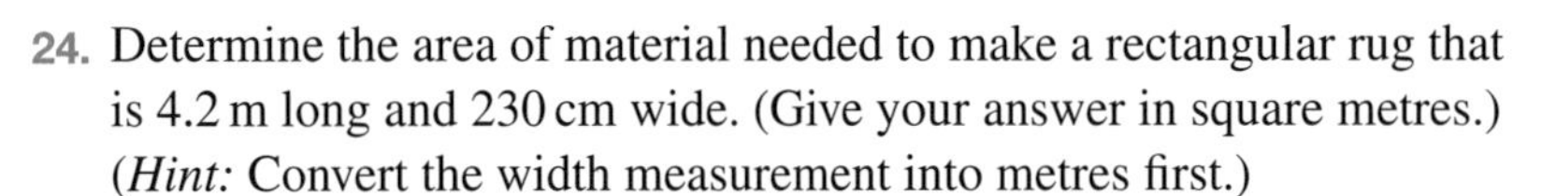

24. Determine the area of material needed to make a rectangular rug that is 4.2 m long and 230 cm wide. (Give your answer in square metres.) (*Hint:* Convert the width measurement into metres first.)

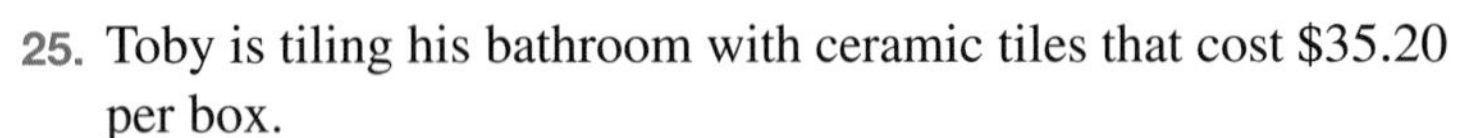

25. Toby is tiling his bathroom with ceramic tiles that cost $35.20 per box.
    a. Calculate how many square metres of tiles he will need if the rectangular room has a width of 2.5 m and a length of 3 m.
    b. Calculate how many boxes of tiles he should order if each box contains enough tiles to cover $0.5\,\text{m}^2$.
    c. Determine the cost of the tiles.

## Reasoning

26. Alana, who works for Fast Glass Replacements, has been asked for a quote to replace three windows. Each window is 1.8 m long and 0.8 m wide. Determine the price Alana should quote if the glass costs $27 per square metre. (Include a delivery cost of $25 in the quote.)

27. A floor tiler charged $640.00 to tile a rectangular room. Her next job is to tile the floor of a rectangular room twice as long and twice as wide. Determine how much she should charge for the larger room. (The answer is not $1280.00.) Justify your answer.

28. Glen wants to pave his back courtyard. His courtyard is $12\,\text{m} \times 12\,\text{m}$. He has a choice of $10\,\text{cm} \times 25\,\text{cm}$ clay pavers that cost $2.30 each, or concrete pavers that are $70\,\text{cm} \times 70\,\text{cm}$ and cost $42 each. Compare the costs of each to determine the cheaper option.

## Problem solving

29. For the sailboat shown, calculate the approximate area of the triangular mainsail when the sail is flat. Use the dimensions shown.

30. Jane is a landscape gardener who is laying a new lawn. The rectangular lawn is 13 m long and 8 m wide. Calculate how many square metres of turf Jane should order.
    Calculate the total cost of the turf if it costs $12.50 per square metre.

31. Calculate the total area of the cattle station shown in the diagram, which has 3 large paddocks. (Give your answer in square kilometres.)

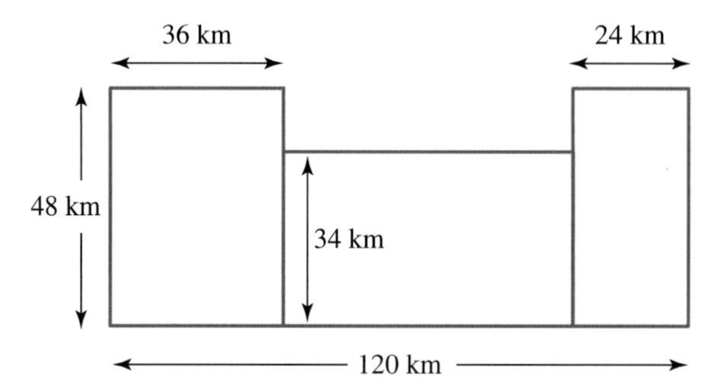


32. Determine the area of cloth required to make the kite shown, given that the length BD is 80 cm and the length AC is 1.5 m.

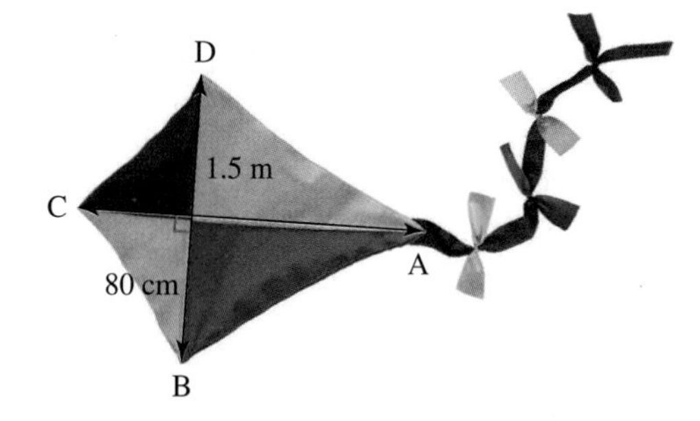


33. Calculate the total floor area of a concrete slab for a house as shown in the following diagram.

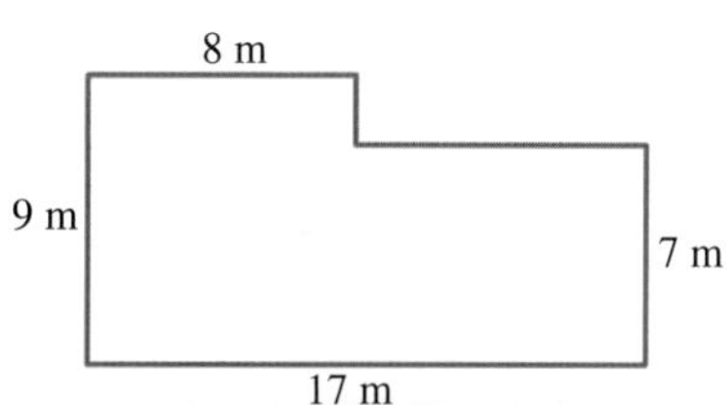


34. A church spire has six identical triangular faces that have the dimensions shown. Determine the area of copper roofing required to cover all six faces of the spire.

35. Calculate the total wing area of the following delta-winged jet aircraft.

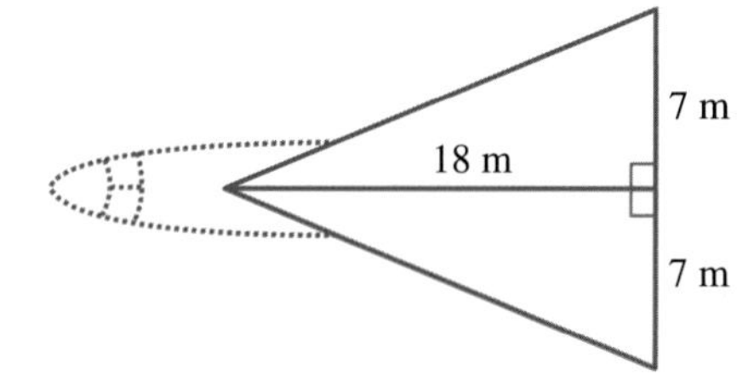


36. A rectangle has an area of 36 $cm^2$ and a perimeter of 26 cm. Each side of the rectangle is a whole number of centimetres. Determine the length and width of this rectangle. Show your working.

37. Geoff wants to establish a rectangular vegetable garden bed on his farm. His work shed is 20 m long, and it will act as one of the boundaries for the garden bed. (*Note:* The garden will not necessarily be this long.) He has 24 metres of fencing, which he plans to use to fence the other three sides. Describe how Geoff could use this fencing to enclose the largest possible area.

# LESSON
# 10.7 Area of composite shapes

## LEARNING INTENTION

At the end of this lesson you should be able to:

- separate composite shapes into simple shapes
- calculate the areas of composite shapes by adding or subtracting areas of simple shapes.

## 10.7.1 Calculating the area of composite shapes

eles-4558

- **Composite shapes** can be separated into parts, each of which is a simple shape (such as, for example, a rectangle or a triangle).
- The area of a composite shape can be found by adding the areas of each of the separate parts.
- For example, the area of the following pentagon can be found by adding the areas of the triangle and rectangle.

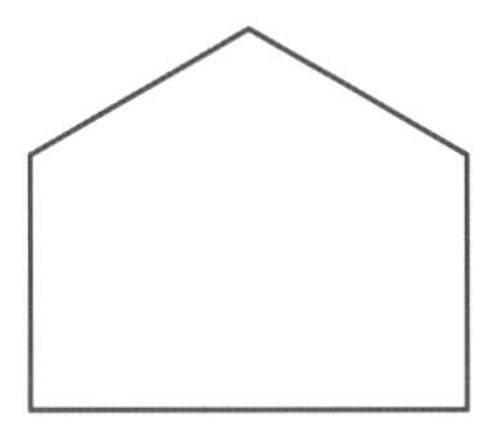

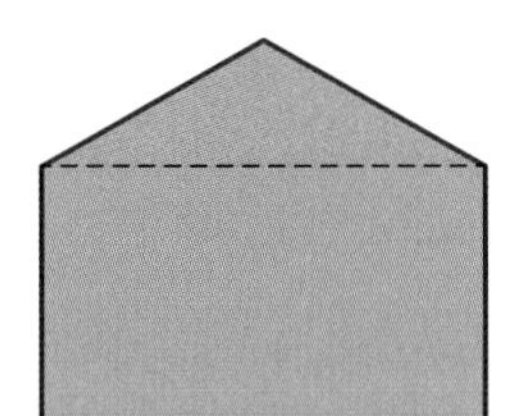

### WORKED EXAMPLE 16 Calculating the area of a composite shape using addition

**Calculate the area of the following shape.**

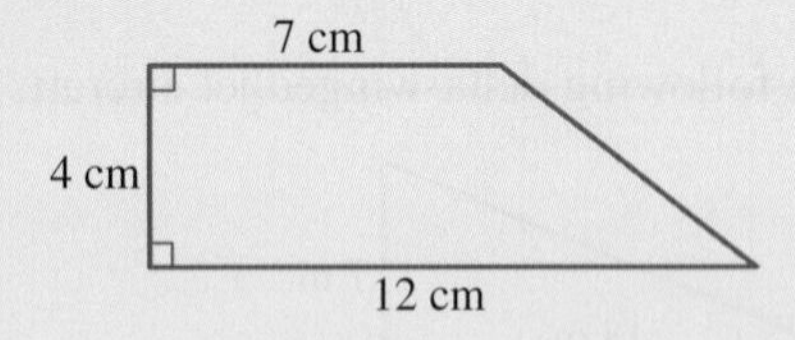


**THINK**

1. Divide the shape into a rectangle and a triangle.

**WRITE**

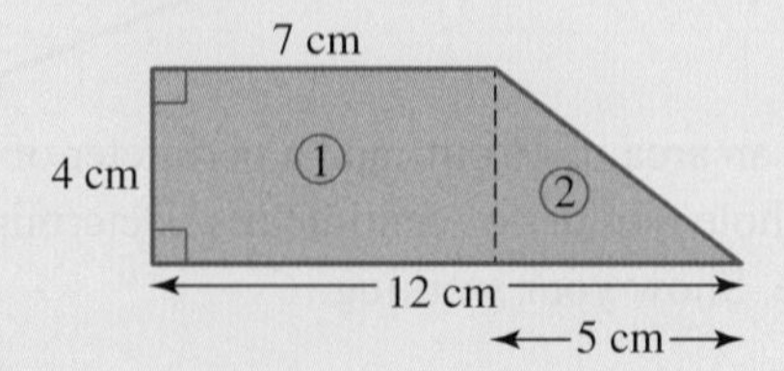

2. Calculate the area of shape 1 using the formula $A = lw$. The value of $l = 7$ cm and $w = 4$ cm.

Area 1 $= lw$ $\quad l = 7\text{ cm}, w = 4\text{ cm}$

3. Substitute the values of $l$ and $w$ in the formula to calculate the area of the rectangle.

$= 7 \times 4$

$= 28\text{ cm}^2$

4. Calculate the area of shape 2 using the formula $A = \frac{1}{2}bh$. The value of $b = 5$ cm and $h = 4$ cm.

Area 2 $= \frac{1}{2}bh$ $\quad b = 5\text{ cm}, h = 4\text{ cm}$

5. Substitute the values of $b$ and $h$ in the formula to calculate the area of the triangle.

$= \frac{1}{2} \times 5 \times 4$

$= 10\text{ cm}^2$

6. Add the two areas to get the total area.

Total area $= A_1 + A_2$

$= 28 + 10$

$= 38\text{ cm}^2$

7. Write the answer with the correct units.

The area of the composite shape is $38\text{ cm}^2$.

- In many situations the area of a given shape can be found by subtracting individual areas from each other.

## WORKED EXAMPLE 17 Calculating the area of a composite shape using subtraction

**Calculate the shaded area in the following figure.**

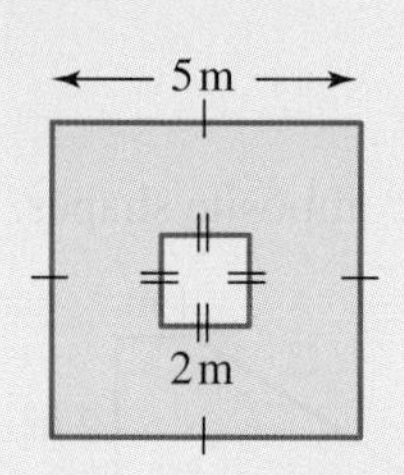


**THINK**

1. The figure shows a large square with the blue shaded area and a small square.

2. Calculate the area of the large square.

3. Calculate the area of the small square.

**WRITE**

Area of a square $= l^2$

Area of the large square: $A = 5^2$

$= 5 \times 5$

$= 25\text{ m}^2$

Area of the small square: $A = 2^2$

$= 2 \times 2$

$= 4\text{ m}^2$

4. To calculate the area of the shaded region, subtract the area of the small square from that of the large one.

$$\text{Shaded area} = 25 - 4$$
$$= 21\ \text{m}^2$$

5. Write the answer with the correct unit.

The shaded area is $21\ \text{m}^2$.

 Resources

 **eWorkbook** Topic 10 Workbook (worksheets, code puzzle and project) (ewbk-1960)

 **Video eLesson** Composite area (eles-1886)

 **Interactivities** Individual pathway interactivity: Area of composite shapes, using addition and subtraction (int-4359)

 Area of composite shapes (int-4020)

## Exercise 10.7 Area of composite shapes

learnon

10.7 Quick quiz on | 10.7 Exercise

### Individual pathways

| ■ PRACTISE | ■ CONSOLIDATE | ■ MASTER |
| --- | --- | --- |
| 1, 5, 7, 9, 11, 14 | 2, 6, 10, 12, 15 | 3, 4, 8, 13, 16 |

### Fluency

1. **WE16** Calculate the area of the following composite shapes.

a. 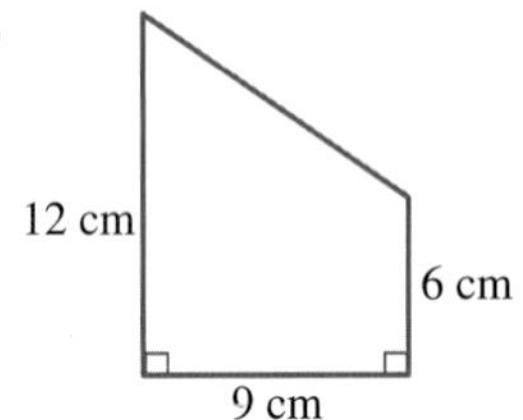


b. 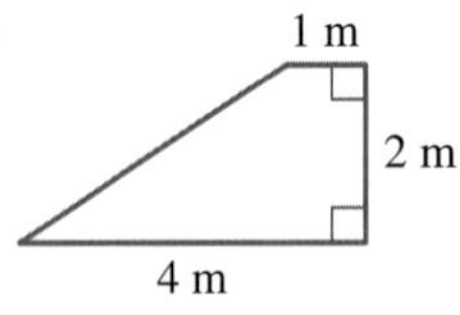


c. 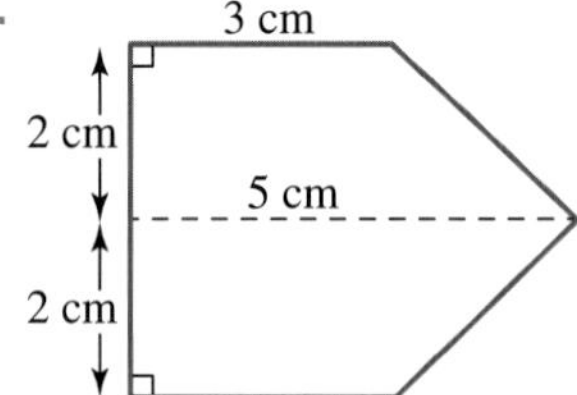


2. Calculate the area of the following composite shapes.

a. 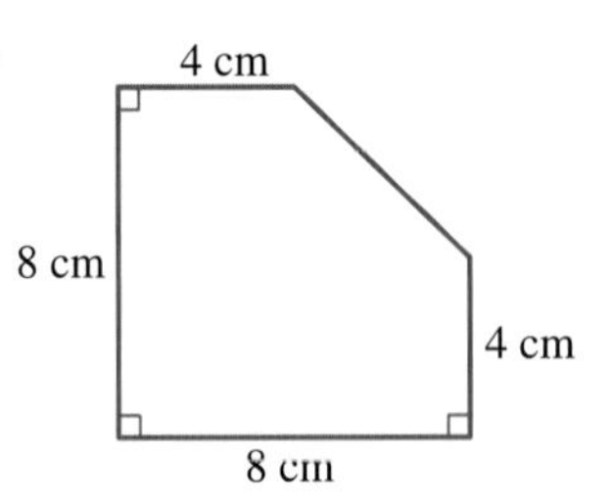


b. 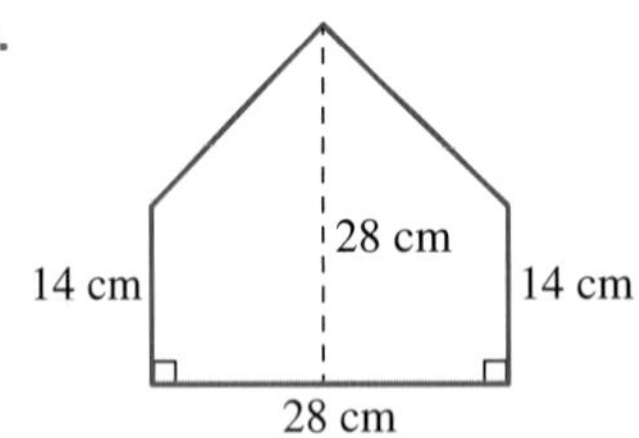


c. 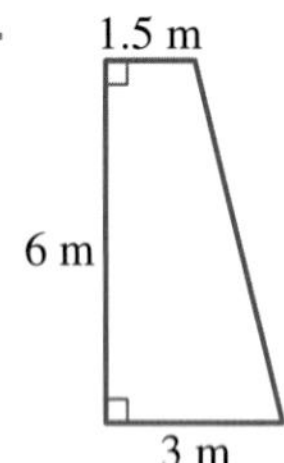

3. **WE17** Calculate the shaded area of the following shapes.

a.

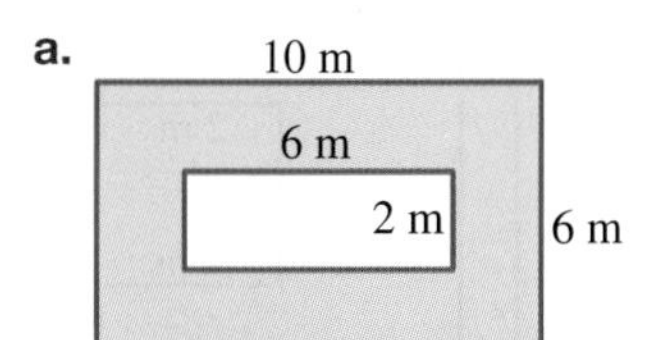


b.

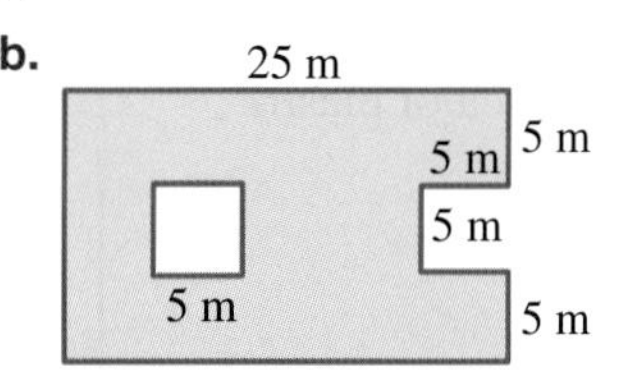


4. Calculate the area of the following garage wall.

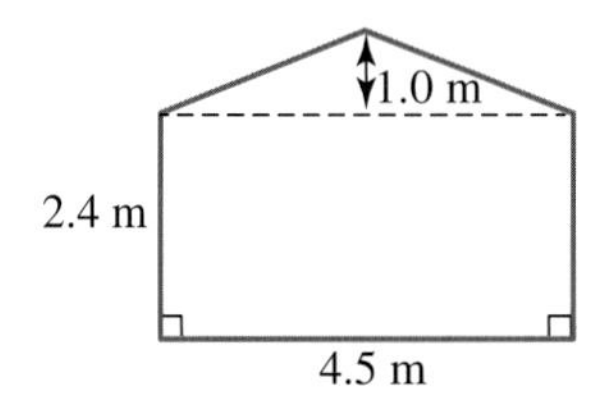


5. Determine the area of bricks needed to cover the following courtyard.

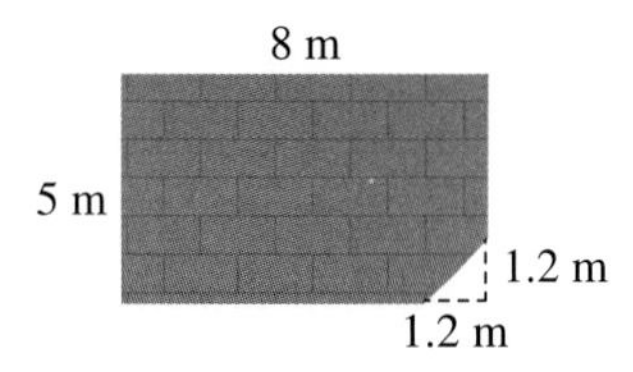


## Understanding

6. Determine the area of the hotel lobby in the following diagram.

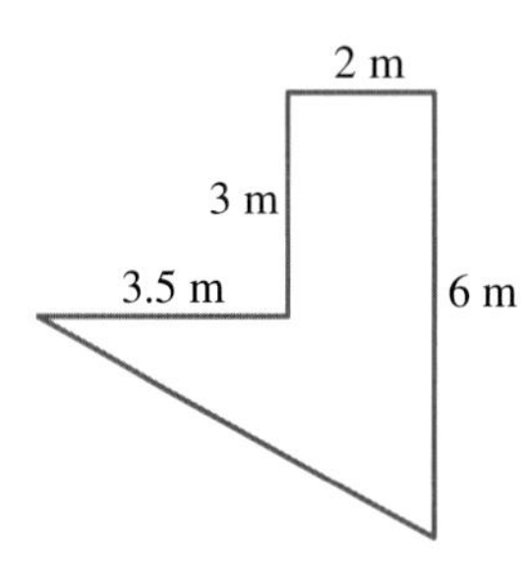


7. a. Calculate the area of carpet needed for the floor plan shown.
   b. Determine the cost of carpeting the room if the carpet costs $25 per square metre.

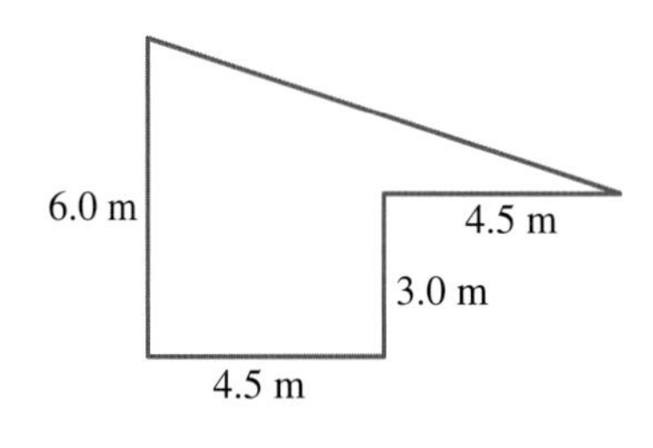


8. Michael is paving a rectangular yard that is 15.5 m long and 8.7 m wide. A square fishpond with side lengths of 3.4 m is in the centre of the yard. Calculate the cost of paving Michael's yard if the paving material costs $17.50 per square metre.

9. Members of the Lee family want to pave the area around their new swimming pool. The pool is set into the corner of the yard as shown in the diagram. Calculate the area of paving (in square metres) required to cover the yard around the pool (shaded in the diagram).

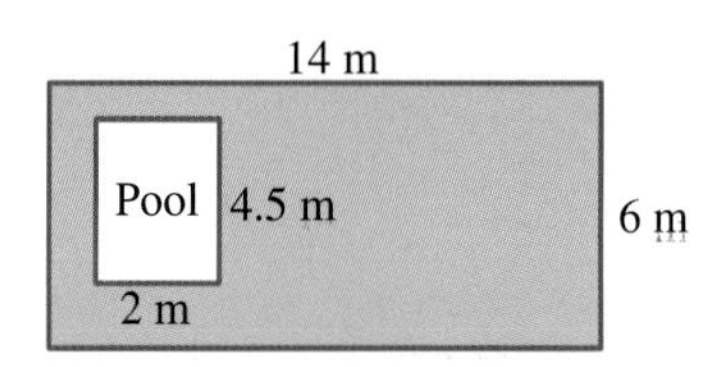

10. In order to construct an advertising sign, the following letters are cut from plastic sheets. Calculate the total area of plastic required to make all 3 letters.

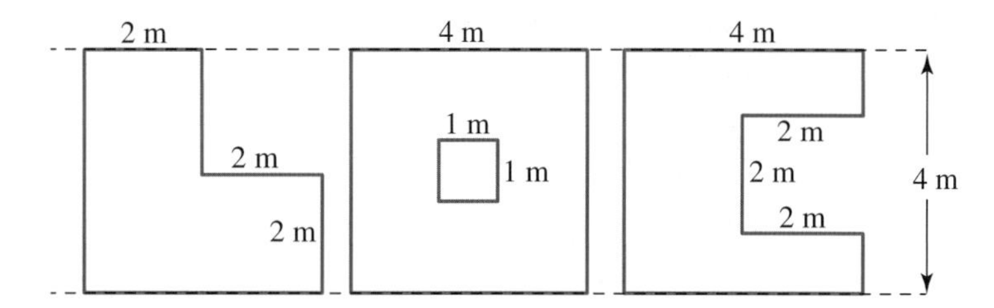


### Reasoning

11. a. Calculate the area of the diagram. Show all your working.
    b. Calculate the area of the figure using an alternative method. Show all your working.
    c. Explain the differences between the two methods.

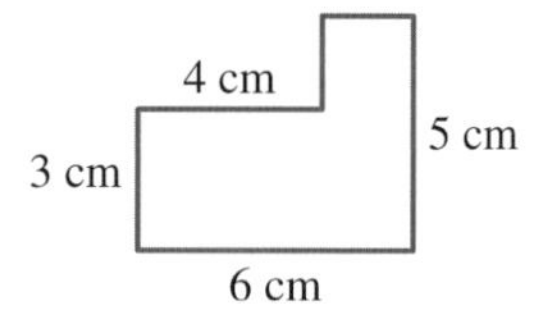


12. When determining the area of a composite shape, explain how you would decide whether to use addition or subtraction.

13. The stylised diagram of a set of traffic lights has been constructed using basic shapes, as shown.
    a. State the shapes used to draw the diagram and their dimensions in cm.
    b. Explain the steps required in calculating the black area surrounding the traffic lights shown.
    c. If the area of one of the large semicircles is closest to 353 $cm^2$, and the area of one of the small coloured circles is closest to 177 $cm^2$, calculate the area of the black surface. State the answer in $cm^2$.
    d. Explain why it is more appropriate to express the area in $cm^2$ rather than $mm^2$.

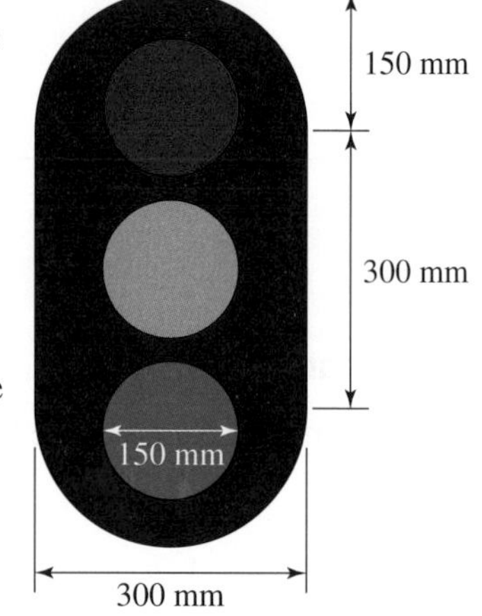


### Problem solving

14. Ellen wants to frame her cross-stitch work. Her cross-stitch work is 90 cm long and 75 cm wide. The frame's border, shown in the diagram, measures 2.5 cm wide. One company charges $120 for the frame. Determine the cost of the frame per square centimetre. Round your answer to 2 decimal places.

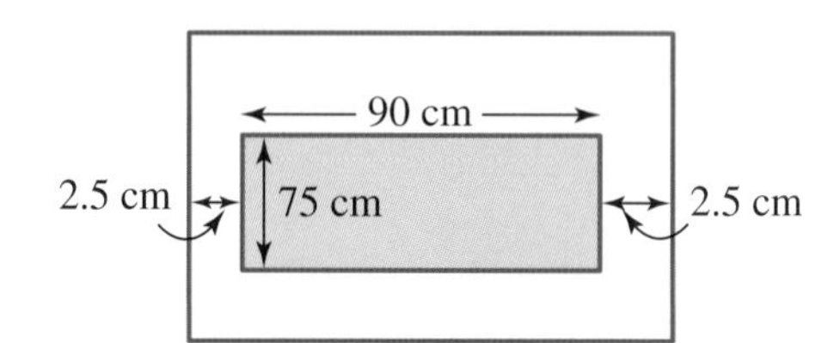


15. A square of side length 1 metre is made up of smaller internal squares that are formed by joining the midpoints of the outer square as shown in the diagram. Determine how much of the square the shaded area represents.

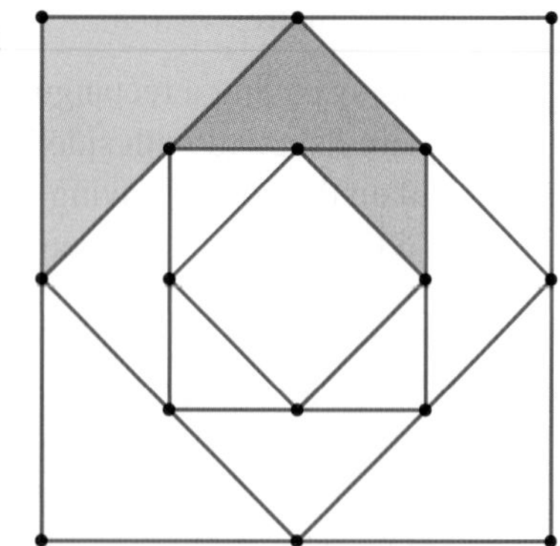

16. The diagram of the letter M shown has been constructed using basic shapes.
 a. Explain two methods of calculating the area of the letter using:
  i. addition  ii. subtraction.
 b. Using your preferred method, calculate the area of the letter.

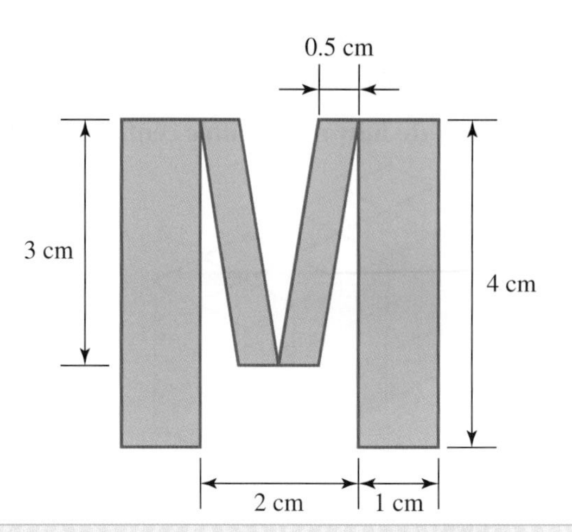


# LESSON
# 10.8 Volume of right prisms

## LEARNING INTENTION

At the end of this lesson you should be able to:
- calculate the volume of right prisms.

## 10.8.1 Volume

eles-4559

- The **volume** of a **3-dimensional object** is the amount of space it occupies.
- Volume is measured in cubic units such as $mm^3$, $cm^3$ and $m^3$.
- A cubic centimetre $\left(cm^3\right)$ is the space occupied by a cube with sides of 1 cm.
- **Right prisms** are three-dimensional solids that are formed by two identical end faces connected by rectangular side faces.

A sugar cube has a volume of about 1 $cm^3$.

A cubic metre $\left(m^3\right)$ is the space occupied by a cube with sides of 1 m.

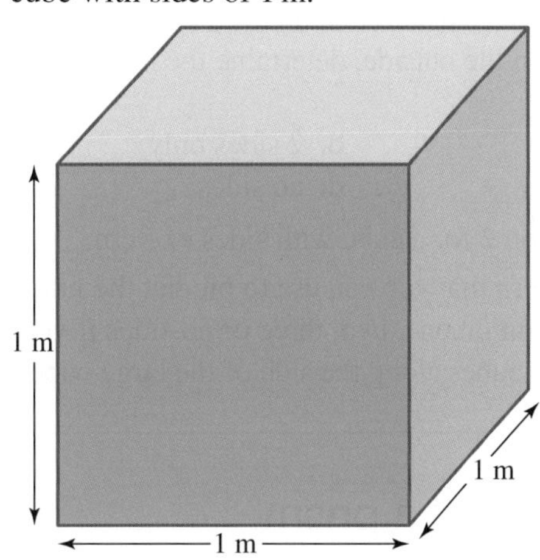

WORKED EXAMPLE 18 Counting cubes in solid shapes

**Calculate how many cubic centimetres are in each solid shape. (Each small cube represents 1 $cm^3$.)**

a.

b.

| THINK | WRITE |
|---|---|
| a. 1. Count the cubes.<br>There are 10 cubes in each layer.<br>There are three layers altogether. | a. Volume $= 10\,cm^3 \times 3$ |
| 2. Give the answer in cubic centimetres. | $= 30\,cm^3$ |
| b. 1. Count the cubes.<br>There are 12 cubes in the first layer.<br>There are 6 cubes in the second layer.<br>There are 3 cubes in the third layer. | b. Volume $= 12\,cm^3 + 6\,cm^3 + 3\,cm^3$ |
| 2. Give the answer in cubic centimetres. | $= 21\,cm^3$ |

COLLABORATIVE TASK: The blue cube

Work in pairs to answer the following questions.

1. The outside of a cube with 3 cm sides is painted blue. The cube is then cut into smaller cubes, each with sides 1 cm long. How many of the 1 cm cubes will have paint on:
   a. 1 side only
   b. 2 sides only
   c. 3 sides only
   d. no sides?
2. For a cube with 4 cm sides, made from 1 cm cube blocks and painted blue on the outside, determine the number of cubes painted on:
   a. 1 side only
   b. 2 sides only
   c. 3 sides only
   d. no sides.
3. Repeat question 2 for a cube with sides of 5 cm.
4. Is there a pattern that you can use to predict the numbers of cubes with paint on one, two, three or no sides if you know the number of cubes along the side of the large cube?

## 10.8.2 Volume of a prism

eles-4560

The volume of a prism can be calculated by the formula $V = A \times H$, where $A$ is the area of the cross-section of the prism and $H$ is the dimension that is at right angles to the cross-section, often referred to as the height.

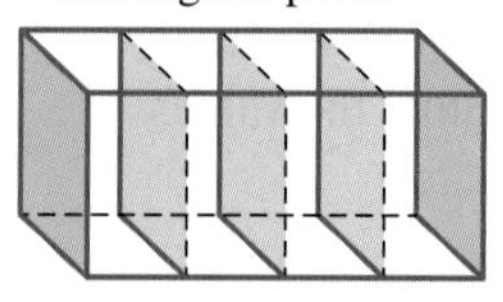


$$V_{prism} = A_{base} \times H$$

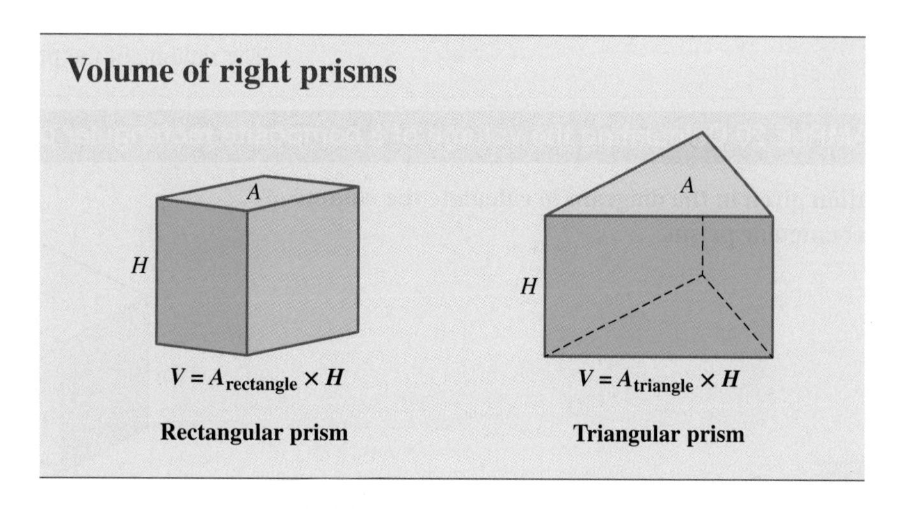


## WORKED EXAMPLE 19 Calculating the volume of right prisms

a. **Calculate the volume of the following rectangular prism.**

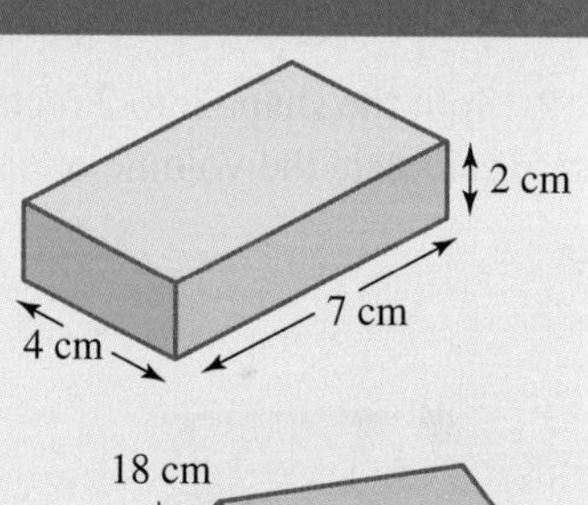


b. **Calculate the volume of the triangular prism shown.**

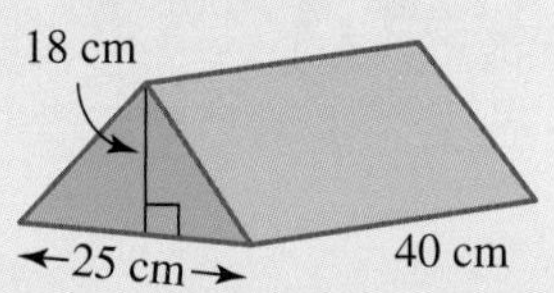


| THINK | WRITE |
|---|---|
| a. 1. State the formula for the volume of a rectangular prism. | $V = lwh$ |
| 2. Identify the length ($l = 7$ cm), width ($w = 4$ cm), and height ($h = 2$ cm) of the prism and substitute the values into the formula. | $= 7\text{ cm} \times 4\text{ cm} \times 2\text{ cm}$ |
| 3. Calculate the volume, stating the answer in the appropriate units ($cm^3$). | $= 56\text{ cm}^3$ |
| b. 1. Write the formula for the volume of a prism. | $V = A \times H$ |
| 2. This object has a triangular cross-section, so $A$ represents the area of the triangle. Write the formula for the area of the triangle. | $A_{triangle} = \frac{1}{2} \times b \times h$ |

3. Identify the values of the pronumerals. (*Note:* Here, $h$ is the height of the triangle, not of the prism.) Substitute the values of the pronumerals into the formula and evaluate.

$b = 25$ cm; $h = 18$ cm

$$A_{\text{triangle}} = \frac{1}{2} \times 25 \times 18$$
$$= 225 \text{ cm}^2$$

4. State the value of $H$. Since the prism is not standing on its triangular end, $H$ represents the prism's length.

$H = 40$ cm

5. Substitute the values of $A$ and $H$ into the volume formula and evaluate. Remember to include units in your answer.

$$V = 225 \times 40$$
$$= 9000 \text{ cm}^3$$

The volume of the prism is 9000 cm$^3$.

## WORKED EXAMPLE 20 Calculating the volume of a composite rectangular prism

**Use the information given in the diagram to calculate the volume of the composite rectangular prism.**

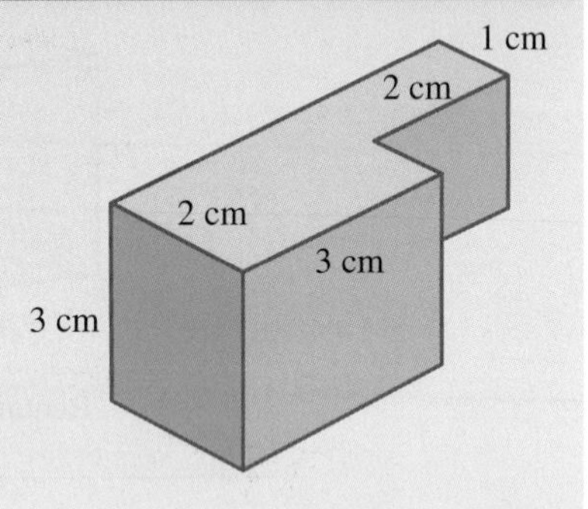


**THINK** | **WRITE**

1. To calculate the volume, first write the formula for the volume of a rectangular prism.

$V = lwh$

2. Split the shape into 2 rectangular prisms and calculate the volume of each rectangular prism.

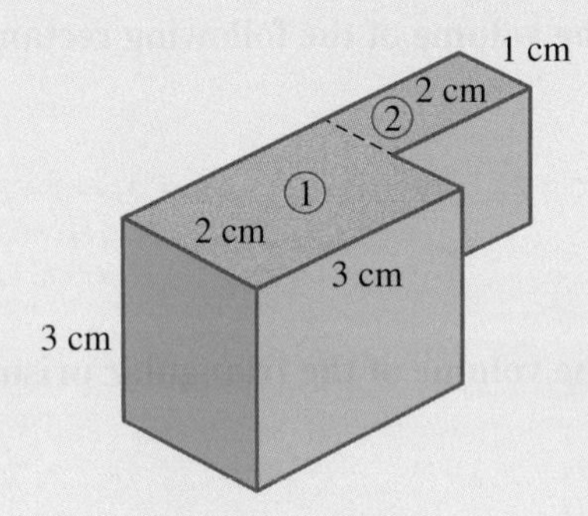


$$\text{Volume of rectangular prism 1} = 3 \times 2 \times 3$$
$$= 18 \text{ cm}^3$$
$$\text{Volume of rectangular prism 2} = 2 \times 1 \times 3$$
$$= 6 \text{ cm}^3$$

3. Add the volumes to determine the volume of the composite rectangular prism.

$$V = 18 \text{ cm}^3 + 6 \text{ cm}^3$$
$$= 24 \text{ cm}^3$$

Resources

eWorkbook Topic 10 Workbook (worksheets, code puzzle and project) (ewbk-1960)

Interactivities Individual pathway interactivity: Volume of rectangular prisms (int-8470)

Volume (int-4021)

Volume of a rectangular prism (int-4022)

# Exercise 10.8 Volume of right prisms

learn**on**

10.8 Quick quiz **on** | 10.8 Exercise

Individual pathways

| ■ PRACTISE | ■ CONSOLIDATE | ■ MASTER |
|---|---|---|
| 1, 2, 6, 9, 13, 16, 21, 24 | 3, 5, 7, 10, 11, 14, 17, 19, 22, 25 | 4, 8, 12, 15, 18, 20, 23, 26 |

## Fluency

1. **WE18a** Calculate how many cubic centimetres are in each solid shape. (Each small cube represents 1 cm$^3$.)

a. 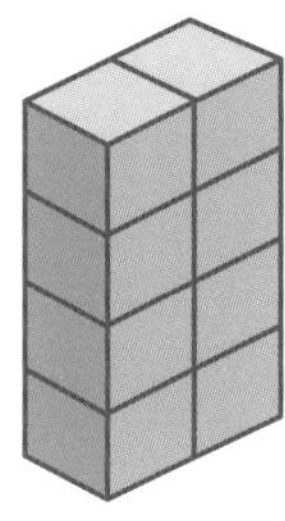

b. 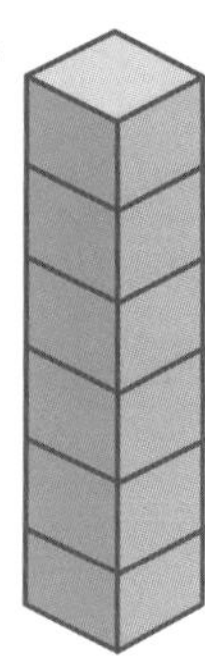

c. 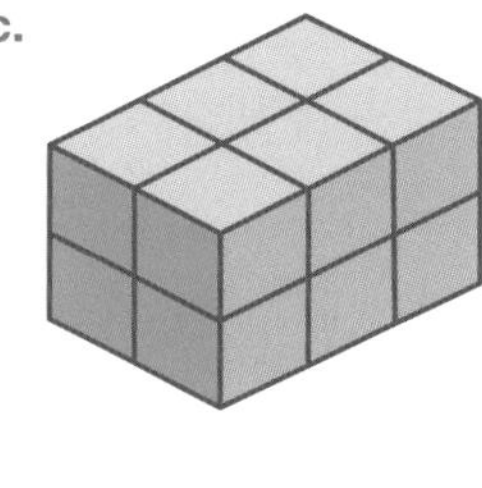

d. 

2. **MC** Select the volume of the prism shown.

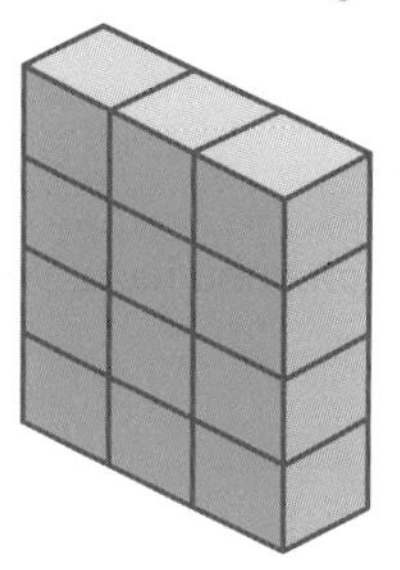

A. 4 cm$^3$
B. 12 cm$^3$
C. 1 cm$^3$
D. 8 cm$^3$
E. 19 cm$^3$

*Note:* Each cube has a volume of 1 cm$^3$.

3. **WE18b** Calculate the volumes of the following solids. (Each small cube represents 1 cm$^3$.)

a. 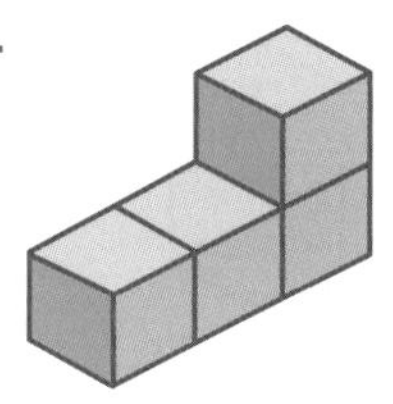

b. 

c. 

d. 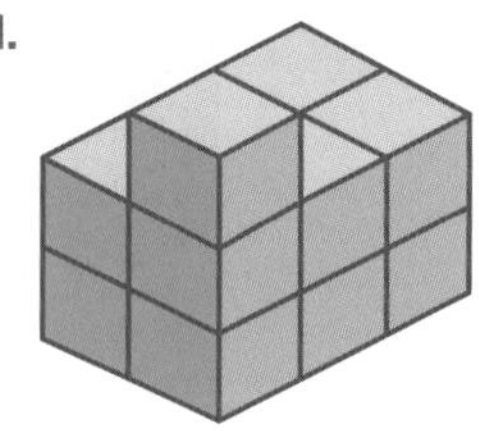

4. Calculate the volumes of the following solids. (Each small cube represents 1 $cm^3$.)

a.

b.

c.

d.

5. **MC** Select the volume of the prism shown.

A. 20 $cm^3$
B. 16 $cm^3$
C. 10 $cm^3$
D. 22 $cm^3$
E. 3 $cm^3$

*Note:* Each cube has a volume of 1 $cm^3$.

6. **WE19a** Calculate the volumes of the following rectangular prisms.

a.

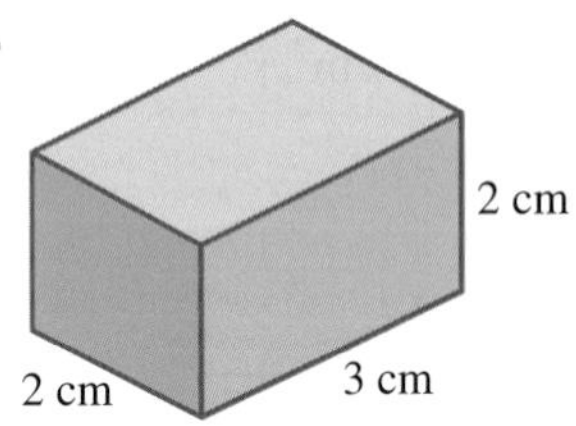


b.

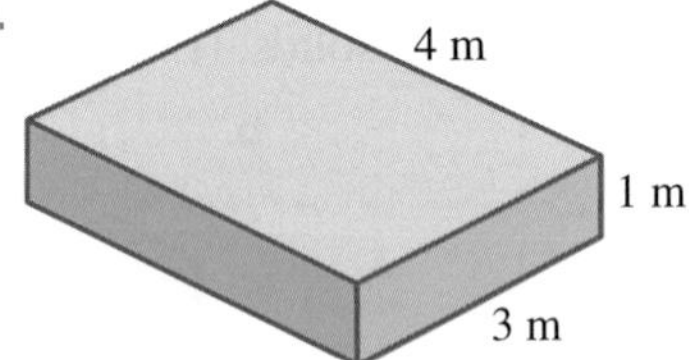


c.

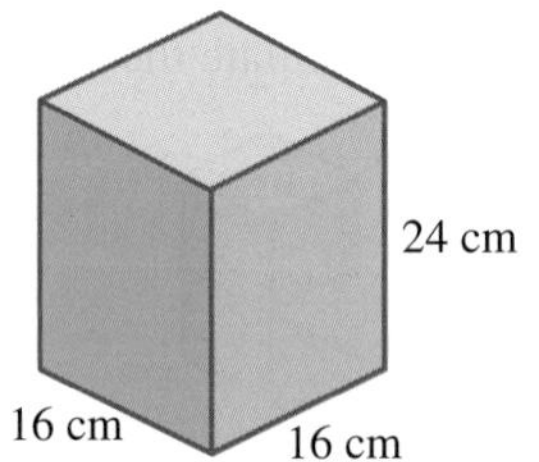


7. Determine the volume of each of the following. (Each cube has a volume of 1 $cm^3$.)

a.

b.

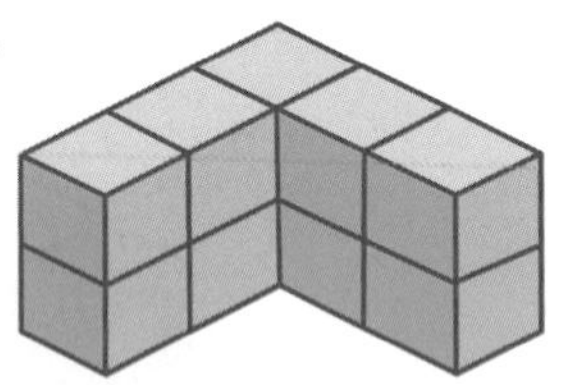

c.

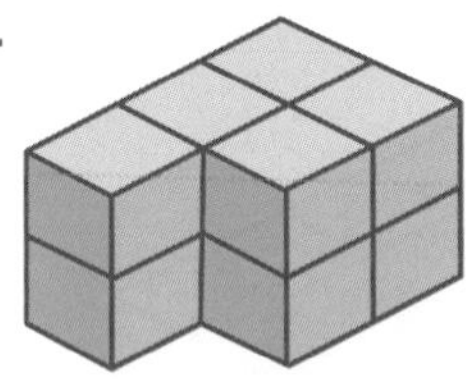

8. **WE19b** Calculate the volumes of the following rectangular prisms.

a.

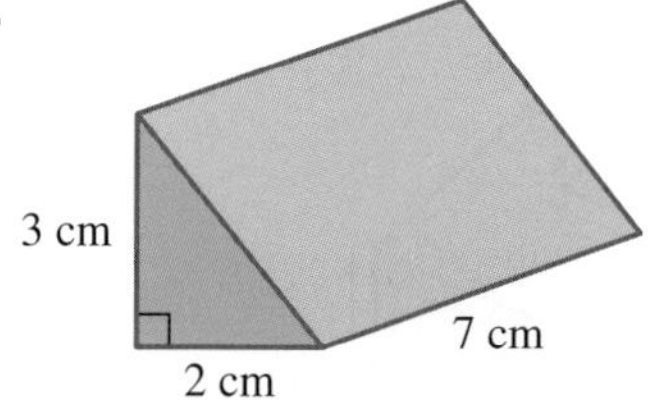


b.

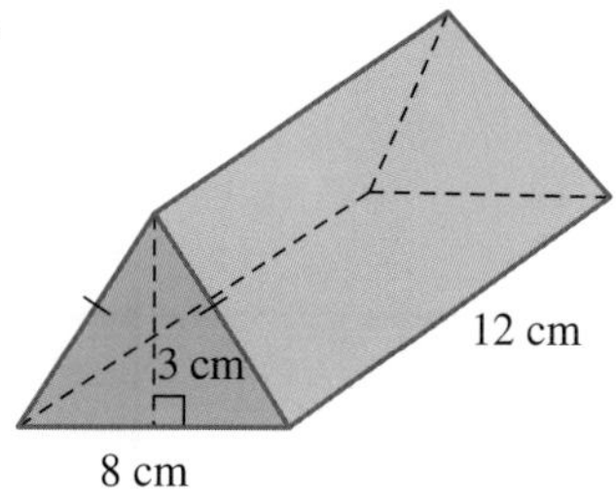


c.

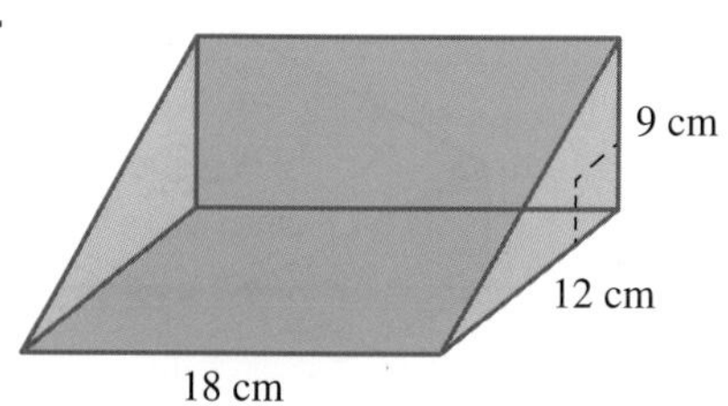


9. **MC** Select the volume of the prism shown.

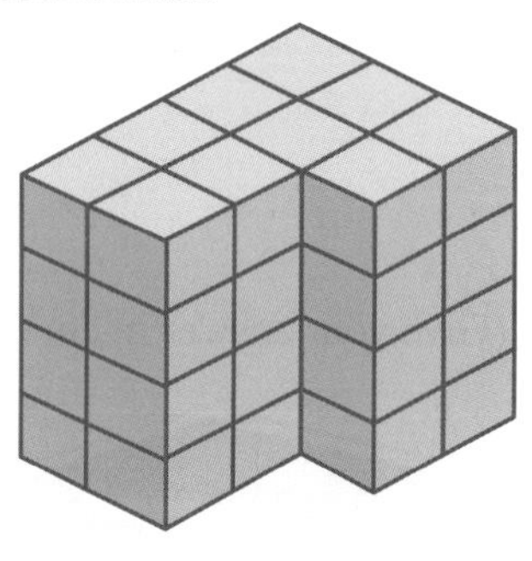

A. $10\,\text{cm}^3$ B. $20\,\text{cm}^3$ C. $30\,\text{cm}^3$ D. $40\,\text{cm}^3$ E. $80\,\text{cm}^3$

*Note:* Each cube has a volume of $1\,\text{cm}^3$.

10. **MC** Select the volume of the prism shown.

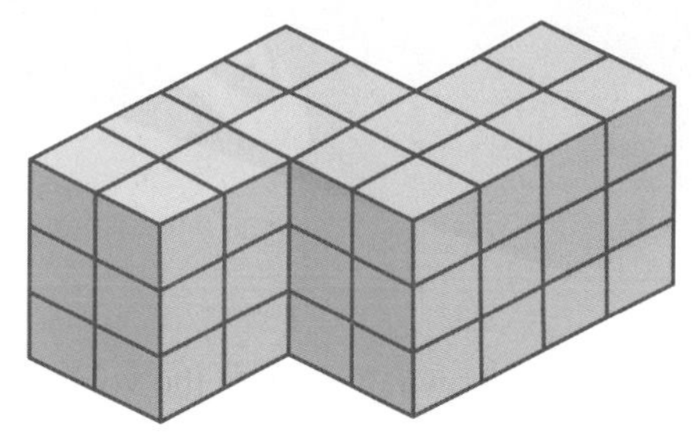

A. $16\,\text{cm}^3$ B. $48\,\text{cm}^3$ C. $32\,\text{cm}^3$ D. $40\,\text{cm}^3$ E. $160\,\text{cm}^3$

*Note:* Each cube has a volume of $1\,\text{cm}^3$.

11. **WE20** Calculate the volumes of the following composite prisms.

a.

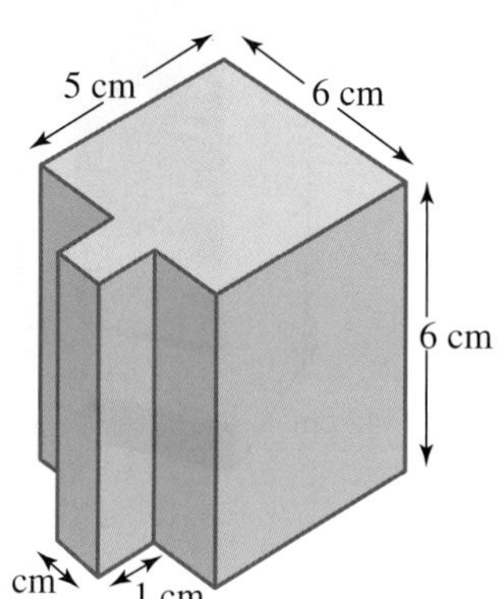


b.

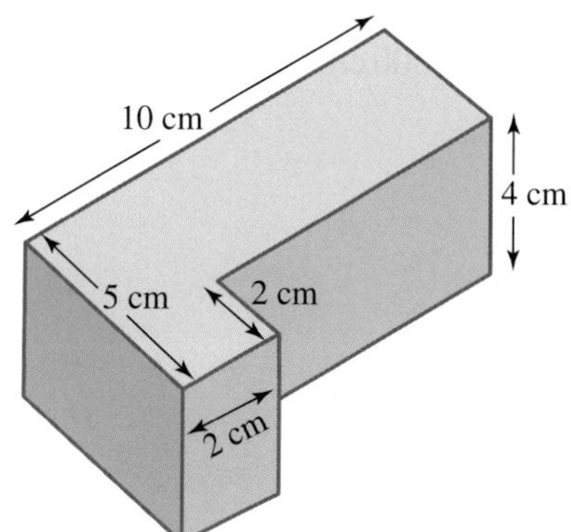

12. Calculate the volumes of the following composite prisms.

a.

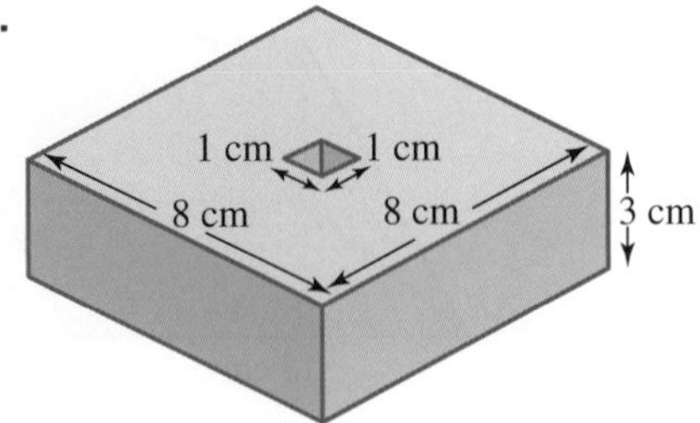


b.

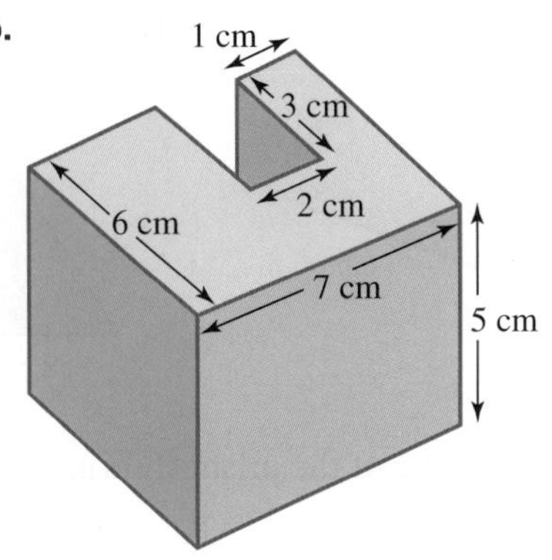


## Understanding

13. Calculate the volume of the shoe box shown. (Give your answer in $cm^3$.)

14. The inside dimensions of a refrigerator are shown. Calculate the volume available for food storage inside the refrigerator.

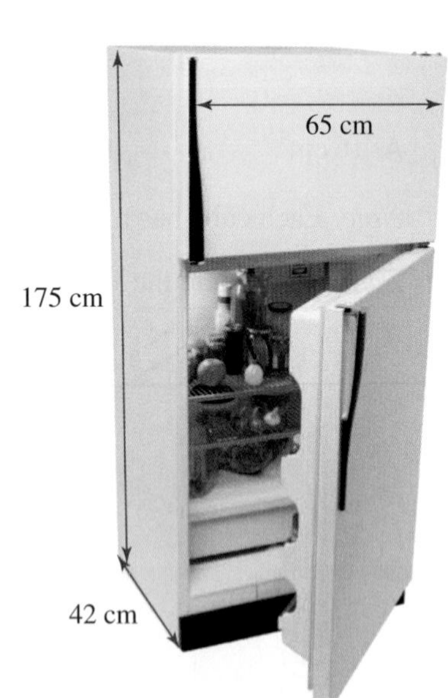


15. a. A rectangular prism has a length of 30 cm and a width of 15 cm. If its volume is 9000 $cm^3$, determine its height.
   b. A rectangular prism has a length of 13 cm and a width of 17 cm. If its volume is 1989 $cm^3$, determine its height.
   c. A triangular prism has a base of 10 cm and a triangle height of 15 cm. If its volume is 1800 $cm^3$, determine its height.

16. The lunch box shown is a rectangular prism.
 a. Calculate the volume of the lunch box in cubic centimetres.
 b. Change each measurement to millimetres and hence calculate the volume of the lunch box in cubic millimetres.

17. Determine the volume of concrete (in cubic metres) that would be needed to make the base for a garage that is 6.5 m wide and 3 m long. The concrete base is 0.25 m deep.

18. a. Calculate the volume (in cubic centimetres) of the matchbox shown.
 b. Change each measurement to millimetres and hence determine the volume of the matchbox in cubic millimetres.
 c. Matches are rectangular prisms of length 44 mm, width 2 mm and height 2 mm. Calculate the volume of a match (in cubic millimetres). (Ignore the red substance on the end of each match.)

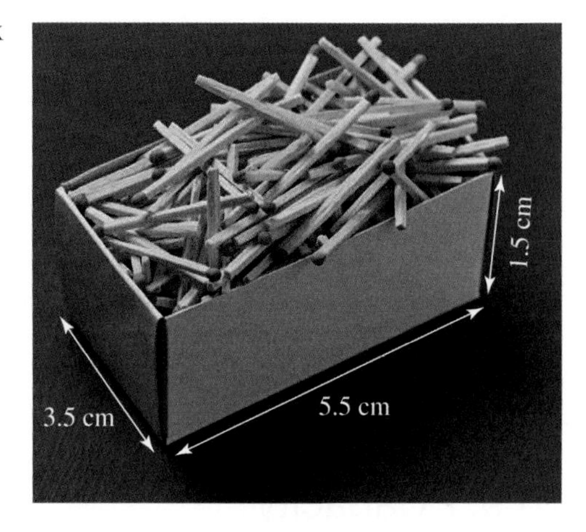


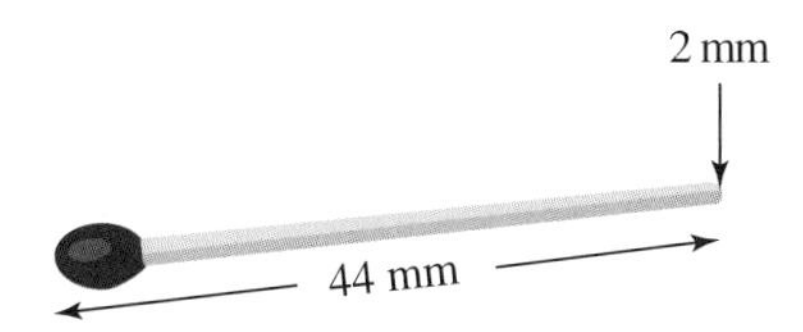


 d. If a matchbox contains 50 matches, determine how much space is left in a full box.

19. Calculate how many cubic metres of water would be needed to fill a diving pool that has a length of 16 m, a width of 12 m and a depth of 4 m.

20. Heather wishes to cover a rectangular lawn with topsoil to a depth of 0.1 m. If the lawn is 24 m long and 17 m wide, calculate the volume of soil (in cubic metres) she needs to order.

## Reasoning

21. The fruitcake shown in the photo is to be divided equally among 100 people at a wedding reception. Determine the volume of cake (in cubic centimetres) each guest will receive.

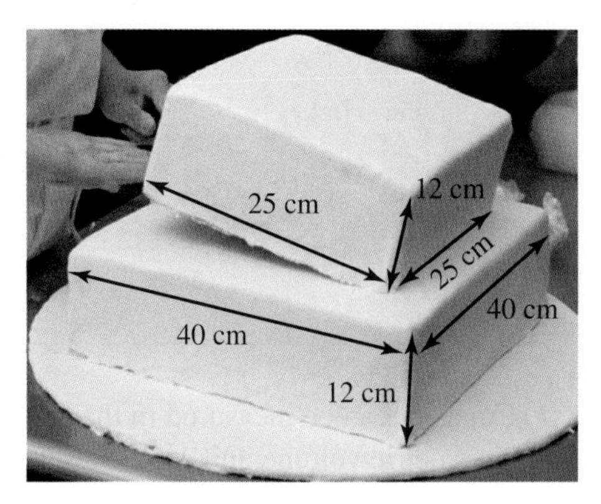


22. A swimming pool is rectangular and its width is exactly half its length. Determine the volume of water needed to fill it if the swimming pool is 50 metres long and has a constant depth of 2 metres.

23. A new rectangular patio has been built on the end of your house. It measures 3.8 m by 1.9 m. You want to plant a garden that is 1.5 m wide around the patio on three sides. The garden beds are to be 0.5 metres deep. If you order 7.3 cubic metres, explain with a diagram whether enough soil has been ordered.

### Problem solving

24. Calculate the volume of chocolate (assume the package is full of chocolate!) in the Toblerone bar shown.

25. The areas of the three sides of a rectangular box are shown in the diagram. Determine the volume of the box, showing all your working.

26. A big cargo box measures 12 m by 6 m by 5 m. Determine how many small boxes of 40 cm by 25 cm by 10 cm can fit in the cargo box.

# LESSON
# 10.9 Capacity

## LEARNING INTENTION

At the end of this lesson you should be able to:
- calculate the capacity of a container
- convert between units of volume and capacity.

## 10.9.1 Capacity

eles-4561

- The **capacity** of a container is the volume of liquid that it can hold.
- Capacity is commonly measured in millilitres (mL), litres (L), kilolitres (kL) and megalitres (ML).
- The following chart can be used to convert between megalitres, kilolitres, litres and millilitres.

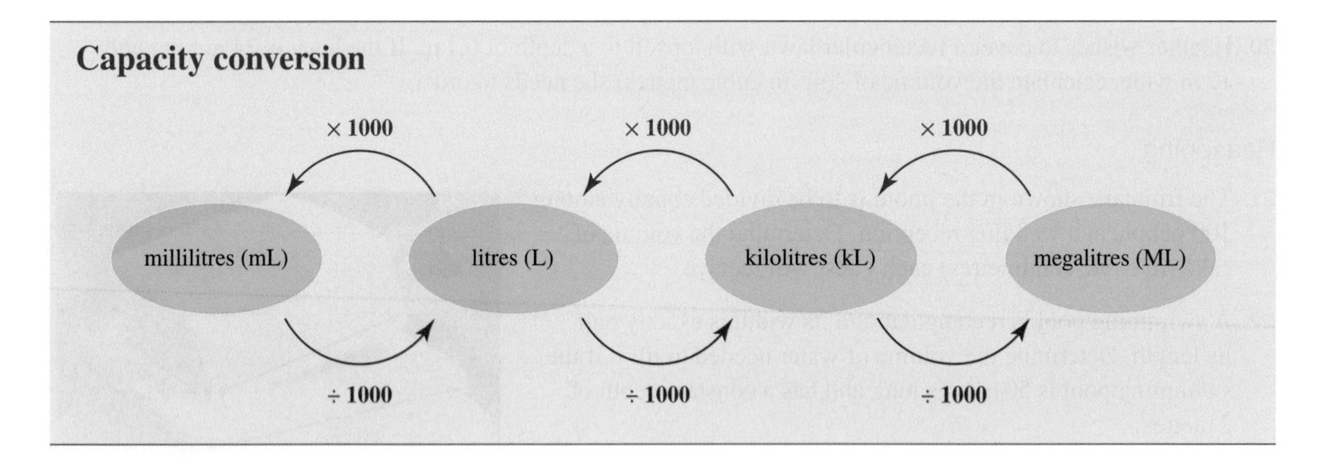


- Capacity can be measured in the same units as volume (e.g. $cm^3$ or $m^3$).
- The metric volume unit 1 $cm^3$ has a capacity of 1 mL. The metric capacity unit 1 L is equivalent to 1000 $cm^3$.

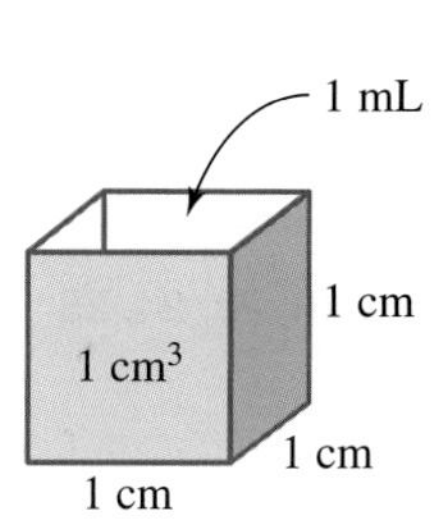


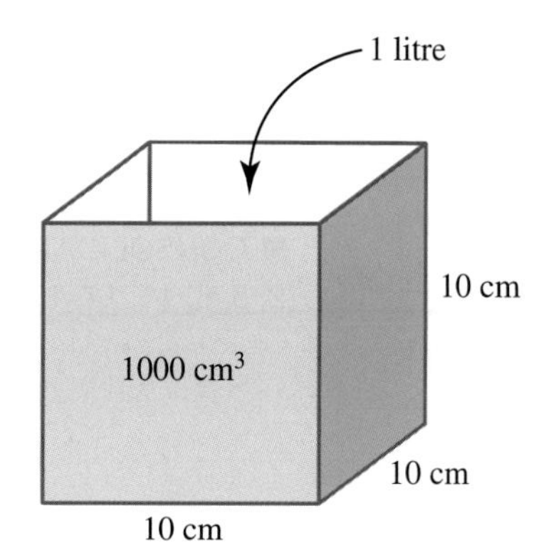


## WORKED EXAMPLE 21 Converting between different units of capacity

**Complete the following unit conversions.**

a. **6 L = ______mL** b. **700 mL = ________L** c. **0.45 L = _______cm³**

| THINK | WRITE |
|---|---|
| a. Check the conversion chart. To convert litres to millilitres, multiply by 1000. Since 6 is a whole number, do this by adding three zeros to the end of the number. | a. $6\text{ L} = 6 \times 1000\text{ mL}$<br>$= 6000\text{ mL}$ |
| b. Check the conversion chart. To convert millilitres to litres, divide by 1000. To do this, move the decimal point 3 places to the left. | b. $700\text{ mL} = 700 \div 1000\text{ L}$<br>$= 0.7\text{ L}$ |
| c. Check the conversion chart. To convert litres to millilitres, multiply by 1000. To do this, move the decimal point 3 places to the right. Note that $1\text{ mL} = 1\text{ cm}^3$. | c. $0.45\text{ L} = 0.45 \times 1000\text{ mL}$<br>$= 450\text{ mL}$<br>$= 450\text{ cm}^3$ |

### DISCUSSION

How would you explain the difference between volume and capacity?

### Resources

**eWorkbook** Topic 10 Workbook (worksheets, code puzzle and project) (ewbk-1960)

**Interactivities** Individual pathway interactivity: Capacity (int-4361)
Capacity (int-4024)

## Exercise 10.9 Capacity

learnon

10.9 Quick quiz on | 10.9 Exercise

### Individual pathways

| ■ PRACTISE | ■ CONSOLIDATE | ■ MASTER |
|---|---|---|
| 1, 4, 7, 10, 13, 15, 17, 20 | 2, 5, 8, 11, 14, 18, 21, 22 | 3, 6, 9, 12, 16, 19, 23, 24, 25 |

### Fluency

1. **WE21a,b** Complete the following unit conversions.

   a. $2\text{ L} =$ __________ mL
   b. $3000\text{ mL} =$ __________ L
   c. $7000\text{ mL} =$ __________ L
   d. $5500\text{ mL} =$ __________ L

2. Complete the following unit conversions.

   a. $2\frac{1}{2}\text{ L} =$ __________ mL
   b. $32\,000\text{ mL} =$ __________ L
   c. $0.035\text{ L} =$ __________ mL
   d. $420\text{ L} =$ __________ mL

3. Complete the following unit conversions.

   a. $1\frac{87}{100}\text{ L} =$ __________ mL
   b. $22\,500\text{ mL} =$ __________ L
   c. $\frac{1}{10}\text{ L} =$ __________ mL
   d. $25\text{ L} =$ __________ kL

4. **WE21c** Complete the following unit conversions.

   a. $750\text{ cm}^3 =$ ______ mL
   b. $2500\text{ m}^3 =$ ______ kL
   c. $2.45\text{ L} =$ ______ $\text{m}^3$
   d. $78\,000\text{ cm}^3 =$ ______ L

5. Complete the following unit conversions.

   a. $40\,000\text{ cm}^3 =$ ______ mL $=$ ______ L
   b. $6\text{ L} =$ ______ mL $=$ ______ $\text{cm}^3$
   c. $5200\text{ L} =$ ______ kL $=$ ______ $\text{m}^3$

6. **MC** A capacity of 25 L is equal to:

   A. 0.025 mL
   B. 250 mL
   C. 0.25 mL
   D. 25 000 mL
   E. 2500 mL

7. **MC** A capacity of 35 400 mL is equal to:

   A. 35 400 000 L
   B. 0.354 L
   C. 3.5400 L
   D. 35.4 L
   E. 35 L

8. Arrange in order from smallest to largest:

   a. $2\frac{1}{2}$ L, 25 000 mL, $\frac{1}{4}$ L, 2.45 L
   b. 760 mL, 0.765 L, 7.65 mL, 7.60 L
   c. 110 mL, 0.1 L, 0.011 L, 1.1 L

### Understanding

9. A water bottle has a capacity of 2 litres. Calculate how many 125 mL bottlefuls are required to fill it.

10. A bottle contains 250 mL of orange juice concentrate. Calculate how much water should be added to make up 2 L of juice from the concentrate.

11. A scientist dilutes (waters down) an acid solution by adding 120 mL of acid to 1.5 L of water. Calculate how much of the diluted solution this will make.

12. Most sports drinks are sold in 500 mL bottles. Calculate how many litres of sports drink are in one dozen (12) bottles.

13. A medicine bottle contains 125 mL of cough syrup. Calculate how many 2.5 mL doses could be administered from this bottle, assuming that none is spilt.

14. Anthea runs a market stall selling detergent. Calculate how many 200 mL bottles of detergent she could fill from a 45 L bulk container.

15. A milk bar sells 55 small bottles of lemon drinks in one week. If each bottle contains 180 mL, calculate how many litres are sold each week.

16. Petrov is working as a school laboratory technician. If there are 12 groups of students and each group requires 400 mL of salt solution for an experiment, calculate how many litres of solution should be prepared.

## Reasoning

17. a. Claire has made her favourite green cordial with 0.25 L of water and 30 mL of cordial. Determine whether a glass with a capacity of 250 mL will hold this volume.
    b. Sharmila has made fresh juice with 0.3 L of apple juice and 40 mL of carrot juice. Determine whether a glass with a capacity of 350 mL will hold this volume.

18. A 185 mL container of hair conditioner is sold at the special price of $3.70. A 0.5 L container of the same conditioner costs $11.00. Compare them to identify the better buy. (*Hint:* Calculate the cost of 1 mL of hair conditioner in each case.)

19. Laurie connected a water tank to his 5 m × 3 m roof in order to collect rain water. In four hours, 2 cm of rain fell. The water tank holds 280 litres. Determine whether the tank overflowed. *Hint:* $1000\,\text{L} = 1\,\text{m}^3$.

## Problem solving

20. Determine the number of millilitres of milk the container in the picture holds.

21. Water is to be poured into this fish tank. How many litres of water are needed to fill the tank to a depth of 28 cm?

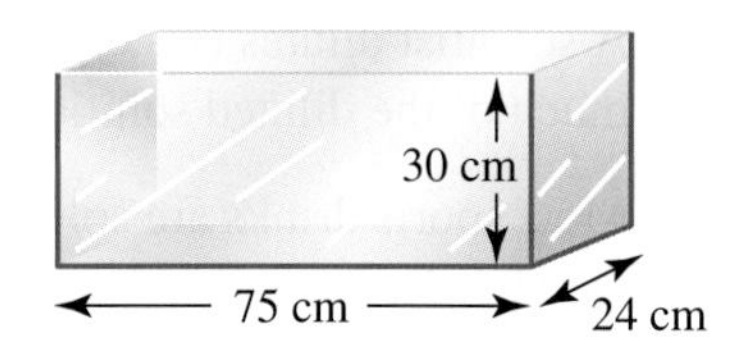


22. Using the measurements shown in the photo of the sinks, estimate how many litres of water each sink will hold if filled to the top. Both sinks have the same dimensions. (*Hint:* First convert the measurements to centimetres.)

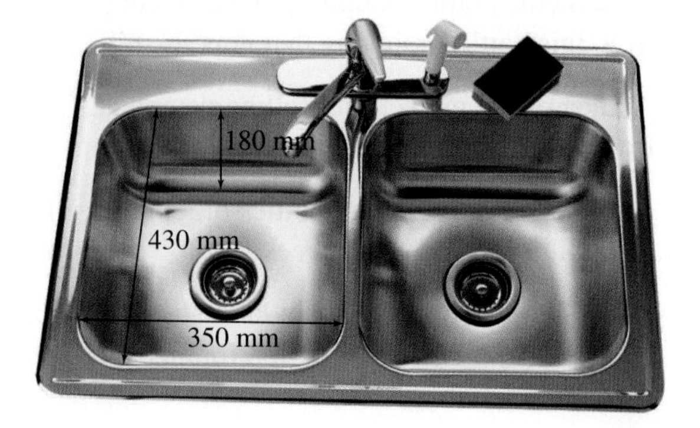


23. One litre of paint covers an area of 20 square metres. Calculate how many millilitres of paint are required to cover an area of 6 square metres. Show your working.

24. A tap drips water at a rate of one drop every three seconds. If it takes 750 drops of water to fill a 100 mL container, determine how many litres of water are lost in one year (365 days).

25. To achieve a world record, a 16.4 kL strawberry milkshake was made in the UK in 1996. To understand how large this is, consider the size of a rectangular prism that would have this capacity. Suggest three possible sets of measurements for this container.

# LESSON
# 10.10 Drawing solids

## LEARNING INTENTION

At the end of this lesson you should be able to:
- draw the cross-section of a given prism
- draw the views, or elevations, of a given solid
- construct a solid given its views or elevations.

## 10.10.1 Prisms

eles-4562

- A prism is a 3-dimensional figure that can be cut into parallel slices or cross-sections that have the same size and shape.

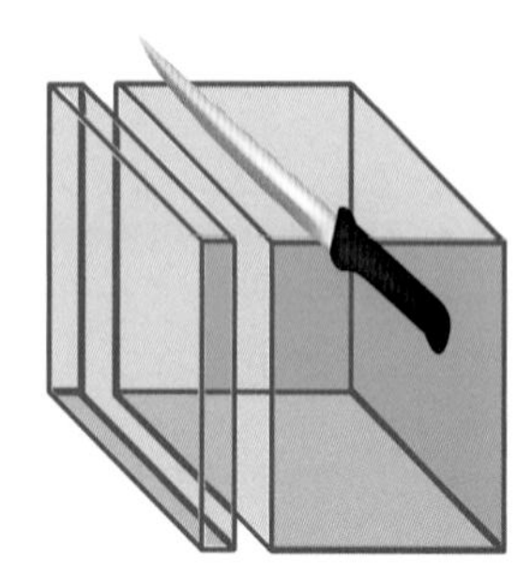

## WORKED EXAMPLE 22 Drawing the cross-section of a prism

**Draw the cross-section of the following prisms.**

**a.**

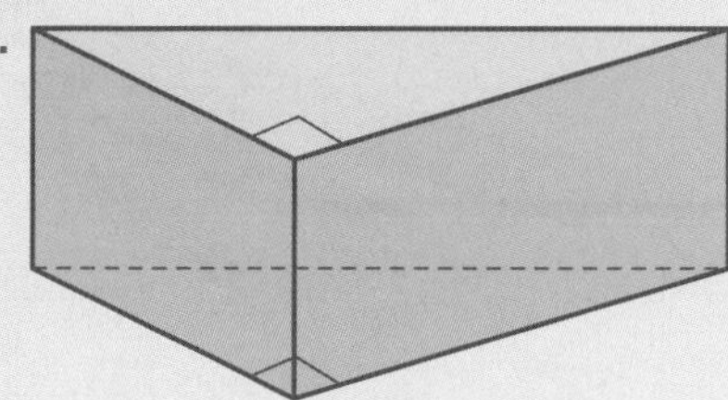

**b.**

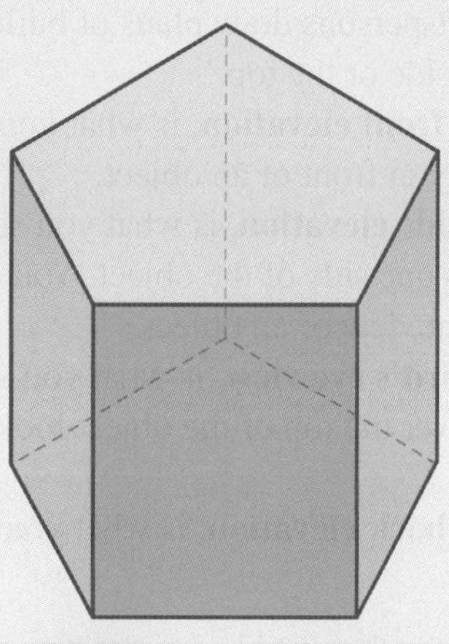

**THINK**

**a. 1.** Identify the cross-section of the prism that is uniform along the length.

**DRAW**

**a.**

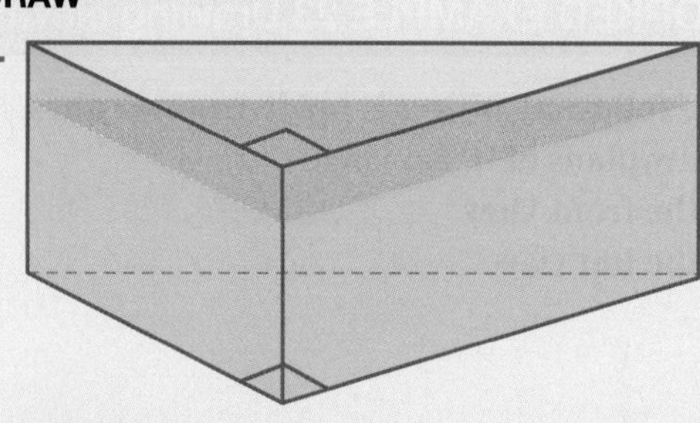

**2.** Draw the cross-section as a 2-dimensional shape.

**b. 1.** Identify the cross-section of the prism that is uniform along the length.

**b.**

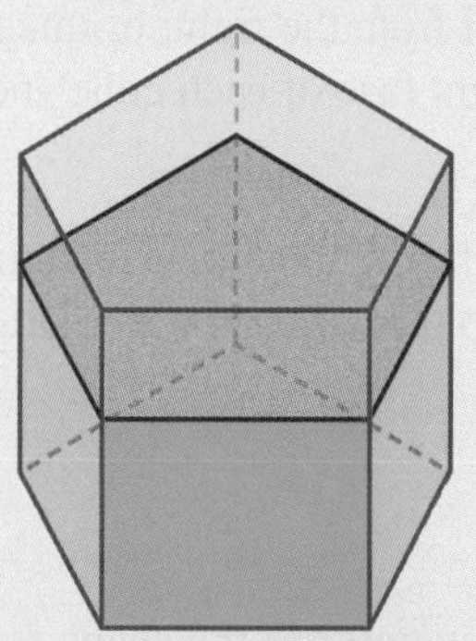

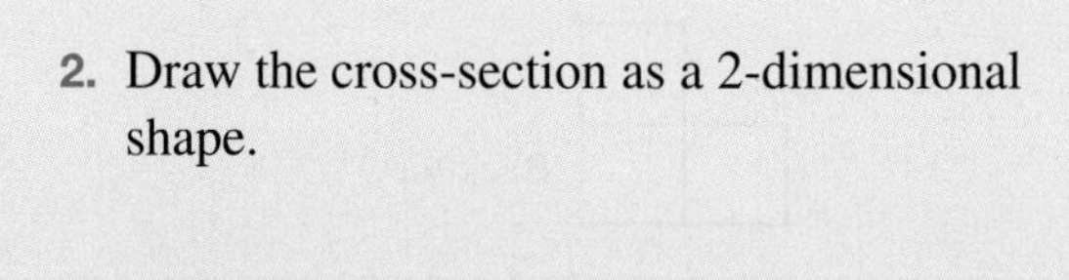

**2.** Draw the cross-section as a 2-dimensional shape.

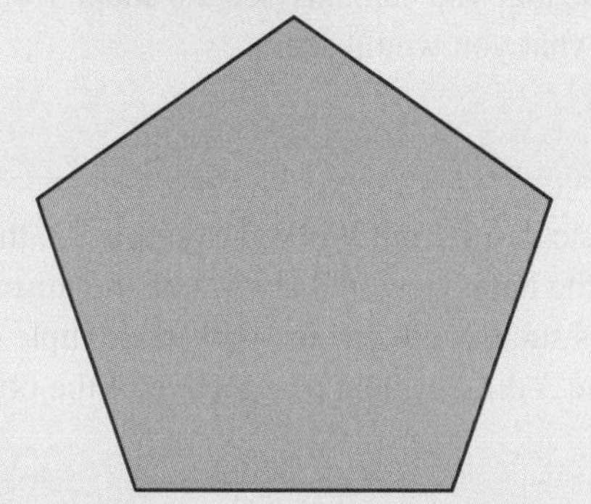

## 10.10.2 Plans and views

eles-4563

- An object can be viewed from different angles.
- Architects and draftspersons draw plans of buildings viewed from the front, the side or the top.
- The **front view**, or **front elevation**, is what you see if you are standing directly in front of an object.
- The **side view**, or **side elevation**, is what you see if you are standing directly to one side of the object. You can draw the left view or the right view of an object.
- The **top view**, or **bird's eye view**, is what you see if you are hovering directly over the top of the object looking down on it.
- The **back view**, or **back elevation**, is what you see if you are standing directly behind an object.

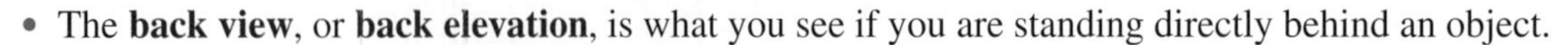

### WORKED EXAMPLE 23 Drawing elevations or views

**The following object is made from 4 cubes.**
**Draw plans of it showing:**

a. **the front view**
b. **the right view**
c. **the top view**
d. **the back view.**

Front

| THINK | DRAW |
|---|---|
| a. Make this shape using cubes. Place the shape at a considerable distance and look at it from the front (this way you can see only the front face of each cube). Draw what you see. (Or simply imagine looking at the shape from the front and draw what you see.) | a. Front view |
| b. Look at your model from the right, or imagine that you can see only the right face of each cube and draw what you see. | b. Right view |
| c. Look at your model from the top, or imagine that you can see only the top face of each cube. Draw what you see. | c. Top view<br>Front |
| d. Imagine that you can only see the shape from behind. Draw what you would see. | Back view |

- As you can see from Worked example 23, the back view is a mirror image of the front view. In fact, for any solid, the back view will always be the mirror image of the front view.
- Figures such as the one in Worked example 23 can be drawn using isometric dot paper. This will help to give the 3-dimensional perspective of the object.

## WORKED EXAMPLE 24 Drawing a view of a solid

**Draw plans of this solid showing:**
a. **the front view**
b. **the right view**
c. **the top view.**

**THINK**

1. Determine an object of similar shape, or visualise the object in your head.
2. Whether viewed from the front, or from the right of the object, the rectangular shaft will appear as a long thin rectangle. The square prisms will also be seen as a pair of identical rectangles. So the front view and the right view are the same.
3. When the object is viewed from above, all we can see is the flat surface of the top square prism: that is, a large and a small square with the same centre.

**DRAW**

Front view

Right view

Top view

## 10.10.3 Isometric drawing

eles-4564

- When working with 3-dimensional models and designs, it is often useful to have the design or model drawn on paper (that is, in 2 dimensions).
- An **isometric drawing** is a 2-dimensional drawing of a 3-dimensional object.
- Isometric dot paper can be used to help with these drawings.

## WORKED EXAMPLE 25 Completing an isometric drawing of an object

**First copy the incomplete figure (Figure 1) onto isometric dot paper. Complete the isometric drawing of the object (Figure 2).**

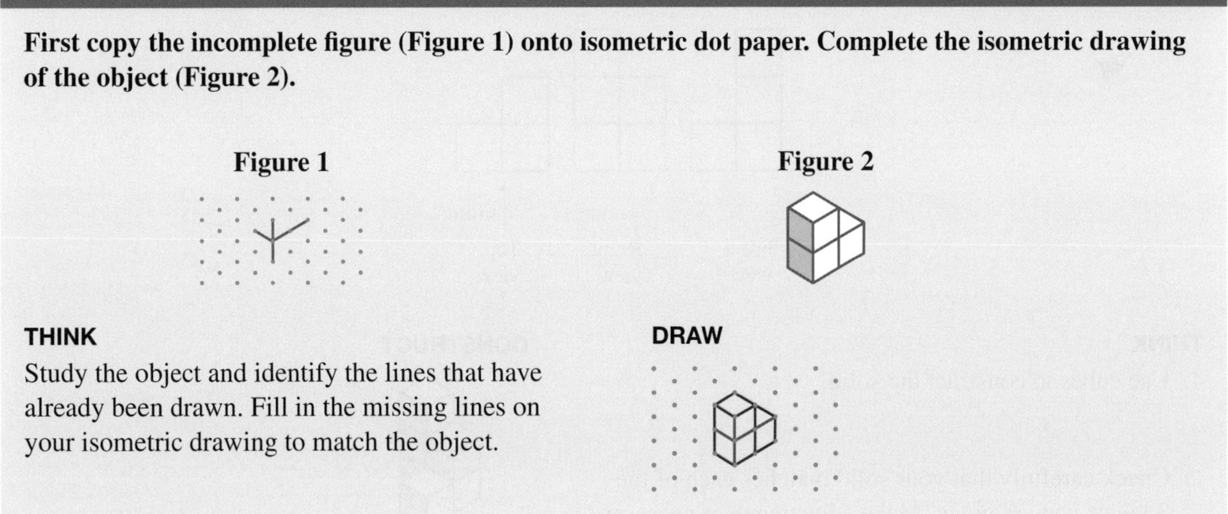


**THINK**

Study the object and identify the lines that have already been drawn. Fill in the missing lines on your isometric drawing to match the object.

**DRAW**

## WORKED EXAMPLE 26 Drawing an object on isometric dot paper

**Draw the following object on isometric dot paper. (You could construct it from a set of cubes.)**

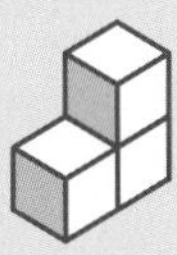

**THINK**

1. Use cubes to make the object shown (optional). Draw the front face of the object. The vertical edges of the 3-dimensional object are shown with vertical lines on the isometric drawing; the horizontal edges are shown with the lines at an angle (by following the dots on the grid paper).
2. Draw the left face of the object.
3. Add the top face to complete the isometric drawing of the object.

**DRAW**

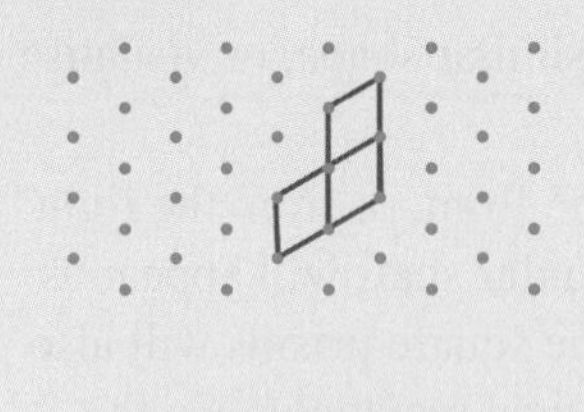

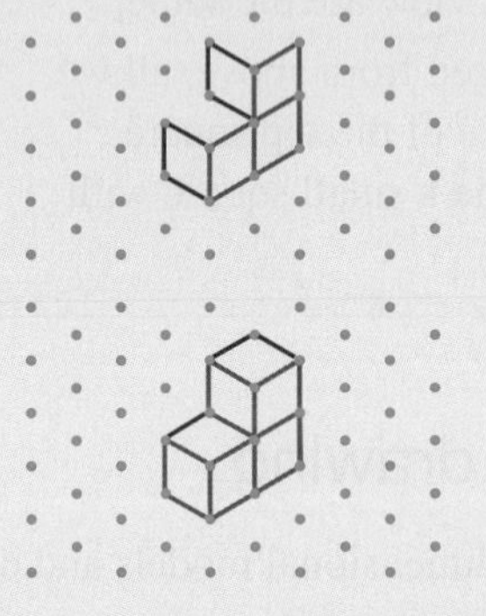

## WORKED EXAMPLE 27 Constructing a solid from given information

**The front, right and top views of a solid are shown. Use cubes to construct the solid, then sketch the solid.**

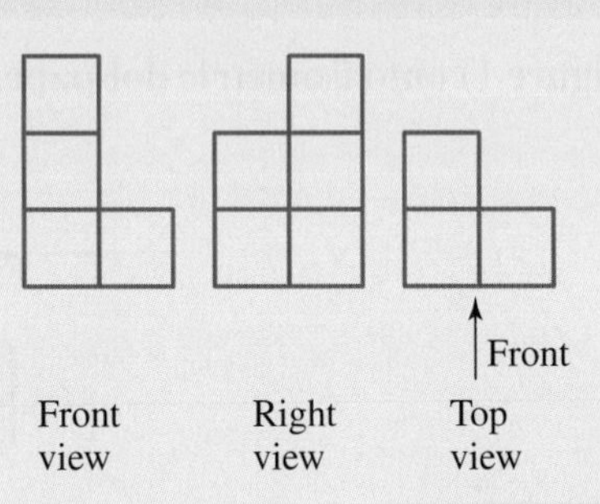


**THINK**

1. Use cubes to construct the solid.
2. Check carefully that your solid matches each of the 3 views you are given. Make adjustments if necessary.

**CONSTRUCT**

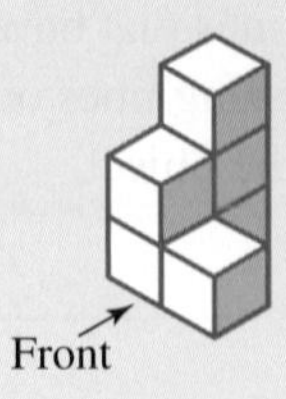

## COLLABORATIVE TASK: Build it!

**Equipment:** building blocks

For this activity, you will work in groups of three or four as directed by your teacher.

1. Stack ten building blocks in any formation. An example is shown.
2. Draw diagrams of the formations you have assembled to show:
   a. the top view
   b. the front view
   c. the side view.
3. Dismantle your formation. Swap your drawings with another group and ask them to build your formation by using your diagrams.

## DISCUSSION

In your small groups from the previous activity, discuss some of the uses of these types of drawings. In which professions might these types of drawings be used?

 Resources

 **eWorkbook** Topic 10 Workbook (worksheets, code puzzle and project) (ewbk-1960)

**Interactivities** Individual pathway interactivity: Plans and views (int-4392)
Prisms (int-4148)
Plans and views (int-4149)

# Exercise 10.10 Drawing solids

learnon

| 10.10 Quick quiz  | 10.10 Exercise |
|---|---|

Individual pathways

| ■ PRACTISE | ■ CONSOLIDATE | ■ MASTER |
|---|---|---|
| 1, 3, 7, 8, 10, 12, 16, 19, 20 | 2, 4, 5, 9, 11, 14, 17, 21 | 6, 13, 15, 18, 22, 23 |

## Fluency

1. WE22 Draw the cross-sections of the following prisms.

a. 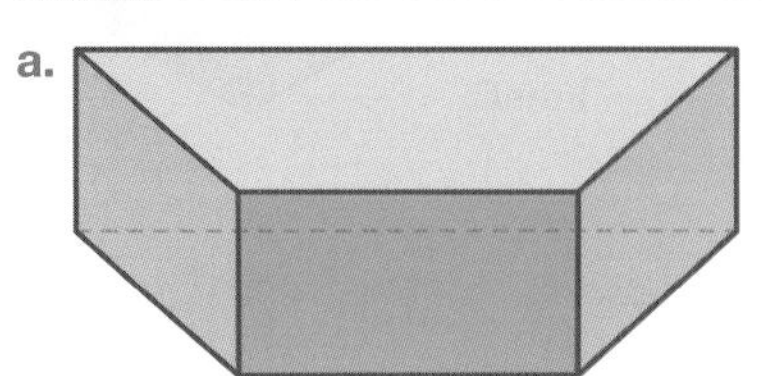

 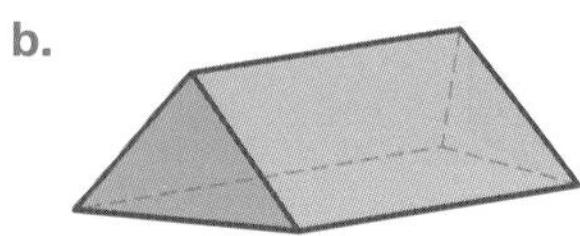

2. Draw the cross-sections of the following prisms.

a.

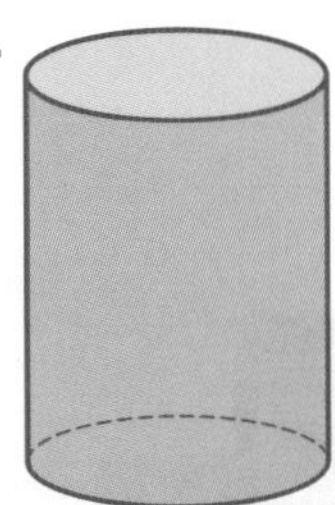

b.

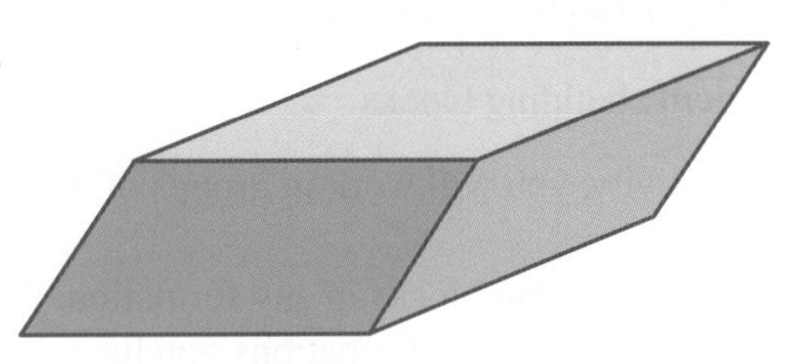

3. WE23 The following objects are made from cubes. For each object draw the plans, showing the front view, the right view, the top view and the back view. (You may wish to use a set of cubes or building blocks to help you.)

a.

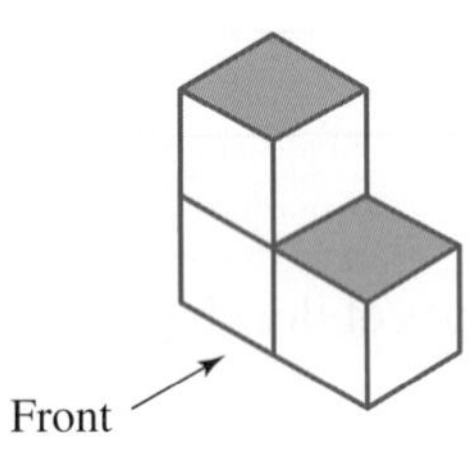


b.

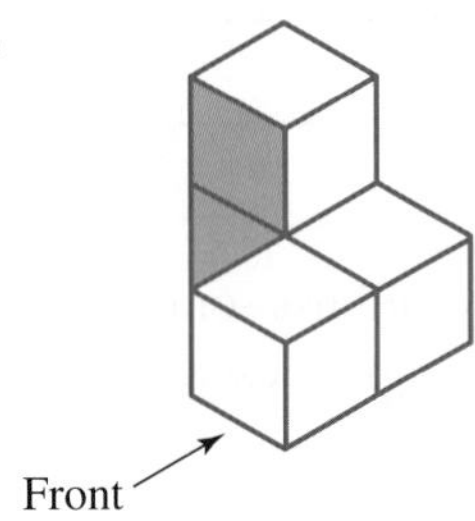


c.

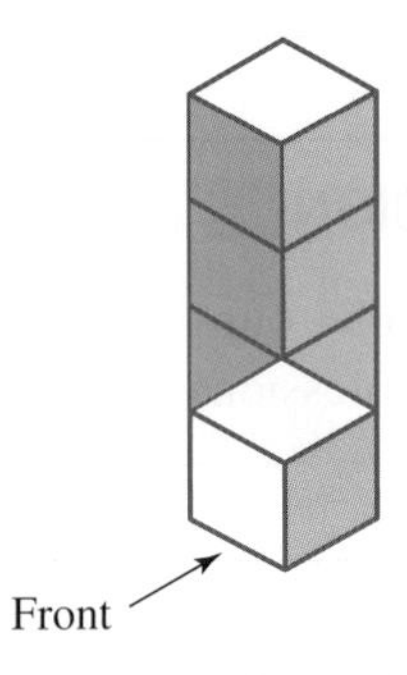


4. The following objects are made from cubes. For each object draw the plans, showing the front view, the right view, the top view and the back view. (You may wish to use a set of cubes or building blocks to help you.)

a.

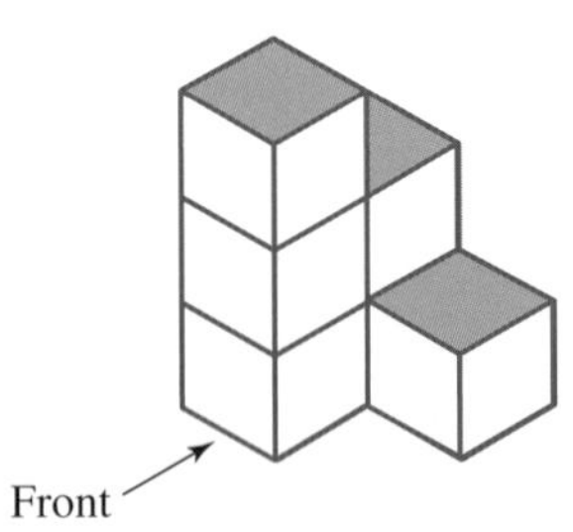


b.

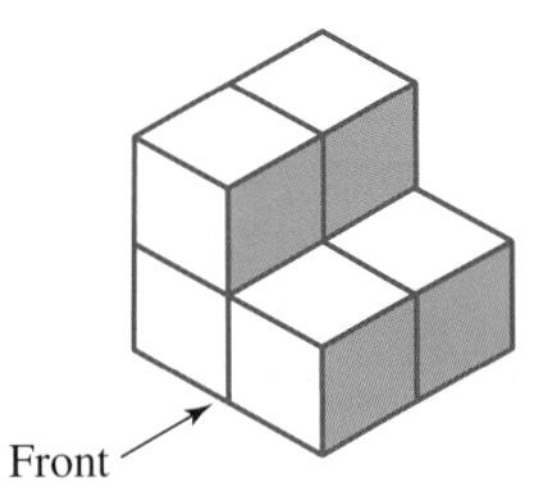


c.

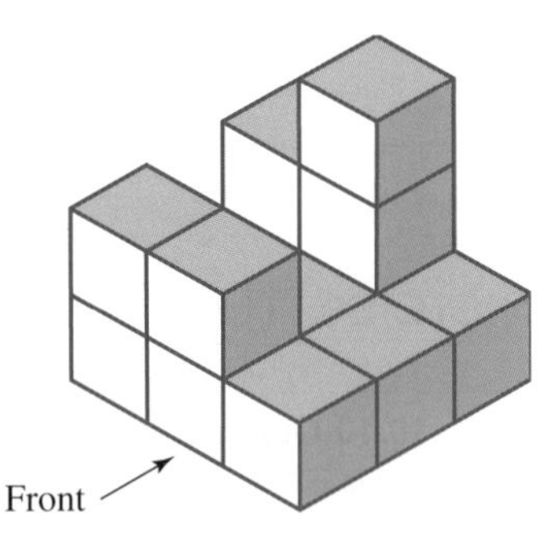


5. Draw the front, right and top views of these objects.

a.

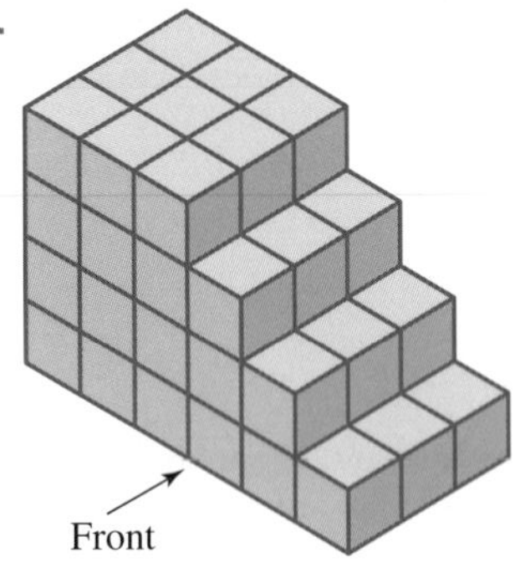


b.

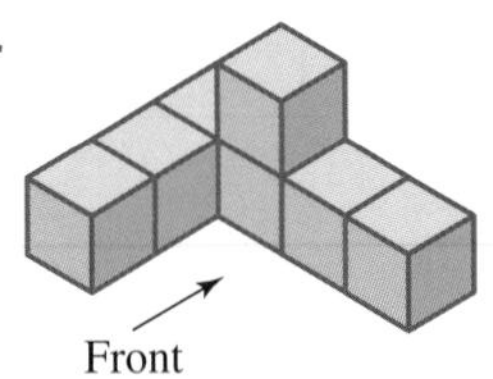


c.

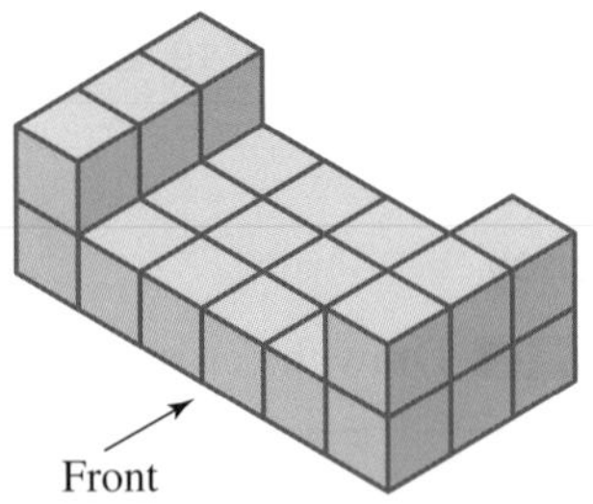

6. Draw the top, front and side views of the object shown in the following isometric drawing.

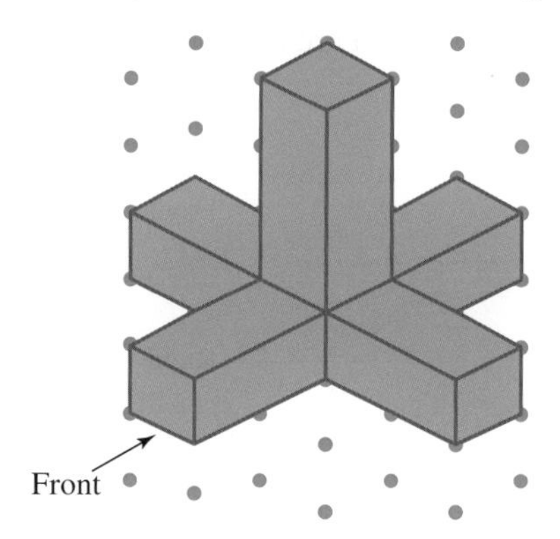


7. WE24 Draw the front, right and top views of each solid shown.

b.

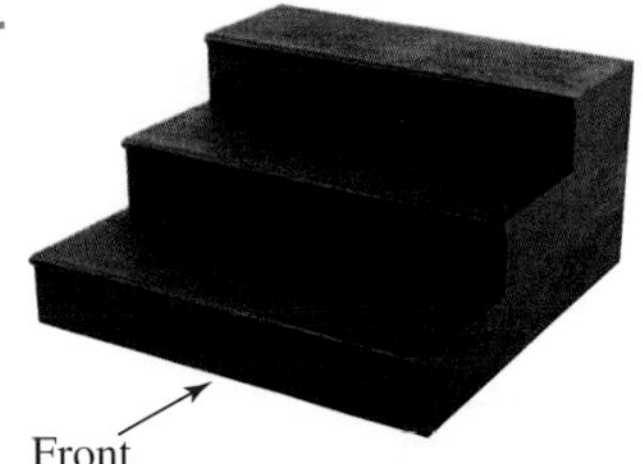


8. WE25 Copy the following figures onto isometric dot paper and complete the isometric drawing of the objects shown.

a.

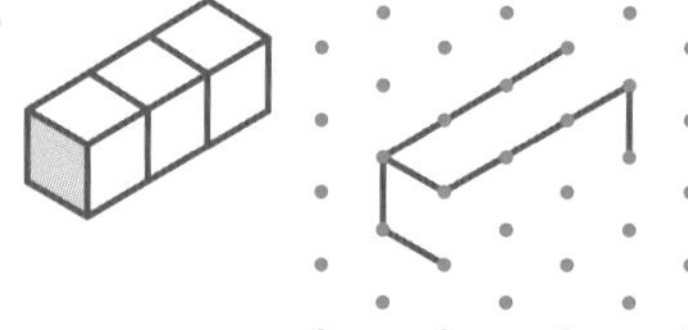

b.

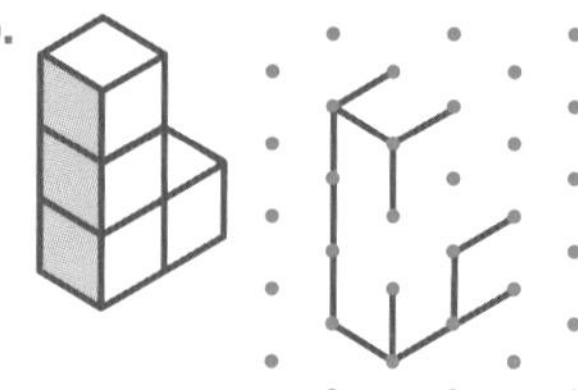

c.

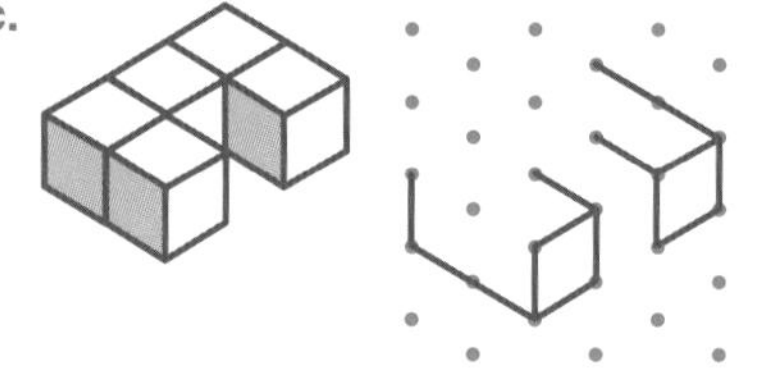

d.

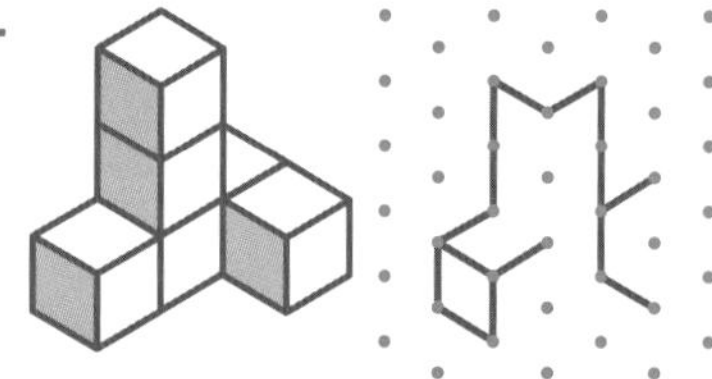

9. WE26 Draw each of the following objects on isometric dot paper. (You might wish to make them first from a set of cubes.)

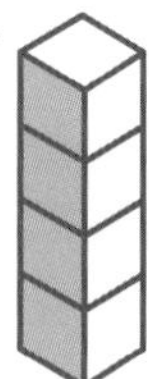

b.

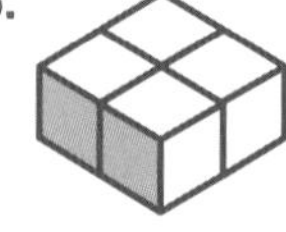

c.

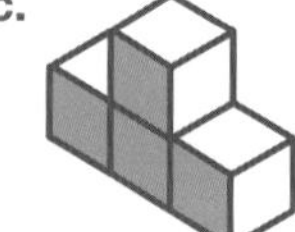

d.

## Understanding

**10.** WE27 The front, right and top views of a solid are shown. In each case, use cubes to construct the solid, then sketch the solid.

**a.**

Front view

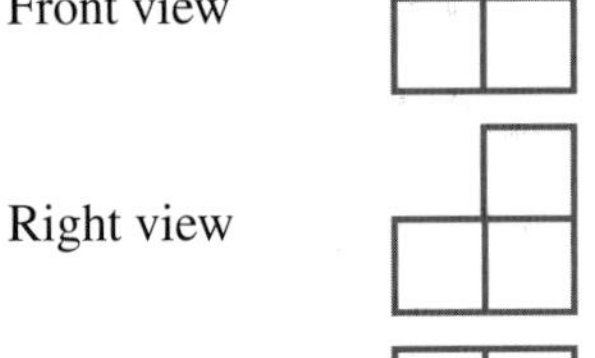

Right view

Top view

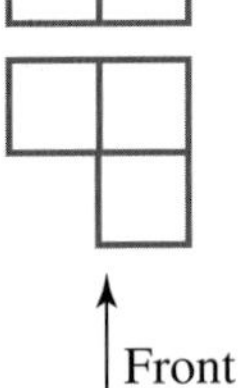

Front

**b.**

Front view

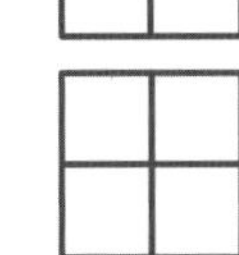

Right view

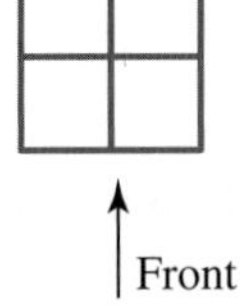

Top view

Front

**11.** The front, right and top views of a solid are shown. In each case, use cubes to construct the solid, then sketch the solid.

**a.**

Front view

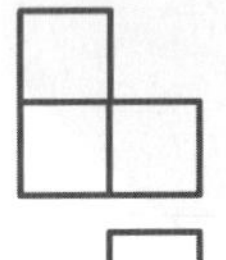

Right view

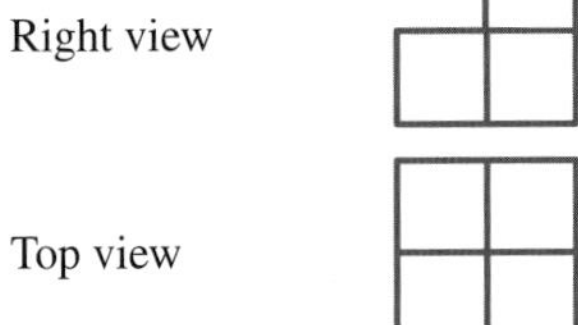

Top view

Front

**b.**

Front view

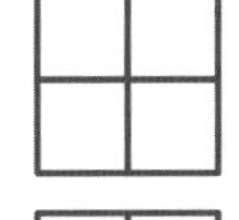

Right view

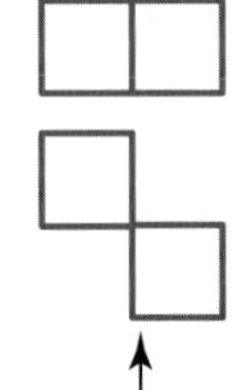

Top view

Front

**12.** MC The front, right and top views of a solid are shown. Select the correct solid represented by these views.

Front view 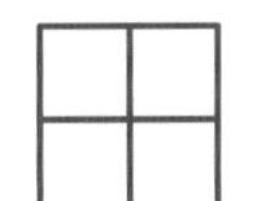 Right view 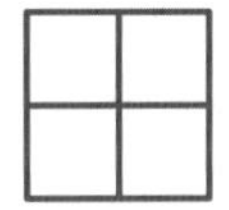 Top view 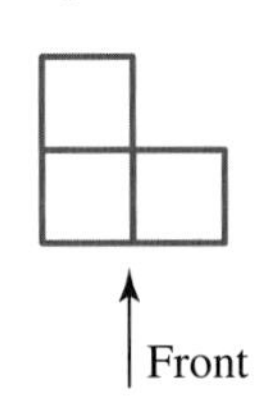

Front

**A.**

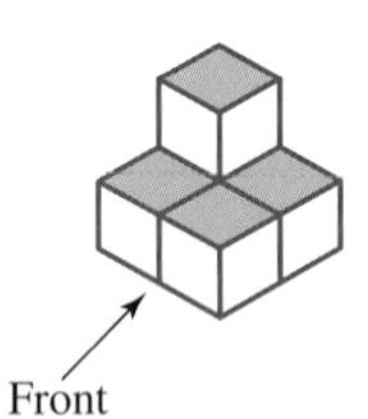


**B.**

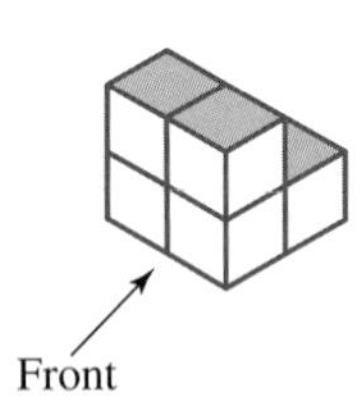


**C.**

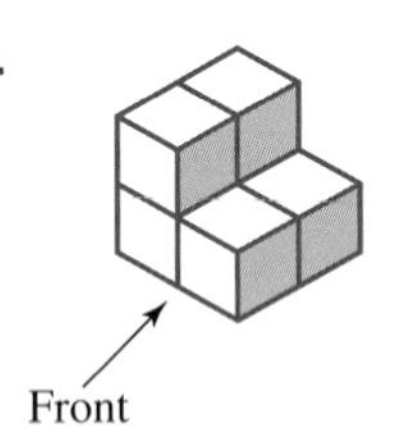


**D.**

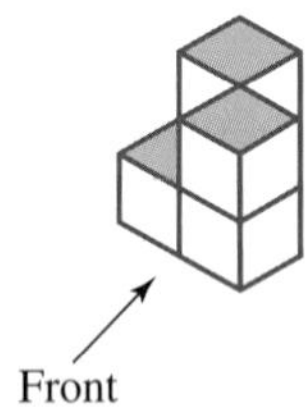


**E.**

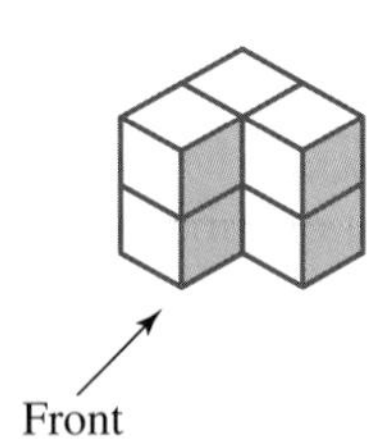

13. Construct the following letters using cubes, and then draw the solids on isometric dot paper:

a. the letter T with 5 cubes
b. the letter L with 7 cubes
c. the letter E with 10 cubes
d. the letter H with 7 cubes.

14. Draw the following figure on isometric dot paper.

15. Draw a selection of buildings from this photograph of a city skyline on isometric dot paper.

## Reasoning

16. Consider the shapes with the following front and right views.
Determine the minimum number of cubes that could be used to construct the shape with the following views.
Determine the maximum number of cubes that could be used. Explain your reasoning.

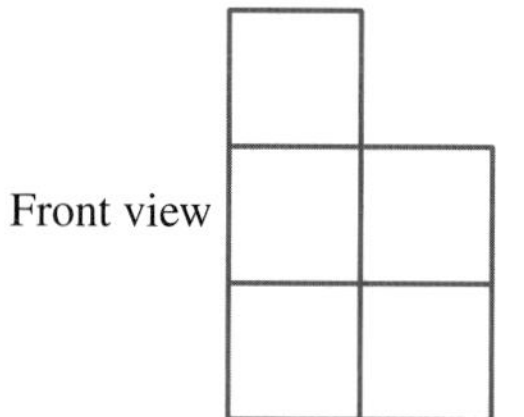


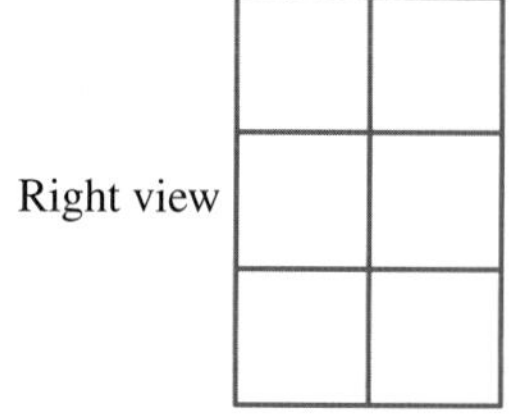


17. A shape is made using only 4 cubes. Its front view, right view and top view are shown.

Front view | Right view | Top view

Front

a. Explain whether it is possible to construct this solid.
b. Describe or draw what this solid would look like.

18. How many objects can you construct using the top view shown? Explain your answer with the use of diagrams.

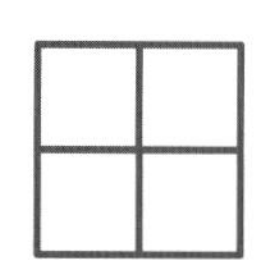

## Problem solving

19. On the same isometric dot paper, draw the two wooden boxes shown.

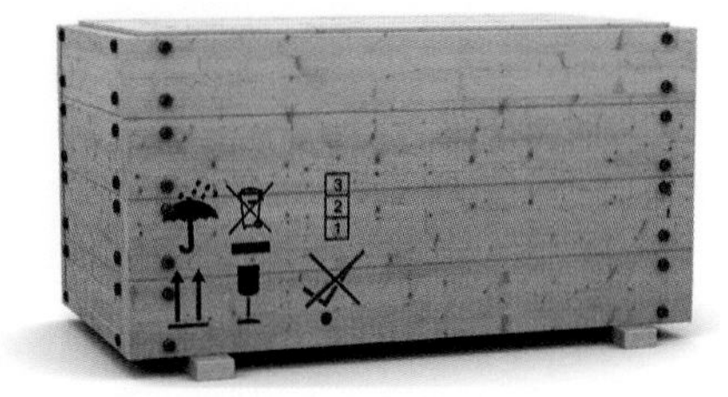

20. Draw the plans of the following diagram, showing the front view, right view and top view.

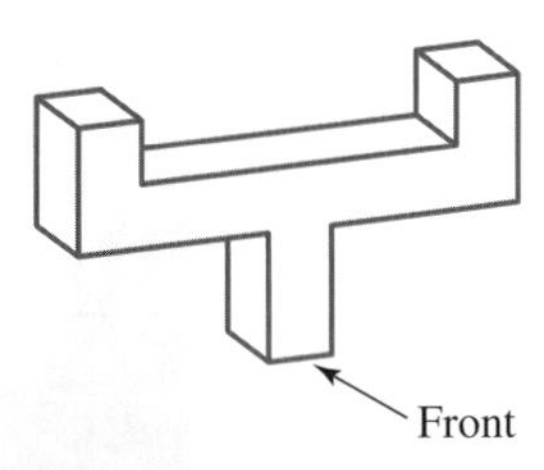


21. Jonah is building a brick wall as shown in the photograph. Draw this part of the wall on isometric dot paper.

22. Draw top, front and side views for the 3D objects shown.

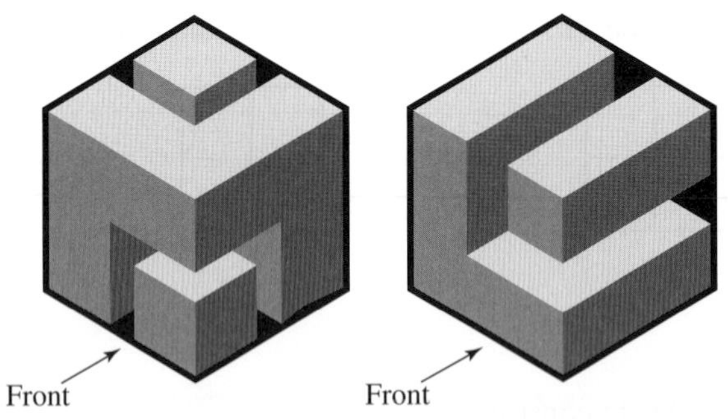


23. Sam has a set of wooden cube blocks he uses to build different shapes. His sister Chris challenged him to see how many different ways he could stack 4 cubes. He can only stack them on top of each other or side by side — not one behind the other. The arrangements must be different and not simply be a mirror image or rotation of another shape. Draw the different arrangements possible.

# LESSON
# 10.11 Review

## 10.11.1 Topic summary

# MEASUREMENT

### Area

- The area of a shape is the amount of 2-dimensional space within it.
- Area is measured in square units: $mm^2$, $cm^2$ $m^2$ and $km^2$.
- The area formulas for some common shapes are listed below.

| Shape | Area formula |
|---|---|
| Square ($l$) | $A = l \times l$<br>$= l^2$ |
| Rectangle ($l$, $w$) | $A = l \times w$ |
| Parallelogram ($b$, $h$) | $A = b \times h$ |
| Triangle ($b$, $h$) | $A = \frac{1}{2} \times b \times h$ |

- The hectare, ha, is a non-SI metric unit of area equal to a square with sides of 100 m.
  1 ha = 10 000 $m^2$

### Units of length

- The modern metric system in Australia is defined by the International System of Units (SI).
- The most common units of length are millimetres (mm), centimetres (cm), metres (m) and kilometres (km).
- You can convert between these units by using the diagram below.

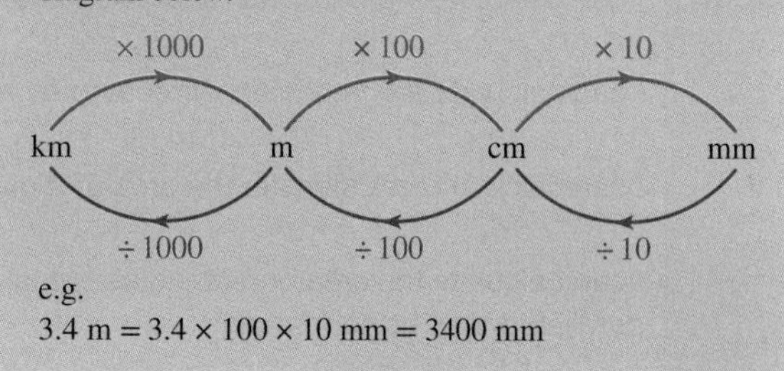


e.g.
3.4 m = 3.4 × 100 × 10 mm = 3400 mm

### Perimeter

- The perimeter, $P$, is the distance around a closed shape.
- When calculating the perimeter of a shape, make sure all lengths are in the same units before adding them together.

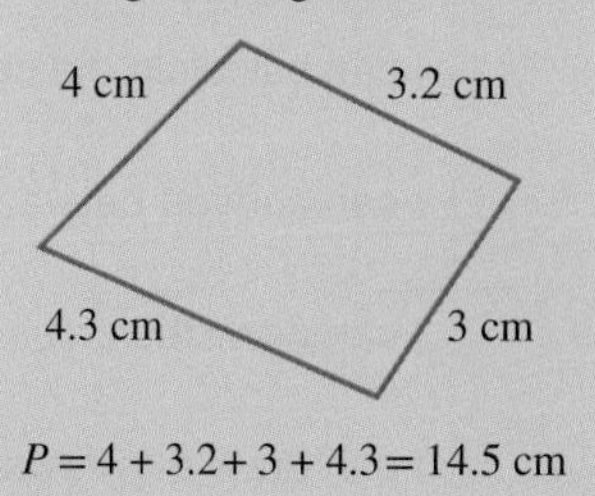


$P = 4 + 3.2 + 3 + 4.3 = 14.5$ cm

### Capacity

- The amount of liquid an object can hold is known as its *capacity*.
- Capacity can be measured in cubic units ($cm^3$ or $m^3$); however, it is more commonly measured in millilitres (mL), litres (L), kilolitres (kL) or megalitres (ML).
- You can convert between these units by using the diagram below.

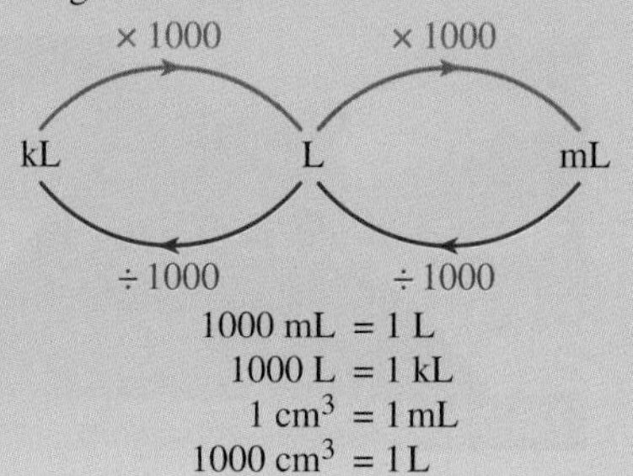


1000 mL = 1 L
1000 L = 1 kL
1 $cm^3$ = 1 mL
1000 $cm^3$ = 1 L

### Volume

- Volume is the amount of 3-dimensional space an object takes up.
- Volume is measured in cubic units: $mm^3$, $cm^3$, $m^3$ and $km^3$.
- A prism is a solid object with identical ends, flat faces and the same cross-section along its length.
- The volume of a right prism is given by:

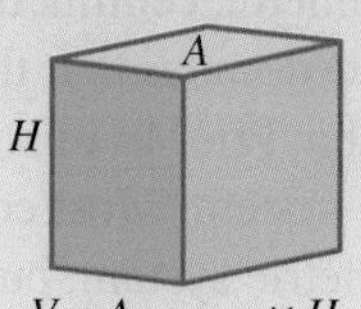


$V = A_{rectangle} \times H$
Rectangular prism

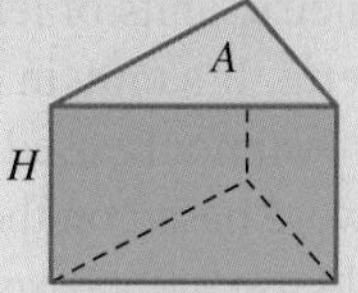


$V = A_{triangle} \times H$
Triangular prism

## 10.11.2 Success criteria

Tick the column to indicate that you have completed the lesson and how well you think you have understood it using the traffic light system.

(**Green:** I understand; **Yellow:** I can do it with help; **Red:** I do not understand)

| Lesson | Success criteria | Green | Yellow | Red |
|---|---|---|---|---|
| **10.2** | I can convert between different units of length. | | | |
| **10.3** | I can read and interpret various scales. | | | |
| **10.4** | I can calculate the perimeter of a given shape. | | | |
| | I can calculate the perimeter of rectangles and squares. | | | |
| **10.5** | I can calculate, or estimate, the area of a given shape. | | | |
| | I can calculate the area of rectangles, squares, triangles and parallelograms using formulas. | | | |
| **10.6** | I can separate composite shapes into simple shapes. | | | |
| | I can calculate the area of composite shapes by adding or subtracting areas of simple shapes. | | | |
| **10.7** | I can calculate the volume of a rectangular prism. | | | |
| **10.8** | I can calculate the capacity of a container. | | | |
| | I can convert between units of volume and capacity. | | | |
| **10.9** | I can draw the cross-section of a given prism. | | | |
| | I can draw the views, or elevations, of a given solid. | | | |
| | I can construct a solid given its views or elevations. | | | |

## 10.11.3 Project

### Old units of length

When studying length and perimeter, you learned about the common metric units of length: millimetre, centimetre, metre and kilometre. It's only in the last few decades that these units have been commonly used in Australia. Prior to this, the units of length were based on the imperial system. Most countries in the world use units based on the metric system, but a few continue to use imperial units.

The origins of the imperial units are far and wide. Many have been adapted from ancient times or translated into English from other languages. Other units, as strange as it may seem, are based on parts of the body.

The ancient Egyptians built the Great Pyramid of Giza over four and a half thousand years ago. It was constructed with side lengths of 440 cubits and was 280 cubits high.

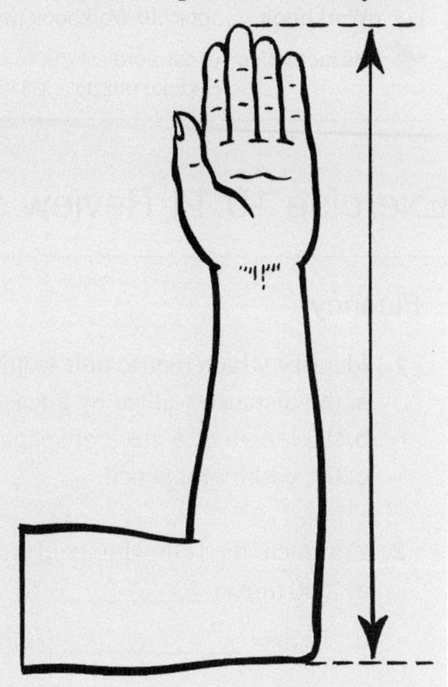

The *cubit* was a unit equal to the 'length of the forearm of the bent elbow to the tip of the extended middle finger'.

1. The metric height of the Great Pyramid of Giza is 146.5 metres; its base length is 230 metres. Use these to determine the value of the cubit to 3 decimal places.
2. Below are some units of length from the imperial system. Select five units, and calculate their equivalent in the metric system. Find out where and how they were used (what sorts of objects they measured, in what situations they were used — for example, for small or large distances, for racetracks, for the depth of water or in land surveying).

| league | hand | line | mile |
|---|---|---|---|
| furlong | chain | inch | link |
| rod | foot | yard | fathom |

Although Australians have used units of length from the metric system for over 3 decades, many institutions have stuck with units common to the imperial system. For example, a cricket pitch is still 22 yards long.

3. Find another sport that still uses lengths associated with the imperial system and state the unit(s) involved.

## Resources

**eWorkbook** Topic 10 Workbook (worksheets, code puzzle and project) (ewbk-1960)

**Interactivities** Crossword (int-2600)
Sudoku puzzle (int-3170)

# Exercise 10.11 Review questions

learnon

### Fluency

1. Identify which metric unit would be most suitable for measuring:
   a. the distance walked by a teacher at school during a week
   b. the length of a piece of spaghetti
   c. the width of a pencil.

2. Complete the following conversions.
   a. 560 mm = ___________ m
   b. 2300 cm = ___________ km
   c. 17 m = _____________ cm
   d. $\frac{3}{4}$ km = ___________ mm
   e. 2.09 m = __________mm

3. Complete the following conversions.
   a. $6\frac{4}{5}$ cm = __________ m
   b. 22.5 mm = _________ cm
   c. $\frac{63}{10000}$ km = ________ m
   d. 82 000 000 m = _______ km
   e. $5\frac{9}{10}$ mm = ________ cm

4. Arrange from smallest to largest: 44.5 m, 455 cm, 455 000 mm, 0.004 45 km.

5. a. Add 45.6 km to 5600 m.
   b. Calculate the difference between 80 m and 4300 cm.

6. Identify the reading in kilograms on this bathroom scale. *Note:* The scale for kilograms is the inner circle.

7. Estimate the height of the front tree in the photograph, given that the person standing is 1.7 m tall.

8. Identify which unit of area would be the most appropriate for measuring the following. (Choose from $mm^2$, $cm^2$, $m^2$, ha and $km^2$.)
   a. The floor area of your classroom
   b. The area of the city of Brisbane
   c. A pin head
   d. A market garden
   e. The area of the continent of Antarctica

9. Calculate the circumference of each of these circles corect to 2 decimal places.

a.

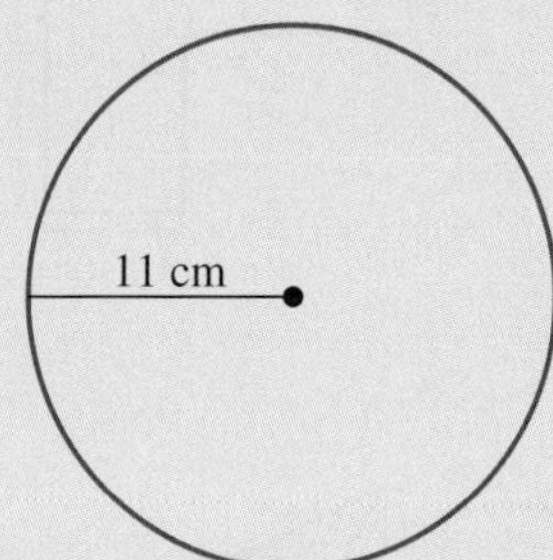


b.

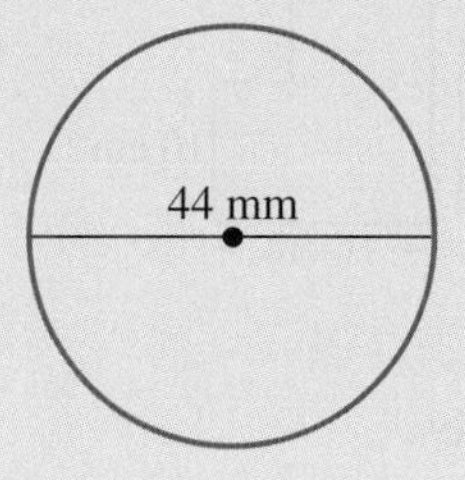


c.

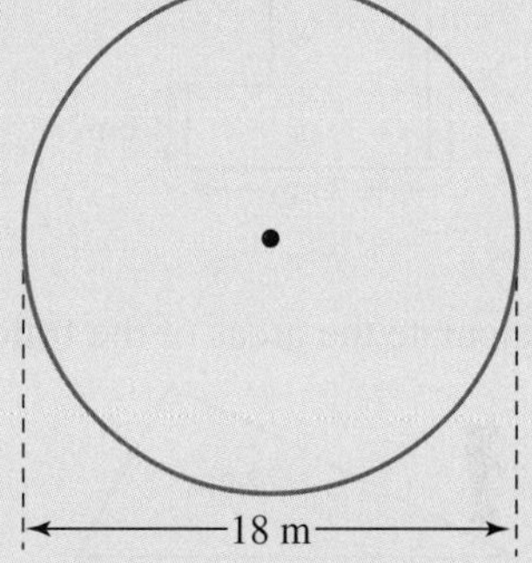


10. Estimate the shaded area in the figure.
*Note:* Each square on the grid has an area of 1 $cm^2$.

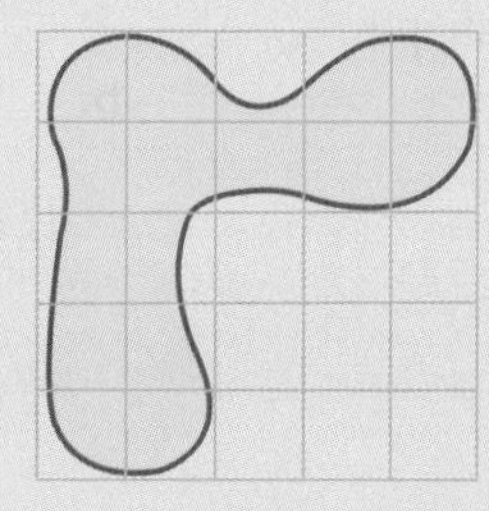

11. Use the formula $A = lw$ to calculate the areas of the following rectangles.

a. 15 m, 10 m

b.

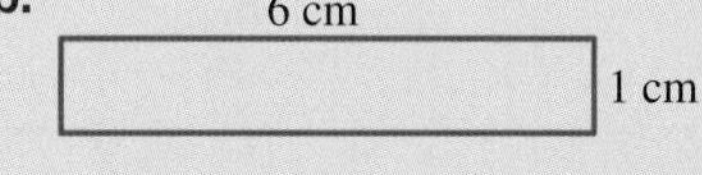


c.

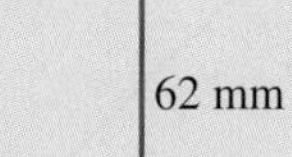

**12.** Calculate the areas of the following shapes by first dividing them into rectangles.

**a.**

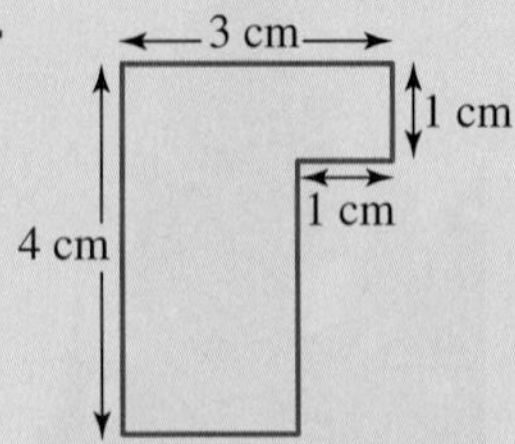


**b.**

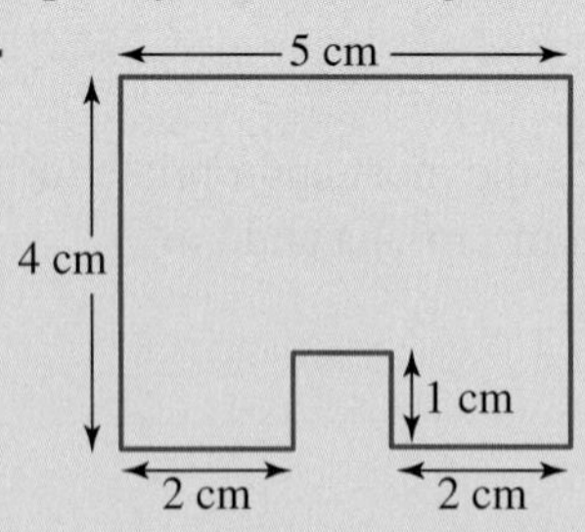


**c.**

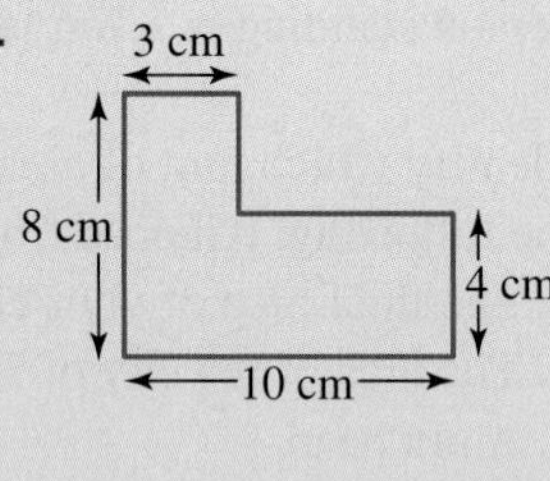


**13.** Calculate the areas of the following shapes by first dividing them into rectangles.

**a.**

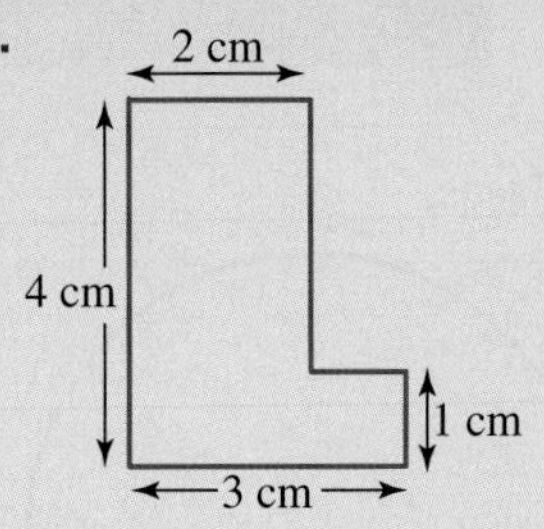


**b.**

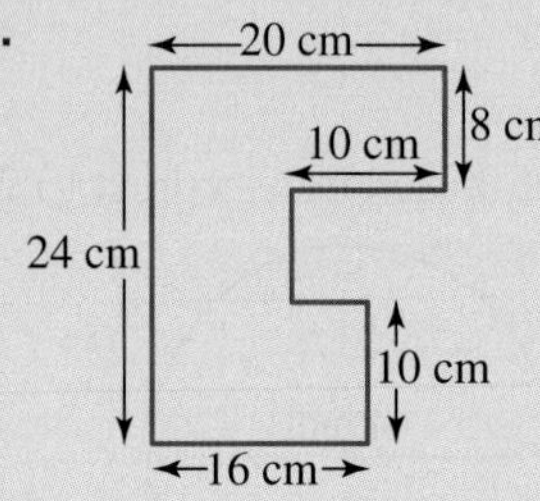


**c.**

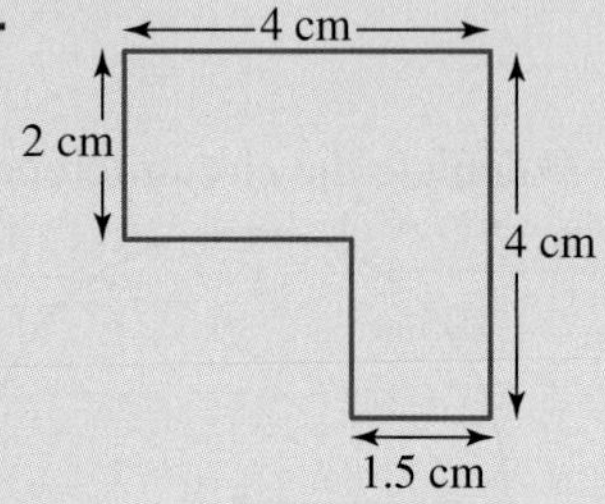


**14.** Calculate the areas of the following triangles.

**a.**

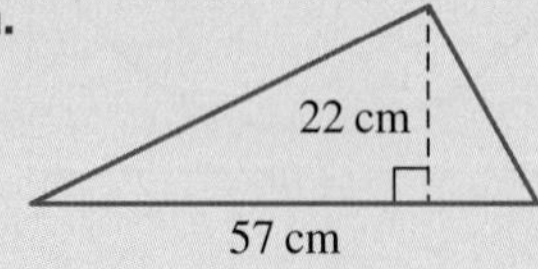


**b.**

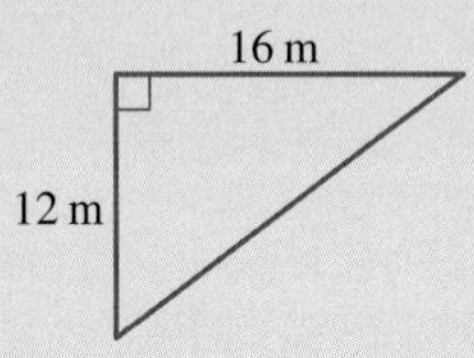


**c.**

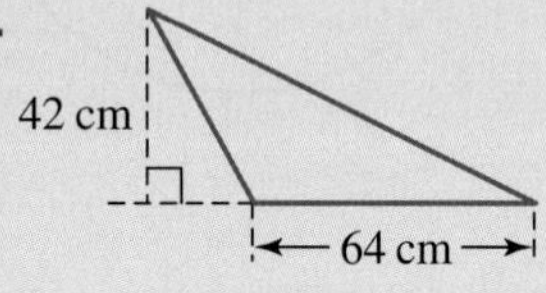


**15.** Calculate the areas of the following shapes.

**a.**

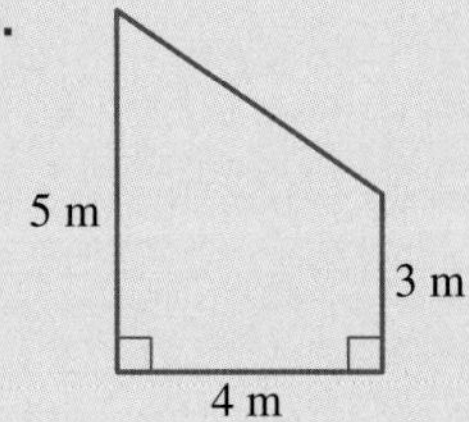


**b.**

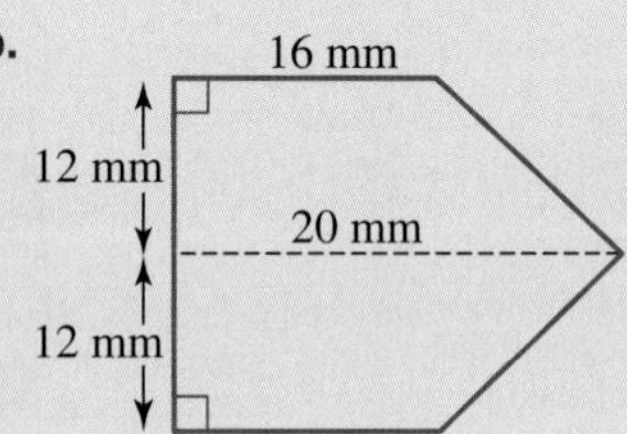


**16.** Calculate the areas of the following shapes.

**a.**

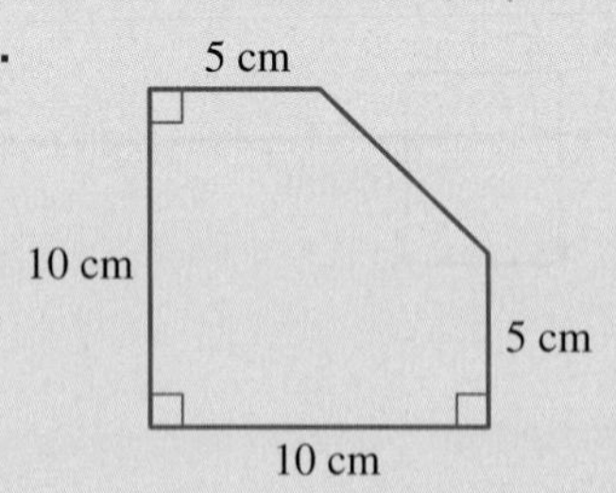


**b.**

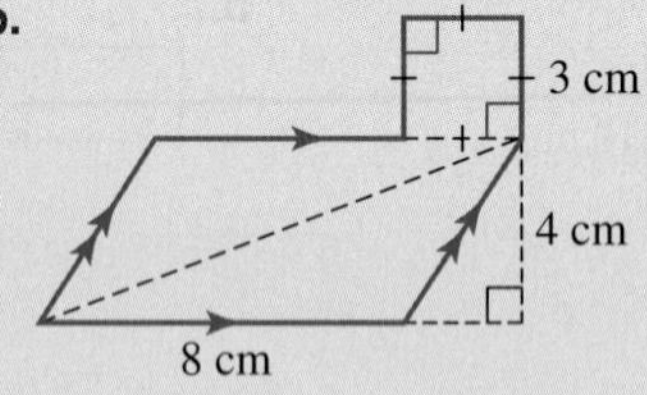

**17.** Calculate the volume of each shape in cubic centimetres ($cm^3$). Each cube has a volume of $1\ cm^3$.

**a.**

**b.**

**c.**

**d.**

**18.** Draw the front, right and top views of these objects.

**a.**

**b.**

**19.** Draw isometric views of the objects, whose front (F), right (R) and top (T) views are given here.

**a.**

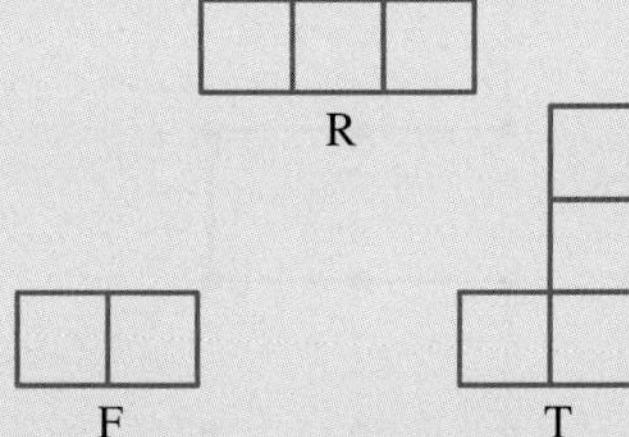


**b.**

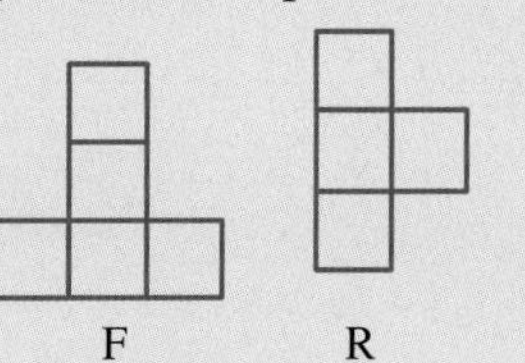


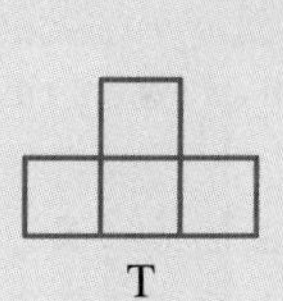


**20.** Use the information given in the diagram to calculate the volume of the shape.

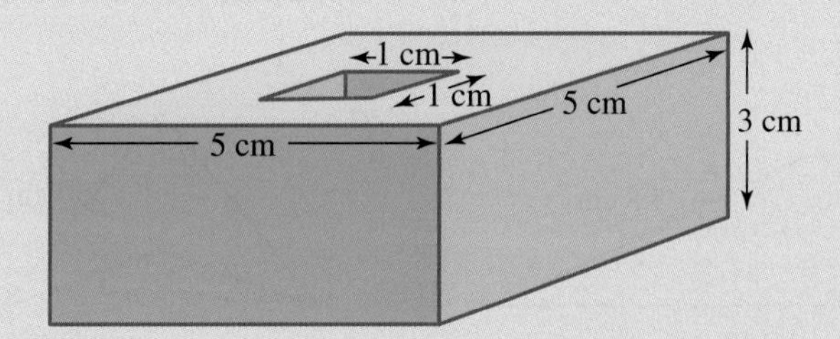

21. Use the information given in the diagram to determine the volume of the shape.

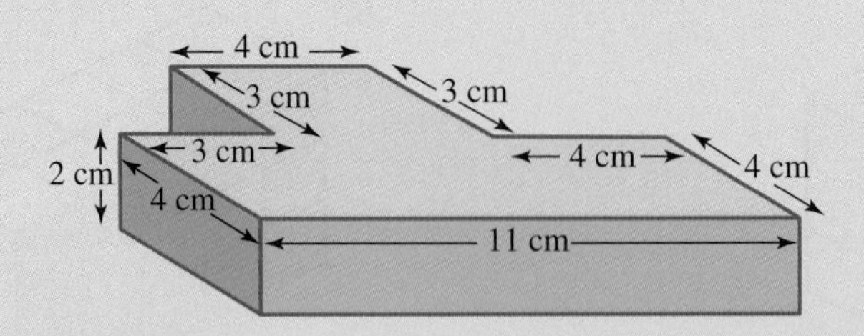


22. Complete the following conversions.

a. $8\,\text{L} =$ __________ mL

b. $\frac{21}{50}\text{L} =$ __________ mL

c. $3300\,\text{mL} =$ __________ L

d. $1\frac{3}{250}\,\text{L} =$ __________ mL

23. Complete the following conversions.

a. $4\frac{3}{10}\,\text{kL} =$ __________ L

b. $0.0034\,\text{kL} =$ __________ L

c. $4755\,\text{L} =$ __________ kL

d. $432\,\text{mL} =$ __________ L

## Problem solving

24. During a rescue operation in calm seas, a 16.5 m rope is dangled from a helicopter hovering 20 m above sea level. A man who is 175 cm tall is standing on the deck of a boat when he reaches 50 cm above his head for the rope. By how much does he fail to reach the rope if the deck is 1 m above sea level?

25. Calculate the perimeter of each shape.

a.

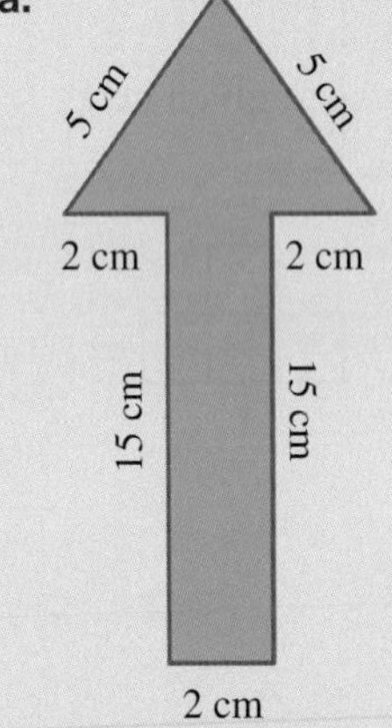


b.

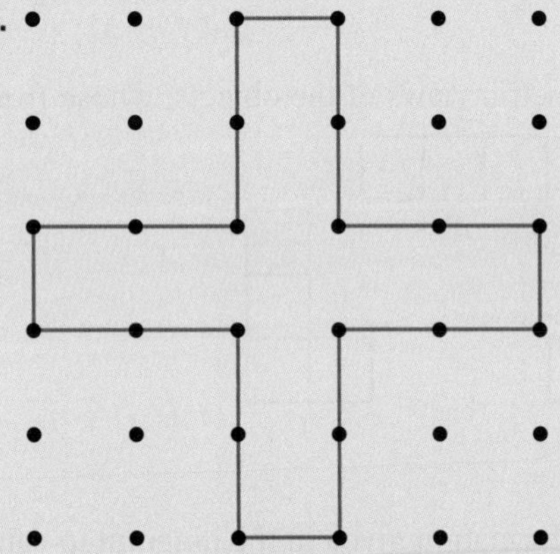

*Note:* The dots are 1 cm apart.

c.

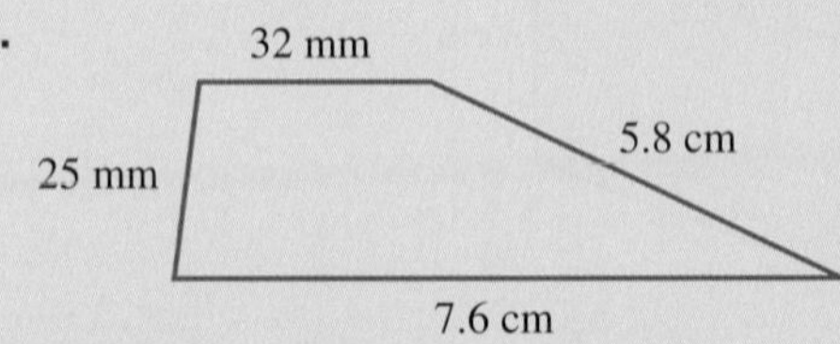


d.

**26.** Michelle rides three laps of the following dirt-bike track. Calculate how far she rides altogether.

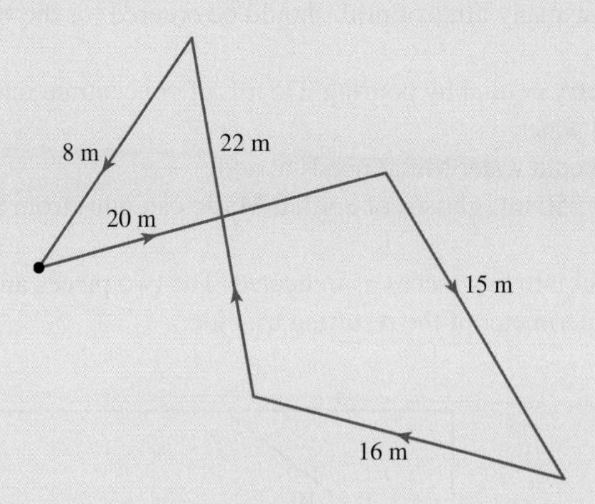


**27.** Determine the length of satin ribbon required to edge a rectangular blanket on all four sides if the blanket is 240 cm long and 195 cm wide. (Assume there is no overlap.)

**28.** Determine the cost of paving a rectangular courtyard that is 6.5 m long and 3.2 m wide. The courtyard is to be paved with concrete paving blocks, which cost $28 per square metre.

**29.** Calculate the area of card needed to make the triangular display sign shown.

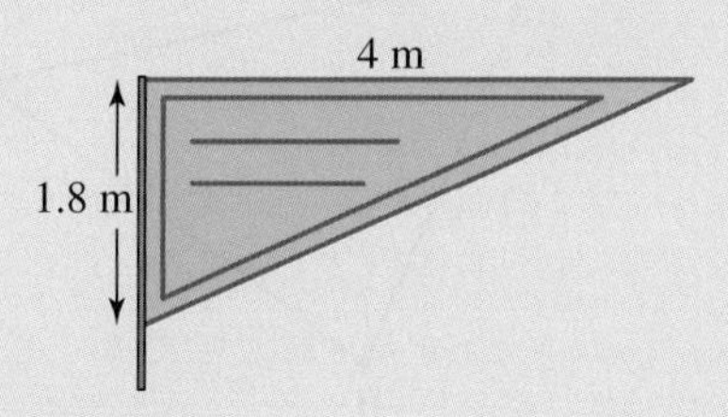


**30.** When making a box kite, Katie cuts a square piece of material diagonally, as shown by the dotted line in the diagram. What is the area of each triangular piece of material?

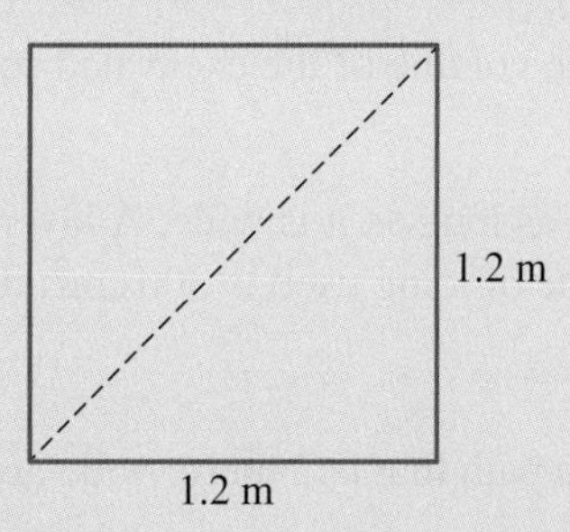


**31.** **a.** A rectangular prism has a length of 40 cm and a width of 26 cm. If its volume is 29 120 $cm^3$, calculate the height of the rectangular prism.

**b.** A rectangular prism has a length of 30 cm and a width of 15 cm. If its volume is 3825 $cm^3$, calculate its height.

**32.** In a food technology class, each of the 14 groups of students uses 350 mL of fresh milk to make pancakes. Calculate how many litres of milk should be ordered for the whole class.

**33.** Mario makes up raspberry cordial by pouring 275 mL of concentrate into a 2-litre container and filling the container with cold water.

**a.** Calculate how much cold water Mario needs to add.

**b.** Calculate how many 250 mL glasses of cordial Mario can pour from the large container.

**34.** A $6 \times 16$ rectangle is cut into two pieces as indicated. The two pieces are rearranged to form a right triangle. Calculate the perimeter of the resulting triangle.

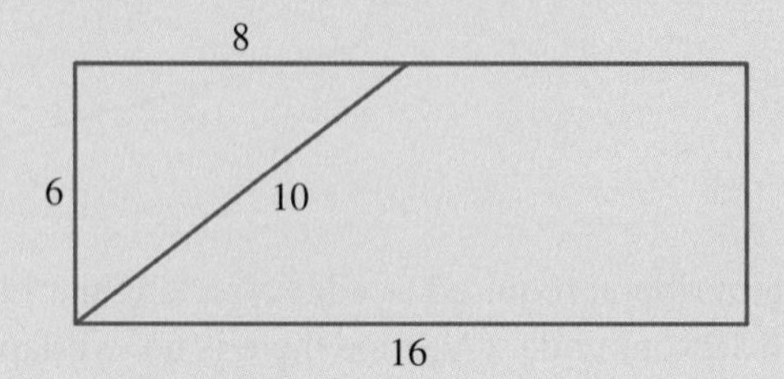


**35.** Each small square in the following diagram measures 1 cm × 1 cm. Calculate the area of triangle ABC (in square cm).

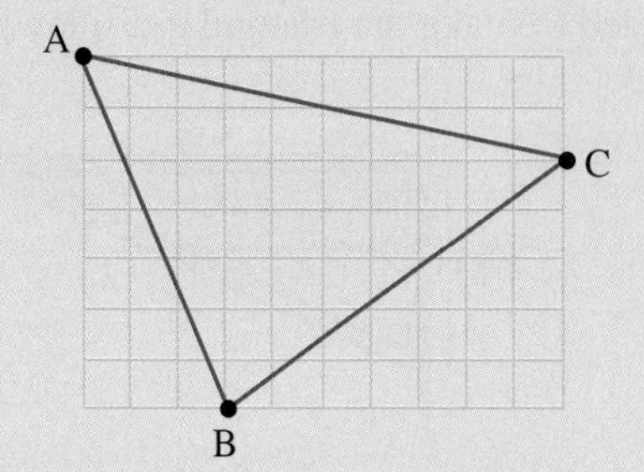


**36.** Stewart is having a swimming pool built in his back yard. The hole for the pool is 6 metres long, 4 metres wide and 1.5 metres deep. The excavated soil becomes aerated as it is dug out, and its volume is increased by $\frac{1}{10}$. Determine the volume of the excavated soil.

**37.** A rectangular cake is three times as long as it is wide. A layer of icing that is 1 cm thick is spread on top of the cake. If the total volume of icing used is 300 cm$^3$, determine the dimensions of the top of the cake.

**38.** A gardener is employed to pave a path that is 2 metre wide around a 10 m × 5.5 m rectangular swimming pool.

**a.** Determine the area of the paved path.

**b.** Determine the minimum number of bricks that should be ordered if there are 60 bricks per square metre.

**39.** The perimeter of a rectangle is 20 cm. Investigate the shape of all such rectangles if the sides are whole numbers. Identify the dimensions of the rectangle with the largest area.

**40.** The following rectangle is divided into three identical smaller rectangles. The length of the large rectangle is 12 cm.
Determine the width of the large rectangle.

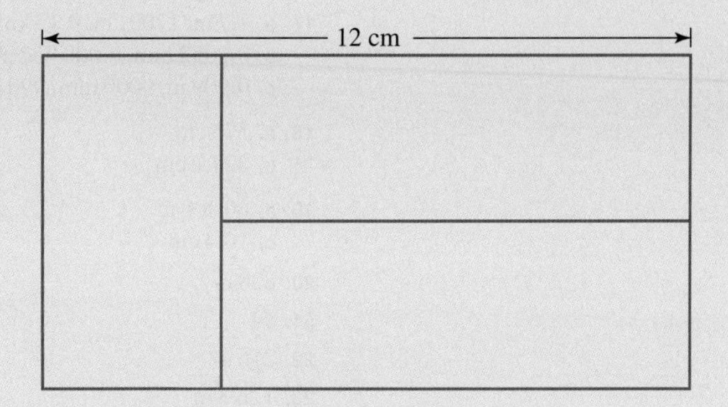


**41.** The front, side and top views of a solid are shown. Construct this solid using blocks, then sketch the solid.

**a.**

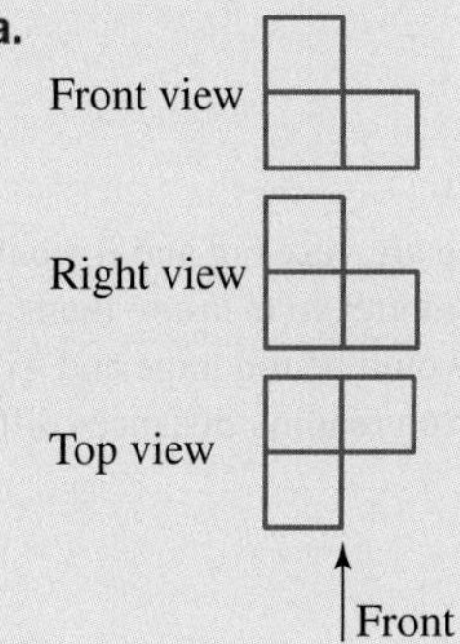


**b.**

Front view

Right view

Top view

Front

**on** To test your understanding and knowledge of this topic, go to your learnON title at www.jacplus.com.au and complete the **post-test**.

# Answers

## Topic 10 Measurement

### 10.1 Pre-test

1. 56.5 cm
2. 0.020 12 km
3. 90 m
4. 1.8 km
5. 46.8 cm
6. 50 m
7. 27 $m^2$
8. $472.50
9. 112 $cm^2$
10. 12.72 $m^2$
11. 48 $m^3$
12. 7 cm
13. 54 L
14. a. ii b. iii c. v d. i e. iv
15. a. B b. A c. C d. D

### 10.2 Units of measurement

1. a. m b. km c. cm d. km e. m
2. a. m b. cm c. km d. mm
3. A
4. a. 2000 m b. 7000 m c. 5300 m d. 660 m
5. a. 25 cm b. 280 mm c. 2000 mm d. 70 000 cm
6. a. 8 km b. 6.5 km c. 0.7 km d. 0.05 km
7. a. 60 m b. 0.57 m c. 4.5 cm d. 2560 cm
8. a. 800 000 cm b. 101 000 mm c. 72 330 mm d. 0.030 35 mm
9. D
10. A
11. a. 0.0452 km b. 0.56 m c. 87.5 mm
12. a. 60 cm b. 309 000 cm c. 4.8 cm
13. Sample responses are given here:
    a. the length of tube required to allow a cat to breathe through while in surgery
    b. the circumference of an actor's head so that an old-fashioned hat of the correct size can be made
    c. the length of a piece of milled hardwood required for the tabletop of a new table for a large family
    d. the height of a hill that requires terracing
    e. the length, in paces, of the runway leading to a long jump pit
14. a. 868 cm b. 10.8 cm
15. Everest: 8.848 km, K2: 8.611 km
16. a. 150 cm, 12.5 m, 0.02 km
    b. 0.445 m, 3000 mm, 350 cm
    c. 50 000 mm, 500 m, 50 km
17. a. 1.7 m, 1700 cm, 0.17 km
    b. 0.0052 mm, 0.000 052 m, 0.052 cm
    c. 0.909 m, 9000 mm, 990 cm
18. a. 375 cm b. 10 200 m c. 224.2 cm d. 155.3 cm
19. a. 66 400 m b. 410.4 m c. 10.4 cm d. 1.7 km
20. 828 m
21. 44
22. 216 m
23. 1.593 m
24. Yes, 36 cm
25. 198 cm
26. 2.4 m
27. a. i. 18 000 000 km
    ii. 1 080 000 000 km
    iii. 25 920 000 000 km
    iv. 9 460 800 000 000 km
    b. Distances in the universe are very big and it would be inconvenient to use kilometres to measure them. The numbers in kilometres would be too large and would make interpreting and even reading distances difficult.
28. 71 428 pins
29. 38
30. 40 cm
31. 215 cm, 105 cm, 155 cm, 170 cm
32. a. i. 50 rotations
    ii. 100 rotations
    b. i. 500 rotations
    ii. 1000 rotations
33. 2 cm × 4 cm, 2 cm × 8 cm, 2 cm × 12 cm, 2 cm × 16 cm
34. 38.5 km

### 10.3 Reading scales and measuring length

1. a. 4 cm b. 7 cm c. 2 cm
2. a. 9.5 cm b. 6.4 cm c. 10.1 cm d. 5.8 cm
3. 2.3 units
4. a. 45 °F b. 34 °C c. 26 °C d. 10 °C
5. a. 45 km/h b. 80 km/h c. 100 km/h
6. a. 1280 mL b. 1800 mL
7. 30 °C
8. a. 130 °C b. 220 °C c. 360 °F
9. B
10. C
11. D

12. D
13. B
14. 1.74 m
15. C
16. A
17. B
18. Sample answers include the following:
    a. Measuring a piece of wood
    b. Drawing straight lines
    c. Determining waist size
    d. Measuring garden dimensions
19. The lamppost is about 4 times the height of the car, so it is about 7.2 m high.
20. Measure the thickness of the book using a ruler, then divide the thickness of the book by the number of pages (two sided) in the book to calculate the thickness of each page.
21. 0.2 km
22. 293 cm

## 10.4 Perimeter

1. a. 8 units b. 18 units c. 18 units d. 22 units e. 12 units
2. a. 40 cm b. 92 mm
3. a. 30 m b. 44 cm
4. 74 cm
5. 68 cm
6. 28 cm
7. a. 173 mm b. 2995 cm
8. a. 408 mm b. 180 mm
9. a. 14 433 m b. 17 579 m
10. B
11. a. 156 cm b. 664.2 cm c. 14.8 cm
12. a. $12\frac{9}{10}$ m or 12.9 m b. $32\frac{4}{5}$ mm or 32.8 mm
13. 48.5 m
14. 252 cm
15. 325.8 m
16. 139.8 m
17. a. 193 m b. $3667
18. a.

23.77 m
3.6 m
10.97 m
6.4 m
6.4 m
3.6 m

b. 109.48 m
c. $4050.76

19. For example, fence construction companies would need to calculate perimeters on a daily basis.
20. 99 cm
21. a. 120 m
    b. $228
    c. $810
    d. 25 posts
    e. $341.25
    f. i. 1600 ii. $1680
    g. $4059.25
22. a. 8 different rectangles
    b. Square sides 8 cm
23. $P = 6a - 2b$
24. 6 cm

## 10.5 Circles and circumference

1. a. $2\pi$ cm b. $10\pi$ cm c. $7\pi$ mm
2. a. 2.58 m b. 23.25 km c. 106.81 m
3. a. $8\pi$ m b. $34\pi$ mm c. $16\pi$ cm
4. a. 8.98 km b. 2.51 m c. 66.60 m
5. a. 25.71 cm b. 82.27 mm c. 61.70 m d. 39.28 mm e. 71.42 cm
6. a. 120.82 cm b. 5.88 m c. 250.80 m d. 252.81 cm
7. B
8. C
9. a. 24.38 cm b. 35.28 mm c. 31.83 cm d. 65.99 cm
10. 119.38 mm
11. 100.53 m
12. 100.53 cm
13. 25.13 m
14. 21.58 m
15. a. 6.00 m b. 20.63 cm c. 23.75 mm
16. a. 2.01 cm b. 7.54 m c. 24.99 mm
17. 19.19 m
18. 15.71 m
19. 80 m
20. a. **iii** b. **i** c. **iv** d. **ii**
21. 41.82 km/h
22. 9.55 cm
23. a. 25.71 cm b. 483.29 mm c. 31.07 m d. 5.71 cm
24. Length of arc = 6.98 cm; perimeter = 13.74 cm.

## 10.6 Area

1. a. $cm^2$ b. $m^2$ c. $mm^2$ d. $km^2$
2. a. $m^2$ b. $km^2$ c. $m^2$ d. ha

3. a. $14\,cm^2$ b. $10\,cm^2$ c. $10\,cm^2$
d. $16\,cm^2$ e. $8\,cm^2$

4. a. $9\,cm^2$ b. $6\,cm^2$ c. $9\frac{1}{2}\,cm^2$
d. $12\,cm^2$ e. $10\,cm^2$

5. A

6. C

7. C

8. a. $15\,cm^2$ b. $14\,km^2$ c. $10\,m^2$

9. a. $20\,cm^2$ b. $48\,m^2$

10. a. $56\,m^2$ b. $400\,mm^2$

11. a. $72\,cm^2$ b. $76\,m^2$

12. a. $135\,mm^2$ b. $686\,mm^2$

13. a. $4\,cm^2$ b. $3\,cm^2$ c. $40.5\,cm^2$ d. $50\,cm^2$

14. C

15. A

16. A

17. a. $416\,cm^2$ b. $3.125\,m^2$ c. $2.4\,m^2$ d. $493\,mm^2$

18. a. $583\,cm^2$ b. $600\,km^2$
c. $337.5\,cm^2$ d. $152\,cm^2$

19. $28.14\,mm^2$

20. C

21. $200\,000\,m^2$

22. a. $16\,000\,000\,000\,m^2$
b. 1 600 000 hectares

23. $17.5\,m^2$

24. $9.66\,m^2$

25. a. $7.5\,m^2$ b. 15 boxes c. $528

26. $141.64

27. $2560

28. It would be cheaper to use the 70 cm × 70 cm concrete pavers.

29. $88.8\,m^2$

30. $104\,m^2$, $1300

31. $4920\,km^2$

32. $6000\,cm^2$ ($0.6\,m^2$)

33. $135\,m^2$

34. $12.6\,m^2$

35. $126\,m^2$

36. Length 9 cm, width 4 cm

37. Two sides of the fencing are 6 m long and the other side is 12 m, providing an area of $72\,cm^2$.

## 10.7 Area of composite shapes

1. a. $81\,cm^2$ b. $5\,m^2$ c. $16\,cm^2$

2. a. $56\,cm^2$ b. $588\,cm^2$ c. $13.5\,m^2$

3. a. $48\,m^2$ b. $325\,m^2$

4. $13.05\,m^2$

5. $39.28\,m^2$

6. $14.25\,m^2$

7. a. $27\,m^2$ b. $675

8. $2157.60 (rounded to the nearest 5 cents)

9. $75\,m^2$

10. $39\,m^2$

11. a. $22\,cm^2$
b. $22\,cm^2$
c. Sample responses can be found in the worked solutions in the online resources.

12. If your composite shape is made up of two or more simple shapes, it is best to determine the area by addition. If the area of a shape has a section not required (a shaded area), it is best to calculate the total area and then subtract the area not required.

13. a. Square of side length 30 cm
Three circles of diameter 15 cm
Two semicircles of diameter 30 cm
b. Using addition of areas: square + circle of diameter 30 cm (the two equal semicircles form one circle).
Using subtraction of areas: area calculated in previous step $-3\times$ area of the smaller (coloured) circle.
c. $1075\,cm^2$
d. The area in $mm^2$ would be $107\,500\,mm^2$. This is more inconvenient to use because it is a larger number, which is harder to interpret and use in further calculations if needed.

14. $\$0.15/cm^2$

15. The shaded area is $\frac{7}{32}$ of the square.

16. a. i. Two rectangles with length 4 cm and width 1 cm and two parallelograms with length 3 cm and width 0.5 cm
ii. One square of side length 4 cm, one rectangle with length 2 cm and width 1 cm, one triangle with height 3 cm and base 1 cm and two right-angled triangles with height 3 cm and base 0.5 cm. The last two triangles form a triangle with base 1 cm and height 3 cm.
b. $11\,cm^2$

## 10.8 Volume of rectangular prisms

1. a. $8\,cm^3$ b. $6\,cm^3$ c. $12\,cm^3$ d. $48\,cm^3$

2. B

3. a. $4\,cm^3$ b. $24\,cm^3$ c. $15\,cm^3$ d. $13\,cm^3$

4. a. $7\,cm^3$ b. $35\,cm^3$ c. $14\,cm^3$ d. $17\,cm^3$

5. A

6. a. $12\,cm^3$ b. $12\,m^3$ c. $6144\,cm^3$

7. a. $12\,cm^3$ b. $10\,cm^3$ c. $10\,cm^3$

8. a. $21\,cm^3$ b. $144\,cm^3$ c. $972\,cm^3$

9. D

10. B

11. a. $186\,cm^3$ b. $136\,cm^3$

12. a. $189\,cm^3$ b. $180\,cm^3$

13. $9690\,cm^3$

14. $477\,750\,cm^3$

15. a. 20 cm  b. 9 cm  c. 24 cm

16. a. $2700\ \text{cm}^3$  b. $2\,700\,000\ \text{mm}^3$

17. $4.875\ \text{m}^3$

18. a. $28.875\ \text{cm}^3$  b. $28\,875\ \text{mm}^3$
 c. $176\ \text{mm}^3$  d. $20\,075\ \text{mm}^3$

19. $768\ \text{m}^3$

20. $40.8\ \text{m}^3$

21. Each guest receives $267\ \text{cm}^3$ of cake.

22. $2500\ \text{m}^3$

23. $0.65\ \text{m}^3$ more soil is needed (or $2.075\ \text{m}^3$).

24. $36\ \text{cm}^3$

25. Length 18 cm, width 15 cm and height 10 cm. Volume is $2700\ \text{cm}^3$.

26. 36 000

## 10.9 Capacity

1. a. 2000 mL  b. 3 L  c. 7 L  d. 5.5 L

2. a. 2500 mL  b. 32 L  c. 35 mL  d. 420 000 mL

3. a. 1870 mL  b. 22.5 L  c. 100 mL  d. 0.025 kL

4. a. 750 mL  b. 0.0025 kL
 c. $2450\ \text{cm}^3$  d. 78 L

5. a. 40 000 mL = 40 L  b. $6000\ \text{mL} = 6000\ \text{cm}^3$
 c. $5.2\ \text{kL} = 5.2\ \text{m}^3$

6. D

7. D

8. a. 0.25 L, 2.45 L, 2.5 L, 25 000 mL
 b. 7.65 mL, 760 mL, 0.765 L, 7.60 L
 c. 0.011 L, 0.1 L, 110 mL, 1.1 L

9. 16 bottles

10. 1750 mL (1.75 L)

11. 1620 mL (1.62 L)

12. 6 L

13. 50 doses

14. 225 bottles

15. 9.9 L

16. 4.8 L

17. a. No  b. Yes

18. The 185 mL container is the better buy.

19. The tank overflowed.

20. 616 mL

21. 50.4 L

22. Approximately 27 L (27.09 L)

23. 300 mL

24. 1401.6 L

25. $16.4\ \text{kL} = 16.4\ \text{m}^3$. An example could be length 2 m, width 2 m, height 4.1 m.

## 10.10 Drawing solids

1. a.

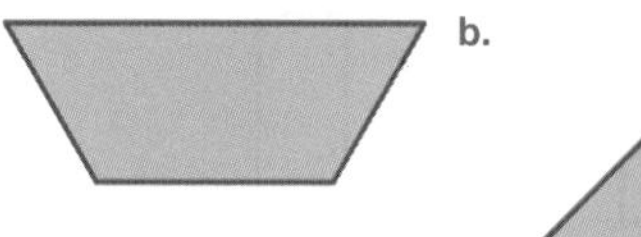

b.

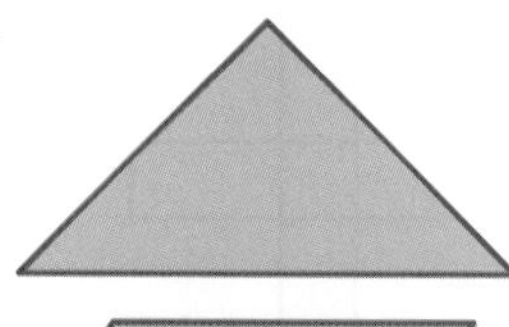

2. a.

b.

3 a.

b.

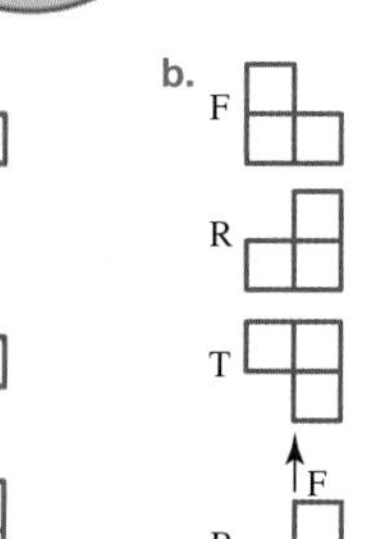


c.

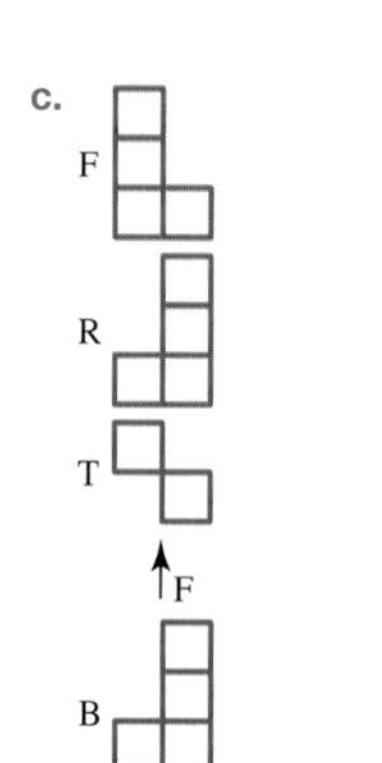


4. a.

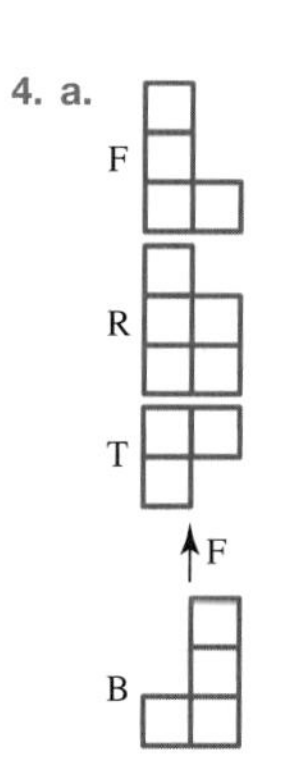


b.

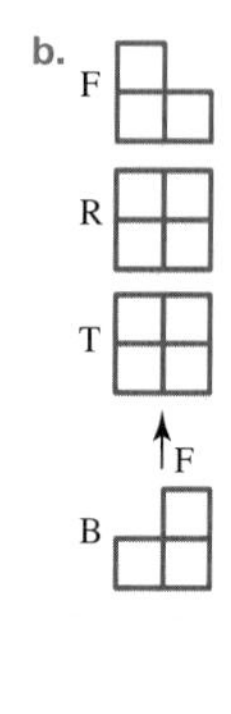


c.

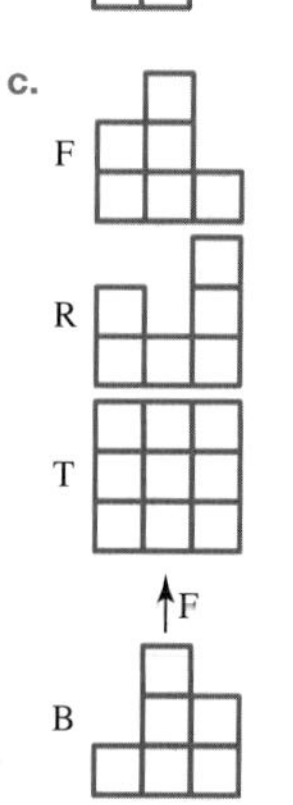


5. a.

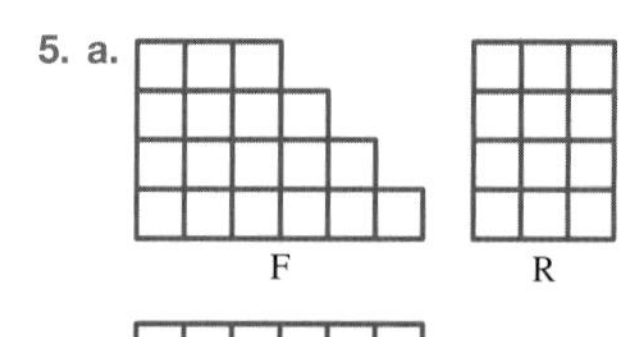


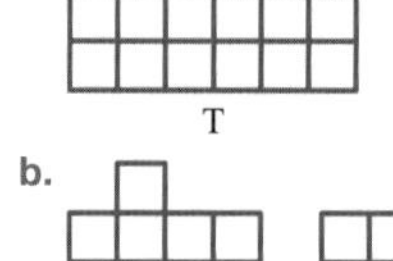


b. c.

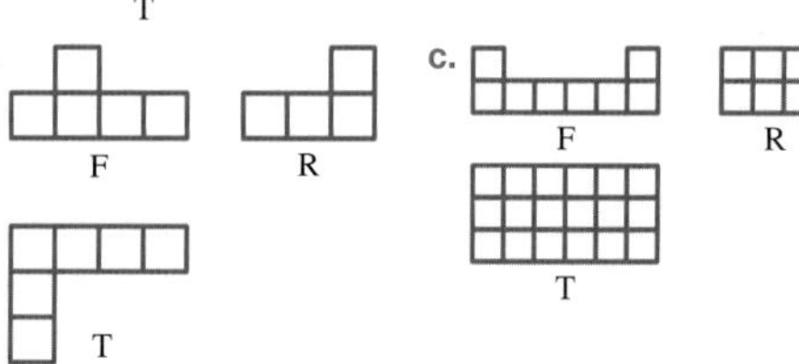

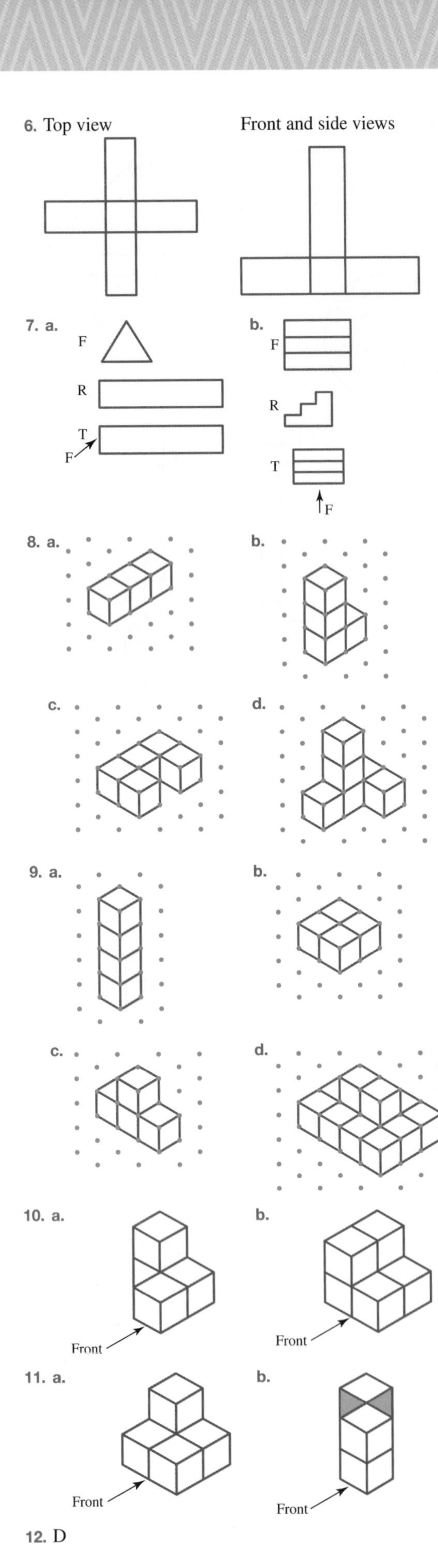


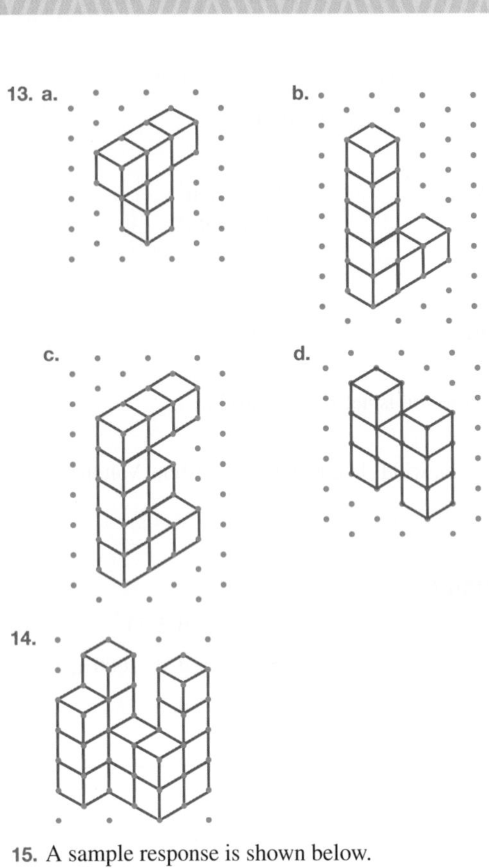


15. A sample response is shown below.

16. The minimum number of cubes that could be used is 8, as shown in blue in the figure. There are two spaces that could be filled with cubes without changing the front or right views (shown in pink); therefore, the maximum number of cubes that could be used is 10.

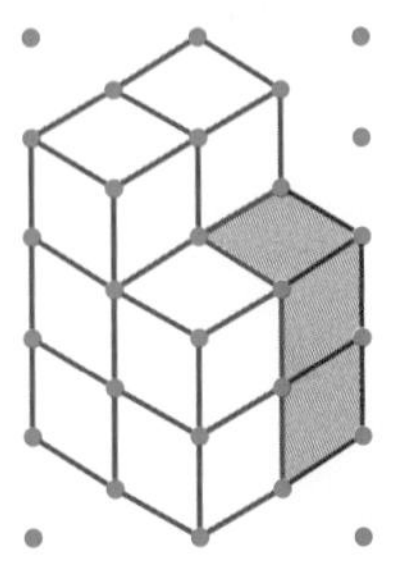

17. a. It would be difficult to construct this solid.

b. 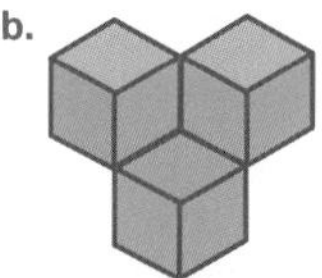

The solid would look like the one shown. The fourth cube is behind the front bottom cube, resting on the surface. The two top cubes have no cubes supporting them.

18. An infinite number of shapes because there are no restrictions on the other two views

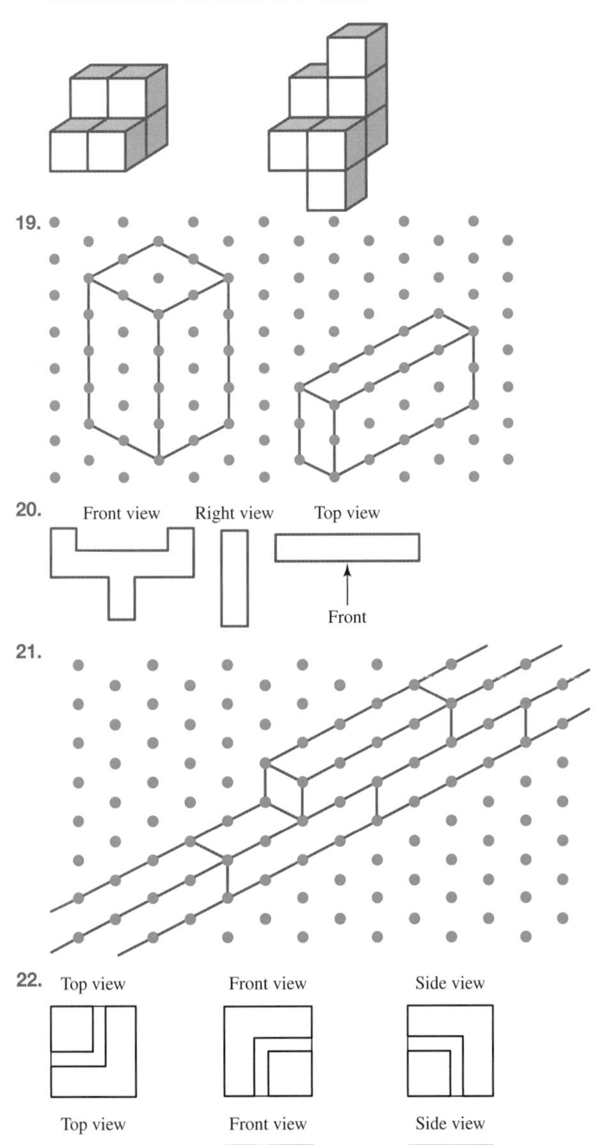


23. The arrangements are shown below.

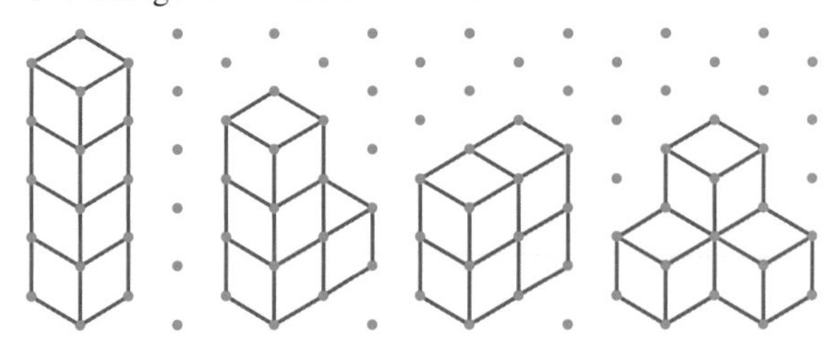

## Project

1. 1 cubit = 0.523 metres
2. A sample response is given here.

   Inch — a unit of length used as a measure for display screens, the height of a person or sometimes the thickness of an object

   Mile — a unit of length or distance used to measure the distance usually seen on road signs in the US

   Furlong — a unit of distance used to measure dimensions of farmland and distance in horse racing

   Foot — a unit of length used to measure the height of a person or an object

   Fathom — a unit of length used to measure the depth of the water, or the length of a fisher's setline

3. A sample response is given here.

**Basketball:**

| | |
|---|---|
| Baskets: | Inches<br>Rings are 18 inches in inside diameter, with white cord 12-mesh nets, 15 to 18 inches in length. |
| Height of basket: | Feet<br>10 feet (upper edge) |
| Circumference of ball: | Inches<br>Not greater than 30 inches and not less than $29\frac{1}{2}$ inches. |

**Soccer:**

| | |
|---|---|
| Goals: | Yards/feet/inches<br>Distance between posts is 8 yards.<br>Distance from crossbar to the ground is 8 feet. |
| Penalty area: | Yards<br>Two lines drawn at right angles to the goal line, 18 yards from the inside of each goalpost. Lines extend into playing field for 18 yards and are joined by a line drawn parallel with the goal line. |

## 10.11 Review questions

1. a. Kilometres or metres

   b. Centimetres

   c. Millimetres

2. a. 0.56 m b. 0.023 km c. 1700 cm
d. 750 000 mm e. 2090 mm

3. a. 0.068 m b. 2.25 cm c. 6.3 m
d. 82 000 km e. 0.59 cm

4. 0.004 45 km, 455 cm, 44.5 m, 455 000 mm

5. a. 51 200 m (51.2 km)
b. 3700 cm (37 m)

6. 80 kg

7. 7 m

8. a. $m^2$ b. $km^2$ c. $mm^2$
d. ha e. $km^2$

9. a. 69.12 cm b. 138.23 mm c. 56.55 m

10. $15 \text{ cm}^2$

11. a. $150 \text{ m}^2$ b. $6 \text{ cm}^2$ c. $2914 \text{ mm}^2$

12. a. $9 \text{ cm}^2$ b. $19 \text{ cm}^2$ c. $52 \text{ cm}^2$

13. a. $9 \text{ cm}^2$ b. $380 \text{ cm}^2$ c. $11 \text{ cm}^2$

14. a. $627 \text{ cm}^2$ b. $96 \text{ m}^2$ c. $1344 \text{ cm}^2$

15. a. $16 \text{ m}^2$ b. $432 \text{ mm}^2$

16. a. $87.5 \text{ cm}^2$ b. $41 \text{ cm}^2$

17. a. $18 \text{ cm}^3$ b. $30 \text{ cm}^3$
c. $20 \text{ cm}^3$ d. $28 \text{ cm}^3$

18. a.

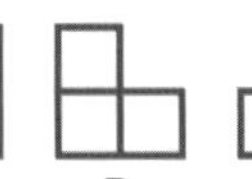

F R T

b.

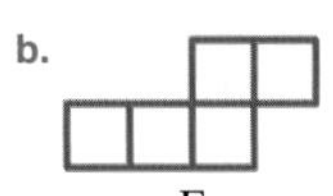

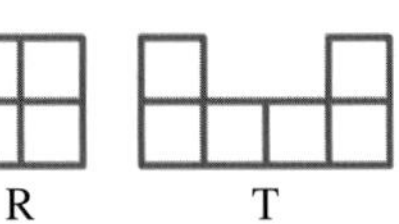

F R T

19. a.

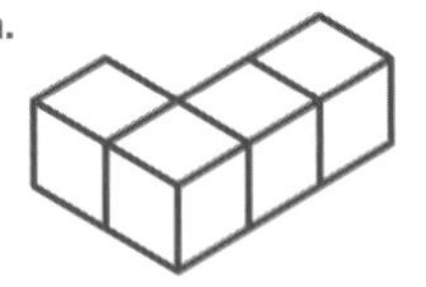

b.

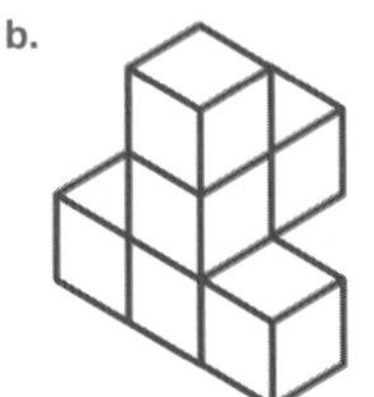

20. $72 \text{ cm}^3$

21. $112 \text{ cm}^3$

22. a. 8000 mL b. 420 mL
c. 3.3 L d. 1012 mL

23. a. 4300 L b. 3.4 L
c. 4.755 kL d. 0.432 L

24. 25 cm

25. a. 46 cm b. 20 cm
c. 191 mm (19.1 cm) d. 21 cm (210 mm)

26. 243 m

27. 870 cm

28. $582.40

29. $3.6 \text{ m}^2$

30. $0.72 \text{ m}^2$

31. a. 28 cm b. 8.5 cm

32. 4.9 L

33. a. 1725 mL (1.725 L)
b. 8 glasses

34. The perimeter is 46.

35. $32 \text{ cm}^2$

36. 39.6 cubic metres

37. 10 cm by 30 cm

38. a. Area of path is $78 \text{ m}^2$.
b. 4680 bricks are needed.

39. Possible side dimensions: $9 \times 1; 8 \times 2; 7 \times 3; 6 \times 4; 5 \times 5$.
Largest area is 25 square units.

40. 8 cm

41. a.

b.

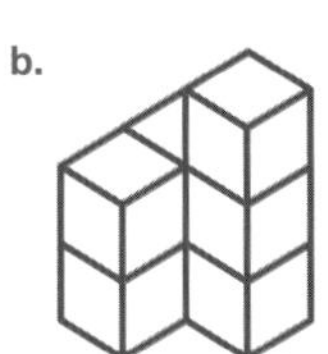

# 11 Coordinates and the Cartesian plane

## LESSON SEQUENCE

11.1 Overview ....614
11.2 The Cartesian plane ....617
11.3 Linear number patterns ....625
11.4 Plotting simple linear relationships ....632
11.5 Interpreting graphs ....639
11.6 Review ....651

# LESSON
## 11.1 Overview

### Why learn this?

Using a coordinate system enables us to pinpoint locations in different settings. When you go to the movies, a musical, a concert or a sporting grand final, your ticket shows where you will be seated. For example, a ticket to a concert may give the location of your seat as Section 47, Row G, Seat 5. A boarding pass on a flight always shows where you will be sitting. For example, a boarding pass that shows seat 30A means you will be seated in row 30, seat A, by the window.

Have you ever wondered where you are on a map, or globe, of the world? Melbourne is 37.81°S, 144.96°E, which means it is in the southern hemisphere, 37.81° south of the equator and 144.96° east of the Royal Observatory in Greenwich, England. New York is 40.71°N, 74.01°W, so it is in the northern hemisphere and west of Greenwich. In coordinate geometry, we define a point by how far it is horizontally and vertically from the centre point. Scientists and geographers often use coordinates when they analyse data. Reading and interpreting maps involves knowing about coordinates. Computer programmers use coordinates when creating artwork. A computer screen is made up of tiny points of light called pixels. Programmers use coordinates to identify which pixels to light up when creating art. Understanding coordinates, and maps, assists you in many of your everyday activities.

Use the following map to answer questions **1** and **2**.

1. Use the grid map to give the grid references for:

   **a.** Mt Kosciuszko **b.** Port Campbell **c.** The Grampians.

2. Using the grid map, name a town that is located within each of the following grid references.

   **a.** C1 **b.** B2 **c.** D3

3. Consider the chessboard shown.

   **a.** One of the black rooks (in chess, rook is another name for castle) is on a8. State the location of the white rooks.

   **b.** One of the white bishops is on c1. State the position of the second white bishop.

   **c.** A bishop can move diagonally any number of (unoccupied) squares at a time. With its next move, the bishop at b3 could capture the pawn at what position?

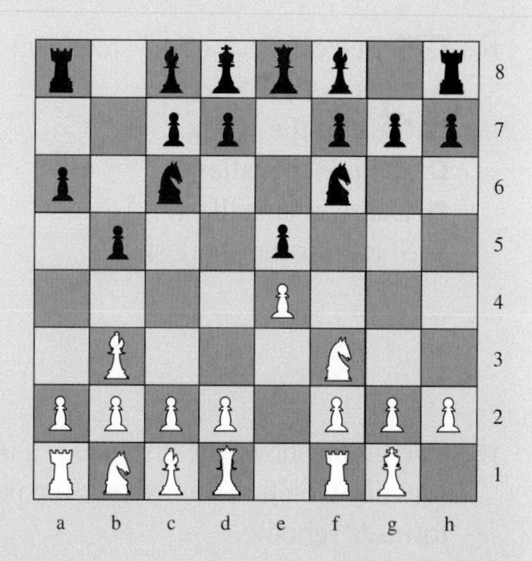


4. Identify the coordinates of point X on the Cartesian plane.

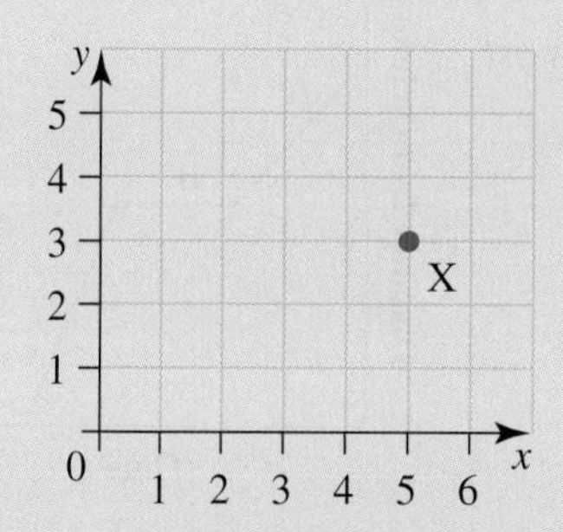

5. Calculate the area of the rectangle formed by connecting the points $(-2, 3), (3, 3), (3, -4)$ and $(-2, -4)$ on a Cartesian plane.

6. **MC** Identify which of the following points is located in the second quadrant.
   A. (5, 1) B. (1, 5) C. (1, –5) D. (–1, –5) E. (–5, 1)

7. Determine the coordinates of the midpoint between (–2, 6) and (0, 2).

8. **MC** Identify the point that sits on the $x$-axis.
   A. (1, 1) B. (1, 0) C. (3, 3) D. (–3, 3) E. (0, 1)

9. Determine the value of point P if the points (2, 5), (4, 11) and (6, P) are to be linear.

10. If the values in the following table are linear, determine the missing value.

| $x$ | 2 | 3 | 5 | 7 |
|---|---|---|---|---|
| $y$ | 5 | 9 | | 25 |

11. **MC** Select which of the following points form a linear pattern when plotted.
    A. (0, 2), (2, 4), (3, 6) B. (–1, 0), (–3, 4), (–4, 6) C. (–1, 1), (1, 1), (2, 2)
    D. (3, 3), (4, –4), (5, 5) E. (0, 0), (–1, –2), (3, 4)

Use the following information to answer questions **12** and **13**.

Teo plants a fast-growing hedge. The plant starts at a height of 45 cm, and for each of the next three weeks he measures the height to be 51 cm, 57 cm and 63 cm.

12. Explain whether the hedge is following a linear growth pattern.

13. If $H$ represents the height of the hedge after $x$ weeks, determine the equation that models the height of the plants.

14. **MC** Select which of the statements is true about the graph shown.
    A. Claire is the oldest.
    B. Heidi is the tallest.
    C. Jodie is the tallest.
    D. Claire is the tallest.
    E. Sharon is the shortest.

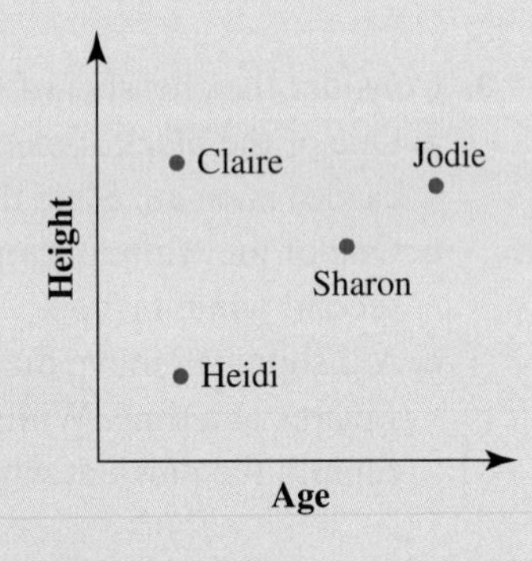


15. The graph shows the distance Frankie is from his school.
    Identify which section shows the period when Frankie is travelling fastest towards school.

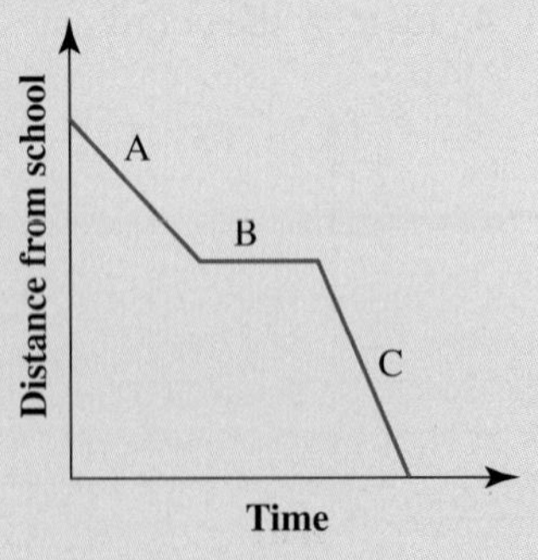

# LESSON
# 11.2 The Cartesian plane

## LEARNING INTENTION

At the end of this lesson you should be able to:
- plot points on a Cartesian plane
- identify the coordinates of points on a Cartesian plane.

## 11.2.1 The Cartesian plane

eles-4635

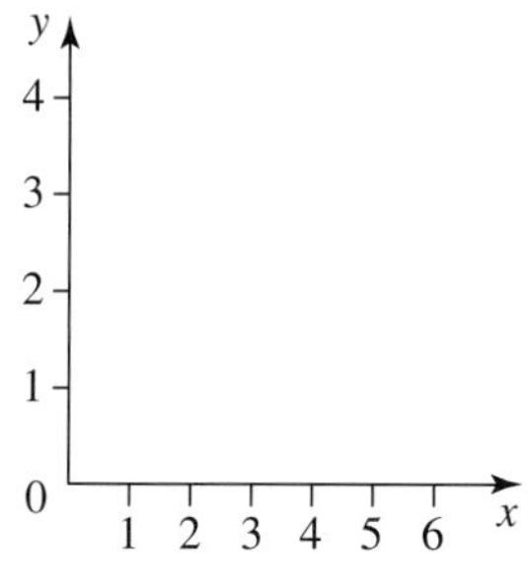


- The **Cartesian plane** is formed from two straight, perpendicular lines called **axes.**
- The horizontal axis is referred to as the ***x*-axis**.
- The vertical axis is referred to as the ***y*-axis**.
- The point where the two axes intersect is called the **origin** and can be labelled with the letter O. The origin has the coordinate $(0, 0)$.
- Both axes must be marked with evenly spaced ticks and numbered with an appropriate scale (usually the distance between each tick is one unit).
- The $x$-axis must be labelled with an $x$ at the far right of the horizontal axis and the $y$-axis must be labelled with a $y$ at the top of the vertical axis.
- A **Cartesian coordinate** is a pair of numbers, written in brackets and separated by a comma, which gives the position of a point on the plane.
- In a coordinate, the first number is the $x$-coordinate and shows how far right (or left) from the origin a point is located. The second number is the $y$-coordinate and shows how far up (or down) from the origin a point is located.

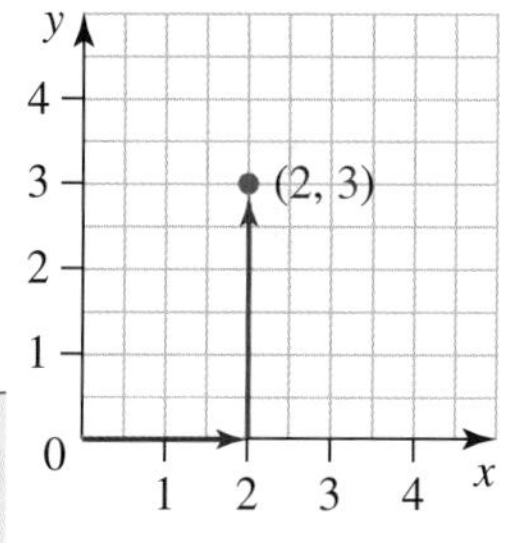


**Cartesian coordinates**

**A Cartesian coordinate is written in the form: $(x, y)$.**

**The Cartesian coordinate (2, 3) can be located by moving two units right from the origin and then three units up.**

### WORKED EXAMPLE 1 Creating a Cartesian plane and plotting points

**Draw a Cartesian plane with axes extending from 0 to 6 units. Mark the following points with a dot, and label them.**

a. $(2, 4)$ b. $(5, 0)$ c. $(0, 2)$ d. $\left(3\frac{1}{2}, 2\right)$

**THINK**

First rule and label the axes.

a. (2, 4) means starting at the origin, go across 2 units, and then up 4 units.

b. (5, 0) means go across 5 units and up 0 units. It lies on the $x$-axis.

c. (0, 2) means go across 0 units and up 2 units. It lies on the $y$-axis.

d. $\left(3\frac{1}{2}, 1\right)$ means go across $3\frac{1}{2}$ units and up 1 unit.

Label each point.

**DRAW**

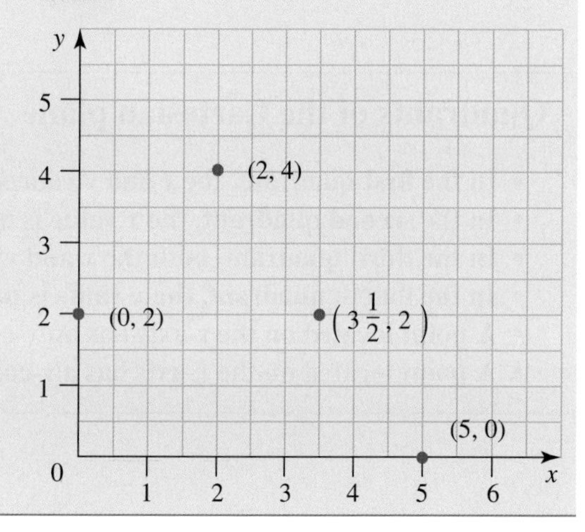

## WORKED EXAMPLE 2 Determining the coordinates of points on the Cartesian plane

**Identify the Cartesian coordinates for each of the points A, B, C and D.**

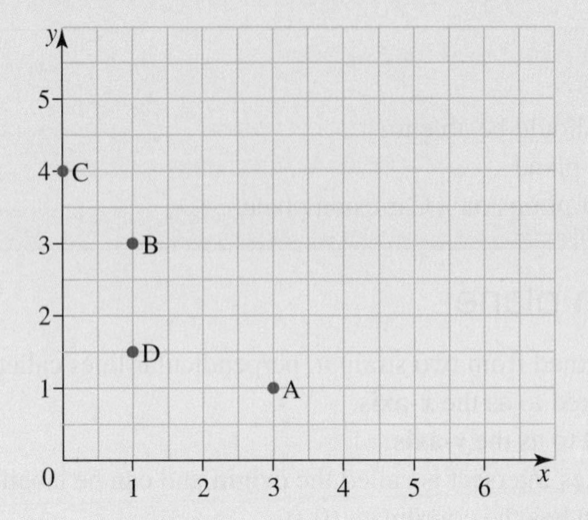


| THINK | WRITE |
|---|---|
| Point A is 3 units across and 1 unit up. | A is at (3, 1) |
| Point B is 1 unit across and 3 units up. | B is at (1, 3) |
| Point C is 0 units across and 4 units up. | C is at (0, 4) |
| Point D is 1 unit across and $1\frac{1}{2}$ units up. | D is at $\left(1, 1\frac{1}{2}\right)$ |

## 11.2.2 The four quadrants of the Cartesian plane

eles-4636

- The axes on the Cartesian plane can extend infinitely in both directions.
- On the $x$-axis, the values to the left of the origin are negative.
- On the $y$-axis, the values below the origin are negative.
- By extending the axes in both directions, the Cartesian plane is divided into four sections called **quadrants**.
- The quadrants are numbered one to four, starting with the top right as quadrant one, then numbering in an anti-clockwise direction.

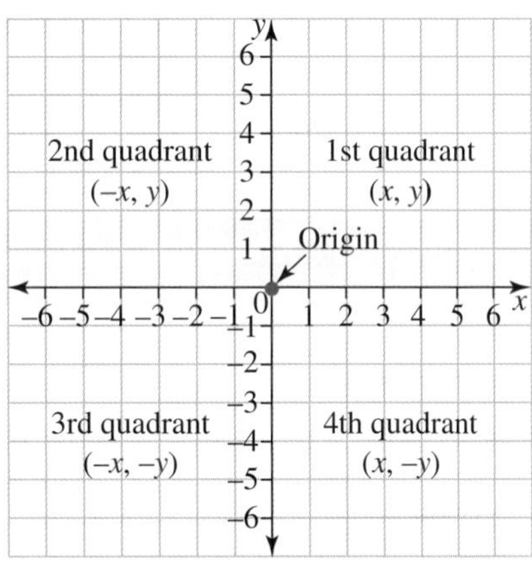


### Quadrants of the Cartesian plane

- **In the first quadrant, the $x$ and $y$ values are both positive.**
- **In the second quadrant, the $x$ value is negative and the $y$ value is positive.**
- **In the third quadrant, both the $x$ and $y$ values are negative.**
- **In the fourth quadrant, the $x$ value is positive and the $y$ value is negative.**
- **A point located on the $y$-axis has an $x$-coordinate of 0.**
- **A point located on the $x$-axis has a $y$-coordinate of 0.**

## WORKED EXAMPLE 3 Locating the quadrant of points on the Cartesian plane

**a. Plot the following points on the Cartesian plane.**

$$A(-1, 2), B(2, -4), C(0, -3), D(4, 0), E(-5, -2)$$

**b. State the location of each point on the plane (that is, the quadrant, or the axis on which it sits).**

**THINK**

a. 1. Draw a set of axes, ensuring that they are long enough to fit all the values.

2. Plot the points. The first point is one unit to the left and two units up from the origin. The second point is two units to the right and four units down from the origin (and so on).

**WRITE**

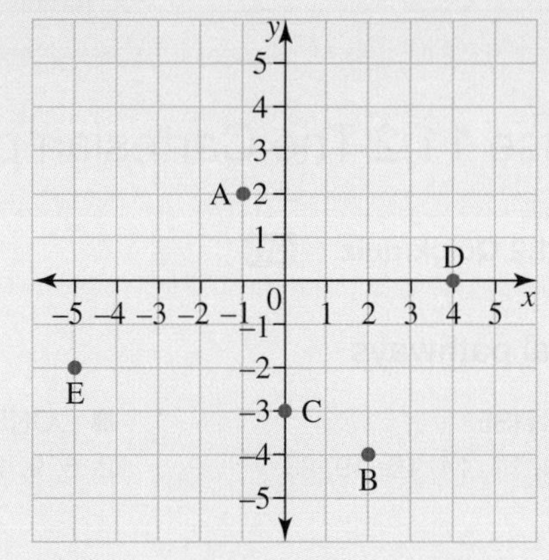


**THINK**

b. Look at the plane and state the location of each point. Remember that the quadrants are numbered in an anti-clockwise direction, starting at the top right. If the point is on the axis, specify which axis it is.

**WRITE**

Point A is in the second quadrant.
Point B is in the fourth quadrant.
Point C is on the $y$-axis.
Point D is on the $x$-axis.
Point E is in the third quadrant.

## COLLABORATIVE TASK: Alphanumeric grid references on a map

- Similar to a Cartesian plane, areas of land can be displayed by using a grid reference map, where the map is divided into squares.
- Because grid references refer to a square region rather than a point on a map, they are often called area references.
- The reference for each square on a map is given by two coordinates — a letter and a number (hence the name alphanumeric).
- Alphanumeric grid references are usually found in tourist maps, street directories or atlases.
- A map of central London with an alphanumeric grid is shown.

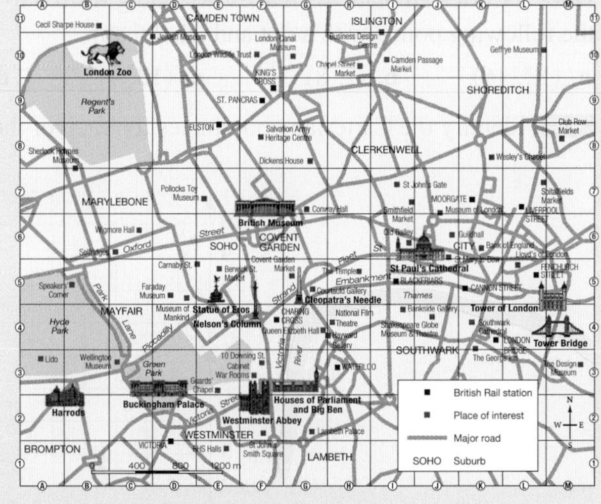


1. Using the map provided, give the grid reference for each of the following locations.
   a. The British Museum
   b. Westminster Abbey
   c. Cleopatra's Needle
2. Name the major feature located at each of the following grid references.
   a. D3
   b. F4
   c. H10
3. Suggest a disadvantage to using a map with alphanumeric grid references.
4. Investigate other ways that locations are given on maps. How do pilots navigate to specific runways in large cities around the world?
5. Discuss with your classmates other uses for alphanumeric grid references.

## Resources

eWorkbook Topic 11 Workbook (worksheets, code puzzle and project) (ewbk-1912)

Interactivities Individual pathway interactivity: The Cartesian plane (int-4384)
The Cartesian plane (int-4136)
Extending the axes (int-4137)

# Exercise 11.2 The Cartesian plane

learnon

11.2 Quick quiz 

11.2 Exercise

### Individual pathways

| ■ PRACTISE | ■ CONSOLIDATE | ■ MASTER |
|---|---|---|
| 1, 4, 5, 8, 11, 14, 16, 19, 23 | 2, 6, 9, 12, 13, 17, 20, 24, 25 | 3, 7, 10, 15, 18, 21, 22, 26, 27 |

### Fluency

1. WE1 Draw a Cartesian plane with axes extending from 0 to 5 units. Mark the following points with a dot, and label them.
   a. (4, 3) b. (1, 4) c. (3, 3) d. (2, 0) e. (0, 4) f. (0, 0)

2. Draw a Cartesian plane with axes extending from 0 to 10 units. Mark the following points with a dot and label them.
   a. (4, 2) b. (3, 8) c. (9,9) d. (5, 0) e. (7, 2) f. (0, 10)

3. Draw a Cartesian plane with axes extending from 0 to 20 units using a scale of 2 units for both axes. Mark the following points with a dot and label them.
   a. (12, 16) b. (2, 8) c. (18, 4) d. (10, 10) e. (6, 0) f. (14, 20)

4. WE2 Identify the Cartesian coordinates for each of the points A to L.

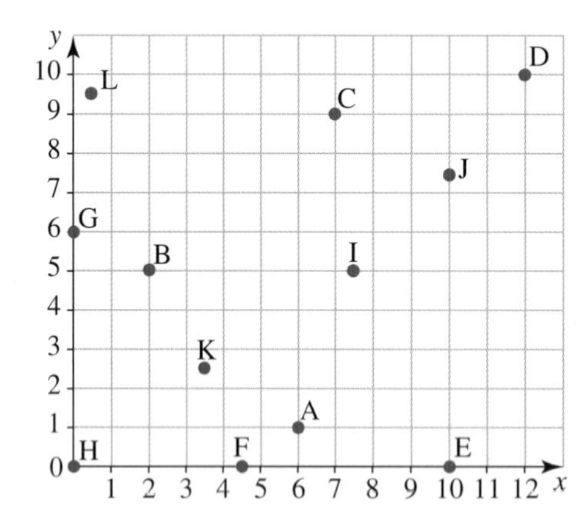


5. WE3 a. Plot the following points on the Cartesian plane.
   A(−1, −3), B(−2, 5), C(3, −3), D(0, −4), E(−2, −2), F(−5, 0), G(3, 1), H(3, 0), I(−4, −2), J(4, −5)
   b. State the location of each point on the plane (that is, the quadrant, or the axis it sits on).

6. a. Plot the following points on the Cartesian plane.
   A(−4, 3), B(5, −7), C(−5, −5), D(0, −7), E(−9, 0), F(−2, 8), G(−3, −6), H(−1, 6), I(2, 8), J(−7, 7)
   b. State the location of each point on the plane (that is, the quadrant or the axis it sits on).

7. a. Plot the following points on the Cartesian plane.
   A(−12, 7), B(−7, 1), C(−10, −10), D(0, 7), E(−3, 0), F(−13, 5), G(−9, −11), H(1, −9), I(2, 8), J(3, −3)
   b. State the location of each point on the plane (that is, the quadrant or the axis it sits on).

8. Identify the Cartesian coordinates for each of the points A to F.

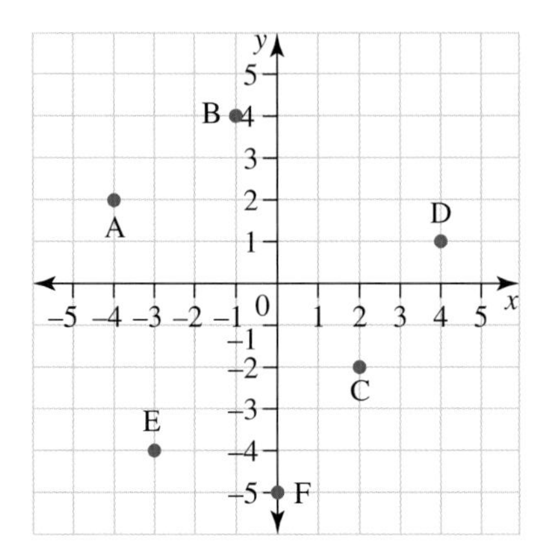


9. Identify the Cartesian coordinates for each of the points A to F.

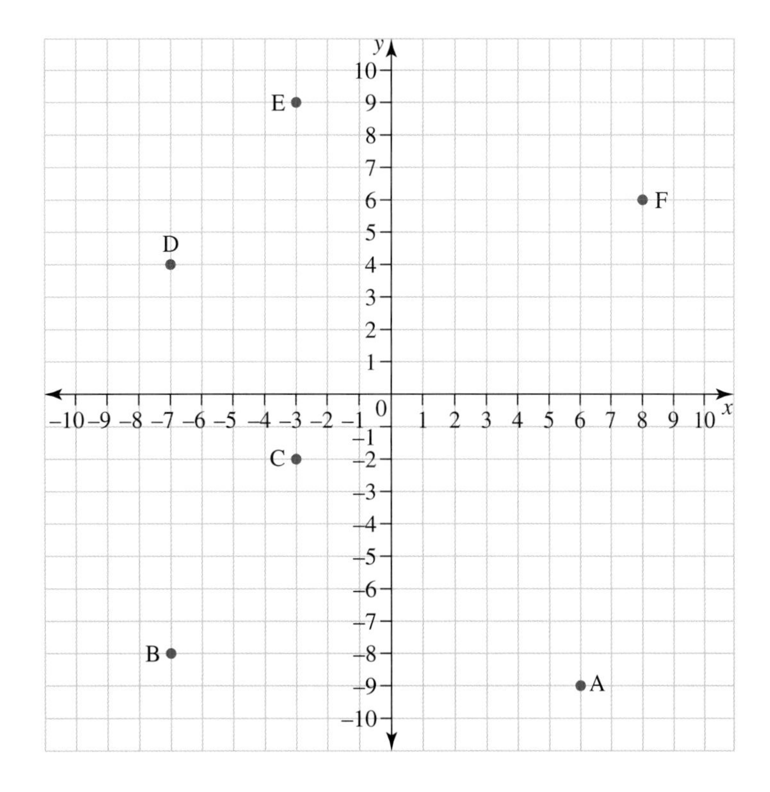

**10.** Identify the Cartesian coordinates for each of the points A to F.

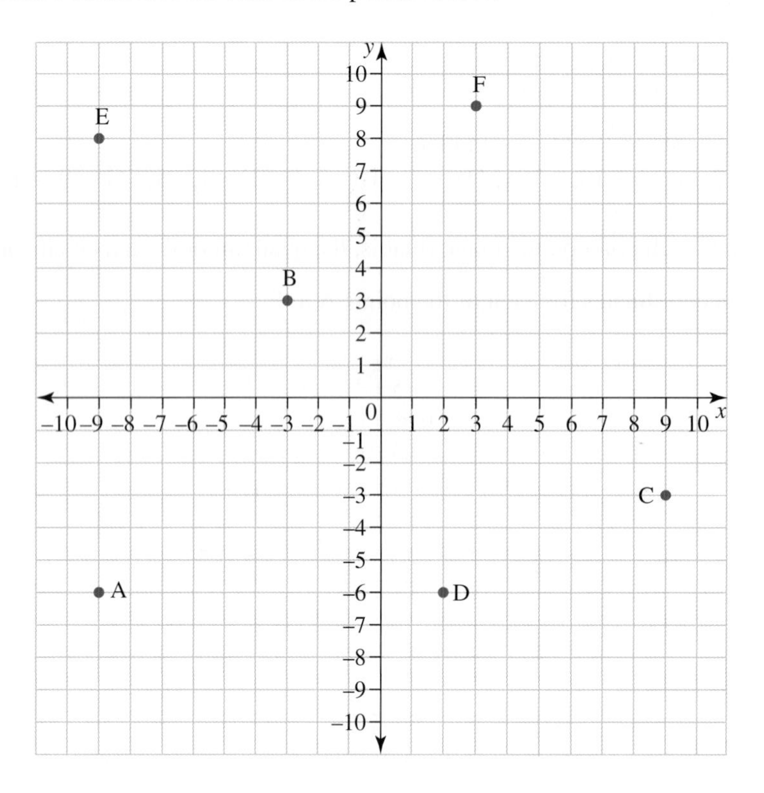


## Understanding

**11.** Each of the following sets of Cartesian axes (except one) has something wrong with it. From the list below, match the mistake in each diagram with one of the sentences.

**A.** The units are not marked evenly.
**B.** The $y$-axis is not vertical.
**C.** The axes are labelled incorrectly.
**D.** The units are not marked on the axes.
**E.** There is nothing wrong.

**a.**

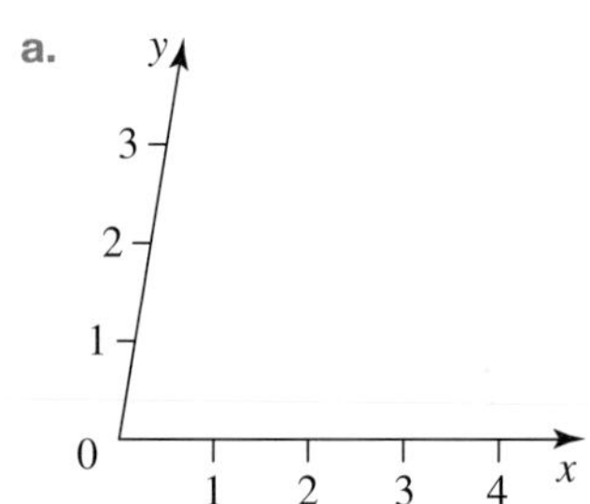

**b.**

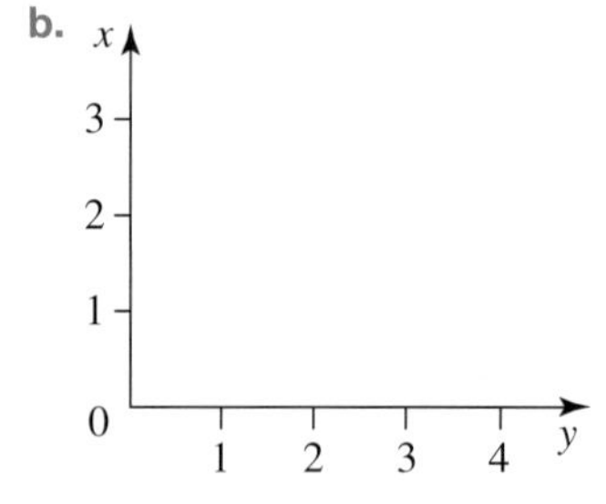

**c.**

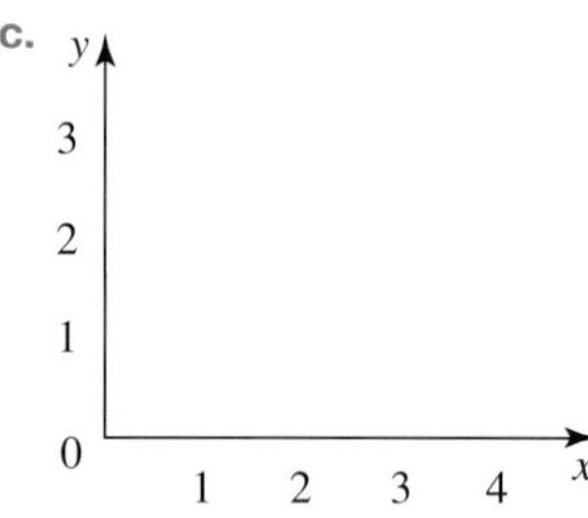

**d.**

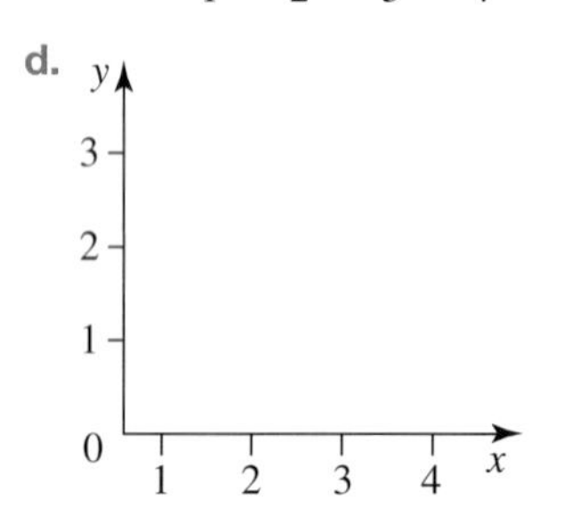

**e.**

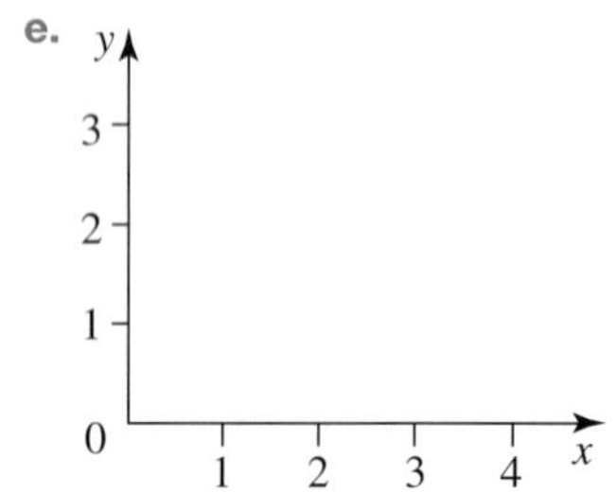

**f.**

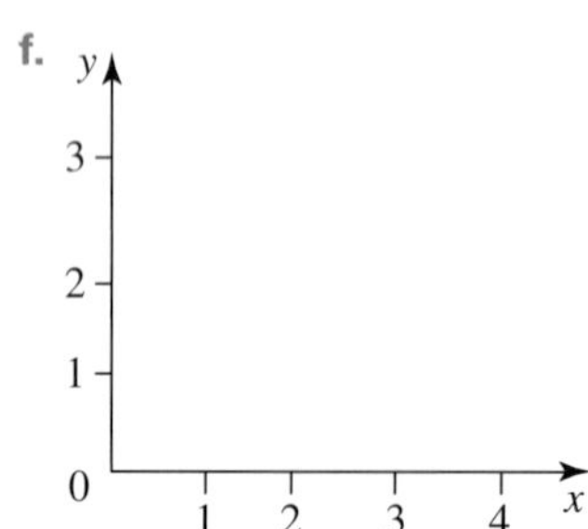

12. From the graph, write down the coordinates of 2 points that:

  a. have the same $x$-coordinate
  b. have the same $y$-coordinate.

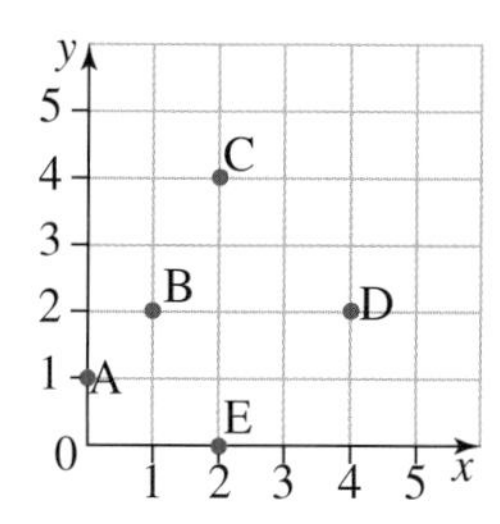


13. Messages can be sent in code using a grid like the one shown in the following graph, where the letter B is represented by the coordinates (2, 1).
Use the graph to decode the answer to the following riddle:
Q: Where did they put the man who was run over by a steamroller?
A: (4, 2) (4, 3), (3, 2) (5, 3) (4, 4) (1, 4) (4, 2) (5, 4) (1, 1) (2, 3), (4, 2) (4, 3), (3, 5) (1, 1) (3, 4) (4, 1) (4, 4) (4, 4) (4, 2) (4, 5), (4, 4) (5, 1) (2, 5) (5, 1) (4, 3), (5, 1) (4, 2) (2, 2) (3, 2) (5, 4), (1, 1) (4, 3) (4, 1), (4, 3) (4, 2) (4, 3) (5, 1)
The commas outside the brackets in the above indicate separate words. Do not use the commas in your answer; use spaces between your decoded words.

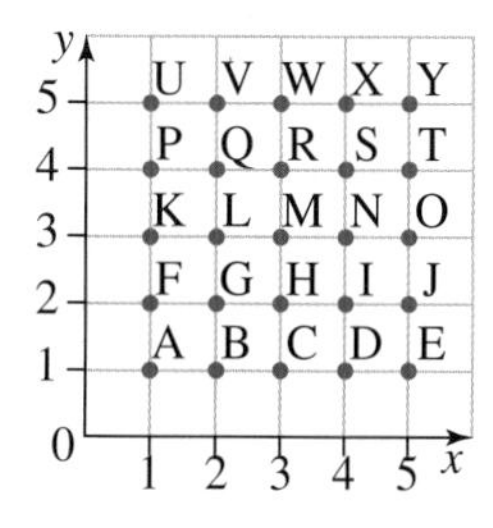


14. Rule a Cartesian plane with both axes extending from 0 to 10 units. Plot the following points and join them in the order given to make a geometric figure. Name each shape.

  a. (2, 2) – (5, 2) – (2, 6) – (2, 2)
  b. (4, 4) – (8, 4) – (6, 8) – (4, 4)
  c. (1, 1) – (10, 1) – (8, 9) – (2, 9) – (1, 1)
  d. (0, 0) – (8, 0) – (10, 10) – (2, 10) – (0, 0)

15. Here is an exercise that may require extra care and concentration. On graph paper or in your exercise book rule a pair of Cartesian axes. The $x$-axis must go from 0 to 26 and the $y$-axis from 0 to 24. Plot the following points and join them in the order given.
(0, 15) – (4, 17) – (9, 22) – (10, 21) – (12, 24) – (16, 22) – (15, 21) – (18, 19) – (20, 24) – (22, 18) – (26, 12)

$$-(26, 10) - (23, 4) - (20, 3) - (18, 4) - (14, 7) - (11, 7) - (4, 6) - (2, 7) - \left(2\frac{1}{2}, 8\right) - (0, 15)$$

Complete the picture by joining (19, 2) – (21, 2) – (20, 0) – (19, 2).

16. Consider a rectangle formed by connecting the points (3, 2), (9, 2), (9, 5) and (3, 5) on a Cartesian plane. Calculate:

  a. the area of the rectangle
  b. the perimeter of the rectangle.

17. Consider the following set of points:
A(2, 5), B(−4, −12), C(3, −7), D(0, −2), E(−10, 0), F(0, 0), G(−8, 15), H(−9, −24), I(18, −18), J(24, 0).

Identify which of the following statements are true.

  a. Points A and J are in the first quadrant.
  b. Points B and H are in the third quadrant.
  c. Only point I is in the fourth quadrant.
  d. Only one point is in the second quadrant.
  e. Point F is at the origin.
  f. Point J is not on the same axis as point E.
  g. Point D is two units to the left of point F.

18. Consider the following triangle ABC.

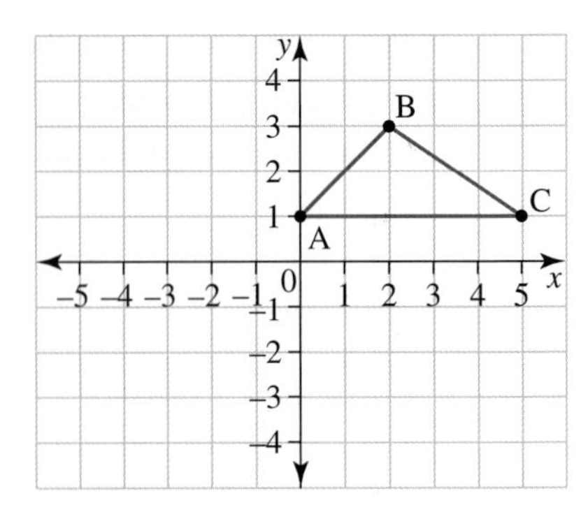


a. State the coordinates of the vertices of the triangle ABC.
b. Calculate the area of the triangle.
c. Reflect the triangle in the $x$-axis. (You need to copy it into your workbook first.) What are the new coordinates of the vertices?
d. Now reflect the triangle you have obtained in part **c** in the $y$-axis, and state the new coordinates of the vertices.

## Reasoning

19. Consider the square ABCD shown.

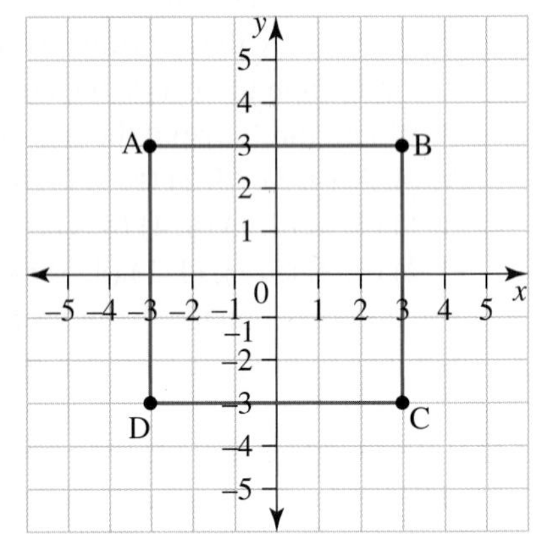


a. State the coordinates of the vertices of the square ABCD.
b. Calculate the area of the square ABCD.
c. Extend sides AB and CD three squares to the right and three squares to the left.
d. Calculate the area of the new shape.
e. Extend sides AD and BC three squares up and three squares down.
f. Calculate the area of the new shape.
g. Compare the three areas calculated in parts b, d and f. Explain the changes of area in relation to the change in side length.

20. a. Explain what would happen to a shape in quadrant 3 if it was reflected in:
   i. the $x$-axis
   ii. the $y$-axis.
   b. Explain what would happen to a shape in quadrant 4 if it was reflected in:
   i. the $x$-axis
   ii. the $y$-axis.
   c. Write a rule about reflections:
   i. in the $x$-axis
   ii. in the $y$-axis.

21. a. Determine the coordinate of the point that lies in the middle of the following pairs of points:
   i. (2, 6) and (4, 10)
   ii. (9, 3) and (3, 1)
   iii. (−7, 4) and (−1, −6)
   iv. (−2, − 3) and (8, −7)
   b. Write a rule that could be used to find the middle of the points $(a, b)$ and $(m, n)$.

22. $(x, y)$ is a point that lies somewhere on the Cartesian plane. State which quadrant/s the point could be located in, given the following conditions:
   a. $x < 0$
   b. $y > 0$
   c. $x < 0$ and $y < 0$
   d. $x > 0$ and $y < 0$

## Problem solving

23. A line AB passes through points A(−5, 6) and B(3, 6).
   a. Draw line AB on a number plane.
   b. Determine the length of AB.
   c. Determine the coordinates of the middle point of the line interval AB.
   d. Draw another horizontal line and determine the coordinates of the middle point.
   e. Is there a formula to calculate the $x$-coordinate of the middle point of a horizontal line?

24. A line CD passes through points C(1, 7) and D(1, −5).
   a. Draw line CD on a number plane.
   b. Determine the length of CD.
   c. Determine the coordinates of the middle point of the line interval CD.
   d. Draw another vertical line and determine the coordinates of the middle point.
   e. Explain whether there is a formula to calculate the $y$-coordinate of the middle point of a vertical line.
25. Consider the die shown. If we consider the centre dot to be the origin, give a possible set of coordinates for the other four dots.

26. Consider the following set of points:
   (0, 3), (2, 7), (4, 11), (6, 15), (8, 19)
   a. Describe the pattern that connects these points to one another.
   b. Write the next three points in the sequence.
   c. Write a rule that connects the $x$-coordinate to the $y$-coordinate for each point in this sequence.
27. Consider a circle with a radius of 5 units that has its centre as the origin.
   a. Identify the coordinates of the four points where the circle cuts each axis.
   b. By drawing this circle on the Cartesian plane, explain whether you can identify any other points with integer coordinates that lie on this circle.

# LESSON
# 11.3 Linear number patterns

**LEARNING INTENTION**

At the end of this lesson you should be able to:
- construct a number pattern from a rule
- identify a rule from a number pattern
- identify a number pattern that can be used to describe a geometric sequence
- express a number pattern algebraically.

## 11.3.1 Describing relationships and patterns

eles-4637

- Mathematicians aim to describe relationships in the world around us. The relationships and rules they find can often be described using number patterns.
- Each number in a number pattern is called a **term**.
- Linear relationships are number patterns that increase or decrease by the same amount from one term to the next. For example, the sequence 1, 4, 7, 10,... increases by 3 from one term to the next, so it is a linear relationship.
- Number patterns can be described by rules. For example, the number pattern above can be described by the rule 'start with 1 and add 3 each time'.

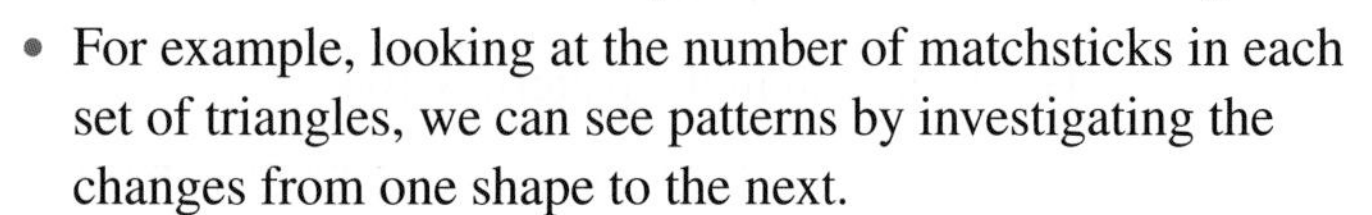

## 11.3.2 Geometric patterns

eles-4638

- We can often find number patterns when we look at geometric shapes.
- For example, looking at the number of matchsticks in each set of triangles, we can see patterns by investigating the changes from one shape to the next.

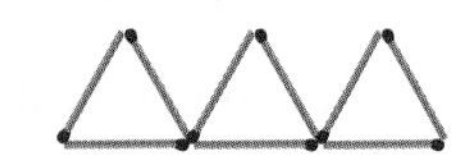

- By using a table of values, we can see a number pattern developing.
- The pattern in the final column is 3, 6, 9, … and we can see that the rule here is 'next number = previous number + 3'.
- We can also look for a relationship between the number of triangles and the number of matchsticks in each shape. If you examine the table, you will see that a relationship can be found. The relationship can be described by the rule: number of matchsticks = 3 × numbers of triangles.

| Number of triangles | Number of matchsticks |
|---|---|
| 1 | 3 |
| 2 | 6 |
| 3 | 9 |
| 4 | 12 |
| 5 | 15 |
| 6 | 18 |

### WORKED EXAMPLE 8 Geometric number patterns

**Consider a set of hexagons constructed according to the following pattern.**

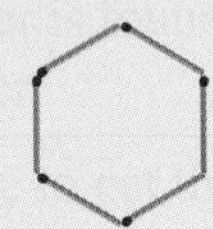

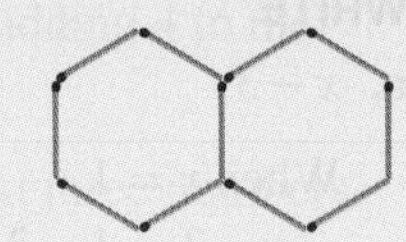

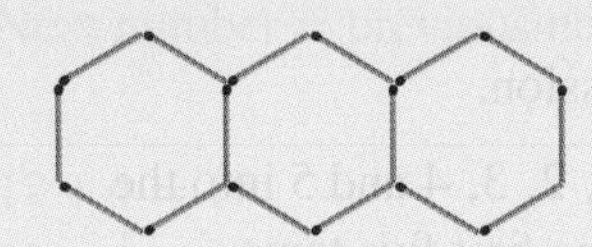

**a. Using matches, pencils or similar objects, construct these figures, then draw the next two figures in the series.**

**b. Draw a table showing the relationship between the number of hexagons in the figure and the number of matches used to construct the figure.**

**c. Devise a rule to describe the number of matches required for each figure in terms of the number of hexagons in the figure.**

**d. Use your rule to determine the number of matches required to make a figure consisting of 15 hexagons.**

| THINK | WRITE |
|---|---|
| a. Construct the given figures with matches, then draw the next two figures. | a. The next two figures are: 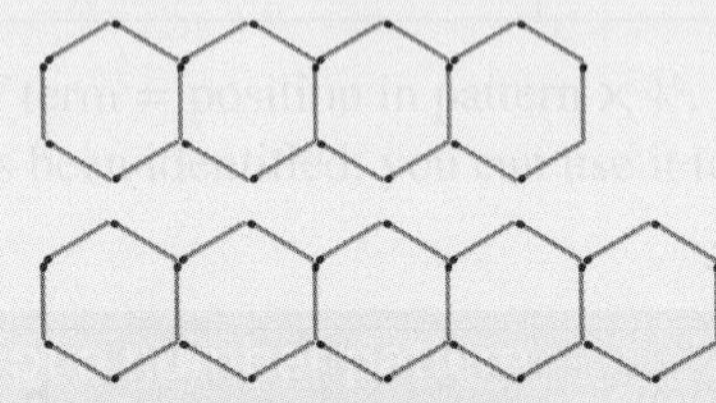 |
| b. Draw a table showing the number of matches needed for each figure in terms of the number of hexagons. Fill it in by looking at the figures. | b. (table below) |
| c. 1. Look at the pattern in the number of matches going from one figure to the next. It is increasing by 5 each time. | c. The number of matches increased by 5 in going from one figure to the next. |
| 2. If we take the number of hexagons and multiply this number by 5, it does not give us the number of matches. However, if we add 1 to this number, it does give us the number of matches in each shape. | Number of matches = number of hexagons × 5 + 1 |

| Number of hexagons | 1 | 2 | 3 | 4 | 5 |
|---|---|---|---|---|---|
| Number of matches | 6 | 11 | 16 | 21 | 26 |

| | |
|---|---|
| **d. 1.** Use the rule to find the number of matches to make a figure with 15 hexagons. | **d.** Number of matches for 15 hexagons $= 15 \times 5 + 1$ $= 76$ |
| **2.** Write the answer. | 76 matches would be required to construct a figure consisting of 15 hexagons. |

## 11.3.3 Using algebra to describe patterns

eles-4639

- Once a pattern has been identified, we can use algebra to describe it.
- If we let $n =$ the number of matches and $h =$ the number of hexagons from the pattern in the previous worked example, the equation $n = 5h + 1$ would describe this pattern.

### WORKED EXAMPLE 9 Using algebra to describe patterns

**Consider the set of shapes constructed according to the following pattern.**

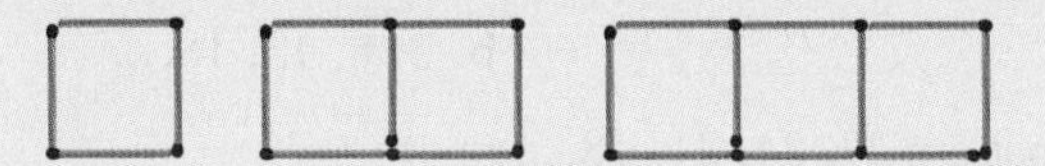

**a. Draw a table of values showing the number of matches needed for each figure in the pattern.**
**b. Express the relationship between the number of matches ($n$) and the number of squares ($s$) in the pattern as an algebraic equation.**

| THINK | WRITE |
|---|---|
| **a.** Look at the figures in the pattern to complete a table of values. | **a.** Number of squares: 1, 2, 3; Number of matches: 4, 7, 10 |
| **b. 1.** Look at the pattern in the number of matches going from one figure to the next. It is increasing by 3 each time. | **b.** The number of matches increased by 3 in going from one figure to the next. |
| **2.** If we take the number of squares and multiply this number by 3, it does not give us the number of matches. However, if we add 1 to this number, it does give us the number of matches in each shape. | Number of matches = number of squares $\times 3 + 1$ |
| **3.** Replace 'number of matches' with '$n$' and 'number of squares' with '$s$' to create your algebraic equation, then simplify by removing the multiplication sign. | $n = s \times 3 + 1$<br>$n = 3s + 1$ |

| Number of squares | 1 | 2 | 3 |
|---|---|---|---|
| Number of matches | 4 | 7 | 10 |

**on** Resources

| | |
|---|---|
| **eWorkbook** | Topic 11 Workbook (worksheets, code puzzle and project) (ewbk-1912) |
| **Interactivities** | Individual pathway interactivity: Identifying patterns (int-4446)<br>Identifying patterns (int-3801)<br>Geometric patterns (int-3802) |

## Exercise 11.3 Linear number patterns

**11.3 Quick quiz** on | **11.3 Exercise**

Individual pathways

| ■ PRACTISE | ■ CONSOLIDATE | ■ MASTER |
|---|---|---|
| 1, 3, 6, 8, 12, 14, 17 | 2, 4, 7, 9, 13, 15, 18, 19 | 5, 10, 11, 16, 20, 21, 22 |

### Fluency

1. **WE4** Using the following rules, write down the first five terms of the number patterns.
   a. Start with 1 and add 4 each time.
   b. Start with 5 and add 9 each time.
2. Using the following rules, write down the first five terms of the number patterns.
   a. Start with 50 and subtract 8 each time.
   b. Start with 64 and subtract 11 each time.
3. Write down the next three terms for the following number patterns.
   a. 2, 4, 6, 8, ...
   b. 3, 8, 13, 18, ...
4. **WE5** Consider the number pattern 27, 24, 21, 18, ... .
   a. Describe the pattern in words.
   b. Write down the next three terms in the pattern.
5. Consider the number pattern 48, 41, 34, 27, ...
   a. Describe the pattern in words.
   b. Write down the next three terms in the pattern.
6. Fill in the missing numbers in the following number patterns.
   a. 3, ___, 9, 12, ___, ___
   b. 8, ___, ___, 14, ___
   c. 4, 8, ___, 32, ___
7. Fill in the missing numbers in the following number patterns.
   a. ___, ___, 13, 15, ___
   b. 66, 77, ___, 99, ___, 121
   c. 100, ___, ___, 85, 80, ___
8. **WE6** Consider the number pattern 5, 15, 25, ...
   a. Determine the rule for the number pattern by relating the value of a term to its position in the number pattern.
   b. Calculate the 15th term in the pattern.
9. Consider the number pattern 44, 41, 38, ...
   a. Determine the rule for the number pattern by relating the value of a term to its position in the number pattern.
   b. Calculate the 15th term in the pattern.
10. Consider the number pattern 102, 108, 114, ...
    a. Determine the rule for the number pattern by relating the value of a term to its position in the number pattern.
    b. Calculate the 35th term in the pattern.
11. Consider thc number pattern 212, 199, 186, ...
    a. Determine the rule for the number pattern by relating the value of a term to its position in the number pattern.
    b. Calculate the 23rd term in the pattern.

### Understanding

12. WE7 The algebraic expression $x + 4$ is used to generate a number pattern, where $x$ represents the position in the pattern.
    a. Generate the first five terms of the number pattern.
    b. Determine the 10th term of the number pattern.
    c. Determine the 100th term of the number pattern.

13. A number pattern is given by the algebraic expression $2x - 2$, where $x$ represents the position in the pattern.
    a. Generate the first five terms of the number pattern.
    b. Generate the 20th term of the number pattern.
    c. Generate the 77th term of the number pattern.

### Reasoning

14. Explain why it is more useful to describe a number pattern in terms of relating the value of a term to its position in the pattern than to describe a number pattern by the rule between terms.

15. Are all patterns that you can describe by written descriptions also describable using algebraic symbols? Explain your answer.

16. Explain why it is useful to describe the rule for a pattern by drawing a table and looking for the connection between the top row and the bottom row of the table.

### Problem solving

17. WE8 a. Construct the following figures using matches, then draw the next two figures in the pattern.

  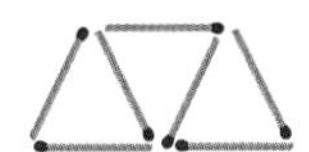

    b. Draw a table showing the relationship between the number of shapes in each figure and the number of matchsticks used to construct it.
    c. Devise a rule that describes the pattern relating the number of shapes in each figure and the number of matchsticks required to construct it.
    d. Use your rule to calculate the number of matchsticks required to construct a figure made up of 7 such shapes.

18. a. Construct the following figures using matches, then draw the next two figures in the pattern.

 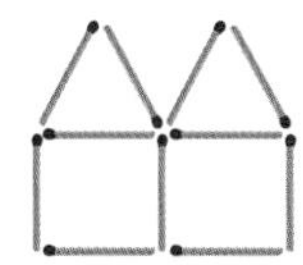 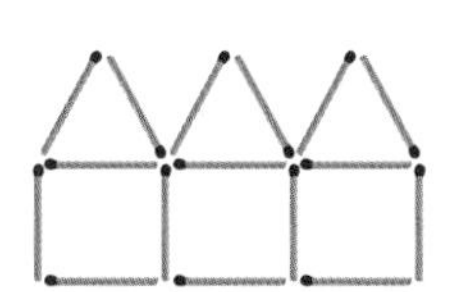

    b. Draw a table showing the relationship between the number of shapes in each figure and the number of matchsticks used to construct it.
    c. Devise a rule that describes the pattern relating the number of shapes in each figure and the number of matchsticks required to construct it.
    d. Use your rule to calculate the number of matchsticks required to construct a figure made up of 7 such shapes.

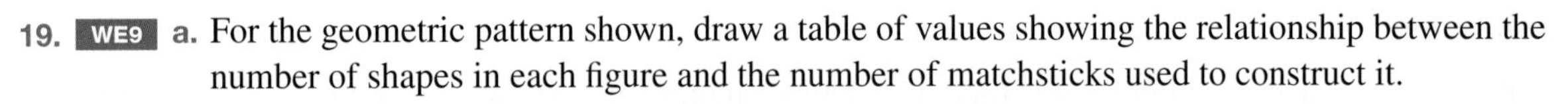

19. WE9 a. For the geometric pattern shown, draw a table of values showing the relationship between the number of shapes in each figure and the number of matchsticks used to construct it.

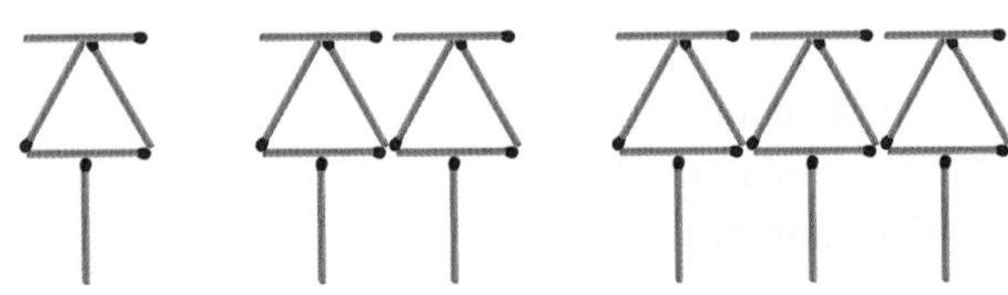

b. Express the relationship between the number of matches ($m$) and the number of shapes ($n$) in the pattern as an algebraic equation.

20. a. Construct the following figures using matches, then draw the next two figures in the pattern.

b. Draw a table showing the relationship between the number of shapes in each figure and the number of matchsticks used to construct it.

c. Devise a rule that describes the pattern relating the number of shapes in each figure and the number of matchsticks required to construct it.

d. Use your rule to calculate the number of matchsticks required to construct a figure made up of 7 such shapes.

21. a. For the geometric pattern shown, draw a table of values showing the relationship between the number of shapes in each figure and the number of matchsticks used to construct it.

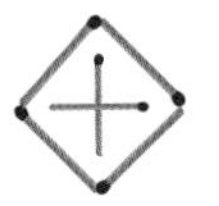
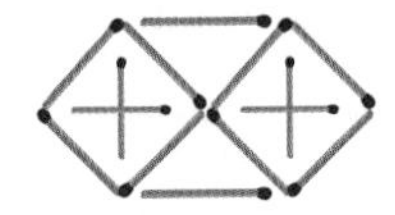
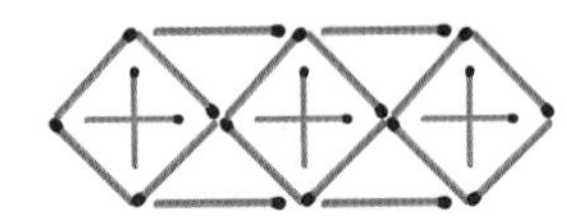

b. Express the relationship between the number of matches ($m$) and the number of shapes ($n$) in the pattern as an algebraic equation.

22. Create a matchstick pattern that can be represented by the rule $m = 5n + 3$.

# LESSON
# 11.4 Plotting simple linear relationships

## LEARNING INTENTION

At the end of this lesson you should be able to:
- plot a simple linear relationship from a table of values
- identify whether a series of points forms a linear relationship.

## 11.4.1 Plotting simple linear relationships

eles-4640

- When a set of points is plotted on the Cartesian plane, a pattern may be formed. If a pattern forms a straight line, we call it a *linear pattern*. If the pattern forms a curve, we call it a *non-linear pattern*.
- The coordinates of the points that form a pattern can be presented in a table of values or as a set of points (as shown in the following table).

- If shown in a table, the coordinates of each point should be read 'in columns'; the top number in each column gives the $x$-coordinate and the bottom number gives the corresponding $y$-coordinate of the point.

| $x$ | 0 | 1 | 2 | 3 |
|---|---|---|---|---|
| $y$ | 8 | 6 | 4 | 2 |

(0, 8), (1, 6), (2, 4), (3, 2)

- The Cartesian coordinates of the points are ordered pairs. That is, the first number always represents the $x$-coordinate of a point and the second always represents the $y$-coordinate. A set of ordered pairs forms a relationship between $x$ and $y$.
- If the points form a linear pattern when plotted, we say that the relation between $x$ and $y$ is linear.

## WORKED EXAMPLE 10 Plotting a set of points to observe a pattern

**Plot the following set of points on the Cartesian plane, and comment on any pattern formed.**
**$(1,3),(2,4),(3,5),(4,6),(5,7)$**

**THINK**

1. Look at the coordinates of the points in the set: the $x$-values range between 1 and 5, while the $y$-values range between 3 and 7. Draw a set of axes, ensuring they are long enough to fit all the values.
2. Plot the points.
3. Comment on the pattern that the points form.

**WRITE**

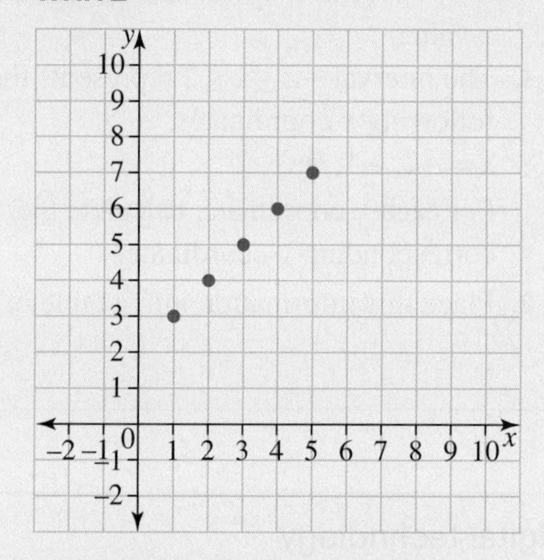


The points form a linear pattern.

## WORKED EXAMPLE 11 Plotting points presented in a table of values

**Plot the following points on a Cartesian plane and decide whether the relationship is linear.**

| $x$ | 0 | 1 | 2 | 3 | 4 |
|---|---|---|---|---|---|
| $y$ | 8 | 7 | 6 | 5 | 4 |

**THINK**

1. The $x$-values range from 0 to 4; the $y$-values range from 4 to 8. Draw a set of axes that will include all these values. (The axes need not be the same length. In this case you can extend the y-axis a bit more than the $x$-axis, as the $y$-values are higher.)
2. The entries in the table can be rewritten as a set of points. The top number in each column gives the $x$-coordinate of the point and the bottom number gives the corresponding $y$-coordinate. Thus the first column gives the point (0, 8), the second (1, 7), and so on. Plot all points.

**WRITE**

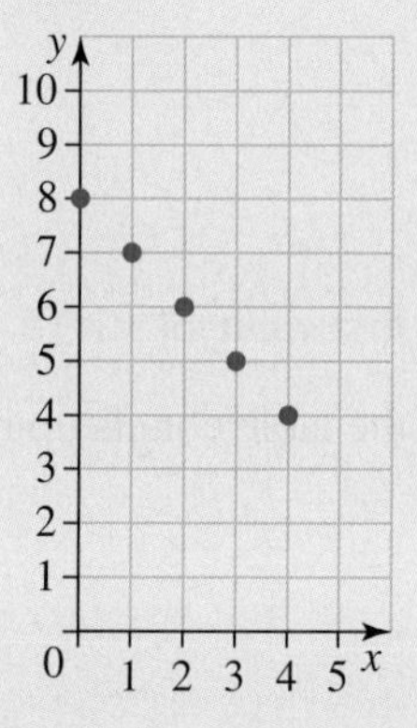

3. Consider the pattern formed by the points. It is a straight line. Draw a suitable conclusion about the relationship between $x$ and $y$. — The relationship between $x$ and $y$ is linear.

## WORKED EXAMPLE 12 Constructing a table of values from a rule

**Consider the relationship between $x$ and $y$ given by the rule $y = 3x - 4$.**

**a. Describe the set of points produced by this relationship between $x$ and $y$.**

**b. Construct a table of values for the integers $-2 \le x \le 2$**

**THINK**

a. When a rule is written in this form, it means that each $y$-coordinate must be found from a corresponding $x$-coordinate. In this case the $y$-value is four less than three times the $x$-value.

**WRITE**

a. The rule describes a set of points such that the $y$-coordinate of each point is four less than three times the value of the $x$-coordinate.

**THINK**

b. 1. The interval $-2 \le x \le 2$ represents the following $x$-coordinates:
$x = -2, -1, 0, 1, 2$
For each $x$-coordinate, calculate the corresponding $y$-coordinate.

**WRITE**

b. $y = 3 \times -2 - 4 = -10$
$y = 3 \times -1 - 4 = -7$
$y = 3 \times 0 - 4 = -4$
$y = 3 \times 1 - 4 = -1$
$y = 3 \times 2 - 4 = 2$

2. Place this information into a table of values.

| $x$ | –2 | –1 | 0 | 1 | 2 |
|---|---|---|---|---|---|
| $y$ | –10 | –7 | –4 | –1 | 2 |

### Digital technology

Scientific calculators have the ability to create a table of values.

Start by typing the rule for the relationship between $x$ and $y$ by pressing the table button then the $x$ button for the equation itself.

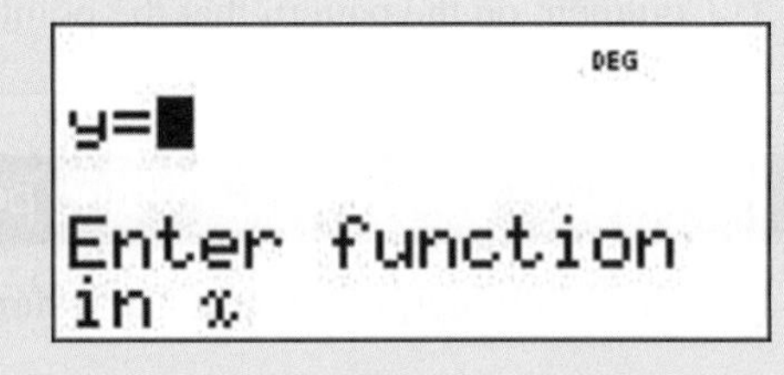

Once the rule is typed in, press Enter. You will be able to change some settings.

DEG
y=2x−3

**Start = 0** means the first value of $x$ is 0.

**Step = 1** means that the table counts up in single units (ones).

DEG
Start=
Step=1
Auto Ask-x
OK

Selecting **Auto** will generate the table.

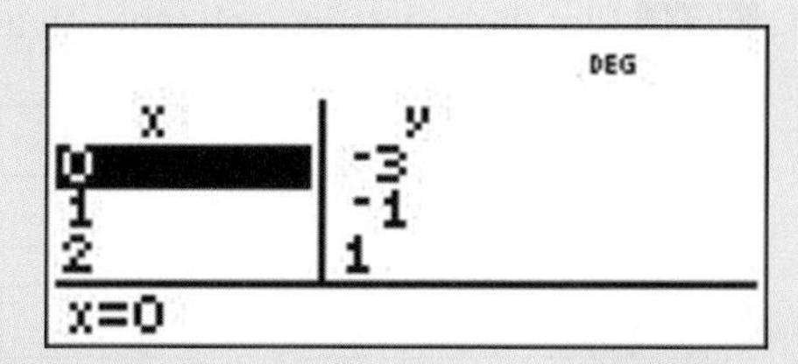


Selecting **Ask-*x*** will let you choose your own values of *x* and allow you to generate your own unique table.

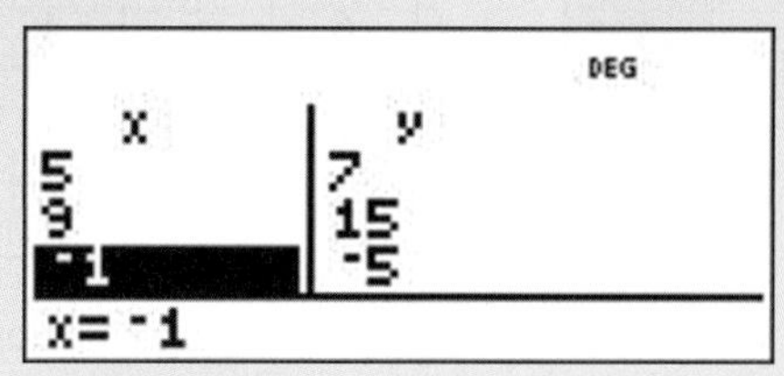


## Resources

 **eWorkbook** Topic 11 Workbook (worksheets, code puzzle and project) (ewbk-1912)

 **Interactivities** Individual pathway interactivity: Plotting simple linear relationships (int-4385)

Plotting simple linear relationships (int-4138)

# Exercise 11.4 Plotting simple linear relationships

learnon

11.4 Quick quiz  | 11.4 Exercise

### Individual pathways

| ■ PRACTISE | ■ CONSOLIDATE | ■ MASTER |
|---|---|---|
| 1, 4, 7, 10, 12, 15 | 2, 5, 8, 9, 13, 16 | 3, 6, 11, 14, 17, 18 |

### Fluency

1. WE10 Plot the following sets of points on the Cartesian plane, and comment on any pattern formed.
   a. (0, 1), (1, 2), (2, 3), (3, 4), (4, 5)
   b. (0, 3), (1, 4), (2, 5), (3, 6), (4, 7)
   c. (2, 2), (3, 3), (4, 4), (5, 5), (6, 6)
   d. (0, 2), (1, 3), (2, 1), (3, 0), (4, 5)
   e. (1, 10), (2, 8), (3, 6), (4, 4), (5, 2)
   f. (2, 3), (3, 3), (4, 3), (5, 3), (6, 3)
2. Plot the following sets of points on the Cartesian plane. (Note that some coordinates are negative, so you will need to use all four quadrants.) Comment on any pattern formed.
   a. (−2, 0), (−1, 2), (0, 4), (1, 6), (2, 8)
   b. (−3, −3), (−2, 0), (−1, 3), (0, 6), (1, 9)
   c. (−2, 7), (−1, 6), (0, 4), (1, 3), (2, 1)
   d. (−6, 8), (−5, 6), (−4, 4), (−3, 2), (−2, 0)
3. Plot the following sets of points on the Cartesian plane. Comment on any pattern formed.
   a. (−3, 8), (−1, 5), (1, 2), (3, −1), (5, −4)
   b. (−5, 0), (−5, 5), (−5, 3), (−5, −2), (−5, −4)
   c. (−4, 4), (−1, 3), (2, 1), (5, −3), (8, −11)
   d. (7, 8), (1, −2), (−2, −7), (4, 3), (10, 13)

4. WE11 Plot the points in each table on a Cartesian plane, and decide whether or not the relationship is linear.

a.

| $x$ | 0 | 1 | 2 | 3 | 4 |
|---|---|---|---|---|---|
| $y$ | 3 | 4 | 5 | 6 | 7 |

b.

| $x$ | 0 | 1 | 2 | 3 | 4 |
|---|---|---|---|---|---|
| $y$ | 0 | 1 | 4 | 9 | 16 |

c.

| $x$ | 1 | 2 | 3 | 4 | 5 |
|---|---|---|---|---|---|
| $y$ | 7 | 6 | 5 | 4 | 3 |

d.

| $x$ | 1 | 2 | 3 | 4 | 6 |
|---|---|---|---|---|---|
| $y$ | 12 | 6 | 4 | 3 | 2 |

e.

| $x$ | 0 | 1 | 2 | 3 | 4 |
|---|---|---|---|---|---|
| $y$ | 0 | 3 | 6 | 9 | 12 |

f.

| $x$ | 2 | 3 | 4 | 5 | 6 |
|---|---|---|---|---|---|
| $y$ | 0 | −2 | −4 | −6 | −8 |

5. Plot the points in each table on a Cartesian plane and decide whether the relationship is linear or not.

a.

| $x$ | 0 | 1 | 2 | 3 | 4 |
|---|---|---|---|---|---|
| $y$ | 1 | 4 | 7 | 10 | 13 |

b.

| $x$ | −2 | −1 | 0 | 1 | 2 |
|---|---|---|---|---|---|
| $y$ | 10 | 5 | 0 | −5 | −10 |

c.

| $x$ | 0 | 1 | 2 | 3 | 4 |
|---|---|---|---|---|---|
| $y$ | −1 | 0 | 2 | 6 | 14 |

d.

| $x$ | −3 | −2 | −1 | 0 | 1 |
|---|---|---|---|---|---|
| $y$ | 10 | −8 | 6 | −4 | 2 |

6. Plot the points in each table on a Cartesian plane, and decide whether or not the relationship is linear.

a.

| $x$ | 0 | 1 | 2 | 3 | 4 |
|---|---|---|---|---|---|
| $y$ | 8 | 5 | 2 | −1 | −4 |

b.

| $x$ | −2 | −1 | 0 | 1 | 2 |
|---|---|---|---|---|---|
| $y$ | 5 | 3 | 1 | −1 | −3 |

c.

| $x$ | 0 | 1 | 2 | 3 | 4 |
|---|---|---|---|---|---|
| $y$ | −2 | 4 | −6 | 8 | −10 |

d.

| $x$ | −2 | −1 | 0 | 1 | 2 |
|---|---|---|---|---|---|
| $y$ | 8 | 7 | 4 | −1 | −8 |

### Understanding

7. a. Plot the following points on the set of axes: (0, 10), (1, 8), (2, 6), (3, 4), (4, 2), (5, 0).
   b. Describe the pattern formed by the points.
   c. Extend the pattern by plotting the next two points.
   d. State the coordinates of the two points you have plotted.

8. The points in the following table form a linear pattern. The y-coordinate of the middle point is missing.

| $x$ | 1 | 2 | 3 | 4 | 5 |
|---|---|---|---|---|---|
| $y$ | 0 | 3 |  | 9 | 12 |

   a. Plot the four points (whose $x$- and $y$-coordinates are both known) on the Cartesian plane.
   b. Use your graph to predict the missing $y$-coordinate of the middle point.
   c. Add the middle point to your graph and check whether it fits the pattern.

9. Which of the following sets of points, when plotted, will form a linear pattern?
   a. (0, 3), (1, 5), (2, 7), (3, 9), (4, 11)
   b. (−1, 8), (−2, 7), (−3, 6), (−4, 5), (−5, 4)
   c. (1, 6), (2, 5), (3, 7), (4, 8), (5, 9)
   d. (2, 3), (3, 5), (4, 7), (5, 9), (6, 11)
   e. (0, 2), (1, 4), (2, 4), (3, 5), (4, 6)
   f. (10, 7), (9, 4), (8, 3), (7, 1), (6, 0)

10. Consider the linear pattern shown.

   a. If the pattern is continued to the right, identify the coordinates of the next point.
   b. If the pattern is extended to the left, identify the coordinates of the previous point.

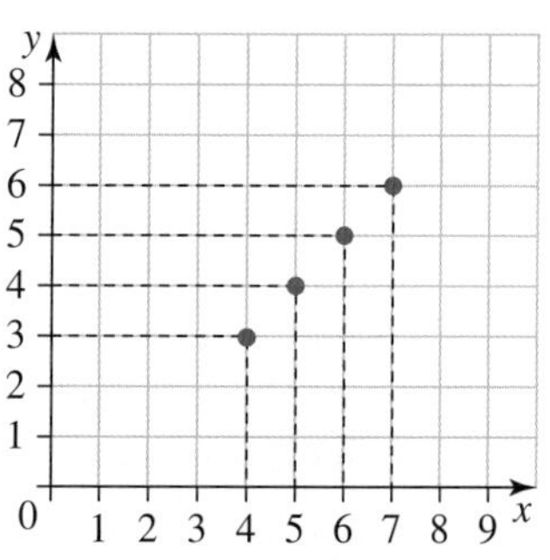


11. Rachel planted a cherry tomato plant in her vegetable patch. She measured the height of her plant every week for four consecutive weeks, obtaining the following data: 20 mm, 24 mm, 28 mm, 32 mm. Rachel forgot to measure her plant in week 5, but the following week she found the height of the cherry tomato to be 40 mm.

   a. Fill in the table of values below.

| Week | | | | | | |
|---|---|---|---|---|---|---|
| Height (mm) | | | | | | |

   b. Plot the heights of the cherry tomato plant on a set of axes.
   c. Explain whether the relationship between the number of the week and the height of the plant is linear.
   d. Assuming the pattern was the same for the first 6 weeks, determine the height of the tomato plant in week 5.
   e. If the pattern continues, predict the height of the cherry tomato plant in week 8.

## Reasoning

12. **WE12** The relationship between the two variables $x$ and $y$ is described by the rule $y = x + 3$.

   a. Complete the following statement: 'The rule describes a set of points such that the $y$-coordinate of each point is …'
   b. Construct a table of values for whole numbers $-2 \le x \le 2$ using the given rule.
   c. Plot the points on a set of axes.
   d. Is the relationship linear? Justify your answer.

13. The relationship between the two variables $x$ and $y$ is described by the rule $y = 2x - 3$.

   a. Complete the following statement: 'The rule describes a set of points such that the $y$-coordinate of each point is three less than …'
   b. Construct a table of values for whole numbers $-2 \le x \le 2$ using the given rule.
   c. Plot the points from the table on a set of axes.
   d. Is the relationship linear? Justify your answer.
   e. The $x$-values of the relationship form the *domain* of that relationship, while the $y$-values form its *range*. The domain of the given relationship is $-2 \le x \le 2$. Identify the range.

14. a. On the same set of axes, plot the following linear relationships: $y = x$, $y = 3x + 1$ and $y = 5x - 2$.
   b. On the same set of axes, plot the following linear relationships: $y = -x$, $y = -3x + 1$ and $y = -5x - 2$.
   c. Explain the differences and similarities between the two sets of graphs.

## Problem solving

**15. a.** Use the following information to complete the table of coordinates.

Point A is two units to the left of the origin and five units below the origin.
Point B is one unit to the left of the origin and two units above point A.
Point C is one unit below the origin on the vertical axis.
Point D is one unit to the right of C and two units above point C.
Point E is two units to the right of the $y$-axis and three points above the $x$-axis.

| $x$ | | | | | |
|---|---|---|---|---|---|
| $y$ | | | | | |
| | A | B | C | D | E |

**b.** Plot these points on a Cartesian plane.
**c.** Determine the value of the horizontal step between two adjacent points.
**d.** Is the relationship linear? Explain your answer.

**16.** The table of values shown displays the relationship between time, in minutes, and parking fee, in dollars.

| **Time, minutes** | 30 | 60 | 90 | 120 | |
|---|---|---|---|---|---|
| **Parking fee, $** | 1.20 | 2.40 | | 4.80 | 6.00 |

**a.** Write a rule for this relationship.
**b.** Use the rule found in part **a** to calculate the missing values.

**17.** The following table of values gives the relationship between the weight of coffee beans (in kg) and the cost per kg from a wholesale distributor.

| **Weight (kg)** | 1 | 5 | | 20 | 25 |
|---|---|---|---|---|---|
| **Cost per kg ($)** | 49.00 | 46.00 | 42.25 | 34.75 | |

**a.** Write a rule for this relationship.
**b.** Use the rule found in part **a** to calculate the missing values.
**c.** Joel orders a 22 kg bag of beans each week for his coffee shop. Calculate how much he has to pay the wholesaler for his coffee.
**d.** The biggest bag that the wholesaler produces is a 30 kg bag of coffee beans. A chain coffee store orders a number of these bags each week. If the total cost of their order was $5722.50, determine the weight of coffee beans that was ordered.

**18.** Each row in my classroom has the same number of desks, and the rows are straight. My desk is fourth from the front and fourth from the back. It has one desk to its left and two desks to its right. Determine how many desks are in the classroom.

# LESSON
# 11.5 Interpreting graphs

## LEARNING INTENTION

At the end of this lesson you should be able to:
- interpret information present on a Cartesian plane
- identify the domain of a set of points or line graph
- identify the range of a set of points or line graph
- read a travel graph and calculate average speed.

## 11.5.1 Interpreting graphs

eles-4641

- As graphs are found in many areas of daily life, it is important to be able to interpret graphs by considering the following points:
  - Look at the overall picture. Where are the points situated? Does the line go up from left to right (an increasing trend) or go down from left to right (a decreasing trend)?
  - Look at what the horizontal and vertical axes represent, the values shown on the scale (if any) and their units of measurement.
  - Note that the values increase as you move up the vertical axis and move to the right along the horizontal axis.
- A scatter plot graph shows individual points of information only. A line graph can show how the information changes.

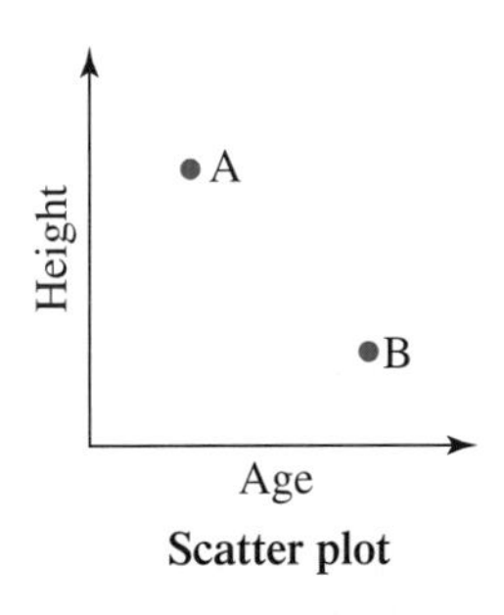

Scatter plot

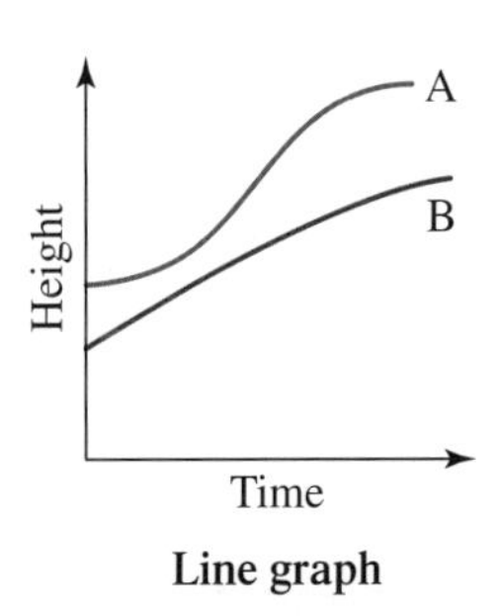

Line graph

### WORKED EXAMPLE 13 Interpreting a scatter plot

**Use the graph shown to answer the following questions about Brendan and Kelly.**

**a. Identify who is taller, Brendan or Kelly.**

**b. Identify who is younger.**

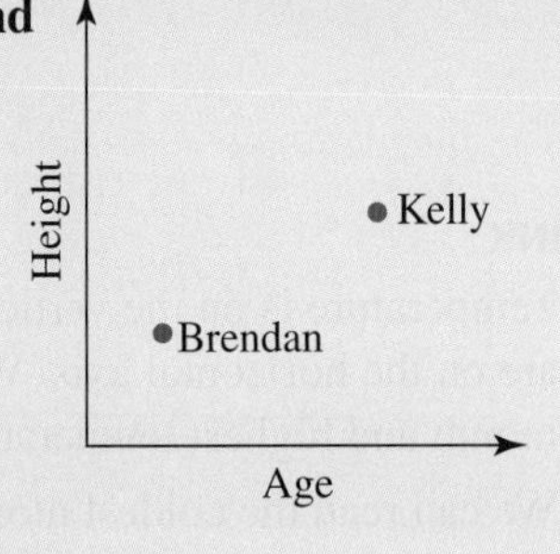


| THINK | WRITE |
|---|---|
| a. Height is on the vertical axis. Kelly is above Brendan, so Kelly is taller. | a. Kelly is taller. |
| b. The horizontal axis shows age. Brendan is to the left of Kelly, so Brendan is younger. | b. Brendan is younger. |

## WORKED EXAMPLE 14 Interpreting a line graph

**This graph shows the distance Conchita is from home when she is out hiking. Describe what each straight line section of the graph shows.**

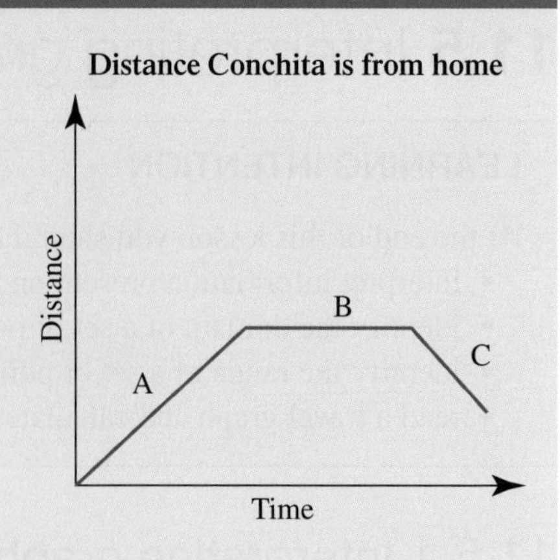


| THINK | WRITE |
|---|---|
| **1.** As you move along section A, the distance is increasing and time is increasing, so Conchita is moving further away from home. | Section A : Conchita is walking away from home. |
| **2.** As you move along section B, the distance is staying the same but time is still increasing. | Section B : Conchita could be resting. |
| **3.** As you move along section C, the distance from home is decreasing and time is increasing. | Section C : Conchita is walking towards home. |

## WORKED EXAMPLE 15 Comparing line graphs

**This graph shows the average maximum and minimum monthly temperatures for Sydney. Use the graph to answer the following questions.**

**a. Identify which month was the hottest and its average maximum temperature.**

**b. Identify which month was the coldest and its average minimum temperature.**

**c. Identify which months had an average minimum temperature of 12 °C.**

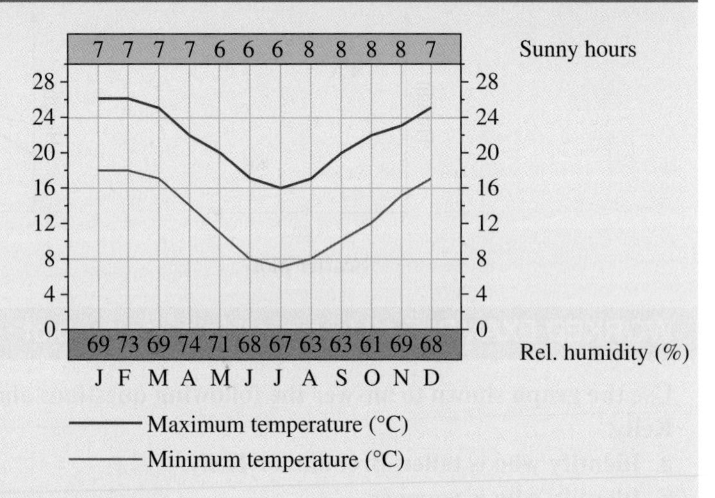


| THINK | WRITE |
|---|---|
| **a.** Temperature is on the vertical axis and months are on the horizontal axis. We can read the hottest month and highest temperature from the graph. | **a.** Hottest month is January and the average maximum temperature is 26 °C. |
| **b.** We can read the coldest month and lowest temperature from the graph. | **b.** Coldest month is July and the average minimum temperature is 7 °C. |
| **c.** We can read the average minimum temperature of 12 °C from the vertical axis. | **c.** Average minimum temperature of 12 °C was in May and October. |

## 11.5.2 Domain and range

eles-4642

- For any relationship between $x$ and $y$, the **domain** is the set of all $x$-coordinates and the **range** is the set of all $y$-coordinates.
- For a collection of points, domain and range can be represented as a set of numbers.
- For a line segment, the domain and range can be represented as an interval between two values.
- Domain and range are always given in ascending order (lowest value to highest value).

**Set of points**

$(1, 2), (3, 6), (-4, -8), (0, 0), (-5, -10)$

Domain: $\{-5, -4, 0, 1, 3\}$
Range: $\{-10, -8, 0, 2, 6\}$

**Table of values**

| $x$ | −2 | −1 | 0 | 1 | 2 | 3 |
|---|---|---|---|---|---|---|
| $y$ | −5 | −2 | 1 | 4 | 7 | 10 |

Domain: $\{-2, -1, 0, 1, 2, 3\}$
Range: $\{-5, -2, 1, 4, 7, 10\}$

### WORKED EXAMPLE 16 Finding domain and range

**State the domain and range for the following:**

**a. $(1, 2), (5, -6), (7, 4), (-3, 3), (0, -5), (2, 4)$**

b.

| $x$ | −3 | −1 | 0 | 2 | 4 |
|---|---|---|---|---|---|
| $y$ | 6 | 2 | 0 | −4 | −8 |

c.

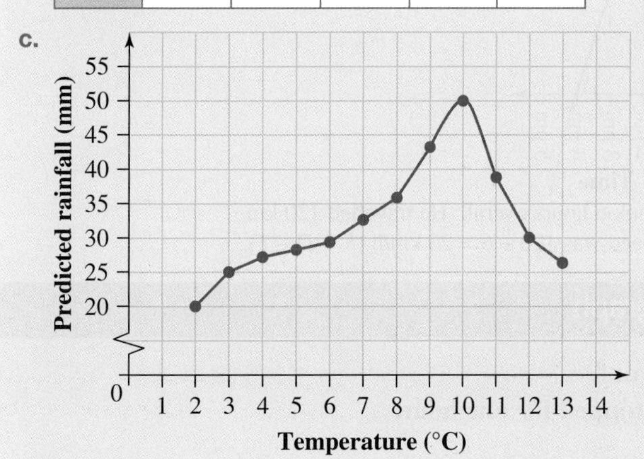


**THINK**

a. The domain is a list of all $x$-coordinates in the set of points and the range is a list of all $y$-coordinates in the set. The domain and range must be listed from lowest values to highest.

The range has one less value as the number 4 appears twice as a $y$-coordinate and we only need to list it once when stating the range.

**WRITE**

a. Domain = $\{-3, 0, 1, 2, 5, 7\}$
Range = $\{-6, -5, 2, 3, 4\}$

**THINK**

b. We can get the domain and range directly from the table. Remember to order the range in ascending order

**WRITE**

b. Domain = $\{-3, 1, 0, 2, 4\}$
Range = $\{-8, -4, 0, 2, 6\}$

**THINK**

c. Since this is a continuous line, the domain and range include every possible value between the lowest and highest value. This means we need to use an interval to represent domain and range.

**WRITE**

c. Domain = $2 \le x \le 13$
Range = $20 \le y \le 50$

## 11.5.3 Travel graphs

eles-4643

- A **travel graph** is a special type of line graph.
- A travel graph is used to represent a journey. Time is shown on the horizontal axis and distance is shown on the vertical axis. The following graph is an example of a travel graph, including all of the information this type of graph can show.
- This graph shows information about Enrico's trip on his scooter. Some of the information we can find out from the graph is already shown.

**Calculating average speed**

**The average speed of a journey can be calculated using the formula:**

$$\text{average speed} = \frac{\text{total distance travelled}}{\text{total time taken}}$$

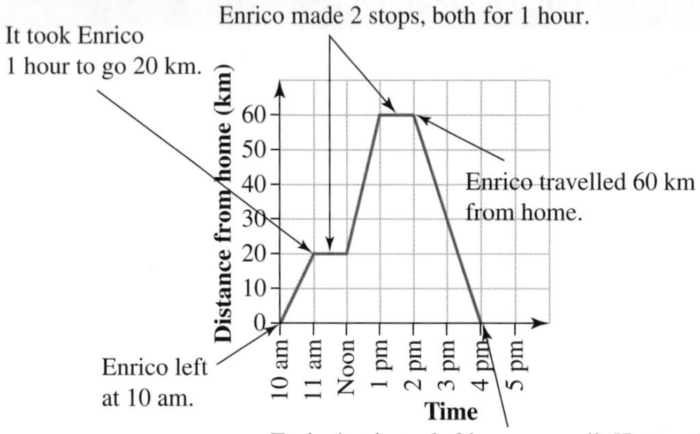


Enrico's trip took 6 hours overall. He travelled 120 km.
His average speed was $120 \div 6 = 20$ km/h $[S = D \div T]$.

### WORKED EXAMPLE 17 Interpreting a travel graph

**The following travel graph shows Todd's afternoon out.**

**a. Identify how far Todd was from home when he stopped for one hour.**

**b. Calculate how far Todd travelled overall.**

**c. Calculate Todd's average speed for the return journey.**

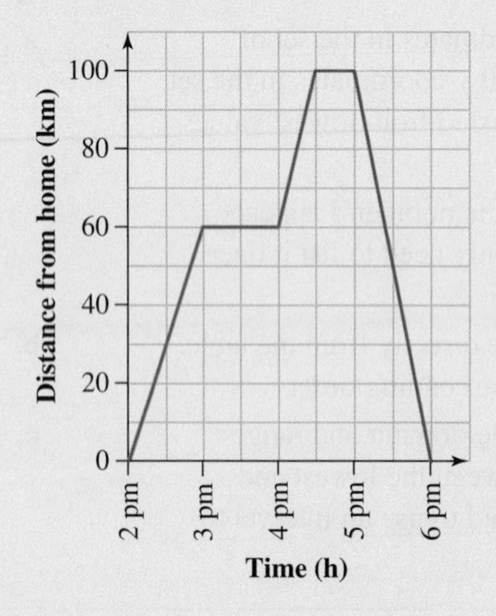

| THINK | WRITE |
|---|---|
| **a.** The line on the travel graph is horizontal for one hour from 3 pm indicating that no distance was covered in that time. On the $y$-axis the distance from home at 3 pm was 60 km. | **a.** Todd was 60 km from home when he stopped for one hour. |
| **b.** The outward journey and return journey are both 100 km. | **b.** Overall, Todd travelled $100 + 100\text{ km} = 200\text{ km}$ |
| **c.** Todd's 200 km trip took from 2 pm to 6 pm, which is 4 hours overall. Divide the distance travelled by the time taken to calculate Todd's average speed. | **c.** $\text{Speed} = \frac{\text{Distance}}{\text{Time}} = \frac{200}{4} = 50$<br>Todd's average speed was 50 km/h. |

## Resources

**eWorkbook** Topic 11 Workbook (worksheets, code puzzle and project) (ewbk-1912)

**Interactivities** Individual pathway interactivity: Interpreting graphs (int-4386)

Interpreting graphs (int-4139)

## Exercise 11.5 Interpreting graphs

learnon

11.5 Quick quiz on | 11.5 Exercise

### Individual pathways

| ■ PRACTISE | ■ CONSOLIDATE | ■ MASTER |
|---|---|---|
| 1, 2, 5, 7, 10, 14, 17, 22, 25, 28 | 3, 6, 8, 11, 13, 15, 18, 20, 23, 27, 29 | 4, 9, 12, 16, 19, 21, 24, 26, 30 |

### Fluency

1. WE13 Use the following graph to answer the questions about Lucas and Selina.
   a. Identify who is taller.
   b. Identify who is younger.

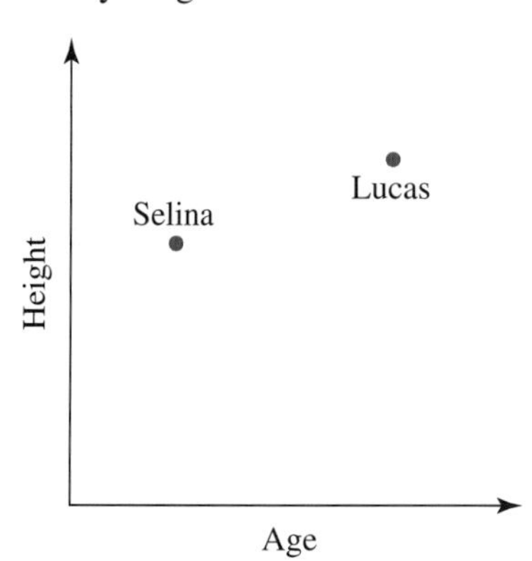

2. Use the graph to answer the following questions.
   a. Identify whether Yelena has a larger shoe size than Andrea.
   b. Identify whether Yelena is older than Andrea.

Shoe size
Yelena
Andrea
Age

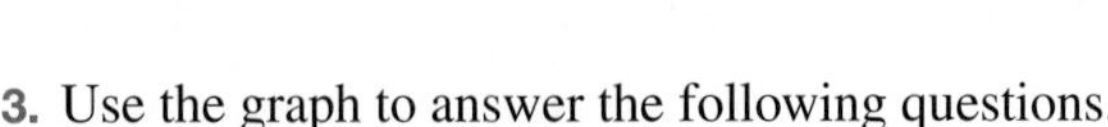

3. Use the graph to answer the following questions.
   a. Identify who is the youngest.
   b. Identify who weighs the most.
   c. Decide whether the oldest person is also the heaviest.

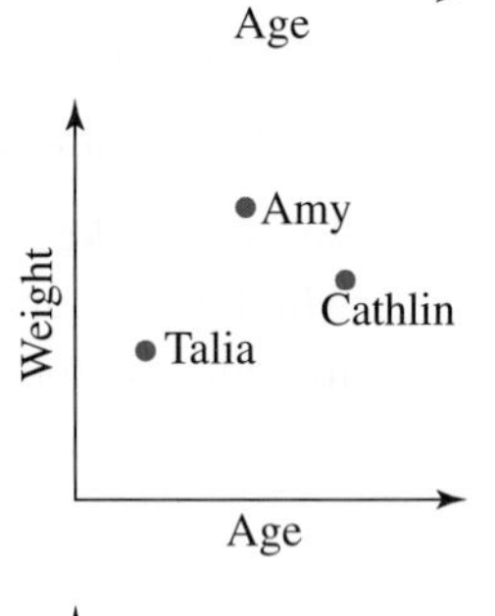


4. Use the graph to answer the following questions.
   a. Identify who is given the most pocket money.
   b. Decide whether the youngest person is given the least amount of pocket money.
   c. Decide whether the oldest person is given the most pocket money.

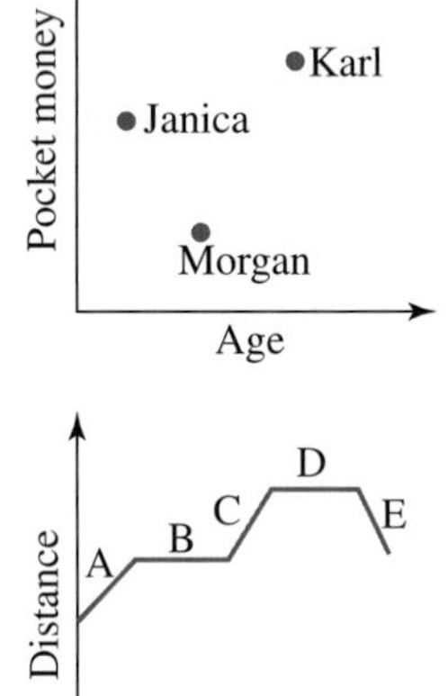


5. **WE14** The following graph shows the distance Claire is from home when she is out rockclimbing. Describe what each straight line section of the graph is showing.

Distance
A
B
C
D
E
Time

## Understanding

6. a. Use the graph to answer these questions.
      i. Identify who weighs the most.
      ii. Identify who is the tallest.
      iii. Decide whether the shortest person weighs the least.
   b. Copy this graph and plot the height and weight of Georgia, who is taller than Linh and weighs less than Hannah but more than James.

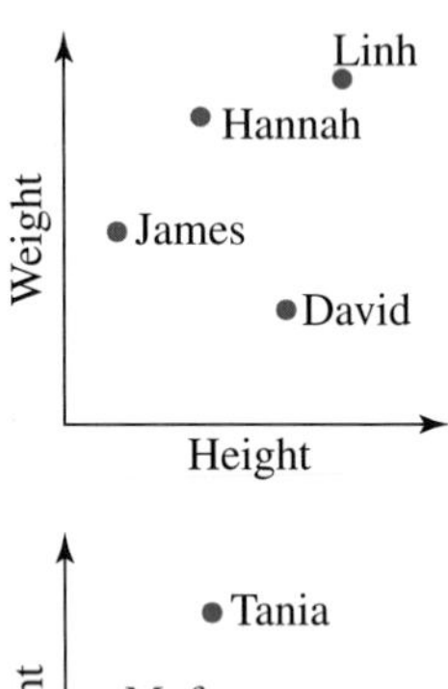


7. **MC** Select which of the following is shown by this graph.
   A. Myfanwy is the lightest and youngest.
   B. Tania is the oldest.
   C. Myfanwy is older than Louise and younger than Tania.
   D. Tania and Louise are the same age.
   E. Louise weighs the least.

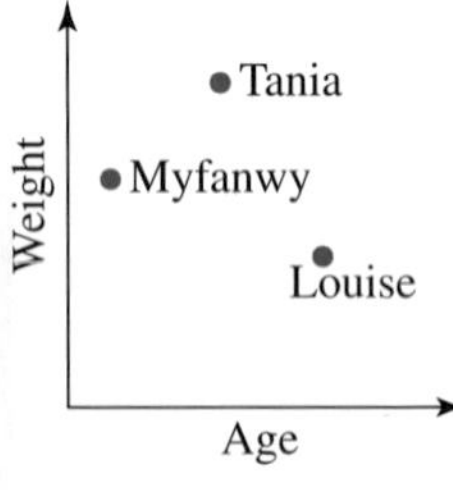

8. **MC** Select which of the following is shown by this graph.

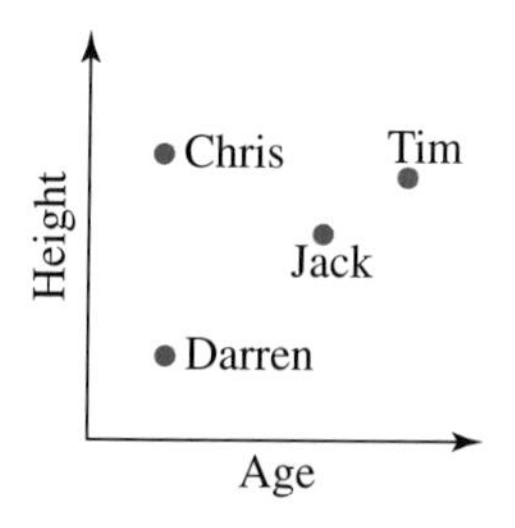


A. Jack is the tallest.
B. Chris is the oldest.
C. Darren and Chris are the same age.
D. Tim is the tallest.
E. Darren is the same height as Chris.

9. The graph below shows Thomas's head circumference for the first 12 weeks after birth.
   a. Explain why the graph doesn't start at zero.
   b. Describe the change in Thomas's head circumference during the first 12 weeks.
   c. Suggest what the graph would look like beyond the first 12 weeks.

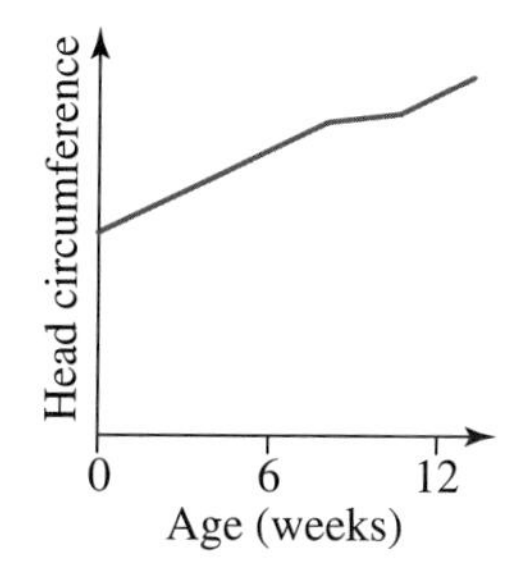


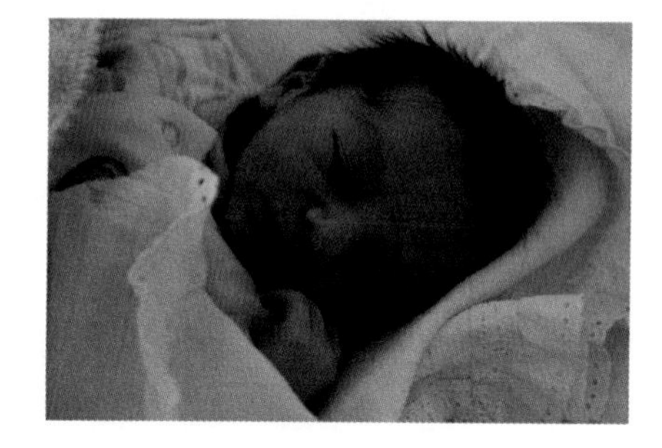

10. The graph shows Monique's journey as she walks to school.

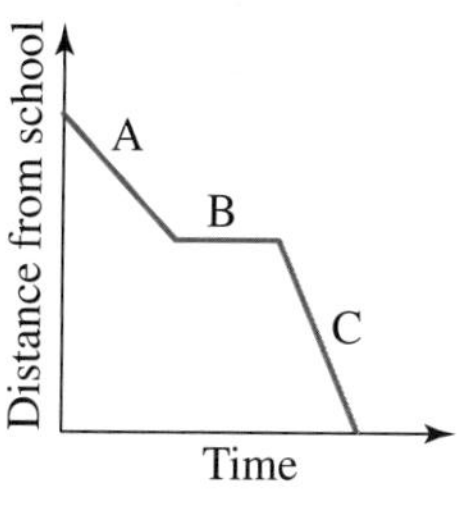


   a. Identify the section of the graph that shows when Monique is not moving.
   b. The line in section C is steeper than the line in section A. Describe what this tells us about section C.
   c. Identify where Monique is at the end of section C.

11. Yupa owns an ice-cream shop. The following graph shows his profit for the first 4 months of the year.

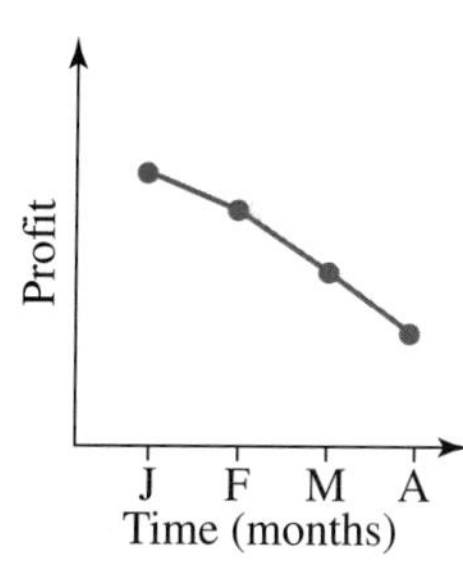


   a. Identify the month when Yupa made the most profit.
   b. Identify the month when Yupa made the least profit.
   c. Describe what has happened to Yupa's profit over the 4 months.

12. Copy the following axes and draw a line graph that you think would show rainfall over one year.

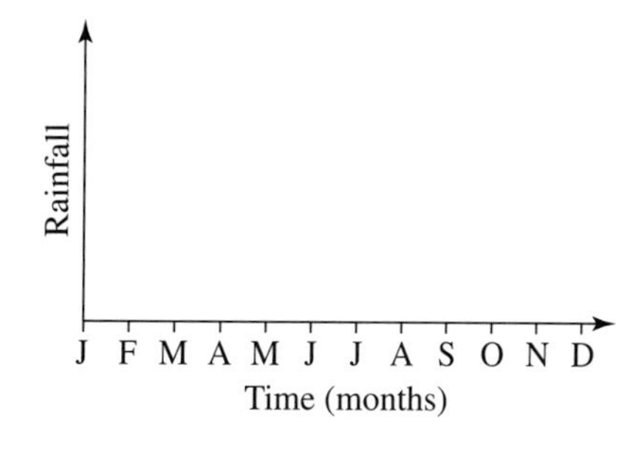

13. Cathy has been training hard for the 100 m sprint. She has kept a record of the fastest time she has run each week. Draw a graph that shows the following sections:
Section A: Cathy increased her speed for the 100 m sprint each week for the first 4 weeks.
Section B: After 4 weeks Cathy's speed stayed the same for the next 2 weeks.
Section C: Cathy had a cold and her speed decreased for 2 weeks.
Section D: Having recovered from her cold, Cathy trained hard and increased her speed again.

14. WE15 The following graph shows the average rainfall and average maximum and minimum temperatures for Muswellbrook and the Upper Hunter Valley.

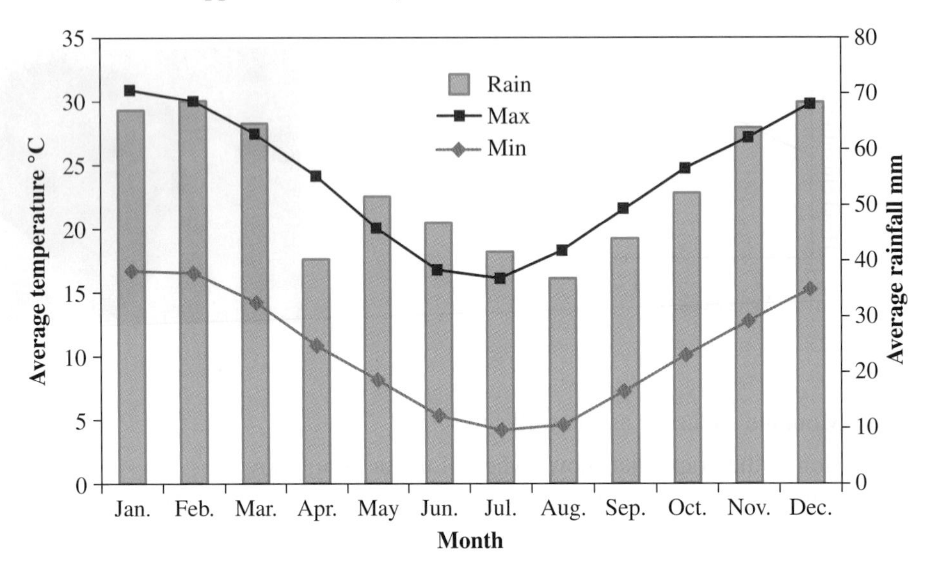


a. Identify which two months had the lowest average rainfall.
b. Identify the lowest average rainfall.
c. Identify which month had the lowest average minimum temperature.

15. Looking at the following graph, comment about the future for wind power in Australia and the world.

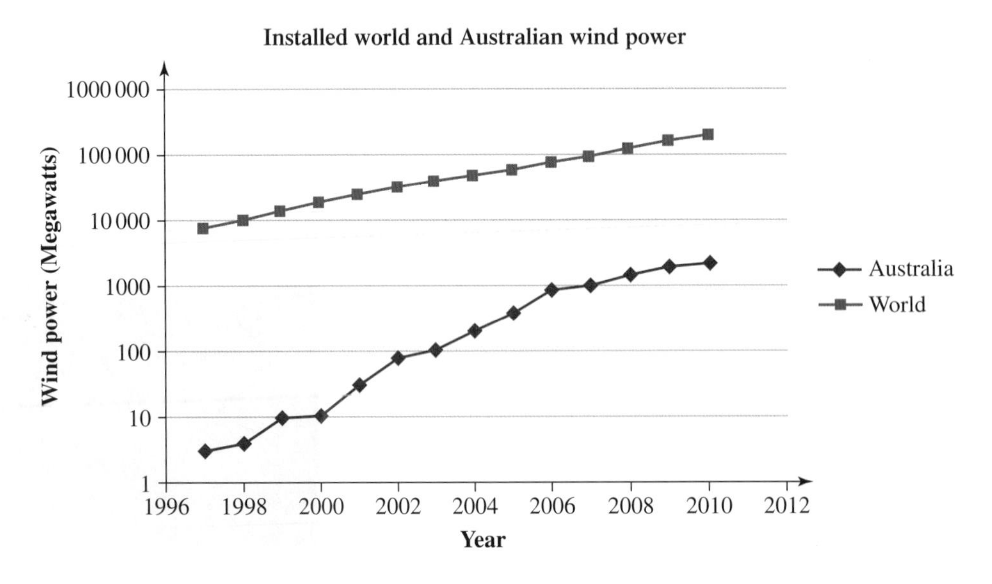

**16.** The following graph shows the percentage of visitors to Parks and Wildlife Group–managed parks in New South Wales for 3 years.

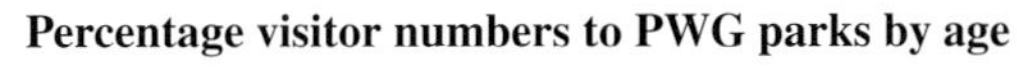


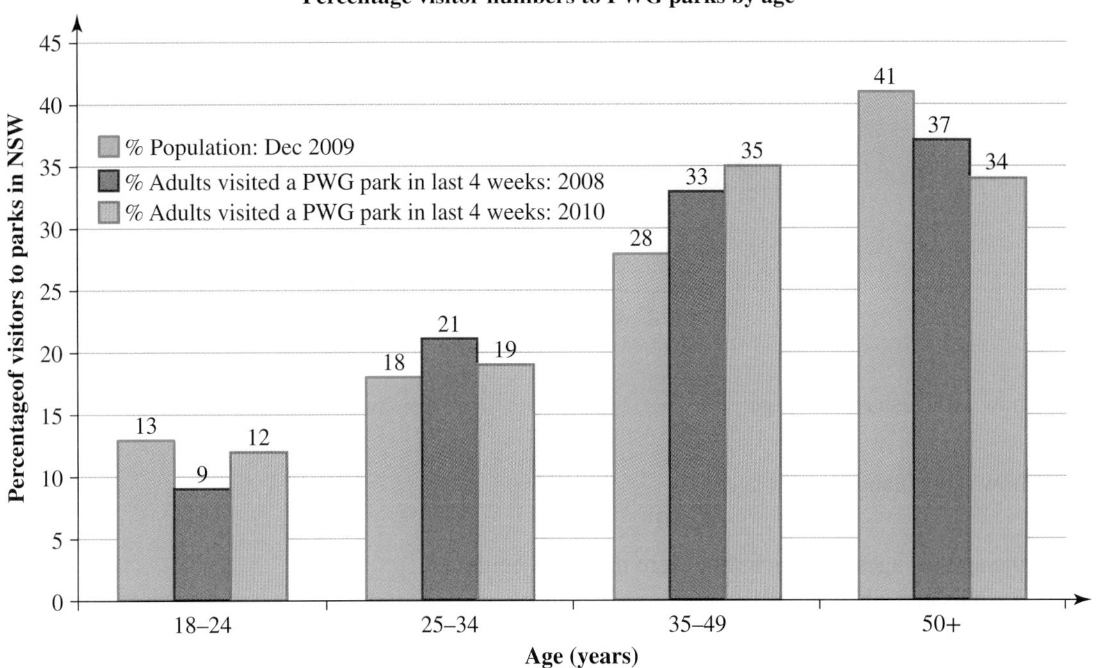


**a.** For the 25–34 age group, identify which year had the lowest percentage of visits.
**b.** Identify which age group visited the parks more frequently than any other group.
**c.** Suggest a reason why this age group visited the parks more often.

**17.** WE16 State the domain and range of each of the following:

**a.** $(3, 4)$, $(5, 6)$, $(7, 8)$, $(9, 10)$, $(11, 12)$

**b.** $(8, 8)$, $(7, 6)$, $(6, 4)$, $(5, 2)$, $(4, 0)$

**c.**

| $x$ | 1 | 2 | 3 | 4 | 5 |
|---|---|---|---|---|---|
| $y$ | 6 | 2 | 0 | −4 | −8 |

**d.**

| $x$ | −4 | −2 | 0 | 2 | 4 |
|---|---|---|---|---|---|
| $y$ | 8 | 4 | 0 | −4 | −8 |

**18.** State the domain and range of each of the following:

**a.** $(1, 2)$, $(2, 4)$, $(3, 6)$, $(4, 8)$, $(5, 10)$

**b.** $(7, 2)$, $(3, 5)$, $(-6, 9)$, $(-5, -2)$, $(0, 1)$

**c.**

| $x$ | −1 | −2 | −3 | −4 | −5 |
|---|---|---|---|---|---|
| $y$ | 3 | 6 | 9 | 12 | 15 |

**d.**

| $x$ | −4 | −2 | 0 | 2 | 4 |
|---|---|---|---|---|---|
| $y$ | 3 | 1 | 0 | 1 | 3 |

**19.** State the domain and range of each of the following:

**a.** $(1, -2)$, $(-2, 4)$, $(-3, 6)$, $(4, -8)$, $(-5, 10)$

**b.** $(2, 2)$, $(3, 2)$, $(-6, 4)$, $(5, -2)$, $(0, 0)$

**c.**

| $x$ | 0 | 3 | 4 | 7 | 9 |
|---|---|---|---|---|---|
| $y$ | 5 | 5 | 5 | 5 | 5 |

**d.**

| $x$ | 1 | 2 | 0 | 2 | 3 |
|---|---|---|---|---|---|
| $y$ | 4 | 6 | 8 | 1 | 3 |

20. The line graph shown compares the heights of Yasha and Yolande over a 20-year period.

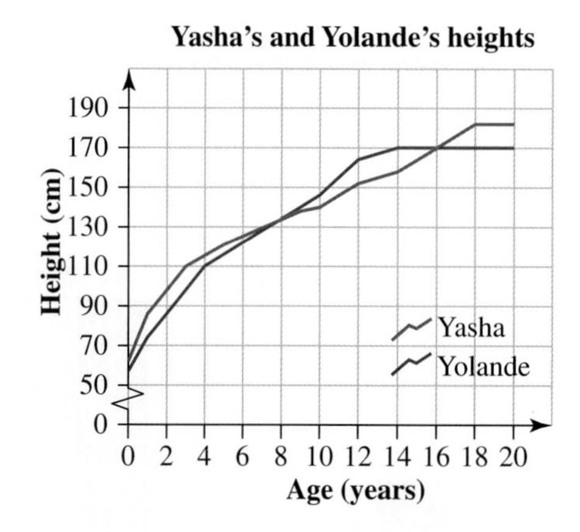


a. State how tall Yasha was at age:
   i. 3   ii. 5   iii. 14.

b. State how tall Yolande was at age:
   i. 4   ii. 10   iii. 14.

c. State the age (or ages) at which they were the same height.

d. State Yasha's age when she was:
   i. 140 cm tall   ii. 176 cm tall.

e. Identify the range of ages when Yolande was taller than Yasha.

f. Between which two birthdays did each person grow the fastest?

21. Weekly sales of song downloads for the pop group The Mathemagics are shown in the following table.

| Week | 1 | 2 | 3 | 4 | 5 | 6 | 7 | 8 |
|---|---|---|---|---|---|---|---|---|
| Sales | 1500 | 2800 | 3750 | 4000 | 3600 | 3000 | 2400 | 1900 |

a. Construct a line graph of the data.   b. Estimate the sales for week 9.

22. WE17 The graph shows the bike trips made by Svetlana and Sam.

a. State who left first.

b. Calculate how far Sam travelled.

c. Calculate how far Svetlana travelled.

d. State who got home first.

e. Identify who stopped, when, and for how long.

f. Calculate Sam's and Svetlana's average speeds for the full trips, correct to two decimals places where appropriate.

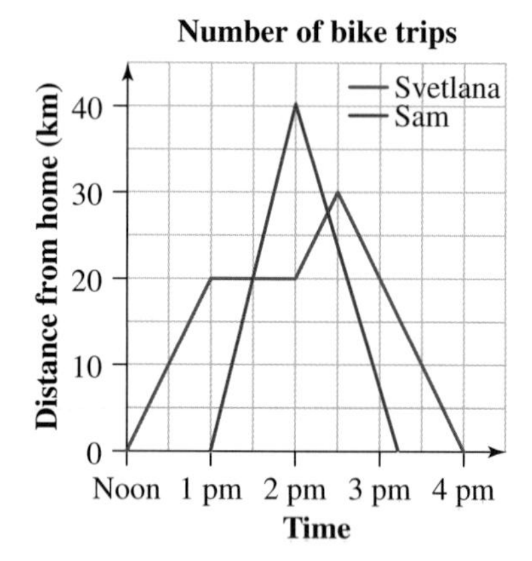

23. The graph at the right shows Steven's return trip from home to the Victoria Market.
   a. Calculate how far he travelled.
   b. Calculate the average speed of his trip.
   c. State how long he was at the market for.
   d. Was he driving faster on his way to the market or on his way home? Justify your answer.

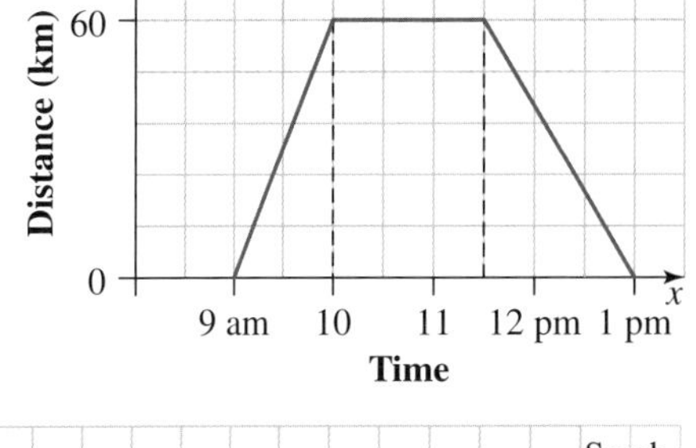


24. Tom and Sarah's journey to school is shown on the graph.
   a. State who lives further from school.
   b. Identify who takes the train to school. Justify your answer.
   c. Calculate the average speed of each of their journeys.
   d. Identify the time at which they were the same distance from school.

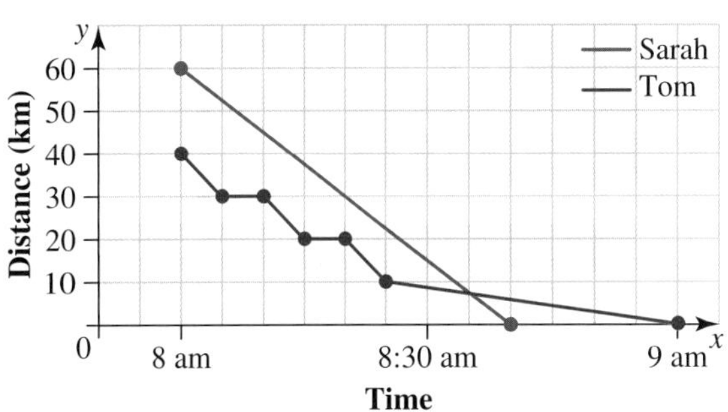


## Reasoning

25. Lisa created a line graph to illustrate the number of glass pendants she has sold for each of the last ten years. She intends to use the graph for a presentation at her bank, in order to obtain a loan.
   a. Identify in what year of operation she sold the most pendants. State how many she sold.
   b. Identify in what year of operation she sold the fewest pendants. State how many she sold.
   c. Describe any trend in the sales of the pendants.
   d. Will this trend help Lisa obtain the loan?
   e. State the domain and range of the graph of glass pendant sales.

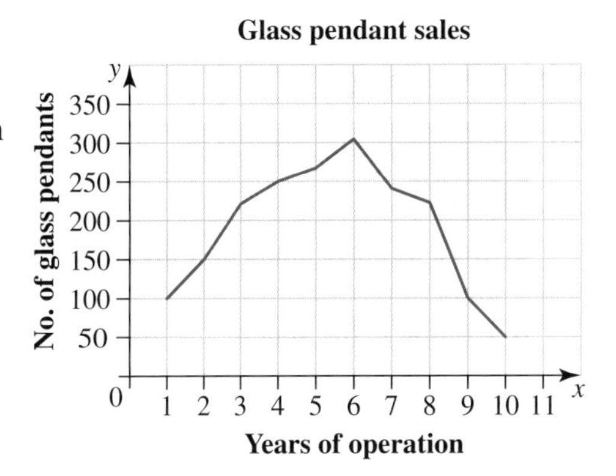


26. Peter and Sonja are walking home from school. A dog started chasing them and Peter started running away from the dog. Sonja noticed that this was her neighbour's dog so she stopped to pat the dog and then started to walk towards Peter who was now waiting for her.
   Draw a graph to represent the relationship between distance travelled and time for both Peter and Sonja.

## 11.6.2 Success criteria

Tick the column to indicate that you have completed the lesson and how well you think you have understood it using the traffic light system.

(**Green:** I understand; **Yellow:** I can do it with help; **Red:** I do not understand)

| Lesson | Success criteria | Green | Yellow | Red |
|---|---|---|---|---|
| 11.2 | I can plot coordinates on a Cartesian plane. | | | |
| | I can identify the coordinates of points on a Cartesian plane. | | | |
| 11.3 | I can construct a number pattern from a rule. | | | |
| | I can identify a rule from a number pattern. | | | |
| | I can identify a number pattern that can be used to describe a geometric sequence. | | | |
| | I can express a number pattern algebraically. | | | |
| 11.4 | I can plot a simple linear relationship from a table of values. | | | |
| | I can identify whether a series of points forms a linear relationship. | | | |
| 11.5 | I can interpret the information presented on a Cartesian plane. | | | |
| | I can identify the domain of a set of points or line graph. | | | |
| | I can identify the range of a set of points or line graph. | | | |
| | I can read a travel graph and calculate average speed. | | | |

## 11.6.3 Project

### Battleships game

The goal of this two-person game is to sink all of your opponent's ships before they do the same to you.

**Equipment:** Each player has two $10 \times 10$ grids. The grids are labelled with letters across the top (A to J) and numbers down the side (1 to 10). One of the grids is labelled 'My grid' and the other is labelled 'Opponent's grid'.

**Set-up:** Each player decides where to place four ships on their 'My grid': a five-square battleship, a four-square cruiser, a three-square submarine and a two-square destroyer. All ships must be placed in straight lines either horizontally or vertically, but not diagonally. It is fine (but not required) for two or more ships to be adjacent to each other.

**How to play:** Players take turns taking shots at each other. A shot is taken by calling out the coordinates of a square on the $10 \times 10$ grid, such as A1 or H4. Each player takes one shot at a time. If the player calls the coordinates of a square where one of the opponent's ships is located, the opponent calls out 'Hit' and says which ship was hit; if the shot misses, the opponent says 'Miss'.

Players should mark the shots they take on their 'Opponent's grid' and whether each shot was a hit or a miss, so that they don't call any square more than once. Players should also mark their 'My grid' with shots taken by their opponent.

A ship is sunk when all of its squares have been hit. When this happens, the player whose ship was sunk says, for example, 'You sank my battleship.'

The first player to sink all of their opponent's ships wins the game.

| My grid | | | | | | | | | | |
|---|---|---|---|---|---|---|---|---|---|---|
| | A | B | C | D | E | F | G | H | I | J |
| 1 | | | | | | | | | | |
| 2 | | | | | | | | | | |
| 3 | | | | | | | | | | |
| 4 | | | | | | | | | | |
| 5 | | | | | | | | | | |
| 6 | | | | | | | | | | |
| 7 | | | | | | | | | | |
| 8 | | | | | | | | | | |
| 9 | | | | | | | | | | |
| 10 | | | | | | | | | | |

| Opponent's grid | | | | | | | | | | |
|---|---|---|---|---|---|---|---|---|---|---|
| | A | B | C | D | E | F | G | H | I | J |
| 1 | | | | | | | | | | |
| 2 | | | | | | | | | | |
| 3 | | | | | | | | | | |
| 4 | | | | | | | | | | |
| 5 | | | | | | | | | | |
| 6 | | | | | | | | | | |
| 7 | | | | | | | | | | |
| 8 | | | | | | | | | | |
| 9 | | | | | | | | | | |
| 10 | | | | | | | | | | |

| | |
|---|---|
| **eWorkbook** | Topic 11 Workbook (worksheets, code puzzle and project) (ewbk-1912) |
| **Interactivities** | Crossword (int-2610) |
| | Sudoku puzzle (int-3175) |

## Exercise 11.6 Review questions

learnon

### Fluency

**1.** The grid shown in part **b** shows a simplified map that uses alphanumeric grid references.

**a.** Fill in the blank cells in the table.

| Location | Grid reference |
|---|---|
| | B3 |
| Alex's house | |
| | D1 |
| Lena's house | |

**b.** A playground is to be built at A3. Mark its location on the grid with an X.

| | | | | |
|---|---|---|---|---|
| D | Post-office | Bus-stop | Star cafe | Library |
| C | | Apple street | | |
| B | Alex's house | Maya's house | Joseph's house | Lena's house |
| A | | | | |
| | 1 | 2 | 3 | 4 |

2. Study the graph to answer these questions.
   a. Identify the letter that represents the point (−3, 4).
   b. Identify the points on the $x$-axis.
   c. Name all points whose $x$-coordinate is zero.
   d. Identify the points in the second quadrant. Give their coordinates.
   e. State the coordinates of C and F.

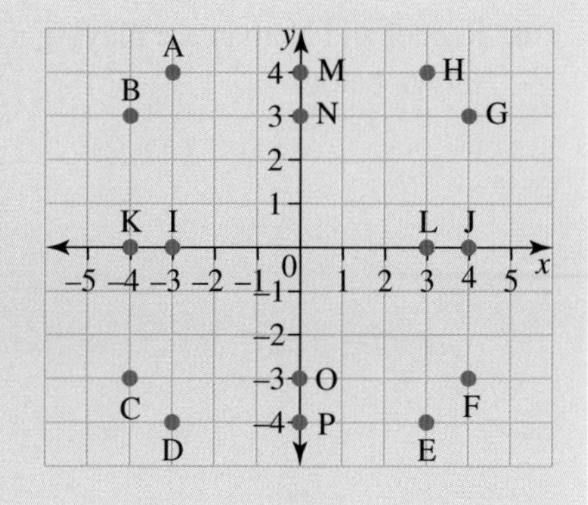


3. Use the graph to answer these questions.
   a. Fill in the blanks in the table.

| Coordinates | Point |
|---|---|
| (0, 2) | |
| | C |
| | D |

   b. Mark E (4, 6) on the graph. Join all the points with a straight line.

4. The following table gives the $x$ and $y$ coordinates of some points.

| $x$ | 0 | 1 | 2 | 3 |
|---|---|---|---|---|
| $y$ | 0 | 2 | 4 | 6 |

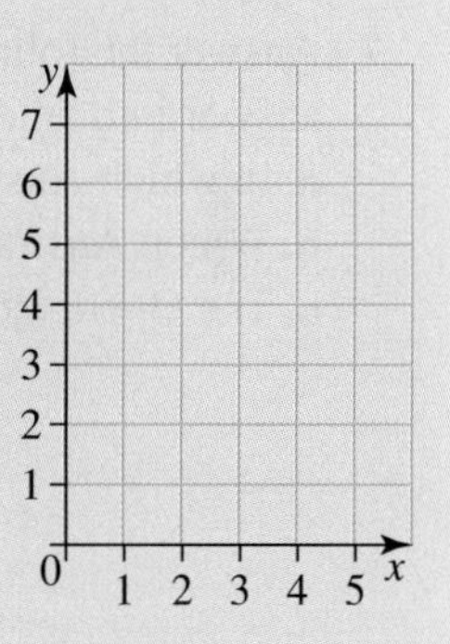

   a. Using the axes shown, plot the points given in the table and join them.
   b. Explain whether the relationship between $x$ and $y$ is linear.

## Problem solving

5. A game consists of a grid, and a dart that is thrown to land on a square of the grid.

| A | B | C | D | E |
|---|---|---|---|---|
| F | G | H | I | J |
| K | L | M | N | O |
| P | Q | R | S | T |
| U | V | W | X | Y |

The dart is thrown and lands on a letter that:
- has horizontal symmetry
- also has vertical symmetry
- and is not a vowel.

What are the possibilities for the landing position of the dart? Give instructions for locating any possibilities on the grid.

6. Rachel and Nathan are playing a battleship game. Rachel's grid (in progress, up to this moment) is shown below.

|  | A | B | C | D | E | F | G | H | I | J |
|---|---|---|---|---|---|---|---|---|---|---|
| 1 | • |  |  |  |  |  |  |  |  | • |
| 2 |  |  |  |  |  |  |  |  |  |  |
| 3 |  |  |  | • |  |  |  |  |  |  |
| 4 |  |  |  |  |  |  | • |  |  |  |
| 5 | ⊠ | • |  |  | • |  |  |  |  |  |
| 6 | ⊠ |  |  |  |  |  |  | ⊠ |  |  |
| 7 |  |  |  |  |  | • |  |  |  |  |
| 8 |  |  |  | • |  |  |  |  |  |  |
| 9 |  |  |  |  |  |  |  |  |  |  |
| 10 |  |  | • |  |  |  |  | • |  |  |

a. For each of the following attempts, state whether Nathan will miss (M), sink (S) or hit (H) one of Rachel's ships.

i. A7 ii. F9 iii. I10 iv. C3

b. Specify the location of all of Rachel's one-berth ships.

c. State the set of grid references that Nathan needs if he is to sink Rachel's largest ship.

7. Answer the following questions, then translate each letter of your answer into an ordered pair by using the grid.

a. In which Australian state is the Great Barrier Reef?

b. Which Australian state is known as the Garden State?

c. The Opera House is in which city?

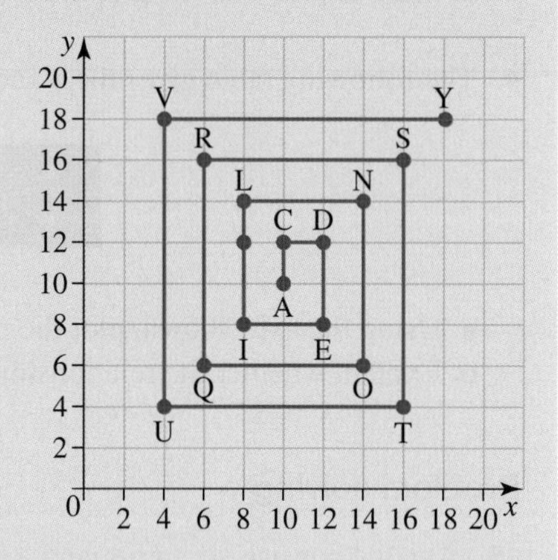


8. The following diagram gives the key to a code by matching letters to coordinates. Use the key to write in code this famous statement by Descartes: 'I think, therefore I am.'

9. On a carefully ruled pair of Cartesian axes, join the following points in the order given. It reveals a symbol used by an ancient mathematical secret society.

$$(1, 0) - (6, 4) - (0, 4) - (5, 0) - (3, 6) - (1, 0)$$

10. ABCD is a square drawn in the first quadrant as shown.
Given A(0, 3) and B(4, 0) and C(7, 4), find the coordinates of D.

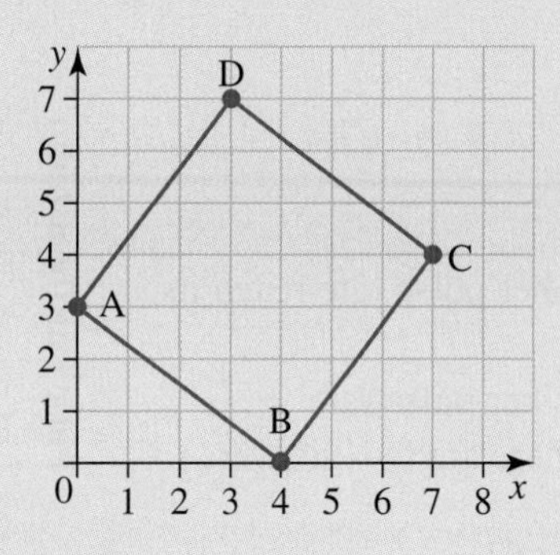


11. a. Plot the points A(3, 0), B(6, 7), C(10, 7), D(7, 0).
    i. What type of quadrilateral is ABCD?
    ii. Calculate its area.
  b. Draw a set of axes and on them plot points so that the encompassing triangle has an area of 20 square units. List the points and show why the area is 20 square units.

12. Four points — A, B, C and D — are drawn on a Cartesian plane and are joined in order to form a square. A is the point (−4, 0), B is (1, 5) and C is (6, 0). State the coordinates of point D.

13. a. State the domain and range of this table.
  b. Does this table represent a relationship between $x$ and $y$ that is linear?

| $x$ | 1 | 2 | 3 | 4 |
|---|---|---|---|---|
| $y$ | 1 | 4 | 9 | 16 |

  Explain why or why not.

14. Consider the following statement: 'A linear relation, when plotted on a Cartesian plane, produces a straight line that cuts the $x$-axis at one point, and the $y$-axis at one point.'
Discuss whether or not this statement is true.

15. The following graph shows Anton's distance from the shore during a surf life-saving drill.

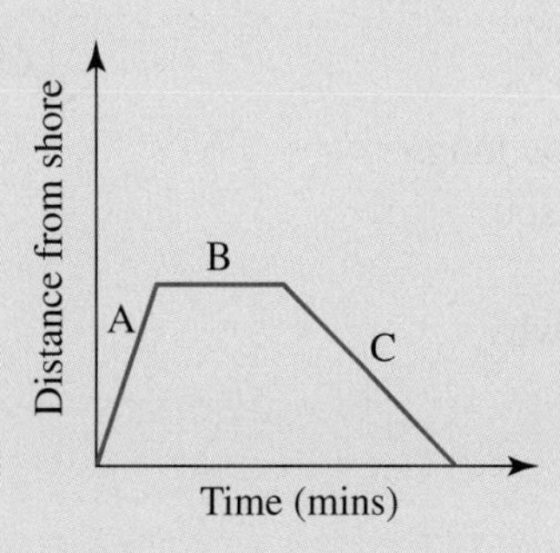


  a. Describe what each section of the graph is showing.
  b. The line in section A is steeper than the line in section C. Describe what this tells you about Anton's swimming speed in section A compared to his speed in section C.

16. A Year 8 student doing an exercise program recorded her pulse rate at one-minute intervals, as shown in the graph.
   a. State her pulse rate after:
      i. 2 minutes
      ii. 4 minutes.
   b. After how many minutes did her pulse rate reach its maximum?
   c. After how many minutes was her pulse rate 80 beats/min? Explain why there are two answers to this question.
   d. State her pulse rate after $3\frac{1}{2}$ minutes.
   e. Calculate for how long her pulse rate was above 80 beats/min.

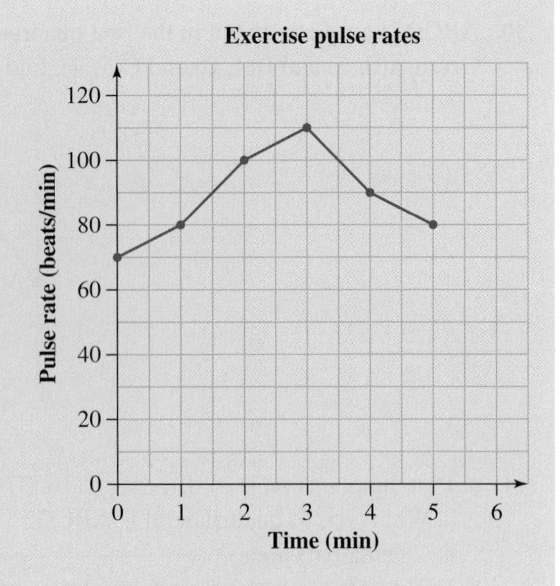


17. The following line graph shows the traffic flow on a particular road on a certain day.
   a. State when the traffic flow hit its peak.
   b. During the period studied, state when the number of vehicles was lowest.
   c. State how many cars used the road at:
      i. 11.00 am
      ii. 9.30 am.
   d. State the times of the day when:
      i. 300 cars were on the road
      ii. 225 cars were on the road.
   e. Discuss why a line graph is used for this information rather than a column or a sector graph.

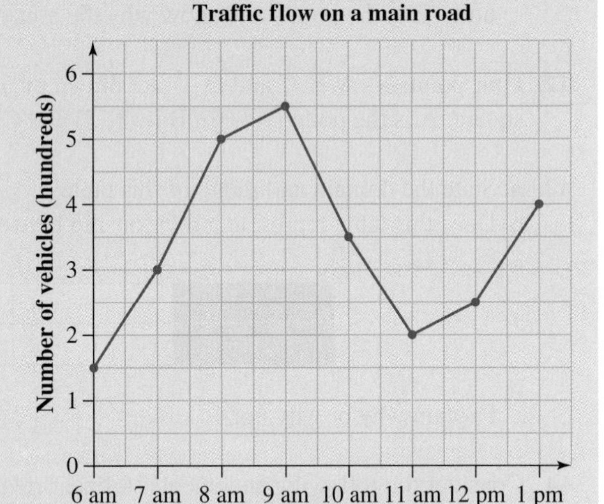


18. The following graph shows Wilma's bike trip for an afternoon. Write a report on Wilma's trip, including the following:
   - when she left home
   - when she stopped and for how long
   - when she arrived at Mt Cranson
   - when she left to go home
   - how fast she was going and why
   - when she got home again.

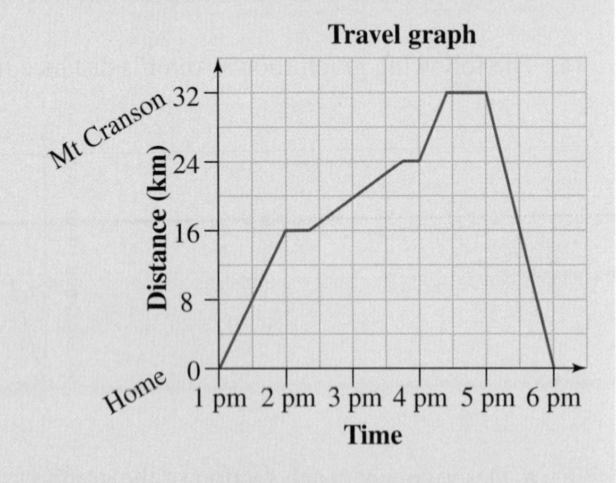


**on** To test your understanding and knowledge of this topic, go to your learnON title at www.jacplus.com.au and complete the **post-test**.

# Answers

## Topic 11 Coordinates and the Cartesian plane

### 11.1 Pre-test

1. a. E2
   b. B1
   c. B2
2. a. Any of the following: Torquay, Geelong, Queenscliff, Sorrento, Wonthaggi, Berwick
   b. Any of the following: Warracknebeal, Dimboola, Wycheproof, Inglewood, Murtoa, Horsham, Navarre, Stawell, Avoca, Ararat, Clunes, Dunkeld, Penshurst, Derrinallum, Skipton, Clunes, Ballarat
   c. Any of the following: Griffith, Narrandera, Wagga Wagga
3. a. a1 and f1 b. b3 c. f7
4. (5, 3)
5. 35 units$^2$
6. E
7. (−1, 4)
8. B
9. 17
10. 17
11. B
12. True
13. $H = 6x + 45$
14. D
15. Section C

### 11.2 The Cartesian plane

1.

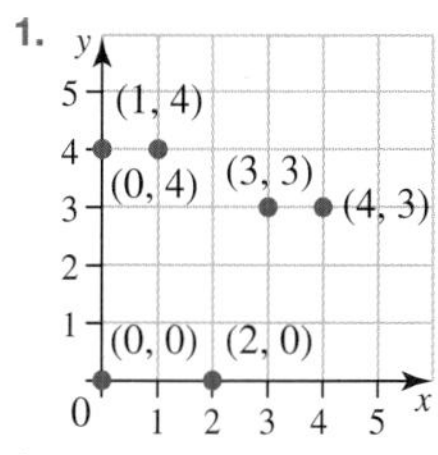


2.

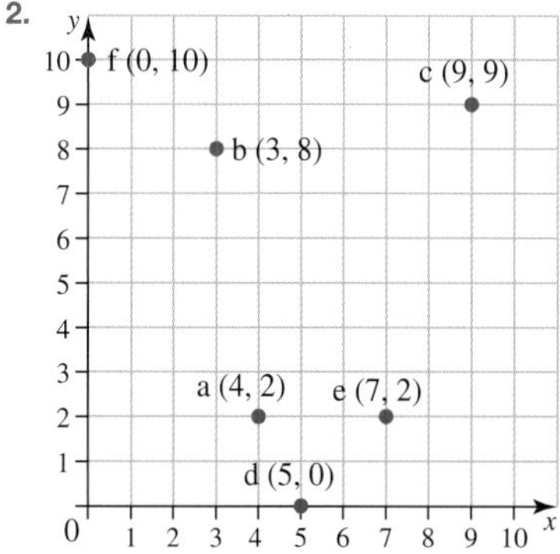


3.

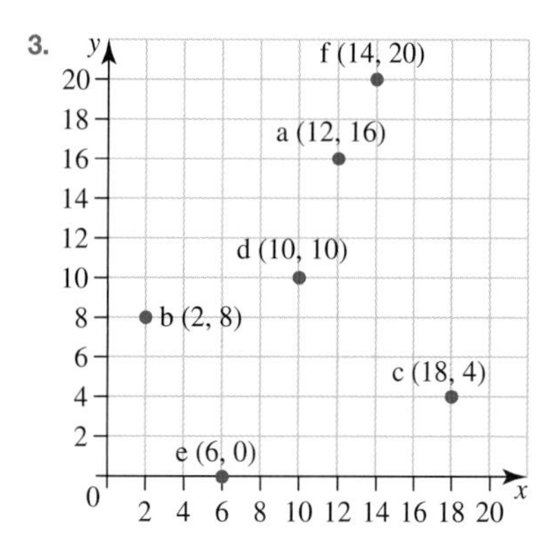


4. A(6, 1) B(2, 5) C(7, 9) D(12, 10) E(10, 0) F$\left(4\frac{1}{2}, 0\right)$ G(0, 6) H(0, 0) I$\left(7\frac{1}{2}, 5\right)$ J$\left(10, 7\frac{1}{2}\right)$ K$(3\frac{1}{2}, 2\frac{1}{2})$ L$(\frac{1}{2}, 9\frac{1}{2})$

5. a.

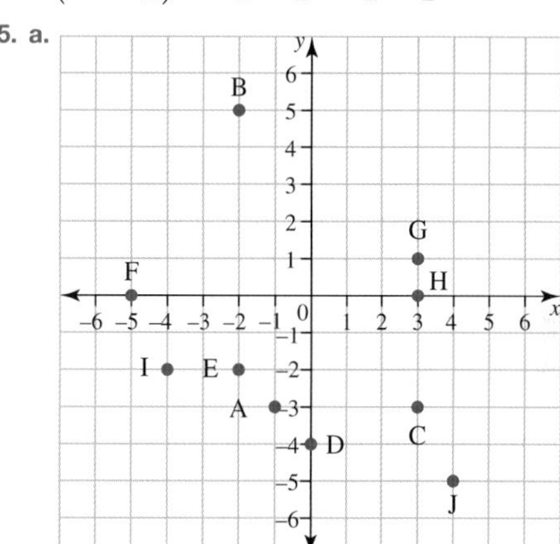


b. Quadrant I: point G; Quadrant II: point B; Quadrant III: points A, E, I; Quadrant IV: points C, J; $x$–axis: points F, H; $y$–axis: point D.

6. a.

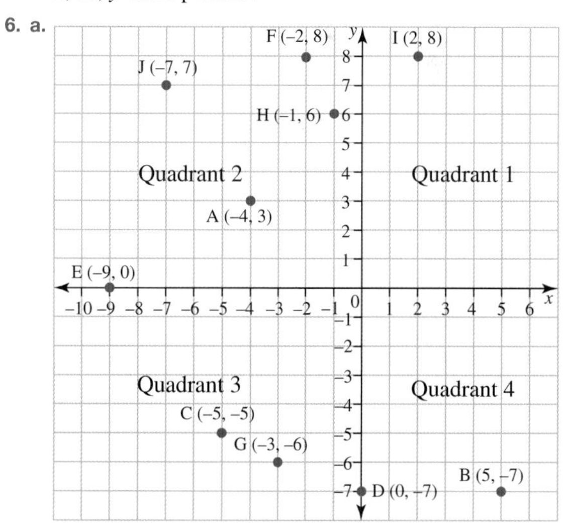


b. Quadrant I: point I; Quadrant II: points A, F, H, J; Quadrant III: point C, G; Quadrant IV: point B; $x$–axis: point E; $y$–axis: point D.

**7. a.**

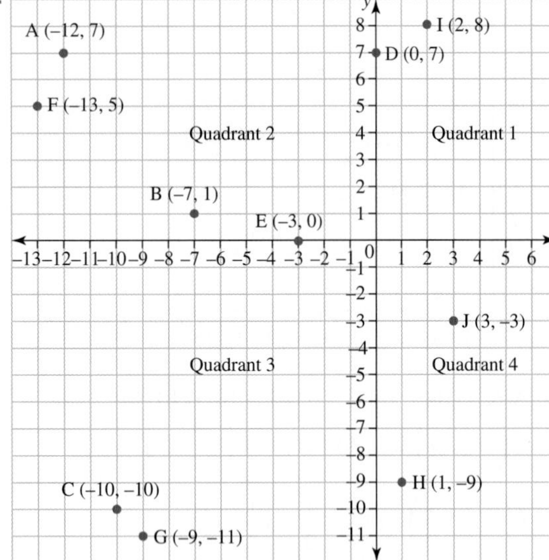


**b.** Quadrant I: point I; Quadrant II: point A, B, F; Quadrant III: points C, G; Quadrant IV: points H, J; $x$–axis: points E; $y$–axis: point D.

**8.** A(–4, 2), B(–1, 4), C(2, –2), D(4, 1), E(–3, –4), F(0, –5)

**9.** A(6, –9), B(–7, –8), C(–3, −2), D (–7, 4) , E (–3, 9), F (8, 6)

**10.** A(–9, –6), B(–3, 3), C(9, –3), D(2, –6), E(–9, 8), F(3, 9)

**11. a.** B **b.** C **c.** D
**d.** A **e.** E **f.** A

**12. a.** E, C **b.** B, D

**13.** In hospital; in wards six, seven, eight and nine

**14.**

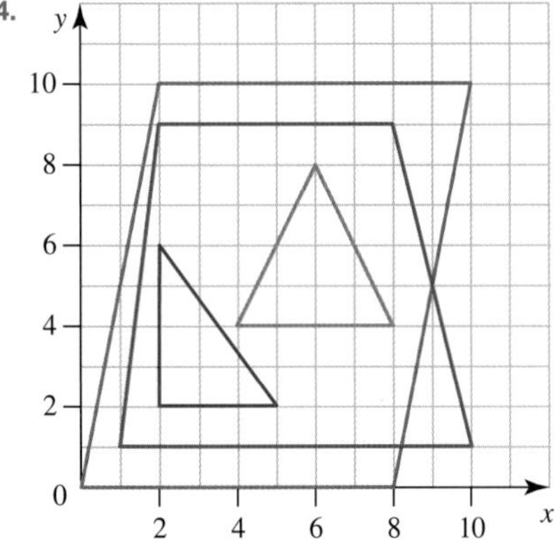

**a.** Right-angled triangle
**b.** Isosceles triangle
**c.** Trapezium
**d.** Parallelogram

**15.**

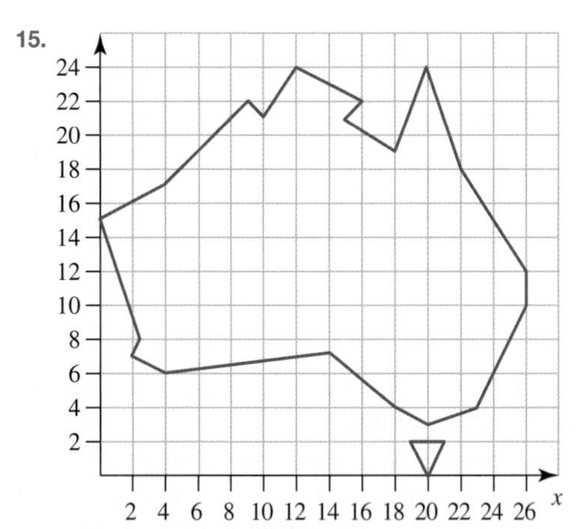

**16. a.** Area is 18 square units
**b.** Perimeter is 18 units

**17.** b, d, e

**18. a.** A(0, 1) B(2, 3) C(5, 1)
**b.** 5 square units
**c.** A(0, −1) B(2, −3) C(5, −1)
**d.** A(0, −1) B(−2, −3) C(−5, −1)

**19. a.** A(−3, 3), B(3, 3), C(3, −3), D(−3, −3)
**b.** 36 square units.
**c.**

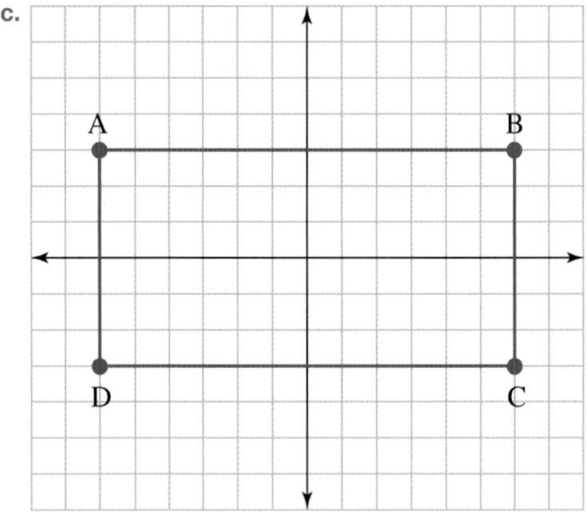


**d.** 72 square units.
**e.**

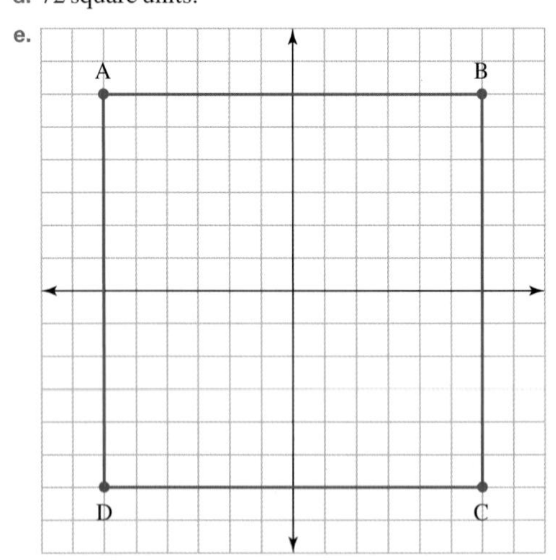


**f.** 144 square units.
**g.** The length doubled, so the area doubled.
When both the length and the width doubled, the area became four times bigger.

20. a. i. It would move in quadrant 2.

ii. It would move in quadrant 4.

b. i It would move in quadrant 1.

ii It would move in quadrant 3.

c. i. Vertical reflection.

ii. Horizontal reflection.

21. a. i. $(3, 8)$

ii. $(6, 2)$

iii. $(-4, -1)$

iv. $(3, -5)$

b. $\left(\dfrac{a+m}{2}, \dfrac{b+n}{2}\right)$

22. a. Quadrant 2 or 3 b. Quadrant 1 or 2
c. Quadrant 3 d. Quadrant 4

23. a. 

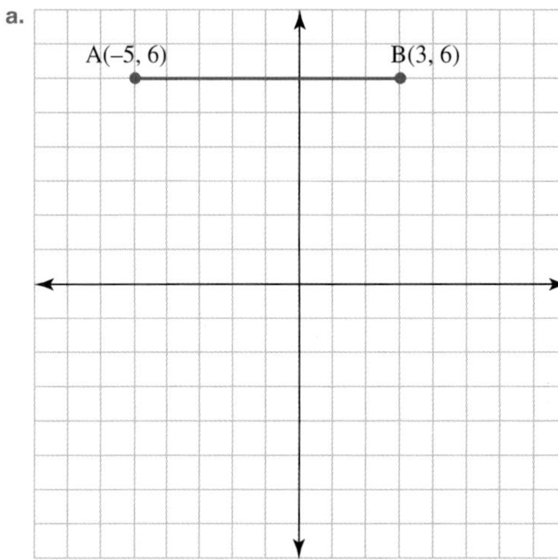


b. 8 units

c. $(-1, 6)$

d. Take care when determining the middle point of your second horizontal line.

e. $\dfrac{x_1 + x_2}{2}$

24. a. 

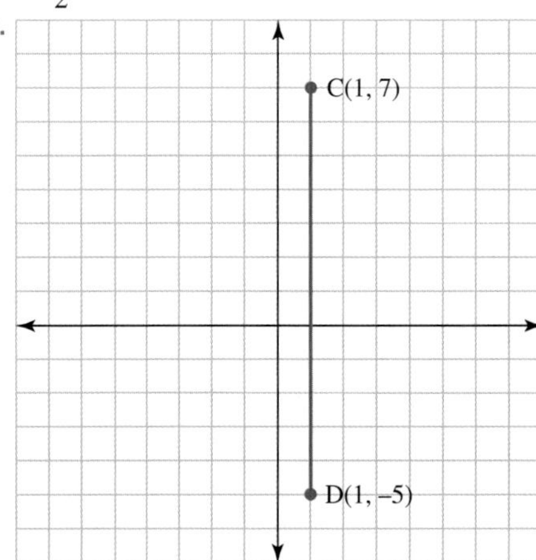


b. 12 units

c. $(1, 1)$

d. Take care when determining the middle point of your second vertical line.

e. $\dfrac{y_1 + y_2}{2}$

25. One possible answer is: $(2, 2)$ $(-2, 2)$ $(-2, -2)$ $(2, -2)$.

26. a. The $x$-coordinate increases by 2 and the $y$-coordinate increases by 4. The $y$-coordinate is always 3 more than double the $x$-coordinate.

b. $(10, 23)$ $(12, 27)$ $(14, 31)$

c. $y = 2x + 3$

27. a. $(0, 5)$ $(5, 0)$ $(-5, 0)$ $(0, -5)$

b. $(3, 4)$ $(4, 3)$ $(-3, 4)$ $(-4, 3)$ $(-3, -4)$ $(-4, -3)$ $(3, -4)$ $(4, -3)$ all lie on a circle of radius 5 units with centre at the origin.

## 11.3 Linear number patterns

1. a. 1, 5, 9, 13, 17

b. 5, 14, 23, 32, 41

2. a. 50, 42, 34, 26, 18

b. 64, 53, 42, 31, 20

3. a. 10, 12, 14 b. 23, 28, 33

4. a. Value of term = 30 – position in pattern × 3

b. 15, 12, 9

5. a. Value of term = 55 – position in pattern × 7

b. 20, 13, 6

6. a. 6, 15, 18 b. 10, 12, 16 c. 16, 64

7. a. 9, 11, 17 b. 88, 110 c. 95, 90, 75

8. a. Value of term = position in pattern × 10 – 5

b. 145

9. a. Value of term = 47 – position in pattern × 3

b. 2

10. a. Value of term = position in pattern × 6 + 96

b. 306

11. a. Value of term = 225 – position in pattern × 13

b. −74

12. a. 5, 6, 7, 8 and 9

b. 14

c. 104

13. a. 0, 2, 4, 6 and 8

b. 38

c. 152

14. When relating the value of a term to its position in the pattern, you can use that description to find any other future term in the number pattern.

15. Yes, you can substitute appropriate algebraic symbols for parts of the written descriptions to create an algebraic statement. For example, let the number of matches = $m$ and the number of shapes = $n$.

16. Once you have identified the connection between the top row and the bottom row in a table of values, you have identified the rule for the pattern.

17. a.

b.

| Number of triangles | 1 | 2 | 3 | 4 | 5 |
|---|---|---|---|---|---|
| Number of matches | 3 | 5 | 7 | 9 | 11 |

c. Number of matches = number of triangles × 2 + 1

d. 15

18. a. 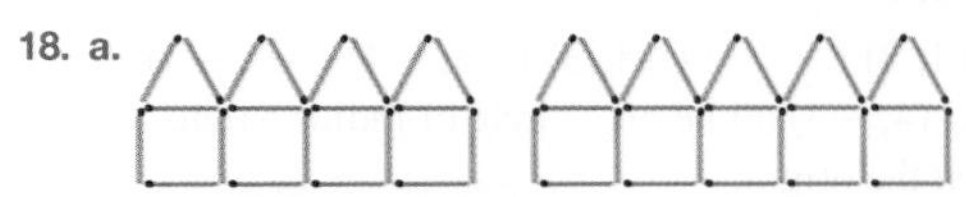

b.

| Number of houses | 1 | 2 | 3 | 4 | 5 |
|---|---|---|---|---|---|
| Number of matches | 6 | 11 | 16 | 21 | 26 |

c. Number of matches = number of houses × 5 + 1

d. 36

19. a.

| Number of shapes | 1 | 2 | 3 |
|---|---|---|---|
| Number of matches | 5 | 10 | 15 |

b. $m = 5n$

20. a. 

b.

| Number of fence units | 1 | 2 | 3 | 4 | 5 |
|---|---|---|---|---|---|
| Number of matches | 4 | 7 | 10 | 13 | 16 |

c. Number of matches = number of fence units × 3 + 1

d. 22

21. a.

| Number of shapes | 1 | 2 | 3 |
|---|---|---|---|
| Number of matches | 6 | 14 | 22 |

b. $m = 8n - 2$

22. See a sample pattern below. Other sample responses can be found in the worked solutions in the online resources.

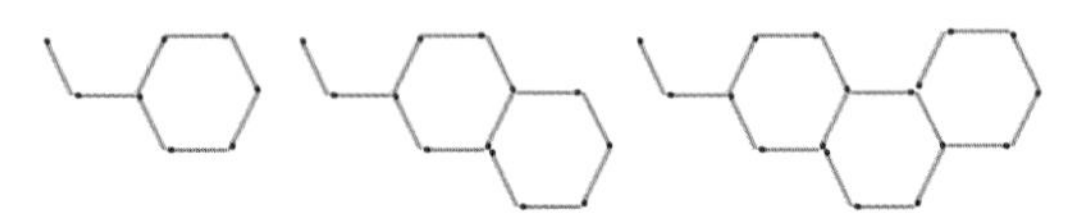

## 11.4 Plotting simple linear relationships

1. a. 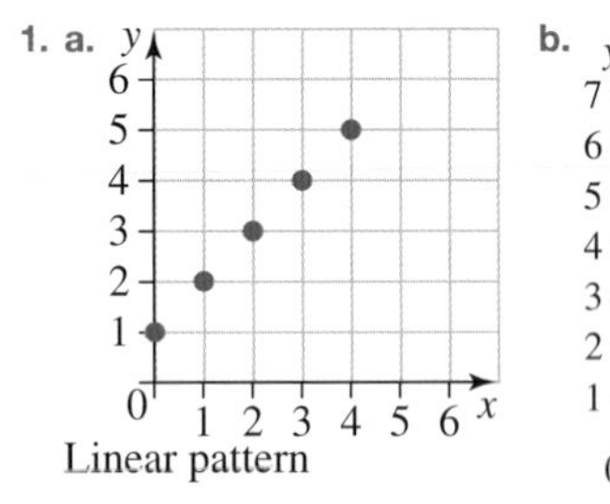

Linear pattern

b. 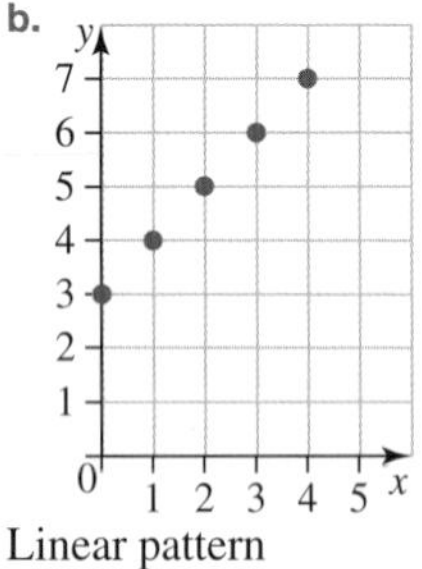

Linear pattern

c. 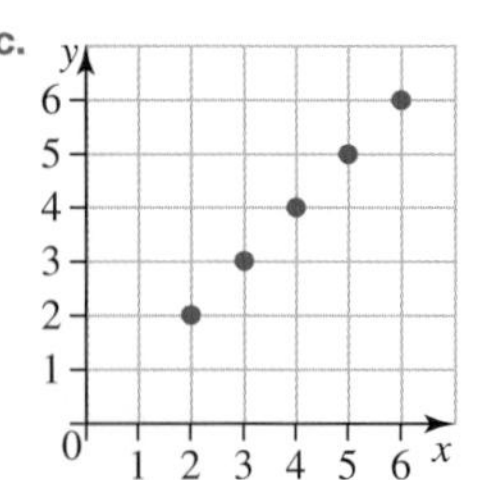

Linear pattern

d. 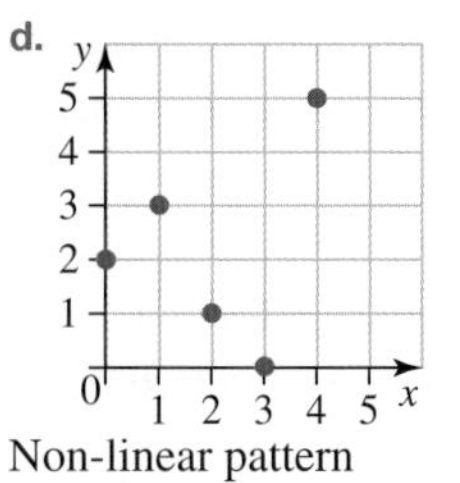

Non-linear pattern

e. 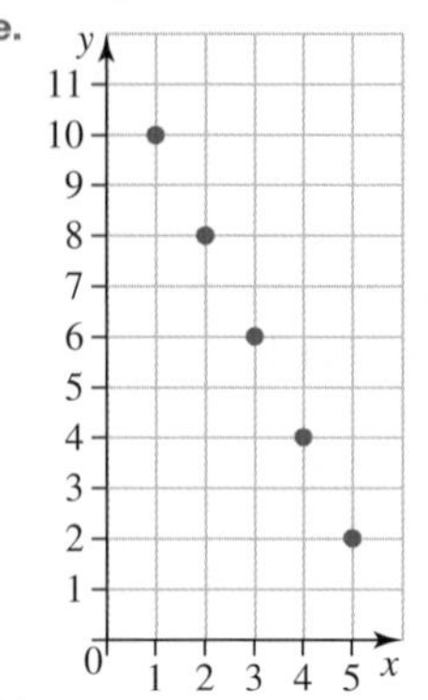

Linear pattern

f. 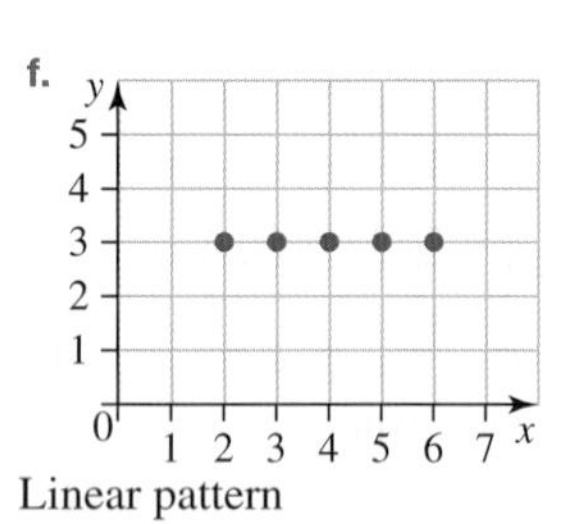

Linear pattern

2. a. b. 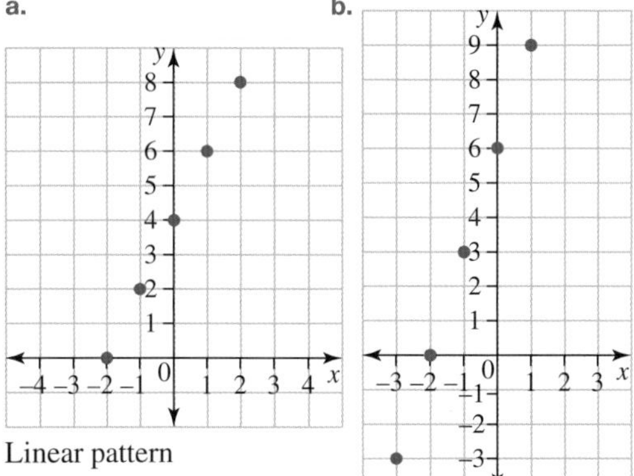

Linear pattern

Linear pattern

c. 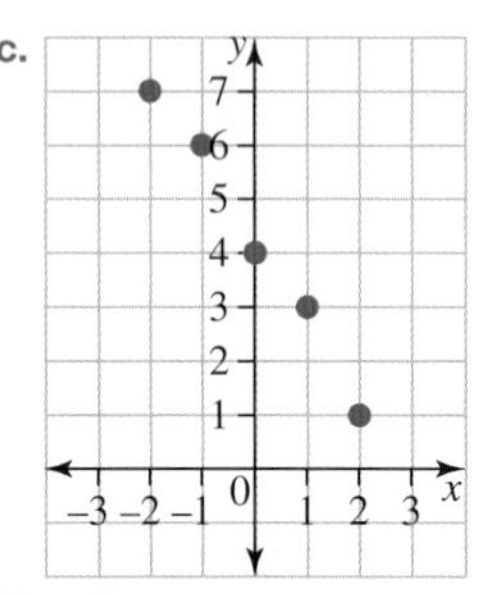

Non-linear pattern

d. 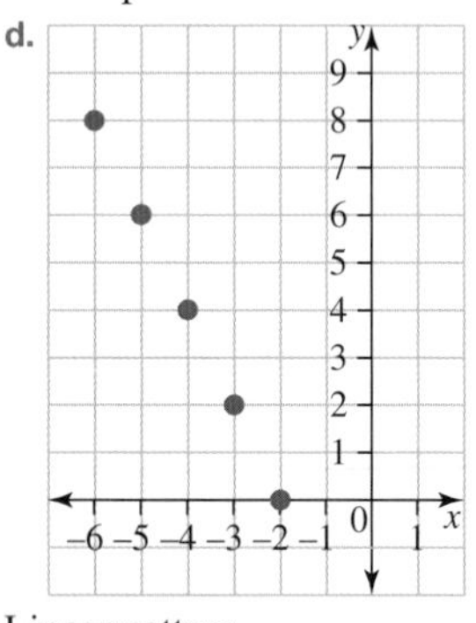

Linear pattern

3. a.

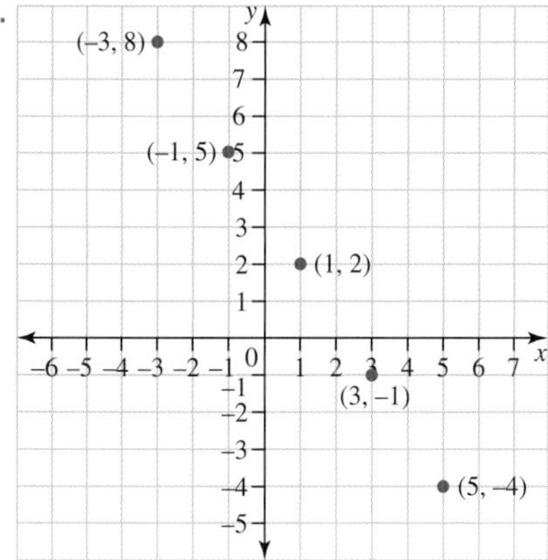


Linear pattern

b.

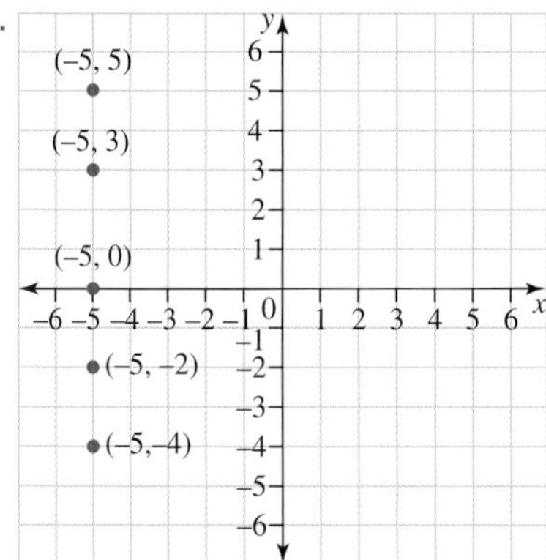


Vertical line

c.

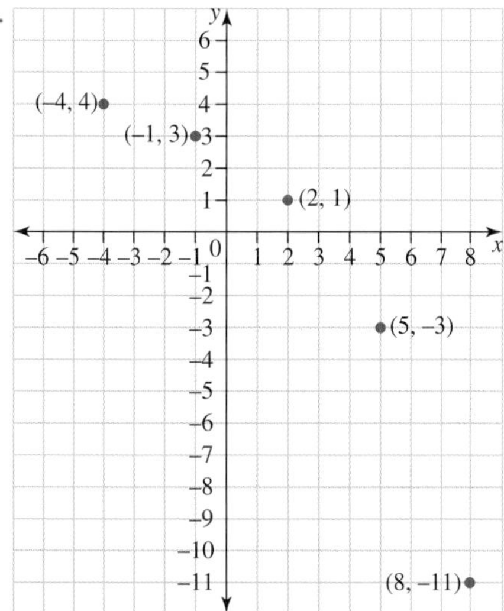


Non-linear pattern

d.

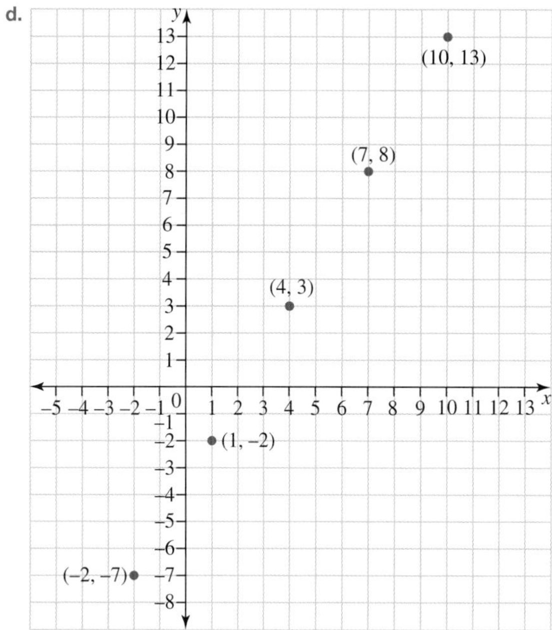


Linear pattern

4. a.

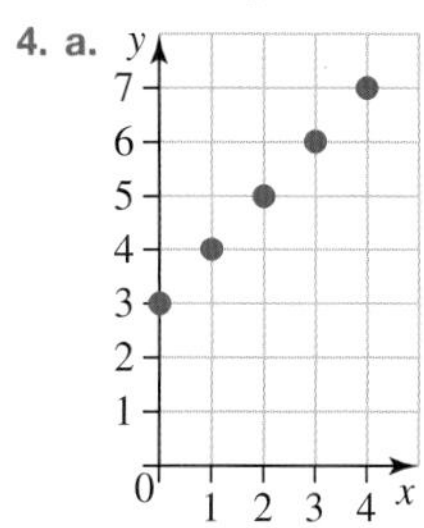

Linear relationship

b.

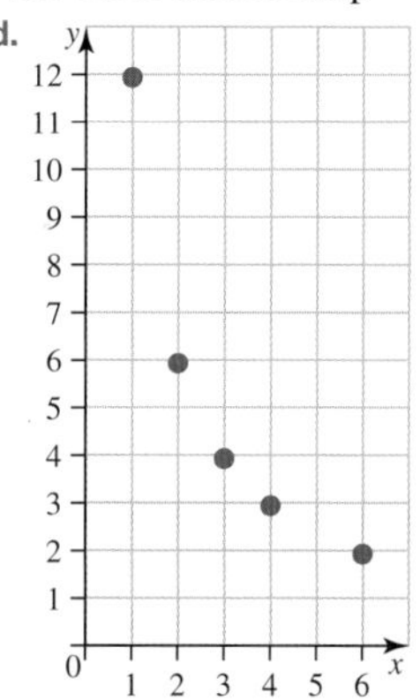

Non-linear relationship

c.

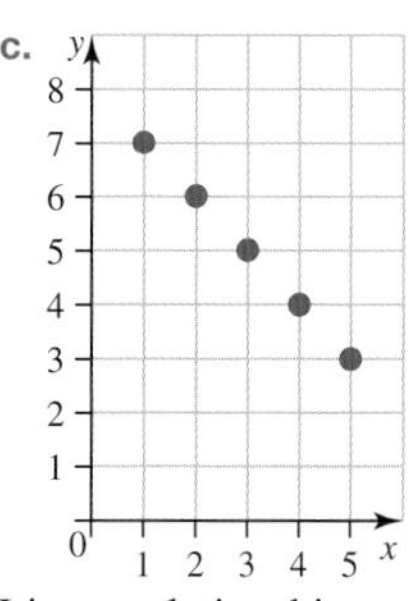

Linear relationship

d.

Non-linear relationship

e.

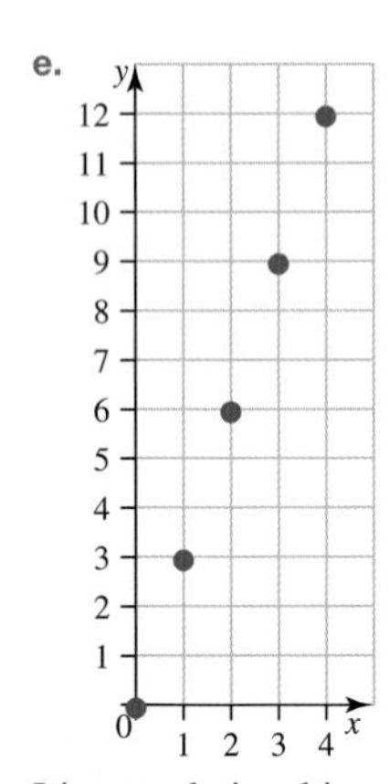


Linear relationship

f.

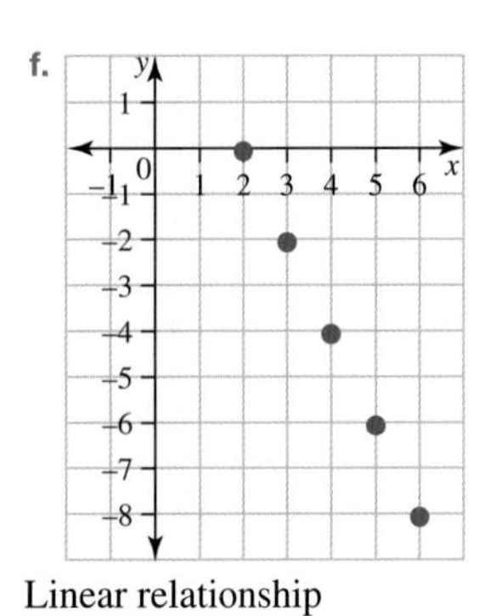


Linear relationship

5. a.

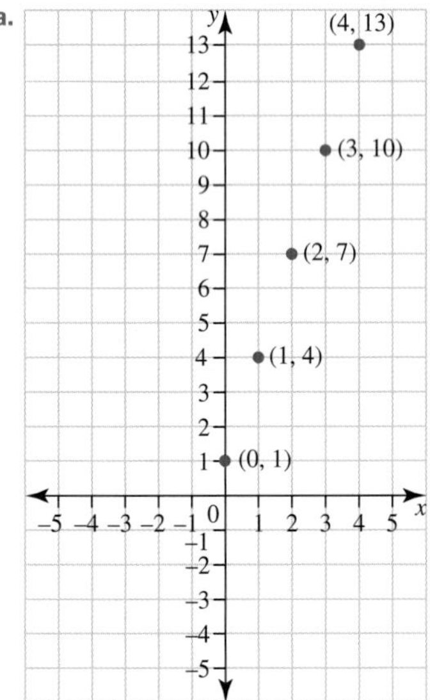


Linear pattern

b.

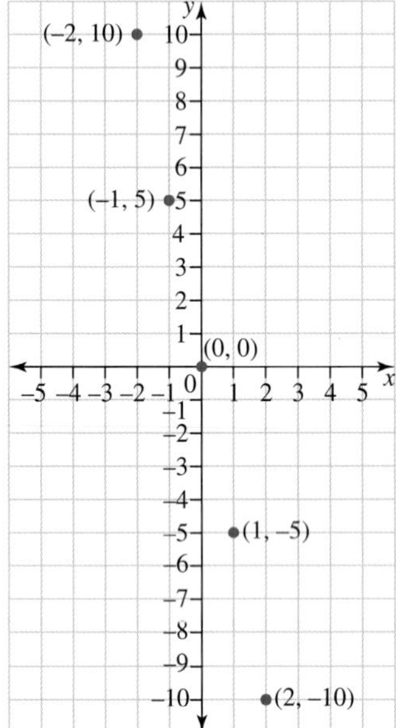


Linear pattern

c.

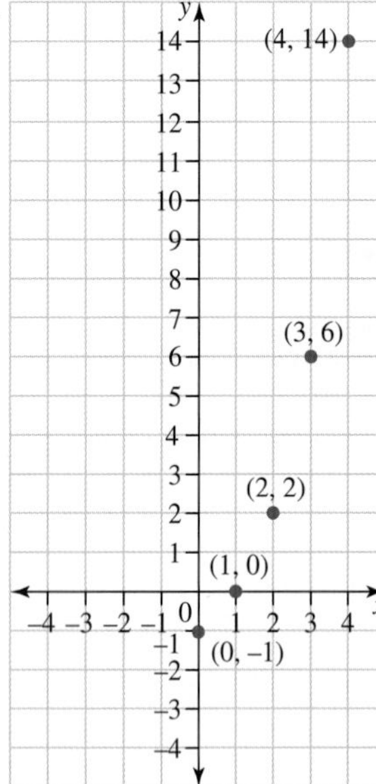


Non-linear pattern

d.

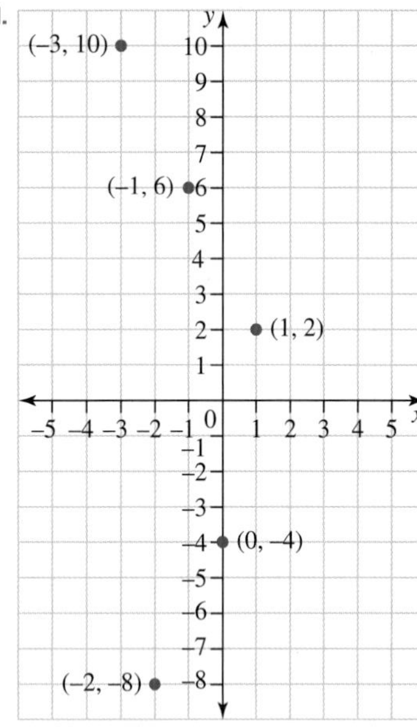


Non-linear pattern

6. a.

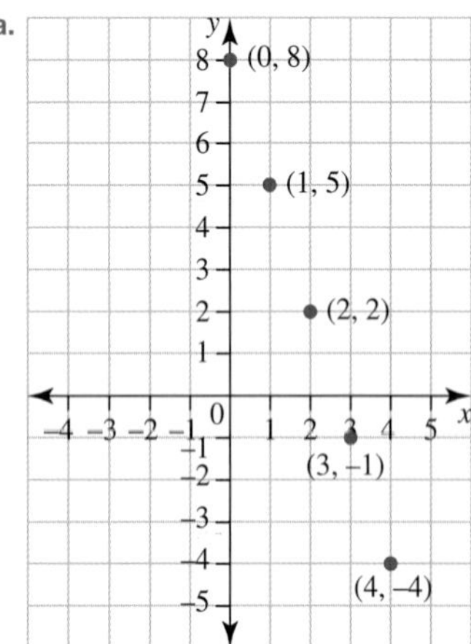


Linear pattern

**b.**

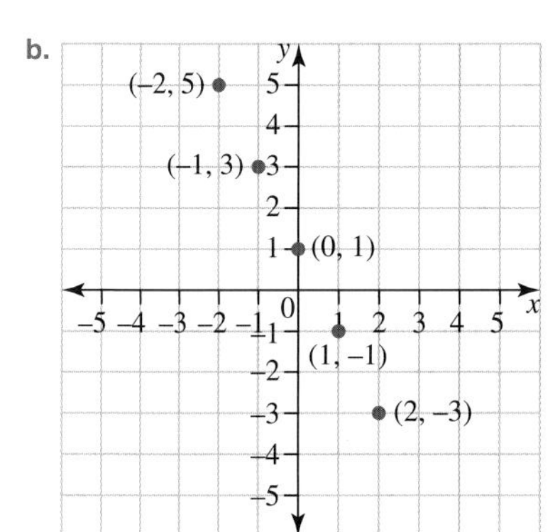


Linear pattern

**c.**

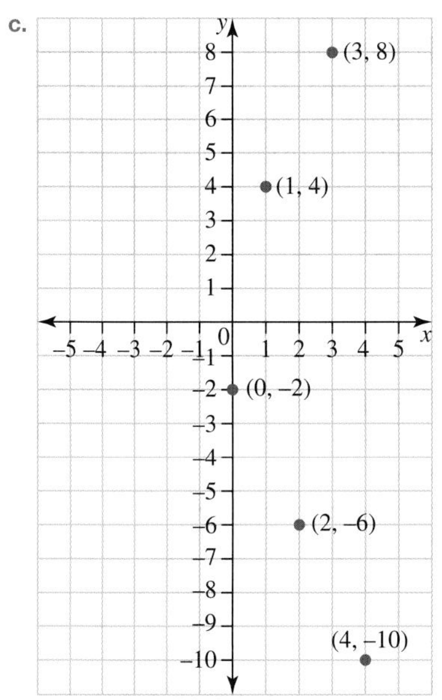


Non-linear pattern

**d.**

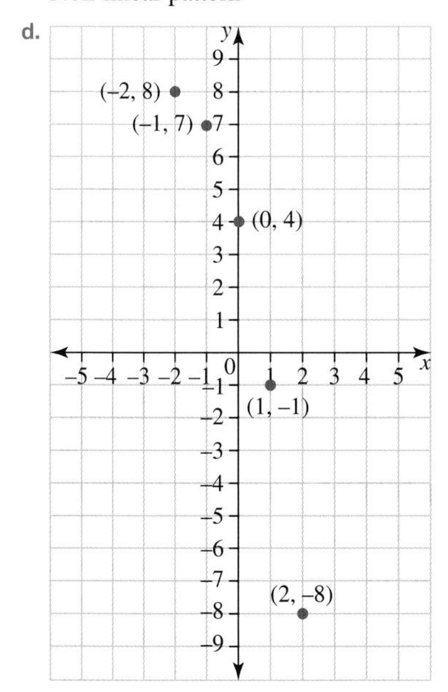


Non-linear pattern

**7. a, c.**

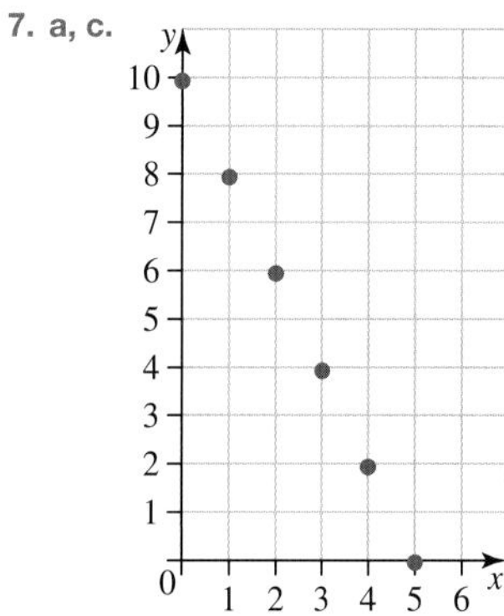


**b.** Linear pattern

**d.** (4, 2), (5, 0)

**8. a, c.**

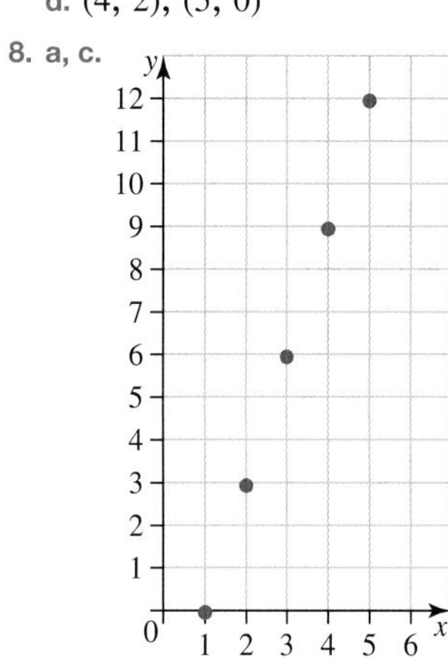


**b.** $y = 6$

**9.** a, b, d

**10. a.** (8, 7) **b.** (3, 2)

**11. a.**

| Week | 1 | 2 | 3 | 4 | 5 | 6 |
|---|---|---|---|---|---|---|
| Height (mm) | 20 | 24 | 28 | 32 |  | 40 |

**b.**

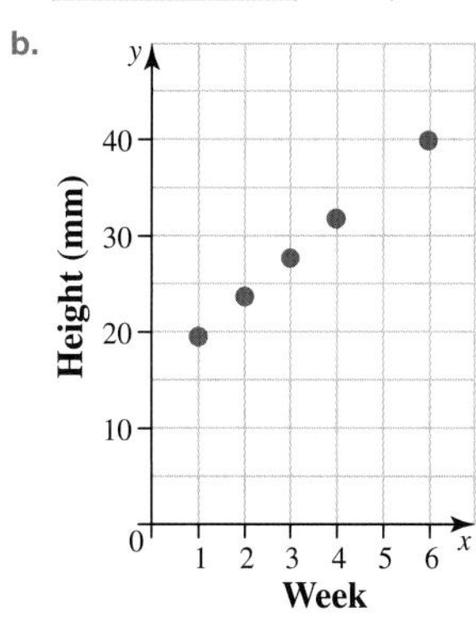


**c.** Yes, the points form a straight line.

**d.** 36 mm

**e.** 48 mm

**12. a.** The rule describes a set of points such that the $y$-coordinate of each point is 3 more than the $x$-coordinate.

**b.**

| $x$ | −2 | −1 | 0 | 1 | 2 |
|---|---|---|---|---|---|
| $y$ | 1 | 2 | 3 | 4 | 5 |

26.

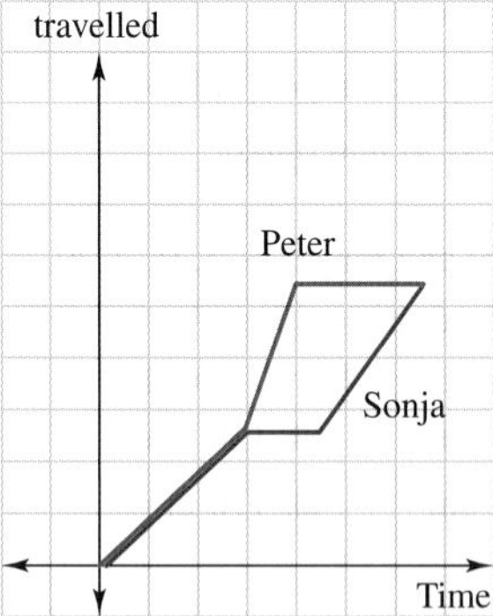


27. It cost 25 cents to make a phone call and then 2 cents for every second of the call.

28. a.

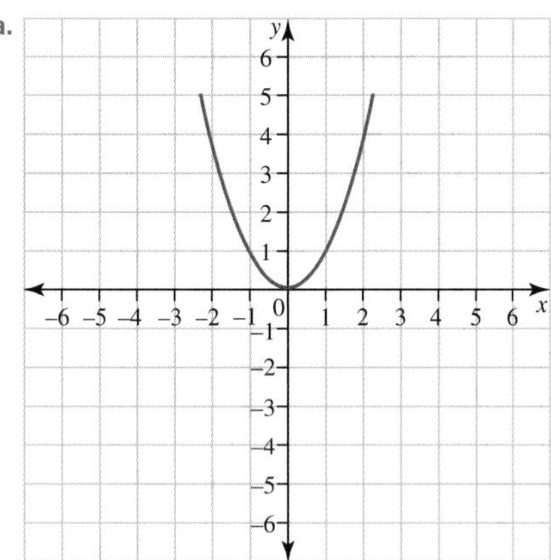

b. As the value of $x$ increases, the value of $y$ decreases to zero and then increases.

29. a.

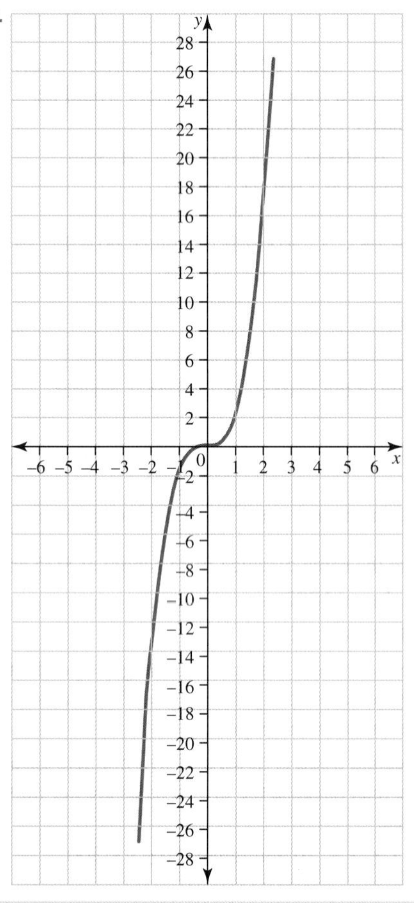

b. As the value of $x$ increases, the value of $y$ increases.

30. 36.12 km per hour

## Project

Players should apply their knowledge of coordinates and the Cartesian plane and play the game to be the first to sink all their opponent's ships.

## 11.6 Review questions

1. a. Joseph's house: B1; Post office: B4

b.

| | | | | |
|---|---|---|---|---|
| D | Post-office | Bus-stop | Star cafe | Library |
| C | | Apple street | | |
| B | Alex's house | Maya's house | Joseph's house | Lena's house |
| A | | | X | |
| | 1 | 2 | 3 | 4 |

2. a. A

b. K, I, L, J

c. M, N, O, P

d. A(−3, 4) B(−4, 3)

e. C(−4, −3) F(4, −3)

3. a. A(2, 4) (3, 5)

b. A B C D E

4. a.

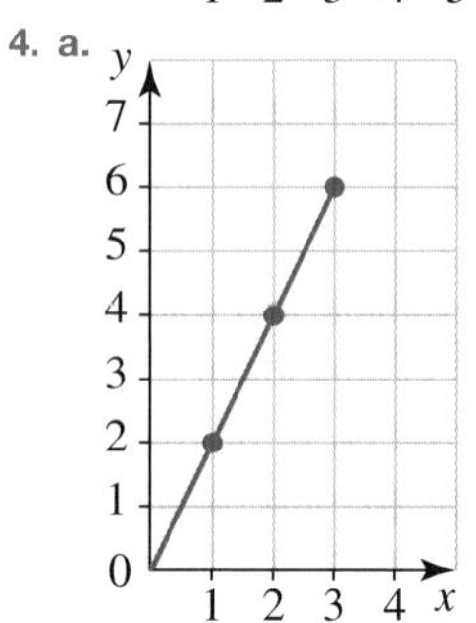

b. Yes, the points form a straight line.

5. H or X
H lies in column 3 and row 2.
X lies in column 4 and row 5.

6. a. i. H ii. H iii. S iv. M

b. B9, C2, H6, I10

c. A5 and A6 are already hit, so he needs A4 and A7 to sink it.

7. a. QUEENSLAND (6, 6), (4, 4), (12, 8), (12, 8), (14, 14), (16, 16), (8, 14), (10, 10), (14, 14), (12, 12)

b. VICTORIA (4, 18), (8, 8), (10, 12), (16, 4), (14, 6), (6, 16), (8, 8), (10, 10)

c. SYDNEY (16, 16), (18, 18), (12, 12), (14, 14), (12, 8), (18, 18)

8. (3, 1) (2, 3) (2, 1) (3, 1) (2, 2) (5, 1) (2, 3) (2, 1) (5, 0) (0, 3) (5, 0) (0, 1) (3, 2) (0, 3) (5, 0) (3, 1) (1, 0) (1, 2)

9. 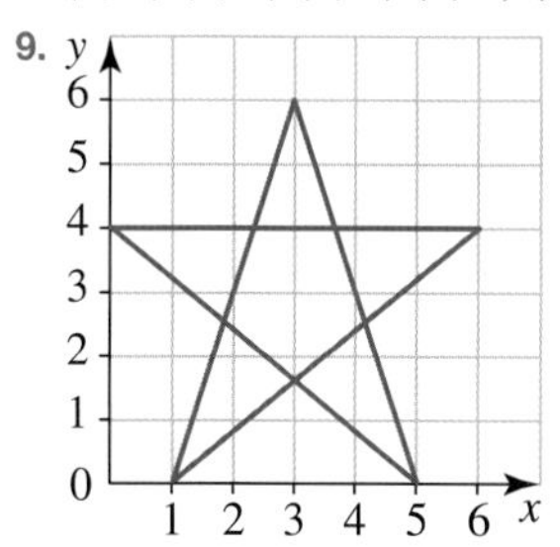


10. D is (3, 7)

11. a. 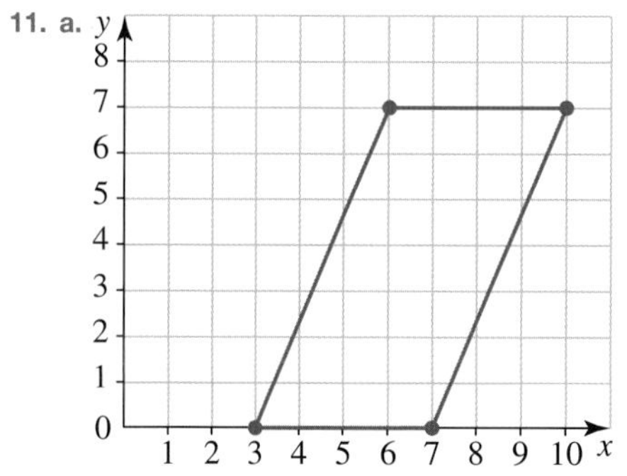


i. Parallelogram

ii. 28 square units

b. 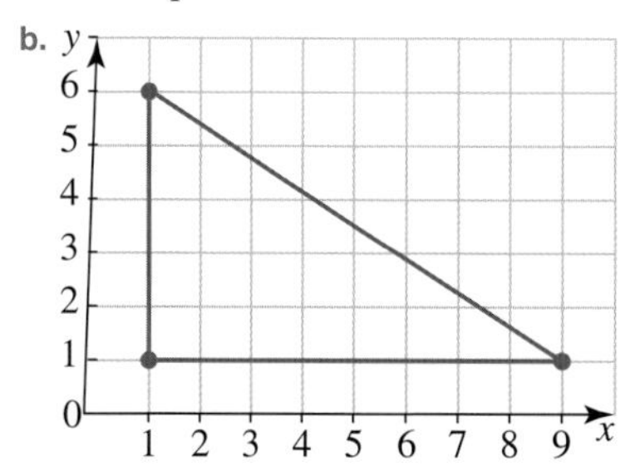


One example could be Triangle ABC where A is (1, 1), B is (1, 6) and C is (9, 1).

The area of this triangle is $\frac{1}{2} \times \text{base} \times \text{height} = \frac{1}{2} \times 8 \times 5 = 20$ square units as required.

12. D(1, −5)

13. a. Domain: {1, 2, 3, 4} and Range {1, 4, 9, 16}

b. No; the relation is $y = x^2$, which is not in the form $y = mx + c$.

14. Not true, as a straight line could:
- be parallel to the $y$-axis and only cut the $x$-axis once
- be parallel to the $x$-axis and only cut the $y$-axis once
- pass through the origin, cutting the axes at only that point.

15. a. A: Anton is swimming out from the shore.
B: Anton is staying the same distance from the shore.
C: Anton is swimming back to the shore.

b. Anton is swimming faster in section A than in section C.

16. a. i. 100 beats/min

ii. 90 beats/min

b. 3 min

c. 1, 5 min. Pulse rate increases and exceeds 80 beats/min while exercising and then begins to decrease and comes back down to 80 beats/min after exercising.

d. 100 beats/min

e. 4 minutes

17. a. 9:00 am

b. 6:00 am

c. i. 200 ii. 450

d. i. 7:00 am, 10:15 am, 12:15 pm

ii. 6:30 am, 10:45 am, 11:30 am

e. The line graph displays how traffic changes over time and we can estimate traffic flow at in-between times.

18. Wilma left home at 1 pm and rode for 16 km (this took 1 hour) before she rested for 15 minutes. She continued for $1\frac{1}{2}$ hours, travelling 8 km, then rested again for 15 minutes. She rode on for 8 km, which took 30 minutes. She reached Mt Cranson at 4:30 pm. She stopped for 30 minutes, then rode home at 5 pm non-stop (32 km) in 1 hour, getting back home at 6 pm. Her speeds between stops were 16 km/h, $5\frac{1}{3}$ km/h, 16 km/h and 32 km/h. These speeds suggest that she rode uphill to Mt Cranson and downhill home. Her average speed for the entire day was 12.8 km/h.

# 12 Transformations

## LESSON SEQUENCE


**12.1** Overview ....... 672
**12.2** Line and rotational symmetry ....... 674
**12.3** Translations ....... 680
**12.4** Reflections ....... 686
**12.5** Rotations and combined transformations ....... 693
**12.6** Review ....... 705

# LESSON
# 12.1 Overview

## Why learn this?

Geometry is not only the study of figures and shapes; it is the study of their movement. *Transformation* is a term used to describe a change to a shape or its position. Shapes can be rotated, reflected or translated (moved to another position). Some shapes, such as isosceles and equilateral triangles, have symmetrical properties. Symmetry is found in nature; some examples are honeycombs, starfish and orchids.

Transformations can be found in many different places from ancient times to the present day. Tiling a floor or wall uses the concept of transformation, where the same shape or shapes are moved to cover the surface. For example, Ancient Roman mosaics and Islamic tilings often feature intricate geometric patterns.

During the twentieth century, the graphic artist Escher used transformations in his paintings and drawings. The verandahs of period homes in Australia are often tessellated or tiled in traditional English patterns involving just a few different tiles that are repeated to cover the space.

The concept of transformation is still used in architecture and building today. It can also be seen in design. If you look carefully, you will find symmetry and geometric patterns in the world around you, all of which are examples of how transformations can be applied in everyday life.

## Exercise 12.1 Pre-test

learnon

1. State the number of axes of symmetry a square has.

2. MC Select which of these times on an analogue clock has a vertical axis of symmetry.
   A. 3:00 B. 3:45 C. 9:15 D. 12:00 E. 6:30

3. MC Select which of the following capital letters does not have an axis of symmetry.
   A. A B. B C. E D. F E. H

4. State which regular polygon has 6 axes of symmetry.

5. State the final position of an object, from its original position, if it undergoes translations of: 2U, 4D, 6L, 4R, 7U and 3R.

6. MC If, relative to its original position, the final position of a point is 2U and 3L, the translations that it could have undergone are:
   A. 3U, 5R, 5D, 2L B. 3U, 5L, 5D, 2R C. 6L, 4U, 5U, 3R, 7D
   D. 7U, 2D, 4R, 6L, 2D E. 4D, 2R, 7L, 2U, 2R

7. State the final position of the point $(-3, 5)$ after it undergoes the following translations: 3R, 4L, 3U, 1L, 5D.

8. State the coordinates of the point $(-2, 3)$ after it undergoes a reflection in the $x$-axis.

9. MC An analogue clock is showing a time of 3:30. If the little hand is reflected about the line from the 12 to the 6, the new time on the clock is:
   A. 3:30 B. 4:30 C. 8:30 D. 9:30 E. 9:00

10. MC Select which of the following letters does not remain the same if reflected either horizontally or vertically.
    A. A B. B C. O D. L E. M

11. State the coordinates of the point $(-4, -7)$ after it is reflected in the $y$-axis.

12. MC In the diagram shown, select which two transformations the shape on the left has undergone to become the shape on the right.
    A. A reflection and then a rotation
    B. A reflection and then a translation
    C. A rotation and then a translation
    D. A rotation and then a reflection
    E. A translation and then a reflection

13. MC In the diagram, point A is located at the origin and each square represents one unit.
    If the shape is rotated at point A, 90° in the anticlockwise direction, identify the new coordinates of point B.
    A. $(0, 0)$ B. $(6, 0)$ C. $(0, -6)$ D. $(-6, 0)$ E. $(-6, -6)$

B
C A

## WORKED EXAMPLE 2 Identifying rotational symmetry

**Identify whether the following shapes have rotational symmetry. State the order of rotational symmetry for each shape.**

a. 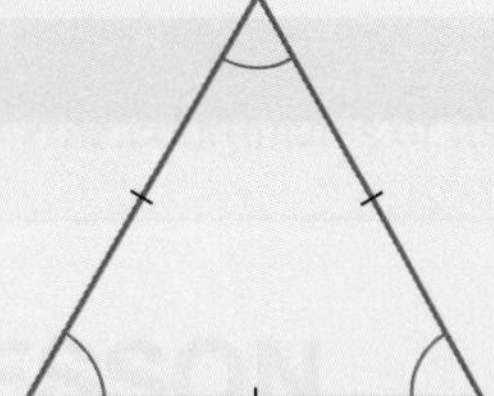

b. 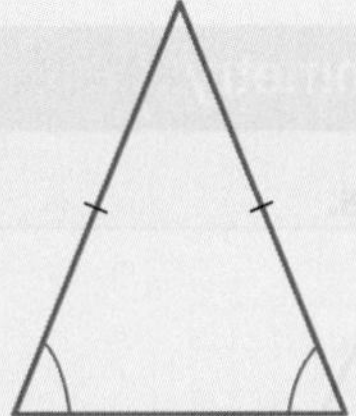

c. 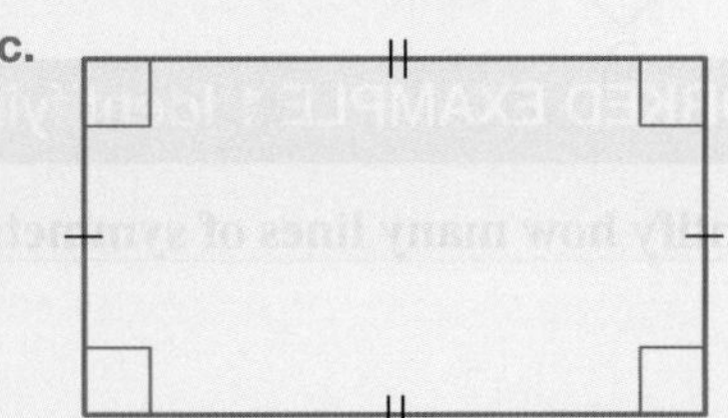

| THINK | WRITE |
|---|---|
| a. 1. An equilateral triangle can be rotated less than 360° to match the original shape, so it has rotational symmetry. | a. Yes, this shape has rotational symmetry. |
| 2. A match occurs three times throughout the rotation (each of the three sides can sit horizontally). | The shape has a rotational symmetry of order 3. |
| b. 1. An isosceles triangle must be rotated 360° to match the original shape, so it does not have rotational symmetry. | b. This shape does not have rotational symmetry. This shape has rotational symmetry of order 1. |
| c. 1. A rectangle can be rotated less than 360° to match the original shape, so it has rotational symmetry. | c. Yes, this shape has rotational symmetry. |
| 2. A match occurs two times throughout the rotation (each of the two long sides can sit horizontally). | The shape has a rotational symmetry of order 2. |

### DISCUSSION

The flags of many countries contain lines of symmetry. For example, the Canadian flag has one line of symmetry. Look at some other countries' flags and identify lines of symmetry or rotational symmetry you find in them.

 Resources

 **eWorkbook** Topic 12 Workbook (worksheets, code puzzle and project) (ewbk-1913)

 **Interactivities** Individual pathway interactivity: Line and rotational symmetry (int-4387)

Axes of symmetry (int-4140)

## Exercise 12.2 Line and rotational symmetry

learnon

12.2 Quick quiz on | 12.2 Exercise

### Individual pathways

| ■ PRACTISE | ■ CONSOLIDATE | ■ MASTER |
|---|---|---|
| 1, 4, 8, 9, 12, 15, 18 | 2, 5, 7, 10, 13, 16, 19 | 3, 6, 11, 14, 17, 20 |

### Fluency

1. **WE1** Identify the number of axes of symmetry for each shape. (Some shapes will have more than one axis of symmetry.) *Hint*: Copy each of the following shapes onto grid or squared paper. Carefully cut out each shape. Fold it to find the axes of symmetry.

   a. 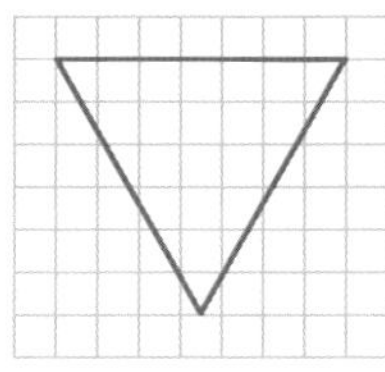
   b. 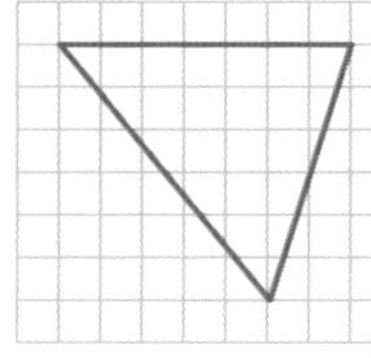
   c. 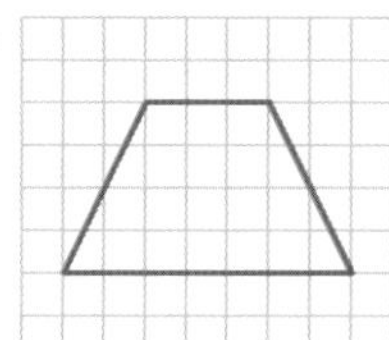
   d. 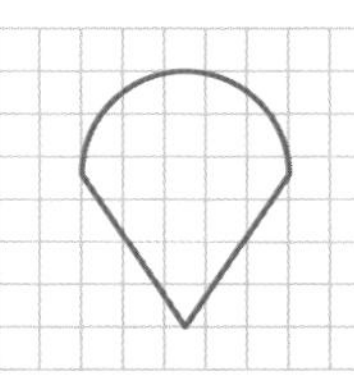
   e. 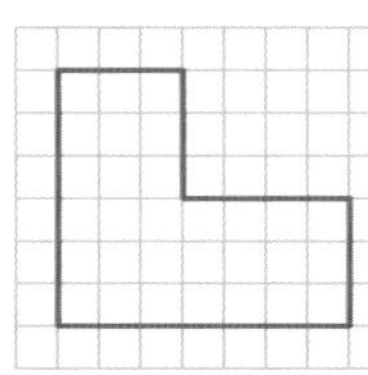

2. Identify the number of axes of symmetry for each shape.

   a. 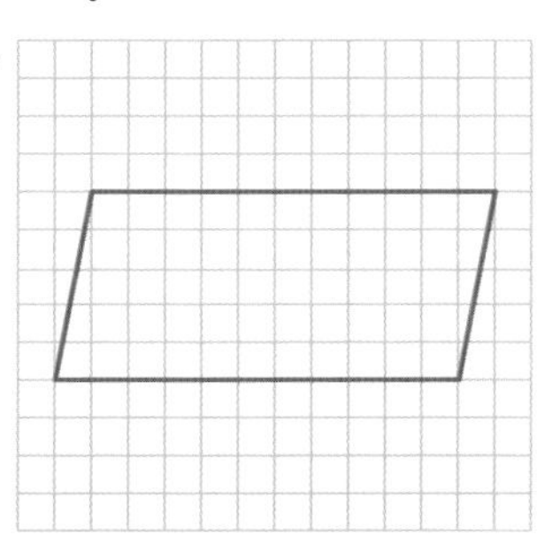
   b. 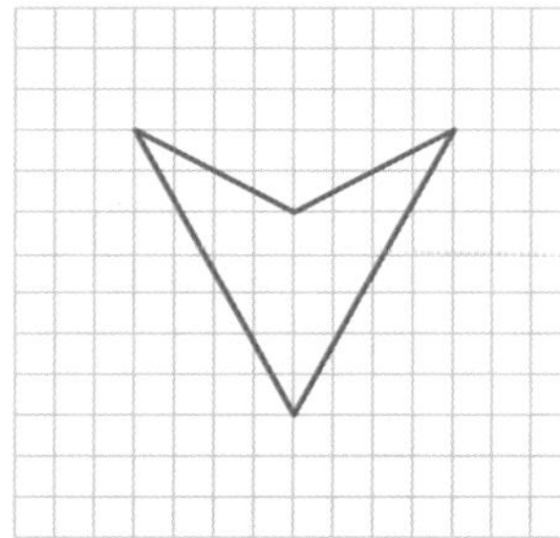
   c. 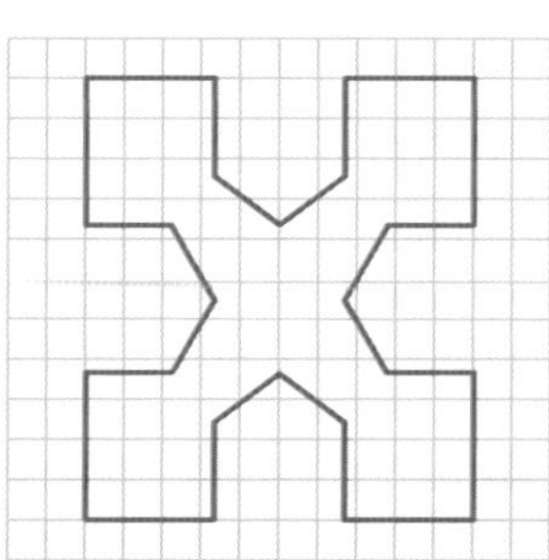
   d. 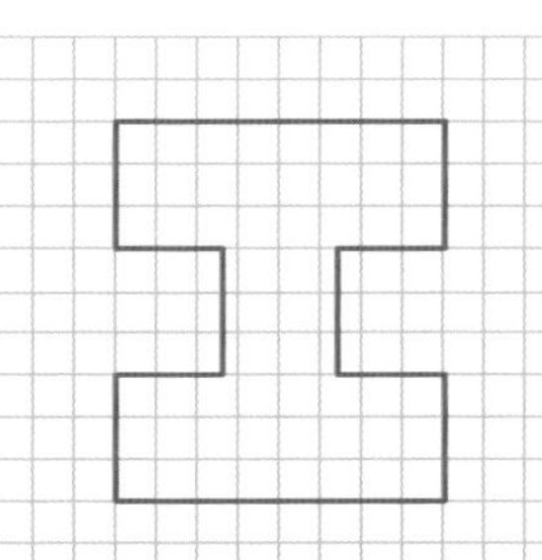
   e. 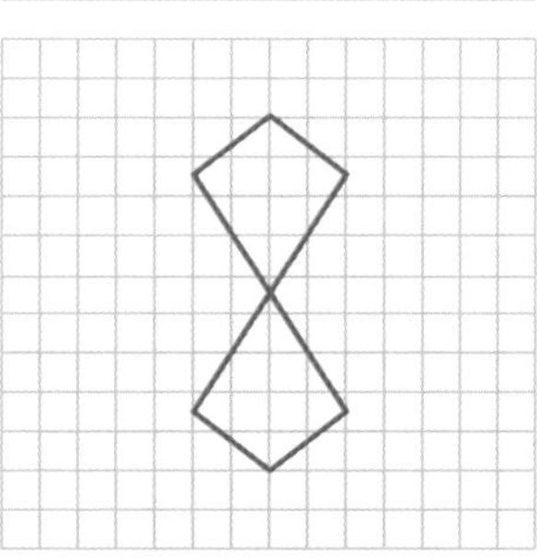

3. Identify how many axes of symmetry each of the following shapes has.

   a. 
   b. 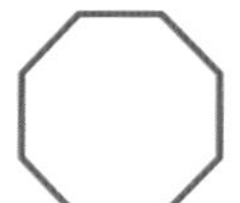
   c. 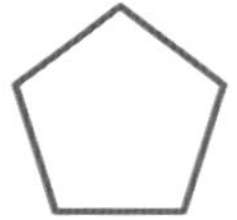
   d. 

4. WE2 Identify which of the shapes in question 1 have rotational symmetry. For each shape, state the order of rotational symmetry.

5. Identify which of the shapes in question 2 have rotational symmetry. For each shape, state the order of rotational symmetry.

6. For each figure in question **3**, state the order of rotational symmetry.

7. MC Identify the number of axes of symmetry in the shape shown.

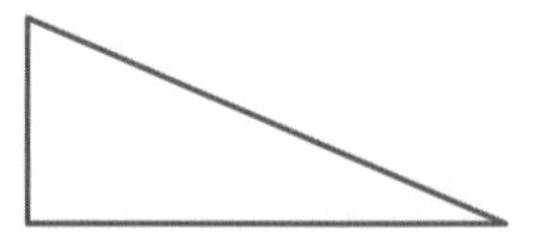

A. 0    B. 1    C. 2    D. 3    E. 4

8. a. For each of the following diagrams, use the dotted line as an axis of symmetry to complete the picture.

i. 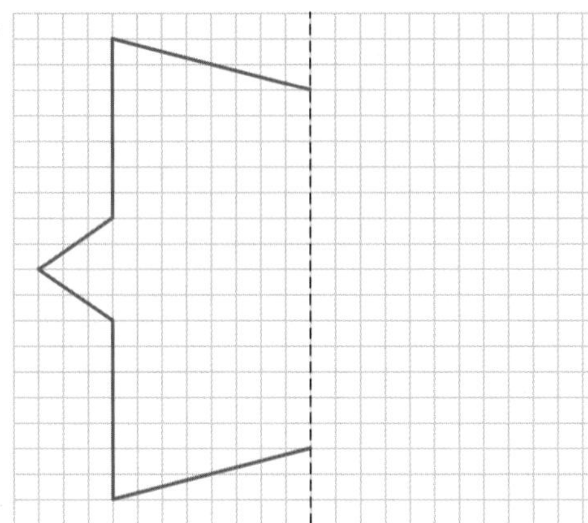

ii. 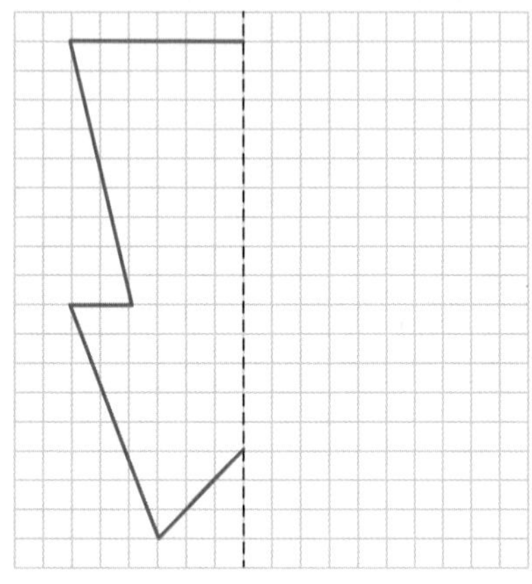

b. State the order of rotational symmetry of the completed shapes in part a.

## Understanding

9. a. Determine how many axes of symmetry an equilateral triangle has.
   b. Determine how many axes of symmetry an isosceles triangle has.
   c. Determine how many axes of symmetry a scalene triangle has.

10. a. State how many lines of symmetry a square has.
    b. State how many lines of symmetry a rectangle has.
    c. State how many lines of symmetry a parallelogram has.

11. a. Determine how many lines of symmetry a rhombus has.
    b. Determine how many lines of symmetry a trapezium has.
    c. Determine how many lines of symmetry a kite has.

12. a. State the order of rotational symmetry of an equilateral triangle.
    b. State the order of rotational symmetry of an isosceles triangle.
    c. State the order of rotational symmetry of a scalene triangle.

13. a. State the order of rotational symmetry of a square.
    b. State the order of rotational symmetry of a rectangle.
    c. State the order of rotational symmetry of a parallelogram.

14. a. State the order of rotational symmetry of a rhombus.
    b. State the order of rotational symmetry of a trapezium.
    c. State the order of rotational symmetry of a kite.

## Reasoning

15. a. Determine how many axes of symmetry a circle has. Justify your answer.
    b. Determine the order of rotational symmetry of a circle. Justify your answer.

16. The following shapes are regular polygons.

    a. Show how many axes of symmetry each shape has.
    b. Determine the order of rotational symmetry of each shape.
    c. Is there a relationship between the number of vertices in each of these shapes and the number of axes of symmetry? Explain your answer.

17. a. Complete the pattern knowing that the horizontal and the vertical dashed lines are both axes of symmetry for this pattern.
    b. How did you know where to draw the pattern? Explain your answer using the concept of symmetry and its meaning.
    c. Determine the order of rotational symmetry of the completed pattern.

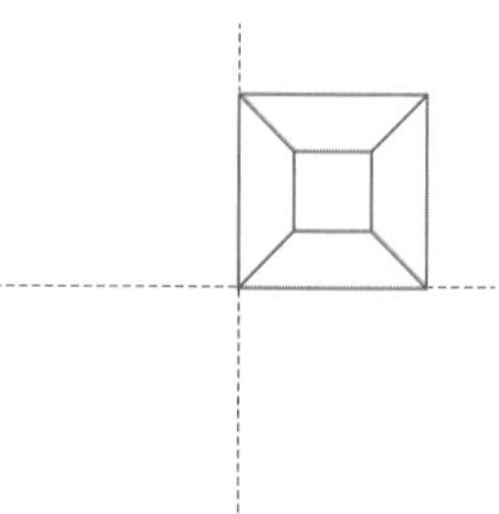

## Problem solving

18. The shape given is a sector of a circle.
    a. Determine how many axes of symmetry this shape has.
    b. Determine the order of rotational symmetry of the shape.

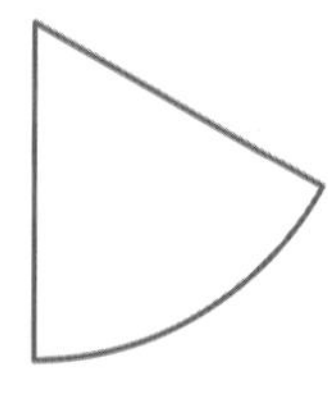

19. The shape given is a circle with a marked chord.
    a. Determine how many axes of symmetry this shape has.
    b. Determine the order of rotational symmetry of the shape.

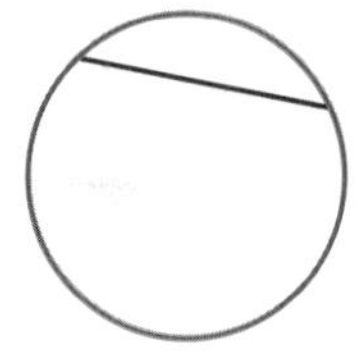

20. Answer the following questions for the given shape.
    a. Copy the following shape and draw in its axes of symmetry.
    b. Determine the order of rotational symmetry of this shape.

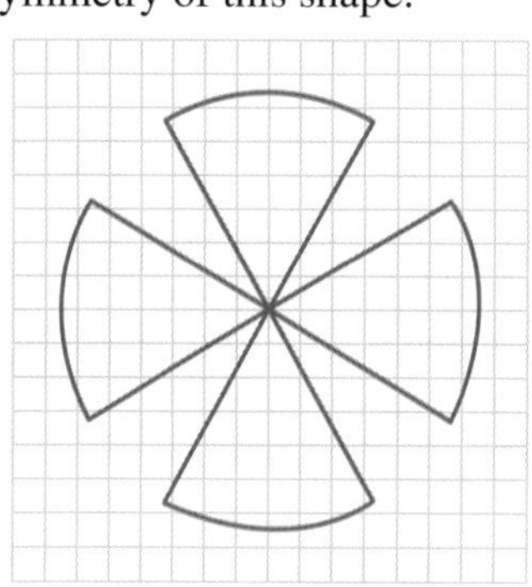

# LESSON
# 12.3 Translations

## LEARNING INTENTION

At the end of this lesson you should be able to:
- translate points and objects in the Cartesian plane
- identify translations in the Cartesian plane.

## 12.3.1 Transformations

eles-4682

- Transformations are operations that move a point P from one position to another.
- The position of the point, P, after a transformation is called the **image** of the point and is denoted P′.
- Shapes can also undergo transformations. In this case, the transformations are applied to all points of the shape. For example, the image of the rectangle ABCD after a transformation is A′B′C′D′.
- There are different types of transformations. In this topic we will cover **translations**, **reflections** and **rotations**.

### Translations

- A translation is the movement of a point or an object up, down, left or right (U, D, L or R) without flipping, turning or changing size.
- After a translation, the object has the same shape and size but is in a different position. For this reason translations are also known as 'slides'.
- The Cartesian plane in the diagram shows some examples of translations.
- The point P has been translated 3 units to the right.
- The triangle ABC has been translated 3 units to the left and 4 units down.

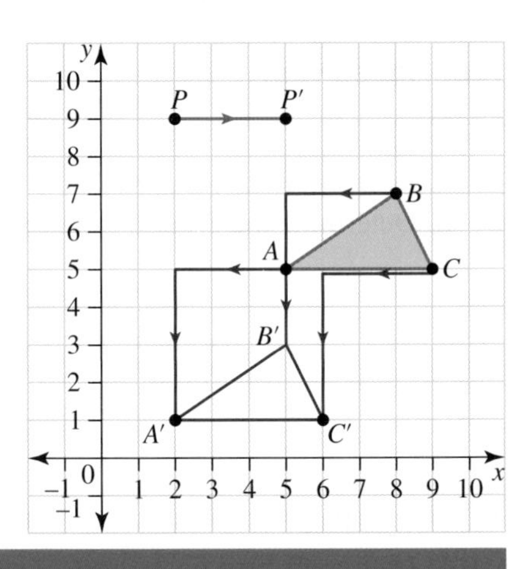


### WORKED EXAMPLE 3 Identifying a translation

**State the translation of shape *a* to *a*′.**

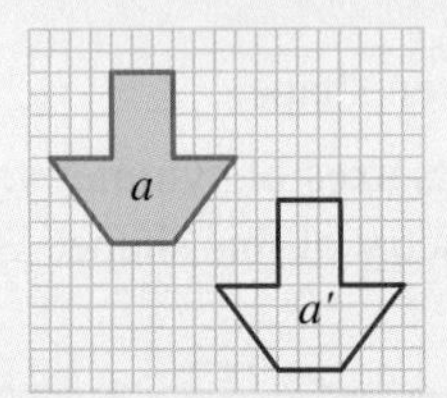


**THINK**

1. Select any point on the boundaries of the original object, *a* (any vertex is good to choose), and name it A. Locate the corresponding point on the image and name it A′. From point A, draw a horizontal line until it is directly above A′; then draw a vertical line so that it meets A′.

2. Count the number of units that the object has moved across (to the right) and down and, hence, record the translation that took place.

**WRITE/DRAW**

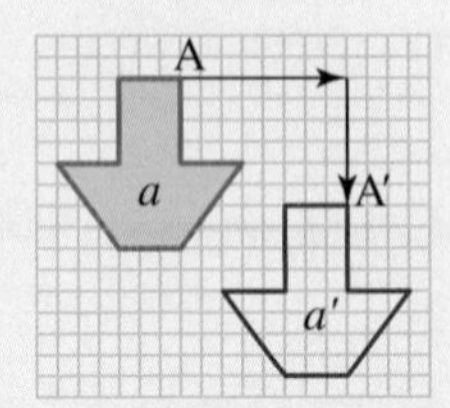


Translation: 8R 6D

## Translating points

- To translate a point horizontally, add or subtract the appropriate number from the $x$-coordinate of the point.
  - Right is the positive direction.
  - Left is the negative direction.
- To translate a point vertically, add or subtract the appropriate number from the $y$-coordinate of the point.
  - Up is the positive direction.
  - Down is the negative direction.

### WORKED EXAMPLE 4 Translating points

**State the coordinates and sketch the image of the point $P(2, -1)$ after:**

a. **it is translated 3 units to the left**

b. **it is translated 5 units up**

c. **it is translated 1 unit to the right and 2 units down.**

| THINK | WRITE |
|---|---|
| a. 1. To translate 3 units to the left, subtract 3 from the $x$-coordinate. | a. $P' = (2-3, -1)$ <br> $= (-1, -1)$ |
| 2. Sketch the image P′. | |

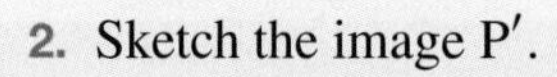

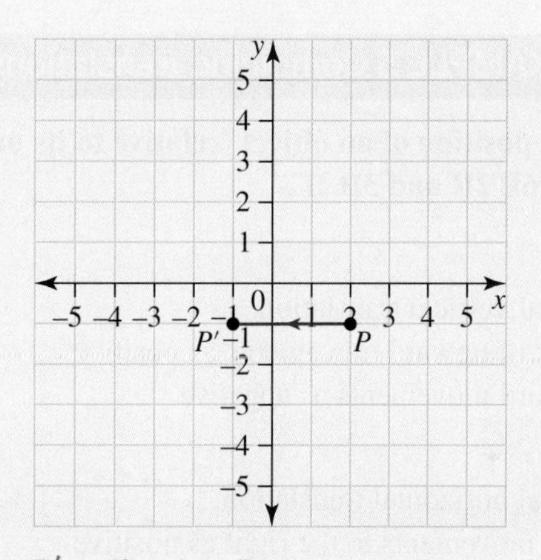


| THINK | WRITE |
|---|---|
| b. 1. To translate 5 units up, add 5 to the $y$-coordinate. | b. $P' = (2, -1+5)$ <br> $= (2, 4)$ |
| 2. Sketch the image P′. | |

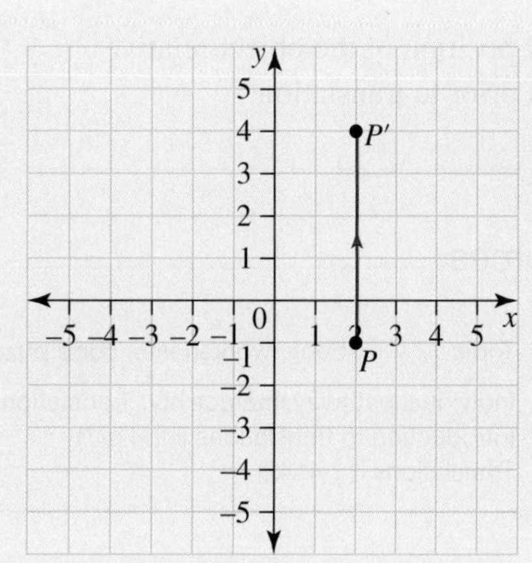

c. 1. To translate 1 unit to the right and 2 units down, add 1 to the $x$-coordinate and subtract 2 from the $y$-coordinate.

2. Sketch the image P′.

c. $P' = (2+1, -1-2)$
$= (3, -3)$

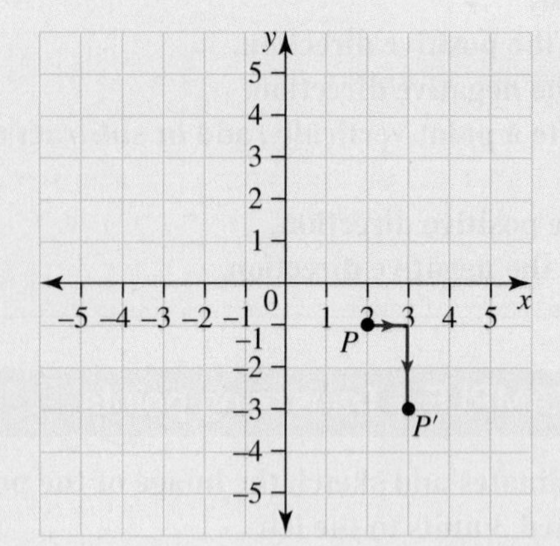

## 12.3.2 Combining translations

eles-4684

- If a point or a shape has been translated several times in both horizontal and vertical directions, its final position (relative to the original one) can be described by the total horizontal and total vertical translations.

### WORKED EXAMPLE 5 Combining translations

**State the final position of an object (relative to its original position) after the translation: 4U 2L, 2D 5L, 6U 2R and 3D 2L.**

| THINK | WRITE |
|---|---|
| 1. Find the total vertical translation. *Note:* Think of upward movements as positive and downward movements as negative. | $4U + 2D + 6U + 3D$ $= 4 - 2 + 6 - 3$ $= 5$ So, the vertical translation is 5U. |
| 2. Find the total horizontal translation. *Note:* Treat movements to the right as positive and movements to the left as negative. | $2L + 5L + 2R + 2L$ $= -2 - 5 + 2 - 2$ $= -7$ So, the horizontal translation is 7L. |
| 3. State the final position of the object relative to its position prior to translation. | The position after translation, relative to the original position, is 5U 7L. |

on Resources

**eWorkbook** Topic 12 Workbook (worksheets, code puzzle and project) (ewbk-1913)

**Interactivities** Individual pathway interactivity: Translations (int-4388)
Introduction to translations (int-4141)
Translations (int-4142)

# Exercise 12.3 Translations

learnon

12.3 Quick quiz on | 12.3 Exercise

### Individual pathways

| ■ PRACTISE | ■ CONSOLIDATE | ■ MASTER |
|---|---|---|
| 1, 4, 7, 10, 14, 17 | 2, 5, 8, 11, 13, 15, 18 | 3, 6, 9, 12, 16, 19 |

### Fluency

1. **WE3** State the translation that has occurred to each of the shapes labelled $a$ to $d$ in the following figure.

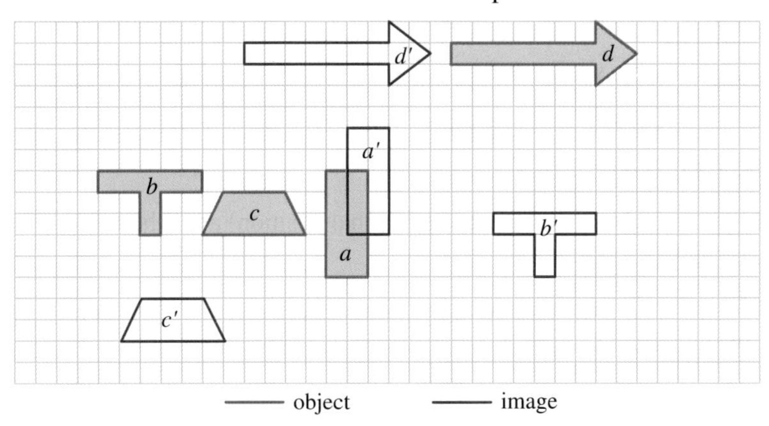


2. State the translation that has occurred to each of the shapes labelled $e$ to $h$ in the following figure.

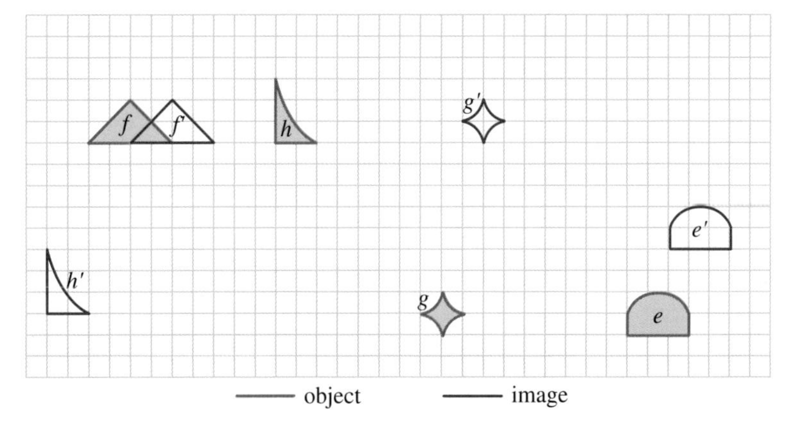


3. State the translation that has occurred to each of the shapes labelled $i$ to $l$ in the following figure.

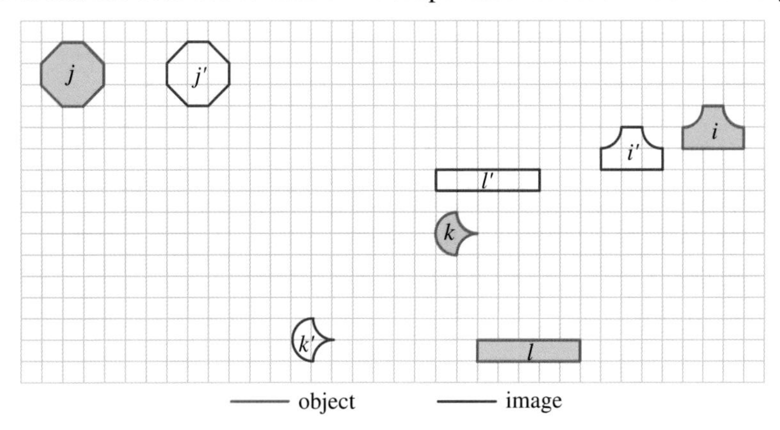

4. **WE4** State the coordinates and sketch the image of the point $P(1, 3)$ after:

a. it is translated 2 units to the left
b. it is translated 4 units up
c. it is translated 1 unit to the right and 2 units down.

5. State the coordinates and sketch the image of the point $P(4, -1)$ after:

a. it is translated 5 units to the right
b. it is translated 2 units down
c. it is translated 4 units to the left and 3 units up.

6. State the coordinates and sketch the image of the point $P(-3, -5)$ after:

a. it is translated 7 units to the right
b. it is translated 3 units up
c. it is translated 4 unit to the right and 4 units down.

7. **WE5** State the final position of an object (relative to its original position) after the translation: 1R 7U, 2R 3D, 5L 2D and 2L 4U.

8. State the final position of an object (relative to its original position) after the translation: 5L 2U, 6L 4U, 8R 3U and 7L 11D.

9. State the final position of an object (relative to its original position) after the translation: 8R 6D, 1R 5U, 3L 9U and 7R 7D.

### Understanding

10. Translate the square ABCD, shown in the diagram, 3 units right and 1 unit up, then translate the resulting image 2 units left and 6 units up.

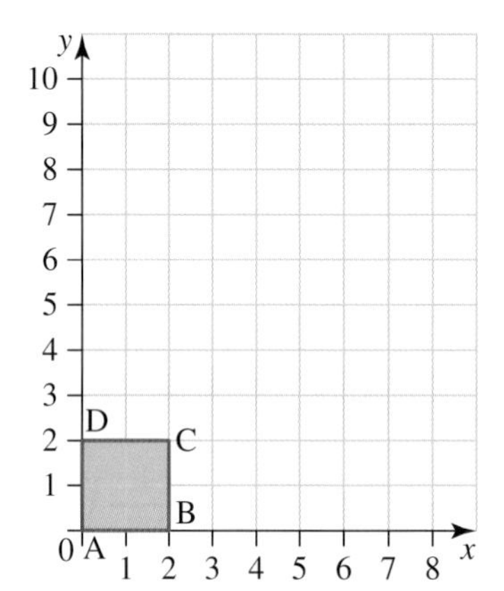


11. Determine each of the following translations of the labelled arrows in the diagram.

a. A to C
b. A to F
c. F to E
d. E to F
e. D to B
f. E to A

12. Using the diagram in question 11, state which arrow is the object and which arrow is the image in the following translations.

a. 2U 5R
b. 2D 5L
c. 2U 9L
d. 2L 8U
e. 4U 10L
f. 14L

13. a. Translate the point P (3, 1) 3 units up and 2 units to the left. Label the new point P′ and give the coordinates of this point.
    b. Translate P′ 2 units down and 3 units to the left. Label the new point P″ and give the coordinates of this point.
    c. Determine what single translation would move P to P″.

## Reasoning

14. A city road map can be thought of using Cartesian coordinates with the north being the positive direction of the *y*-axis. Lucia's home is located at (3, 11) and her work is located at (−6, −2).
    a. Describe the translation that describes Lucia's trip home from work.
    b. Lucia does not always take the same route home from work. Sometimes she drives past the supermarket to buy groceries on her way home. If the supermarket is located at (1, 5) describe the transformations that describe Lucia's trip home.
    c. Show that the transformations in parts a and b are equivalent.

15. In a game of draughts, a counter is removed from the board when another counter jumps over it. Four counters are arranged on the corner of a draughts board as shown. Use three horizontal or vertical jumps to remove all three red counters, leaving the single black counter. Copy and complete the table to show your working.

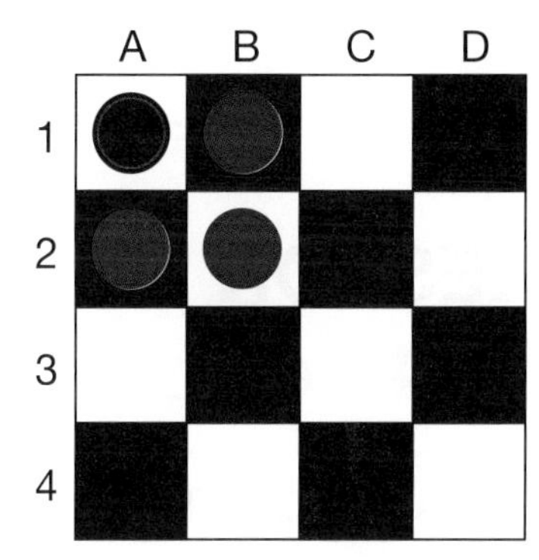


| Jump | From | To |
|---|---|---|
| 1 | | |
| 2 | | |
| 3 | | |

16. In another game, any counter can jump any other counter. Use seven horizontal or vertical jumps to remove all seven red counters, leaving the single black counter. Copy and complete the table to show your working.

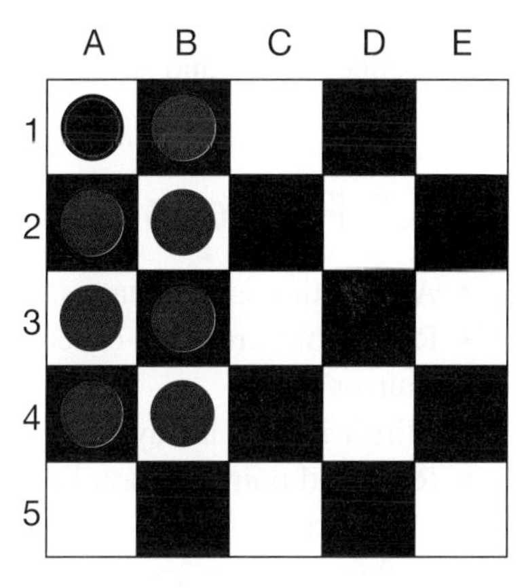


| Jump | From | To |
|---|---|---|
| 1 | | |
| 2 | | |
| 3 | | |
| 4 | | |
| 5 | | |
| 6 | | |
| 7 | | |

## Problem solving

17. A crane is moving a girder needed in the construction of a building 5 m up from an initial position of 3 m above the ground. Determine how far above ground level the girder will be at the end of the translation.

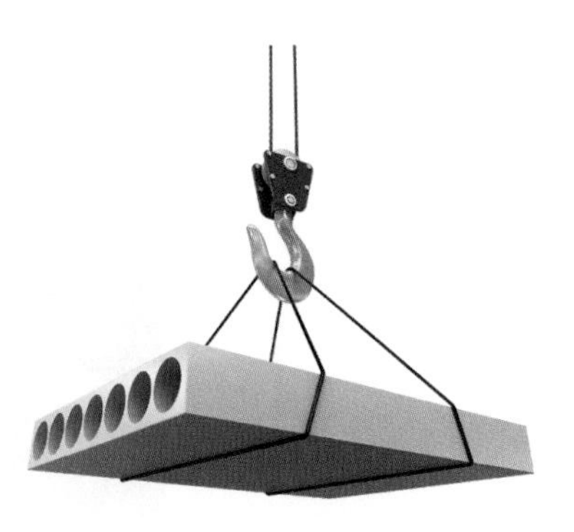

18. Alja likes to skip around her backyard.
   a. If her starting position on a number plane is at point A (1, 1) and she skips two steps forward and one step to the right, draw her next four locations (on the following graph) after every set of three steps.
   b. Do the five points form a pattern?
   c. From her last location, Alja decides to skip one step backwards and two steps to the left. She skips in this pattern four times.
   d. Do these five points form a pattern?

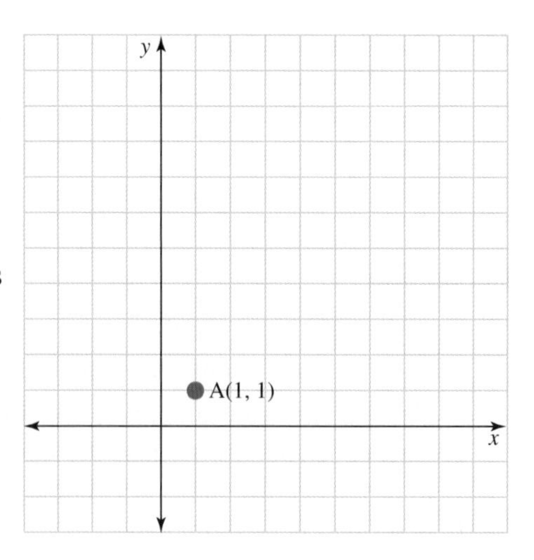


19. Sketch the quadrilateral ABCD that is formed by the vertices that are described by the following translations from the origin.
   A: 3R 2D
   B: 5R 1U
   C: 1R 5U
   D: 4L 3D

# LESSON
# 12.4 Reflections

**LEARNING INTENTION**

At the end of this lesson you should be able to:
- reflect points and shapes in mirror lines
- reflect points and shapes in the Cartesian plane.

## 12.4.1 Reflections in mirror lines

eles-4685

- A reflection is the image of a point or an object, as seen in the mirror.
- Reflections are often called 'mirror images' and the lines in which the objects are reflected are called 'mirror' lines.
- Mirror images always have reverse orientation; that is, left appears to be right and vice versa.
- Reflected points always lie on the line that is perpendicular to the mirror line that passes through the original point.
- Reflected points are the same distance from the mirror line as the original point on the other side of the mirror.
- Reflections are also known as 'flips'.

- When reflecting shapes in a given line, the following steps can be of assistance:

  Step 1: Select some key points on the original object (the vertices are usually a good choice).

  Step 2: From each point, draw a line perpendicular (at a right angle) to the mirror line. Extend each line beyond the mirror line.

  Step 3: For each selected point, measure its distance along the line from the mirror. The image of the point will be the same distance from the mirror line, on the other side of it.

  Step 4: Complete the image using the reflections of the key points.

## WORKED EXAMPLE 6 Reflections in a mirror line

**Sketch the image of the triangle after it is reflected in the mirror line.**

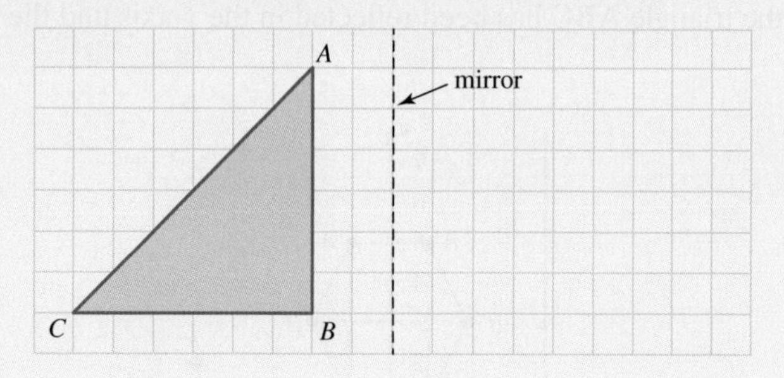

| THINK | WRITE |
|---|---|
| 1. From each vertex of the given triangle, draw the lines perpendicular to and extending beyond the mirror line. | 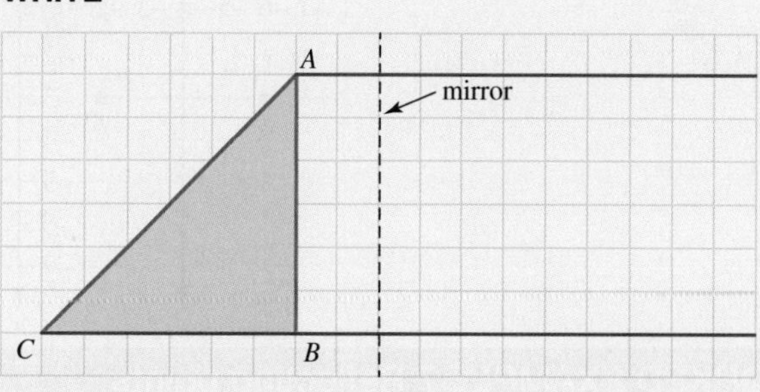  |
| 2. Points A and B are both 2 units to the left of the mirror line. Therefore their images A′ and B′ will be 2 units to the right of the mirror line.<br><br>Point C is 8 units to the left of the mirror line. Therefore C′ will be 8 units to the right of the mirror line. | 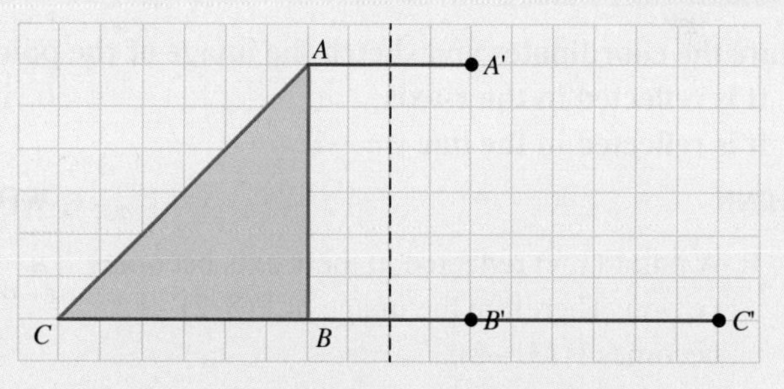  |
| 3. Join the vertices A′, B′ and C′ to complete the image. | 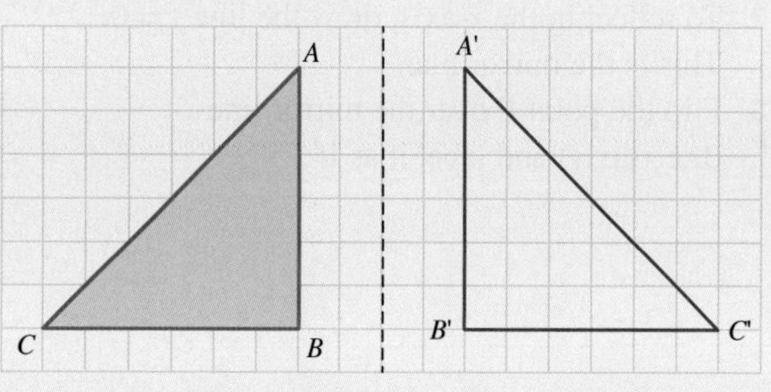  |

## 12.4.2 Reflections in the Cartesian plane

eles-4686

- Points and objects in the Cartesian plane can be reflected in lines and axes.

### Reflecting points in an axis

**If a point $P = (x, y)$ is reflected in the $x$-axis, the $x$-coordinate stays the same and the $y$-coordinate switches sign. Therefore $P' = (x, -y)$.**

**If a point $P = (x, y)$ is reflected in the $y$-axis, the $y$-coordinate stays the same and the $x$-coordinate switches sign. Therefore $P' = (-x, y)$.**

- The Cartesian plane in the following diagram shows some examples of reflections. The point P has been reflected in the $x$-axis, the triangle ABC has been reflected in the $y$-axis and the point S has been reflected in the line $x = -4$.

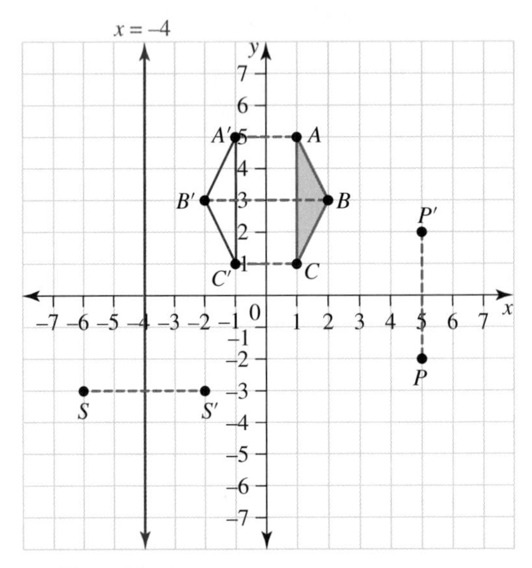


### WORKED EXAMPLE 7 Reflecting points

**State the coordinates and sketch the image of the point P(3, 2) after:**

**a. it is reflected in the $x$-axis**

**b. it is reflected in the $y$-axis**

**c. it is reflected in the line $y = -1$.**

**THINK**

a. 1. A point $(x, y)$ reflected in the $x$-axis becomes $(x, -y)$. Therefore the image of $P(3, 2)$ becomes $P'(3, -2)$.

2. To reflect in the $x$-axis, draw the line $y = 0$. This is the mirror line.

3. Flip the point P over the mirror line (the $x$-axis) and label it as P′.

**WRITE**

a. $P' = (3, -2)$

b. 1. A point $(x, y)$ reflected in the $y$-axis becomes $(-x, y)$. Therefore the image of P(3, 2) becomes P′(−3, 2).

2. To reflect in the $y$-axis, draw the line $x = 0$. This is the mirror line.

3. Flip the point P over the mirror line (the $y$-axis) and label as P′.

b. P′ = (−3, 2)

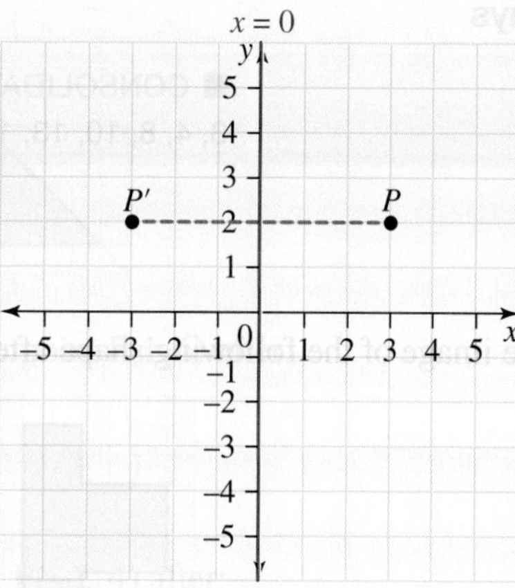


c. 1. Draw the line $y = -1$. This is the mirror line.

2. Draw a second line that is perpendicular to $y = -1$ and passes through P.

c.

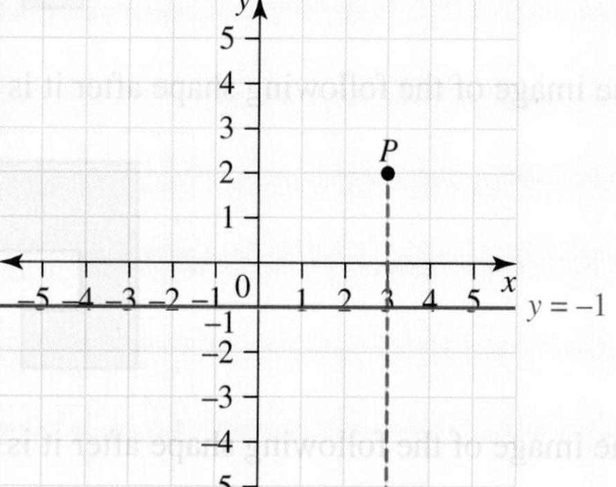


3. Place the image P′ on the dashed line so that the distance from the line $y = -1$ is equal for P and P′.

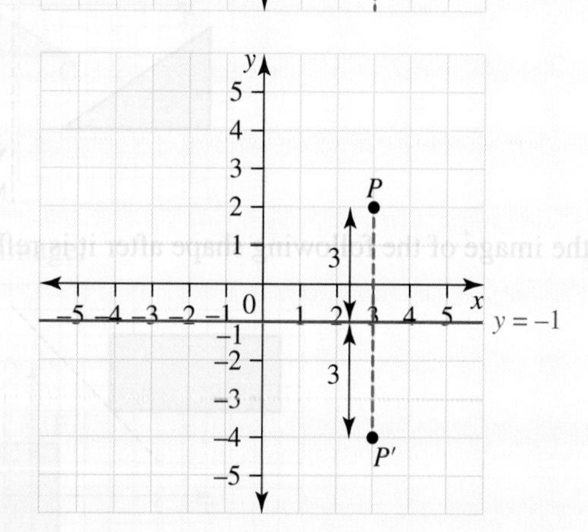


4. State the coordinates of P′.

P′ = (3, −4)

## Resources

**eWorkbook** Topic 12 Workbook (worksheets, code puzzle and project) (ewbk-1913)

**Interactivities** Individual pathway interactivity: Reflections (int-4389)

Reflections (int-4141)

**11.** Consider the following shape.

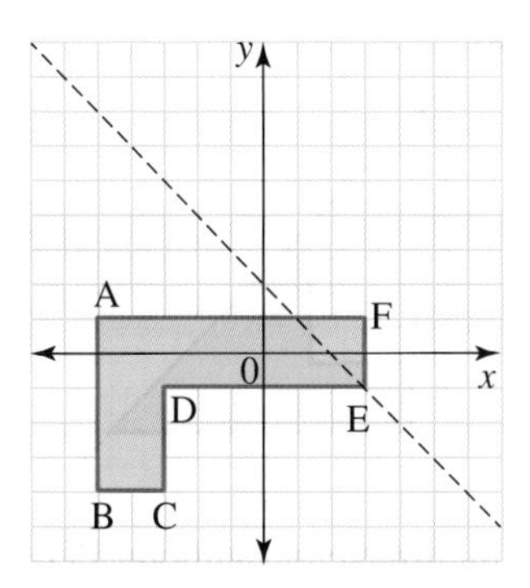


**a.** Draw the image reflected in the mirror.
**b.** Determine the coordinates of the vertices of the image.

## Reasoning

12. Explain the statement: In a reflection, the axis of reflection becomes the axis of symmetry if the object and its image are next to each other.

13. Explain the difference between reflection in a horizontal axis and reflection in a vertical axis.

**14.** Explain which one of the four strawberries is a reflection in the dotted line of the strawberry marked asthe original.

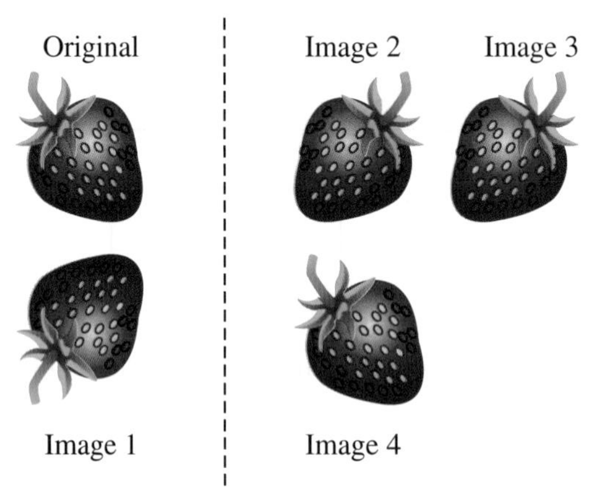


## Problem solving

15. Reflections in mirrors transform objects. Complete each of the patterns by determining images of the objects reflected in a single mirror, as shown in the following.

a.

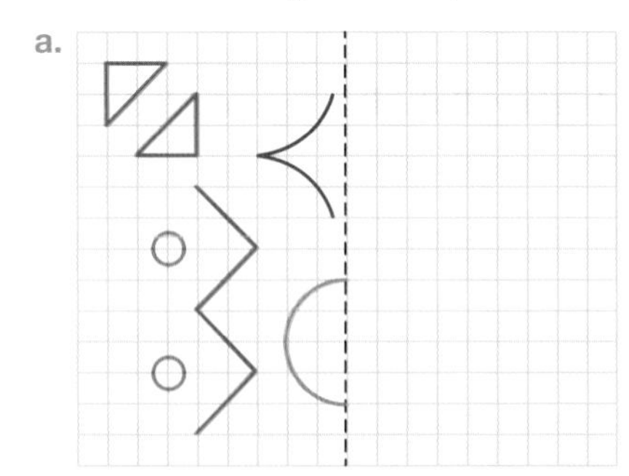

b.

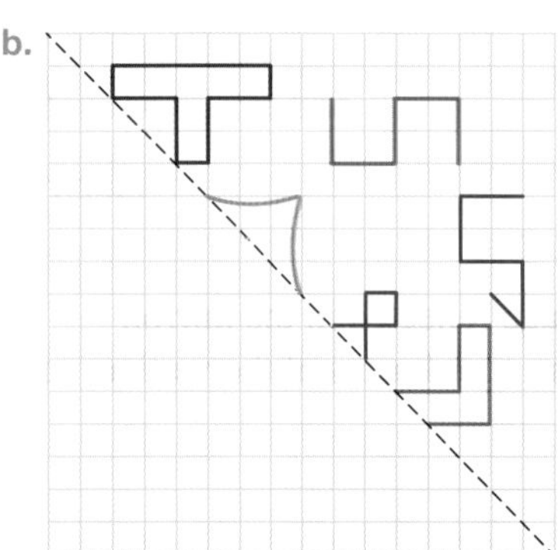

16. This is a backward clock. It has the numbers in reverse order, and the hands sweep anticlockwise. Determine the time shown by the clock.

17. A kaleidoscope uses more than one mirror to create wonderful patterns. To help design a kaleidoscope, reflect each object in each of the mirrors to complete the pattern in each of these figures.

a. 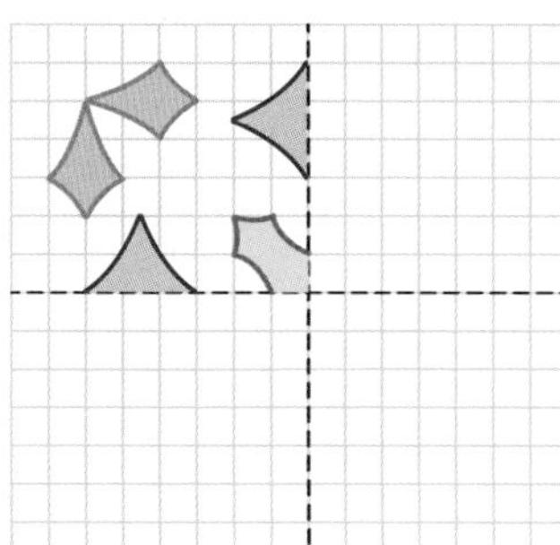

b. 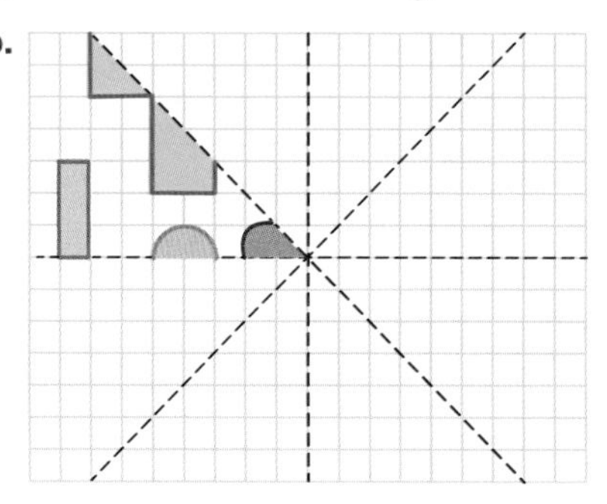

# LESSON
# 12.5 Rotations and combined transformations

## LEARNING INTENTION

At the end of this lesson you should be able to:
- rotate a point or object in the Cartesian plane about the origin
- determine the image of a point or object after it has undergone multiple transformations.

## 12.5.1 Rotations

eles-4687

- A rotation is simply a turn. By turning your phone from portrait to landscape you have performed a rotation.
- A point or a shape can be rotated about a point called the **centre of rotation**. To specify the rotation, we need to give the angle through which the object is to be turned and the direction of the rotation (clockwise or anticlockwise).
- Consider a centre of rotation at the origin (0, 0). Any point may be rotated about the origin in either direction, clockwise or anticlockwise.

### Rotation by 90° about the origin

- To understand what happens when a point is rotated by 90° about the origin, consider the following graphs.

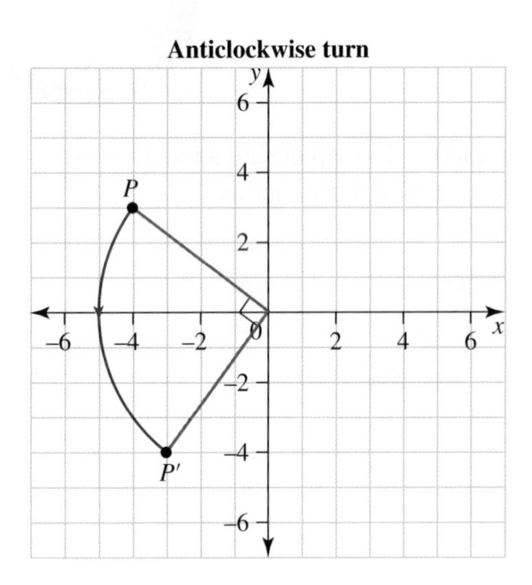


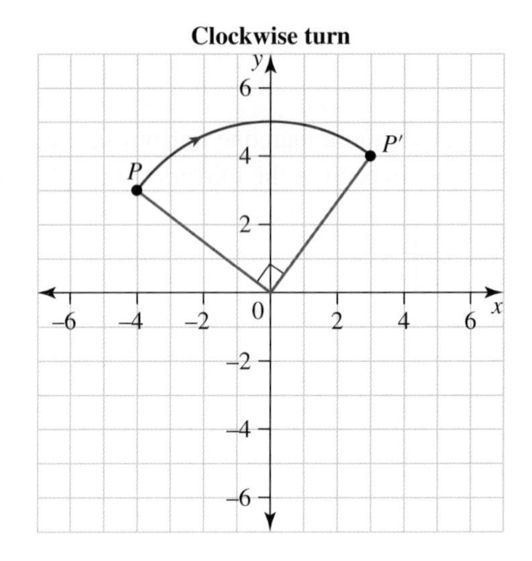


- Rotating a point 90° about the origin simply means that the angle at the origin between a point and its image must be 90°.
- By applying your knowledge of rotating one point, you can rotate multiple points at once. Consider the following graphs.

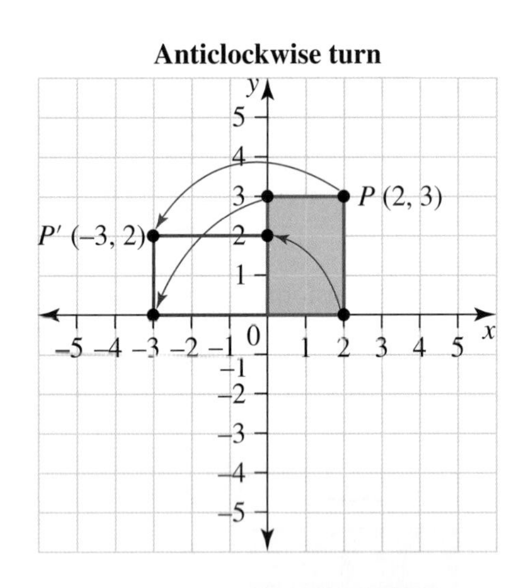


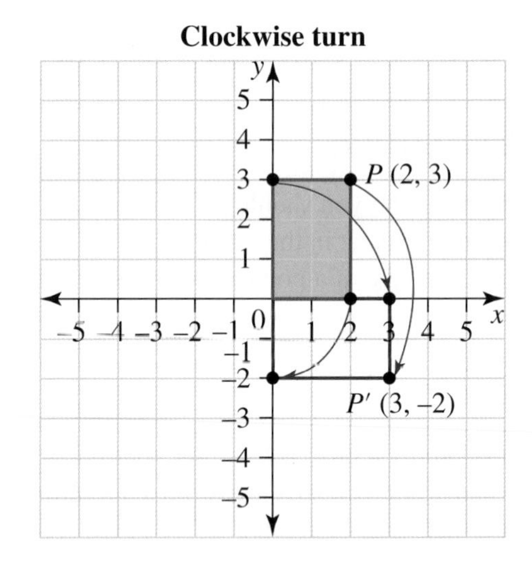


- By analysing these graphs we can see the rules that hold true for rotations by 90° about the origin.

**Rotating points by 90° about the origin**

**Anticlockwise: The image of a point P$(x, y)$ that is rotated by 90° anticlockwise about the origin is P$'(-y, x)$.**

**Clockwise: The image of a point P$(x, y)$ that is rotated by 90° clockwise about the origin is P$'(y, -x)$.**

## Rotation by 180° about the origin

- A similar method can be used to understand what happens when a point is rotated by 180° about the origin.

**Anticlockwise turn** **Clockwise turn**

- An alternate method is to draw a straight line through the point and the centre of rotation. The image will be on the line, at the same distance from the centre of rotation.

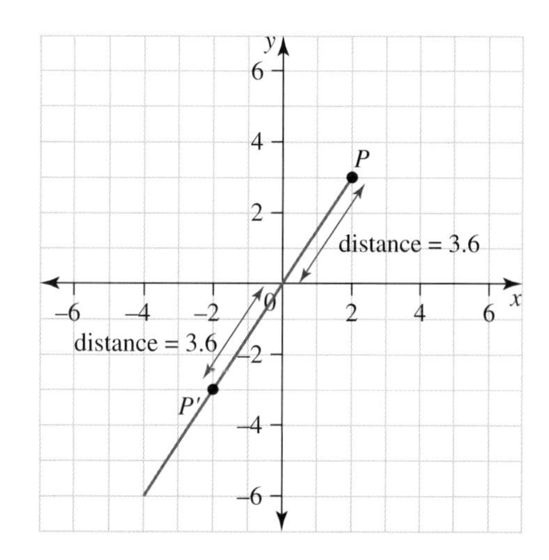


### Rotating points by 180° about the origin

**Regardless of the direction of rotation, the image of a point P($x$, $y$) that is rotated by 180° about the origin is P′($-x$, $-y$).**

### WORKED EXAMPLE 8 Rotating points

**State the coordinates and sketch the image of the point P(−4, −2) after:**
**a. it is rotated by 90° anticlockwise about the origin**
**b. it is rotated by 90° clockwise about the origin**
**c. it is rotated by 180° anticlockwise about the origin.**

| THINK | WRITE |
|---|---|
| a. 1. To rotate by 90° anticlockwise about the origin, the point $(x, y)$ becomes $(-y, x)$. | a. $(x, y) \rightarrow (-y, x)$ |

2. To determine the image of P, swap the $x$-coordinate with the $y$-coordinate then change the sign of the new $x$-coordinate.

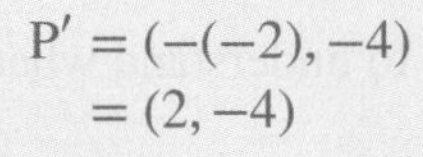
$P' = (-(-2), -4)$
$= (2, -4)$

3. Sketch the image P′.

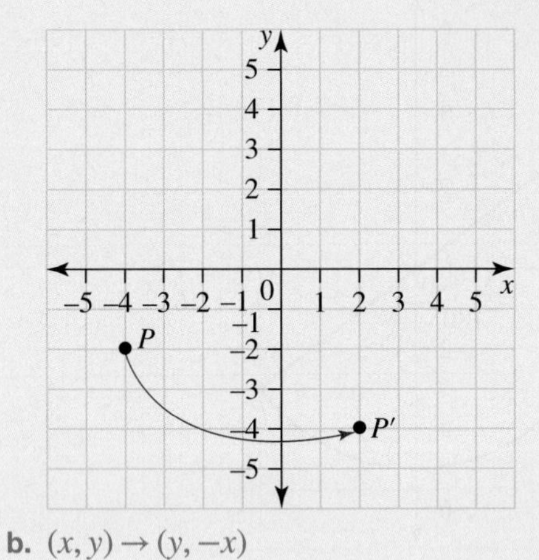


**b.** 1. To rotate by 90° clockwise about the origin, the point $(x, y)$ becomes $(y, -x)$.

**b.** $(x, y) \rightarrow (y, -x)$

2. To determine the image of P, swap the $x$-coordinate with the $y$-coordinate then change the sign of the new $y$-coordinate.

$P' = (-2, -(-4))$
$= (-2, 4)$

3. Sketch the image P′.

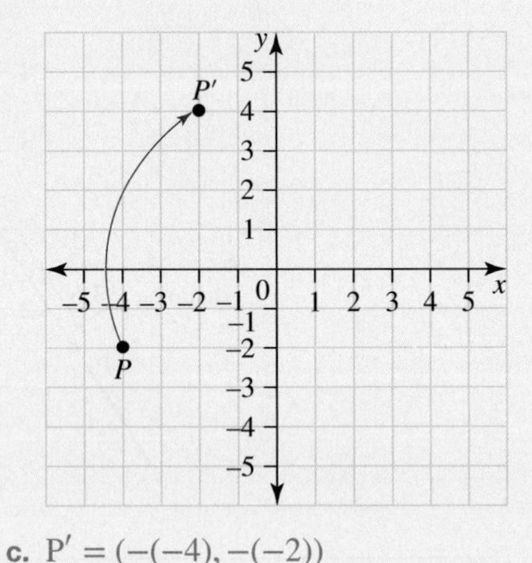


**c.** 1. To rotate by 180° anticlockwise about the origin, the point $(x, y)$ becomes $(-x, -y)$. We can simply change the signs of the $x$- and $y$-coordinates.

**c.** $P' = (-(-4), -(-2))$
$= (4, 2)$

2. Sketch the image P′.

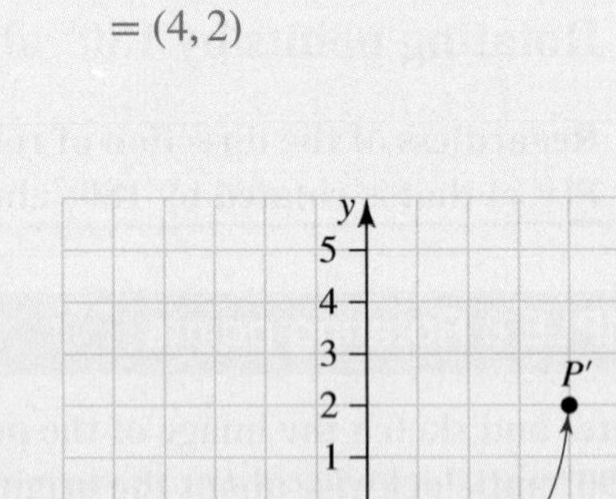

## WORKED EXAMPLE 9 Rotating shapes

**State the coordinates and sketch the image of the following shape after it is rotated by 180° clockwise about the origin.**

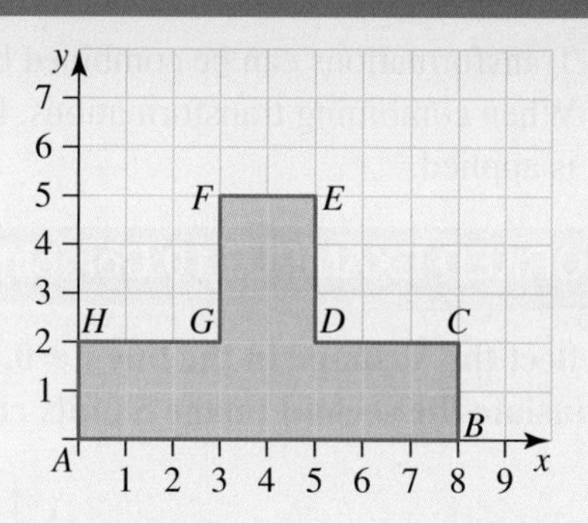


| THINK | WRITE |
|---|---|
| 1. Write the coordinates of the original shape. | The vertices of the original image are: A(0, 0), B(8, 0), C(8, 2), D(5, 2), E(5, 5), F(3, 5), G(3, 2) and H(0, 2) |
| 2. A rotation of 180° clockwise about the origin changes the signs of the $x$- and $y$-coordinates. That is $(x, y) \rightarrow (-x, -y)$. | The vertices of the image are: A′(0, 0), B′(−8, 0), C′(−8, −2), D′(−5, −2), E′(−5, −5), F′(−3, −5), G′(−3, −2) and H′(0, −2) |
| 3. Sketch the image using the coordinates of the image that were determined in step 2. | 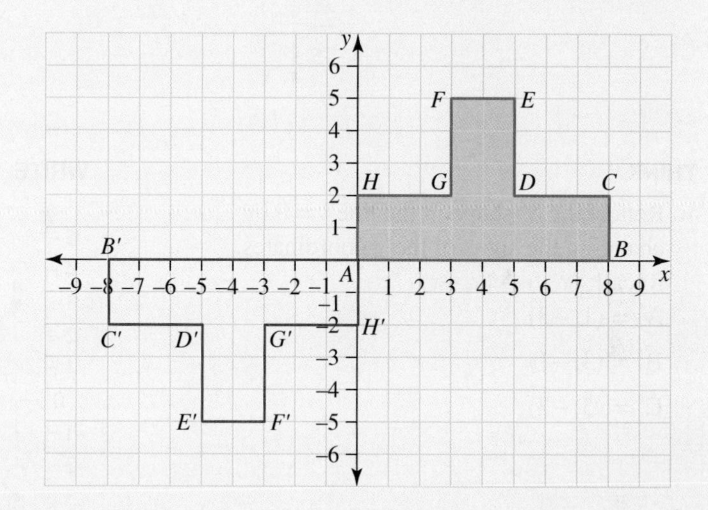 |



## Digital technology

Desmos is a free, interactive web application that contains a graphing calculator and a geometry tool. The geometry tool can be used to draw points, circles, polygons, angles, rays and lines. It can also be used to perform and visualise all types of transformations.

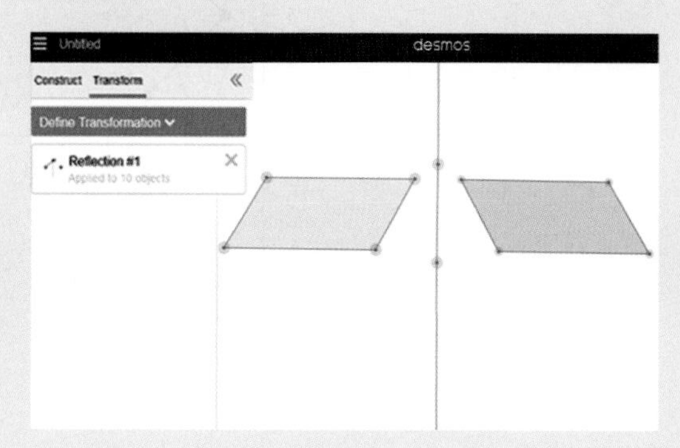

## 12.5.2 Combined transformations in the Cartesian plane

eles-4688

- Transformations can be combined by performing separate transformations, one after the other.
- When combining transformations, keep track of the coordinates of the point/points as each transformation is applied.

**WORKED EXAMPLE 10 Combining transformations**

**Reflect the V-shape in the line $y = 0$, then rotate the image 90° anticlockwise about the origin, then translate the second image 5 units right and 3 units down.**

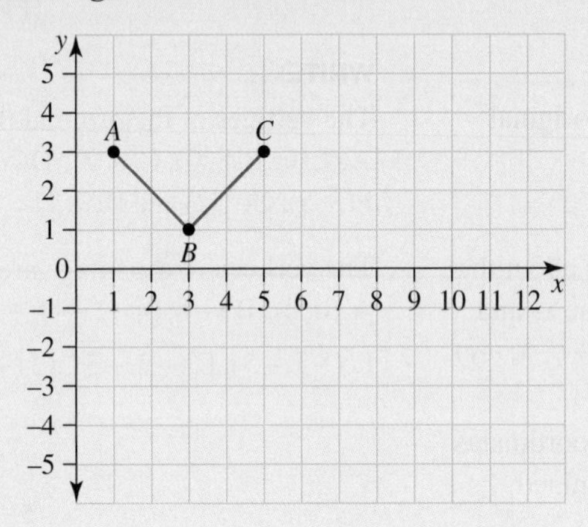

**THINK**

1. Reflect the V-shape in the line $y = 0$ by changing the signs of the $y$-coordinates.
   Recall: $(x, y) \rightarrow (x, -y)$
   $A' = (1, -3)$
   $B' = (3, -1)$
   $C' = (5, -3)$

**WRITE**

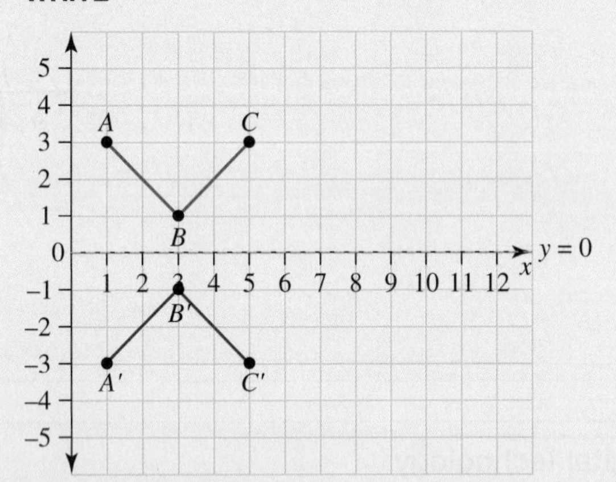

2. Rotate the image by 90° anticlockwise about the origin by swapping the $x$-coordinates with the $y$-coordinates then changing the signs of the new $x$-coordinates.
   Recall: $(x, y) \rightarrow (-y, x)$
   $A'' = (3, 1)$
   $B'' = (1, 3)$
   $C'' = (3, 5)$

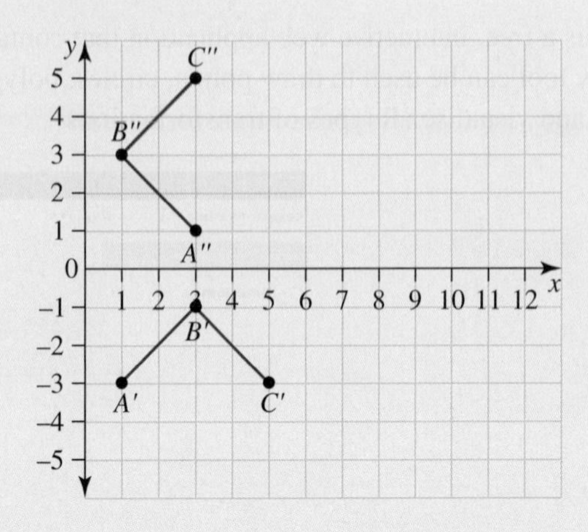

3. Translate the second image 5 units right and 3 units down by adding 5 to the $x$-coordinates and subtracting 3 from the $y$-coordinates.
$A''' = (8, -2)$
$B''' = (6, 0)$
$C''' = (8, 2)$

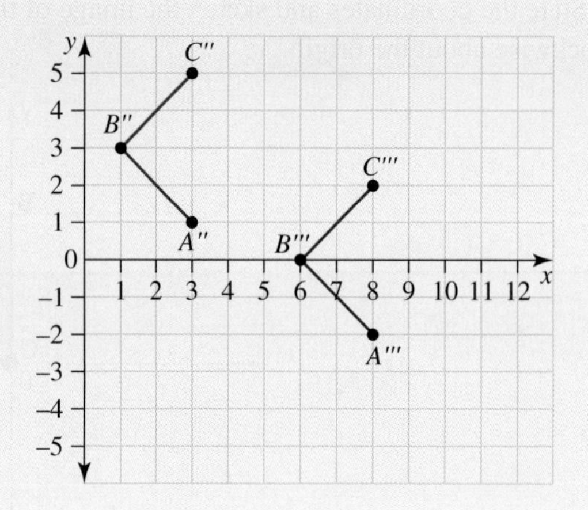

## Resources

eWorkbook Topic 12 Workbook (worksheets, code puzzle and project) (ewbk-1913)

Interactivities Individual pathway interactivity: Rotations and combined transformations (int-4390)
Rotations (int-4144)
Combined transformations (int-4145)

## Exercise 12.5 Rotations and combined transformations

learnon

12.5 Quick quiz  | 12.5 Exercise

### Individual pathways

| ■ PRACTISE | ■ CONSOLIDATE | ■ MASTER |
|---|---|---|
| 1, 4, 7, 10, 14, 18, 19 | 2, 5, 8, 11, 13, 15, 16, 20 | 3, 6, 9, 12, 17, 21 |

### Fluency

1. WE8 State the coordinates and sketch the image of the point P(1, 5) after:
   a. it is rotated by 90° anticlockwise about the origin
   b. it is rotated by 90° clockwise about the origin
   c. it is rotated by 180° anticlockwise about the origin.

2. State the coordinates and sketch the image of the point Q(−2, 4) after:
   a. it is rotated 90° anticlockwise about the origin
   b. it is rotated 90° clockwise about the origin
   c. it is rotated 180° anticlockwise about the origin.

3. State the coordinates and sketch the image of the point R(−3, −1) after:
   a. it is rotated 90° anticlockwise about the origin
   b. it is rotated 90° clockwise about the origin
   c. it is rotated 180° clockwise about the origin.

4. **WE9** State the coordinates and sketch the image of the following shape after it has been rotated by 180° anticlockwise about the origin.

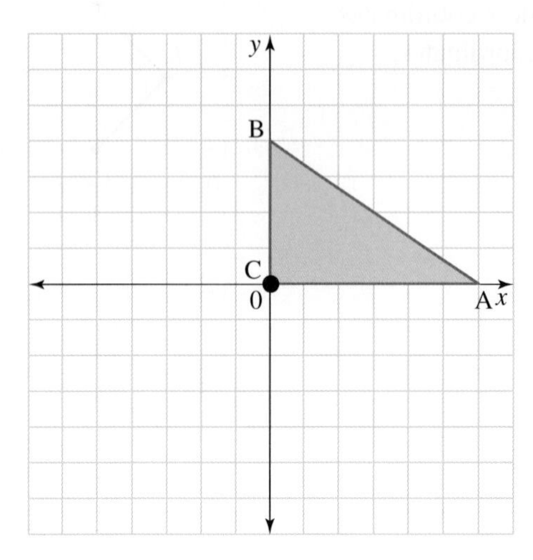


5. State the coordinates and sketch the image of the following shape after it has been rotated by 90° clockwise about the origin.

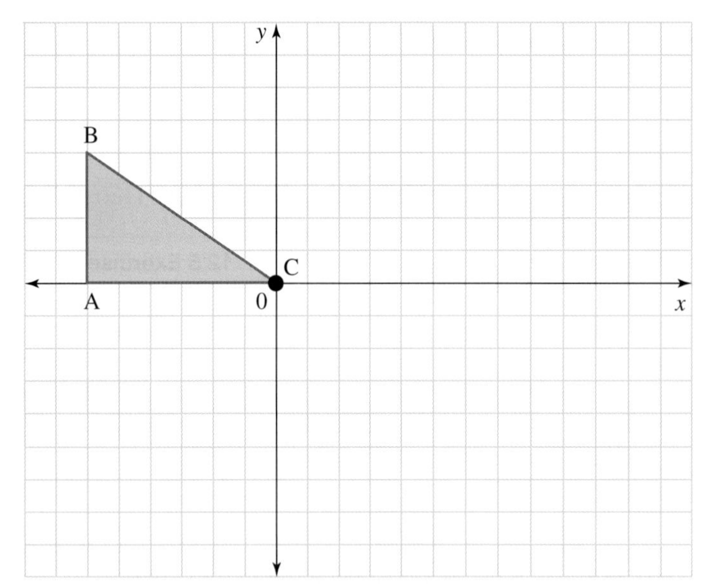

6. State the coordinates and sketch the image of the following shape after it has been rotated by 90° anticlockwise about the origin.

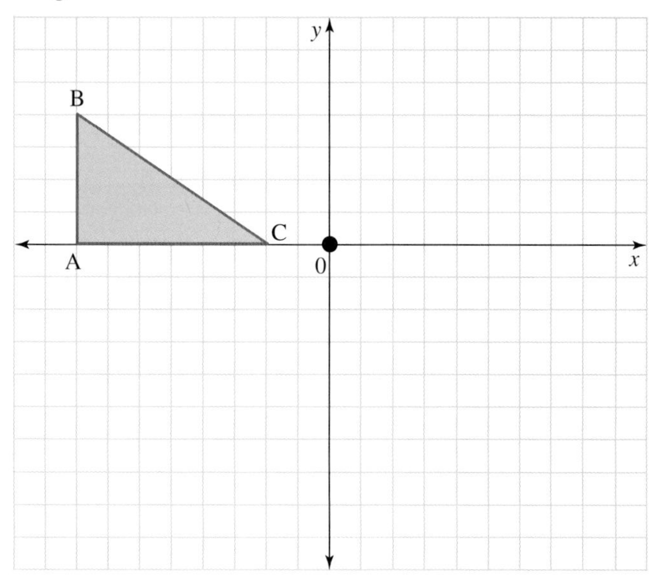


7. State the coordinates and sketch the image of the following shape after it has been rotated by 90° anticlockwise about the origin.

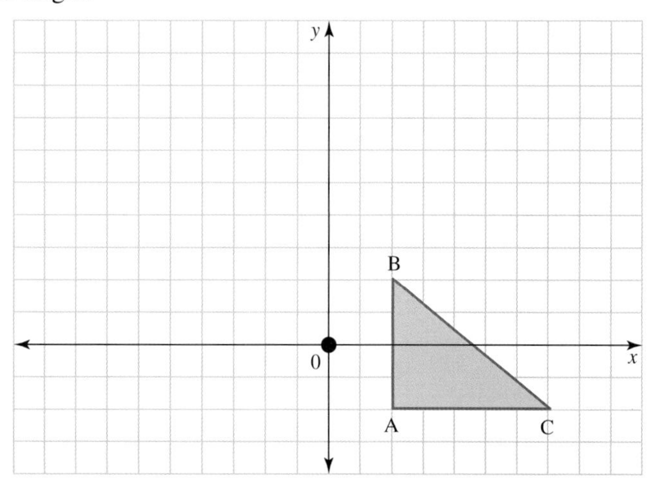


8. State the coordinates and sketch the image of the following shape after it has been rotated by 90° clockwise about the origin.

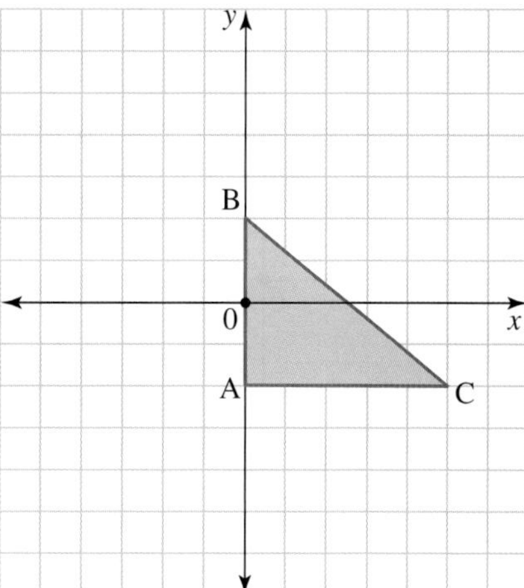

9. State the coordinates and sketch the image of the following shape after it has been rotated by 90° clockwise about the origin.

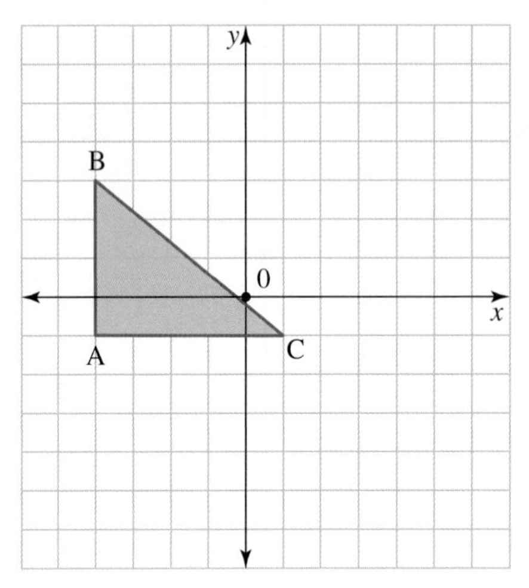


## Understanding

10. **WE10** Reflect the triangle ABC in the $x$-axis, then rotate the image 90° clockwise about the origin and then translate the second image 4 units right and 6 units down.

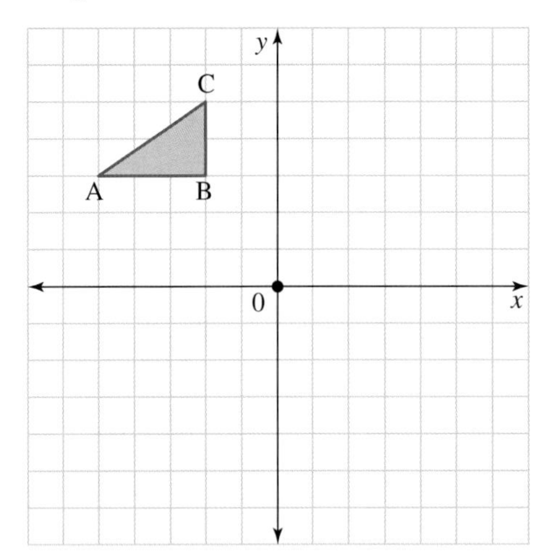


11. Translate the V-shape 3 units left and 2 units up, then rotate the image 90° anticlockwise about the origin and then reflect the second image in the $y$-axis.

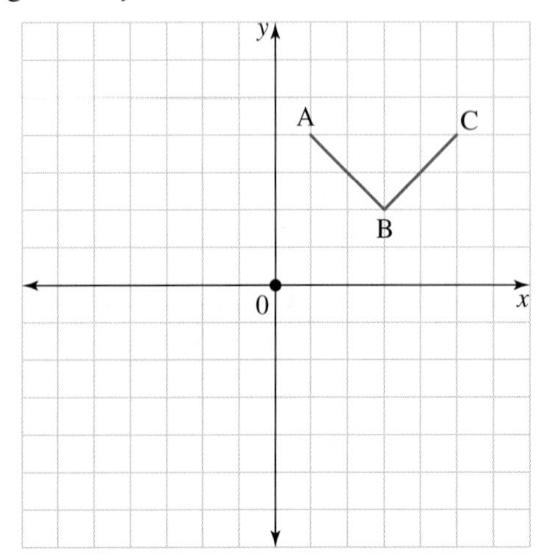

**12.** Rotate the tick ABC 90° clockwise about the origin, then translate the image 7 units right and 1 unit down and then reflect the second image in the $x$-axis.

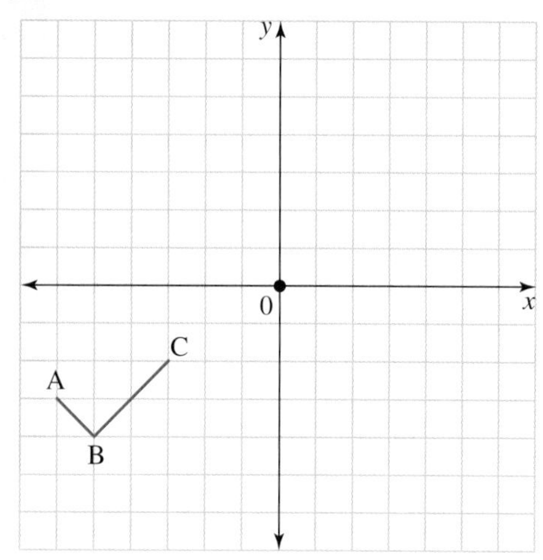


**13.** **MC** The following diagram shows an object and its image after a certain set of transformations. The object was:

**A.** reflected in the $y$-axis, then translated 3R 2D.
**B.** reflected in the $x$-axis, then translated 4R 2D.
**C.** translated 3R 2D, then reflected in the $x$-axis.
**D.** translated 4R 2D, then reflected in the $x$-axis.
**E.** translated 3R 2U, then reflected in the $y$-axis.

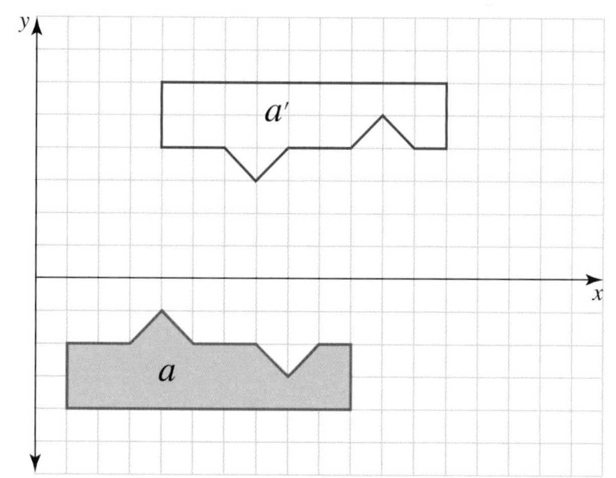


## Reasoning

**14.** The following object has been rotated 180° about point A. Is there another transformation that would produce the same image? Explain your response.

**15.** Using the point $(x, -y)$, prove that a rotation of 180° anticlockwise gives an identical result to a rotation of 180° clockwise.

**16.** The point A $(a, b)$ is rotated anticlockwise through 90° about the origin, then reflected in the $x$-axis, translated by 4 units in the negative $x$ direction and finally reflected in the $y$-axis. Determine the final coordinates of A.

**17.** Using the point R $(-r, -q)$ and showing all working, prove that a rotation of 180° followed by a reflection in the $y$-axis is equivalent to a reflection in the $x$-axis.

## Problem solving

18. Earth completes a full rotation around the Sun in approximately one year. Determine the approximate angle of rotation of a point on Earth in one day.

19. The point P $(a, -b)$ is reflected in the $y$-axis and then rotated through $180°$ about the origin. Determine the coordinates of the new point. Show your working.

20. a. Translate the point P (3, 1) 3 units up and 2 units to the left. Label the new point P′ and give the coordinates of this point.
    b. Translate P′ 2 units down and 3 units to the left. Label the new point P″ and give the coordinates of this point.
    c. Identify the single translation that would move P to P″.

21. A point is translated seven units to the right and four units down, rotated 180° clockwise about the initial position, then reflected in a vertical line passing through the initial position, and then translated seven units to the left and four units down. What are the coordinates of the point at the end of these transformations? Draw a diagram to represent the transformations.

# LESSON
# 12.6 Review

## 12.6.1 Topic summary

### TRANSFORMATIONS

#### Rotational symmetry

- A shape has rotational symmetry if rotating it less than 360° matches the original figure.
- The number of times a match occurs in a 360° rotation is the order (or degree) of rotational symmetry.
- The flower below has rotational symmetry of order 5, and the Mercedes-Benz logo has rotational symmetry of order 3.

- All shapes have rotational symmetry of at least order 1, since a rotation of 360° always creates a match.

#### Translations

- Translations move points up, down, left or right.
- The directions are often abbreviated by their first letter. For example, a translation 1 unit down and 5 units right is often expressed as D1 5R.

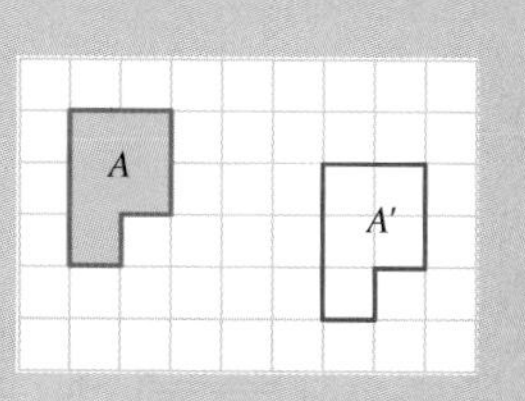


#### Reflections

- Reflections are images of a point or object as seen in a mirror.
- If a point $P(x, y)$ is reflected in the $x$-axis, $P' = (x, -y)$.
- If a point $P(x, y)$ is reflected in the $y$-axis, $P' = (-x, y)$.

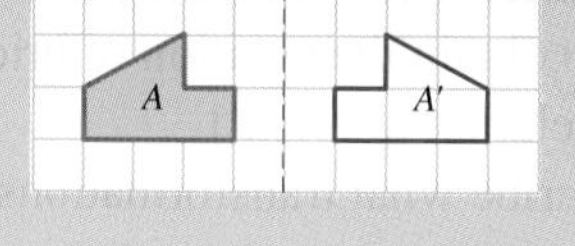


#### Line symmetry

- A shape is symmetrical along a line if it can be folded along that line to create two identical parts that overlap perfectly.
- Shapes can have zero, one or multiple lines of symmetry.

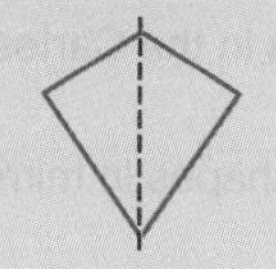

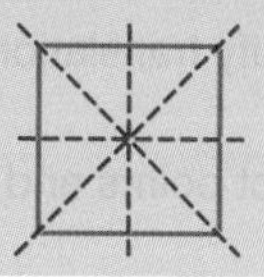

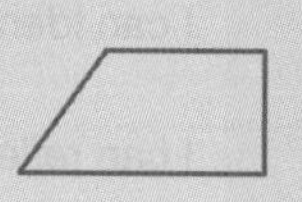

#### Images

- The position of a point or object after transformation is called the image of that point or object.
- The notation for the image is to place an apostrophe after the name of the point or object.
  - The image of a point $P$ is denoted as $P'$.
  - The image of an image $P'$ is denoted as $P''$.

#### Rotations

- A point or object in the Cartesian plane can be rotated about the origin.
- If a point $P(x, y)$ is rotated 90° clockwise about the origin, $P' = (y, -x)$.
- If a point $P(x, y)$ is rotated 90° anticlockwise about the origin, $P' = (-y, x)$.
- If a point $P(x, y)$ is rotated 180° in either direction about the origin, $P' = (-x, -y)$.

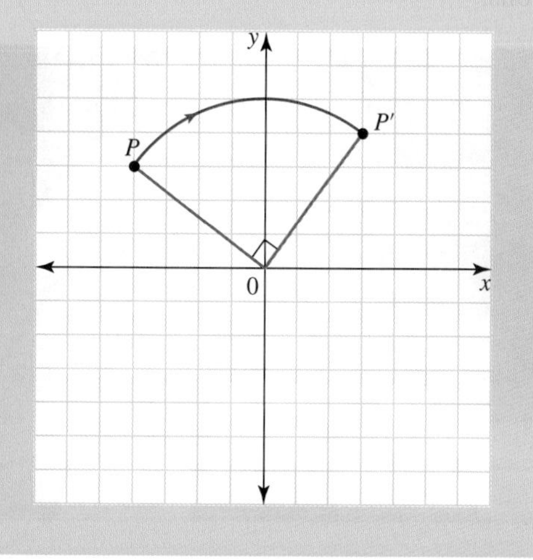


#### Combined transformations

- Multiple transformations can be combined by performing them one after the other.
- It is a good idea to keep track of the coordinates of the point(s) as each transformation is performed.

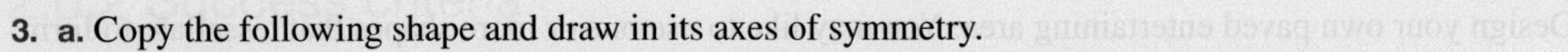

3. a. Copy the following shape and draw in its axes of symmetry.
   b. Determine the order of rotational symmetry of this shape.

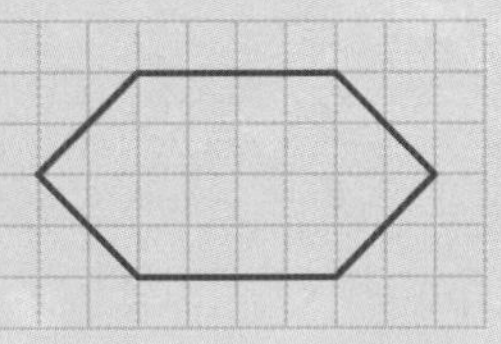

4. Show the following translations by copying the shapes and producing an image on a grid.

a.

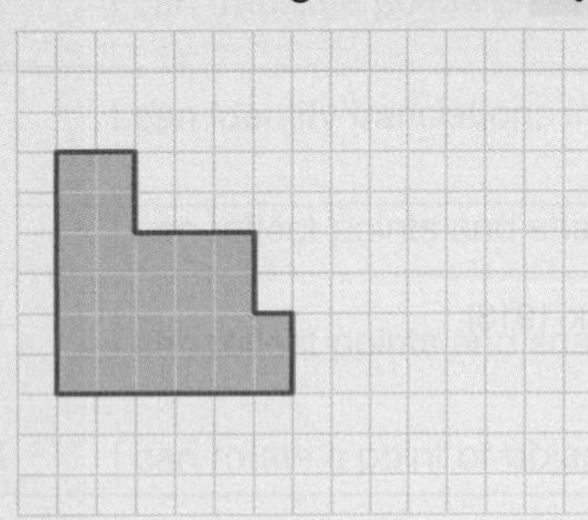

7R 2U

b.

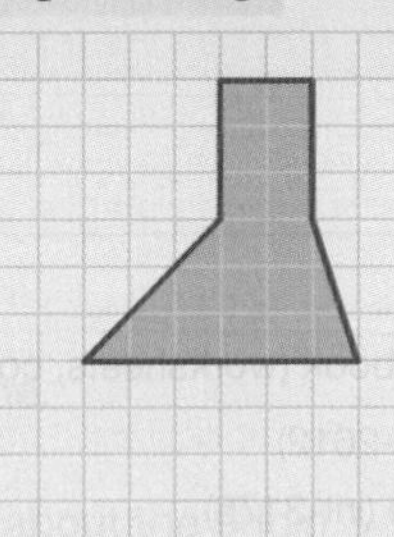

2L 3D

c.

5R 2D

5. State the final position after each set of the following translations.
   a. 3U 2R, 4D 6R, 2U 3L
   b. 2D 3L, 4U 5R, 2D 6L, 3D 2R
   c. 3L 2U, 9R 5D, 2L 3D, 4R 7U
   d. 12L 3U, 4R 2D, 6L 5U, 2R 2D

6. For each of thc following shapes, sketch the reflected image in the mirror line shown.

a.

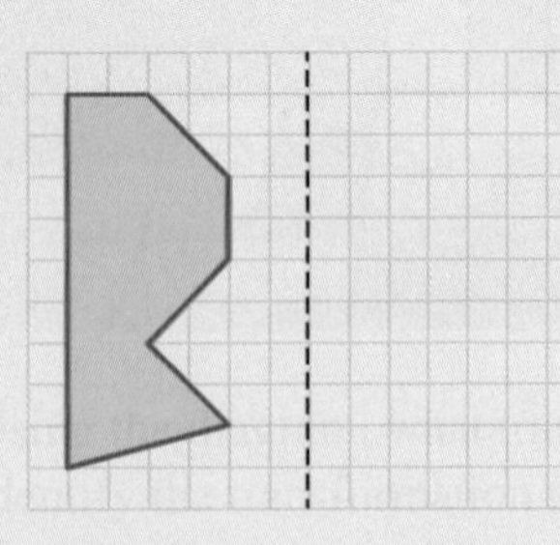

b.

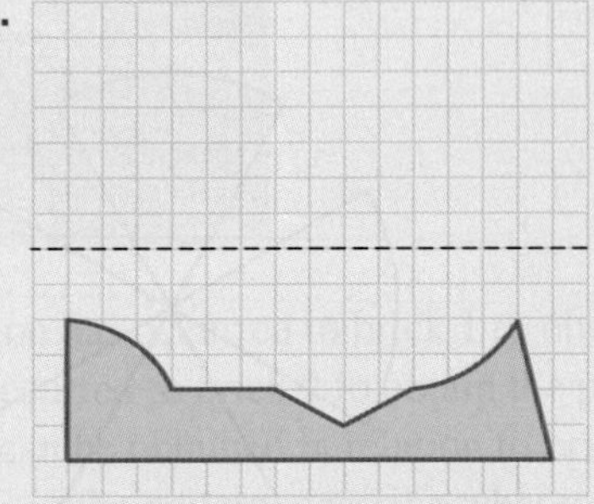

c.

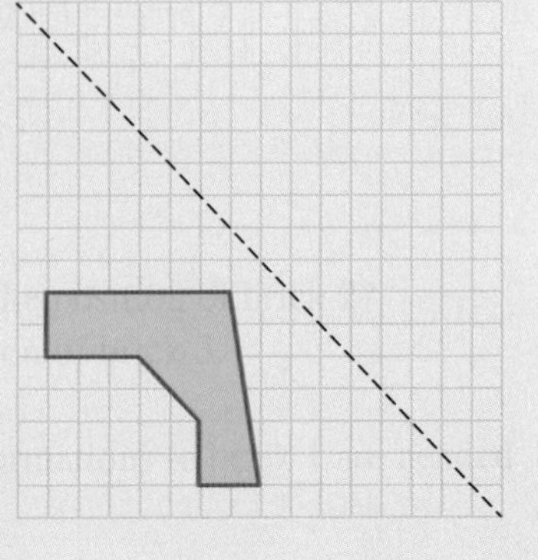

7. **a.** Sketch the image after the sets of transformations shown in these objects.

**i.**

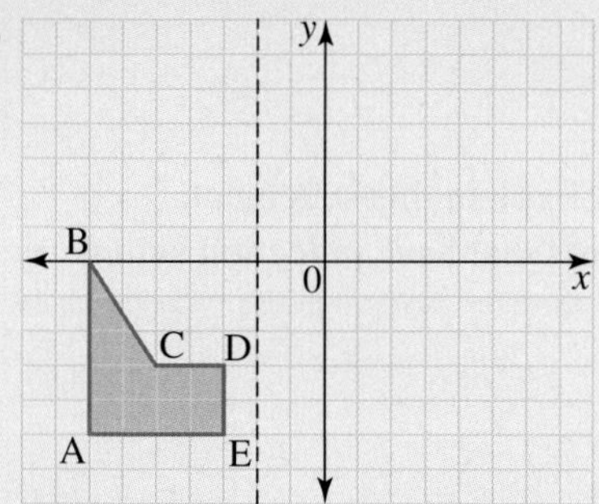


Translated 3L 4U; then reflected in the mirror

**ii.**

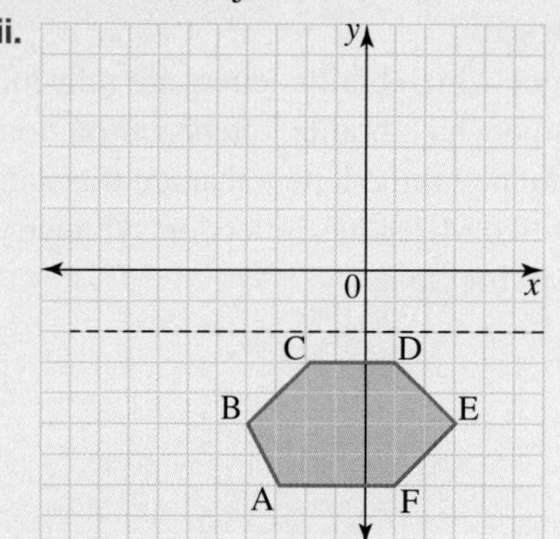


Reflected in the mirror; then translated 3U 5L

**b.** Determine the coordinates of the vertices of the object and its image.

8. For each of the following shapes, show the image after the rotation about the origin as specified in the following figures.

**a.**

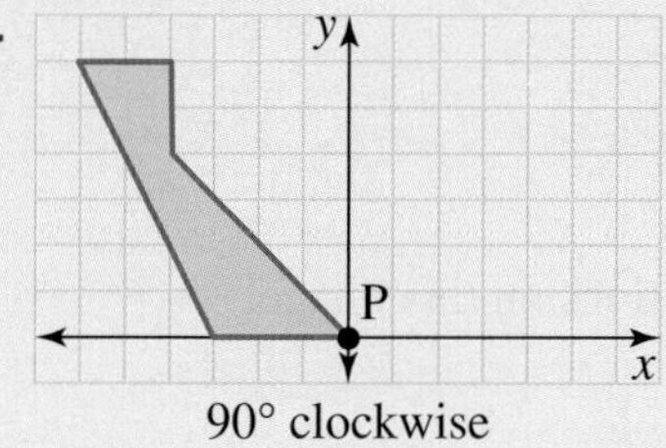


90° clockwise

**b.**

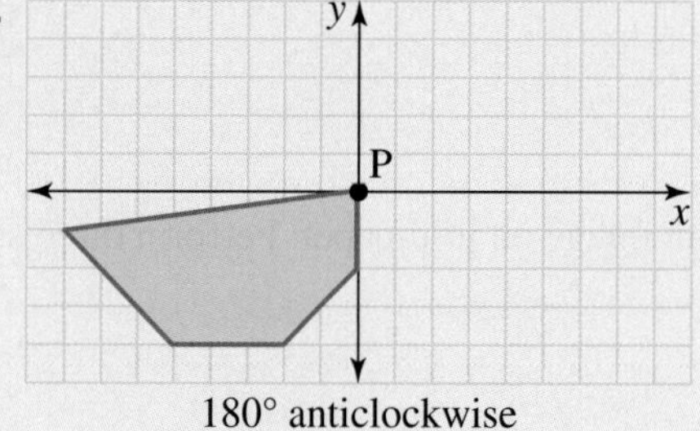


180° anticlockwise

**c.**

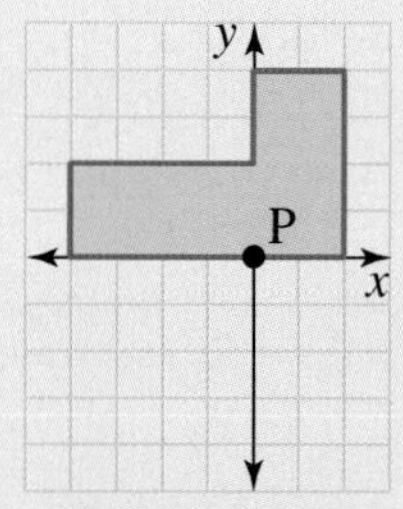


270° clockwise

## Problem solving

9. Braille is a code of raised dots that can be read by touch. It was developed by a 15-year-old blind French student named Louis Braille. The braille alphabet is based on a cell three dots high and two dots wide.

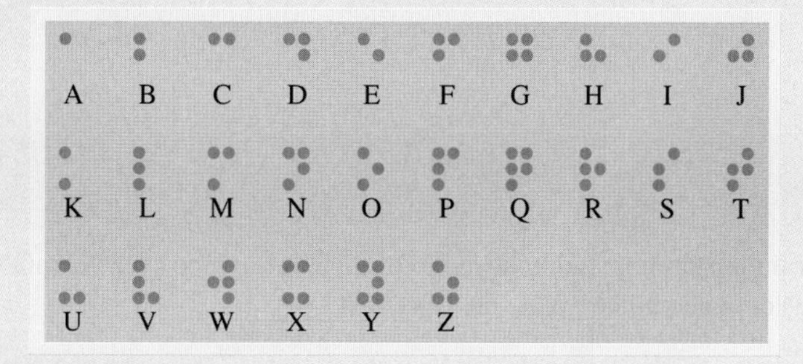

**a.** Compare the first 10 letters of the braille alphabet with the second 10 letters. Describe any patterns that you see.
**b.** Determine which braille letters are reflections of each other.
**c.** Determine which braille letters are rotations of each other.
**d.** There are no braille letters that are translations of each other. Explain this statement.
**e.** A word like MUM has reflection symmetry. Identify two words that have reflection symmetry when written in braille.

**f.** A word like SWIMS has rotational symmetry. Identify a two-letter combination that has rotational symmetry when written in braille.

**10.** Design an image on grid paper. Perform three successive transformations on this image.

on To test your understanding and knowledge of this topic, go to your learnON title at www.jacplus.com.au and complete the **post-test**.

# Answers

## Topic 12 Transformations

### 12.1 Pre-test

1. 4
2. D
3. C
4. Hexagon
5. 1R 5U
6. C
7. $(-5, 3)$
8. $(-2, -3)$
9. C
10. D
11. $(4, -7)$
12. C
13. D
14. $(2, 4)$
15. $(5, 8)$

### 12.2 Line and rotational symmetry

| | | | | |
|---|---|---|---|---|
| 1. a. 3 | b. 0 | c. 1 | d. 1 | e. 0 |
| 2. a. 0 | b. 1 | c. 4 | d. 2 | e. 2 |
| 3. a. 6 | b. 8 | c. 5 | d. 0 | |
| 4. a. 3 | b. 1 | c. 1 | d. 1 | e. 1 |
| 5. a. 2 | b. 1 | c. 4 | d. 2 | e. 2 |
| 6. a. 6 | b. 8 | c. 5 | d. 3 | |

7. A
8. a. i.

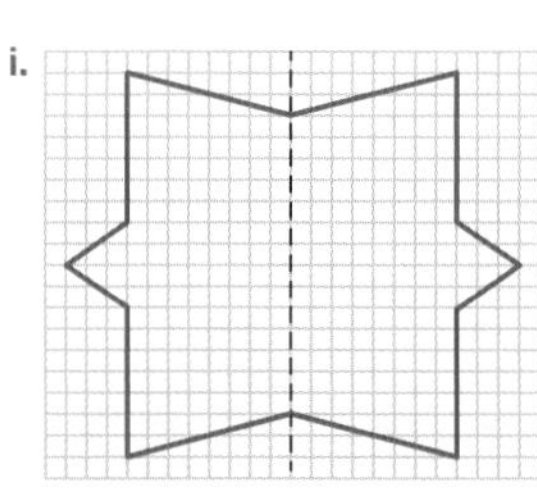

ii.

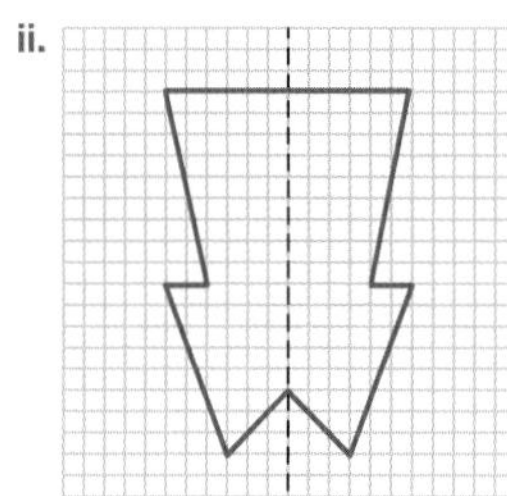

b. i. 2
ii. 1

| | | |
|---|---|---|
| 9. a. 3 | b. 1 | c. 0 |
| 10. a. 4 | b. 2 | c. 0 |
| 11. a. 2 | b. 0 or 1 | c. 1 |
| 12. a. 3 | b. 1 | c. 1 |
| 13. a. 4 | b. 2 | c. 2 |
| 14. a. 2 | b. 1 | c. 1 |

15. a. An infinite amount
b. An infinite amount
16. a. 4, 5, 6, 7
b. 4, 5, 6, 7
c. Yes, the number of vertices is double the number of axes of symmetry, because each axis of symmetry goes through two vertices.
17. a.

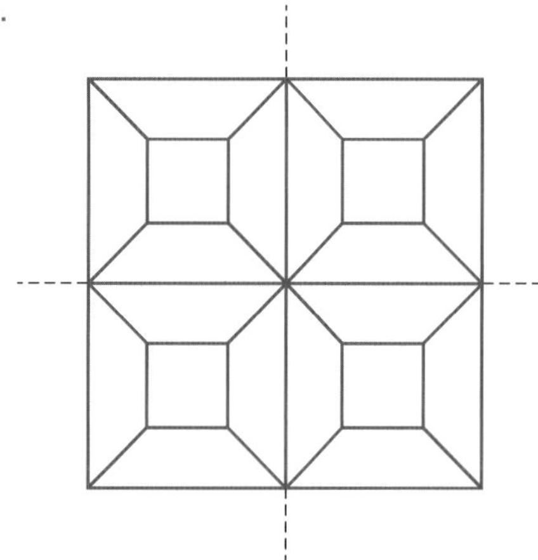

b. Every point on the pattern has to be equally distant from the axis of symmetry on either side of the axis.
c. 4
18. a. 1 b. 1
19. a. 1 b. 1
20. a.

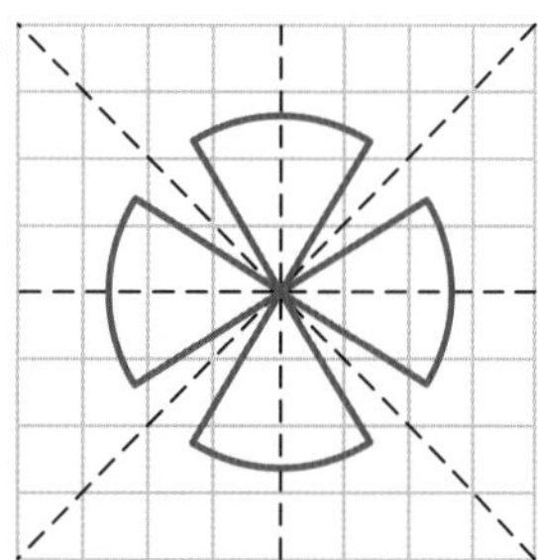

b. 4

### 12.3 Translations

| | | | |
|---|---|---|---|
| 1. a. 1R 2U | b. 19R 2D | c. 4L 5D | d. 10L |
| 2. e. 2R 4U | f. 2R | g. 2R 9U | h. 11L 8D |
| 3. i. 4L 1D | j. 6R | k. 7L 5D | l. 2L 8U |

4. a. $P' = (-1, 3)$

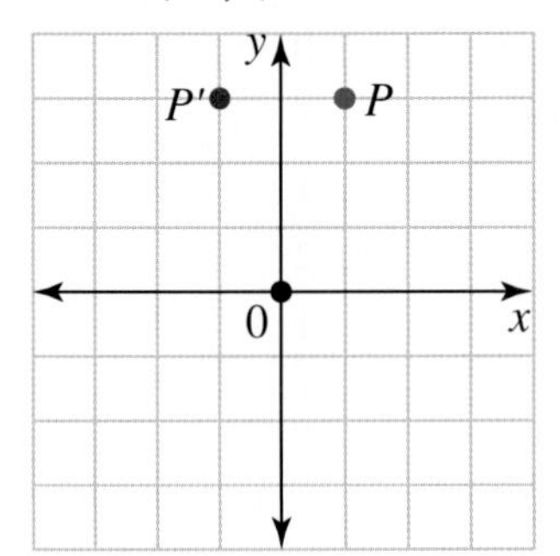


b. $P' = (1, 7)$

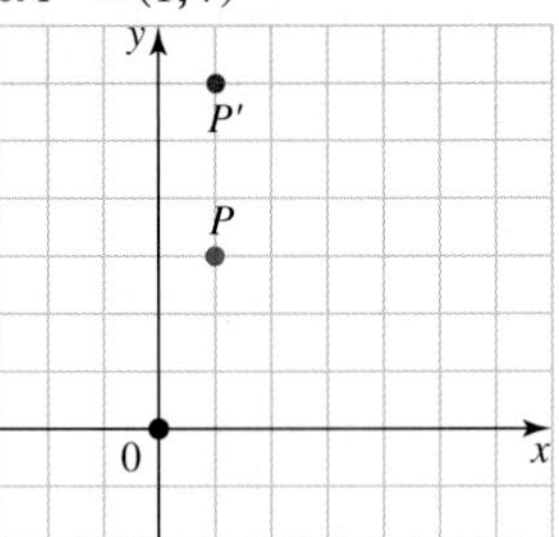


c. $P' = (2, 1)$

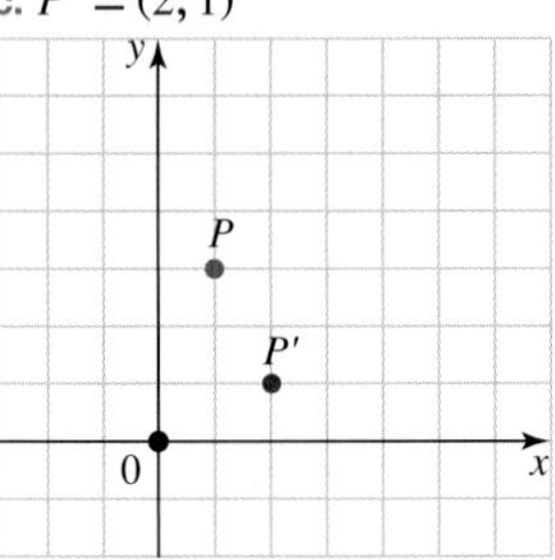


5. a. $P' = (9, -1)$

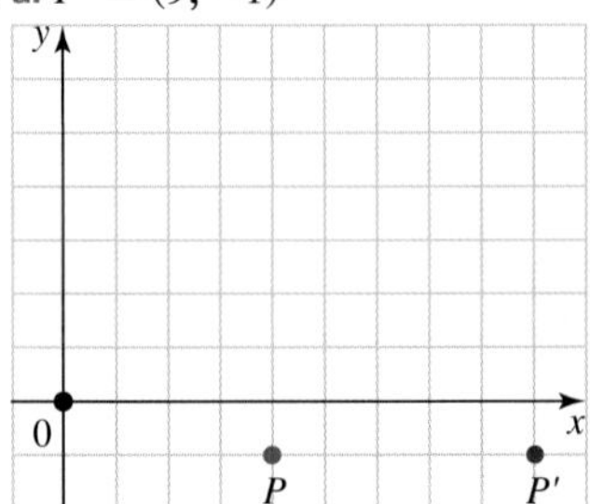


b. $P' = (4, -3)$

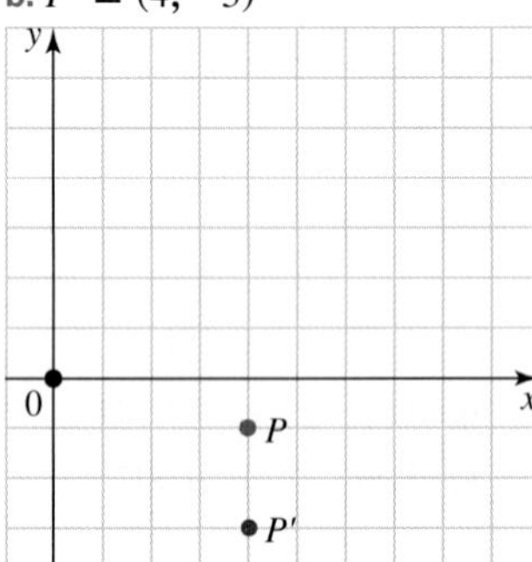


c. $P' = (0, 2)$

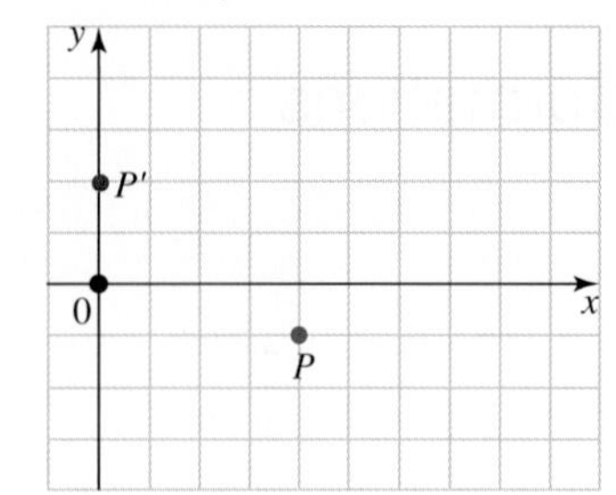


6. a. $P' = (4, -5)$

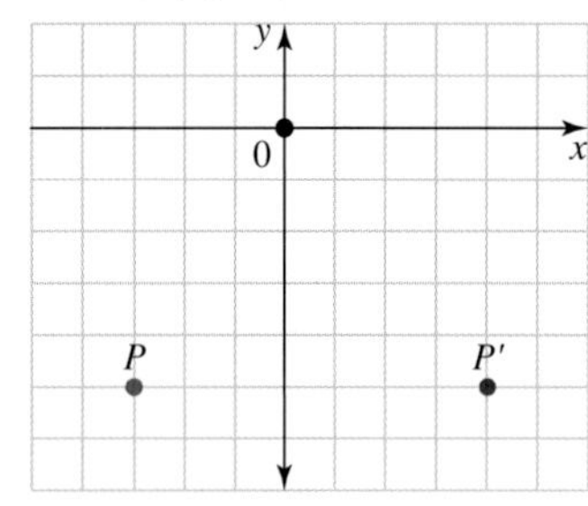


b. $P' = (-3, -2)$

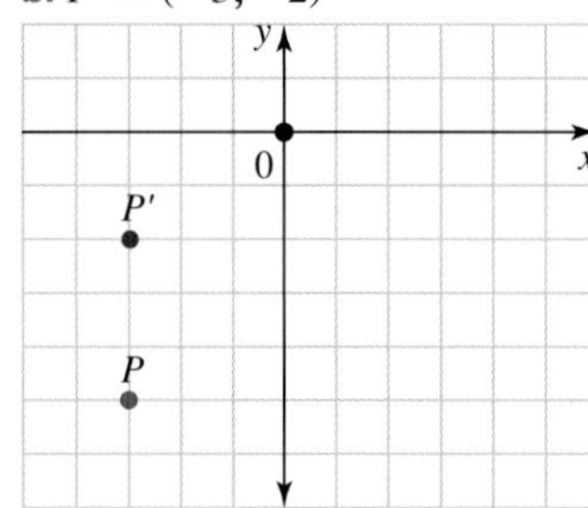


c. $P' = (1, -9)$

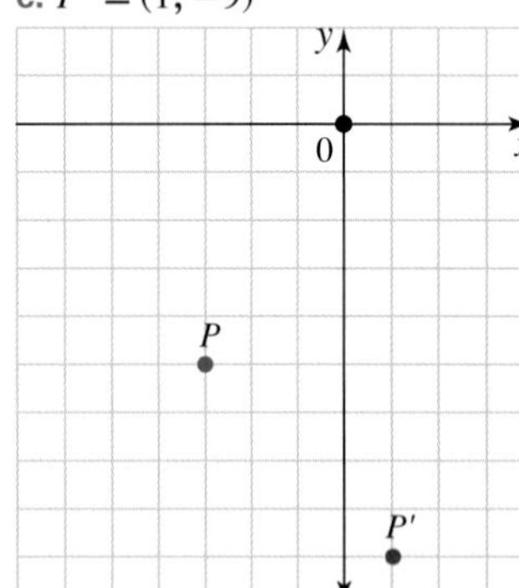


7. 4L 6U

8. 10L 2D

9. 13R 1U

10.

y
10
9
8
7
6
5
4
3
2
1
0
1 2 3 4 5 6 7 8 x
D″ C″
A″ B″
D′ C′
A′ B′
D C
A B

11. a. 5R 6D b. 21R 5D c. 2L 1D
d. 2R 1U e. 7R 6U f. 19L 6U

12. a. C to D b. D to C c. E to D
d. E to B e. D to A f. E to C

13. a. $P'(1, 4)$
b. $P''(-2, 2)$
c. A translation of 1 unit up and 5 units to the left.

14. a. 9R 13U b. 7R 7U then 2R 6U
c. $7R + 2R = 9R$
$7U + 6U = 13U$

15.

| Jump | From | To |
|---|---|---|
| 1 | A2 | C2 |
| 2 | A1 | C1 |
| 3 | C1 | C3 |

16.

| Jump | From | To |
|---|---|---|
| 1 | A4 | C4 |
| 2 | A3 | C3 |
| 3 | C4 | C2 |
| 4 | A1 | A3 |
| 5 | B1 | B3 |
| 6 | A3 | C3 |
| 7 | C3 | C1 |

17. 8 m

18. a.

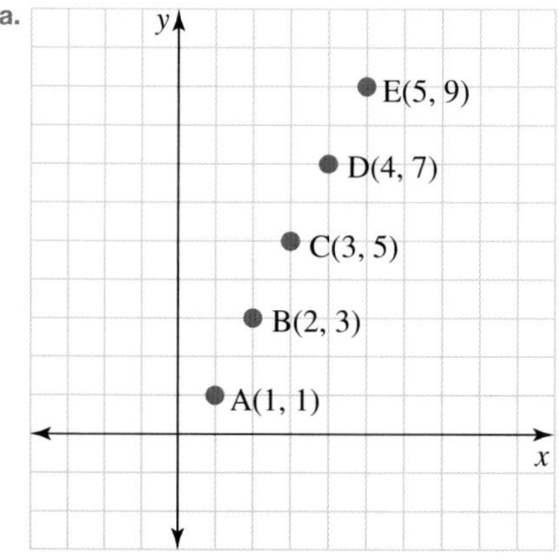


b. Form a straight line.

c.

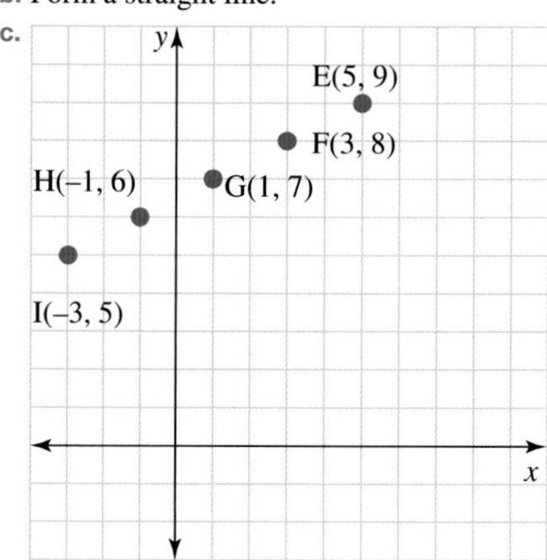


d. Form a straight line.

19.

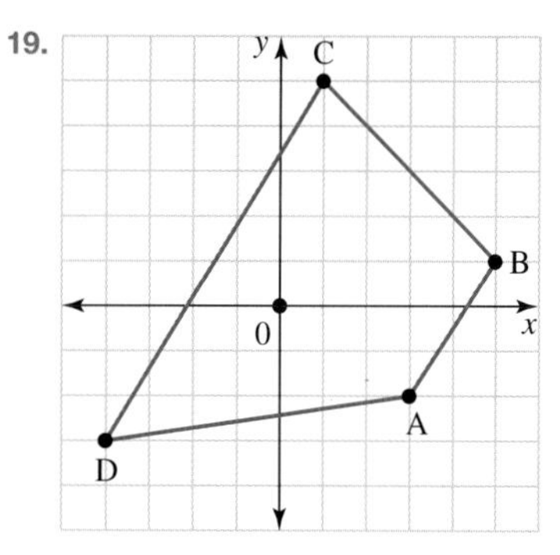


## 12.4 Reflections

1.

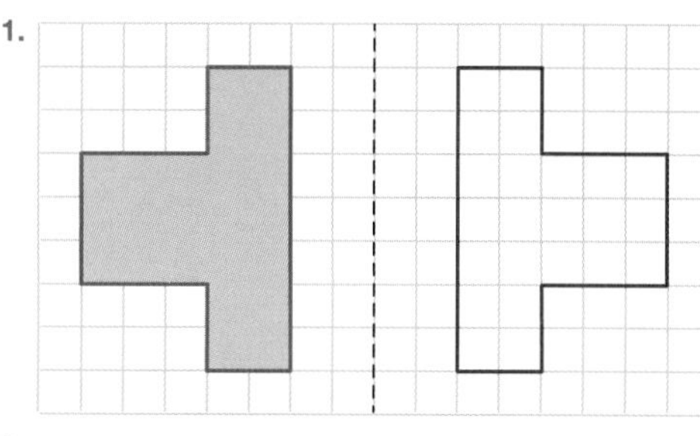

2.

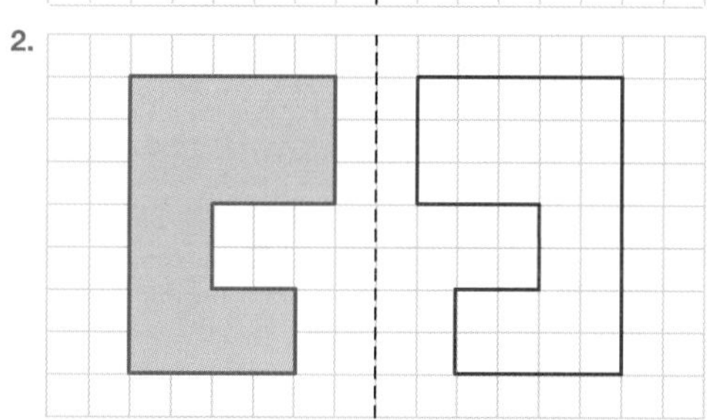

3.

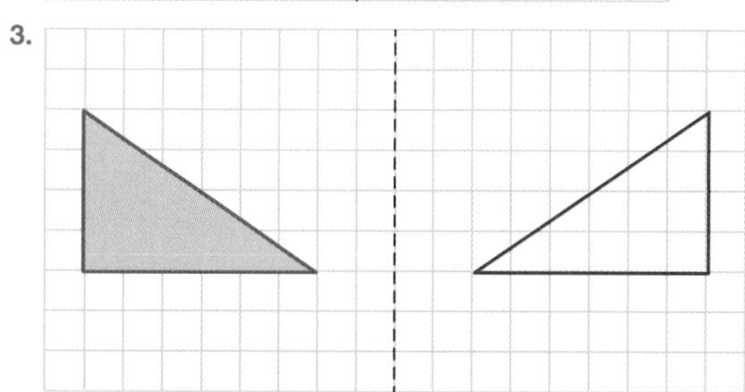

4.

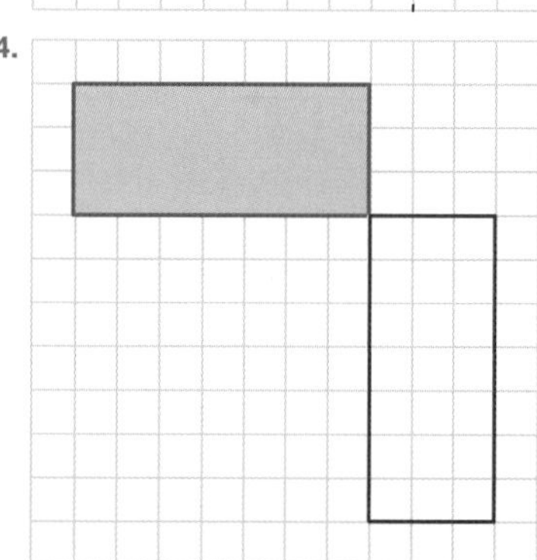

5.

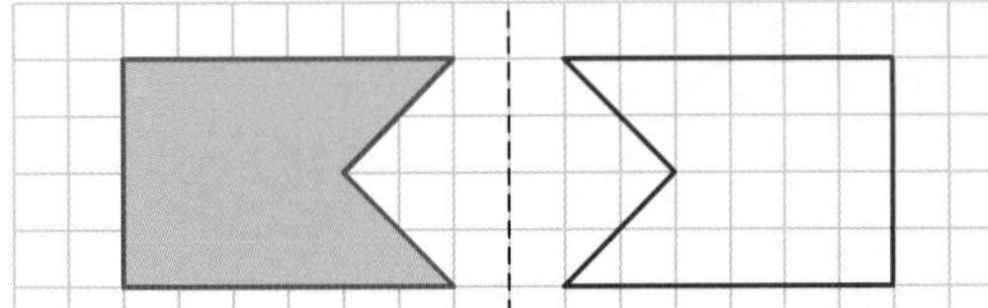

6.

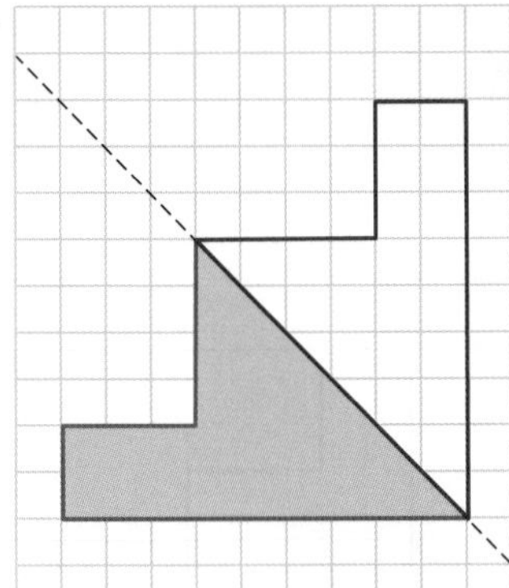

7. a. $P' = (1, -6)$

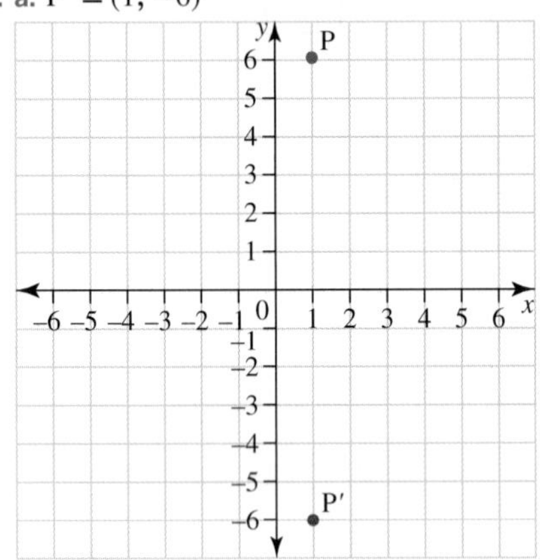


b. $P' = (-1, 6)$

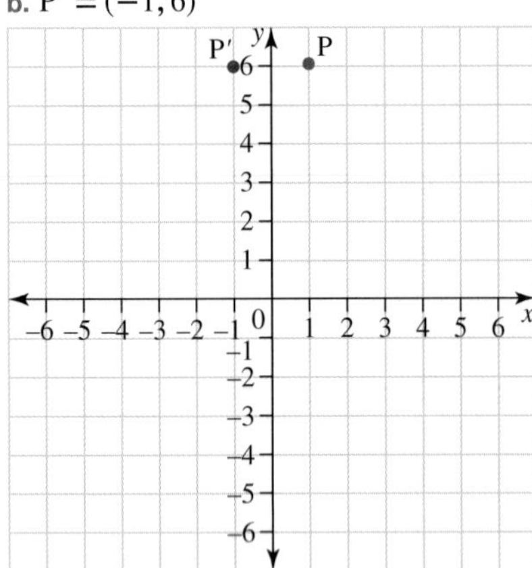


c. $P' = (1, 0)$

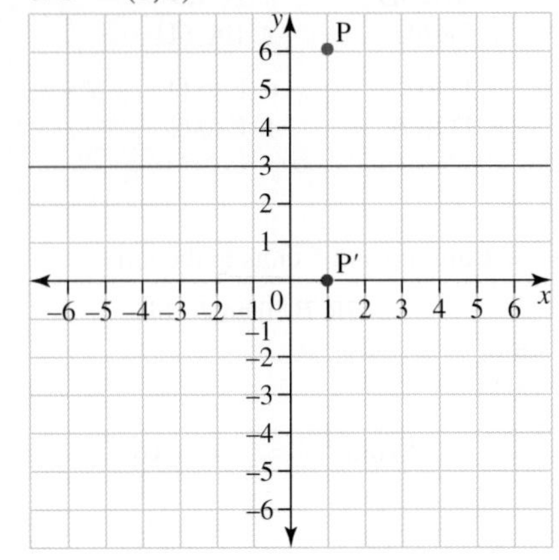


8. a. $P' = (-2, -4)$

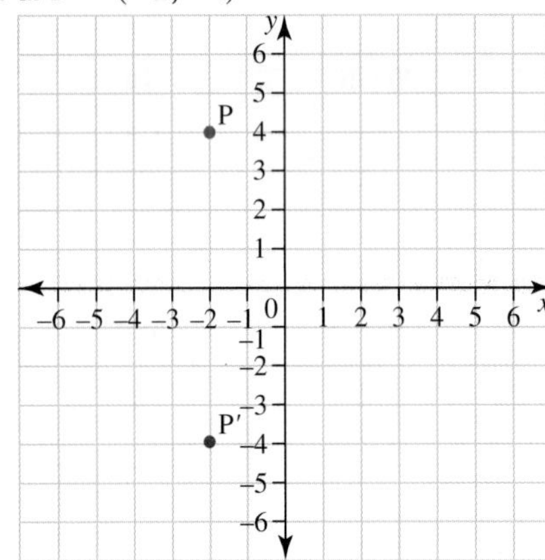


b. $P' = (2, 4)$

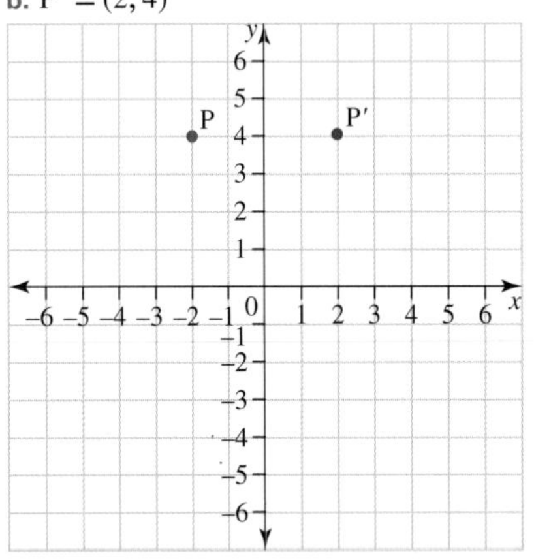

**c.** $P' = (0, 4)$

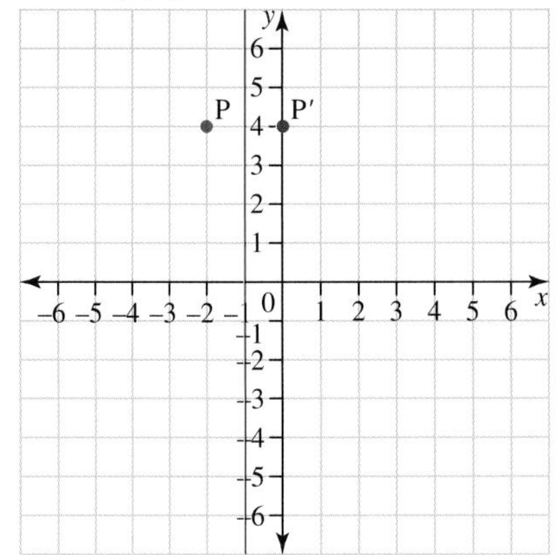


**9. a.**

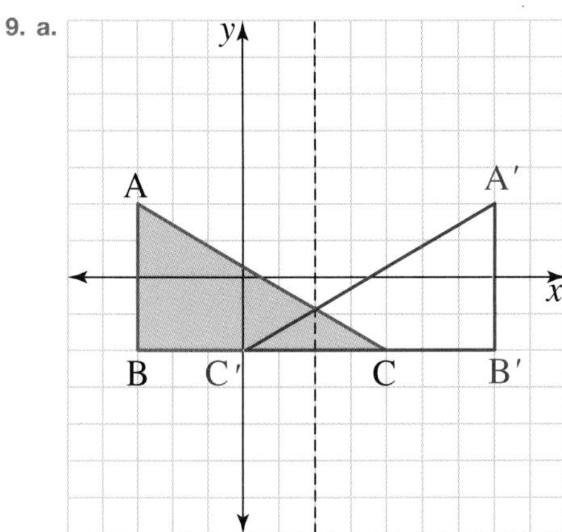


**b.** $A'(7, 2), B'(7, -2), C'(0, -2)$

**10. a.**

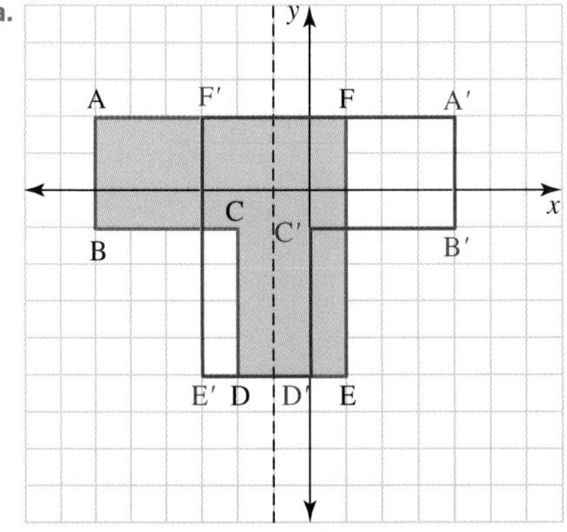


**b.** $A'(4, 2), B'(4, -1), C'(0, -1),$
$D'(0, -5), E'(-3, -5), F'(-3, 2)$

**11. a.**

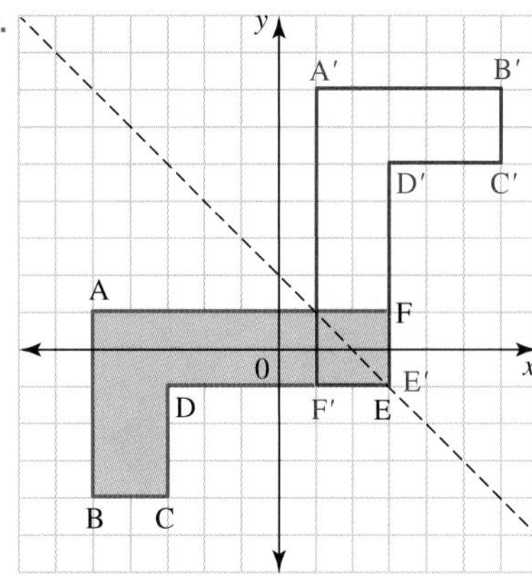


**b.** $A'(1, 7), B'(6, 7), C'(6, 5),$
$D'(3, 5), E'(3, -1), F'(1, -1)$

**12.** When reflected in an axis, the image is the mirror representation of the original object. If the object and its image are next to each other, they touch each other so they could form one object together. As they are both identical, the axis of reflection becomes the axis of symmetry of the newly formed object.

**13.** A reflection in a horizontal axis requires the object to be reflected in a vertical direction: up or down.

**14.** Image 2. It is equally distant from the reflection line as the original strawberry, it is the mirror image of the original strawberry and it is placed on the line perpendicular to the reflection line passing through the original strawberry.

**15. a.**

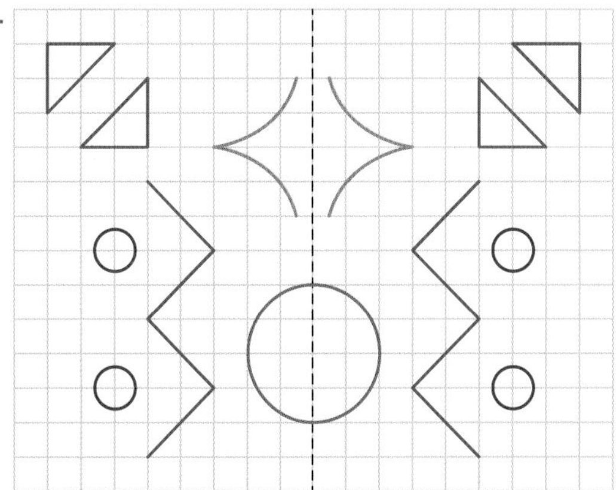

**b.**

**16.** 10:15

**17. a.**

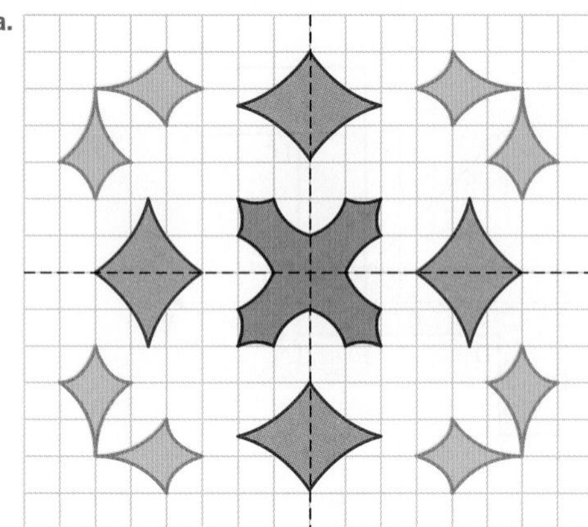

**b.**

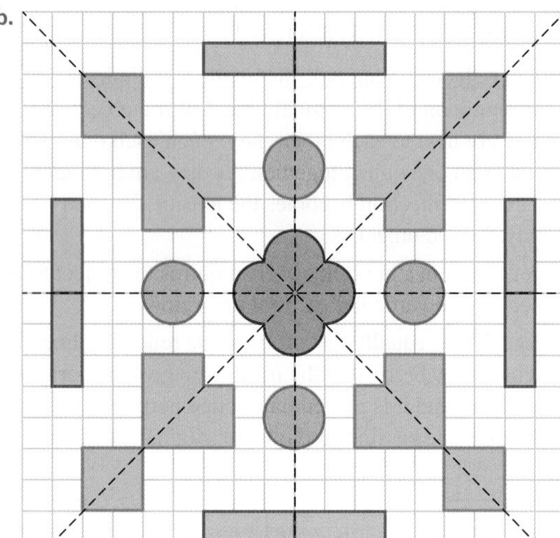

## 12.5 Rotations and combined transformations

**1. a.** $P' = (-5, 1)$

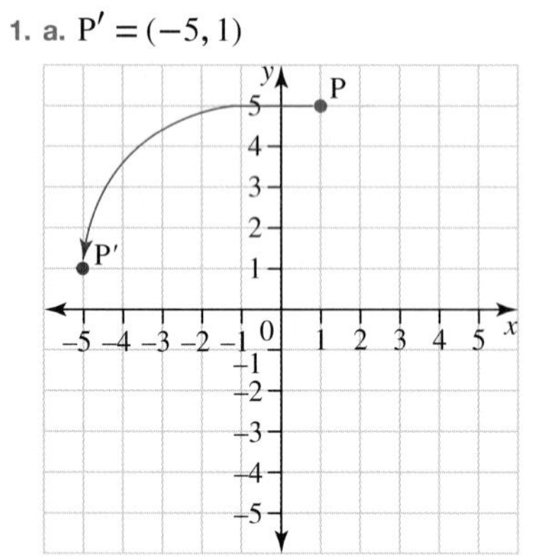


**b.** $P' = (5, -1)$

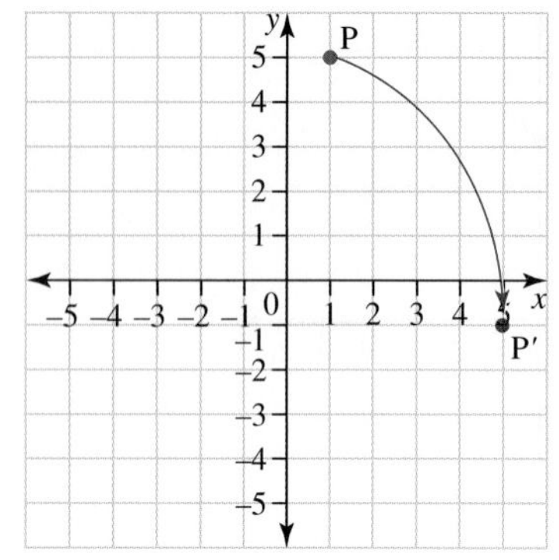


**c.** $P' = (-1, -5)$

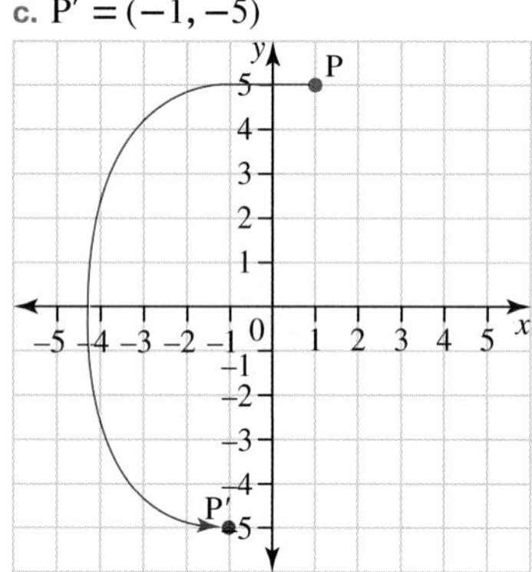


**2. a.** $Q' = (-4, -2)$

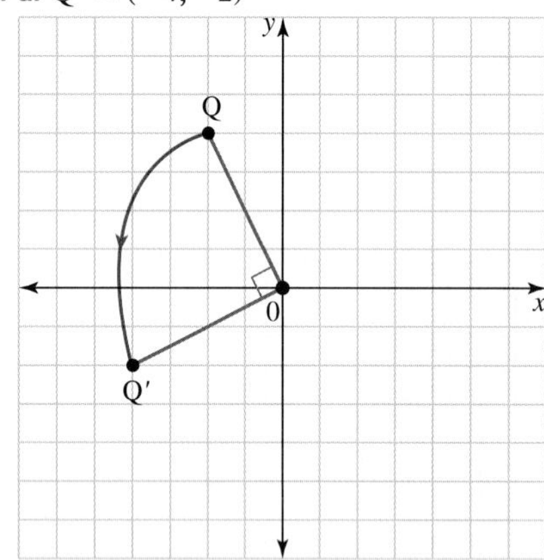


**b.** $Q' = (4, 2)$

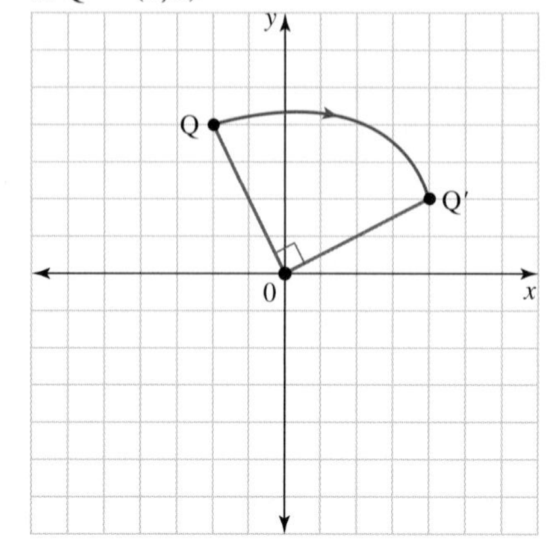

c. $Q' = (2, -4)$

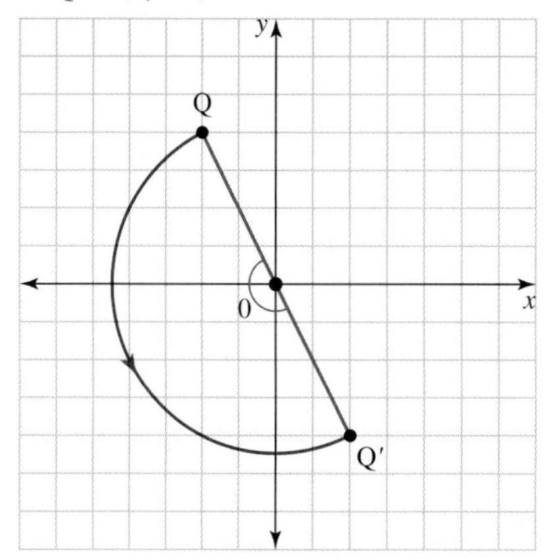


c. $R' = (3, 1)$

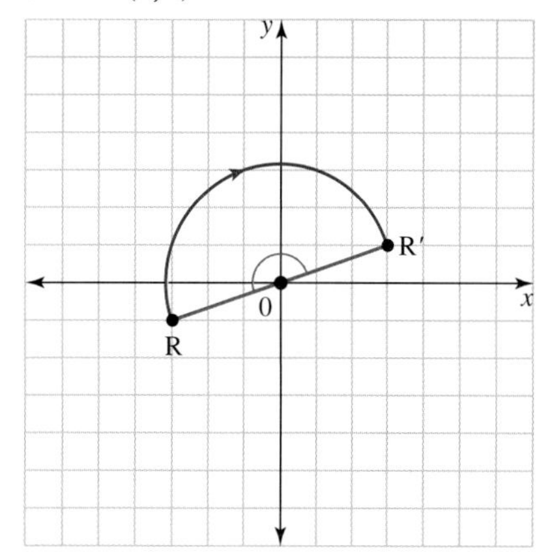


**3.** a. $R' = (1, -3)$

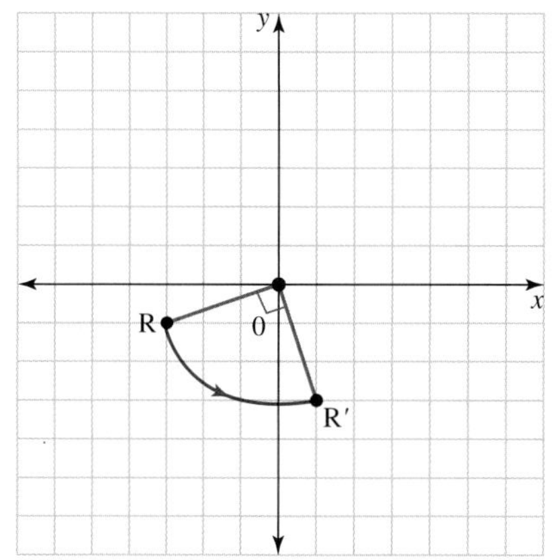


b. $R' = (-1, 3)$

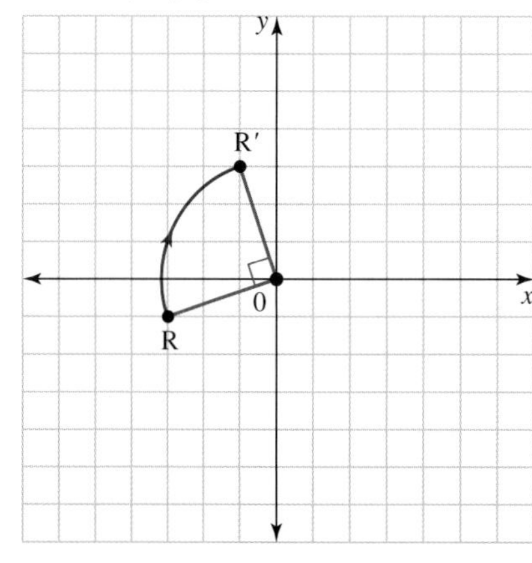


**4.** $A' = (-6, 0)$, $B' = (0, -4)$, $C' = (0, 0)$

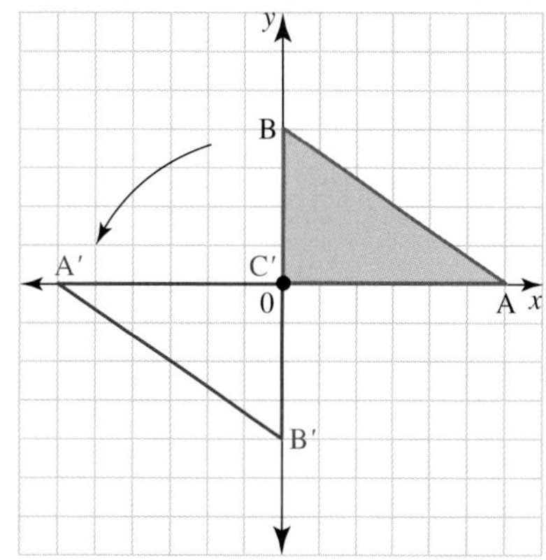


**5.** $A' = (6, 0)$, $B' = (4, 6)$, $C' = (0, 0)$

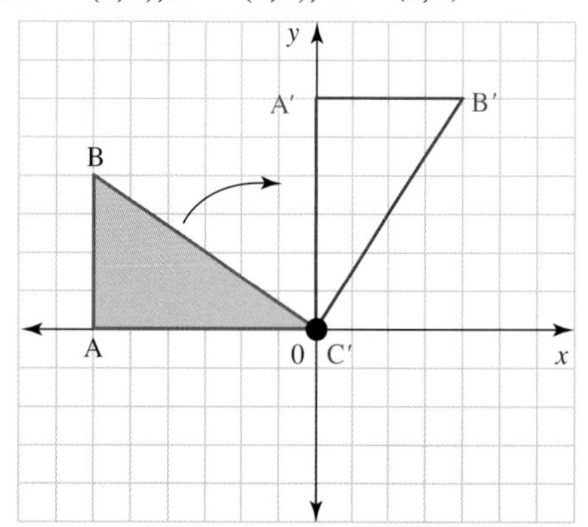

6. $A' = (0, -8)$, $B' = (-4, -8)$, $C' = (0, -2)$

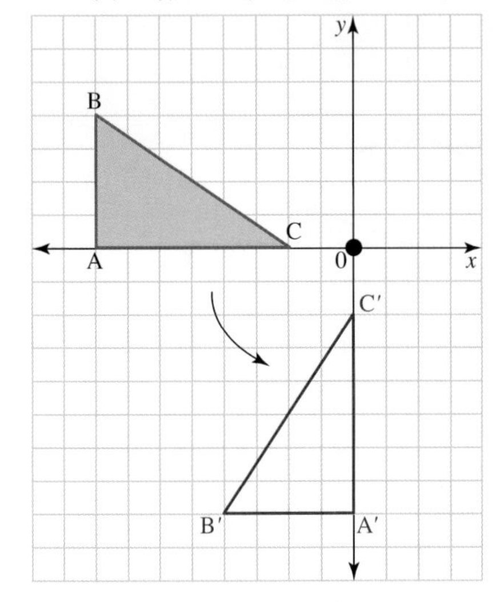


7. $A' = (2, 2)$, $B' = (-2, 2)$, $C' = (2, 7)$

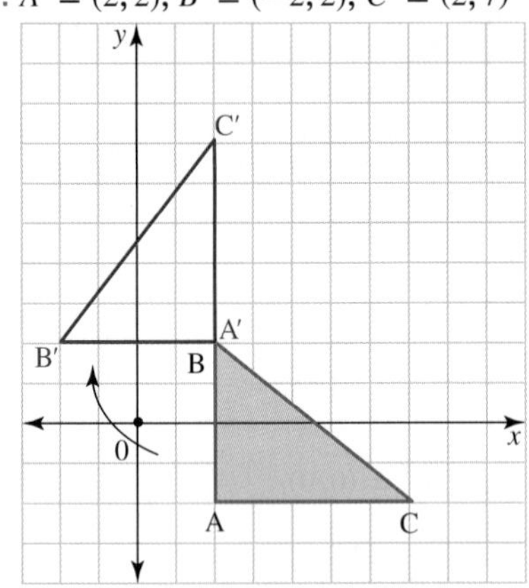


8. $A' = (-2, 0)$, $B' = (2, 0)$, $C' = (-2, -5)$

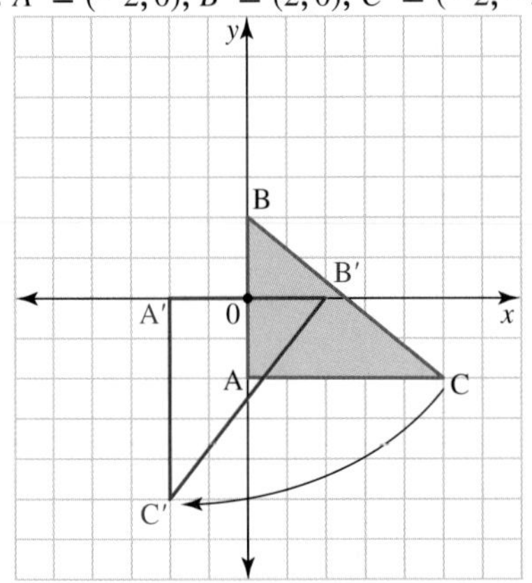


9. $A' = (-1, 4)$, $B' = (3, 4)$, $C' = (-1, -1)$

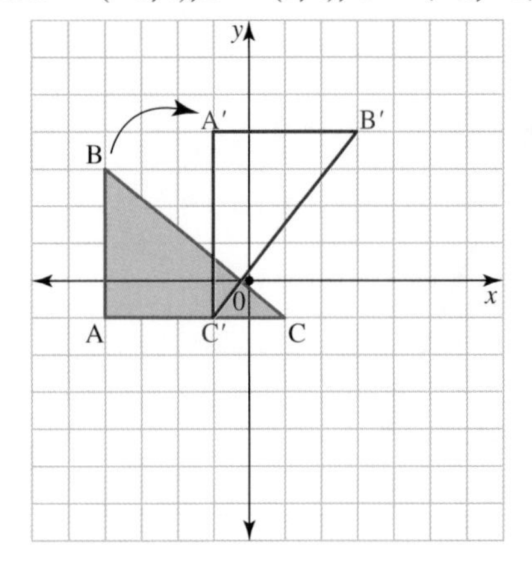


10.

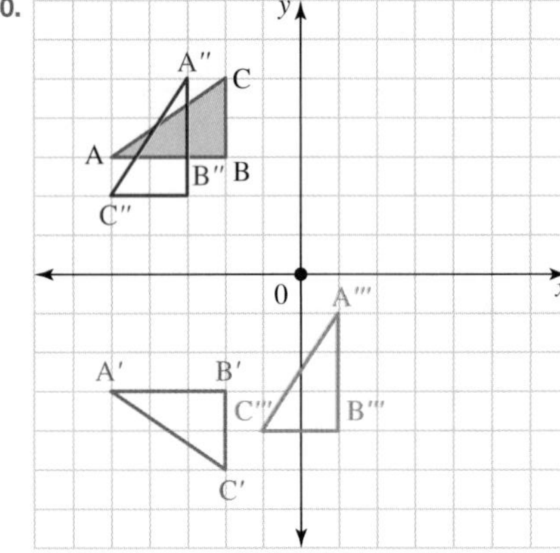


11.

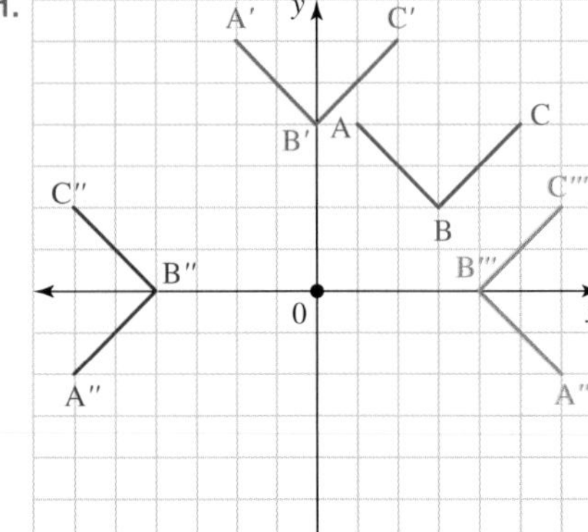


12.

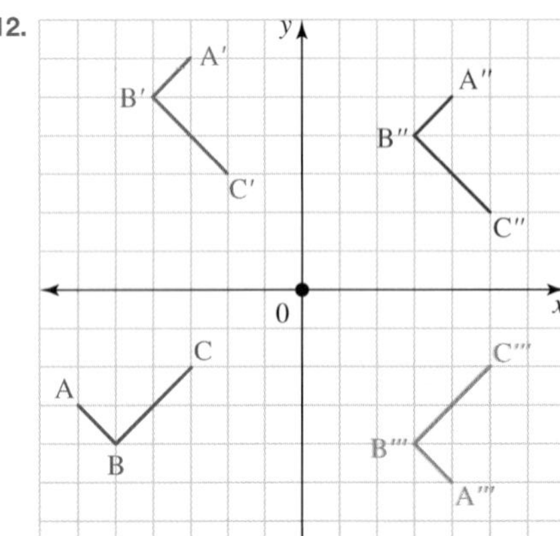


13. C

14. A reflection in the dashed line shown.

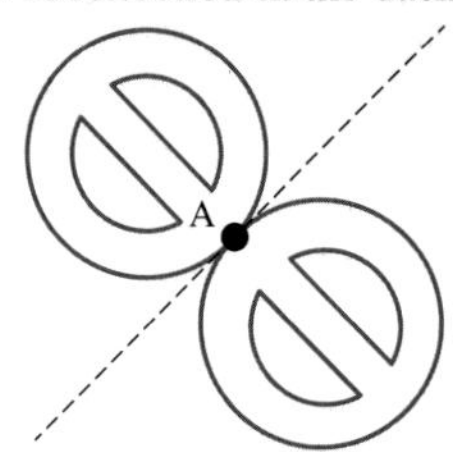


15. A full circle is 360°. 180° is half of a circle. It does not matter which direction you move along a circle; if you travel 180°, you will reach the opposite side. This can be seen in the following diagram.

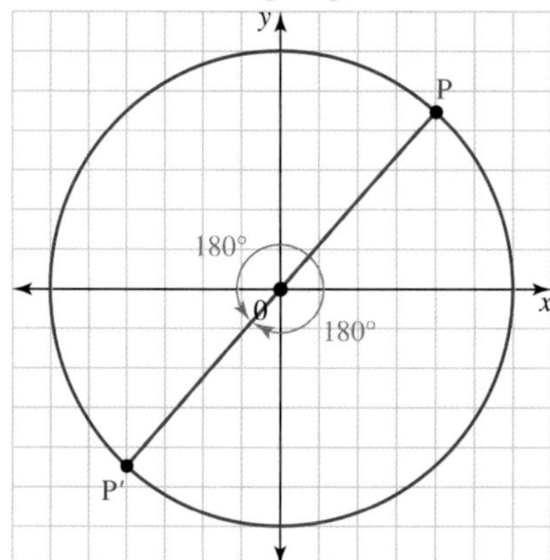


16. $(4 - b, -a)$

17. $(-r, -q) \Rightarrow (r, q) \Rightarrow (-r, q); (-r, -q) \Rightarrow (-r, q)$

18. Approximately 1°.

19. $\text{P}(a, -b) \Rightarrow \text{P}'(-a, -b) \Rightarrow \text{P}''(a, b)$

20. a., b.

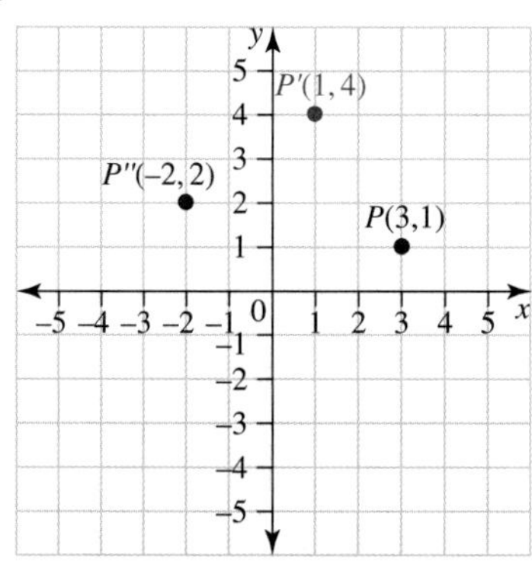


c. A translation of 1 unit up and 5 units to the left.

21. The coordinates of the point remain unchanged.

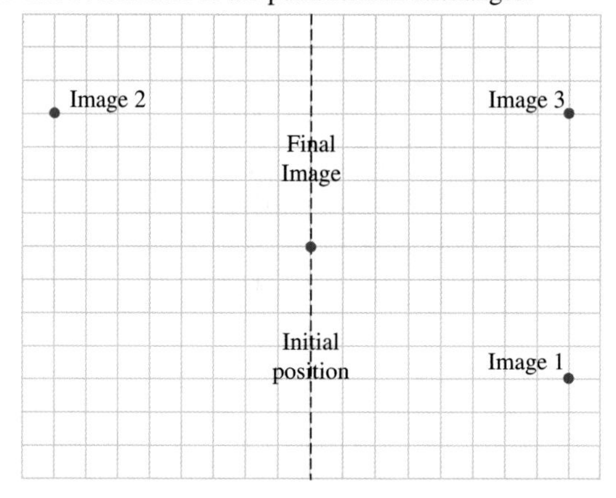


## Project

1. 90° clockwise rotation around the bottom right corner of brick 1
2. Translation of one brick-width to the right
3. 90°clockwise rotation around a point halfway between the bases of bricks 1 and 3
4. Brick 4 : translate one brick-width up
   Brick 5 : rotate 180° around the point halfway between the closest points connecting bricks 1 and 5
   Brick 6: rotate 90° clockwise around the bottom right corner of brick 3
   Brick 7: rotate 90° clockwise around the bottom right corner of brick 1 and then translate it four brick-widths to the right
   Brick 8: rotate 90° clockwise around the point at the bottom left corner of brick 7
5. This is an example of a tessellation because the pattern repeats itself and there are no gaps.
6. Students could use one or more shapes that look like a tessellation for their entertaining area. After creating the pattern, students should describe what transformations are involved.

## 12.6 Review questions

1. a. 3 b. 3

2. a.

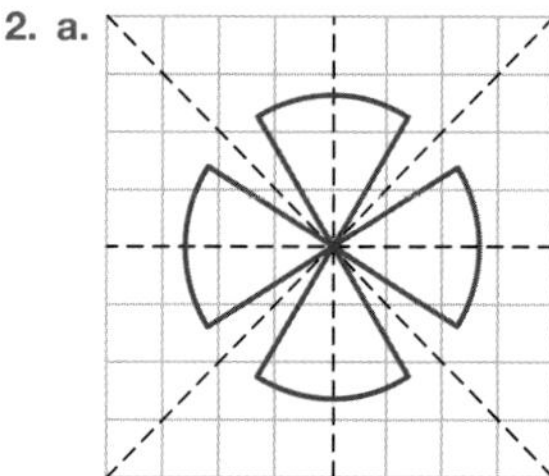

b. 4

3. a.

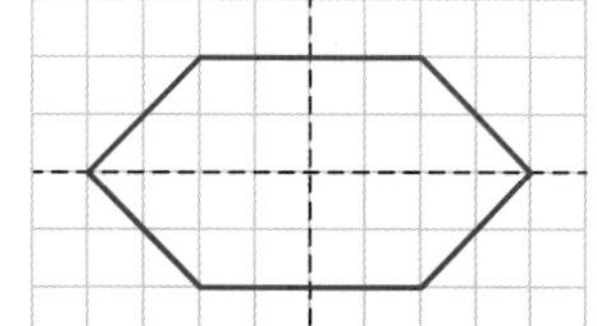

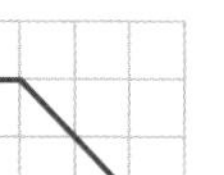

b. 2

4. a.

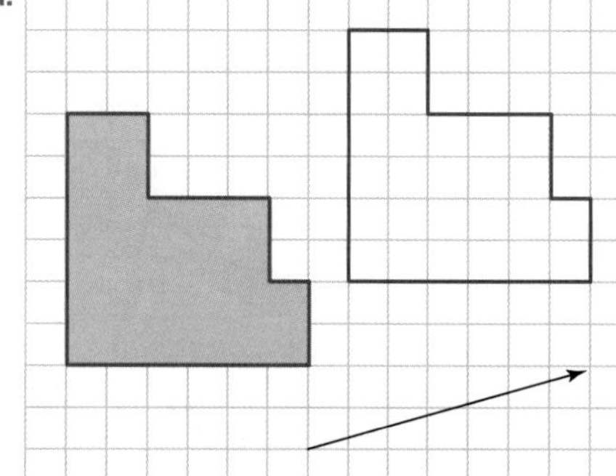

**b.**

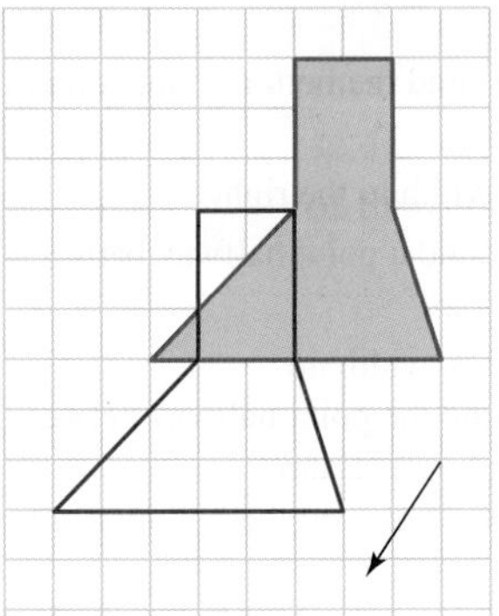

**c.**

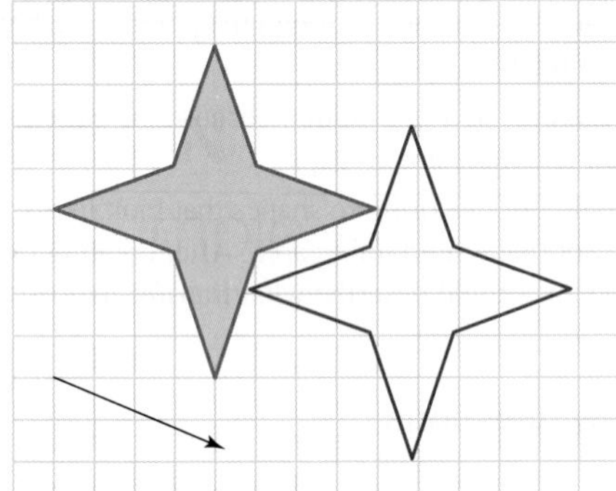

**5. a.** 1U 5R **b.** 3D 2L **c.** 8R 1U **d.** 12L 4U

**6. a.**

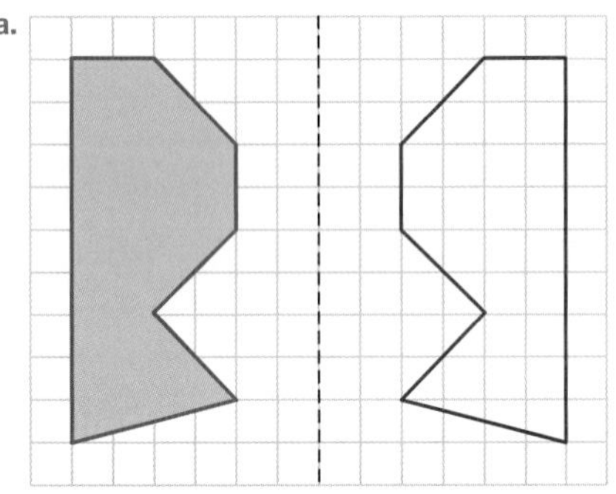

**b.**

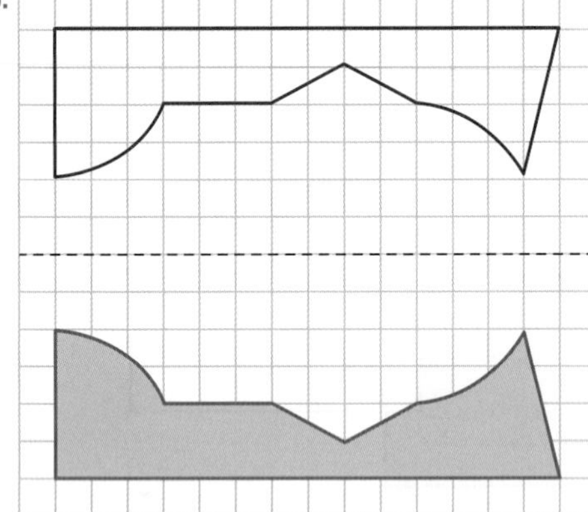

**c.**

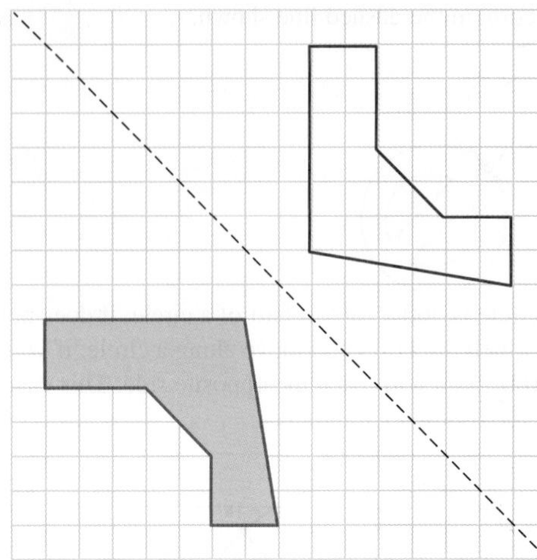

**7. a. i.**

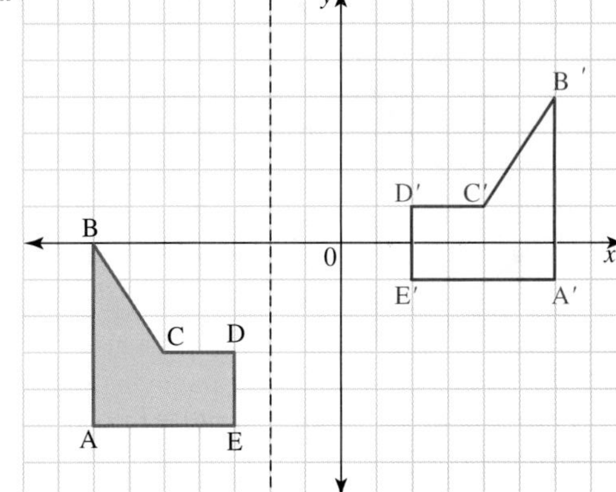


**ii.**

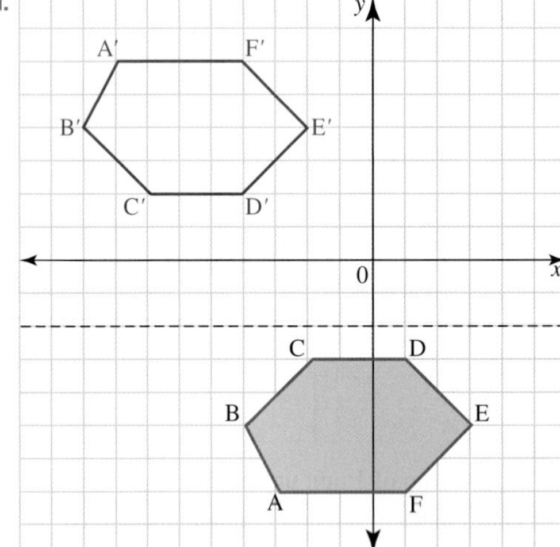


**b. i.** $A(-7, -5) \rightarrow A'(6, -1)$,

$B(-7, 0) \rightarrow B'(6, 4)$,

$C(-5, -3) \rightarrow C'(4, 1)$,

$D(-3, -3) \rightarrow D'(2, 1)$,

$E(-3, -5) \rightarrow E'(2, -1)$,

ii. $A(-3, -7) \rightarrow A'(-8, 6)$,
$B(-4, -5) \rightarrow B'(-9, 4)$,
$C(-2, -3) \rightarrow C'(-7, 2)$,
$D(1, -3) \rightarrow D'(-4, 2)$,
$E(3, -5) \rightarrow E'(-2, 4)$,
$F(1, -7) \rightarrow F'(-4, 6)$

8. a. 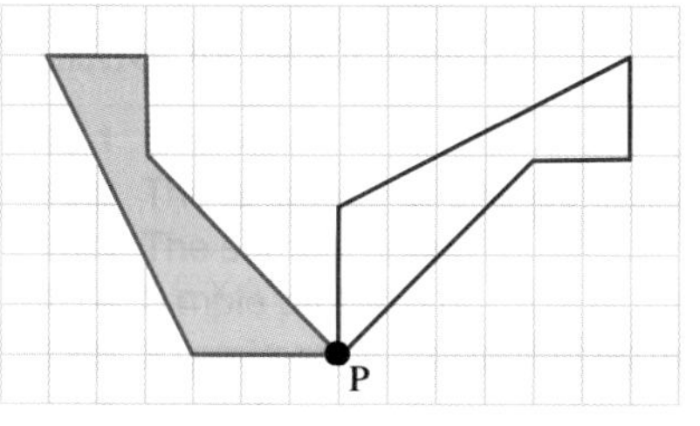


b. 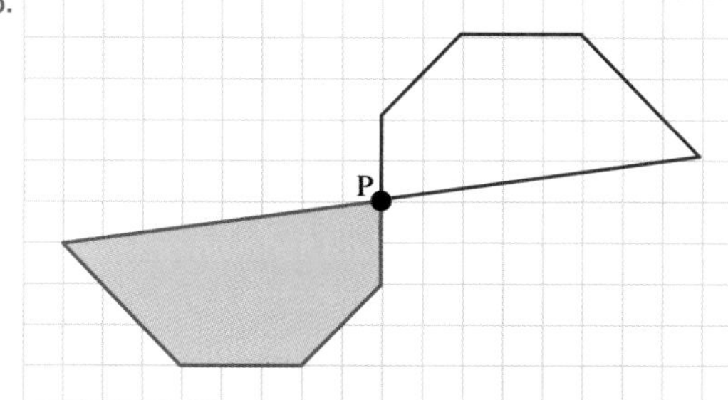


c. 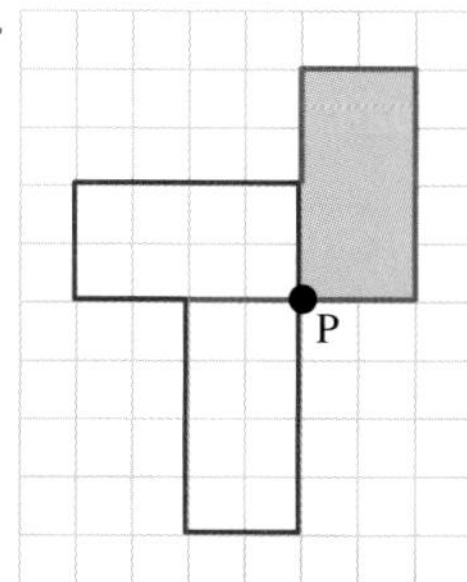


9. a. Each of the second group of 10 letters has the same pattern as the corresponding letter but with an extra dot underneath.
b. D and F, E and I, H and J, M and U, N and Z, P and V, R and W
c. B, C, E and I
d. Braille is read by touch and translations would not be determined by touch.
e. The word *fixed* has rotational symmetry when written in braille:

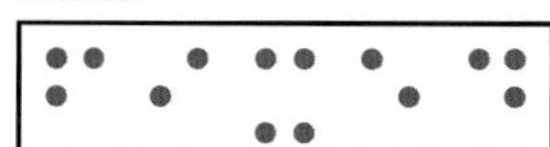

f. The letters W and R form a rotational symmetrical pair in braille:

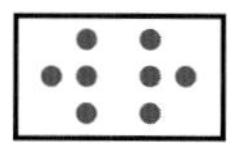

10. Sample responses can be found in the worked solutions in the online resources.

# LESSON
# 13.1 Overview

## Why learn this?

Probability describes the chances of different events occurring. As with many different areas of mathematics, probability has its own terminology or language. To understand the chances of an event happening, it is important to understand the language of probability.

Whether you realise it or not, you use and consider probabilities every single day. For example, the weather report may say that there is a high chance of rain tomorrow. When you decide what to wear each day, you are taking into account the likelihood of it raining, or the likelihood of the sun shining all day.

Many board games depend on probability. Have you ever played a board game with 2 dice where you wanted to throw a double 6? The chance, or probability, of that occurring is 1 in 36. Lotteries use probability to determine the likelihood of a winning ticket being bought, and set the prize money based on these calculations. To win first prize in TattsLotto, you need to have selected the correct 6 balls from the 45 balls in the barrel. The chance of this happening is 1 in 8 145 060. This means that you would be likely to win TattsLotto approximately once per 8 million tickets you buy! In Oz Lotto you need to select 7 balls from a 45 ball barrel, with the chance of winning division 1 dropping lower, to 1 in 45 379 620!

Probability is also widely used in the finance industry. Insurance brokers look at the chance, or likelihood, of an event occurring, and set their rates accordingly. Many different professions require knowledge of probability, including science, market research analysis, meteorology, financial analysis, statistics and many others.

## Exercise 13.1 Pre-test

learnon

1. MC Select the option that best describes the probability of rolling an odd number with a die.
   A. Impossible  B. Unlikely  C. Even  D. Likely  E. Certain

2. MC Select the probability that is best linked to a likely probability.
   A. 0  B. 0.1  C. 0.3  D. 0.5  E. 0.7

3. A bag contains 3 blue balls, 5 red balls and 2 yellow balls. Determine the probability of randomly selecting a yellow ball from the bag.

4. MC A ball is selected from a bag. Determine the sample space of the bag that contains 5 blue balls, 8 red balls and 2 yellow balls. (B = blue, R = red, Y = yellow)
   A. $S = \{5B, 8R, 2Y\}$  B. $S = \{B, R, Y\}$
   C. $S = \{B, R, W, Y\}$  D. $S = \{BBBBB, RRRRRRRR, YY\}$
   E. $S = \{B, G, R, Y\}$

5. A card is selected from a standard deck of cards and the suit is noted. Determine the number of elements in the sample space.

6. Two dice are rolled and the sum of the two dice is 5. Determine the number of ways the sum of 5 can be obtained.

7. MC If a fair die is rolled, the probability of rolling a number less than 5 is:
   A. $\frac{2}{3}$  B. $\frac{1}{6}$  C. $\frac{1}{3}$  D. $\frac{2}{5}$  E. $\frac{1}{2}$

8. MC If a 12-sided die is rolled, the probability of rolling a multiple of 4 is:
   A. $\frac{1}{2}$  B. $\frac{1}{4}$  C. $\frac{1}{3}$  D. $\frac{5}{12}$  E. $\frac{7}{12}$

9. MC A card is chosen at random from a standard deck of cards. The probability of selecting a red picture card is:
   A. $\frac{2}{13}$  B. $\frac{3}{26}$  C. $\frac{3}{28}$  D. $\frac{3}{13}$  E. $\frac{1}{4}$

10. A bag contains 5 blue balls, 7 red balls and 3 green balls. If a ball is selected at random, calculate the probability that neither a red or blue ball was selected.

11. MC Lauren does not know when Emily's birthday is. Determine which of the following months it is most likely to be in.
    A. April  B. June  C. September
    D. November  E. All options are equally likely.

12. Australia Post delivers parcels during business hours (9 am to 5 pm). Sally has ordered a parcel that will arrive next week. Determine the probability that the parcel arrives before midday on Tuesday.

13. MC Identify which of the following options best describes the probability of randomly choosing a prime number out of the first 100 whole numbers.
    A. Impossible  B. Unlikely  C. Even chance  D. Likely  E. Certain

▶

14. **MC** Identify which of the following options would best simulate selecting a ball from a bag of 6 differently coloured balls.
   A. Tossing a coin
   B. Spinning a 4-sector circular spinner
   C. Spinning a 5-sector circular spinner
   D. Rolling a standard die
   E. Spinning a square spinner

15. A box contains 3 red dice and 4 blue dice. Determine the smallest number of dice needed to be taken from the box to be *sure* you have a die of each colour.

# LESSON
# 13.2 The language of chance

## LEARNING INTENTION

At the end of this lesson you should be able to:
- understand that a probability is a number between 0 and 1 that describes the possibility of an event
- classify the chance of an event occurring using words such as *certain, likely, unlikely, even chance* or *impossible*.

## 13.2.1 The language of chance

eles-4710

- An **event** is a result that may occur.
  - When classifying the chance of an event occurring, we use words such as *certain, likely, even chance, unlikely* and *impossible.*
  - Probabilities can be written as decimals, fractions and percentages. For example, $0.5 = \frac{1}{2} = 50\%$.

- The following **probability** scale associates important words used to describe probability with their approximate corresponding numerical values.

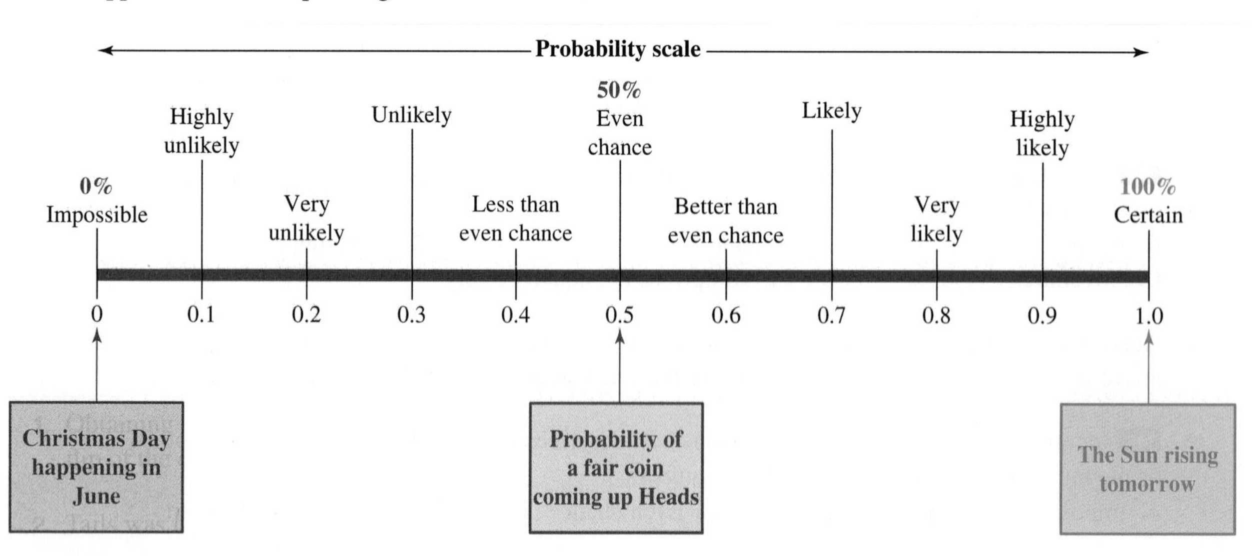

- An **outcome** in a chance experiment refers to any possible result.
- An event can describe either one outcome (for example, rolling a 1 on a die) or a collection of outcomes (for example, rolling an even number on a die).

### DISCUSSION

Discuss other examples of events that have a probability of 0, 0.5 or 1.

### WORKED EXAMPLE 1 Classifying an event occurring

**For each of the given statements, specify whether the chance of the following events occurring is *certain, likely, even chance, unlikely* or *impossible*.**

**a. You will compete in the next Olympics.**
**b. Every student in Year 7 will obtain 100% in their next mathematics test.**
**c. Each person in your class has been to the zoo.**
**d. You flip a coin and Tails comes up.**
**e. March is followed by April.**

| THINK | WRITE |
|---|---|
| a. The chance of a person competing in the next Olympics is very small; however, it could happen. | a. It is *unlikely* that this event will occur. |
| b. Due to each student having different capabilities and the number of students involved, this situation will almost never happen. | b. The chance of this event occurring is *impossible.* |
| c. The chance that each student in your class has been to the zoo, either with their family or primary school, is probable. | c. It is *likely* this event has occurred. |
| d. When you flip a coin in a chance experiment, there are only two possibilities, a Head or a Tail. So there is a 50% chance of Tails coming up. | d. There is an *even chance* this event will occur. |
| e. April always follows the month of March. This is a true statement. | e. It is *certain* this event will occur. |

## WORKED EXAMPLE 2 Estimating the probability of an event occurring

**Assign a number between and including 0 and 1 to represent the estimated probability of the following events, depending on how likely they are.**

a. **One of two evenly matched tennis players will win the next game.**

b. **You will guess the correct answer on a multiple choice question with four options.**

c. **Rolling a fair die and obtaining a number less than 6.**

| THINK | WRITE |
|---|---|
| a. 1. Determine the likelihood of an event occurring, with reasoning. | a. Since the two players are evenly matched, one does not have an advantage over the other. Therefore, they each have an equal chance of winning the next game. |
| 2. Express the answer as a decimal. | The probability that one player wins the game is $\frac{1}{2}$ or 0.5. |
| b. 1. Determine the likelihood of an event occurring, with reasoning. | b. When guessing an answer on a multiple choice question with 4 options, 1 out of the 4 possibilities will be correct. One out of 4 may be expressed as a fraction. |
| 2. Express the answer as a decimal. | The probability of guessing the correct answer is $\frac{1}{4}$ or 0.25. |
| c. 1. Determine the likelihood of an event occurring, with reasoning. | c. In the chance experiment of rolling a die, there are six possibilities. They are 1, 2, 3, 4, 5, 6. A number less than 6 includes 1, 2, 3, 4, 5. Therefore, five out of the six possibilities may be rolled. Five out of six may be expressed as a fraction. |
| 2. Express the answer as a decimal, correct to 2 decimal places. | The probability of obtaining a number less than six is $\frac{5}{6}$ or approximately 0.83. |

## COLLABORATIVE TASK: Draw a spinner

Create spinners with the following probabilities.

a. $P(\text{blue}) = \frac{1}{3}$ and $P(\text{white}) = \frac{2}{3}$

b. $P(\text{blue}) = \frac{1}{2}$, $P(\text{white}) = \frac{1}{4}$, $P(\text{green}) = \frac{1}{8}$ and $P(\text{red}) = \frac{1}{8}$

c. $P(\text{blue}) = 0.75$ and $P(\text{white}) = 0.25$

Once you have created your spinners, test them out with your classmates. How accurately do your results match the probabilities for which the spinners were designed?

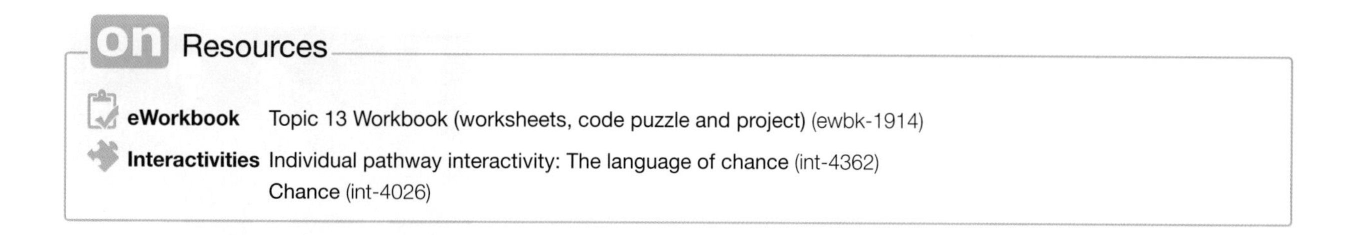
Resources

eWorkbook Topic 13 Workbook (worksheets, code puzzle and project) (ewbk-1914)

Interactivities Individual pathway interactivity: The language of chance (int-4362)
Chance (int-4026)

## Exercise 13.2 The language of chance

learnon

13.2 Quick quiz on | 13.2 Exercise

### Individual pathways

| ■ PRACTISE | ■ CONSOLIDATE | ■ MASTER |
|---|---|---|
| 1, 3, 5, 9, 12, 15 | 2, 6, 7, 10, 13, 16 | 4, 8, 11, 14, 17 |

### Fluency

1. WE1 For each of the given statements, specify whether the chance of the following events occurring is certain, likely, even chance, unlikely or impossible.

   a. New Year's Day will be on 1 January next year.
   b. You roll a fair die and obtain a number less than 5.
   c. Water will boil in the fridge.
   d. There will be snow on the ski fields this winter.
   e. You will grow 18 cm taller this year.

2. For each of the given statements, specify whether the chance of the following events occurring is certain, likely, even chance, unlikely or impossible.
   a. You will win first prize in Tattslotto.
   b. You roll a fair die and obtain an odd number.
   c. The year 2024 will be a leap year.
   d. You choose a white ball from a bag that contains only white balls.
   e. You choose a yellow ball from a bag containing 4 red balls and 4 yellow balls.

3. WE2 Assign a number between and including 0 and 1 to represent the estimated probability of the following events, depending on how likely they are.
   a. You flip a coin and obtain a Tail.
   b. You choose a red ball from a bag containing only 8 white balls.
   c. You guess the correct answer in a multiple choice question with 5 options.
   d. You roll a die and obtain a number greater than 4 on a fair die.

4. Assign a number between and including 0 and 1 to represent the estimated probability of the following events, depending on how likely they are.
   a. You flip a coin and obtain a Head.
   b. You choose a green ball from a bag containing only four green balls.
   c. You have science classes this year.
   d. Australia will win the Boxing Day cricket test.

5. **MC** The word that has the same meaning as *improbable* is the word:
   A. unlikely  B. impossible  C. uncertain  D. certain  E. likely

6. **MC** The word that has the same meaning as *certain* is the word:
   A. definite  B. possible  C. likely  D. probable  E. unlikely

### Understanding

7. Compare the given events A, B, C, D and order them from least to most likely.
   a. It will be sunny in Queensland most of the time when you visit.
   b. Melbourne Cup Day will be on the first Tuesday in November next year.
   c. You will be the next Australian Eurovision contestant.
   d. Saturn will be populated next year.

8. List five events that are:
   a. impossible  b. unlikely to happen
   c. likely to happen  d. sure to happen.

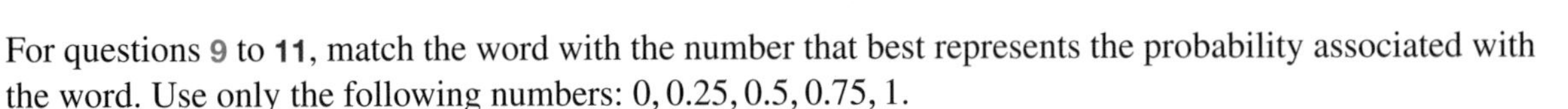

For questions 9 to 11, match the word with the number that best represents the probability associated with the word. Use only the following numbers: $0, 0.25, 0.5, 0.75, 1$.

9. a. Certain  b. Likely  c. Unlikely  d. Probable  e. Improbable

10. a. Slim chance  b. Sure  c. Doubtful
    d. Not able to occur  e. More than likely

11. a. Definite  b. Impossible  c. Fifty-fifty  d. Fair chance

### Reasoning

12. Explain your answers to the following questions using the language learned in this subtopic.
    a. If today is Monday, what is the chance that tomorrow is Thursday?
    b. If today is the 28th of the month, what is the chance for tomorrow to be the 29th?
    c. If you toss a coin, what is the chance it will land Heads up?

13. 'Fifty-fifty' is an expression commonly used in probability. Explain the meaning of this expression, giving its fractional value and its decimal number form, as well as expressing it as a percentage.

14. Five balls numbered 1, 2, 3, 4 and 5 are placed in a bag. You draw a ball out of the bag at random and check whether it is odd or even. Explain why the probability of drawing an odd ball from the bag is not 0.5.

### Problem solving

15. Anthony has 10 scrabble pieces, as shown. His friend Lian is blindfolded and is asked to pick a piece at random during a chance experiment. Determine the chance she will pick:

a. an I
b. an A
c. a U
d. an E.

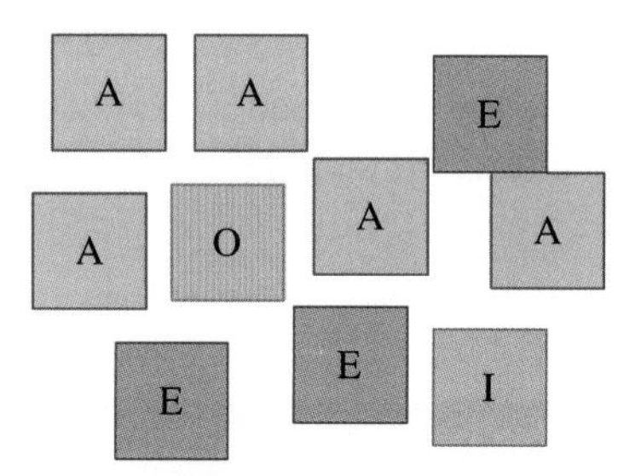


16. On 1 January, two friends, Sharmela and Marcela, were chatting with each other. Sharmela commented, 'It is very likely that tomorrow the temperature will be around 35 °C.' Marcela replied, 'It is very likely that tomorrow it is going to snow.' They were both correct.

a. Explain how this is possible.
b. Think and discuss any other situations like the one described in this question.

17. You roll a regular 6-sided die. Identify an event that could be described by the following words:

a. Impossible
b. Very unlikely
c. Even chance
d. Very likely
e. Certain

# LESSON
# 13.3 The sample space

**LEARNING INTENTION**

At the end of this lesson you should be able to:
- determine the outcomes and sample space of a chance experiment
- identify the sample space of a two-step chance experiment using a two-way table.

## 13.3.1 The sample space

eles-4711

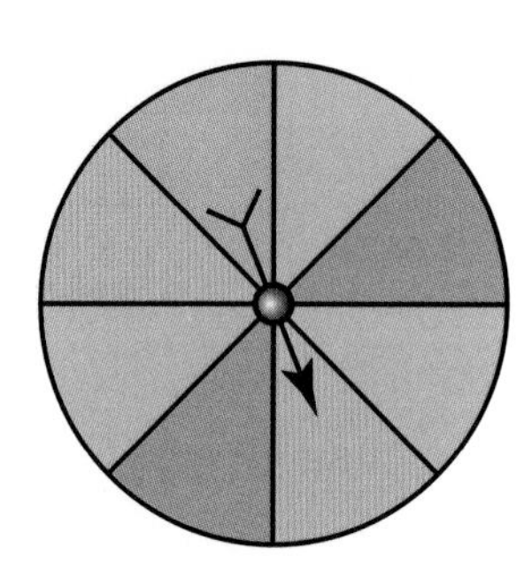

- The **sample space** refers to the list of all possible outcomes of a chance experiment. It is represented by the letter $S$.
  For example, the sample space for the spinner shown is $S = \{\text{blue, pink, green, orange}\}$.
  *Note:* Each outcome is only listed once in the sample space (for example, even if there are multiple sectors with red on the spinner, red is only listed once).
- Each outcome in the sample space is called an element of the sample space.
  For example, green is an outcome of the chance experiment and an element of the sample space.
- The number of elements in the sample space is denoted $n(S)$.
- The number of elements in the sample space of the spinner shown is 4: $n(S) = 4$.

## WORKED EXAMPLE 3 Listing the sample space

**A card is drawn from a standard deck. The suit of the card is then noted. List the sample space for this chance experiment.**

| THINK | WRITE |
|---|---|
| Although there are 52 cards in the deck, we are concerned only with the suit. List each of the four suits as the sample space for this chance experiment. | $S = \{\text{clubs, spades, diamonds, hearts}\}$ |

## WORKED EXAMPLE 4 Identifying the number of elements in a sample space

**A die is rolled and the number on the uppermost face is noted. Identify the number of elements in the sample space.**

| THINK | WRITE |
|---|---|
| A die has six outcomes: 1, 2, 3, 4, 5 or 6. | $n(S) = 6$ |

## 13.3.2 Two-way tables

eles-4712

- **Two-way tables** are used to show the sample space of two consecutive chance experiments.
- Suppose we want to spin a spinner and then toss a coin. The outcomes of each experiment are listed in the first column and the first row of the following table.

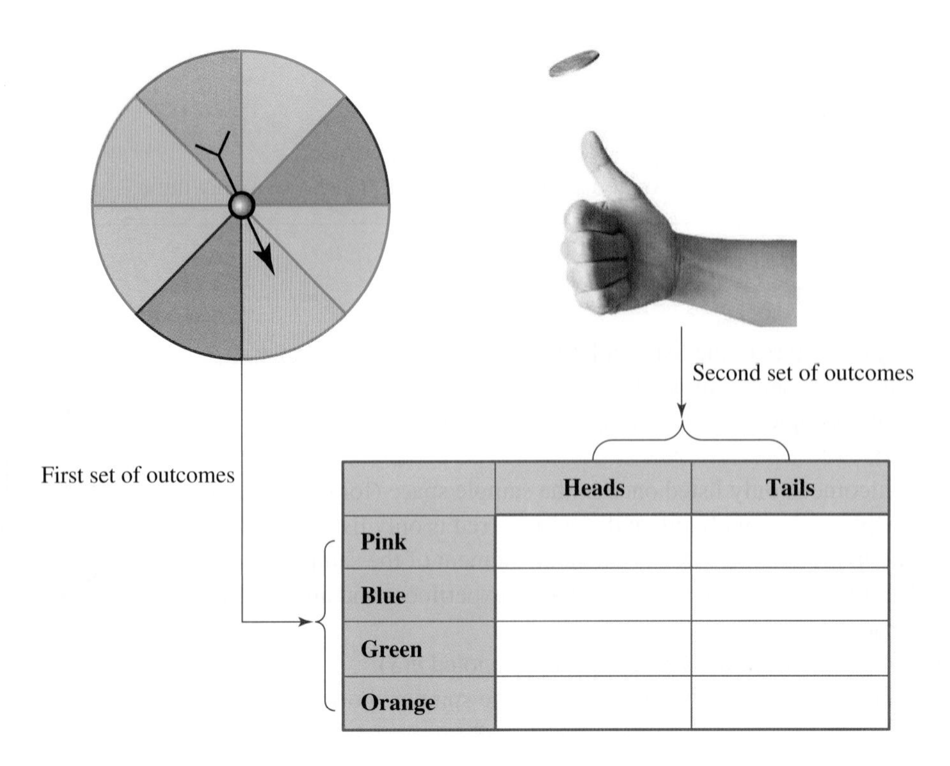


| | Heads | Tails |
|---|---|---|
| Pink | | |
| Blue | | |
| Green | | |
| Orange | | |

- The sample space for the **two-step experiment** is then listed in the remaining cells.

| | Heads | Tails |
|---|---|---|
| **Pink** | Pink, Heads | Pink, Tails |
| **Blue** | Blue, Heads | Blue, Tails |
| **Green** | Green, Heads | Green, Tails |
| **Orange** | Orange, Heads | Orange, Tails |

Sample space

$$S = \begin{Bmatrix} \text{(Pink, Heads), (Pink, Tails), (Blue, Heads), (Blue, Tails),} \\ \text{(Green, Heads), (Green, Tails), (Orange, Heads), (Orange, Tails)} \end{Bmatrix}$$

## WORKED EXAMPLE 5 Creating a two-way table and listing the sample space

**a.** **Draw a two-way table and list the sample space for the experiment 'tossing a coin and rolling a die'.**
**b.** **State how many different outcomes or results are possible.**

**THINK**

a. 1. Rule a table consisting of 7 rows and 3 columns. Leave the first cell blank.
2. Label the second and third cells of the first row as H and T respectively.
3. Label cells 2 to 7 of the first column as 1, 2, 3, 4, 5, 6 respectively.

**WRITE**

a.

| | H | T |
|---|---|---|
| 1 | H1 | T1 |
| 2 | H2 | T2 |
| 3 | H3 | T3 |
| 4 | H4 | T4 |
| 5 | H5 | T5 |
| 6 | H6 | T6 |

4. Answer the question by combining the outcome pairs in the order in which they occur in each of the remaining cells, that is, the first event result followed by the second event result.

The sample space for the experiment 'tossing a coin and rolling a die' is:
$\{(H, 1), (H, 2), (H, 3), (H, 4), (H, 5), (H, 6), (T, 1), (T, 2), (T, 3), (T, 4), (T, 5), (T, 6)\}$.

b. Count the number of different outcomes and answer the question.

b. There are 12 different outcomes.

**on** Resources

**eWorkbook** Topic 13 Workbook (worksheets, code puzzle and project) (ewbk-1914)

**Interactivities** Individual pathway interactivity: The sample space (int-4363)
The sample space (int-4027)
Tables and sample spaces (int-4029)

# Exercise 13.3 The sample space

learnon

13.3 Quick quiz on | 13.3 Exercise

## Individual pathways

| ■ PRACTISE | ■ CONSOLIDATE | ■ MASTER |
|---|---|---|
| 1, 2, 3, 9, 10, 13, 16, 19 | 4, 6, 7, 11, 14, 17, 20 | 5, 8, 12, 15, 18, 21, 22 |

### Fluency

**WE3** For each of the chance experiments in questions 1 to 6, list the sample space.

1. A spinner with equal sectors labelled 1 to 10 is spun.
2. A coin is tossed.
3. A multiple choice question has five alternative answers: A, B, C, D and E.
4. A soccer team plays a match and the result is noted.

5. A card is selected from the four aces in a deck.
6. An exam paper is given the grade $A$ to $F$.
7. A card is selected from a standard deck. List the sample space if we are interested in:
   a. the suit of the card chosen
   b. the colour of the card chosen.
8. A bag contains 8 red marbles, 9 green marbles and 2 orange marbles. A marble is selected from the bag. List the sample space.

### Understanding

9. **WE4** A coin is tossed. Identify the number of elements in the sample space.
10. In each of the following, state the number of elements in the sample space.
    a. A card selected from a standard deck.
    b. The first ball drawn in the Tattslotto draw. (Balls are numbered from 1 to 45.)
    c. The winner of the AFL premiership. (There are 18 teams.)
    d. A day of the year is selected.

11. In each of the following, state the number of elements in the sample space.
    a. A letter from the alphabet is selected at random.
    b. The first prize in the lottery is chosen from tickets numbered 1 to 180 000.
    c. A term is selected from a school year.
    d. You win a medal at your chosen event at the world swimming championships.

12. MC From the following list, select the event that has the most elements in the sample space.

A. Selecting a card from a standard deck
B. Selecting a page at random from this book
C. Selecting an exercise book from your school bag
D. Selecting a student at random from your class
E. Selecting a page at random from the phone directory

13. WE5 The spinner shown is spun and a coin is tossed.

a. Construct a two-way table to list the sample space of this two-step experiment.
b. State how many different outcomes or results are possible.

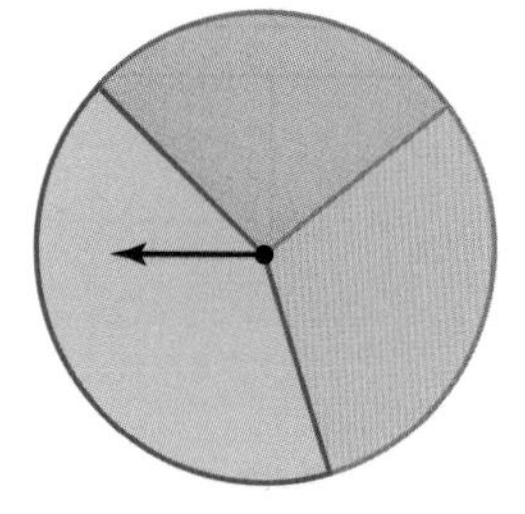

14. a. Draw a two-way table and list the sample space for the experiment 'spinning a circular spinner divided into 3 equal sectors labelled A, B, C and rolling a die'.
b. State the number of different outcomes or results.

15. A five-sided die and a 3-sided die are rolled simultaneously.

a. Use a two-way table to list the sample space of this two-step experiment.
b. Use your two-way table to help you construct a sample space for 'the sum of the two dice rolled'.

### Reasoning

16. Explain the difference between a chance experiment, an outcome, an event and the sample space.

17. Are all elements in a sample space equally likely to occur? Justify your answer.

18. Determine the number of different ways in which change can be given for a 50 cent coin using only 20 cent, 10 cent and 5 cent coins. Justify your answer.

### Problem solving

19. Michelle studies elective music. Her assignment this term is to compose a piece of music using as many instruments as she chooses, but only those that she can play. Michelle plays the acoustic guitar, the piano, the double bass and the electric bass. Determine the different combinations Michelle could choose.

20. Alex has one brother, one sister, a mother and a father. The family owns a 5-seat car. When the family goes out, the parents always sit in the front two seats. There are three seats behind that. Determine the different seating arrangements that are there.

21. If you had any number of ordinary six-sided dice, determine the number of different ways you could roll the dice and obtain a total of 6.

22. Four students, Aimee, Ben, Carla and Donald, are standing in a line waiting to enter the classroom. Determine the number of possible ways these 4 students can be arranged if Ben refuses to stand at the front of the line.

# LESSON
# 13.4 Simple probability

## LEARNING INTENTION

At the end of this subtopic you should be able to:
- calculate theoretical probabilities.

## 13.4.1 Simple probability

eles-4713

- An outcome is a particular result of a chance experiment.
- A **favourable outcome** is one that we are looking for.

### Theoretical probability

- **The theoretical probability of a particular result or event is defined as:**

$$\textbf{Pr(event)} = \frac{\textbf{number of favourable outcomes}}{\textbf{total number of outcomes}}$$

- **Equally likely outcomes** have an equal chance of occurring.
- For the spinner shown:
  - there are three blue sectors out of eight possible outcomes (sectors), so $\text{Pr(blue)} = \frac{3}{8}$
  - pink and green are equally likely outcomes as they both are two out of the eight sectors: $\text{Pr(pink)} = \frac{2}{8} = \frac{1}{4}$ and $\text{Pr(green)} = \frac{2}{8} = \frac{1}{4}$
  - orange is the least likely outcome as it is one sector out of the eight, so $\text{Pr(orange)} = \frac{1}{8}$.

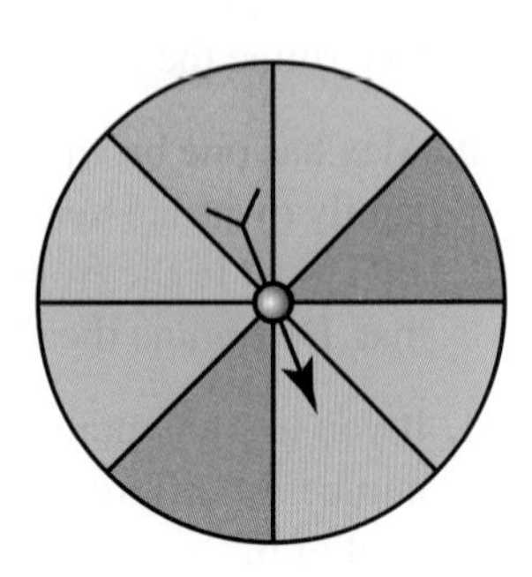

## DISCUSSION

Design a spinner that has four colours: red, green, blue and yellow.

Red should be twice as likely to occur as green and three times as likely to occur as blue. Yellow should be half as likely to occur as green.

Are there multiple designs that meet the description above?

Discuss with your classmates and try to come up with multiple designs.

### WORKED EXAMPLE 6 Determining outcomes of chance experiments

**State how many possible outcomes there are for each of the following chance experiments and specify what they are. Discuss whether the outcomes are equally likely or not equally likely to occur.**

a. **Tossing a coin**

b. **Spinning a circular spinner with 9 equal sectors labelled from *a* to *i***

c. **Drawing a picture card (jack, queen, king) from a standard pack of cards**

| THINK | WRITE |
|---|---|
| a. 1. Make a note of how many sides the coin has and what each side represents. | a. The coin has 2 sides, a Head and a Tail. |
| 2. State the outcomes and specify whether they are equally likely or not equally likely to occur. | When tossing a coin, there are two possible outcomes: Head or Tail. These outcomes are equally likely to occur. |
| b. 1. Make a note of how many sectors are in the circular spinner and what each one represents. | b. The circular spinner has 9 sectors labelled $a$ to $i$. |
| 2. State the outcomes and specify whether they are equally likely or not equally likely to occur. | When spinning the circular spinner, there are 9 possible outcomes; they are $a$, $b$, $c$, $d$, $e$, $f$, $g$, $h$ or $i$. These outcomes are equally likely to occur. |
| c. 1. State the possible outcomes. | c. There are 2 possible outcomes, drawing a picture card or drawing a non-picture card. |
| 2. Consider the possibility of the outcomes and specify whether they are equally likely or not equally likely to occur. | There are 52 cards in a standard pack. There are 3 picture cards in each of the four suits: $3 \times 4 = 12$ picture cards in a pack. Therefore the probabilities are: $\Pr(\text{picture card}) = \frac{12}{52}$ $\Pr(\text{non-picture card}) = \frac{52-12}{52} = \frac{40}{52}$ The probability of drawing a picture card is less than the probability of drawing a non-picture card. |

### WORKED EXAMPLE 7 Calculating probability

**Christopher rolls a fair 6-sided die.**

**a. State all the possible results that could be obtained.**

**b. Calculate the probability of obtaining:**

**i. a 4**

**ii. a number greater than 2**

**iii. an odd number.**

| THINK | WRITE |
|---|---|
| a. Write all the possible outcomes. | a. There are six possible outcomes: $1, 2, 3, 4, 5, 6$. |
| b. i. 1. The number 4 occurs once. | b. i. Number of favourable outcomes $= 1$<br>Total number of outcomes $= 6$ |
| 2. Write the rule for probability. | $\text{Pr(event)} = \dfrac{\text{number of favourable outcomes}}{\text{Total number of outcomes}}$ |
| 3. Substitute the known values into the probability formula and evaluate the probability of obtaining a 4. | $\text{Pr}(4) = \dfrac{1}{6}$ |
| 4. Write the answer. | The probability of obtaining a 4 is $\dfrac{1}{6}$. |
| ii. 1. 'A number greater than 2' implies: $3, 4, 5, 6$. | ii. Number of favourable outcomes $= 4$<br>Total number of outcomes $= 6$ |
| 2. Write the rule for probability. | |
| 3. Substitute the known values into the probability formula and evaluate the probability of obtaining a number greater than 2. | $\text{Pr(greater than 2)} = \dfrac{4}{6}$ |
| 4. Simplify the fraction. | $= \dfrac{2}{3}$ |
| 5. Write the answer. | The probability of obtaining a number greater than two is $\dfrac{2}{3}$. |
| iii. 1. 'An odd number' implies $1, 3, 5$. | iii. Number of favourable outcomes $= 3$<br>Total number of outcomes $= 6$ |
| 2. Write the rule for probability. | |
| 3. Substitute the known values into the probability formula and evaluate the probability of obtaining an odd number. | $\text{Pr(an odd number)} = \dfrac{3}{6}$ |
| 4. Simplify the fraction. | $= \dfrac{1}{2}$ |
| 5. Write the answer. | The probability of obtaining an odd number is $\dfrac{1}{2}$. |

## COLLABORATIVE TASK: Experimental versus theoretical probability!

1. Roll a die 20 times and record how many even numbers you roll.
2. Calculate the number of even numbers you should have rolled. How does your number compare to the theoretical number you would expect from 20 rolls?
3. Compare your results with your classmates' results. How do your classmates' results compare to the expected number?
4. Collate all of your classmates' results together. How does the total number of evens compare to the theoretical expected number? Are the class results closer to the expected results than each student's individual results?

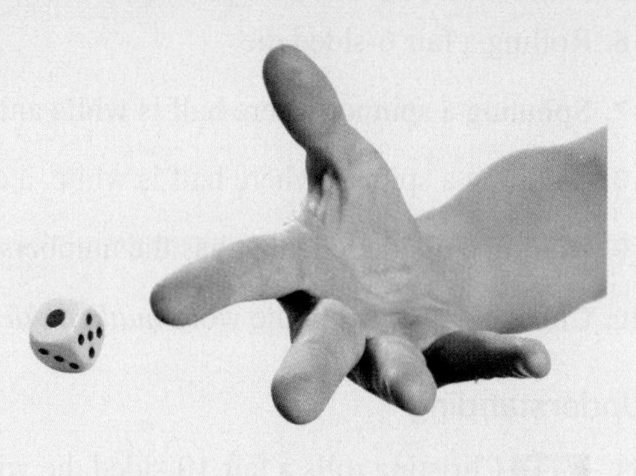

 Resources

 **eWorkbook** Topic 13 Workbook (worksheets, code puzzle and project) (ewbk-1914)

**Interactivities** Individual pathway interactivity: Simple probability (int-4364)
Simple probability (int-4028)

## Exercise 13.4 Simple probability

learnon

13.4 Quick quiz 

13.4 Exercise

### Individual pathways

| ■ PRACTISE | ■ CONSOLIDATE | ■ MASTER |
|---|---|---|
| 1, 4, 7, 11, 14, 18, 21 | 2, 5, 8, 10, 13, 16, 19, 22 | 3, 6, 9, 12, 15, 17, 20, 23 |

### Fluency

For each of the chance experiments in questions 1 to 5, state the number of possible outcomes and specify what they are.

1. Rolling a 12-sided die, numbered 1 to 12 inclusively
2. Spinning a spinner for a game that has 5 equal-sized sections, numbered 1 to 5 inclusively
3. Choosing a consonant from the word *cool*
4. Rolling an even number greater than 2 on a fair 6-sided die
5. Choosing an odd number from the first 20 counting numbers

**WE6** For each of the chance experiments in questions **6** to **10**, state the possible outcomes and specify whether each outcome is equally likely to occur or not.

**6.** Rolling a fair 6-sided die

**7.** Spinning a spinner where half is white and half is black

**8.** Spinning a spinner where half is white, a quarter is blue and a quarter is red

**9.** Rolling a 6-sided die that has the numbers 1, 2, 3, 4, 5, 5 on it

**10.** Choosing a vowel in the word *mathematics*

### Understanding

**11.** **WE7** Christina rolls a fair 10-sided die with faces numbered from 1 to 10.

a. State all the possible results that could be obtained.

b. Calculate the probability of obtaining:

i. a 9  ii. a number less than 7  iii. a prime number

iv. a number greater than 3  v. a multiple of 3.

**12.** Leo has been given a bag of marbles to play with. Inside the bag, there are 3 blue, 6 red, 4 green and 7 black marbles. If Leo takes out one marble from the bag, calculate:

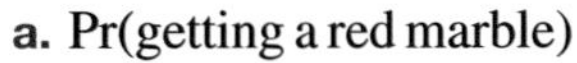

a. Pr(getting a red marble)

b. Pr(getting a green marble)

c. Pr(getting a black marble)

d. Pr(getting a blue or black marble)

e. Pr(getting a green, red or blue marble)

f. Pr(getting a green, red, blue or black marble).

**13.** There is a valuable prize behind 2 of the 5 doors in a TV game show. Determine the probability that a player choosing any door will win a valuable prize.

**14.** **MC** A circular spinner is shown. Select the probability of obtaining an orange sector.

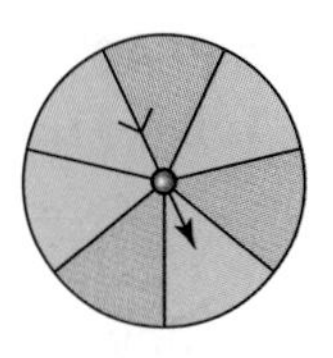

A. $\frac{4}{7}$  B. $\frac{1}{7}$  C. 75%  D. $\frac{1}{2}$  E. $\frac{3}{7}$

**15.** **MC** For an octagonal spinner with equal sectors numbered 1 to 8, select the chance of getting a number *between* 2 and 7.

A. $\frac{5}{8}$  B. $\frac{3}{8}$  C. $\frac{3}{4}$  D. 0.5  E. 25%

**16.** A pack of playing cards is shuffled and a card is chosen at random (in no particular order or pattern). Calculate the probability that the card chosen is:

a. a black card (that is, spades or clubs)

b. an ace

c. a diamond

d. a picture card (that is, jack, queen or king)

e. the queen of hearts.

17. A pack of playing cards is shuffled and a card is chosen at random (in no particular order or pattern). Calculate the probability that the card chosen is:

a. a diamond or a black card
b. not a king
c. a club, diamond, heart or spade
d. not a spade
e. red and a ten.

## Reasoning

18. If we know the probability of an event occurring, explain how we can work out the probability of it not occurring.

19. In a raffle where there is only 1 prize (a car), 100 000 tickets have been sold, at a cost of $5.00 each.

a. Determine the chance of winning the prize for a buyer who:
   i. purchases only 1 ticket
   ii. purchases 20 tickets
   iii. purchases 50 tickets
   iv. purchases all the tickets.

b. If someone bought all the tickets in the raffle, explain whether they had made a wise purchase.

20. Answer the following questions with full working.

a. If you had only one pair of shoes, determine what the probability would be that you would wear that pair of shoes on any given day.
b. If you had two pairs of shoes, determine what the probability would be that you would wear a certain pair of shoes on any given day.
c. If you had seven pairs of shoes, determine what the probability would be that you would wear a certain pair of shoes on any given day.
d. If you had seven pairs of shoes but two pairs were identical, state what the probability would be that you would wear one of the two identical pairs of shoes on any given day.
e. Explain what happens to the probability when the number of pairs of shoes increases.
f. Explain what happens to the probability when the number of identical pairs of shoes increases.

## Problem solving

21. Melbourne City FC is a soccer team in the Australian A-league. Over its history the team has won 60% of its matches, and lost half as many as it has won. If one of the team's past matches is selected at random, determine the probability that the match was a draw.

22. At Jaca college there are 220 Year 7 students, with a gender split of 55% girls and 45% boys. Every morning, 59 Year 7 students catch a bus to school. If one third of Year 7 boys catch a bus to school every morning, determine the probability that a randomly selected Year 7 student is a girl who does not catch a bus to school.

23. An apple, a banana, an orange, a peach and a bunch of grapes are in May's fruit basket. If she chooses two fruits to take to school with her lunch, determine the probability that one of the fruits is a banana.

# LESSON
# 13.5 Experimental probability

**LEARNING INTENTION**

At the end of this lesson you should be able to:
- calculate experimental probabilities from data and by performing repeated trials of a chance experiment.

## 13.5.1 Experimental probability

eles-4714

- In real life, the chance of something occurring may be based on factors other than the number of favourable and possible outcomes.
  For example, the chances of you beating your friend in a game of tennis could theoretically be $\frac{1}{2}$, as you are one of 2 possible winners. In practice, however, there are other factors (such as experience and skill) that would influence your chance of winning.
- A **trial** is one performance of an experiment to collect a result.
- An **experiment** is a process that allows us to collect data by performing trials. In experiments with repeated trials, it is important to keep the conditions for each trial the same.
- A **successful trial** is one that results in the desired outcome.
- The **experimental probability** of an event is found by conducting an experiment and counting the number of times the event occurs.

**Experimental probability**

**The experimental probability of a particular result or event is defined as:**

$$\textbf{Pr(event)} = \frac{\textbf{number of successful trials}}{\textbf{total number of trials}}$$

**WORKED EXAMPLE 8 Calculating experimental probability**

**A coin is flipped 10 times and the results are seven Heads and three Tails. Calculate the experimental probability of obtaining a Tail.**

| THINK | WRITE |
|---|---|
| 1. Obtaining a Tail is considered a success. Each flip of the coin is a trial. | $\text{Pr(success)} = \frac{\text{number of successful trials}}{\text{total number of trials}}$ |
| 2. Tails was flipped three times, so there were three successful trials out of a total of 10 trials. | $\text{Pr(Tail)} = \frac{3}{10}$ <br> $= 0.3$ |

## 13.5.2 Long-term trends

eles-4715

- The more times an experiment is performed, the more accurate the experimental probability becomes.
- In fair experiments, the **long-term trend** (that is, the trend observed for results from a very large number of trials) shows that the results obtained through experimental probability will match those of theoretical probability.

### WORKED EXAMPLE 9 Investigating experimental probability

**a. Copy the following table. Toss a coin 10 times and record the result in row 1 of the table.**

| Experiment | Heads | | Tails | |
|---|---|---|---|---|
| number | Tally | Count | Tally | Count |
| 1 | | | | |
| 2 | | | | |
| 3 | | | | |
| 4 | | | | |
| 5 | | | | |
| 6 | | | | |
| | Total | | Total | |

**b. Calculate the probability of obtaining a Head from your experiment.**
**c. Calculate the probability of obtaining a Tail from your experiment.**
**d. Discuss how these values compare with the theoretical results.**
**e. Repeat step a another 5 times and combine all of your results.**
**f. Explain how the combined result compare with the theoretical results.**

**THINK**

a. Toss a coin 10 times and record the results in the first row of the table.
*Notes:* (a) Place a stroke in the appropriate tally column each time an outcome is obtained. Five is denoted by a 'gatepost': that is, 4 vertical strokes and 1 diagonal stroke (卌).
(b) The same coin must be used throughout the experiment. The style of the toss and the surface the coin lands on must be the same.

**WRITE**

a.

| | Heads | | Tails | |
|---|---|---|---|---|
| Exp. no. | Tally | Count | Tally | Count |
| 1 | \|\|\|\| | 4 | 卌 \| | 6 |
| 2 | \|\|\|\| | 4 | 卌 \| | 6 |
| 3 | 卌 \|\| | 7 | \|\|\| | 3 |
| 4 | \|\|\| | 3 | 卌 \|\| | 7 |
| 5 | \|\|\|\| | 4 | 卌 \| | 6 |
| 6 | 卌 \|\|\| | 8 | \|\| | 2 |
| | Total | 30 | Total | 30 |

b. 1. For experiment 1, calculate the probability of obtaining a Head using the rule.

b. $\Pr(\text{event}) = \dfrac{\text{number of favourable outcomes}}{\text{number of possible outcomes}}$

$\Pr(\text{Heads}) = \dfrac{\text{number of Heads obtained}}{\text{total number of tosses}}$

2. Substitute the given values into the rule.

$\Pr(\text{Heads}) = \dfrac{4}{10}$

3. Evaluate and simplify.

$= \dfrac{2}{5}$

| THINK | WRITE |
| --- | --- |
| 4. Convert the fraction to a percentage by multiplying by 100%. | As a percentage $\frac{2}{5} = \frac{2}{5} \times 100\%$ <br> $= \frac{200}{5}\%$ <br> $= 40\%$ |
| 5. Answer the question. | The probability of obtaining a Head in this experiment is $\frac{2}{5}$ or 40% |
| **c.** 1. For experiment 1, calculate the probability of obtaining a Tail for this experiment. | **c.** $\Pr(\text{Tails}) = \frac{\text{number of Tails obtained}}{\text{total number of tosses}}$ |
| 2. Substitute the given values into the rule and simplify. | $\Pr(\text{Tails}) = \frac{6}{10}$ <br> $= \frac{3}{5}$ |
| 3. Convert the fraction to a percentage by multiplying by 100%. | As a percentage $\frac{3}{5} = \frac{3}{5} \times 100\%$ <br> $= \frac{300}{5}\%$ <br> $= 60\%$ <br> The probability of obtaining a Tail in this experiment is $\frac{3}{5}$ or 60%. |
| **d.** Compare the results obtained in parts **b** and **c** with the theoretical results. | **d.** The experimental value obtained for the Pr(Heads) is $\frac{2}{5}$ (or 40%) and Pr(Tails) is $\frac{3}{5}$ (or 60%). The theoretical value of these probabilities is $\frac{1}{2}$ (or 50%). Therefore, the experimental probabilities differ from the theoretical probabilities by 10%. |
| **e.** 1. Repeat the procedure of part **a** 5 times. <br> 2. Calculate the total number of Heads and Tails obtained and enter the results in the table. | **e.** Refer to the results in the table in part a. |
| **f.** 1. Calculate the probability of obtaining a Head for all 10 experiments combined. | **f.** $\Pr(\text{Heads}) = \frac{\text{number of Heads obtained}}{\text{total number of tosses}}$ |
| 2. Substitute the given values into the rule and simplify. | $\Pr(\text{Heads}) = \frac{30}{60}$ <br> $= \frac{1}{2}$ |
| 3. Convert the fraction to a percentage by multiplying by 100%. | As a percentage $\frac{1}{2} = \frac{1}{2} \times 100\%$ <br> $= \frac{100}{2}\%$ <br> $= 50\%$ |

The probability of obtaining a Head in all 10 experiments is $\frac{1}{2}$ or 50%.

4. Calculate the probability of obtaining a Tail for all 10 experiments combined.

$$\Pr(\text{Tails}) = \frac{\text{number of Tails obtained}}{\text{total number of tosses}}$$

$$\Pr(\text{Tails}) = \frac{30}{60}$$

$$= \frac{1}{2}$$

As a percentage $\frac{1}{2} = \frac{1}{2} \times 100\%$

$$= \frac{100}{2}\%$$

$$= 50\%$$

The probability of obtaining a Tail in all 10 experiments is $\frac{1}{2}$ or 50%.

5. Compare the combined result obtained with the theoretical results.

The combined results in this experiment produced probability values that were equal to the theoretical probability values.

Therefore, the long-term trend of obtaining a Head or Tail when tossing a coin is equal to $\frac{1}{2}$.

## Digital technology

Chance experiments can be simulated using simple devices (such as coins or dice) or by using technology. Examples include random number generators in Excel spreadsheets or online adjustable spinner simulators, as shown. These simulators allow you to select the number of sectors and the number of spins (trials) and let the simulation run. You will notice that the experimental probabilities get closer to the theoretical probabilities as you increase the number of spins.

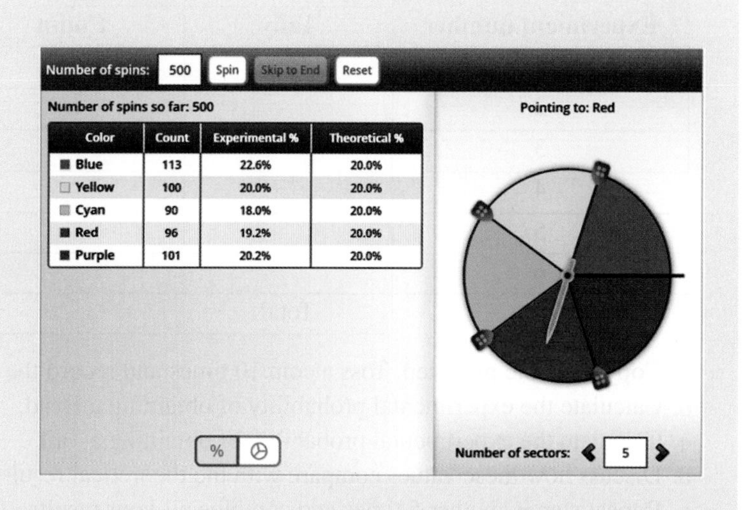

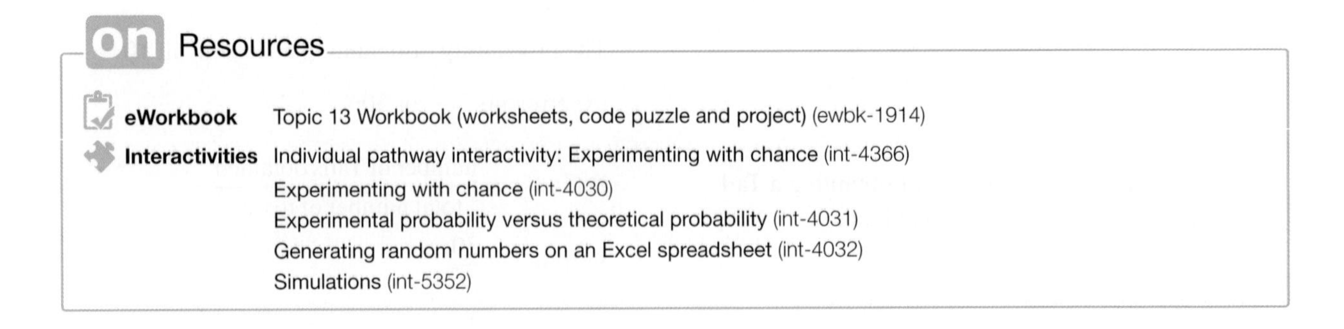

Resources

eWorkbook Topic 13 Workbook (worksheets, code puzzle and project) (ewbk-1914)

Interactivities Individual pathway interactivity: Experimenting with chance (int-4366)
Experimenting with chance (int-4030)
Experimental probability versus theoretical probability (int-4031)
Generating random numbers on an Excel spreadsheet (int-4032)
Simulations (int-5352)

## Exercise 13.5 Experimental probability

learnon

13.5 Quick quiz on | 13.5 Exercise

### Individual pathways

| ■ PRACTISE | ■ CONSOLIDATE | ■ MASTER |
|---|---|---|
| 1, 3, 4, 9, 10, 13 | 2, 5, 7, 11, 14 | 6, 8, 12, 15 |

### Fluency

1. **WE8** Teagan was playing the game Trouble and recorded the number of times she rolled a 6. During the game, she was successful 5 times out of the 25 times she tried. Calculate the experimental probability of rolling a 6 in the game.
2. In the 2020 AFL season, the Brisbane Lions won 14 out of 17 games. Determine the experimental probability of the Lions winning a game in 2020.
3. **WE9**

| | Heads | | Tails | |
|---|---|---|---|---|
| **Experiment number** | **Tally** | **Count** | **Tally** | **Count** |
| 1 | | | | |
| 2 | | | | |
| 3 | | | | |
| 4 | | | | |
| 5 | | | | |
| 6 | | | | |
| | Total | | Total | |

   a. Copy the table provided. Toss a coin 10 times and record the results in the first row of the table.
   b. Calculate the experimental probability of obtaining a Head.
   c. Calculate the experimental probability of obtaining a Tail.
   d. Discuss how these values compare with the theoretical results.
   e. Repeat step a another 5 times and combine all your results.
   f. Explain how the combined results compare with the theoretical results.

4. If you wanted to create a device that would give a theoretical probability of achieving a particular result as $\frac{1}{4}$, state how many sections a spinner such as this would need to be divided into.

5. Determine how you would divide or colour a spinner if you wanted to achieve the probability of a success equal to $\frac{3}{10}$.

6. Two students conducted a chance experiment using a standard 6-sided die. The results were recorded in the following table.

| Experiment number | Even number | | Odd number | |
|---|---|---|---|---|
| | Tally | Count | Tally | Count |
| 1 | ||| | 3 | || | 2 |
| 2 | |||| | 4 | | | 1 |
| 3 | || | 2 | ||| | 3 |

a. Describe in words the experiment that was conducted by the two students.
b. State the number of times they conducted the experiment.
c. Calculate the theoretical probability of rolling an even number.
d. Compare the experimental probability with the theoretical probability of rolling an even number.
e. Compare the experimental probability with the theoretical probability of rolling an odd number.

## Understanding

7. Toss a coin 60 times, tallying up the number of Heads and Tails that you toss in trials of 10 tosses at a time. (If you have already completed question 3, you may use the results you obtained from that experiment).
   a. The long-term trend of the probability of obtaining a Head on the toss of a coin is the Pr(Heads) from your experiment.
   Determine the long-term trend of the probability after:
      i. 10 tosses of the coin
      ii. 20 tosses of the coin
      iii. 30 tosses of the coin
      iv. 60 tosses of the coin.
   b. Obtain a classmate's 60 results. Combine these with yours. State the long-term trend of Pr(Heads) obtained.
   c. Combine your pair's 120 results with those of another pair. State the long-term trend of Pr(Heads) obtained.
   d. Finally, count the results obtained by the whole class for this experiment. (Make sure nobody's results are counted twice.) You should have 60 tosses per person. State the long-term trend of Pr(Heads) obtained.
   e. Copy and complete the following table.

| Number of tosses | Heads | | Tails | |
|---|---|---|---|---|
| | Pr(Heads) | Pr(Heads) as percentage | Pr(Tails) | Pr(Tails) as percentage |
| 10 | | | | |
| 20 | | | | |
| 30 | | | | |
| 60 | | | | |
| 120 | | | | |
| 240 | | | | |
| Whole class (specify number of tosses) | | | | |

   f. Comment on the changes of the long-term trend value of Pr(Heads) as you toss the coin more times.

8. Use a random number generator to simulate a 5-colour spinner.
   **a.** Determine the chance of getting any one of the five colours when you spin the spinner (theoretically).
   **b.** Spin the spinner 10 times and, using a table such as the following one, record your results.

| Colour | 1 | 2 | 3 | 4 | 5 |
|---|---|---|---|---|---|
| Number of times it occurs | | | | | |

   **c.** From your results, list the probabilities of obtaining each colour. For example, divide the number of times a particular colour was obtained by the total number of spins (that is, 10).
   **d.** Explain why these probabilities might not be the same as the theoretical probability would suggest.
   **e.** Spin your spinner and record the results for another 10 spins.
   **f.** Spin your spinner so that you have 100 results. Is the experimental probability closer to the pure probability? Why might this be? Discuss.

### Reasoning

9. Cory records the fact that it has rained on 65 out of 365 days in a year.
   **a.** Write the number of days that it has rained as a simple fraction.
   **b.** Karen says that since any day can be wet or dry, the probability of rain on any day is $\frac{1}{2}$. Explain whether Karen is correct.
   **c.** Determine the experimental probability of rain on any given day, expressed as a decimal, correct to 2 decimal places.

10. **a.** Conduct the following experiments:
      i. Toss a coin 10 times and record the number of Heads and Tails that occur.
      ii. Toss a coin 25 times and record the number of Heads and Tails that occur.
      iii. Toss a coin 50 times and record the number of Heads and Tails that occur.
      iv. Toss a coin 100 times and record the number of Heads and Tails that occur.
   **b.** Calculate the experimental probability for each experiment.
   **c.** Compare these values and explain your findings.

11. A die is rolled 30 times, giving the following results.

4 3 5 4 3 5 2 1 1 5
3 2 2 4 1 3 1 6 1 3
2 1 6 6 3 5 1 3 5 3

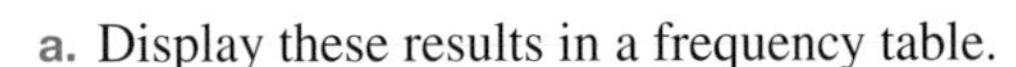

   **a.** Display these results in a frequency table.
   **b.** Calculate the probability of obtaining a 6 when you roll a die.
   **c.** Calculate how many times you would expect to obtain a 6 in 30 rolls of a die.
   **d.** Explain the difference between your expected results and the actual results shown in this question.

12. You have calculated previously that the chance of getting any particular number on a 6-sided die is $\frac{1}{6}$. You have 2 different coloured dice. Explain whether there is any difference in your dice apart from colour and whether one could be biased (more likely to give a particular result than theory says it should).
   **a.** Design a test to determine whether the dice you have are fair. Write down what you are going to do.
   **b.** Perform your test and record results.
   **c.** Determine the probability of getting each of the numbers on each of your dice, based on your tests and on the long-term trend you have observed.
   **d.** State what your test says about your two dice. Explain whether there are any things that need to be considered before giving your answer. (Perhaps your dice have slightly uneven shapes or something that might cause them to lean towards one result more than others.)

## Problem solving

13. Inside a bag are 36 shapes that are either squares or triangles. One shape is taken out at random, its shape noted and put back in the bag. After this is repeated 72 times, it is found that a triangle was taken out 24 times. Estimate how many triangles and how many squares there are in the bag.

14. In your desk drawer, there are 5 identical red pens and 6 identical black pens. Determine the smallest number of pens you have to remove from the drawer in the dark so that you will be absolutely sure of having:
   a. 2 black pens
   b. 2 red pens
   c. 1 black pen and 1 red pen.
   d. Explain your answers to parts **a**, **b** and **c**.

15. Jim operates a parachute school. Being a man who is interested in statistics, he keeps a record of the landing position of each jump from first-time parachutists. With experience, parachutists are able to land on a particular spot with great accuracy. However, first-time parachutists do not possess this ability. Jim has marked out his landing field with squares, as shown here.

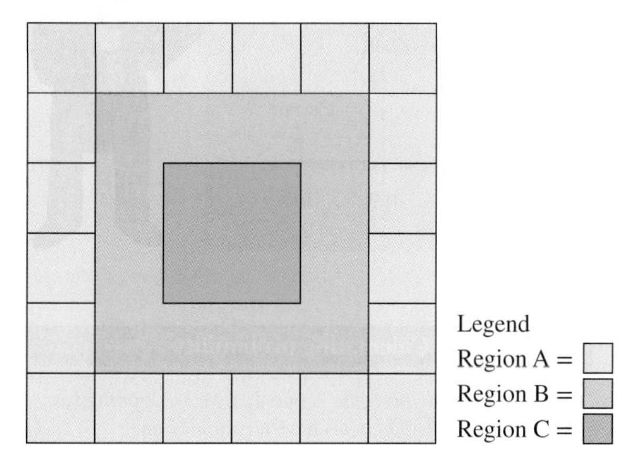

We are going to look at the areas of each of the regions A, B and C. To do this, we will determine each of the areas in terms of one of the small squares in region A. We will say that each small square has an area of 1 square unit.

   **a.** Determine the area of Jim's whole landing field (in square units).
   **b.** Determine the areas of regions A, B and C (in square units).
   **c.** Assuming that the parachutist lands in the field, calculate the probability that the landing will occur in:
      **i.** region A
      **ii.** region B
      **iii.** region C.

   These represent theoretical probabilities.

   **d.** Jim's records indicate that, from 5616 jumps of first-time parachutists, the landing positions were:
      **i.** 592 in region C
      **ii.** 1893 in region B
      **iii.** 3131 in region A.

   Comment on these results in comparison with the probabilities you calculated in part **c**.

# LESSON
# 13.6 Review

## 13.6.1 Topic summary

### Understanding probability

- Probability looks at how **likely** something is to happen.
- Probabilities are stated as a number between 0 and 1 inclusive, where 0 corresponds to *impossible* and 1 corresponds to *certain*. That is,

$$\mathbf{0 \leq Probability \leq 1}$$

- Different words or phrases in the English language can be associated with different probabilities, as indicated on the probability scale below.

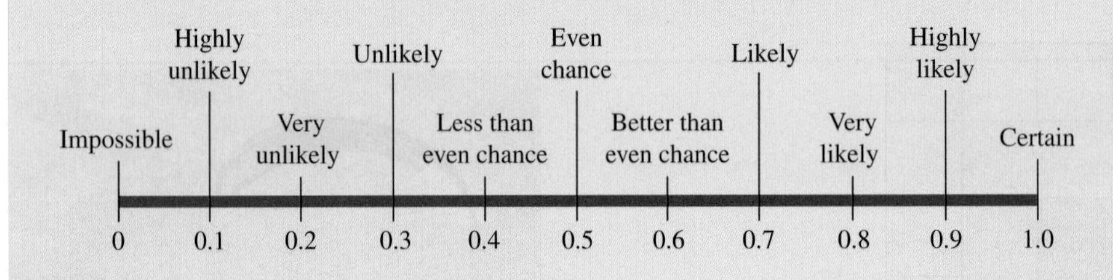

# INTRODUCTION TO PROBABILITY

### The language of chance

- A **chance experiment** is an experiment in which there are several different possible outcomes, each having a defined chance of occurring.
  e.g. Rolling a die is a chance experiment.
- An **outcome** is a possible result of a chance experiment, such as rolling a 2.
- An **event** is either a single outcome (e.g. rolling a 4) or a collection of outcomes (e.g. rolling a 1, 2 or 3).
- The **sample space**, $S$, is the set of *all* possible outcomes of a chance experiment.
  e.g. The sample space of rolling a fair six-sided die is $S = \{1, 2, 3, 4, 5, 6\}$.
- The number of elements in the sample space is often important in probability. It is denoted by the symbol $n(S)$.

### Calculating probability

- A *favourable outcome* is one that we are looking for.
- *Equally likely outcomes* have an equal chance of occurring.
- The **theoretical probability** of an event occurring is defined as:

$$\mathbf{Pr(event)} = \frac{\textbf{number of favourable outcomes}}{\textbf{total number of outcomes}}.$$

e.g. Consider the following chance experiments.

Rolling a die: $\Pr(\text{rolling a } 1) = \frac{1}{6}$

Flipping a coin: $\Pr(\text{Tail}) = \frac{1}{2}$ (or 0.5)

Picking a card from a deck of cards:
$\Pr(\text{diamond}) = \frac{13}{52} = \frac{1}{4}$ (or 0.25)

### Experimental probability

- The **experimental probability** of an event is determined by the results of repeated trials of the experiment.
- $\mathbf{Pr(event)} = \dfrac{\textbf{number of successful trials}}{\textbf{total number of trials}}$

  e.g. If a team has won 42 out of their last 70 matches, the experimental probability that they will win a match is $\frac{42}{70} = \frac{3}{5} = 60\%$.
- The experimental probability of an event becomes more accurate as the number of trials increases, and approaches the theoretical probability.

## 13.6.2 Success criteria

Tick the column to indicate that you have completed the lesson and how well you think you have understood it using the traffic light system.

(**Green:** I understand; **Yellow:** I can do it with help; **Red:** I do not understand)

| Lesson | Success criteria | | | |
|---|---|---|---|---|
| 13.2 | I understand that a probability is a number between 0 and 1 that describes the possibility of an event. | | | |
| | I can classify the chance of an event occurring using words such as *certain*, *likely*, *unlikely*, *even chance* or *impossible*. | | | |
| 13.3 | I can determine the outcomes and sample space of a chance experiment. | | | |
| | I can identify the sample space of a two-step chance experiment using a two-way table. | | | |
| 13.4 | I can calculate theoretical probabilities. | | | |
| 13.5 | I can calculate experimental probabilities from data and by performing repeated trials of a chance experiment. | | | |

## 13.6.3 Project

### Snakes, ladders and probability!

In the game of snakes and ladders, you roll 2 dice, add the 2 numbers that appear on the uppermost faces to get a total, and move a marker the total number around a 100-square board. If you are lucky and your marker lands on the base of a ladder, you can advance more quickly in the game by 'climbing' to the top of the ladder. However, if your marker lands on the head of a snake, you must 'slide down' the length of the snake.

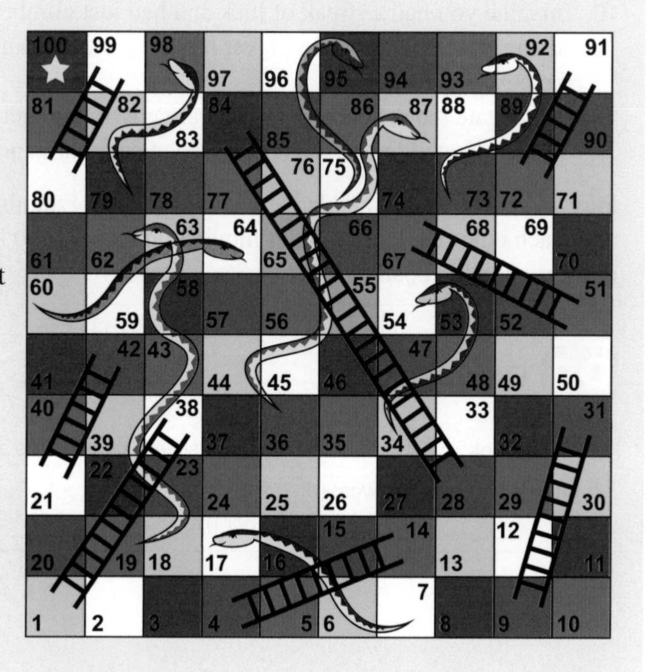


1. Complete the following table to show the possible totals when 2 dice are rolled. Some of the values have been included for you.
2. List the sample space for the possible totals when 2 dice are rolled.
3. Determine which total appears the most.
4. Determine which total appears the least.
5. Calculate the probability of getting a total of 3 when 2 dice are rolled.

| Die 2 \ Die 1 | 1 | 2 | 3 | 4 | 5 | 6 |
|---|---|---|---|---|---|---|
| 1 | 2 | 3 | 4 | | | |
| 2 | 3 | 4 | 5 | | | |
| 3 | 4 | | | | | |
| 4 | | | | | | |
| 5 | | | | | | |
| 6 | | | | | | |

Using the table that shows the frequency of the totals, we can investigate the probabilities involved in moving around the snakes and ladders board. The following situations will enable you to investigate some of the possibilities that occur in snakes and ladders.

6. Imagine you landed on square 95 and slid down the snake to square 75. What total would you need to get to go up the ladder at square 80 on your next move? Determine the number of ways in which this total can be achieved in one turn.
7. If you slid down the snake at square 87, is it possible to move up the next ladder with your next turn? Explain.
8. Explain what would happen if you were on square 89 and rolled two 1s and rolled two 1s again with your next turn. Discuss what would be the likelihood of this happening in a game.
9. Describe how you could get from square 71 to square 78 in one turn. Work out the probability of this happening.
10. Imagine you had a streak of luck and had just climbed a ladder to square 91. Your opponent is on square 89. Explain which player has the greater chance of sliding down the snake at square 95 during the next turn.
11. Investigate the different paths that are possible in getting from start to finish in the fewest turns. For each case, explain the totals required at each turn and discuss the probability of obtaining these totals.

Play a game of snakes and ladders with a partner. Examine your possibilities after each turn, and discuss with each other the likelihood of moving up ladders and keeping away from the snakes' heads as you both move around the board.

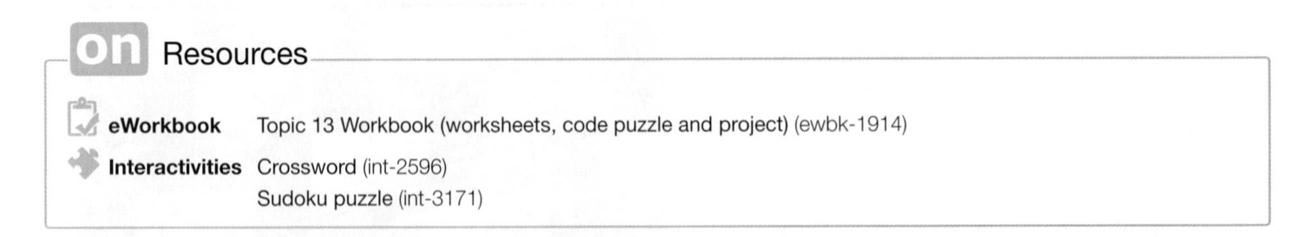

## Exercise 13.6 Review questions

learnon

### Fluency

1. For each of the given statements, specify whether the chance of the following events occurring is certain, likely, even chance, unlikely or impossible.
   a. Australia is in the Southern Hemisphere.
   b. You will still be alive in the next century.
   c. You obtain an even number on a circular spinner numbered from 1 to 16.

2. For each of the given statements, specify whether the chance of the following events occurring is certain, likely, even chance, unlikely or impossible.
   a. England is in the Southern Hemisphere.
   b. You roll a fair die and obtain a number less than or equal to 4.
   c. Humans can survive without water indefinitely.

3. List one event for which the chance of it occurring would be:
   a. impossible
   b. certain
   c. unlikely

4. Copy the number line and place the following words on it to indicate what sort of chance each number would represent: *certain, very unlikely, unlikely, likely, highly likely, highly unlikely, even chance, impossible, very likely*

   0 0.1 0.2 0.3 0.4 0.5 0.6 0.7 0.8 0.9 1.0

5. List the sample space for the following chance experiments.
   a. Tossing a coin
   b. Drawing a marble out of a bag containing red, green and blue marbles
   c. Answering a true/false question in a test

6. List the sample space for the following chance experiments.
   a. Rolling a 6-sided die
   b. Spinning a circular spinner numbered from 1 to 10
   c. Competing in a race in which there are 9 runners

7. For each of the following, state the number of elements in the sample space, $n(S)$.
   a. The first prize in a raffle is chosen from tickets numbered from 1 to 5000.
   b. A card is selected from the red cards in a standard deck.
   c. A circular spinner numbered 1 to 5 is spun.
   d. A day is selected in the month of July.

**8.** A six-sided die labelled 1, 2, 3, 3, 4, 5 is rolled.
State:
**a.** Pr(getting a 5)
**b.** Pr(getting a 3)
**c.** Pr(getting an even number)
**d.** Pr(getting an odd number).

**9.** A six-sided die labelled 1, 2, 3, 3, 4, 5 is rolled.
State:
**a.** Pr(getting a number greater than or equal to 3)
**b.** Pr(getting a number less than 3)
**c.** Pr(getting a 0)
**d.** Pr(getting a 6).

**10.** A person has a normal pack of cards and draws one out.
Calculate:
**a.** Pr(picking a heart)
**b.** Pr(picking a spade)
**c.** Pr(picking a picture card)
**d.** Pr(picking a card with a number less than 5 on it). Do not count an ace as a number.

**11.** Ten cards are numbered from 1 to 10, shuffled and placed face down on a table. If a card is selected at random, calculate the probability that the card selected is:
**a.** 4
**b.** an even number
**c.** divisible by 3
**d.** an even number and divisible by 3.

**12.** Ten cards are numbered from 1 to 10, shuffled and placed face down on a table. If a card is selected at random, calculate the probability that the card selected is:
**a.** an even number or divisible by 3
**b.** not divisible by 3
**c.** greater than 8
**d.** 12.

**13.** A sample of 250 students at a particular school found that 225 of them had access to the internet at home. Given this sample is a good representation for the entire school, calculate the probability that a student selected at random in the school will have internet access at home.

**14.** Use a table to show the sample space for the experiment 'tossing a coin and rolling a die'.
Calculate:
**a.** Pr(a Tail and a number less than 3)
**b.** Pr(a Head and a number greater than 3)
**c.** Pr(a Tail and an even number)
**d.** Pr(a Head and a prime number).

**15.** Use a table to show the sample space for the experiment 'spinning a spinner with 10 sectors and rolling a die'.
Determine:
**a.** Pr(the sum totals to an even number)
**b.** Pr(odd number on the spinner and an even number on the die)
**c.** Pr(both digits having a value less than 5)
**d.** Pr(the sum totals to a value between 7 and 10)
**e.** Pr(the sum totals to a value greater than or equal to 13 but less than 16)
**f.** Pr(the sum totals to a prime number).

## Problem solving

**16.** In your sock drawer there are 4 identical blue socks and 5 identical black socks. Determine the smallest number of socks you can take from the drawer in the dark so that you will be absolutely sure of having:

**a.** a pair of black socks
**b.** a pair of blue socks
**c.** a black pair and a blue pair of socks.

**17.** The arrows on Spinner A and Spinner B are spun. (If an arrow lands on a line, the spinner is spun again.) The two numbers are added to get a score. For example, in the diagram the score is 10.

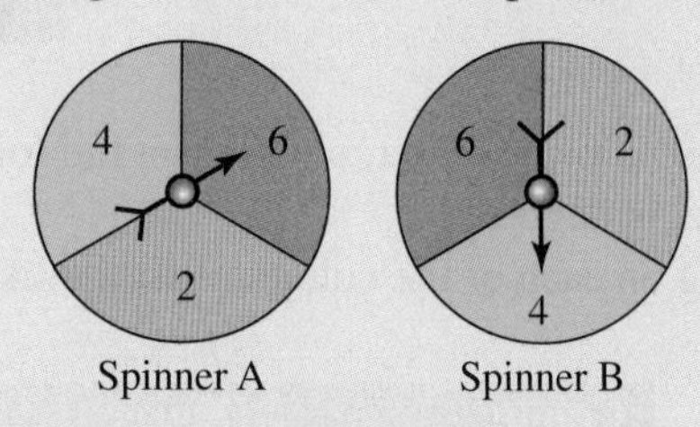


**a.** Determine the highest total score possible.
**b.** Determine the possible total scores.
**c.** List all the ways to get a total score of 8.
**d.** Determine the probability of getting a total of 9.
**e.** Determine the probability of getting a total score of 10.
**f.** Draw a grid showing the probabilities of getting all possible totals.

**18.** Charlotte and Rhianna have a flower garden. They have 17 red flowers, 12 pink flowers and 13 yellow flowers. Charlotte picks some flowers for her vase. She selects 6 red flowers, 3 pink and 3 yellow flowers. If Rhianna picks a flower at random for her hair, determine the probability that it is yellow.

**19.** A bowl contains blue marbles and white marbles. If there are twice as many blue marbles as white marbles, determine the probability that a blue marble is selected.

**20.** Chloë is a contestant on a game show. There are five sealed cases on the podium and each of the cases contains one of the following amounts: \$1, \$50, \$500, \$1000 and \$5000. The game show host offers her a deal of \$400, or she can choose a case and keep the amount of money in it, instead. Determine the probability she will win more than the \$400 the host is offering.

21. Rebecca plays a sideshow game where she puts four balls into a clown's mouth, and the balls then fall into slots numbered 1 to 6. To win, one of her balls must go into slot 6.
She has noticed that the numbers 2 and 5 come up 4 times as often as 1 and 6, and the numbers 3 and 4 come up 5 times as often as 1 and 6. Determine the probability that she will get a 6.

22. You have a spinner. Draw a pie or circle graph to represent the following list of colours and their associated sectors of the circle, in degrees.
Discuss the chances of landing on each of the colours. List the likelihoods from smallest to largest.

| Colour | Degree |
|---|---|
| Yellow | 25° |
| Orange | 40° |
| Green | 60° |
| Pink | 100° |
| Purple | 15° |
| Blue | 120° |

on To test your understanding and knowledge of this topic, go to your learnON title at www.jacplus.com.au and complete the **post-test**.

# Answers

## Topic 13 Introduction to probability

### 13.1 Pre-test

1. C
2. E
3. $\frac{1}{5}$
4. B
5. 4
6. 4
7. A
8. B
9. B
10. $\frac{1}{5}$
11. E
12. $\frac{11}{40}$
13. B
14. D
15. 5

### 13.2 The language of chance

1. a. Certain b. Likely c. Impossible d. Likely e. Unlikely
2. a. Unlikely b. Even chance c. Certain d. Certain e. Even chance
3. a. $\frac{1}{2}$ (or 0.5) b. 0 c. $\frac{1}{5}$ (or 0.2) d. $\frac{1}{3}$
4. a. $\frac{1}{2}$ (or 0.5)
   b. 1
   c. 1
   d. 0.3–0.7. Any value within this range is reasonable.
5. A
6. A
7. D, C, A, B
8. Sample responses can be found in the worked solutions in the online resources.
9. a. 1 b. 0.75 c. 0.25 d. 0.75 e. 0.25
10. a. 0.25 b. 1 c. 0.25 d. 0 e. 0.75
11. a. 1 b. 0 c. 0.5 d. 0.5
12. a. It is impossible because Tuesday always follows Monday.
    b. It is certain because the 29th always follows the 28th day of the month. However, if the month is February, depending on whether the year is a leap year or not, the chance is either impossible or certain.
    c. It has an equal chance or a fifty-fifty chance if it is a coin with a Head on one side and a Tail on the other side.
13. Equal chance: 50% or $\frac{1}{2}$ or 0.5.
14. There are 3 odd-numbered balls (1, 3, 5) and only 2 even-numbered balls (2, 4). Therefore you are more likely to draw an odd-numbered ball from the bag than an even-numbered ball.
15. a. Highly unlikely because there is only one piece with the letter I. The chance is $\frac{1}{10}$ or 0.1.
    b. Equal chance because there are five pieces with the letter A and five pieces with other vowels.
    c. Impossible because there are no pieces with the letter U.
    d. Unlikely because the chance is $\frac{3}{10}$ or 0.3.
16. a. The two girls are living in two different countries.
    b. Time zones in different places, night-time and daytime in different places in the world.
17. a. Rolling a 7
    b. Rolling a 1
    c. Rolling an even number
    d. Rolling a number that is greater than 1
    e. Rolling a number that is less than 10

### 13.3 The sample space

1. $S = \{1, 2, 3, 4, 5, 6, 7, 8, 9, 10\}$
2. $S = \{\text{Heads, Tails}\}$
3. $S = \{A, B, C, D, E\}$
4. $S = \{\text{win, loss, draw}\}$
5. $S = \{\text{ace of clubs, ace of spades, ace of hearts, ace of diamonds}\}$
6. $S = \{A, B, C, D, E, F\}$
7. a. $S = \{\text{clubs, spades, diamonds, hearts}\}$
   b. $S = \{\text{red, black}\}$
8. $S = \{\text{red, green, orange}\}$
9. 2
10. a. 52 b. 45 c. 18 d. 365 (or 366 in a leap year)
11. a. 26 b. 180 000 c. 4 d. 3
12. E
13. a.

| | Heads | Tails |
|---|---|---|
| Blue | Blue, H | Blue, T |
| Green | Green, H | Green, T |
| Orange | Orange, H | Orange, T |

$S = \{(\text{Blue, H}), (\text{Blue, T}), (\text{Green, H}), (\text{Green, T}), (\text{Orange, H}), (\text{Orange, T})\}$

b. $n(S) = 6$

14. a.

| Die | Spinner A | Spinner B | Spinner C |
|---|---|---|---|
| 1 | A1 | B1 | C1 |
| 2 | A2 | B2 | C2 |
| 3 | A3 | B3 | C3 |
| 4 | A4 | B4 | C4 |
| 5 | A5 | B5 | C5 |
| 6 | A6 | B6 | C6 |

$\{(A,1),(A,2),(A,3),(A,4),(A,5),(A,6),$
$(B,1),(B,2),(B,3),(B,4),(B,5),(B,6),$
$(C,1),(C,2),(C,3),(C,4),(C,5),(C,6)\}$

b. 18

15. a.

| | 1 | 2 | 3 |
|---|---|---|---|
| 1 | (1, 1) | (1, 2) | (1, 3) |
| 2 | (2, 1) | (2, 2) | (2, 3) |
| 3 | (3, 1) | (3, 2) | (3, 3) |
| 4 | (4, 1) | (4, 2) | (4, 3) |
| 5 | (5, 1) | (5, 2) | (5, 3) |

$S = \{(1,1), (2,1), (3,1), (4,1), (5,1), (1,2), (2,2), (3,2), (4,2), (5,2), (1,3), (2,3), (3,3), (4,3), (5,3)\}$

b. $S = \{2, 3, 4, 5, 6, 7, 8\}$

16. A chance experiment is an experiment in which the outcome is left to chance. An outcome is any possible result of the chance experiment. An event can describe either one outcome or a collection of outcomes. The sample space is a list of all possible outcomes.

17. It depends on whether each element has an equal chance of occurring. In the sample space for rolling a die, each element is equally likely to occur. In more complex experiments, each event may not be likely to occur.

18. 12 different ways:
20, 20, 10; 20, 20, 5, 5; 20, 10, 10, 10;
20, 10, 10, 5, 5; 20, 10, 5, 5, 5, 5; 20, 5, 5, 5, 5, 5, 5;
10, 10, 10, 10, 10; 10, 10, 10, 10, 5, 5;
10, 10, 10, 5, 5, 5, 5; 10, 10, 5, 5, 5, 5, 5, 5;
10, 5, 5, 5, 5, 5, 5, 5, 5; 5, 5, 5, 5, 5, 5, 5, 5, 5, 5;

19. 15 choices

20. 12

21. 11

22. 18 ways

## 13.4 Simple probability

1. 12 : 1, 2, 3, 4, 5, 6, 7, 8, 9, 10, 11, 12
2. 5 : 1, 2, 3, 4, 5
3. 2 : C, L
4. 2 : 4, 6
5. 10 : 1, 3, 5, 7, 9, 11, 13, 15, 17, 19
6. 1, 2, 3, 4, 5, 6. All equally likely.
7. White, black. Each is equally likely.
8. White, blue, red. It is more likely you will get white rather than red or blue, because white takes up a bigger area.
9. 1, 2, 3, 4, 5. It is more likely that a five will be rolled because it appears twice on the die.
10. a, e, i. It is more likely that a consonant would be obtained as there are 7 consonants and 4 vowels.
11. a. 1, 2, 3, 4, 5, 6, 7, 8, 9, 10
    b. i. $\frac{1}{10}$ ii. $\frac{3}{5}$ iii. $\frac{2}{5}$ iv. $\frac{7}{10}$ v. $\frac{3}{10}$
12. a. $\frac{3}{10}$ b. $\frac{1}{5}$ c. $\frac{7}{20}$
    d. $\frac{1}{2}$ e. $\frac{13}{20}$ f. 1
13. $\frac{2}{5}$
14. E
15. D
16. a. $\frac{1}{2}$ b. $\frac{1}{13}$ c. $\frac{1}{4}$ d. $\frac{3}{13}$ e. $\frac{1}{52}$
17. a. $\frac{3}{4}$ b. $\frac{12}{13}$ c. 1 d. $\frac{3}{4}$ e. $\frac{1}{26}$
18. An event must either occur or not occur — there is no other option. If we know the probability of an event occurring, the probability it will not occur is 1 − P(event).
19. a. i. $\frac{1}{100\,000}$
    ii. $\frac{1}{5000}$
    iii. $\frac{1}{2000}$
    iv. 1
    b. It would not be a very wise purchase because the total cost would be \$500 000 unless of course the car was worth more than \$500 000.
20. a. 100% b. 50%
    c. $\frac{1}{7}$ d. $\frac{2}{7}$
    e. Decrease f. Increase
21. 0.1 = 10%
22. $\frac{95}{220} = \frac{19}{44} \approx 43\%$
23. The probability of taking a banana is $\frac{2}{5}$.

## 13.5 Experimental probability

1. $\frac{1}{5}$
2. $\frac{14}{17} \approx 82\%$
3. a.

| | Heads | | Tails | |
|---|---|---|---|---|
| Exp. no. | Tally | Count | Tally | Count |
| 1 | \|\|\|\| | 4 | 卌 \| | 6 |
| 2 | \| | 1 | 卌 \|\|\|\| | 9 |
| 3 | 卌 \| | 6 | \|\|\|\| | 4 |
| 4 | 卌 \|\|\| | 8 | \|\| | 2 |
| 5 | \|\|\|\| | 4 | 卌 \| | 6 |
| 6 | 卌 \|\| | 7 | \|\|\| | 3 |
| | Total | 30 | Total | 30 |

*Note:* This is only one possible solution. Answers will differ each time.

b. $\frac{2}{5}(40\%)$

c. $\frac{3}{5}(60\%)$

d. The theoretical value for both results is $\frac{1}{2}(50\%)$. They differ by 10%.

e. Refer to table.

f. The combined result equals the theoretical value.

4. 4
5. Divide the spinner into 10 sections; 3 of these sections will be shaded in 1 colour and 7 sections will be shaded in another colour.
6. a. The students rolled a die 5 times and recorded whether the result was an odd number or an even number.

b. 3 times

c. 0.5

d. The experimental probability was 0.6; this is 0.1 higher than the theoretical probability.

e. The experimental probability was 0.4, this is 0.1 lower than the theoretical probability.

7. Values will differ for each group. As you complete more trials, you will probably notice the values you get experimentally are closer to those you would expect from theoretical probability $\left(\frac{1}{2} \text{ or } 50\%\right)$. If this is not occurring, you will probably need more trials.
8. a. $\frac{1}{5}$

b.

| Colour | 1 | 2 | 3 | 4 | 5 |
|---|---|---|---|---|---|
| Number of times it occurs | | \|\| | \|\| | \|\|\|\| | \|\| |

c. $P(1) = 0$, $P(2) = \frac{1}{5}$, $P(3) = \frac{1}{5}$, $P(4) = \frac{2}{5}$, $P(5) = \frac{1}{5}$

*Note:* This is only one possible solution. Answers will differ each time.

d. Your spinner may not be evenly balanced and this may lead to an increased likelihood of getting one result or another.

e-f. Values will differ for each group. As you complete more trials, you will probably notice the values you get experimentally are closer to those you would expect from theoretical probability $\left(\frac{1}{5}\right)$ or 20%). If this is not occurring, you will probably need more trials.

9. a. $\frac{13}{73}$

b. Karen is not correct, as a wet day (65 out of 365) and a dry day (300 out of 365) are not equally likely.

c. 0.18

10. The general trend as the number of trials increases should show that the probability of getting a Head or a Tail gets closer to 50%.
11. a.

| Score | Frequency |
|---|---|
| 1 | 7 |
| 2 | 4 |
| 3 | 8 |
| 4 | 3 |
| 5 | 5 |
| 6 | 3 |

b. $P(6) = \frac{1}{6}$

c. 5

d. The result differs by pure chance. The more often we roll the die, the closer we can expect the results to match the expected number of sixes.

12. a. Each die would need to be thrown about 120 times for us to get some impression of whether it is biased or not. The same person would need to roll the die each time in the same manner and onto the same surface.

b.

| Number on die | Black | White |
|---|---|---|
| 1 | 20 | 19 |
| 2 | 19 | 21 |
| 3 | 20 | 20 |
| 4 | 21 | 20 |
| 5 | 19 | 20 |
| 6 | 21 | 20 |
| Total | 120 | 120 |

*Note:* These values will differ each time the experiment is conducted.

c.

| Number on die | Probability of black | Probability of white |
|---|---|---|
| 1 | $\frac{20}{120} = \frac{1}{6}$ | $\frac{19}{120}$ |
| 2 | $\frac{19}{120}$ | $\frac{21}{120} = \frac{7}{40}$ |
| 3 | $\frac{20}{120} = \frac{1}{6}$ | $\frac{20}{120} = \frac{1}{6}$ |
| 4 | $\frac{21}{120} = \frac{7}{40}$ | $\frac{20}{120} = \frac{1}{6}$ |
| 5 | $\frac{19}{120}$ | $\frac{20}{120} = \frac{1}{6}$ |
| 6 | $\frac{21}{120} = \frac{7}{40}$ | $\frac{20}{120} = \frac{1}{6}$ |
| Total | 1 | 1 |

The long-term trend suggests that the probability of obtaining each value on either die will be $\frac{1}{6}$.

d. The 2 dice appear to be fair as each value occurred approximately 20 times, which is what we would expect in 120 throws. More trials however could be conducted. It is very important when conducting an experiment such as this that the devices used are even in shape and size and that one doesn't have an advantage over the other.

13. 12 triangles, 24 squares

14. a. 7

b. 8

c. 7

d. In part **a** there are 5 red pens, so, to make sure that you pick 2 black pens, you have to pick $5 + 2 = 7$.
In part **b** there are 6 black pens, so, to make sure that you pick 2 red pens, you have to pick $6 + 2 = 8$.
In part **c** there are 5 red pens and 6 black pens, so, to make sure that you pick 1 black pen and 1 red pen, you have to pick $6 + 1 = 7$.

15. a. 36 square units

b. A = 20 square units, B = 12 square units, C = 4 square units

c. i. $\frac{5}{9}$

ii. $\frac{1}{3}$

iii. $\frac{1}{9}$

d. i. $\frac{37}{351}$, which is approximately 0.11. This is very close to the theoretical probability.

ii. $\frac{631}{1872}$, which is approximately 0.34. This is very close to the theoretical probability.

iii. $\frac{3131}{5616}$, which is approximately 0.56. This is very close to the theoretical probability.

## Project

1.

| | | Die 1 | | | | | |
|---|---|---|---|---|---|---|---|
| | | 1 | 2 | 3 | 4 | 5 | 6 |
| Die 2 | 1 | 2 | 3 | 4 | 5 | 6 | 7 |
| | 2 | 3 | 4 | 5 | 6 | 7 | 8 |
| | 3 | 4 | 5 | 6 | 7 | 8 | 9 |
| | 4 | 5 | 6 | 7 | 8 | 9 | 10 |
| | 5 | 6 | 7 | 8 | 9 | 10 | 11 |
| | 6 | 7 | 8 | 9 | 10 | 11 | 12 |

2. {2, 3, 4, 5, 6, 7, 8, 9, 10, 11, 12}

3. 7

4. 2, 12

5. $\frac{1}{18}$

6. 5, four different ways.

7. You end up at square 36. The next ladder is at square 51. It is not possible to obtain a 15 when you roll 2 dice.

8. At the end of these two turns, you would end up at square 73. P (two ones in a row) $= \frac{1}{36} \times \frac{1}{36} = \frac{1}{1296}$.

9. You need to roll a total of 7. P (7) $= \frac{1}{6}$.

10. Your opponent has a better chance, because the chance of scoring a 6 is higher than that of scoring a 4.

11. There are many different paths that get from start to finish in 5 turns.

## 13.6 Review questions

1. a. Certain b. Unlikely c. Even chance

2. a. Impossible b. Likely c. Impossible

3. a. The month of January having 29 days

b. Tuesday coming after Monday

c. Rolling a 3 on a regular die

4.

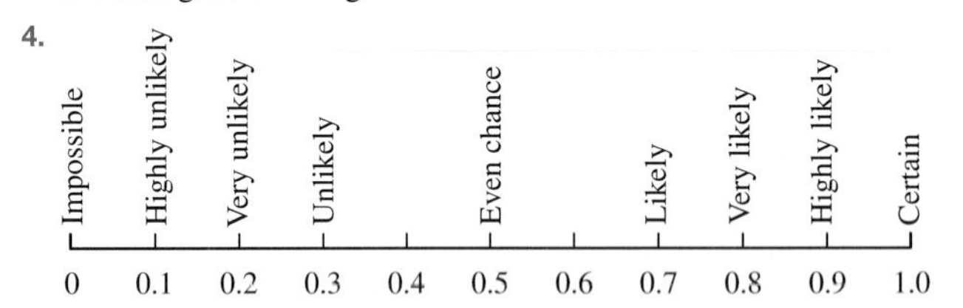


5. a. {Head, Tail}

b. {red marble, green marble, blue marble}.

c. {true, false}

6. a. {1, 2, 3, 4, 5, 6}

b. {1, 2, 3, 4, 5, 6, 7, 8, 9, 10}

c. {1, 2, 3, 4, 5, 6, 7, 8, 9}

7. a. 5000 b. 26 c. 5 d. 31

8. a. $\frac{1}{6}$ b. $\frac{1}{3}$ c. $\frac{1}{3}$ d. $\frac{2}{3}$

9. a. $\frac{2}{3}$ b. $\frac{1}{3}$ c. 0 d. 0

10. a $\frac{1}{4}$ b $\frac{1}{4}$ c $\frac{3}{13}$ d $\frac{3}{13}$

11. a. $\frac{1}{10}$ b. $\frac{1}{2}$ c. $\frac{3}{10}$ d. $\frac{1}{10}$

12. a. $\frac{7}{10}$ b. $\frac{7}{10}$ c. $\frac{1}{5}$ d. 0

13. $\frac{9}{10}$

14.

| Die | Coin: Head | Tail |
|---|---|---|
| 1 | H1 | T1 |
| 2 | H2 | T2 |
| 3 | H3 | T3 |
| 4 | H4 | T4 |
| 5 | H5 | T5 |
| 6 | H6 | T6 |

a. $\frac{1}{6}$ b. $\frac{1}{4}$ c. $\frac{1}{4}$ d. $\frac{1}{4}$

15. See the table at the foot of the page.*

a. $\frac{1}{2}$ b. $\frac{1}{4}$ c. $\frac{4}{15}$

d. $\frac{1}{5}$ e. $\frac{3}{20}$ f. $\frac{23}{60}$

16. a. 6 socks. Worst case scenario is first picking 4 blue socks, then 2 black socks.
    b. 7 socks. Worst case scenario is first picking 5 black socks then 2 blue socks.
    c. 7 socks. Minimum needed to guarantee a pair of black socks is 6 (from part a). Picking one more guarantees a pair of blue socks will be included (from part b).

17. a. 12
    b. $4, 6, 8, 10, 12$
    c. $2 + 6, 6 + 2, 4 + 4$
    d. 0
    e. $\frac{2}{9}$
    f.

| Spinner B \ Spinner A | 2 | 4 | 6 |
|---|---|---|---|
| 2 | 4 | 6 | 8 |
| 4 | 6 | 8 | 10 |
| 6 | 8 | 10 | 12 |

**Total**

| 4 | 6 | 8 | 10 | 12 |
|---|---|---|---|---|
| $\frac{1}{9}$ | $\frac{2}{9}$ | $\frac{3}{9}=\frac{1}{3}$ | $\frac{2}{9}$ | $\frac{1}{9}$ |

18. $\frac{1}{3}$

19. $\frac{2}{3}$

20. $\frac{3}{5}$

21. $\frac{29\,679}{160\,000}$

22. Likelihood smallest to largest: purple, yellow, orange, green, pink, blue

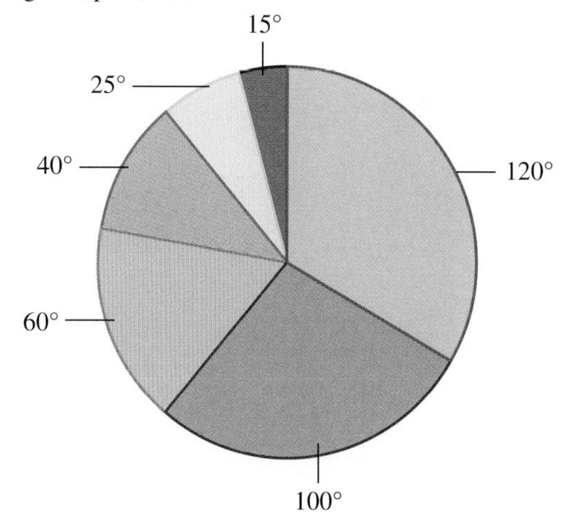


*15.

| Die \ Spinner | 1 | 2 | 3 | 4 | 5 | 6 | 7 | 8 | 9 | 10 |
|---|---|---|---|---|---|---|---|---|---|---|
| 1 | 11 | 21 | 31 | 41 | 51 | 61 | 71 | 81 | 91 | 101 |
| 2 | 12 | 22 | 32 | 42 | 52 | 62 | 72 | 82 | 92 | 102 |
| 3 | 13 | 23 | 33 | 43 | 53 | 63 | 73 | 83 | 93 | 103 |
| 4 | 14 | 24 | 34 | 44 | 54 | 64 | 74 | 84 | 94 | 104 |
| 5 | 15 | 25 | 35 | 45 | 55 | 65 | 75 | 85 | 95 | 105 |
| 6 | 16 | 26 | 36 | 46 | 56 | 66 | 76 | 86 | 96 | 106 |

# 14 Representing and interpreting data

## LESSON SEQUENCE


**14.1** Overview ... 764
**14.2** Collecting and classifying data ... 767
**14.3** Displaying data in tables ... 774
**14.4** Measures of centre and spread ... 780
**14.5** Column graphs and dot plots ... 791
**14.6** Stem-and-leaf plots ... 798
**14.7** Pie charts and divided bar graphs (extending) ... 804
**14.8** Comparing data ... 814
**14.9** Review ... 820

# LESSON
# 14.1 Overview

## Why learn this?

Every day we see graphs, charts and statistics in the media. These graphs, charts and statistics all serve to condense massive amounts of information into a few simple numerical facts. This is the beauty of statistics. It turns complicated data into simpler forms that, with the knowledge gained throughout this topic, you will be able to interpret and understand.

You may have seen graphs or charts of housing prices in the news or online. One house price report stated that in September 2020 the median house price was \$1 154 406 in Sydney, \$875 980 in Melbourne and \$596 316 in Brisbane. What is the median house price? Why are house prices discussed in terms of the median and not the mean? This topic will help you to understand the answers to these questions, and apply your understanding to other contexts.

In every business it is vital to keep a close eye on finances. Statistics can be used to determine the economic health of a business by looking at changes in revenue, costs, customer satisfaction, market share and many other factors over time. Line graphs are often used to show changes over time, and pie charts are often used to display market share or customer satisfaction.

Data is also extremely important in the medical field. Statistics are used to collate data from around the world to better understand the rates of disease, effectiveness of different medicines or treatments, and many other things. In 2020, when the Covid-19 pandemic hit the world, graphs and charts started appearing in the news and on social media displaying the case numbers for different states and countries. The data collected by each state and country was displayed graphically, enabling residents to be informed and educated on the latest developments.

## Exercise 14.1 Pre-test

learnon

1. **MC** Select the data category that suits the number of players in a netball team.
   - **A.** Categorical and discrete
   - **B.** Numerical and discrete
   - **C.** Numerical and continuous
   - **D.** Categorical and continuous
   - **E.** Categorical and numerical

2. **MC** Select the data category that suits the amount of rainfall.
   - **A.** Numerical and continuous
   - **B.** Categorical and discrete
   - **C.** Numerical and discrete
   - **D.** Categorical and continuous
   - **E.** Numerical and categorical

Use the following information to answer questions **3** and **4**.

The following frequency table represents the heights of 20 students in a Mathematics class.

| Height group (cm) | Frequency |
|---|---|
| 120–129 | 2 |
| 130–139 | 8 |
| 140–149 | 7 |
| 150–159 | |

3. Determine the number of students in the 150–159 cm group.

4. **MC** Choose the most common height group in the class.
   - **A.** 120–129 cm
   - **B.** 130–139 cm
   - **C.** 140–149 cm
   - **D.** 150–159 cm
   - **E.** 160–169 cm

5. A book store sold the following number of books each day for the last week. Calculate the daily average number of books sold.

   Book sales: 23, 34, 18, 28, 36, 45, 33

6. Calculate the median of the following data set.

   3, 3, 7, 8, 9, 9, 9, 11, 12, 13, 13, 17, 18, 21

7. Buddy kicks a ball 7 times. His kicks are 49, 52, 57, 43, 55, 62 and 59 metres in distance. Calculate the range of Buddy's 7 kicks.

8. The following data represents the sizes of Nike runners sold over a week. State the most common shoe size.

   Nike shoe sizes: 7, 7, 8, 9, 9, 9, 9, 10, 10, 11, 11, 11, 12, 12

**9.** The following graph shows the preferred sports of a group of students.

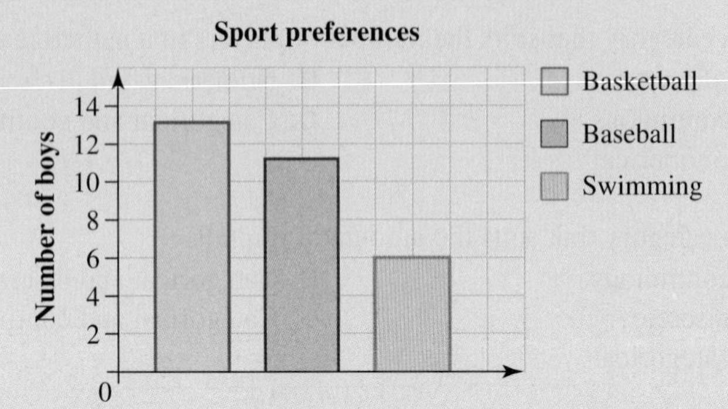


To the nearest per cent, determine the percentage of students who preferred swimming.

**10.** Determine the median value in the following dot plot.

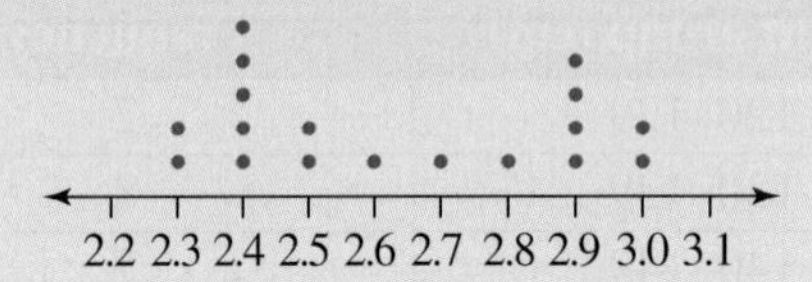


**11.** The following stem-and-leaf plot represents the ages of a group of golfers. Determine the number of golfers that are older than 20 years of age.

Key: 2|4 = 24

| *Stem* | *Leaf* |
|---|---|
| 1 | 78899 |
| 2 | 2479 |
| 3 | 1338 |
| 4 | 022266 |
| 5 | 57 |
| 6 | 4 |

**12.** The following back-to-back stem-and-leaf plot shows the heights of Year 7 students. Determine the difference between the height of the tallest boy and the shortest girl.

Key: 13|7 = 137 cm

| *Leaf (boys)* | *Stem* | *Leaf (girls)* |
|---|---|---|
| 98 | 13 | 78 |
| 98876 | 14 | 356 |
| 988 | 15 | 1237 |
| 7665 | 16 | 356 |
| 876 | 17 | 1 |

**13.** A survey asked people for their favourite brand of car. The following results were found:

Toyota = 34, Ford = 28, Holden = 30, Mazda = 18

If a sector graph was constructed using these results, determine the angle used for Toyota. Give your answer to the nearest degree.

14. The data presented in a divided bar graph shows the preferred Australian holiday destinations.

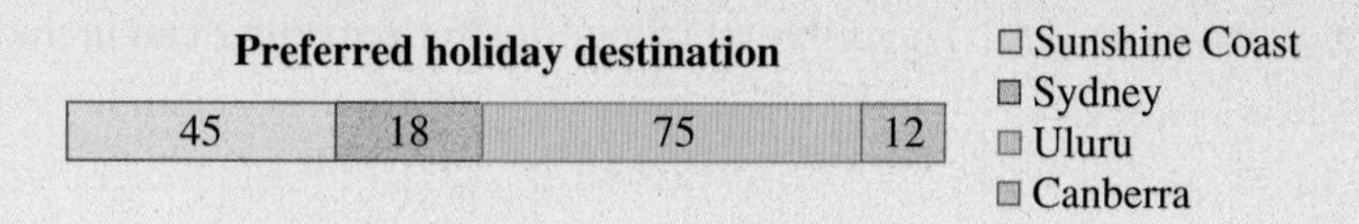


a. Based on this graph, calculate the number of people who were interviewed about their favourite Australian holiday destination.
b. Identify the number of people who preferred the Sunshine Coast.
c. Calculate the percentage of the total number of people who preferred Canberra.
d. Calculate the percentage of the total number of people who preferred Uluru.

15. The following sector graph is based on the way students travel to school. Determine the percentage, to the nearest whole per cent, of students who get to school by bus.

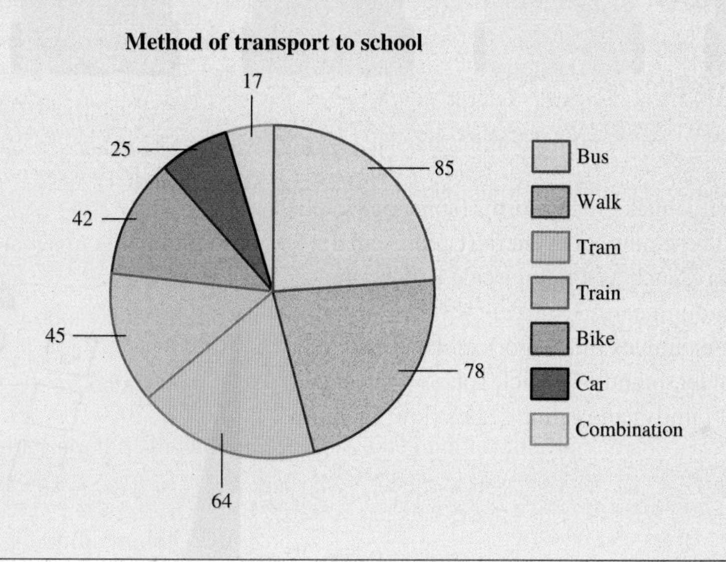


# LESSON
# 14.2 Collecting and classifying data

## LEARNING INTENTION

At the end of this lesson you should be able to:

- describe types of data using the key words 'categorical' (discrete) or 'numerical' (continuous)
- understand the difference between primary and secondary data
- explain why some data is more reliable that other data.

## 14.2.1 How to classify data

eles-4716

- Statistics is the study of collecting, organising, presenting, analysing and interpreting data.
- **Data** is information that is collected about a certain group.
- A **variable** is a particular characteristic that can be observed, counted or measured, such as eye colour or height. A variable can change over time or between individual observations.
- Data can be classified as **categorical data** or **numerical data**.

## Categorical data

- Categorical data provides information about variables that can be *grouped into categories*.
- Take, for example, recycling bins. The different categories for recycling can include organic, paper, plastic, glass and metal.

- Categorical data can take many forms (some of which are numbers). These numbers generally represent a group or category, such as a rating scale (for example, 1 = dislike, 2 = don't care, 3 = like).
- Other common examples of categorical data include hair colour, gender, brand of vehicle, pizza size (small, medium, large, family) and rating scales (low, medium, high).

## Numerical data

- Numerical data provides information about variables that can be *counted or measured*.
  For example, the number of text messages you send in a day, the number of puppies in a litter, the height of a tree at your school, and the daily pollution levels on a particular highway are all examples of data that can be counted or measured.
- Numerical data can be further classified as **discrete** (able to be counted), or **continuous** (able to be measured). For example, *number of goals scored* is a discrete variable, while *height of tree* is a continuous variable.

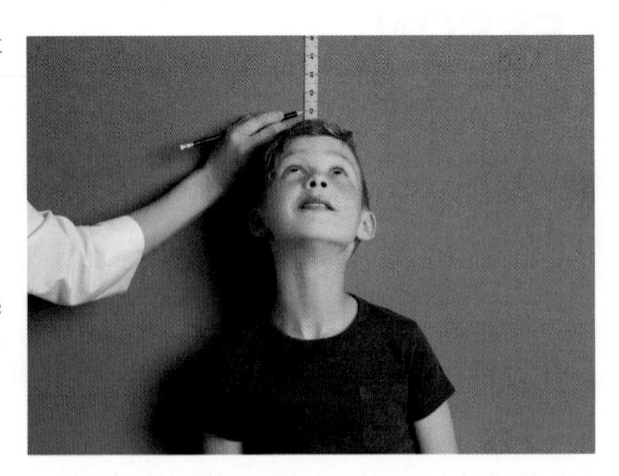

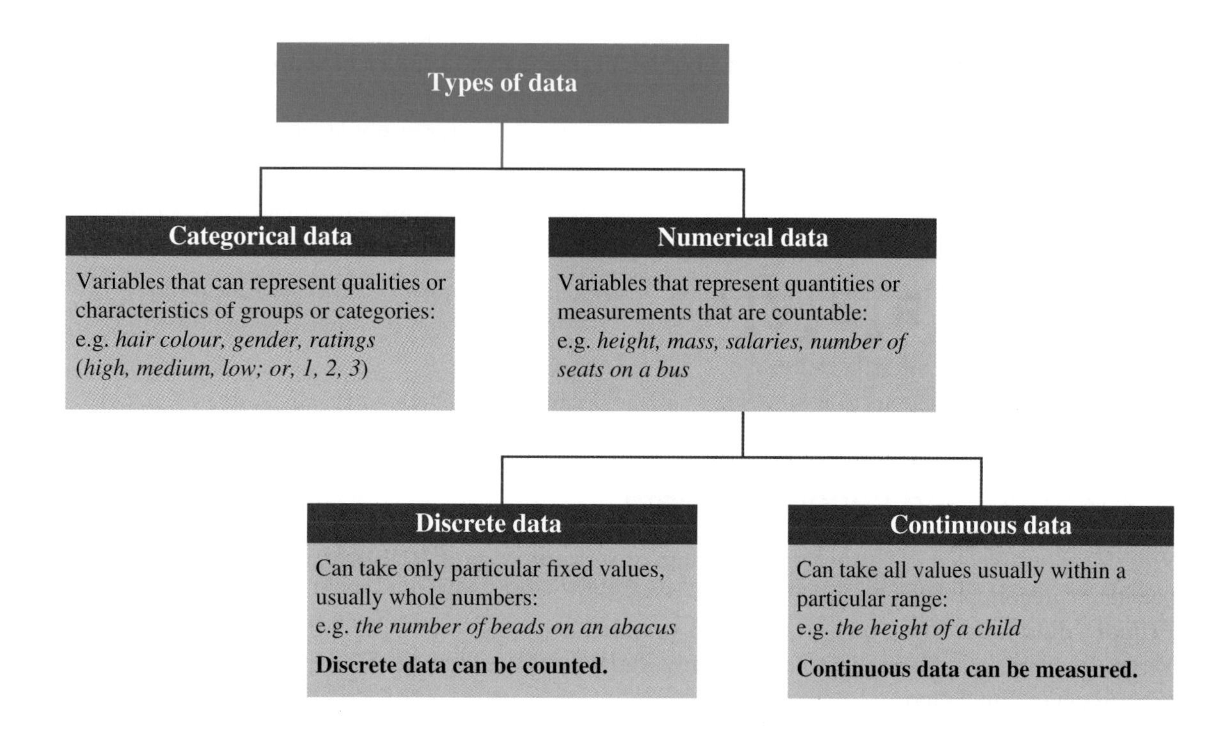


## Numerical or categorical?

**Deciding whether data is numerical or categorical is not always easy. Two things that can help your decision-making are the following.**

1. **Numerical data can always be used to perform arithmetic computations. This is not the case with categorical data. This is a good test to apply when in doubt.**
   **For example, it makes sense to calculate the average weight of a group of individuals, but not the average hair colour of a class of students.**
2. **It is not the variable alone that determines whether data is numerical or categorical — it is also the way the data is recorded.**
   **For example, if the data for the variable *weight* is recorded in kilograms, it is numerical. However, if the data is recorded using labels such as 'underweight', 'normal weight' and 'overweight', it is categorical.**

## WORKED EXAMPLE 1 Classifying data

**Classify the following data using the following descriptive words: 'categorical' or 'numerical' (discrete or continuous).**

a. **The number of students absent from school**
b. **The types of vehicle using a certain road**
c. **The eye colour of each student in your class**
d. **The room temperature at various times during a particular day**

| THINK | WRITE |
|---|---|
| a. 1. Determine whether the data is categorical or numerical. | a. The data is *numerical* because the number of students can be counted or measured. |
| 2. Determine whether the data is discrete (countable) or continuous (measurable). | The data is *discrete* because the number of absences can be counted rather than measured. |

| | |
|---|---|
| **b.** Determine whether the data is categorical or numerical. | **b.** The data is *categorical* because the types of vehicles can be placed into groups or categories (for example, car, truck, motorbike). |
| **c.** Determine whether the data is categorical or numerical. | **c.** The data is *categorical* because the colour of a person's eyes fits into groups (for example, brown, blue, green). |
| **d. 1.** Determine whether the data is categorical or numerical. | **d.** The data is *numerical* because the room temperature can be counted or measured. |
| **2.** Determine whether the data is discrete (countable) or continuous (measurable). | The data is *continuous* because temperature can be measured rather than counted. |

## 14.2.2 Primary and secondary data

eles-4717

- When data is collected it is either **primary data** or **secondary data**. Both primary and secondary data can be either categorical data or numerical data.
- **Primary data** is data that you have collected yourself.
- Various methods of collecting primary data are available. These include observation, measurement, survey, experiment or simulation.
- **Secondary data** is data that has already been collected by someone else.
- The data can come from a variety of sources:
  - books, journals, magazines, company reports
  - online databases, broadcasts, videos
  - government sources — the Australian Bureau of Statistics (ABS) provides a wealth of statistical data
  - general business sources — academic institutions, stockbroking firms, sporting clubs
  - media — newspapers, TV reports.
- Secondary data sources often provide data that it would not be possible for an individual to collect.

### WORKED EXAMPLE 2 Classifying primary and secondary data

**Classify the following as primary data or secondary data.**
**a. The results of a class survey that you ran about favourite football teams**
**b. The Bureau of Meteorology's rainfall data from the last 10 years**

| THINK | WRITE |
|---|---|
| **a. 1.** Identify the person/people who collected the data. | **a.** I collected the data. |
| **2.** I collected the data; therefore, it is primary data. | It is primary data. |
| **b. 1.** Identify the person/people who collected the data. | **b.** The Bureau of Meteorology collected the data. |
| **2.** Someone else collected the data; therefore, it is secondary data. | It is secondary data. |

### Reliability of data

- When using data it is important to identify whether or not the data is reliable.

## Determining the reliability of data

**To determine the reliability of data, consider the following factors:**

- **Data source**
  - **Was the data collected by a reputable organisation or individual?**
  - **Did the people who collected the data have a motivation to find certain results? That is, does the data contain bias? (For example, a cigarette company might wish to fund a survey that finds no link between cigarettes and lung cancer.)**
- **Sample size**
  - **Was the sample of data collected small? This is not very reliable.**
  - **Was the sample of data collected large? This is generally more reliable.**

DISCUSSION

How are graphs and statistics used to summarise data?

 Resources

 **eWorkbook** Topic 14 Workbook (worksheets, code puzzle and project) (ewbk-1915)

 **Interactivity** Individual pathway interactivity: Classifying data (int-7045)

## Exercise 14.2 Collecting and classifying data

learnon

**14.2 Quick quiz** 

**14.2 Exercise**

Individual pathways

| ■ PRACTISE | ■ CONSOLIDATE | ■ MASTER |
|---|---|---|
| 1, 2, 5, 6, 10, 13 | 3, 7, 8, 11, 14 | 4, 9, 12, 15, 16 |

### Fluency

1. Match each word with its correct meaning.

| | |
|---|---|
| a. discrete | i. placed in categories or classes |
| b. categorical | ii. data collected by someone else |
| c. continuous | iii. data in the form of numbers |
| d. numerical | iv. used for measurements |
| e. primary data | v. counted in exact values |
| f. secondary data | vi. data collected by me |

2. WE1 Classify each of the following data as 'categorical' or 'numerical' (discrete or continuous).
   a. The population of your town or city
   b. The types of motorbikes in a car park
   c. The heights of people in an identification line-up
   d. The weights of babies in a group
   e. The languages spoken at home by students in your class
   f. The time spent watching TV

3. Classify each of the following numerical data as either 'discrete' or 'continuous'.
   a. The number of children in the families in your suburb
   b. The air pressure in your car's tyres
   c. The number of puppies in a litter
   d. The times for swimming 50 metres
   e. The quantity of fish caught in a net
4. Classify each of the following data as 'categorical' or 'numerical' (discrete or continuous).
   a. The types of shops in a shopping centre
   b. The football competition ladder at the end of each round
   c. The lifetime of torch batteries
   d. The number of people attending the Big Day Out
   e. Final Year 12 exam grades
   f. Hotel accommodation ratings
5. WE2 Classify the following as primary data or secondary data.
   a. The number of cars you counted going past your house in the last hour
   b. The Australian government's data about the number of students enrolled in Australian schools

### Understanding

6. Data representing shoe (or rollerblade) sizes can be classified as categorical. Identify whether this statement is true or false.
7. Data representing the points scored in a basketball game can best be described as discrete. Identify whether this statement is true or false.
8. MC Select an example of categorical data.
   A. Heights of buildings in Sydney
   B. Number of pets in households
   C. Types of pets in households
   D. Birthday year of students in Year 7
   E. Number of hours spent playing sport
9. a. Write a sentence explaining the difference between discrete and continuous data. Give an example of each.
   b. Write a sentence explaining the difference between primary and secondary data. Give an example of each.

### Reasoning

10. A Likert scale is used to represent people's attitudes to a topic. A five-level Likert scale could be as follows:
    1. Strongly disagree
    2. Disagree
    3. Neither agree nor disagree
    4. Agree
    5. Strongly agree

The data collected from a Likert scale is recorded as a numerical value. Explain why this data is categorical instead of numerical.

11. Explain why data such as postal codes, phone numbers and drivers' licence numbers is not numerical data.
12. The Fizzy Drink company surveyed 50 people about the health benefits of soft drinks. The results showed that soft drink is healthy. Explain why this data is unreliable giving at least two reasons.

### Problem solving

13. The following questions would collect categorical data. Rewrite the questions so that you could collect numerical data.
    a. Do you read every day?
    b. Do you play sport every day?
    c. Do you play computer games every day?
14. The following questions would collect numerical data. Rewrite the questions so that you could collect categorical data.
    a. On average, how many minutes per week do you spend on Maths homework?
    b. How many books have you read this year?
    c. How long does it take you to travel to school?
15. A fisheries and wildlife officer released 200 tagged trout into a lake. A week later, the officer took a sample of 50 trout and found that 8 of them were tagged. The officer can use this information to estimate the population of trout in the lake. Calculate the number of trout that are in the lake. Explain how you got this answer.

16. Bill is a soccer coach who trains 150 players. Bill surveys 5 players and the results say that Bill is the best soccer coach in Australia.
    a. Discuss whether Bill has collected primary or secondary data.
    b. Explain whether this data is reliable.
    c. Bill wants people to think he is the best soccer coach. Explain how this affect the reliability of the data.
    d. If the survey was done by the Australian Institute of Soccer, explain whether the data would be more or less reliable.

# LESSON
# 14.3 Displaying data in tables

## LEARNING INTENTION

At the end of this lesson you should be able to:
- construct a frequency distribution table for a set of data
- analyse and report on the information in a frequency distribution table.

## 14.3.1 Frequency distribution tables

eles-4718

- Data can be displayed in a variety of forms.
- In general, data is first organised into a table, and then a suitable graph is chosen as a visual representation.
- One way of presenting data is by using a frequency distribution table.
- **Frequency** is the number of times a result or piece of data occurs.
- A **frequency distribution table** consists of three columns: score, tally and frequency.
- The following frequency table is the result of a survey of 30 students, and it shows the number of family members who live at home with them.

| Score (number of family members living at home) | Tally | Frequency |
|---|---|---|
| 2 | 𝍷𝍷𝍷 | 3 |
| 3 | 𝍸 | 5 |
| 4 | 𝍸𝍷𝍷𝍷𝍷 | 9 |
| 5 | 𝍸𝍷𝍷 | 7 |
| 6 | 𝍷𝍷𝍷𝍷 | 4 |
| 7 | 𝍷 | 1 |
| 8 | 𝍷 | 1 |
| | Total | 30 |

### WORKED EXAMPLE 3 Constructing and interpreting a frequency table

**A particular class was surveyed to determine the number of pets per household, and the data was recorded. The raw data was:**
**0, 3, 1, 2, 0, 1, 0, 1, 2, 4, 0, 6, 1, 1, 0, 2, 2, 0, 1, 3, 0, 1, 2, 1, 1, 2**

**a. Organise the data into a frequency distribution table.**
**b. Calculate the number of households that were included in the survey.**
**c. Determine the number of households that have fewer than 2 pets.**
**d. State the most common number of pets.**
**e. Determine the number of households that have 3 or more pets.**
**f. Determine the fraction of households surveyed that had no pets.**

**THINK**

a. 1. Draw a frequency distribution table comprising three columns, headed Number of pets, Tally and Frequency.
2. In the first column list the possible number of pets per household (from 0 to 6).
3. Place a stroke in the tally column each time a particular score is noted.
*Hint:* A score of 5 is shown as a 'gate post' (four vertical strokes and one diagonal stroke: 𝍸).

**WRITE**

a.

| Number of pets | Tally | Frequency |
|---|---|---|
| 0 | 𝍸𝍷𝍷 | 7 |
| 1 | 𝍸𝍷𝍷𝍷𝍷 | 9 |
| 2 | 𝍸𝍷 | 6 |
| 3 | 𝍷𝍷 | 2 |
| 4 | 𝍷 | 1 |
| 5 | | 0 |
| 6 | 𝍷 | 1 |
| | Total | 26 |

| | |
|---|---|
| 4. Write the total number of tally strokes for each pet in the frequency column. | |
| 5. Calculate the total of the frequency column. | |
| b. The total of the frequency column gives the number of households surveyed. | b. 26 households were surveyed. |
| c. 1. Calculate the number of households that have fewer than two pets. *Hint:* Fewer than two means zero pets or one pet. | c. Fewer than two pets $= 7 + 9$ $= 16$ |
| 2. Write the answer. | 16 households have fewer than two pets. |
| d. 1. Make a note of the highest value in the frequency column and check which score it corresponds to. | d. The score with the highest frequency (9) corresponds to one pet. |
| 2. Write the answer. | The most common number of pets is one. |
| e. 1. Calculate the number of households that have 3 or more pets. *Hint:* 3 or more means 3, 4, 5 or 6. | e. 3 or more pets $= 2 + 1 + 0 + 1$ $= 4$ |
| 2. Write the answer. | Four households have 3 or more pets. |
| f. 1. Write the number of households with no pets. | f. Households with no pets $= 7$ |
| 2. Write the total number of households surveyed. | Total number of households surveyed $= 26$ |
| 3. Define the fraction and substitute the known values into the rule. | $\frac{\text{Households with no pets}}{\text{Total number of households surveyed}} = \frac{7}{26}$ |
| 4. Write the answer. | Of the households surveyed, $\frac{7}{26}$ have no pets. |

## Class intervals

- Sometimes data may contain too many numerical values to list them all individually in the 'score' column. In this case, we use a range of values, called a *class interval*, as our category. For example, the range 100–104 may be used to cater for all the values that lie within the range, including 100 and 104.

### WORKED EXAMPLE 4 Using class intervals in a frequency table

**The data below shows the ages of a number of mobile phone owners.**

**12, 11, 21, 12, 30, 26, 13, 15, 29, 16, 17, 17, 17, 21, 19, 12, 14, 16, 43, 18, 51, 25, 30, 28, 33, 62, 39, 40, 30, 18, 19, 41, 22, 21, 48, 31, 33, 33, 34, 41, 18, 17, 31, 43, 42, 17, 46, 23, 24, 33, 27, 31, 53, 52, 25**

a. **Construct a frequency table to classify the given data. Use a class interval of 10, that is, ages 11–20, 21–30 and so on, as each category.**
b. **Calculate the number of people who were surveyed.**
c. **State the age group that had the most people with mobile phones.**
d. **State the age group that had the fewest people with mobile phones.**
e. **Determine the number of people in the 21–30 age group who own a mobile phone.**

| THINK | WRITE |
|---|---|
| **a. 1.** Draw a frequency distribution table.<br>**2.** In the first column, list the possible age groups (11–20, 21–30 and so on).<br>**3.** Systematically go through the results and place a stroke in the tally column each time a particular age group is noted.<br>**4.** Write the total tally of strokes for each age group in the frequency column.<br>**5.** Calculate the total of the frequency column. | **a.** (see table below) |
| **b.** The total of the frequency column gives us the number of people surveyed. | **b.** A total of 55 people were surveyed. |
| **c. 1.** Make note of the highest value in the frequency column and check which age group it corresponds to. | **c.** The 11–20 age group has the highest frequency: that is, a value of 19. |
| **2.** Write the answer. | The 11–20 age group has the highest number of people with mobile phones. |
| **d. 1.** Make note of the lowest value in the frequency column and check which age group it corresponds to.<br>*Note:* There may be more than one answer. | **d.** The over-60 age group has the lowest frequency: that is, a value of 1. |
| **2.** Write the answer. | The over-60 age group has the lowest number of people with mobile phones. |
| **e. 1.** Check the 21–30 age group in the table to see which frequency value corresponds to this age group. | **e.** The 21–30 age group has a corresponding frequency of 15. |
| **2.** Write the answer. | 15 people in the 21–30 age group own a mobile phone. |

| Age group | Tally | Frequency |
|---|---|---|
| 11–20 | 𝍸 𝍸 𝍸 \|\|\|\| | 19 |
| 21–30 | 𝍸 𝍸 𝍸 | 15 |
| 31–40 | 𝍸 𝍸 | 10 |
| 41–50 | 𝍸 \|\| | 7 |
| 51–60 | \|\|\| | 3 |
| over 60 | \| | 1 |
| | **Total** | 55 |

**on Resources**

**eWorkbook** Topic 14 Workbook (worksheets, code puzzle and project) (ewbk-1915)

**Interactivities** Individual pathway interactivity: Displaying data in tables (int-4379)
Frequency distribution tables (int-4051)

# Exercise 14.3 Displaying data in tables

learnon

14.3 Quick quiz on | 14.3 Exercise

### Individual pathways

| ■ PRACTISE | ■ CONSOLIDATE | ■ MASTER |
|---|---|---|
| 1, 3, 6, 7, 10, 13 | 2, 4, 8, 11, 14 | 5, 9, 12, 15 |

### Fluency

1. WE3 The number of children per household in a particular street is surveyed and the data recorded as follows.

   0, 8, 6, 4, 0, 0, 0, 2, 1, 3, 3, 3, 1, 2, 3,
   2, 3, 2, 1, 2, 1, 3, 0, 2, 2, 4, 2, 3, 5, 2

   a. Organise the data into a frequency distribution table.
   b. Calculate the number of households that are included in the survey.
   c. Determine the number of households that have no children.
   d. Determine the number of households that have at least 3 children.
   e. State the most common number of children.
   f. Determine the fraction of households surveyed that have 4 children.

2. WE4 The following data shows the incomes of a sample of doctors.

   \$100 000, \$105 000, \$110 000, \$150 000, \$155 000, \$106 000, \$165 000, \$148 000, \$165 000, \$200 000, \$195 000, \$138 000, \$142 000, \$153 000, \$173 000, \$149 000, \$182 000, \$186 000

   a. Construct a frequency table to classify the given data. Use a class interval of 10 000 to organise incomes into brackets of \$100 000 to \$109 999, \$110 000 to \$119 999, and so on.
   b. Calculate the number of people who were surveyed.
   c. State the income bracket that was most common.
   d. State the income bracket that was least common.
   e. Determine the number of doctors who earned \$140 000 to \$149 999.

3. Use this frequency distribution table to answer the following questions.

| Score | Tally | Frequency |
|---|---|---|
| 0 | \|\| | 2 |
| 1 | 𝍸 | 5 |
| 2 | \|\|\| | 3 |
| 3 | 𝍸 𝍸 \| | 11 |
| 4 | 𝍸 \|\|\| | 8 |
| 5 | \|\|\|\| | 4 |
| | Total | |

   a. Calculate the number of people who participated in the survey.
   b. State the most frequent score.
   c. Determine the number of people who scored less than 3.
   d. Determine the number of people who scored 3 or more.
   e. Calculate the fraction of the people surveyed who scored 3.

4. Draw a frequency distribution table for each of the following data sets.
   a. Andrew's scores in Mathematics tests this semester are:
   6, 9, 7, 9, 10, 7, 6, 5, 8, 9.
   b. The number of children in each household of a particular street are:
   2, 0, 6, 1, 1, 2, 1, 3, 0, 4, 3, 2, 4, 1, 0, 2, 1, 0, 2, 0.
   c. The weights (in kilograms) of students in a Year 7 class are:
   46, 60, 48, 52, 49, 51, 60, 45, 54, 54, 52, 58, 53, 51,
   54, 50, 50, 56, 53, 57, 55, 48, 56, 53, 58, 53, 59, 57.
   d. The heights of students in a Year 7 class are:
   145, 147, 150, 150, 148, 145, 144, 144, 147, 149, 144, 150, 150, 152,
   145, 149, 144, 145, 147, 143, 144, 145, 148, 144, 149, 146, 148, 143.

5. Rosemary decided to survey the participants of her local gym about their preferred sport. She asked each participant to name one preferred sport. Her results are below:

   hockey, cricket, cricket, tennis, scuba diving, netball, tennis, netball, swimming, netball, tennis, hockey, cricket, lacrosse, lawn bowls, hockey, swimming, netball, tennis, netball, cricket, tennis, hockey, lacrosse, swimming, lawn bowls, swimming, swimming, netball, netball, tennis, golf, hockey, hockey, lacrosse, swimming, golf, hockey, netball, swimming, scuba diving, scuba diving, golf, tennis, cricket, cricket, hockey, lacrosse, netball, golf

   a. State any problems with the way Rosemary has displayed the data.
   b. Organise Rosemary's results into a frequency table to show the participants' preferred sports.
   c. From the frequency table, determine:
      i. the most preferred sport
      ii. the least preferred sport.
   d. State whether any sport(s) have the same frequency.

### Understanding

6. A random sample of 30 families was surveyed to determine the number of children in high school in each family. The following is the raw data collected.
   2, 1, 1, 0, 2, 0, 1, 0, 2, 0, 3, 1, 1, 0, 0,
   0, 1, 4, 1, 0, 0, 1, 2, 1, 2, 0, 3, 2, 0, 1

   a. Organise the data into a frequency distribution table.
   b. Determine the number of families that have no children of high school age.
   c. Determine the number of families that have 2 or more children of high school age.
   d. State the score that has the highest frequency.
   e. From the 30 families surveyed, state the highest number of children in high school from one family.
   f. Calculate the fraction of families that had 2 children of high school age.

7. Draw a frequency table to classify the following data on students' heights. Use a range of values (for example, 140–144, 145–149, and so on) as each category. The values are listed below.

   168 cm, 143 cm, 145 cm, 151 cm, 153 cm, 148 cm, 166 cm, 147 cm, 160 cm, 162 cm,
   175 cm, 168 cm, 143 cm, 150 cm, 160 cm, 180 cm, 146 cm, 158 cm, 149 cm, 169 cm,
   167 cm, 167 cm, 163 cm, 172 cm, 148 cm, 151 cm, 170 cm, 160 cm

8. A real estate agent has listed all the properties sold in the area in the last month. She wants to know what has been the most popular type of property from the list of sold properties:

   2-bedroom house, 4-bedroom house, 3-bedroom house, 2-bedroom unit, 4-bedroom house, 1-bedroom unit, 3-bedroom house, 2-bedroom unit, 3-bedroom house, 1-bedroom unit, 2-bedroom unit, 3-bedroom house, 3-bedroom house, 3-bedroom house, 2-bedroom unit, 1-bedroom unit

   a. Complete a frequency table for the list of properties and work out which type of property was most popular.

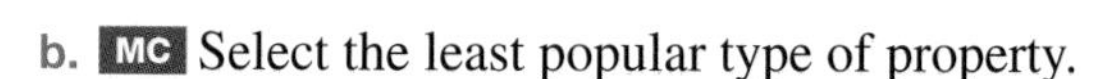

   b. **MC** Select the least popular type of property.

      A. 1-bedroom unit
      B. 2-bedroom unit
      C. 2-bedroom house
      D. 3-bedroom house
      E. 4-bedroom house

   c. **MC** Choose the type of property that is half as popular as a 2 bedroom unit.

      A. 4-bedroom house
      B. 3-bedroom house
      C. 2-bedroom house
      D. 1-bedroom unit
      E. none of these

9. **MC** Select the statement(s) about the frequency column of a frequency table that are true.

   A. It adds up to the total number of categories.
   B. It adds up to the total number of results given.
   C. It adds up to the total of the category values.
   D. It displays the tally.
   E. None of these statements are true.

## Reasoning

10. Explain why tallies are drawn in batches of four vertical lines crossed by a fifth one.

11. Discuss what you need to consider when selecting a class interval for a frequency distribution table.

12. Explain the reasoning behind using class intervals and state the circumstances in which they should be used.

## Problem solving

13. The following frequency distribution table displays the number of pets in the families of the students of a class.

| Number of pets | Frequency | % Frequency |
|---|---|---|
| 0 | 8 | |
| 1 | 11 | |
| 2 | 3 | |
| 3 | 2 | |
| 4 | 0 | |
| 5 | 1 | |

   a. Determine the number of students in the class.
   b. Calculate the percentage frequency for each row of the table.

   Another class containing 25 students creates their own version of this frequency distribution table. When the data in both tables is combined, the percentage frequency of having 1 pet is 34%.

   c. Determine the number of students in the other class with one pet in their family.

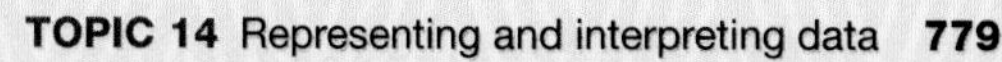

14. This frequency table shows the percentage occurrence of the vowels in a piece of writing. Two pieces of data are missing — the percentage occurrence of the letters O and U. The occurrence of O is 2.6 times that of U. Determine the two missing values in this table. Show your working.

| Vowel | Percentage frequency |
|---|---|
| A | 22.7 |
| E | 27.6 |
| I | 19.9 |
| O | |
| U | |

15. A survey was conducted on 45 households to identify the number of children living in each household. Use the following information to determine the results and complete the table.
    - The number of households with no children is equal to the number of households with 3 children.
    - The number of households with 1 child was 4 times the number of households with no children.
    - The number of households with 2 children was 5 times the number of households with no children.
    - There was one household with 4 children.

| Score | Frequency |
|---|---|
| 0 | |
| 1 | |
| 2 | |
| 3 | |
| 4 | |
| **Total** | |

# LESSON
# 14.4 Measures of centre and spread

## LEARNING INTENTION

At the end of this lesson you should be able to:
- calculate the mean, median and mode of a set of data
- calculate the range of a set of data.

## 14.4.1 Measures of centre

eles-4719

- Measures of centre are statistical values that give us an idea about the centre of a data set. The measures of centre are **mean**, **median** and **mode**.

### The mean

- The mean of a set of data is the average of the values.
- The mean is denoted by the symbol $\bar{x}$ (a lower-case $x$ with a bar on top).
- The average is found by adding up all of the values in the data set, then dividing by the number of values in the set.
- The mean is not necessarily a whole number, or a number that was in the original set of data.

## Calculating the mean

- **The formula for the mean is:**

$$\bar{x} = \frac{\text{sum of data values}}{\text{total number of data values}}$$

- **In most cases the mean is thought of as the centre of a set of data.**

### WORKED EXAMPLE 5 Calculating the mean

**For each of the following sets of data, calculate the mean, $\bar{x}$.**
**Give your answers correct to two decimal places where appropriate.**

a. **5, 5, 6, 4, 8, 3, 4**

b. **0, 0, 0, 1, 1, 1, 1, 1, 4, 4, 4, 4, 5, 5, 5, 7, 7**

| THINK | WRITE |
|---|---|
| a. 1. Calculate the sum of the given values. | a. Sum of data values $= 5+5+6+4+8+3+4$ $= 35$ |
| 2. Count the number of values. | Number of data values $= 7$ |
| 3. Define the rule for the mean. | $\bar{x} = \frac{\text{sum of data values}}{\text{total number of data values}}$ |
| 4. Substitute the known values into the rule and evaluate. | $= \frac{35}{7}$ $= 5$ |
| 5. Write the answer. | The mean (or average) of the set of data is 5. |
| b. 1. Calculate the sum of the given values. Take note of the number of times each value occurs. 0 occurs 3 times $(3 \times 0)$, 1 occurs 5 times $(1 \times 5)$, 4 occurs 4 times $(4 \times 4)$, 5 occurs 3 times $(5 \times 3)$, 7 occurs 2 times $(7 \times 2)$ | b. Sum of data values $= 3\times0+5\times1+4\times4+3\times5+2\times7$ $= 0+5+16+15+14$ $= 50$ |
| 2. Count the number of values. | Number of data values $= 17$ |
| 3. Define the rule for the mean. | $\bar{x} = \frac{\text{sum of data values}}{\text{total number of data values}}$ |
| 4. Substitute the known values into the rule and evaluate. | $= \frac{50}{17}$ $= 2.94117647...$ |
| 5. Round the answer to 2 decimal places. | $= 2.94$ |
| 6. State the answer. | The mean (or average) of the set of data is 2.94. |

## The median

- The median represents the middle score when the data values are ordered from **smallest** to **largest** (also known as 'numerical order' or 'ascending order').

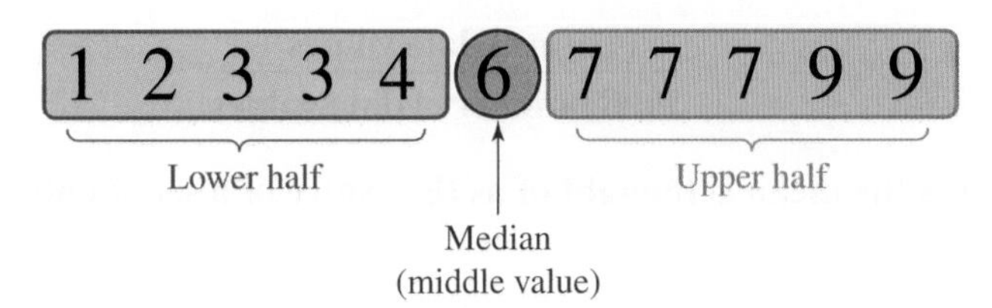


- An equal number of data values can be found below the median and above it.

### Calculating the median

**When determining the median, follow these steps.**

1. **Arrange the data values in ascending order (smallest to largest).**
2. **Locate the middle value (ensure that there are as many values above the median as there are below it).**
3. **If the data set contains an even number of data values there will be two 'middle' numbers. The median is the average of these numbers.**

**Alternatively:**

**The *position* of the median is the $\left(\frac{n+1}{2}\right)$ th data value, where $n$ is the total number of data values.**

### WORKED EXAMPLE 6 Determining the median

**Determine the middle value (median) for the following sets of data.**

**a. 5, 4, 2, 6, 3, 4, 5, 7, 4, 8, 5, 5, 6, 7, 5**

**b. 8, 2, 5, 4, 9, 9, 7, 3, 2, 9, 3, 7, 6, 8**

| THINK | WRITE |
|---|---|
| a. 1. Arrange the data values in ascending order (smallest to largest). | a. 2, 3, 4, 4, 4, 5, 5, 5, 5, 5, 6, 6, 7, 7, 8 |
| 2. Locate the middle value by inserting a vertical line that splits the data into two equal sections. The pink line inserted through the third number 5 splits the data into two equal sections that each have 7 data values. | 2, 3, 4, 4, 4, 5, 5, \|5, 5, 5, 6, 6, 7, 7, 8 |
| 3. Write the answer. | The median of the data set is 5. |
| b. 1. Arrange the data values in ascending order (smallest to largest). | b. 2, 2, 3, 3, 4, 5, 6, 7, 7, 8, 8, 9, 9, 9 |
| 2. Locate the middle value by inserting a vertical line that splits the data into two equal sections. The pink line inserted in between the 6 and 7 splits the data into two equal sections that each have 6 data values. | 2, 2, 3, 3, 4, 5, 6, \| 7, 7, 8, 8, 9, 9, 9 |

3. Since the median (or middle position) occurs between 6 and 7, we need to determine the average of these two numbers. The median of the data lies directly between 6 and 7.

$$\text{Median} = \frac{6+7}{2} = \frac{13}{2} = 6.5$$

4. The number directly between 6 and 7 is 6.5. Write the answer.

The median of the data set is 6.5.

**Alternative method:**

**a.** 1. Arrange the data values in ascending order (smallest to largest).

**a.** 2, 3, 4, 4, 4, 5, 5, 5, 5, 5, 6, 6, 7, 7, 8

2. Count the number of data values in the set.

$n = 15$

3. Determine the $\left(\frac{n+1}{2}\right)^{\text{th}}$ data value.

$$\left(\frac{n+1}{2}\right) = \frac{15+1}{2} = 8$$

4. The median is located at the $\left(\frac{n+1}{2}\right)^{\text{th}}$ data value.

The median is the 8th value in the set of data. The 8th number in the set of data is 5.

5. Write the answer.

The median of the data set is 5.

**b.** 1. Arrange the data values in ascending order (smallest to largest).

**b.** 2, 2, 3, 3, 4, 5, 6, 7, 7, 8, 8, 9, 9, 9

2. Count the number of data values in the set.

$n = 14$

3. Determine the $\left(\frac{n+1}{2}\right)^{\text{th}}$ data value.

$$\left(\frac{n+1}{2}\right) = \frac{14+1}{2} = 7.5$$

4. The two middle values are the 7th and 8th values.

The middle values are 6 and 7

5. Calculate the average of the middle values.

$$\frac{6+7}{2} = \frac{13}{2} = 6.5$$

6. Write the answer.

The median of the data set is 6.5.

- As the median is the middle value, it is a common measure of centre, especially for large data sets or data sets where the average is skewed by a small number of extreme (outlier) values.

## The mode

- The mode is the most common value in a set of data. It is the value that occurs most frequently.
- Some sets of data have more than one mode.
- Some sets of data have no mode at all.

### Calculating the mode

**When determining the mode:**

1. **arrange the data values in ascending order (smallest to largest) — this step is optional but does help**
2. **look for the value that occurs most often (the one that has the highest frequency).**

### Important notes about the mode

- **If a set of data has no mode, your answer should be the words 'no mode', not '0'. This is what would happen when no value occurs more than once in a set of data.**
  **For example, in the set of data 1, 2, 3, 4, 6, 7, 10, 12, 13, no value is repeated, so there is no mode.**
- **If a set of data has one mode, write this value as your answer.**
  **For example, for the set of data 15, 31, 12, 47, 21, 65, 12, the mode is 12 because it appears twice — each other value only appears once.**
- **If a set of data has more than one mode, write all of the values that occur the most as your answer.**
  **For example, for the set of data 1, 1, 2, 2, 2, 4, 5, 6, 6, 6, 9, 9, 12, 12, 12, the modes are 2, 6 and 12.**

### WORKED EXAMPLE 7 Identifying the mode

**Identify the mode of the following scores.**

a. **2, 3, 4, 5, 5, 6, 6, 7, 8, 8, 8, 9**
b. **12, 18, 5, 17, 3, 5, 2, 10, 12**
c. **42, 29, 11, 28, 21**

| THINK | WRITE |
|---|---|
| a. 1. Look at the set of data and highlight any values that have been repeated. | a. 2, 3, 4, 5, 5, 6, 6, 7, 8, 8, 8, 9 |
| 2. Choose the value that has been repeated the most. | The numbers 5 and 6 occur twice. However, the number 8 occurs three times. |
| 3. Write the answer. | The mode for the given set of values is 8. |
| b. 1. Look at the set of data and highlight any values that have been repeated. | b. 12, 18, 5, 17, 3, 5, 2, 10, 12 |
| 2. Choose the value(s) that have been repeated the most. | The number 5 occurs twice. The number 12 occurs twice. |
| 3. Write the answer. | The modes for the given set of values are 5 and 12. |
| c. 1. Look at the set of data and highlight any value(s) that have been repeated. | c. 42, 29, 11, 28, 21 |
| 2. Write the answer. | No values have been repeated. The set of data has no mode, since each of the numbers occurs once only. |

- The mode is generally not used as a measure of centre, rather as a measure of the most common, or popular, value.

### COLLABORATIVE TASK: Which average?

In this activity you will investigate how the mean, the median and the mode are affected by the size of a data set. You will also determine which of these is the most useful value.

1. Working in groups of 4 or more, list the number of pets each student has. This will be the data set for your group.
2. Calculate the mean, the median and the mode of your data set. Explain which measure of centre best describes the number of pets that people in your group own.

3. Compare your results with the results found by another group.
4. If you were to combine your data set with the other group's data set, predict what would happen to the mean, the median and the mode.
5. Combine the two sets of data and find the mean, the median and the mode of this new data set.
   a. Did the new values you calculated agree with your prediction? Discuss your findings.
   b. Explain which of the new measures of centre best describes the number of pets that people in the combined group own.
6. Combine your latest set of data with another group's data and repeat your calculations for the mean, the median and the mode.

7. Your teacher will now record the data for the whole class on the board. Calculate the mean, the median and the mode for the whole class.
8. Discuss the results. Compare the values you found for the mean of each different-sized data set. Do the same for the median and the mode.
9. If you were to rely on just one of these measures of centre to describe the number of pets that people in your class own, determine which one you would choose. Discuss your reasons.

## 14.4.2 Measures of spread

eles-4720

- A measure of spread indicates how far data values are spread from the centre, or from each other.
- Looking at the following two graphical displays, the pink graph on the left is more 'spread out' compared to the green graph on the right. That is, the pink graph has a larger spread (from 0 to 22) compared to the green graph (from 6 to 18).
- There are several measures of spread, but at this stage we will only discuss the **range**.
- The range of a set of values is the difference between the largest and smallest values.

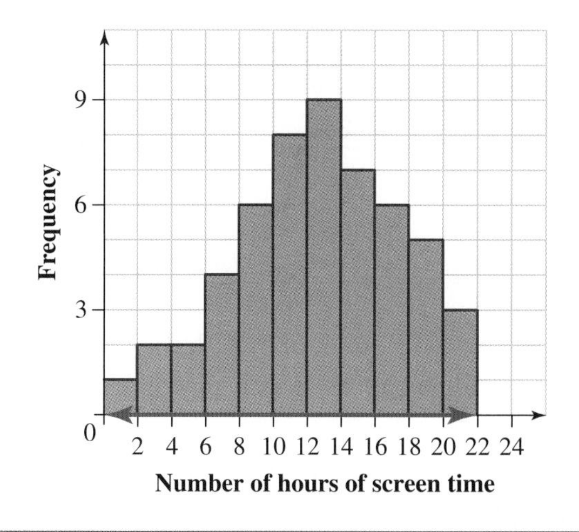


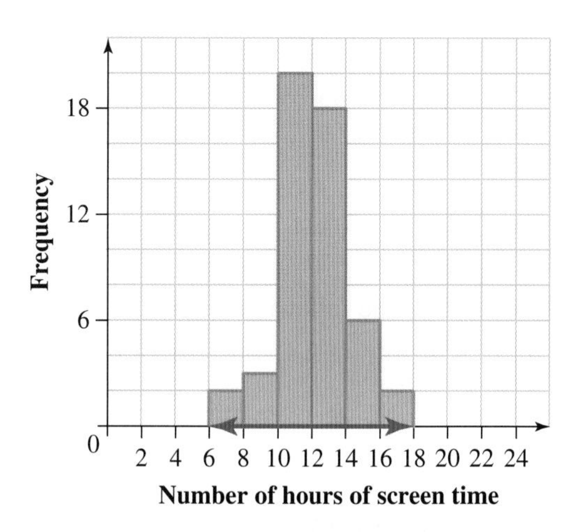


### Calculating the range

**The range of a set of data is the difference between the largest and smallest values in that set.**

**Range = (largest data value) – (smallest data value)**

## WORKED EXAMPLE 8 Determining the range

**Determine the range of the following data.**

**12, 76, 35, 29, 16, 45, 56**

| THINK | WRITE |
|---|---|
| 1. Identify the largest and smallest values. | Largest value = 76<br>Smallest value = 12 |
| 2. State the formula for the range and substitute the values in to evaluate it. | Range = largest value − smallest value<br>= 76 − 12<br>= 64 |
| 3. Write the answer. | The range of the data set is 64. |

### Resources

**eWorkbook** Topic 14 Workbook (worksheets, code puzzle and project) (ewbk-1915)

**Interactivities** Individual pathway interactivity: Measures of centre and spread (int-4380)
Measures of centre (int-4052)
Mean (int-4053)
Median (int-4054)
Mode (int-4055)
Range (int-4056)

## Exercise 14.4 Measures of centre and spread

**14.4 Quick quiz**  **14.4 Exercise**

### Individual pathways

| ■ PRACTISE | ■ CONSOLIDATE | ■ MASTER |
|---|---|---|
| 1, 4, 8, 10, 11, 15, 16, 21, 22, 27, 28, 29 | 2, 5, 6, 9, 12, 17, 18, 23, 24, 30, 31, 32 | 3, 7, 13, 14, 19, 20, 25, 26, 33, 34, 35 |

### Fluency

1. **WE5** For each of the following sets of data, calculate the mean.
   a. 3, 4, 5, 5, 6, 7
   b. 5, 6, 7, 5, 5, 8
   c. 4, 6, 5, 4, 2, 3
   d. 3, 5, 6, 8, 7, 7
   e. 5, 4, 4, 6, 2, 3
2. For each of the following sets of data, calculate the mean.
   a. 2, 2, 2, 4, 3, 5
   b. 12, 10, 13, 12, 11, 14
   c. 11, 12, 15, 17, 18, 11
   d. 12, 15, 16, 17, 15, 15
   e. 10, 14, 12, 12, 16, 14
3. For each of the following sets of data, calculate the mean.
   *Hint:* Use the grouping of values to help you.
   a. 2, 2, 2, 4, 4, 4
   b. 1, 2, 2, 4, 4, 5
   c. 1, 2, 2, 5, 5, 6, 7
   d. 9, 9, 8, 8, 7, 1, 1, 1, 1
   e. 3, 3, 3, 1, 1, 1, 2, 2, 2
   f. 2, 2, 2, 3, 3, 3, 3, 6

4. WE6a Determine the median (middle value) for the following sets of data.
   a. 3, 3, 4, 5, 5, 6, 7
   b. 1, 2, 2, 3, 4, 8, 9
   c. 2, 2, 2, 3, 3, 4, 5
   d. 5, 5, 6, 6, 7, 7, 8, 9, 9
5. Determine the median (middle value) for the following sets of data.
   a. 7, 7, 7, 10, 11, 12, 15, 15, 16
   b. 4, 3, 5, 3, 4, 4, 3, 5, 4
   c. 1, 2.5, 5, 3.4, 1, 2.4, 5
   d. 1.2, 1.5, 1.4, 1.8, 1.9
6. WE6b Determine the median (middle value) for the following sets of data.
   a. 1, 1, 2, 2, 4, 4
   b. 4, 5, 5, 5, 6, 7
   c. 4, 5, 7, 7, 8, 9
   d. 1, 2, 2, 3, 3, 4
7. Determine the median (middle value) for the following sets of data.
   a. 2, 4, 4, 6, 8, 9
   b. 1, 5, 7, 8
   c. 2, 4, 5, 7, 8, 8, 9, 9
   d. 1, 5, 7, 8, 10, 15
8. WE7 Identify the mode for each of the following sets of data.
   a. 3, 3, 4, 4, 4, 5, 6
   b. 2, 9, 8, 8, 4, 5
   c. 1, 1, 2, 2, 2, 3
   d. 4, 6, 4, 2, 7, 3
   e. 2, 4, 3, 6, 2, 4, 2
9. Identify the mode for each of the following sets of data.
   a. 4, 8, 8, 3, 3, 4, 3, 3
   b. 6, 2, 12, 10, 14, 14, 10, 12, 10, 12, 10, 12, 10, 12
   c. 7, 9, 4, 6, 26, 71, 3, 3, 3, 2, 4, 6, 4, 25, 4
   d. 2, 2, 3, 4, 4, 9, 9, 9, 6
   e. 3, 7, 4, 5, 22
10. WE8 Determine the range of the following sets of data:
   a. 15, 26, 6, 38, 10, 17
   b. 12.8, 21.5, 1.9, 12.0, 25.4, 2.8, 1.3

### Understanding

11. MC Select the method(s) that can be used to calculate the mean for the data 5, 5, 6, 7, 2.
   A. Adding all the results and multiplying by the number of results
   B. Adding all the results and dividing by the number of results
   C. Adding all the results
   D. Choosing the middle result
   E. Ordering the results, then choosing the middle result
12. MC Select statement(s) that are true about calculating the mean of a set of data.
   A. Zeroes do not matter at all.
   B. Zeroes must be counted in the number of results.
   C. Zeroes must be added to the total as they will change it.
   D. Zeroes will make the mean zero.
   E. None of these are true.
13. MC Choose the statement that is true about the set of data 2.6, 2.8, 3.1, 3.7, 4.0, 4.2.
   A. The mean value for the data will be above 4.2.
   B. The mean value for the data will be below 2.6.
   C. The mean value for the data will be between 2.6 and 3.0.
   D. The mean value for the data will be between 3.0 and 4.0.
   E. The mean value for the data will be between 4.0 and 4.2.

14. MC Choose the correct statement.

A. The mean, median and mode for any set of data will always be the same value.
B. The mean, median and mode for any set of data will never be the same value.
C. The mean, median and mode for any set of data must always be close in value.
D. The mean, median and mode for any set of data are sometimes close in value.
E. None of these statements are true.

15. MC Select the range of the set of numbers: 16, 33, 24, 48, 11, 30, 15

A. 48 B. 59 C. 37 D. 20 E. 11

16. Eleanor wanted to know what her Mathematics test average was. Eleanor's teacher told her that she had used the mean of Eleanor's test results to calculate the end-of-year mark. Eleanor's test results (written as percentages) are:

89, 87, 78, 75, 89, 94, 82, 93, 78

Calculate Eleanor's mean Mathematics test score.

17. A cricketer had scores of 14, 52, 35, 42 and 47 in her last 5 innings. Determine her mean score.

18. Eliza works in a shoe factory as a shoe inspector. The number of shoes Eliza inspected in an hour was counted over a number of days' work. The following list contains the results.

105, 102, 105, 106, 103, 105, 105,
102, 108, 110, 102, 103, 106, 107,
108, 102, 105, 106, 105, 104, 102,
99, 98, 105, 102, 101, 97, 100

Calculate the mean number of shoes checked by Eliza in one hour. Round your answer to the nearest whole number.

19. The number of students in the canteen each lunchtime was surveyed for 2 weeks. The following are the results.

52, 45, 41, 42, 53, 45, 47, 32, 52, 56

Calculate the mean number of students in the canteen at lunchtime in that fortnight. Round your answer to the nearest whole number.

20. Answer the following questions.

a. Explain what 'the mean number of goals scored in a soccer match' means.
b. Explain what 'the median house price' means.
c. Explain what 'the mode of shoe-size data' means.

### Reasoning

21. Suggest why we summarise data by calculating measures of centre and spread.

22. The local football team has been doing very well. They want to advertise their average score to attract new club members. You suggest that they use the mean of their past season's game scores. They ask you to find it out for them. Their results are listed below (scores are given as total points for each game).

110, 112, 141, 114, 112, 114, 95, 75, 58, 115, 116, 115, 75,
114, 78, 96, 78, 115, 112, 115, 102, 75, 79, 154, 117, 62

a. Calculate the team's mean score. Show your working.
b. Identify whether the mode or median is the 'better' value to use to attract new club members. Justify your answer.

23. A clothing company wanted to know the size of jeans that should be manufactured in the largest quantities. A number of shoppers were surveyed and asked their jeans size. The results are given below.

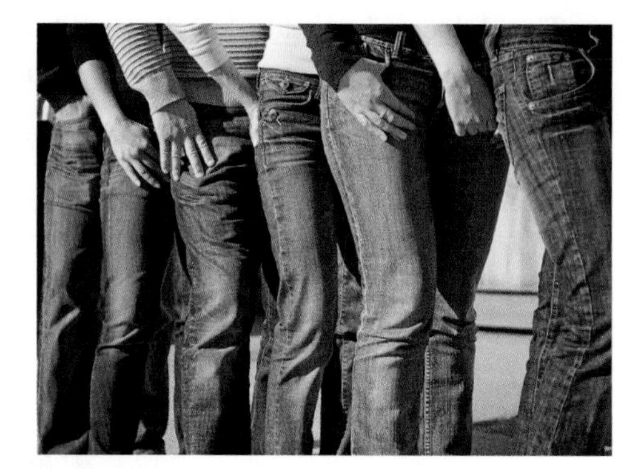

13, 12, 14, 12, 15, 16, 14, 12, 15, 14, 12,
14, 13, 14, 11, 10, 12, 13, 14, 14, 10, 12,
14, 12, 12, 10, 8, 16, 17, 12, 11, 13, 12,
15, 14, 12, 17, 8, 16, 11, 12, 13, 12, 12

a. Determine the mode of the data.
b. Explain why the company would be more interested in the mode than the mean or median values.

24. Explain whether we are able to calculate the mean, median, mode and range of categorical data.

25. Tom thinks that the petrol station where he buys his petrol is cheaper than the one where his friend Sarah buys her petrol. They begin to keep a daily watch on the prices for 4 weeks and record the following prices (in dollars per litre).

Tom: 1.32, 1.37, 1.39, 1.39, 1.40, 1.42, 1.41, 1.38, 1.34, 1.30, 1.29, 1.28, 1.27, 1.26, 1.25, 1.25, 1.24, 1.25, 1.24, 1.24, 1.25, 1.25, 1.26, 1.27, 1.28, 1.28, 1.30, 1.32

Sarah: 1.41, 1.37, 1.32, 1.31, 1.29, 1.27, 1.24, 1.22, 1.21, 1.21, 1.20, 1.19, 1.20, 1.21, 1.22, 1.23, 1.24, 1.24, 1.25, 1.26, 1.27, 1.28, 1.28, 1.28, 1.29, 1.30, 1.32, 1.31

a. Calculate the mean petrol prices for Tom and Sarah.
b. Identify the petrol station that sells cheaper petrol on average. Justify your answer.
c. Explain why Tom might have been wrong about thinking his petrol station was cheaper.

26. Jennifer wants to make sure the mean height of her jump in the high jump for 10 jumps is over 1.80 metres.
a. Calculate her current mean if her jumps so far have been (in metres) 1.53, 1.78, 1.89, 1.82, 1.53, 1.81, 1.75, 1.86, 1.82. Show your working.
b. Determine what height she needs to jump on the tenth jump to achieve a mean of 1.80. Show your working.
c. Discuss whether this is likely, given her past results.

### Problem solving

27. Kim has an average (mean) score of 72 in Scrabble. He has played 6 games. Determine what he must score in the next game to keep the same average.

28. Peter has calculated his mean score for history to be 89%, based on 5 tests. If he scores 92% in the sixth test, calculate his new mean score.

29. The club coach at a local cycling track was overheard saying that he thought at least half of the cyclists using the track were cycling at speeds of 30 km/h or more. The speeds (in km/h) of the club cyclists are listed as follows.

31, 22, 40, 12, 26, 39, 49, 23, 24, 38, 27, 16, 25, 37, 19, 25, 45, 23, 17, 20, 34, 19, 24, 15, 40, 39, 11, 29, 33, 44, 29, 50, 18, 22, 51, 24, 19, 20, 30, 40, 49, 29, 17, 25, 37, 25, 18, 34, 21, 20, 18

Discuss whether the coach's statement is correct.

30. Identify five whole numbers that have a mean of 10 and a median of 12.

31. The mean of five different test scores is 15. All test scores are whole numbers. Determine the largest and smallest possible test scores, given that the median test score is 12.

32. Gavin records the amount of rainfall in millimetres each day over a two-week period. Gavin's results are:

11, 24, 0, 6, 15, 0, 0, 0, 12, 0, 0, 127, 15, 0.

a. Determine the mean rainfall for the two-week period.
b. Determine the median rainfall.
c. Identify the mode of the rainfall.
d. State which of the mean, median and mode is the best measure of the typical rainfall. Explain your choice.

33. A group of 3 children have a mean height of 142 cm. The middle height is the same as the mean. The tallest child leaves the group, and is replaced by a child with the same height as the shortest child.
The mean height of this group of three children is now 136 cm. Calculate the heights of the 4 children. Explain how you got your answer.

34. The mean of 5 different test scores is 10. All test scores are whole numbers. Determine the largest and smallest possible values for the median.

35. The mean of 9 different test scores ranging from 0 to 100 is 85. The median is 80. All test scores are whole numbers. Determine the greatest possible range between the highest and lowest possible test scores.

# LESSON
# 14.5 Column graphs and dot plots

## LEARNING INTENTION

At the end of this lesson you should be able to:
- display data as a column or bar graph
- display data as a dot plot
- analyse and interpret column and bar graphs and dot plots.

## 14.5.1 Column and bar graphs

eles-4721

### Features of graphs

- Graphs are a useful way of displaying data, or numerical information. TV, websites, newspapers and magazines frequently display data as graphs.
- All graphs should have the following features:
  - a *title* — to tell us what the graph is about
  - clear *labels* for the axes — to explain what is being shown
  - *evenly scaled axes* — if the graph has numerical axes, they must have a scale that is constant for the length of the axes, with a clear indication of the units being used
  - **legends** — these are not always necessary, but they are necessary when any symbols or colours are used to illustrate some element of the graph.
- Different graphical displays are used depending whether the data is categorical or numerical.

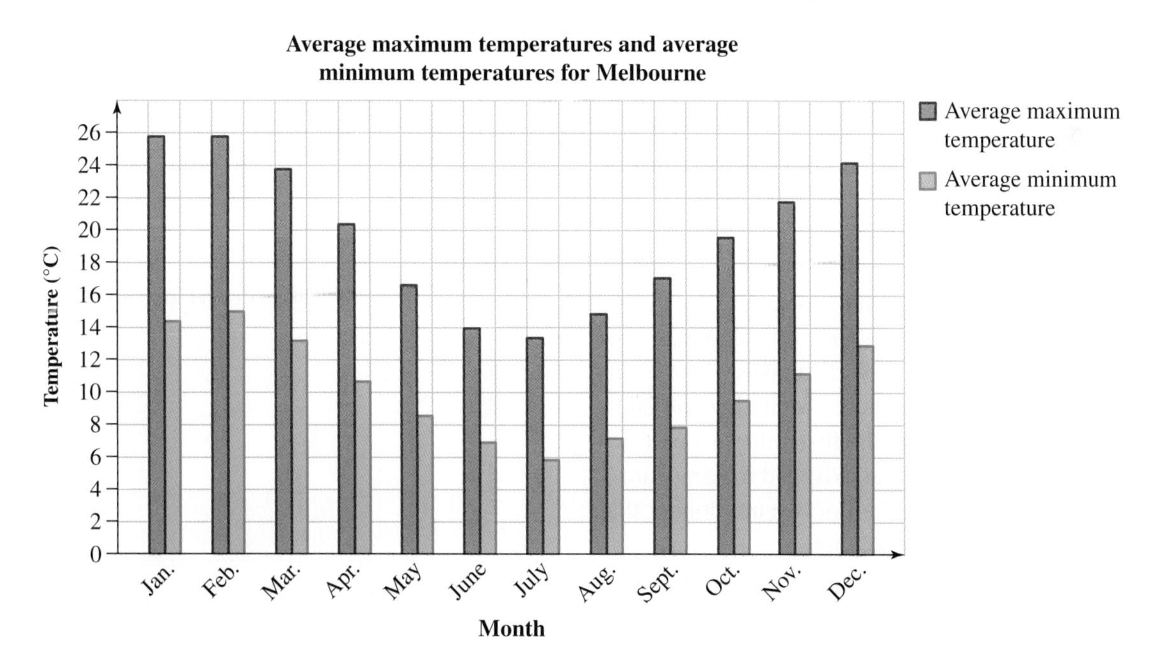


### Using column graphs and bar graphs

- Column and bar graphs can be used to represent categorical data.
- The frequency for each category determines the length of the bar or the height of the column.
- It is easiest to graph the data from a frequency table.
- The advantage of column and bar graphs is that they are usually easy to prepare and interpret.

- The disadvantages of column and bar graphs are:
  - if they are displaying a large number of categories they become difficult to understand
  - they don't provide a visualisation of what proportion each category is of the whole.

## Column graphs

- **Column graphs** should be presented on graph paper and have:
  - a title
  - labelled axes that are clearly and evenly scaled
  - columns of the same width
  - an even gap between each column
  - a space between the vertical axis and the first column.

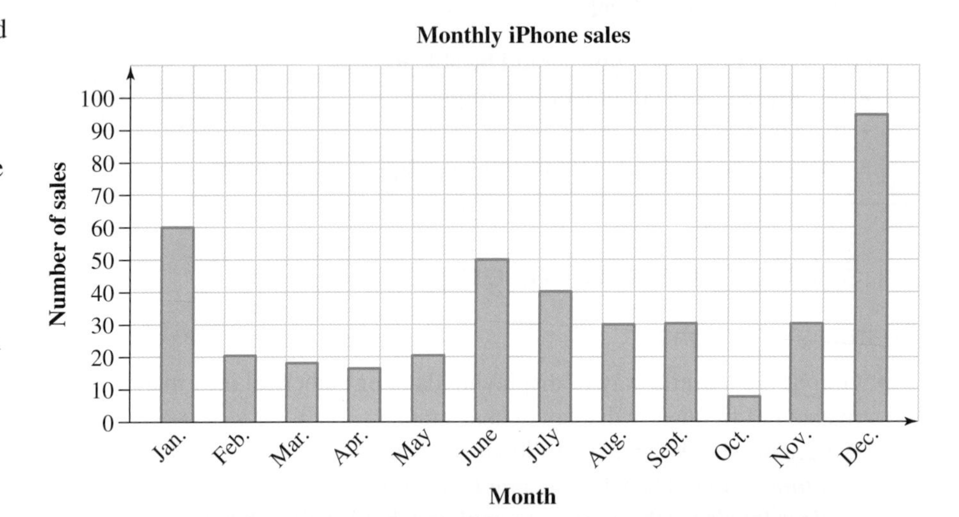


### WORKED EXAMPLE 9 Constructing a column graph from a frequency table

**Beth surveyed the students in her class to find out their preferences for the school uniform. Her results are shown in the following table. Construct a column graph to display the results.**

| Type of uniform | Tally | Frequency |
|---|---|---|
| White shirt and black skirt/trousers | 𝍸 III | 8 |
| Blue shirt and black skirt/trousers | IIII | 4 |
| Blue shirt and navy skirt/trousers | 𝍸 𝍸 II | 12 |
| White shirt and navy skirt/trousers | 𝍸 | 5 |
| | Total | 29 |

**THINK**

1. Rule a set of axes on graph paper. Give the graph a title. Label the horizontal and vertical axes.
2. Scale the horizontal and vertical axes.
   *Hint:* Leave an interval at the beginning and end of the graph by placing the first column a half or one unit from the vertical axis, and making sure the horizontal axis is a half or one unit longer than needed to fit all of the columns.
3. Draw the first column so that it reaches a vertical height corresponding to 8 people. Label the section of the axis below the column as 'White shirt and black skirt/trousers'.
4. Leave a gap (measuring one column width) between the first column and the second column.
5. Repeat steps **3** and **4** for each of the remaining uniform types.

**DRAW**

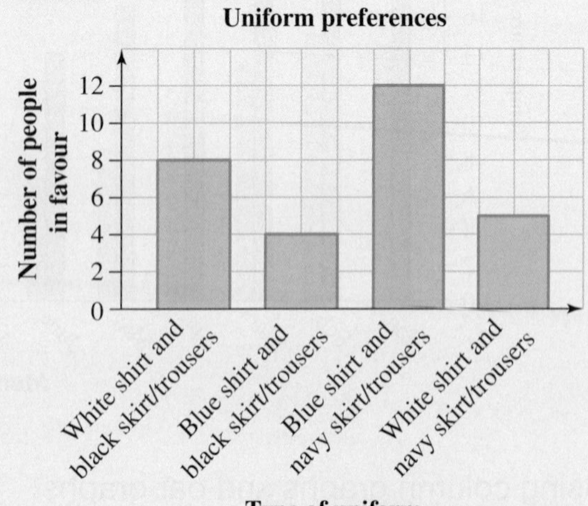

## Bar graphs

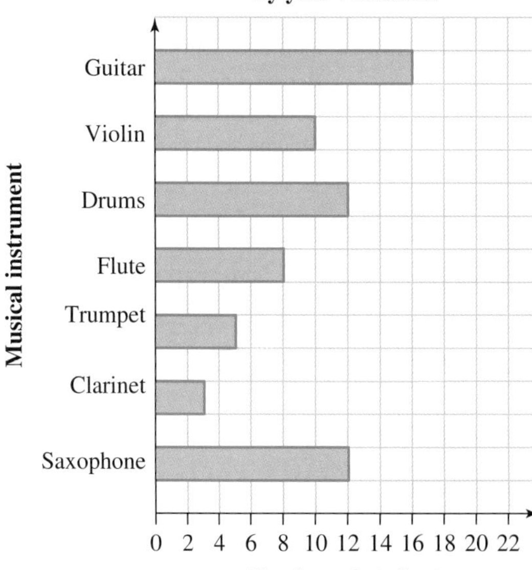


- Column graphs that are drawn horizontally are known as **bar graphs**. Bar graphs can be used when category names are too long to fit at the bottom of a column.
- Bar graphs should have the same features as a column graph; that is:
  - a title
  - labelled and evenly scaled axes
  - bars of the same width
  - even gaps between bars.

## 14.5.2 Dot plots

eles-4723

- **Dot plots** are usually used to represent numerical data. They use dots on a number line to show individual data values.
- If a data value appears more than once, the dots are stacked vertically.
- Dot plots give a quick overview of distribution. They show **clustering** (groups of dots) and **outliers** (values that are extremely small or large compared to the majority of values).
- The dot plot shown here displays the number of hours of screen time each day for a group of people. Each dot represents the number of hours of screen time per day for each person. Two people had 6 hours of screen time on a particular day.

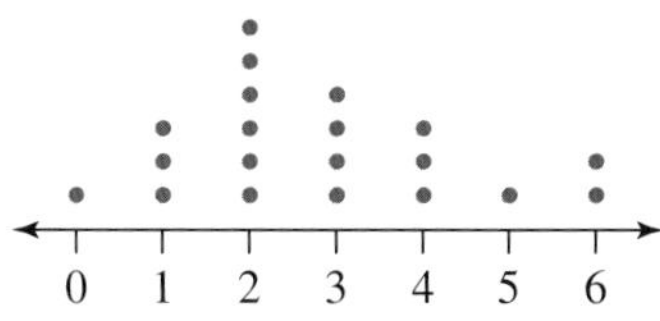


### WORKED EXAMPLE 10 Constructing a dot plot

**Over a 2-week period, the number of packets of potato chips sold from a vending machine each day was recorded. The results are listed below.**

**10, 8, 12, 11, 12, 18, 13, 11, 12, 11, 12, 12, 13, 14**

a. **Draw a dot plot of the data.**
b. **Comment on the distribution.**

| THINK | WRITE |
|---|---|
| a. 1. Use a number line to include the full range of data recorded.<br>2. Place a dot above the number for each value recorded. | a. 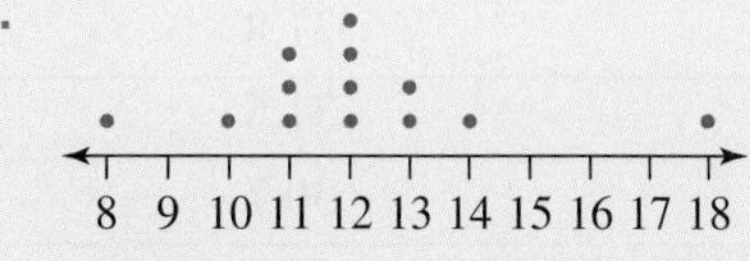  |
| b. Comment on interesting features of the dot plot, for example the range, clustering, extreme values or any practical information about the situation. | b. The scores on this dot plot extend from 8 to 18. That gives a range of 10. On most days between 11 to 13 packets of chips were sold. There were very few days on which 8 and 18 packets of chips were sold. Making sure that there are enough chips to sell 20 packets per day should cover even the most extreme demand for chips from this machine. |

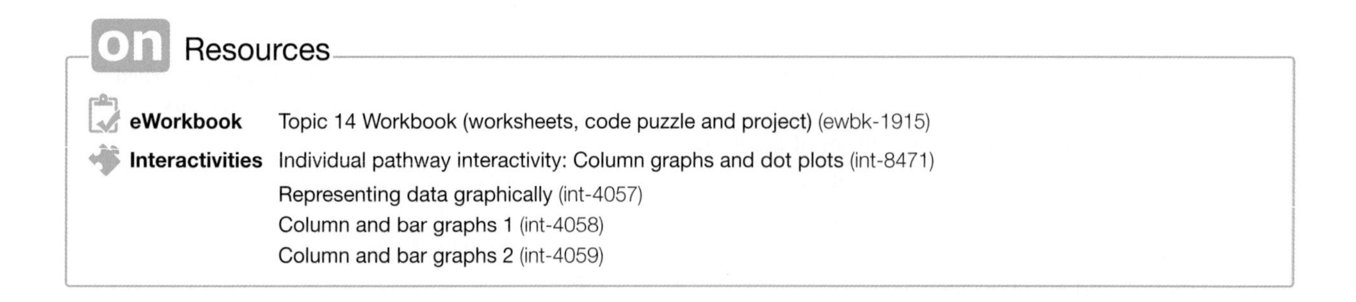


## Exercise 14.5 Column graphs and dot plots

learnon

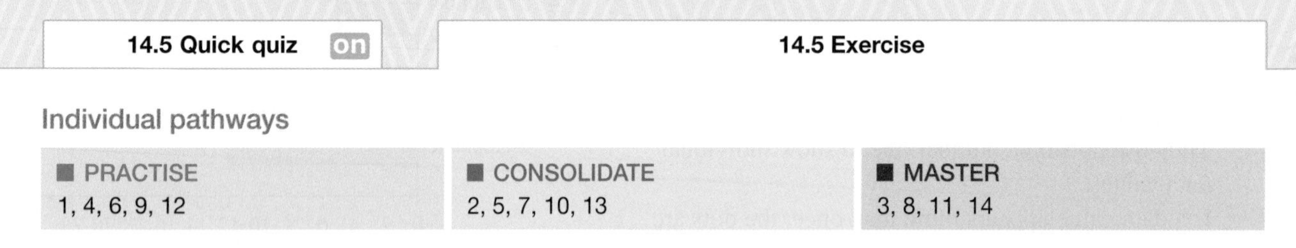


### Fluency

1. **WE9** Beth surveyed the students in her class to find out their method of travelling to school. Her results are shown in the following table. Construct a column graph to display the data.

| Transport | Tally | Frequency |
|---|---|---|
| Car | 𝍸 𝍸 𝍸 | 15 |
| Tram | 𝍸 IIII | 9 |
| Train | 𝍸 𝍸 𝍸 III | 18 |
| Bus | 𝍸 III | 8 |
| Bicycle | III | 3 |
| | **Total** | 53 |

2. Construct a column graph to display the data in the table shown, displaying the mean daily maximum temperatures for each month in Cairns, Queensland.

| Month | Mean daily maximum temperature (°C) |
|---|---|
| January | 31.8 |
| February | 31.5 |
| March | 30.7 |
| April | 29.3 |
| May | 27.6 |
| June | 25.9 |
| July | 25.6 |
| August | 26.4 |
| September | 27.9 |
| October | 29.7 |
| November | 30.8 |
| December | 31.8 |

3. The data in the table displays the number of students absent from school each day in a fortnight. Construct a bar graph to display the data.

| Day | Number of students absent |
|---|---|
| Monday | 15 |
| Tuesday | 17 |
| Wednesday | 20 |
| Thursday | 10 |
| Friday | 14 |
| Monday | 16 |
| Tuesday | 14 |
| Wednesday | 12 |
| Thursday | 5 |
| Friday | 14 |

4. WE10 Over a 2-week period, the number of packets of potato chips sold from a vending machine each day was recorded as follows: 15, 17, 18, 18, 14, 16, 17, 6, 16, 18, 16, 16, 20, 18.

a. Draw a dot plot of the data.
b. Comment on the distribution.

5. Draw a dot plot for each of the following sets of data.

a. 2, 0, 5, 1, 3, 3, 2, 1, 2, 3
b. 18, 22, 20, 19, 20, 21, 19, 20, 21
c. 5.2, 5.5, 5.0, 5.8, 5.3, 5.2, 5.6, 5.3, 6.0, 5.5, 5.6
d. 49, 52, 60, 55, 57, 60, 52, 66, 49, 53, 61, 57, 66, 62, 64, 48, 51, 60

## Understanding

6. Phone bills often include a graph showing your previous bill totals. Use the phone bills column graph shown to answer the following questions.

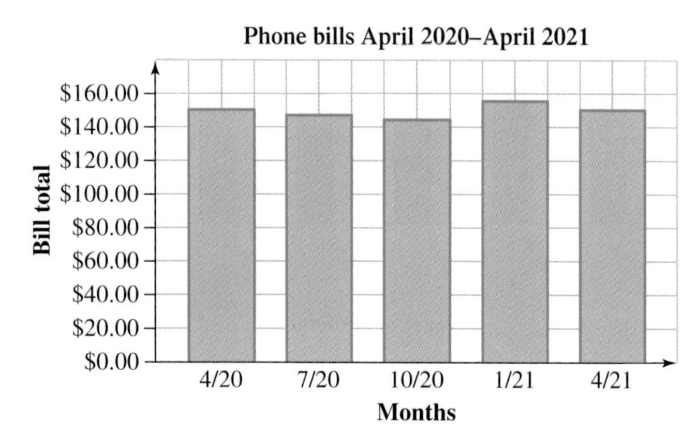


a. State the horizontal axis label.
b. State the vertical axis label.
c. Determine how often this person receives a phone bill.
d. State in which month their phone bill was the highest.
e. Explain why it would be useful to receive a graph like this with your phone bill.
f. If the next bill was for $240.09, explain whether this would be expected.

7. An apple grower records his sales for a 12-week period.

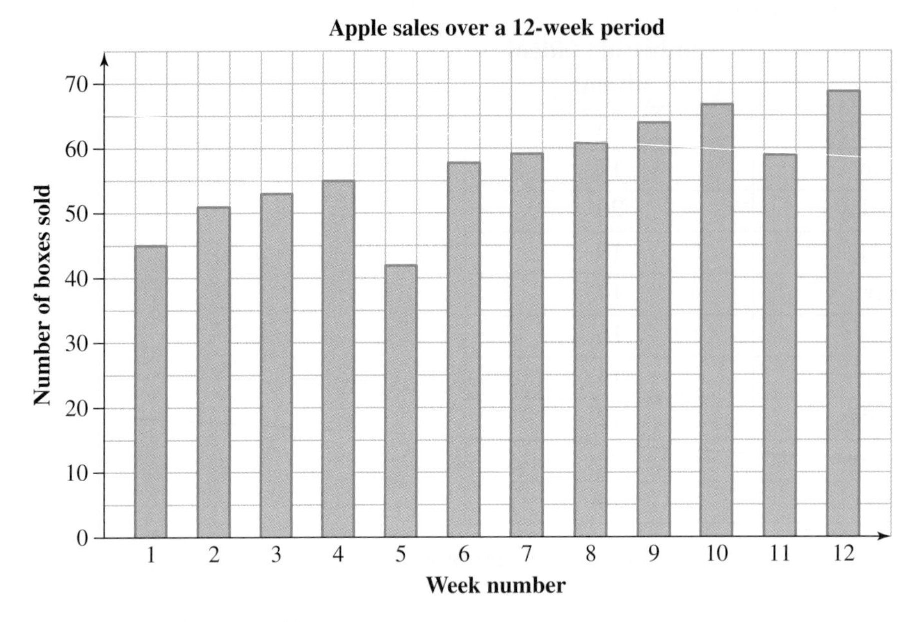


a. Determine the number of boxes sold:

i. in the first week  ii. in the fifth week  iii. in the eighth week.

b. The values for some weeks might be unusual. Determine which values these ones are.

c. Does the graph indicate that apple sales are improving? Explain your answer.

8. The students in a Year 7 physical education class were asked to sprint for 10 seconds. The teacher recorded their results on 2 different days. The graph shown displays the results collected.

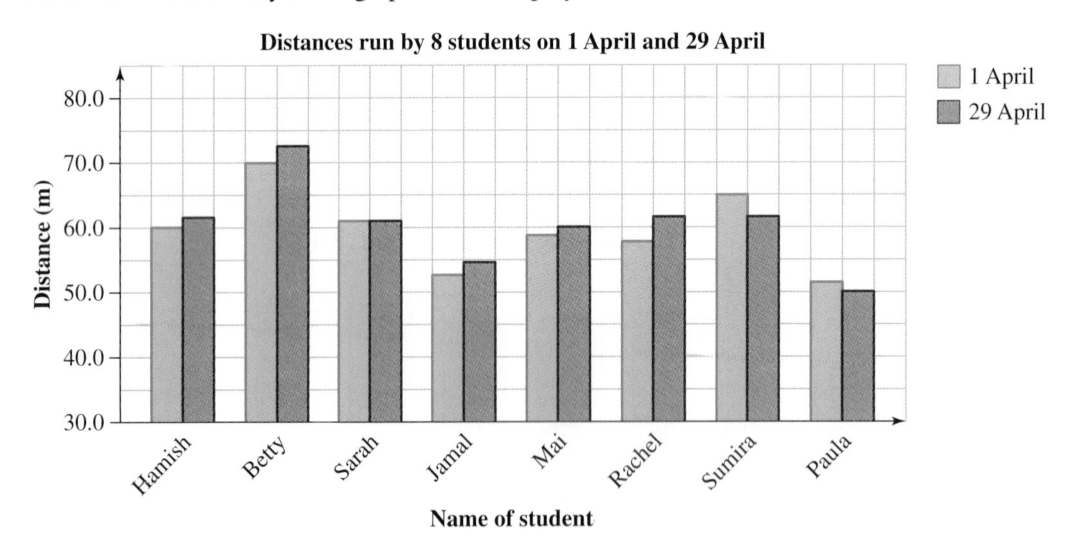


a. State who was the fastest student on each day.

b. Identify how far the fastest student ran on each day.

c. Determine who improved the most.

d. Determine which students did not improve, if any.

e. Explain whether this graph could be misleading in any way.

f. Why might the graph's vertical axis start at 30 m?

## Reasoning

9. Explain why it is important for the data points in a dot plot to be neatly aligned above the value they are representing.

10. Explain why column and bar graphs are more often used to represent categorical data rather than numerical data.

11. Explain and give an example of the effect that outliers in a set of data have on the:

    a. mean
    b. median
    c. mode
    d. range.

## Problem solving

12. Loic surveys his class to determine which eye colour is most common and obtains the following results.
    Blue, Brown, Brown, Green, Brown, Blue, Hazel, Brown, Brown, Hazel, Brown, Brown,
    Blue, Brown,Brown, Brown, Hazel, Brown, Brown, Blue, Brown, Blue, Blue

    a. Represent this data as a bar graph.
    b. Determine the median of this sample.
    c. Identify the mode of this sample.
    d. If a student is selected at random from the class, calculate the probability that they don't have brown eyes.

13. Ten randomly chosen students from Class A and Class B each sit for a test in which the highest possible mark is 10. The results of the ten students from the two classes are:

| Class A | 1 | 2 | 3 | 4 | 5 | 6 | 7 | 8 | 9 | 10 |
|---|---|---|---|---|---|---|---|---|---|---|
| Class B | 1 | 2 | 2 | 3 | 3 | 4 | 4 | 5 | 9 | 10 |

    a. Display the data on a dot plot.
    b. Calculate measures of centre and spread.
    c. Explain any similarities or differences between the results of the two classes.

14. A new whitegoods store opened up in town five weeks ago. The column graph shows the number of fridges they have sold over their first five weeks.

    a. Identify the number of fridges they sold in week 2.
    b. Describe how the number of fridges sold changed from week to week.
    c. Write a formula to describe the number of fridges sold each week over the first five weeks. Let $F =$ the number of fridges sold, and $w =$ the week number.
    d. Using your answer to part **b** or **c**, predict the number of fridges they will sell in week 10.

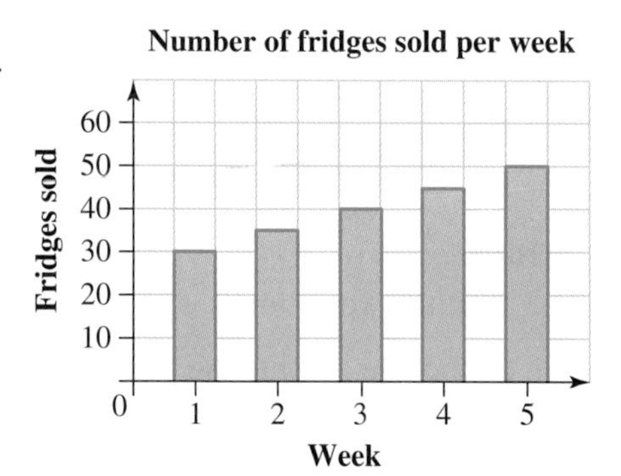

# LESSON
# 14.6 Stem-and-leaf plots

## LEARNING INTENTION

At the end of this lesson you should be able to:
- display data as a stem-and-leaf plot
- analyse and interpret stem-and-leaf plots.

## 14.6.1 Stem-and-leaf plots

eles-4724

- A **stem-and-leaf plot** may be used as an alternative to a frequency distribution table.
- Stem-and-leaf plots are used to display numerical data.
- Sometimes the term 'stem-and-leaf plot' is shortened to 'stem plot'.
- Each piece of data in a stem plot is made up of two components: a stem and a leaf.
  For example, the value 28 is made up of 20, a tens component (the stem), and 8, the units component (the leaf).
  In a stem plot, 28 would be written as:
  Key: $2|8 = 28$

| *Stem* | *Leaf* |
|---|---|
| 2 | 8 |

- It is important to provide a key when drawing up stem plots to make sure the data can be read and interpreted correctly.
- **Ordered stem plots** have the leaf part written in ascending order from left to right.

### WORKED EXAMPLE 11 Constructing an ordered stem-and-leaf plot

**Prepare an ordered stem-and-leaf plot for each of the following sets of data.**
**a. 129, 148, 137, 125, 148, 163, 152, 158, 172, 139, 162, 121, 134**
**b. 1.6, 0.8, 0.7, 1.2, 1.9, 2.3, 2.8, 2.1, 1.6, 3.1, 2.9, 0.1, 4.3, 3.7, 2.6**

**THINK**

a. 1. Rule two columns with the headings 'Stem' and 'Leaf'.
2. Include a key to the plot that tells the reader the meaning of each entry.
3. Make a note of the smallest and largest values of the data (that is, 121 and 172 respectively). List the stems in ascending order in the first column (that is, 12, 13, 14, 15, 16, 17). *Note:* The hundreds and tens components of the number represent the stem.
4. Systematically work through the given data and enter the leaf (unit component) of each value in a row beside the appropriate stem. *Note:* The first row represents the interval 120–129, the second row represents the interval 130–139 and so on.

**WRITE**

a. Key: $12|1 = 121$

| *Stem* | *Leaf* |
|---|---|
| 12 | 9 5 1 |
| 13 | 7 9 4 |
| 14 | 8 8 |
| 15 | 2 8 |
| 16 | 3 2 |
| 17 | 2 |

5. Redraw the stem plot so that the numbers in each row of the leaf column are in ascending order.

Key: 12|1 = 121

| *Stem* | *Leaf* |
|---|---|
| 12 | 159 |
| 13 | 479 |
| 14 | 88 |
| 15 | 28 |
| 16 | 23 |
| 17 | 2 |

**b.** 1. Rule the stem and leaf columns and include a key.

2. Include a key to the plot that tells the reader the meaning of each entry.

3. Make a note of the smallest and largest values of the data (that is, 0.1 and 4.3 respectively). List the stems in ascending order in the first column (that is, 0, 1, 2, 3, 4). *Note:* The units components of the decimal represent the stem.

4. Systematically work through the given data and enter the leaf (tenth component) of each decimal in a row beside the appropriate stem. *Note:* The first row represents the interval 0.1–0.9, the second row represents the interval 1.0–1.9 and so on.

**b.** Key: 0|1 = 0.1

| *Stem* | *Leaf* |
|---|---|
| 0 | 871 |
| 1 | 6296 |
| 2 | 38196 |
| 3 | 17 |
| 4 | 3 |

5. Redraw the stem plot so that the numbers in each row of the leaf column are in ascending order.

Key: 0|1 = 0.1

| *Stem* | *Leaf* |
|---|---|
| 0 | 178 |
| 1 | 2669 |
| 2 | 13689 |
| 3 | 17 |
| 4 | 3 |

- The advantage of using a stem plot instead of a grouped frequency distribution table is that all the original pieces of data are retained. This means it is possible to identify the smallest and largest values. It is also possible to identify any repeated values. This makes it possible to calculate measures of centre (such as mean, median and mode) and spread (range). These identifications and calculations cannot be done when values are grouped in class intervals.
- The disadvantages of stem plots are:
  - they are not very informative for small sets of data or data that is tightly clustered together
  - they are not very manageable for data that has a large range.

## Back-to-back stem-and-leaf plots

- Two sets of data can be displayed using just one stem, by placing the leaf parts for each set on either side of the stem. This is called a **back-to-back stem-and-leaf plot**.

- Back-to-back stem-and-leaf plots are used to display numerical data for two different groups. For example, test scores for two classes, Class A and Class B, can be displayed as shown here:

Key: 5|6 = 56%

| *Leaf (Class A)* | *Stem* | *Leaf (Class B)* |
|---|---|---|
| 4 3 2 | 5 | 6 |
| 8 5 3 4 1 | 6 | 0 0 2 4 |
| 9 6 4 4 3 1 0 | 7 | 3 5 6 7 8 8 8 9 |
| 8 8 7 5 3 | 8 | 0 1 4 6 |
| 8 6 3 | 9 | 2 3 5 6 7 8 |

- In a back-to-back stem-and-leaf plot, the values in the leaf are written in ascending order, working outwards from the stem.

### WORKED EXAMPLE 12 Constructing a back-to-back stem-and-leaf plot

**The ages of male and female groups using a ten-pin bowling centre are listed.**

**Males: 65, 15, 50, 15, 54, 16, 57, 16, 16, 21, 17, 28, 17, 27, 17, 22, 35, 18, 19, 22, 30, 34, 22, 31, 43, 23, 48, 23, 46, 25, 30, 21**

**Females: 16, 60, 16, 52, 17, 38, 38, 43, 20, 17, 45, 18, 45, 36, 21, 34, 19, 32, 29, 21, 23, 32, 23, 22, 23, 31, 25, 28**

**Display the data as a back-to-back stem plot and comment on the distribution.**

**THINK**

1. Rule three columns, headed 'Leaf (female)', 'Stem' and 'Leaf (male)'.
2. Make a note of the smallest and largest values across both sets of data (15 and 65). List the stems in ascending order in the middle column.
3. Beginning with the males, work through the given data and enter the leaf (unit component) of each value in a row beside the appropriate stem.
4. Repeat step 3 for the females' set of data.

**WRITE**

Key: 1|5 = 15

| *Leaf (female)* | *Stem* | *Leaf (male)* |
|---|---|---|
| 9 8 7 7 6 6 | 1 | 5 5 6 6 6 7 7 7 8 9 |
| 8 5 3 2 3 3 1 9 1 0 | 2 | 1 8 7 2 2 2 3 3 5 1 |
| 1 2 2 4 6 8 8 | 3 | 5 0 4 1 0 |
| 5 5 3 | 4 | 3 8 6 |
| 2 | 5 | 0 4 7 |
| 0 | 6 | 5 |

5. Include a key that tells the reader the meaning of each entry.
6. Redraw the stem plot so that the numbers in each row of the leaf columns are in ascending order.
*Note:* The smallest values are closest to the stem column and increase as they move away from the stem.

Key: 1 | 5 = 15

| *Leaf (female)* | *Stem* | *Leaf (male)* |
|---|---|---|
| 9 8 7 7 6 6 | 1 | 5 5 6 6 6 7 7 7 8 9 |
| 9 8 5 3 3 3 2 1 1 0 | 2 | 1 1 2 2 2 3 3 5 7 8 |
| 8 8 6 4 2 2 1 | 3 | 0 0 1 4 5 |
| 5 5 3 | 4 | 3 6 8 |
| 2 | 5 | 0 4 7 |
| 0 | 6 | 5 |

7. Comment on any interesting features.

The youngest male attending the ten-pin bowling centre is 15 and the oldest is 65. The youngest female attending the centre is 16 and the oldest is 60. Ten-pin bowling is most popular for men in their teens and 20s, and for females in their 20s and 30s.

## COLLABORATIVE TASK: Secret numbers

As a class, construct a stem-and-leaf plot using data supplied by each student.

1. Write a whole number between 210 and 290 on a piece of paper.
2. As a class, decide what the stems will be for this data and come up with a key that matches this.
3. One student should volunteer to draw two columns headed 'Stem' and 'Leaf' on the left-hand side of the board and list the stems in ascending order in the first column. Remember to also include the key that you came up with in step 2.
4. Each student should then come to the board one at a time to write the leaf part of their own number in the appropriate row of the stem-and-leaf plot. Show a classmate your number and have them check that you have written your piece of data in the correct row.
5. Is the stem-and-leaf plot drawn on the board an ordered stem-and-leaf plot? How can you tell? If it is not, how can it be redrawn to make it ordered?
6. Several students should volunteer to use the answers from step 5 to draw the final version of the stem-and-leaf plot on the right-hand side of the board.
7. As a class, make some observations about the distribution of the data. Did any students have the same secret number?
8. Explain whether there is a type of graph that looks similar to a stem-and-leaf plot? Discuss the advantages and disadvantages of using this kind of graph compared with using a stem-and-leaf plot for displaying data.

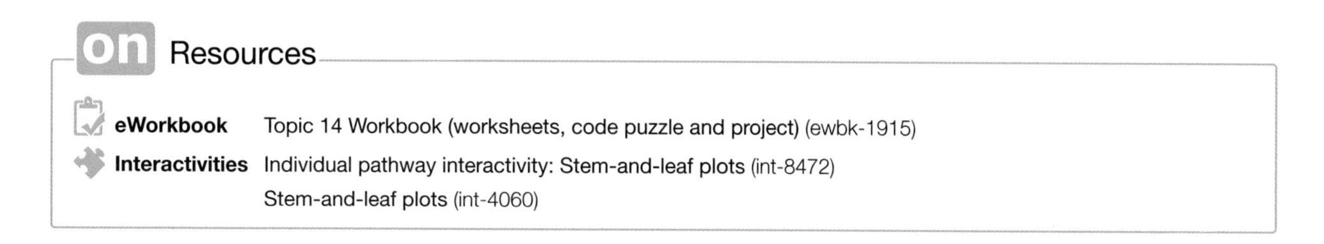

on Resources

**eWorkbook** Topic 14 Workbook (worksheets, code puzzle and project) (ewbk-1915)

**Interactivities** Individual pathway interactivity: Stem-and-leaf plots (int-8472)
Stem-and-leaf plots (int-4060)

# Exercise 14.6 Stem-and-leaf plots

learn on

**14.6 Quick quiz** on | **14.6 Exercise**

Individual pathways

| ■ PRACTISE | ■ CONSOLIDATE | ■ MASTER |
|---|---|---|
| 1, 5, 6, 9, 12 | 2, 3, 7, 10, 13 | 4, 8, 11, 14 |

### Fluency

1. **WE11** The following data gives the number of fruit that have formed on each of 40 trees in an orchard.

   29, 37, 25, 62, 73, 41, 58, 62, 73, 67, 47, 21, 33, 71,
   92, 41, 62, 54, 31, 82, 93, 28, 31, 67, 29, 53, 62, 21,
   78, 81, 51, 25, 93, 68, 72, 46, 53, 39, 28, 40

   Prepare an ordered stem plot that displays the data.

2. The number of mistakes made each week by 30 machine operators is recorded below.

12, 2, 0, 10, 8, 16, 27, 12, 6, 1, 40, 16, 25, 3, 12,
31, 19, 22, 15, 7, 17, 21, 18, 32, 33, 12, 28, 31, 32, 14

Prepare an ordered stem plot that displays the data.

3. Prepare an ordered stem plot for each of the following sets of data.
   a. 132, 117, 108, 129, 165, 172, 145, 189, 137, 116, 152, 164, 118
   b. 131, 173, 152, 146, 150, 171, 130, 124, 114
   c. 207, 205, 255, 190, 248, 248, 248, 237, 225, 239, 208, 244
   d. 748, 662, 685, 675, 645, 647, 647, 708, 736, 691, 641, 735

4. Prepare an ordered stem plot for each of the following sets of data.
   a. 1.2, 3.9, 5.8, 4.6, 4.1, 2.2, 2.8, 1.7, 5.4, 2.3, 1.9
   b. 7.7, 6.0, 9.3, 8.3, 6.5, 9.2, 7.4, 6.9, 8.8, 8.4, 7.5, 9.8
   c. 14.8, 15.2, 13.8, 13.0, 14.5, 16.2, 15.7, 14.7, 14.3, 15.6, 14.6, 13.9, 14.7, 15.1, 15.9, 13.9, 14.5
   d. 0.18, 0.51, 0.15, 0.02, 0.37, 0.44, 0.67, 0.07

5. **WE12** The number of goals scored in football matches by Mitch and Yani were recorded as follows.

| **Mitch** | 0 | 3 | 1 | 0 | 1 | 2 | 1 | 0 | 0 | 1 |
|---|---|---|---|---|---|---|---|---|---|---|
| **Yani** | 1 | 2 | 0 | 1 | 0 | 1 | 2 | 2 | 1 | 1 |

Display the data as a back-to-back stem plot and comment on the distribution.

### Understanding

6. Answer the following questions for the back-to-back stem plot in question 5.
   a. Determine the number of times each player scored more than 1 goal.
   b. State who scored the greatest number of goals in a match.
   c. State who scored the greatest number of goals overall.
   d. State who is the more consistent performer.

7. The stem plot shown gives the ages of members of a theatrical group.
   a. Determine the number of people that are in the theatrical group.
   b. Identify the age of the youngest member of the group.
   c. Identify the age of the oldest member of the group.
   d. Determine the number of people who are over 30 years of age.
   e. State the most common age in the group.
   f. Determine the number of people who are over 65 years of age.

Key: 2 | 4 = 24

| *Stem* | *Leaf* |
|---|---|
| 1 | 7 8 8 9 9 |
| 2 | 2 4 7 9 |
| 3 | 1 3 3 8 |
| 4 | 0 2 2 2 6 6 |
| 5 | 5 7 |
| 6 | 4 |

8. Sprint times, in seconds, over 100 metres were recorded for a random sample of 20 runners as follows.

10.8, 11.0, 12.0, 13.2, 12.4, 13.9, 11.8, 12.8, 14.0, 15.0,
11.2, 12.6, 12.5, 12.8, 13.6, 11.5, 13.6, 10.9, 14.1, 13.9

   a. Show the data as a stem-and-leaf plot.
   b. Comment on the range of performance and other interesting points.
   c. Draw conclusions about the runners' performances.

### Reasoning

9. Explain why it is important to use a key with all stem-and-leaf plots. Give an example to illustrate your answer.

10. Explain why it is important to order and align data values when constructing a stem-and-leaf plot.

**11.** The scores in a Mathematics test for two classes are shown in the following table.

| 7F | 91 | 77 | 85 | 82 | 43 | 84 | 77 | 79 | 78 | 92 | 81 | 80 | 41 | 88 |
|---|---|---|---|---|---|---|---|---|---|---|---|---|---|---|
| 7G | 76 | 85 | 82 | 74 | 89 | 83 | 68 | 66 | 78 | 82 | 80 | 78 | 75 | 76 |

**a.** Display the data as a back-to-back stem-and-leaf plot.
**b.** Calculate the mean and median for each class.
**c.** Explain which of these measures of centre is the most appropriate in this situation.

## Problem solving

**12.** The following stem-and-leaf plot displays the ages of a sample of people who were in a grocery store on a particular day.

**a.** Determine the range of their ages.
**b.** Determine the mode of their ages.
**c.** One more person in the store is added to the sample. The mean is now equal to the mode. Determine the age of this additional person. Show your working.

Key: 2 | 1 = 21

| *Stem* | *Leaf* |
|---|---|
| 1 | 8 |
| 2 | 2 4 |
| 3 | 0 9 |
| 4 | 1 3 3 5 8 |
| 5 | 7 |
| 6 | 5 |
| 7 | 1 3 |

**13.** The following data shows the maximum temperature for 7 straight days in two different cities. The temperatures were recorded in the month of Feburary.

| City A | 25 | 23 | 21 | 24 | 29 | 36 | 30 |
|---|---|---|---|---|---|---|---|
| City B | 20 | 17 | 13 | 13 | 12 | 12 | 13 |

**a.** Display the data as a back-to-back stem-and-leaf plot.
**b.** Calculate the range, median, mean and mode for each city.
**c.** One of the cities is in the northern hemisphere and the other is in the southern hemisphere. Use your answers to part **b** to determine which city is in the northern hemisphere and which is in the southern hemisphere.

**14.** Percentages in a Mathematics exam for two classes were as follows:

| 9A | 32 | 65 | 60 | 54 | 85 | 73 | 67 | 65 | 49 | 96 | 57 | 68 |
|---|---|---|---|---|---|---|---|---|---|---|---|---|
| 9B | 46 | 74 | 62 | 78 | 55 | 73 | 60 | 75 | 73 | 77 | 68 | 81 |

**a.** Construct a back-to-back stem plot of the data.
**b.** Determine the percentage of each group who scored above 50.
**c.** State which group had more scores over 80.
**d.** Compare the clustering for each group and comment on any extreme values.
**e.** Parallel dot plots are obtained by constructing individual dot plots for each set of data and positioning them on a common scale. Construct a parallel dot plot of the data (use colour).
**f.** Compare class performances by reference to both graphs.

# LESSON
# 14.7 Pie charts and divided bar graphs

**LEARNING INTENTION**

At the end of this lesson you should be able to:
- display data as a sector graph (pie chart)
- display data as a divided bar graph
- analyse and interpret sector and divided bar graphs.

## 14.7.1 Sector graphs and divided bar graphs

eles-4725

- **Sector graphs** or pie charts, are made up of sectors of a circle. **Divided bar graphs** are made up of a rectangle divided into bars.
- Both of these types of graphs are used to represent categorical data — they are not suitable for numerical data.
- The sector graph and divided bar graph shown have six different sectors.

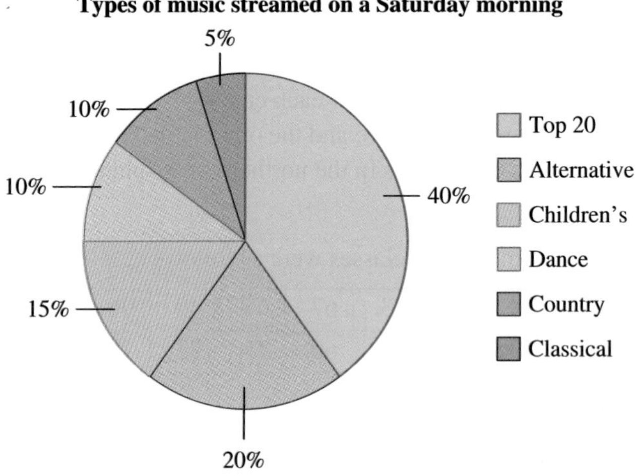


- In a sector graph the sectors should be ordered from largest to smallest in a clockwise direction (starting at 12 o'clock).
- In a divided bar graph the bars should be ordered largest to smallest from left to right.

**Types of music streamed on a Saturday morning**

40% | 20% | 15% | 10% | 10% | 5%

Top 20, Alternative, Children's, Dance, Country, Classical

- Each sector or bar must be labelled appropriately either on the graph or using a legend.

### Drawing sector graphs and divided bar graphs by hand

- To draw sector graphs and divided bar graphs you must determine the fraction (or percentage) of the total that is represented by each category.
- The fraction of the whole that each category represents should correspond to the fraction/percentage of the sector/divided bar graph.

- The following table shows how to draw a sector graph and a divided bar graph from data.

| Category | Frequency | Fraction of total | Angle at centre of circle | Length of bar |
|---|---|---|---|---|
| A | 3 | $\frac{3}{12}=\frac{1}{4}$ | $\frac{1}{4}\times 360° = 90°$ | $\frac{1}{4}\times 12 = 3\,\text{cm}$ |
| B | 4 | $\frac{4}{12}=\frac{1}{3}$ | $\frac{1}{3}\times 360° = 120°$ | $\frac{1}{3}\times 12 = 4\,\text{cm}$ |
| C | 5 | $\frac{5}{12}$ | $\frac{5}{12}\times 360° = 150°$ | $\frac{5}{12}\times 12 = 5\,\text{cm}$ |
| **Total** | 12 | | | |

- A protractor and a ruler are required to draw these graphs by hand.

**A sector graph**

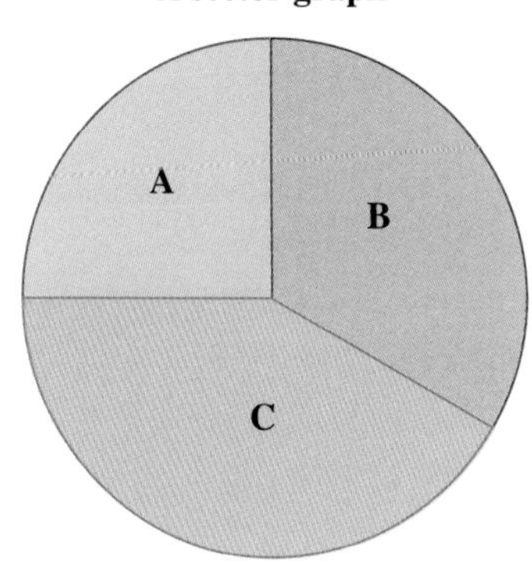


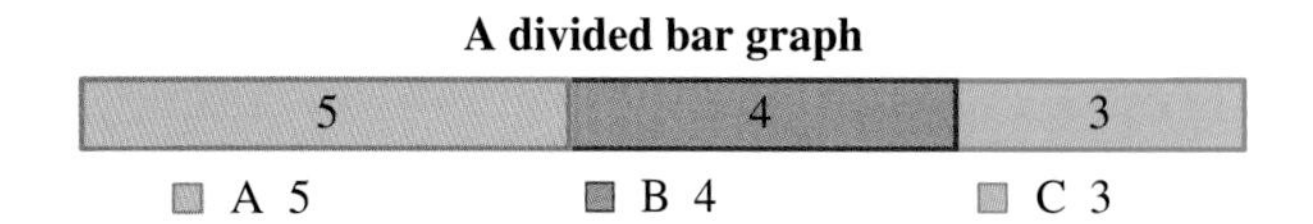


## Drawing sector graphs and bar graphs using technology

- The quickest and easiest way to create sector graphs and divided bar graphs is to use a spreadsheet such as Microsoft Excel.
- The following worked examples provide a detailed explanation of how to create sector graphs and divided bar graphs in Microsoft Excel.

### WORKED EXAMPLE 13 Constructing a sector graph

**Of 120 people surveyed about where they would prefer to spend their holidays this year, 54 preferred to holiday in Australia, 41 preferred to travel overseas and 25 preferred to stay at home. Use a spreadsheet program such as Microsoft Excel to represent the data as a sector graph.**

**THINK**

1. Input the data into a spreadsheet in order from largest category to smallest category. Put labels along the top row and numbers along the second row.

**WRITE**

| | A | B | C | D |
|---|---|---|---|---|
| 1 | Australia | Overseas | At home | |
| 2 | 54 | 41 | 25 | |
| 3 | | | | |

2. Select the data.

| | A | B | C | D |
|---|---|---|---|---|
| 1 | Australia | Overseas | At home | |
| 2 | 54 | 41 | 25 | |
| 3 | | | | |

3. Click **insert**, select the **Insert Pie or Doughnut Chart** icon then click **Pie.**

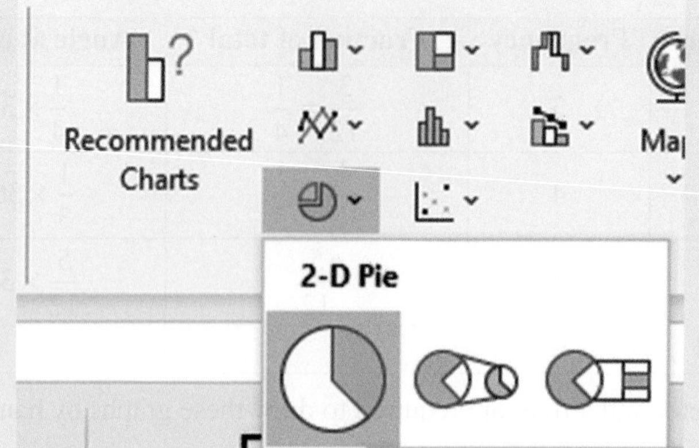


4. The chart will appear.

Chart Title

□ Australia □ Overseas □ At home

5. Rename the chart by clicking **Chart Title** and writing a name for your chart.

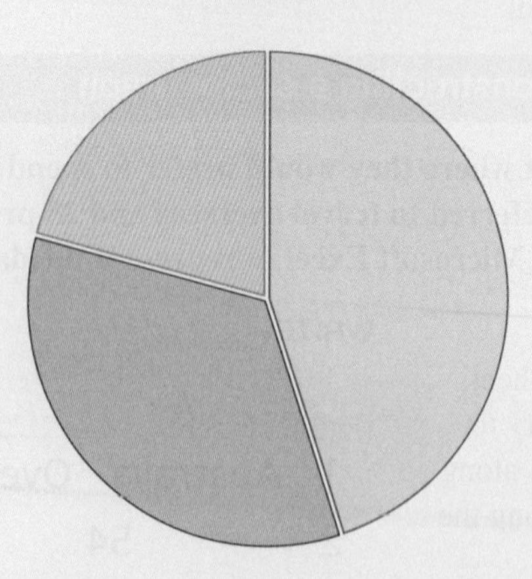

6. Click on the **Chart Elements** in the top right corner and select **Data Labels**.

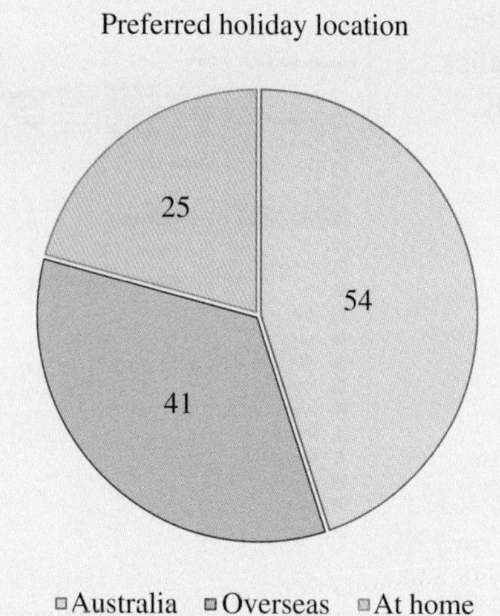


## WORKED EXAMPLE 14 Constructing a divided bar graph

**From a community group, 90 people were surveyed about their favourite weekend activities. Of these, 43 people preferred to go to the beach, 25 preferred to watch sports and 22 preferred to go shopping. Use a spreadsheet program such as Microsoft Excel to represent the data as a divided bar graph.**

**THINK**

**WRITE**

1. Input the data into a spreadsheet in order from largest category to smallest category. Put labels along the top row and numbers along the second row.

| | A | B | C | D |
|---|---|---|---|---|
| 1 | Beach | Sports | Shopping | |
| 2 | 43 | 25 | 22 | |
| 3 | | | | |

2. Select the data.

| | A | B | C | D |
|---|---|---|---|---|
| 1 | Beach | Sports | Shopping | |
| 2 | 43 | 25 | 22 | |
| 3 | | | | |

3. Click **insert**, select the **Insert Column or Bar Chart** icon then click **More Column Charts**.

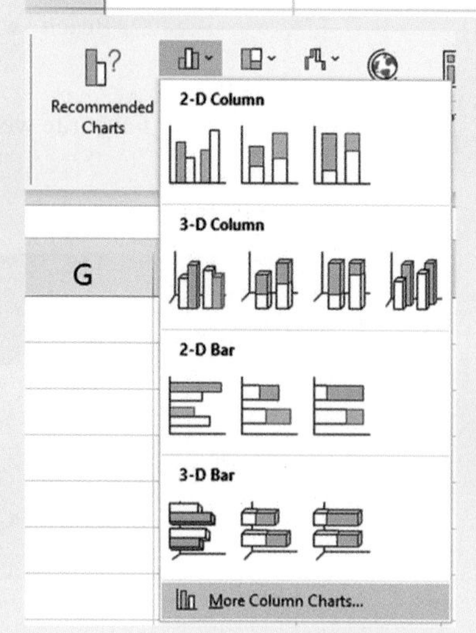

4. From the **Bar** options, select the second option at the top then click the second graph and click **OK**.

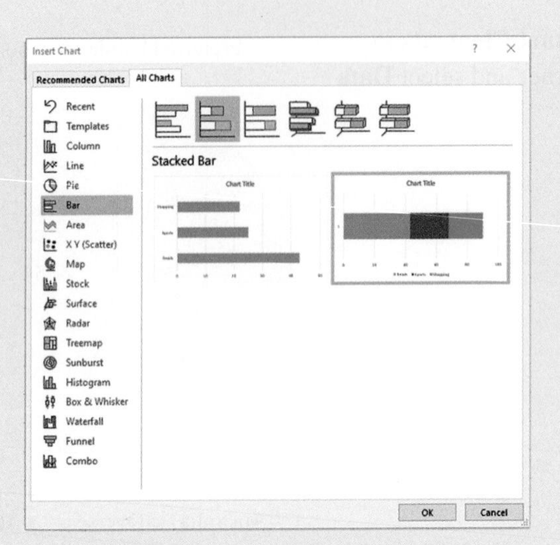


5. The chart will appear.

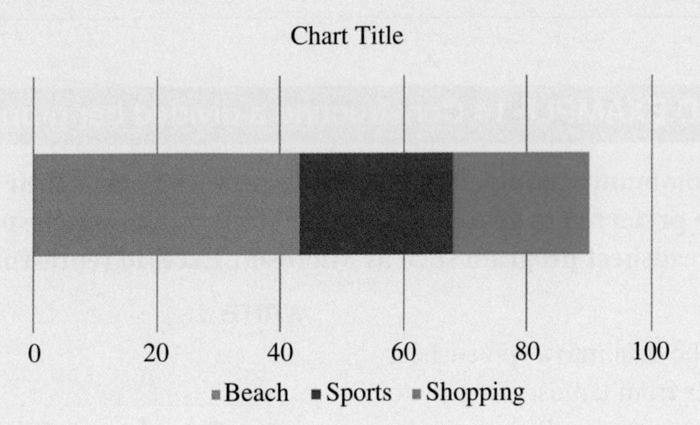


6. Rename the chart by clicking **Chart Title** and writing a name for your chart.

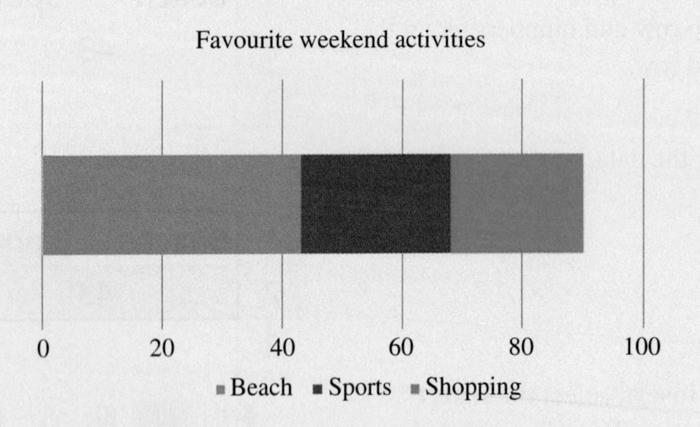


7. Click on the **Chart Elements** in the top right corner and select **Data Labels**.

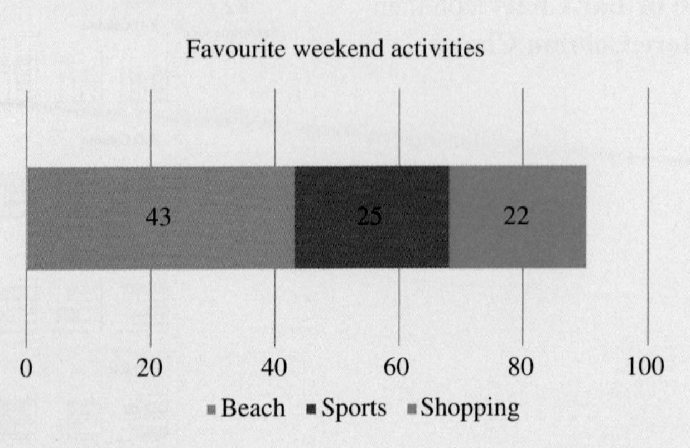

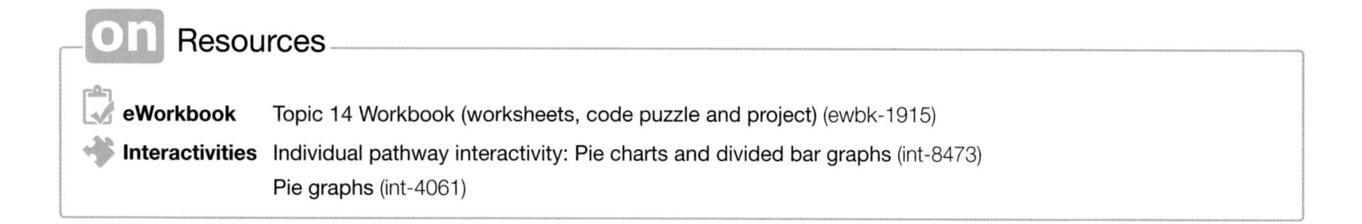

on Resources

eWorkbook Topic 14 Workbook (worksheets, code puzzle and project) (ewbk-1915)

Interactivities Individual pathway interactivity: Pie charts and divided bar graphs (int-8473)

Pie graphs (int-4061)

## Exercise 14.7 Pie charts and divided bar graphs

learnon

14.7 Quick quiz on | 14.7 Exercise

### Individual pathways

| ■ PRACTISE | ■ CONSOLIDATE | ■ MASTER |
|---|---|---|
| 1, 2, 6, 9, 12, 13, 16 | 4, 5, 7, 10, 14, 17 | 3, 8, 11, 15, 18 |

### Fluency

1. A survey was conducted of a group of students to determine their method of transport to school each day. The following sector graph displays the survey's results.

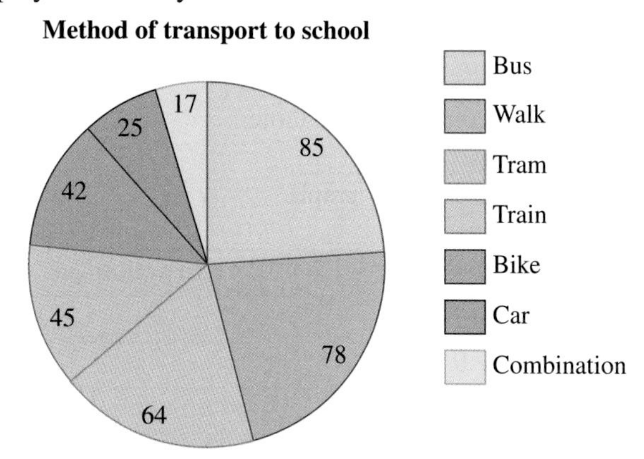


a. State the number of students surveyed.
b. State the most common method of transport to school.

2. WE13 The following table shows the different sports played by a group of Year 7 students.
a. Complete the table.
b. Draw a sector graph to display the data.
c. Use technology to represent the data as a sector graph.

| Sport | Number of students | Fraction of students | Angle at centre of circle |
|---|---|---|---|
| Basketball | 55 | $\frac{55}{180}$ | $\frac{55}{\not{180}^{1}} \times \frac{\not{360}^{\circ 2}}{1} = 110^\circ$ |
| Netball | 35 | $\frac{35}{180}$ | |
| Soccer | 30 | | |
| Football | 60 | | |
| **Total** | 180 | | |

3. The sector graph shown displays the candidates and the percentage of votes received in a community organisation's recent election.

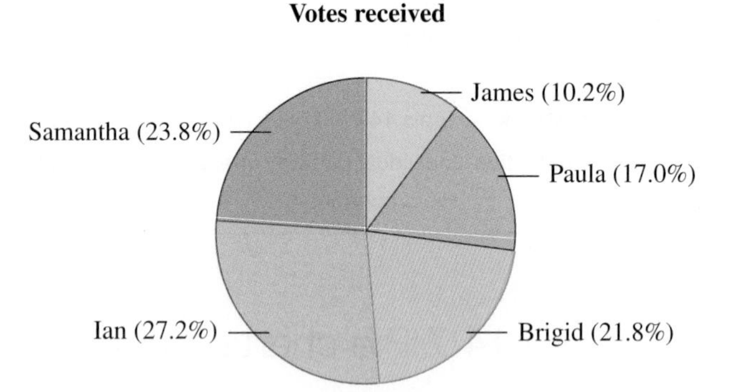


   a. State who received the highest number of votes.
   b. State the percentage of votes Paula received.
   c. Can this sector graph tell us how many people voted in the election? Explain your answer.
   d. Calculate the size of the angle for the sector representing Brigid's votes. Give your answer to 2 decimal places.

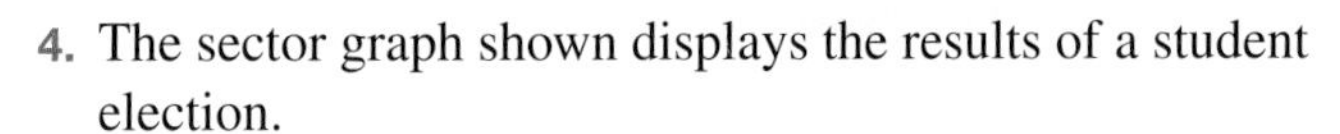

4. The sector graph shown displays the results of a student election.

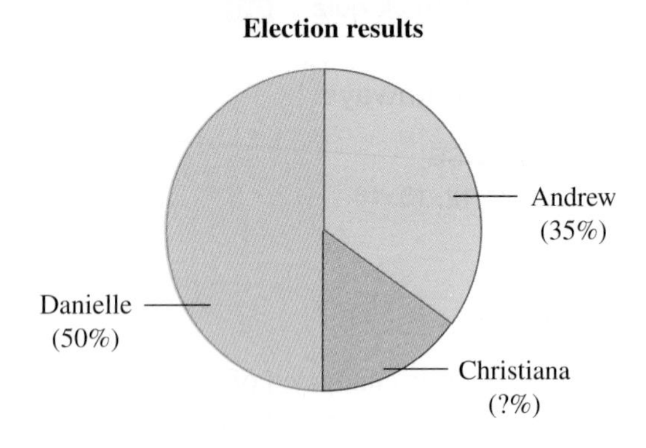


   a. Identify the percentage of votes that went to Danielle.
   b. Identify the percentage of votes that went to Andrew.
   c. Calculate the percentage of votes that went to Christiana.
   d. Evaluate the total of the percentages.
   e. Using calculations, determine the size of the angle of each sectors showing the votes for each of the 3 candidates.

5. For the following table of values showing the type of roofing material used in 36 houses, answer the following questions.
   a. Complete the fraction and angle size columns in the table.
   b. Draw a sector graph to display the data.
   c. Use technology to represent the data as a sector graph.

| Roofing material | Number of houses | Fraction | Angle size (°) |
|---|---|---|---|
| Concrete tiles | 12 | | |
| Terracotta tiles | 9 | | |
| Galvanised iron | 6 | | |
| Colourbond | 3 | | |
| Slate | 2 | | |
| Shingles | 4 | | |
| **Total** | 36 | | |

6. The following divided bar graph displays the types of dogs in a dog rescue centre.

**Types of dogs in the rescue centre**

18 | 12 | 6 | 6 | 3

Terrier
Rottweiler
Labrador
Schnauzer
German Shepherd

   a. Based on this graph, calculate the total number of dogs in the rescue centre.
   b. Identify the number of Labradors that are in the centre.
   c. Identify the number of Rottweilers that are in the centre.
   d. Calculate the percentage of the total number of dogs that are terriers.
   e. State which type of dog there are the fewest of.

7. WE14 The following table of values shows the number of kilograms of meat of various types that a butcher sold in a day.

a. Copy and complete the table of values.
b. Construct a divided bar graph to display the data.
c. Use technology to represent the data as a divided bar graph.

| Type of meat | Amount sold (kg) | Fraction |
|---|---|---|
| Lamb | 10 | |
| Beef | 45 | |
| Pork | 5 | |
| Chicken | 15 | |
| Turkey | 10 | |
| Rabbit | 5 | |
| Total | 90 | |

8. The students at Mount Birdie Secondary College counted the number of birds that landed on their school oval on one day. The results are in the following table.

| Type of bird | Number of birds | Fraction |
|---|---|---|
| Magpie | 24 | |
| Crow | 12 | |
| Cockatoo | 12 | |
| Wattlebird | 6 | |
| Rosella | 6 | |
| **Total** | 60 | |

a. Copy and complete this table.
b. Use technology to represent the data as:
   i. a sector graph
   ii. a divided bar graph.

## Understanding

9. The following table shows the different forms of transport that 100 students use to get to school.

a. Copy the table and complete the *Fraction* column (the first one has been done for you).
b. Using a total bar length of 10 cm, complete the *Length of bar* column (the first one has been done for you).
c. Represent the data as a divided bar graph.
d. If the total bar length was 5 cm (instead of 10 cm) calculate how long the *Walk* section would be.

| Transport | Number of students | Fraction | Length of bar |
|---|---|---|---|
| Walk | 50 | $\frac{50}{100} = \frac{1}{2}$ | $\frac{1}{2} \times 10 = 5\text{ cm}$ |
| Car | 20 | | |
| Bus | 15 | | |
| Bike | 15 | | |
| **Total** | 100 | | |

10. The sector graph shown represents the favourite colours of 200 students. The angle of each section is shown.

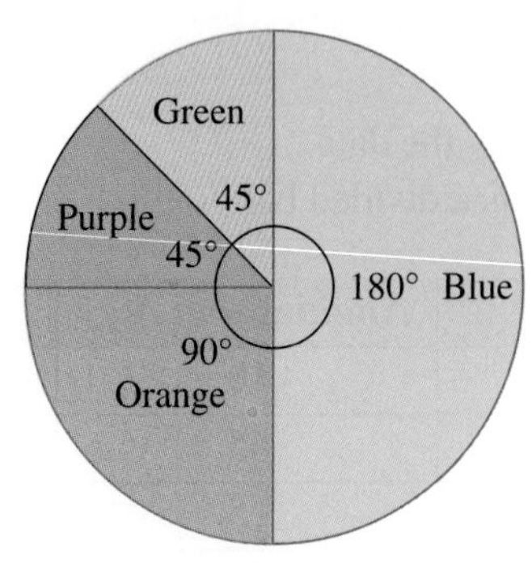


a. Calculate the number of students who said their favourite colour was blue.
b. Calculate the number of students who said their favourite colour was orange.
c. Calculate the number of students who said their favourite colour was purple.

11. The sector graph shown represents the furniture sold at Devin's Furniture Land over the period of one week. The total number of sales was 90 items. Calculate the number of pieces of each furniture item that were sold. Use a protractor to measure the angles in the sector graph.

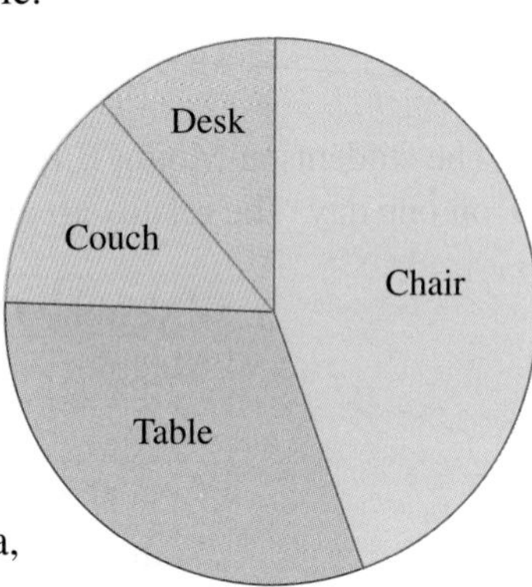


12. State which type of data is represented by the following graphs: column and bar graphs, dot plots, stem-and-leaf plots and sector graphs.

## Reasoning

13. Discuss the advantage of using a divided bar graph to display categorical data, rather than a column graph.

14. The following two sector graphs represent the sales at Charlie's Kiosk. One sector graph is from summer and one sector graph is from winter.

**Graph 1**

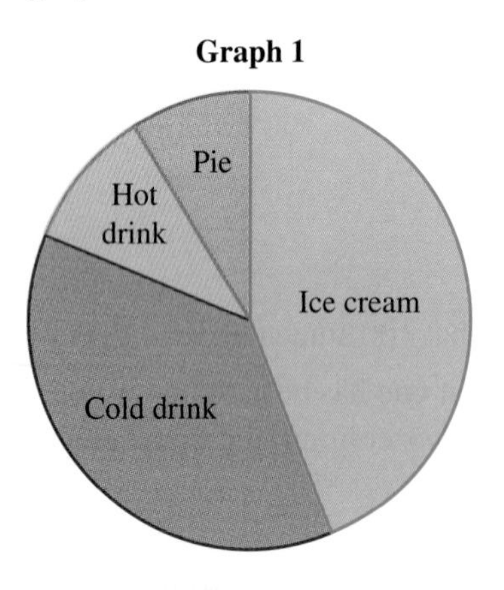


**Graph 2**

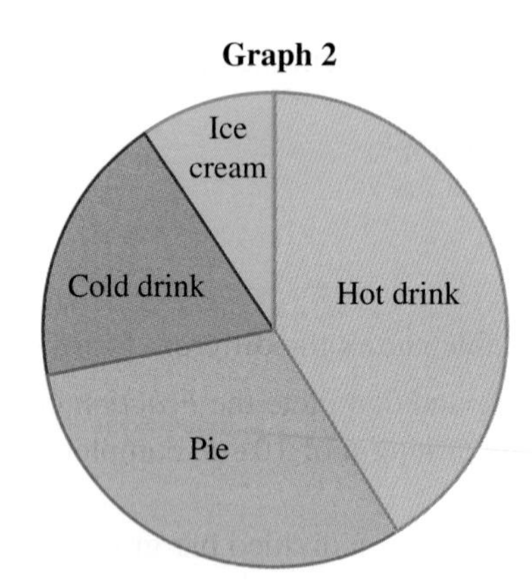


a. State which graph represents the sales from summer.
b. Explain your answer to part a.

15. For the table of values showing the sports played by a group of students, answer the following questions.

**a.** Complete the angle size column before drawing a sector graph to display the data.
**b.** In this data we are not told how many students were surveyed.

i. Explain why the data could not represent ten students.
ii. Explain why the data could represent twenty students.

| Sport | Percentage of total | Angle size (°) |
|---|---|---|
| Cricket | 20 | |
| Hockey | 25 | |
| Netball | 40 | |
| Soccer | 10 | |
| Tennis | 5 | |
| Total | 100 | |

### Problem solving

16. Omar is working on his budget. His income is $2000 per month and the following divided bar graph shows how he is currently spending his money. Omar needs your help to increase the amount he saves.

a. Omar does not want to move house and he cannot change his utility bills. Determine the sections of the bar graph that must stay the same.
b. Determine how much Omar is currently spending on food and entertainment in total.
c. If Omar halves the amount he spends on food and entertainment and puts that money into savings, determine how much he would:
   i. spend on food
   ii. spend on entertainment
   iii. put into savings in total each month.
d. Use technology to draw a new divided bar graph showing the changes to Omar's budget you made in part c.

17. Beach Brellas Inc. is designing a new beach umbrella using the design shown.
   a. Determine the percentage of the umbrella that is blue.
   b. Determine the angle for one section of the umbrella.
   c. If it takes 1.5 $m^2$ of fabric to make the whole design, determine how much of each colour fabric is needed to make the design. Give your answer to three decimal places and include units.

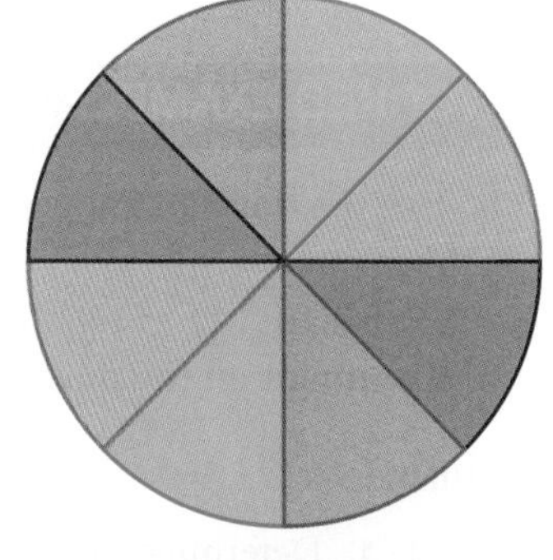

18. The sector graph shown has four sectors representing different amounts. Starting at the smallest sector, each new sector is double the size of the previous one. Without using a protractor, calculate the third sector as a percentage. Show your working.

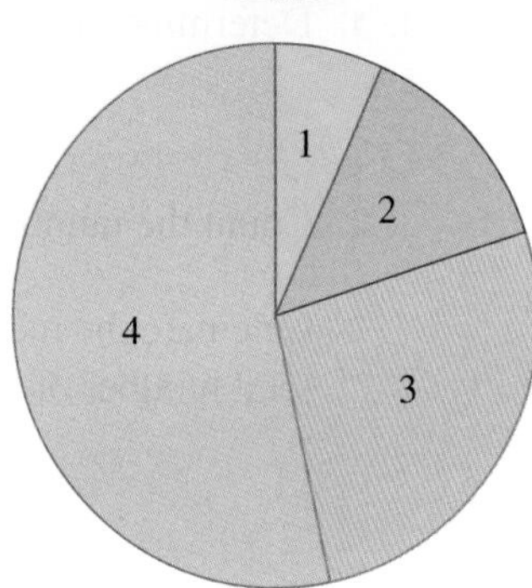

# LESSON
# 14.8 Comparing data

## LEARNING INTENTION

At the end of this lesson you should be able to:
- calculate measures of centre and spread from a graphical display
- identify the most appropriate measure of centre for a graphical display.

## 14.8.1 Determining measures of centre and spread from graphs

eles-4726

- We have considered the calculation of measures of centre and spread from lists of data.
- We also need to know how to calculate these measures from graphs of individual pieces of data.
- We can then make comparisons between data presented in listed form and data presented in graphical form.
- When data is displayed graphically, the spread may be obvious, but we often need to calculate the data's measures of centre so that we can understand it better.
- We also need to be able to determine which measure of centre best represents the data.

### Determining the most appropriate measure of centre

- If a set of data contains outliers, the median is the most appropriate measure of centre because outliers do not affect it, whereas outliers can significantly affect the value of the mean.
- If a set of data does not contain outliers, the mean or median could be used as the measure of centre. Since the concept of the mean (or average) is more familiar to most people, it is generally used in preference to the median.
- The mode is not normally considered a good measure of centre because it tells us only about the most common score.

### WORKED EXAMPLE 15 Determining measures of centre and spread from a dot plot

**a. Use the dot plot shown to determine the following statistical measures.**

**i. Mean ii. Median iii. Mode iv. Range**

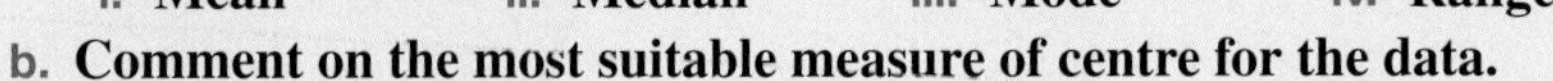

**b. Comment on the most suitable measure of centre for the data.**

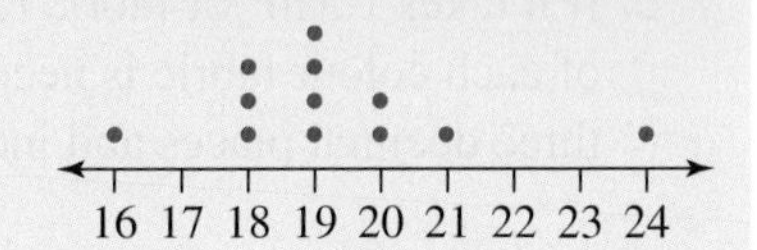


| THINK | WRITE |
|---|---|
| a. i. 1. Determine the sum of the values. | a. Sum of values $= 16 + 3 \times 18 + 4 \times 19 + 2 \times 20 + 21 + 24 = 231$ |
| 2. Count the number of values. | There are 12 values. |
| 3. Calculate the mean by dividing the sum by the total number of values. | $\text{Mean} = \dfrac{\text{sum of data values}}{\text{total number of data values}} = \dfrac{231}{12} = 19.25$ |

ii. 1. The values are already in order. The median is the middle value. There are 12 values, so the middle one is the average of the 6th and 7th values. — The 6th and 7th values are both 19.

2. Calculate the average of these. — The median value is 19.

iii. The mode is the most common value. Look for the one that occurs most frequently. — The mode is 19.

iv. The range is the difference between the largest value and the smallest value. — $\text{Range} = 24 - 16$
$= 8$

b. The scores 16 and 24 are possible outliers. When a set of data contains outliers the best measure of centre is the median, as it is not affected by outliers as much as the mean. — b. Since there may be outliers in the data, the median is the most suitable measure of centre in this case.

## WORKED EXAMPLE 16 Determining measures of centre and spread from a stem plot

**a. Use the stem plot to determine the following measures of the data.**

**i. Mean**

**ii. Median**

**iii. Mode**

**iv. Range**

**b. Comment on the most suitable measure of centre for the data.**

**Key: 1|8 = 18**

| *Stem* | *Leaf* |
|---|---|
| 1 | 8 9 |
| 2 | 2 2 5 7 7 8 |
| 3 | 0 1 4 6 7 |
| 4 | 0 5 |

**THINK** / **WRITE**

a. i. 1. Determine the sum of the values.

a. i. $\text{Sum of values} = 18 + 19 + 22 + 22 + 25 + 27 + 27 + 28 + 30 + 31 + 34 + 36 + 37 + 40 + 45$
$= 441$

2. Count the number of values. — There are 15 values.

3. Calculate the mean by dividing the sum by the total number of values.

$$\text{Mean} = \frac{\text{sum of data values}}{\text{total number of data values}} = \frac{441}{15} = 29.4$$

ii. The values are already in order. The median is the middle value. There are 15 values, so the middle one is the 8th value. Locate this. — The middle position of the 15 values is the 8th value. This is 28. The median value is 28.

iii. The mode is the most common value. Look for the one that occurs most frequently. — There are two modes (the data set is bimodal): 22 and 27.

iv. The range is the difference between the largest value and the smallest value. — $\text{Range} = 45 - 18$
$= 27$

b. Look at the measures of mean, median and mode to see which best represents the values in terms of their closeness to the centre.

b. Because there does not appear to be any outliers, the mean or median could be used to represent the centre of this set of data.

Resources

**eWorkbook** Topic 14 Workbook (worksheets, code puzzle and project) (ewbk-1915)

**Interactivities** Individual pathway interactivity: Comparing data (int-4382)

Determining measures of centre and spread from graphs (int-4062)

## Exercise 14.8 Comparing data

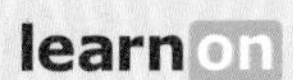

14.8 Quick quiz 

14.8 Exercise

### Individual pathways

| ■ PRACTISE | ■ CONSOLIDATE | ■ MASTER |
|---|---|---|
| 1, 4, 6, 7, 13, 16 | 2, 5, 8, 9, 10, 14, 17 | 3, 11, 12, 15, 18 |

### Fluency

1. **WE15** Consider the dot plot shown.

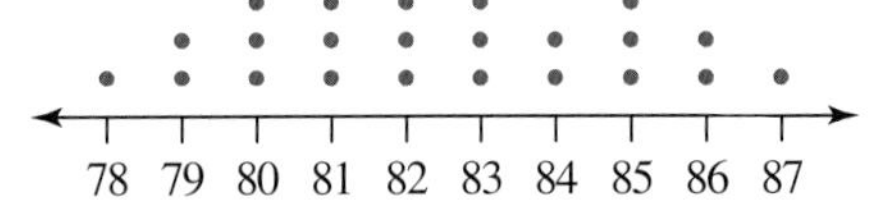


   a. Use the dot plot to determine the following measures of the data.
      i. mean ii. median iii. mode iv. range
   b. Comment on the most suitable measure of centre for the data.

2. Consider the dot plot shown.

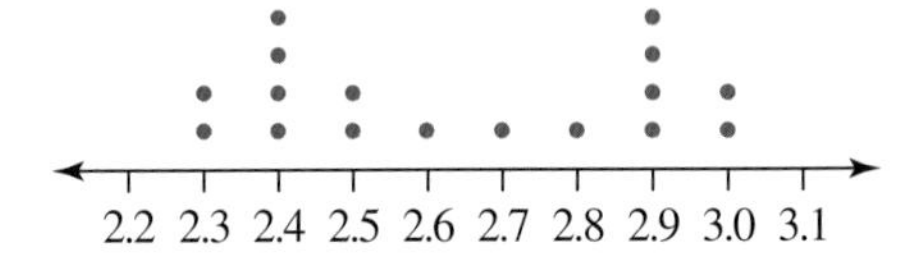


   a. Use the dot plot to determine the following measures of the data.
      i. mean ii. median iii. mode iv. range
   b. Comment on the most suitable measure of centre for the data.

3. Consider the dot plot shown.

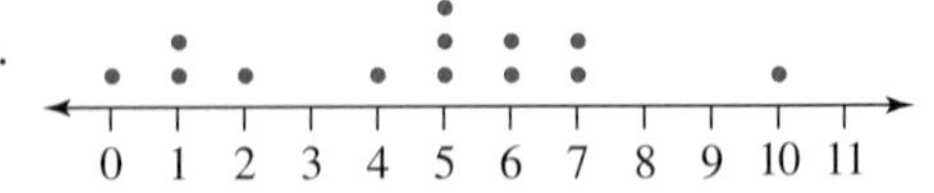


   a. Use the dot plot to determine the following measures of the data.
      i. mean ii. median iii. mode iv. range
   b. Disregard the score of 10 and recalculate each of these values.
   c. Discuss the differences/similarities in your two sets of results.

4. **WE16** Consider the stem-and-leaf plot shown.

Key: $6.1|8 = 6.18$

| *Stem* | *Leaf* |
|---|---|
| 6.1 | 8 8 9 |
| 6.2 | 0 5 6 8 |
| 6.3 | 0 1 2 4 4 |

   a. Use the stem plot to determine the following measures of the data.
      i. mean
      ii. median
      iii. mode
      iv. range
   b. Comment on the most suitable measure of centre for the data.

5. Consider the stem-and-leaf plot shown.

Key: 1|10 = 10

| *Stem* | *Leaf* |
|---|---|
| 1 | 0 2 |
| 2 | 1 3 3 5 |
| 3 | |
| 4 | 4 |

a. Use the stem plot to determine the following measures of the data.
   i. mean
   ii. median
   iii. mode
   iv. range

b. Disregard the score of 44 and recalculate each of these values.

c. Discuss the differences/similarities in your two sets of results.

## Understanding

6. MC Select the score that shows up most often.

A. median B. mean C. mode D. average E. frequency

7. MC Select the best term for *average* in everyday use.

A. the mean B. the mode C. the median D. the total E. none of these

8. MC Choose the measure affected by outliers (extreme values).

A. the middle
B. the mode
C. the median
D. the mean
E. none of these

The back-to-back stem plot shown displays the heights of a group of Year 7 students. Use this plot to answer questions 9 and 10.

Key: 13 | 7 = 137cm

| *Leaf (boys)* | *Stem* | *Leaf (girls)* |
|---|---|---|
| 9 8 | 13 | 7 8 |
| 9 8 8 7 6 | 14 | 3 5 6 |
| 9 8 8 | 15 | 1 2 3 7 |
| 7 6 6 5 | 16 | 3 5 6 |
| 8 7 6 | 17 | 1 |

9. MC Select the total number of Year 7 students.

A. 13 B. 17 C. 30 D. 36 E. 27

10. MC Choose the tallest male and shortest female heights.

A. 186 cm and 137 cm
B. 171 cm and 148 cm
C. 137 cm and 188 cm
D. 178 cm and 137 cm
E. None of these

11. A survey of the number of people in each house in a street produced the following data:

2, 5, 1, 6, 2, 3, 2, 1, 4, 3, 4, 3, 1, 2, 2, 0, 2, 4

a. Prepare a frequency distribution table with an $f \times x$ column and use it to calculate the average (mean) number of people per household.
b. Draw a dot plot of the data and use it to find the median number per household.
c. Calculate the modal number per household.
d. State which measure of centre (mean, median or mode) would be most useful to:
   i. real estate agents renting out houses
   ii. a government population survey
   iii. an ice-cream truck owner.

12. The mean of 12 scores is 6.3. Calculate the total of the scores.

## Reasoning

13. A class of 26 students had a median mark of 54 in Mathematics, even though no-one actually scored this result.
   a. Explain how this is possible.
   b. Explain how many students must have scored below 54.
14. A clothing store records the dress sizes sold during one day in order to work out the most popular sizes. The results for a particular day are given below.

12, 14, 10, 12, 8, 12, 16, 10, 8, 12, 10, 12, 18, 10, 12, 14
16, 10, 12, 12, 12, 14, 18, 10, 14, 12, 12, 14, 14, 10

Rebecca is in charge of marketing and sales. She uses these figures when ordering future stock. From these figures she decides on the following ordering strategy. She will order:
- the same number of size 8, 16 and 18 dresses
- three times this number of size 10 and size 14 dresses
- five times as many size 12 dresses as size 8, 16 and 18.

Comment on Rebecca's strategy.

15. A small business pays these salaries listed below (in thousands of dollars) to its employees.

18, 18, 18, 18, 26, 26, 26, 35

The business owner's salary is 80 thousand dollars.
   a. Determine the wage earned by most workers. Make sure to include the business owner's salary in this calculation.
   b. Calculate the average wage. Show your working.
   c. Calculate the median of the distribution.
   d. Explain which measure of centre might be used in salary negotiations by:
      i. the union that represents the business's employees
      ii. the business owner.

### Problem solving

16. Five scores have an average of 8.2. Four of those scores are 10, 9, 8 and 7. Determine the fifth score.

17. A tyre manufacturer selects 48 tyres at random from their production line for testing. The total distance travelled during the safe life of each tyre is shown in the following table.

| Distance in km ('000) | 46 | 50 | 52 | 56 | 78 | 82 |
|---|---|---|---|---|---|---|
| Number of tyres | 4 | 12 | 16 | 10 | 4 | 2 |

a. Calculate the mean, median and mode.
b. Explain which measure best describes the average safe life of these tyres.
c. Recalculate the mean with the 6 longest-lasting tyres removed. Determine how much it is lowered by.
d. If you selected a tyre at random, determine how long it would most likely last.
e. In a production run of 10 000 tyres, determine how many could be expected to last for a maximum of 50 000 km.
f. If you were the tyre manufacturer, explain what kind of distance guarantee you would be prepared to offer on your tyres.

18. A soccer team had averaged 2.6 goals per match after 5 matches. After their 6th match the average dropped to 2.5 goals per match. Determine the number of goals they scored in their 6th match. Show your working.

# LESSON
# 14.9 Review

## 14.9.1 Topic summary

### Types of data

- There are two main types of data: *categorical* data and *numerical* data.

| Categorical data | Numerical data |
|---|---|
| Data that can be placed into a group or category e.g. Eye colour, gender, shoe size, rating scale (high, medium, low) | Data that can be counted or measured<br>• **Discrete** numerical data can be counted. e.g. The number of people in a room<br>• **Continuous** numerical data can be measured. e.g. A person's height |

## REPRESENTING AND INTERPRETING DATA

### Frequency distribution tables

- Frequency is the number of times a result or piece of data occurs.
- A frequency table shows the score, tally and frequency in a tabular form.

e.g.

| Age group (score) | Tally | Frequency |
|---|---|---|
| 11–20 | \|\|\|\| | 4 |
| 21–30 | \|\| | 2 |
| 31–40 | \| | 1 |
| | Total | 7 |

### Representing data graphically

- There are many types of graphical displays to represent data.
- For *categorical* data the most appropriate graphical displays are:
  - frequency table
  - column or bar graph
  - sector graph (pie chart)
  - divided bar graph.
- For *numerical* data the most appropriate graphical displays are:
  - frequency table
  - dot plot
  - stem-and-leaf plot.

### Measures of centre

- A measure of centre gives us some idea about the location of the centre of the data.
- The three measures of centre are *mean*, *median* and *mode*.
- The **mean** is commonly referred to as the *average*. The symbol used to represent the mean is $\bar{x}$. The formula used to calculate the mean is:

$$\bar{x} = \frac{\text{sum of data values}}{\text{total number of data values}}.$$

- The **median** represents the middle score when the data is ordered from smallest to largest.

e.g. 1, 1, 3, 4, 4, 10, 12, 25, 41
Median = 4

- The **mode** is the most commonly occurring value in a data set. Data sets might have one mode, no mode or more than one mode.

### Comparing data

- By calculating measures of centre, we can make comparisons between data sets. e.g. Mean battery life is higher for brand A batteries than for brand B so, on average, brand A batteries last longer.
- When extreme values (outliers) are present in a data set, the best measure of centre is the median as it is least affected by outliers.

### Measures of spread

- A measure of spread tells us how spread out the data is.
- The **range** is one measure of spread; it is the difference between the largest and smallest values in that set.

$$\text{Range} = \text{largest value} - \text{smallest value}$$

## 14.9.2 Success criteria

Tick the column to indicate that you have completed the lesson and how well you think you have understood it using the traffic light system.
(**Green:** I understand; **Yellow:** I can do it with help; **Red:** I do not understand)

| Lesson | Success criteria | Green | Yellow | Red |
|---|---|---|---|---|
| 14.2 | I can describe types of data using the key words 'categorical' (discrete) or 'numerical' (continuous). | | | |
| | I understand the difference between primary and secondary data. | | | |
| | I can explain why some data is more reliable than other data. | | | |
| 14.3 | I can construct a frequency distribution table for a set of data. | | | |
| | I can analyse and report on the information in a frequency distribution table. | | | |
| 14.4 | I can calculate the mean, median and mode of a set of data. | | | |
| | I can calculate the range of a set of data. | | | |
| 14.5 | I can display data as a column or bar graph. | | | |
| | I can display data as a dot plot. | | | |
| | I can analyse and interpret column and bar graphs and dot plots. | | | |
| 14.6 | I can display data as a stem-and-leaf plot. | | | |
| | I can analyse and interpret stem-and-leaf plots. | | | |
| 14.7 | I can display data as a sector graph (pie chart). | | | |
| | I can display data as a divided bar graph. | | | |
| | I can analyse and interpret sector and divided bar graphs. | | | |
| 14.8 | I can calculate measures of centre and spread from a graphical display. | | | |
| | I can identify the most appropriate measure of centre for a graphical display of data. | | | |

## 14.9.3 Project

### Families with children

Surveys help us to find out about lots of different things. They can be conducted by collecting information from an entire population or smaller groups. The information collected in a survey, known as data, can be examined and presented in many different ways. Graphs are commonly used to present the results of a survey.

The graph shown displays the percentage of families in Australia with children under the age of 15. This data is taken from the 2011 census.

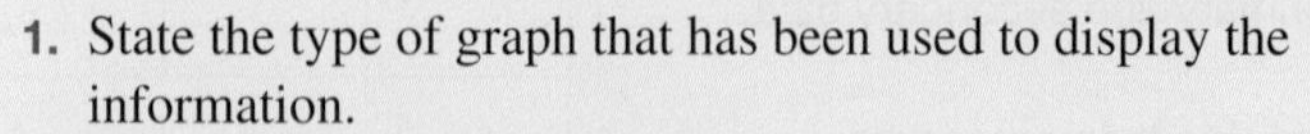


1. State the type of graph that has been used to display the information.
2. State the most common category of Australian families with children under 15.
3. Give an example of another type of graph that could be used to present this information.

Conduct your own survey of the number of children under 15 in your classmates' families. Compare your results with the results of the 2011 census.

4. Record your survey results in the following frequency distribution table.

| Number of children | Tally | Frequency | Percentage |
|---|---|---|---|
| 1 | | | |
| 2 | | | |
| 3 | | | |
| 4 or more | | | |
| | **Total** | | |

5. Is the survey you conducted an example of a *census* (contains the entire population) or a *sample* (contains only a portion of the entire population)? Explain your answer.
6. Explain whether the data you collected is classified as categorical data or numerical data.
7. Present the information from your survey as a column graph. Use the percentage values on the vertical axis and number of children on the horizontal axis.
8. Explain how the results of your class compare with the results obtained in the 2011 census. Discuss the major difference between these two sets of results.
9. Design and conduct a new survey on a topic of interest (choose the topic as a class). You can survey members of your class on this topic or expand it to include a larger target audience. Present your data as a poster that displays the findings of your survey using an appropriate style of graph.

**on** Resources

**eWorkbook** Topic 14 Workbook (worksheets, code puzzle and project) (ewbk-1915)

**Interactivities** Crossword (int-2608)

Sudoku puzzle (int-3174)

## Exercise 14.9 Review questions

learnon

### Fluency

1. Identify whether the following statement is true or false.
   The classification of data describing the number of iPhones sold during the year is continuous.

2. State whether the data that describes a person's mass (in kilograms) is categorical, discrete or continuous.

3. State whether the data that describes sandwich types at a takeaway outlet is categorical, discrete or continuous.

4. Eighty students in a school of 800 participated in a survey to find their favourite drinks. The results were: Coke 20, Pepsi 14, other soft drink 10, milk/milk-based 16, coffee 9, tea 4 and fruit juice 7.
   a. Identify the most popular beverage.
   b. Identify the least popular beverage.

5. Use the frequency distribution table shown to answer the following questions.

| Score | Frequency |
|---|---|
| 13 | 2 |
| 14 | 9 |
| 15 | 3 |
| 16 | 5 |
| 17 | 6 |
| 18 | 1 |

   a. Calculate the number of students in the class.
   b. State the most frequent score.
   c. State the least frequent score.
   d. Determine the number of students who scored 15 or less.
   e. Determine the number of students who scored at least 16.

6. A random sample of 24 families was surveyed to determine the number of vehicles in each household. The following list is the raw data.

   2, 0, 3, 2, 1, 0, 2, 3, 4, 2, 2, 1, 0, 1, 3, 2, 1, 0, 0, 0, 2, 2, 3, 3

   a. Organise the data into a frequency distribution table.
   b. Identify the number of families who have no vehicles in their household.
   c. Determine the number of families who have two or more vehicles in their household.
   d. Identify the score that has the highest frequency.
   e. State the highest score.
   f. Calculate the fraction of families who had two vehicles in their household.

7. For the following sets of data, determine the following.

i. the mean ii. the mode iii. the median

a. 4, 4, 4, 3, 2, 2, 3
b. 2, 4, 2, 4, 3, 7, 4
c. 4, 5, 6, 6, 5, 5, 3
d. 1, 2, 1, 4, 1, 5, 1
e. 5, 8, 1, 7, 7, 5, 2, 7, 5, 5
f. 1, 5, 8, 7, 4, 8, 5, 6, 8

8. The graph shown displays the test marks of a group of students.

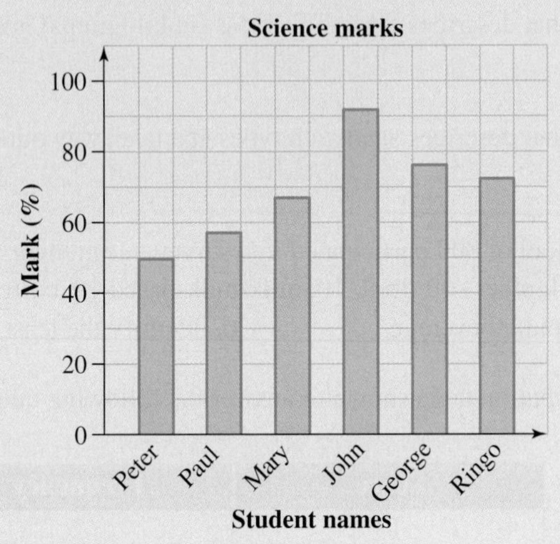


a. State the label on the horizontal axis.
b. State the scale used on the vertical axis.
c. State the name of the student with the highest mark.
d. Determine which student(s) would have failed if the pass mark was 50%.
e. Identify Paul's mark.
f. Identify George's mark.

9. Look at the following graph, showing air quality.

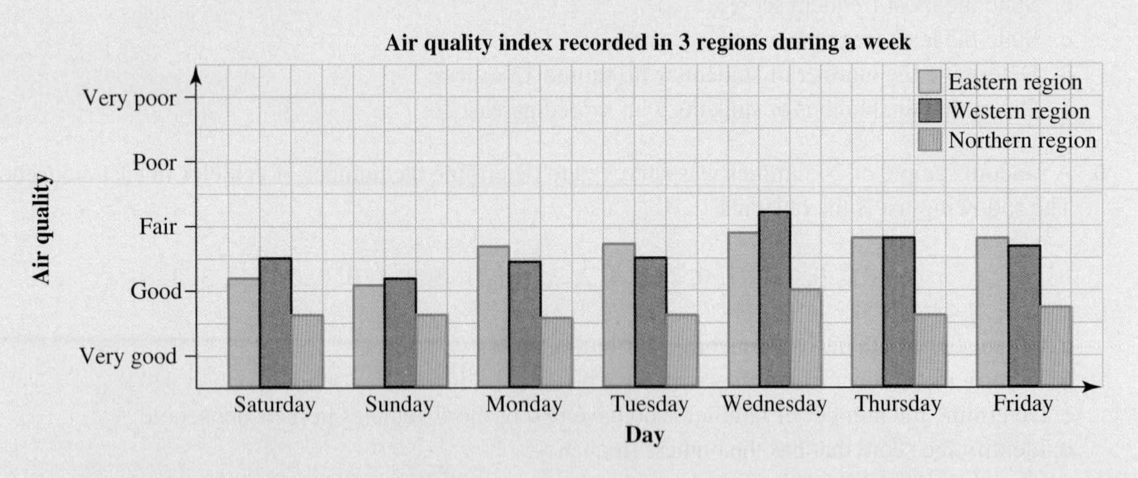


a. State what the horizontal axis represents.
b. State what the vertical axis represents.
c. Identify the region with a rating that ranged from very good to good every day.
d. Determine the day when the Western region had a fair to poor air quality rating.
e. Determine the region with a rating that ranged from good to fair every day.

10. **a.** The following data gives the speed of 30 cars recorded by a roadside speed camera along a stretch of road where the speed limit is 80 km/h.

75, 90, 83, 92, 103, 96, 110, 92, 102, 93, 78, 94, 104, 85, 88,
82, 81, 115, 94, 84, 87, 86, 96, 71, 91, 91, 92, 104, 88, 97

Present the data as an ordered stem-and-leaf plot.
**b.** Comment on the data recorded by the roadside speed camera.

11. Complete the specified tasks for the given sets of data.

| **Set A** | 64 | 30 | 59 | 1 | 57 | 13 | 45 | 28 |
|---|---|---|---|---|---|---|---|---|
| **Set B** | 3 | 41 | 38 | 68 | 29 | 32 | 30 | 31 |

**a.** Construct a back-to-back stem plot of the two sets of data.
**b.** For each data set, determine the:
**i.** lowest score **ii.** highest score **iii.** mean **iv.** median **v.** range.
**c.** Compare the two data sets.

12. Consider the dot plot shown, which represents the number of soft drinks sold at lunchtime over a 20-day period from a vending machine in a school cafeteria in the USA.

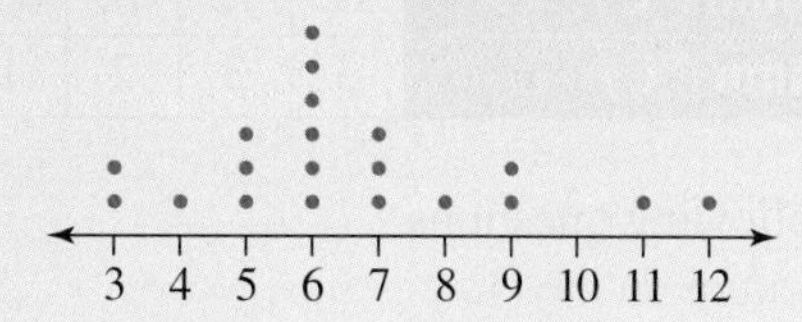


**a.** Use the dot plot to determine the following.
**i.** mean **ii.** median **iii.** mode **iv.** range
**b.** Comment on the distribution of the data.

13. Consider the stem plot shown, which represents the ages of participants in an aerobics class at a gym.

Key: 1|7 = 17

| *Stem* | *Leaf* |
|---|---|
| 1 | 4 7 8 8 9 |
| 2 | 0 1 3 5 6 7 9 |
| 3 | 2 4 5 6 7 8 |
| 4 | 2 5 |
| 5 | 3 |

**a.** Use the graph to determine the:
**i.** mean age
**ii.** median age
**iii.** modal age
**iv.** range of ages.
**b.** Comment on the distribution of the data.

14. A group of 150 people were surveyed about what they had for breakfast that morning. Of the people surveyed, 62 had cereal, 70 had toast, 12 had a cooked breakfast and 6 had no breakfast. Use technology to represent the data as a sector graph.

15. Use technology to draw a divided bar chart to display the following data.

| Hobby | Percentage of total |
|---|---|
| Stamp collecting | 20 |
| Photography | 15 |
| Sailboarding | 20 |
| Skateboarding | 30 |
| Model making | 15 |
| Total | 100 |

## Problem solving

16. This table shows the maximum and minimum daily temperatures in a city over a one-week period.

| Day | 1 | 2 | 3 | 4 | 5 | 6 | 7 |
|---|---|---|---|---|---|---|---|
| Maximum (°C) | 12 | 13 | 10 | 11 | 9 | 10 | 8 |
| Minimum (°C) | 3 | 3 | 2 | 1 | 0 | 4 | 2 |

Use the table to answer the following questions.

**a.** State the maximum temperature on day 3.
**b.** State the day that had the lowest minimum temperature.
**c.** Determine the coldest day.
**d.** Identify the day that had the warmest overnight temperature.
**e.** Calculate the temperature range (variation) on day 2.
**f.** Determine the day that had the smallest range of temperatures.

The table shown is part of a teacher's mark book. Use it to answer questions **17** and **18.**

| Name | Test 1 | Test 2 |
|---|---|---|
| John | 85 | 94 |
| Peter | 85 | 63 |
| Mark | 95 | 58 |
| James | 82 | 67 |
| David | 76 | 95 |
| Rachel | 62 | 85 |
| Mary | 87 | 75 |
| Eve | 94 | 63 |
| Esther | 68 | 68 |

17. **a.** Calculate the class mean for test 1.
**b.** Calculate the class mean for test 2.
**c.** Identify the test in which the class received better marks.
**d.** Determine the number of students who scored above the mean in test 1.
**e.** Determine the number of students who scored above the mean in test 2.

18. a. Determine the mode score for test 1.
b. Determine the mode score for test 2.
c. Determine the median score for test 1.
d. Determine the median score for test 2.
e. Calculate the mean score for the two tests for each student and list them.
f. Determine the range of values for these means.

19. The results of height measurements (in cm) for the students in class 9A are given below.

145, 152, 148, 152, 163, 148, 165, 158, 159, 162, 145, 153, 156, 158, 157,
159, 169, 156, 156, 156, 152, 154, 128, 141, 154, 153, 156, 156, 165, 168

a. Calculate the range of heights.
b. Calculate the mean height of the class.
c. Determine the mode for the heights.
d. Determine the median height of the class.
e. Explain whether these three values (mean, mode, median) are similar.

20. Use the table of values shown to answer the following questions.

| Month | Number of flies (thousands) |
|---|---|
| January | 24 |
| February | 28 |
| March | 26 |
| April | 20 |
| May | 18 |
| June | 15 |
| July | 12 |
| August | 10 |
| September | 11 |
| October | 12 |
| November | 15 |
| December | 19 |

a. State what would be on the horizontal axis in drawing a column graph using this data.
b. State what would be on the vertical axis in drawing a column graph.
c. Identify the minimum value required on the vertical axis.
d. Identify the maximum value required on the vertical axis.
e. Determine the scale that should be used for the vertical axis.
f. Draw a column graph to display the data.

**21.** The number of pets cared for by each of 20 families was surveyed, giving the following data.

1, 2, 3, 2, 0, 1, 0, 2, 5, 3, 2, 1, 2, 0, 2, 0, 1, 3, 2, 1

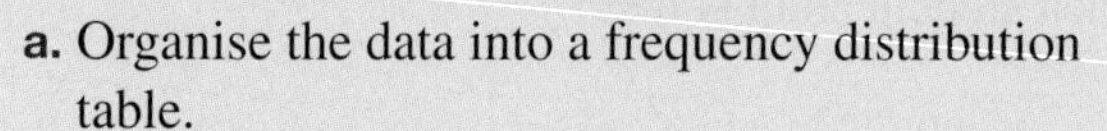

**a.** Organise the data into a frequency distribution table.
**b.** Draw a dot plot of the data and comment on the distribution.
**c.** Calculate the mode, median and mean.

**22. a.** For the given set of scores 9, 3, 8, 5, 6, 9, 4, 5, 5, 25, determine the following.
**i.** mode
**ii.** median
**iii.** mean
**iv.** outliers (if any)
**v.** mean when outliers are omitted
**vi.** mean and median when 10 is added to *each* of the given scores
**b.** Comment on any discoveries from part **a**.

**23.** This back-to-back stem-and-leaf plot shows the heights of a group of boys and girls.

Key: 13 | 7 = 137 cm

| *Leaf (boys)* | *Stem* | *Leaf (girls)* |
|---:|---|---|
| 9 8 | 13 | 7 8 |
| 9 8 8 7 6 | 14 | 3 5 6 |
| 9 8 8 | 15 | 1 2 3 7 |
| 7 6 6 5 | 16 | 3 5 6 |
| 8 7 6 | 17 | 1 |

Give a full description of the heights of the boys and the girls.

**on** To test your understanding and knowledge of this topic, go to your learnON title at www.jacplus.com.au and complete the **post-test**.

# Answers

## Topic 14 Representing and interpreting data

### 14.1 Pre-test

1. B
2. A
3. 3
4. B
5. 31
6. 10
7. 19 m
8. 9
9. 20%
10. 2.55
11. 17
12. 41 cm
13. 111
14. a. 150 b. 45
    c. 8% d. 50%
15. 24%

### 14.2 Collecting and classifying data

1. a. v b. i
   c. iv d. iii
   e. vi f. ii
2. a. Numerical (discrete)
   b. Categorical
   c. Numerical (continuous)
   d. Numerical (continuous)
   e. Categorical
   f. Numerical (continuous)
3. a. Discrete b. Continuous c. Discrete
   d. Continuous e. Discrete
4. a. Categorical
   b. Categorical
   c. Numerical (continuous)
   d. Numerical (discrete)
   e. Categorical
   f. Categorical
5. a. primary data b. secondary data
6. True
7. True
8. C
9. a. Discrete data deals with values that are exact and must be counted, for example the number of people at a football match. Continuous data deals with values that are measured and may be written in decimal form, for example the length of each football quarter.
   b. Primary data is data collected by me, for example a survey I conduct, and secondary data is data collected by someone else, for example AFL/NRL data.
10. In a Likert scale the data is split into categories, so even though these categories can be represented by numbers, the data is still categorical.
11. Although represented by digits, these types of data are categorical because they do not represent numerical ordered values.
12. Many possible answers. Some ideas: small amount of data; the company is motivated to find that soft drink is healthy.
13. Possible answers:
    a. How many minutes, on average, do you read per day?
    b. How many times a week do you play sport?
    c. How many times a week do you play computer games?
14. Possible answers:
    a. Do you do Maths homework every night?
    b. What books have you read this year?
    c. What suburb do you live in?
15. 1250
16. a. Primary data
    b. 5 is a small amount of data
    c. Data unreliable because Bill is motivated to get a certain result.
    d. More reliable because there is more data and there is no motivation to get a certain result.

### 14.3 Displaying data in tables

1. a.

| Score | Tally | Frequency |
|---|---|---|
| 0 | 卌 | 5 |
| 1 | \|\|\|\| | 4 |
| 2 | 卌\|\|\|\| | 9 |
| 3 | 卌\|\| | 7 |
| 4 | \|\| | 2 |
| 5 | \| | 1 |
| 6 | \| | 1 |
| 7 | | 0 |
| 8 | \| | 1 |
| | Total | 30 |

b. 30
c. 5
d. 12
e. 2
f. $\frac{2}{30}=\frac{1}{15}$

2. a.

<table><tr><th>Value (thousand dollars)</th><th>Tally</th><th>Frequency</th></tr>
<tr><td>100–109</td><td>|||</td><td>3</td></tr>
<tr><td>110–119</td><td>|</td><td>1</td></tr>
<tr><td>120–129</td><td></td><td>0</td></tr>
<tr><td>130–139</td><td>|</td><td>1</td></tr>
<tr><td>140–149</td><td>|||</td><td>3</td></tr>
<tr><td>150–159</td><td>|||</td><td>3</td></tr>
<tr><td>160–169</td><td>||</td><td>2</td></tr>
<tr><td>170–179</td><td>|</td><td>1</td></tr>
<tr><td>180–189</td><td>||</td><td>2</td></tr>
<tr><td>190–199</td><td>|</td><td>1</td></tr>
<tr><td>200–209</td><td>|</td><td>1</td></tr>
<tr><td></td><td>Total</td><td>18</td></tr></table>

b. 18

c. \$100 000 – \$109 999, \$140 000 – \$149 999 and \$150 000 – \$159 999

d. \$120 000 – \$129 999

e. 3

3. a. 33 b. 3 c. 10

d. 23 e. $\frac{1}{3}$

4. a.

<table><tr><th>Score</th><th>Tally</th><th>Frequency</th></tr>
<tr><td>5</td><td>|</td><td>1</td></tr>
<tr><td>6</td><td>||</td><td>2</td></tr>
<tr><td>7</td><td>||</td><td>2</td></tr>
<tr><td>8</td><td>|</td><td>1</td></tr>
<tr><td>9</td><td>|||</td><td>3</td></tr>
<tr><td>10</td><td>|</td><td>1</td></tr>
<tr><td></td><td>Total</td><td>10</td></tr></table>

b.

<table><tr><th>Score</th><th>Tally</th><th>Frequency</th></tr>
<tr><td>0</td><td>~~||||~~</td><td>5</td></tr>
<tr><td>1</td><td>~~||||~~</td><td>5</td></tr>
<tr><td>2</td><td>~~||||~~</td><td>5</td></tr>
<tr><td>3</td><td>||</td><td>2</td></tr>
<tr><td>4</td><td>||</td><td>2</td></tr>
<tr><td>5</td><td></td><td>0</td></tr>
<tr><td>6</td><td>|</td><td>1</td></tr>
<tr><td></td><td>Total</td><td>20</td></tr></table>

c.

<table><tr><th>Score</th><th>Tally</th><th>Frequency</th></tr>
<tr><td>45</td><td>|</td><td>1</td></tr>
<tr><td>46</td><td>|</td><td>1</td></tr>
<tr><td>47</td><td></td><td>0</td></tr>
<tr><td>48</td><td>||</td><td>2</td></tr>
<tr><td>49</td><td>|</td><td>1</td></tr>
<tr><td>50</td><td>||</td><td>2</td></tr>
<tr><td>51</td><td>||</td><td>2</td></tr>
<tr><td>52</td><td>||</td><td>2</td></tr>
<tr><td>53</td><td>||||</td><td>4</td></tr>
<tr><td>54</td><td>|||</td><td>3</td></tr>
<tr><td>55</td><td>|</td><td>1</td></tr>
<tr><td>56</td><td>||</td><td>2</td></tr>
<tr><td>57</td><td>||</td><td>2</td></tr>
<tr><td>58</td><td>||</td><td>2</td></tr>
<tr><td>59</td><td>|</td><td>1</td></tr>
<tr><td>60</td><td>||</td><td>2</td></tr>
<tr><td></td><td>Total</td><td>28</td></tr></table>

d.

<table><tr><th>Score</th><th>Tally</th><th>Frequency</th></tr>
<tr><td>143</td><td>||</td><td>2</td></tr>
<tr><td>144</td><td>~~||||~~ |</td><td>6</td></tr>
<tr><td>145</td><td>~~||||~~</td><td>5</td></tr>
<tr><td>146</td><td>|</td><td>1</td></tr>
<tr><td>147</td><td>|||</td><td>3</td></tr>
<tr><td>148</td><td>|||</td><td>3</td></tr>
<tr><td>149</td><td>|||</td><td>3</td></tr>
<tr><td>150</td><td>||||</td><td>4</td></tr>
<tr><td>151</td><td></td><td>0</td></tr>
<tr><td>152</td><td>|</td><td>1</td></tr>
<tr><td></td><td>Total</td><td>28</td></tr></table>

5. a. The list is messy. It is difficult to see how many different sports there are or to tell how many people prefer a particular sport.

b.

| Score | Tally | Frequency |
|---|---|---|
| Hockey | 𝍸𝍷𝍷𝍷 | 8 |
| Cricket | 𝍸𝍷 | 6 |
| Tennis | 𝍸𝍷𝍷 | 7 |
| Netball | 𝍸𝍷𝍷𝍷𝍷 | 9 |
| Swimming | 𝍸𝍷𝍷 | 7 |
| Golf | 𝍷𝍷𝍷𝍷 | 4 |
| Scuba diving | 𝍷𝍷𝍷 | 3 |
| Lacrosse | 𝍷𝍷𝍷𝍷 | 4 |
| Lawn bowls | 𝍷𝍷 | 2 |
| | Total | 50 |

c. i. Netball

ii. Lawn bowls

d. Yes, Tennis and swimming had a frequency of 7, and golf and lacrosse had a frequency of 4.

6. a.

| Score | Tally | Frequency |
|---|---|---|
| 0 | 𝍸 𝍸 𝍷 | 11 |
| 1 | 𝍸 𝍸 | 10 |
| 2 | 𝍸𝍷 | 6 |
| 3 | 𝍷𝍷 | 2 |
| 4 | 𝍷 | 1 |
| | Total | 30 |

b. 11

c. 9

d. 0

e. 4

f. $\frac{1}{5}$

7.

| Height | Tally | Frequency |
|---|---|---|
| 140–144 | 𝍷𝍷 | 2 |
| 145–149 | 𝍸𝍷 | 6 |
| 150–154 | 𝍷𝍷𝍷𝍷 | 4 |
| 155–159 | 𝍷 | 1 |
| 160–164 | 𝍸 | 5 |
| 165–169 | 𝍸𝍷 | 6 |
| 170–174 | 𝍷𝍷 | 2 |
| 175–179 | 𝍷 | 1 |
| 180–184 | 𝍷 | 1 |
| | Total | 28 |

8. a.

| Type | Tally | Frequency |
|---|---|---|
| 1-bedroom unit | 𝍷𝍷𝍷 | 3 |
| 2-bedroom unit | 𝍷𝍷𝍷𝍷 | 4 |
| 2-bedroom house | 𝍷 | 1 |
| 3-bedroom house | 𝍸𝍷 | 6 |
| 4-bedroom house | 𝍷𝍷 | 2 |
| | Total | 16 |

A 3-bedroom house was the most popular.

b. C

c. A

9. B

10. Because it makes it easier to collect and read the data.

11. When selecting a class interval we need to select ranges that are narrow enough to highlight variations in the data, and that are also wide enough to ensure the amount of data we are looking at is manageable.

12. Class intervals can be used to group continuous numerical data so that it can be represented in tables and other graphical forms.

13. a. 25

b.

| Number of pets | Frequency | % Frequency |
|---|---|---|
| 0 | 8 | 32 |
| 1 | 11 | 44 |
| 2 | 3 | 12 |
| 3 | 2 | 8 |
| 4 | 0 | 0 |
| 5 | 1 | 4 |

c. 6

14. O: 21.5%, U: 8.3%

15.

| Score | Frequency |
|---|---|
| 0 | 4 |
| 1 | 16 |
| 2 | 20 |
| 3 | 4 |
| 4 | 1 |
| Total | 45 |

## 14.4 Measures of centre and spread

1. a. 5 b. 6 c. 4
   d. 6 e. 4
2. a. 3 b. 12 c. 14
   d. 15 e. 13
3. a. 3 b. 3 c. 4
   d. 5 e. 2 f. 3
4. a. 5 b. 3 c. 3 d. 7
5. a. 11 b. 4 c. 2.5 d. 1.5
6. a. 2 b. 5 c. 7 d. 2.5
7. a. 5 b. 6 c. 7.5 d. 7.5
8. a. 4 b. 8 c. 2
   d. 4 e. 2
9. a. 3 b. 10 and 12 c. 4
   d. 9 e. no mode
10. a. 32 b. 24.1
11. B
12. B
13. D
14. D
15. C
16. 85%
17. 38
18. 104 shoes
19. 47 students
20. a. The average number of goals scored in each match, calculated by summing the total number of goals scored and dividing by the number of matches played.
    b. The average house price, calculated by listing all of the house prices and finding the value that has the same number of houses being more expensive and less expensive than it.
    c. The most popular shoe size.
21. By calculating measures of centre and spread we can see how data is grouped around a central point. This allows us to see any patterns in the data and interpret those patterns.
22. a. 101.9
    b. Both the mode (115) and median (112) give a better impression of how the team has performed, even though they could give a 'misleading' impression of the team's performance.
23. a. Mode = 12
    b. The mode shows which size to order more of. The mean and median would not show the more common sizes and would give only an indication of the middle of the range.
24. We cannot calculate the mean, median and range of categorical data, because these values are all calculated using the numerical values of the variables. We can calculate the mode of categorical data, which is the most popular or frequency category.
25. a. Tom: $1.30, Sarah: $1.27
    b. Sarah's sells cheaper petrol.
    c. Tom may have seen the price on a very expensive day. There's a greater range of values at Sarah's petrol station.
26. a. 1.75 m
    b. 2.21 m
    c. Based on her past performances and her past range of values, she cannot jump this high.
27. 72
28. 89.5%
29. The coach's statement was not correct.
30. One possible answer is 5, 6, 12, 13, 14. If the 5 numbers are in ascending order, the 3rd number must be 12 and the other 4 numbers must add up to 38.
31. The highest score is 49 and the lowest score is 0. The scores would be 0, 1, 12, 13, 49
32. a. 15
    b. 3
    c. 0
    d. The median is the best measure because the mode is the lowest of all scores and the mean is inflated by one much larger figure.
33. 133 cm, 133 cm, 142 cm, 151 cm
34. Largest value for the median is 15. Scores would be 0, 2, 15, 16, 17 or 0, 1, 15, 16, 18. Smallest value for the median is 2. Scores would be 0, 1, 2, $a$, $b$, where $a+b=47$
35. 43; highest score: 100; lowest score: 57. The scores would be 57, 77, 78, 79, 80, 97, 98, 99, 100

## 14.5 Column graphs and dot plots

1.

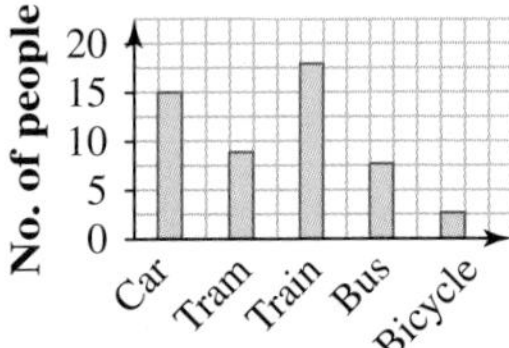


2.

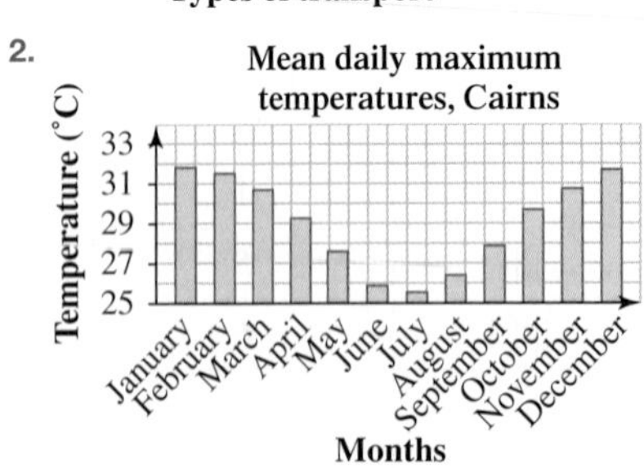

3.

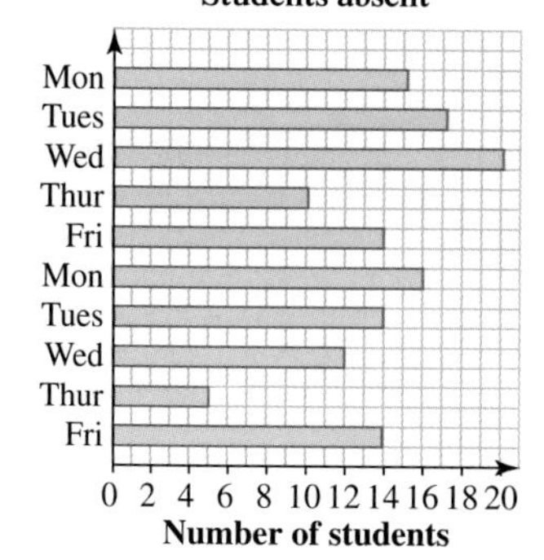


4. a.

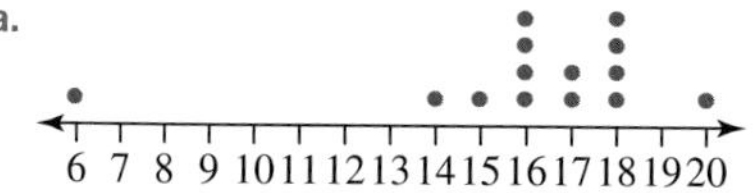


b. The scores lie between and include 6 to 20; that is, a range of fourteen.
Mostly 16 to 18 packets were sold. Sales of 6 and 20 packets of chips were extremely low. A provision of 20 packets of chips each day should cover the most extreme demands.

5. a.

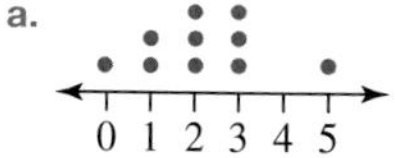


b.

18 19 20 21 22

c.

5.0 5.2 5.4 5.6 5.8 6.0
5.1 5.3 5.5 5.7 5.9

d.

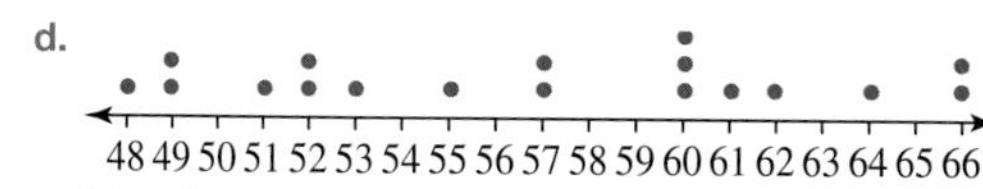


6. a. Months
b. Bill total
c. Every 3 months (quarterly)
d. January 2021
e. A graph like this is useful to monitor the spending pattern over time and to decide whether there have been any unusual increases.
f. It would not be normal, but there might be reasons to explain why it was so much higher, for example making a lot of overseas phone calls.

7. a. i. 45
ii. 42
iii. 61
b. 5 and 11
c. It does indicate an improvement, despite these two low weeks, because there is an overall upward trend throughout the time period.

8. a. Betty
b. 1 April: 70 m, 29 April: 73 m
c. Rachel
d. Yes. Sarah remained the same. Sumira and Paula sprinted at a slower pace.
e. Yes, the graph could be misleading because the vertical scale begins at 30 m and not at 0 m. If the vertical scale was not read properly, it might appear that the students covered a greater distance in a shorter period of time.
f. The graph's vertical axis starts at 30 m because no-one ran less than that distance and the teacher wanted to use all the space to show small differences in the distances.

9. By aligning the data points neatly above the value they are representing, these data points cannot be mistakenly seen as representing any other value. A dot plot is also neater and easier to interpret.

10. Column and bar graphs have discrete columns and bars, which are perfect for representing categorical data. There are often better graphs for displaying numerical data.

11. a–d. Generally, outliers affect the mean and range, have little effect on the median, and have no effect on the mode.

12. a.

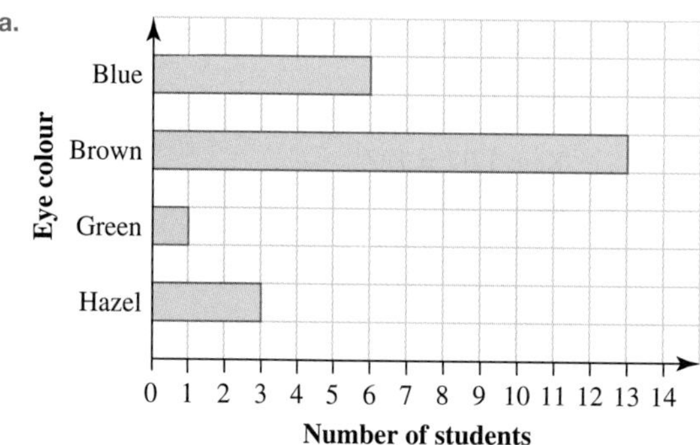


b. The data is categorical, not numerical. Therefore there is no median.
c. Brown
d. $\frac{10}{23} \approx 43\%$

13. a.

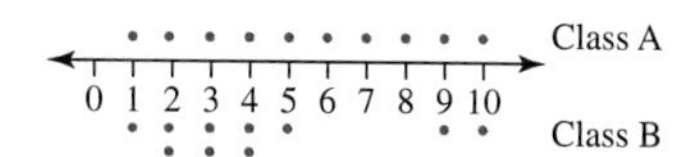


b. Class A: Mean 5.5, median 5.5, no mode, range 9; Class B: Mean 4.3, median 3.5, modes 2, 3 and 4, range 9
c. Class A has a higher mean and median than Class B, while the range is the same for both classes. The results for Class A are scattered throughout the whole range, while those for Class B are concentrated more towards the lower end of the range.

14. a. 35
b. Each week the number of fridges sold increases by 5.
c. $F = 25 + 5w$
d. $F = 25 + 5(10)$
$= 25 + 50$
$= 75$

## 14.6 Stem-and-leaf plots

**1.** Key: 2|7 = 27

| *Stem* | *Leaf* |
|---|---|
| 2 | 11558899 |
| 3 | 11379 |
| 4 | 01167 |
| 5 | 13348 |
| 6 | 2222778 |
| 7 | 12338 |
| 8 | 12 |
| 9 | 233 |

**2.** Key: 3|6 = 36

| *Stem* | *Leaf* |
|---|---|
| 0 | 0123678 |
| 1 | 022224566789 |
| 2 | 12578 |
| 3 | 11223 |
| 4 | 0 |

**3. a.** Key: 12|7 = 127

| *Stem* | *Leaf* |
|---|---|
| 10 | 8 |
| 11 | 678 |
| 12 | 9 |
| 13 | 27 |
| 14 | 5 |
| 15 | 2 |
| 16 | 45 |
| 17 | 2 |
| 18 | 9 |

**b.** Key: 13|2 = 132

| *Stem* | *Leaf* |
|---|---|
| 11 | 4 |
| 12 | 4 |
| 13 | 01 |
| 14 | 6 |
| 15 | 02 |
| 16 | |
| 17 | 13 |

**c.** Key: 23|7 = 237

| *Stem* | *Leaf* |
|---|---|
| 19 | 0 |
| 20 | 578 |
| 21 | |
| 22 | 5 |
| 23 | 79 |
| 24 | 4888 |
| 25 | 5 |

**d.** Key: 65|2 = 652

| *Stem* | *Leaf* |
|---|---|
| 64 | 1577 |
| 65 | |
| 66 | 2 |
| 67 | 5 |
| 68 | 5 |
| 69 | 1 |
| 70 | 8 |
| 71 | |
| 72 | |
| 73 | 56 |
| 74 | 8 |

**4. a.** Key: 1|7 = 1.7

| *Stem* | *Leaf* |
|---|---|
| 1 | 279 |
| 2 | 238 |
| 3 | 9 |
| 4 | 16 |
| 5 | 48 |

**b.** Key: 6|7 = 6.7

| *Stem* | *Leaf* |
|---|---|
| 6 | 059 |
| 7 | 457 |
| 8 | 348 |
| 9 | 238 |

**c.** Key: 13|7 = 13.7

| *Stem* | *Leaf* |
|---|---|
| 13 | 0899 |
| 14 | 3556778 |
| 15 | 12679 |
| 16 | 2 |

d. Key: 1|5 = 0.15

| Stem | Leaf |
|---|---|
| 0 | 2 7 |
| 1 | 5 8 |
| 2 | |
| 3 | 7 |
| 4 | 4 |
| 5 | 1 |
| 6 | 7 |

5.

Key: 0|2 = 2

| *Leaf (Mitch)* | *Stem* | *Leaf (Yani)* |
|---|---|---|
| 3 2 1 1 1 1 0 0 0 0 | 0 | 0 0 1 1 1 1 1 2 2 2 |

Mitch scored between 0 and 3 goals inclusive.
Yani scored between 0 and 2 goals inclusive.

6. a. Mitch: 2, Yani: 3
b. Mitch
c. Yani: 11
d. Yani, as he scored goals in more games.

7. a. 22 b. 17
c. 64 d. 13
e. 42 f. 0

8. a. Key: 10|8 = 10.8

| *Stem* | *Leaf* |
|---|---|
| 10 | 8 9 |
| 11 | 0 2 5 8 |
| 12 | 0 4 5 6 8 8 |
| 13 | 2 6 6 9 9 |
| 14 | 0 1 |
| 15 | 0 |

b. Range = 4.2 There is a reasonably large range of sprint times in this sample.
c. The majority of these runners completed the 100 m sprint between 11 and 13.9 seconds. The average sprint time was 12.68 s and the distribution is approximately evenly spread around the middle.

9. Stem plots need a key because these plots can show a variety of data, ranging from whole numbers to decimals. A key helps to show what the data represents. For example, a stem plot key showing data could be 15|1 = 151 while a stem plot key showing decimals could be 11|1 = 11.1. In these examples, the number on the leaf represents units (in the first example) and decimal values (in the second example).

10. By ordering a dot plot it is easy to identify key values, for example the median, the lowest value and the highest value. By aligning the data values we can easily see how many values lie in each class interval. We can also see the shape of the data.

11. a.

**Test scores (%)**
Key: 5|6 = 56%

| *Leaf (Class 7F)* | *Stem* | *Leaf (Class 7G)* |
|---|---|---|
| 3 1 | 4 | |
| | 5 | |
| | 6 | 6 8 |
| 9 8 7 7 | 7 | 4 5 6 6 8 8 |
| 8 5 4 2 1 0 | 8 | 0 2 2 3 5 9 |
| 2 1 | 9 | |

b. 7F: mean = 77, median = 80.5
7G: mean = 78, median = 78
c. Since class 7F has two outliers, the best measure of centre is the median, not the mean.

12. a. 55 b. 43 c. 26

13. a.

**Temperatures (°C)**
Key: 2 | 1 = 21°C

| *Leaf (City A)* | *Stem* | *Leaf (City B)* |
|---|---|---|
| | 1 | 2 2 3 3 3 7 |
| 9 5 4 3 1 | 2 | 0 |
| 6 0 | 3 | |

b. City A: range = 15, median = 25, mean = 26.9, there is no mode.
City B: range = 8, median = 13, mean = 14.3, mode = 13.
c. In February the northern hemisphere is in winter and the southern hemisphere is in summer. Looking at the answers to part **b** City A has a higher mean and median temperature, so is most likely in the southern hemisphere. City B has a lower mean and median, so is in the northern hemisphere.

14. a.

Key: 3|2 = 32

| *Leaf* (9*A*) | *Stem* | *Leaf* (9*B*) |
|---|---|---|
| 2 | 3 | |
| 9 | 4 | 6 |
| 7 4 | 5 | 5 |
| 8 7 5 5 0 | 6 | 0 2 8 |
| 3 | 7 | 3 3 4 5 7 8 |
| 5 | 8 | 1 |
| 6 | 9 | |

b. 9A: 83%; 9B: 92%
c. 9A: 2
d. For class 9A, exam marks are quite spread out overall, but there is a cluster of students scoring in the 60% range. For class 9B, the spread of exam marks is smaller — they are generally clustered in the 70% range.
e. See the dot plot at the foot of the page.*

*14. e.

32 34 36 38 40 42 44 46 48 50 52 54 56 58 60 62 64 66 68 70 72 74 76 78 80 82 84 86 88 90 92 94 96
33 35 37 39 41 43 45 47 49 51 53 55 57 59 61 63 65 67 69 71 73 75 77 79 81 83 85 87 89 91 93 95

Percentages

f. The dot plot tells a slightly different story than the stem-and-leaf plot. For class 9A there appears to be 3 outliers (32%, 85% and 96%). This was not apparent in the stem-and-leaf plot. The results do not seem clustered around any particular percentage range. For class 9B there is one clear outlier (there are possibly more) at 46%. The results still appear clustered in the 70% range.

## 14.7 Pie charts and divided bar graphs

1. a. 356 b. Bus

2. a.

| Sport | Number of students | Fraction of students | Angle at centre of circle |
|---|---|---|---|
| Basketball | 55 | $\frac{55}{180}$ | $\frac{55}{^{1}\cancel{180}} \times \frac{^{2}\cancel{360°}}{1} = 110°$ |
| Netball | 35 | $\frac{35}{180}$ | $\frac{35}{\cancel{180}} \times \frac{^{2}\cancel{360}}{1} = 70°$ |
| Soccer | 30 | $\frac{30}{180}$ | $\frac{30}{\cancel{180}} \times \frac{^{2}\cancel{360}}{1} = 60°$ |
| Football | 60 | $\frac{60}{180}$ | $\frac{60}{\cancel{180}} \times \frac{^{2}\cancel{360}}{1} = 120°$ |
| Total | 180 | | |

b, c. Sports played by Year 7 students

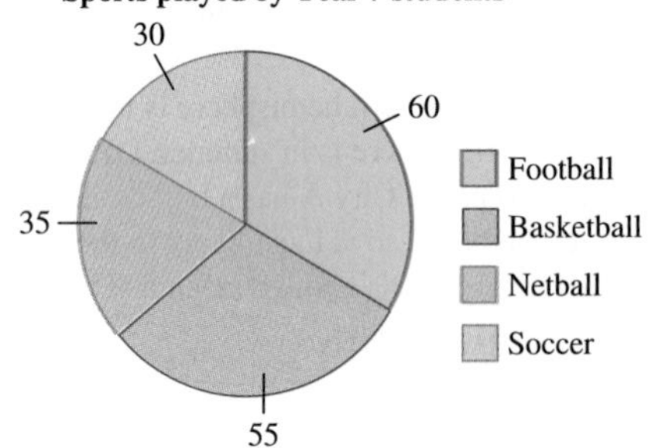


3. a. Ian
b. 17%
c. No, this sector graph gives percentages only, not actual values or numbers of people who voted.
d. 78.48°

4. a. 50%
b. 35%
c. 15%
d. 100%
e. i. 180° ii. 126° iii. 54°

5. a.

| Roofing material | Number of houses | Fraction | Angle size (°) |
|---|---|---|---|
| Concrete tiles | 12 | $\frac{1}{3}$ | 120 |
| Terracotta tiles | 9 | $\frac{1}{4}$ | 90 |
| Galvanised iron | 6 | $\frac{1}{6}$ | 60 |
| Colourbond | 3 | $\frac{1}{12}$ | 30 |
| Slate | 2 | $\frac{1}{18}$ | 20 |
| Shingles | 4 | $\frac{1}{9}$ | 40 |
| Total | 36 | 1 | 360 |

b. Roofing materials

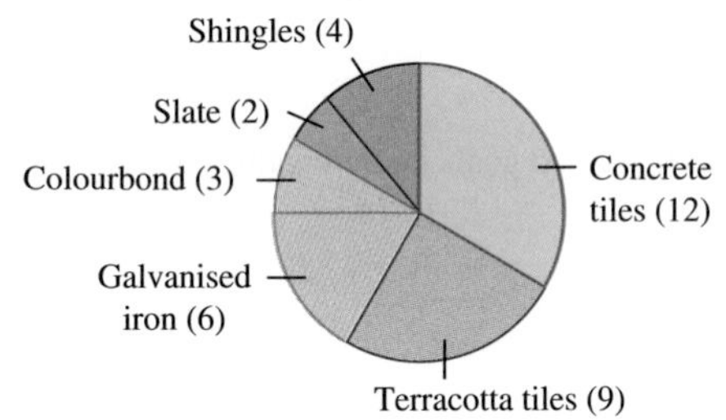


6. a. 45
b. 6
c. 12
d. 40%
e. German Shepherd

7. a.

| Type of meat | Amount sold (kg) | Fraction |
|---|---|---|
| Lamb | 10 | $\frac{1}{9}$ |
| Beef | 45 | $\frac{1}{2}$ |
| Pork | 5 | $\frac{1}{18}$ |
| Chicken | 15 | $\frac{1}{6}$ |
| Turkey | 10 | $\frac{1}{9}$ |
| Rabbit | 5 | $\frac{1}{18}$ |
| Total | 90 | 1 |

b, c. See the figure at the foot of the page.*

*7b, c.

Type of meat sold

45 | 15 | 10 | 10 | 5 | 5

Beef
Chicken
Lamb
Turkey
Pork
Rabbit

8. a.

| Type of bird | Number of birds | Fraction |
|---|---|---|
| Magpie | 24 | $\frac{2}{5}$ |
| Crow | 12 | $\frac{1}{5}$ |
| Cockatoo | 12 | $\frac{1}{5}$ |
| Wattlebird | 6 | $\frac{1}{10}$ |
| Rosella | 6 | $\frac{1}{10}$ |
| **Total** | 60 | |

b. i.

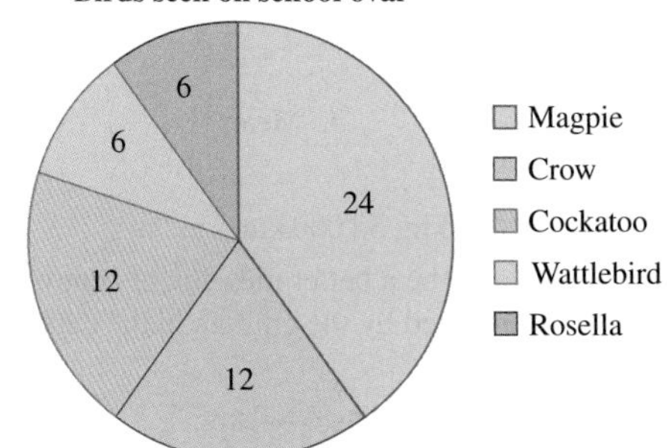


ii.

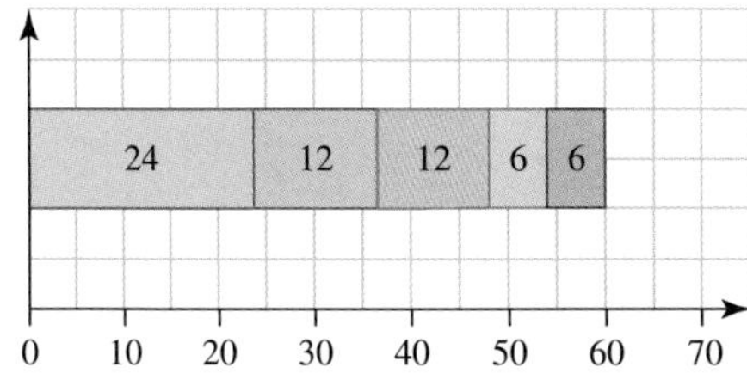


9. a, b.

| Transport | Number of students | Fraction | Length of bar |
|---|---|---|---|
| Walk | 50 | $\frac{50}{100}=\frac{1}{2}$ | $\frac{1}{2}\times 10 = 5\,\text{cm}$ |
| Car | 20 | $\frac{20}{100}=\frac{1}{5}$ | $\frac{1}{5}\times 10 = 2\,\text{cm}$ |
| Bus | 15 | $\frac{15}{100}=\frac{3}{20}$ | $\frac{3}{20}\times 10 =$ 1.5 cm |
| Bike | 15 | $\frac{15}{100}=\frac{3}{20}$ | $\frac{3}{20}\times 10 =$ 1.5 cm |
| **Total** | 100 | | |

c.

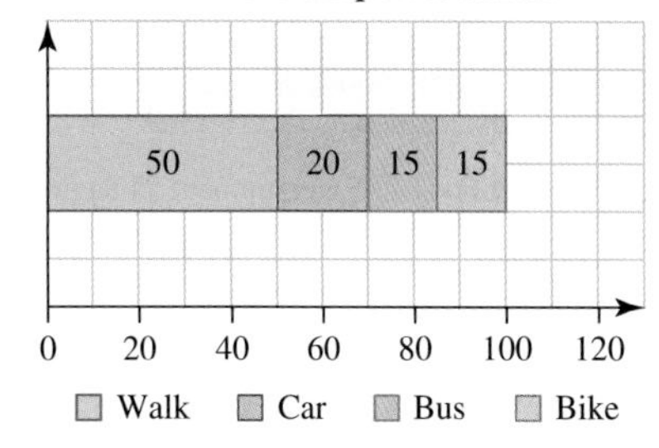


d. 2.5 cm

10. a. 100 b. 50 c. 25

11. Chair: 40; Table: 28; Couch: 12; Desk: 10.

12. Categorical data: column and bar graphs and sector graphs
Numerical data: dot plots, stem-and-leaf plots

13. A divided bar graph gives a graphical representation of the proportion of each category as a percentage of the whole.

14. a. Graph 1

b. In graph 1 there was a much greater proportion of ice creams and cold drinks sold than pies and hot drinks. Ice creams and cold drinks would be much more popular in summer than pies and hot drinks.

15. a.

| Sport | Percentage of total | Angle size (°) |
|---|---|---|
| Cricket | 20 | 72 |
| Hockey | 25 | 90 |
| Netball | 40 | 144 |
| Soccer | 10 | 36 |
| Tennis | 5 | 18 |
| Total | 100 | 360 |

**Sports**

Tennis (5)
Soccer (10)
Cricket (20)
Hockey (25)
Netball (40)

b. i. If there were only 10 students in the class, each student would represent 10% of the class total. It would therefore be impossible for only 5% of the class to play tennis.

ii. If there were 20 students in the class, each student would represent 5% of the class total. Since all of the percentages for the sports are multiples of 5%, this is a plausible scenario.

16. a. Rent and utilities

b. $800

c. i. $200 ii. $200 iii. $500

d. **Omar's new budget**

700 | 400 | 200 | 200 | 500

Rent, Utilities, Food, Entertainment, Savings

17. a. 25% b. 45° c. $0.375\ m^2$

18. $26\frac{2}{3}\%$

## 14.8 Comparing data

1. a. i. 82.4 ii. 82 iii. 81 iv. 9
   b. The mean, median and mode are all quite close, so any could be used as a measure of centre.
2. a. i. 2.63 ii. 2.55 iii. 2.4 iv. 0.7
   b. This distribution is quite spread out, with a significant number at the lower end and towards the top. For this reason the mode is probably the best measure of centre.
3. a. i. 4.5 ii. 5 iii. 5 iv. 10
   b. i. 4.1 ii. 5 iii. 5 iv. 7
   c. The mean and range are affected by the outlier. The median and mode have not been affected.
4. a. i. 6.26 ii. 6.27
   iii. 6.18 and 6.34 iv. 0.16
   b. The mean or median would probably be the best measures of centre in this case.
5. a. i. 22.6 ii. 23 iii. 23 iv. 34
   b. i. 19 ii. 22 iii. 23 iv. 15
   c. The mean and range are most affected by the outlier. The median is affected slightly, while the mode is not affected.
6. C
7. A
8. D
9. C
10. D
11. a.

| Score $x$ | Frequency $f$ | Freq. × score $f \times x$ |
|---|---|---|
| 0 | 1 | 0 |
| 1 | 3 | 3 |
| 2 | 6 | 12 |
| 3 | 3 | 9 |
| 4 | 3 | 12 |
| 5 | 1 | 5 |
| 6 | 1 | 6 |
| | $n = 18$ | $\sum fx = 47$ |

Mean $\approx 2.6$

b.

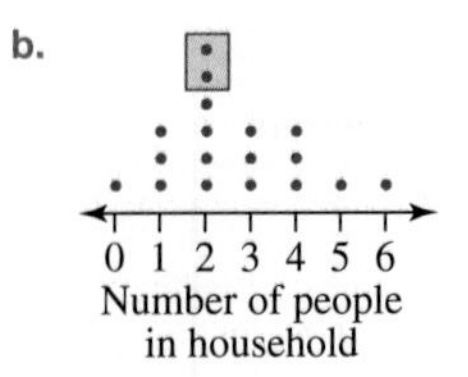


c. 2
d. i. Median ii. Mean iii. Mode

12. 75.6
13. a. The median was calculated by taking the average of the 2 middle scores.
    b. 13
14. Rebecca's strategy seems reasonable. Sample responses can be found in the worked solutions in your online resources.
15. a. \$18 000
    b. \$29 444
    c. \$26 000
    d. i. Mode ii. Mean
16. 7
17. a. 55 250 km, 52 000 km, 52 000 km
    b. The median would be a better measure of centre because it will be less affected by the outliers in the data than the mean.
    c. 51 810 km; It is reduced by 3440 km
    d. 52 000 km
    e. 3333
    f. 50 000 km; 92% last that distance or more.
18. 2

## Project

1. Sector graph
2. Families with no children in this age range is the most common category.
3. Column or bar graph
4. Responses will vary depending on the students surveyed. Here is a sample response.

| Number of children | Tally | Frequency | Percentage |
|---|---|---|---|
| 1 | \|\|\|\|\| | 6 | 24% |
| 2 | \|\|\|\|\|\|\|\|\|\| | 11 | 44% |
| 2 | \|\|\|\|\| | 5 | 20% |
| 4 or more | \|\|\| | 3 | 12% |
| | Total | 25 | |

5. Sample (the entire population was not surveyed)
6. Numerical data

7. Responses will vary depending on the students surveyed. The following is a sample response.

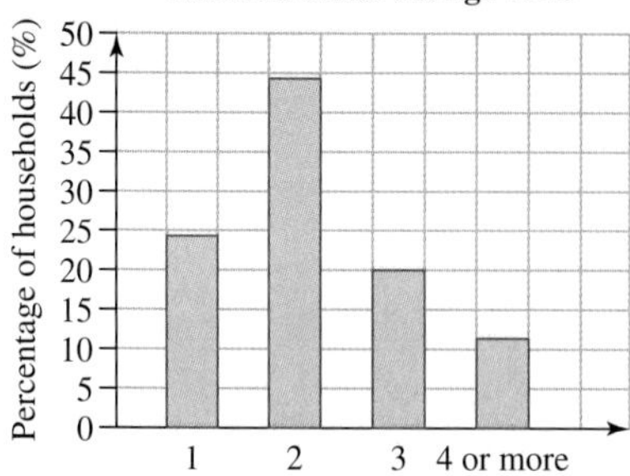


8. Responses will vary depending on the students surveyed. One major difference will be that none of the households surveyed have 0 children under 15 years as they all contain at least 1 Year 7 student. This is a major variation from the census data.

9. Responses will vary depending on the survey undertaken by the student.

## 14.9 Review questions

1. False
2. Continuous
3. Categorical
4. a. Coke b. Tea
5. a. 26 b. 14 c. 18 d. 14 e. 12
6. a.

| Cars | Frequency |
|---|---|
| 0 | 6 |
| 1 | 4 |
| 2 | 8 |
| 3 | 5 |
| 4 | 1 |
| Total | 24 |

b. 6
c. 14
d. 2
e. 4
f. $\frac{1}{3}$

7. a. i. 3.14 ii. 4 iii. 3
b. i. 3.71 ii. 4 iii. 4
c. i. 4.86 ii. 5 iii. 5
d. i. 2.14 ii. 1 iii. 1
e. i. 5.2 ii. 5 iii. 5
f. i. 5.78 ii. 8 iii. 6

8. a. Names b. 1 unit = 10%
c. John d. Peter
e. 58% f. 77%

9. a. Days of week b. Air quality rating
c. Northern region d. Wednesday
e. Eastern region

10. a. Key: $3|6 = 36$

| *Stem* | *Leaf* |
|---|---|
| 7 | 1 5 8 |
| 8 | 1 2 3 4 5 6 7 8 8 |
| 9 | 0 1 1 2 2 2 3 4 4 6 6 7 |
| 10 | 2 3 4 4 |
| 11 | 0 5 |

b. Only 3 cars were travelling under the 80 km/h speed limit. The slowest speed recorded was 71 km/h, while the fastest speed recorded was 115 km/h. Most of the recorded speeds were in the 90–99 km/h class interval. The most common (modal) speed recorded was 92 km/h.

11. a. Key: $3|1 = 31$

| *Leaf Set A* | Stem | *Leaf Set B* |
|---|---|---|
| 1 | 0 | 3 |
| 3 | 1 | |
| 8 | 2 | 9 |
| 0 | 3 | 0 1 2 8 |
| 5 | 4 | 1 |
| 9 | 5 | |
| 4 | 6 | 8 |

b. Set A
i. 1
ii. 64
iii. 37.1
iv. 37.5
v. 63

Set B
i. 3
ii. 68
iii. 34
iv. 31.5
v. 65

c. The two sets have similar minimum and maximum values, giving them a similar range. The mean of Set A is slightly higher than that of Set B. The scores of Set A are spread out more evenly than those in Set B, which tend to be clustered towards the middle.

12. a. i. 6.55 ii. 6 iii. 6 iv. 9
b. There appear to be two outliers (11 and 12). The mean, median and mode values are quite close, indicating that it is common for 6 soft drinks to be sold quite frequently. The minimum number sold over the period is 3, while the maximum is 12.

13. a. i. 29 years    ii. 27 years
iii. 18 years    iv. 39 years

b. The ages of the group range from 14 years to 53 years, with most of the group being under 40. The mean or median values would be a good representation of the age of the group.

14. **People who ate different types of breakfast**

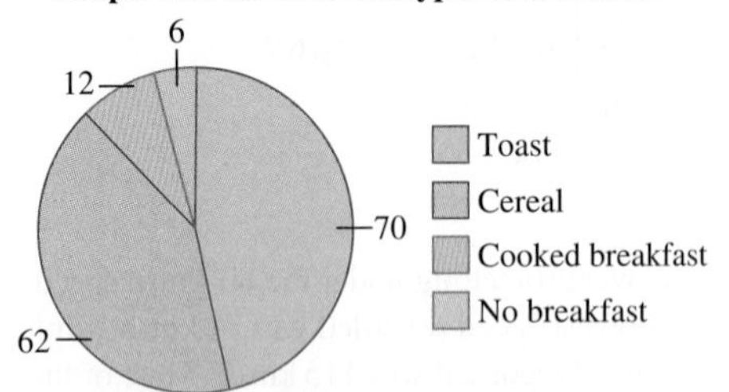


15. See the figure at the foot of the page.*

16. a. 10 °C    b. Day 5    c. Day 7
d. Day 6    e. 10 °C    f. Days 6 and 7

17. a. 81.56    b. 74.22    c. Test 1
d. 6    e. 4

18. a. 85
b. 63
c. 85
d. 68
e. John 89.5, Peter 74, Mark 76.5, James 74.5, David 85.5, Rachel 73.5, Mary 81, Eve 78.5, Esther 68
f. 21.5

19. a. 41 cm
b. 154.8 cm
c. 156 cm
d. 156 cm
e. Yes

20. a. Months
b. Number of flies
c. 10 000
d. 28 000
e. 1 cm = 5000 flies (if beginning from 0 flies) or
1 cm = 2000 flies (if beginning from 10 000 flies)

f. **Fly population**

Number of flies (thousands): 0, 5, 10, 15, 20, 25, 30

Months: January, February, March, April, May, June, July, August, September, October, November, December

21. a.

| $x$ | $f$ |
|---|---|
| 0 | 4 |
| 1 | 5 |
| 2 | 7 |
| 3 | 3 |
| 4 | 0 |
| 5 | 1 |
| Total | 20 |

b.

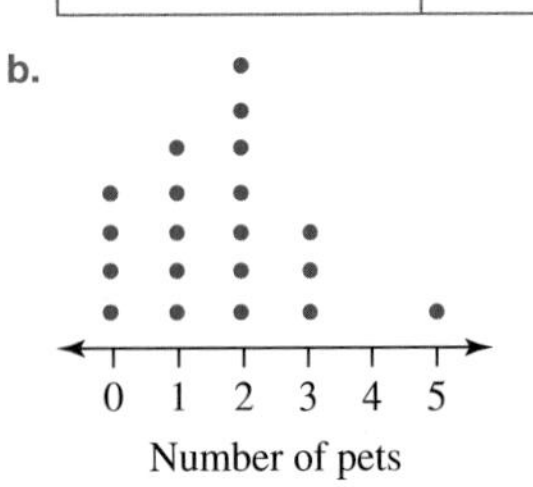


c. 2, 2, 1.65

22. a. i. 5
ii. 5.5
iii. 7.9
iv. 25
v. 6
vi. 16.9, 15.5

b. When 10 was added to each score, both the mean and median increased by exactly 10.

23. The boys have a mean height of 158 cm, with a median height of 158 cm and three modes (148 cm, 158 cm, and 166 cm). They range in height from 138 cm to 178 cm.

The girls have a mean height of 153 cm, with a median height of 152 cm and no mode. They range in height from 137 cm to 171 cm.

The shortest person is a girl, while the tallest person is a boy. On the whole, the boys are taller than the girls.

*15 **Hobbies**

30 | 20 | 20 | 15 | 15

Skateboarding
Stamp collecting
Sailboarding
Photography
Model making

# Semester review 2

The learnON platform is a powerful tool that enables students to complete revision independently and allows teachers to set mixed and spaced practice with ease.

## Student self-study

Review the **Course Content** to determine which topics and lessons you studied throughout the year. Notice the green bubbles showing which elements were covered.

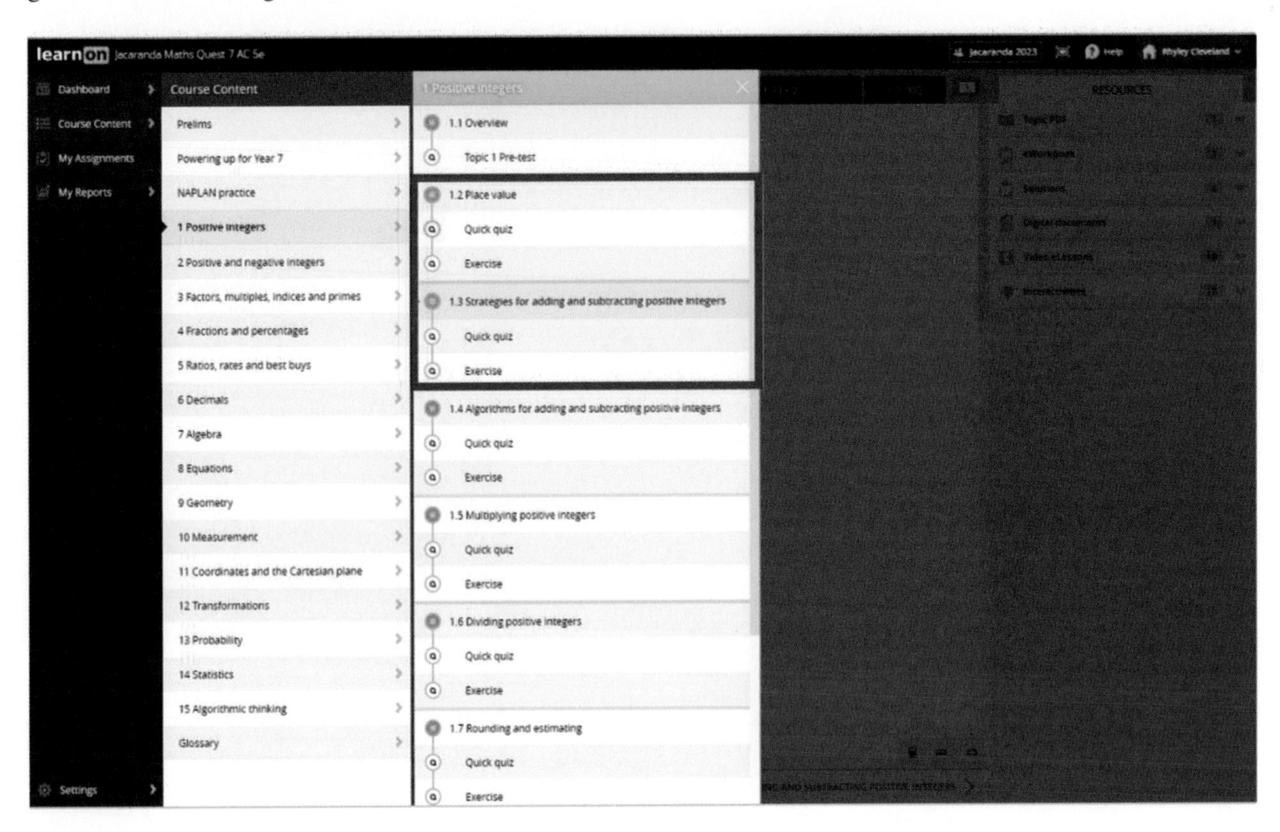


Review your results in **My Reports** and highlight the areas where you may need additional practice.

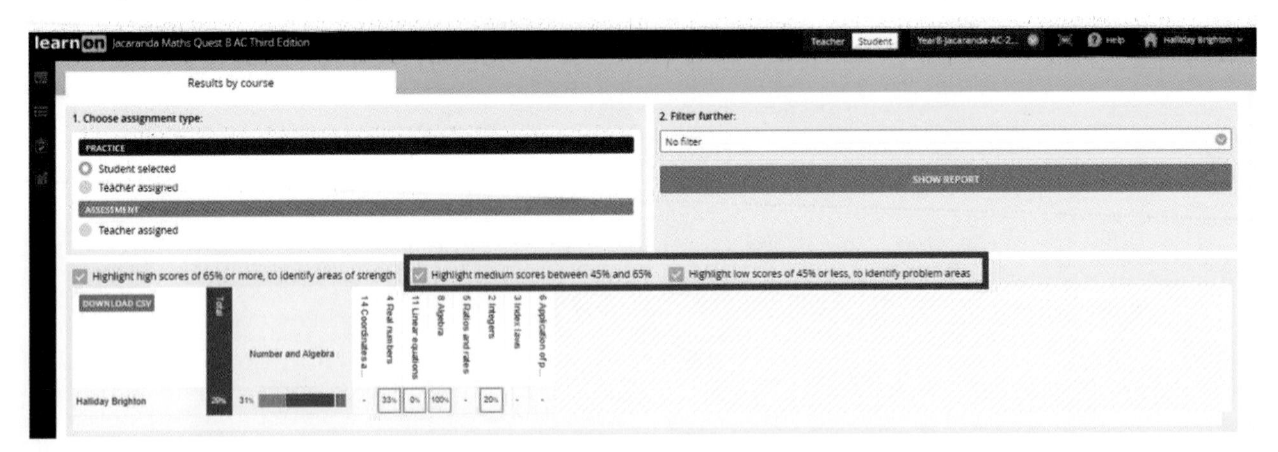


Use these and other tools to help identify areas of strengths and weakness and target those areas for improvement.

## Teachers

It is possible to set questions that span multiple topics. These assignments can be given to individual students, to groups or to the whole class in a few easy steps.

Go to **Menu** and select **Assignments** and then **Create Assignment**. You can select questions from one or many topics simply by ticking the boxes as shown below.

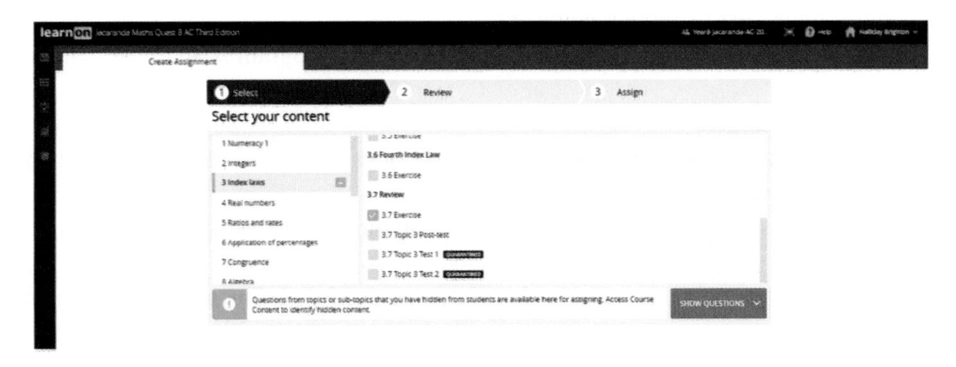


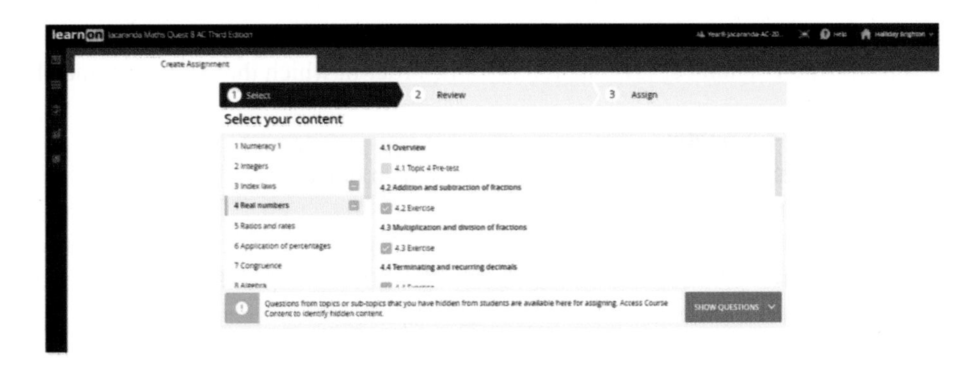


Once your selections are made, you can assign to your whole class or subsets of your class, with individualised start and finish times. You can also share with other teachers.

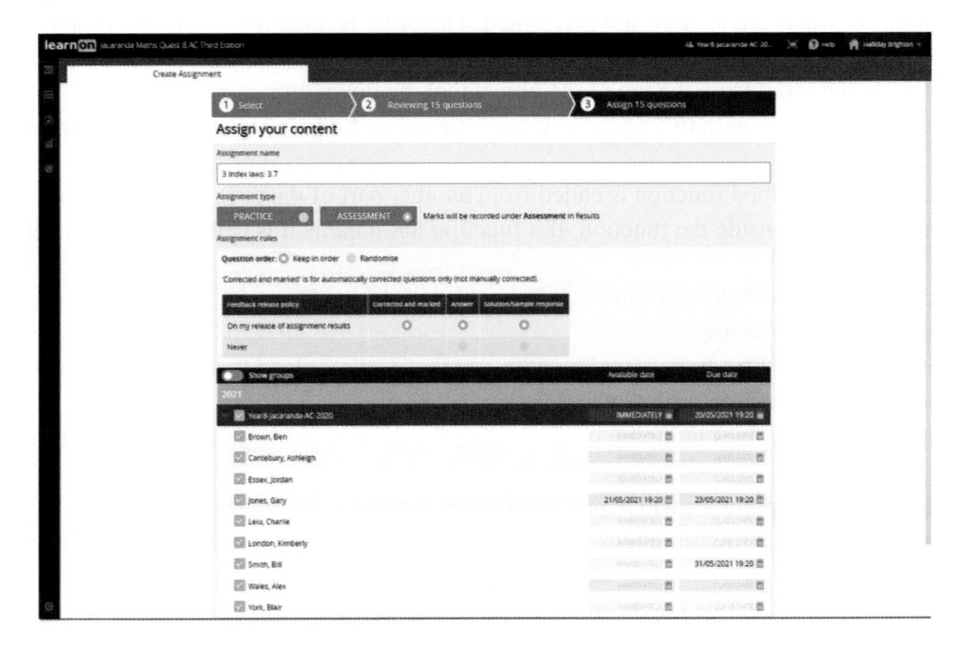


More instructions and helpful hints are available at www.jacplus.com.au.

# GLOSSARY

**3-dimensional object** shape that has depth, as well as length and width; also called a polyhedron
**acute angle** angle greater than 0° but less than 90°
**acute-angled triangle** triangle where each angle is acute; that is, less than 90°
**adjacent angles** angles lying next to another angle
**algorithm** a step-by-step set of tasks to solve a particular problem. A program is an implementation of an algorithm.
**allied angles** *see* **co-interior angles**
**alternate angles** angles 'inside' the parallel lines, on alternate sides of a transversal; think of them as Z-shaped
**arc** a section of the circumference of a circle
**area** amount of surface enclosed by a shape; measured in square units, such as square metres, $m^2$, and square kilometres, $km^2$
**arithmagon** triangular figure in which the two numbers at the ends of each line add to the number along the line
**ascending order** in order from smallest to largest
**Associative Law** a number law that refers to the order in which three numbers may be added, subtracted, multiplied or divided, taking two at a time
**Associative Law for addition** a number law that refers to the order in which three numbers may be added or subtracted
**axis of symmetry** *see* **line of symmetry**
**back elevation** the view observed from standing directly behind an object
**back view** *see* **back elevation**
**back-to-back stem-and-leaf plot** a type of table used to organise, display and compare two sets of data
**backtracking** process of working backwards through a flow chart. Each step is the inverse operation of the corresponding step in the flow chart. It can be used to solve equations.
**bar graph** graph drawn in a similar way to a column graph, with horizontal bars instead of vertical columns. Categories are graphed on the vertical axis and frequencies (numbers) on the horizontal axis.
**base** the number being repeatedly multiplied in a power term. The power is the number of times the base is written.
**BIDMAS** the order in which calculations are performed. The order is: brackets; index (or power); division; multiplication from left to right; addition; and subtraction from left to right.
**bird's eye view** view observed from the top of the object looking down on it
**Boolean** a JavaScript data type with two possible values: `true` or `false`. JavaScript Booleans are used to make logical decisions.
**called** in programming, a defined function is called from another part of the program. Calling a function will run the statements defined inside the function. If a function has a name it is called with the statement `name();`.
**capacity** maximum amount of fluid that can be contained in an object. The units may be the same as for volume or litres and millilitres may be used.
**Cartesian plane** area formed by a horizontal line with a scale ($x$-axis) joined to a vertical line with a scale ($y$-axis). The point of intersection of the lines is called the origin.
**categorical data** data that involves grouping or classifying things, not numbers; for example, grouping hair colour
**centre of rotation** a specific point around which an object is rotated
**character** in programming, a string of length `1`. A JavaScript character is used to represent a letter, digit or symbol.
**chord** a straight line joining any two points on the circumference of a circle
**circumference** the distance around a circle
**clustering** a grouping of values around a particular value

**coefficient** the number written in front of a term
**co-interior angles** angles on the same side of the transversal positioned between the parallel lines. These angles are supplementary.
**column graph** graph in which equal width columns are used to represent the frequencies (numbers) of different categories
**common factor** a factor common to more than one number
**common multiple** a number that is a multiple of two or more numbers
**Commutative Law** a rule for a particular operation showing that the result is the same regardless of the order of the numbers
**Commutative Law for addition** a number law that refers to the order in which two numbers may be added or subtracted
**complementary angles** angles that sum (add) to 90°
**complete revolution** an angle of 360° (a full circle)
**composite number** number with more than two factors
**composite shape** figure made up of more than one basic shape
**console** a special region in a web browser for monitoring the running of JavaScript programs
**constant** a value that is not a variable
**continuous** describes data that take on any value that is measurable; for example, the heights of people in a class
**conversion factor** used to convert one unit of measurement to another
**convert** to change a measurement or value to different units
**convex** a shape with all its interior angles less than 180°
**coordinate** a pair of numbers that gives the position of a point on the Cartesian plane. The first number is the $x$-coordinate and the second number is the $y$-coordinate.
**corresponding angles** angles on the same side of the transversal that are both either above or below the parallel lines. The angles are equal in size.
**cross-section** the two-dimensional shape we can see when we cut straight through a three-dimensional object
**cube number** a number that can be arranged in a cube
**cube root** number that multiplies by itself three times to equal the original number. Finding a cube root is the opposite of cubing a number.
**data** information collected from surveys
**decimal places** digits after the decimal point
**decimal point** a dot, placed after a whole number, used to represent parts of a decimal fraction
**decomposition method** method of subtraction. The larger number is decomposed by taking 10 from the tens column and adding it to the units column. It may also involve taking 100 from the hundreds column and adding it to the tens column.
**denominator** bottom term of a fraction. It shows the total number of parts into which the whole has been divided.
**descending order** numbers in order from largest to smallest
**diameter** the straight-line distance across a circle through its centre. The diameter is twice the radius.
**difference** the answer obtained when two numbers are subtracted
**directed number** a number with a positive or negative sign indicating that it is a certain direction from the origin along a number line
**discrete** describes data that takes on exact values that may be counted; for example, the number of televisions sold each week
**Distributive Law** a rule stating that each term inside a pair of brackets is to be multiplied by the term outside the brackets
**divided bar graph** a representation of data whereby the different sectors of a bar represent different groups of data
**dividend** the quantity in a division that is to be divided
**divisibility test** a series of rules that tests for factors of a number
**division** the sharing of a number into equal parts; the opposite of multiplication

**divisor** the quantity in a division by which the dividend is divided
**domain** the set of all $x$-values in a relationship between $x$ and $y$
**dot plot** consists of a horizontal axis labelled and evenly scaled, with each data value represented by a dot
**equally likely outcomes** two or more outcomes that have an identical chance of occurring
**equation** mathematical statement containing a left-hand side, a right-hand side and an equals sign between them
**equilateral triangle** triangle with all sides equal in length. All angles will be equal too.
**equivalent expressions** two or more expressions that are identical but displayed in different ways
**equivalent fractions** two or more fractions that are equal in value but have different numerators and denominators; for example, $\frac{1}{2} = \frac{2}{4}$
**event** a possible outcome of a trial in a probability experiment
**experiment** the process in probability of performing repeated trials of an activity for obtaining data in order to be able to predict the chances of certain things happening
**experimental probability** the probability determined by conducting an experiment and gathering data
**exponent** *see* **power**
**expression** mathematical statement made up of pronumerals (letters) and numbers. An expression does not contain an equals sign.
**exterior angle** angle formed when any side of a closed shape is extended outwards. The exterior angle and the interior angle adjacent to it are supplementary (add up to 180°).
**factor** a whole number that divides exactly into another whole number, without any remainder
**factor tree** diagram that displays the prime factors of a composite number. Each branch shows a factor of all numbers above it. The last numbers at the bottom of the tree are the prime factors of the original number.
**favourable outcome** the outcome we are looking for
**fraction** used to describe a part of a whole number
**frequency** number of times a result or piece of data occurs.
**frequency distribution table** table used to organise data by recording the number of times each data value occurs
**front elevation** view observed from directly in front of an object
**front view** *see* **front elevation**
**highest common factor (HCF)** largest factor that is common to all the numbers given
**image** a point or shape after it has been transformed
**improper fraction** fraction whose numerator is larger than its denominator; for example, $\frac{7}{4}$
**index** *see* **power**
**input number** the first number in a flow chart
**integers** positive whole numbers, negative whole numbers and zero
**interior angle** an angle on the inside edges of a shape
**inverse operation** an operation that 'undoes' a previous operation. Addition and subtraction are inverse operations; and multiplication and division are inverse operations.
**is greater than** more or higher in value, expressed by the symbol >
**is less than** less or lower in value, expressed by the symbol <
**isometric drawing** a 2-dimensional drawing of a 3-dimensional object
**isosceles triangle** triangle that has two sides of equal length. The two angles opposite the equal sides will also be equal.
**legend** lists definitions of different symbols or colours used in a graph
**like terms** terms that contain exactly the same pronumeral (letter) part; for example, $5a$ and $33a$ are like terms but $4ab$ is not
**line of symmetry** line that divides a shape into two identical parts
**long-term trend** in probability, the trend observed for results from a very large number of trials. This tends to match the theoretical probability.
**lowest common multiple (LCM)** lowest multiple that two or more numbers have in common
**magnitude** the size of a number, relative to its distance from zero; also called absolute value

**mean** a measure of the centre of a set of data. If the total value of all the observations is shared equally among all of the values in the data set, the mean is the amount that each observation would receive.
**median** the middle value of a set of data when the data is arranged in order from smallest to largest
**method** in JavaScript, a defined function applied to an object. For example, the method indexOf can be applied to a string, `"string".indexOf("in")`, to find the index of the substring `"in"`.
**mixed number** number that is made up of a whole number and a fraction; for example, $2\frac{3}{4}$
**mode** the most common value in a set of data
**multiple** the result of multiplying a number by another whole number
**multiplication** a quick, shortened process of adding a number to itself a specific number of times
**negative integers** whole numbers less than zero
**number** in programming, a JavaScript data type that represents a numerical value
**number plane** *see* **Cartesian plane**
**numerator** top term of a fraction. It shows how many parts there are.
**numerical data** data that can be measured with a numerical value
**obtuse angle** angle greater than 90° but less than 180°
**obtuse-angled triangle** triangle that has one angle greater than 90°
**opposite integers** numbers equidistant from zero, but one is positive and one is negative
**opposite operations** *see* **inverse operations**
**order of operations** set of rules that determines the order in which mathematical operations are performed
**order of rotational symmetry** the number of times that a shape matches the original figure when being rotated through 360°
**ordered stem plot** a type of table used to organise and display data, with the leaf part written in ascending order from left to right
**outcome** result of an experiment
**outlier** an extreme value in a data set
**output number** the last number in a flow chart
**parallel lines** lines that run in the same direction and never meet
**parallelogram** quadrilateral with both pairs of opposite sides parallel to each other. Rectangles, squares and rhombuses are parallelograms.
**per** used to indicate 'for each' or 'for every' in a rate; for example, 20 km per hour or 20 km/h. It indicates that you are dividing the first quantity by the second quantity.
**per cent** out of 100
**perfect square** a number that can be arranged in a square; a number that is the square of another number
**perigon** an angle of 360° (a full circle)
**perimeter** distance around the outside (border) of a shape
**perpendicular** describes two objects with a right angle (90°) between them
**place value** the position of a digit in a number
**positive integers** whole numbers greater than zero
**power** the number of times that the base is written in the expanded form of a power term; also known as an exponent or index
**primary data** data that has been collected first-hand (by you)
**prime factors** factors that are only prime numbers
**prime number** counting number that has exactly two factors, itself and 1
**prism** a solid object with identical parallel ends, and with the same cross-section along its length
**probability** likelihood or chance of a particular event (result) occurring
**product** result of a multiplication
**pronumeral** letter used in place of a number
**proper fraction** fraction whose numerator is smaller than its denominator; for example, $\frac{3}{4}$
**protractor** instrument used to measure angles in degrees
**quadrant** one of four regions of the Cartesian plane produced by the intersection of the $x$- and $y$-axes
**quadrilateral** 2-dimensional, closed shape formed by four straight lines

**quotient** result of a division
**radius** the straight-line distance from the centre of a circle to its circumference. The radius is half the diameter.
**range** the set of all $y$-values in a relationship between $x$ and $y$ (algebra); the difference between the largest and smallest data value (measure of spread)
**rate** a ratio that compares two measurements with different units; for example, km/h or $/g
**ratio** a comparison of two or more quantities of the same kind
**reciprocal** the reciprocal of a number is obtained by first expressing it as a fraction then tipping the fraction upside down; for example, $\frac{4}{3}$ is the reciprocal of $\frac{3}{4}$. A mixed number must be converted to an improper fraction first and then tipped.
**recurring decimal** decimals that have one or more digits repeated continuously; for example, 0.999… They can be expressed exactly by placing a dot or horizontal line over the repeating digits, as in this example: $8.343\,434\ldots = 8.\dot{3}\dot{4}$ or $8.\overline{34}$.
**reflection** exact image of an object, as seen in a mirror, as far behind the mirror as the object is in front of it. Reflections always have reversed orientations; right appears left and left appears right.
**reflex angle** an angle greater than 180° but less than 360°
**remainder** number left over after numbers have been divided
**right angle** angle that equals 90°
**right-angled triangle** triangle that has one of its angles equal to 90° (a right angle). A small square is placed in that angle to mark the right angle.
**right prisms** a three-dimensional solid that is formed by two identical end faces connected by rectangular side faces
**rotation** the turning of an object about a certain point — the centre of rotation
**sample space** in probability, the complete set of outcomes or results obtained from an experiment. It is shown as a list enclosed in a pair of braces (curled brackets) { }.
**scale** a series of marks indicating measurement increasing in equal quantities
**scalene triangle** triangle that has no equal sides. The angles will all be unequal too.
**secondary data** data that has been collected second-hand (by someone else)
**sector** a region of a circle bounded by two radii and the arc joining them
**sector graph** a representation of data whereby the different sectors of a circle represent different groups of data; also known as a pie graph
**segment** a section of a circle bounded by a chord and an arc
**side elevation** view observed when standing directly to one side of an object. The left view or the right view of the object can be drawn.
**side view** *see* **side elevation**
**square number** a numbers that can be arranged in a square
**square root** number that multiplies by itself to equal the original number. Finding a square root is the opposite of squaring a number; for example, $\sqrt{49} = 7$.
**squaring a number** multiplying a number by itself; for example, $2 \times 2 = 2^2 = 4$
**stem-and-leaf plot** a type of table used to organise and display data
**straight angle** an angle that equals exactly 180°
**string** a JavaScript data type that represents text
**substitution** process by which a number replaces a pronumeral in a formula
**successful trial** a trial that results in the outcome you wanted
**sum** answer obtained when numbers are added together
**supplementary angles** angles that sum (add) to 180°
**term** part of an equation or expression. A term can be expressed as one or more than one pronumeral, or a number only. Examples include $5x$, $3ab$, $xyz$, 8 and so on.
**terminating decimal** a decimal number that has a fixed number of decimal places; for example, 0.6 and 2.54
**tessellation** a repeated pattern that covers a flat surface without any gaps
**theoretical probability** probability determined by predicting outcomes rather than by gathering data
**top view** *see* **bird's eye view**

**translation** horizontal (left/right) or vertical (up/down) movement
**transversal** a line that intersects a pair (a set) of parallel lines
**travel graph** a line graph that represents a journey. The horizontal axis represents time and the vertical axis represents distance.
**trial** number of times a probability experiment is conducted
**triangle** 2-dimensional closed shape formed by three straight lines
**two-step experiment** in probability, the process of performing repeated trials of two consecutive chance experiments to predict the chances of two events occurring; this data is recorded in a two-way table
**two-way table** a diagram that represents the relationship between two non-mutually exclusive attributes
**unit price** the price for a single unit of the item in question. This can be the price of a specified quantity (e.g. 1 kg or 100 mL) or of a single unit (e.g. 1 banana).
**unitary method** a method comparing the unit cost of varying quantities of items to help determine the best buy
**unlike terms** terms that do not contain exactly the same pronumeral (letter) part; for example, $5a$ and $33a$ are like terms but $4ab$ is an unlike term
**value** the worth of a number
**variable** another word for a pronumeral (algebra); a symbol in an equation or expression that may take many different values (data)
**vertex** the point at which two line s meet
**vertically opposite angles** special angles formed when two straight lines intersect. The two non-adjacent angles are called vertically opposite angles. These angles are equal in size.
**vinculum** horizontal bar separating the numerator from the denominator
**volume** amount of space that a 3-dimensional object occupies. The units used are cubic units, such as cubic metres, $m^3$, and cubic centimetres, $cm^3$. A volume of $1\ m^3$ is the space occupied by a cube of side length 1 m.
**whole numbers** counting numbers: 0, 1, 2, 3, 4, …

# INDEX


**A**

acute angle 459
acute-angled
triangle 466
addition
decimals 311–2
fractions 192–7
integers 80–3, 85
mixed numbers 213–5
positive integers
algorithms 16–17
compensation strategy 11
jump strategy 10
rearrange strategy 11–12
split strategy 10–11
adjacent angles 487
algebra
division terms 368–9
evaluation 358–62
expanding brackets 375–9
like terms 362–7
multiplication terms 367–8
number laws 371–5
substitution 358–60
allied angles 490
alphabet sizes 157
alternate angles 490
angles 445–6
classification 459
drawing 460
estimation 445–6
measurement 446–8
naming 460–1
protractor 453–4
quadrilateral 479–82
triangle 468–70
types of 459–60
area
composite shapes 566–71
metric units 553–4
parallelogram 559–60
rectangle 555–7
triangle 557–9
arithmagon 23
Associative Law 372–4
for addition 17
axes 617

**B**

back elevation, drawing solids 585
back view, drawing solids 585
backtracking 403–5
inverse operations 415–24
solve equations 408–12
base 128
battleships game 652
BIDMAS, positive and negative
integers 97–9
bird's eye view, drawing solids 585
Brackets, Indices or roots, Division
and Multiplication, Addition and
Subtraction (BIDMAS) 45–8

**C**

capacity 579–83
Cartesian coordinates 617
Cartesian plane 617–8, 620–5
creating 617
determining coordinates of points
on 618
four quadrants 618–20
reflections in 688–90
categorical data 767
chance 726–9
clustering 793
co-interior angles 490
column graphs 792
common factors 123
common multiples 123
common percentages, short cuts
236–8
Commutative Law 371–2
for addition 17
comparing numbers 5–7
compensation strategy
addition 11
subtraction 14–5
complementary angles 488
complete revolution 459
composite number 133
factor trees 135–7
composite rectangular prism 573
composite shapes 566–71
constructing angles
greater than 180°, semicircular
protractor 454–6
protractor 453–4
continuous numerical data 768
conversion
decimals to fractions 300–2
fractions to decimals 302–3
conversion factors 522
convert 272
convex quadrilaterals 477
coordinate 617
coordinate geometry 614
coordinate system 614
corresponding angles 490
Covid-19 pandemic and data 764
cube numbers 150–1
cube roots 151–3

**D**

daily life, positive and negative
numbers 72–3
data 764, 767
collecting and classifying 767–74
column graphs and dot plots
791–8
comparison 814–20
displaying data, in tables 774–80
measures of spread 785–6
numerical 767
pie charts and divided bar graphs
(extending) 804–14
stem-and-leaf plots 798–804
decimals
adding and subtracting 311–7
checking by estimation 313
comparing 294–7
converting to fractions 300–2
division
another decimal number 325–7
multiple of 10 324–5
whole number 323–4
multiplication 317–8
another decimal number 325–7
multiples of 10 319–21
short cuts 320–1
squaring 318–9
percentages 330–1
place value 292–300
places 292
whole numbers 292–4
decomposition method 19
denominator 171
descending order 5
dice fractions 196
difference result 18
directed numbers 72
discount 280
discrete numerical data 768
Distributive Law 29, 375–7

divided bar graphs 804
dividend (numerator) 37
divisibility tests 147–9
division
algebra terms 368–9
decimals
another decimal number 325–7
multiple of 10 324–5
whole number 323–4
fraction 208–11
integers 92–4
positive integers 34
numbers, multiples of 10 37–8
short division algorithm 35–7
using expanded form 35
divisor (denominator) 37
dollars 273
domain and range 641–2
drawing solids
isometric drawing 586–8
plans and views 585–6
prisms 583–5

**E**

Egyptian numbers 53–4
equally likely outcomes 736
equations 396–8
backtracking 403–5, 408–12
balanced 415–6
checking solutions 424–9
expressions 402–8
flowcharts 402–3
guess, check and improve 398–400
inspection 398
inverse operations 415–24
overview 394–5
pronumeral positive 420–2
writing, from words 396–8
equilateral triangle 466
equivalent algebraic expressions 364
equivalent fractions 172–4
equivalent ratios 266
equivalent repeating fractions 308
Eratosthenes 134
estimate angle measures 445–6
estimation integers 42–3
event 726
experimental probability of 742
expanding brackets, algebra
Distributive Law 375–7
experiment 742
experimental probability 742
defining 742–3
long-term trends 743–6
vs. theoretical probability 739
expressions
algebra 351–4
exterior angle 470

**F**

factors 119–20
factor tree 135
favourable outcome 736
fractions 170
adding and subtracting 192–200
comparing 175–8
converting to decimals 302–3
dice 196
division 208–11
equivalent 172–4
mixed numbers 186–92
multiplication 200–3
number line 174–5
percentages 220–5
simplifying 182–4
frequency 774
frequency distribution tables 774–7
front elevation, drawing solids 585
front view, drawing solids 585

**G**

geometry 672
angles 445–53
constructing angles, protractor 453–9
overview 442–5
parallel and perpendicular lines 487–98
triangles 465–77
graphs
interpretation of
domain and range 641–2
travel graphs 642–3
graphs, interpretation of 639–41
Greek numbers 54–5

**H**

highest common factor (HCF) 123–4

**I**

image 680
improper fractions 171
to mixed numbers 186–7
index 128
index notation
indices 128–30
place values 130–1
indices 128–30
place values 130–1
input flowchart 402
inspection equations 398
integers 70–1
adding integers 80–5
division of 92–4
multiplication 91–2
number line 70–5
number plane 75–7, 80
order of operations 97–102
subtraction 85–7
interior angles 470
inverse operations equations 415–24
isometric drawing 586–8
isosceles triangle 466

**J**

jump strategy
addition 10
subtraction 12–13

**L**

language of chance 726–31
legends 791
like terms, algebra 362–4
line graph, interpretation 640
line symmetry 674–5
linear number patterns 625–32
geometric patterns 628–9
relationships and patterns, description of 625–8
linear pattern 632
long-term trends 743–6
lowest common denominator (LCD) 193–7
lowest common multiple (LCM) 123–4, 193

**M**

measurement
area 553–66
capacity 579–83
drawing solids 583–94
perimeter 536–44
reading scales and measuring length 528–31
units of 521–8
volume of rectangular prisms 571–9
measures of centre 780–5
determination from graphs 814–6
measures of spread 785–6
determination from graphs 814–6
memory game 244
metric units, area 553–4
mirror lines, reflections in 686–8
mixed numbers 171
adding and subtracting 213–5

mixed numbers (*cont.*)
  improper fractions 187–9
  multiplication 215–6
  working with 213–20
multiples 30, 118–9
multiplication
  algebra terms 367–8
  decimals 317–8
    multiples of 10 319–21
    short cuts 320–1
    squaring 318–9
  fractions 200–3
    word of 203–6
  integers 91–2
  mixed numbers 215–6
  positive integers 25
    algorithm 27–8
    area mode 26–7
    long multiplication algorithm 28–9
    mental strategies 29–31
    using diagrams 25–6
    using expanded form 26

**N**

naming angles 460–1
naming quadrilaterals 478
naming triangles 465–8
negative integers 70
  BIDMAS 97–9
negative numbers 72–3
non-linear pattern 632
number laws, algebra
  Associative Law 372–4
  Commutative Law 371–2
  Distributive Law 375–7
number line 71–2
  fractions 174–5
  integers 70–1
number plane
  integers 76–7
  positive integers and zero 75–6
numerator 171
numerical data 767

**O**

obtuse angle 459
obtuse-angled triangle 466
Olympic Games 430
opposite integers 71
order of operations 45, 144–5
  BIDMAS 45–8
  integers 97–102
order of rotational symmetry 675
ordered stem plots 798
ordering numbers 5–7
origin 617
outcome 727, 736
outliers 793
output flowchart 402

**P**

parallel lines 489–90
  calculation angles 490–92
parallelograms 477
parallelograms, area 559–60
per, rates 271
percentages
  amount using fractions 225–7
  decimal 330–31
  definition 220–22
  short cuts 236–8
  unitary method 277–9
  whole number 236
perfect squares 141
perigon 459
perimeter 536–44
  rectangle and square 538–9
perpendicular lines 492–3
pixels 614
place values
  decimals 292–300
  indices 130–1
  place-holding zeros 4–5
  reading numbers 5
plotting simple linear relationships 632–9
positive integers 70
  addition 10–12
  BIDMAS 97–9
  division 34–5
  multiplication 25–7
  number plane 75–6
  order of operations 45–52
  ordering and comparing numbers 5–7
  overview 2–4
  place value 4–5
  rounding and estimating 41–5
  subtraction 12–5
positive numbers 72–3
power 128
primary data 770
prime factors 135
prime numbers 133–5
prisms 583–5
probability 724, 726
  experimental probability 742–50
  language of chance 726–31
  sample space 731–6
  simple probability 736–42
product 25
pronumerals 350–1
proper fractions 171
protractor 446
  constructing angles 453–4
  greater than 180°, angles 454–6

**Q**

quadrants 618
  of Cartesian plane 618–20
quadrilaterals 477
  angles 479–82
  properties 482–7
  types of 477–9
quotient 37

**R**

range 641, 785
rates 271
  applications of 273–4
  identification 271–2
  simplification 272–3
ratios 265
  simplification 266–8
reading scales 528–31
rearrange strategy, addition 11–12
reciprocal of fraction 208
rectangle, area 555–7
recurring decimal 307
reflections 680, 686–93
  Cartesian plane 688–90
  mirror lines 686–8
reflex angle 459
remainder 35
repeating decimals 307–9
right angle 459
right-angled triangle 466
rotating points 627
rotational symmetry 675–7
rotations 680
rounding 305–6
rounding integers
  first or leading digit 41–2
  nearest 10 41
  nearest 100 41
rule
  creating sequence from 626
  identifying 626
  number pattern, determination of 626

**S**

sample space 731–2
scale, measuring length 528–31
scalene triangle 466

scatter plot, interpretation 639
secret numbers 405
short division algorithm 35–7
side elevation, drawing solids 585
side view, drawing solids 585
simplifying fractions 182–4
sketching quadrilaterals 481
solving equations
  backtracking 408–12
  balancing 416–8
  guess, check and improve 398–400
  inverse operations 415–20, 424
  pronumeral positive 420–2
sorting, multiples and factors 120
split strategy, addition 10–11
square numbers 141–2
square roots 142–5
squaring decimals 318–9
stem-and-leaf plots 798–804
straight angle 459
substitution
  algebra 358
  equations 424–6
subtraction
  decimals 312–3
  fractions 192–7
  integers 85–7
  mixed numbers 213–5
  positive integers
    algorithms 18–21
    compensation strategy 14–5
    decomposition 19–21
    find the difference 13
    jump strategy 12–3
    with borrowing 19
    with no borrowing 18
successful trial 742
successive discounts 282
sum 16
supplementary angles 488
symbol means greater than 6
symbol means less than 6


**T**


tangrams 499
term 625
terminating decimal 307
theoretical probability 736
  vs. experimental probability 739
top view, drawing solids 585
transformations 672, 680–2
  line and rotational symmetry 674–80
  reflections 686–90
  translations 680–6
translations 680–6
  combining 682–3
transversal parallel lines 489
travel graphs 642–3
trial 742
  successful 742
triangles 465
  area 557–9
  classification 466–8
  interior and exterior angles 470–1
  naming triangles 465–8
two-step experiment 733
two-way tables 732–4


**U**


unit price 276–7
unitary method 276
  best buys 276–81
  percentages 277–9
units of length
  converting 522–4
  metric 521–2
units of measurement
  converting units of length 522–4
  metric units of length 521–2
unlike terms, algebra 362


**V**


value 272
variables 767
vertically opposite angles 487, 490
vinculum 171
volume 571–2
  rectangular prisms 571–4, 579


**W**


whole numbers 70
  decimals 292–300
  dividing a decimal 323–4
  short cut, percentages 236
  subtraction fraction 195
writing number sentences 6


**X**


x-axis 617


**Y**


y-axis 617